2022 24th European Conference on Power Electronics and Applications (EPE'22 ECCE Europe)

Hanover, Germany
5-9 September 2022

Pages 1350-2020

IEEE Catalog Number: CFP22850-POD
ISBN: 978-1-6654-8700-9

Copyright © 2022, The European Power Electronics and Drives Association
All Rights Reserved

*** *This is a print representation of what appears in the IEEE Digital Library. Some format issues inherent in the e-media version may also appear in this print version.*

IEEE Catalog Number: CFP22850-POD
ISBN (Print-On-Demand):   978-1-6654-8700-9
ISBN (Online):   978-9-0758-1539-9

Additional Copies of This Publication Are Available From:

Curran Associates, Inc
57 Morehouse Lane
Red Hook, NY 12571 USA
Phone: (845) 758-0400
Fax: (845) 758-2633
E-mail: curran@proceedings.com
Web: www.proceedings.com

2022 24th European Conference on Power Electronics and Applications (EPE'22 ECCE Europe)

Hanover, Germany
5-9 September 2022

Pages 1350-2020

IEEE Catalog Number: CFP22850-POD
ISBN: 978-1-6654-8700-9

TABLE OF CONTENTS

Dynamic Power Analysis of Inverter-Fed Drives Based on the Switching Period of the Power Electronics 1
Alexander Stock

Stability Analysis in an Inverter-Dominant Microgrid Facing In-Rush Current of an Induction Machine 11
Nastaran Fazli, David Hammes, Sidney Gierschner, Hans-Gunter Eckel

Self-Oscillating Capacitive Power Transfer with Multiple Receiver Capability and Coupling Path Adaption 22
Norbert Seliger

An Electrically Driven Gas Compressor for Hydrogen Refueling Stations with Active Power Smoothing 30
Alfred Rufer

Unsymmetrical Fault Behavior of PLL Based Grid-Connected Converters 39
Philipp Hackl, Ziqian Zhang, Robert Schuerhuber

Stability Assessment and Optimization of MMC Energy Balancing for Drive Applications at Standstill using an Averaging Approach 49
Qiuye Gui, Hendrik Fehr, Albrecht Gensior

Turn-On Losses Optimization for Medium Power SiC MOSFET Half-Bridge Module 59
Pham Ha Trieu To, Felix Kayser, Hans-Günter Eckel

Oscillation Damping in a 500kW Hybrid Si/SiC Three-Level ANPC Inverter with Decoupling Capacitor 70
Pham Ha Trieu To, Hans-Günter Eckel

Multi Busbar Sub-Module Modular Multilevel STATCOM with Partially Rated Energy Storage Configured in Sub-Stacks 80
Chuantong Hao, Wenhao Ma, Michael Merlin, Paul Judge, Stephen Finney

Three-Phase ZVS Inverter with Variable and Fixed Frequency Operation Based on GaN Semiconductors 88
Benedikt Kohlhepp, Michael Lutsch, Thomas Dürbaum

Influences of Conductor Positions and Fast Rising Impulse Voltages on the Line-End Coil Based on a Three-Phase High-Frequency Model 97
Ting Helmholdt-Zhu, Volker Grabs

Simulation Tool for Optimization of Digital Active Gate Drive Sequence using Genetic Algorithm 108
Hajime Takayama, Shuhei Fukunaga, Takashi Hikihara

Analysis of Balancing Algorithms for Quasi- Two/Three-Level Single Phase Operation of a Flying Capacitor Converter 115
Stefan Mersche, Markus Bayer, Kai Rickert, Marc Hiller

Instability in Active Balancing Control of Dc Bus Voltages in VSC Converters Interconnected via Multi-Winding Transformers 125
Duro Basic, Sami Siala

Online Learning-Based Islanding Detection Scheme for Grid-Connected Systems............................ 135
Mohammed Ali Khan, V S Bharath Kurukuru, Rupam Singh

Difference in the Design Process of LCL Filters for Grid Connected VSI When using SiC/GaN
Instead of Si Semiconductors 145
Dennis Kampen, Lukas Fräger, Niklas Badenhop, Arthur Mambetow

Analysis and Design of a Resonant DC/DC Transformer in Modular Operation............................ 152
Abraham López, Manuel Arias, Pablo F. Miaja, Arturo Fernández

Predictive Braking Algorithm for Soft Starter Driven Induction Motors....................................... 160
Hauke Nannen, Heiko Zatocil, Gerd Griepentrog

Ambient Electromagnetic Energy Harvesting Circuit using Rectennas Manufactured with
Stereolithography Resin 169
Xuan Viet Linh Nguyen, Tony Gerges, Jacques Verdier, Philippe Lombard, Michel Cabrera,
Bruno Allard, Jean-Marc Duchamp, Philippe Benech

Boost/Buck-Boost Based Grid Connected Solar PV Micro-Inverter with Reduced Number of
Switches and Having Power Decoupling Capability 178
Arup Ratan Paul, Arghyadip Bhattacharya, Kishore Chatterjee

Operation and Selection of Multilevel Power Converters for Doubly Fed Induction Generator-
Based Wind Turbines 187
Kapil Jha, Joseph Banda, Hridya I, Arvind Tiwari

A Detailed View on the Trapezoidal Operation for MMC Type Braking Chopper in Medium
Voltage Application.. 195
Patrick Hofstetter, Viktor Hofmann, Dennis Karwatzki

Influence of Operating Frequency on High-Power Medium-Voltage Medium-Frequency
Transformers 203
Thomas B. Gradinger, Ralph M. Burkart, Marko Mogorovic

Output Power Characteristics of Isolated Secondary-Resonant SAB DC-DC Converter for Output
Voltage Variation 213
Shota Yamashita, Kohei Budo, Takaharu Takeshita

Hardware and Control Design of a High Precision Modular Power Converter Based on GaN
Technology for Particle Accelerator Magnets 223
Thomas Margreiter, Ivan De Cesaris, Maurizio Incurvati, Sebastien Pelletier, Martin
Schiestl, Ronald Stärz

Battery Cycler to Generate Open Li-Ion Cell Aging Data and Models.. 232
Matthias Luh, Thomas Blank

Function Blocks of a Highly-Integrated All-In-GaN Power IC for DC-DC Conversion 242
Michael Basler, Richard Reiner, Stefan Moench, Patrick Waltereit, Rüdiger Quay

Comparison of Redundancy Requirements for Modular Multilevel Converter Considering
Manufacturer Reliability Inputs and Mission Profile 251
Diego Velazco, Guy Clerc, Emmanuel Boutleux, Francois Wallart

Impact of Insulation and Cooling on Performance Due to Reliability-Oriented Design of Electrical
Machines 261
Lucas Vincent Hanisch, Jonas Franzki, Markus Henke

Long Switching Horizon Model Predictive Controller for High-Speed Integrated Modular Motor Drives 268

    Martin Schiestl, Maurizio Incurvati, Ronald Starz, Markus Schmid

Standalone Power Management System for Flexible Piezo Electric Nano Generators (PENG) Based on the Co-Polymer P(VDF:TrFE) 279

    Alexander Wölk, Mahmoud Shousha, Shashank Shekhawat Singh, Martin Haug, Lorandt Fölkel, Michael Brooks, Asier Alvarez, Andreas Petritz, Philipp Schäffner, Jonas Groten, Andreas Tschepp, Barbara Stadlober

Analysis and Estimation of Neutral-Point Voltage Balancing Ability of an Optimized Balancing Algorithm for Grid Connected Active-NPC Converter 289

    Joseph Banda, Kapil Jha, Hridya Ittamveettil, Arvind Kumar Tiwari, Fernando Ramirez

A Direct Model Predictive Control Strategy of Back-To-Back Modular Multilevel Converters using Arm Energy Estimation 297

    Akseli Hakkila, Antonios Antonopoulos, Petros Karamanakos

Study on Commutation Loop Inductance and Current Distribution to DC-Link Capacitors in a GaN Half-Bridge 307

    Benedikt Kohlhepp, Samuel Faber, Jeremias Kaiser, Thomas Dürbaum

Cooperative Control of Online Impedance Spectroscopy Monitoring Method and Maximum Power Point Tracking Method for Photovoltaic Panels 315

    Xin Wang, Zhixue Zheng, Michel Aillerie, Alexandre De Bernardinis, Jean–paul Sawicki, Marie-Cécile Péra, Daniel Hissel

Benefits of Switching from Si to SiC Modules with Further Converter Optimization 325

    Antxon Arrizabalaga, Mikel Mazuela, Iosu Aizpuru, June Urkizu, Jon Aztiria

On the Reduction of Output Capacitance in Two-Level Three Phase PFC Boost Rectifier for Pulsating Loads 335

    Tania C. Cano, Douglas Pedroso, Alberto Rodríguez, Ignacio Castro, Diego G. Lamar

Cognitive Insights into Metaheuristic Digital Twin Based Health Monitoring of DC-DC Converters 344

    Abdul Basit Mirza, Kushan Choksi, Sama Salehi Vala, Krishna Moorthy Radha, Madhu Sudhan Chinthavali, Fang Luo

A Three-Phase Isolated Secondary-Resonant Single-Active-Bridge DC-DC Converter with a Delta-Star Connected Transformer 351

    Atsushi Nishio, Kohei Budo, Mai Van Tuan, Takaharu Takeshita

A Novel Concept to Optimize Core Loss in Planar Magnetic Based on an Unbalanced-Flux-Approach 361

    Sobhi Barg, Kent Bertilsson, Grover Torrico

Model Reduction using Singular Perturbation Methods for a Microgrid Application 370

    Lasse Gnärig, Albrecht Gensior, Saioa Burutxaga Laza, Miguel Carrasco, Carsten Reincke-Collon

Drive Level Parameter Identification of an Induction Motor 380

    Andreas Bünte, Alex Hald, Andreas Kirsch

Impedance Stability of Single-Phase LCL Grid-Connected Voltage Source Inverters with Wideband Gap Devices Under Different Control Approaches 390

    Ramy Ali, Terence O'Donnell

Design and Modulation Optimization of an MMC Based Braking Chopper.. 400
    Viktor Hofmann, Patrick Hofstetter

Modeling the Arrangement of Drill Holes for Orthogonal Biasing in Controllable Inductors for
Power Electronic Converters.. 411
    Jonas Pfeiffer, Christoph Drexler, Pierre Küster, Peter Zacharias, Michael Schmidhuber

A Sectorized FCS-MPC Transformerless SST for Power Transmission Application 421
    Gabriel Gaburro Bacheti, Renner Sartório Camargo, Emilio José Bueno, Marco Liserre,
    Lucas Frizera Encarnação

Inductance Estimation for Square-Shaped Multilayer Planar Windings 432
    Theofilos Papadopoulos, Antonios Antonopoulos

Cost and Efficiency Considerations in On-Board Chargers 442
    Marija Jankovic, Christian Felgemacher, Kevin Lenz, Aly Mashaly, Abdelmouneim
    Charkaoui

A Novel Combined Control of Ground Current and DC-Pole-To-Ground Voltage in Symmetrical
Monopole Modular Multilevel Converters for HVDC Applications... 451
    Pablo Briff, Amit Kumar

A PFC Boost Converter with Reduced Switching Losses Operating at a Fixed Switching Frequency............ 459
    Burkhard Ulrich

Predictive Control of Power Electronics Autotransformer for Mitigating Three-Phase Grid Current
Unbalance in Railway Supply Systems... 468
    Tabish Nazir Mir, Faysal Hardan, Masood Hajian, Tamer Kamel, Pietro Tricoli

Parameter Sensitivity of a MRAS-Based Sensorless Control for AFPMSM Considering Speed
Accuracy and Dynamic Response at Multiple Parameter Variations.. 474
    Michael Brüns, Christian Rudolph, Tankred Müller

Synchronization Stability of a Grid Forming Converter Under the Effect of Current Limit in
Voltage Dips with VI Based Current Limiting Method: Analysis and Solution 484
    Siam Hasan Khan, Markel Zubiaga Lazkano, Pedro Izurza, Alain Sanchez-Ruiz, Javier Cañas
    Aceña, Joseba Arza

Analytic Calculation of Touch and Leakage Currents of Non-Isolated EV Chargers using a Fast
Common Mode Calculation Method and Non-Ideal Passive Component Models 493
    Christian Stutz, Sebastian Nielebock, Martin März

Triple-Phase-Shift Controlled Dual Active Bridge Converter with Variable Input Voltage in
Auxiliary Railway Supply 504
    Martin Scohier, Olivier Deblecker, Carlos Valderrama

Loss Characterization Methodology for Soft Magnetic Nano-Crystalline Tape Materials in Coupled
Inductors.. 514
    David Bohne, Valentin Wagner, Patrick Deck, Christian P. Dick

Substitution of Nanocrystalline Toroid by Laminated Ferrite Toroid in the Application of a
Common-Mode Choke 525
    Lukas Reißenweber, Fritz Wohlrath, Alexander Stadler

Direct Active Stabilization of the DC-Link in Voltage-Source Converters 534
    Matthieu Bertin, Mohamad Koteich

Hardware-In-The-Loop Control of a Modular Induction Motor Drive in Power Electronics Education.. 544
Jens Peter Kaerst

Design and Efficiency Analysis of an LCL Capacitive Power Transfer System with Load-Independent ZPA... 554
Francesco Musolino, Ahmed Abdullah, Mario Pavone, Fabio Ferreyra, Paolo Crovetti

A Pulse Generator Based on Transmission Line Transformer for Insulation Aging Test............... 562
Xiao Yu, Khanh-Hung Nguyen, Peter Zacharias

Design of a Single-Phase Common Mode and Differential Mode Inductor for Interleaved Converters 572
Jonathan Robinson, Gopal Mondal, Stefan Hänsel, Matthias Neumeister

Steady-State Analysis and Comparison of SSFB, SDFB and DSFB MMC-Based STATCOM 582
Mohamed Moez Belhaouane, Pierre Vermeersch, François Gruson, Pierre Rault, Sébastien Dennetiere, Xavier Guillaud

Current Distribution Control in Parallel Connected Power Converters with Continuous Output Voltage 593
Sabrina Ulmer, Andreas Brunner, Philipp Czerwenka, Gernot Schullerus, Ertugrul Sönmez

Optimized Pulse Pattern with Half-Wave Symmetry for 5-Level Converter 604
Jonas Weires, Pedro Leal Dos Santos, Steven Liu

Characterization of Si-IGBT Crosstalk with a Concentration on Power Circuit Parasitic Elements and the Device Operation Point... 614
Amir Azam Rajabian, Sadegh Mohsenzade, Javad Naghibi, Kamyar Mehran

Impact of Higher Current Harmonics on Component Current Stress and Conduction Losses of Half-Bridge-Series-Resonant-Converters in Discontinuous Conduction Mode for High-Power Applications... 624
Daniel Haake, Anton Grodnichev, Fabian Schnabel, Marco Jung

Control of a Zero-Voltage Switching Isolated Series-Resonant Power Circuit for Direct 3-Phase AC to DC Conversion 634
Yusuf Kosesoy, Remco Bonten, Henk Huisman, Jan Schellekens

Design of a Robust Voltage Control for Inverters with LC Filter Based on the Internal Model Control.. 641
Frederik Stallmann, Axel Mertens, Lukas Fräger

Influence of Power Semiconductor Device Variations on Pulse Shape of Nanosecond Pulses in a Solid-State Linear Transformer Driver.. 651
Raffael Risch, Anliang Hu, Jürgen Biela

Optimal Design of Integrated Motor Drives - Comparison of Topologies (2L/3L/Modular), PWM Variants, and Switch Technologies (Si/SiC/GaN).. 662
Thilo Bringezu, Jürgen Biela

Distribution Transformer Voltage Control using a Single-Phase Matrix Converter 673
Rui Wang, Henk Huisman, Korneel Wijnands

Influence of Carrier-Based PWM Techniques on the Common-Mode Voltage and Common-Mode Current of Six-Phase Full-Bridge Inverters.. 681
Juris Arrozy, Esin Ilhan Caarls, Henk Huisman, Jorge L. Duarte, Lorenzo Ceccarelli

Mitigation of Dead-Time Effects on Transient DC Bias Elimination in Dual Active Bridge Link Current 689

    MK Kharabela Mohanta, Dipankar De, Silpashree Sahu, Alberto Castellazzi

Generalized Automated Tool for Analysis and Design of Multiphase Coupled Inductor Buck Converters 698

    Rana Asad Ali, Mahmoud Shousha, Martin Haug

Experimental Study of a Directly Oil-Cooled Electrical Machine for a Full-Electric Vehicle by using Low Viscosity Oil 709

    Huihui Xu, Georg Tobias Götz, Shimin Zhang, Rik W. De Doncker

Development of a Family of High Voltage Gain Step-Up Multi-Port DC-DC Converters for Fuel Cell-Based Hybrid Vehicular Power Systems 719

    Pouya Zolfi, Sina Vahid, Ayman El-Refaie

Bidirectional DC Circuit Breaker with Improved Performance During Commissioning and Reclosing 730

    Aditya Pogulaguntla, Venkata Raghavendra I, Satish Naik Banavath, Andrii Chub, T Sreekanth, Harish Sarma Krishnamoorthy

Modeling Method for Conducted Noise Flowing in Power Lines of DC/DC Converter 739

    Takato Hattori, Wataru Kitagawa, Takaharu Takeshita

High-Bandwidth Power Hardware-In-The-Loop for Motor and Battery Emulation at High Voltage Levels 749

    Manuel Fischer, Philipp Kemper, Johannes Herbold, Daniel Epping, Frank Puschmann

Analysis and Discussion of Different Three-Phase dv/dt Filter Topologies and the Influences of Their Filter Parameters on Losses and EMC 758

    Eric Fritze, Michael Meissner, Klaus F. Hoffmann, Kai-Uwe Rathjen, Stefan Dickmann, Oliver Woywode

State of Charge Prediction of Lithium-Ion Batteries Based on Artificial Neural Networks and Reduced Data 767

    Sebastian Pohlmann, Ali Mashayekh, Dominic Karnehm, Manuel Kuder, Antje Gieraths, Thomas Weyh

Investigation for Condensation Test Condition of HVIGBT Modules 777

    Kenji Hatori, Keiichi Nakamura, Wakana Noboru, Nils Soltau, Eugen Wiesner

Three Phase PV Inverter LCOE Optimization Considering Technological Choice 787

    Morteza Tadbiri Nooshabadi, Jean-Luc Schanen, Shahrokh Farhangi, Hossein Iman-Eini

Square Wave Operation to Reduce Pulsating Power in Isolated MMC-Based Ultrafast Chargers 798

    Ygor Pereira Marca, Maurice G. L. Roes, Jorge L. Duarte, Korneel Wijnands

Surge Current Protection for Railway Traction Applications 805

    Michael Gleissner, Mark-M. Bakran

Impedance-Based Analysis of HVDC Converter Control for Robust Stability in AC Power Systems 814

    André Schön, Andreas Lorenz, Rodrigo Alonso Alvarez Valenzuela

Class-E Push-Pull Resonance Converter with Load Variation Robustness for Industrial Induction Heating 825

    Janus Dybdahl Meinert, Benjamin Futtrup Kjærsgaard, Thore Stig Aunsborg, Asger Bjorn Jorgensen, Stig Munk-Nielsen, Sune Bro Duun

Review of Power Converter Topologies for Electrochemical Impedance Spectroscopy of Lithium-Ion Batteries 833

Hamzeh Beiranvand, Julius M. Placzek, Marco Liserre, Giorgia Zampardi, Doriano Constantino Brogioli, Fabio La Mantia

Design and Experimental Validation of a Voltage Sensing-Current Cancellation Common Mode Linear Active Filter 843

B. Mohamed Nassurdine, PE Lévy, D. Labrousse, JL Schanen, X. Maynard, S. Carcouet

Partial Discharges of Insulated Wires Under Impulses from Wide Bandgap Power Electronics 854

Ting Helmholdt-Zhu, Vivien Grau, Urs Obernolte

Analysis of a Droop-Based Power Controller for Three-Phase Microgrids 865

Andrea Lauri, Hossein Abedini, Davide Biadene, Tommaso Caldognetto, Paolo Mattavelli

Efficiently Paralleling GaN-Transistors for High Current and High Frequency Applications using a Butterfly Layout 873

Martin Wattenberg, Oscar Lorenz, Juan Sanchez

Data-Driven Decentralized Volt/Var Control for Smart PV Inverters in Distribution Systems 883

Yizhou Lu, Qianwen Xu, Lars Nordström

Study of Current Ripple Generators for Accelerated Ageing of Capacitors 891

Robert Keilmann, Hendrik Schefer, Regine Mallwitz

Intra-Arm Balancing Control of Cascaded Multi-Port Converter for Whole Power Unbalance Conditions 902

Takumi Yasuda, Jun-Ichi Itoh

Investigation of Creepage Distances on Printed Circuit Boards for Avionic Applications 912

Hendrik Schefer, Zhongqing Xu, Tobias Kopp, Regine Mallwitz, Michael Kurrat

A 20 kW, 3-Level Flying Capacitor 1500 V Inverter with Characterized GaN Devices for Grid-Tie Applications 922

Van Sang Nguyen, Anthony Bier, Hajar Es-Seghier, Ulrich Soupremanien, Gérard Delette, Stephane Catellani

New Analytical Model for Calculating HF-Losses in Litz Wire Regions Located Outside the E/U-CoreWindow of Transformers 933

Qingchao Meng, Jürgen Biela

Fast and Accurate Soft-Switching and Hard-Switching Losses Estimation for Power Converter, Application to the Dual Active Bridge (DAB) Converter 944

Francois Boige, Nicolas Videau, Adel Ziani, Bruno Guerrero, Julien Laclaverie

Influence of an Electrical Machine on the Dimension and Packaging of Multi-Machine Systems 952

Thomas Stöckl, Hans-Georg Herzog

Design of a Serial Impingement Cooling Heatsink for a 30 kW PV String Inverter 960

Paul Bruyere, Guillaume Piquet Boisson, Gaëtan Perez

Online Junction Temperature Measurement of SiC-MOSFETs via Gate Impedance using the Gate-Signal Injection Method 971

David Hirning, Luca Bauer, Johannes Ruthardt, Jörg Haarer, Philipp Ziegler, Jörg Roth-Stielow

Powercycling Test Bench with Realistic Loss Distribution and Temperature Ripples 980
Till-Mathis Plötz, Jan Fuhrmann, Hans-Günter Eckel

Design, Implementation and Characterization of an Integrated Current Sensing in GaN HEMT
Device by using the Current-Mirroring Technique 990
*Van-Sang Nguyen, René Escoffier, Stéphane Catellani, Murielle Fayolle-Lecocq, Jérémy
Martin*

GaN-Based Modular Multilevel Converter for Low-Voltage Grid Enables High Efficiency 999
Philip Kiehnle, Patrick Himmelmann, Marc Hiller

Energy Management of Smart Homes with Electric Vehicles using Deep Reinforcement Learning............. 1006
Xavier Weiss, Qianwen Xu, Lars Nordström

Simple and Low-Computational Losses Modeling for Efficiency Enhancement of Differential
Inverters with High Accuracy at Different Modulation Schemes.. 1015
Ahmed Shawky, Mokhtar Aly, Emad M. Ahmed, Samir Kouro, José Rodriguez

Estimation of Battery Parameters in Cascaded Half-Bridge Converters with Reduced Voltage
Sensors 1025
Nima Tashakor, Bita Arabsalmanabadi, Elham Hosseini, Kamal Al-Haddad, Stefan Goetz

Method to Analyze the Influence of Switching Behavior in Hard Switching Half Bridge Topologies
for Traction Application.. 1036
Dominik Nehmer, Michael Gleissner, Lukas Bergmann, Mark-M. Bakran

Impact of Aluminum Casing on High-Frequency Transformer Leakage Inductance and AC
Resistance.. 1046
*Reda Bakri, Xavier Margueron, Wendell Da Cunha Alves, Xavier Cimetiere, Frédéric Gillon,
Antoine Bruyere, Lucian Vatamanu*

Neural Networks-Generalized Predictive Control for MIMO Grid-Connected Z-Source Inverter
Model 1056
Navid Salehi, Herminio Martinez-Garcia, Guillermo Velasco-Quesada

Voltage Estimation for Diode-Clamped MMCs Based on a Simplified Neural Network 1064
Nima Tashakor, Davood Keshavarzi, Shady Banana, Stefan Goetz

A Non-Cooperative Game-Theoretic Distributed Control Approach for Power Quality
Compensators 1074
*Claudio Burgos-Mellado, Victor Bucarey, Helmo K. Morales-Paredes, Diego Muñoz-
Carpintero*

A Comparative Analysis of Power Converter Topologies for Integration of Modular Batteries in
Electric Vehicles... 1083
*Alberto Cárcamo, Aitor Vázquez, Alberto Rodriguez, Diego G. Lamar, Marta M. Hernando,
Daniel Remón*

Design of a High-Dynamic Test Bench for Accelerated Dielectric Lifetime Testing with Adjustable
Voltage Slopes and Temperatures 1094
Hendrik Schefer, Lucas Hanisch, Tim-Hendrik Dietrich, Regine Mallwitz, Markus Henke

Novel Modulation Method for Common-Mode Noise Reduction in Solid-State Transformer Based
on ISOP Configuration.. 1104
Naoto Kikuchi, Hiroki Watanabe, Keisuke Kusaka, Jun-Ichi Itoh

Modular STATCOM for Compensation of Reactive Power and Voltage Asymmetry in Medium-Voltage Distribution Power Grids 1114
Josef Štengl, Tomáš Kormska, Jakub Talla, Zdenek Peroutka

Novel Method for Active Short Circuit (ASC) Tests of Power Module in Automotive Traction Application 1121
Tobias Appel, Arne Bieler

Short Circuit Performance and Current Limiting Mode of a Monolithically Integrated SiC Circuit Breaker for DC Applications Up to 800 V 1128
Norman Boettcher, Taro Takamori, Keiji Wada, Wataru Saito, Shin-Ichi Nishizawa, Tobias Erlbacher

Application of a HV Bipolar Square-Wave Voltage Generator for Qualification and Assessment of Energy Equipment 1137
Rico Fischer-Baeumer, Kai Gohrmann, Konrad Domes, Benjamin Sahan, Christian Staubach

A Decentralized and Communication-Free Control Algorithm of DC Microgrids for the Electrification of Rural Africa 1147
Lucas Richard, David Frey, Marie-Cecile Alvarez-Herault, Bertrand Raison

Universal Real-Time Model for Active Rectifiers in Versatile Totem-Pole PFC Configurations 1157
Axel Kiffe, Thorben Hoffstadt

Investigation of Core-Loss Mechanisms in Large-Scale Ferrite Cores for High-Frequency Applications 1167
Michael Baumann, Christoph Drexler, Jonas Pfeiffer, Jens Schueltzke, Erwin Lorenz, Michael Schmidhuber

Generation of Methodology for Making Benchmark Microgrids and Application in ESUSCON Microgrid 1177
Oscar Dorner, Patricio Mendoza-Araya

An Overview of Grid-Connection Requirements for Converters and Their Impact on Grid-Forming Control 1187
Paul Imgart, Mebtu Beza, Massimo Bongiorno, Jan R. Svensson

Modular Battery-Integrated Power Electronics-Modelling, Advantages, and Challenges 1197
Nima Tashakor, Jan Kacetl, Tomas Kacetl, Stefan Goetz

Design of Triple-Active Bridge Converter with Inherently Decoupled Power Flows 1207
Dong-Uk Kim, Byengjoo Byen, Byunghwang Jeong, Sungmin Kim

Application of a Multi-Winding Magnetic Component Characterization Method to Optimize Cross-Regulation Performances in DCM Flyback Converters 1216
Denis Motte-Michellon, Brahim Ramdane, Yves Lembeye, Bruno Cogitore

Application of an Electrostatic Machine in a Low-Voltage Microgrid 1226
Gabriel Ramos Huerta, Patricio Mendoza-Araya

Influences of Parasitic Capacitances in Wide Bandwidth Rogowski Coils for Commutation Current Measurement 1237
Philipp Ziegler, Tobias Festerling, Jorg Haarer, Philipp Marx, David Hirning, Jorg Roth-Stielow

Systematic Analysis of Oscillations in DC-Links of Fast Switching Power Electronics 1247
Tobias Fricke, Regine Mallwitz

EMI Mitigation Induced by an IGBT Driver Based on a Controlled Gate Current Profile 1256
Daniel S. Martinez-Padron, Nicolas Patin, Eric Monmasson

An Accurate and Fast Model of Three-Level Three-Phase Dual-Active Bridge Converters in Real-
Time Simulation 1266
Ming Jia, Philipp Joebges, Rik W. De Doncker

A Calorimetric and Electrical Method for Measuring Loss Energies of Half-Bridges 1277
Jörg Haarer, Mattea Eckstein, Philipp Ziegler, Philipp Marx, David Hirning, Jörg Roth-
Stielow

Condition Monitoring Approach of a SiC Power Semiconductor using Turn-Off Delay with an
Integration in a SiC Driver 1286
Victor Golev, Ulf Schümann, Rando Raßmann, Jan Bockholt

Measurement Results of Multilevel Hysteresis Control for Paralleled Two-Level Converters 1294
Magdalena Gierschner, Yves Hein, Hans-Günter Eckel, Christian Heien

Design and Development of a Short-Circuit Test Bench for Low-Voltage Direct Current Protection
Devices 1300
Simon Ravyts, Thomas Vandenbussche, Koen Stul, Jan Cappelle

A Novel Modified-TOGI Based PLL for the Three-Phase Unbalanced and Distorted Grid
Conditions 1309
Khanh-Hung Nguyen, Ahmad Ali Nazeri, Xiao Yu, Peter Zacharias

Comparison of Two and Three-Level AC-DC Rectifier Semiconductor Losses with SiC MOSFETs
Considering Reverse Conduction 1319
Guangyao Yu, Thiago Batista Soeiro, Jianning Dong, Pavol Bauer

Measurement Method for Simple Determination of Sinusoidal Large Signal Losses in Inductive
Components 1328
Peter Zacharias, Alejandro Aganza-Torres

A Novel Technique for the Suppression of the Displacement Current Through Power Module Base-
Plate Capacitance 1336
Mahmoud Saeidi, Ahmad Ali Nazeri, Rufad Zilic, Peter Zacharias

Analysis and Implementation of Effective Placement of EMC Capacitors for WBG Modules 1343
Mahmoud Saeidi, Ahmad Ali Nazeri, Firas Jenhani, Peter Zacharias

Power Hardware-In-The-Loop Verification of a Cold Load Pickup Scenario for a Bottom-Up Black
Start of an Inverter-Dominated Microgrid 1350
Mina Mirzadeh, Robin Strunk, Tobias Erckrath, Axel Mertens

Detection of Incipient Inter-Turn Short-Circuit Faults by Artificial Intelligence Classifiers 1361
Osman Örgüt, Ilker Sahin, Ece Olcay Günes

Modeling the Impact of Grid-Forming E-STATCOMs on Inter-Area System Oscillations 1371
A. Bolzoni, N. Johansson, J. P. Hasler

Combining Schwarz-Christoffel Mappings and Biot-Savart Law to Calculate the High-Frequency
Current Distribution Inside a Single Slot 1381
Torben Fricke, Phil Leon Pickert, Babette Schwarz, Bernd Ponick

Standardised Switching Cell Building Block for Converter Design Optimisation with Detailed Electro-Thermal Model 1391

Georgios Papadopoulos, Jürgen Biela

Design Procedure for Transformer-Based Solid-State Pulse Modulators with Damping Network 1402

Spyridon Stathis, Juergen Biela

DC Bias Impact on Magnetic Core Losses at High Frequency 1413

Bima Nugraha Sanusi, Ziwei Ouyang

Investigation of the Short-Circuit Type II Safe Operating Area of IGBTs 1424

Madhu Lakshman Mysore, Mohamed Alaluss, Abhishek Maitra, Thomas Basler, Roman Baburske, Franz-Josef Niedernostheide, Hans-Joachim Schulze

Single Transformer, MMC Based MV Power Electronic Traction Transformer 1434

Simon Fuchs, Simon Beck, Jürgen Biela

A New Power MOSFET Technology Achieves a Further Milestone in Efficiency 1445

Ralf Siemieniec, Michael Hutzler, Cesar Braz, Tomasz Naeve, Elias Pree, Heimo Hofer, Ingmar Neumann, David Laforet

Experimental Evaluation of Battery Impedance and Submodule Loss Distribution for Battery Integrated Modular Multilevel Converters 1456

Arvind Balachandran, Tomas Jonsson, Lars Eriksson, Anders Larsson

Constant DC Power Infeed Grid Forming with Improved Ability to Ride-Through Unbalanced Low-Voltage Faults 1466

Tayssir Hassan, Malte Eggers, Huoming Yang, Peter Teske, Sibylle Dieckerhoff

Constrained Long-Horizon Direct Model Predictive Control for Grid-Connected Converters with LCL Filters 1476

Mattia Rossi, Petros Karamanakos, Francesco Castelli-Dezza

Performance Evaluation of SiC-Based Isolated Bidirectional DC/DC Converters for Electric Vehicle Charging 1486

Kaushik Naresh Kumar, Rafal Miskiewicz, Przemyslaw Trochimiuk, Jacek Rabkowski, Dimosthenis Peftitsis

Impact of Threshold Voltage Shifting on Junction Temperature Sensing in GaN HEMTs 1497

Burhan Etoz, Jose Ortiz Gonzalez, Arkadeep Deb, Saeed Jahdi, Olayiwola Alatise

Comparison of Power Cycling Results of Discrete GaN Cascodes for Automotive Power Electronics with High Temperature Swings 1506

Florian Lippold, Philipp Hauenschild, Regine Mallwitz

Current Distortion Study for Hybrid Multi-Level Grid Inverter with Active Neutral-Point-Clamped 4-Leg Topology 1515

Jonas Steffen, Matthias Klee, Fabian Schnabel, Axel Seibel, Marco Jung

Dynamic Maximum Power Point Tracking Method Including Detection of Varying Partial Shading Conditions for Photovoltaic Systems 1525

Rosalie Rouphael, Nezha Maamri, Jean-Paul Gaubert

Novel Operation Mode of the Modular Multilevel Matrix Converter Based on a Dimensioning Algorithm 1533

Rebecca Dierks, Axel Mertens

On the Cosmic Ray Influence on the Electronics Design of a High Altitude Electric Aircraft 1543
Philippe Morey, Mauro Carpita

DC-Bus Control Considerations of Asymmetrical Multilevel Inverters with Embedded Buck-Boost
Converter 1551
Theodoros P. Mouselinos, Emmanuel C. Tatakis

A Seamless Modulation Strategy for Step-Up/Down Partial Power Processing Converter (SUD-
P3C) 1561
Chao Liu, Zhe Zhang, Ziwei Ouyang, Jiasheng Huang, Michael A. E. Andersen, Tiberiu
Gabriel Zsurzsan

Performances Analysis of Non-Model-Based Speed Estimation Algorithms for Motor Drives 1569
Gaetano Turrisi, Luigi Danilo Tornello, Giacomo Scelba, Giulio De Donato, Giuseppe
Scarcella

A Method to Design Power Control System of Wayside Energy Storage System for Energy Saving
in DC-Electrified Railway 1580
Kota Sato, Keiichiro Kondo, Hiroyasu Kobayashi, Makoto Chida

A Reconfigurable Single-Stage Three-Phase Electric Vehicle DC Fast Charger Compatible with
Both 400V and 800V Automotive Battery Packs 1590
Mojtaba Forouzesh, Yan-Fei Liu, Paresh C. Sen

Efficiency Improvement of Single-Stage AC-DC LLC Converter using a Line Cycle Synchronous
Rectifier (SR) Driving Strategy 1601
Mojtaba Forouzesh, Yan-Fei Liu, Paresh C. Sen

Influence of DC Supply Voltage Unbalances on the Performance of ARCP Inverters 1611
Gholamreza Tabrizi, Sebastian Sprunck, Marco Jung

Grid-Forming Control for Enhanced Microgrid Interconnection 1620
Tobias Erckrath, Christian Bendfeld, Peter Unruh, Axel Seibel, Marco Jung

Low Phase Shift Filter for Current Sensing Based on the Difference Between AC Machine Models
with and Without Iron Losses 1631
Niklas Himker, Marcel Krümpelmann, Axel Mertens

Design and Analysis of a Voltage Clamping Active Delay Control Method for Series Connected
SiC MOSFETs 1641
Rui Wang, Asger Bjørn Jørgensen, Hongbo Zhao, Stig Munk-Nielsen

Practical Implementation of a Concept for In-Situ Detection of Humidity-Related Degradation of
IGBT Modules 1649
Benedikt Kostka, Axel Mertens

Design for Enhanced Noise Immunity of PCB Coils Used for Sensing Current Through Power
Devices 1658
Aamir Rafiq, Sumit Pramanick

Measurement Principle for Measuring High Frequency Bearing Currents in Electric Machines and
Drive Systems 1665
Benjamin Knebusch, Lennart Junemann, Pauline Holtje, Axel Mertens, Bernd Ponick

Climatically Induced Insulation Degradation in Power Semiconductor Modules of Wind Turbines 1674
Timo Lichtenstein, Sören Fröhling, Bernd Tegtmeier, Katharina Fischer

Comparison of Magnetic Noise Compensation Techniques for Dual Three-Phase Electrically Excited Synchronous Machines.. 1684
Jonas Henkenjohann, Jan Andresen, Axel Mertens

PCB Technology Comparison Enabling a 900V SiC MOSFET Half Bridge Design for Automotive Traction Inverters 1692
Matthias Spieler, Che-Wei Chang, Ayman El-Refaie, Muhammad H Alvi, Dong Dong, Rolando Burgos

Desaturated Turn-Off of Low-Saturation IGBTs with Clamping Method to Reduce Turn-Off Energy Losses.. 1703
Vishwas Acharya Nayampalli, Hans-Günter Eckel

Impact of Bond Wire Configuration on the Power Cycling Capability of Discrete SiC-MOSFET Devices 1713
Patrick Heimler, Nick Thönelt, Josef Lutz, Thomas Basler

A Low-Leakage, Low-Loss Magnetic Transformer Structure for High-Frequency Applications................. 1722
Allen Nguyen, Ajinkya Phanse, Michael Solomentsev, Alex J. Hanson

Temperature Distribution of an IGBT Chip During Repetitive Switching Events Under Consideration of Front-Side Ageing.. 1733
Christian Bäumler, Bo Zhang, Maximilian Goller, Xing Liu, Thomas Basler

Boosting Pilot-Diode Reverse-Conducting IGBTs Turn-ON and Reverse-Recovery Losses with a Simple Gate-Control Technique 1744
Daniel Lexow, Hans-Günter Eckel

Modeling of an Interleaved DC-DC Boost Converter for a Direct Model Predictive Control Strategy... 1754
Thomas Effenberger, Hannes Böorngen, Eyke Liegmann, Michael Hoerner, Petros Karamanakos, Ralph Kennel

Static Analysis and Control Strategies of the Single Active Bridge Converter 1765
Alexis A. Gómez, Alberto Rodríguez, Marta M. Hernando, Diego G. Lamar, Javier Sebastián, Ibán Ayarzaguena, Jose Manuel Bermejo, Igor Larrazabal, David Ortega, Francisco Vázquez

Multi-Port Inductive Power Transfer System Considering Charging Auxiliary Battery in EVs..................... 1776
Zhuoqi Zhang, Ryosuke Ota, Ryohei Okada, Nobukazu Hoshi

Influence of IGBT and Diode Parameters on the Current Sharing and Switching-Waveform Characteristics of Parallel-Connected Power Modules.. 1785
Y. Ando, J. Sakai, K. Hatori, N. Soltau, E. Wiesner

Innovative Driving Scheme for Electrical Generators in More Electric Aircrafts Employing Series Active Filtering.. 1796
Nena Apostolidou, Nick Papanikolaou

Field-Measurement Based Hygrothermal Modelling of the Converter-Cabinet Climate in Wind Turbines... 1804
Katharina Fischer, Katherina Gohler

A Multi-Mode Control Based Asymmetrical Dual-Active-Bridge Series-Resonant DC-DC Converter (DABSRC) 1815
M. Yaqoob, Grover Torrico, Wang Shuqin

Extended Balancing and Dimensioning of Capacitors in MMC Double Submodules 1824
Ali Sharaf Addin, Christopher Dahmen, Thomas Brückner

Saliency Extraction and Torque Sharing Estimation of Dual Motor Drive using Special Current Sensor Configuration.. 1834
E. Rodriguez Montero, M. Vogelsberger, T. Wolbank

Soft-Switching Converter for Inductive Power Transfer System with Double-Sided LCC Resonant Network 1844
Ryohei Okada, Ryosuke Ota, Nobukazu Hoshi

Ultra Low Loss - MMC Submodules Favorable for SiC-FET Enabling High Functional Safety 1855
Christopher Dahmen, Rainer Marquardt

Control of an Active Gate Driver for an Electric Vehicle Traction Inverter using Artificial Neural Networks 1865
Julius Wiesemann, Jacob Dumtzlaff, Axel Mertens

Cascaded H-Bridge Converter Designs for Future Short-Range All-Electric Aircraft Propulsion 1875
Maximilian Hagedorn, Malte Lorenz, Axel Mertens

Overview and Evaluation of Energy Balancing Techniques for MMCs with Various Input and Output Frequencies.. 1885
Gyanendra Kumar Sah, Michael Schütt, Hans-Günter Eckel

Comparative Lifetime Estimations for IGBT Modules in Wind Turbine Converters 1895
Christian Neumann, Hans-Gunter Eckel

Single-Phase, Five-Level Inverter with SPWM-Based Neutral Point Voltage Balancing Scheme 1906
Dmytro Kondratenko, Arkadiusz Lewicki, Charles Odeh

Magnetic Core Evaluation Kit for the Comparison of Core Losses 1914
Wilmar Martinez, Xiaobing Shen, Siqi Lin, Jens Friebe

Multi-Objective Optimization of Modular Multilevel Converter Systems.. 1923
Nikolaus Patzelt, Christian Schlegel, Michail Vasiladiotis

Sizing of Hybrid Energy Storage System for Residential PV Applications 1933
Xiangqiang Wu, Zhongting Tang, Tamas Kerekes

DC Bias Currents in Full-Bridge DC-DC Converters in Context of WBG Semiconductors and High Switching Frequencies.. 1939
Niklas Badenhop, Lukas Fräger, Dennis Kampen, Sascha Langfermann, Michael Owzareck

Parameter Tuning Method for Class Φ_2 Converters for High-Frequency Wireless Power Transfer Applications.. 1947
Yining Liu, Prasad Jayathurathnage, Jorma Kyyrä

Inductor Design Optimization using FEA Supervised Machine Learning 1955
D. Cajander, I. Viarouge, P. Viarouge, D. Aguglia

Enabling Large-Scaled MMC EMT-RMS Co-Simulation by Data Exchange in the Loop (DXiL)............... 1966
Xiong Xiao, Soham Choudhury, Martin Coumont, Jutta Hanson

Advanced Low-Voltage System-In-Package Half-Bridge MOSFET with Added Protection Features.......... 1975
S. Musumeci, V. Barba, F. Scrimizzi, C. Mistretta

Evaluation of Common-Mode Leakage Current of Aalborg-Type Transformerless PV Inverters 1985
Georgios I. Orfanoudakis, Eftychios Koutroulis, Georgios Foteinopoulos, Weimin Wu

Multi-Frequency Traction-To-Auxiliary Integrated EV Drivetrain: Eliminating the Need for an
Auxiliary Power Module 1995
Caniggia Viana, Mehanathan Pathmanathan, Peter W. Lehn

Potentials to Improve the Post-Fault Performance of a Fault-Tolerant Inverter System in Electrified
Aircraft Propulsion System 2003
Yongtao Cao, Leon Fauth, Jens Friebe, Axel Mertens

Model Predictive Control-Enabled Fault Ride Through Operation Strategy for High Power Wind
Turbine 2011
Pedro Catalán, Yanbo Wang, Zhe Chen, Joseba Arza

A Theoretical Comparison of Different Virtual Synchronous Generator Implementations on
Inverters... 2021
Patrick Körner, Andrea Reindl, Hans Meier, Michael Niemetz

Linear Flux-Switching Machine Design - A Multiobjective Optimization 2030
Hendrik Marks, Henning Schillingmann, Sridhar Balasubramanian, Markus Henke

Single-Arm MMC-Based Converter for Transformerless Rail Interties... 2038
Simon Beck, Simon Fuchs, Jürgen Biela

Medium Voltage Diode Rectifier Design for High Step-Up DC-DC Converter 2049
Pierre Le Métayer, Cyril Buttay, Drazen Dujic, Piotr Dworakowski

Fast Switching Planar Inductance Current Source ZETA Converter with Integrated Common Mode
Filter 2058
Benjamin H. Zacher, Christian Schumann

System Level Simulation of Moisture Propagation and Effects in Wind Power Converters......................... 2066
Johannes C. Wenzel, Axel Mertens

PWM-Based Optimization-Free Active Voltage-Balancing Control of 7-Level Active Neutral-
Point-Clamped Flying-Capacitor Multicell Inverters 2073
Vahid Dargahi

Model Predictive Power Sharing Algorithm for Fuel Cell Integration in a Dual Inverter Electric
Vehicle Drivetrain 2084
Mehanathan Pathmanathan, Caniggia Viana, Sukhjit Singh, Peter W. Lehn

Comparative Evaluation of the 5-Phase Vienna and the 5-Phase PWM Rectifiers Under DC
Voltage Control 2092
A. Dieng

Modelling and Control of a 50kW SiC-Based Isolated DAB Converter for Off-Board Chargers of
Electric Vehicles.. 2101
*Haaris Rasool, Manh Tuan Tran, Sajib Chakraborty, Joeri Van Mierlo, Thomas Geury,
Mohamed El Baghdadi, Omar Hegazy*

Impact of Cyber Attacks on Cost Oriented Power Routing Schemes in Microgrids..................................... 2110
Kirti Gupta, Subham Sahoo, Bijaya Ketan Panigrahi, Frede Blaabjerg

Response of IGBT Chip Characteristics Due to Critical Stress... 2119
Kohei Yamauchi, Rik W. De Doncker

Mega-Hertz High-Power WPT System with Parallel-Connected Inverters using Current Balance Circuit................2127
Masamichi Yamaguchi, Keisuke Kusaka, Jun-Ichi Itoh

Investigation and Mitigation of Common-Mode Voltage in Four-Level NPC Converters Modulated by Redundant Level Modulation................2136
Jun Wang, Wei Xu, Xibo Yuan, Lihong Xie

Ferrite Optimization for a Three-Phase Wireless Power Transfer System for Electric Vehicles................2145
Shuang Nie, Mehanathan Pathmanathan, Peter W. Lehn

Frequency and Modulation Index Related Effects in Continuous and Discontinuous Modulated Y-Inverter for Motor-Drive Applications................2156
Hamzeh J. Jaber, Alberto Castellazzi

Performance Evaluation of Sinusoidal-Flux Reluctance Machine for Improving Power Density with Reduced Torque and Input-Current Ripples................2164
Kiwa Nagayasu, Masaki Iida, Kazuhiro Umetani, Mastaka Ishihara, Eiji Hiraki

Power Hardware-In-The-Loop Test of Low-Voltage Battery for a Plug-In Hybrid Electric Vehicle................2175
Ronan German, Florian Tournez, Alain Bouscayrol, Aurelien Lievre, Betty Lemaire-Semail

Stability Analysis of DFIG System Connected with High-Frequency Capacitive Grid Based on Closed-Loop Current Control and Direct Power Control................2182
Bin Hu, Heng Nian, Subham Sahoo, Frede Blaabjerg, Yaqian Zhang, Zixiao Xu

Full-Bridge Modular Multilevel Converter for the Four-Quadrant Supply of High Power Magnets in Particle Accelerators................2189
Manuel Colmenero, Ricardo Vidal-Albalate, Francisco R. Blanquez, Ramon Blasco-Gimenez

Deep Neural Network for Magnetic Core Loss Estimation using the MagNet Experimental Database................2197
Xiaobing Shen, Hans Wouters, Wilmar Martinez

Hybrid Circuit Board Structure for Power Electronics................2205
Gerrit Braun, Deniz-Heinz Moldenhauer

Active Control of Gear Mesh Vibration using a Permanent-Magnet Synchronous Motor and Simultaneous Equation Method................2211
Dominik Reitmeier

Research Laboratory for Testing Grid Connected Devices Under Grid Voltage / Grid Impedance Variations and Microgrid Conditions................2219
Swen Bosch, Jochen Staiger, Heinrich Steinhart

Reducing the Impact of Skin Effect Induced Measurement Errors in M-Shunts by Deliberate Field Coupling................2230
Hauke Lutzen, Jonas Müller, Vladimir Polezhaev, Till Huesgen, Nando Kaminski

Grid Forming Control for HVDC Systems: Opportunities and Challenges................2241
Adil Abdalrahman, Ying-Jiang Häfner, Malaya Kumar Sahu, Khirod Kumar Nayak, Ashkan Nami

A Highly Integrated and Modular High Speed Electric Drive for Lightweight Electric Mountain Bikes................2251
Matthias Hofer, Mario Nikowitz, Manfred Schrödl

Performance Enhancement of Power Conditioning Systems in More Electric Aircrafts 2257
Nick Rigogiannis, Nick Papanikolaou, Yongheng Yang

Steady State Simulations of a Hybrid HVAC/HVDC Network using OS Based ARM Devices 2266
Ioan Catalin Damian, Mircea Eremia

Experimental Comparison of FPGA-Implemented Model Predictive Voltage Control to Cascaded
Proportional Resonant Control for a Three-Phase Four-Wire Three-Level Grid-Forming Inverter of
250 kVA 2276
Jarren Lange, Dominik Schmies, Karl Stephan Stille, Joachim Böcker, Oliver Wallscheid

Experimental Study of Interleaved Y-Inverter Performance 2285
Yusuke Endo, Masataka Minami, Hamzeh J. Jaber, Alberto Castellazzi

Design of a GaN-Based Reconfigurable Resonant Converter for High Frequency On-Board
Charger of Battery Electric Vehicles 2293
*Manh Tuan Tran, Haaris Rasool, Dai Duong Tran, Mohamed El Baghdadi, Philippe Lataire,
Omar Hegazy*

Transient Liquid Phase Bond Reliability Evaluation of Die-Attach for Power Module Packaging 2301
Laxma R. Billa, Yangang Wang, Thomas Grant, Xiang Li, Harley Neal, Muhammad Morshed

Experimental Evaluation on Observer-Based Delay-Compensating Active Damping for LC-Filters 2308
Michael Schütt, Hans-Günter Eckel

Influence of Static Rotor Imbalance on the Roller Bearing Damage Due to Inverter-Induced
Bearing Currents 2316
Martin Weicker, Omid Safdarzadeh, Andreas Binder

Novel Current Balancing Method for HF Interleaved Converters with Reduced Control Effort 2327
Christian Beckemeier, Jens Friebe

dV/dt-Based Filter Design for Motor Inverters with Continuous Output Voltage 2334
Sabrina Ulmer, Stevan Bugarski, Gernot Schullerus, Ertugrul Sönmez

Evaluation of Core Losses in Transformers for Three-Phase Multi-Level DAB Converters 2344
Babak Khanzadeh, Yuriy Serdyuk, Torbjörn Thiringer

A Quasi-Offline Condition Monitoring Method of DC-Link Capacitor Banks in Accelerator Power
Converters 2355
*Timm Felix Baumann, Konstantinos Papastergiou, Raul Murillo Garcia, Dimosthenis
Peftitsis*

Minimizing Voltage Stress in Auxiliary Resonant Commutated Pole Inverters using Saturable
Inductors 2366
Markus Zocher, Norbert Grass, Ralph Kennel

Adaptive Dead-Time Control in a Resonant Wireless Power Transfer System 2375
Tim Krigar, Martin Pfost

Multilevel Battery Converter with Cascaded H-Bridges on Cell Level-Battery Management System
Or a Renewed Attempt for Power Electronic Building Blocks? 2383
*Max Rothenburger, Markus Horn, Xiao Yu, Gerold Schulze, Koenraad Muyllaert, Peter
Zacharias, Ludwig Brabetz, Hartmut Hillmer*

Design and Potential of EMI cm Chokes with Integrated DM Inductance 2392
Mohammad Ali, Rehnuma Bushra, Jens Friebe, Axel Mertens

Implementation Options of a Fully SiC Buck-CSI for Advanced Motor Drive Application......................... 2402
Yonghwa Lee, Alberto Castellazzi

Optimized Control Scheme to Achieve ZVS for the Complete Pre-Charging Phase of
Supercapacitors with a 500 kHz SiC- And GaN-Based Dual Active Bridge 2413
Patrick Lenzen, Martin Pfost

Fault Blocking Capability in the DC-MMC with Reduced Number of Sub-Modules................................... 2422
J. D. Páez, F. Morel, S. Bacha, P. Dworakowski

An Open-Source FEM Magnetic Toolbox for Calculating Electric and Thermal Behavior of Power
Electronic Magnetic Components 2432
Nikolas Förster, Jonas Hölscher, Till Piepenbrock, Philipp Rehlaender, Oliver Wallscheid,
Frank Schafmeister, Joachim Böcker

Comparison of Dual-Active-Bridge-Based Topologies for Single-Phase Single-Stage EV On-Board
Chargers 2441
Daniel Gaona, Denis Pauls, Eduardo Facanha De Oliveira

Design Concepts for Medium Voltage DC Networks Supplying the Future Circular Collider (FCC).......... 2451
Manuel Colmenero, Francisco R. Blanquez, Ramon Blasco-Gimenez

A Novel Dual CC-CV Output Wireless EV Charger with Minimal Dependency on Both Coil
Coupling and Load Variation 2462
Subhranil Barman, Kishore Chatterjee

A High-Performance EMI Filter Based on Laminated Ferrite Ring Cores 2470
Marcin Kacki, Marek S. Rylko, John G. Hayes, Charles R. Sullivan

Investigation of the Static Performance and Avalanche Reliability of High Voltage 4H-SiC
Merged-PiN-Schottky Diodes 2477
Chengjun Shen, Saeed Jahdi, Phil Mellor, Juefei Yang, Erfan Bashar, Jose Ortiz-Gonzalez,
Olayiwola Alatise

On Chain-Link Based Multi-Port Converters Able to Connect HVDC and MVDC to AC
Transmission Network... 2486
Daniele Falchi, Oriol Gomis-Bellmunt, Eduardo Prieto-Araujo, Olivier Despouys

Voltage Control Scheme for Multilevel Interfacing PV Application: Real-Time MRAC-Based
Approach 2496
Mohammad Sadegh Orfi Yeganeh, Mehdi Rahmani, Nenad Mijatovic, Tomislav Dragicevic,
Frede Blaabjerg, Pooya Davari

Control Principles for Island Operation and Black Start by Offshore Wind Farms Integrating Grid-
Forming Converters... 2504
Daniela Pagnani, Lukasz Kocewiak, Jesper Hjerrild, Frede Blaabjerg, Claus Leth Bak

Experimental Study of the Reduction and Removal of Turn-On Snubber for IGCT Based MMC
Submodule using Fast Silicon Diodes 2515
Arthur Boutry, Cyril Buttay, Besar Asllani, Bruno Lefebvre, Eric Vagnon, Dong Dong

Characterisation of a Ferrite-Polymer Based Magnetic Material 2526
Johan Le Leslé, Guillaume Lefevre, Julien Morand, Rémi Perrin, Pierre-Yves Pichon,
Guillaume Regnat

Model Predictive-Based Control Technique for Fault Ride-Through Capability of VSG-Based Grid-Forming Converter 2537
  Mobina Pouresmaeil, Amir Sepehr, Basit Ali Khan, Jafar Adabi, Edris Pouresmaeil

Grounding Points in HV/MV Hybrid Transformer Auxiliary Converters 2544
  Adrian Wiemer, Jürgen Biela

Non-Parasitic Induced Transient Overvoltage in ANPC Topology Due to Critical Switching Sequences 2554
  Michael Geiss, Robert Kragl, Jürgen Thoma, Benjamin Volzer

Open-Delta SBC: A New Converter Topology with Low Number of Sub-Modules for MV Applications 2564
  D. Lanzarotto, P. B Steckler, K. Vershinin, F. Morel

Characterising the Effect of an Inverter on the Regulation of the AC Voltage using a Frequency Response Identification Technique 2574
  Mohamed Aldarmon, Joan Marc Rodriguez, Adria Junyent-Ferre

Artificial-Intelligence Based DC-DC Converter Efficiency Modelling and Parameters Optimization 2581
  Fanghao Tian, Diego Bernal Cobaleda, Wilmar Martinez

Analysis of the Loss Distribution of a 6 kW Two Stage Power Supply for 600 V DC Applications 2588
  Lukas Fräger, Sascha Langfermann, Michael Owzareck, Dennis Kampen, Jens Friebe

Study on the Gate Loop Design and Its Impact on Switching Characteristics of GaN Transistors 2596
  Xiaomeng Geng, Carsten Kuring, Oliver Hilt, Mihaela Wolf, Joachim Würfl, Sibylle Dieckerhoff

Analysis of Current Sharing in the Parallel Connection of GaN Transistors 2607
  Frederik Stalleicken, Sibylle Dieckerhoff, Karsten Handt, Sebastian Nielebock

Verification of GaN-HEMT Spice Models using an S-Parameters Approach 2618
  Alonso Gutierrez, Nasri Said, Emmanuel Marcault, Mathieu Gavelle

Power Loss Modelling of GaN HEMT-Based 3L-ANPC Three-Phase Inverter for Different PWM Techniques 2628
  Salvatore Mita, Arjun Sujeeth, Giuseppe Aiello, Dario Patti, Francesco Gennaro, Giacomo Scelba, Mario Cacciato

Generalized Core and Winding Area Ratio - Trends for Inductors and Transformers in Power Electronics with High Switching Frequencies 2638
  Siqi Lin, Leon Fauth, Wilmar Martnez, Jens Friebe

Active Substrate Termination of Discrete and Monolithic Bidirectional GaN HEMTs in a T-Type Inverter 2644
  Carsten Kuring, Yannic Lange, Xiaomeng Geng, Oliver Hilt, Mihaela Wolf, Joachim Würfl, Sibylle Dieckerhoff

Transformer Design Optimization and Comparison for a DC-DC Converter Used in PV Micro-Inverters 2655
  Tobias Manthey, Meriem Khader, Jens Friebe

Automated Gate Impedance Network Design for SiC MOSFETs using SPICE Solver Interfaced with MATLAB Environment 2661
  Pawel Piotr Kubulus, Szymon Michal Beczkowski, Stig Munk-Nielsen, Asger Bjørn Jørgensen

An Improved Multi-Loop Resonant and Plug-In Repetitive Control Schemes for Three-Phase
Stand-Alone PWM Inverter Supplying Non-Linear Loads 2670
Ahmad Ali Nazeri, Peter Zacharias

High Switching Frequency Operation of a Single-Phase Five-Level Hybrid Active Neutral Point
Clamped Inverter with a Model Predictive Control Approach 2682
Mohammad Najjar, Mahdi Shahparasti, Rasool Heydari, Morten Nymand

Design of Planar Coupled Inductor Applied to Zero-Current Switching Clamped Current Converter 2689
Vinicius Freire Bezerra, Tobias Manthey, Montiê Alves Vitorino, Jens Friebe

Characterization of Online Junction Temperature of the SiC Power MOSFET by Combination of
Four TSEPs using Neural Network 2698
Kanuj Sharma, Simon Kamm, Kevin Muñoz Barón, Ingmar Kallfass

Novel Extended Robust Disturbance Observer for Improved Cogging Force Compensation in
Permanent Magnet Linear Motors 2706
Franz Luckert, Axel Mertens

Improvement of a Self-Powered Gate Driver Power Supply 2715
*Mariana Raya, Oriol Aviñó, Sergio Busquets-Monge, Xavier Perpiñá, Miquel Vellvehi, Xavier
Jordà*

Optimization and Scaling of a Compact High-Power IGCT Capacitor Charger Based on Simulation
and Measurements with a 300 kW/3.3 kV Demonstrator 2726
Felix Haag, Fabian Albrecht, Volker Brommer, Oliver Liebfried, Klaus F. Hoffmann

Multilayer Busbars for Medium Voltage ANPC Converter Dedicated to Battery Energy Storage
Systems 2736
Mamadou Lamine Beye, Luc Bimmel, Anthony Bier, Jérémy Martin

A Simulation Model for SiC MOSFET Switching Transients Controlled by an Adaptive Gate
Driver with the Capability of Reducing Switching Losses and EMI Across the Full Operating
Range 2744
Zheming Li, Robert W. Maier, Mark-M. Bakran, Franz-J. Niedernostheide, Daniel Domes

Phase-Shift Modulation for Flying-Capacitor DC-DC Converters 2754
Philipp Rehlaender, Frank Schafmeister, Joachim Böcker

An EV Integrated Isolated DC Charger using a Six-Phase Synchronous Machine 2763
Sukhjit S Ghumman, Mehanathan Pathmanathan, Peter W Lehn

Configurable ISOP-IPOP DC-DC Converter for Universal Solid-State Transformer 2773
Pramod Apte, Jens Friebe, Lukas Fräger

Using System-On-Chip Boards for the Deployment of Controller for Verification and Prototyping 2780
Adeel Jamal, Gerd Griepentrog

Utilizing the Reactive Current Control Capability of an MMC-Fed AC/DC Converter for Volt-
Second Balancing in Medium Frequency Transformers 2788
*Kaveh Pouresmaeil, Maurice Roes, Jorge Duarte, Korneel Wijnands, Nico Baars, George
Papafotiou*

Cost Comparison for Different PV-Battery System Architectures Including Power Converter
Reliability 2795
*Martijn Deckers, Leander Van Cappellen, Glenn Emmers, Fereshteh Poormohammadi, Johan
Driesen*

Insulation Design and Analysis of a Medium Voltage Planar PCB-Based Power Bus Considering Interconnects and Ancillary Circuit Integration 2806
Joshua Stewart, Rolando Burgos, Dushan Boroyevich

Modular Multilevel Converter Control with using a General Space Vector PWM Method in Medium Voltage Hydro Power Application.. 2813
Chengjun Tang, Torbjörn Thiringer

A Technical Overview of Single-Stage Three-Port DC-DC-AC Converters 2824
Sebastian Neira, Zoe Blatsi, Michael M. C. Merlin, Javier Pereda

Common-Mode EMI Noise Modeling of Three-Level T-Type Inverter for Adjustable Speed Drive Systems.. 2835
Vefa Karakasli, Abdelmoumin Allioua, Gerd Griepentrog

A Condition Monitoring Scheme for Semiconductor Devices in Modular Multilevel Converters with Cascaded H-Bridge Submodules... 2843
Mohsen Asoodar, Mehrdad Nahalparvari, Christer Danielsson, Hans-Peter Nee

Particular Requirements on Drive Inverters for Safe and Robust Operation on an Open Industrial DC Grid... 2852
Simon Puls, Jan-Niklas Koch, Martin Ehlich, Holger Borcherding

Investigation About Operation and Performance of Gate Drivers for Power Electronics Converters for Cryogenic Temperatures... 2860
Mustafeez-Ul-Hassan, Yuxuan Wu, Vyacheslav Solovyov, Fang Luo

Synchronization Angle Determination in DVCSFO of DFIM Naval Propulsion......................... 2869
Youssef Drimizi, Maria Pietrzak-David, Pascal Maussion

Power Control of LCR-DAB Converter with Phase Shift in Fixed Switching Frequency 2877
Seung-Hyuk Baek, Jaehong Lee, Seung-Hwan Lee, Sungmin Kim

A Simplified Braking Method for Direct Matrix Converter-Fed PMSM Drives with Consideration of Avoiding Regenerative Energy 2885
Jun Xie, Dustin Henneberg, Martin Suberski, Thomas Ellinger, Uwe Radel, Jürgen Petzoldt

Inverter-Machine Parametric Co-Design for Energy Efficient Electric Drives............................ 2893
Jaedon Kwak, Alberto Castellazzi

Bidirectional Cuk Converter in Partial-Power Architecture with Current Mode Control for Battery Energy Storage System in Electric Vehicles 2903
J. S. Artal-Sevil, J. Anzola, V. Ballestín-Bernad, I. Aizpuru

Design Space Exploration for a Capacitive 36V, 4A, 4:1 DCDC Converter with GaN Switches using a Performance-Cost-Matrix Including Uncommon Topologies... 2912
Adrian Gehl, Malte Kempchen, Simon Disselkamp, Markus Olbrich, Bernhard Wicht

A Fast Control for a Three-Switch Multi-Input DC-DC Converter... 2919
Simone Cosso, Andrea Formentini, Mario Marchesoni, Massimiliano Passalacqua, Luis Vaccaro

Impact on the Torque and on the Copper Losses Under Fault-Tolerant Control of 5-Phase PMSG.............. 2930
A. Dieng

Weighting Factor Design for FS-MPC in VSCs: A Brain Emotional Learning-Based Approach 2939
Mohammad Sadegh Orfi Yeganeh, Arman Oshnoei, Saeed Peyghami, Nenad Mijatovic,
Tomislav Dragicevic, Frede Blaabjerg

A Strategy for Smooth Microgrid Transitions Without Phase Misalignment and Voltage Mismatch 2948
Gabriel Silva Rocha, Amiron Wolff Dos Santos Serra, Cesar Augusto Santana Castelo
Branco, Hercules Araujo Oliveira, Jose Gomes De Matos, Luiz Antonio De Souza Ribeiro

Subtle Design and Performance Comparison of WF-FSM and DC-VRM for Large-Scale Direct-
Drive Wind Power Generation 2958
Udochukwu B. Akuru, Maarten J. Kamper, Zi-Qiang Zhu

Analysis and Implementation of Different Non-Isolated Partial-Power Processing Architectures
Based on the Cuk Converter.. 2967
J. S. Artal-Sevil, J. Anzola, V. Ballestín-Bernad, J. L. Bernal-Agustín

GaN HEMT and SiC Diode Commutation Cell Based Dual-Buck Single-Phase Inverter with
Premagnetized Inductors and Negative Gate Driver Turn-Off Voltage 2977
Tobias Brinker, Hendrik Gräber, Jens Friebe

Determination of Optimal Associated Discrete Circuit Switch Model Parameters for Real-Time
Simulation of Dual-Active Bridge Converters 2985
Marija Stevic, Ravinder Venugopal

Integrated Motor Drive: A Multidisciplinary Approach.. 2996
Betty Lemaire-Semail, Nadir Idir, Eric Semail, Souad Harmand

Hardware in the Loop Test of an Electric Aircraft Powertrain.. 3005
Sebastian Mönninghoff, Moritz Scholjegerdes, Kay Hameyer

A Multi-Port Smart Transformer for Green Airport Electrification 3014
Giampaolo Buticchi, Giovanni De Carne, Thiago Pereira, Kangan Wang, Xiang Gao, Jiajun
Yang, Youngjong Ko, Zhixiang Zou, Marco Liserre

Improvement of EMI Filter Attenuation using Shielding... 3022
Mohammad Ali, Rehnuma Bushra, Jens Friebe, Axel Mertens

Implementation of Onsite Junction Temperature Estimation for a SiC MOSFET Module for
Condition Monitoring.. 3031
Farzad Hosseinabadi, Shahid Jaman, Sachin Kumar Bhoi, Md. Mahamudul Hasan, Sajib
Chakraborty, Mohamed El Baghdadi, Omar Hegazy

Energy Storage Systems for Airborne Wind Generators.. 3037
Bakr Bagaber, Axel Mertens

Design Interactions of AC- And DC-Side Filters for Traction Drives with SiC Inverters 3048
Hedieh Movagharnejad, Benjamin Knebusch, Axel Mertens, Bernd Ponick

Investigation of an Interleaved Current-Fed Single Active Bridge DC-DC Converter for PV
Applications.. 3059
Lucas Vinícius De Araújo Gomes, Tobias Manthey, Montiê Alves Vitorino, Jens Friebe

Real-Time Thermal Characterization of Power Semiconductors using a PSO-Based Digital Twin
Approach 3067
Johannes Kuprat, Yoann Pascal, Marco Liserre

Self-Sensing Design and Control for an Induction Machine with an Additional Short-Circuited Rotor Coil 3075
Stefan Luecke, Axel Mertens

Calculating the Tractive Power and Power Conversion Efficiency of Battery Electric Vehicles using a Global Navigation Satellite System and a Road Elevation Database 3084
Shinichi Domae, Alberto Castellazzi, Hamzeh J. Jaber, Tenghui Dong, Taketsune Nakamura

PCB Layer Optimization of Planar Medium Frequency Transformer for On-Board EV Chargers 3092
Fabian Groon, Hamzeh Beiranvand, Thiago Pereira, Görkem Can, Marco Liserre

Fault Current Capability Assessment of Low-Voltage Side Inverters in Smart-Transformers 3101
Thiago Pereira, Luis Camurca, Francisco Santos, Marco Liserre

Adaptive Resonant-Valley Switching for a GaN HEMT Direct AC-AC Auxiliary Resonant Commutated Pole Converter 3112
Kyle Steyn, Johan Beukes

The Variation of Core Loss in High-Frequency Transformers Under Different Load Conditions 3120
Navid Rasekh, Jun Wang, Xibo Yuan

A Complete PFC Inductor Design for Lighting Equipment Applications 3130
Wai Keung Mo, Kasper M. Paasch, Thomas Ebel

Automatic Generation Control-Based Charging/Discharging Strategy for EV Fleets to Enhance the Stability of a Vehicle-To-Weak Grid System 3140
Majid Mehrasa, Mehrdad Gholami, Reza Razi, Khaled Hajar, Antoine Labonne, Ahmad Hably, Seddik Bacha

Model-Based Converter Control for the Emulation of a Wind Turbine Drive Train 3149
Alexander Ernst, Wilfried Holzke, Dawid Koczy, Nando Kaminski, Bernd Orlik

A Novel Grid-Demanded Power Point Tracking (GPPT) Control Method for Wind Turbines to Preserve Grid Stability with High Wind Energy Penetration 3159
David Matthies, Alexander Ernst, Henning Sauerland, René Reimann, Wilfried Holzke, Bernd Orlik

Extension and Implementation of a Model-Based Lifetime Monitoring System with Parallel Calculation of Multiple Power Semiconductors 3169
Steffen Menzel, Wilfried Holzke, Michael Hanf, Holger Groke, Bernd Orlik, Nando Kaminski

Smart Charging Strategy for Electric Vehicles using an Optimized Fuzzy Logic System 3179
M. Gholami, M. Mehrasa, R. Razi, K. Hajar, A. Hably, S. Bacha, A. Labonne

Analysis and Discussion of a Concept for an Adjustable Inductance Based on an Impact of an Orthogonal Magnetic Field 3188
Guido Schierle, Michael Meissner, Klaus F. Hoffmann

A Field Programmable and Dynamic Configurable Power Electronic Converter Concept 3198
Bjarte Hoff

DAB Converter Discrete ADRC Control into Real-Time CHIL Simulation of a MVDC/LVDC Power Grid 3206
Alessio Clerici, Riccardo Chiumeo, Diego Raggini, Alessandro Veroni

SNNFT: Sequential Neural Network-Fuzzy Thermal Early Warning System for Lithium-Ion Batteries.. 3215
Marui Li, Chaoyu Dong, Yunfei Mu, Qian Xiao, Jingming Cao, Hongjie Jia

Fine-Grained Dynamics Representation and Stability Analysis for MMC-Based Hybrid AC/DC Power Systems 3225
Jingming Cao, Chaoyu Dong, Qian Xiao, Marui Li, Xiaodan Yu, Hongjie Jia

Adaptive Pontryagin's Minimum Principle-Inspired Supervised-Learning-Based Energy Management for Hybrid Trains Powered by Fuel Cells and Batteries 3235
Hujun Peng, Feifei Li, Zhu Chen, Kai Deng, Sebina Jeschke, Kay Hameyer

A Case Study of Pole-Phase Changing Induction Machine Performance 3246
Konstantina Bitsi, Sjoerd G. Bosga

New Topology of Superconducting Fault Current Limiter with Bypass Resistor 3254
D. Baimel, Eli Barbi, S. Bronstein, N. Baimel, A. Kuperman

A Pre- And Discharge Unit for Capacitive DC-Links Based on a Dual-Switch Bidirectional Flyback Converter 3262
Madlen Hoffmann, Martin März

Control and Integration of a Multiphase Brushless Wounded Synchronous Motor Drive 3272
Remi Perrin, Guilherme Bueno-Mariani

A Way Forward to Achieve Interoperability in Multi-Vendor HVDC Systems 3282
Adil Abdalrahman, Ying-Jiang Häfner, Philippe Maibach, Christoph Haederli

Model Predicitve Position Control of Electrical Drives on an Industrial PC 3292
Fabian Karau, Michael Leuer

Bidirectional Active EMC Filter for Industrial Power Converters.. 3301
Bernhard Wunsch, Stanislav Skibin, Ville Forsstrom

A General Method to Measure Parasitic Capacitance of Transformer using Guarding Technique............... 3309
Shaokang Luan, Stig Munk-Nielsen, Bruce Wakelin, Magnus Hortans, Jan Schupp, Hongbo Zhao

Inductance Analysis of Electric Machines by Classical and Numerical Methods........................ 3318
J. J. Germishuizen, T. J. E. Miller

Dynamic Wireless Power Transfer DWPT Time Domain Model: Xyz Position and Speed Coupling Effect... 3327
Iosu Aizpuru, Eneko Agirrezabala, Mikel Mazuela, Unai Iraola, Estanis Oyarbide, Carlos Bernal

Dynamic Average Small Signal Model of the SAB Converter 3336
Alexis A. Gómez, Alberto Rodríguez, Marta M. Hernando, Diego G. Lamar, Javier Sebastián, Ibán Ayarzaguena, Jose Manuel Bermejo, Igor Larrazabal, David Ortega, Francisco Vázquez

Algorithm for Optimal Selection of Drive Motor Transmission Combination............................ 3344
Santiago Ramos Garces, Dries Jacques, Stijn Derammelaere, Simon Houwen, Nick Van Oosterwyck, Bart Vanwalleghem

Evaluation of Drain-Source Voltage in Switch Transient Time Intervals as Gate Oxide Degradation Precursor of SiC Power MOSFETs 3353
Javad Naghibi, Sadegh Mohsenzade, Kamyar Mehran, Martin P. Foster

Active Output LLC Converter Topology 3362
Hannes Börngen, Eyke Liegmann, Sriram Jagannath, Ralph Kennel

Short Circuit Type II and III Behavior of 1.2 kV Power SiC-MOSFETs 3373
Xing Liu, Xupeng Li, Thomas Basler

Analog MPPT Comparison for Interplanetary Small Satellites Missions 3382
C. Torres, A. Garrigós, J. M. Blanes, P. Casado, D. Marroquí, C. Orts

Feasibility Assessment of Variable-Speed Generator Set Concepts with Focus on Rating of Power
Electronic Equipment 3391
Hendrik Fehr, Albrecht Gensior, Andreas Möckel, Frank Atzler, Tilo Roß, Carsten Reincke-
Collon

Bus Voltage Regulation using Sequentially Switched ZVZCS Converters for Spacecraft Power
Systems 3401
A. Garrigós, C. Orts, D. Marroquí, J. M. Blanes, C. Torres, P. Casado

A Standardized and Modular Power Electronics Platform for Academic Research on Advanced
Grid-Connected Converter Control and Microgrids 3411
Frank S. R., Schulz D., Stefanski L., Schwendemann R., Hiller M.

Gate Input Capacitance Characterization for Power MOSFETs using Turn-On and Turn-Off
Switching Waveforms 3420
Yota Nishitani, Michiko Inoue, Takashi Sato, Michihiro Shintani

AC Battery: Modular Layout with Cell-Level Degradation Control 3429
Claudio Burgos-Mellado, Marcos Orchard, Diego Muñoz-Carpintero, Tomislav Dragicevic,
Lorenzo Reyes-Chamorro, Jacqueline Llanos

Analysis of Test Methods for Measurement of Leakage and Magnetising Inductances in Integrated
Transformers 3440
Sajad A. Ansari, Jonathan N. Davidson, Martin P. Foster, David A. Stone

A Topology-Morphing Series Resonant Converter for Photovoltaic Module Applications 3450
Grigorios Sergentanis, Liliana De Lillo, Lee Empringham, C. Mark Johnson

A Novel Parameter for the Evaluation of Protective Circuits for IGBT Explosion Protection in
Submodules of MMC 3460
Christoph Junghans, Hans-Guenter Eckel

Sub-Modules Switching Algorithms for Dual Active Bridge Modular Multilevel Converters to
Optimize Capacitor Voltage Deviation Versus Power Efficiency 3470
Peizhou Xia, Chuantong Hao, Stephen Finney, Michael Merlin

Systematic Adaptive Robust State Feedback Control for Active Front-End Rectifiers 3480
Aidar Zhetessov, Giri Venkataramanan

An Optimized Compensation Strategy of Direct Matrix Converter-Fed PMSM Drives with Field
Weakening Under Unbalanced Supply Conditions 3491
Jun Xie, Dustin Henneberg, Martin Suberski, Manuel Kusebauch, Uwe Rädel, Jürgen
Petzoldt

Double Inverter Concept for High-Speed Drives Without Motor Filters 3501
Henning Kasten, Stephan Beineke, Matthias Bachmann

A Universal Single Stage Current-Fed Bidirectional Converter with Both AC and DC Input Power Source Compatibility 3511
    Manish Kumar, Sumit Pramanick, Bijaya Ketan Panigrahi

Optimization of Electric Vehicle Charge Scheduling with Consideration of Battery Degradation 3518
    Raka Jovanovic, Sertac Bayhan, Islam Safak Bayram

Onboard ESU Sizing and Dynamic IPT Charging Scenarios for a Tramway Application 3529
    Endika Bilbao Muruaga, Irma Villar, Florian Legay, Pierre Prenleloup, Jean-François Reynaud

Investigations on the Active Reduction of Common Mode Noise with Opposing Noise Sources 3536
    Philipp Marx, Felix Seybold, Philipp Ziegler, David Hirning, Jörg Roth-Stielow

Knowledge Based Grey Box Modeling of Inaccessible Circuits for System EMC-Simulation in Time Domain 3545
    Jan-Philipp Roche, Jens Friebe, Oliver Niggemann

Novel Quasi-Direct Rotor Position Estimator for Permanent Magnet Synchronous Machines Based on the Back-Electromotive Force using Current Oversampling 3555
    Georg Lindemann, Viktor Willich, Axel Mertens

Design Considerations for Fast On-State Voltage Measurement Circuits 3565
    Mathias C. J. Weiser, Manuel Rueß, Ingmar Kallfass

Analytical, FEM and Experimental Study of the Influence of the Airgap Size in Different Types of Ferrite Cores 3574
    Asier Arruti, Francisco Jose Perez-Cebolla, Jon Anzola, Iosu Aizpuru, Mikel Mazuela

Design Method of a High Frequency GaN-Based Half-Bridge with Bottom-Side Cooled Transistors using Multi-PCB Assembly 3582
    Loris Pace, Florian Chevalier, Thierry Duquesne, Nadir Idir

A 30 kW Dynamic Wireless Inductive Charging System for EVs 3590
    Zariff Meira Gomes, José Renes Pinheiro, Gilney Damm, Karim Kadem, Hassan Moussa

Dynamic Control of the Switching Behavior of SiC MOSFETs in Converter Operation 3599
    Jochen Henn, Laurids Schmitz, Rik W. De Doncker

A Series Resonant Balancing Converter for Bipolar DC Grids on Ships 3607
    Sachin Yadav, Zian Qin, Pavol Bauer

A V2G-Enabled Seven-Level Buck PFC Rectifier for EV Charging Application 3615
    Anekant Jain, Ritika Agarwal, Krishna Kumar Gupta, Sanjay K. Jain

Experimental Demonstration of a 2.2kW Active-Clamp Converter for High-Current Wide-Voltage-Transfer Ratio Applications 3625
    Philipp Rehlaender, Bastian Korthauer, Frank Schafmeister, Joachim Böcker

A Simplified Model for the Battery Ageing Potential Under Highly Rippled Load 3636
    Tomáš Kacetl, Jan Kacetl, Nima Tashakor, Stefan Goetz

System Modeling and Design of a Hybrid Renewable Energy System for a Cable Network Head-End Station in Rural Area 3646
    Tobias Schillinger, Thomas Schuhmann, Martin Eckart

Comparison of System-Level Availability in Industrial Grids 3655
G. Emmers, J. Driesen

Ageing Mitigation and Loss Control in Reconfigurable Batteries in Series-Level Setups 3665
Tomáš Kacetl, Jan Kacetl, Nima Tashakor, Stefan Goetz

Characterization of Conventional and Advanced Current Measurement Techniques Suitable for
WBG Semiconductor Devices 3676
Severin Klever, André Thönnessen, Rik W. De Doncker

Zero-Sequence Voltage Reduces DC-Link Capacitor Demand in Cascaded H-Bridge Converters for
Large-Scale Electrolyzers by 40% 3686
Roland Unruh, Frank Schafmeister, Joachim Böcker

Thermal Behavior Impact on the Electric Motor Shape Multi-Objective Optimization 3696
Aissam Riad Meddour, Anthony Babin, Nassim Rizoug, Christopher Vagg, Richard Burke,
Laid Degaa

Modelling Approaches of Power Systems Considering Grid-Connected Converters and Renewable
Generation Dynamics 3704
Jaume Girona-Badia, Vinícius Albernaz Lacerda, Eduardo Prieto-Araujo, Oriol Gomis-
Bellmunt, Stephan Kusche, Florian Pöschke, Horst Schulte

Efficiency and Lifetime Analysis of Several Airborne Wind Energy Electrical Drive Concepts 3711
Bakr Bagaber, Daniel Heide, Bernd Ponick, Axel Mertens

Design and Performance Analysis of Single-Phase Axial Flux Permanent Magnet Motor for
Coaxial Cascade 3722
Chu Wang, Xiaowei Hu, Xiaoya Wang, Weiwei Geng, Qiang Li, Jingning Hou

Comparison of Pulse Current Capability of Different Switches for Modular Multilevel Converter-
Based Arbitrary Wave Shape Generator Used for Dielectric Testing of High Voltage Grid Assets 3729
Dhanashree Ashok Ganeshpure, Ajeeth Phrassanna Soundararajan, Thiago Batista Soeiro,
Mohamad Ghaffarian Niasar, Peter Vaessen, Pavol Bauer

Accurate Modeling of IGBT-Based Converters in PLECS 3740
Anne Von Hoegen, Philipp Tillmann, Tetsuya Kojima, Rik W. De Doncker

Novel Analytical Method for Estimating the Junction-To-Top Thermal Resistance of Power
MOSFETs 3750
José Miguel Sanz-Alcaine, Francisco Jose Perez-Cebolla, Carlos Bernal-Ruiz, Asier Arruti,
Iosu Aizpuru

DC-Side Impedance for Handling Interoperability of Multi-Vendor Multi-Terminal HVDC
Systems 3757
Ashkan Nami, Adil Abdalrahman, Ying-Jiang Häfner, Malaya Kumar Sahu, Khirod Kumar
Nayak

Utilizing the Electroluminescence of SiC MOSFETs as Degradation Sensitive Optical Parameter 3766
Lukas A. Ruppert, Michael Laumen, Rik W. De Doncker

Characterization of GaN-On-AlN/SiC Transistors Towards Monolithic Integrability 3775
Nick Wieczorek, Xiaomeng Geng, Carsten Kuring, Oliver Hilt, Frank Brunner, Mihaela Wolf,
Joachim Würfl, Sibylle Dieckerhoff

Optimal Frequency for Dynamic Wireless Power Transfer 3786
Mincui Liang, Khalil El Khamlichi Drissi, Christophe Pasquier

A Wide-Input-Voltage-Range 50W Series-Capacitor Buck Converter with Ancillary Voltage Bus for Fast Transient Response in 48V PoL Applications.. 3796
Nameer Khan, James Xu, Gerard Villar Piqué, John Pigott, Henk Jan Bergveld, Alaa El Sherif, Olivier Trescases

Four-Level Boost Inverter Based on ANPC Topology with Switched-Capacitor Branch............................ 3804
Robert Stala, Adam Penczek, Stanislaw Piróg, Aleksander Skala, Andrzej Mondzik, Zbigniew Waradzyn, Krishna Kumar Gupta, Pallavee Bhatnagar, Sanjay K. Jain, Kasinath Jena

Comparative Evaluation of Partially-Rated Energy Storage Integration Topologies for High Voltage Modular Multilevel Converters... 3813
Zoe Blatsi, Sebastian Neira, Stephen Finney, Michael M. C. Merlin

Influence of Current Collapse Due to V_{ds} Bias Effect on GaN-HEMTs I_d-V_{ds} Characteristics in Saturation Region 3822
Xuyang Lu, Arnaud Videt, Ke Li, Soroush Faramehr, Petar Igic, Nadir Idir

Deep-Learning Fault Detection and Classification on a UAV Propulsion System 3831
Pierre-Yves Brulin, Fouad Khenfri, Nassim Rizoug

A Compact Solid State Transformer for Replacing Conventional Medium Power Transformer in Weight-Critical Applications... 3838
Leon Fauth, Felix Willer, Jens Friebe

Comparative Study of Single-Phase and Three-Phase DAB for EV Charging Application........................... 3846
Nicola Blasuttigh, Hamzeh Beiranvand, Thiago Pereira, Marco Liserre

Dynamic Load Emulation for Automotive Power IC Robustness Validation 3855
Alexander Ulbing, Daniel Kostynski, Markus Sievers

DAB Frequency Decoupling Control with Current Minimization 3862
Simon Uicich, Jean-Yves Gauthier, Xuefang Lin-Shi, Bruno Allard, Arnaud Plat

Design and Performance Analysis of a Modified Proportional Multi-Resonant (PMR) Controller for Three-Phase Voltage-Source Inverters 3871
Ahmad Ali Nazeri, Mahmoud Saeidi, Peter Zacharias

Proposition and Comparison of Several Solutions for High Induced Voltage Across Inactive Transmitting Coils in a Series-Series Compensation DIPT System.. 3883
Wassim Kabbara, Tanguy Phulpin, Mohamed Bensetti, Antoine Caillierez, Serge Loudot, Daniel Sadarnac

Modeling and Measuring the Bearing Capacitance of Radially Loaded Bearings 3893
Stefan Quabeck, Daniel C. Rodriguez, Rik W. De Doncker

Comprehensive Control of Matrix Converters in On-Board Electric Drive Applications............................ 3903
Galina Mirzaeva

Power System Simulation Tool for Quick Benchmarking of Innovative MVDC Grids in E-Mobility Applications... 3910
Daniel Siemaszko, Philippe Noisette

An Artificial Intelligence Pipeline for Critical Equipment Thermal Conditioning System Design 3920
Raik Orbay, Athanasios Tzanakis, Inko Marcaide, Jonas Löfgren, Torbjörn Thiringer, Thomas Bernichon

Aspects of Stability Issues of HVAC/HVDC Coupled Grids.. 3928
Gianni Bakhos, Kosei Shinoda, Juan-Carlos Gonzalez-Torres, Abdelkrim Benchaib, Luigi Vanfretti, Seddik Bacha

Measurement of Coss-V Characteristic of the 1.7kV/900A SiC Power Module and Estimation of the Channel Current.. 3938
Jacek Rabkowski, Fernando Gonzalez-Hernando, Mariusz Zdanowski, Irma Villar, Uxue Larrañaga

In-Slot Cooling of Electrical Machines using Traditional Techniques and Additive Manufacturing 3947
Ahmed Hembel, Gokhan Cakal, Bulent Sarlioglu

Comparison of High-Power 2-Level and 3-Level Converters in Terms of Power Density, Costs and Performance.. 3957
Ludwig Schlegel, Wilfried Hofmann

Autonomous Characterization of Lithium-Ion Battery Model Parameters Utilizing a Mathematical Optimization Methodology 3966
Hamzeh Beiranvand, Helge Krüger, Sandra Hansen, Marco Liserre, Christian Werlig, Andreas Würsig

SOC Governed Algorithm for an EV Cascaded H-Bridge Connected to a DC Charger 3975
Giulia Tresca, Andrea Formentini, Filippo Gemma, Federico Lusardi, Riccardo Leuzzi, Pericle Zanchetta

Shaping the Transition from Si-Based Power Devices to SiC MOSFETs and GaN HEMTs 3984
Gerald Deboy

Reinventing Batteries Through Nanotechnology 3986
Yi Cui

Advancing GaN Power ICs: Efficiency, Reliability & Autonomy.. 3987
Dan Kinzer

Electrification Strategy of Volkswagen Group.. 3989
Alexander Krick

Make it Fly — the Future of Sustainable Aviation.. 3991
Tanja Neuland

The Instrumental but Extremely Challenging Role of Hydrogen Towards a Decarbonized Society 3992
Stefan Linder

Short Circuit Behavior of Dual Three-Phase Permanent Magnet Synchronous Motors with Different Mutual Inductance in Electric Propulsion Application 3993
Yinghui Yang, Georg Möhlenkamp

Hybrid Silicon-SiC Inverter – Combining the Best of Both Worlds 4003
Hans-Günter Eckel, Felix Kayser, Pham Ha Trieu To

Robustness of SiC Trench MOSFETs 4004
Christian Felgemacher

3D Predictive Fatigue Modeling of Power Modules 4005
Ben Samples, Brandon Passmore

Heterogeneous Integration of Power Conversion using Power Supply on Chip and Power Supply in Package... 4006
Cian Ó Mathúna, Seamus O'Driscoll

Driving Innovations for Power Electronics with Integratable and Sustainable Magnetics........................... 4008
Matt Wilkowski

Impact of Package Technology on the Switching Behavior of High-Voltage GaN FETs............................. 4011
Sebastian Klötzer

Impact of Power Electronics on Battery Operation 4012
Dirk Uwe Sauer

Trends in Power Electronics and Batteries for Electrified Vehicle Infrastructure.. 4013
Torsten Leifert

Impact of High Frequency Current Pulses on Battery Ageing 4014
Julia Kowal

Aircraft Electrification – System-Level Potentials for Aviation Decarbonization 4015
Kathrin Ebner, Antoine Habersetzer, Arne Seitz

About Power Electronics Challenges in Aviation 4016
Marco Bohllaender

Development of Electric Motors for Aircraft Applications.. 4017
Simon Wolfstädter

Powertrain Trends in Electric Trucks.. 4018
Luciana C. Afonso

Modulation Strategy Impact of BEV Inverters on the Voltage Ripple and the High-Voltage Traction System Stability 4019
Cornelius Rettner

Zero Emission Trucks & Bodies 4020
Martin Glaser

Integrating Offshore Wind & Hydrogen - An Operator's View 4021
Florian Gremme

Status Quo and Future Prospects of Power Electronic Solutions for Electrolysis Plants 4022
Sven Schumann

Modular Power Supply System for Large Scale Water Electrolyzers... 4023
Ralf Juchem, Klaus Rigbers

Properties of a Lithium-Ion Battery as a Partner of Power Electronics.. 4025
Alexander Blömeke, Katharina Lilith Quade, Dominik Jöst, Weihan Li, Florian Ringbeck, Dirk Uwe Sauer

Author Index

Power Hardware-in-the-Loop Verification of a Cold Load Pickup Scenario for a Bottom-up Black Start of an Inverter-dominated Microgrid

Mina Mirzadeh[*,1], Robin Strunk[1], Tobias Erckrath[2] and Axel Mertens[1]

1) Institute for Drive Systems and Power Electronics, Leibniz University Hannover
Welfengarten 1, 30163 Hannover, Germany
[*]Phone: +49 (0)511 762-2848
[*]Email: mina.mirzadeh@ial.uni-hannover.de
URL: www.ial.uni-hannover.de
2) Fraunhofer Institute for Energy Economics and Energy System Technology (IEE)
Joseph-Beuys-Straße 8, 34117 Kassel, Germany

Acknowledgements

This work was supported by Federal Ministry for Economic Affairs and Climate Action on the basis of a decision by the German Bundestag. Project RuBICon, Funding number: 03EI4003A. The authors would like to acknowledge the technical support from OPAL-RT Germany GmbH for real-time simulations.

Keywords

≪Smart microgrids≫, ≪Grid-forming converters≫, ≪Power Hardware-in-the-Loop≫, ≪Non-linear loads≫, ≪Demand response≫, ≪Smart meters≫, ≪Grid restoration≫

Abstract

Black start capability is one of the challenges in the future grid, to be dominated by distributed power electronic converter systems. A bottom-up multi-master black start scenario based on droop-controlled grid-forming inverters was earlier introduced as a response to this challenge. With reduced reliance on the availability of smart loads, this paper adapts this scenario so that a minimum change in the current configuration of the low voltage residential load sector is required. Moreover, with a focus on the cold load pick-up response, the power-sharing among two grid-forming inverters in the presence of high inrush currents is investigated using Power Hardware-in-the-Loop tests and detailed load models. The results confirm the formation of a laboratory-scale islanded Microgrid through grid-forming inverters where smart meters coordinate an autonomous dynamic partial loading only based on local measurements.

1 Introduction

The conventional approach for a black start after a large-scale power outage is to restore the supply from the transmission grid with the help of bulk generation units [1]. With the increasing penetration of small distributed generation (DG) units, however, such concepts need to be adapted to allow for a bottom-up black start procedure in the distribution network when a fault has happened at higher voltage levels. Followed by a global power outage in a low voltage (LV) residential sector, small-scale inverter-based generation units could not only be responsible for the local demand but also a quick black start of an islanded Microgrid (MG).

On this track, a new concept was already introduced in [2] to enable the black start capability of DG in the LV grid with no dependency on the availability of communication channels. This concept counts on droop-based grid-forming (GFm) inverters and a demand-side participatory approach different from the typical remote central control schemes found in the literature. Similar to the droop behavior of inverters, local measurement of voltage and frequency through smart meters is used as an indicator for the available power to allow a priority-based switching of individual smart household appliances. This scenario remains the base of this paper. However, in the transition towards the dominance of smart loads in the residential sector, this paper adapts the loading sequence for non-intelligent loads. Automated operation of circuit breakers for different rooms are used for demand-side management. In this way, each smart meter-controlled residential household turns into a self-switching unit contributing to a bottom-up black start.

Recent developments in addressing challenges in the integration of smart grid technologies have led to the creation of new validation methods based on real-time simulation (RTS) and Hardware-in-the-Loop (HIL) tests [3]. Correspondingly, this paper uses Power Hardware-in-the-Loop (PHIL) laboratory tests for the validation of the proposed restoration scheme. In these tests, a power amplifier emulates the load-side behavior through RTS while the DG units are represented with physical inverter systems.

As listed by [4], it is a challenge for the grid operators to test network restoration schemes under realistic conditions, especially feasible load characteristics. This is well in line with the findings in [5], which identify inadequate load models contributing to unsuccessful grid reconstruction. Therefore, it is important to count for appropriate representation of LV residential loads in black start studies. Counting for the laboratory limitation of hardware representation of LV residential loads, the analytical load models in [6] are used in this work to effectively reduce the gap between simulation and real-life recorded system responses. These component-level models are based on the startup current measurements of real household appliances and include their non-linearity and inrush behavior.

This paper is structured as follows: Section 2 provides the background description of the considered black start sequence (BSS) including the requirements in the operation of GFm inverters and residential loads. The PHIL setup is then specified in detail in Section 3 and the measurement results are shown in Section 4. Finally, the conclusion is drawn in Section 5.

2 Black Start Sequence

The BSS follows the specification in [2] for a blackout caused at a higher voltage level. The main objective in such a sequence is to initially build up an LV islanded Microgrid through the contribution of multiple DGs (multi-master approach). This MG will (partially) supply local loads and later will be able to resynchronize to the medium voltage (MV) level once it is restored. As shown in Fig. 1, the BSS is initiated by a leading GFm inverter, called *Master Seed*, that ramps up its output voltage to the nominal voltage V_N. Other GFm inverters in the MG, called *Master Fellows*, synchronize to this voltage, connect to the MG and take over a share of the required power. For simplicity, this paper considers a static assignment of the *Master Seed*. The next sections shortly provide the working principle of the supply-side, consisting of GFm inverters, and the requirements from the LV residential load-side based on a novel operation of smart meters (SM) and circuit breakers (CB).

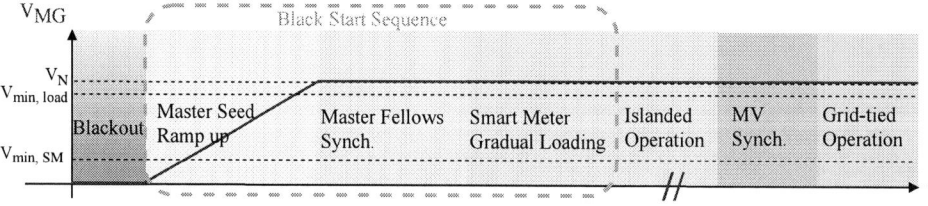

Fig. 1: Proposed BSS depicting the Microgrid voltage reference value (V_{MG}) with respect to the minimum required voltages for the smart meter ($V_{min,SM}$) and controlled loading ($V_{min,loading}$).

2.1 Grid-forming Inverter Operation

The inverters that restore the grid voltage after a blackout need to be controlled in a grid-forming mode as opposed to the grid-feeding mode, where the objective is only to deliver power without controlling the voltage [7]. Here, inverters with a constant DC-link voltage and fixed available power are considered. This is a valid approximation of a battery storage system if only a short period is considered. Generally, inverters equipped with considerable energy storage are preferred for *Master* operation. Alternatively, inverters with fluctuating available power, e.g. PV-coupled systems, can also be controlled in a grid-forming mode. However, they need to curtail their output power during normal operation to enable flexible reactions to any load changes [8].

In the scope of this paper, a standard droop control with two cascaded inner control loops $G_V(s)$ and $G_I(s)$, respectively for the output voltage and current, are applied to all inverters, according to the block diagram shown in Fig. 2 [9]. $G_{LP}(s)$ is a low pass filter with the corner frequency ω_{LP} that adds virtual inertia to the droop control. G_V and G_I are both proportional-resonant (PR) controllers in the $\alpha\beta$-frame, as indicated by the indices of the input and output quantities. The resonance frequency is always adjusted to match the reference frequency ω^* that is determined by the droop control. The transfer functions are

$$G_{LP}(s) = \frac{\omega_{LP}}{\omega_{LP} + s} \quad (1) \quad G_V(s) = K_{p,V} + K_{r,V}\frac{s}{s^2 + \omega^{*2}} \quad (2) \quad G_I(s) = K_{p,I} + K_{r,I}\frac{s}{s^2 + \omega^{*2}} \quad (3)$$

which are discretized according to the method proposed in [10]. Besides, a proportional controller $K_{p,I0}$ with a reference value of zero for the zero-sequence current is added in order to reduce the corresponding distortions in the laboratory setup. Furthermore, a feedforward path for the voltage reference, a virtual inductance L_v and an active damping resistance R_{ad} are included in the control. The control parameters are tuned in dependence of the parameters of the LC filter and the required control bandwidth, as listed in Table I. The reference values for active power P^* and reactive power Q^* are set to zero in the absence of a secondary control. The grid forming inverters are equipped with an LC filter, where $\underline{v}_{C,\alpha\beta}$ is the voltage across the capacitor, $\underline{i}_{L,\alpha\beta}$ is the current through the inductance, $\underline{i}_{C,\alpha\beta}$ is the current in the capacitor and $\underline{i}_{out,\alpha\beta}$ is the output current. The reference voltage for the pulse-width modulation (PWM) is $\underline{v}^*_{PWM,\alpha\beta}$. The sinusoidal reference for the output capacitor voltage $\underline{v}^*_{C,\alpha\beta}$ is calculated from the voltage reference amplitude $\hat{V}^*$ and angle θ^*. The amplitude of the current reference vector $\underline{i}^*_{L,\alpha\beta}$ is limited to the rated current. However, the protection from the transient overcurrents with shorter time constants than the dynamics of the current control require the blocking of PWM pulses. Since operation in current limiting generally causes disturbance in voltage control, this work avoids triggering the current limiting through load-side management.

The GFm inverters share the load according to their P-ω and Q-V droop coefficients: $m_P = \frac{2\pi \cdot 1\,\mathrm{Hz}}{S_N}$ and $m_Q = \frac{0.1V_N}{S_N}$, respectively, where S_N is the rated power of the inverter and V_N is the nominal grid voltage. The chosen droop parameters are a compromise between smaller transient power oscillations and a wider range of frequency and voltage for easier detection of thresholds. The reference values for active power P^* and reactive power Q^* are set to zero as a secondary control is not considered. The *Master Fellows* synchronize to the voltage provided by the *Master Seed* by using a phase-locked loop (PLL) before connection. A proportional controller with gain $K_{p,PLL}$ and a low-pass filter with corner frequency $\omega_{LP,PLL}$ are used in the PLL. At the time of connection, the voltage and frequency measured by the PLL are taken as initial reference values for the droop control. Consequently, the power provided by the *Master Fellows* will ideally be zero at first. Then, the reference values are gradually changed to the nominal voltage V_N and frequency ω_N. As a result, the master fellows take over an equal share of power according to the droop coefficients. Any existing grid-feeding inverters would reconnect to the grid once the voltage is reestablished according to [2]. However, they are not considered in the PHIL setup to keep the focus on the interaction between the Gfm inverters and the load-side.

2.2 Load-side Requirements

Although there is an accelerating trend for manufacturers towards smart household appliances, the residential load profile still dominantly consists of conventional loads with no overlaid or remote control. Apart from current protection schemes via conventional CBs inside households, it is the responsibility of the grid provider to assure power quality to the loads in the allowed range (According to VDE-AR-N 4100: $0.8V_N < V < 1.1V_N$ and $47.5\,\mathrm{Hz} < f < 51.5\,\mathrm{Hz}$). Consequently, in the case of a blackout, since

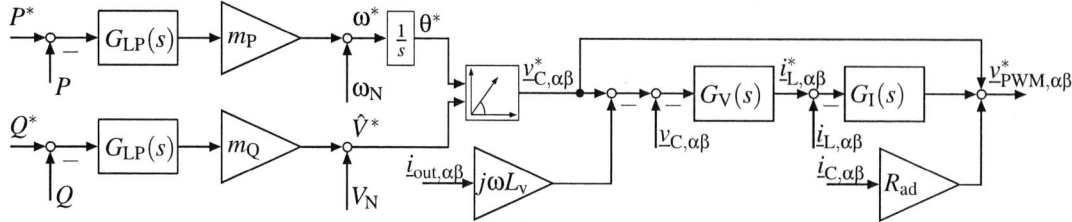

Fig. 2: Droop control with inner voltage and current control loops, virtual inductance and active damping for grid-forming inverters

Table I: Control parameters

Parameter	Symbol	Value
Corner frequency of the low-pass filter for the droop control	ω_{LP}	$314\,\frac{rad}{s}$
P-ω droop coefficient	m_P	$0.393\,\frac{rad}{kWs}$
Q-V droop coefficient	m_Q	$2.04\,\frac{V}{kVAr}$
Proportional gain of the voltage control	$K_{p,V}$	$7.540\,\frac{mA}{V}$
Resonant gain of the voltage control	$K_{r,V}$	$94.75\,\frac{A}{Vs}$
Proportional gain of the current control	$K_{p,I}$	$2.67\,\frac{V}{A}$
Resonant gain of the current control	$K_{r,I}$	$4195\,\frac{V}{As}$
Proportional gain of the zero-sequence current control	$K_{p,I0}$	$0.534\,\frac{V}{A}$
Virtual inductance	L_v	$3\,mH$
Active damping resistance	R_{ad}	$1\,\Omega$
Proportional gain for PLL	$K_{p,PLL}$	$266.7\,\frac{rad}{s}$
Corner frequency of the low-pass filter for the PLL	$\omega_{LP,PLL}$	$158\,\frac{rad}{s}$

it is conventionally expected that only quality-assured power is restored from the MV level, the loads remain connected through the closed CBs and are then directly exposed to the BSS.

For a bottom-up black start, however, controllable loads are required to ensure the protection of loads and GFm units, and therefore increase the possibility of a successful black start. This is the case especially during the ramp-up by the *Master Seed* to avoid overloading, and later in the *Masters* Synchronization, where a low-load operation significantly reduces circulating transients among inverters. Until successful completion of *Masters* synchronization, the loads, on the other hand, should be protected from a grid, possibly far from its nominal operation values. Moreover, to increase the success of the black start even in low supply scenarios, a fair partial loading should be enabled at the household level. Consequently, smart units are required inside households to take care of switching on/off (*add/shed*) the loads.

2.2.1 Smart Meter-Controlled Circuit Breaker Operation

A demand-side response to black start capability is proposed in this paper with minimum implied requirements compared to the current residential configuration, as opposed to the device-level control proposed in [2]. As shown in Fig. 3 (right), a household is equipped with a smart meter that operates smart circuit breakers with normally-open operation, each assigned to one or two rooms, i.e load categories. The more smart circuit breakers configured to be controlled by the smart meter, the more distinctive the partial loading will be. The device-level switching in [2] can be therefore seen as the equivalent to the case where the number of CBs is equal to the number of loads. For a partial loading to be fair among different consumers, at least two smart circuit breakers are necessary per household. As a sample configuration, 3 smart circuit breakers are considered in Fig. 3, each controlling one phase. Assuming random assignment of phases to rooms in the LV grid, the gradual closure of the smart circuit breakers is not expected to introduce additional asymmetry to the overall system.

In the absence of external communication or a central control unit, the smart meter measures the input voltage magnitude and frequency and decides when each controllable circuit breaker should be opened

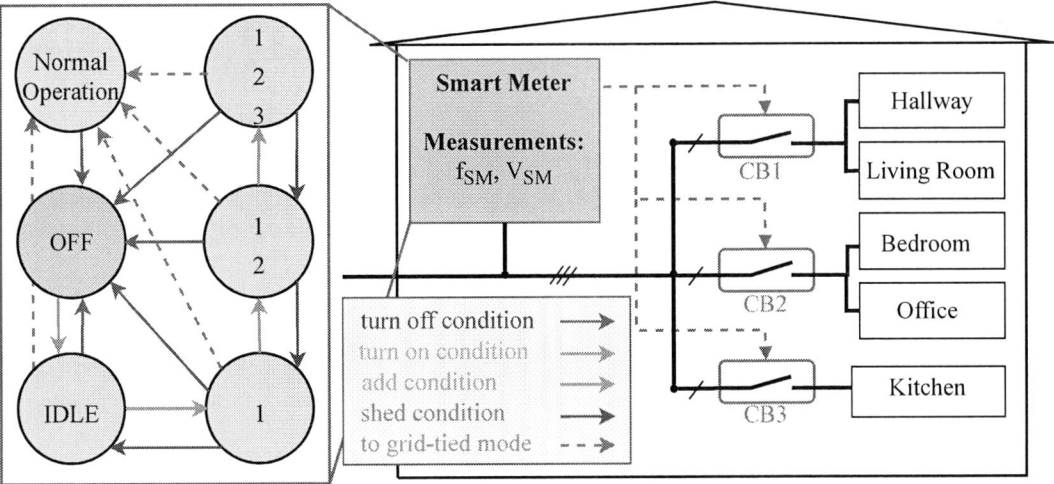

Fig. 3: Schematic of a representative household with a smart meter: sample configuration of 3 circuit breakers (right) and smart meter state machine during BSS (left)

or closed. Fig. 3 (left) shows a simplified state machine of the smart meter for 3 circuit breakers which can be further applied to more circuit breakers. The working principle is based on a *first in, last out* approach, where the highest priority is given to CB1, meaning it will be the first to be *added* and the last to be *shed*. Similarly, CB2 is given a higher priority than CB3. According to Fig. 3, the key conditions for smart meter state change can be defined as follows:

- The smart meter turns on once a minimum supply voltage $V_{min,SM}$ is restored (turn on condition). It turns off at any point when voltage goes below this limit (turn off condition).

- The smart meter would only go back to its normal operation when a successful BSS ends (to grid-tied mode) through typical communications signals.

- The criterion to *add* or *shed* load categories is in line with the concept of under-frequency relays for non-critical loads in real MGs [11]. Generally, adding a load category follows an observation of a minimum voltage $V_{min,loading}$ and a maximum frequency drop Δf (*add* condition). Furthermore, a category is *shed* when either the frequency drop exceeds Δf or the voltage drops below $V_{min,loading}$ (*shed* condition).

Counting in for the droop behavior of GFm inverters, Δf should be predefined as $\Delta f = P_{max,load} \cdot m_P$ where $P_{max,load}$ is the maximum allowed loading in per unit active power. Therefore, according to the chosen value of m_P in this work, Δf should be kept smaller than $1\,Hz$. This ensures that loads are shed before the GFm inverters reach their current limit, counting for the reactive power demand in the grid and the transient power peaks due to inrush currents. A waiting time is considered between state changes to allow a stable frequency measurement and in the case of the first add, a successful synchronization of the *Masters*. Additionally, an internal flag is used to track the connection of each circuit breaker. This flag is used to avoid an immediate *add* followed by a *shed*. Table II shows the parameterization of the Smart Meter state machine, where the minimum voltage values for the smart meter and loading are chosen according to available smart meter data sheets and German grid codes, respectively. The maximum allowed loading $P_{max,load}$, consequently defining the maximum allowed frequency drop, is the authors' choice for a conservative supply in the given testbench with the considered overcurrent protection. This marginal supply counts for the possible high-load transients between an *add* and a *shed*. Lastly, the waiting time is chosen based on the speed of the applied PLL and for better visibility in the RT tests.

Table II: Smart Meter State Machine Parameters

Parameter	Symbol	Value
Minimum supply voltage for the smart meter	$V_{\min,\text{SM}}$	$80\,\text{V}$
Minimum supply voltage for loading	$V_{\min,\text{loading}}$	$0.9\,\text{pu}$
Maximum allowed loading	$P_{\max,\text{load}}$	$\frac{2}{3}\,\text{pu}$
Maximum frequency drop before *shed*	Δf	$0.67\,\text{Hz}$
Waiting time between state change	T_{wait}	$2\,\text{s}$

2.2.2 Residential Load Classification

This work considers a selection of balanced 3-phase loads representing typical residential loads in the Microgrid. Assuming a complementary assignment of phases to CBs, each appliance is considered as a symmetric three-phase device representing three identical single-phase devices in the Microgrid. A single smart meter is then sufficient to control all the controllable loads according to Fig. 3. Similar three-phase consideration is applied to the non-controllable loads, resulting in a significant reduction of computational burden for RTS.

As extensively discussed in [6], most residential household appliances show high levels of inrush currents when they startup, known in grid restoration literature as *cold load pickup*. For some Switch-Mode-Power-Supplies (SMPS) and Electric Machines (EM), the inrush current exceeds ten times the steady-state rated current. The aggregation of these inrush currents at the time of the black start could consequently exceed the current limits of the supplying inverters. Although this shows the necessity of fast overcurrent protection for all GFm inverters, optimized distribution of the aggregated inrush currents in a simulated black start event in [6] was shown effective in not only reducing the peak current values but also the duration of overcurrent operation. In this approach, low-power high-inrush current loads of class SMPS are initially switched on, followed by EMs with mid-power and mid-inrush current levels, and lastly the high-power resistive loads. Generally, the mentioned principle, *SMPS-first resistive-last*, should be used to prioritize loads in a one-time configuration of the smart meter, based on the number of CBs and the profile of the connected loads. It can be assumed that apart from a few portable devices like a charger or a vacuum cleaner, most of the devices are stationary within a household and therefore have a fixed CB connected.

In this paper, the smart meter is configured for a selection of typical appliances listed in Table III as follows: The rooms with the highest inrush currents (SMPS loads) were assigned to CB1. CB2 controls the loads in the bedroom/office with medium levels of inrush current. Last but not least, the kitchen with mostly resistive loads is assigned to CB3. The non-controllable load is kept smaller than the controllable load to represent a high penetration of smart meters in the Microgrid.

3 Power Hardware-in-the-Loop Laboratory Setup

The verification of the BSS is carried out in a PHIL laboratory setup, as shown in Fig. 4. Two 3-phase inverter systems, one as *Master Seed* and one as *Master Fellow*, are considered with the GFm control explained in section 2.1. Through different lengths of LV lines, the two *Masters* are physically connected to a power amplifier that emulates the behavior of the load with the characteristics explained in section 2.2. According to the current-type ideal transformer model [12], the control loop of the power amplifier is closed as shown in Fig. 5. The physical measurement of the voltage at the terminal of the power amplifier is passed via *Power Amplifier Interface* to an ideal 3-phase controllable voltage source in the RTS *Model*. A low pass filter with a corner frequency of $500\,\text{Hz}$ is added to the voltage measurements in the simulation to increase the stability of the PHIL setup [3]. The RTS current of the aggregated load model is then passed as the current setpoint to *Power Amplifier Interface*. Additionally, the RT control is enabled through a console that provides RT monitoring and data logging. More details on the hardware characteristics and the RTS environment are given in the following sections.

Table III: Residential load configuration and component-level characteristics including inrush current I_{inrush}, and steady-state current I_{ss} according to measurements in [6].

Section	Appliances	Type	I_{inrush}(peak A)	I_{ss}(peak A)	P_{rated}(W)
Controllable Loads: CB1 (Hallway/ Living room)	Vacuum Cleaner	EM	30.7	8.3	1400
	TV (stand-by)	SMPS	4.8	0.3	2
	Chiller	SMPS	31.7	0.2	2100
Controllable Loads: CB2 (Bedroom/ Office)	Vacuum Cleaner	EM	13.6	9.3	850
	Fan	EM	0.4	0.3	45
	Printer (stand-by)	SMPS	19.6	0.6	1850
	Residual	R	NA	8.1	2640
Controllable Loads: CB3 (Kitchen)	Fridge	EM	12.6	1	90
	Toaster	R	NA	5.3	900
	Microwave	SMPS	5.5	9.1	1150
	Electric Kettle	R	NA	14.1	2400
Non-controllable Loads	Rice Cooker	R	NA	2.5	350
	LED	SMPS	0.2	0.1	2.3

3.1 Hardware Parameters

The power lines in the setup are modeled by physical passive components. One line section has a resistance of $70.3\,m\Omega$ and an inductance of $79.29\,\mu H$ and resembles a cable of 340 m length and the type NAYY $4 \cdot 150\,mm^2$. As shown in Fig. 4, the *Master Seed* inverter is connected to the busbar via one line section while the *Master Seed* is connected via three line sections. The two *Masters* share a common DC-link. The DC-link voltage is set to a constant value of 730 V and it is controlled by an active front end. The switching frequency is $f_s = 16\,kHz$ and the rated power of each inverter is $S_N = 16\,kVA$. Furthermore, both inverters comprise an LC filter with the parameters $L_f = 850\,\mu H$ and $C_f = 10\,\mu F$. Corresponding to the rated power, the rated peak current of each phase is 33 A.

The power amplifier has a rated power of $S_N = 100\,kVA$ and a switching frequency of $f_s = 125\,kHz$. The DC-link voltage is 770 V and is controlled by an active front end. The details of the filter and control design of the power amplifier are not known. However, it shall be noted here that resonance interactions between the filter of the power amplifier and the LC filters of the *Master* inverters might occur. A set of control parameters that leads to acceptable damping of any high-frequency oscillations is chosen. Further information and a picture of the laboratory setup can also be found in [13].

3.2 Real-time Residential Load Emulation

The Opal RT real-time simulator OP5707 is used in this CPU-based setup with a sampling time of $t_S = 50\,\mu s$ that corresponds to the expected transients in the system and the switching frequency of the power amplifier while maintaining an RT execution. The coupled RT Lab software implements the ARTEMiS solver add-on in SimPowerSystems (SPS) of Matlab Simulink. This solver has a higher order (order 5) than SPS (order 2) which aims for higher accuracy in RTS. Additionally, since ARTEMiS precomputes and discretizes the typical state-space matrices of SPS in advance, it significantly reduces the computational time so that the real-time limit is not crossed [14]. As the electrical system expands, the size of the state-space matrice exceeds, causing memory flows for the precomputation of ARTEMiS. To overcome this challenge, a system decoupling is achieved with the State Space Nodal (SSN) methods [15] which divides the electrical system into nodal groups specified by the user. Each group is then solved by a state-space method while the nodal method computes the interface between the groups through SSN Nodal Interface Blocks (SSN NIB). In this way, the total size of the state-space matrices in the memory can be significantly reduced. However, not every SPS block is compatible with SSN NIB. The SPS-based models of resistive and SMPS loads of [6] can be further used in ARTEMiS RTS connected to V-type (virtual voltage source) SSN NIB . The SPS Asynchronous Machines, on the other hand, should be replaced by SSN Induction Machines (SSN IM) connected to X-type (external group) SSN NIB.

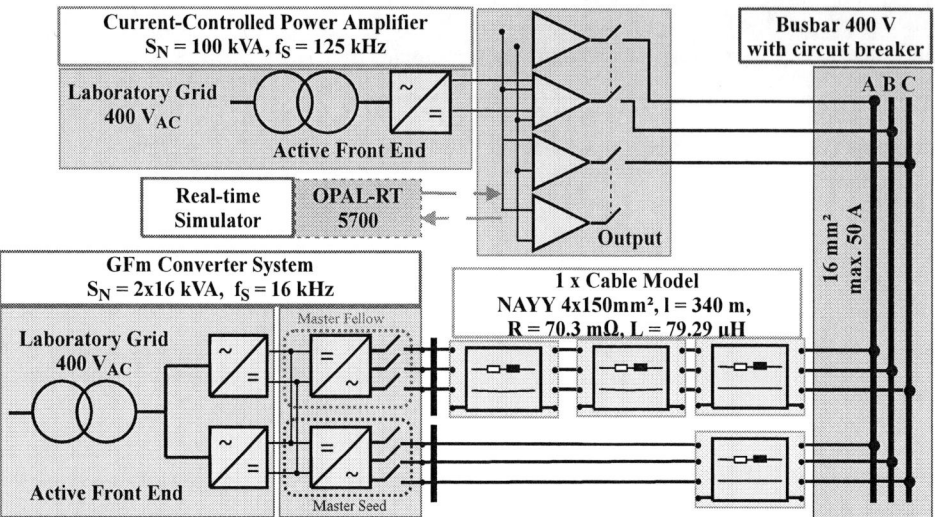

Fig. 4: System under test: MG consisting of GFm converters, a power amplifier, LV cable boxes, and an RTS considering an aggregated load model of multiple households in a symmetric operation.

The main challenge here is to intelligently introduce nodes to keep the system running in real-time. Generally, a minimum number of nodes should be introduced, while each group should be kept with a reasonable number of states and switches. The indicator of the size of the electrical system, i.e. the size of the corresponding state-space matrix, is the number of switches in each SSN group, the maximum of which is 15. For the residential configuration given in Table III, such optimization is carried out according to Fig. 5 as follows: Each EM with a 3-phase SSN IM model forms a nodal group through a separate 3-phase SSN NIB. For the SMPS load LED with 4 switches per phase, one nodal group is introduced for its 3-phase model (12 switches) through a 3-phase SSN NIB. It is important to note that since the current levels (both inrush and steady-state) of n number of LEDs are equal to n times the current levels of a single LED [16], the parameters of the circuit model of an LED are accordingly scaled to represent 20 instances with no additional switches needed in the circuit. In the case of the Microwave and the Printer, two additional breakers were used inside the SMPS model to count for the inrush behavior. Therefore, each phase of the device would unavoidably need a separate grouping (6 switches). Alternatively, for the TV and the Chiller in the same room, one 1-phase SSN NIB can mutually group each phase of the two SMPS loads (12 switches). Last but not least, the three 3-phase Circuit Breakers are grouped into a 3-phase SSN NIB (9 switches). In total, the residential load in this setup is grouped into 15 SSN nodes with moderate execution time and CPU usage, resulting in no overruns during operation. A resistor of $10\,\mathrm{m\Omega}$ per phase is assumed as the equivalent of the Smart Meter hardware. The voltage magnitude and frequency at the terminal of the Smart Meter are measured by a DSOGI-PLL, based on SPS blocks. The measurements are then fed to the Smart Meter state-machine which is implemented as a Moore-Machine in state-flow Simulink according to Fig. 3. The line model represents a service connection including the electricity meter and over-current protection elements with a standard length of 30 m and $6\,\mathrm{m}^2$ cross-section ($R = 3.690\,\frac{\Omega}{\mathrm{km}}$, $X = 0.094\,\frac{\Omega}{\mathrm{km}}$), as specified in [17].

4   Measurement Results

All measurements results are based on the internal sensors of the power amplifier and the inverter systems. The PHIL demonstration of the proposed BSS is shown in Fig. 6. For a clear illustration of power-sharing between the GFm inverters, the harmonics are reduced via the application of a low pass filter with a corner frequency of 50 Hz to the active and reactive power measurements of the *Master Seed* and *Master Fellow* (P_{MS}, P_{MF}, Q_{MS} and Q_{MF}, respectively). The frequency f_{SM} and the voltage V_{SM} are measured with the PLL of the smart meter which is located at the load connection point in the RTS. The

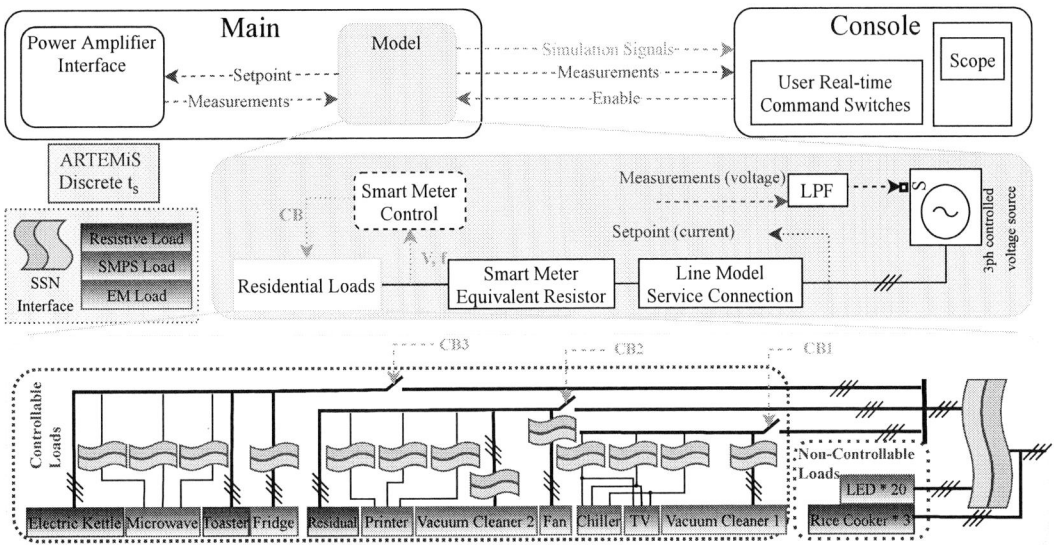

Fig. 5: Real-Time simulation environment and the optimized SSN grouping.

same PLL design as described in section 2.1 is also used for the smart meters.

The sequence starts with the voltage ramp up by the *Master Seed* that readily supplies the directly-connected base load as well as the significant demand of negative reactive power of the filter capacitor of the connected *Master Fellow*. The *Master Fellow* synchronizes to the sensed voltage and is enabled at $t = 6$ s, which leads to oscillations in the power. Due to the gradual change of the voltage and frequency setpoints of the *Master Fellow* to the nominal values, an equal active power-sharing of the two GFm inverters is achieved after approximately 3 s. At $t = 13$ s the smart meter starts to add load categories according to its state machine. After category C is added, the frequency drops below the predefined threshold $\Delta f = 0.67$ Hz, corresponding to a power of $P_{\text{max,load}} = \frac{2}{3}$ pu. Therefore, the smart meter sheds category C and the steady-state operation of the Microgrid is reached. While the frequency drop follows the active power of the GFm inverters according to the droop characteristic, the voltage drop not only depends on the reactive power but additionally on the virtual impedance in the control and the physical impedance of the power lines. Therefore, in the absence of an integral behavior in the Q-V droop and different lengths of the line of the two GFM inverters to the load, the reactive power is not equally shared.

In Fig. 7, the measured currents of the power amplifier i_{Load} are shown. The inrush currents at the connection of each load category are mainly caused by non-linear loads including diode rectifiers and induction machines. Since the inverters in this work are not equipped with a fast current limiting, the inrush currents become a bottleneck in the BSS. Although the inverters do not operate at nominal power in the steady-state, the peak current is already close to the maximum value of 33 A per inverter. Besides, the fast change of the current leads to high-frequency oscillations that are quickly damped. This can be explained by the interactions between the power amplifier and the inverter system and their respective filters. During steady-state operation with all load categories connected, the total harmonic distortion of the current is 2.7 %, well below the threshold 8 % specified by EN 61000-2-4.

5 Conclusion

In this paper, a bottom-up sequence is verified in Power Hardware-in-the-Loop laboratory tests to establish an islanded Microgrid after a blackout. The sequence requires a minimum hardware change of household circuit breakers but no external communication infrastructure. Following predefined rules, power is restored in the Microgrid based on a multi-master operation of droop-controlled inverters while the smart meters govern a partial loading. Cold load pickup is handled by smart meters through prioritized gradual adding of residential load groups according to their start-up behavior. On the other hand,

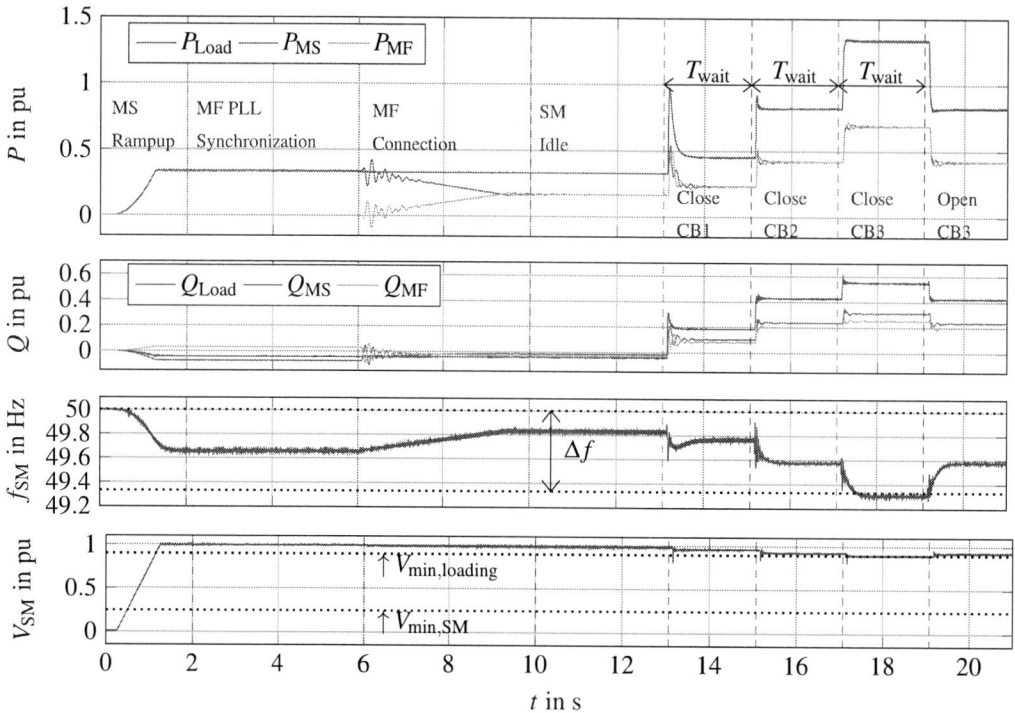

Fig. 6: PHIL measurements of the inverters and the SM during the proposed BSS. The base values of the per-unit system are the rated power S_N of each GFm inverter and the nominal grid voltage V_N.

an overload operation is reduced through the early shedding of the last added load groups. An aggregated household model comprising a variety of loads is successfully integrated into the test bench by real-time simulation. Withstanding peak inrush currents of the loads without triggering the protection of the grid-forming inverters is found as the bottleneck of the operation. Although such optimized distribution of load inrush currents minimizes the duration and level of overcurrent operation, further study into compatible fast current limiting methods is identified as a key requirement for higher reliability in the black start sequence and better protection of the grid-forming inverters. Furthermore, in case of moderate penetration of smart meters in the Microgrid, a lower voltage level and hence an earlier synchronization of grid-forming inverters should be further investigated to reduce the loading of the leading inverter that is responsible for the supply of uncontrollable base loads.

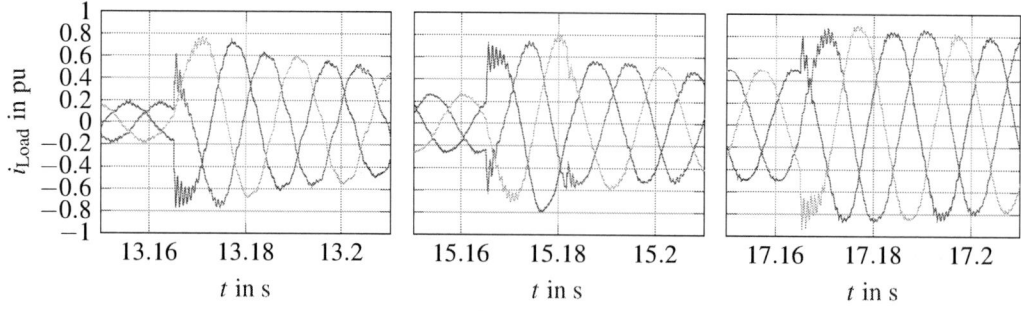

Fig. 7: Measured inrush currents at the connection time of CB1, CB2 and CB3. The base value of the per-unit system 1 pu = 66 A peak is the accumulated rated current amplitude of both GFm inverters.

References

[1] F. O. Resende, N. J. Gil, and J. A. P. Lopes, "Service restoration on distribution systems using Multi-MicroGrids," Eur. Trans. Electr. Power, vol. 21, no. 2, pp. 1327–1342, Mar. 2011, doi: 10.1002/etep.404.

[2] M. Mirzadeh, et al, "A Rule-based Concept for a Bottom-up Multi-Master Black Start of an Inverter-Dominated Low-Voltage Cell," to be published in Proc. IEEE 13th International Symposium on Power Electronics for Distributed Generation Systems (PEDG), 2022.

[3] J. Montoya et al., "Advanced Laboratory Testing Methods Using Real-Time Simulation and Hardware-in-the-Loop Techniques: A Survey of Smart Grid International Research Facility Network Activities," Energies, vol. 13, no. 12, Art. no. 12, Jan. 2020, doi: 10.3390/en13123267.

[4] H. Haes Alhelou, M. Hamedani-Golshan, T. Njenda, and P. Siano, "A Survey on Power System Blackout and Cascading Events: Research Motivations and Challenges," Energies, vol. 12, no. 4, p. 682, Feb. 2019, doi: 10.3390/en12040682.

[5] J. V. Milanovic, K. Yamashita, S. Martínez Villanueva, S. Ž. Djokic and L. M. Korunović, "International Industry Practice on Power System Load Modeling," in IEEE Transactions on Power Systems, vol. 28, no. 3, pp. 3038-3046, Aug. 2013, doi: 10.1109/TPWRS.2012.2231969.

[6] M. Mirzadeh and A. Mertens, "Measurement-based Component-level Load Modeling for Evaluation of a Current-suppressing Loading Scenario for Microgrid Black Start Events," to be published in Proc. 13th IEEE PES ISGT North America 2022, Apr. 2022.

[7] J. Rocabert, A. Luna, F. Blaabjerg and P. Rodríguez, "Control of Power Converters in AC Microgrids," in IEEE Transactions on Power Electronics, vol. 27, no. 11, pp. 4734-4749, Nov. 2012, doi: 10.1109/TPEL.2012.2199334.

[8] M. E. Elkhatib, W. Du and R. H. Lasseter, "Evaluation of Inverter-based Grid Frequency Support using Frequency-Watt and Grid-Forming PV Inverters," 2018 IEEE Power & Energy Society General Meeting (PESGM), 2018, pp. 1-5, doi: 10.1109/PESGM.2018.8585958.

[9] J. C. Vasquez, J. M. Guerrero, M. Savaghebi, J. Eloy-Garcia and R. Teodorescu, "Modeling, Analysis, and Design of Stationary-Reference-Frame Droop-Controlled Parallel Three-Phase Voltage Source Inverters," in IEEE Transactions on Industrial Electronics, vol. 60, no. 4, pp. 1271-1280, Apr. 2013, doi: 10.1109/TIE.2012.2194951.

[10] S. A. Richter and R. W. De Doncker, "Digital proportional-resonant (PR) control with anti-windup applied to a voltage-source inverter," Proceedings of the 2011 14th European Conference on Power Electronics and Applications, 2011, pp. 1-10.

[11] M. Barnes et al., "Real-World MicroGrids-An Overview," in 2007 IEEE International Conference on Systemof Systems Engineering, Apr. 2007, pp. 1–8

[12] W. Ren, M. Steurer and T. L. Baldwin, "Improve the Stability and the Accuracy of Power Hardware-in-the-Loop Simulation by Selecting Appropriate Interface Algorithms," in IEEE Transactions on Industry Applications, vol. 44, no. 4, pp. 1286-1294, Jul-Aug. 2008, doi: 10.1109/TIA.2008.926240.

[13] M. Dokus and A. Mertens, "On the Coupling of Power-Related and Inner Inverter Control Loops of Grid-Forming Converter Systems," in IEEE Access, vol. 9, pp. 16173-16192, 2021, doi: 10.1109/ACCESS.2021.3053060.

[14] RT-LAB Help, ARTEMiS User's Guide, Version 2019.1.0.140, OPAL-RT Technologies.

[15] C. Dufour, J. Mahseredjian and J. Belanger, "A combined state-space nodal method for the simulation of power system transients," 2011 IEEE Power and Energy Society General Meeting, 2011, pp. 1-1, doi: 10.1109/PES.2011.6038887.

[16] J. F. G. Cobben and N. Arumdati, "Inrush Related Problems Caused by Lamps and Electric Vehicle," in Conference on Applied Electromagnetic Technology (AEMT), Lombok, Jul. 2011, p. 7.

[17] S. Papathanassiou, N. Hatziargyriou and K. Strunz, "A Benchmark Low Voltage Microgrid Network", CIGRE TF 6.04.02, Jan. 2005.

Detection of Incipient Inter-Turn Short-Circuit Faults by Artificial Intelligence Classifiers

Osman Örgüt[12], İlker Şahin[2], Ece Olcay Güneş[3]

[1] Istanbul Technical University
İTÜ Ayazağa Campus, Building of Graduate School, 34469, Maslak
Istanbul, Türkiye
Phone: +90 (545) 843-1845
Email: orgut20@itu.edu.tr

[2] ASELSAN Inc.
Mehmet Akif Ersoy, İstiklal Marşı Caddesi No:16, 06200
Ankara, Türkiye
Email: oorgut@aselsan.com.tr

[3] Istanbul Technical University
İTÜ Ayazağa Campus, Faculty of Electrical and Electronics Engineering, 34469, Maslak
İstanbul, Türkiye

Keywords

≪Fault detection≫, ≪Induction motor≫, ≪Machine learning≫, ≪Neural network≫, ≪Model predictive control≫

Abstract

This paper presents two artificial intelligence (AI) based identification methods for inter-turn short-circuit fault (ISCF) detection in induction motors (IMs), driven by voltage source inverters (VSIs). It was previously observed that, for an IM driven by finite control set model predictive control (FCS-MPC), the ISCF occurrence disturbs the balanced distribution of the resultant switching vectors, which are merely the control outputs of the FCS-MPC scheme. This effect of the ISCFs is utilized for fault detection purposes. The proposed method successfully detects the ISCF using AI methods which are fed by histograms of switching vectors along with torque and speed. This is especially convenient from the motor driver's perspective since no additional sensor or hardware is required for fault detection. The dataset utilized in this paper was obtained from an experimental IM drive test setup, on which intentional ISCFs can be created. The test results proved that the average fault detection rate is 95.8%, for an ISCF of 2-turns in a 104-turns phase winding.

Introduction

The squirrel cage induction motors are the most widely used electrical machine type due to their lower cost, ruggedness, robust structure, lower maintenance requirements, market availability, and ability to work in harsh working environments [1], [2], [3]. About 85% of all motors used in industrial equipment are IMs [4], [5]. According to [6], the stator-related faults constitute 54.8% of all IM faults, based on surveys conducted in different repair workshops. It is reported in [7] that, stator faults generally start as ISCF and if they cannot be diagnosed in the early stage, the fault can lead to phase to phase or phase to ground short circuits. Thus, early detection of the ISCF can prevent most of the faults which occur in electrical machines. Therefore, this study considers the early detection of ISCFs for IMs.

There are many studies in the literature to diagnose ISCFs by using several different methods, which can be broadly categorized into three groups. The first group is the signal-based methods, where fault traces are detected based on measured quantities. In [8], [9] and [10], stator currents are used as the main detection element. The motor current signature analysis (MCSA) is employed and fault traces are analyzed. The second most utilized quantity is vibration. The ISCF fault detection based on vibration data is studied in [11], [12], [13] and reported promising results. However, in signal-based techniques, the need for an external sensor is a major disadvantage. The second group of fault detection methods is the model-based approach. In this approach, the knowledge of mathematical and physical features of the system is utilized so that any significant change from the healthy state helps diagnose the fault. In [14], an extended Kalman filter is designed for a permanent magnet synchronous generator (PMSG) based on the model of PMSG. In [15], an error vector is calculated based on the phase currents and the model of the permanent magnet synchronous motor (PMSM). The study reported the method as a good monitoring technique in transient conditions. However, the necessity of the model parameters is a serious drawback for model-based approaches. The third group is the artificial intelligence methods, which rely on processing the data. Based on the faulty and healthy data, a classification structure can be trained and the trained structure can be used as a classifier [16]. In [17] and [18], similar neural network (NN) architectures are designed for detection. However, the inputs of the neural network classifiers originated from phase currents. In [19], the output torque waveform of the electrical machine is fed to the NN classifier. In [20], the phase current information is converted into images, and a traditional convolutional neural network (CNN) architecture is utilized as the classifier. Up to 99% accuracy of detection is reported. The problem of these diagnosis techniques is the same as the signal-based techniques, which is the necessity of a quantity that is measured by an external sensor. In the proposed method, the quantities that are generated in the control algorithms are utilized as the base of the input feature vector of the classifier. In addition, using AI methods for classification eliminates the human decision for fault threshold determination.

It is previously shown in [21] that, for an IM driven by a VSI via model predictive control (MPC), the occurrence of ISCF results in an unexpected unbalance in the switching vector distribution over full electrical periods of the motor. The unbalance in the switching vector statistics was evaluated to yield a "fault indicating score", and the healthy/faulty decision was made by comparison to a predefined threshold. The selection of the predefined threshold is subject to change for different torque and speed. Thus, this drawback addresses the consideration of more intelligent classifiers. In this study, the switching vectors generated by the motor drive setup of [23], under various torque-speed values, for both healthy and faulty modes are utilized for ISCF diagnosis. The recorded data is first prepared to be fed into the machine learning (ML) algorithms. By observing different types of ML algorithms Quadratic Support Vector Machines (QSVM) and NN structures are selected as classifiers. The training phase of SVM and NN classifiers with the recorded switching vector data was realized in MATLAB Classification Learner Toolbox and Neural Net Pattern Recognition Toolbox, respectively. Trained classifiers are tested and it is proven that the proposed methods can diagnose ISCF with an average accuracy of 93.52% and 95.88% respectively for a wide range of operating conditions.

A fundamental difference between the proposed method from the previous literature is the quantity that is used to diagnose the fault. In the previous literature, the main quantities which are used to diagnose a fault are; current, voltage, and vibration. These quantities would often require external sensors. However, switching sequences are already generated by the controller. Thus, diagnosing a fault through the examination of switching sequences requires no additional hardware, and it is very convenient from the motor driver's viewpoint.

As the ISCF fault of an electrical machine might be fatal in terms of safety and costs, the condition monitoring of this fault is a significant field of research. For that reason, a method that does not require an additional sensor is proposed for fault detection. The proposed method utilizes the voltage vector data which are generated in the motor driver. In addition, since the methods are based on artificial intelligence, there is no human decision for fault thresholds. As an outcome, the proposed methods detect the incipient ISCF (2 out of 104 turns of a phase winding) with an average accuracy of 95.8%.

The Proposed Fault Detection Method

In FCS-MPC, the controller evaluates all possible switching vectors at each control cycle and applies the optimum switching vector at the next switching instant [23]. Thus, the FCS-MPC-based controller's outputs are discrete voltage vectors which makes any statistical analysis more convenient. For illustration purposes, the raw data, which corresponds to the switching sequences of 3750 rpm and 0.3 Nm operation, is depicted for both healthy and faulty motor in Fig. 1. The faulty data correspond to the 2-turns ISCF (out of 104-turns in a phase winding), as detailed in [21]. As can be seen from Fig. 1, raw data of healthy and faulty cases seem almost similar. However, the difference between faulty and healthy cases is more obvious in the histogram in Fig. 2. The occurrence of an ISCF manifests itself in the distribution of switching vectors as an unexpected unbalance. Our main motivation is to detect this change with the use of ML algorithms. Since ML algorithms are utilized for classification, there is no need for manual threshold selection. This feature of the method provides a more general solution for fault detection.

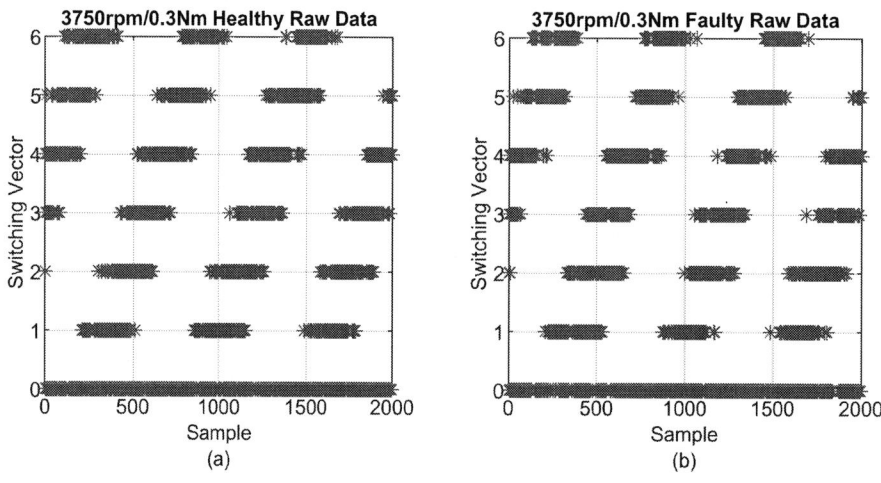

Fig. 1: Raw data of 3750 rpm and 0.3 Nm operation for (a) healthy and (b) faulty motor.

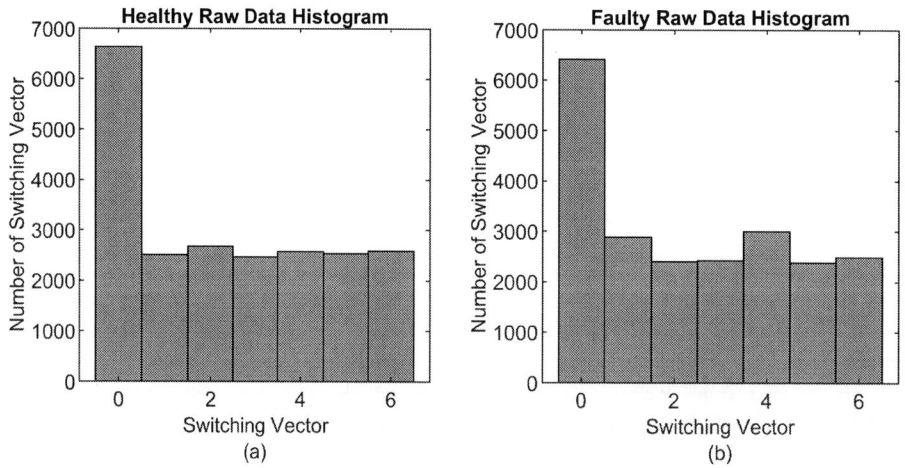

Fig. 2: Histogram of raw data of 3750 rpm and 0.3 Nm operation for (a) healthy and (b) faulty motor.

The proposed method includes four main units, which are data acquisition, data pre-processing, feature vector generation, and classification unit. The block diagram of the proposed method is shown in Fig. 3.

The methodology is as follows: five periods of MPC-generated switching vectors, along with the torque and speed values of the operating point, are taken in the data acquisition unit with a sampling frequency of 40 kHz. The experimental setup which the data are taken from is shown in Fig. 4. The speed information of the motor is already available in the motor driver and the torque of the induction motor is measured from an external torque sensor. Although the torque values from the external sensor are utilized in the experimental dataset, torque is also estimated in the standard FCS-MPC routine. Thus, the proposed methods are valid in the absence of a torque sensor. The switching vectors of five electrical periods are fed into the data pre-processing unit, in which all switching vectors are counted and the ratio of each vector to the total number of vectors is found. This operation yields seven feature vectors. Torque and speed data are added to the feature vector in the feature vector generation unit. Then, the feature vector generation unit outputs nine features vector to be fed to the classification unit. Finally, the classification unit takes the nine-features vector and outputs the machine status as healthy and faulty.

The occurrence of ISCF in an IM controlled by a VSI via MPC results in an unanticipated unbalance in the switching vector distribution during whole electrical periods of the motor, as previously shown in [21]. The unbalance in the switching vector data was assessed to produce a "fault indicating score," and the healthy/faulty choice was made by comparing the score to a predetermined threshold. For variable torque and speed, the specified threshold may need to be adjusted. Thus, this drawback addresses the consideration of more intelligent classifiers. Hence, In the classification unit, artificial intelligence-based classifiers are utilized. In this work, two methods, namely QSVM and single-layer NN are implemented as classifiers and the test results are presented.

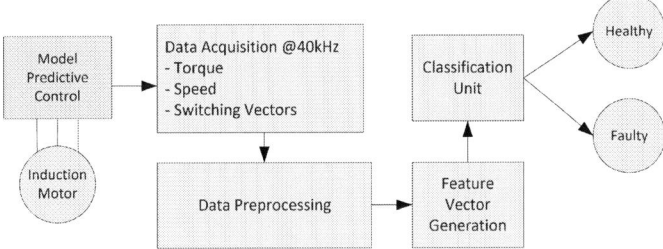

Fig. 3: Block diagram of proposed method.

Training

The data which is utilized in this study for both training and testing is collected from the test setup in Fig. 4 and the details of the setup can be found in [23]. The dataset is divided into train and test splits with a ratio of 0.52. Although the generally accepted training-test splits ratio is about 0.7 since the samples are acquired for various speed and torque pairs, to test every torque and speed that are trained, the ratio had to be close to 0.5. Training-test splits can be seen from Table I. The base of the dataset that is used in this work can be found in [22]. However, the data are slightly different from [22] in terms of train/test splits.

MATLAB Classification Toolbox is used to train the classifier. There are 23 classifiers available in the toolbox. Each classification algorithm in the toolbox is trained, validated, and tested with the default options of the Classifier Toolbox. As training data, 186 feature vectors which are composed of histograms of voltage vectors, torque, and speed are utilized. In the training phase, MATLAB splits this training data as 0.7 training, 0.15 validation, and 0.15 test data. These ratios are altered to increase the performance of the classifiers and the mentioned ratios are determined. To avoid over-fitting MATLAB toolbox benefits from the 5-fold Cross-Validation technique. The training and test accuracy are calculated as the sum of true positives and true negatives over all predictions. As test data, a total of 170 feature vectors are used. Training and test accuracy results of all classifiers are presented in Table II. As can be seen from the Table II, Support Vector Machine classifiers are better in terms of the test accuracy. However, to implement the algorithm to a microcontroller, K-Nearest Neighbors' (KNN) algorithms can be more

Table I: Data information for training and test

Speed (rpm)	Torque (Nm)	Motor Status	Train Samples	Test Samples
1500	0.3	Healthy	2	2
1500	0.3	Faulty	2	2
1500	0.9	Healthy	2	2
1500	0.9	Faulty	3	2
2250	0.3	Healthy	3	0
2250	0.3	Faulty	3	3
2250	0.3	Healthy	4	4
2250	0.3	Faulty	4	4
2250	1.25	Faulty	4	4
2250	1.25	Healthy	8	4
2250	1.25	Faulty	4	4
3000	0.3	Healthy	5	5
3000	0.3	Faulty	5	5
3000	0.3	Healthy	5	5
3000	0.3	Faulty	5	5
3000	1.2	Healthy	5	5
3000	1.2	Healthy	5	5
3000	1.3	Faulty	5	5
3000	1.3	Faulty	5	6
3000	1.35	Healthy	5	5
3000	1.35	Faulty	5	5
3750	0.3	Healthy	6	0
3750	0.3	Faulty	6	6
3750	0.3	Healthy	6	6
3750	0.3	Faulty	6	6
3750	1.25	Healthy	0	7
3750	1.25	Faulty	7	7
3750	1.25	Healthy	7	7
3750	1.25	Faulty	7	7
4500	0.3	Healthy	5	2
4500	0.3	Faulty	5	2
4500	0.3	Healthy	8	8
4500	0.3	Faulty	8	8
4500	1.15	Healthy	5	3
4500	1.15	Faulty	5	3
4500	1.15	Healthy	8	8
4500	1.15	Faulty	8	8
		Total	186	170

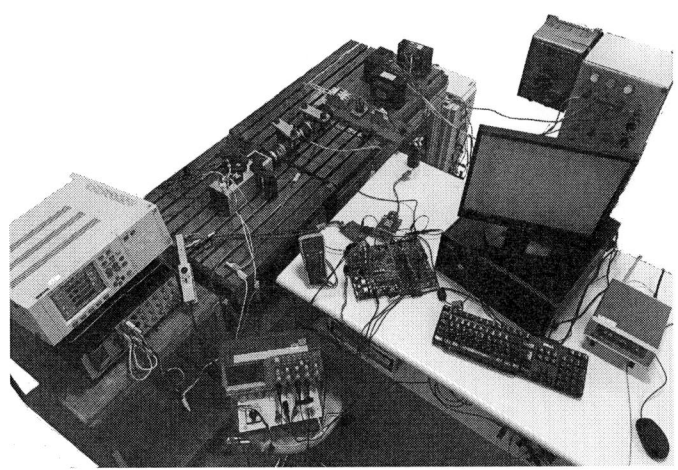

Fig. 4: Experimental setup which the data are acquired [23],[21].

useful. Since the implementation is not considered in this work, the QSVM classifier is selected.

Table II: Comparison of different classifiers in Classifier Learner Toolbox

Classifier	Train Accuracy (%)	Test Accuracy (%)
Quadratic SVM	91.4	93.53
Cubic SVM	90.9	91.76
Medium Gaussian SVM	86.6	91.18
Fine Gaussian SVM	90.9	90.59
Cosine KNN	86.6	90
Weighted KNN	91.4	89.41
Fine Tree	81.2	88.82
Medium Tree	81.2	88.82
Medium KNN	80.6	88.24
Subspace KNN	87.1	88.24
Linear SVM	86	87.65
Fine KNN	88.7	87.65
RUSBoosted Trees	84.9	87.65
Cubic KNN	83.9	87.06
Bagged Trees	89.2	87.06
Boosted Trees	60.8	84.71
Subspace Discriminant	83.9	84.12
Linear Discriminant	79	80
Logistic Regression	79.6	77.06
Coarse Gaussian SVM	72	75.29
Coarse Tree	73.7	74.7
Coarse KNN	54.8	60

As the second classifier, a neural network architecture is considered. For this structure, MATLAB Neural Pattern Recognition Toolbox is used. In this toolbox, training-validation-test sets are randomly divided as 0.7, 0.15, 0.15, respectively. The optimization algorithm to find optimum parameters is declared as scaled-conjugant gradient back-propagation by the toolbox. As a loss function, cross-entropy loss function is utilized. For the sake of simplicity and implementation purposes, a single-layer network that includes 39 hidden neurons is trained. Block diagram of trained NN structure can be seen in Fig. 5. To choose a neuron number, a training loop structure is configured and the neuron number is generated randomly.

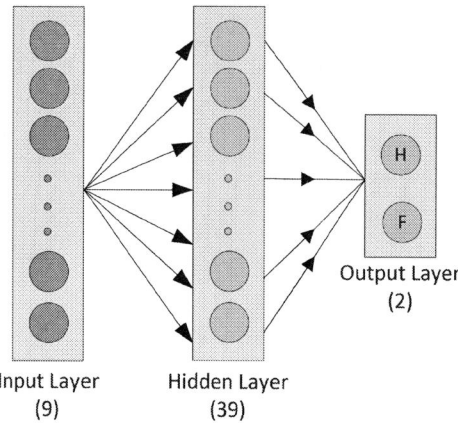

Fig. 5: Block diagram neural network classifier.

A comparison in terms of the accuracy is conducted and 39 is selected as a hidden neuron number. Training performance of the neural network is presented in Fig. 6. The confusion matrices of the training phase are also illustrated in Fig. 7. As can be seen from the Fig. 7, there is no over-fitting or under-fitting problem in the NN classifier.

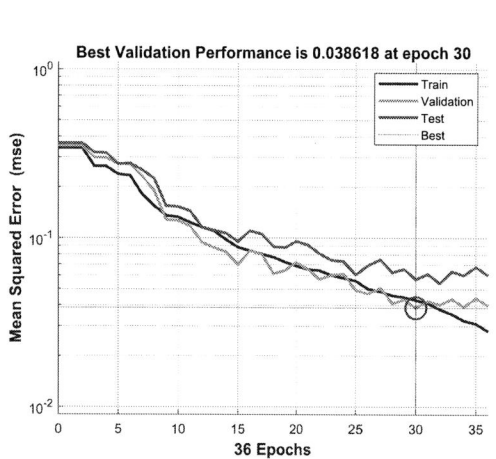

Fig. 6: Training performance of the proposed NN classifier.

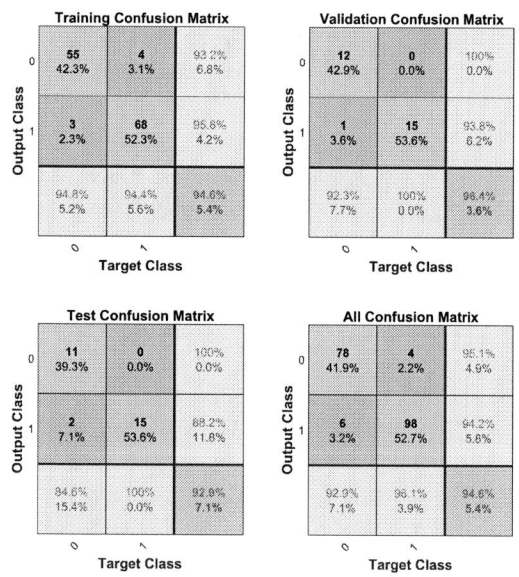

Fig. 7: Confusion matrix of train, validation and test for training phase.

Test Results

To measure the performance of the proposed condition monitoring techniques, accuracy metrics are calculated for the test dataset. This dataset is fully independent of the training dataset to measure the performance of the classifiers on unseen data. Test data includes 170 samples of the feature vector.

This feature vector includes periodical histograms of the voltage switching vectors, torque and speed information. Since the histograms are calculated period by period, sample sizes are relatively small. However, the information in these histograms is derived from the average of 500 samples in each period.

The accuracy metric is simply the ratio of the sum of the true positives and true negatives over all predictions. The test accuracy of QSVM was calculated as 93.53% based on 170 test samples for various torque and speeds. The second classification method is a single-layer neural network that includes 39 hidden neurons. The test accuracy of NN is up to 95.88%. The results of these two methods are presented in Table III. The confusion matrix of the NN classifier for the test dataset is presented in Fig. 8.

As can be seen from Table III, Quadratic SVM diagnoses the IFSC faults with the accuracy of 93.52%, and NN diagnoses the fault with the accuracy of 95.88%. For both classifiers, 4500 rpm and 1.15 Nm is the worst-case scenario, and the accuracy of diagnosis decreases down to 50% in this operating condition. Full torque at high-speed operation (where the inverter is running close to the overmodulation region, i.e. the utilization of zero vectors starts to diminish) was also observed to yield significantly lower "fault indicating scores" in [21].

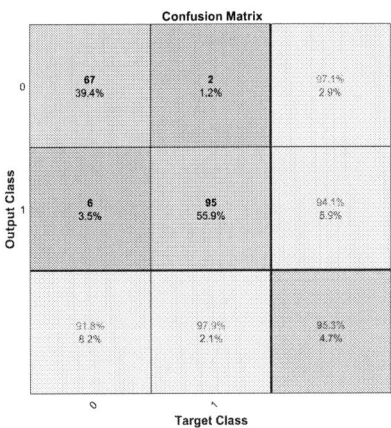

Fig. 8: Confusion matrix of test dataset.

Conclusion

In this paper, two types of artificial intelligence classifiers, namely quadratic support vector machines and single-layer neural networks, are utilized to detect incipient ISCF (2 out of 104 windings) in an IM. In the proposed detection methods, no external sensors or additional hardware are required to diagnose faults. Internally generated switching vectors in the inverter, torque, and speed of the motor are used as the indicators, which is also convenient for implementation purposes. All of the quantities which are used as inputs for the feature vector are already calculated or measured in the motor driver for the control algorithm. The training and testing data are collected from a real induction motor drive test setup. The accuracy of the classifiers is calculated and test results are promising for both methods. In the Quadratic SVM method, accuracy of 93.52% is achieved and in the NN method, the accuracy of 95.88% is measured for overall test data. This work aims to be introductory research for ISCF detection in induction motors by using AI techniques. The result from the comparison of the proposed two classifiers is that the neural network architectures might be better in this field. Thus, more sophisticated neural network architectures can be designed for improved performance. In addition, a real-time implementation of the proposed methods to the actual motor drive setup and considering transient conditions such as torque and speed transitions are being planned as future work.

Table III: Test Results of Quadratic SVM and NN classifiers

Speed (rpm)	Torque (N.m)	Sample	Motor Status	Quadratic SVM (%)	Neural Network (%)
3750	1.25	7	Healthy	57	100
2250	1.25	4	Healthy	100	100
2250	0.3	3	Faulty	100	100
2250	1.25	4	Faulty	100	100
3750	0.3	6	Faulty	100	100
3750	1.25	7	Faulty	100	100
3000	0.3	5	Healthy	100	100
3000	1.2	5	Healthy	100	100
1500	0.3	2	Healthy	100	100
1500	0.9	2	Healthy	100	100
4500	0.3	8	Healthy	100	100
4500	1.15	8	Healthy	62.50	50
2250	0.3	4	Healthy	100	100
3750	0.3	6	Healthy	100	100
3750	1.25	7	Healthy	100	100
3000	0.3	5	Faulty	80	100
3000	1.3	6	Faulty	100	100
1500	0.3	2	Faulty	100	100
1500	0.9	2	Faulty	100	100
4500	0.3	8	Faulty	100	100
4500	1.15	8	Faulty	50	62.50
2250	0.3	4	Faulty	100	100
2250	1.25	4	Faulty	100	100
3750	0.3	6	Faulty	100	100
3750	1.25	7	Faulty	100	100
3000	1.35	5	Faulty	100	100
3000	0.3	5	Faulty	100	100
3000	1.35	5	Healthy	100	100
3000	0.3	5	Healthy	100	100
3000	1.25	5	Faulty	100	100
3000	1.25	5	Faulty	100	100
4500	0.3	2	Healthy	100	100
4500	0.3	2	Faulty	100	100
4500	1.15	3	Healthy	100	100
4500	1.15	3	Faulty	100	100
			Total Accuracy	93.52	95.88

References

[1] Marino, R., Tomei, P., & Verrelli, C. M. (2010). Induction motor control design. Springer Science & Business Media.

[2] Tze Fun Chan; Keli Shi, "Philosophy of Induction Motor Control," in Applied Intelligent Control of Induction Motor Drives , IEEE, 2011, pp.9-30, doi: 10.1002/9780470825587.ch2.

[3] G. Pellegrino, A. Vagati, B. Boazzo and P. Guglielmi, "Comparison of Induction and PM Synchronous Motor Drives for EV Application Including Design Examples," in IEEE Transactions on Industry Applications, vol. 48, no. 6, pp. 2322-2332, Nov.-Dec. 2012, doi: 10.1109/TIA.2012.2227092.

[4] M. Riera-Guasp, J. A. Antonino-Daviu and G. Capolino, "Advances in Electrical Machine, Power Electronic, and Drive Condition Monitoring and Fault Detection: State of the Art," in IEEE Transactions on Industrial Electronics, vol. 62, no. 3, pp. 1746-1759, March 2015, doi: 10.1109/TIE.2014.2375853.

[5] S. Kumar et al., "A Comprehensive Review of Condition Based Prognostic Maintenance (CBPM) for Induction Motor," in IEEE Access, vol. 7, pp. 90690-90704, 2019, doi: 10.1109/ACCESS.2019.2926527.

[6] N. Bessous, "Reliability Surveys of Fault Distributions in Rotating Electrical Machines : – Case Study of Fault Detections in IMs –," 2020 1st International Conference on Communications, Control Systems and Signal Processing (CCSSP), 2020, pp. 535-543, doi: 10.1109/CCSSP49278.2020.9151672.

[7] A. K. Bonnett and G. C. Soukup, "Cause and analysis of stator and rotor failures in 3-phase squirrel cage induction motors," Conference Record of 1991 Annual Pulp and Paper Industry Technical Conference, 1991, pp. 22-42, doi: 10.1109/PAPCON.1991.239667.

[8] J. -H. Jung, J. -J. Lee and B. -H. Kwon, "Online Diagnosis of Induction Motors Using MCSA," in IEEE Transactions on Industrial Electronics, vol. 53, no. 6, pp. 1842-1852, Dec. 2006, doi: 10.1109/TIE.2006.885131.

[9] F. Çira, M. Arkan, B. Gümüş and T. Goktas, "Analysis of stator inter-turn short-circuit fault signatures for inverter-fed permanent magnet synchronous motors," IECON 2016 - 42nd Annual Conference of the IEEE Industrial Electronics Society, 2016, pp. 1453-1457, doi: 10.1109/IECON.2016.7793717.

[10] S. E. Zouzou, M. Sahraoui, A. Ghoggal and S. Guedidi, "Detection of inter-turn short-circuit and broken rotor bars in induction motors using the Partial Relative Indexes: Application on the MCSA," The XIX International Conference on Electrical Machines - ICEM 2010, 2010, pp. 1-6, doi: 10.1109/ICEL-MACH.2010.5607874.

[11] J. Seshadrinath, B. Singh and B. K. Panigrahi, "Investigation of Vibration Signatures for Multiple Fault Diagnosis in Variable Frequency Drives Using Complex Wavelets," in IEEE Transactions on Power Electronics, vol. 29, no. 2, pp. 936-945, Feb. 2014, doi: 10.1109/TPEL.2013.2257869.

[12] Z. Yang, X. Shi and M. Krishnamurthy, "Vibration monitoring of PM synchronous machine with partial demagnetization and inter-turn short circuit faults," 2014 IEEE Transportation Electrification Conference and Expo (ITEC), 2014, pp. 1-6, doi: 10.1109/ITEC.2014.6861774.

[13] H. Bilal, N. Heraud and E. J. R. Sambatra, "Comparison of quadrature currents and vibration signal to detect inter-turn short circuit on a Doubly Fed Induction Machine (DFIM)," 2021 IEEE 62nd International Scientific Conference on Power and Electrical Engineering of Riga Technical University (RTUCON), 2021, pp. 1-6, doi: 10.1109/RTUCON53541.2021.9711678.

[14] B. Aubert, J. Régnier, S. Caux and D. Alejo, "Kalman-Filter-Based Indicator for Online Interturn Short Circuits Detection in Permanent-Magnet Synchronous Generators," in IEEE Transactions on Industrial Electronics, vol. 62, no. 3, pp. 1921-1930, March 2015, doi: 10.1109/TIE.2014.2348934.

[15] A. Kiselev, A. Kuznietsov and R. Leidhold, "Model based online detection of inter-turn short circuit faults in PMSM drives under non-stationary conditions," 2017 11th IEEE International Conference on Compatibility, Power Electronics and Power Engineering (CPE-POWERENG), 2017, pp. 370-374, doi: 10.1109/CPE.2017.7915199.

[16] Z. Xu et al., "Data-Driven Inter-Turn Short Circuit Fault Detection in Induction Machines," in IEEE Access, vol. 5, pp. 25055-25068, 2017, doi: 10.1109/ACCESS.2017.2764474.

[17] P. P. Tun, P. S. Kumar, R. A. Pratama and L. Shuyong, "Brushless Synchronous Generator Turn-to-Turn Short Circuit Fault Detection Using Multilayer Neural Network," 2018 Asian Conference on Energy, Power and Transportation Electrification (ACEPT), 2018, pp. 1-8, doi: 10.1109/ACEPT.2018.8610686.

[18] M. Mohamed, E. Mohamed, A. -A. Mohamed, M. Abdel-Nasser and M. M. Hassan, "Detection of Inter Turn Short Circuit Faults in Induction Motor using Artificial Neural Network," 2020 26th Conference of Open Innovations Association (FRUCT), 2020, pp. 297-304, doi: 10.23919/FRUCT48808.2020.9087535.

[19] W. Pietrowski and K. Górny, "Wavelet torque analysis and neural network in detection of induction motor inter-turn short-circuit," 2017 18th International Symposium on Electromagnetic Fields in Mechatronics, Electrical and Electronic Engineering (ISEF) Book of Abstracts, 2017, pp. 1-2, doi: 10.1109/ISEF.2017.8090751.

[20] K. -J. Shih, M. -F. Hsieh, B. -J. Chen and S. -F. Huang, "Machine Learning for Inter-turn Short-circuit Fault Diagnosis in Permanent Magnet Synchronous Motors," in IEEE Transactions on Magnetics, doi: 10.1109/TMAG.2022.3169173.

[21] İ. Şahin and O. Keysan, "Model Predictive Controller Utilized as an Observer for Inter-Turn Short Circuit Detection in Induction Motors," in IEEE Transactions on Energy Conversion, vol. 36, no. 2, pp. 1449-1458, June 2021, doi: 10.1109/TEC.2020.3048071.

[22] Mustafa Umit Oner, İlker Şahin, Ozan Keysan, February 9, 2022, "Inverter Switching Vectors Data of An FCS-MPC Driven Induction Machine for Healthy and Faulty Conditions", IEEE Dataport, doi: https://dx.doi.org/10.21227/chp0-5x97.

[23] İ. Şahin, "Model predictive torque control of an induction motor enhanced with an inter-turn short circuit fault detection feature," Ph.D. Dissertation, Middle East Technical University, 2021.

Modeling the impact of grid-forming E-STATCOMs on inter-area system oscillations.

A. Bolzoni[*], N. Johansson[±], J. P. Hasler[°]
HITACHI ENERGY
[*]Hitachi Energy Research, Segelhofstrasse, Baden Dättwil (CH)
[±]Hitachi Energy Research, Vaesteras (SE)
[°] Grid and Power Quality Solutions, Hitachi Energy, Vaesteras (SE).
E-Mail: alberto.bolzoni@hitachienergy.com
URL: https://www.hitachienergy.com/

Keywords

«Frequency Dynamics», «Inertia support», «Converter control», «STATCOM», «FACTS».

Abstract

The reduction of physical inertia in power systems represents one of the major trends affecting public grids operations. Under this scenario, it becomes crucial to assess the positive contribution achievable through the application of advanced control strategies to converter-based units at the transmission and distribution levels. In this perspective, this paper analyzes how the introduction of grid-forming control functionalities in STATCOM devices could help toward the stabilization of the network transients and the reduction of inter-area phenomena.

Introduction

Low-inertia operations for power networks are becoming increasingly common, due to the replacement of synchronous machines supplied by fossil fuel power plants, and their substitution with renewable generators. A typical issue related to the reduced inertia is the amplification of inter-area oscillations: according to the definition proposed by the IEEE Power System Dynamics and CIGRE 38 study committees [1], these phenomena are defined as the relative swings between the phase angles of distant generators and may potentially lead to uncontrolled large-scale stability disruptions and to a reduction of the quality of the power supply. These oscillations are excited by a local and temporary mismatch between generation / consumption in one of the areas of the system and are characterized by a resonance angular frequency of around $0.2 - 2.5$ rad/s [2]-[3].

In recent years, several authors analyzed the possibility to mitigate inter-area phenomena through diverse technologies like decentralized [4] or coordinated [5] control architectures for wind production plants, and Wide Area Monitoring infrastructures [6]. Another alternative consists in equipping the STATCOM devices installed at the transmission level with grid-forming control functionalities and with a supercapacitor storage placed at the DC side of the converter; this leads to the so called Enhanced-STATCOM (E-STATCOM) architecture.

This paper deals with the impact of E-STATCOM technology towards inter-area phenomena damping. More specifically, the goals are the following: to show how these devices mange to improve the stability of the network during transients; to derive an analytical model of the E-STATCOMs effect on inter-area modes; to assess the impact of the grid-forming scheme toward network stabilization; to provide some insights in the E-STATCOM best placement expressed as a function of the installation area.

To this purpose, a simple two-area configuration derived from [7] (and often referred as Kundur network) is considered; although simple, the architecture has been successfully exploited to test the positive effect of novel electronic devices on network system stability and inter-area phenomena [8]. DigSilent PowerFactory is used as simulation software for the analysis.

System modeling

Simplified transients' modeling

Consider the simple four-generator, two-area Kundur network typically exploited to study the impact of inter-area phenomena (Figure 1a). A simple representation of the active exchanges between the constituent areas is firstly introduced; this hypothesis will be later relaxed and a more general model able to include active / reactive dynamics and their inherent couplings will be presented in the next section.

The graphical representation in Figure 1b provides a simple representation of the dynamics occurring within and between the areas. Whenever a small local power mismatch is experienced in any of the two areas (Δp_{A1} or Δp_{A2}), the entire network undergoes a transitory condition: the regulation provided by local synchronous generators is combined with the power component Δp_{12} exchanged between the areas. Each of the areas $\{A1; A2\}$ is modelled as an equivalent synchronous machine with starting time T_a (equal to twice the inertia constant H) and steady-state primary frequency control determined by the inverse droop coefficient r_{droop}. A common per-unit system (V_b: phase-to-phase per-unit base voltage 400 kV, S_b: per-unit base power 1800 MVA, ω_b: base angular frequency $2\pi \cdot 50$ rad /s) is adopted to enable homogenous calculations, and the actual characteristics of the specific areas can be referred to the base system as $T_a^{Ai} = T_a^{Ai\prime} \cdot S_b'/S_b$, $r_{droop}^{Ai} = r_{droop}^{Ai\,\prime} \cdot S_b'/S_b$, $i \in \{1,2\}$ (primed parameters stand for the original quantities, before they are referred to the common base system).

In addition to the inertial (T_a) and frequency regulation (r_{droop}) effects, the model includes the dynamics of the steam turbine and machine governors for the two areas through the following time constants: τ_1, internal governor delay; τ_2, derivative compensation of steam turbine delay; τ_3, internal delay associated to the turbine control. The full set of parameters for the setup is provided in Tables 1-3.

Table 1: Ratings of synchronous generators and network

Synchronous generators ratings (same for each synchronous gen.)	Numeric value
Rated apparent power [MVA]	900
Rated transformer ratio [kV / kV] and short-circuit voltage [%]	400 / 20 − 15%
Connecting impedance between the two areas $\overline{Z}_{12}$ [Ω]	13.25 + j 132.5

Table 2: Frequency regulation parameters for synchronous generators

Synchronous generators characteristics	Area 1 (G1 − G2)	Area 1 (G3 − G4)
Machine starting time T_a ($T_a = 2H$) [s]	6	10
Turbine gov. time constant τ_1 [s]	0.5	2
Turbine derivative time constant τ_2 [s]	1	5
Turbine delay time constant τ_3 [s]	3	5
Inverse droop coefficient r_{droop} [pu]	10	10

Table 3: E-STATCOM parameters

E-STATCOM characteristics	E-STAT 1 / E-STAT 2
Rated apparent power S_c [MVA]	500 MVA
Additional inertia H_{stat} [s] and additional damping term D_{stat} [s]	20 s, 0.3 s
Virtual admittance terms X_v [pu] - R_v [pu]	0.44 - 0.1
E-STATCOM supercapacitor parameters	1.78 F − 40 kV

In case each single area $i \in \{1,2\}$ operates independently of the other, the frequency dynamics under a load power increase Δp_{Ai} are expressed by:

$$\frac{\Delta \omega_{Ai}}{\Delta p_{Ai}} = K_{\omega P}^{Ai}(s) = -\frac{\left(s\tau_1^{Ai} + 1\right)\left(s\tau_2^{Ai} + 1\right)}{sT_a^{Ai}\left(s\tau_1^{Ai} + 1\right)\left(s\tau_2^{Ai} + 1\right) + \left(s\tau_3^{Ai} + 1\right)r_{droop}^{Ai}}, \qquad i \in \{1,2\}. \qquad (1)$$

Once the areas are connected through the impedance $\overline{Z}_{12}$, the exchanged power Δp_{12} should also be included. Figure 1b shows the equivalent feedbacks (in blue, in the picture) associated to inter-area power exchange Δp_{12} proportional to the phase angle difference between the areas. For simplicity, here it is assumed a predominantly inductive impedance $\overline{Z}_{12} \simeq jX_{12}$: this hypothesis will be later removed in the complete model. Phase angles deviations (from steady-state values) $\Delta\vartheta_{A1}$ / $\Delta\vartheta_{A2}$ are introduced for the two areas, defined as the integral of the local angular frequencies $\Delta\omega_{A1}$ / $\Delta\omega_{A2}$.

$$\Delta p_{12} = \frac{V_1 V_2}{X_{12}} \sin(\Delta\vartheta_{A1} - \Delta\vartheta_{A2}) \simeq \frac{V_1 V_2}{X_{12}} (\Delta\vartheta_{A1} - \Delta\vartheta_{A2}) \qquad (2)$$

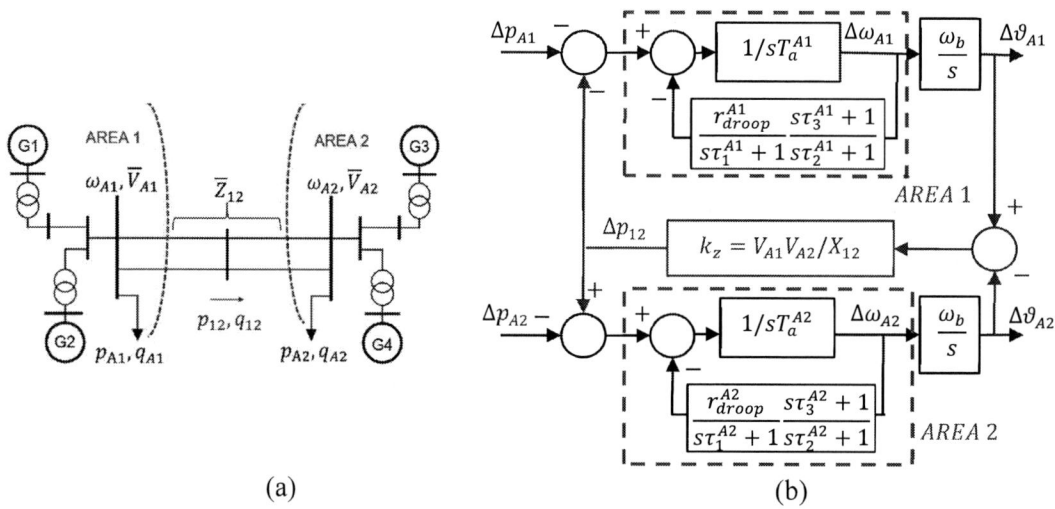

(a)   (b)

Fig. 1: (a) equivalent circuit for the two-area network setup and (b) simplified dynamic system for the modeling of active power exchanges in the test network. Constant-power models are used for the loads.

The dynamics of the interconnected system under possible active load variations at the two sides are expressed by the MIMO (Multi-Input Multi-Output) transfer matrix $\underline{G}(s)$ in (3), whose elements account for the active power interactions between the two areas; the term k_z exploited in (3) is equal to $k_z = V_{A1} V_{A2}/ X_{12}$ and it is derived from (2).

$$\begin{bmatrix} \Delta\omega_{A1} \\ \Delta\omega_{A2} \end{bmatrix} = -\underline{G}_{2\times2}(s) \begin{bmatrix} \Delta p_{A1} \\ \Delta p_{A2} \end{bmatrix} \qquad (3)$$

$$G_{ii}(s) = K_{\omega P}^{Ai}(s) \left(1 + K_{\omega P}^{Ai}(s)\frac{k_z \omega_b}{s}\left(1 + K_{\omega P}^{Aj}(s) \cdot \frac{k_z \omega_b}{s}\right)^{-1}\right)^{-1} \qquad i = \{1,2\}, \qquad j \neq i \quad (3.b)$$

$$G_{ij}(s) = \frac{K_{\omega P}^{Ai}(s)K_{\omega P}^{Aj}(s)k_z \omega_b}{s + K_{\omega P}^{Aj}(s) k_z \omega_b}\left(1 + \frac{\frac{K_{\omega P}^{Ai}(s)K_{\omega P}^{Aj}(s)k_z \omega_b^2}{s^2}}{1 + K_{\omega P}^{Aj}(s)\frac{k_z \omega_b}{s}}\right)^{-1} \qquad i = \{1,2\}, \qquad j \neq i \quad (3.c)$$

The comparison between the Bode diagrams of the separate areas functions $K_{\omega P}^{A1}(s)$ and $K_{\omega P}^{A2}(s)$ from (1) against the elements of the transfer matrix $\underline{G}(s)$ (like $G_{11}(s)$ in Figure 2.a) shows the amplification of a resonance point in the equivalent regulation curve of the system. This resonance condition is the source of the inter-area oscillations, and its natural frequency T_2^{-1} (in Hz) is marked in Fig. 2.a (where $T_2 \simeq 1.4$ s from Fig. 2.b). The local maxima in the magnitude profile of $G_{11}(s)$ matches with the dominant harmonic components identifiable in the time profiles of $\Delta\omega_{A1}/\Delta\omega_{A2}$, confirming the correctness of the proposed modeling.

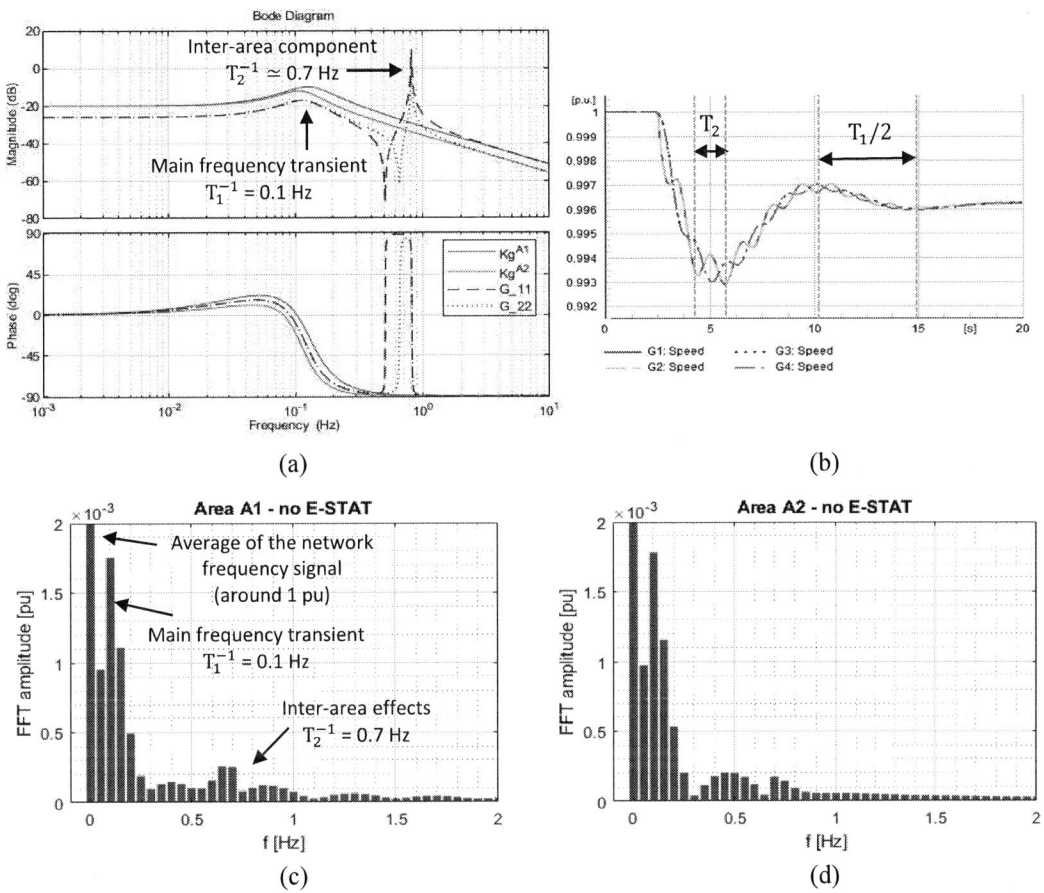

(a)

(b)

(c)

(d)

Fig. 2: (a) comparison between the equivalent regulation characteristics of the independent areas $K_g^{A1}(s)$ / $K_g^{A2}(s)$ and the ones resulting from the areas interconnection (represented by the term $G_{11}(s)$ and $G_{22}(s)$ in the MIMO-system); (b) the time profiles from DigSilent PowerFactory for $\Delta\omega_{A1}$ and $\Delta\omega_{A2}$ under a step change Δp_{A1}. The spectral decomposition of the local frequency signals at the two areas are also reported in (c) and (d), as an additional verification.

Extended MIMO model for assessment of the E-STATCOM effect.

Although the simple configuration reported in the previous subsection provides interesting insights in the core dynamical principles of inter-area phenomena, the assessment of the impact of the E-STATCOM technology requires an extensive modeling of the active / reactive power interaction between the converter-based units and the network. A full MIMO extension of the system is thus developed for this purpose; in this context, the grid-forming E-STATCOM is controlled to provide just inertial support, with no proportional frequency droop.

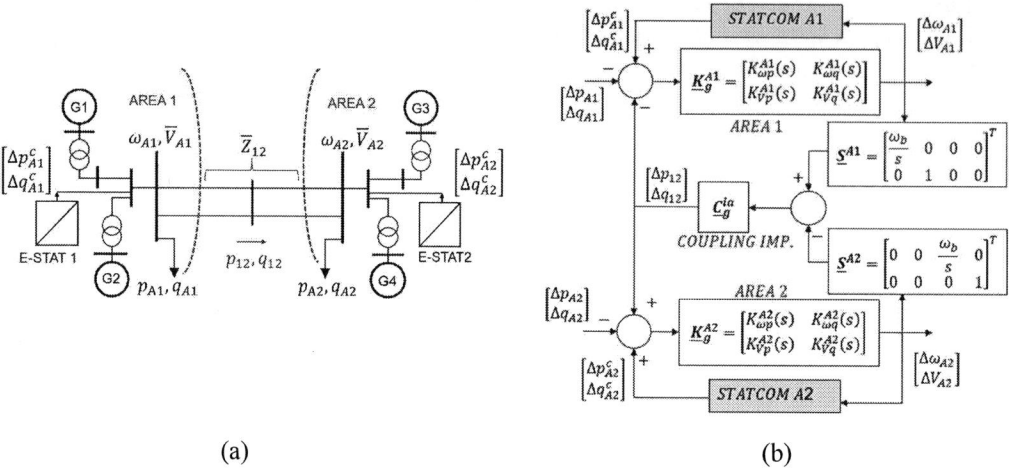

(a)   (b)

Fig. 3: MIMO model of the network for the assessment of E-STATCOM impact.

Figure 3 extends the approach introduced in Figure 1 to the generalized case of non-perfect decoupling between active and reactive power dynamics. The first step for the model development consists of deriving the natural response of the areas when just the loads and synchronous generators are connected (no E-STATCOMs). Extending the simple modeling introduced in (1), the area characteristics are now expressed through the 2x2 transfer matrix $\underline{K}_g$ in (4).

The elements of $\underline{K}_g$ can be derived either through analytical considerations or through system identification tests performed on the separate areas. In the DigSilent environment, the two areas are disconnected and characterized separately looking at their dynamical response following a local load step. The analytical formulation (1) is exploited to model the element $K_{\omega p}^{Ai}(s)$ in $\underline{K}_g$, while the remaining elements of the matrix are obtained through system identification. Figure 4 shows profiles of the simulated transients and their reconstructed analytical behavior ones under separate active / reactive steps for each of the two areas. As an alternative to the system identification, a more detailed modeling of the generators damper windings and PSS (Power-System Stabilizers) available in the time-domain simulation could be introduced.

$$\begin{bmatrix} \Delta\omega_{Ai} \\ \Delta V_{Ai} \end{bmatrix} = -\underline{K}_{g\,2\times2}^{Ai}(s) \begin{bmatrix} \Delta p_{Ai} \\ \Delta q_{Ai} \end{bmatrix} = - \begin{bmatrix} K_{\omega p}^{Ai}(s) & K_{\omega q}^{Ai}(s) \\ K_{Vp}^{Ai}(s) & K_{Vq}^{Ai}(s) \end{bmatrix} \begin{bmatrix} \Delta p_{Ai} \\ \Delta q_{Ai} \end{bmatrix} \qquad i \in \{1,2\} \qquad (4)$$

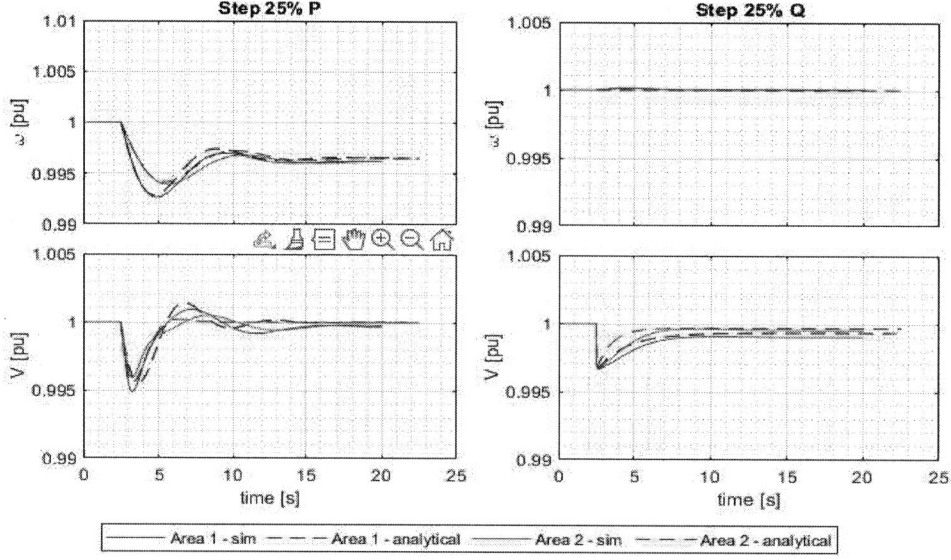

Fig. 4: Characterization of the two areas.

The local characteristics for each area are later combined to account for the inter-area exchanges $\Delta p_{12}/\Delta q_{12}$. This can be through the sensitivity matrixes $\underline{S}^{A1}/\underline{S}^{A2}$ (as defined in Figure 3.b) and the coupling matrix $\underline{C}_g^{ia}$: this last element includes the linearized effects of the power flow equations [7] along the inter-area tie line which, for the generic case of ohmic-inductive impedance $\overline{Z}_{12} = R_{12} + jX_{12}$, is:

$$
\underline{C}_g^{ia} = \begin{bmatrix} \dfrac{V_1 V_2 \, X_{12}}{X_{12}^2 + R_{12}^2} & \dfrac{(2V_1 - V_2)R_{12}}{X_{12}^2 + R_{12}^2} & \dfrac{V_1 V_2 \, X_{12}}{X_{12}^2 + R_{12}^2} & \dfrac{V_1 R_{12}}{X_{12}^2 + R_{12}^2} \\[2mm] -\dfrac{V_1 V_2 \, R_{12}}{X_{12}^2 + R_{12}^2} & \dfrac{(2V_1 - V_2)X_{12}}{X_{12}^2 + R_{12}^2} & -\dfrac{V_1 V_2 \, R_{12}}{X_{12}^2 + R_{12}^2} & \dfrac{V_1 X_{12}}{X_{12}^2 + R_{12}^2} \end{bmatrix}. \tag{5}
$$

An active / reactive power increase in one of the two areas impacts the voltage / frequency transients in each of the parts of the interconnected system. This case can be described through the MIMO system (6), assuming, as a base case, no contribution from the E-STATCOMs: the transfer matrix $\underline{G}_{4\times4}^{\text{no STAT}}$ describes the dynamics generated by any active / reactive load variation on the frequency / voltage dynamics of each area.

The frequency response of $\underline{G}_{4\times4}^{\text{no STAT}}$ provides insights in which area is potentially more affected by the inter-area resonance. In the specific case study, the shape of $\underline{G}_{4\times4}^{\text{no STAT}}$ indicates that area A1 is potentially affected by larger swings of the voltage phase-angle compared to area A2 (Figure 5).

$$
\begin{bmatrix} \Delta\omega_{A1} \\ \Delta V_{A1} \\ \Delta\omega_{A2} \\ \Delta V_{A2} \end{bmatrix} = -\,\underline{G}_{4\times4}^{\text{no STAT}}(s) \begin{bmatrix} \Delta p_{A1} \\ \Delta q_{A1} \\ \Delta p_{A2} \\ \Delta q_{A2} \end{bmatrix} \tag{6}
$$

$$
\underline{G}_{(1:2\,;1:2)}^{\text{no STAT}} = \underline{K}_g^{A1} \cdot \left(\underline{I}_2 - \underline{C}_g^{ia} \cdot \left(\underline{I}_4 + \underline{S}^{A1}\underline{K}_g^{A1}\,\underline{C}_g^{ia} + \underline{S}^{A2}\underline{K}_g^{A2}\,\underline{C}_g^{ia} \right)^{-1} \underline{S}^{A1}\underline{K}_g^{A1} \right) \tag{7}
$$

$$
\underline{G}_{(1:2\,;3:4)}^{\text{no STAT}} = \underline{K}_g^{A1}\,\underline{C}_g^{ia} \cdot \left(\underline{I}_4 + \underline{S}^{A1}\underline{K}_g^{A1}\,\underline{C}_g^{ia} + \underline{S}^{A2}\underline{K}_g^{A2}\,\underline{C}_g^{ia} \right)^{-1} \cdot \underline{S}^{A2}\underline{K}_g^{A2}
$$

$$
\underline{G}_{(3:4\,;1:2)}^{\text{no STAT}} = \underline{K}_g^{A2}\,\underline{C}_g^{ia} \cdot \left(\underline{I}_4 + \underline{S}^{A1}\underline{K}_g^{A1}\,\underline{C}_g^{ia} + \underline{S}^{A2}\underline{K}_g^{A2}\,\underline{C}_g^{ia} \right)^{-1} \cdot \underline{S}^{A1}\underline{K}_g^{A1}
$$

$$
\underline{G}_{(3:4\,;3:4)}^{\text{no STAT}} = \underline{K}_g^{A2} \cdot \left(\underline{I}_2 - \underline{C}_g^{ia} \cdot \left(\underline{I}_4 + \underline{S}^{A1}\underline{K}_g^{A1}\,\underline{C}_g^{ia} + \underline{S}^{A2}\underline{K}_g^{A2}\,\underline{C}_g^{ia} \right)^{-1} \underline{S}^{A2}\underline{K}_g^{A2} \right)
$$

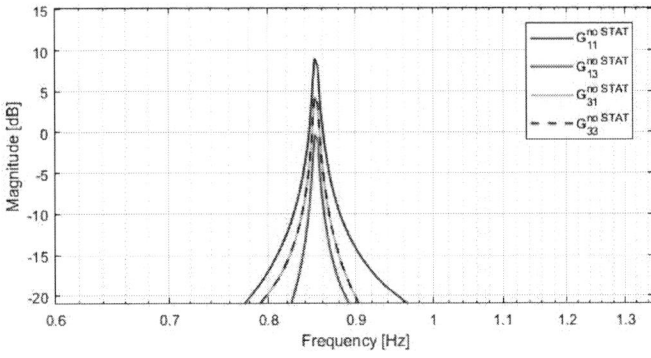

Fig. 5: Analytical assessment of the areas potentially more subjected to larger phase-angle variations.

Consider now the extension of the base case through the E-STATCOM installation. The impact can be modelled as additional feedbacks between the local area quantities $\Delta\omega_{Ai}$ / ΔV_{Ai} and the converter active / reactive power contributions Δp_{Ai}^c / Δq_{Ai}^c. For this analysis, the well-known Virtual Synchronous Machine approach (VSM) will be deployed for the E-STATCOMs grid-forming control, whose synchronization mechanism is reported in Figure 6-a. Still, it is important to highlight that the positive effect associated to the E-STATCOM installation is independent with respect to the chosen grid-forming scheme and any alternative control configuration can be assessed in the light of the same proposed methodology.

By applying the linearization of the converter synchronization according to [9], it is possible to derive the equivalent MIMO system associated to the E-STATCOM (Figure 6-b). In this specific case, no explicit AC-voltage controller is included in the modeling; still, the proposed approach can be further extended to account also for the additional loops in the control.

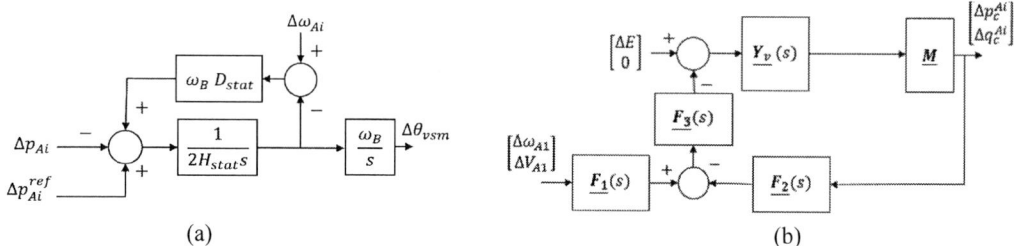

(a)   (b)

Fig. 6: Synchronization mechanism for the VSM system (a) and E-STATCOM control system MIMO representation for the assessment of its impact on inter-area phenomena (b); ΔE in (b) represents the variation of the back-EMF magnitude, assumed constant in this case.

The full E-STATCOM behavior is thus expressed in (8): the definition of the transfer matrix $\underline{F}_1, \underline{F}_2$ and $\underline{F}_3$ can be derived combining the VSM synchronization system in Fig.6-a with the linearization approach from reference [9]; $\Delta\theta_{vsm}$ in Fig.6.a represents the converter synchronization angle used for the application of the linearization approach from [9].

$$\begin{bmatrix} \Delta p_{Ai}^c \\ \Delta q_{Ai}^c \end{bmatrix} = \underline{F}^{stat} \begin{bmatrix} \Delta\omega_{Ai} \\ \Delta V_{Ai} \end{bmatrix} = -\left(\underline{I}_{2\times2} - \underline{M}\,\underline{Y}_v\,\underline{F}_3\,\underline{F}_2\right)^{-1} \left(\underline{M}\,\underline{Y}_v\,\underline{F}_3\,\underline{F}_1\right) \begin{bmatrix} \Delta\omega_{Ai} \\ \Delta V_{Ai} \end{bmatrix} \quad (8)$$

$$\underline{Y}_v(s) = \begin{bmatrix} R_v + \dfrac{s}{\omega_B}L_v & -X_v \\ X_v & R_v + \dfrac{s}{\omega_B}L_v \end{bmatrix}^{-1} \qquad \underline{M} = \dfrac{S_c}{S_b}\begin{bmatrix} 1 & 0 \\ 0 & -1 \end{bmatrix} \quad (9)$$

The introduction of the E-STATCOM effects $\underline{F}^{stat}$ changes the characteristics of the local areas in Figure 3-b: the transfer matrixes $\underline{K}_g^{Ai}$ modify in $\underline{K}_g^{Ai\,mod}$ to account for the E-STATCOM effects:

$$\underline{K}_{g\,2\times2}^{Ai\,mod}(s) = \left(\underline{I}_2 - \underline{K}_g^{Ai}\,\underline{F}^{stat}\right)^{-1} \cdot \underline{K}_g^{Ai}(s) \quad (10)$$

The global network characteristics change according to the new transfer matrix $\underline{G}_{4\times4}^{STAT}$, which is formally equal to (7) but is now calculated deploying the modified area characteristics $\underline{K}_g^{Ai\,mod}(s)$ instead of the original ones $\underline{K}_g^{Ai}(s)$, to account for the effects of the E-STATCOMs installation in the two areas.

E-STATCOM impact assessment: Oscillation damping.

Figure 7 and Figure 8 show respectively the time profiles associated to the network quantities in the base case where no E-STATCOMs are considered (Fig.7), and once E-STATCOM devices are connected at both sides of the two-area network (Fig.8). At a first glance, it is possible to appreciate the substantial reduction of the inter-area oscillations following a 50% variation of the local load connected to area A1. The presence of the E-STATCOMs helps toward the damping of the oscillatory phenomena in the measured local frequency profiles both for the area A1 and the area A2, improves the nadir of the transients (both for the frequencies and local voltages) and smoothens the profiles of the active / reactive power injections associated to the nearby synchronous machines.

Figure 9 shows the comparison of the theoretical model developed in the previous section against the spectral analysis of the time series from Figure 7 – Figure 8 (without and with E-STATCOM respectively). According to the theoretical model, the installation of the E-STATCOM devices is expected to induce a positive damping effect on the inter-area modes due to the reduction of the resonance peak around 0.7 Hz (see Fig. 9 - a); this result is confirmed by the reduction of the spectral components in the same frequency range as obtained from the simulations.

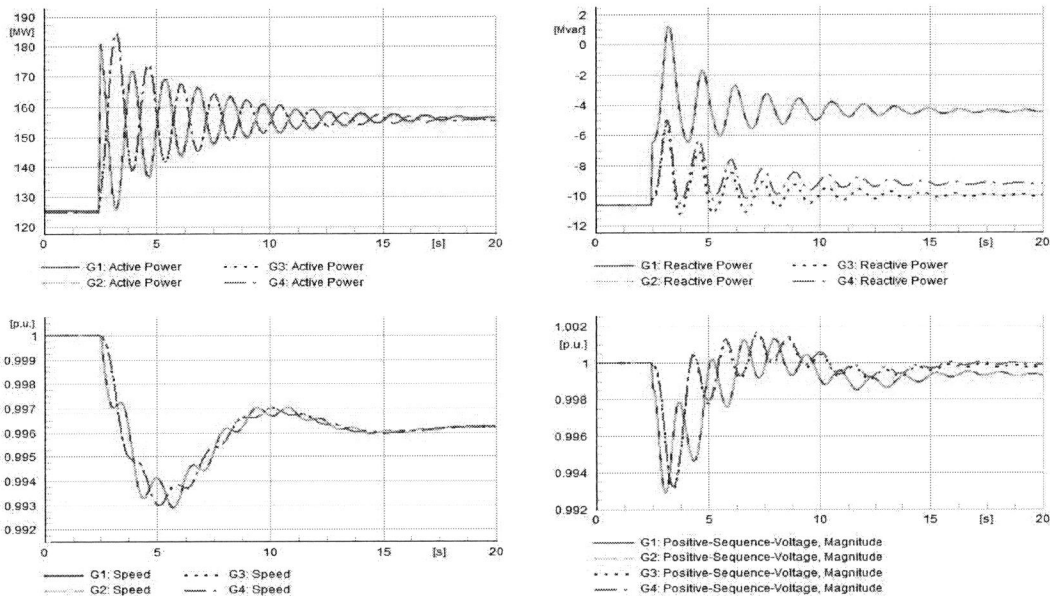

Fig. 7: Time profiles of the main generators' quantities (active / reactive power, angular frequency, and voltage magnitude) for the **BASE CASE**.

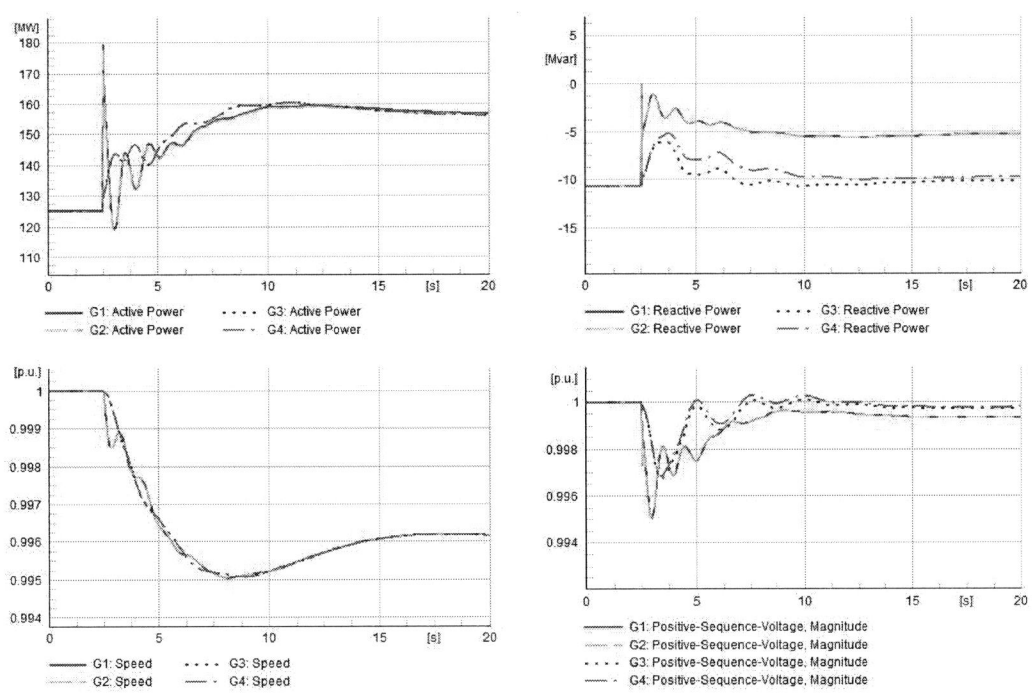

Fig. 8: Time profiles of the main generators' quantities (active / reactive power, angular frequency, and voltage magnitude, STATCOM active / reactive power) for the case with **both E-STATCOM 1 / E-STATCOM 2** connected.

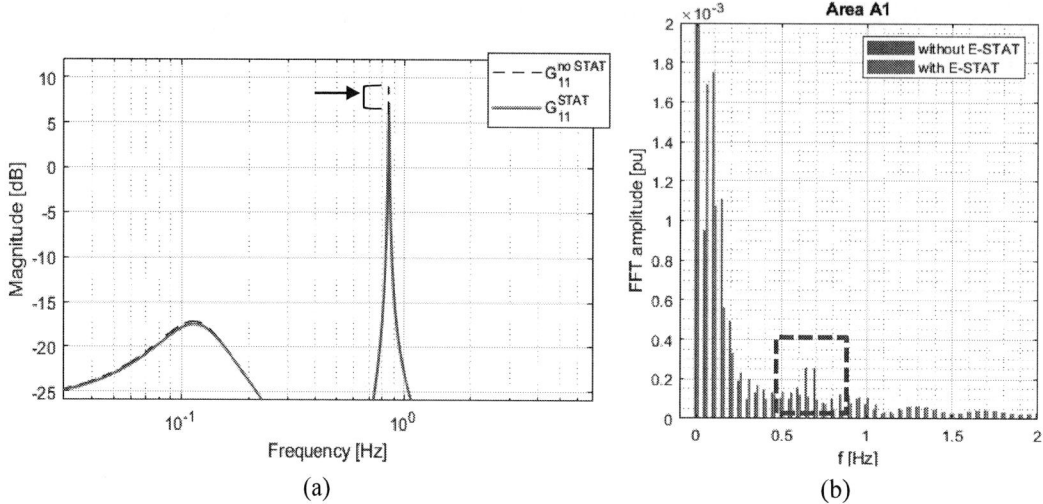

(a)   (b)

Fig. 9: (a) theoretical damping effect of the inter-area dynamics obtained from the analytical model and (b) measured impact of this effect in terms on the local frequency spectrum for area A1.

E-STATCOM impact assessment: Placement.

Figures 10 and 11 show the effect of the E-STATCOM placement, assuming that just one unit may be placed into the network (either E-STATCOM 1 or E-STATCOM 2). This allows to derive insights in the best location for these elements. Figure 10 shows the frequency profiles once a 50% increase of the active load connected to area A1 is assumed as operating scenario; this scenario is evaluated both for the single-unit placement in area A1 (Figure 10-a) or in area A2 (Figure 10-b). A similar approach is considered in Figure 11, but in this case the load variation is applied to area A2.

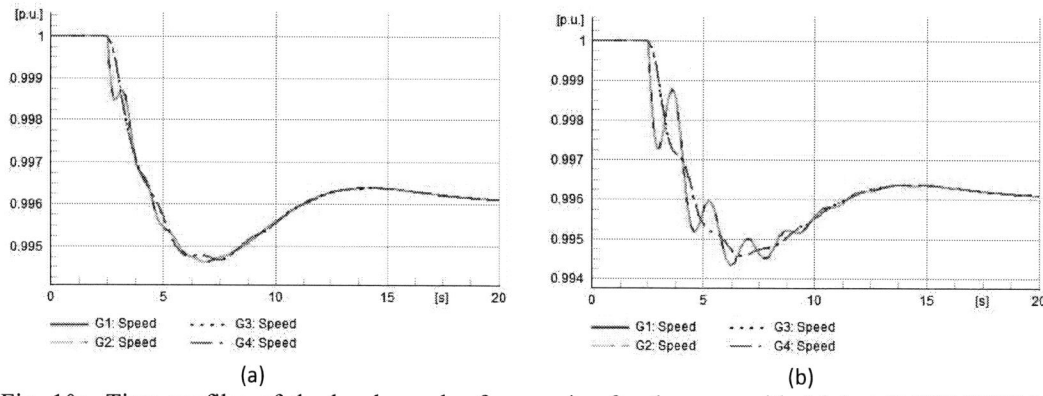

(a)   (b)

Fig. 10: Time profiles of the local angular frequencies for the case with (a) just E-STATCOM 1 connected and (b) just E-STATCOM 2 connected. Both cases are tested assuming the load variation in area A1 as the source of the inter-area event.

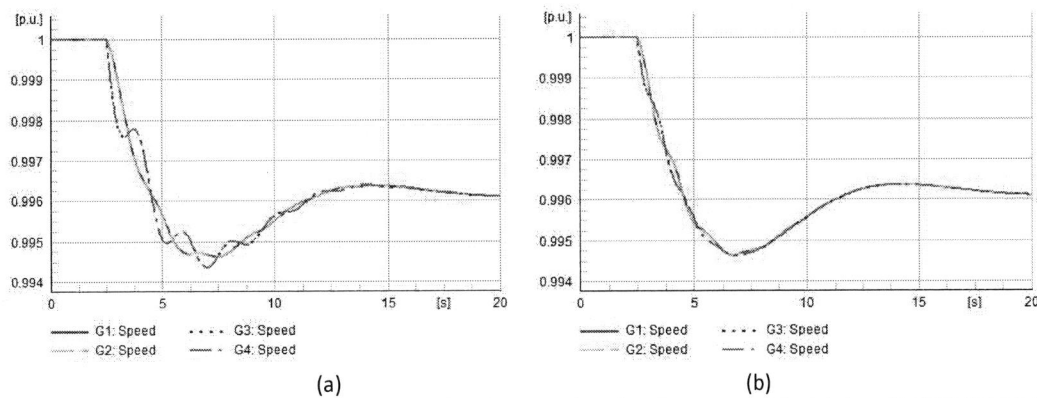

(a)   (b)

Fig. 11: Time profiles of the local angular frequencies for the case with (a) just E-STATCOM 1 connected and (b) just E-STATCOM 2 connected. Both cases are tested assuming load variation in area A2 as the source of the inter-area event.

The following considerations can be derived:

- For each of the considered load scenarios (Figure 10 and Figure 11 separately), the location of the E-STATCOM device close to the source of the disturbance provides the better results; that is, case (a) for Figure 10 and case (b) for Figure 11. If the source of the disturbance is known, the optimal location of the E-STATCOM is as close as possible to the origin of the oscillation.
- In case the source of the event is unknown (which is normally the case) and assuming no statistic difference in the size / likelihood of the disturbance events at the two sides of the network, the worst-case condition in Fig. 10 – b (considering both Figure 10 and Figure 11 together) indicates A1 as the more sensitive location for inter-area phase-angle fluctuations. This result is consistent with the analytical derivation in Fig. 5, confirms the correctness of the proposed approach, and provides useful insights on the areas to strengthen through larger E-STATCOM installations.

Conclusion

In this paper, the positive effect associated to E-STATCOM devices installation toward the damping of inter-area phenomena has been demonstrated and an analytical methodology based on MIMO characterization is proposed to predict the main dynamics of the system. The results (both at an analytical and simulated level) confirm the positive impact of E-STATCOM devices toward the stabilization of inter-area dynamics and provide useful insights in the optimal placement of these units.

References

[1] P. Kundur et al., "Definition and classification of power system stability IEEE/CIGRE joint task force on stability terms and definitions," in IEEE Trans. on Power Systems, vol. 19, no. 3, pp. 1387-1401, Aug. 2004

[2] S. A. N. Sarmadi and V. Venkatasubramanian, "Inter-Area Resonance in Power Systems from Forced Oscillations," in IEEE Transactions on Power Systems, vol. 31, no. 1, pp. 378-386, Jan. 2016

[3] Y. Yu, and oth., "Oscillation Energy Analysis of Inter-Area Low-Frequency Oscillations in Power Systems," in IEEE Transactions on Power Systems, vol. 31, no. 2, pp. 1195-1203, March 2016.

[4] M. J. Morshed and A. Fekih, "A Coordinated Controller Design for DFIG-Based Multi-Machine Power Systems," in IEEE Systems Journal, vol. 13, no. 3, pp. 3211-3222, Sept. 2019.

[5] S. Yari and M. Khatibi, "Damping Improvement of Inter-Area Oscillations Using Large-Scale Wind Farms," 7th Iran Wind Energy Conference (IWEC2021), 2021, pp. 1-5.

[6] I. Zenelis, X. Wang and I. Kamwa, "Online PMU-Based Wide-Area Damping Control for Multiple Inter-Area Modes," in IEEE Transactions on Smart Grid, vol. 11, no. 6, pp. 5451-5461, Nov. 2020.

[7] P. Kundur, "Power System stability and control", Mc. Graw Hill, 1993, pp. 810.

[8] A. Venkatraman, U. Markovic, et al., "Improving Dynamic Performance of Low-Inertia Systems Through Eigensensitivity Optimization," in IEEE Trans. on Power Systems, vol. 36, no. 5, pp. 4075-4088, Sept. 2021.

[9] A. Bolzoni, "Generalized Nyquist MIMO Stability of Frequency Regulation Services in Power Networks," 2020 IEEE 21st Workshop on Control and Modeling for Power Electr. (COMPEL), 2020, pp. 1-7.

Combining Schwarz-Christoffel Mappings and Biot-Savart Law to Calculate the High-Frequency Current Distribution Inside a Single Slot

Torben Fricke[1], Phil Leon Pickert[1], Babette Schwarz[2], Bernd Ponick[1]
LEIBNIZ UNIVERSITY HANNOVER[1], VOITH HYDRO HOLDING GMBH & CO. KG[2]
Institute for Drive Systems and Power Electronics[1]
Hannover, Germany[1] and Heidenheim an der Brenz, Germany[2]
Email: torben.fricke@ial.uni-hannover.de
URL: https://ial.uni-hannover.de[1] and https://voith.com[2]

Keywords

≪Converter machine interactions≫, ≪Electrical machine≫, ≪Adjustable speed drive≫, ≪Software≫

Abstract

A novel calculation approach for the current and field distribution inside conductors in a slot at high frequencies based on a combination of a Schwarz-Christoffel Mapping and Biot-Savart Law is presented. The stator and rotor laminations are modeled using a surface impedance boundary condition and the results are validated against 2D FEA. The proposed method is overall less capable and mature than widely available 2D FEA software but has some appeal in niche applications, where licensing issues with FEA software are of concern or a high degree of integration into a broader tool chain is desirable.

Introduction

The proliferation of wide-bandgap semiconductors makes higher efficiencies and smaller form factors possible in power electronics [1]. These technological advances pose challenges in the design of electrical machines, as the increased switching speed exaggerates problems related to high frequency (HF) leakage currents [2]. Designing electrical machines that are resilient to adverse effects from HF voltage components requires advanced calculation approaches to predict high frequency leakage currents. One important component of such a calculation approach is an HF model of the stator winding. The inductances, resistances and capacitances of the stator winding HF model are commonly calculated using 2D FEA [3]. In this paper, an alternative calculation approach based on a combination of a Schwarz-Christoffel (SC) mapping and Biot-Savart (BS) law is presented and evaluated. The idea of combining Biot-Savart law with a conformal mapping dates back to 1979, when Reppe [4] used a numerically solved SC mapping in combination with BS law to calculate the air-gap field in a salient-pole synchronous machine. This combination of methods has since been used in geophysics [5], the current distribution calculation in PCB traces [6], superconductor research [7] and, of course, field problems in electrical machines [8]. In this paper, the idea of combining an SC mapping with BS law will be extended to new HF field problems by introducing a surface impedance boundary condition.

Method

The algorithm used to calculate the current distribution inside an arbitrary number of conductors in a slot can be summarized by the following steps, which will be explored in the subsequent subsections.
1. Solve the Schwarz-Christoffel mapping for the desired geometry.
2. Discretize each conductor into a number of partial conductors with a constant current distribution.
3. Generate surface conductors that represent a surface impedance boundary condition (SIBC).

4. Generate the surface conductor inductance matrix and use it to calculate the SIBC-aware vector potential.
5. Calculate the partial conductor inductance matrix.
6. Solve the partial conductor inductance matrix for the current distribution, taking into account the voltage induced by other distributed conductors. Repeat this step until the current distribution is constant.

Schwarz-Christoffel Mapping

A conformal mapping is a technique which maps one bound area (or domain) to another while preserving angles locally [9], as seen in Fig. 1. This allows us to solve certain kinds of field problems in one domain and transform the solution to another domain, which is usually more complex. The Schwarz-Christoffel (SC) mapping is one kind of conformal mapping, which maps from an infinite half plane (canonical domain) to an arbitrary polygon (physical domain), as shown in Fig. 4. For simple geometries, this mapping can be calculated analytically. A collection of analytically solved SC mappings is provided by Gibbs [10]. More complex geometries have to be approached numerically, as pioneered by Reppe [4] and later expanded upon by Driscoll, who open sourced his implementation [11]. Further explorations into numerical SC mappings unfortunately cannot be provided in this paper due to space constraints. However, Driscoll's book [9] is a great resource on this matter.

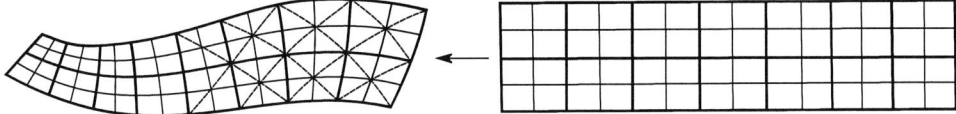

Fig. 1: Example of a conformal mapping. Note how the size and shape of the squares changes, but the angles remain unchanged.

Discretization

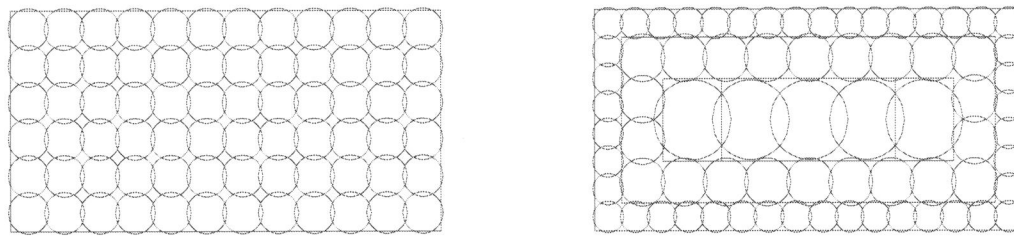

(a) Uniform mesh containing 72 partial conductors

(b) Skin mesh containing 71 partial conductors

Fig. 2: Two ways to discretize a rectangular distributed conductor into circular partial conductors.

Any distributed conductor – be it a hairpin, round wire or Roebel bar subconductor – has to first be discretized into multiple circular partial conductors with a constant current distribution. Fig. 2 shows two ways in which a distributed conductor can be discretized. The total area of all partial conductors equals the area of the distributed conductor. The field of a partial conductor can now be calculated using the geometric vector potential

$$\text{GVP} = \frac{1}{4\pi} \cdot \begin{cases} \left(\frac{r^2}{R^2} + \ln(R^2) - 1 \right) & \text{inside the conductor } (r \leq R) \\ \ln(r^2) & \text{outside of the conductor } (r > R), \end{cases} \tag{1}$$

which has been defined such that the vector potential outside a partial conductor is independent of its radius (this property will become important later on). The geometric vector potential GVP (shown in Fig. 3) can easily be turned into the vector potential

$$\underline{A} = \text{GVP} \cdot \underline{I} \cdot \mu_r \mu_0, \tag{2}$$

where the current $\underline{I}$ is defined as an RMS phasor.

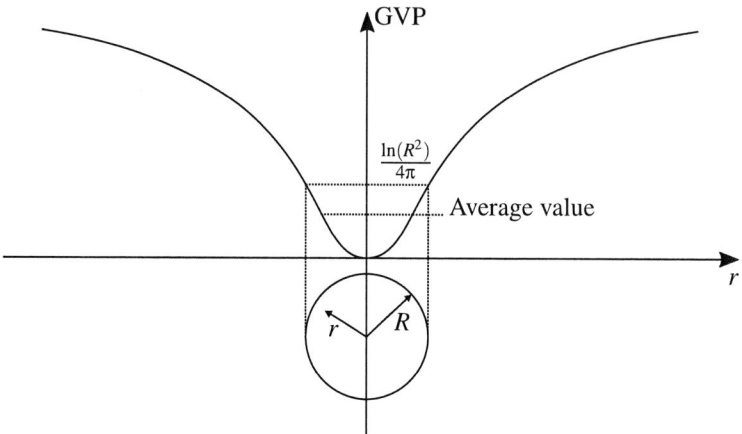

Fig. 3: Geometric vector potential (GVP) of a circular conductor carrying a constant current

The vector potential is calculated in the canonical domain using the method of image currents. When using BS law in combination with a conformal mapping, the image current

$$I' = \begin{cases} I & \mu \to \infty \text{ field lines are normal to the surface} \\ -I & \mu = 0 \text{ field lines are tangential to the surface} \end{cases} \tag{3}$$

can only take one of two values. Mapping a field solution obtained using any other image current produces an incorrect result. Fig. 4 shows the field produced by a line conductor arbitrarily placed in an arbitrary physical mapping. The field solution has been calculated in the canonical domain and transformed into the physical domain. All Biot-Savart calculations, even those of the surface currents introduced in the next section, are performed in the canonical domain, assuming $\mu \to \infty$, and mapped into the physical domain.

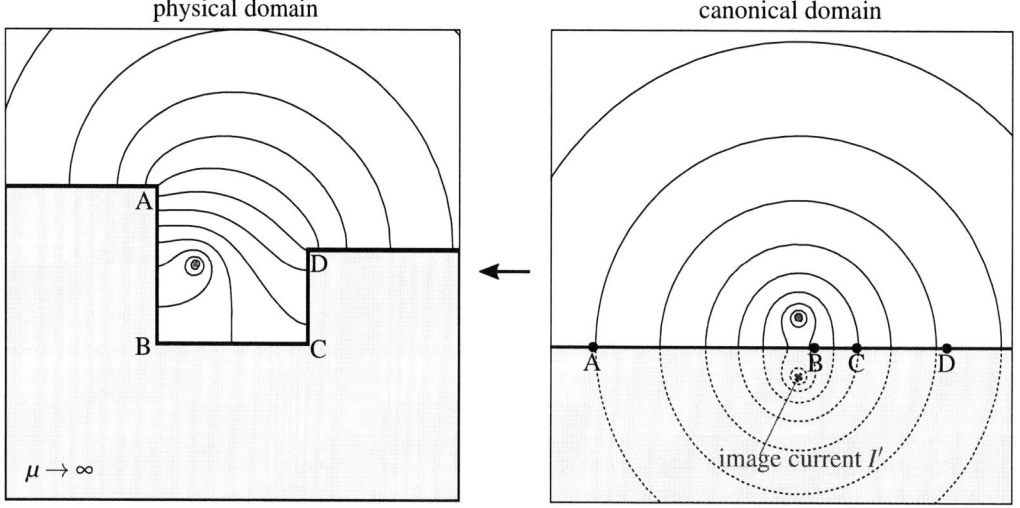

Fig. 4: Method of image currents in an arbitrary SC mapping

Surface Conductors

Next, the surface impedance boundary condition (SIBC) can be applied. In this work, the simplest surface impedance boundary condition, often referred to as the skin effect [12], is used. Using this surface impedance boundary condition requires some assumptions. First, the stator and rotor material has to be assumed to have a constant permeability μ_r and conductivity σ. All geometric features at the boundary must be small compared to the skin depth

$$\delta = \frac{1}{\sqrt{\pi f \sigma \mu_r \mu_0}}. \tag{4}$$

Assuming a typical conductivity for M330-35A electrical steel of $\sigma = 2.2$ MS/m [13] and a permeability of $\mu_r = 2000$, this requires the frequency to be higher than about 5000 Hz for the skin depth $\delta(f = 5000\,\text{Hz}) = 0.11$ mm to become sufficiently small compared to a common lamination thickness of 0.35 mm. In cases where the teeth are saturated from the fundamental harmonic, the frequency needs to be higher for this assumption to remain valid. The surface impedance can be written as

$$\underline{Z}_s'' = \frac{1+j}{\sigma \delta} \tag{5}$$

and has the unit $[\underline{Z}_s''] = \Omega/\text{m}^2$.

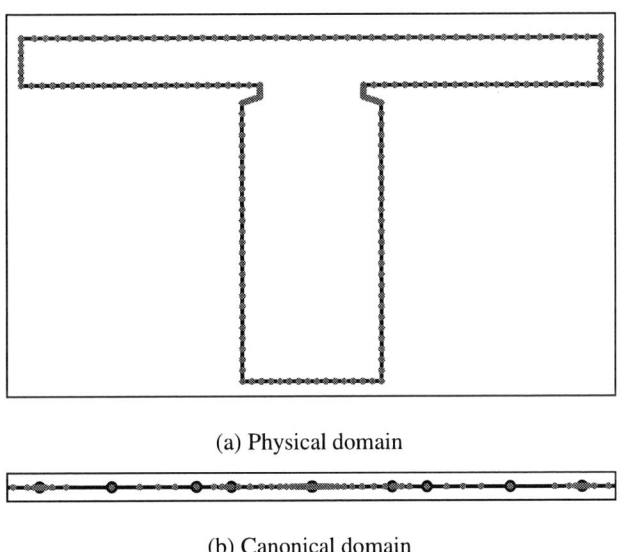

(a) Physical domain

(b) Canonical domain

Fig. 5: Locations of surface conductors in both domains of the SC mapping.

In order to discretize the SIBC for use in the proposed calculation approach, surface conductors are applied along the edges of the SC mapping polygon as shown in Fig. 5. The surface conductors are spaced uniformly along each edge, with shorter edges having a higher conductor density. Now the surface impedance per unit depth (all impedances and voltages are per unit depth) can be calculated as

$$\underline{Z}_k' = \underline{Z}_s'' \cdot \frac{d_{k-1} + d_k}{2}, \tag{6}$$

where d_k and d_{k-1} are the distances to the neighboring surface conductors, measured in the physical domain as shown in Fig. 6.

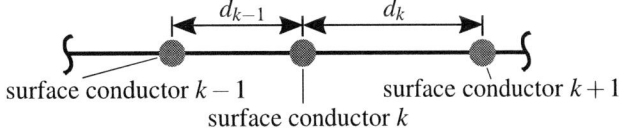

Fig. 6: Distances to neighboring surface conductors measured in the physical domain.

SIBC-aware Vector Potential

Now we can assemble the *surface impedance matrix* that links all surface conductors to each other. The surface impedance matrix need only be created once and can be written as

$$\underline{\mathbf{Z}}'_{\text{surf}} = \begin{pmatrix} \underline{Z}'_k & \underline{Z}'_{k,k+1} & \underline{Z}'_{k,k+2} & \cdots \\ \underline{Z}'_{k,k+1} & \underline{Z}_{k+1}' & \underline{Z}'_{k+1,k+2} & \cdots \\ \underline{Z}'_{k,k+2} & \underline{Z}'_{k+1,k+2} & \underline{Z}_{k+2}' & \cdots \\ \vdots & \vdots & \vdots & \ddots \end{pmatrix}, \tag{7}$$

where each impedance is, as always, defined per unit depth.

The mutual impedance between two arbitrary conductors a and b

$$\underline{Z}'_{a,b} = j\omega \left(A_{a,\infty} - A_{a,b} \right) / I_a \tag{8}$$

can be calculated from the vector potential of conductor a at infinity, carrying a current I_a, $A_{a,\infty} = 0$ and the vector potential of the conductor a at the other conductor $A_{a,b}$. The vector potential at infinity equals $A_{p,\infty} = 0$ because we assume the current loop of the partial conductor to close at infinity. Therefore, the sum of the vector potentials of the partial conductor and the returning conductor (at infinity) equals zero. This equation is used for every non-diagonal element of the matrix. The diagonal elements of the matrix represent the self-inductance of each surface conductor and are calculated using equation (6).

It is worth inverting the surface impedance matrix $\underline{\mathbf{Z}}'_{\text{surf}}$ as it is used repeatedly. Afterwards, the surface currents can be calculated as

$$\underline{\mathbf{I}}'_{\text{surf}} = \underline{\mathbf{U}}'_p \cdot \underline{\mathbf{Z}}'^{-1}_{\text{surf}}. \tag{9}$$

The voltage per length, induced by a partial conductor with the index p,

$$\underline{\mathbf{U}}'_p = \underline{I}_p \cdot \begin{pmatrix} \underline{Z}'_{p,k} \\ \underline{Z}'_{p,k+1} \\ \underline{Z}'_{p,k+2} \\ \vdots \end{pmatrix} \tag{10}$$

is calculated from its current $\hat{I}_p$ and the mutual impedance between the partial conductor and each surface conductor, which in turn can be calculated from equation (8).

Now we can introduce the SIBC-aware vector potential

$$\underline{A}^* = \underline{A}_p + \sum_k \underline{A}_{\text{surf},k}, \tag{11}$$

which contains the contribution of a partial conductor itself $\underline{A}_p$ and of the surface currents it causes $\underline{A}_{\text{surf},k} = A(\underline{I}_{\text{surf},k})$. The *SIBC-aware vector potential* allows us to perform Biot-Savart calculations that take the surface impedance boundary condition into account, while abstracting the surface impedance matrix away.

Partial Conductor Inductance Matrix

Next, the *partial conductor inductance matrix* that describes how the partial conductors of one distributed conductor are linked to each other needs to be generated. This matrix has to be created for each distributed conductor and can be written as

$$\underline{\mathbf{Z}}'_{\text{part}} = \begin{pmatrix} \underline{Z}_p{}' & \underline{Z}'_{p,p+1} & \underline{Z}'_{p,p+2} & \cdots \\ \underline{Z}'_{p,p+1} & \underline{Z}_{p+1}{}' & \underline{Z}'_{p+1,p+2} & \cdots \\ \underline{Z}'_{p,p+2} & \underline{Z}'_{p+1,p+2} & \underline{Z}_{p+2}{}' & \cdots \\ \vdots & \vdots & \vdots & \ddots \end{pmatrix}, \tag{12}$$

with the nomenclature shown in Fig. 7.

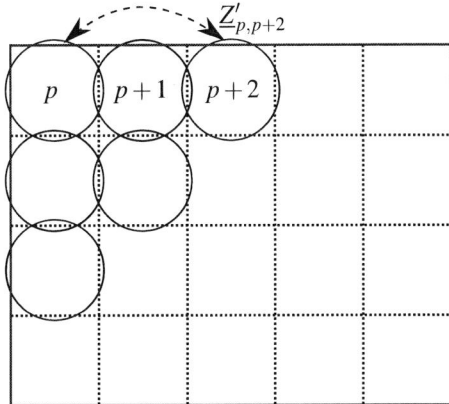

Fig. 7: Nomenclature for the partial conductor inductance matrix. The blue rectangle represents the outline of a distributed conductor.

The non-diagonal elements are obtained similarly to equation (8), only now using the SIBC-aware vector potential:

$$\underline{Z}'_{a,b} = j\omega \left(\underline{A}^*_{a,\infty} - \underline{A}^*_{a,b} \right) / I_{\text{a}} \tag{13}$$

The diagonal elements contain the self-impedance of each partial conductor

$$\underline{Z}'_p = \frac{1}{\sigma \cdot \pi R^2} + j\omega \left(\underline{A}^*_{p,\infty} - \underline{A}^*_p \right) / I_{\text{p}}, \tag{14}$$

where the first term accounts for the partial conductor's resistance, with σ being the conductivity of the corresponding distributed conductor. The second term represents the self-inductance of the partial conductor, where $\underline{A}^*_{p,\infty} = 0$ is the vector potential at infinity and $\underline{A}^*_p$ is the vector potential caused by the partial conductor in question, at its center.

Current Distribution

Finally, we can calculate the actual current distribution inside a distributed conductor. The current carried by the distributed conductor is set as an input value, while its voltage is a result of the current distribution. Fig. 8 gives an overview of how the current distribution is calculated.

First, the terminal voltage along the distributed conductor is set to an initial, virtually infinite value $U'_{\text{terminal}} = 10^{10}$ V/m.

Next, the voltage induced by other distributed conductors is determined by calculating the mutual inductance between each external partial conductor and each partial conductor of the distributed conductor in question using equation (13), as shown in Fig. 9. This requires a large number of calculations to be performed. If we assume $n_{\text{distrib}} = 6$ distributed conductors made up of $n_{\text{part}} = 70$ partial con-

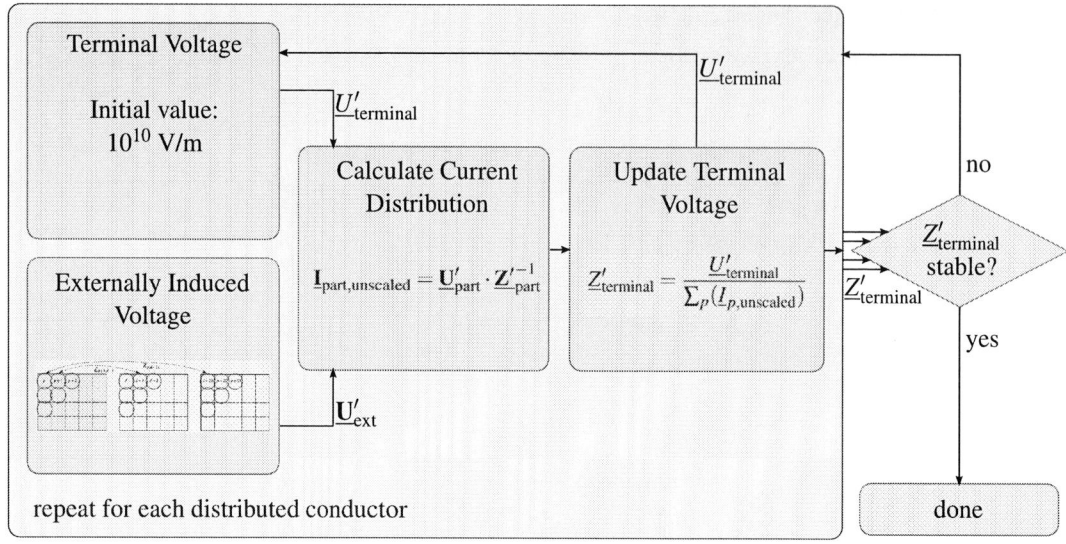

Fig. 8: Flowchart describing how the current distribution is calculated.

ductors each, this involves a total of $n_{\text{part}} \cdot n_{\text{part}} \cdot (n_{\text{distrib}} - 1) = 24500$ impedance calculations to calculate the externally induced voltage of one distributed conductor. If external distributed conductors are approximated using one effective conductor each, the number of calculations can be reduced to $n_{\text{part}} \cdot 1 \cdot (n_{\text{distrib}} - 1) \cdot n_{\text{distrib}} = 350$. This approximation will be used in the validation section, as it provides a worthwhile trade-off between accuracy and calculation time. The voltage induced by other distributed conductors can be written as

$$\underline{\mathbf{U}}'_{\text{ext}} = \begin{pmatrix} \underline{Z}'_{p,d} & \underline{Z}'_{p,d+1} & \underline{Z}'_{p,d+2} & \underline{Z}'_{p,d+3} & \cdots \\ \underline{Z}'_{p+1,d} & \underline{Z}'_{p+1,d+1} & \underline{Z}'_{p+1,d+2} & \underline{Z}'_{p+1,d+3} & \cdots \\ \underline{Z}'_{p+2,d} & \underline{Z}'_{p+2,d+1} & \underline{Z}'_{p+2,d+2} & \underline{Z}'_{p+2,d+3} & \cdots \\ \vdots & \vdots & \vdots & \vdots & \ddots \end{pmatrix} \cdot \begin{pmatrix} \underline{I}_d \\ \underline{I}_{d+1} \\ \underline{I}_{d+2} \\ \underline{I}_{d+3} \\ \vdots \end{pmatrix}, \tag{15}$$

where $\underline{I}_d, \underline{I}_{d+1}, \ldots$ refer to the partial conductor currents in the external distributed conductors (or the effective conductor which it has been approximated by), while the indices $p, p+1, \ldots$ refer to the partial conductors of the distributed conductor in question, as shown in Fig. 9.

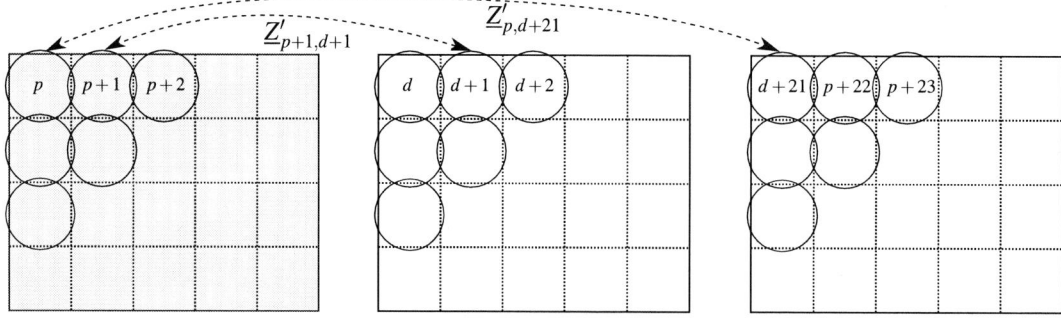

Fig. 9: Nomenclature indicating how the mutual inductances between the partial conductors of the distributed conductor in question and those of external distributed conductors relate.

The partial conductor voltage vector

$$\underline{\mathbf{U}}'_{\text{part}} = U'_{\text{terminal}} + \left(\underline{\mathbf{U}}'_{\text{ext}} - \text{mean}(\underline{\mathbf{U}}'_{\text{ext}})\right) \tag{16}$$

is a result of the terminal voltage and the voltage induced by external distributed conductors. Now, the current distribution can be calculated as

$$\underline{\mathbf{I}}_{\text{part,unscaled}} = \underline{\mathbf{U}}'_{\text{part}} \cdot \underline{\mathbf{Z}}'^{-1}_{\text{part}}, \tag{17}$$

where the partial conductor inductance matrix $\underline{\mathbf{Z}}'_{\text{part}}$ has been inverted to avoid solving a system of equations in each iteration. Next, the current distribution is scaled

$$\underline{\mathbf{I}}_{\text{part}} = \underline{\mathbf{I}}_{\text{part,unscaled}} \cdot \frac{\underline{I}_{\text{terminal}}}{\sum_p (\underline{I}_{p,\text{unscaled}})} \tag{18}$$

so that the sum of all partial currents equals the terminal current $\underline{I}_{\text{terminal}}$.

Lastly, the apparent terminal impedance

$$\underline{Z}'_{\text{terminal}} = \frac{\underline{U}'_{\text{terminal}}}{\sum_p (\underline{I}_{p,\text{unscaled}})} \tag{19}$$

is calculated, and the terminal voltage is updated for the next iteration

$$\underline{U}'_{\text{terminal}} = \underline{Z}'_{\text{terminal}} \cdot \underline{I}_{\text{terminal}}. \tag{20}$$

These steps are performed for each distributed conductor. Afterwards, the whole process is repeated until the change in terminal impedance $\underline{Z}'_{\text{terminal}}$ between two iterations is below a threshold value in every distributed conductor.

Validation

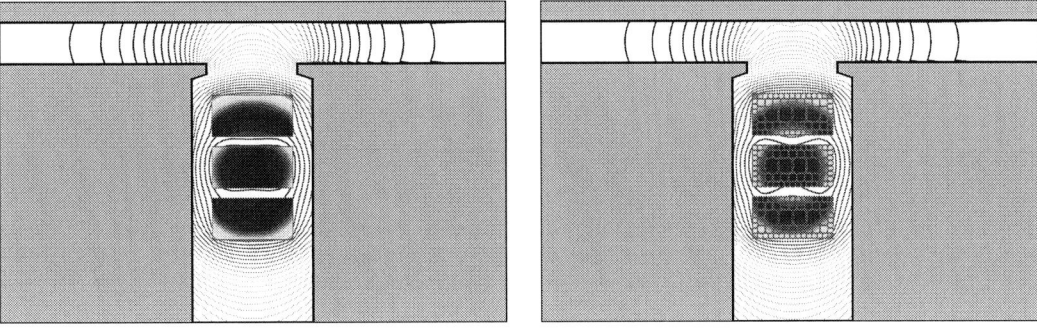

(a) FEMM (2D FEA)   (b) Schwarz-Christoffel and Biot-Savart

Fig. 10: Flux lines and current distribution in a slot.

The proposed calculation approach was validated by comparing it to 2D FEA simulations using the open-source FEMM software [14]. The geometry, material properties, and boundary conditions were kept the same in both calculations. In FEMM, the corresponding boundary condition is called *Small Skin Depth* and was assigned to every edge enclosing the simulation domain. Fig. 10 shows the current distribution and the field lines (lines of constant vector potential) for the case where every conductor carries the same current. For the Schwarz-Christoffel and Biot-Savart calculation, external distributed conductors were approximated using an effective conductor, as described earlier.

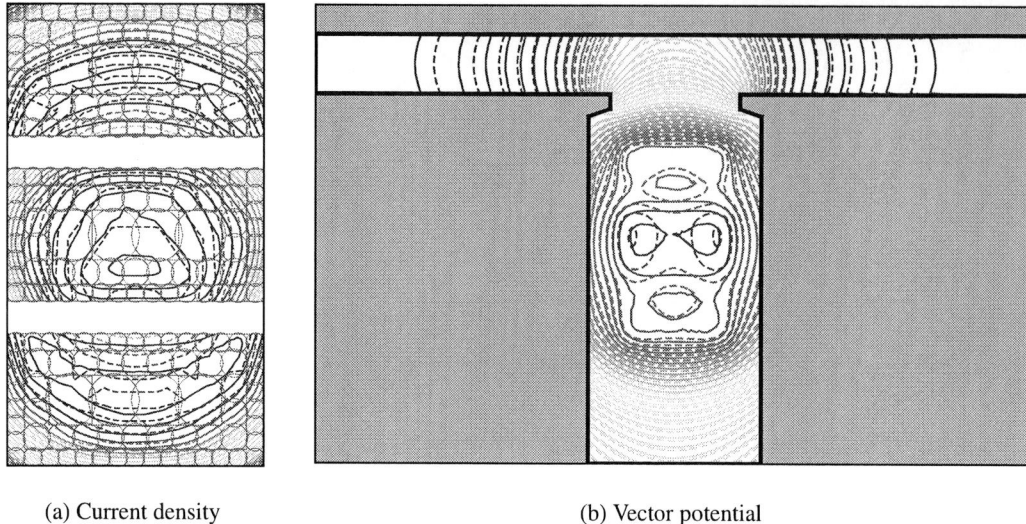

(a) Current density   (b) Vector potential

Fig. 11: Comparison between the 2D FEA simulation (solid lines) and the Schwarz-Christoffel and Biot-Savart based approach (dotted lines).

Plotting the current distribution and the field lines from these calculations together, as shown in Fig. 11 allows us to assess the differences between the calculations more clearly. While there are easily noticeable differences in both the current distribution and field lines, they are overall acceptable given the likely much higher error, compared to reality, due to the simplifications made to allow for the SIBC and only considering one slot. Fig. 12 shows a more complex example of what the proposed method is capable of, with twelve conductors within one slot being modeled.

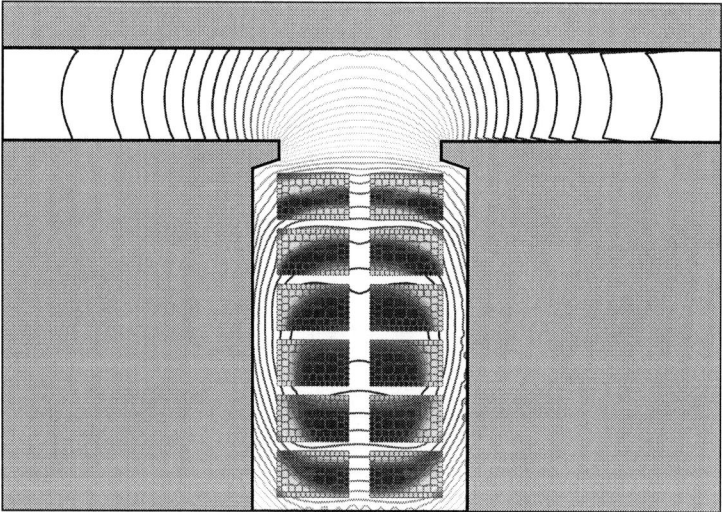

Fig. 12: A more complex example of a current distribution, calculated using the Schwarz-Christoffel and Biot-Savart based approach.

Conclusion

The proposed Schwarz-Christoffel and Biot-Savart based approach is able to accurately calculate the current distribution of conductors within a single slot. A key difference from 2D FEA is the fact that the field can be calculated wherever it is needed instead of meshing the entire domain.

Using the proposed method currently requires a deep understanding of how Schwarz-Christoffel mappings are solved numerically in order to circumvent numerical challenges such as crowding and inverse mapping convergence issues [9]. Given how mature 2D FEA software has become, it is hard to make the case that a combination of Schwarz-Christoffel mappings and Biot-Savart law should become a general-purpose field and loss calculation tool. However, the proposed method is appealing in niche applications where licensing costs or restrictions of FEA software are an issue or where a high level of integration into a broader toolchain and the possible benefits in calculation time resulting from only having to calculate the field where it is needed, are worth the implementation effort.

References

[1] F. D. Giovanni. *Wide Bandgap Semiconductors: to EV and beyond*. Power Electronics News. 2020. URL: https://www.powerelectronicsnews.com/wide-bandgap-semiconductors-to-ev-and-beyond/ (visited on 11/2021).

[2] G. Grandi, D. Casadei, and U. Reggiani. "Common- and differential-mode HF current components in AC motors supplied by voltage source inverters". In: *IEEE Transactions on Power Electronics* 19.1 (2004). Conference Name: IEEE Transactions on Power Electronics, pp. 16–24. DOI: 10.1109/TPEL.2003.820564.

[3] B. Heidler, K. Brune, and M. Doppelbauer. "High-frequency model and parameter identification of electrical machines using numerical simulations". In: *2015 IEEE International Electric Machines Drives Conference (IEMDC)*. 2015, pp. 1221–1227. DOI: 10.1109/IEMDC.2015.7409217.

[4] K. Reppe. *Berechnung von Magnetfeldern mit Hilfe der konformen Abbildung durch numerische Integration der Abbildungsfunktion von Schwarz–Christoffel*. Vol. 8. Siemens Forschungs- und Entwicklungsberichte. Heidelberg, Germany: Springer., 1979.

[5] P. Janhunen and A. Viljanen. "Application of conformal mapping to 2-D conductivity structures with non-uniform primary sources". In: *Geophysical Journal International* 105.1 (1991), pp. 185–190. DOI: 10.1111/j.1365-246X.1991.tb03454.x.

[6] V. V. Amelichev et al. "Conformal transformation method as applied to finding the current density distribution and induced magnetic field in a strip conductor with a rectangular cut". In: *Computational Mathematics and Mathematical Physics* 54.10 (2014), pp. 1618–1625. DOI: 10.1134/S0965542514100017.

[7] E. Costamagna, P. Di Barba, and R. Palka. "Field models of high-temperature superconductor devices for magnetic levitation". In: *Engineering Computations: Int J for Computer-Aided Engineering* 29 (2012). DOI: 10.1108/02644401211246328.

[8] D. C. J. Krop, E. A. Lomonova, and A. J. A. Vandenput. "Application of Schwarz-Christoffel Mapping to Permanent-Magnet Linear Motor Analysis". In: *IEEE Transactions on Magnetics* 44.3 (2008). Conference Name: IEEE Transactions on Magnetics, pp. 352–359. DOI: 10.1109/TMAG.2007.914513.

[9] T. Driscoll and L. Trefethen. *Schwarz-Christoffel Mapping*. Cambridge Monographs on Applied and Computational Mathematics. Cambridge University Press, 2002.

[10] W. J. Gibbs. *Conformal transformations in electrical engineering*. Chapman & Hall, 1958.

[11] T. A. Driscoll. "An improved Schwarz-Christoffel Toolbox for Matlab". In: *Department of Mathematical Sciences, University of Delaware* (2003).

[12] S. V. Yuferev and N. Ida. *Surface Impedance Boundary Conditions: A Comprehensive Approach*. CRC Press, 2009.

[13] *isovac® - Elektroband, Elektroblech, Generatorenbau*. URL: https://www.voestalpine.com/isovac (visited on 11/2021).

[14] D. Meeker. *Finite Element Method Magnetics (FEMM)*. URL: http://www.femm.info/wiki/HomePage (visited on 11/2019).

Standardised switching cell building block for converter design optimisation with detailed electro-thermal model

Georgios Papadopoulos and Jürgen Biela
Laboratory for High Power Electronic Systems, ETH Zürich, Switzerland
Email: papadopoulos@hpe.ee.ethz.ch

Abstract

The design of power electronic systems is typically performed based on optimisation procedures. In this paper, the switching cell of the converter is introduced as a new building block in the optimisation procedure of the converter systems. A standardised half-bridge arrangement is modelled and optimised in terms of efficiency, volume, parasitics, and thermal performance.

1 Introduction

For the design of power electronics systems, various optimisation procedures have been presented for identifying the optimal converter design. Often multiple requirements like high power density, high efficiency, low failure rate, and low cost should be fulfilled at the same time by the design. Typically, many of these goals are competing, as for example, the demand for lower losses, which results in higher efficiency, is usually contradictory to designing converters with a lower volume/weight what results in a higher power density. Therefore, multi-objective and multi-domain optimisation procedures are used in order to evaluate the converter system in various domains, as e.g. electric, electromagnetic, thermal, etc., and to derive the best possible overall design compromise for each application considering the given specifications and constraints [1–3].

To evaluate the performance of the converter and to compare different designs/topologies, accurate and comprehensive models of the system behaviour and its components are required. The combination of these models basically represents a virtual prototype/digital twin of the system and allows to analyse the behaviour and the performance with respect to various aspects [4]. The models can either be based on analytical equations, numerical simulations, or a combination of these. By using virtual prototypes and simulation tools, the design of converters and the converter performance can be evaluated and compared in a comprehensive and time efficient manner.

Virtual prototypes of the converter systems are typically based on system level models and individual component models for semiconductor devices [5, 6], cooling systems [7], gate drivers [8], capacitors [9, 10], magnetics [11], and temperature distribution [12], which all have been extensively modelled and investigated in the literature. Concerning the geometrical component/converter design in the existing multi-objective models and optimisation procedures, the focus has been on optimising the mechanical design of magnetics [11], capacitors [9], and the cooling system for dissipating the semiconductors' losses [7].

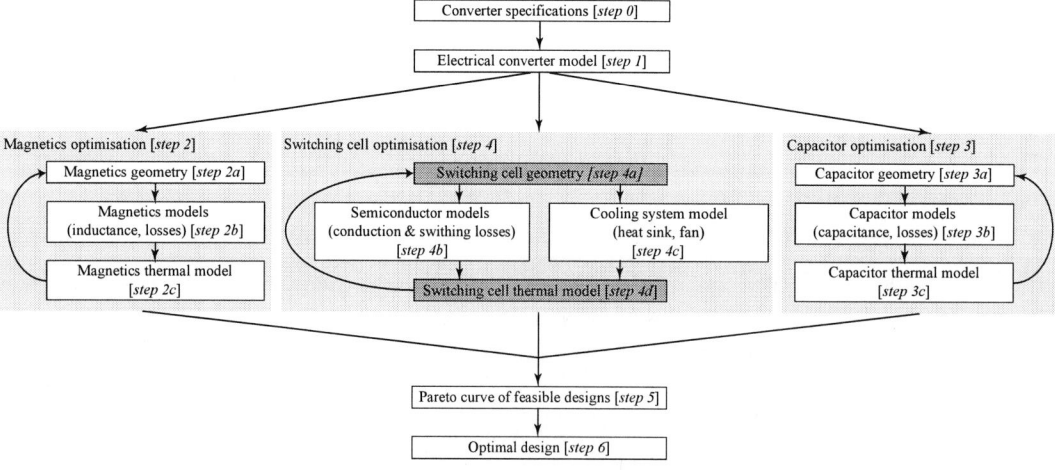

Fig. 1: Proposed converter optimisation procedure including model and optimisation of the converter switching cell.

In the converter optimisation procedures, the geometrical converter design can be divided into four major building blocks, which are the magnetics, the DC-link capacitors, the semiconductors, and the cooling system from a hardware point of view. There, the DC-link capacitors can be split up into the energy and the decoupling capacitors. The energy capacitors are used as energy storage device responsible for reducing/minimising the effects of voltage/load variations on the converter output (e.g. hold-up time capacitor), while the decoupling capacitors are responsible for improving the transient switching behaviour of the semiconductor devices (i.e minimising the parasitic inductance in the commutation loop). In the considered power range of a few kW, the power semiconductor devices, the decoupling DC-link capacitors, and the gate driver circuits are typically mounted in close vicinity on a printed circuit board (PCB), which is usually placed close to the heat sink of the cooling system. Thus, from a geometrical point of view, these components can be grouped in a unified component building block, the so called switching cell building block (SCBB), as illustrated in fig. 2. There, the DC-link energy capacitors are typically excluded from the SCBB because the required energy storage of the converter's DC-link capacitors, which is correlated to their volume, strongly depends on the application, and by excluding the energy storage capacitors, the complexity for standardising the mechanical layout of the switching cell (SC) is significantly reduced.

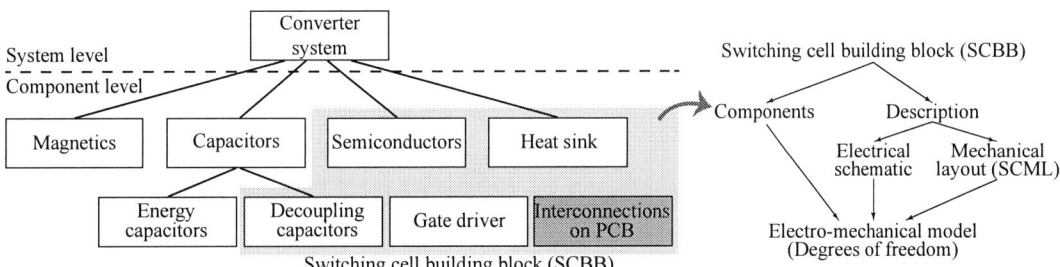

Fig. 2: "Building blocks" considered in the converter optimisation.

So far, the geometrical arrangement/mechanical design of the SCBB has not been modelled and integrated in the optimisation routine, although its design significantly impacts the overall converter performance. Therefore, in the proposed converter optimisation procedure depicted in fig. 1, steps for optimising the switching cells mechanical design and thermal modelling on the switching cell level are introduced. The modelling principle is similar to the magnetics building block, which is composed of the individual magnetic components (e.g. core, windings, etc.) and described by their electro-magnetic behaviour and their mechanical arrangement. In case of the SCBB, it is described by an electrical schematic (fig. 3), and a mechanical layout as presented in section 2. The individual models of the SCBB components together with the electrical and geometrical configuration are combined to an electro-mechanical model of the SCBB, which can be used to optimise the SCBB design based on predefined degrees of freedom.

For demonstrating the benefits of integrating the SCBB in the converter optimisation procedure, a half-bridge buck converter for a battery interface application is examined as example. For simplicity, and to further narrow down the design space, the focus is on standard TO-247 packages, SOIC-8 integrated gate driver circuits, ceramic decoupling capacitors, and heat sink with a cooling fan. However, the basic concept could be also extended to other components.

In section 2, first standard geometrical arrangements of the switching cell building block components of a half-bridge are defined and qualitatively compared in terms of cooling performance, efficiency, and power density. In section 3, models of the switching cell building block are presented. In section 4, simulation results of the optimisation of a switching cell mechanical layout are presented, and afterwards in section 5, these are validated by FEM simulations. Finally, in section 6, the benefits of standardising the switching cell as a building block for the overall converter optimisation are discussed.

2 Standardised half-bridge switching cell mechanical layout

In order to include the half-bridge switching cell as a building block in the converter optimisation procedure, its components, electrical behaviour, and mechanical layouts need to be modelled. The heat sink of the cooling system for dissipating the power losses of the switching cell components, constitutes the bulkiest component of the switching cell and thus it is used as kind of "base/anchor" element in the standardisation of the switching cell mechanical layout (SCML).

As depicted in fig. 4, the SCMLs typically can be categorised based on two main geometrical parameters. The first one is the face of the heat sink (top/side) on which the semiconductor devices are mounted and the second one is

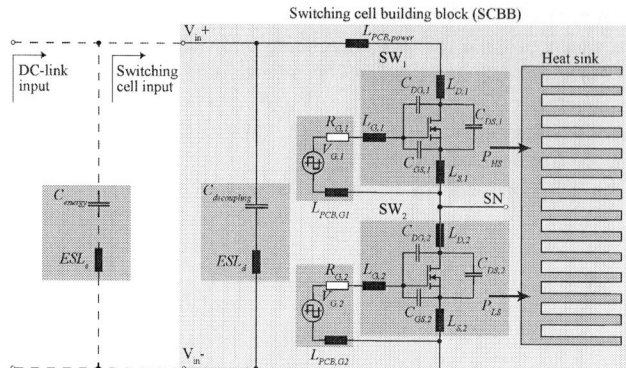

Fig. 3: Basic electrical schematic of a half-bridge switching cell building block (SCBB).

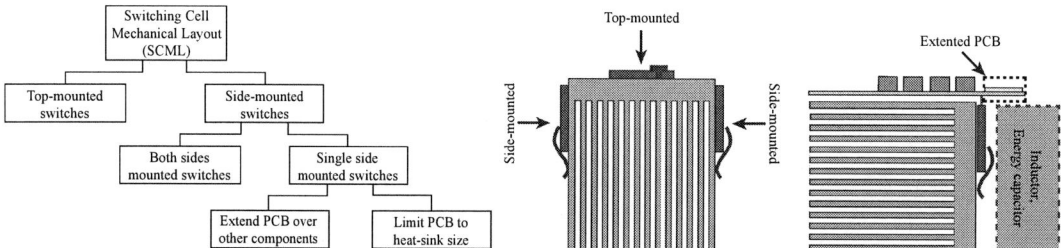

Fig. 4: Standardisation of switching cell mechanical layout (SCML) using the heat sink as "base/anchor" element.

whether the PCB of the switching cell is extended beyond the heat sink on the side of the switch, and over other components, as e.g. the magnetics (inductor) of the converter. Based on the assumed standard components given in Table I, 10 standard half-bridge SCMLs can be derived as illustrated in fig. 5. In order to simplify the modelling process, only one package type is selected for each component as exampmle, but the basic concept could easily be adapted to other package types. Ceramic capacitors with a 2220 SMD package provide the required capacitance for decoupling in case e.g. SiC MOSFETs are used and at the same time, there are so compact to be placed as close as possible to the semiconductor switches in order to minimise the power loop parasitic inductance.

Component	Package/type
Semiconductor switch	TO-247
Gate driver IC	SOIC-8
Decoupling capacitor	Ceramic 2220
SMD driver resistors/capacitors	1210
Mounting device (side-mounted)	Spring
Mounting device (top-mounted)	Screw
Forced cooling fan	Quadratic fan 60x60 mm

Table I: Considered components, packages, and dimensions for standardising SCMLs of half bridge converters.

The different SCMLs are qualitatively compared in fig. 7 in terms of volume, parasitic elements, and complexity. The inductances $L_{PCB,power}$, $L_{PCB,G}$, L_S, L_D, and L_G refer to the power loop, the gate loop, and the packaging pin (source, drain, gate) inductances, the variable $Area_{capacitors}$ to the available space for the DC-link decoupling capacitors and the variable $Volume$ to the required effort/"complexity" for assembling the switching cell (e.g. mounting the semiconductor device with a screw or a spring to the heat sink). The variable $Volume$ refers to the total boxed volume of the switching cell's outer dimensions including the cooling fan, the heat sink, the semiconductors, the mounting devices, the PCB, the gate drivers and the decoupling capacitors. In case of SCMLs with extended PCBs over other components, the boxed volume could overestimated the system volume since the space "under the PCB" can be used for other components (e.g. inductor as shown in fig. 4) on the system level.

In order to compare the different SCMLs in terms of power and gate loop parasitic inductances, the power and the gate loops need to be determined. Fig. 6 illustrates a basic 2-layer PCB layout (top and bottom view) for two of the SCMLs: $SCML_h$ and $SCML_j$ as example. The power loop parasitic inductance results from the closed loop which is defined by the semiconductor switches and the DC-link decoupling capacitors. Thus, it is the sum of the semiconductors' drain and source pins' inductances (L_D and L_S), the PCB trace power loop inductance, and the

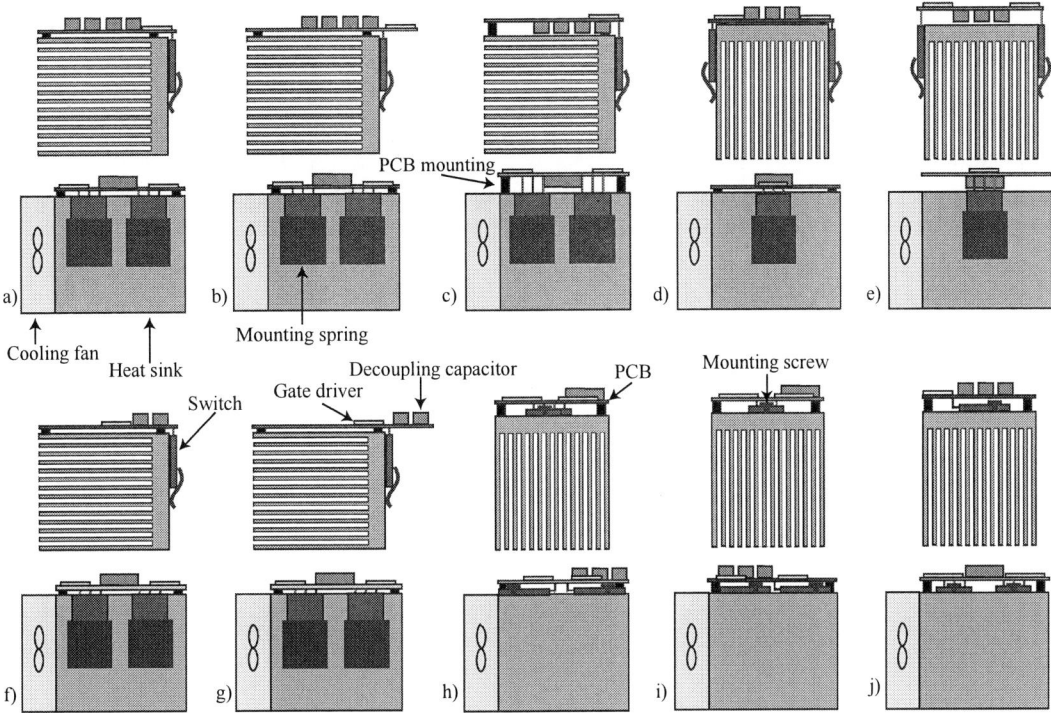

Fig. 5: Standardised switching cell mechanical layouts (SCMLs) based on the standard components given in Table I.

decoupling capacitor's ESL. So the total PCB power loop inductance is given by: $L_{PCB,power} = L_{PCB,D} + L_{PCB,DS} + L_{PCB,S} + ESL_d$. The gate loop parasitic inductance results from the closed loop defined from the semiconductor switch, its gate driver IC, the gate driver's supply capacitor and the gate resistor. Thus, it is the sum of the semiconductors' gate and source pins' inductances (L_G and L_S), the PCB trace gate loop inductance ($L_{PCB,G1}$ and $L_{PCB,G2}$), and the ESL of the supply capacitor.

Regarding the parasitic inductances in the various SCMLs, it should be noted that the parasitic inductances of the PCB loops are proportional to the loops' length and inversely proportional to their widths. Consequently, it can be concluded that the power loop inductance of $SCML_h$ (fig. 6a) is typically higher than the one of $SCML_j$ (fig. 6b) since the PCB power planes in this case are longer and thinner in the considered example layout. However, the difference in the parasitic power loop inductance of those SCMLs is usually relatively small and depends on the specific layout and the number of PCB layers. Concerning the gate loop inductance both SCMLs indicate similar inductances with short paths between the driver IC, the driver's SMD gate resistor, the drivers SMD supply capacitor and the semiconductor switch.

The parasitic inductances are a key parameter for evaluating the SCMLs since they have a significant impact on the semiconductor switching losses/converter's efficiency and also on the transient overvoltages. Therefore, SCMLs which basically result in relatively high power loop inductances ($SCML_a$, $SCML_c$, $SCML_d$, $SCML_e$ and $SCML_i$) are neglected from the following consideration regarding the optimisation of the switching cell. Such SCMLs are typically only considered if they are beneficial due to other design constraints.

SCMLs in which the PCB board is extended beyond the heat sink for placing either the gate driver or the decoupling capacitors ($SCML_b$ and $SCML_g$) typically result in smaller parasitics as well as improving the electrical and thermal performance. However, such designs are not the most compact ones and increase the overall volume of the switching cell, if the space under the extended PCB is not used for other components.

Regarding the volume of the heat sink, it is typically directly linked with the efficiency of the switching cell, since a bigger heat sink typically results in a better cooling for the semiconductors, and eventually a lower junction temperature, and thus, lower conduction losses and power losses of the fan. The size of the heat sink as well as the placement of the switching cell's components on the heat sink/PCB are basically free design parameters (degrees of freedom) for optimising the switching cell design, which is presented in the following section.

The switching cell mechanical layout $SCML_j$ shown in fig. 8 is used as an example for deriving the switching cell models in the following section, since it is one of the best solutions in terms of parasitic inductances, volume, available space for additional decoupling capacitors based on the qualitative analysis. These benefits come with a

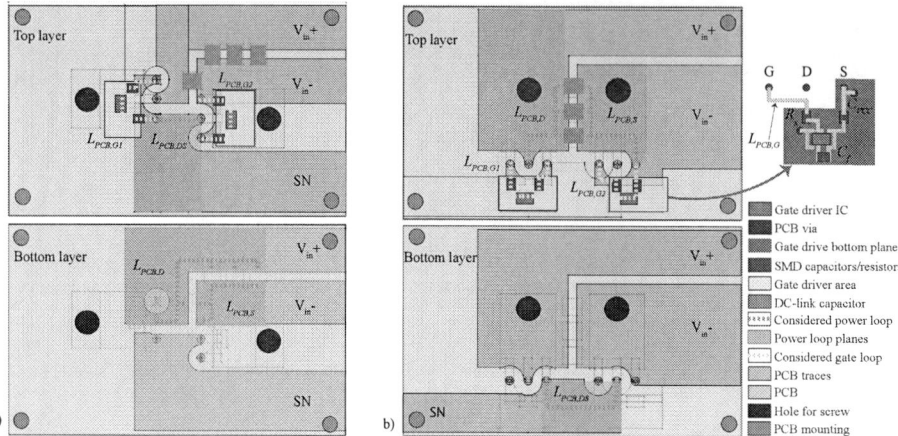

Fig. 6: Considered simplified sketch of the power and the gate loop of $SCML_h$ and $SCML_j$ on a 2-layer PCB for calculating the parasitic loop inductances (step 8 in fig. 9).

	$SCML_a$	$SCML_b$	$SCML_c$	$SCML_d$	$SCML_e$	$SCML_f$	$SCML_g$	$SCML_h$	$SCML_i$	$SCML_j$
$L_{PCB,G}$	↓↓	↓↓	↓	↓↓	↑	↑↑	↓↓	↓↓	↓↓	↓↓
$L_{PCB,D}+L_{PCB,DS}+L_{PCB,S}$	↑	↓↓	↓↓	↑↑	↑↑	↓↓↓	↓↓	↓↓↓	↑↑	↓↓↓
L_G, L_D, L_S	↓	↓	↑↑	↓	↑↑	↓	↓	↑		↑
$Area_{capacitors}$	↑↑↑	↑↑↑	↑↑	↑↑↑	↑↑↑	↓↓	↓↓	↑↑	↑	↑↑
$Assembly$	↑↑↑	↑↑	↑↑	↑↑↑	↑↑	↑↑↑	↑↑	↓	↓	↓
$Volume$	↓	↑↑↑	↑	↓	↑	↓	↑↑↑	↓	↓	↓

Fig. 7: Qualitative comparison of standardised switching cell mechanical layouts given in fig. 5.

cost of an increased difficulty in assembling the switching cell, since the TO semiconductors need to be soldered on the PCB with bended pins and then mounted on the heat sink with a screw. In applications where easy assembly is important, side-mounted TO packages are preferable, since in those cases the screw/spring is not covered by the PCB.

3 Degrees of freedom/model of the selected mechanical layout standard

Based on the standardised SCML building block and the identified best SCML shown in fig. 8 as example, the basic degrees of freedom (DOFs) of the design are identified in a next step and the models for the optimisation of the building block are presented. As shown in fig. 8, the DOFs for the geometrical arrangement optimisation of the switching cell are its outer dimensions, i.e. the width w_{sc}, the length l_{sc}, and the height h_{sc} of the switching cell, as well as the distance between the semiconductor switches d_{semi}. The height and the width of the switching cell are mainly determined by the dimensions of the selected fan. But the length of the heat sink and the operating point of the fan which affect the thermal resistance of the heat sink, and thus, the operating temperature of the switches are DOFs to optimise. The switches are usually placed in or close to the middle of the heat sink in order to maximise the heat spreading and the dissipation capability of the cooling system. The power loop parasitics depend on the placement of the DC-link capacitors and the switches on the PCB and they have an impact on the switching losses.

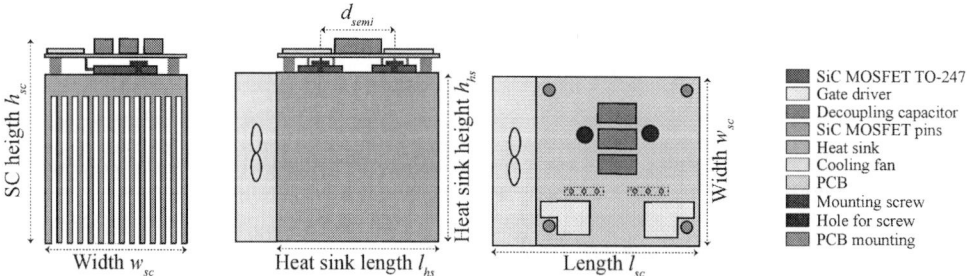

Fig. 8: Degrees of freedom (h_{sc}, w_{sc}, l_{sc}, h_{hs}, l_{hs}, d_{semi}) of the selected SCML - $SCML_j$ given in fig. 5.

Typically, the decoupling capacitors are placed as close as possible to the semiconductor switches. However, the minimum distance and the routing of the PCB traces/planes are limited by the clearance distance (shown in fig. 6), which is given as input in the optimisation procedure based on the converter's specifications (i.e. voltage level, pollution degree etc.).

The flow chart of the proposed procedure for optimising the mechanical layout of the half-bridge switching cell is illustrated in fig. 9. The detailed steps for evaluating the SCMLs in terms of parasitics, losses, and cooling system design are explained in the following.

- **Step 1**: Select a SCML standard based on the system requirements and the qualitative evaluation presented in section 2. In the following, the SCML illustrated in fig. 8 is considered as example.

- **Step 2**: Derive the mechanical constraints for the optimisation variables based on the selected SCML. For example, in the selected SCML (fig. 8) the length of the heat sink l_{hs} cannot be smaller than the sum of the switches' width plus the distance of the centres d_{semi}, since otherwise they cannot be mounted on the heat sink.

- **Step 3**: Select a semiconductor switch based on the converter specifications (voltage, current) and cost.

- **Step 4**: Select decoupling capacitors based on the analysis presented in [10]. The minimum recommended capacitance value should be higher than at least 50 times the output capacitance of the selected semiconductor switch at DC-link voltage. This is the lower boundary for the decoupling capacitor: $C_{decouple} \geq 50 \cdot C_{OSS}$. The upper limit of parallel capacitors is given by the available space on the PCB. By connecting more capacitors in parallel, the equivalent series inductance of the capacitor ESL_d is divided by the number of capacitors $n_{decouple}$ and the total trace inductance decreases also. In the considered optimisation procedure, the maximum number of parallel capacitors fitting on the available PCB space is selected in step 4 in order to minimize the power loop parasitic inductance.

- **Step 5**: Select gate driver IC and gate resistors based on the specifications of the selected semiconductor device.

- **Step 6**: Mechanically "arrange" the components in the virtual prototype and the selected SCML based on the geometrical parameters illustrated in fig. 8. The width w_{sc} and the height h_{sc} of the switching cell are determined based on the dimensions of the selected cooling fan, which is given as input in the optimisation routine. The gate driver is mounted on the PCB as close as possible to the semiconductor switch to minimised the parasitic gate loop inductance. The PCB traces and planes illustrated in fig. 6b) are routed according to the electrical schematics (fig. 3), the clearance distance and the number of PCB layers. The length l_{sc} of the switching cell and the distance between the semiconductors d_{semi} vary in order to optimise the SCML.

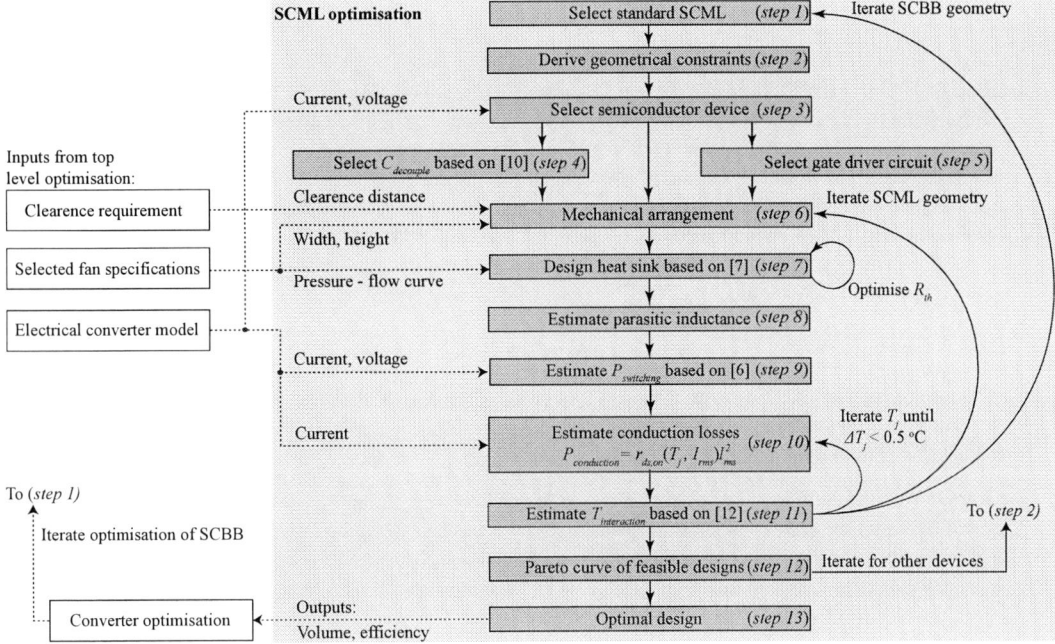

Fig. 9: Switching cell optimisation procedure based on an electro-thermal model with forced air cooling.

- **Step 7**: For given from step 6 heat sink's outer dimensions l_{hs}, w_{hs} and h_{hs}, locally optimise the rest heat sink geometrical parameters, i.e. base plate thickness d_{hs}, channel width s_{hs}, fin length c_{hs}, fin width t_{hs}, the number of channels n_{fins}, and the cooling fan operational point in order to minimise the thermal resistance of the heat sink $R_{th,hs}$ and the cooling fan power losses as described in [7].

$$R_{th,hs} = R_{th,d} + R_{th,fin} \tag{1}$$

There the thermal resistance of the base plate of the heat sink considering the aluminium thermal conductivity λ_{al} is:

$$R_{th,d} = \frac{d_{hs}}{w_{sc} l_{sc} \lambda_{al}} \tag{2}$$

And the thermal resistance of the fins is given as a function of the average air volume flow $\dot{V}$ as derived from the fan's operation point based on the flow-pressure characteristics for the air density ρ_{air} and the thermal air capacity c_{air}.

$$R_{th,fin}(\dot{V}) = \frac{1}{\rho_{air} c_{air} \dot{V} \left(1 - e^{-\frac{hA_{eff}}{\rho_{air} c_{air} \dot{V}}}\right)} \tag{3}$$

There h is the heat transfer coefficient determining the thermal resistance of the effective fin surface area assuming a uniform wall temperature and considering the cooling air flow and the thermal conductivity of the air λ_{air}.

$$h = \frac{Nu_{\sqrt{A}} \lambda_{air}}{d_h} \tag{4}$$

The Nusselt number $Nu_{\sqrt{A}}$ is calculated based on the Prandtl number and the Reynolds number as described in [7] and the hydraulic diameter is given by:

$$d_h = \frac{2 s_{hs} c_{hs}}{s_{hs} + c_{hs}} \tag{5}$$

The effective fin surface is given by:

$$A_{eff} = (n_{fins} - 1)(2 c_{hs} a_{fin} + s_{hs}) l_{hs} \tag{6}$$

Where: $a_{fin} = \dfrac{tanh(f)}{f}$ and $f = \sqrt{\dfrac{2h(t_{hs} + l_{sc})}{\lambda_{al} t_{hs} l_{sc}}}$

- **Step 8**: Estimate the PCB power and gate loop inductances based on the PCB layout given in fig. 6b). For simplifying the calculation, the single trace/plane inductance formula: $L_{dc} = \mu_r \cdot \frac{d_t l_t}{w_t}$ is used as an approximation, where $\mu_r = 1$ is the assumed relative permeability of copper and d_t, l_t, as well as w_t are the thickness, the length and the width of the single trace/plane. To improve the accuracy of the estimated inductances, refined parasitic inductance models can be used. The gate loop is defined by the closed loop connecting the gate and the source of each semiconductor with the driver IC, the gate resistor, and the gate driver's supply capacitor. The power loop is the closed loop connecting the decoupling capacitors with the drain and the source pins of the semiconductors. The PCB power loop parasitic inductance is approximately given by:

$$L_{PCB,power} = L_{PCB,D} + L_{PCB,S} + L_{PCB,DS} + ESL_d / n_{decouple} \tag{7}$$

The approximation of the parasitic inductances is based on the PCB layout given in fig. 6 b) for a 2-layer PCB. In case of a 4 or more layer PCB, further layout optimisation can be performed to minimise the parasitics, as for example parallel connected power planes.

- **Step 9**: The switching losses of each semiconductor switch are estimated as described in the model presented in [6] based on the semiconductors' typical capacitances, the selected gate driver (gate-source voltage and gate resistors), the parasitic inductances estimated on step 8 as well as the operational voltage and current, whose waveforms are given as inputs, in e.g. a piecewise linear (PWL) format, to the optimisation routine from the electrical model of the converter.

- **Step 10**: Calculate the conduction losses with $P_{conduction} = r_{ds,on}(T_j, I_{rms}) \cdot I_{rms}^2$ based on an assumed operation junction temperature T_j and the RMS current of each switch, which is calculated based on the current's waveform (in e.g. PWL format) given as input based on the electrical model of the converter. The semiconductor's on-resistance $r_{ds,on}$ is derived as function of the operational temperature and current from the respective conductance curves given in the semiconductor's datasheet. The PCB traces/planes conduction losses are also calculated in this step based on the copper resistivity ρ_{cu} and the PCB layout presented in fig. 6 by:

$$r_{dc,pcb} = \rho_{cu} \cdot \frac{l_t}{d_t \cdot w_t} \cdot \left(1 + \left(\theta_{cu} \cdot (T_j - T_{amb})\right)\right) \tag{8}$$

- **Step 11**: The temperature interaction between the two switches is estimated based on the temperature spreading model for the isotropic base plate as described in [12]. The temperature at any point of the heat sink plate and also for the semiconductors is given by the equation:

$$\begin{aligned}
\theta(x,y,z) = A_0 + B_0 z &+ \sum_{m=1}^{\inf} cos(\lambda x)[A_m cosh(\lambda z) + B_m sinh(\lambda z)] \\
&+ \sum_{n=1}^{\inf} cos(\delta y)[A_n cosh(\delta z) + B_n sinh(\delta z)] \\
&+ \sum_{m=1}^{\inf}\sum_{n=1}^{\inf} cos(\lambda x)cos(\delta y)[A_{mn} cosh(\beta z) + B_{mn} sinh(\beta z)]
\end{aligned} \tag{9}$$

where $\lambda = m\pi/l_{sc}$, $\delta = n\pi/w_{sc}$ and $\beta = \sqrt{\lambda^2 + \delta^2}$.

The Fourier coefficients A_0, A_m, A_n, A_{mn}, B_0, B_m, B_n, B_{mn} are given in [12] as a function of the heat sink geometry, the heat transfer coefficient due to the cooling air flow, the semiconductor's package dimensions and the total losses of each semiconductor (conduction + switching losses).

Assuming a linear thermal model, the junction temperature of each switch is derived based on:

$$T_{j,i} = (P_{conduction,i} + P_{switching,i})R_{th,hs} + T_{interaction,k \to i} + T_{amb} + T_{cj} \tag{10}$$

There, $T_{interaction,k \to i} = \theta(X_{Ci}, Y_{Ci}, 0)$ is the superimposed temperature rise of i-th semiconductor switch, with the centre position X_{Ci} and Y_{Ci}, caused by the conduction and switching losses of the k-th semiconductor switch. The variable T_{amb} is the ambient temperature and the variable T_{cj} is the case to junction temperature. The calculated junction temperature is iterated in step 10 and 11 until the relative error between the assumed junction temperature at step 10 and the calculated one is smaller than $0.5\,^{\circ}$C.

- **Step 12**: Derive the pareto-front curves for the feasible geometries where the junction temperature does not exceed the maximum allowed junction temperature: $T_{j,max} = 150\,^{\circ}$C, minus a margin, which is chosen to be $10\,^{\circ}$C in the considered case. At this point additional optimisation iterations could be performed for other devices/components.

- In the final step, the best design is chosen based on the converter system's volume and efficiency requirements. The output of the suggested routine can be linked to a converter system optimisation, where the SCBB is optimised based on the proposed models for several SCMLs, cooling fans or topologies.

For demonstrating the operation of the presented routine and models, optimisation results of the selected SCML given in fig. 8 are presented in the following section.

4 Optimisation results for the selected switching cell mechanical layout standard

In order to demonstrate the application and performance of the presented modelling and optimisation of the SCML, a DC-DC converter for a charger in an electric vehicle with the specifications given in Table II is used as an example. The results of the optimisation for the selected components presented in Table III are illustrated in figs. 10 and 11. Fig. 10 shows the pareto front power losses vs. volume and the selected design point for the SC. In fig. 11 it is illustrated, that there is an optimal distance between the two semiconductor switches for the selected standard SCML, where the power losses of both semiconductor switches are minimised. The power losses initially decrease with an increasing distance between the switches due to two main reasons. The first reason is the initial decrease of the switching losses due to the impact of the clearance distances, which results in thinner PCB traces/planes, and thus, higher parasitic inductances for short distances d_{semi}. The second reason is because of the thermal interaction between the two switches, which exponentially reduces with increasing distances.

At larger distances d_{semi}, the increasing distance between the semiconductors results again in higher power losses due to increasing power loop parasitic inductances, which results in higher switching losses. Consequently, there is an optimal distance for minimal losses.

Parameter	Symbol	Value
Topology		Buck
Input voltage	V_{in}	800 V
Output voltage	V_{out}	540 V
Output current	I_{out}	40 A
Ripple voltage	ΔV_{pp}	< 5 %

Table II: Converter specifications with output power $P_{out} = 540\,\text{V} \cdot 40\,\text{A} = 21.6\,\text{kW}$.

Component	Manufacturer	Product number
Switch	Infineon	IMW120R030M1H
Gate driver IC	Infineon	1EDC40I12AHXUMA1
Decoupling cap.	Knowles Syfer	2220Y1K50154KXTWS3
Cooling fan	Sanyo Denki	109P0612T7H122

Table III: Chosen components for optimising the selected SCML - $SCML_j$ given in fig. 5.

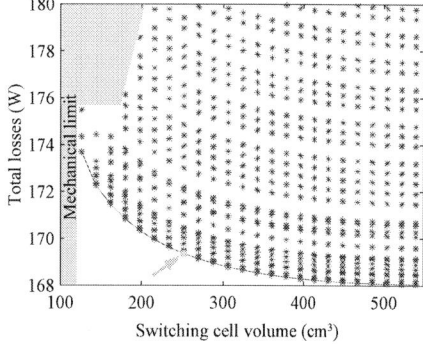

Fig. 10: Pareto front losses-volume of the half-bridge SC with a 60x60 mm cooling fan.

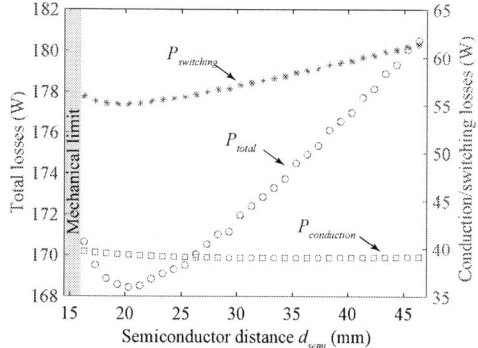

Fig. 11: Total as well as high-side switch losses in the half-bridge SC with a 60x60 mm cooling fan and a length of 70 mm.

5 Validation of thermal model by FEM

In this section, simulation results based on an electro-thermal FEM simulation for the optimised SCML (marked as green point in fig. 10) are presented for validating the electro-mechanical model described in section 3.

Initially, a universal geometrical model of a switching cell depending on the selected cooling fan is implemented in FEM. The geometry and the variables of the selected and optimised switching cell are illustrated in fig. 12. This switching cell is simulated for various distances between the switches and lengths of the SC in order to validate the proposed analytical model.

In the FEM model, the on-resistance of the selected semiconductor switch is implemented as a function of the operating current and temperature. Additionally, the switching losses are implemented as a function of the estimated parasitics based on the designed geometry of the SC. The cooling performance of the fan is calculated based on the channel surface shown in red fig. 12 for one channel as heat dissipating surface. The effective heat transfer coefficient of this surface due to the cooling fan is determined based on the average air flow velocity and pressure calculated with the analytical model described in [7]. This results in $h = 134.35\,\text{W}/(\text{m}^2\text{K})$.

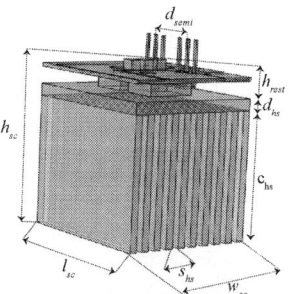

Fig. 12: Mechanical model of the optimised switching cell. The considered heat dissipating channel surface shown in red for one channel is used for calculating the cooling performance of the fan.

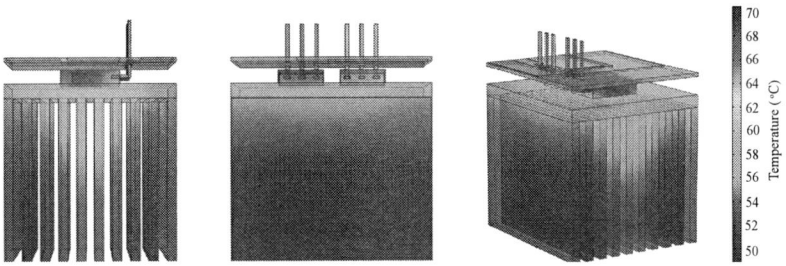

Fig. 13: Temperature distribution of the optimised half-bridge SC with 60x60 mm cooling fan and length of 70 mm.

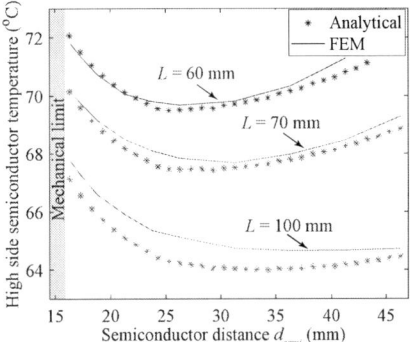

Fig. 14: Operating temperature of a switch of the half-bridge SCs as a function of the distance and the length of the SC at an ambient temperature $T_{amb} = 40\,^{\circ}$C. With the observed $\Delta T \approx 30\,^{\circ}$C, the considered design could also operate in applications with higher ambient temperatures.

The temperature distribution of the optimised half-bridge SC with 60x60 mm cooling fan and length of 70 mm is illustrated in fig. 13. The validation of the thermal model (thermal resistance and thermal interaction) is performed by comparing the operating temperatures of the semiconductor switches resulting from the analytical and the FEM model for several SC lengths and semiconductor distances.

The resulting operating temperature of a semiconductor switch of the half-bridge SC as a function of the SC's length and distance between the semiconductors is illustrated in fig. 14 as example. The reasons for resulting in relatively low junction temperatures are, first, the fact that the design is optimised mostly towards higher efficiency and, second, the selected cooling fan provides a high air flow volume.

As illustrated in the figure, the results from the analytical model are very close to the results derived from FEM. The error in the case of $L = 100$ mm is caused by the assumption of a uniform wall temperature in the analytical model resulting in a better cooling performance compared to the FEM model. As can be seen in fig. 13, the fin wall's temperature is lower at the bottom of the heat sink and higher at the parts of the fins being closer to the mounting points of the switches.

6 Conclusion

In this paper, the concept of a standardised switching cell is introduced as an additional building block in converter optimisation procedures. The SCBB is composed of individual component models (power semiconductors, heat sink, parasitics, DC-link decoupling capacitors, gate drivers). It is described by an electrical schematic and a mechanical layout. The mechanical layout of the switching cell is standardised in order to be included in an electro-mechanical model, which is used to optimise the switching cell building block design. An example of a buck converter design is used to demonstrate the benefits of modelling and optimising the switching cell mechanical layout. Finally, it is shown that there is an optimal design of the switching cell with minimal losses.

The suggested electro-mechanical model of the switching cell is universal and the basis for more complex designs. The proposed optimisation routine of the SCBB could be also applied to designs with other packages, topologies or converter system requirements. The proposed standardised switching cell mechanical layouts enable to move one step closer towards a comprehensive virtual prototyping optimisation procedure by integrating one additional automated design step and to increase the number of optimised building blocks in the design procedures of the converter systems.

References

[1] A. Stupar, T. McRae, N. Vukadinović, A. Prodić, and J. A. Taylor, "Multi-objective optimization of multi-level DC–DC converters using geometric programming," *IEEE Trans. on Power Electronics*, vol. 34, no. 12, 2019.

[2] S. Waffler, M. Preindl, and J. W. Kolar, "Multi-objective optimization and comparative evaluation of Si soft-switched and SiC hard-switched automotive DC–DC converters," in *Industrial Electronics Annual Conference (IECON)*, 2009.

[3] R. Burkart and J. W. Kolar, "Component cost models for multi-objective optimizations of switched-mode power converters," in *IEEE Energy Conversion Congress and Exposition (ECCE USA)*, 2013.

[4] J. Biela, J. W. Kolar, A. Stupar, U. Drofenik, and A. Müsing, "Towards virtual prototyping and comprehensive multi-objective optimisation in power electronics," in *International Exhibition & Conference for Power Electronics Intelligent Motion Power Quality (PCIM)*, 2010.

[5] D. Christen and J. Biela, "Analytical switching loss modeling based on datasheet parameters for SiC MOSFETs in a half-bridge," *IEEE Trans. on Power Electronics*, vol. 34, 2019.

[6] A. Hu and J. Biela, "An analytical switching loss model for a SiC MOSFET and schottky diode half-bridge based on nonlinear differential equations," in *European Conference on Power Electronics and Applications (EPE - ECCE Europe)*, 2021.

[7] D. Christen, M. Stojadinovic, and J. Biela, "Energy efficient heat sink design: Natural versus forced convection cooling," *IEEE Trans. on Power Electronics*, vol. 32, 2017.

[8] M. Moradpour, A. Lai, A. Serpi, and G. Gatto, "Multi-objective optimization of gate driver circuit for GaN HEMT in electric vehicles," in *Annual Conference of the IEEE Industrial Electronics Society (IECON)*, 2017.

[9] D. Menzi, D. Bortis, G. Zulauf, M. Heller, and J. W. Kolar, "Novel iGSE-C loss modeling of X7R ceramic capacitors," *IEEE Trans. on Power Electronics*, vol. 35, 2020.

[10] Z. Chen, D. Boroyevich, P. Mattavelli, and K. Ngo, "A frequency-domain study on the effect of DC-link decoupling capacitors," in *IEEE Energy Conversion Congress and Exposition (ECCE USA)*, 2013.

[11] M. Stojadinović and J. Biela, "Modelling and design of a medium frequency transformer for high power DC–DC converters," in *International Power Electronics Conference (IPEC - ECCE Asia)*, 2018.

[12] Y. Muzychka, J. Culham, and M. Yovanovich, "Thermal spreading resistance of eccentric heat sources on rectangular flux channels," *Journal of Electronic Packaging*, vol. 125, 2003.

Design Procedure for Transformer-based Solid-State Pulse Modulators with Damping Network

Spyridon Stathis and Juergen Biela

Laboratory for High Power Electronic Systems, ETH Zürich

stathis@hpe.ee.ethz.ch

Keywords

"Pulse Transformer", "Solid-State Modulator", "Damping Network", "Optimization", "Varistor", "Mineral oil"

Abstract

This paper presents a systematic procedure for designing transformer-based solid-state pulse modulators, which include a damping network at the load side in order to minimize the rise time and the overshoot of the pulse. The design procedure is applied to the specifications of the CARM modulator system to evaluate its performance.

1 Introduction

In many particle accelerator facilities ultra-precise voltage/current pulses with fast rise and settling times with a high flat-top stability and a low overshoot are required. To generate such pulses, often solid-state pulse modulators are used and pulse transformer-based modulator concepts have proven to be able to meet strict pulse requirements [1–3]. For the design of such systems, optimization procedures have been developed [3–5] and as the specifications of the pulses become more and more demanding, a detailed design procedure is inevitable for a reliable prediction of the pulse performance, before the modulator is built.

An example of such a modulator with challenging requirements is the CARM modulator, a mm-wavelength source, which is currently under development at ENEA research center with the specifications listed in Table I [6]. The definitions of the main transient characteristics of the pulse are also given in Fig. 1a). In [7], a first feasibility study is presented for this modulator based on an optimization procedure. This analysis showed that a damping network (DN) arrangement at the load side is mandatory for a good pulse performance as the transformer parasitics result in unwanted resonances. The optimization routine allowed a comparison between different DN types and one DN type proved to be a promising configuration for the specifications of the CARM modulator (Fig. 1b). The optimization routine combined the design of the pulse transformer and the damping network, and it resulted in a preliminary set of transformer parasitics and damping network parameters for an optimal pulse performance.

However, in [7] an ideal behaviour for the operation of the metal-oxide varistors (MOVs) in the DN has been assumed and only a very small parameter range has been considered in the optimization. In order to improve the accuracy of the optimization procedure, a detailed model of the varistor branches is presented in this paper, so that the full potential of the DN and its effect on the transient characteristics of the pulse can be examined. Many

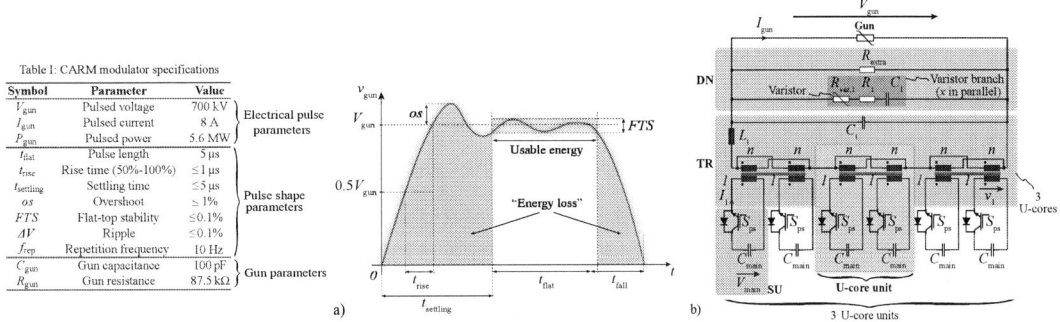

Table I: CARM modulator specifications

Symbol	Parameter	Value	
V_{gun}	Pulsed voltage	700 kV	Electrical pulse parameters
I_{gun}	Pulsed current	8 A	
P_{gun}	Pulsed power	5.6 MW	
t_{flat}	Pulse length	5 μs	
t_{rise}	Rise time (50%-100%)	≤1 μs	
$t_{settling}$	Settling time	≤5 μs	Pulse shape parameters
os	Overshoot	≤ 1%	
FTS	Flat-top stability	≤0.1%	
ΔV	Ripple	≤0.1%	
f_{rep}	Repetition frequency	10 Hz	
C_{gun}	Gun capacitance	100 pF	Gun parameters
R_{gun}	Gun resistance	87.5 kΩ	

Fig. 1: Table I lists the main specifications of the CARM modulator. a) Indicative voltage pulse shape and definitions of the transient quantities. The energy during $t_{settling}$ and t_{fall} can not be used by the application ("energy loss"). Once the pulse lies within FTS the flat-top interval t_{flat} starts. b) Electrical equivalent circuit of the CARM modulator consisting of 3 U-core units, the pulse transformer (TR) with its total parasitic inductance L_t and capacitance C_t, the damping network (DN) and the gun.

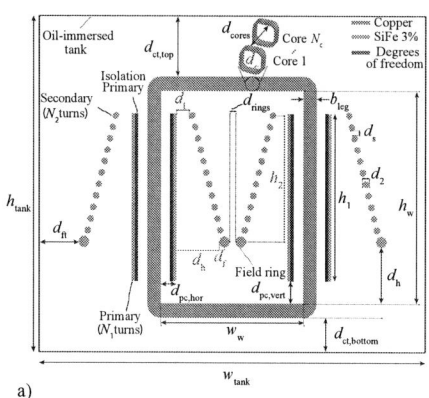

Table II: Considered degrees of freedom & parameters for the transformer optimization

Symbol	Parameter	Value	
d_h	HV side distance	$90\,\text{mm} \leq d_h \leq 190\,\text{mm}$	
d_l	LV side distance	$5\,\text{mm} \leq d_l \leq 20\,\text{mm}$	5 degrees of
d_c	Core depth	$140\,\text{mm} \leq d_c \leq 180\,\text{mm}$	freedom
$h_{2,\text{add}}$	Additional secondary winding height	$10\,\text{mm} \leq h_{2,\text{add}} \leq 100\,\text{mm}$	
d_f	Field ring radius	$8\,\text{mm} \leq d_f \leq 15\,\text{mm}$	
B_{ap}	Applied flux density	$1.5\,\text{T}$	
$d_{\text{pc,hor}}$	Horizontal distance core & primary winding	$4.5\,\text{mm}$	
$d_{\text{pc,vert}}$	Vertical distance core & primary winding	$15\,\text{mm}$	
$d_{\text{ct,bottom}}$	Bottom distance core & tank	$15\,\text{mm}$	
$d_{\text{ct,top}}$	Top distance core & tank	$50\,\text{mm}$	
d_{rings}	Distance between rings	$45\,\text{mm}$	
d_{ft}	Distance field ring & tank	$d_h + 40\,\text{mm}$	Considered
d_2	Thickness of secondary winding turn	$2\,\text{mm}$	parameters
d_s	Secondary winding turns distance	$1\,\text{mm}$	
d_{cores}	Distance between cores	$30\,\text{mm}$	
$\varepsilon_{r,\text{oil}}$	MIDEL oil relative permittivity	3.2	
t_r/t_f	IGBT rise/fall time	$200\,\text{ns}$	
C_{divider}	HV divider capacitance	$50\,\text{pF}$	
L_{gen}	SU parasitic inductance	$100\,\text{nH}$	

Fig. 2: a) 2D representation of the oil immersed split-U-core transformer and tank. A tilted secondary winding is realised for minimum leakage inductance [11]. Table II summarizes the degrees of freedom and additional considered parameters for the CARM transformer optimization. The degrees of freedom are shown with red in Fig. 2a).

MOVs models have been proposed in literature. For instance, in [8], the MOV is modelled as a non-linear resistor, which results only in a static model. More realistic models include the parasitic inductances and capacitances [9] as well as the hysteresis of the MOV [10], what results in dynamic models. In this work, a dynamic model for the MOV, including its parasitics, is integrated in the optimization procedure. For validating the accuracy of the transformer models, 3D FEM simulations of the pulse transformer are also performed. The FEM results are used to validate the scaling lengths, which are used for scaling the per unit length parasitic inductance and capacitance in the optimization. In addition, the analytical calculation of the maximum E-field of the transformer geometry is validated with FEM simulations.

The structure of the paper is as follows: In section 2, the optimization routine is discussed with focus on the DN modeling and the suggested improvements on the transformer models. In section 3, results for the CARM transformer design and the enhanced DN modelling are presented. Finally, section 4 summarizes the main outcomes of this work.

2   System Optimization Procedure

The optimization procedure shown in Fig. 3, is based on the procedure presented in [7]. The red frames illustrate the additional improvements and extensions proposed in this paper. The main extension of the routine is the enhanced DN model in the damping network design section. In step S3, the maximum E-field outside of the core window is also taken into account besides the maximum E-field inside the core window. An additional extension is the transformer 3D FEM simulations, which improve the modeling of the parasitics (S5). A representation of the considered 2D transformer geometry is shown in Fig. 2a). Table II provides the degrees of freedom and additional parameters, which were considered during the optimization of the CARM modulator.

2.1   Transformer Design

The procedure starts (S1) by setting the ranges of the 5 degrees of freedom, the pulse specifications as well as some predefined geometrical quantities (Table II). Next, the core dimensions are calculated, which result in the tank and the transformer volume, and in a minimum height for the secondary winding $h_{2,\text{min}}$ (S2). Finally, the secondary winding height h_2 is controlled by the relation $h_2 = h_{2,\text{min}} + h_{2,\text{add}}$. For determining the cross-sectional area of the core, the volts-seconds of the pulse and the allowed saturation flux density are used.

The procedure continues with the calculation of the maximum E-field inside the core window $E_{\text{max,IW}}$ and the maximum E-field outside of the window $E_{\text{max,OW}}$ (S3). Both E-field values are considered because the maximum E-field outside of the core window might be even higher than the E-field inside the core window as FEM simulations revealed. For the analytical calculation of the E-field, the charge simulation method (CSM) is applied [12]. According to [13], a limit of $E_{\text{th}} = 20\,\text{kV}\,\text{mm}^{-1}$ is suitable for short pulse lengths and for mineral oil as an insulating material. However, in particle accelerator facilities, there is a rising demand to replace the mineral oil with more eco-friendly oils such as natural esters [3,14]. Therefore, the synthetic ester oil Midel 7131 has been selected for the design of the ENEA transformer. In [15], it has been shown that the insulating strength of Midel 7131 is lower than mineral oil under standard impulse conditions (1.2/50 µs). The aging of the oil also affects its insulating strength. Consequently, a more conservative threshold value of $18\,\text{kV}\,\text{mm}^{-1}$ is used in the optimization presented in this paper. For evaluating the impact of the oil on the transformer design, two more cases are considered with a mineral oil and a $E_{th} = 20\,\text{kV}\,\text{mm}^{-1}$ and $E_{th} = 25\,\text{kV}\,\text{mm}^{-1}$.

The extraction of the total secondary-referred winding resistance R_{wdg} follows in step S4. The winding losses due to the skin and the proximity effect have been studied extensively in [16] for foil and round conductors. Here, it is assumed that both windings cover the entire core window height. In order to simplify the model, a trapezoidal current shape is assumed with a rise and fall time time of $1\,\mu s$, a flat-top of $5\,\mu s$ and a repetition frequency of $f_{rep} = 10\,\text{Hz}$ [7]. Using Fourier series, this current is analyzed with 100 harmonics in order to calculate the skin depth. The secondary round turns are transformed to a sheet conductor for calculating the proximity effect losses as described in [16].

Afterwards, the transformer leakage inductance L_σ and the distributed capacitance C_d, referred to the secondary side, are calculated in step S5. For analyzing the transformer parasitics, the geometry is simplified by approximating the primary winding with N_2 circular conductors equally distributed along the height of the primary winding. Also, a homogeneous current and voltage distribution is assumed in the parasitics calculation. The influence of the grounded tank and the grounded core is taken into account. For calculating the distributed capacitance per unit length inside and outside the core window the charge simulation method (CSM) is used [12]. For determining the parasitic inductance per unit length inside and outside the core window, the mirror current method is applied according to [17]. In [18], a more accurate model is proposed where an additional elementary cross section between the cores was introduced and this is used to improve the leakage inductance model. The per unit length quantities are then scaled with the appropriate scaling lengths for the inside and the outside core window areas as suggested in [18]. Finally, the total parasitic inductance L_t and capacitance C_t are determined by

$$L_t = L_{gen,s} + L_\sigma \qquad C_t = C_d + C_{extra}$$

There, $L_{gen,s}$ is the parasitic inductance of the switching unit S_{ps} referred to the secondary side (Fig. 1c) and C_{extra} is the sum of the gun capacitance C_{gun} and the capacitance of the HV divider $C_{divider}$, which is used for measuring the gun voltage. In step S6, the total parasitic inductance and capacitance, the resistance of the windings, the maximum E-field and the respective values of the degrees of freedom are stored. With the stored values, the pareto front of the transformer parasitics is obtained and a design point with minimal transformer parasitics on the knee of the pareto is selected, which results in the fastest rise time. This method is used to drastically reduce the computational effort of the complete optimization procedure as it allows for a more extensive parametric sweep across the DN parameters of the DN design section [7]. For validating the transformer routine, the models have been also applied to the specifications of the SwissFEL modulator at Paul Scherrer Institute (PSI) [2] and to the CLIC modulator for European Organization for Nuclear Research (CERN) [19] and the measured values have been compared to the calculated ones.

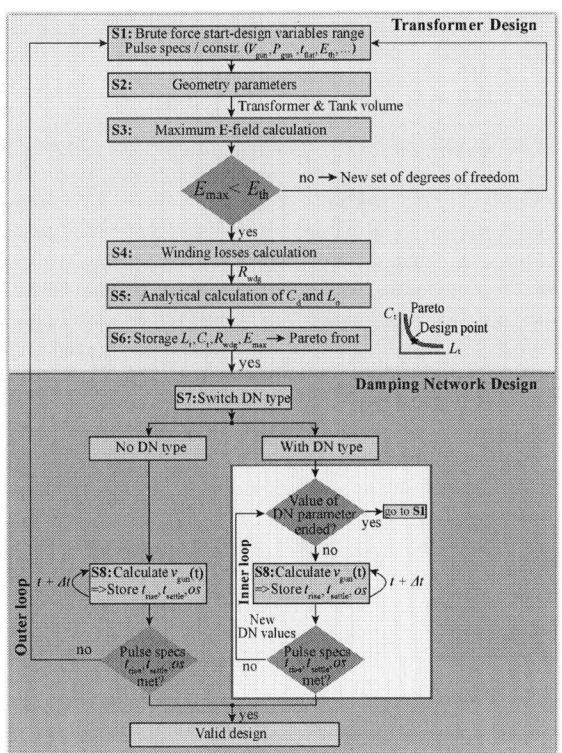

Fig. 3: Flowchart of the optimization procedure including the transformer design and the damping network design. The red frames indicate the improvements of the routine.

2.2 Transformer Design Check

For the validation of the models for the transformer parasitics as well as the maximum E-field inside and outside the core window several design points have been examined. The validation is performed with automated 3D FEM simulations, which allow to compare L_σ, C_d, and E_{max} with the results from the analytical calculations resulting from steps S3 & S5. For the analytical calculation of the transformer parasitics in the optimization, 2D models are used resulting in per unit length quantities and they are multiplied by the associated scaling lengths to get the final 3D values. Therefore, the 3D FEM simulations serve as a reference point in order to verify whether the per unit length leakage inductance and distributed capacitance have been scaled with the appropriate scaling lengths inside

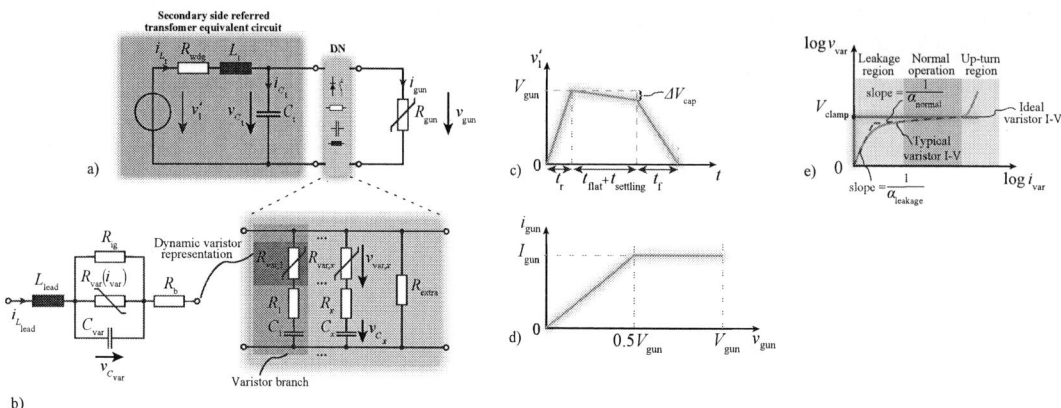

Fig. 4: a) Simplified representation of the transformer and the switching unit including a damping network referred to the secondary side of the transformer. b) Damping network for the CARM modulator. Varistors are placed in series with RC components forming a varistor branch. On the left side, the dynamic varistor representation including its parasitics is provided. c) Voltage signal generated by the switching unit referred to the secondary side, which emulates the switching and the steady behaviour of the switching unit. The IGBT has rise time of t_r and a fall time t_f as given in Table II. d) Current-voltage characteristic of the gun. e) Ideal and typical varistor current-voltage characteristic. The typical I-V characteristic can be approximated by straight lines with different slopes for each operating region.

the optimization and how large the deviation between analytical and 3D FEM values is. With this, the validity of the pareto curve is ensured.

2.3 Damping Network Modeling

With the validated transformer design, the DN is optimized in the next stage (S7 & S8) with a model for the transient behaviour. The dynamic behaviour of the transformer and the switching unit is modeled by the network given in Fig. 4a), which is a simplified representation referred to the secondary side of the transformer. There, v_1' is the primary winding voltage v_1 referred to the secondary side (Fig. 4c). The voltage v_1 it is a linear approximation of the switching and steady state of the IGBT with a rise time t_r and a fall time t_f. An additional slope ΔV_{cap} during the time interval $t_{flat} + t_{settling}$ is considered, which emulates the voltage droop of the main capacitors C_{main} during the pulse [7]. This voltage droop results in a more limited flat-top stability FTS_{lim}, which is given as $FTS_{lim} = FTS - \Delta V_{cap}\%$, with FTS being defined in the specifications of Table I. The value of C_{main} should be chosen, so that $\Delta V_{cap}\% \ll FTS$ to limit the influence of the droop on the pulse shape. A possible droop compensation is not considered in this paper in order to simplify the analysis. Moreover, the magnetizing inductance as well as the iron losses are not taken into account as they have insignificant impact due to the short pulse length [11]. The gun has a non-linear current-voltage characteristic, which is depicted in Fig. 4d). The network for the pulse modulator is solved analytically for different types of DN and also for the case without DN (S8), in order to compute the gun voltage v_{gun} during the pulse (S9). A first comparison between different DN types has been made in [7] and, for the CARM modulator, the DN arrangement shown in Fig. 4b) was proved to be promising. In the considered DN type, varistors are placed in series to a RC circuit forming a varistor branch. Multiple varistor branches can be connected in parallel. In addition, an extra resistor R_{extra} should be added to increase the damping as the gun by itself presents a quite high impedance [7].

For achieving a reliable design optimization, a detailed model of the DN is required. In the following, it is considered that the varistors of the parallel branches could have different clamping voltages so that the varistor branches kick in at different voltage levels. There, a dynamic model of the varistor is crucial in order to examine its effect on the transient voltage pulse. For deriving the dynamic varistor model, the typical I-V characteristic of a varistor is used (Fig. 4e), where three operating regions can be identified: the leakage, the normal operation, and the up-turn region [20]. The equivalent circuit of a varistor, which includes the parasitics, is depicted in Fig. 4b). In the varistor model, L_{lead} models the parasitic inductance of the leads, C_{var} is the capacitance of the varistor, $R_{var}(i_{var})$ is the non-linear resistance, which depends on the current that flows through the varistor, R_{ig} models the resistance of the intergranular boundary, and R_b is the bulk resistance of the zinc oxide [20].

For the leakage region, R_{ig} dominates ($R_{ig} \approx 10^9\,\Omega$) over $R_{var}(i_{var})$. Consequently, L_{lead} and the parallel combination of C_{var} and R_{off} form essentially the equivalent circuit. In the normal operation region, the varistor conducts and $R_{var}(i_{var})$ dominates now over R_{ig} so that L_{lead} and R_{var} form the equivalent circuit of the varistor. Although the bulk resistance of the zinc oxide is very low ($R_b \approx 1\,\Omega$), it is included in the model as many varistors are expected to be connected in series, which will, eventually, increase the resistance. The current that flows through the non-linear resistor $R_{\mathrm{var}}(i_{\mathrm{var}})$ in the normal operating region follows a power law equation $i_{\mathrm{var,normal}} = k_{\mathrm{normal}} v_{\mathrm{var}}^{\alpha_{\mathrm{normal}}}$, where k_{normal} is a constant and α_{normal} is also a constant, which defines the degree of non-linearity, and which is basically the slope of the I-V characteristic. It can be determined by choosing two operating points on the I-V characteristic of the selected varistor component. By performing measurements at varistor components, the authors of [21] showed, that the current in the leakage region can also be approximated by a similar power law relation as for the normal operating region, namely bx $i_{\mathrm{var,leakage}} = k_{\mathrm{leakage}} v_{\mathrm{var}}^{\alpha_{\mathrm{leakage}}}$. By adding these two currents a combined model results, which models the behaviour of the varistor in the leakage and the normal region. Hence,

$$i_{\mathrm{var}} = i_{\mathrm{var,leakage}} + i_{\mathrm{var,normal}} = k_{\mathrm{leakage}} v_{\mathrm{var}}^{\alpha_{\mathrm{leakage}}} + k_{\mathrm{normal}} v_{\mathrm{var}}^{\alpha_{\mathrm{normal}}} \tag{1}$$

where v_{var} is the varistor voltage, which is equal to the voltage across the varistor capacitance C_{var}. The four constant parameters of the power law functions are determined through curve fitting. At high currents, in the kA range, the varistor is in the up-turn region. However, in the considered CARM modulator, the varistors are not intended to operate at such high currents so that this region is not considered in the model. It should also be noted that this model ignores the hysteresis loop of varistors [22]. When the surge across the varistor vanishes, the current rapidly drops, but it follows a different path on the I-V curve creating a hysteresis loop. This phenomenon is not modeled as the focus is on the rising edge and there the hysteresis does not play any role.

The capacitance C_{var} is usually provided in the data-sheets at a specific frequency, e.g at 1 MHz. Also, R_{ig} is known to be in the range of $R_{\mathrm{ig}} \approx 10^9 \Omega$ and remains in the MΩ range even at elevated temperatures whereas R_{b} is usually in the range of approximately 1 Ω. However, the parasitic inductance L_{lead} is typically not provided in the data-sheets. As typically the focus is on the normal operating region, many data-sheets do not provide the I-V characteristic during the leakage mode of operation. This data limitation becomes more significant for high voltage varistors where the leakage region is often completely neglected. To lift these limitations, and in order to show the modeling procedure, a specific radial-lead varistor is chosen for the damping network design, which offers data for a broader current range from μA to kA. The lead inductance of a radial-lead varistor, which is in fact the self-inductance of a loop of wire, can be calculated by [21]

$$L_{\mathrm{lead}} = \frac{\mu_0 h}{\pi} \ln\left(\frac{d - t_{\mathrm{w}}}{t_{\mathrm{w}}}\right) \tag{2}$$

where μ_0 is the relative permeability of free space, h is the height of the radial leads, d is the distance between the wires and t_{w} is the thickness of the wire. The geometrical quantities of the considered varistor are extracted through the data-sheet. On a system level, multiple varistors are connected in series in each varistor branch as the estimated clamping voltages of each branch are in the range of half of the gun voltage (350kV) and above. The number of the varistors $N_{\mathrm{var},x}$, which should be connected in series at the branch x is determined by the relation

$$N_{\mathrm{var},x} = \mathrm{ceil}\left(\frac{V_{\mathrm{var},x}}{V_{\mathrm{clamp}}}\right) \qquad x \in \mathbb{Z}^+ \tag{3}$$

where V_{clamp} is the maximum nominal clamping voltage of the selected varistor, $V_{\mathrm{var},x}$ is the clamping voltage of the x varistor branch, i.e the voltage where all the varistors in series enter the normal operating region. Consequently, the resistances and inductances for the equivalent varistor network have to be multiplied by $N_{\mathrm{var},x}$ and the capacitance has to be divided by the same number. However, radial-lead varistors do not show high clamping voltages and therefore, hundreds of varistor components must be connected in series for the damping network, what results in a reduced reliability of the system. For the real CARM modulator system, varistors with higher voltage ratings will be used and the parasitics could be measured, for example, with an impedance analyzer.

The clamping voltage of the varistor depends on the rise time and the decay time of the applied current waveform [20]. The value of V_{clamp} is usually given in the datasheets for an applied current pulse with a rise time of 8 μs and a decay time of 20 μs (8/20 μs), which has quite a different slope compared to the desired 1 μs rise time waveform for the considered ENEA modulator. According to [20], the clamping voltage tends to increase with a shorter rise and decay time of the applied current pulse. Although this increase of V_{clamp} becomes more severe in the kA range, it also influences the operation at lower currents in the normal operating region. The dependency of the clamping voltage on the wavefront of the applied current is typically not provided in the data-sheets. However, in [20], I-V curves for applied current impulses of 1/3 μs, 0.5/1.5 μs and 8/20 μs for a radial-lead varistor in the normal operating region are given. The leakage region remains approximately unchanged by the different current slopes. If one compares the maximum clamping voltage at 100 A (at this current the clamping voltage is typically provided in the data-sheet) for the 1/3 μs pulse, which has approximately the same slope as the desired voltage pulse of the modulator, with the reference 8/20 μs curve, it can be found that there is an increase of around 6% of the clamping voltage. Then, the parameter α_{normal} of equation (1) should show a decrease of around 6% as the slope of the I-V characteristic is in fact the inverse value of α_{normal} (Fig. 4e). This value has been considered in the modeling for the additional increase of the V_{clamp}. The response time of a varistor is in the range of a few ns [20], which is not comparable to the required rise time of the voltage pulse, and consequently, it does not affect the rise time of the gun voltage.

One should set the differential equations of the system, which is used to solve with respect to the gun voltage v_{gun}, which is equal to v_{C_t}. In order to explain the modeling procedure, the case of a single varistor branch with a clamping voltage $V_{var,1}$ is used as a reference. Applying the voltage/current equations to the circuit of Fig. 4a) and taking into account the equivalent circuit of the varistor in Fig. 4b), one can derive the set of differential equations for the respective state vector $\bar{x}_1$

$$\bar{x}_1 = \left[i_{L_t},\ v_{C_t},\ i_{L_{lead}},\ v_{C_{var}},\ v_{C_1}\right]^T \tag{4}$$

$$\frac{di_{L_t}}{dt} = \frac{1}{L_t}\left[v'_1 - i_{L_t}R_{wdg} - v_{C_t}\right] \tag{5}$$

$$\frac{dv_{C_t}}{dt} = \frac{1}{C_t}\left[i_{L_t} - i_{L_{lead}} - \frac{v_{C_t}}{R_{extra}} - i_{gun}\right] \tag{6}$$

$$\frac{di_{L_{lead}}}{dt} = \frac{1}{N_{var,1}L_{lead}}\left[v_{C_t} - v_{C_{var}} - i_{L_{lead}}(R_1 + N_{var,1}\cdot R_b) - v_{C_1}\right] \tag{7}$$

$$\frac{dv_{C_{var}}}{dt} = \frac{1}{N_{var,1}^{-1}C_{var}}\left[i_{L_{lead}} - \frac{v_{C_{var}}}{N_{var,1}R_{ig}} - \left(k_{leakage}v_{var}^{\alpha_{leakage}} + k_{normal}v_{var}^{\alpha_{normal}}\right)\right] \tag{8}$$

$$\frac{dv_{C_1}}{dt} = \frac{1}{C_1}i_{L_{lead}} \tag{9}$$

with $v_{var} = v_{C_{var}}/N_{var,1}$ due to the serial connection of the varistors.

After calculating v_{gun}, the transient characteristics of the pulse, such as rise time, overshoot and settling time can be computed and stored (S8). The rise time t_{rise} and the overshoot os are defined as

$$t_{rise} = t_{100} - t_{50} \qquad os = \left(\frac{V_{gun,max}}{V_{gun}} - 1\right)\cdot 100\% \tag{10}$$

To calculate the rise time (50%-100%) of the pulse, the points in time t_{50} and t_{100} are used, which indicate the time instances where the voltage pulse reaches 50% and 100%, respectively, of its rated value. In order to compute the overshoot, the maximum gun voltage $V_{gun,max}$ is determined and the os is compared to the acceptable overshoot given in the specifications. The settling time ends when the voltage pulse lies completely within the FTS band. By stepping through v_{gun} in the time domain, the settling point in time is obtained. To limit the calculation time of the optimization, the non-linear set of differential equations (5)-(9) is solved only for the time interval of $t_{control} = t_{settle} + t_{add}$, where $t_{settle} = 5\,\mu s$ is set by the application and $t_{add} = 2\,\mu s$ is chosen, which is a margin to ensure that the pulse will remain safely within the FTS limits.

The dependencies between the transient characteristics are investigated via a parametric sweep across the quantities R_1, C_1, R_{extra}, $V_{var,1}$, and at the end, the designer can choose the best set of DN parameters for an optimized pulse performance. The same procedure is basically applied in the case where more varistor branches are connected in parallel to the load. There, the set of differential equations is expanded accordingly depending on the number of the parallel varistor branches.

3   CARM Modulator Feasibility Study

The procedure presented in section 2 is applied to the specifications of the CARM modulator, which are listed in Table I, to verify the effectiveness of the optimization routine.

3.1   Transformer Design

The optimization procedure is set to run with the considered degrees of freedom and parameters as shown in Table II. In [7], different number of cores were examined, and a 3-core transformer was found to give a good compromise between the transformer parasitics, the transient performance of the output pulse and the complexity of the system. This, still holds true for the improved optimization routine presented in this paper. Therefore, 3-core transformers are considered in the following. The resulting optimal points and the pareto front of the transformer parasitics are shown in Fig. 5a) in dependency of the maximum E-field. From the pareto front a design point is selected at the knee of the curve, for minimum parasitics, which result in a minimum rise time. The chosen design point results in

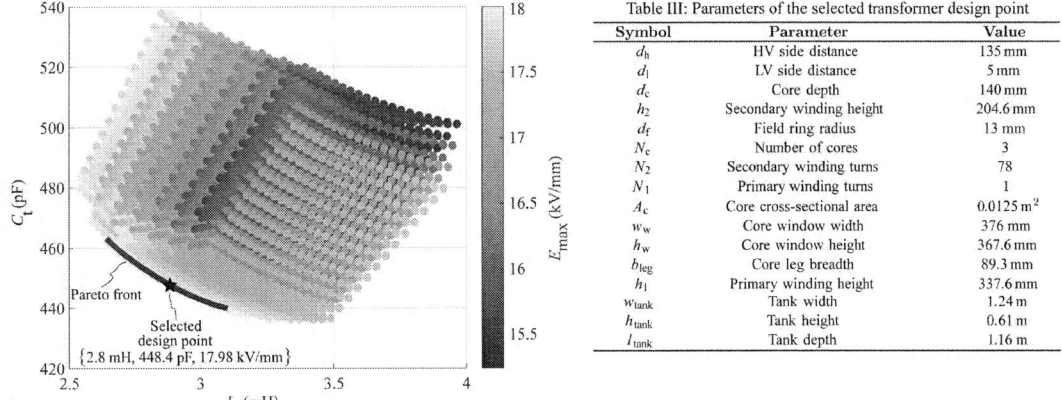

Table III: Parameters of the selected transformer design point

Symbol	Parameter	Value
d_h	HV side distance	135 mm
d_l	LV side distance	5 mm
d_c	Core depth	140 mm
h_2	Secondary winding height	204.6 mm
d_f	Field ring radius	13 mm
N_c	Number of cores	3
N_2	Secondary winding turns	78
N_1	Primary winding turns	1
A_c	Core cross-sectional area	$0.0125\,\mathrm{m}^2$
w_w	Core window width	376 mm
h_w	Core window height	367.6 mm
b_{leg}	Core leg breadth	89.3 mm
h_1	Primary winding height	337.6 mm
w_{tank}	Tank width	1.24 m
h_{tank}	Tank height	0.61 m
l_{tank}	Tank depth	1.16 m

Fig. 5: a) Total parasitic capacitance C_t versus total parasitic inductance L_t in relation to the maximum E-field for a transformer with $N_c = 3$ cores. The red curve indicates the pareto front, where a design point ("★") is selected. The lower the L_t the higher the E_{max} becomes. Table III lists some of the main parameters of the selected design point.

a tank and a transformer geometry with the parameters listed in Table III. It is worth mentioning that the maximum E-field occurs on the surface of the field ring outside the core window, a fact which proves the importance of including the calculation of the maximum E-field outside the core window area also in the optimization. Fig. 5a) also clearly demonstrates that the lower the parasitic inductance, the higher E_{max} becomes given that the distances between the windings tend to be shorter.

3.2 Transformer Design Check

For validating the transformer design, the parasitics and the maximum E-field for the chosen design point are checked via electrostatic and magnetostatic 3D FEM simulations, and the results are compared to the analytical ones of the optimization procedure. All the calculated values are listed in Table IV. For the simulations half of the transformer and the tank geometry is simulated. This helps to investigate the electric and magnetic energy contribution of the region between the cores as well as potential interactions between both secondary windings inside the core window. The FEM simulations are set in such a way so that the calculation of the parasitics inside the core window is feasible. With the 3D simulations as a reference, it is clear that the deviation of the parasitic capacitance C_d is kept below 5% between the 3D simulations and analytical calculations. However, the calculation of L_σ shows a higher value. For the analytical leakage inductance calculation, the model proposed in [18] was used, which increases the accuracy compared to the model used in [3]. The main contributor to the error is the leakage inductance outside of the core window area where an error of 11% error results for the simulated design. For the inside window area the error of the leakage inductance is around 7%. The electrostatic simulations confirm that the maximum E-field lies outside the core window and not inside the window as it is the case for the designs presented in [2,3]. The maximum E-field is around $18.53\,\mathrm{kV\,mm^{-1}}$. The maximum E-field inside the core window tends to be lower as the electric field of the two field shape rings superpose each other resulting in a lower total field. This indicates that the optimization procedure has to take into account the maximum E-field outside the window as well and not only the inside window as proposed in [3]. In addition, in Fig. 6, the norm of the E-field and H-field are visualized with three 2D cuts.

3.3 Damping Network Design

in the next step, the damping network is designed following the modeling procedure described in section 2.3. For the DN model, the radial-lead varistor V25S750P of UltraMOV25S series from Littelfuse is selected as example, since data on the different operating regions is provided for this device. The main parameters of the varistor are given in Table V. According to (2), the exact number of the varistor components, which must be connected in series, depends on the chosen varistor clamping voltage $V_{var,x}$. The considered ranges of all the parameters of the

Table IV: Comparison of the parasitics & peak E-field values
between 3D FEM and analytical calculation for the selected transformer design

Parameter	Analytical	3D-FEM	Error: Analytical vs 3D
C_d(pF)	298.4	285.6	-4.49%
L_σ(mH)	1.9	1.61	-18.1%
E_{max} (kV/mm)	17.98	18.53	2.97%

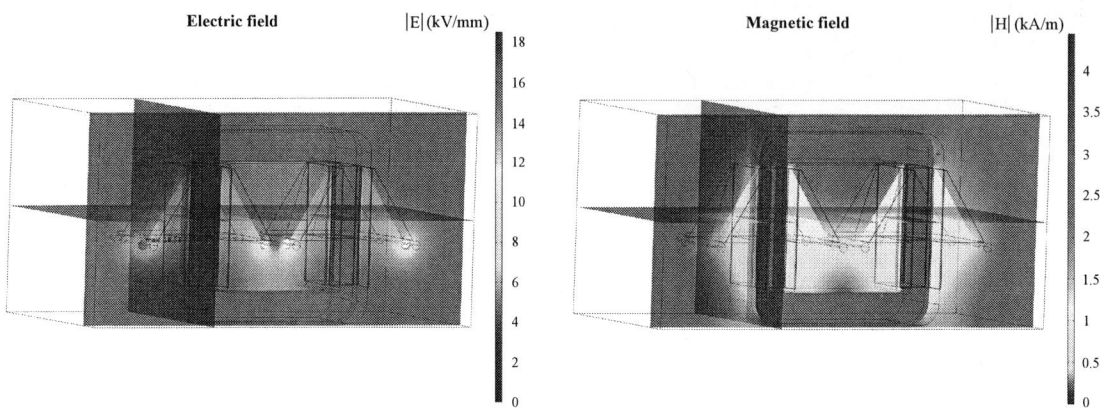

Fig. 6: 3D FEM results for the selected transformer design point with three 2D cuts: a) Norm of the E-field in $kV\,mm^{-1}$. The maximum E-field appears on the surface of the field ring outside of the core window. b) Norm of the magnetic field in $kA\,m^{-1}$.

DN are provided in Table VI. With a parametric sweep, the best set of these parameters is identified for an optimal pulse performance.

Based on the parameter ranges and the parasitics for the selected design point, the dependency between the rise time and the overshoot can be determined for different number of varistor branches N_b in parallel and clamping voltages $V_{var,x}$ as depicted in Fig. 7. The results indicate that a single varistor branch ($N_b = 1$) does not provide sufficient damping to the output voltage pulse at the 1 µs rise time limit. The damping is significantly improved by adding one additional branch in parallel ($N_b = 2$) as it can be seen in Fig. 7b). There, the overshoot levels are lower at rise times less than 1 µs. Another important conclusion resulting from these curves is that varistor clamping voltages above 600 kV provide an ultra-fast rise time and at the same time very low overshoot levels. This is explained from the fact that the varistor branches operate in the leakage region for voltages lower than the clamping voltages $V_{var,x}$ and, as a result, the network in Fig. 4a) mainly consists of the parasitic inductance L_t and capacitance C_t as well as the resistor R_{extra} at the output and the gun. The damping network improves the overshoot, but it also inserts a delay (i.e it increases the rise time) as it is essentially a RC circuit in parallel to the transformer secondary winding and therefore, the optimal clamping voltages $V_{var,x}$, when each varistor branch would be "activated", need to be identified. For this, different curves are plotted in Fig. 4b) with a focus on clamping voltages higher than 600 kV. There, it can be seen that there is an optimal set of varistor clamping voltages (red dashed line) for which low rise times and low overshoots can be achieved at the same time. However, even at the 1 µs rise time limit, the overshoot is around 3%, which remains outside of the value required by the application (1%). Consequently, three and four varistor branches are also investigated in Fig. 7c) and d). By adding a third varistor and a fourth branch significant lower overshoot levels are observed and the 1% overshoot is feasible for both of these cases. The red dashed curves denote again the clamping voltages where minimum rise times and overshoots can be achieved.

By the inspection of how the DN parameters, i.e the quantities R_x, C_x and R_{extra} relate to each other around the rise time limit for the cases $N_b = 1$ and $N_b = 2$, one can further reduce the ranges of the parameters of interest. This helps to reduce the intervals of the parametric sweep, especially when a third and a fourth varistor branch are inserted, which increases the computational time significantly.

Although the obtained curves given in Fig. 7 do provide a relation between two important transient characteristics of the pulse, i.e the rise time and the overshoot, they do not give an insight regarding the settling time of the pulse, a characteristic which is equally or even more significant than the rise time itself. This is because a high settling time means a long exposure of the gun to a high voltage, which could potentially lead to a breakdown of

Table V: Varistor parameters

Parameter	Value
C_{var}	250 pF
R_{off}	1000 MΩ
L_{lead}	32 nH
V_{clamp} (8/20 µs)	1.65 kV
h	25.4 mm
d	10 mm
t_w	0.81 mm

Table VI: Considered boundaries for the DN parameters

Parameter	Value
R_x	$0.02\,k\Omega \leq R_x \leq 5\,k\Omega$
C_x	$50\,nF \leq C_x \leq 1\,\mu F$
$V_{var,x}$	$350\,kV \leq V_{var,x} < 700\,kV$
R_{extra}	$0.087\,R_{gun} \leq R_{extra} \leq 0.2\,R_{gun}$

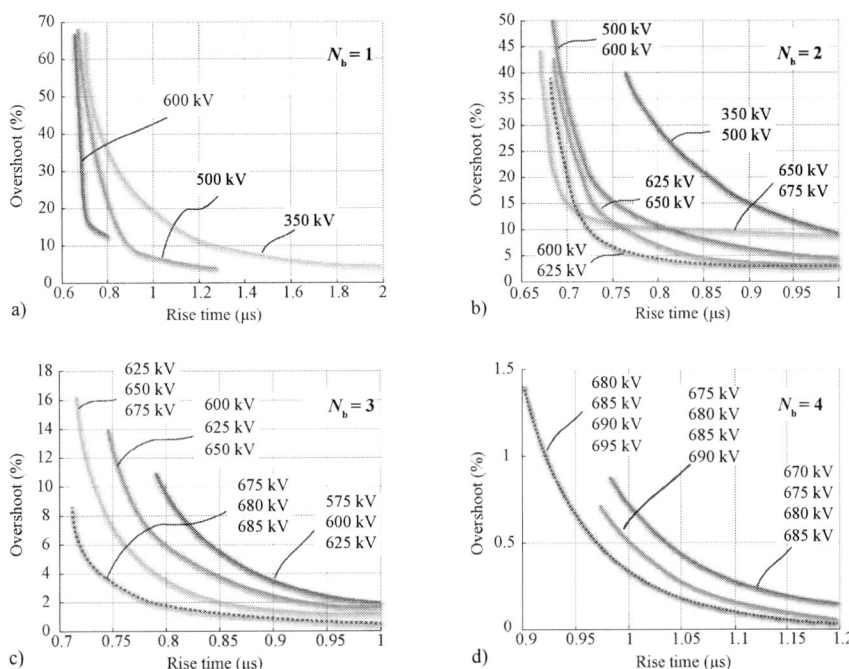

Fig. 7: Dependency between rise time & overshoot for different number N_b of varistor branches in parallel: a) $N_b = 1$, b) $N_b = 2$, c) $N_b = 3$, d) $N_b = 4$. The red dashed lines indicate the clamping voltages, which offer minimum rise times and overshoots at the same time for each number of varistor branches.

the gun. Furthermore, even though the rise time and the overshoot might fulfill the requirements, the pulse might not lie within the FTS band in an acceptable settling time. Therefore, it is interesting to discuss the dependencies between the settling time and the other two transient characteristics, i.e the rise time and the overshoot, for damping networks with two, three and four branches in parallel. For this, design points are selected on the red dashed curves of Fig. 7b) and c), which are found to be the clamping voltages offering optimal combinations of rise times and overshoots. So far, the red curves are limited until the rise time limit of 1 μs, but now they are extended further in order to find the optimal settling times as well. The optimal design points are summarized in Table VII for a 0.1% FTS. Of course, there are numerous designs that lead to settling times less than 5 μs with different combinations of R_x, C_x and R_{extra} even outside the considered DN parameters boundaries. However, the values of the DN parameters should be small and therefore, designs with minimum settling times and minimum DN parameters values at the same time are listed in Table VII. From this table, it is clearly visible that a settling time of less than 5 μs can already be achieved with 2 varistor branches, but the rise time remains higher than 1 μs. However, three varistor branches in parallel result in an improved pulse performance with a shorter settling time of 2.44 μs, a rise time of 1.07 μs and a negligible overshoot. In case four branches are inserted, the settling time drops to 2.33 μs with a rise time of less than 1 μs, bringing the pulse to completely meet the imposed specifications. The resulting pulse shape (dark blue) for the case of four varistor branches is depicted in Fig. 8a) whereas in Fig. 8b) a zoomed illustration of the pulse after its settling and until the end of the flat-top is shown. In Fig. 8c), a CAD illustration of the transformer geometry for the selected design point is shown.

In the same table, the pulse shape and the DN parameters for different oil types and maximum allowed field values E_{th} are shown, in order to investigate the influence of these two quantities. For this, the optimization procedure is set to assume mineral oil, having a relative permittivity of 2.2 and two different threshold E_{th} values of $20\,\text{kV}\,\text{mm}^{-1}$ and $25\,\text{kV}\,\text{mm}^{-1}$. For these cases, only the optimal $V_{var,x}$, which have been identified above, for two and three branches are considered. Also, a design point at the knee of the respective Pareto fronts is chosen. With a higher E_{th} and using mineral oil, which has a lower permittivity compared to the synthetic ester oil Midel 7131, a Pareto front with lower leakage inductance/parasitic capacitance sets results. Due to the smaller parasitics, shorter rise and settling times occur as it is shown in Table VII for two and three branches. In case a higher E_{th} and mineral oil are allowed, significant improvements can be achieved in terms of the rise and settling time, and with three varistor branches the pulse has transient characteristics, which comply with the specifications. The best achievable pulse shape is in the case of mineral oil with an $E_{th} = 25\,\text{kV}\,\text{mm}^{-1}$, where a 0.93 μs rise time is achieved with an ultra-fast settling time of 2.1 μs. The pulse shape for this case (light blue) is given in Fig. 8a) and Fig. 8b).

Table VII: DN & pulse shape parameters for different number of varistor branches and for different oil types and E_{th} values

Oil type/E_{th} (kV/mm)	$V_{var,1}$ (kV)	$V_{var,2}$ (kV)	$V_{var,3}$ (kV)	$V_{var,4}$ (kV)	R_1 (Ω)	R_2 (Ω)	R_3 (Ω)	R_4 (Ω)	C_1 (nF)	C_2 (nF)	C_3 (nF)	C_4 (nF)	R_{extra} (kΩ)	t_{rise} (µs)	$t_{settling}$ (µs)	os (%)
Midel /18	600	625	—	—	440	430	—	—	900	850	—	—	$0.1R_{gun}$	1.49	2.86	0.04
Mineral /20	600	625	—	—	530	540	—	—	700	700	—	—	$0.1R_{gun}$	1.42	2.6	0.049
Mineral /20	600	625	—	—	510	520	—	—	700	700	—	—	$0.1R_{gun}$	1.38	2.55	0.049
Midel /18	675	680	685	—	8	6.5	5.5	—	900	950	950	—	$0.1R_{gun}$	1.07	2.44	0.05
Mineral /20	675	680	685	—	80	81	81	—	800	800	750	—	$0.1R_{gun}$	0.94	2.13	0.048
Mineral /25	675	680	685	—	60	65	65	—	600	700	600	—	$0.1R_{gun}$	0.93	2.1	0.049
Midel /18	680	685	690	695	50	34	32	35	900	900	900	900	$0.15R_{gun}$	0.98	2.33	0.049

3.4 Computational Effort of the Design Procedure

After the results section, the computational effort that is needed for the complete design procedure is discussed. All the steps of the design procedure are implemented on a server-based computer with a 32-core processor at 2.95 GHz (2 processors). For instance, for the case of Midel oil with $E_{max} = 18\,kV\,mm^{-1}$ the elapsed time of the transformer design part was around 4.5 hours and the optimization operates in a brute-force mode. The 3D FEM simulations (electrostatic and magnetostatic) need around 3 hours to finish. This computational time can be further decreased if someone is not interested to investigate the interaction between the 2 secondary windings inside the core window and, as a result, additional symmetry planes can be used to reduce the transformer and tank geometry. The most computationally demanding step is the damping network design. There, hundreds of thousands of iterations are needed to produce the rise time versus overshoot curves and identify, at the end, the optimal clamping voltages of each varistor branch. The elapsed time for two varistor branches and one set of clamping voltages was approximately 24 hours. For the case of three and four varistor branches, where the computational effort rises dramatically for the chosen parameter range, the designer should observe the relation between R_x, C_x in one and two varistor branches, and shrink the intervals of the DN parameters.

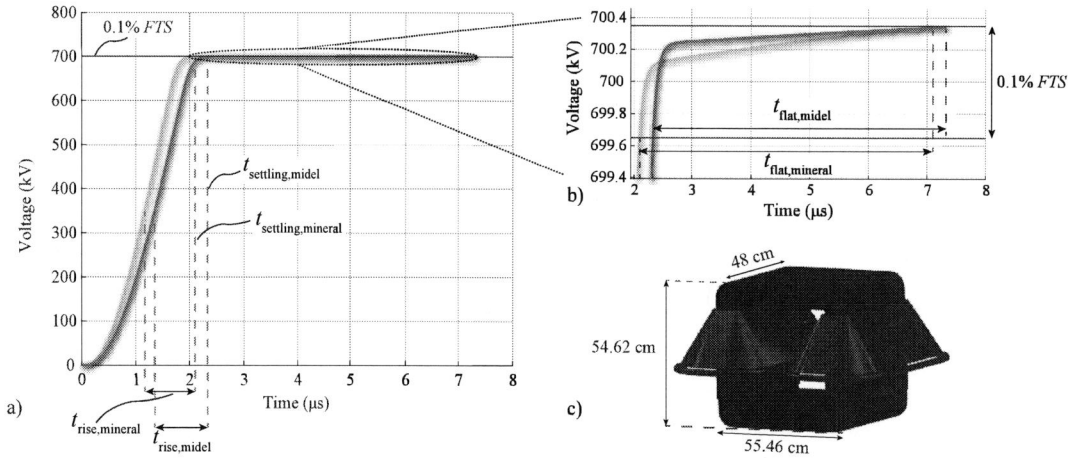

Fig. 8: a) Pulse shapes for Midel 7131 transformer oil, $E_{th} = 18\,kV\,mm^{-1}$ and four varistor branches (dark blue), and mineral oil with $E_{th} = 25\,kV\,mm^{-1}$ and three varistor branches (light blue). For the case of MIDEL 7131 oil, the pulse has a rise time of 0.98 µs and an overshoot of 0.049%. The 0.1% FTS limit can be reached within a settling time of 2.33 µs. For the case of mineral oil, the pulse shows a rise time of 0.93 µs and an overshoot of 0.049%. The 0.1% FTS limit can be met within a settling time of 2.1 µs. b) Zoomed view of the pulses after their settling and until the 5 µs flat-top interval finishes. The pulse lies safely inside the FTS limits until the 5 µs flat-top interval ends. c) CAD illustration of the transformer geometry for the selected design point with MIDEL 7131 transformer oil and an $E_{th} = 18\,kV\,mm^{-1}$.

4 Conclusion

In this paper, an extended systematic design procedure is presented, which combines the optimal design of the transformer and the optimal design of a damping network for transformer-based solid-state modulators. To validate the performance of the optimization routine, it is applied to the specifications of the CARM modulator system. The procedure has improved models regarding the insulation design of the transformer and the analytical calculation of the transformer parasitics as well as a detailed damping network modeling. By applying the optimization procedure, different transformer and damping network designs are proposed for the CARM modulator system, which manage to comply with the strict specifications imposed by the application. The optimization procedure is general and it can be adopted to arbitrary rise time ranges and various pulse lengths leading to a transformer with a compact volume.

References

[1] J. Holma and M. J. Barnes, "The prototype inductive adder with droop compensation for the CLIC kicker systems," *IEEE Trans. on Plasma Science*, vol. 42, no. 10, pp. 2899–2908, 2014.

[2] D. Gerber and J. Biela, "Design of an ultraprecise 127-MW/3-μs solid-state modulator with split-core transformer," *IEEE Trans. on Plasma Science*, vol. 44, no. 5, pp. 829–838, May 2016.

[3] S. Blume, M. Jaritz, and J. Biela, "Design and optimization procedure for high-voltage pulse power transformers," *IEEE Trans. on Plasma Science*, vol. 43, no. 10, pp. 3385–3391, Oct. 2015.

[4] S. Candolfi, P. Viarouge, D. Aguglia, and J. Cros, "Hybrid design optimization of high voltage pulse transformers for klystron modulators," *IEEE Trans. on Dielectrics and Electrical Insulation*, vol. 22, no. 6, pp. 3617–3624, 2015.

[5] M. Jaritz and J. Biela, "System design and measurements of a 115-kV/3.5-ms solid-state long-pulse modulator for the european spallation source," *IEEE Trans. on Plasma Science*, vol. 46, no. 10, pp. 3232–3239, Oct. 2018.

[6] I. Spassovsky, "From research and design work toward the realization of CARM source at ENEA," in *44th Intern. Conf. on Infrared, Millimeter, and Terahertz Waves (IRMMW-THz)*, 2019.

[7] S. Stathis and J. Biela, "Optimal design of a transformer-based solid-state pulse modulator with a damping network for ultra-fast rise times," in *European Conf. on Power Electronics and Applications (EPE ECCE Europe)*, 2021.

[8] G. V. N. Bezerra *et al.*, "Evaluation of surge arrester models for overvoltage studies," in *ICHVE International Conf. on High Voltage Engineering and Application*, 2014.

[9] CIGRE, "Guide for the development of models for HVDC converters in a HVDC grid," 2014.

[10] T. Hagiwara *et al.*, "A metal-oxide surge arrester model with active V-I characteristics," in *Electrical Engineering in Japan*, vol. 121, 1997.

[11] D. Bortis, J. Biela, G. Ortiz, and J. W. Kolar, "Design procedure for compact pulse transformers with rectangular pulse shape and fast rise times," in *IEEE Intern. Power Modulator and High Voltage Conf.*, 2010, pp. 298–302.

[12] H. Singer, H. Steinbigler, and P. Weiss, "A charge simulation method for the calculation of high voltage fields," *IEEE Trans. on Power Apparatus and Systems*, vol. PAS-93, no. 5, pp. 1660–1668, 1974.

[13] D. Bortis, "20 MW halbleiter-leistungsmodulator-system," Ph.D. dissertation, ETH Zürich, 2009.

[14] S. Candolfi, S. Blume, D. Aguglia, P. Viarouge, J. Biela, and J. Cros, "Evaluation of insulation systems for the optimal design of high voltage pulse transformers," in *IEEE Intern. Power Modulator and High Voltage Conf. (IPMHVC)*, 2014, pp. 557–560.

[15] Q. Liu, "Electrical performance of ester liquids under impulse voltage for application in power transformers," Ph.D. dissertation, The University of Manchester, 2011.

[16] J. Biela, "Optimierung des electromagnetisch integrierten serien-parallel resonanzkonverters mit eingeprägtem ausgangsstrom," Ph.D. dissertation, ETH Zürich, 2005.

[17] M. S. Blume, "Highly efficient pulse modulator system with active droop compensation for linear colliders," Ph.D. dissertation, ETH Zürich, 2016.

[18] R. Schlesinger and J. Biela, "Analytical triple-2D leakage inductance model of cone winding matrix transformers," in *23rd European Conf. on Power Electronics and Applications (EPE ECCE Europe)*, 2021.

[19] S. Blume and J. Biela, "Optimal transformer design for ultraprecise solid state modulators," *IEEE Trans. on Plasma Science*, vol. 41, no. 10, pp. 2691–2700, 2013.

[20] Littelfuse, "Littelfuse varistors-basic properties, terminology and theory," 1999.

[21] N. Kularatna *et al.*, "Design of protection systems," 2018.

[22] F. Hohmann and M.-M. Bakran, "Impacts on the current distribution of metal oxide varistors for overvoltage protection in IGBT modules," in *European Conf. on Power Electronics and Applications (EPE ECCE Europe)*, 2017.

DC Bias Impact on Magnetic Core Losses at High Frequency

Bima Nugraha Sanusi and Ziwei Ouyang
DTU Electro, Technical University of Denmark (DTU)
Kongens Lyngby, Denmark
Email: bnusa@elektro.dtu.dk
URL: https://www.ele.elektro.dtu.dk

Keywords

≪Core loss≫, ≪Core Loss Modeling≫, ≪Magnetic Device≫.

Abstract

This paper aims to provide more insight into core losses at high frequency (MHz range) with superimposed DC bias in ferrite. Inductive cancellation method is employed to reduce measurement sensitivity to phase errors. The tested MnZn ferrite has a nominal relative permeability (μ_r) of 1500 (core A) and 800 (core B), which was tested at frequency from 500 kHz to 3 MHz. The DC bias appears to create a shift in the Steinmetz parameters, in particular the k_i and β are affected the most. Higher B_{DC} creates higher losses gain and higher f_{sw} makes the relative losses increase lower. This is a new finding. Furthermore, the Steinmetz Premagnetization Graph (SPG) with iGSE method ise used to create a core losses prediction model. A maximum error of 25% were found. The measurement data and the built model will be published online for use by other magnetics designer.

1 Introduction

It is increasingly necessary to know core losses with reasonable accuracy in the intended application. This is to avoid overheating on the one hand and oversizing on the other. There are 2 steps in knowing the core losses. The first is measuring the core losses to know the behaviour of the core. The second is modeling the core losses to predict what losses are going to occur under a certain condition. The designer may skip the first step by relying on the published core losses data from the manufacturer. However, the intended application may have different core geometry, excitation waveform, and operating condition than the published data. This will have an impact on the core losses prediction accuracy. The modeling stage aims to predict the core losses based on a set of measurement data. With accurate measurement and modeling, the designer can find the optimal point in the magnetic component design.

There are numerous ways to model the magnetic core losses. They can be classified as the following.

- Hysteresis loop based: the dissipated energy per cycle is determined from the hysteresis loop area. The Preisach [1] and Jiles-Atherton models are examples of this. The approach is accurate but requires many parameters and accurate hysteresis loop measurement.
- Steinmetz Equation (SE) based: core losses per volume is modeled by the power equation. From there stems various models [2, 3, 4], which attempt to improve the loss prediction accuracy under different excitation.
- Loss separation based: the total power losses are constructed by calculating separately the hysteresis loss and eddy current loss and summing them. The calculation uses basic core geometry and material properties, such as conductivity and coercivity. This approach was demonstrated in [5, 6]. It is useful when extensive measurement data is not available.

The SE based model is the most adopted approach by magnetics designer. This is due to the practicality of power formulae. However, measurement data is necessary for this approach.

Core loss measurement techniques can be classified into calorimetric and electric method. Calorimetric method works based on the energy conservation law. The core losses creates heat. The heat is measured in the form of temperature rise. From temperature rise and thermal properties, the core losses can be estimated. This method is well known. A recent work [7] uses the transient principle to reduce measurement time. The major drawback of this method is the difficulty to separate winding loss. Electric method does not have this drawback. It works by measuring induced voltage and excitation current. The induced voltage measurement automatically excludes the winding losses. Therefore, this method is chosen in this paper.

Methods such as [8, 4, 9] can be used to predict core losses with any excitation waveform. However, their validity needs to be re-investigated. Those models work under the assumption of non-changing Steinmetz parameters. Unfortunately, the Steinmetz parameters can change at higher frequency [10] or in the presence of DC bias [11, 12]. There is also a missing experimental verification for core losses under square wave excitation at higher frequency. Recent reports in [7, 13] measure losses only under sine wave excitation. The authors in [14, 15] measure losses with square wave excitation only up to 500 kHz. Therefore, this paper aims to provide more insight to core losses by providing more measurement results.

The focus of the present work is to investigate magnetic core losses under square wave excitation with DC bias. The frequency of interest is up to 3 MHz. The DC bias impact will be quantified and the current core losses model will be extended or modified to incorporate the impact. Behaviour in different core material and geometry is also investigated. Through the end, we will see if the current losses model still matches experimental measurement. The parameter shift will also be analyzed.

2 Core Losses Measurement Setup

Measurement Principle

The main difficulty in measuring core losses with electric method is that ferrites generally have low losses compared to the stored energy. This creates the phase angle ϕ between v_{DUT} and i (illustrated in Fig. 1) close to $90°$, since the impedance of magnetizing inductor $L_{\text{m,DUT}}$ is much smaller than the equivalent core loss resistor $R_{\text{Fe,DUT}}$. The relative power error caused by the phase discrepancy $\Delta\phi$ for sinusoidal excitation is given by (1), which was derived in [16]. It is virtually impossible to eliminate $\Delta\phi$ in an actual measurement setup. Therefore, a more effective solution to limit the power error is to control $tan(\phi)$.

$$\left| \frac{\Delta P}{P} \right| = |tan(\phi)| \cdot |\Delta\phi| \tag{1}$$

The mutual inductance neutralization [17] or inductive cancellation [18] technique can bring $tan(\phi)$ to a more feasible region. However, this method requires a very accurate compensating inductance or capacitance. This is not easy to design. A more recent improvement of this cancellation concept [14] allows less strict compensation requirement. This is called partial cancellation concept and is done by introducing the cancellation factor k. The same concept is adopted in this work.

The core losses measurement circuit is shown in Fig. 1. R_{p} and L_{p} represent the winding resistance and stray inductance, respectively, on the primary side. On the secondary side, they are represented by R_{s} and L_{s}. L_{m} is the magnetizing inductance. A DC blocking capacitance C_{DC} is necessary to avoid saturation in the magnetic core. Current sensing is done by measuring voltage across the resistor R_{sense}. However, due to component nonideality, a parasitic inductance L_{sense} will also be present. As the focus of this work is square wave excitation, a half bridge is used to generate it. A controllable DC current source I_{DC} is also used to generate the DC bias, which is also an important part of this work. Parasitic capacitances from measurement devices are shown by C_{probe} and C_{diff}. It is worth mentioning that for measuring low

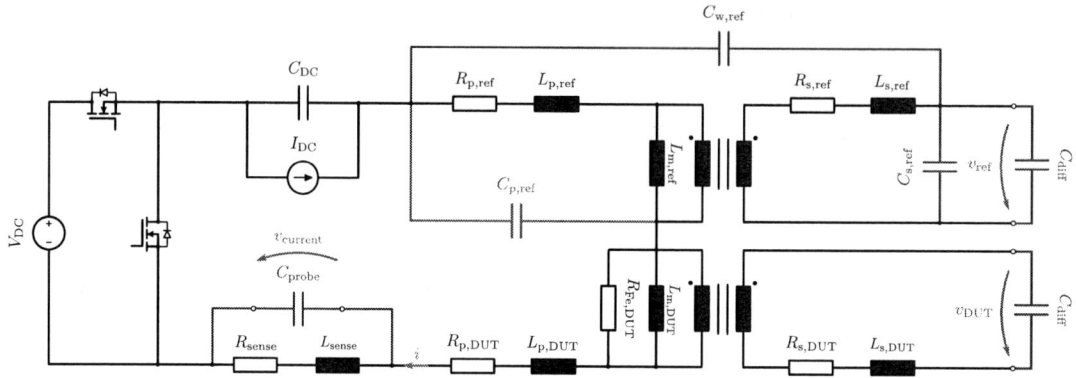

Fig. 1: Electrical circuit of the core losses measurement system. *DUT* means the magnetic device under test and *ref* means the reference / compensation element. Magnetic core losses is represented by $R_{\text{Fe,DUT}}$

losses core, it is difficult to find a magnetic core with much negligible core losses, compared to *DUT*. Therefore, we chose to make the reference element by air core inductor which does not have core losses.

Based on the measurement circuit, the core losses can be calculated by (2), where T_{sw} is the waveform period. For a square wave excitation, the cancellation factor k is calculated by (3). This factor represents the percentage of cancelled reactive voltage to the total reactive voltage. The partial cancellation mechanism creates a virtual voltage $v_{\text{comp}} = v_{\text{DUT}} - v_{\text{ref}}/k$ to replace the initial v_{DUT}. As a result, the phase shift between v_{comp} and measured current can be kept minimum, even when the reference element value $L_{\text{m,ref}}$ has a mismatch with $L_{\text{m,DUT}}$. Therefore, the measurement error can be minimized.

$$P_{\text{Fe}} = \frac{1}{T_{\text{sw}}} \left(\int_0^{T_{\text{sw}}} v_{\text{DUT}} \cdot i_{\text{meas}} \, dt - \frac{1}{k} \int_0^{T_{\text{sw}}} v_{\text{ref}} \cdot i_{\text{meas}} \, dt \right) \tag{2}$$

$$k = \frac{V_{\text{ref,pp}}}{V_{\text{DUT,pp}}} \tag{3}$$

Voltage Measurement

Differential voltage probes are used for the voltage measurement because of the floating measuring position. We chose 25 MHz active differential probe from Testec, which should provide enough bandwidth for 3 MHz square waveform. An additional benefit of the differential probe is the low parasitic capacitance. In this case, it has an equivalent capacitance of 2.75 pF, which is lower compared to 10 pF of common passive probes. One typical drawback is the measurement delay and rise time. Therefore, the differential probes are calibrated to generate identical waveform as measured by a 200 MHz passive probe, before any measurements are performed. A comparison of the waveforms is shown in Fig. 2. The post-processing step will compensate the time delay based on the calibration data.

Voltage measurements (v_{ref} and v_{DUT}) are influenced by the parasitic capacitances. In the DUT, intra and inter-winding capacitances are minimized by using only 1 turn for primary and secondary winding, and keeping enough distance between both windings. Therefore, DUT's parasitic capacitances are neglected in this work. The reference air core needs more winding turns and close distance between primary and secondary. It makes the parasitic capacitances non-negligible. In particular, $C_{\text{w,ref}}$ and $C_{\text{s,ref}}$ will influence the voltage reading. Those capacitances cause extra current flow which makes a voltage drop on secondary $R_{\text{s,ref}}$. This voltage drop is included in the voltage measurement and affect the power calculation.

The formulas given in [14] are used to quantify to potential error. The errors from reference core inter-winding and intra-winding capacitances are given in (4) and (5) respectively. The error from DUT side, caused by voltage probe capacitance, is given in (6). These formulas are mainly derived by calculating the power error caused by the current flowing in the capacitors. The worst case values of the relevant parameters in this work are listed in Table I together with the expected error in Table II. No significant error is expected in this work.

$$\left|\frac{\Delta P}{P}\right|_{C_{w,ref}} \approx \frac{R_{Fe,DUT}C_{w,ref}}{kL_{m,DUT}^2}\left(R_{s,ref}L_{p,ref} + R_{p,ref}L_{s,ref}\right) \tag{4}$$

$$\left|\frac{\Delta P}{P}\right|_{C_{s,ref}} \approx -\frac{C_{s,ref} + C_{diff}}{k}\left(\frac{R_{p,ref}R_{Fe,DUT}}{L_{m,DUT}} + \omega^2 L_{p,ref}\right) \tag{5}$$

$$\left|\frac{\Delta P}{P}\right|_{C_{diff}} \approx C_{diff}\left(\frac{R_{s,DUT}R_{Fe,DUT}}{L_{m,DUT}} + \omega^2 L_{s,DUT}\right) \tag{6}$$

Table I: Parasitic components considered for error calculation

Parameters	Value
C_{diff}	2.75 pF
$C_{w,ref}$	65 pF
$C_{p,ref}, C_{s,ref}$	8 pF
$R_{p,ref}, R_{s,ref}$	200 mΩ
$L_{p,ref}, L_{s,ref}$	38 nH
$R_{p,DUT}, R_{s,DUT}$	200 mΩ
$L_{p,DUT}, L_{s,DUT}$	35 nH
$L_{m,DUT}$	486 nH
$R_{Fe,DUT}$	250 Ω
ω	$2 \cdot \pi \cdot 30$ MHz
k	0.8

Table II: Calculated errors due to parasitic capacitances

Parameters	Value
$\left|\frac{\Delta P}{P}\right|_{C_{w,ref}}$	1.3 %
$\left|\frac{\Delta P}{P}\right|_{C_{s,ref}}$	1.6 %
$\left|\frac{\Delta P}{P}\right|_{C_{diff}}$	3 %

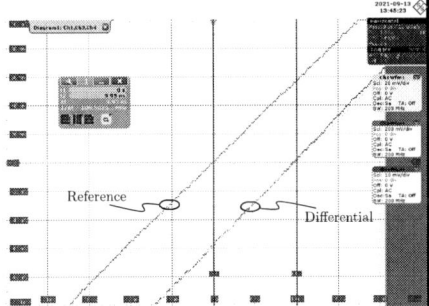

Fig. 2: Voltage measurement comparison: passive probe (reference) and two differential probes. About 10 ns time delay is observed while the gradient is identical.

Current Measurement

Current sensing is a critical part of the measurement system. Shunt resistor is chosen over current probe mainly due to the DC current measuring capability. The shunt resistor needs to have high precision with low stray inductance. Metal foil chip resistor from Susumu PRL series is used in this work. The shunt resistor has $R_{\text{sense}} = 0.1\ \Omega$ and $L_{\text{sense}} = 1\ nH$ as the nominal values, which is shown in Fig. 3. The shunt voltage is measured by the voltage probe, which brings the parasitic capacitance C_{probe}. Its nominal value is 9.5 pF for the RT-ZP10 from Rohde & Schwarz, which is used in this work. Passive probe can be used because the negative side is connected to ground. In addition, it provides high enough bandwidth, 200 MHz in this case.

The last step in current measurement is to transform the measured voltage into current. The stray inductance L_{sense} and voltage probe capacitance C_{probe} make the transformation less straightforward. The measured voltage v_{current} needs to be filtered by transfer function as shown in (7). The difference in measured voltage and calculated current is shown in Fig. 4. It can be seen that much smaller current is flowing through C_{probe}. This approach also anticipates the phase discrepancy brought by the shunt stray inductance.

$$i(s) = v_{\text{current}}(s) \left(\frac{1}{s \cdot L_{\text{sense}} + R_{\text{sense}}} + s \cdot C_{\text{probe}} \right) \tag{7}$$

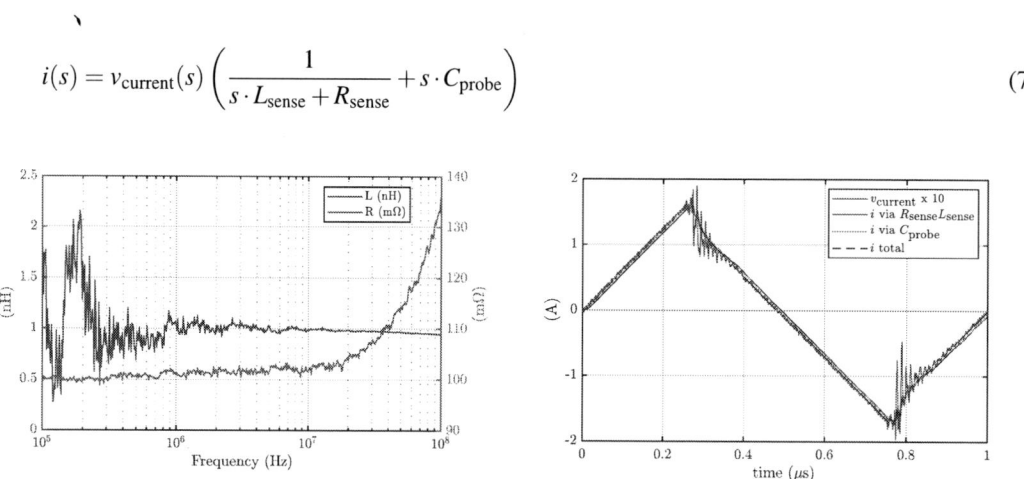

Fig. 3: Current shunt equivalent resistance and inductance

Fig. 4: Current measurement comparison: directly measured voltage and calculated current

Setup Verification

The setup is verified by testing toroid core under sine wave excitation. Then, the measurement results are compared with datasheet values. The measured cores are PC50 R22.1/13.7/7.9 and PC200 R15.8/8.9/4.7 from TDK. Fig. 5 shows the comparison result. The measurement point is given together with its uncertainty range, which is derived from 10 measurement captures. Measurement results match the datasheet values up to 1 MHz for PC50 material. However, for PC200 material, there is a discrepancy and the measurement gives higher losses than datasheet values. This discrepancy can happen in real practice [19], especially as the frequency goes higher. Nevertheless, the sine wave measurement results here can still be the basis for comparison in the next chapters.

3 Experimental Investigation

This work aims at measuring and analyzing core losses under square wave excitation from frequency of 500 kHz up to 3 MHz. The frequency range is motivated by state-of-the-art commercially available MnZn ferrite for power electronics application. The measurements were performed on two magnetic cores with different relative permeability (μ_r) level. They are listed in Table III along with their detailed properties. V_e, A_e, and l_e mean the magnetic core volume, effective cross section, and mean magnetic path length, respectively.

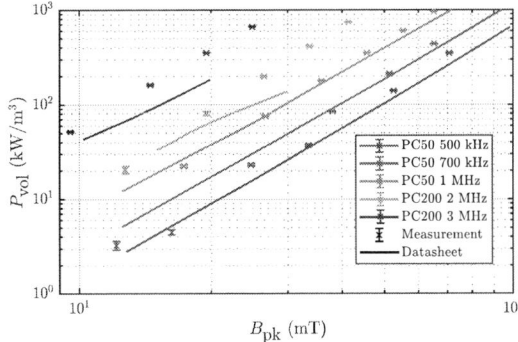

Fig. 5: Comparison between datasheet values and measurement results with sinusoid excitation

The measurement setup is discussed in the previous section and, in the end (2) and (3) are used to calculate the core losses. The magnetic flux density (B) is defined in Fig. 6 and can be calculated using (8) for the AC component and (9) for the DC component. Here N_s means the number of turns of sensing winding in the DUT and v_{DUT} is the sensed voltage. Care must be taken when calculating (3), due to the voltage overshoot at the edge of square wave. Voltage at the flat area of square wave is used to calculate (3) in this work.

$$B_{\text{DUT}}(t) = \frac{1}{N_s \cdot A_e} \int_0^t v_{\text{DUT}}(t)\, dt \qquad (8) \qquad\qquad H_{\text{DC}} = \frac{N_{\text{exc}} \cdot I_{\text{DC}}}{l_e} \qquad (10)$$

$$B_{\text{DC}} = \mu_0 \cdot \mu_r \cdot H_{\text{DC}} \qquad (9)$$

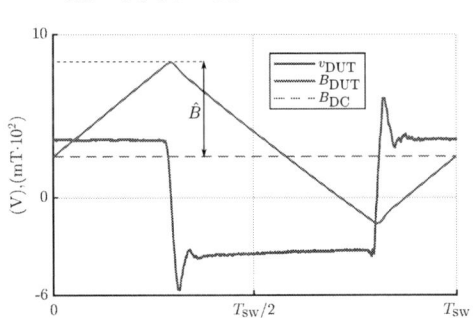

Fig. 6: Sensed DUT voltage (v_{DUT}) and the calculated flux density (B_{DUT}). This example is taken from core sample B at 1 MHz switching frequency

Table III: Core Under Test

Parameters	Core A	Core B
Core Material	PC50	PC200
Core Geometry	R 22.1/13.7/7.9	ER 14.5/6
V_e	1763 mm^3	334 mm^3
A_e	32.6 mm^2	17.6 mm^2
l_e	54.2 mm	19 mm
Initial μ_r	1500	800
B_{sat} at 25°C	490 mT	480 mT
Intended f_{sw}	0.3 - 1 Mhz	1 - 3 Mhz

Impact on Loss Curve

The first step to understand the effect of DC bias is to look at the loss curve. Fig. 7 and Fig. 8 show the losses density against peak AC flux density ($\hat{B}$) for core A and B, while Fig. 9 and Fig. 10 show the losses density against switching frequency. The diamond points in the figures are measurement points while the line is the result of regression.

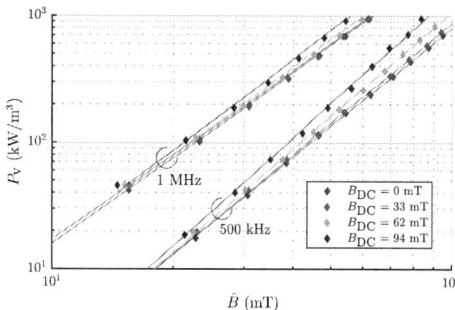
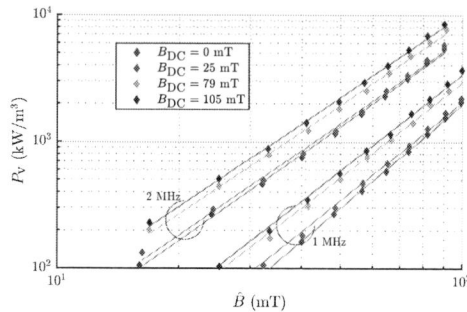

Fig. 7: Core losses density (P_v) versus flux density ($\hat{B}$) measured on test core A with different DC flux bias (B_{DC}) and frequency

Fig. 8: Core losses (P_v) versus flux density ($\hat{B}$) measured on test core B with different DC flux bias (B_{DC}) and frequency

The influence of DC bias (B_{DC}) on P_v vs $\hat{B}$ can be seen in Fig. 7 and Fig. 8. At low DC bias the losses increase is relatively small. However, after the DC bias reaches a certain value, the losses increase starts to be significant. For example, at 500 kHz and $\hat{B} = 50mT$ in core A the losses increase at 33 mT bias is only 5.3 % but, at 62 mT bias the increase becomes 20 % and at 94 mT 51.4 %. For core B at 1 MHz and $\hat{B} = 50mT$ the trend is similar but at a different level. With 25 mT bias, the increase is 15.1 % but with 79 mT it becomes 75 %.

At a single frequency, the measured losses points still follow a straight regression line in logarithmic scale, even after DC bias is applied. This suggests that $P_v \propto \hat{B}^{\beta}$ relation still hold with DC bias. It is also observed that there is a slight shift in the regression line slope. This may suggest a shift in the β parameter. We also see this slope shift is less pronounced in core B, compared to core A. A more detailed modeling is presented in the later section.

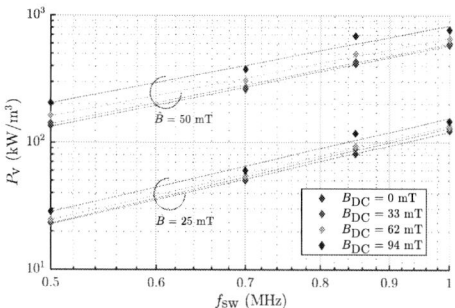
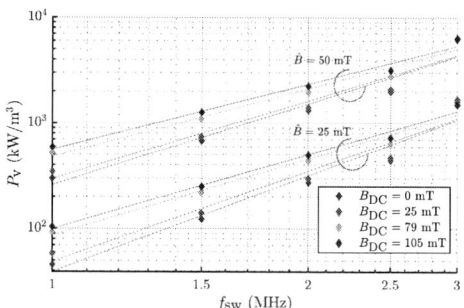

Fig. 9: Core losses (P_v) versus operating (switching) frequency (f_{sw}) measured on test core A with different DC flux bias (B_{DC}) and AC peak flux density ($\hat{B}$)

Fig. 10: Core losses (P_v) versus operating (switching) frequency (f_{sw}) measured on test core B with different DC flux bias (B_{DC}) and AC peak flux density ($\hat{B}$)

Fig. 9 and Fig. 10 show the influence of DC bias (B_{DC}) on P_v vs f_{sw} curve. A similar observation can be found where, again, at low DC bias the loss density increase is small. For test core A, the DC bias seems to offset the loss density curve, as in Fig. 9. Although not so clear, we can also see the $P_v \propto f_{sw}^{\alpha}$ relation and the shift in α parameter is not likely for core A. Meanwhile for core B, the relationship is less straightforward. The DC bias not only offsets the curve, but also changes the slope. In particular at $f_{sw} > 2MHz$, the effect becomes less predictable. At 3 MHz, the loss increase is almost unnoticeable. This can be caused by an error in the measurement system or it is a newfound characteristic of the magnetic core.

Relative Core Loss Increase

This subsection tried to answer when the DC bias starts to have important effect to the core losses. It is also chosen to quantify the DC bias using B field instead of H field. This selection enables magnetic core performance comparison across different permeability value. The relative core losses increase ($P_{v,rel}$) is found by dividing the losses at a certain DC bias with the losses at no bias condition.

Figure 11 shows for core A when the DC bias impact becomes significant. The sum of AC and DC component of B field is taken as the X axis to show if there is relation between peak flux density and losses increase. The increase starts to become important, e.g more than 20 %, when the total B is above 100 mT. This can be seen by looking at the measurement points upper envelope. However, for $B_{DC} < 50$ mT, the increase stays low even when the sum is above 100 mT. It means the bias magnitude also plays a role in determining the increase. Furthermore, different operating frequency brings different impact. The higher frequency seems to create lower relative increase, as can be seen by comparing 1 MHz and 500 kHz points.

In test core B the trend and relationship is less straightforward. Figure 12 shows the relation between total B field and the relative losses increase. The measurement points are more widespread than core A. Contrary to core A behaviour, the relative loss increase is lower as the total flux increases, for a certain B_{DC}. This trend applies to all tested frequency and DC bias for core B. The DC bias magnitude also determines the relative increase and here the loss gain is more pronounced, e.g. at 1 MHz the $P_{v,rel}$ is close to two when $\hat{B} + B_{DC}$ is around 100 mT and $B_{DC} = 78$ mT. This is a steep increase and designers should take care of it when designing the magnetic components. Meanwhile, the impact of frequency is quite similar to core A. The higher frequency creates a lower relative loss increase.

The same data in Fig. 12 can also be presented in a different way as in Fig. 13. The B ratio of AC and DC components is plotted against loss increase. This is done to check if a different characteristics of the core losses can be seen. Nevertheless, a similar observation is found. The B_{DC} is the main controlling factor for $P_{v,rel}$. The B ratio plays a smaller role, although still has an impact, i.e. higher ratio makes lower loss increase. This observation reflects a complex physics behind the core losses phenomenon.

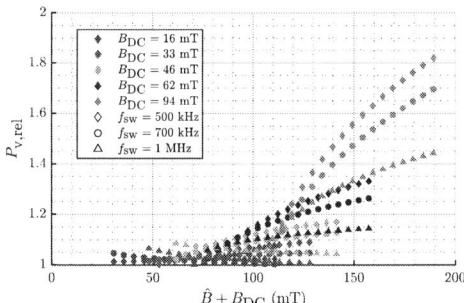

Fig. 11: Relative core losses ($P_{v,rel}$) versus peak total flux density ($\hat{B} + B_{DC}$) at different DC bias and frequency for test core A

Fig. 12: Relative core losses ($P_{v,rel}$) versus peak total flux density ($\hat{B} + B_{DC}$) at different DC bias and frequency for test core B

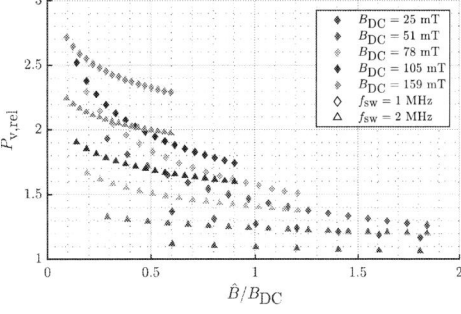

Fig. 13: Relative core losses ($P_{v,rel}$) versus flux density ratio ($\hat{B}/B_{DC}$) at different DC bias and frequency for test core B

4 Core Losses Modeling

This section does not try to propose a new model to predict core losses. Instead, we will see if the available models can still accurately predict core losses from the measurement results generated in this work, particularly this means incorporating the impact of DC bias on core losses. Out of many previously proposed core losses modeling [2, 4, 12, 15], the method in [12] will be used in the following analysis.

Analytical: iGSE with SPG

The main advantage of using analytical formula in modeling core losses is getting the insight into the physics behind the losses mechanism. The explicit formula shows the user which parameters are affecting the core losses. Hence, giving some understanding of the core losses behaviour.

Since in power electronics the flux (B) waveform is commonly not sinusoidal, the established improved Generalized Steinmetz Equation (*iGSE*) method [4] is adopted to calculate the core losses. However, this modeling does not include DC bias effect. Therefore, it needs to be modified with other techniques. The Steinmetz Premagnetization Graph (*SPG*) method [12] can incorporate the DC bias effect without changing the original equation so, it is used in this work. The iGSE formula when applied to triangular flux waveform is reduced to the form in (11). Then, with SPG, the α, β, and k_i will be modified following (12) which incorporates the dependency on DC bias B_{DC}.

The appropriate coefficients of SPG matrix (first term of right hand side in (12)) needs to be found in order to extract the dependency. This is done by solving an optimization or curve fitting problem. A least square algorithm has been implemented that fits calculated curve with measured data by minimizing the relative error. Despite the 4th order polynomial used here, the order can be varied if the optimization / curve fitting result does not generate a good prediction accuracy. In this work, the 4th order proves to be sufficient.

$$P_{v,\text{iGSE,tri}} = k_i \cdot (2f)^\alpha \cdot (\Delta B)^\beta \tag{11}$$

$$\begin{bmatrix} \alpha \\ \beta \\ k_i \end{bmatrix} = \begin{bmatrix} \alpha_0 & 0 & 0 & 0 & 0 \\ \beta_0 & p_{\beta 1} & p_{\beta 2} & p_{\beta 3} & p_{\beta 4} \\ k_{i0} & p_{ki1} & p_{ki2} & p_{ki3} & p_{ki4} \end{bmatrix} \cdot \begin{bmatrix} 1 \\ B_{DC} \\ B_{DC}^2 \\ B_{DC}^3 \\ B_{DC}^4 \end{bmatrix} \tag{12}$$

Following the mentioned procedure, the SPG dependency graph can be obtained. This is shown in Fig. 14 and Fig. 15 for core A and B, respectively. The data points are measurement result and the line is curve-fitting result. For core A, the α and β parameters stay relatively constant over the whole B_{DC} range. Meanwhile, k_i fluctuates smoothly at low B_{DC} and increases quite steeply after $B_{DC} > 50$ mT. This observation aligns with the analysis in section 3, where the relative losses gain rises quickly after a certain B_{DC} values. For core B, the k_i increases steadily from low B_{DC} and shows an almost linear relation to B_{DC}. The β parameters has a slight decline from low to mid B_{DC}, while α shows a decreasing trend towards higher B_{DC}. The SPG prediction cannot capture the α shift because (12) assumes a constant α value. This assumption may not hold true anymore after seeing the measurement result in Fig. 12 and 13. Therefore, the modification of (12) is foreseen. Nevertheless, here the original SPG formulation is kept for the sake on consistency with previous works.

After having the SPG coefficients matrix, (11) is used to calculate the core losses in the presence of DC bias. The results are shown in Fig. 16 and Fig. 17 for core A and B, respectively, with several different $\hat{B}$ and f_{sw} values. The measurement points are also shown to compare the loss prediction and measurement. The iGSE + SPG method appears to be reasonably accurate in the whole B_{DC} range for core A. The accuracy at low $\hat{B}$ (30 mT) is better than at high $\hat{B}$ (60 mT). For core B, this method loses accuracy at high $\hat{B}$ and high f_{sw}, as can be seen in Fig. 17. Nonetheless, the SPG enables losses prediction with DC bias.

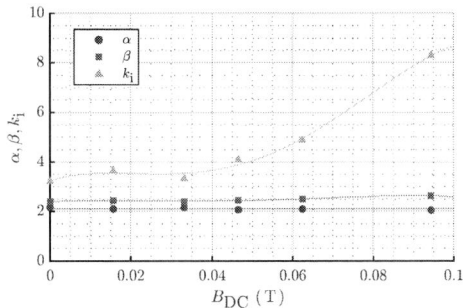

Fig. 14: Steinmetz Pre-magnetization Graph for test core A. The k_{i} value is scaled down by 10^4 to increase graph readability. The data points are measurement result and the line is curve-fitting result.

Fig. 15: Steinmetz Pre-magnetization Graph for test core B. The k_{i} value is scaled down by 10^4 to increase graph readability. The data points are measurement result and the line is curve-fitting result.

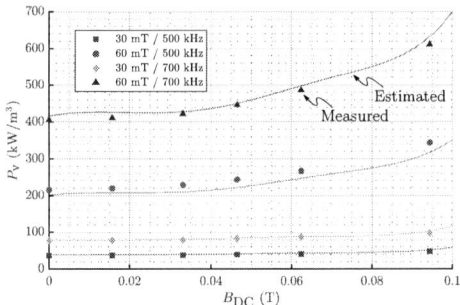
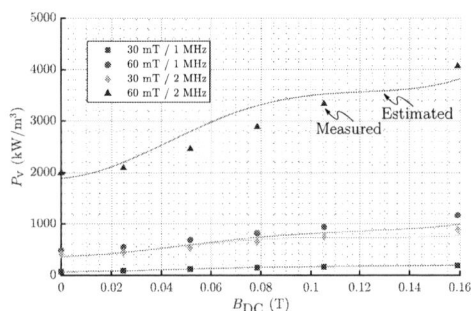

Fig. 16: Core loss density (P_{v}) estimation of test core A. The scatter points are measurement data while the curves are prediction result.

Fig. 17: Core loss density (P_{v}) estimation of test core B. The scatter points are measurement data while the curves are prediction result.

5 Conclusion and Future Work

This paper investigates magnetic core losses under square wave excitation with superimposed DC bias. The frequency of interest is from 500 kHz up to 3 MHz and the tested material is MnZn ferrite with nominal relative permeability (μ_{r}) of 1500 and 800 which are the state-of-the-art for power electronics application. Chapter 2 describes the core losses measurement setup. The mutual inductance neutralization method is used to minimize the measurement error. The measurement results are presented in chapter 3. The DC bias basically generates an offset to the loss curve, creating higher losses. It was also seen that there is a shift in Steinmetz parameter, especially β, after bias is applied. Nevertheless, the proportionality $P_{\mathrm{v}} \propto \hat{B}^{\beta}$ and $P_{\mathrm{v}} \propto f_{\mathrm{sw}}^{\alpha}$ still hold. Chapter 4 attempts to model the core losses behaviour by using the combined approach: Steinmetz Premagnetization Graph (SPG) with iGSE. This method modifies the Steinmetz parameter using a polynomial function of the DC bias and the model generates a maximum error percentage of 25% in our test. In summary, this paper clarifies ferrite core losses behaviour in the mentioned frequency range under DC bias, and quantitatively analyzes the possible models. As a further step, the models and measurement data can be expanded and shared to be used with other magnetic designers.

REFERENCES

[1] I. D. Mayergoyz and G. Friedman, "Generalized Preisach model of hysteresis," *IEEE Transactions on Magnetics*, vol. 24, no. 1, pp. 212–217, 1988.

[2] M. Albach, T. Durbaum, and A. Brockmeyer, "Calculating core losses in transformers for arbitrary magnetizing currents a comparison of different approaches," in *PESC Record. 27th Annual IEEE Power Electronics Specialists Conference.* IEEE, 1996.

[3] J. Reinert, A. Brockmeyer, and R. De Doncker, "Calculation of losses in ferro- and ferrimagnetic materials based on the modified Steinmetz equation," *IEEE Transactions on Industry Applications*, vol. 37, no. 4, 2001.

[4] K. Venkatachalam, C. Sullivan, T. Abdallah, and H. Tacca, "Accurate prediction of ferrite core loss with nonsinusoidal waveforms using only Steinmetz parameters," in *2002 IEEE Workshop on Computers in Power Electronics, 2002. Proceedings.* IEEE, 2002.

[5] W. Roshen, "Ferrite core loss for power magnetic components design," *IEEE Transactions on Magnetics*, vol. 27, no. 6, 11 1991.

[6] W. A. Roshen, "A Practical, Accurate and Very General Core Loss Model for Nonsinusoidal Waveforms," *IEEE Transactions on Power Electronics*, vol. 22, no. 1, pp. 30–40, 2007.

[7] P. Papamanolis, T. Guillod, F. Krismer, and J. W. Kolar, "Transient Calorimetric Measurement of Ferrite Core Losses up to 50 MHz," *IEEE Transactions on Power Electronics*, vol. 36, no. 3, 3 2021.

[8] Jieli Li, T. Abdallah, and C. Sullivan, "Improved calculation of core loss with nonsinusoidal waveforms," in *Conference Record of the 2001 IEEE Industry Applications Conference. 36th IAS Annual Meeting (Cat. No.01CH37248).* IEEE, 2001.

[9] A. Van den Bossche, V. Valchev, and G. Georgiev, "Measurement and loss model of ferrites with non-sinusoidal waveforms," in *2004 IEEE 35th Annual Power Electronics Specialists Conference (IEEE Cat. No.04CH37551).* IEEE, 2004.

[10] W. G. Hurley, T. Merkin, and M. Duffy, "The Performance Factor for Magnetic Materials Revisited: The Effect of Core Losses on the Selection of Core Size in Transformers," *IEEE Power Electronics Magazine*, vol. 5, no. 3, 9 2018.

[11] A. Brockmeyer, "Experimental evaluation of the influence of DC-premagnetization on the properties of power electronic ferrites," in *Proceedings of Applied Power Electronics Conference. APEC '96.* IEEE, 1996.

[12] J. Muhlethaler, J. Biela, J. W. Kolar, and A. Ecklebe, "Core Losses Under the DC Bias Condition Based on Steinmetz Parameters," *IEEE Transactions on Power Electronics*, vol. 27, no. 2, 2 2012.

[13] A. J. Hanson, J. A. Belk, S. Lim, C. R. Sullivan, and D. J. Perreault, "Measurements and Performance Factor Comparisons of Magnetic Materials at High Frequency," *IEEE Transactions on Power Electronics*, vol. 31, no. 11, 11 2016.

[14] D. Hou, M. Mu, F. C. Lee, and Q. Li, "New high-frequency core loss measurement method with partial cancellation concept," *IEEE Transactions on Power Electronics*, vol. 32, no. 4, pp. 2987–2994, 4 2017.

[15] E. Stenglein and T. Durbaum, "Core Loss Model for Arbitrary Excitations With DC Bias Covering a Wide Frequency Range," *IEEE Transactions on Magnetics*, vol. 57, no. 6, 6 2021.

[16] F. Dong Tan, J. Vollin, and S. Cuk, "A practical approach for magnetic core-loss characterization," *IEEE Transactions on Power Electronics*, vol. 10, no. 2, 3 1995.

[17] C. Baguley, U. Madawala, and B. Carsten, "A New Technique for Measuring Ferrite Core Loss Under DC Bias Conditions," *IEEE Transactions on Magnetics*, vol. 44, no. 11, 11 2008.

[18] M. Mu, Q. Li, D. J. Gilham, F. C. Lee, and K. D. Ngo, "New core loss measurement method for high-frequency magnetic materials," *IEEE Transactions on Power Electronics*, vol. 29, no. 8, pp. 4374–4381, 2014.

[19] P. Papamanolis, T. Guillod, F. Krismer, and J. W. Kolar, "Minimum Loss Operation and Optimal Design of High-Frequency Inductors for Defined Core and Litz Wire," *IEEE Open Journal of Power Electronics*, vol. 1, 2020.

Investigation of the short-circuit type II safe operating area of IGBTs

[1]Madhu Lakshman Mysore, [1]Mohamed Alaluss, [1]Abhishek Maitra, [1]Thomas Basler,
[2]Roman Baburske, [2]Franz-Josef Niedernostheide, [2]Hans-Joachim Schulze

[1]Chemnitz University of Technology, Chair of Power Electronics, Chemnitz, Germany.
Tel.: +49 371/531-36091.
Fax: +49 371/531-836091.
[2]Infineon Technologies AG, Neubiberg, Germany
E-Mail: madhu-lakshman.mysore@etit.tu-chemnitz.de
URL: https://www.tu-chemnitz.de/etit/le/ and https://www.infineon.com/

Acknowledgements

This investigation was funded by the Power2Power project. Power2Power is a European, co-funded innovation project involving the semiconductor industry. The project receives grants from the European H2020 research and innovation program, ECSEL Joint Undertaking, National Funding Authorities, and from eight countries involved in the project including the German Federal Ministry of Education and Research (BMBF) under grant agreement No. 826417. The authors would like to thank Karen Hanson for critically reading the manuscript. Additionally, the authors would like to thank Sven Mauerberger from the faculty of mechanical engineering for his help with the sample preparation for failure analysis.

Keywords

«IGBT», «Short circuit», «Current filaments», «Device simulation», «Ruggedness».

Abstract

This study focuses on the short-circuit type II safe operating area (SC-II SOA) with and without gate-emitter voltage clamping. The SC-II measurements without gate-emitter voltage (V_{GE}) clamping show a reduced SC-II SOA at higher DC-link voltages induced by transient gate-emitter voltages that are far beyond the allowed level. These high transient gate voltages result in correspondingly high peak currents. As a consequence, they cause device failure during the negative di_C/dt phase, which is induced by the inductive overvoltage. However, the SC-II SOA can be completely recovered to the level of the SC-I SOA by applying an appropriate V_{GE} clamping circuit, although the IGBT will be subjected to harsher conditions in the SC-II event compared to the SC-I event. To understand the failure types observed in SC-II measurements with and without V_{GE} clamping, computer-aided TCAD simulations were performed using a real front-side, trench-gate IGBT structure.

Introduction

The insulated gate bipolar transistor (IGBT) is one of the most frequently used power semiconductor devices in the field of power electronics. A very important property for many applications is short-circuit (SC) ruggedness. However, IGBTs can be exposed to different short-circuit types (SC-I, SC-II, SC-III or more) in the application described in [1-6]. SC type I (SC-I), or hard switching faults (HSF), occur if the IGBT turns on into an existing short-circuit [4]. The short-circuit type I safe operating area (SC-I SOA) for different voltage classes ranging from 600 V to 6500 V were investigated in [7-14]. The results shown in [10, 11] indicate that an IGBT SC failure can be due to the formation of current filaments occurring at the collector side. Typically, these current failures occur far beyond the safe operating area (SOA) of the IGBT [10, 11].

Short-circuit type II (SC-II), or fault under load (FUL), occur during the conduction phase of the IGBT [2, 3]. In real applications, it is more likely for the IGBT to experience SC-II than SC-I. Hence, it is important to study the ruggedness of the IGBT under SC-II conditions and to understand the root cause of the limitation in the SC-II SOA of the IGBT.

In the present work, SC-I measurements were initially performed on single-chip, 1700 V IGBTs with trench technology, using chips soldered on a direct copper bonded (DCB) substrate. The measured SC-I results show a U-shaped destructive boundary line in the $I_C - V_{DC\text{-link}}$ phase space, where I_C and $V_{DC\text{-link}}$ denote the collector current of the IGBT and the applied DC-link voltage. To study the SC-II ruggedness of the 1700 V single chip, SC-II measurements were carried out initially without gate-emitter voltage (V_{GE}) clamping and without common-emitter inductance (L_{CE}). The influence of the L_{CE} without V_{GE} clamping was also taken into account in the SC-II measurements. Furthermore, V_{GE} clamping was applied in SC-II measurements. Using a technology computer-aided design (TCAD) tool, 2D electro-thermal SC simulations were performed for a 1.7 kV IGBT trench-gate cell structure to study the internal behavior of the IGBT during SC-II events.

Simulation model

The measurements performed in this work cannot provide an explanation for the internal behavior of the IGBT. However, the electro-thermal simulation results will help to understand the different failure modes that occur during SC-I and SC-II. The following device simulation using the *Synopsys TCAD* tool was carried out to study the SC turn-off failure, the pulse failure and the negative di_C/dt failure [15]. A multi-cell, trench-gate 1.7 kV IGBT was designed and used to investigate the current filament behavior under different $V_{DC\text{-link}}$ and collector current (I_C) conditions. Also, simulations were carried out with and without V_{GE} clamping. The simulated IGBT structure was 192 µm wide. A time-dependent, electro-thermal SC-II pulse was simulated with a starting temperature (T_{start}) of 300 K, parasitic inductance (L_{par}) of 30 nH, for a fixed gate-emitter voltage (V_{GE}) and with an applied DC-link voltage. The University of Bologna mobility model, together with the carrier-carrier scattering model were utilized as well as the University of Bologna avalanche model and the Shockley-Read-Hall (SRH) and Auger recombination models, and the Slotboom model for effective intrinsic density. For the SC simulation, self-heating was considered with a thermal boundary condition at the collector contact and a thermal resistance of 0.505 K/W. The SC pulse width was set to 5 µs.

SC-I SOA of the IGBT

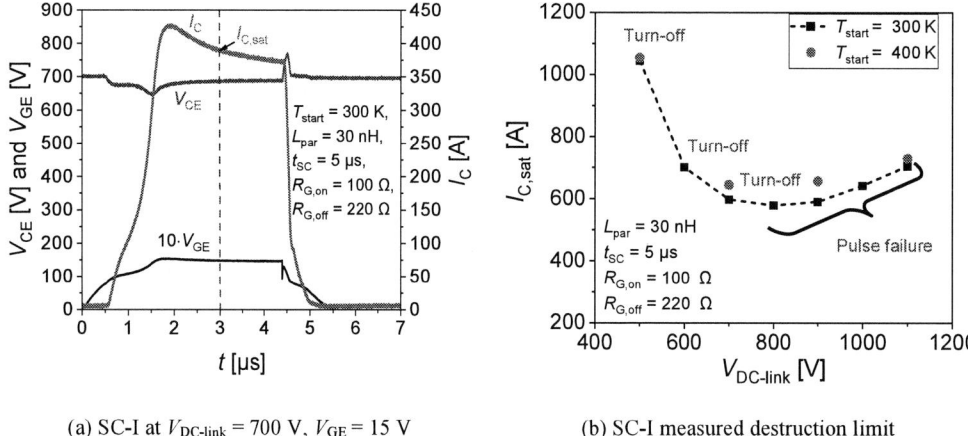

(a) SC-I at $V_{DC\text{-link}}$ = 700 V, V_{GE} = 15 V   (b) SC-I measured destruction limit

Fig. 1: (a) A typical measured SC I waveform for a non-destructive pulse and (b) measured destruction limit far above the SC-I SOA in I_C-$V_{DC\text{-link}}$ phase space for 1.7 kV single-chip IGBT at different temperatures. The text indicated at the different points describes the failure modes.

For comparison with the following SC-II investigations, a typical SC-I waveform is shown in Fig. 1(a). The black and red points show the critical saturation collector current ($I_{C,sat}$) as a function of the DC-link voltage ($V_{DC\text{-link}}$) of the last non-destructive SC-I pulse at two different temperatures in Fig. 1(b). For a fixed V_{CE}, the destruction current limit was measured by increasing the gate voltage in increments of 0.25 V. A short-circuit pulse width of 5 µs was set for all measurements. The minimum destruction indicated occurs at 800 V for the 1.7 kV IGBT during a SC-I occurrence. For both temperatures, as shown in Fig. 1(b), the decrease in SC ruggedness from 500 V to 800 V is due to the formation of destructive current filaments [10, 11]. It is important to note that the respective applied gate voltage for this experiment was clearly above the datasheet limits. Above 800 V, the SC capability increases again due to the recovered homogenous current distribution in the IGBT, as the space-charge region covers the entire drift region at a higher applied DC voltage and higher collector current [11]. For DC-link voltages from 500 V to 700 V, the destruction of the IGBT occurred during the SC turn-off and for DC-link voltages from 800 V to 1100 V during the SC pulse. A similar type of failure occurred at higher temperatures with a slightly higher last-pass saturation collector current.

SC-II SOA of the IGBT without V_{GE} clamping

SC-II measurement setup

The schematic measurement setup used to measure the SC-II of an IGBT is shown in Fig. 2(a). The protection IGBT (PIGBT) and the device under test (DUT) are connected in series close to the DC-link capacitors to reduce the parasitic inductance (L_{par}) of the SC path. A parasitic inductance of the entire SC loop is 30 nH. The PIGBT was used to limit the failure current in the event of DUT failure to the level of its own SC current of the PIGBT. Two gate drive units (GDU) were used to control the gate voltages of the DUT and the PIGBT.

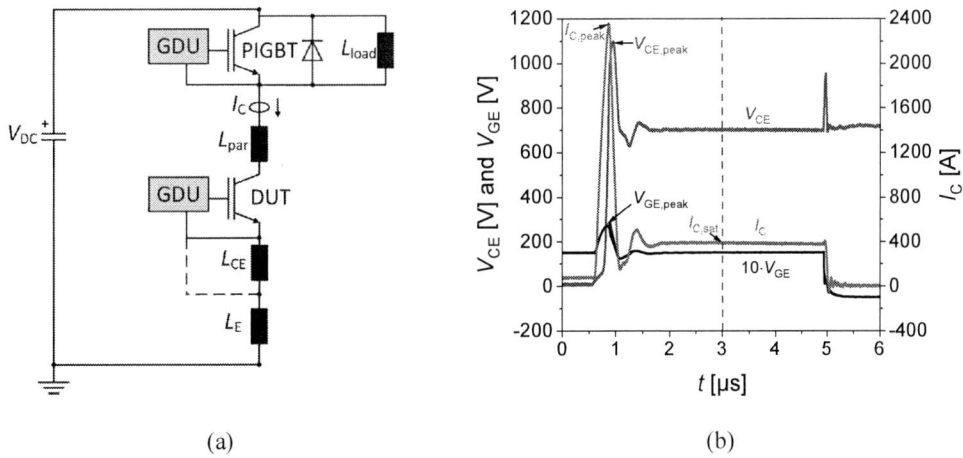

| (a) | (b) |

Fig. 2: (a) Schematic diagram of a SC-II measurement setup without V_{GE} clamping, and (b) measured SC-II pulse without V_{GE} clamping and without L_{CE} influence at given conditions T_{start} = 300 K, L_{par} = 30 nH, L_{load} = 82 µH, R_G = 10 Ω, t_{SC} = 5 µs, $V_{DC\text{-}link}$ = 700 V, V_{GE} = 15 V.

The measurements were carried out without gate-emitter voltage (V_{GE}) clamping. If the gate driver unit (GDU) is connected across the gate and sense emitter of the IGBT as displayed in Fig. 2(a), then measurements were performed without common-emitter inductance (L_{CE}) influence. On the other hand, if the GDU is connected across the gate and load emitter of the IGBT, then the measurements are influenced by L_{CE}. The measured SC-II pulses without V_{GE} clamping and without L_{CE} influence are displayed in Fig. 2(b). Due to the missing V_{GE} clamping, a high current peak is reached in the range of 6 times the nominal SC current, at V_{GE} = 15 V. This is due to the feedback effect across the Miller capacitance (C_{GC}) during voltage desaturation and the removal of remaining plasma from the prior on-phase.

SC-II measurements without V_{GE} clamping and without L_{CE} influence

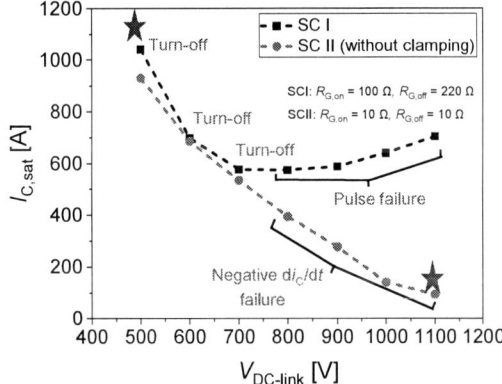

Fig. 3: Measured destruction limit for SC-I and SC-II in the I_C-$V_{DC\text{-}link}$ phase space for 1.7kV single chip at given conditions T_{start} = 300 K, L_{par} = 30 nH, L_{load} = 82 µH, t_{SC} = 5 µs.

To investigate the SC-II ruggedness of the 1700 V IGBT chip at a given DC-link voltage, the gate-emitter voltage was increased in increments of 0.2 V in a very low inductive setup with L_{par} = 30 nH. The measured SC-II destruction limits without V_{GE} clamping are compared with the SC-I destruction boundary (Fig. 3). For the SC- II measurements without V_{GE} clamping and without L_{CE}, the SC-II critical current level $I_{C,sat}$ continuously decreases with increasing DC-link voltage. For DC-link voltages from 500 V to 700 V, the destruction boundary line is approximately the same for both SC-I and SC-II measurements. Furthermore, the failure mode (turn-off failure) is the same for all conditions up to 700 V. The turn-off failure at 500 V is shown in Fig. 4(a). However, SC-II measurements without V_{GE} clamping show a further decrease of critical $I_{C,sat}$ for DC-link voltages above 800 V compared to SC-I. In this voltage range, negative di_C/dt failures were observed after reaching $I_{C,peak}$ [Fig. 4(b)]. The negative di_C/dt failure occurs after the SC-II turn-on collector current peak, when the collector-emitter voltage reaches a maximum peak value. Contrary to the SC-II behavior without clamping, the SC-I measurements show pulse failures above 800 V DC-link voltage. To study the internal mechanisms for the different failure types, SC-II simulations were performed at DC-link voltages of 500 V and 1100 V, as marked with blue stars in Fig. 3.

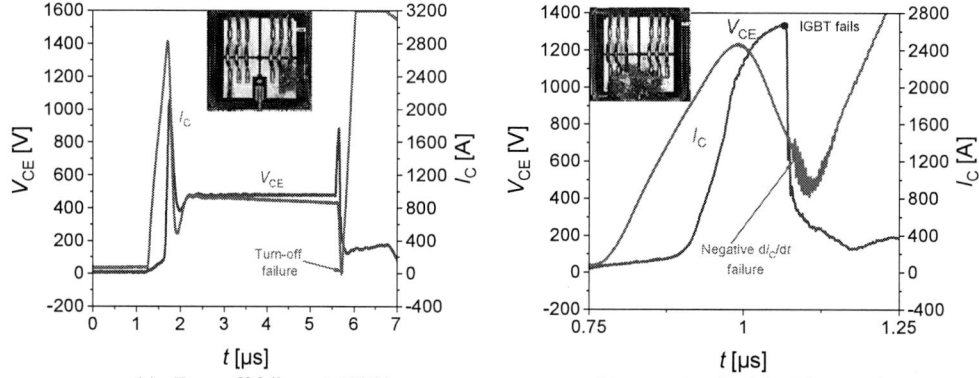

(a) Turn-off failure at 500 V   (b) negative di_C/dt failure at 900 V

Fig. 4: Measured SC-II failure types at given conditions T_{start} = 300 K, L_{par} = 30 nH, L_{load} = 82 µH, R_G = 10 Ω, t_{SC} = 5 µs. Inset: destruction picture of the chip.

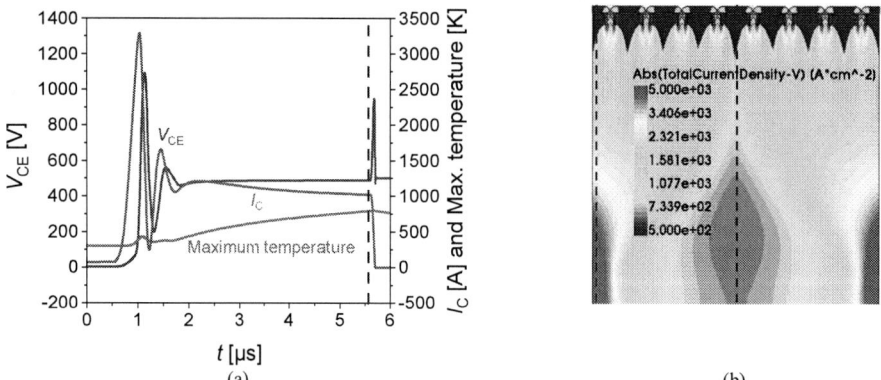

(a)   (b)

Fig. 5: (a) Simulated SC-II at V_{CE} = 500 V with conditions: T_{start} = 300 K, L_{par} = 30 nH, L_{load} = 82 µH, R_G = 10 Ω, t_{SC} = 5 µs (b) current-density distribution at the beginning of the SC-II turn-off for a 1700 V IGBT.

Fig. 5(a) shows the transients of the current, voltage and maximum temperature at $V_{DC\text{-}link}$ = 500 V, $I_{C,sat}$ = 1135 A. Fig. 5(b) shows the distribution of the absolute total current density in the IGBT shortly before the SC-II turn-off. The current flow in the IGBT becomes inhomogeneous. The maximum temperature in the current filaments in the IGBT reaches a value of 795 K. For the simulated structure width, there are two half current filaments at the edges of the IGBT. The edge-to-edge lateral distance between the current filaments is 160 µm. The formation of current filaments depends on the width of the structure, with more current filaments expected in wider structures, as explained in [13].

Fig. 6 shows the vertical cross-section of the absolute value of the electric-field strength, electron density and hole density in the IGBT structure at the time 5.6 µs for the given SC-II conditions. The positions of the vertical cross-sections are marked inside and outside the current filament in Fig. 5(b). The electric-field peak shifts from the emitter side to the collector side [7]. As the current constricts in the current filament, the electron and hole densities are 1.5 orders of magnitude higher in comparison to the outside filament cut. The simulation shows a correlation between the lateral distance between the filaments and the width of the quasi-plasma region. The width of the high-field region defines the diameter of the current filament [13, 14], see Fig. 5(b).

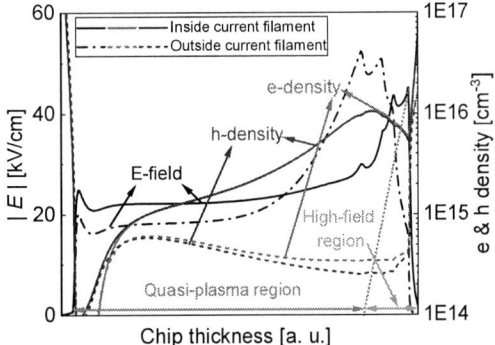

Fig. 6: Absolute value of the electric-field strength, electron and hole density distribution in the IGBT during SC-II simulation at $V_{DC\text{-link}}$ = 500 V, $I_{C,sat}$ = 1135 A, T_{start} = 300 K at 5.6 µs; vertical cuts are marked in Fig. 5(b).

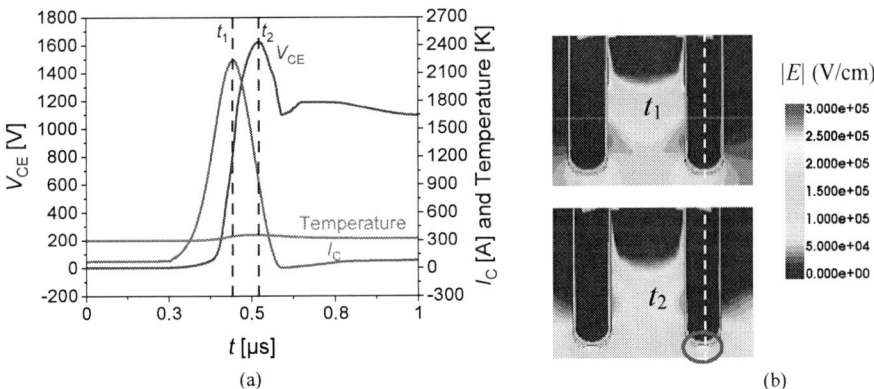

(a)   (b)

Fig. 7: (a) Simulated SC-II pulse at $V_{DC\text{-link}}$ = 1100 V, $I_{C,peak}$ = 2235 A, T_{start} = 300 K, L_{par} = 30 nH, L_{load} = 82 µH, R_G = 10 Ω, t_{SC} = 5 µs. (b) Absolute electric-field strength at time t_1 ($I_{C,peak}$) and t_2 ($V_{CE,peak}$).

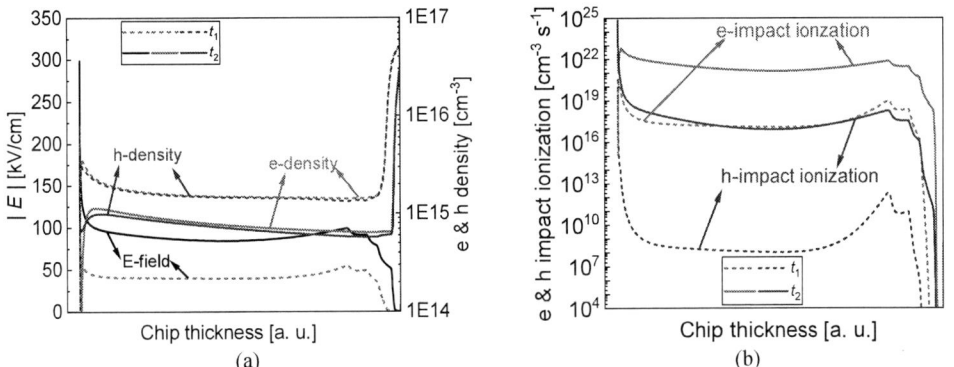

(a)   (b)

Fig. 8: Absolute value of the electric-field strength, electron and hole density distribution in IGBTs obtained by an SC-II simulation at $V_{DC\text{-link}}$ = 1100 V, T_{start} = 300 K at time point t_1 and t_2, (b) Electron and hole impact ionization rates at time point t_1 and t_2; vertical cuts are marked in Fig. 7(b).

The simulated SC-II transients for $V_{DC\text{-link}} = 1100$ V, $I_{C,peak} = 2235$ A, are shown in Fig. 7(a). At time t_1 during the $I_{C,peak}$, the maximum electric field can be found in the channel region, with a value of about 0.3 MV/cm. The vertical distributions of the absolute value of the electric-field strength, electron density and hole density in the IGBT structure at the time t_1 and t_2 for the given SC-II conditions are shown in Fig. 8(a) and the corresponding impact ionization rates of electrons and holes are plotted in Fig 8(b). The positions of the vertical cross-sections through the gate trench are marked in Fig. 7(b). During the time t_1, the IGBT undergoes only a very weak dynamic avalanche.

At time t_2 with $V_{CE} = V_{CE,peak}$, the plasma is reduced, and the electric field expands further into the drift region. The charge carrier concentration in the space-charge region decreases accordingly, as shown in Fig. 8(a). Strong dynamic avalanche occurs, and the maximum electric field now moves below the gate trench, showing a value of approximately 0.3 MV/cm. This exceeds the critical electric field value for impact ionization [Fig. 8(b)] as a consequence of the higher current and voltage stress, resulting eventually in s failure during the negative di_C/dt phase. Furthermore, the appearance of emitter-side filaments is expected during the negative di_C/dt phase, as explained in [16].

SC-II measurements without V_{GE} clamping and with L_{CE} influence

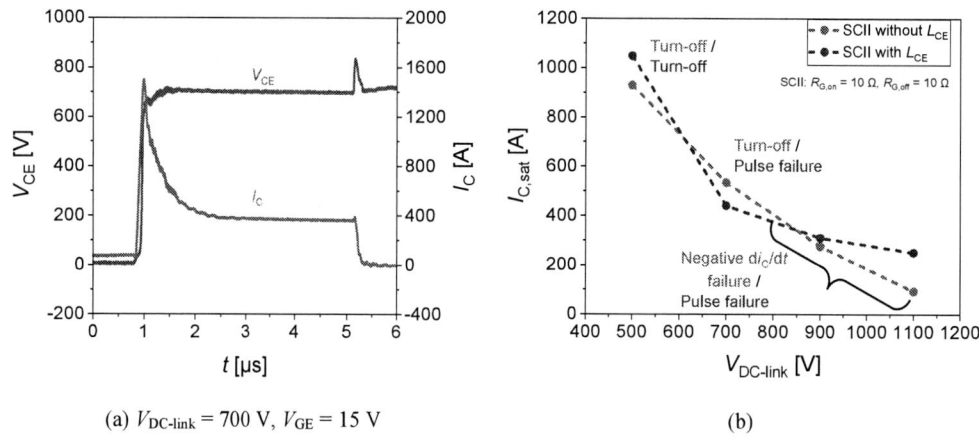

(a) $V_{DC\text{-link}} = 700$ V, $V_{GE} = 15$ V   (b)

Fig. 9: (a) Measured SC-II pulse without V_{GE} clamping and with L_{CE} influence and (b) comparison of the SC-II capability as a function of V_{CE} with and without L_{CE} at the following measurement conditions: $T_{start} = 300$ K, $L_{par} = 30$ nH, $L_{load} = 82$ µH, $R_G = 10$ Ω, $t_{SC} = 5$ µs.

Typical SC-II transients measured with L_{CE} influence and without V_{GE} clamping are shown in Fig. 9(a). During SC-II turn-on, the $V_{CE,peak}$ and $I_{C,peak}$ are significantly reduced by utilizing the negative feedback of the common-emitter inductance. The peak current is just 4 times the SC current at $V_{GE} = 15$ V (compare with Fig. 2b). The last-pass SC-II $I_{C,sat}$ values, with and without L_{CE}, are plotted as a function of $V_{DC\text{-link}}$ in Fig. 9(b). Even though the current peaks and voltage peaks are notably lower with L_{CE} influence, there is only a slightly increased SC-II ruggedness for DC-link voltages 900 V and 1100 V with L_{CE} influence. Further, the failure type for 900 V and 1100 V has changed from negative di_C/dt failure to pulse failure. At $V_{CE} = 700$ V, SC-II capability is reduced slightly for L_{CE}, and the failure type has changed from a turn-off failure to a pulse failure for L_{CE}.

Overall, under the SC II conditions without V_{GE} clamping and with common-emitter inductance influence, there is no significant improvement regarding the SC-II SOA. Therefore, in the next step, we consider an approach for reducing the current and gate-voltage peak by means of a suitable clamping circuit.

SC-II SOA of the IGBT with V_{GE} clamping

Clamping circuit in the gate driver unit

The clamping circuit provides a solution by utilizing the negative feedback generated from the load emitter inductance, which is the result of the high di_C/dt of the short-circuit current. The clamping circuit used in this work was adapted from [17]. The clamping circuit connected between the gate and load emitter is shown in Fig. 10(a). The circuit consists of two fast Schottky diodes (D_1 and D_2) placed in opposite directions. A clamping capacitor

(C_{Clamp}) and a discharge resistor (R_2) in parallel are connected between the two Schottky diodes. Once the driver is powered, the clamping capacitor is charged to a voltage V_{Clamp}, that is approximately equal to the applied gate-emitter voltage. In normal operation, V_{Clamp} does not influence the output voltage V_{GE} due to the Schottky diode D_1.

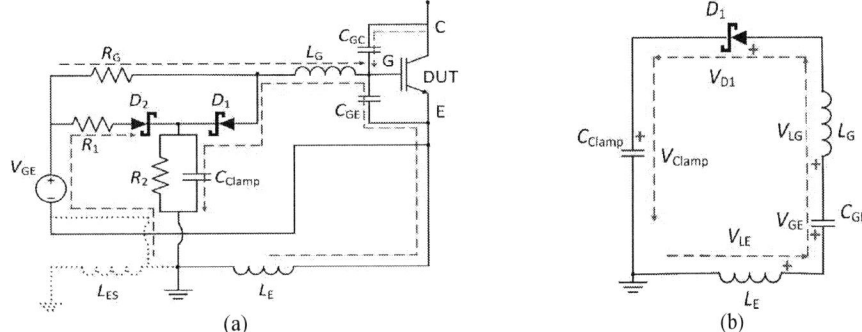

(a) (b)

Fig. 10: (a) Schematic diagram of a clamping circuit used in the GDU for SC-II measurements and (b) schematic diagram of a gate-emitter loop during the onset of a short-circuit II event.

The voltage drop across all the components in the gate-emitter loop during the SC-II event is displayed in Fig. 10(b). The voltage drop V_{LE} across the load emitter inductance L_E acts as negative feedback on the gate voltage, and is responsible for the clamping effect. The resistance R_1 limits the current from the output of the voltage regulator. The D_1 diode blocks any current, which is caused by the emitter stray inductance L_{ES} (for packaged modules), from flowing back to the gate driver unit (GDU). Due to the induced voltage, current passes through the D_1 into the clamping capacitor (C_{Clamp}). The D_2 diode blocks any from current flowing back into the voltage regulator [17].

$$V_{GE} = V_{Clamp} + V_{LG} + V_{D1} - V_{LE} \qquad (1)$$

The relationship between V_{GE} and the voltage drops across the components in the gate-emitter loop, including the clamping components, as seen in Fig. 10(b), is described in Eq. 1. The voltage across the clamping capacitor V_{Clamp} is constant, and is very small across the Schottky diode V_{D1}, since it is in the conducting state. Hence, the voltage drop across the load-emitter inductance (V_{LE}) pulls the gate-emitter voltage V_{GE} down, resulting in a clamping effect. The voltage drop across the V_{LE} is a function of the value of L_E and di_C/dt [17].

SC-II measurements with V_{GE} clamping

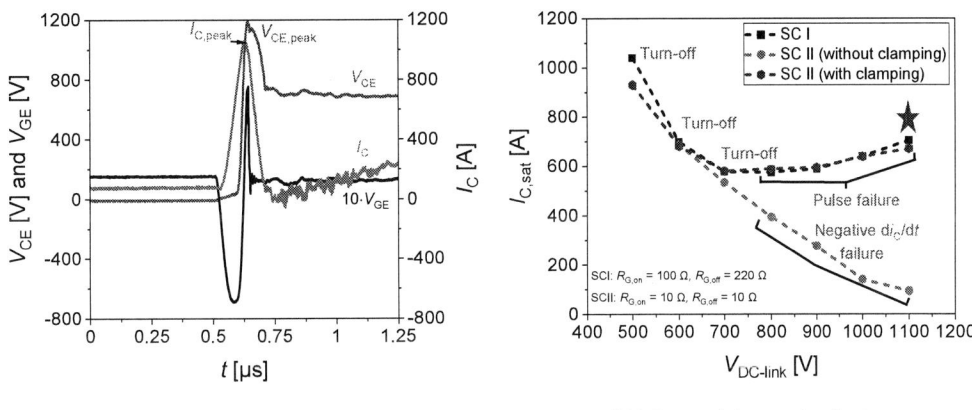

(a) With clamping at V_{GE} = 15 V and V_{CE} = 700 V (b) Measured destruction limit

Fig. 11: (a) A typical SC-II measurement with clamping and (b) measured destruction limit in I_C-$V_{DC-link}$ phase space for SC-II measurements with and without V_{GE} clamping for a 1.7 kV single chip IGBT under the following test conditions: T_{start} = 300 K, L_{par} = 30 nH, L_{load} = 82 µH, R_G = 10 Ω, t_{SC} = 5 µs.

In Fig. 11(a), a typical SC-II measurement with V_{GE} clamping is plotted for $V_{GE} = 15$ V and $V_{CE} = 700$ V. The SC-II measurement with V_{GE} clamping [Fig. 11(a)] shows a significantly reduced collector-current peak ($I_{C,peak}$) during turn-on, with a magnitude of ~2.7 times lower compared to the SC-II measurements without V_{GE} clamping [Fig. 2(b)] under the same test conditions. The clamping is not perfect, as some parasitics are still present between the clamping capacitor and the chip gate (usually internal gate resistor, gate inductance). The $V_{CE,peak}$ is higher with V_{GE} clamping due to the steeper negative di_C/dt but still acceptable. Additionally, the SC-II measurement with V_{GE} clamping shows a self-turn-off of the IGBT after the negative di_C/dt [18]. It should be noted that the plotted V_{GE} is the measured signal of the outer gate terminal, and does not correspond to the internal gate-emitter voltage. Therefore, the current peak value $I_{C,peak}$ is the better characteristic value to assess the effectiveness of the clamping circuit.

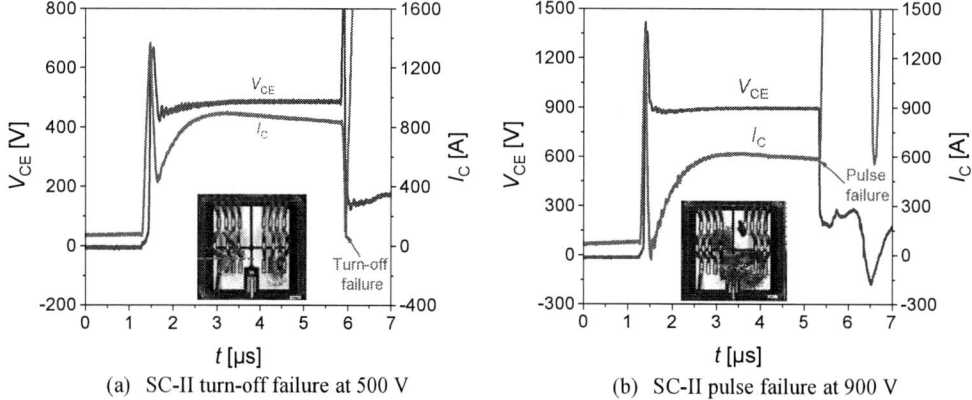

(a) SC-II turn-off failure at 500 V   (b) SC-II pulse failure at 900 V

Fig. 12: Measured SC-II failure types with V_{GE} clamping at given conditions $T_{start} = 300$ K, $L_{par} = 30$ nH, $L_{load} = 82$ µH, $R_G = 10$ Ω, $t_{SC} = 5$ µs. Inset: microscopic image of the destroyed chip.

The SC-II SOA measurements with and without V_{GE} clamping are compared in Fig. 11(b). There is a significant improvement in the SC-II ruggedness of the IGBT when the clamping circuit is used at the GDU, especially at DC-link voltages exceeding 700 V. The destructive boundary line for the SC-II with V_{GE} clamping is almost identical to that of SC-I SOA. The minima can be seen at a V_{CE} of 700 V with a corresponding last-pass saturation current of 589 A for the 1700 V IGBT used. For a DC-link voltage range from 500 V to 700 V, the IGBT fails during the SC-II turn-off. However, instead of a negative di_C/dt failure, a pulse failure is observed for a V_{CE} range of 800 V to 1100 V with V_{GE} clamping in SC-II measurements. The two insets in Fig. 12(a) and (b) show the corresponding chip microscopic images after the SC-II failure.

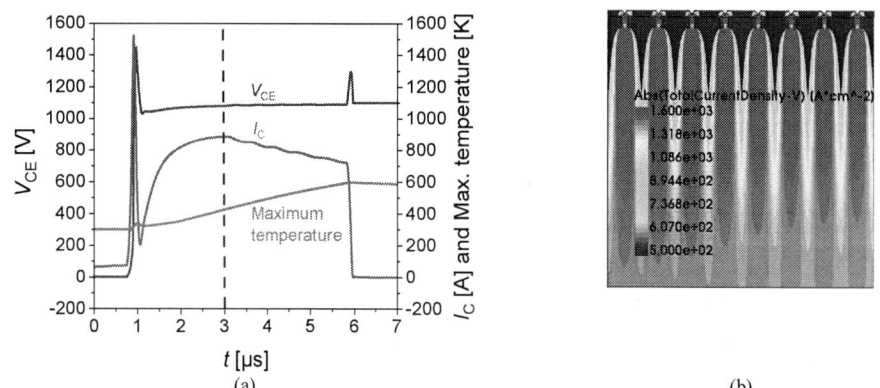

(a)   (b)

Fig. 13: (a) Simulated SC-II at $V_{CE} = 1100$ V with simulated conditions: $T_{start} = 300$ K, $L_{par} = 30$ nH, $L_{load} = 82$ µH, $R_G = 10$ Ω, $t_{SC} = 5$ µs (b) current density distribution at 3 µs for a 1700 V IGBT.

For DC-link voltages from 500 V to 700 V, the IGBT fails during the SC-II turn-off due to the formation of destructive current filaments. At a V_{CE} range from 900 V to 1100 V, the clamped SC-II measurements show pulse failures in contrast to a failure during the negative di_C/dt in SC-II without V_{GE} clamping. The pulse failures above

DC-link voltages of 800 V [Fig. 11(b)] are consistent with the failure signature of the SC-I tests. The change in failure type can be due to the reduced magnitude of the collector current peak and the self-turn-off in the IGBT. Additionally, the duration of very high V_{CE} across the IGBT shortly after the $I_{C,peak}$ is reduced to 140 ns for clamped V_{GE} measurements compared to the 500 ns for the SC-II measurements without V_{GE} clamping (see Fig. 2b).

To gain a better understanding of the observed pulse failure, an SC II simulation with V_{GE} clamping was performed at a $V_{DC\text{-}link} = 1100$ V, $I_{C,sat} = 800$ A; the position is marked with the blue star in Fig. 11(b). The simulated V_{CE}, I_C and maximum temperature transients are shown in Fig. 13(a). Fig. 13(b) shows the absolute total current density in the IGBT during the SC-II pulse at 3 µs. As the space-charge region at higher V_{CE} voltages expands further towards the collector-side, the current distribution becomes more uniform in the IGBT compared to the conditions with lower DC-link voltages. Hence, a higher current is necessary to generate a plasma in the drift region. Consequently, the short-circuit ruggedness increases above 800 V as shown in Fig. 11(b) [7]. For the SC II simulated condition at $I_{C,sat} = 800$ A, which is above the measured destructive boundary, $I_{C,sat}$ shows only a slight inhomogeneous current flow at the anode side in the IGBT, as shown in Fig. 13(b).

Failure analysis

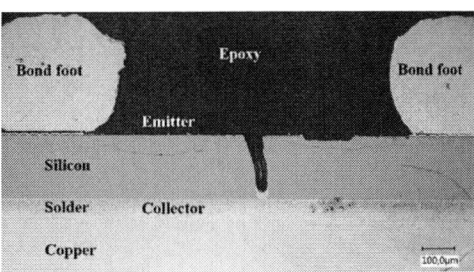

Fig. 14: IGBT cross-section during destructive SC-II turn-off event at $V_{CE} = 500$ V, $T_{start} = 300$ K, $L_{par} = 30$ nH, $L_{load} = 82$ µH, $R_G = 10$ Ω, $t_{SC} = 5$ µs [cut position: as shown in inset Fig. 12(a)].

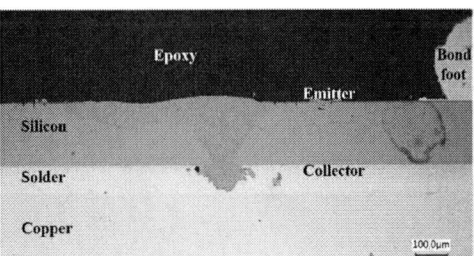

Fig. 15: IGBT cross-section during destructive SC-II negative di_C/dt events at $V_{CE} = 900$ V, $T_{start} = 300$ K, $L_{par} = 30$ nH, $L_{load} = 82$ µH, $R_G = 10$ Ω, $t_{SC} = 5$ µs [cut position: as shown in inset Fig. 4(b)].

In order to analyze the failures due to turn-off and negative di_C/dt, two of the damaged IGBT devices were cut at the precise position where they were destroyed. Fig. 14 and Fig. 15 show the cross-section of the devices after a single destructive SC-II pulse at 500 V and 900 V, respectively. The cross-section in Fig. 14 corresponds to the destroyed chip and plot shown in Fig. 12(a). The IGBT was cut at the position marked by red dashed-line. Similarly, the IGBT cross-section illustrated in Fig. 15, cut at the position where the destructive SC-II negative di_C/dt event occurred corresponds to the destroyed chip and plot shown in Fig. 4(b). The figures displayed in Fig. 14 and Fig. 15 correspond to the filamentation process failures at the collector side and at the emitter side respectively. Also, the destruction on the chip depends on how fast the protection IGBT (PIGBT) reacts after the device under test fails during short-circuit.

Conclusion

The measured SC-II SOA with and without V_{GE} clamping was compared to that of the SC-I SOA. For the SC-II measurements without V_{GE} clamping, the critical saturation collector current decreases continuously with increased

DC-link voltage. For a V_{CE} ranging from 500 V to 700 V, the destruction of the IGBT occurs during the SC turn-off, which is the same as for SC-I. Here, the formation of collector-side current filaments triggers the failures. For DC-link voltages above 700 V, the IGBTs fail during the negative di_C/dt phase. In these cases, the IGBTs undergo strong dynamic avalanche due to the high $V_{CE,peak}$ and $I_{C,peak}$ values occurring during SC-II. Furthermore, it could be shown that the SC-II SOA could be significantly improved by using V_{GE} clamping. As a result of V_{GE} clamping, the SC-II SOA is approximately same as the SC-I SOA.

References

1] J. Lutz, H. Schlangenotto, U. Scheuermann, and R. De Doncker, "Semiconductor Power Devices: Physics, Characteristics, Semiconductor Power Devices: Physics, Characteristics, Reliability," 2nd ed. Berlin, Germany: Springer-Verlag, 2018.

[2] H.-G. Eckel, L. Sack, "Experimental Investigation on the Behaviour of IGBT at Short-Circuit during the On-State," 20th International Conference on Industrial Electronics, Control and Instrumentation, IECON, pp. 118-123, 1994.

[3] J. Lutz, T. Basler, "Short-Circuit Ruggedness of the High-Voltage IGBTs" in Proc. 28th International Conference on Microelectronics (MIEL), pp. 243-250, May, 2012.

[4] R. Letor, C.G. Aniceto, "Short circuit behaviour of IGBTs correlated to the intrinsic device structure and on the application circuit," IEEE Industry Applications Society Annual Meeting, 1992.

[5] D. Hammes, et al. "High-inductive short-circuit Type IV in multi-level converter protection schemes," 42nd Annual Conference of the IEEE Industrial Electronics Society, IECON 2016.

[6] J. Fuhrmann, S. Klauke, H.G. Eckel, "Passive IGBT turn-off during short-circuit type V" in Proc. Conf. International Exhibition and Conference for Power Electronics Intelligent Motion, Power Quality, Nuremberg, 2016.

[7] A. Kopta, M. Rahimo, U. Schlapbach, et al., "Limitation of the Short-Circuit Ruggedness of High Voltage IGBTs," in Proc. ISPSD, 2008, pp. 33-36.

[8] R. Baburske, V. Treek, F. Pfirsch, F. J. Niedernostheide, et al., "Comparison of Critical Current Filaments in IGBT Short Circuit and during Diode Turn-off," in Proc. ISPSD, 2014, pp. 47-50.

[9] M. Tanaka, A. Nakagawa, "Simulation studies for Avalanche induced short-circuit current crowding of MOSFET-Mode IGBT," in Proc. ISPSD, 2015, pp. 121-124.

[10] R. Bhojani, J. Kowalsky, J. Lutz, et al., "Observation of Current Filaments in IGBTs with Thermo-Reflectance Microscopy," in Proc. ISPSD, May 2018, pp. 164-167.

[11] R. Baburske, F. J. Niedernostheide, H.-J. Schulze, et al., "Unified view on energy and electrical failure of the short-circuit operation of IGBTs," Microelectronics Reliability, 2018, Vol. 88-90, pp. 236-241.

[12] M. L. Mysore, R. Bhojani, J. Kowalsky, et al., "Al Modification as Indicator of Current Filaments in IGBTs under repetitive SC operation," Transactions on IET power electronics, ISSN 1755-4535, pp. 1-10, 2019.

[13] R. Bhojani, M. L. Mysore, R. Raihan et al, "Current Filament Behavior in Different Voltage Class IGBTs using Measurements and Simulations," in Proc. ISPSD, Sep. 2020, pp. 446-449.

[14] M. L. Mysore, T. Basler, J. Lutz et al., "Aluminum Modification as Indicator for Current Filaments under Repetitive Short-Circuit in 650V IGBTs," in Proc. of 15th ISPS, pp. 21-29, 2021.

[15] SYNOPSYS, *Sentaurus Device User Guide*, 2020.

[16] T. Basler, "Ruggedness of high-voltage IGBTs and protection solutions," PhD thesis, Technische Universität Chemnitz, 2014, page 94.

[17] X. Liu, J. Kowalsky, C. Herrmann, et al., "Influence of the gate driver and common-source inductance on the SC behavior of IGBT modules," IEEE transactions on power electronics, vol. 35, no. 10, October 2020.

[18] T. Basler, J. Lutz, T. Brückner, et al., "IGBT Self-Turn-Off Under Short-Circuit Condition," in Proc. Int. Conf. 10th International Seminar on Power Semiconductors (ISPS), Prague, 2010.

Single Transformer, MMC based MV Power Electronic Traction Transformer

Simon Fuchs, Simon Beck, Jürgen Biela
Laboratory for High Power Electronic Systems (HPE)
ETH Zürich, Switzerland
Email: becksi@ethz.ch, jbiela@ethz.ch

Keywords

≪Solid-State Transformers≫, ≪Modular Multilevel Converters (MMC)≫, ≪Traction Application≫

Abstract

To reduce the number of modules and the number of transformers of existing power electronic traction transformers (PETTs), a novel PETT topology based on the principle of the modular multilevel converter (MMC) and the dual active bridge (DAB) is introduced. The paper explains the operation principle of the proposed PETT and compares the PETT to two existing concepts.

1 Introduction

Distributed traction drives in modern passenger train concepts benefit from low volume and weight alternatives to classical concepts requiring a transformer operating at grid frequency (16 2/3 or 50 Hz). Such a transformer is inherently large in volume and often has a relatively efficiency, especially at partial loading of the traction system. In contrast, medium frequency transformers (MFT) can be designed to achieve a much lower volume with even higher efficiencies [1, 2]. However, a power electronic converter is required to generate the medium frequency AC input voltage for any MFT. In the following, such a combination of one or more MFT with a power electronic converter is called power electronic traction transformer (PETT) [3, 4].

The grid voltages in the railway power system are typically in the range of 15-25 kV, such that multilevel converters are required on the grid side of the PETT. A promising candidate to realize such a multilevel converter is the concept of the modular multilevel converter (MMC). MMCs are based on modules which are typically implementing an H-bridge in case of bidirectional/AC voltage output. Each of these modules has its own DC-link capacitor and the modules are connected in series to compose a so called MMC arm. Because malfunctioning modules can simply be bypassed during operation, the MMC also offers a high availability, because the $(n-1)$ principle can be applied without extensively over-dimensioning the converter [5, 6].

In literature, there are two main concepts for PETTs based on MMCs [7]:

- In [8], an MMC with four arms is used to feed the primary side of an MFT. The secondary side is used to supply a traditional traction drive inverter. The four arm structure results in a rather high control effort (relatively complicated modulation, energy balancing control, etc.), which is generally not beneficial.

- With the concept presented in [3, 4], only a single MMC arm connected between the line and ground is required. The module capacitor of each of the MMC modules is used as a DC link of an isolated LLC resonant DC/DC converter. The outputs of the LLC resonant converters are connected in parallel to feed a traditional traction drive inverter. This means, that there is one resonant

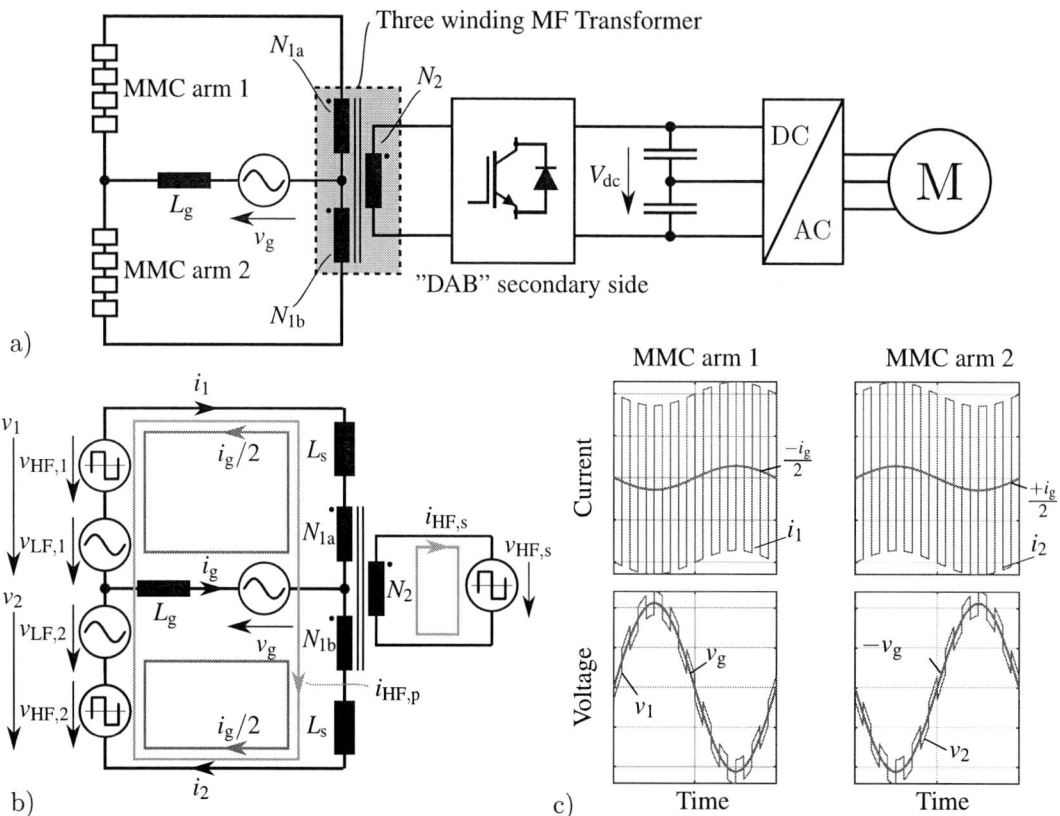

Fig. 1: a) Basic layout of the proposed PETT topology. b) Equivalent circuit of the PETT with the used currents and voltages. c) Basic waveforms for the proposed PETT if the amplitude of $i_{\mathrm{HF,p}}$ is larger than half the grid current amplitude.

converter (including an MFT) for each MMC module. Therefore, the isolation requirements have to be fulfilled for every single module/MFT, which can result in a comparably high volume per module. On the one hand, this enables to use standard control and design schemes for the resonant converters in each module. On the other hand, it also significantly increases the system complexity, if high line voltages and therefore a high number of modules are considered.

In this paper, a concept for a PETT based on the MMC concept as shown in [8] is presented. In the presented concept, an MMC structure is used to control the grid current and to generate the primary voltage for a single three winding MFT operated as a dual-active bridge (DAB). However, only two MMC arms (instead of four in the concept from [8]) are required for the proposed concept, such that the number of required modules is lower as will be shown in the paper. Moreover, the converter can be operated in such a way, that ZVS soft-switching is always achieved. This potentially increases the efficiency compared to [3, 4], where soft-switching can only be achieved in the LLC resonant converters, while the MMC front-end operates with hard-switching.

The paper is structured as follows: First, the basic structure and operation principles of the proposed PETT are introduced and verified with simulation results. Thereafter, equations for the installed semiconductor power as well as the required energy storage in the module capacitance are derived. Finally, the proposed concept is compared with the PETT concepts presented in [3, 4] and [8].

2 Basic Converter Structure and Operation Principle

The proposed MMC based PETT consists of two MMC arms with N H-bridge modules, a medium frequency transformer (MFT) with three windings (N_{1a}, N_{1b} and N_2) as well as a full-bridge operating at the DC-link voltage of the drive inverter connected to the third winding of the MFT as shown in Fig. 1a. The MMC arms are used to implement two voltages as shown in Fig. 1b: First, a low frequency voltage v_{LF} to compensate the grid voltage with the output voltage of each of the MMC arms and control the grid current. Second, each of the two MMC arms additionally implements the primary side of a classical dual-active bridge converter (DAB) with one or more modules providing a high frequency voltage v_{HF} to the windings N_{1a} and N_{1b} of the MFT. The secondary side of the DAB is realised with a full-bridge connected to winding N_2 of the MFT (Fig. 1b).

Basic waveforms for the proposed topology are shown in Fig. 1c). In the considered design, the amplitude of the HF part of the MMC arm currents i_1 and i_2 is larger than the amplitude of their low frequency component that corresponds to half of the grid current. Therefore, soft-switching can be achieved in all modules.

The voltage across the windings N_{1a} and N_{1b} must be synchronous without a phase-shift, such that $v_{HF,1} = v_{HF,2}$ is required. This avoids a power flow from N_{1a} to N_{1b} or vice versa.

Note that the windings N_{1a} and N_{1b} of the MFT also conduct half of the grid current i_g (Fig. 1c). However, because the generated magnetomotive force (MMF) $N_{1a} \cdot i_g$ is negative and $N_{1b} \cdot i_g$ is positive, the grid current i_g does not generate a magnetic flux in the transformer core and also no power transfer to the secondary winding N_2.

In the following, the equations for the MMC arm currents and voltages as well as the currents and voltages of the secondary side as defined in Fig. 1b) are derived. The DAB is operated with 50 % duty cycle, such that the power flow control is done via the phase shift φ_{HF} between the primary voltage $v_{HF,1/2}$ and the secondary voltage V_{dc} of the MFT [9].

2.1 Converter Voltages

The output voltages of MMC arm 1 and 2 are composed of the low frequency grid voltage oscillating with ω_g plus the high frequency rectangular transformer voltage toggling with f_{HF} for the DAB as shown in Fig. 1b:

$$v_1 = \underbrace{-V_g \cdot \sin(\omega_g t)}_{v_{LF,1}} + \underbrace{V_{HF} \cdot \text{rect}(f_{HF} \cdot t)}_{v_{HF,1}}, \qquad v_2 = \underbrace{+V_g \cdot \sin(\omega_g t)}_{v_{LF,2}} + \underbrace{V_{HF} \cdot \text{rect}(f_{HF} \cdot t)}_{v_{HF,2}}, \qquad (1)$$

$$\text{where} \quad \text{rect}(f_{HF} \cdot t) = \begin{cases} +1 & \text{if } (f_{HF} \cdot t \text{ mod } 1.0) \in [0, 0.5[\\ -1 & \text{if } (f_{HF} \cdot t \text{ mod } 1.0) \in [0.5, 1.0[\end{cases} \quad \text{with 50\% duty cycle.}$$

v_1 and v_2 are plotted in the lower part of Fig. 1c. For a given secondary side voltage amplitude V_{dc} and a specific transformer turns ration n, the amplitudes of the HF part of the MMC arms' voltages are $V_{HF} = n \cdot V_{dc}$. The secondary side voltage $v_{HF,s}$ is expressed as

$$v_{HF,s} = V_{dc} \cdot \text{rect}(f_{HF} \cdot t - \varphi_{HF}), \qquad (2)$$

where φ_{HF} is the phase shift to control the power flow as shown in Fig. 2.

2.2 Converter Currents

The MMC arm currents are also composed of a low frequency grid current component and a high frequency DAB component (cf. Fig. 1b):

$$i_1 = -i_{LF} + i_{HF,p}, \qquad i_2 = +i_{LF} + i_{HF,p}, \qquad (3)$$

where $i_{LF} = i_g/2 = P/V_g \cdot \sin(\omega_g t)$ with the transferred real power P.

The waveform of HF part of the current in general strongly varies with the type of the chosen modulation of the DAB [10, 11, 12]. However, assuming that the DAB is operated with 50 % duty cycle and that the ratio of the applied rectangular voltages $V_{HF,p}/V_{dc}$ are close to the transformer turns ratio n, the DAB transformer currents have a trapezoidal shape as shown in Fig. 2. A phase shift φ_{HF} is applied between the two primary voltages and the secondary side voltage in order to control the power transfer

$$P_{\mathrm{HF}} = \frac{n^2 \cdot V_{\mathrm{dc}} \cdot \varphi_{\mathrm{HF}} \cdot (\pi - |\varphi_{\mathrm{HF}}|)}{\pi \cdot f_{\mathrm{HF}} \cdot L_{\mathrm{s}}}. \tag{4}$$

At rated power P_{r}, this phase shift is commonly chosen to be lower than its absolute maximum $\pi/2$ - e.g. around $\varphi_{\mathrm{HF,r}} = \pi/4$. Note that with (4) the choice of $\varphi_{\mathrm{HF,r}}$ also defines the required series inductance

$$L_{\mathrm{s}} = \frac{n^2 V_{\mathrm{dc}}^2 \cdot \varphi_{\mathrm{HF}} \cdot (\pi - \varphi_{\mathrm{HF}})}{2 P_{\mathrm{r}} \pi^2 f_{\mathrm{HF}}} = \frac{3 n^2 V_{\mathrm{dc}}^2}{16 P_{\mathrm{r}} f_{\mathrm{HF}}}. \tag{5}$$

This series inductance includes the transformer leakage inductance plus an optional dedicated inductor. The phase shift also defines the time interval $\varphi_{\mathrm{HF}}/2\pi f_{\mathrm{HF}}$ where the primary and secondary voltages are in opposite directions, such that the voltage $2nV_{\mathrm{dc}}$ is applied across the inductance L_{s} (cf. Fig. 2). The primary side HF current amplitude therefore compiles to

$$I_{\mathrm{HF,p}} = \frac{\varphi_{\mathrm{HF}}}{2\pi f_{\mathrm{HF}}} \cdot \frac{n V_{\mathrm{dc}}}{L_{\mathrm{s}}} = \frac{8 P_{\mathrm{r}}}{3 n V_{\mathrm{dc}}} \cdot \frac{\varphi_{\mathrm{HF}}}{\pi}. \tag{6}$$

At rated power with $\varphi = \pi/4$ this results in $I_{\mathrm{HF,p}} = 2P_{\mathrm{r}}/3nV_{\mathrm{dc}}$. The amplitude of the secondary current $i_{\mathrm{HF,s}}$ is given with $I_{\mathrm{HF,s}} = 2n \cdot I_{\mathrm{HF,p}}$ due to the two primary windings.

2.3 Transformer turns ratio

The transformer turns ratio $n = N_{1a}/N_2 = N_{1b}/N_2$ has a big impact on the operation and the design of the proposed converter: For low turns ratios, the amplitude of the HF part $I_{\mathrm{HF,p}}$ of the current in the MMC arms is relatively high. This results in a zero crossing of the currents in the MMC arms which can be used to achieve zero-voltage-switching in the MMC arms to significantly reduce the switching losses especially if SiC MOSFETs are used to implement the MMC modules.

To obtain this current zero crossing, the primary side HF current amplitude $I_{\mathrm{HF,p}}$ must be higher than the low frequency current amplitude $I_{\mathrm{g}}/2$:

$$k \cdot I_{\mathrm{HF,p}} = k \cdot \frac{2P}{3n V_{\mathrm{dc}}} > \frac{I_{\mathrm{g}}}{2} = \frac{P}{V_{\mathrm{g}}} \qquad \Leftrightarrow \qquad n_{\mathrm{ZVS}} < k \cdot \frac{2 V_{\mathrm{g}}}{3 V_{\mathrm{dc}}} \tag{7}$$

Here, $k < 1$ is a safety factor to ensure enough current though L_{s} to safely fulfil the ZVS conditions [13]. Note that this is only true at rated power, because the phase shift φ_{HF} and therefore also the current amplitude does not scale linearly with the transmitted power P. Therefore, to achieve ZVS also at lower power outputs, e.g. the switching frequency [10] or the modulation scheme [11, 12] can be adopted.

2.4 Semiconductor Requirements

As explained before, both MMC arms must allow a bidirectional output voltage. Therefore, only full-bridge modules are feasible. The required number of modules N depends on the module voltage V_{m} and can be calculated based on the maximum arm voltage (1) to

$$N = 2 \cdot \lceil (V_{\mathrm{g}} + n \cdot V_{\mathrm{dc}})/V_{\mathrm{m}} \rceil \tag{8}$$

including both MMC arms which results in the factor of two. The total number of switches is $4N + 4$, if full-bridge modules are assumed including the secondary side converter.

The installed semiconductor power for the MMC arms P_{sc} results from the number of modules N, the module voltage V_{m} and the maximum MMC arm current:

$$P_{\mathrm{sc,p}} = 4N \cdot V_{\mathrm{m}} \cdot (I_{\mathrm{HF,p}} + P/V_{\mathrm{g}}). \tag{9}$$

The full-bridge on the secondary side requires four switches operating at V_{dc} with a maximum current

Table I: PETT parameters

Variable	Parameter	Value(s)
P	Transmitted power	$3\,\mathrm{MW}$
V_{g}	Grid voltage	$\sqrt{2} \cdot 15\,\mathrm{kV}$
V_{dc}	Secondary side voltage of the transformer(s)	$3\,\mathrm{kV}$
V_{m}	Rated module voltage	$1\ldots5\,\mathrm{kV}$
n	Transformer turns ratio	$0.95 \cdot 2\,V_{\mathrm{g}}/3V_{\mathrm{dc}}$
L_{s}	Series inductance	$2.8\,\mathrm{mH}$
β	Steinmetz core loss parameter	2.5

given by $2n \cdot I_{\mathrm{HF,p}}$ in (6). This results in in

$$P_{\mathrm{sc,s}} = 4 \cdot V_{\mathrm{dc}} \cdot 2n \cdot I_{\mathrm{HF,p}} = 4 \cdot \frac{4P}{3}. \tag{10}$$

2.5 Required Energy Storage (Module Capacitors)

Because the proposed PETT topology basically implements a single phase AC to DC converter, the pulsating power from the single phase grid must be intermediately stored by the converter to allow a constant DC power output. Therefore, the MMC arms require a certain module capacitance value in order to provide an energy buffer. Here, the power drawn by the DAB converter on the secondary side can be assumed to be a DC power, because the operating frequency is typically two orders of magnitude higher than the AC grid frequency. The instantaneous energy fluctuation $e(t)$ of one of the MMC arms can be calculated as follows:

$$e_{\mathrm{arm}}(t) = \int_0^t P_{\mathrm{LF}}(t) - P_{\mathrm{HF}} dt = \int_0^t \frac{V_{\mathrm{g}} \cdot I_{\mathrm{g}}}{2} \sin^2(\omega_{\mathrm{g}}t) - \frac{P}{2} dt = \int_0^t P \sin^2(\omega_{\mathrm{g}}t) - \frac{P}{2} dt \tag{11}$$

$$= -\frac{P}{4\omega_{\mathrm{g}}} \sin(2\omega_{\mathrm{g}}t) \tag{12}$$

The peak-to-peak value of the instantaneous energy fluctuation $e_{\mathrm{arm}}(t)$ defines energy variation $\Delta e_{\mathrm{arm}} = P/2\omega_{\mathrm{g}}$ of the MMC arms. This peak-to-peak energy variation can be used to derive the required module capacitance value C_{m} with

$$\frac{\Delta e_{\mathrm{arm}}}{N} \leq \frac{1}{2}C_{\mathrm{m}}(V_{\mathrm{m}} + \Delta v_{\mathrm{m}})^2 - \frac{1}{2}C_{\mathrm{m}}(V_{\mathrm{m}} - \Delta v_{\mathrm{m}})^2 \quad \Leftrightarrow \quad C_{\mathrm{m}} \geq \frac{\Delta e_{\mathrm{arm}}}{2N \cdot V_{\mathrm{m}} \cdot \Delta v_{\mathrm{m}}} = \frac{P}{4N \cdot \omega_{\mathrm{g}} \cdot V_{\mathrm{m}} \cdot \Delta v_{\mathrm{m}}}, \tag{13}$$

where Δv_{m} is the allowed module voltage deviation from the rated module voltage V_{m}. The total required average energy storage E_{cap} is therefore

$$E_{\mathrm{cap}} = N \cdot C_{\mathrm{m}} \cdot V_{\mathrm{m}}^2 = \frac{V_{\mathrm{m}} \cdot P}{4 \cdot \omega_{\mathrm{g}} \cdot \Delta v_{\mathrm{m}}} \tag{14}$$

for all $2N$ modules of the converter. Note that for MMCs, more sophisticated methods to derive the minimum required module capacitance exist and the operation of the MMC has great impact on the required capacitance value [14, 15, 16]. However, the complexity of the control and modulation of the DAB part would strongly increase with a high module voltage fluctuation, such that this topic exceeds the scope of this paper.

2.6 Simulation results

Figure 2 displays simulation results for the proposed PETT concept. Table I contains the associated parameters used for the simulation model. The simulation is based on average models for the MMC arms as introduced in [17], such that the current ripple in the grid current is not present. It can be seen,

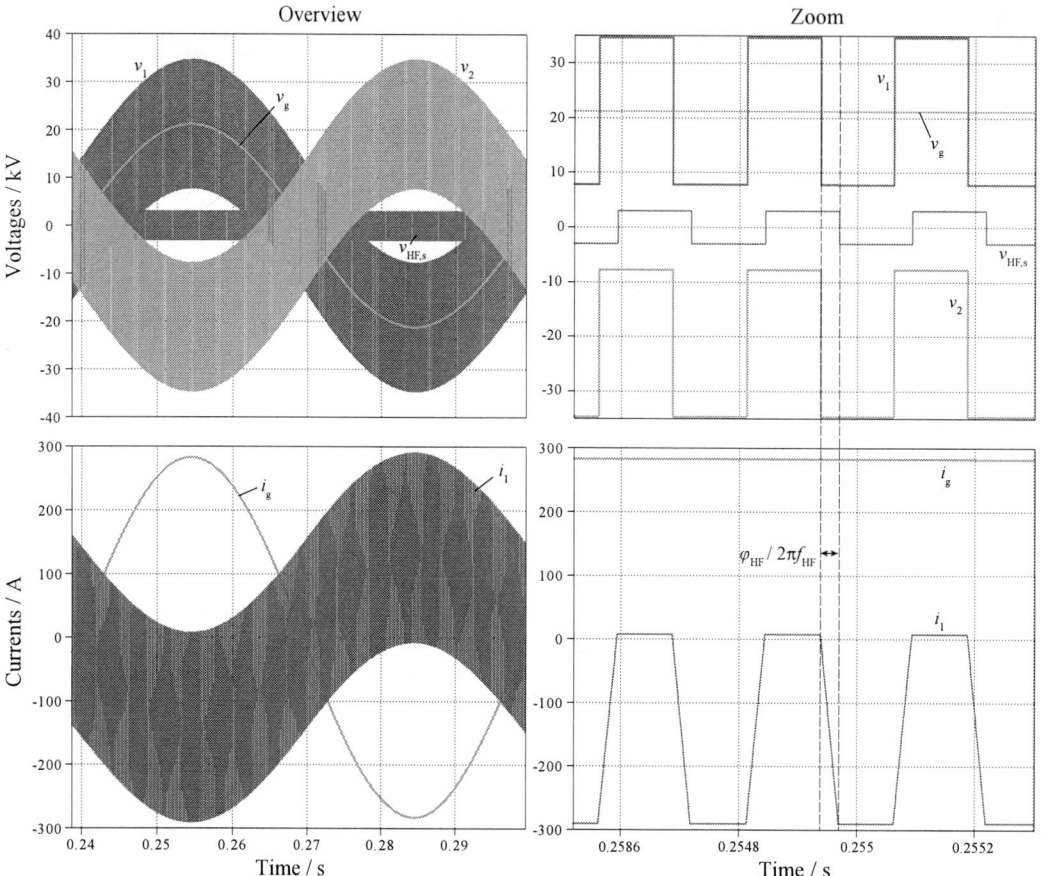

Fig. 2: Simulation results for the operation with $V_{dc} = 3\,\text{kV}$, $n = 4.5$, $f_{HF} = 4\,\text{kHz}$, $L_s = 2.8\,\text{mH}$ and $L_g = 10\,\text{mH}$. The phase shift φ_{HF} is $\pi/4$. Overview shown in the left, zoom-in shown in the right. The remaining system parameters are listed in Tab. I.

that the arm current i_1 has a zero crossing in each switching period throughout the complete grid period, such that ZVS is always achieved. This confirms the considerations in (7). Moreover, the HF current amplitudes also match with the equations presented before.

3 Comparison with existing PETT concepts

In the following, a comparison of the presented PETT concept with the concepts proposed in [3, 4] as depicted in Fig. 4 and [8] as depicted in Fig. 3 is conducted. The equations for the most important system parameters regarding cost and efficiency are derived and presented in Tab. II. Finally, a comparison is conducted for an exemplary parameter set as given in Tab. I.

3.1 Derivations for Concept proposed in [8]

The MMC arm voltage is

$$v_1(t) = V_g/2 \cdot \sin(\omega_g t) - n \cdot V_{dc}/2 \cdot \text{rect}(f_{HF} \cdot t). \tag{15}$$

Therefore the maximum MMC arm voltage results in $(V_g + V_{dc})/2$. The MMC arm current is composed of half the grid current and half the MFT current:

$$i_1(t) = i_g(t)/2 + i_{hf,s}(t)/2n \tag{16}$$

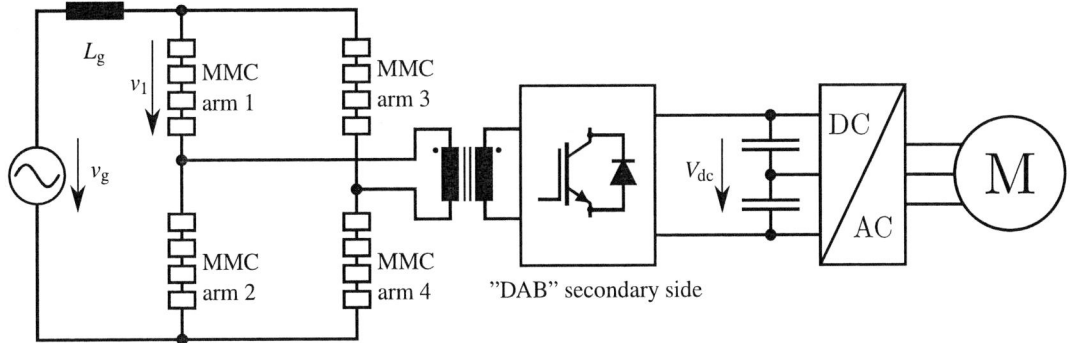

Fig. 3: MMC based PETT topology proposed in [8].

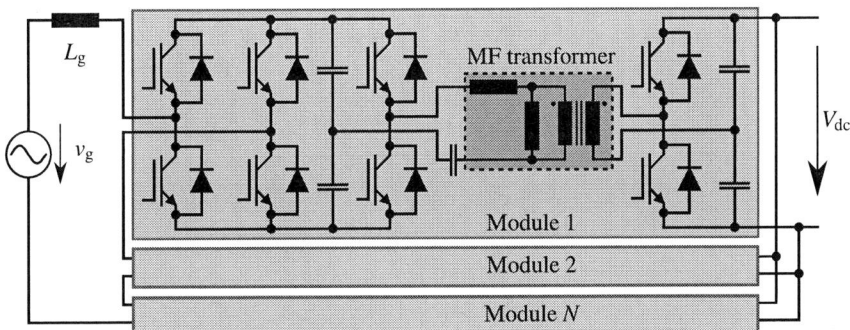

Fig. 4: MMC based PETT topology proposed in [3, 4].

Similarly, as explained for the PETT topology proposed in this paper (cf. (6)), the maximum arm current results in $I = P/V_\text{g} + 2P/3nV_\text{dc}$.

The required number of modules, switches and the required semiconductor power are simple to calculate from here. Derivations are omitted for space reasons. The formulas are given in Tab. II.

To achieve a zero crossing of the current in all arms and therefore ZVS, the amplitude of the HF part of the current must be greater than the grid current part in the MMC arms. Because the arm currents are half of the transformer HF current and half of the grid current, the same equations as for the topology proposed in this paper hold for the transformer turns ratio n_ZVS. Note that this is not mentioned in [8].

Concerning the required energy storage in the module capacitors, similar considerations as for the topology proposed in this paper can be applied: Since the power from the grid and to the MFT is shared equally among the four individual MMC arms, e_arm and its peak-to-peak value Δe_arm compile to

$$e_\text{arm} = \int_0^t P_\text{LF}(t) - P_\text{HF}dt = \int_0^t \frac{P}{2}\sin^2(\omega_\text{g}t) - \frac{P}{4}dt = -\frac{P}{8\omega_\text{g}} \quad \rightarrow \quad \Delta e_\text{arm} = \frac{P}{4\omega_\text{g}}. \tag{17}$$

Therefore, the total required average energy storage is the same as for the topology proposed in this paper:

$$C \geq \frac{P}{8N \cdot \omega_\text{g} \cdot V_\text{m} \cdot \Delta v_\text{m}} \quad \rightarrow \quad E_\text{cap} = 2 \cdot N \cdot C \cdot V_\text{m}^2 = \frac{V_\text{m} \cdot P}{4 \cdot \omega_\text{g} \cdot \Delta v_\text{m}}. \tag{18}$$

3.2 Derivations for Concept proposed in [3, 4]

With the concept proposed in [3, 4], the MMC arm voltage is equal to the grid voltage $V_\text{g} \cdot \sin(\omega_\text{g}t)$, such that the maximum arm voltage is V_g. Furthermore, the arm current is also equal to the grid current, such that the maximum arm current is $I_\text{g} = 2P/V_\text{g}$.

Table II: Comparison of basic PETT concepts

Parameter	This paper	[8]	[3, 4]
MMC arm max. voltage	$V_g + n \cdot V_{dc}$	$(V_g + n \cdot V_{dc})/2$	V_g
MMC arm max. current I	$P/V_g + 4P/3nV_{dc}$	$P/V_g + 4P/3nV_{dc}$	$2P/V_g$
Min. number of modules N_{tot}	$2 \cdot \lceil (V_g + n \cdot V_{dc})/V_m \rceil$	$4 \cdot \lceil (V_g + n \cdot V_{dc})/2V_m \rceil$	$\lceil V_g/V_m \rceil$
Number of switches	$4N_{tot} + 4$	$4N_{tot} + 4$	$N_{tot} \cdot (4+2+2)$
Min. semiconductor power	$4N_{tot} \cdot I \cdot V_m + 4 \cdot 4P/3$	$4N_{tot} \cdot I \cdot V_m + 4 \cdot 4P/3$	$4N_{tot} \cdot I \cdot V_m + 4\pi \cdot P$
Energy storage	$(V_m \cdot P)/(4 \cdot \omega_g \cdot \Delta v)$	$(V_m \cdot P)/(4 \cdot \omega_g \cdot \Delta v)$	$(V_m \cdot P)/(4 \cdot \omega_g \cdot \Delta v)$
Number of MFTs	1	1	N_{tot}
Volume MFTs	$\propto P^{3/8}$	$\propto P^{3/8}$	$\propto N_{tot} \cdot (P/N_{tot})^{3/8}$

The minimum required number of modules $N = \lceil V_g/V_m \rceil$ with the module voltage V_m also defines the minimum number of switches because there are 6 switches required for each module.

Because full-bridge modules with four switches are required, the installed semiconductor power for the MMC part is given by $4N \cdot I_g \cdot V_m$. For the half bride LLC resonant part, the current amplitude is given by $\pi/V_{dc} \cdot P/N$ when neglecting the magnetizing current and assuming an operation at resonance frequency (sinusoidal current, rectangular voltage). Therefore, the installed semiconductor power for the LLC resonant part of the PETT is $4\pi \cdot P$ because there are 4 switches per module that have to transfer the power P/N.

The required energy storage in the module capacitors can be derived on the same basis as shown for the other two topologies. Following similar considerations as in the previous section, the capacitive energy storage is the same as for the other two topologies.

3.3 Medium Frequency Transformer Scaling Laws

The scaling laws for MFTs [18] regarding a constant efficiency, the achievable apparent power of a transformer S_{trafo} per volume V_{trafo} (power density) is

$$\frac{S_{trafo}}{V_{trafo}} \propto x^{\frac{\beta}{\beta-2}} \tag{19}$$

with its linear dimension x and the Steinmetz parameter $\beta \in [2,3]$ for typical MFT core materials [18]. With $\beta = 2.5$, the power density scales with x^5, such that

$$V_{trafo} \propto \frac{S_{trafo}}{x^5} = \frac{S_{trafo}}{V_{trafo}^{5/3}} \quad \Rightarrow \quad V_{trafo} \propto S_{trafo}^{3/8}, \tag{20}$$

because $V_{trafo} = x^3$.

Considering the thermal limitation of a transformer design and therefore assuming a constant temperature rise, the achievable apparent power of a transformer S_{trafo} per volume V_{trafo} (power density) is

$$\frac{S_{trafo}}{V_{trafo}} \propto x^{-\frac{1}{\beta}} = x^{-\frac{1}{2.5}}. \tag{21}$$

Concerning the achievable volume this results in

$$\frac{S_{trafo}}{V_{trafo}} \propto (V_{trafo}^{\frac{1}{3}})^{-\frac{1}{2.5}} = V_{trafo}^{-\frac{2}{15}} \quad \Rightarrow \quad V_{trafo} \propto S_{trafo}^{15/13}, \tag{22}$$

because $x = V_{trafo}^{1/3}$.

3.4 Evaluation with Exemplary Parameter Set

In Fig. 5, the minimum number of modules, the number of switches, and the installed semiconductor power (operating voltage times maximum switch current) and the transformer volume ratios for the

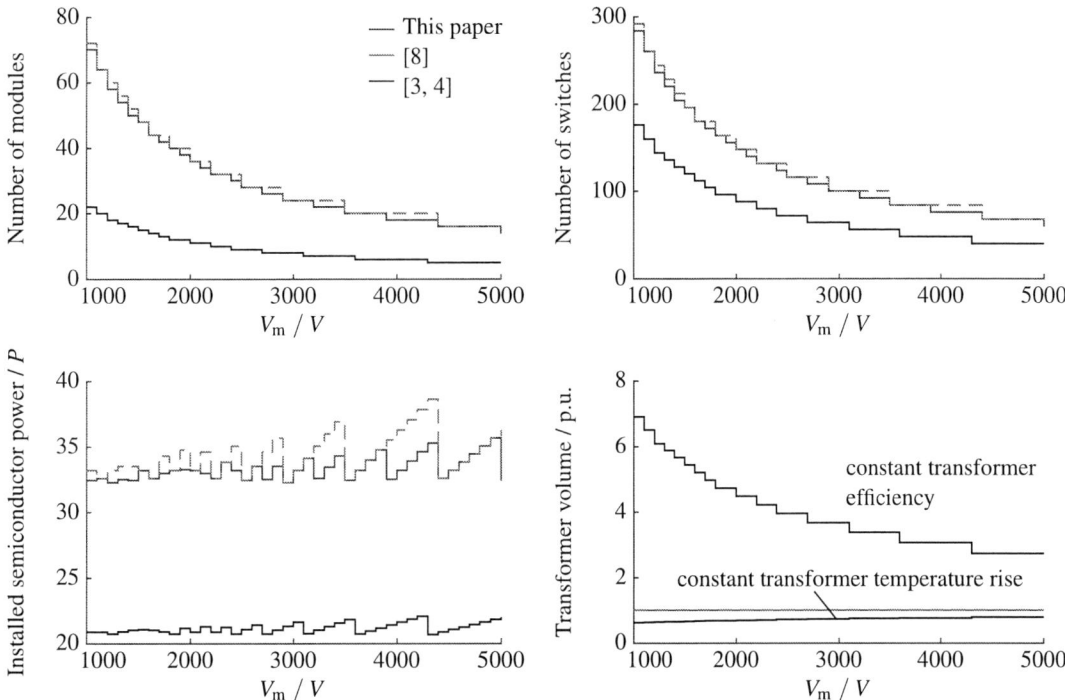

Fig. 5: Comparison of the implementation effort for the proposed concept, as well as the concepts presented in [8] and [3, 4] for a varying module voltage V_m. System parameters are given in Table I. For the concepts proposed in this paper and in [8], the transformer turns ratio n is set to n_{ZVS} (cf. (7)). Note that with the concept from [3, 4], the turns ratio is defined by the ratio between the module voltage and the secondary voltage amplitude V_{dc}. The transformer volume is given as a relation between the many small transformers required with the concept presented in [3, 4] and the single big transformer required for the concept presented in this paper or in [8] (set to 1 p.u.). Here, it is distinguished between the minimum achievable volume given a constant temperature rise and the necessary volume to achieve a constant efficiency.

considered concepts are compared for different module voltages. The parameters used for the comparison are summarized in Tab. I.

One can see, that regarding the number of modules as well as the required to be installed semiconductor power, the concept proposed in [3, 4] is always better than the concepts proposed in this paper and in [8].

The big drawback of the concept presented in [3, 4] is the fact that it needs an MFT in each and every module, such that the implementation effort per module is much higher than for the other concepts. Both MFT scaling laws presented in the previous section are considered in Fig. 5 to compare the transformer volume resulting from the different PETT concepts. Conceptually, if the same transformer efficiency should be achieved, the high number of required MFTs for the concept presented in [3, 4], results in a considerably higher total volume of the transformers of the PETT compared to the concept presented in this paper or [8]. If the efficiency is not first priority and maximum power density should be achieved, the high number of employed MFTs of [3, 4], results in a lower total volume of the transformers of the PETT compared to the concept presented in this paper or [8]. This is because many small transformers simplify the cooling compared to one single big transformer.

However, also an isolation voltage of at least the grid voltage also needs to be provided by all transformers, which is not represented by the scaling laws (20) and (22). Therefore, it would additionally increase the total volume of the many small transformers required with the concept of [3, 4].

When comparing the proposed concept and the one proposed in [8], the main parameters are very similar. This is especially true for the expressions as given in Tab. II. However, one needs to consider, that the number of modules per MMC arm needs to be an integer number, such that four arms can lead to a higher total number of modules in the end. Therefore, the number of modules, the number of switches, as well as the installed semiconductor power with the proposed concept is always lower or equal to [8]. A drawback of the concept proposed in this paper is that the grid current needs to flow though the primary windings of the transformer. However, because of its very low frequency, the related skin and proximity losses are typically negligible and only the DC resistance of the primary windings is of relevance for the related transformer losses.

4 Conclusion

In this paper, a novel MMC based power electronic traction transformer concept is proposed. It can be implemented with only two standard MMC arms using full-bridge modules. In comparison to a similar concept existing in literature, it uses less MMC arms resulting in a lower control and modulation effort. Moreover, the total number of MMC modules can be lower for the same module voltage, such that the installed semiconductor power can be reduced. Other concepts from literature require multiple medium frequency transformers. This is avoided by the proposed concept, at the cost of a higher number of modules and increased installed semiconductor power. However, the proposed concept allows zero voltage switching for all employed switches.

References

[1] S. Farnesi, M. Marchesoni, M. Passalacqua, and L. Vaccaro, "Solid-state transformers in locomotives fed through AC lines: A review and future developments," *Energies*, vol. 12, no. 24, 2019.

[2] T. Guillod and J. W. Kolar, "Medium-frequency transformer scaling laws: Derivation, verif ication, and critical analysis," *CPSS Trans. on Power Electron. and Appl.*, vol. 5, no. 1, 2020.

[3] D. Dujic, C. Zhao, A. Mester, J. K. Steinke, M. Weiss, S. Lewdeni-Schmid, T. Chaudhuri, and P. Stefanutti, "Power electronic traction transformer-low voltage prototype," *IEEE Trans. on Power Electron.*, vol. 28, no. 12, 2013.

[4] C. Zhao, D. Dujic, A. Mester, J. K. Steinke, M. Weiss, S. Lewdeni-Schmid, T. Chaudhuri, and P. Stefanutti, "Power electronic traction transformer—medium voltage prototype," *IEEE Trans. on Ind. Electron.*, vol. 61, no. 7, 2014.

[5] J. E. Huber and J. W. Kolar, "Optimum number of cascaded cells for high-power medium-voltage AC–DC converters," *IEEE Journal of Emerging and Sel. Topics in Power Electron.*, vol. 5, no. 1, 2017.

[6] J. E. Huber, "Conceptualization and multi-objective analysis of multi-cell solid-state transformers," PhD Thesis, ETH Zürich, 2016.

[7] P. Simiyu and I. E. Davidson, "MVDC railway traction power systems, state-of-the art, opportunities, and challenges," *Energies*, vol. 14, no. 14, 2021.

[8] M. Glinka and R. Marquardt, "A new AC/AC multilevel converter family," *IEEE Trans. on Power Electron.*, vol. 52, no. 3, 2005.

[9] R. DeDoncker, D. Divan, and M. Kheraluwala, "A three-phase soft-switched high-power-density DC/DC converter for high-power applications," *IEEE Trans. on Ind. Appl.*, vol. 27, no. 1, 1991.

[10] M. Kheraluwala, R. Gascoigne, D. Divan, and E. Baumann, "Performance characterization of a high-power dual active bridge DC-to-DC converter," *IEEE Trans. on Ind. Appl.*, vol. 28, no. 6, 1992.

[11] F. Krismer and J. W. Kolar, "Closed form solution for minimum conduction loss modulation of DAB converters," *IEEE Trans. on Power Electron.*, vol. 27, no. 1, 2012.

[12] M. Stojadinovic, E. Kalkounis, F. Jauch, and J. Biela, "Generalized PWM generator with transformer flux balancing for dual active bridge converter," in *18th Eur. Conf. on Power Electron. and Appl. (EPE, ECCE Europe)*, 2017.

[13] M. Kasper, R. Burkat, F. Deboy, and J. Kolar, "ZVS of power MOSFETs revisited," *IEEE Trans. on Power Electron.*, 2016.

[14] K. Ilves, S. Norrga, L. Harnefors, and H.-P. Nee, "On Energy Storage Requirements in Modular Multilevel Converters," *IEEE Trans. on Power Electron.*, vol. 29, no. 1, 2014.

[15] S. Fuchs, M. Jeong, and J. Biela, "Reducing the energy storage requirements of modular multilevel converters with optimal capacitor voltage trajectory shaping," in *22nd Eur. Conf. on Power Electron. and Appl. (EPE, ECCE Europe)*, 2020.

[16] S. Fuchs, "MMCs Made Compact and Fast - Towards Modular Multilevel Converters with Minimized Module Capacitance and Fast Transient Operation Utilizing Advanced Control Methods," PhD Thesis, ETH Zürich, 2022.

[17] H. Bärnklau, A. Gensior, and S. Bernet, "Derivation of an equivalent submodule per arm for modular multilevel converters," in *15th Int. Power Electron. and Motion Control Conf. (EPE/PEMC)*, 2012.

[18] C. R. Sullivan, B. A. Reese, A. L. F. Stein, and P. A. Kyaw, "On size and magnetics: Why small efficient power inductors are rare," in *Int. Symposium on 3D Power Electron. Integration and Manufacturing*, 2016.

A new power MOSFET technology achieves a further milestone in efficiency

Ralf Siemieniec, Michael Hutzler, Cesar Braz, Tomasz Naeve, Elias Pree, Heimo Hofer, Ingmar Neumann, David Laforet

INFINEON TECHNOLOGIES AUSTRIA AG
Siemensstrasse 2, 9500 Villach, Austria
ralf.siemieniec@infineon.com
http://www.infineon.com

Keywords

power semiconductor device, MOSFET, new switching devices, robustness, efficiency

Acknowledgements

We thank Marco Kuenstel for performing the various efficiency measurements in the IBC test board and Simone Mazzer for his supportive help providing several drawings and charts. We also want to thank Andrew Wood for carefully editing this article.

Abstract

This work introduces the characteristics and properties of the latest trench MOSFET technology released to the market. Based on the advantages of a revolutionary new cell design combined with the benefits of an advanced manufacturing technology, this new device family brings together the benefits of exceptionally low conduction losses, superior switching performances, improved SOA and good ruggedness. These features make it the best fit for high switching frequency applications, supporting the trend towards significantly higher efficiency while enabling designs for higher power densities and cost effectiveness. Typical applications for these MOSFETs include telecom, server, datacom as well as solar applications, drones, e-bikes, power tools and many other battery-powered applications (BPA). Results presented in this work focus on the 100 V voltage class intended for use in fast switching DC/DC telecom switch-mode power supplies (SMPS), and compares the new technology with its predecessor generation.

1. Introduction

Since their introduction, MOSFET technologies have been noted as excellent candidates to be used as switches in power management circuits. Vertical diffused MOSFET (VDMOS) structures, commercially available since the late seventies, first addressed the needs of a power switch [1]. The superior switching performance and the high input impedance placed the MOSFET as an appealing alternative to bipolar technologies. Still, the high on-state resistance limited the current-handling capabilities of the VDMOS and its applications in the power electronics industry. In a medium-voltage VDMOS, the major contributors to the total on-state resistance between drain and source are represented by the intrinsic channel resistance and by the JFET region limiting the channel current flow into the epitaxially-grown drift region (Fig. 1a). Overcoming this limitation required more than a decade of development in device design and process engineering, which culminated in the late 1980s with the commercialization of the first trench gate MOSFETs. The development of trench power MOSFETs was a milestone for the broad adoption of the field-effect transistors in the power electronics industry [1-3]. By moving the channel to the vertical direction, this device concept allowed a reduction in cell pitch without adversely affecting the current spreading by virtually removing the JFET region, hence dramatically reducing the on-state resistance (Fig. 1b). The achieved ultralow specific channel resistance no longer prevented low on-state resistances, although as a result the substrate and package resistances became significant contributors. However, the remarkable increase in cell density, besides finally establishing the trench MOSFET as a competitive alternative to the planar technology, has also brought to light significant disadvantages. The gate-drain capacitance (related to trench gate penetration in the epi drift region) and gate-source capacitance (overall capacitance between trench gate and source diffusion) increase linearly with the number of trenches, i.e. with the cell density. Together with a sublinear scaling in the on-resistance $R_{DS(on)}$, this significantly impacts the technology figure-of-merit $FOM_g = R_{DS(on)} \times Q_g$. Since the MOSFET is uniquely controlled through its gate terminal, the gate driver circuitry has to provide the total gate charge Q_g required to turn on the transistor. In the case of high switching frequency applications, like SMPS, the lowest gate charge is desirable since it proportionally affects the driving losses. A part of the total gate charge is associated with the gate-to-drain charge Q_{gd}, which governs the drain voltage

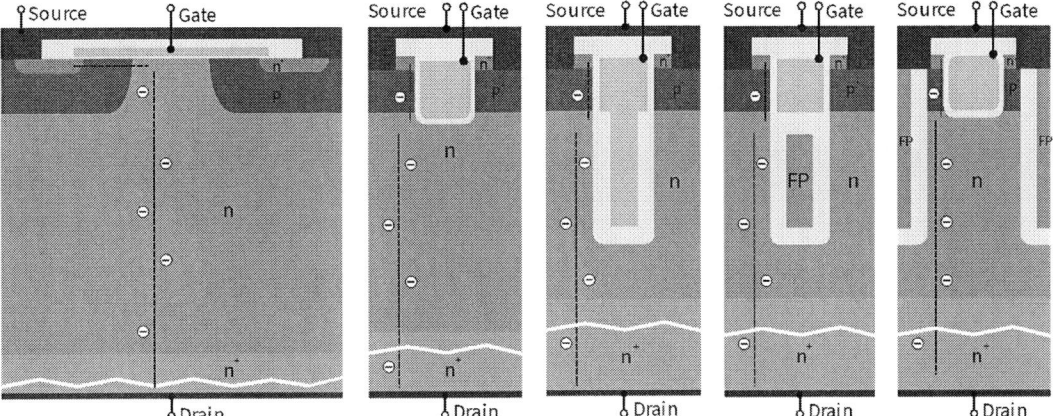

Fig. 1a-e: Exemplary device structures depicting the evolution of power MOSFET:
a) VDMOS structure with lateral channel and planar gate
b) Trench MOSFET structure with vertical channel
c) Trench MOSFET with lateral charge-compensation by a gate-connected field-plate
d) Trench MOSFET with lateral charge-compensation by an insulated field-plate connected to source
e) Trench MOSFET with lateral charge-compensation by an insulated field-plate and separated gate trench

transient. A higher Q_{gd} impacts the transient speed, increases the switching losses, and forces the use of longer dead-times. It became evident that specific measures were needed to reduce the overall gate and gate-drain charge. Additionally, another constraint is imposed by the Miller charge ratio: $Q_{gd}/Q_{gs(th)}$ must be lower than one if the intrinsic robustness against parasitic turn-on of the MOSFET under fast drain voltage transients is to be guaranteed [4].

A new era started with the introduction of charge-compensated structures, exploiting the same principle as super junction devices. The introduction of devices which use an insulated deep field-plate as an extension of the gate electrode enabled the lateral depletion of the drift region in the off state (Fig. 1c) [5]. The lateral depletion alters the electric field distribution throughout the structure, allowing the same voltage to be blocked within a shorter length. Since the electric field can be supported by a thinner and more heavily doped drift region, a substantial reduction in the on-state resistance can be achieved. It is worth noticing that the field-plate (as an extension of the gate electrode) leads to both a significant increase of the reverse-transfer capacitance C_{gd} (hence also Q_{gd} and Q_g) and a nonlinear dependence on the drain voltage. In fact, the transfer capacitance drops abruptly as soon as the mesa region completely depletes. These disadvantages were soon overcome by the use of a field-plate, which was isolated from the gate electrode and instead electrically connected to the source potential (Fig. 1d). While the charge compensation principle operates as before, the buried field-plate does not introduce any additional contributions to the gate-drain capacitance. Instead, the field plate shields the gate electrode from the drain potential, which reduces the gate-drain capacitance C_{gd} and related charges. These devices, at the time of their introduction to the market, showed best-in-class performance with low gate-charge and gate-drain charge characteristics, high switching speeds and good avalanche ruggedness [6]. Still, the presence of the field plate comes with the disadvantage of an increased output capacitance C_{oss} and output charge Q_{oss} - a consequence of the lateral charge-compensation. However, careful device optimization enabled field-plate based power technologies with FOM_{oss} comparable to those of the standard trench MOSFET [7]. These attributes made them even more suitable for a wide variety of switched-mode power supply (SMPS) applications.

2. Features and advantages of the new device technology

2.1 Novel cell design approach

In the development of a new silicon technology, special care must be taken in the definition of its specifications, in order to bring significant system-level advantages and to add value for the customer. The new device was required to provide improvements across all figures of merit, as this is needed to enable high-frequency SMPS operation where losses are associated both with charges (switching) and on-state resistance (conduction).

To meet these more demanding requirements, a novel cell-design approach, which explores a true three-dimensional charge compensation, has been developed and implemented. To follow this path, one first needs to enable a direct connection to the field-plate electrodes from above as illustrated in Fig. 1e. Second, the device layout must move away from the stripe layout, which is commonly used with charge-compensated MOSFETs

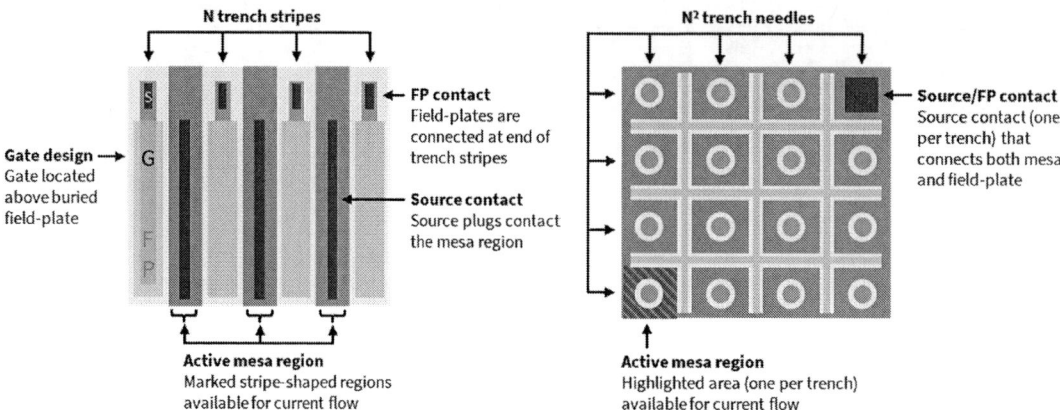

Fig. 2: Comparison of the commonly used stripe layout with the new grid-like layout approach (top-view)

employing a field plate, to a grid-like layout structure as depicted in Fig. 2. In this way the silicon area for current conduction is increased compared to a structure with stripes, allowing a further reduction of the overall on-resistance. In order to not only maintain but also further reduce the $FOM_g = R_{DS(on)} \times Q_g$ and $FOM_{gd} = R_{DS(on)} \times Q_{gd}$ values, the gate trench underwent a complete redesign to minimize its lateral extension.

2.2 Introduction of a metal gate for the trench power MOSFET

A drawback of the design measure to minimize the lateral extension of the gate trench is given by the dramatically reduced conduction area of the gate electrode itself. The use of a common polysilicon gate material results in an unacceptably large gate sheet resistance, linked to a clear increase of the internal gate resistance R_G. Large values for the gate resistance lead to a slowed-down switching behavior and increased switching losses, hence it is mandatory to avoid such an increase.

The introduction of gate fingers represents a common way to reduce the chip's internal gate resistance. The disadvantage of this measure is the reduction of the available active area due the space consumed. This results in an increase of the product on-resistance, especially in cases of smaller dies and the use of several gate fingers. Fig. 3 illustrates this correlation between gate resistance and active area loss for a best-in-class chip in an S3O8 package. The introduction of a metal gate system solves this problem, thanks to the much lower metal sheet resistance, as also indicated in Fig. 3.

Moreover, the use of a metal gate also enables a so-far unmatched switching uniformity over the die area. Simulations of the distributed local gate resistance impressively illustrate the much-improved gate resistance uniformity across the chip compared to the common use of polysilicon gates, see Fig. 4. Together with a direct connection of the field plates to the source metal, a device set-up is realized that not only ensures a very fast and homogeneous transition at turn-on and turn-off that minimizes switching losses, but also reduces the risk of an unwanted, dv/dt induced parasitic turn-on of the MOSFET.

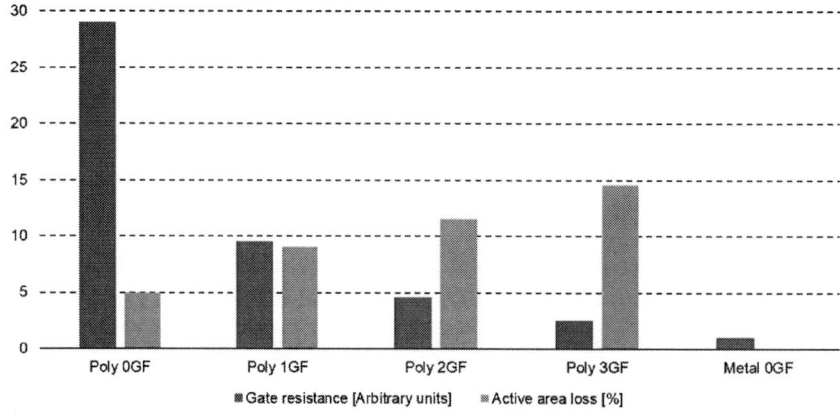

Fig. 3: Reduction of gate resistance depending on the number of gate fingers (GF) and active area loss for a best-in-class chip in an S3O8 package with the use of polysilicon and metal gate

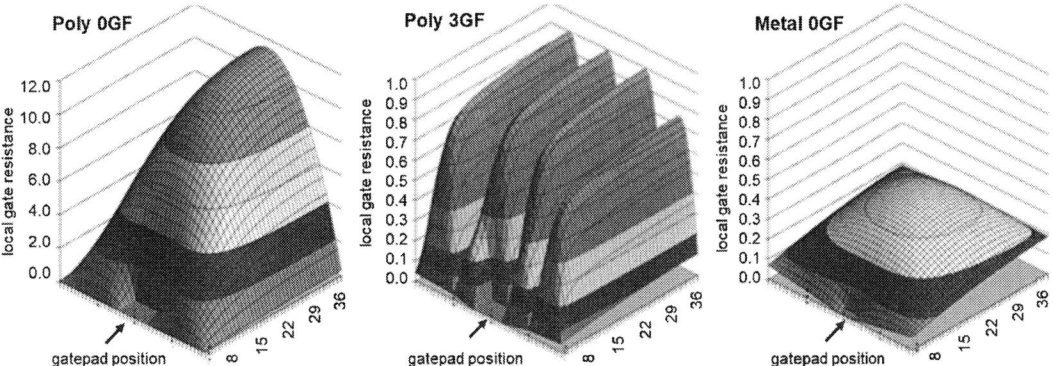

Fig. 4: Comparison of gate resistance uniformity for a best-in-class chip in an S3O8 package using different gate set-ups

3. Device properties

Fig. 5 summarizes the main device parameters of this technology with respect to its predecessor, indicating impressive improvements in all of the relevant parameters. The innovative gate trench engineering of the new device technology results in a remarkable reduction of both gate-source and gate-drain specific capacitances, which is reflected in the respective figures-of-merit FOM_g and FOM_{gd}.

These improvements are also clearly visible in the comparison of gate charge characteristics shown in Fig. 6. The devices have a comparable on-resistance $R_{DS(on)} \sim 3$ mΩ in an SSO8 package. The Q_g reduction helps to achieve better efficiencies especially at light load conditions, due to the much reduced driving losses $P_{AUX} = Q_g \times V_{GS} \times f_{sw}$. This is of even higher importance for SMPS operated at high switching frequencies, as well as in applications where a larger number of MOSFETs are paralleled. In this case, the low gate charge also relaxes the requirements towards the gate driver's current capability. On the other hand, the low Q_{gd} enables fast switching transients, lowering the switching losses.

Fig. 7 compares the typical capacitance characteristics for both technologies. The output capacitance of the MOSFET is charged in every switching cycle. In hard-switched topologies, the energy that is stored in the output capacitance cannot be recovered and contributes significantly to the turn-on losses, as it is dissipated across the channel during the turn-on of the device. Losses associated with the output charge Q_{oss} are increasingly important at higher drain voltages, and scale linearly with the switching frequency. In general, in hard-switched SMPS operated at high switching frequencies, it is desirable to have a low value for the output charge in order to improve the efficiency. The previous technology generation was optimized with a strong focus on achieving an outstanding output charge figure-of-merit [8]. As can be concluded from Fig. 7, the new technology delivers only a minor improvement in this parameter, having opted instead for a compromise with the ease of use of the device. Besides, the clear reduction of the input and reverse transfer capacitance of the newly introduced technology reflect the improvements in the gate charge as just discussed.

The new cell design does influence the transfer characteristics that are important when the MOSFET operates in the linear region, e.g. when the MOSFET operates as a voltage controlled current source. This is a consequence of

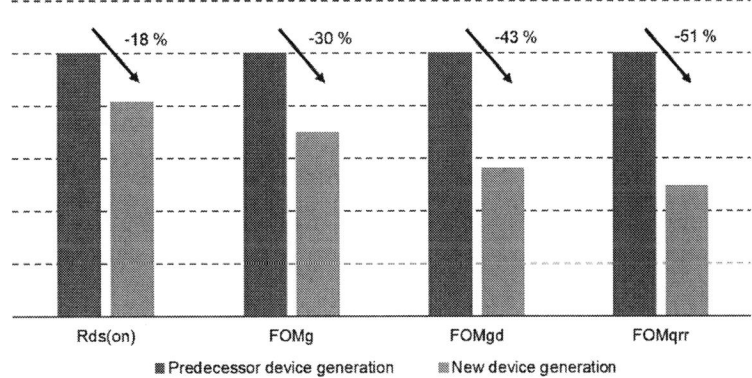

Fig. 5: Gained improvements in device performance for a best-in-class 100 V device in SSO8 package

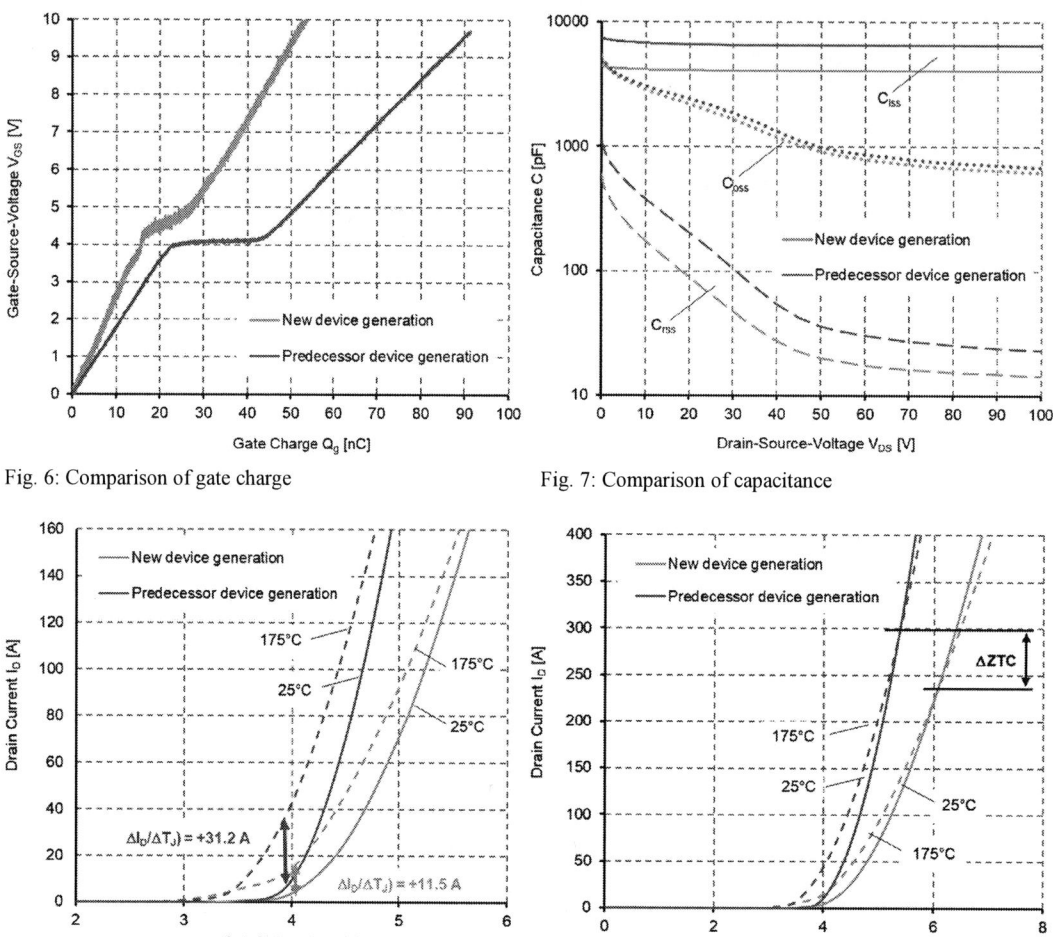

Fig. 6: Comparison of gate charge

Fig. 7: Comparison of capacitance

Fig. 8: Transfer characteristics of the new and predecessor technology for best-in-class devices in SSO8

the reduction in threshold voltage with rising temperature which is not completely compensated by a reducing bulk mobility, as current densities are still comparatively low [9]. The hottest part of the chip will draw more and more current, hence increasing again the temperature of the already hot spot (thermal runaway). At higher current densities, the effect of bulk mobility reduction starts to dominate the device behavior. At the zero-temperature-coefficient (ZTC) point, both effects balance out and the drain current is independent of temperature. Above the ZTC point, the drain current reduces with temperature and the device stabilizes itself.

Shifting the ZTC point to lower drain currents leaves the device less susceptible to thermal runway. Unfortunately, the strong increase of the channel density in low-voltage trench MOSFETs required to minimize on-resistance has shifted the ZTC point towards higher current densities, which has limited the device capabilities in linear mode operation. Fig. 8 compares the transfer characteristics of the new technology with the preceding one, revealing a more gradual turn-on behavior of the new device and consequently a move of the ZTC point towards lower currents. This means that the new technology comes with a lower transconductance in the linear region of operation, improving its robustness against thermal runaway. Along with an excellent thermal resistance from junction to case, the new device technology delivers a significant SOA improvement, as depicted in Fig 9. Such a behavior is very beneficial for applications besides SMPS, such as in hot-swap protection circuits, and widens the range of applications that this new technology generation is capable of addressing.

Hot swap is the procedure of replacing a faulty DC/DC converter by a good one, in a system that is up and running. This operation ought to be carried out without disturbing neighboring loads and without damaging the converter power connector contacts. The hot swap circuitry charges the new converter input capacitors with a charging current that has its peak value limited, avoiding big disturbances in the system bus voltage. MOSFETs employed in such applications must withstand the full system bus voltage and the bulk capacitors' charging current during its charging time: they need a wide SOA area in the region of the charging time pulse.

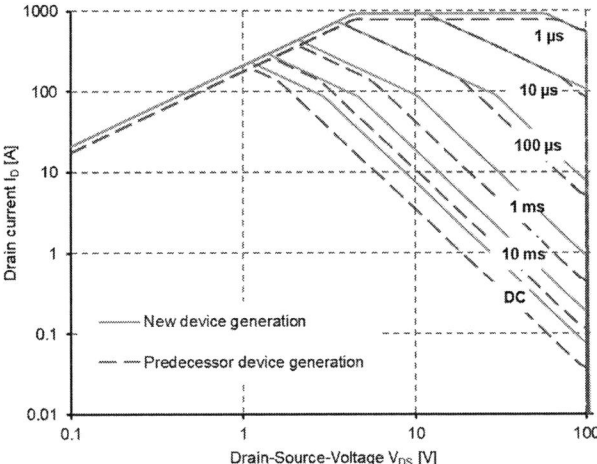

Fig. 9: SOA comparison of the introduced device technology with the predecessor device technology (best-in-class devices in SSO8 package)

4. Performance in target applications

4.1 Device operation in a DC/DC Intermediate Bus Converter for telecom applications

A telecom DC/DC converter serves as a test vehicle to compare the performance of the new-technology device with that of the predecessor technology. Such units are widely used in telecom and datacom power systems, typically as isolated DC/DC Intermediate Bus Converters (IBC) in the overall conversion chain with a nominal 48 V input and a 12 V output voltage bus for the downstream point-of-load converters [10]. The continuous improvement in MOSFET technology has allowed for a power density increase that enables the realization of the IBC in a standard quarter-brick form factor to typically deliver 600 W and more. Fig. 10 shows views of the realized board. For the measurement of the efficiency with the tested devices on the primary side, the test board developed employs a full-bridge topology on both the primary and secondary side, as depicted in Fig. 11. The full-bridge configuration on the secondary side is the topology of choice if the converter power increases, as the utilization of the transformer winding on the secondary side is better in this case. Although this configuration causes the use of two devices in series on the secondary side, the topology achieves a higher efficiency overall. This higher efficiency is also supported by the lower blocking voltage required of the MOSFET (due to the series connection), which is linked to a lower on-resistance. Typically, 40 V devices are used along with a full-bridge topology on the secondary side. To enable an assessment of the efficiency when the 100 V devices are used as synchronous rectifiers on the secondary side, a center-tapped topology needs to be used as shown in Fig. 12. This center-tapped configuration requires devices with a higher blocking voltage of 80 V ... 100 V, and as such, it is the topology of choice in this part of the performance evaluation.

The board operates with a switching frequency of 250 kHz, the operating input voltage is allowed to vary between 36 V and 75 V. Due to the relatively large range of the allowed input voltage, voltage regulation is required [11].

Fig. 13 shows the comparison of the converter efficiencies achieved using the devices to be tested on the primary side with respect to the two technology generations. Here the synchronous rectifier on the secondary side is

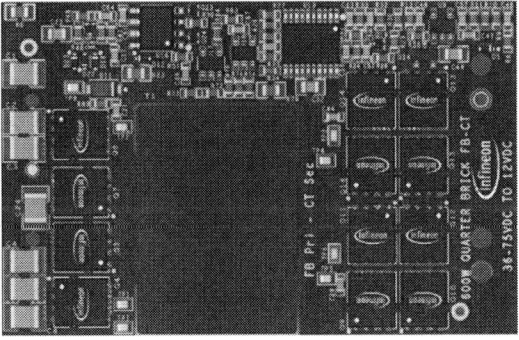

Fig. 10: Views of the 600 W isolated DC/DC IBC test board

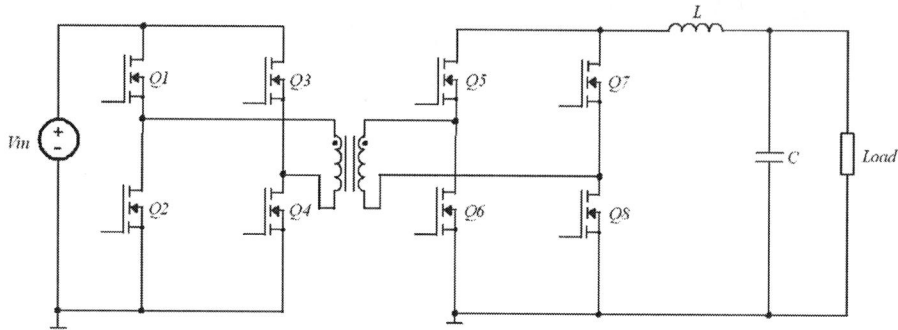

Fig. 11: Basic schematic of the 600 W isolated DC/DC IBC test board in FB-FB configuration

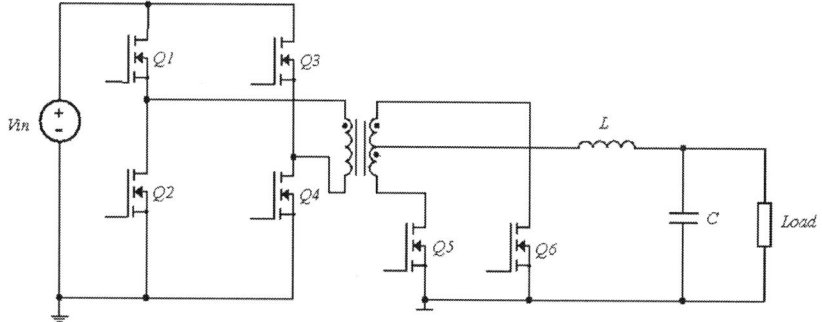

Fig. 12: Basic schematic of the 600 W isolated DC/DC IBC test board in FB-CT configuration

equipped with a fixed set of devices from the previous generation. The on-resistance of both devices was matched as closely as possible, using 6 mΩ devices of the new and 5 mΩ devices of the older technology. Compared to the predecessor solution, a valuable gain of up to 0.7 % is recorded at light-load conditions, which is a direct consequence of the improved switching performance of the new devices. The clearly reduced gate charge lowers the power consumption from the auxiliary bias supply, whereas the gate-drain charge which is less than half of the value of the predecessor technology enables ultra-fast switching, resulting in lowered turn-off losses. This high efficiency gain that is found at light load progressively reduces to a remarkable 0.2 % improvement at half load, where the IBC converter spends most of its operating time, and reaches 0.14 % at full load, despite the increased on-resistance. The reduction in the charges of the new generation device enables the improvements in efficiency even with the higher on-resistance. In the low load range, the reduction of the drain-to-source charge has more impact on the efficiency, whereas from half to full load the reduction of the gate-to-drain capacitance has a higher impact.

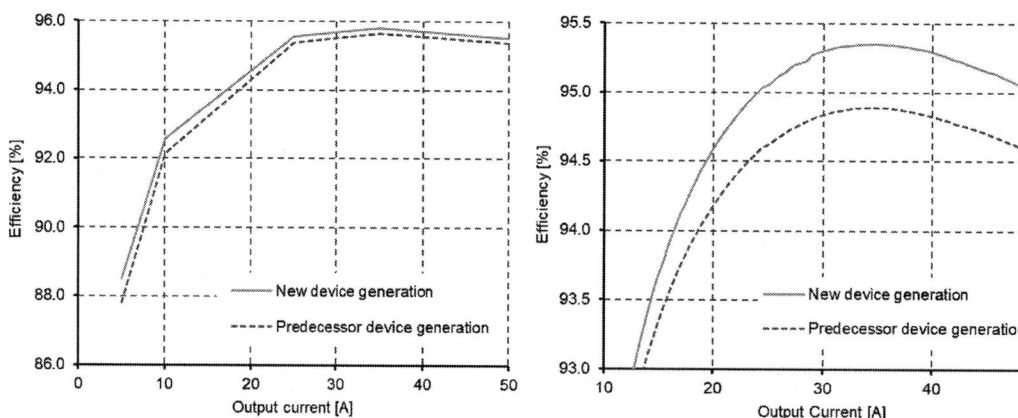

Fig. 13: Converter efficiency comparison of the devices tested in the primary side full-bridge stage of the IBC shown in Fig. 11

Fig. 14: Converter efficiency comparison of the devices tested on the secondary side synchronous rectifier stage of the IBC shown in Fig. 12

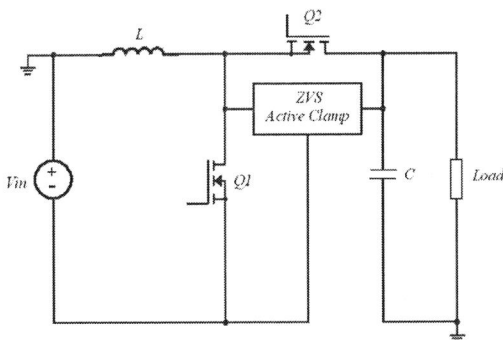

Fig. 15: Basic schematic of one phase of the inverting buck-boost DC/DC converter

Fig. 16: 3D view of the realized evaluation board

For the efficiency investigation on the secondary side, the primary side was equipped with a fixed set of devices from the predecessor technology. Fig. 14 shows the measured efficiency values, again using devices of comparable on-resistance. Here, the efficiency improves by 0.46 % over the entire range from half to full load. This remarkable improvement is mainly attributed to the lower reverse-recovery charge and the lower gate charge. As can be seen in Fig. 2, the new device structure inherently offers a larger area for the current flow when compared to the predecessor device structure. As a consequence, the body diode current density reduces for the same current level in the application, and with it the reverse-recovery charge.

4.2 Device operation in an inverting buck-boost DC/DC ZVS converter for telecom applications

This example covers the efficiency of a telecom DC/DC converter based on a zero-voltage switching (ZVS) inverting buck-boost topology. Ordinary inverting buck-boost topologies find interest in telecom power systems as DC-DC converters that supply the RF power amplifiers (RFPAs). These amplifiers require the highest possible efficiencies for supply voltages ranging from +28 V (for use in LDMOS RFPAs) up to +50 V (for use in GaN RFPAs). The use of a novel active clamp auxiliary circuitry enables the transfer of the reverse recovery charge Q_{rr} from the Synchronous Rectifier MOSFET Q2 toward the output in a non-dissipative way, practically achieving ZVS turn-on for the control switch Q1. As such, the active clamp circuit effectively drives down the overall switching losses in the unit. This enables the use of best-in-class devices with lowest on-resistance, supporting a dramatic increase in the power density [12].

The evaluation board used in this investigation has been adapted to use an output voltage of only 12 V instead of the usually required 28 V, which is suitable for telecom equipment other than RFPA that does not require functional isolation. This configuration enables the use of MOSFETs with a blocking voltage of 100 V, which makes it an excellent platform to study the performance of the discussed devices in a soft-switching topology. The board, based on an interleaved (two-phase) inverting buck-boost, can deliver up to 600 W from an input voltage range of -36 V... -60 V. The basic schematic for one phase of the topology is given in Fig. 15, while the realized demonstrator board is shown in Fig. 16.

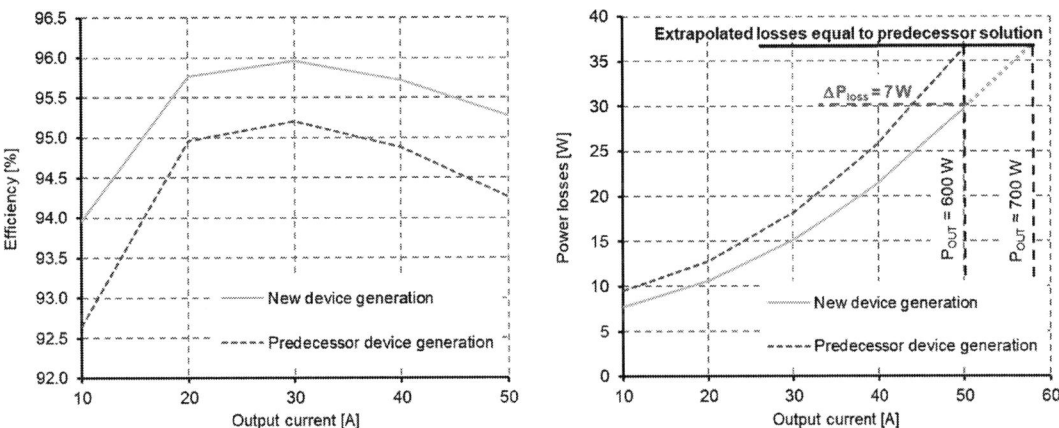

Fig. 17: Efficiency and extrapolated losses for the 600 W ZVS Inverting Buck-Boost Evaluation Board (V_{IN} = -48 V, V_{OUT} = 12 V, f_{SW} = 200 kHz)

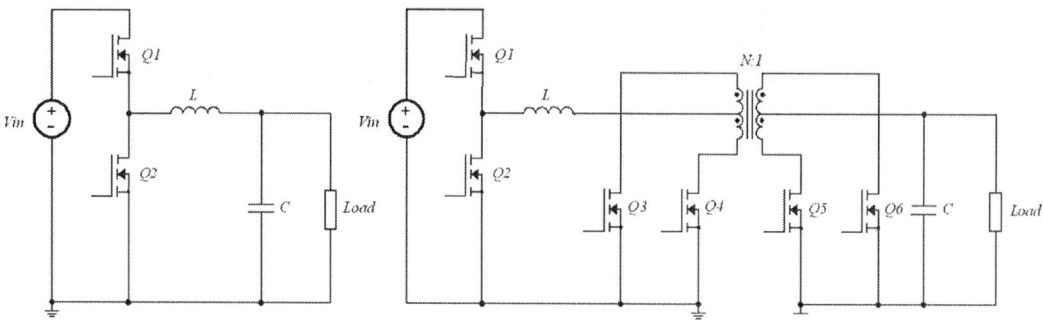

Fig. 18: Basic schematic of the buck converter (left) and the buck current-fed push-pull cascaded converter (right)

Fig. 17 shows the measured efficiencies for the new and predecessor device technology. The results clearly illustrate the huge impact of the effort invested in the device and technology development with an impressive efficiency gain of ~ 1 %. The higher efficiency results in lower losses of 7 W, which enables a power density increase of up to 15 % (see Fig. 17). This reduction in losses is mainly achieved due to the reductions in the on-resistance, the gate charge and the reverse-recovery charge that is transferred to the output by the active ZVS clamp circuit.

4.3   Device operation in a buck converter

The second application example examines the efficiency of operating the devices in a buck converter. This topology, among others, is often used as a front-end stage of current-fed cascaded push-pull converters (see Fig. 18). However, only the buck stage is covered in this investigation. The converter is operated at a switching frequency of 300 kHz, with a 60 V input, a 26 V output voltage, and an output power equal to 450 W. The selected input voltage represents the high voltage level in a 48 V telecom system and the output voltage is the same as the one used as the input for the push-pull stage in the cascaded converter. Devices realized in the new technology are compared with devices based on the previous technology. All devices have identical on-resistance and employ the chip layout without gate fingers; see also Fig. 4 (new technology: Metal 0 GF; previous technology: Poly 0 GF). Fig. 19 gives views of the board design.

Fig. 20 shows a comparison of the drain-source voltages for the devices of the two technologies at turn-off: the new generation device shows a clean drain-source voltage waveform, whereas the device from the previous generation shows a kink in the drain-source voltage at a value of about 15 V. This kink is related to the high dv/dt of the drain-source voltage, which induces a displacement current into the gate through the gate-drain capacitance C_{gd}. As a consequence the MOSFET turns-on despite the external gate driver circuit trying to pull-down the gate of the device. Such a parasitic turn-on occurs for the predecessor technology device, although both MOSFETs come with a similar value of internal gate resistance. The reason is the much less homogeneous distribution of the internal gate resistance. As can be seen from Fig. 4, the effective gate resistance in the region opposite to the gate pad is several times larger than for the area close to the gate pad. Due to this locally increased gate resistance, this part of the MOSFET chip is prone to suffering from an induced turn-on. Further comparing the drain-source voltage waveforms for the two devices it can be seen that the new technology device shows less damping of the

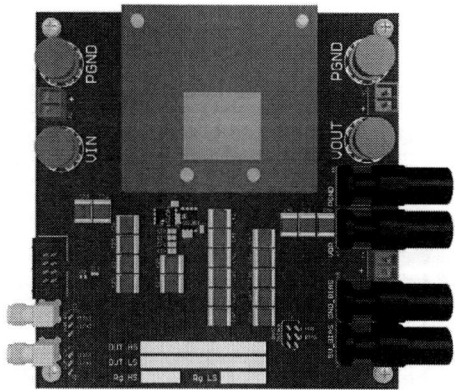 

Fig. 19: Views of the buck converter test board

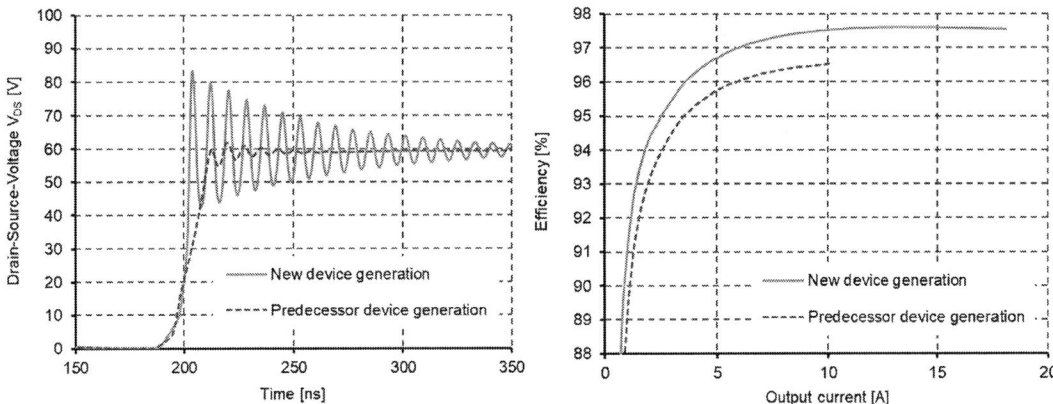

Fig. 20: Comparison of the drain-source-voltages for the two device generations in the buck converter

Fig. 21: Comparison of the efficiency for devices of both technologies in the buck converter

ringing. This effect is due to the low equivalent resistance of the field-plate for the new technology, a consequence of the high density of trenches per area.

In order to enable a valid efficiency comparison, avalanche on the low-side MOSFET needs to be avoided during the buck converter operation. Therefore, the high-side MOSFET needs to be slowed-down to limit the dv/dt over the low-side device, which required the use of a four times larger value for the external gate resistor for the previous technology device compared to the new technology generation.

Fig. 21 compares the achieved efficiencies for the two device generations. The device of the new technology delivers the full power providing a high efficiency, whereas the predecessor device can only deliver half of the required power. The experimental results show that gate fingers ought to be used in the previous-technology device chip layout in order to reduce the peak value of the internal gate resistance to acceptable values and avoid the MOSFET induced turn on. However, the insertion of such gate fingers in the MOSFET chip consumes chip area that would otherwise be used for current conduction and, therefore, a bigger chip ought to be used for the same MOSFET channel on-resistance, which increases the device cost. On the other hand, the use of a metal gate within the introduced new technology, with its more homogeneous distribution of internal gate resistance, provides a good immunity against parasitic induced turn-on without the need for gate fingers.

5. Conclusion

The new power MOSFET technology introduced in this work shows improvements in all important device parameters and combines the benefits of low on-state resistance with a superior switching performance. The new technology is specifically optimized for high switching frequency applications such as telecom SMPS and solar.

The remarkable jump in the overall device performance is enabled by substantial improvements at the device technology level. This has culminated in a unique device structure, which is the first to employ a three-dimensional charge compensation combined with the first-time use of a metal gate in a trench power MOSFET.

The reduction achieved in the on-resistance, the dramatically-lowered gate charge and gate-drain charge, together with the low output charge and improved switching homogeneity across the device area, enhance the system efficiency in the application across all load conditions. Additionally, the new technology offers an optimized transfer characteristic with a low temperature coefficient. This feature enables a safe operating area (SOA) that is significantly enhanced with respect to the predecessor technology, leaving this latest trench power MOSFET technology also as an ideal candidate for battery disconnect switches in battery-powered applications, battery management systems, and further enabling its use with motor drives.

The new device structure is also beneficial for the internal body diode of the MOSFET. While the silicon area conducting current is increased, the body diode current density is decreased which, for the same current level, decreases the reverse recovery charge.

Efficiency measurements in targeted SMPS applications confirm the findings at the semiconductor device level and yield efficiency improvements of up to 1 %, depending on the topology and load condition. These improvements provide a significant margin to meet the demanding requirements of the telecom power arena and other application fields.

6. References

[1] R.K. Williams, M.N. Darwish, R.A. Blanchard, R. Siemieniec, P. Rutter and Y. Kawaguchi, "The Trench Power MOSFET: Part I - History, Technology, and Prospects", IEEE Transactions on Electron Devices, pp. 674-691, Vol. 64, No. 3, 2017

[2] R.A. Blanchard, "Method for making planar vertical channel DMOS structures", U.S. Patent 4767722, 1986

[3] H.-R. Chang, R.D. Black, V.A.K. Temple, W. Tantraporn, and B.J. Baliga, "Self-aligned UMOSFET's with a specific on-resistance of 1 mΩ cm^2", IEEE Trans. Electron Devices, pp. 2329–2334, Vol. ED-34, No. 11, 1987

[4] P. Singh, "Power MOSFET Failure Mechanisms", pp. 499-502, Proc. INTELEC 2004, Chicago, USA, 2004

[5] J. Ejury, F. Hirler and J. Larik, "New P-Channel MOSFET Achieves Conventional N-Channel MOSFET Performance", Proc. PCIM, Nuremberg, Germany, 2001

[6] A. Schlögl, F. Hirler, J. Ropohl, U. Hiller, M. Rösch, N. Soufi-Amlashi and R. Siemieniec, "A new robust power MOSFET family in the voltage range 80 V – 150 V with superior low R_{DSon}, excellent switching properties and improved body diode", Proc. EPE, Dresden, Germany, 2005

[7] R. Siemieniec, C. Mößlacher, O. Blank, M. Rösch, M. Frank and M. Hutzler, "A new Power MOSFET Generation designed for Synchronous Rectification", Proc. EPE, Birmingham, UK, 2011

[8] R. Siemieniec, M. Hutzler, D. Laforet, O. Blank, L.-J. Yip, A. Huang and R. Walter, "Development of low-voltage power MOSFET based on application requirement analysis", Facta Universitatis (Niš), Ser.: Elec. Energ., Vol.28, No.3, 2015

[9] P. Spirito, G. Breglio, V. D'Alessandro and N. Rinaldi, "Thermal Instabilities in High Current Power MOS Devices: experimental Evidence, Electro-thermal Simulations and Analytical Modeling", Proc. MIEL, Niš, Serbia, 2002

[10] S. Li, "Intermediate Bus Converters for High Efficiency Power Conversion: A Review", IEEE Texas Power and Energy Conference (TPEC), 2020

[11] ETSI (2016.10), "Power supply interface at the input to telecommunications and Datacom (IST) equipment; Part 2: Operated by -48V direct current (DC)", EN 300 132-2

[12] Infineon Technologies AG (2021, 8.), Application note: "XDPP1100 two-phase interleaved buck-boost"

Experimental Evaluation of Battery Impedance and Submodule Loss Distribution for Battery Integrated Modular Multilevel Converters

Arvind Balachandran[1], Tomas Jonsson[1,2], Lars Eriksson[1], and Anders Larsson[2]

[1]Linköping University, Linköping, Sweden
[2]Scania, Södertalje, Sweden
Email: arvind.balachandran@liu.se, tomas.u.jonsson@liu.se,
lars.eriksson@liu.se, anders.larsson@scania.com

Abstract

Greenhouse gas emissions and the increase in average global temperature are growing concerns now more so than ever. Therefore it is of importance to increase the use of alternative energy sources, especially in the automotive industry. Battery electric vehicles (BEV) have gained popularity over the past several years. However, the performance of a BEV is limited by the battery pack, in particular, the weakest cell in the pack. Therefore, improved cell controllability and high efficiency are seen as important directions for research and development and one direction where it can be achieved is through using battery-integrated modular multilevel converters (BI-MMC). The battery current in BI-MMCs contains additional harmonics and the frequency dependent losses of these harmonics are determined by the resonance between the battery and the DC-link capacitor bank. The paper presents an experimental validation of previously published theoretical results for both harmonic allocations and loss distribution at the switching frequency within the BI-MMC submodule. Furthermore, a methodology for measuring the battery impedance using the full-load converter switching currents is presented.

Keywords

≪Modular Multilevel Converters (MMC)≫, ≪Power converters for EV≫, ≪Batteries≫, ≪DC-AC converters≫, ≪Automotive application≫.

Introduction

The average global temperature has risen considerably due to greenhouse gas emissions and the automotive sector is responsible for about 15% of the greenhouse gas emissions [1, 2]. Therefore, it is of importance to increase the utilization of alternative energy carriers, that can replace fossil fuels, especially in this aforementioned automotive applications. Automotive battery packs are generally made up of several modules where each module consists of a number of parallel and/or series connected cells [3]. Differences in leakage currents and cell inhomogeneities cause individual cell voltage and state-of-charge (SOC) distribution among the cells to become non-homogeneous [4]. Therefore, the available energy and power is determined not only by the cell type and size, but to a large extent also by the configuration and battery management system (BMS) [5,6]. By restructuring the cell interconnections and introducing more electronics in the pack, more precise control and better utilization of the energy in the individual modules can increase the available energy and provide additional benefits such as improved battery life and increased usable capacity of the battery pack [7,8].

Modular multilevel converters (MMC), have proven to be highly reliable in power grids with HVDC system because of their modularity and improved EMC [9, 10]. Over the past several years, battery-integrated modular multilevel converters (BI-MMC) have gained popularity for EV powertrains because of their added benefits such as high efficiency, greater cell-level control, and better fault isolation capabilities [11, 12]. Several pieces of research also indicate that there is a great benefit in terms of battery life-time and battery utilization by increasing the controllability of the cells [13, 14]. Some articles also suggested that pulsed current charging improves the lifetime of Li-ion batteries [15, 16]. Furthermore, [17] reported that pulsed charging ($\approx 2\,\mathrm{kHz}$ pulse current) significantly improved the battery lifetime when compared to a constant current charging. These articles suggests that frequency of the battery current is an important factor that influences the battery lifetime. The battery current in BI-MMCs are inherently pulsating thus resulting in higher battery losses than a 2-level inverter [18–20].

This paper extends the results in [21] where the potential benefit of introducing MMCs into hybrid and electric vehicles was presented clarifying the advantages and disadvantages of different MMC topologies for a $400\,\mathrm{kW}$ 40 ton commercial vehicle. Furthermore, [21] presented the principle optimization of the submodule DC-link capacitor and MOSFET switching frequency, minimizing the total power losses. It was found that the the resonance between the battery and the DC-link capacitors determined the allocation of harmonics and in its turn the loss distribution between the battery and the DC-link capacitors. The article also showed that a crucial parameter was the battery impedance. Electrochemical impedance spectroscopy (EIS) is a method used to characterize the properties of electrochemical devices like batteries and capacitors. The EIS signature is often determined using low-amplitude perturbations [22, 23]. However, the EIS and impedance characterizations using pulse tests are not identical [24, 25].

The first contribution in this paper is a methodology to characterize the battery impedance utilizing the full load converter switching currents (switched pulse-test). The second contribution is the experimental validation of the DC-current harmonics allocation and the loss distribution at the switching frequency between the battery and DC-link capacitors on a single-phase half-bridge converter.

Topology overview

A BI-MMC topology consists of either one or two arms per phase and each arm is made up of a number of cascaded stages of DC/AC converters that are commonly referred to as submodules (SM). Fig. 1, presents the schematics of 3-phase single-star half-bridge (SSHB) (Fig. 1(a)) and single-star full-bridge (SSFB) (Fig. 1(b)) topologies with $N_{s(cells)}$ series and $N_{p(cells)}$ parallel cells per SM. Each SM consists of $N_{p(mos)}$ parallel MOSFETs and a DC-link capacitor bank. The DC-link capacitor bank consists of $N_{p(cap)}$ parallel capacitors, here represented by its ideal capacitance C, ESR, and ESL. Fig. 1(a) and Fig. 1(b) also show the half-bridge (HB) and full-bridge (FB) submodules, respectively. A single HB-SM connected to an RL load is used for experimental validations and is referred to as an HB-converter.

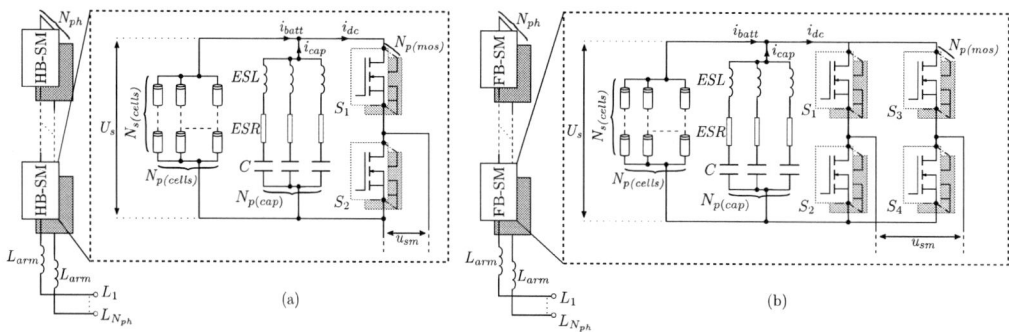

Fig. 1: Battery-integrated modular multilevel inverter topologies. (a) single-star half-bridge and (b) single-star full-bridge topologies.

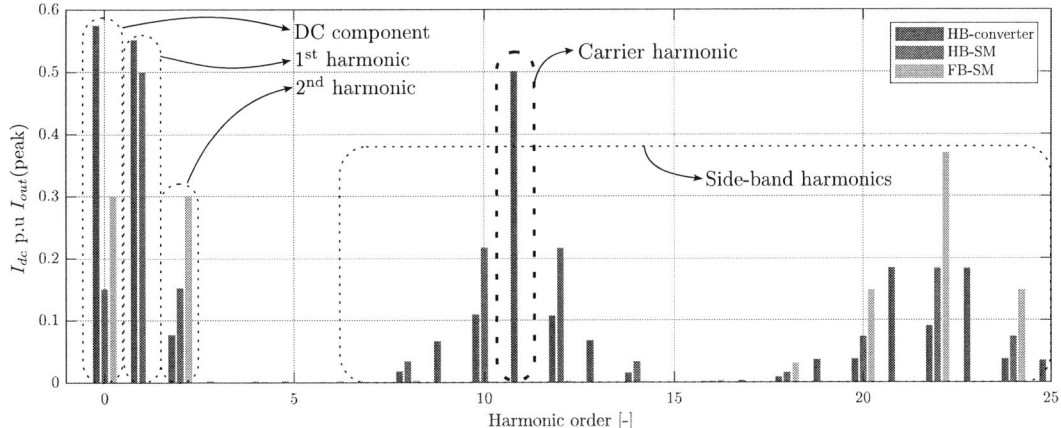

Fig. 2: The DC-side currents per unit output current for HB and FB submodules, and HB-converter considering a sinusoidal carrier based modulation scheme with $f_{sw} = 11 f_1$ with a modulation index of 0.6 and $\cos\phi = 0.9$.

Submodule DC-current

The instantaneous DC-side SM current ($i_{dc}(t)$) is calculated as follows:

$$i_{dc}(t) = S_{conv}(t)\, i_{arm}(t), \qquad\qquad i_{dc(HBc)}(t) = S_{conv}(t)\, i_{out}(t), \qquad\qquad (1)$$

where, $i_{arm}(t)$ represents the instantaneous arm current of the BI-MMC, $S_{conv}(t)$ is the converter/SM switching function, $i_{dc(HBc)}$ is the DC-side current of the HB-converter, and i_{out} is the output current of the HB-converter. In order to capture the high frequency characteristics of the battery, a sinusoidal carrier based modulation is selected. Fig. 2 shows the DC-side currents for the HB and FB SMs, and the HB-converter with a switching frequency, $f_{sw} = 11 f_1$ (where f_1 is the fundamental frequency) considering a modulation index of 0.6 and power factor of 0.9. The modulation index and power factor values correspond to those used in the battery impedance characterization and the validation of the distribution of harmonics between the battery and the DC-link capacitor bank. From the figure it is clear that the DC-side currents for the HB-SM contains DC, 1st, 2nd, and all the side-band harmonic components. However, the DC-side currents for the FB-SM consists of DC, 2nd, and only the even carrier multiples as the side-band harmonic components. However, the HB-converter DC-side current contains not only DC, 1st, 2nd, and side-band harmonics but also the carrier harmonic, i.e, the switching frequency component. The battery current and voltage components at the switching frequency are used to characterize the battery impedance through a set of measurements by varying the switching frequency.

Loss-distribution

The principle distribution of harmonics and losses between the DC-link capacitor bank and the battery is presented in this section.

[21] presented a detailed calculation of the battery and the DC-link capacitor currents, and losses. Two key parameters pointed out in [21] that determine the loss distribution between the battery and the DC-link capacitor bank are the capacitor energy rating and the converter switching frequency. There exists a resonance, at the frequency f_{res}, between the battery and the DC-link capacitor given by the following:

$$f_{res} = \frac{1}{2\pi\,\sqrt{L_{dc}\,C_{cap}}}, \qquad\qquad L_{dc} = L_s + L_p, \qquad\qquad (2)$$

where L_{dc} is the total DC-side inductance, L_s is the inductance of the battery cell, L_p is the total parasitic inductance of the cables connecting the battery and the converter, and C_{cap} is the capacitance of the DC-

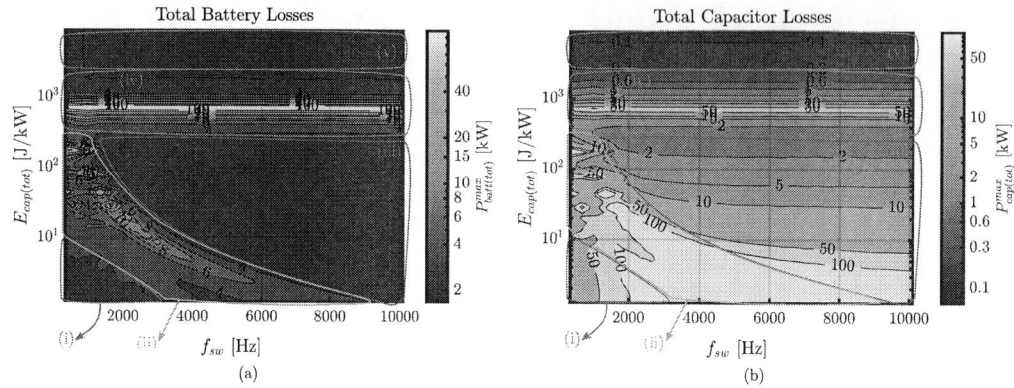

Fig. 3: Power losses as a function of the normalized total energy stored in the capacitors ($E_{cap(tot)}$), and MOSFET switching frequency (f_{sw}) for 3-phase 1-$N_{s(cells)}$ SSFB BI-MMC topology at rated power (P_{tot}^{max}) of 400 kW. (a) total battery losses ($P_{batt(tot)}^{max}$) and (b) total capacitor losses ($P_{cap(tot)}^{max}$).

link capacitor bank. f_{res} defines the allocation of harmonics between the battery and the DC-link capacitor bank. An example of a 3-phase SSFB topology with 1 series battery cell per converter sub-module is presented to illustrate these principles. Fig. 3 shows the calculated total battery losses ($P_{batt(tot)}^{max}$) and capacitor losses ($P_{cap(tot)}^{max}$) as a function of total energy stored in the DC-link capacitor bank normalized to 1 C discharge power of the battery ($E_{cap(tot)}$), and MOSFET switching frequency (f_{sw}) for the converter operating at rated power of 400 kW. The loss distribution can be divided into five regions as shown in the figure. In region (i), $f_{sw} < f_{res}$, as a result, the switching harmonic components of I_{dc} flows through the battery, therefore $P_{batt(tot)}^{max}$ is high. In region (ii), $f_{sw} \approx f_{res}$, i.e, a parallel resonance resulting in low impedance in the loop between L_{dc} and C_{cap} resulting in high circulating currents increases the total losses. In region (iii), $f_{sw} > f_{res}$ and the battery current only consists of the DC-component and the 2nd harmonic (1st harmonic also for HB-SM). As a result, the battery losses are low. Although the capacitor current consists of all the switching frequency components, ESR is low due to large capacitor energy rating, thus resulting in lower capacitor losses. A detailed description of regions (iv) and (v) is presented in [21].

Experimental determination of battery and DC-link capacitor bank impedances

This section describes in detail the battery and DC-link capacitor bank impedance characterization procedure and results. Furthermore, the experimental validation of the principle harmonic allocation and the loss distribution between the battery and the DC-link capacitor bank are presented.

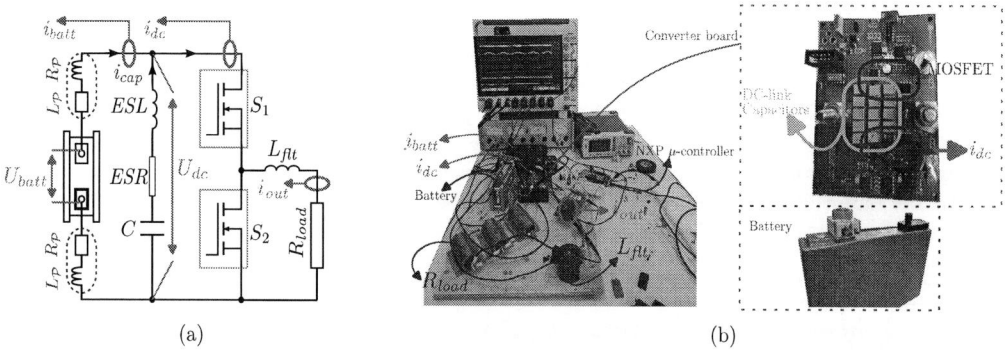

Fig. 4: Battery and DC-link capacitor impedance measurement setup.

Experimental setup

Fig. 4 presents the experimental setup developed and used to determine both the battery and the DC-link capacitor impedances. Fig. 4(a) shows the schematic of the experimental setup to characterize the battery and the DC-link capacitor bank impedances. In the figure, i_{batt}, i_{dc}, and i_{out} are the battery, DC, and output currents measurements, respectively. i_{cap} is the calculated capacitor current (in blue). In order to minimize the drain-source voltage step corresponding to the rate of change of drain current (i.e. minimize ESL in Fig. 4(a)), the trace lengths between the DC-link capacitors and the MOSFETs are short. As a result, i_{cap} cannot be measured using neither the Rogowski transducers nor the LEM transducers. U_{batt} is the battery voltage measurement and U_{dc} is the the DC-link capacitor voltage measurement, L_{flt} is the output filter, R_{load} is the output load resistor, L_P and R_P are the parasitic inductance and resistance of the cables connecting the battery and the converter board, respectively. In order to accurately determine i_{cap}, i_{batt} and i_{dc} are measured using the same type of current sensor. This is done to ensure that both current transducers have identical phase and magnitude. i_{dc} and i_{batt} are measured using T3RC0300-UM Rogowski current waveform transducers [26]. With the use of Rogowski transducers, the DC current measurement is sacrificed. However, a more accurate measurement of the switching currents is achieved. This is an advantage when characterizing the high frequency battery impedance. i_{out} was measured using HAIS 100-P LEM current transducer and the voltage measurements were measured using TPP0500B passive voltage probes with a bandwidth of 500 MHz [27, 28]. U_{batt} measurement is the calculated differential voltage using two passive probes. The measurements were recorded using a Tektronix MSO58, 500 MHz oscilloscope [29]. Fig. 4(b) shows the experimental setup. The figure also shows the Samsung 24 A h Li-ion NMC battery cell connected to a converter board with 14 parallel units of 1 mF, 6.3 V TPSV108M006V0050 AVX Tantalum capacitors [30]. These capacitor units each have an internal resistance (ESR) of 50 mΩ and a maximum current ripple of 2.2 A RMS at 25 °C and 100 kHz. The total energy stored in the DC-link capacitor bank on the converter board normalized to 1 C discharge power of the battery is about $1.2\,\mathrm{J\,kW^{-1}}$. The converter board is a FB inverter with 4 SQJQ142E Vishay MOSFETs. The converter board is operated as an HB-converter and connected to three parallel load resistors of 100 mΩ each (R_{load}) and a filter inductor (L_{flt}) of 48 μH.

The battery discharge rate was set to 1 C and the half bridge converter with a sinusoidal carrier based modulation scheme is utilized to determine the impedances. i_{dc} contains the DC, 1st, 2nd, side-band, and carrier harmonics, and i_{batt} and i_{cap} harmonics depend on the battery, DC-link capacitor bank and cable-parasitic impedances. The current and voltage at the carrier harmonic (i.e. switching frequency component) is used to determine the impedances. By varying f_{sw}, the switching frequency component (or carrier harmonic) is varied and i_{batt}, U_{batt}, i_{dc}, and U_{dc} are measured for every f_{sw}. An impedance spectra is thus obtained. Only the switching frequency components (or carrier harmonics) are used to determine the battery impedance because the amplitude of this component is approximately equal to the 1st harmonic component. The side-band harmonics are not utilized to determine the battery impedance because the amplitude of these harmonics are significantly lower. The fundamental frequency of the output voltage and current is 50 Hz, thus the 1st and 2nd harmonic components, although high in amplitude, are close to the lower bandwidth of the current transducers, thus also not used to determine the impedances.

The battery, DC link capacitor bank, and the parasitic impedances are determined using the following relation:

$$Z_{batt}(\omega_{sw}) = \frac{U_{batt}(\omega_{sw})}{I_{batt}(\omega_{sw})}, \qquad\qquad Z_{cap}(\omega_{sw}) = \frac{U_{dc}(\omega_{sw})}{I_{cap}(\omega_{sw})},$$

$$Z_{dc}(\omega_{sw}) = \frac{U_{dc}(\omega_{sw})}{I_{batt}(\omega_{sw})}, \qquad\qquad Z_{P}(\omega_{sw}) = Z_{dc}(\omega_{sw}) - Z_{batt}(\omega_{sw}), \qquad (3)$$

where Z_{batt} and Z_{cap} are the battery and the DC-link capacitor bank complex impedances, respectively, U_{batt} and U_{dc} are the measured complex battery and DC-link capacitor bank voltages, respectively. I_{batt} is the measured complex battery current, I_{cap} is the calculated complex capacitor current, and ω_{sw} is the angular switching frequency ($\omega_{sw} = 2\pi f_{sw}$). Z_{dc} is the total DC-side impedance and Z_P is the parasitic impedance of the cables connecting the converter board and the battery. It is important to mention that

Experimental Evaluation of Battery Impedance and Submodule Loss Distribution for
Battery Integrated Modular Multilevel Converters

BALACHANDRAN Arvind

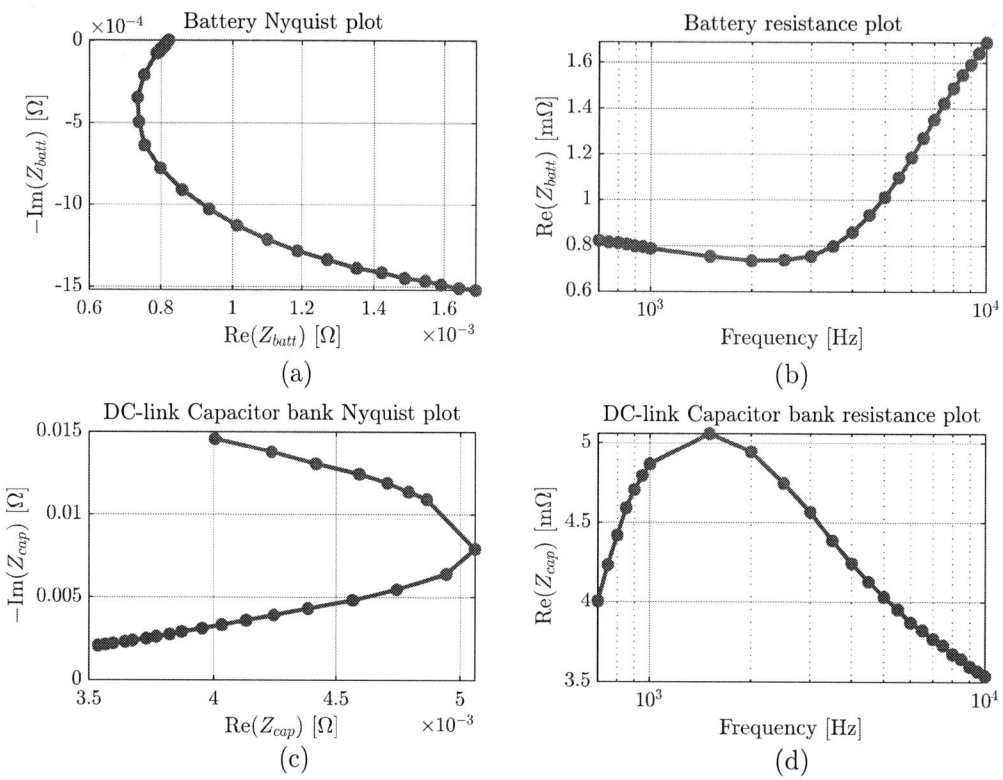

Fig. 5: Experimentally determined battery and DC-link capacitor bank impedances. (a) battery impedance Nyquit plot and (b) resistance plot, and (c) Nyquist plot of the DC-link capacitor bank impedance and (d) resistance plot.

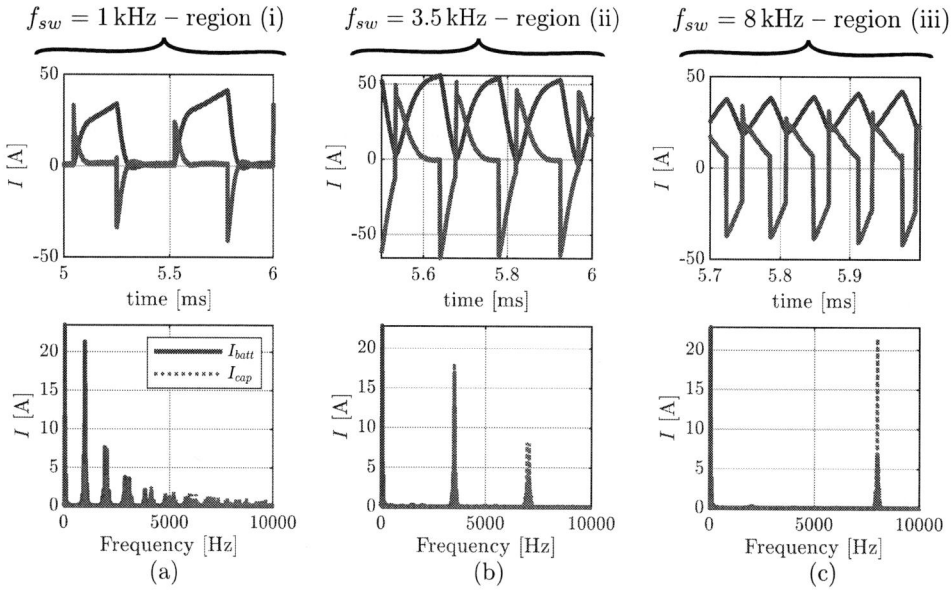

Fig. 6: Measured battery and capacitor currents different converter switching frequencies (f_{sw}). (a) Battery and capacitor current time plot and frequency spectra with $f_{sw} = 1$ kHz, in region (i), (b) $f_{sw} = 3.5$ kHz, in region (ii), and (c) $f_{sw} = 8$ kHz, in region (iii).

the battery state of charge during the measurement period is fairly constant. This is because during the measurement interval, the battery was discharged with 24 A for only a few seconds.

The HB-converter switching frequency was varied from 700 Hz to 10 kHz and Z_{batt}, Z_{cap}, and Z_{dc} were determined at these frequencies. The recording time for each frequency was about 100 ms with a sample rate of 12.25 MS/s. Fig. 5 presents the battery and the DC-link capacitor bank impedances for different f_{sw}. Fig. 5(a) presents the battery impedance Nyquist plot and Fig. 5(b) shows the resistance plot. Fig. 5(c) and Fig. 5(d) shows the DC-link capacitor bank Nyquist and resistance plots, respectively.

Loss Distribution

Fig. 6 shows the measured battery and capacitor current, time and frequency domain plots at three different switching frequencies illustrating the three different regions of harmonic allocations. In the figure, I_{batt} is the measured battery current and I_{cap} is the calculated DC-link capacitor bank current. Fig. 6(a) presents the HB-converter operating in region (i) (see Fig. 3) and it is clear that I_{batt} contains majority of the switching frequency components. Fig. 6(b) presents the HB-converter operating in region (ii) and it is clear that the magnitude of I_{batt} and I_{cap} components at f_{sw} are similar, indicating a resonance between the DC-link capacitor bank and the battery. Fig. 6(c) presents the HB-converter operating in region (iii) and it is clear that I_{cap} contains the majority of the switching frequency components. In all the three cases presented in the figure, it is clear that the resonance defines not only the allocation of the switching frequency component but also all the side bands and the low-frequency harmonic components. Therefore, it is sufficient to concentrate on the switching frequency component to illustrate the distribution of losses between the battery and the DC-link capacitor bank.

Fig. 7 shows the loss distribution between the battery and the DC-link capacitors at different f_{sw}. The figure shows the total DC-side impedance ($Z_{dc} = Z_{batt} + Z_{\mathcal{P}}$, where $Z_{\mathcal{P}}$ is the cable impedance) and the DC-link capacitor bank impedance (Z_{cap}). These impedances determine the allocation of harmonics between the battery the DC-link capacitor bank and is divided into three different regions. The area shaded in blue corresponds to region (i) and here $Z_{dc} > Z_{cap}$. The area shaded in orange represents the region (ii) and here $Z_{dc} \approx Z_{cap}$, indicating the resonance. The resonance frequency was found to be about 3.3 kHz. Finally, the shaded area in yellow represents region (iii), where $Z_{dc} < Z_{cap}$. It is important to mention that the losses presented in the figure are the losses determined at f_{sw}, i.e:

$$P_{batt}(\omega_{sw}) = (I_{batt}(\omega_{sw}))^2 \, \mathrm{Re}(Z_{batt}(\omega_{sw})), \qquad P_{\mathcal{P}}(\omega_{sw}) = (I_{batt}(\omega_{sw}))^2 \, \mathrm{Re}(Z_{\mathcal{P}}(\omega_{sw})),$$

$$P_{cap}(\omega_{sw}) = (I_{cap}(\omega_{sw}))^2 \, \mathrm{Re}(Z_{cap}(\omega_{sw})), \qquad \omega_{sw} \in [1400\pi,\ 20000\pi]. \tag{4}$$

As mentioned previously, the magnitudes of the side-band harmonics are significantly lower. Due to the amplitude and bandwidth limitations of the current transducers and the voltage probes, only the switching frequency components are used determine the loss distribution between the battery and DC-link capacitance. Fig. 7(a) shows the battery losses (P_{batt}) at different f_{sw} and it is evident that P_{batt} has its maximum at region (i), this is due to the fact that the battery current contains majority of the switching-frequency harmonic components. As f_{sw} increases, Z_{batt} increases and Z_{cap} decreases. As a result, the switching harmonics are bypassed through the DC-link capacitor bank. Thus, P_{batt} is low. Fig. 7(b) shows the losses in the cables ($P_{\mathcal{P}}$) between the battery and the converter board and it follows P_{batt} for the same reasons as mentioned earlier. Fig. 7(c) shows the DC-link capacitor bank losses (P_{cap}) and it is clear that it has its minimum in region (i). This is due to the fact that I_{cap} has little to no harmonic components. With an increased f_{sw}, i.e, moving towards region (iii), P_{cap} increases because I_{cap} now contains all the high-frequency harmonic components. In region (ii), P_{cap} has its maximum because of the resonance. It is clear that P_{cap} is subjected to the capacitor design, the ESR related to capacitor type and number of parallel units. An improved capacitor design with lower ESR will reduce P_{cap}. Fig. 7(d) and Fig. 7(e) presents I_{batt} and I_{cap} switching frequency components at different f_{sw}, respectively. From the figure it is clear that in region (i) $I_{batt} > I_{cap}$, region (ii) $I_{batt} \approx I_{cap}$, and in region (iii) $I_{batt} < I_{cap}$. From the figure, it is also clear that the DC-link capacitor bank current is lower than the ripple current limit from the capacitor data-sheet, i.e. $I_{cap} < 18.5$ A at 700 Hz and $I_{cap} < 25$ A at 10 kHz [30, 31].

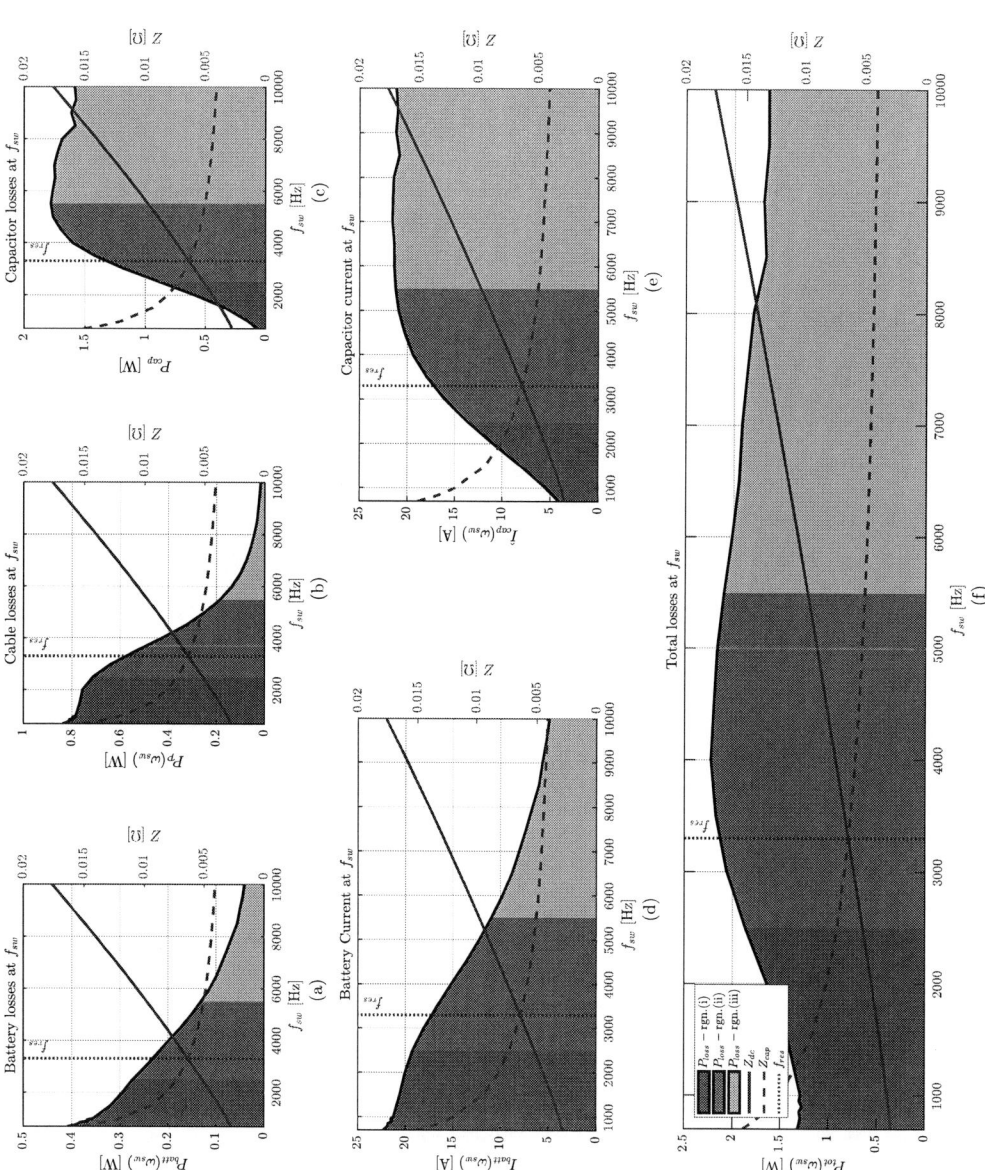

Fig. 7: Loss distribution between the battery and DC-link capacitor bank at different converter switching frequencies (f_{sw}). (a) battery loses (P_{batt}), (b) cable losses, (c) DC-link capacitor bank losses (P_{cap}), (d) battery current (I_{batt}) component at f_{sw}, (e) DC-link capacitor bank current (I_{cap}) component at f_{sw}, and (f) total losses (P_{tot}).

Fig. 7(f) presents the total losses ($P_{tot} = P_{batt} + P_{cap} + P_{\mathcal{P}}$) at different f_{sw}. From the figure it is clear that P_{tot} has its minimum in region (i), the maximum losses occurs in region (ii), and P_{tot} gradually reduces in region (iii).

Conclusion

A methodology to characterize the battery impedance at high frequencies using the full-load converter current is presented. The effect of battery and capacitor impedances on the battery and the DC-link capacitor bank currents' harmonics are experimentally validated. The allocation of harmonics can be divided into three regions given by the selection of switching frequency in relation to the resonance between battery inductance, cable inductance and DC-link capacitance. In region (i), the switching frequency (f_{sw}) is below the resonance (f_{res}), the battery current contains the major share of the switching frequency component and therefore increasing the battery losses. The total losses at the switching frequency are lowest in this region. The total losses peak in region (ii) because of the resonance between the battery and DC-link capacitor bank. Finally, in region (iii), f_{sw} is above the resonance, the battery current has little to no switching frequency harmonics and therefore reducing the battery losses. Although the capacitor currents are high in this region, the low resistance of the capacitors results in lower total losses in region (iii) when compared to region (ii).

References

[1] James Hansena, Makiko Satoa, Reto Ruedyb, Gavin Schmidtc, and Ken Lob. Global temperature in 2019. 2020.

[2] Céline JW Bonfils, Benjamin D Santer, John C Fyfe, Kate Marvel, Thomas J Phillips, and Susan RH Zimmerman. Human influence on joint changes in temperature, rainfall and continental aridity. *Nature Climate Change*, 10(8):726–731, 2020.

[3] Jiuchun Jiang and Caiping Zhang. *Fundamentals and applications of lithium-ion batteries in electric drive vehicles*. John Wiley & Sons, 2015.

[4] Thorsten Baumhöfer, Manuel Brühl, Susanne Rothgang, and Dirk Uwe Sauer. Production caused variation in capacity aging trend and correlation to initial cell performance. *Journal of Power Sources*, 247:332–338, 2014.

[5] Gregory L Plett. *Battery management systems, Volume II: Equivalent-circuit methods*. Artech House, 2015.

[6] Helena Berg. *Batteries for electric vehicles: materials and electrochemistry*. Cambridge university press, 2015.

[7] Ye Li and Yehui Han. A module-integrated distributed battery energy storage and management system. *IEEE Transactions on Power Electronics*, 31(12):8260–8270, 2016.

[8] Chang-Hua Lin, Hsuan-Yi Chao, Chien-Ming Wang, and Min-Hsuan Hung. Battery management system with dual-balancing mechanism for LiFePO$_4$ battery module. In *TENCON 2011-2011 IEEE Region 10 Conference*, pages 863–867. IEEE, 2011.

[9] Anton Lesnicar and Rainer Marquardt. An innovative modular multilevel converter topology suitable for a wide power range. In *2003 IEEE Bologna Power Tech Conference Proceedings,*, volume 3, pages 6–pp. IEEE, 2003.

[10] Kamran Sharifabadi, Lennart Harnefors, Hans-Peter Nee, Staffan Norrga, and Remus Teodorescu. *Design, control, and application of modular multilevel converters for HVDC transmission systems*. John Wiley & Sons, 2016.

[11] Lennart Baruschka and Axel Mertens. Comparison of cascaded H-bridge and modular multilevel converters for BESS application. In *2011 IEEE Energy Conversion Congress and Exposition*, pages 909–916. IEEE, 2011.

[12] Faisal Altaf. *On Modeling and Optimal Control of Modular Batteries*. PhD thesis, Ph. D. dissertation, Chalmers University of Technology, 2016.

[13] Jorn M Reniers, Grietus Mulder, Sina Ober-Blöbaum, and David A Howey. Improving optimal control of grid-connected lithium-ion batteries through more accurate battery and degradation modelling. *Journal of Power Sources*, 379:91–102, 2018.

[14] Nejmeddine Bouchhima, Matthias Gossen, and Kai Peter Birke. Fundamental aspects of reconfigurable batteries: Efficiency enhancement and lifetime extension. *Modern Battery Engineering: A Comprehensive Introduction*, page 101, 2019.

[15] Xinrong Huang, Wenjie Liu, Jinhao Meng, Yuanyuan Li, Siyu Jin, Remus Teodorescu, and Daniel-Ioan Stroe. Lifetime extension of lithium-ion batteries with low-frequency pulsed current charging. *IEEE Journal of Emerging and Selected Topics in Power Electronics*, 2021.

[16] Xinrong Huang, Siyu Jin, Jinhao Meng, Remus Teodorescu, and Daniel-Ioan Stroe. The effect of pulsed current on the lifetime of lithium-ion batteries. In *2021 IEEE Energy Conversion Congress and Exposition (ECCE)*, pages 1724–1729. IEEE, 2021.

[17] Xinrong Huang. The effects of pulsed charging current on the performance and lifetime of lithium-ion batteries. 2021.

[18] Mahran Quraan, Taejung Yeo, and Pietro Tricoli. Design and control of modular multilevel converters for battery electric vehicles. *IEEE Transactions on Power Electronics*, 31(1):507–517, 2015.

[19] Anton Kersten, Manuel Kuder, Emma Grunditz, Zeyang Geng, Evelina Wikner, Torbjörn Thiringer, Thomas Weyh, and Richard Eckerle. Inverter and battery drive cycle efficiency comparisons of CHB and MMSP traction inverters for electric vehicles. In *2019 21st European Conference on Power Electronics and Applications (EPE'19 ECCE Europe)*, pages P–1. IEEE, 2019.

[20] Oskar Josefsson. *Investigation of a multilevel inverter for electric vehicle applications*. PhD thesis, 2015.

[21] Arvind Balachandran, Tomas Jonsson, and Lars Eriksson. Design and analysis of battery-integrated modular multilevel converters for automotive powertrain applications. In *2021 23rd European Conference on Power Electronics and Applications (EPE'21 ECCE Europe)*, pages P–1. IEEE, 2021.

[22] Thomas F Landinger, Guenter Schwarzberger, and Andreas Jossen. A novel method for high frequency battery impedance measurements. In *2019 IEEE International Symposium on Electromagnetic Compatibility, Signal & Power Integrity (EMC+ SIPI)*, pages 106–110. IEEE, 2019.

[23] Oskar Theliander, Anton Kersten, Manuel Kuder, Weiji Han, Emma Arfa Grunditz, and Torbjorn Thiringer. Battery modeling and parameter extraction for drive cycle loss evaluation of a modular battery system for vehicles based on a cascaded h-bridge multilevel inverter. *IEEE Transactions on Industry Applications*, 56(6):6968–6977, 2020.

[24] Awais Chaudhry. Investigation of lithium-ion battery parameters using pulses and EIS. 2018.

[25] Andreas Blidberg. Correlation between different impedancemeasurement methods for battery cells, 2012.

[26] T3RC current probes datasheets – rogowski current probes. https://cdn.teledynelecroy.com/files/pdf/t3rc-current-probes-datasheet.pdf. (Accessed on 06/09/2022).

[27] Current transducer HAIS 50 .. 400–P and 50 .. 150–TP. https://www.lem.com/sites/default/files/products_datasheets/hais_50_400-p_and_50_150-tp.pdf. (Accessed on 06/09/2022).

[28] Passive voltage probes TPP1000 TPP0500B TPP0502 TPP0250 datasheet. https://download.tek.com/datasheet/TPP1000-TPP0500B-TPP0502-TPP0250-Passive-Voltage-Probe-Datasheet-51W261519.pdf. (Accessed on 06/09/2022).

[29] 5 series MSO specifications and performance verification. https://download.tek.com/manual/5-Series-MSO-Specifications-Performance-Verification-Manual-RevA-077130605.pdf. (Accessed on 06/09/2022).

[30] TPS series – low esr. https://datasheets.kyocera-avx.com/TPS.pdf. (Accessed on 06/09/2022).

[31] Kyocera AVX SpiCAT online simulator. https://spicat.kyocera-avx.com/product/tan/chartview/TPSV108M006R0050/dist/mouser. (Accessed on 06/15/2022).

Constant DC power infeed grid forming with improved ability to ride-through unbalanced low-voltage faults

Tayssir Hassan, Malte Eggers, Huoming Yang, Peter Teske, Sibylle Dieckerhoff
TECHNISCHE UNIVERSITÄT BERLIN
Chair of Power Electronics
Einsteinufer 19
10587 Berlin, Germany
Tel: +49 (0)30 314-25513
Email: t.hassan@tu-berlin.de
URL: http://www.pe.tu-berlin.de

Keywords

≪Grid-connected converter≫, ≪Grid-forming converters≫, ≪Converter control≫, ≪Small signal stability≫, ≪Real-time simulation≫.

Abstract

A new optimized grid-forming control strategy based on matching control is presented, which can be used in an unbalanced system and provides minimum frequency support by utilizing the energy stored in the dc-link capacitor. Simulation and measurements in different scenarios and under asymmetrical grid faults evaluate the proposed strategy.

Introduction

In today's power grids, the contribution of renewable energy sources such as wind turbines and photovoltaic cells is increasing rapidly [8]. These sources are mainly connected to the power grid via power electronics converters, which usually operate in a grid-following manner. In this way, they feed the maximum available active power into the grid and act as a constant power source, independent of grid frequency fluctuations, which reduces the ability of the power grid to cope with frequency deviations[9]. To improve stability in grids with high penetration of power electronics, many grid-forming control strategies have been presented. One of these is the matching control, which is introduced in [1]. The matching control creates an equivalence between the energy stored in the dc-link capacitor to the rotational energy of a synchronous generator. This means that the dc-link voltage varies proportional to frequency variations. The authors of [2] modified this strategy to eliminate the steady-state dc-link voltage deviation and performed a stability analysis. In [3] the impact of reaching DC- and AC-current limitation has been investigated in addition to the interaction between the fast grid-forming converters and the slow synchronous machine dynamics.

This paper presents an optimized grid-forming control strategy based on the matching control, focusing on improving the performance of this strategy in the presence of unbalanced grid faults. Moreover, the effectiveness of using the dc-link capacitor to provide frequency support in a limited range is investigated, and the impact of this small inertia contribution to the distribution grid is studied.

A small-signal model helps to study the system stability and to determine the control parameters. For experimental verification, the control strategy is implemented in a test bench with three converters in parallel connected to a hardware-in-the-loop system. Several grid faults are applied, and their effects on the strategy are investigated.

The paper is organized as follows: The original matching control is discussed, highlighting its weaknesses. Then, the proposed changes to this strategy are presented. Finally, the measurement results are introduced and evaluated.

The matching control strategy

The basic idea of the matching control strategy is to establish equivalence between the electrical energy stored in the dc-link capacitor to the rotational energy stored in the rotor of a synchronous generator. Considering the mathematical similarity between the differential equation of the DC capacitor and the swing equation of the synchronous generator. Equation (1) shows the swing equation

$$J_{SG} \cdot \omega_n \cdot \frac{d\omega}{dt} + D_{SG} \cdot \omega_n \cdot (\omega - \omega_n) = P_T - P_{ESG} \tag{1}$$

where J_{SG} is the inertia of the generator, ω is the rotational speed, ω_n is the synchronous rotational speed, D_{SG} is the damping constant, P_T is the mechanical power, and P_{ESG} is the electrical output power of the generator. The angular momentum at the synchronous speed is $M = J_{SG} \cdot \omega_n$ and $D = D_{SG} \cdot \omega_n$ is the damping constant at this speed.

The dynamic equation of the dc-link capacitor is presented in equation (2)

$$C_{DC} \cdot \frac{dV_{DC}}{dt} = i_S - i_E = \frac{P_s}{V_{DC}} - \frac{P_E}{V_{DC}} \qquad \Longrightarrow \qquad \frac{C_{DC}}{2} \cdot \frac{dV_{DC}^2}{dt} = P_S - P_E \tag{2}$$

where P_s is the power delivered by the power source, i_S is the associated DC-current. P_E is the active power output of the converter with i_E as the corresponding DC-current.

According to the definition of the matching control, equation (1) and (2) are equal, by integrating equation (3) and applying the Laplace transform on it equation 4 is obtained.

$$M \cdot \frac{d\omega}{dt} + D \cdot (\omega_{SG} - \omega_n) = \frac{C_{DC}}{2} \cdot \frac{dV_{DC}^2}{dt} \tag{3}$$

$$\omega = \omega_n + \frac{s}{k_j \cdot s + k_d} \cdot (V_{DC}^2 - V_{DC,n}^2) \qquad k_j = \frac{2 \cdot M}{C_{DC}} \qquad k_d = \frac{2 \cdot D}{C_{DC}} \tag{4}$$

Equation (4) represents the relation between the rotational speed and the dc-link voltage. This relation is defined by a differential-controller which reacts to frequency deviation by controlling the dc-link voltage. In [2] equation (4) is modified to a proportional-differential (PD) controller as shown in equation (5),

$$\omega = \omega_{ref} + \frac{s + k_t}{k_j \cdot s + k_d} \cdot (V_{DC}^2 - V_{DC,n}^2) \qquad \omega_{ref} = m \cdot \omega_{PLL} + (1 - m) \cdot \omega_n \tag{5}$$

where ω_{ref} is the reference angular frequency, ω_{PLL} is the angular frequency measured by the phased-locked loop (PLL), m is a weighting coefficient between 0 and 1 and k_t, k_d, k_j are the control parameters. The reference angular frequency ω_{ref}, is not a constant value, which helps to reduce the deviation of the dc-link voltage from its nominal value during frequency deviation. By using the new controller presented in [2], the dynamic equation of the dc-link capacitor becomes as follows:

$$(k_j C_{DC}/2) \cdot s \cdot \omega + (k_d C_{DC}/2) \cdot (\omega - \omega_{ref}) - (P_S - P_E) = (k_t C_{DC}/2) \cdot (V_{DC}^2 - V_{DC,n}^2) \tag{6}$$

If the rotational frequency equals its reference value, then the right side of equation (6) is zero and the dynamic equation of the dc-link capacitor matches the swing equation of the synchronous generator. However, this can only be achieved when the weighting coefficient m is equal to 1, which increases the influence of the PLL on the system. Therefore, the performance of the system under weak grid conditions is limited. Moreover, asymmetrical faults cause 100 Hz power oscillation and thus lead to 100 Hz output frequency oscillation [7]. Under this condition, the PLL may also suffer from stability issues that can cause the whole system to become unstable. The model presented in [2] has been extended and optimized to solve these problems and improve the behavior of the system under such conditions.

The optimized model

The proposed strategy provides a way to eliminate the dc-link voltage deviation without requiring a PLL, resulting in an overall better stability of the system. To achieve this, the PD-controller shown in equation (5) is replaced by a proportional-integral-differential (PID) controller, which ensures zero steady state error of the dc-link voltage. In addition to the new controller, a notch filter is integrated into the control system to eliminate the aforementioned frequency oscillation during unbalanced faults.

$$\omega = \omega_n + \frac{s^2 + k_1 \cdot s + k_2}{k_3 \cdot s \cdot (s + k_4)} \cdot (V_{DC}^2 - V_{DC,n}^2) \tag{7}$$

where k_1, k_2, k_3 and k_4 are the control parameters. With the new control strategy, the dynamic equation of the dc-link capacitor is shown in equation (8).

$$\underbrace{(1/k_2) \cdot s \cdot \omega - (k_1/k_2) \cdot (\omega - \omega_n) - (P_S - P_E)}_{\text{Grid synchronization unit}} = \underbrace{(1/k_2) \cdot (k_3 + k_4 \cdot \frac{1}{s}) \cdot (V_{DC}^2 - V_{DC,n}^2)}_{\text{dc-link voltage controller}} \tag{8}$$

During steady-state, the right side of equation (8) is always zero, which ensures a zero deviation of the dc-link voltage even during frequency deviation.

Fig. 1 shows the block diagram of the control system. The dc-link voltage controller block combined with the grid synchronization unit, represent the dynamics of the dc-link capacitor shown in equation (8). The reactive current droop block, consists of the $-i_Q \sim V$-droop as proposed in [4], and a PI-

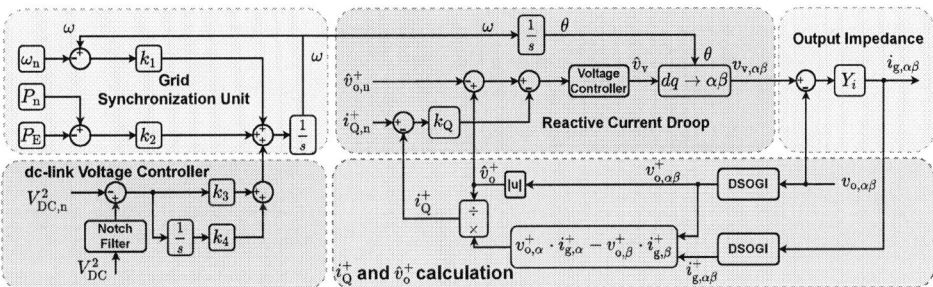

Fig. 1: Block diagram of the control strategy

voltage controller that sets the virtual internal voltage to regulate the reactive current. The output impedance block, is a combination between the grid impedance and a virtual impedance, and is used to calculate the output current based on the virtual voltage and the output voltage. Finally, the last block, calculates the positive and negative sequence of the output voltage and grid current by using a dual second-order generalized integrator (DSOGI), where these quantities are used to calculate the positive sequence voltage amplitude of the converter output voltage $\hat{v}_o^+$, and of the reactive current $\hat{i}_Q^+$ as shown in the figure.

Small-signal Model

To optimize the control parameters, the small-signal model of Fig. 1 is derived. The model consists of two parts, the plant part represents the physical components of the system, while the control part represents the active and reactive power control. The output impedance and the DSOGI build up the plant model, while the remaining blocks of Fig. 1 represent the controller part. For the small-signal model the converter is considered as an ideal part, i.e., the internal current and voltage control in addition to the semiconductor losses were not evaluated. Therefore, the converter is represented by an ideal AC voltage source which is connected to the grid through an impedance. In Fig. 2 a single-phase equivalent circuit of the system is presented in addition to the plant model. Fig. 3 shows the block diagram of the DSOGI, based on this a state space model of the DSOGI is implemented where the inputs are the grid current i_g and the converter output voltage v_o while the system exports

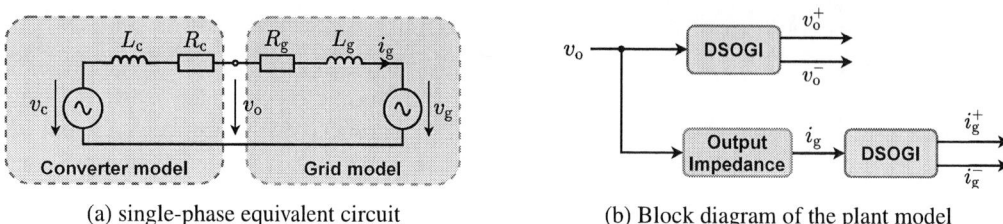

(a) single-phase equivalent circuit   (b) Block diagram of the plant model

Fig. 2: Block diagram of the physical part of the small-signal model

the positive i_g^+, v_o^+ and negative i_g^-, v_o^- sequence of the input quantities. This system is presented in equations (9) and (10), where ω is the fundamental frequency of the input signal and $k = \sqrt{2}$.

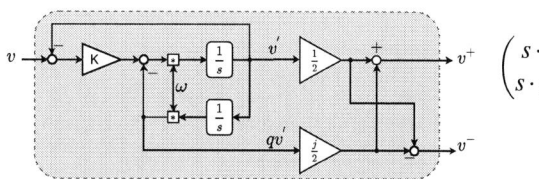

Fig. 3: Block diagram of the DSOGI

$$\begin{pmatrix} s \cdot v'(s) \\ s \cdot qv'(s) \end{pmatrix} = \begin{pmatrix} -\omega k & -\omega \\ \omega & 0 \end{pmatrix} \cdot \begin{pmatrix} v'(s) \\ qv'(s) \end{pmatrix} + \begin{pmatrix} \omega \\ 0 \end{pmatrix} \cdot v(s) \quad (9)$$

$$\begin{pmatrix} v^+(s) \\ v^-(s) \end{pmatrix} = \begin{pmatrix} 1/2 & j/2 \\ 1/2 & -j/2 \end{pmatrix} \cdot \begin{pmatrix} v'(s) \\ qv'(s) \end{pmatrix} \quad (10)$$

The grid impedance together with the virtual impedance used by the reactive power control represent the output impedance of the system. From the differential equation of the output impedance shown in equation (11), the small-signal model of the output impedance can be derived.

$$v_o(s) = s \cdot L_i \cdot i_g(s) + R_i \cdot i_g(s) + v_g(s) \implies s \cdot i_g(s) = \frac{R_i}{L_i} \cdot i_g(s) + \begin{pmatrix} -1/L_i & 1/L_i \end{pmatrix} \cdot \begin{pmatrix} v_g(s) \\ v_o(s) \end{pmatrix} \quad (11)$$

The system of the plant model is then obtained by combining the models of the DSOGI and the output impedance. The acquired system consist of AC quantities which are represented in the complex domain. To apply the control strategies, the Park transform should be performed on the system. Moreover, the system must be transformed in to the real domain to describe the real physical system.

Besides the plant model which represents the physical part, the control model describes the applied active and reactive power control. For the reactive power, the strategy presented in [4] is used. This algorithm replaces the $Q \sim V$-droop with a $-i_Q \sim V$-droop, where i_Q is the reactive current. As a result, the control strategy provides voltage support similar to that required by grid-connected inverters [6] in existing grid codes. In addition to that, the $-i_Q \sim V$-droop allows normal operation for deeper voltage sags at similar voltage support characteristic to the $Q \sim V$-droop in the nominal range. Also, this strategy improves the linearity of the system due to the linear relationship between i_Q and V. According to Fig. 1 the positive sequence of the converter output voltage v_o^+ and grid current i_g^+ are used to calculate the positive sequence amplitude of the converter output voltage $\hat{v}_o^+$ and reactive current i_Q^+. The difference between the calculated values and their reference values represent the input signals of the voltage controller, where a proportional-integral (PI) controller is used. The controller outputs the virtual voltage amplitude $\hat{v}_v$. Equation (12) shows the virtual voltage amplitude in addition to the PI-controller, k_{Qp} and k_{Qi} are the control parameters while k_Q defines the $-i_Q \sim V$-droop.

$$\hat{v}_v = \left(k_{Qp} + \frac{k_{Qi}}{s}\right) \cdot \left(\hat{v}_o^+ - \hat{v}_{o,n}^+ - k_Q \cdot \left(i_Q^+ - i_{Q,n}^+\right)\right) \quad \text{with} \quad k_Q = 0.5 \cdot \frac{V_{o,n}}{I_n} \quad (12)$$

The active power control is ensured by applying the presented algorithm in equation (7). The power-to-frequency characteristic is shown in equation (8). During steady state, the PI-controller ensures zero steady-state deviation of the dc-link voltage. Based on that, equation (8) becomes similar to equation (1) which means that in steady-state, the behaviour of the converter matches that of the synchronous generator. As it can be seen from the control equations the control system is not linear, therefore to build a small-signal model the equations need to be linearized around an operating point to calculate the system matrices. This operating point is chosen to be the nominal operating point of the converter (table II). After the linearization, the control model can be obtained by combining the active and reactive power control systems. Finally, the combination of the plant and control model produces a model that describes the system.

To perform stability analysis, the eigenvalues of the system matrix A of the obtained model must be calculated, where these eigenvalues are related to the control parameters. The matrix A is of order 15, which means that there are 15 eigenvalues.

For the stability analysis, a parameter sweep was performed, starting from a certain point, the parameters were changed to cover a specific range, and at each point, the eigenvalues were calculated and analysed. Due to the six control parameters and the high order of matrix A, the parameter sweep method is complex and the analysis of the results is fraught with difficulties. To reduce the complexity of the problem, the parameter k_4 is not swept. This parameter is the gain of the integrator to keep the dc-link voltage at its nominal value in the steady-state. The mentioned parameter has no significant influence on the stability of the system. It only affects the time needed for the dc-link voltage to reach its nominal value after a disturbance of the system. From the simulation results shown later, a value of $k_4 = 0.08 \, \text{A}^2 \, \text{s} \, \text{J}^{-2}$ gives a good compromise between the control time and the overshoot of the dc-link voltage. For further reduction of complexity, the sweep is performed for each parameter individually after selecting the optimal value, the next parameter will be swept. The sweep start from the following parameters:

$$k_1 = -90 \, \text{s}^{-1}; \quad k_2 = 1.2 \, \text{J}^{-1} \, \text{s}^{-1}; \quad k_3 = 0.02 \, \text{A}^2 \, \text{J}^{-2}; \quad k_{Qi} = -17; \quad k_{Qp} = 0.4 \tag{13}$$

The first parameter that swept is k_2. The movements of the eigenvalues is shown in Fig. 4b. The analysis shows that the eigenvalues 1,2,3,4,7 and 8 are not changed by the shift Whereas, the eigenvalues 5,6,9 and 10 are slightly affected by the variation of k_2. These poles move away from the origin of the coordinate system with a decreasing imaginary part. This improves the damping and response of the system. On the other hand, 11 and 12 shift to the left in the coordinate system when k_2 is increased. This shift enhances the stability of the system. However, eigenvalues 13, 14 and 15 become unstable when k_2 becomes too large. Equation (8) shows that the parameter k_2 is inversely proportional to the emulated inertia. Hence, if k_2 is too large, the emulated inertia will be minimal. As a result, the system response becomes too fast, and damping decreases. Conversely, if the value of k_2 is small, the emulated inertia is large, and cannot be realized due to the significant drop of the dc-link voltage, which will cause large-signal instability.

The optimal value is chosen so that the system has the best damping and fast response. This is achieved by placing most of the eigenvalues on the real axis. For $k_2 = 1.3 \, \text{J}^{-1} \, \text{s}^{-1}$, eigenvalues 13, 14, and 15 have an imaginary part of zero. The remaining poles are well-damped and fast.

Secondly, k_1 will be swept, with a range from 0 to $-150 \, \text{s}^{-1}$. The loci of the eigenvalues are shown in Fig.4a. Similar to k_2, the poles 1,2,3,4,7 and 8 are not affected by the shift. On the other hand, eigenvalues 5,6,9 and 10 tend to shift toward the origin of the coordinate system when k_1 is decreased. However, these poles become better damped. Moreover, poles 11 and 12 start near the origin and move to the left when k_1 decreases. Nevertheless, both poles move back to the origin when $k_1 \leq -95 \, \text{s}^{-1}$. Furthermore, eigenvalues 13, 14 and 15 remain on the real axis when k_1 decreases. A comparison between equation (8) and (1) shows k_1 is proportional to the damping parameter D_{SG}. This means that with a higher value of k_1, the system is better damped. This behavior is also shown in Fig.4a, where the imaginary part of the eigenvalues becomes smaller with larger value of k_1. Even so, when the damping factor is too high, the system becomes unstable. As with k_2, the optimal value is chosen to achieve the best damping and fast response. A value of $k_1 = -110 \, \text{s}^{-1}$ satisfies the requirements. At this value, eigenvalues 11 and 12 occupy the most stable point. Where poles 13,14 and 15 settle on the real axis.

The parameters k_1 and k_2 are responsible for the synchronization of the grid and the control of the active power. After tuning these parameters, the reactive power control is now optimized by finding the optimal values for k_{Qp} and k_{Qi}. Fig. 5a shows the loci of the eigenvalues when k_{Qp} is modified. The eigenvalues 1,2,3,4,7 and 8 are not affected by the sweep of the parameter. However, as can be seen in Fig. 5a, the

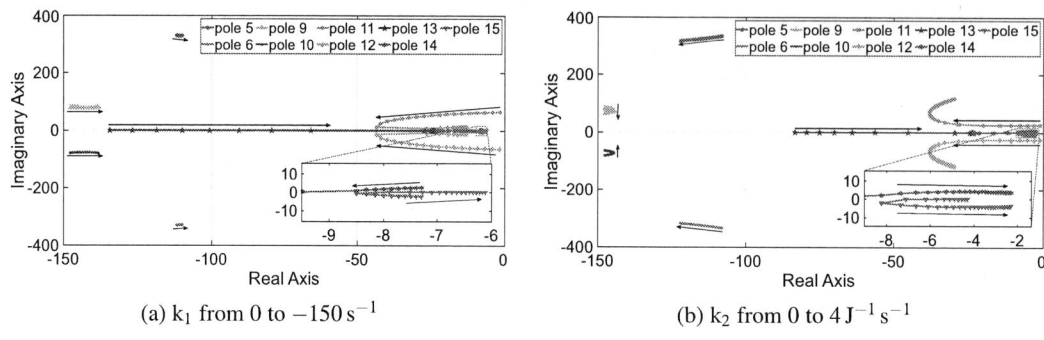

(a) k_1 from 0 to $-150\,\mathrm{s}^{-1}$

(b) k_2 from 0 to $4\,\mathrm{J}^{-1}\,\mathrm{s}^{-1}$

Fig. 4: Small signal model stability analysis active power controller

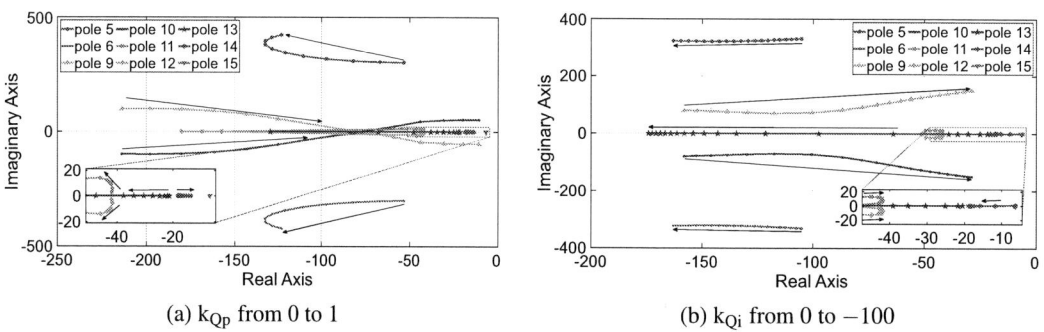

(a) k_{Qp} from 0 to 1

(b) k_{Qi} from 0 to -100

Fig. 5: Small signal model stability analysis reactive power controller

parameter k_{Qp} has a significant influence on the stability of the system. The eigenvalues 5 and 6 shift to the left of the origin at higher values of the parameter, which improves the system's response. However, the imaginary part of these two poles becomes more significant, and thus these eigenvalues become less damped. In addition, eigenvalues 11, 12 and 13 become faster when k_{Qp} is increased, the damping of these poles is not significantly affected. On the other hand, the remaining eigenvalues tend to become unstable when k_{Qp} becomes larger.

The parameter k_{Qp} is the gain of the proportional part of the virtual voltage controller. Through this parameter, the system's response to voltage sags is defined. The optimal value for this parameter should be a good compromise between fast response, and a good damping. A value of $k_{Qp} = 0.4$ achieves this requirement. At this value, the poles 9,10 and 13 are on the real axis, and the remaining eigenvalues are well damped.

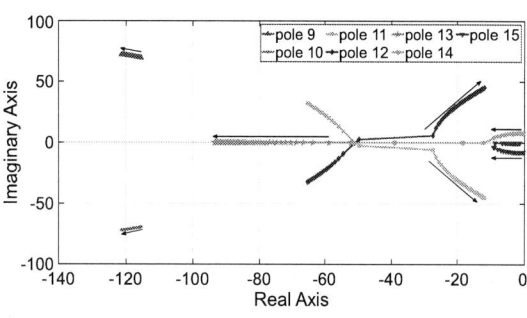

Fig. 6: k_3 from 0 to $0.03\,\mathrm{A}^2\,\mathrm{J}^{-2}$

Besides k_{Qp}, the parameter k_{Qi} also influences the system behavior in case of voltage sag. This parameter is the integrator part of the virtual voltage controller. Figure 5b shows the locus of eigenvalues. Also, for this parameter, eigenvalues 1,2,3,4,7 and 8 are not affected. However, poles 9 and 10 become slower and less damped by higher values of k_{Qi}. In contrast, the remaining poles improve their behaviour as the parameter is increased. Just as for k_{Qp}, an optimum is chosen to obtain the best damping and good system performance. The chosen value is $k_{Qi} = -30$.

Finally, the last parameter swept is k_3. This parameter is responsible for the tracking of the dc-link voltage. In Fig. 6, the loci of the eigenvalues are presented. Only the poles from 9 to 15 are affected by the modified parameter. As shown in Fig. 6, the poles 9, 10, 13 and 14 improve the stability of the system when the parameter increases. However, the other eigenvalues move towards unstable positions.

The optimum value for this parameter is $k_3 = 0.02\,\text{A}^2\,\text{J}^{-2}$. At this value, the poles from 13 to 15 are on the real axis, which improves the damping of the system. The obtained control parameters are used in a large-scale simulation and for the measurements, the corresponding results are presented in the next sections. In summery the parameters determined by the above procedure are:

$$k_1 = -110\,\text{s}^{-1}; \quad k_2 = 1.3\,\text{J}^{-1}\,\text{s}^{-1}; \quad k_3 = 0.02\,\text{A}^2\,\text{J}^{-2}; \quad k_{Qi} = -30; \quad k_{Qp} = 0.4 \tag{14}$$

Simulation results

To show the benefits of the optimized control, a large-signal model is built in Matlab/Simulink which compares the two control strategies. A line-to-line fault with line impedance to fault impedance ratio of 1:4 is simulated, which shows the performance of both strategies for an unbalanced fault. As expected, the dc-link voltage of both control schemes shows the 100 Hz oscillation that is caused by the unbalanced fault (Fig.7b). Additionally, the optimized control leads to a higher overshoot of the dc-link voltage, which is caused by the integral part of the controller. However, in the standard control strategy, the dc-link voltage deviates from its reference value. On the other hand, comparing the frequency response shows the benefits of the optimized control, which eliminates most of the frequency oscillations that are present in the standard strategy (Fig.7a). To validate the simulation results, the proposed strategy is

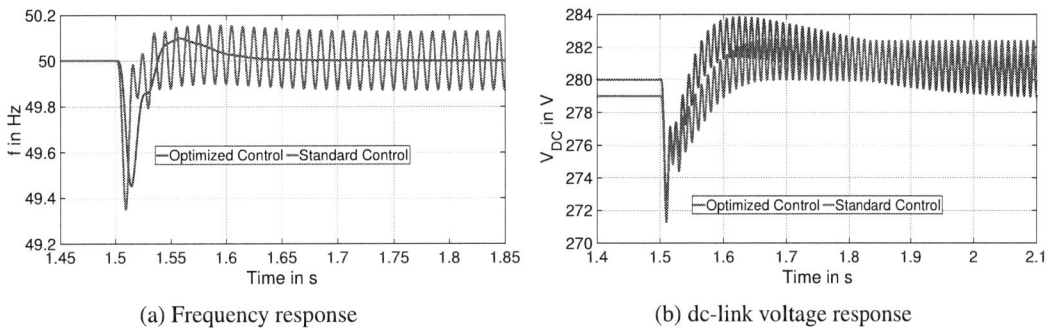

(a) Frequency response   (b) dc-link voltage response

Fig. 7: Comparison of the two control schemes during a line-to-line grid fault

tested under different grid conditions. The measurement results will be shown in the following section.

Measurement results

The test bench used to perform the experiments consists of three two-level converters, which can operate independently or in a parallel circuit. In addition, these converters can feed a passive load or be connected to a grid emulator. Fig. 8 depicts the setup of the test bench, and the system parameters are listed in table II. The devices used to perform the measurements are listed in table I.

Nominal dc-link Voltage	$V_{DC,n} = 309\,V$	dc-link Capacitor	$C_{DC} = 2.2\,mF$
Nominal RMS Output Voltage	$V_{o,n} = 70\,V$	Filter Capacitor	$C_f = 16\,\mu F$
Nominal RMS Current	$I_n = 9.5\,A$	Converter side Inductance	$L_{fC} = 5.6\,mH$
Nominal Power	$P_n = 2000\,W$	Grid side Inductance A	$L_{fg,A} = 2.93\,mH$
Nominal Frequency	$f_n = 50\,Hz$	Grid side Inductance B	$L_{fg,B} = 0\,mH$
Switching Frequency	$f_s = 3200\,Hz$	Grid side Inductance C	$L_{fg,C} = 0.73\,mH$

Table II: List of the converter's nominal values

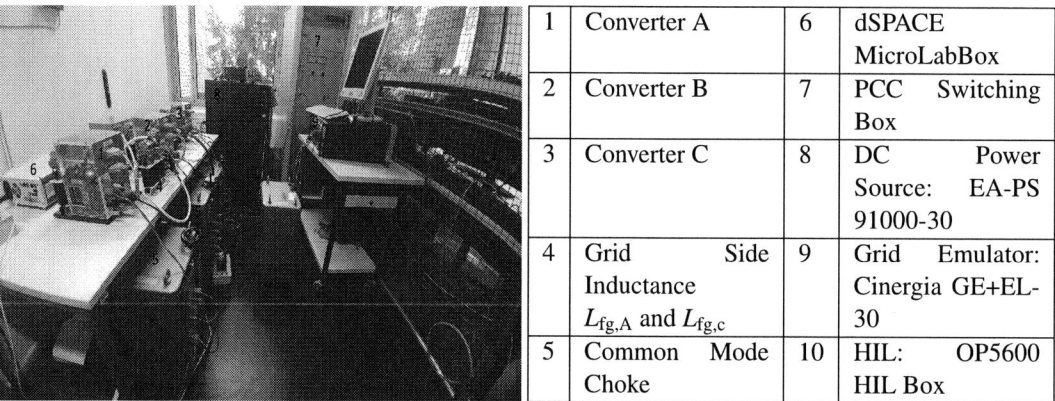

Fig. 8: Photo of the measurement setup

1	Converter A	6	dSPACE MicroLabBox
2	Converter B	7	PCC Switching Box
3	Converter C	8	DC Power Source: EA-PS 91000-30
4	Grid Side Inductance $L_{fg,A}$ and $L_{fg,c}$	9	Grid Emulator: Cinergia GE+EL-30
5	Common Mode Choke	10	HIL: OP5600 HIL Box

Table I: List of the used hardware

For a better understanding of the test bench, Fig. 9 shows a block diagram of the whole system, in addition to the structure of one converter. Each converter is connected to the grid emulator via an **LCL** filter. Except for the grid side filter inductor L_{fg}, the three inverters and filters are identical. The control software is implemented in MATLAB/Simulink, and the hardware is controlled via a dSPACE MicroLabBox.

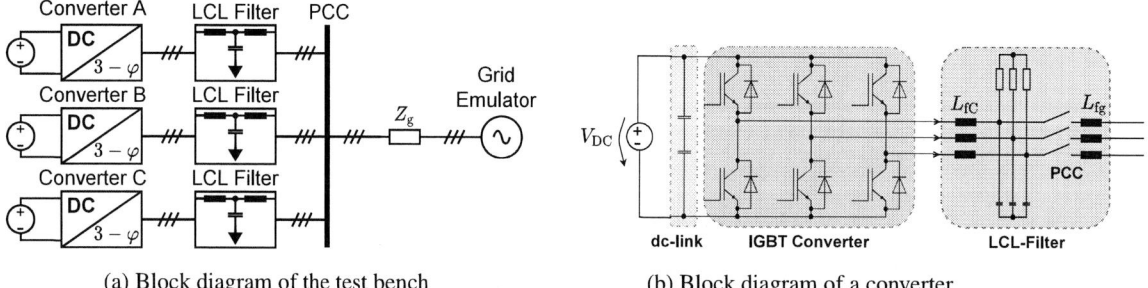

(a) Block diagram of the test bench

(b) Block diagram of a converter

Fig. 9: Test bench setup

Unbalanced grid faults

To validate the simulation results, a line-to-line fault with a line impedance to fault impedance ratio of 1:4 is simulated. During this experiment, converter A is connected to the grid emulator (Cinergia GE+EL-30) that simulates the fault. The experimental results are shown in Fig. 10. The dc-link voltage shows the expected 100 Hz oscillation during the grid fault. However, this oscillation is not present in the frequency response due to the notch filter. Moreover, the new PID-controller regulates the dc-link voltage to its reference value, and the system shows a stable and well-damped behavior. This test shows that the control strategy can work under unbalanced grid conditions without the frequency suffering from oscillations.

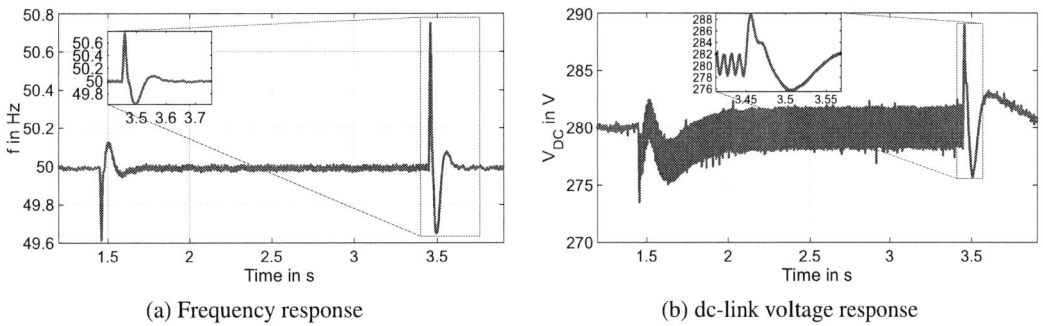

(a) Frequency response

(b) dc-link voltage response

Fig. 10: Response of the proposed control strategy during grid fault

Islanded operation mode

In this test, the focus is on the ability of the proposed control strategy to operate within networks of converters with different control algorithms and the interaction between these control strategies. For this purpose, the three converters were connected in parallel to create an islanded grid. In this grid, converters A and B operate with the proposed control strategy, while converter C performs a grid-forming (GFM) droop control. It is important to note that, both converters A and B have no DC power source and balance power fluctuation using the dc-link only, which will provide minimum frequency support to the grid. To investigate the effects of this minimum support on the grid, load steps were simulated, and the results are shown in Fig. 11a. To demonstrate the effect of the control strategy, converter C was tested separately

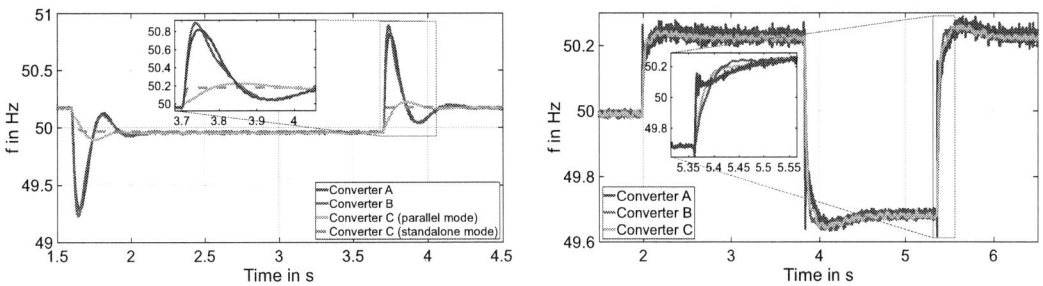

(a) Frequency response during a parallel connection with a grid-forming converter

(b) Frequency response during a parallel connection with a grid-following converter

Fig. 11: System behavior during parallel connection

with the same load step (green curve in Fig. 11a). As can be seen in the figure, the system operates stably during the load change. To evaluate the effects of the minimum inertia support provided by converters A and B, the frequency of converter C is considered in parallel and standalone mode. The frequency change of converter C in parallel operation is slower than in standalone operation, indicating that the rate of change of frequency (RoCoF) is reduced, due to the increase in total system inertia. On the other hand, a small overshoot of the frequency during parallel connection is also present. This overshoot comes from the fact that the matching control adapts the dc-link voltage, which forces the droop control to briefly provide more or less power than is required by the load. The inverter inertia constant provides an indication of the inertia added to the system by the control strategy used. Per definition, the inertia constant is calculated as follows:

$$H = \frac{E}{S_n} \qquad E = \frac{1}{2} \cdot C_{DC} \cdot U^2 \qquad U = V_{DC,min} + \Delta V_{DC} \qquad S_n = 2000\,\mathrm{V\,A} \qquad (15)$$

where E is the available energy stored in the dc-link capacitor, and S_n is the nominal apparent power of the converter. For the used setup, the minimal dc-link voltage is $V_{DC,min} = 220\,\mathrm{V}$ and the acceptable deviation is $\Delta V_{DC} = 100\,\mathrm{V}$. With these values, the converter inertia constant is $H = 30\,\mathrm{ms}$.

Another test was performed in parallel mode, where converters B and C apply the proposed strategy, while converter A operates in grid-following (GFL) mode. The objective of this test is to prove the ability of the control strategy to establish a stable grid where GFL control strategies can operate well. The same load step as in the previous test was performed, the results are shown in Fig. 11b. The GFL strategy operates stably in this parallel connection, with the frequency of converter A following that of converters B and C. However, it can also be seen that the frequency of converter A fluctuates slightly at the moment of load jump, which is caused by the weak grid conditions. Likewise, it is noticeable that the frequency of converter A changes faster than the frequencies of the other converters at the moment of the step, but slows down later and follows the frequency of converters B and C. This behavior is due to a phase jump associated with the load step, which causes some stability problems for the PLL.

Conclusion

This paper proposes an optimized grid-forming control strategy that can successfully operate under unbalanced faults and weak grid conditions. Moreover, the benefits of utilizing the dc-link capacitor as an energy source to deliver minimal inertia support have been investigated. A small-signal model of the system has been derived and used to optimize the control parameters. The system is then tested under different conditions. These tests have shown that the strategy remains stable during unbalanced faults. In addition, the minimum inertial support added by the proposed control strategy contributes to the reduction of RoCoF.

References

[1] I. Cvetkovic, D. Boroyevich, R. Burgos, C. Li and P. Mattavelli, "Modeling and control of grid-connected voltage-source converters emulating isotropic and anisotropic synchronous machines," 2015 IEEE 16th Workshop on Control and Modeling for Power Electronics (COMPEL), 2015, pp. 1-5, doi: 10.1109/COMPEL.2015.7236454.

[2] L. Huang et al., "A Virtual Synchronous Control for Voltage-Source Converters Utilizing Dynamics of DC-Link Capacitor to Realize Self-Synchronization," in IEEE Journal of Emerging and Selected Topics in Power Electronics, vol. 5, no. 4, pp. 1565-1577, Dec. 2017, doi: 10.1109/JESTPE.2017.2740424.

[3] A. Tayyebi, D. Gro, A. Anta, F. Kupzog and F. Drfler, "Frequency Stability of Synchronous Machines and Grid-Forming Power Converters," in IEEE Journal of Emerging and Selected Topics in Power Electronics, vol. 8, no. 2, pp. 1004-1018, June 2020, doi: 10.1109/JESTPE.2020.2966524.

[4] M. Eggers, H. Yang, H. Just and S. Dieckerhoff, "Virtual-Impedance-Based Droop Control for Grid-Forming Inverters with Fast Response to Unbalanced Grid Faults," 2020 IEEE 11th International Symposium on Power Electronics for Distributed Generation Systems (PEDG), 2020, pp. 122-129, doi: 10.1109/PEDG48541.2020.9244440.

[5] Peter Unruh et al."Overview on Grid-Forming Inverter Control MethodsİIn: Energies 13.10 (2020). issn: 1996-1073. doi: 10.3390/en13102589. url: https://www.mdpi.com/1996-1073/13/10/2589.

[6] Summary of the draft VDE-AR-N 4110:2017-02, 2017.

[7] H. Just, H. Yang and S. Dieckerhoff, "Evaluation of Advanced PLL Concepts for Enhanced Fault Ride Through Response," 2018 IEEE Energy Conversion Congress and Exposition (ECCE), 2018, pp. 5684-5691, doi: 10.1109/ECCE.2018.8558444.

[8] Global wind statistics 2017, Global Wind Energy Council, Tech. Rep., 2018.

[9] F. Milano, F. Drfler, G. Hug, D. J. Hill and G. Verbi, "Foundations and Challenges of Low-Inertia Systems (Invited Paper)," 2018 Power Systems Computation Conference (PSCC), 2018, pp. 1-25, doi: 10.23919/PSCC.2018.8450880.

Constrained Long-Horizon Direct Model Predictive Control for Grid-Connected Converters with LCL Filters

Mattia Rossi*, Petros Karamanakos*, Francesco Castelli-Dezza[†]
*Tampere University, Faculty of Information Technology and Communication Sciences
33101 Tampere, Finland
[†]Politecnico di Milano, Department of Mechanical Engineering
20156 Milan, Italy
Email: mattia.rossi@tuni.fi, p.karamanakos@ieee.org, francesco.castellidezza@polimi.it

Keywords

≪Model predictive control≫, ≪Grid-connected converter≫, ≪Optimal control≫, ≪Multi-objective optimization≫, ≪Control methods for electrical systems≫, ≪LCL≫, ≪Converter control≫.

Abstract

This paper presents a direct model predictive control algorithm for a three-level neutral point clamped converter connected to the grid via an LCL filter. The proposed controller simultaneously controls the grid and converter currents as well as the filter capacitor voltage, while meeting the relevant grid standards. Moreover, output constraints are included to ensure operation of the system within its safe operating limits. This is achieved by formulating the direct MPC problem as a constrained integer least-squares optimization problem, wherein the output constraints are mapped into input constraints. The presented results verify the effectiveness of the proposed method.

Introduction

Medium-voltage (MV) grid-tied converters are used for the integration of renewable energy sources, scalable loads, and high-performance variable speed drives into the electrical grid. To ensure smooth operation, grid standards—e.g., the IEEE 519 [1] and the IEC 61000-2-4 [2] standards—are imposed to the point of common coupling (PCC). Such standards set tight limits on the current and voltage harmonics injected by the power electronic systems into the grid. To mitigate the current harmonics, LCL filters are commonly placed between the converter and the PCC. The addition of such a filter, however, gives rise to a third-order system, implying that its control is a nontrivial task. More specifically, not only the grid current needs to be controlled—as is the case with all grid-tied systems—but also the converter current and the capacitor voltage. Moreover, adequate damping of the filter resonance is required. Finally, large overshoots during transients that may harm the hardware components due to the correlated dynamics of the system need to be avoided to ensure safe system operation.

Meeting the above-mentioned tasks with conventional control techniques—that rely on linear control principles—is challenging, while the controller design can become complicated [3]. Furthermore, the most favorable dynamic operation is not achieved due to the cascaded control loops that tend to limit the bandwidth of the controller, especially when operation at low switching frequency is required [4]. As an alternative to linear control techniques, model predictive control (MPC) can be employed. Thanks to its multiple-input multiple-output (MIMO) nature as well as its ability to handle explicit constraints and provide active damping, superior steady-state and dynamic performance of the grid-connected power electronic system can be achieved [5, 6].

The most popular MPC-based method in academia is direct MPC, i.e., a control strategy that directly computes and applies the switching signals. Direct MPC with output reference tracking, also referred to as finite control set MPC (FCS-MPC), fully exploits the advantages that MPC can offer, but, alas, it comes with pronounced computational load [5, 7]. To address this—at least to some extent—the direct

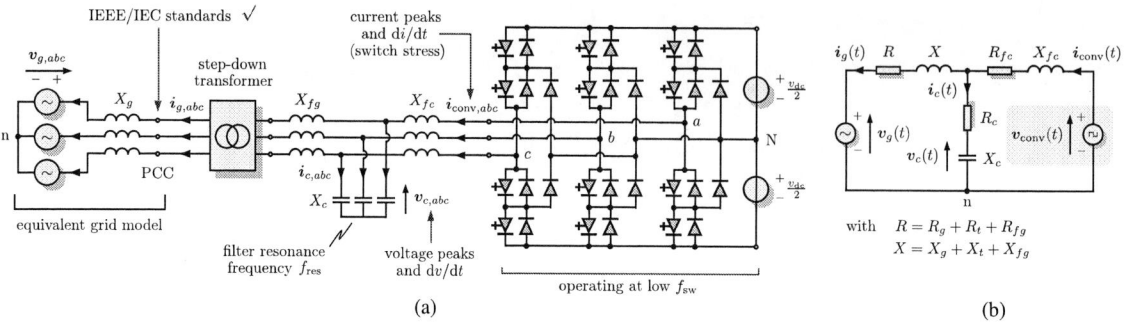

Fig. 1: (a) MV grid-tied three-phase 3L-NPC converter with an LCL-filter placed between the converter and the step-down transformer. (b) Equivalent circuit in the $\alpha\beta$-plane.

MPC problem can be designed as an integer least-squares (ILS) one and solved with a sphere decoding algorithm [8]. Such an approach, however, cannot easily handle constraints that will ensure operation of the system within its safe operating area. Thus, to avoid overcurrents and/or overvoltages that may damage the hardware of the converter and/or the filter, the feasible set of the optimization problem underlying direct MPC needs to be carefully redesigned [9].

Given the above, this paper proposes a constrained direct MPC for an MV three-level neutral point clamped (3L-NPC) converter connected to the grid via an LCL filter. The optimization problem, formulated as an ILS one, accounts for constraints on all the controlled variables to increase the system reliability. To this aim, the output constraints are mapped into constraints on the control input. In doing so, candidate solutions that may lead to large overshoots are already excluded in a preprocessing stage. Moreover, a long horizon is employed, which combined with full-state information, provides damping of the filter resonance without requiring a dedicated active damping loop. As a result, the MV converter can be operated at a very low switching frequency (and close to the resonance), while still meeting the grid standards. The effectiveness of the adopted method is demonstrated with the presented results.

Formulation of the Constrained Direct MPC Optimization Problem

Consider the MV grid-tied 3LNPC converter with an LCL-filter shown in Fig. 1(a). To keep the demonstration of the proposed control method simple, the dc-link voltage V_{dc} is assumed to be constant[1] and the neutral point potential N fixed. All variables given in the abc-plane $\boldsymbol{\xi}_{abc} = [\xi_a\,\xi_b\,\xi_c]^T$ are mapped into two-dimensional vectors $\boldsymbol{\xi}_{\alpha\beta} = [\xi_\alpha\,\xi_\beta]^T$ via the reduced Clarke transformation matrix $\boldsymbol{K}$, i.e., $\boldsymbol{\xi}_{\alpha\beta} = \boldsymbol{K}\boldsymbol{\xi}_{abc}$, see [5].

Controller Model

The equivalent circuit of the system under consideration in the $\alpha\beta$-reference frame[2] is shown in Fig. 1(b). This paper considers the same MV conversion system as the case study described in [10], from which the system parameters are taken (see Table I in [10]). A strong grid is assumed, and the (dominant) LCL-filter resonance frequency is $f_{\text{res}} = 304\,\text{Hz}$. The passive components have very small (series) resistive parts, i.e., they essentially do not provide any passive damping. All SI variables are normalized based on the rated values of the step-down transformer (secondary side).

Depending on the *single-phase* switch position $u_z \in \mathcal{U} = \{-1, 0, 1\}$, with $z \in \{a, b, c\}$, the single-phase output voltage of a 3L-NPC converter can assume three possible discrete voltage levels, i.e., $-\frac{V_{\text{dc}}}{2}$, 0, $\frac{V_{\text{dc}}}{2}$, respectively. By introducing the *three-phase* switch position[3] $\boldsymbol{u}_{abc} = [u_a\,u_b\,u_c]^T \in \boldsymbol{\mathcal{U}} = \mathcal{U}^3$, the converter output voltage $\boldsymbol{v}_{\text{conv}}$ is given by

$$\boldsymbol{v}_{\text{conv}}(t) = \frac{V_{\text{dc}}}{2}\boldsymbol{K}\boldsymbol{u}_{abc}(t) = \frac{V_{\text{dc}}}{2}\boldsymbol{u}(t)\,. \tag{1}$$

---

[1]The outer layer which regulates the dc-link voltage and generates the power references is out of the scope of this paper.

[2]Hereafter, to simplify the notation, variables in the $\alpha\beta$-frame are not indicated by the corresponding subscript, unless otherwise stated.

[3]Note that, given a 3L-NPC topology, the integer input set $\boldsymbol{\mathcal{U}}$ comprises $3^3 = 27$ combinations of $\boldsymbol{u}_{abc}$

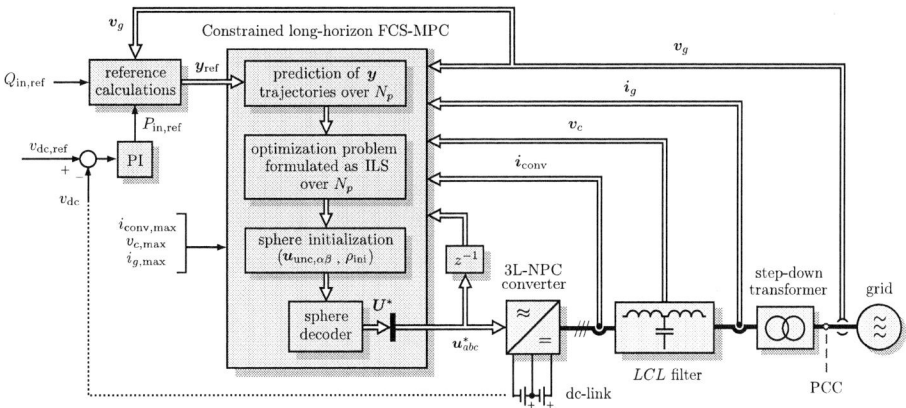

Fig. 2: Structure of the direct MPC (formulated as an ILS problem) to control the system shown in Fig. 1. The preprocessing procedure to compute ρ_{ini} is executed before calling the sphere decoder, which computes then $\boldsymbol{u}_{abc}^*$.

By choosing $\boldsymbol{x} = [\boldsymbol{i}_{\text{conv}}^T \, \boldsymbol{v}_c^T \, \boldsymbol{i}_g^T \, \boldsymbol{v}_g^T]^T \in \mathbb{R}^8$ and $\boldsymbol{y} = [\boldsymbol{i}_{\text{conv}}^T \, \boldsymbol{v}_c^T \, \boldsymbol{i}_g^T]^T \in \mathbb{R}^6$ as the state and output vectors, respectively, and $\boldsymbol{u} \equiv \boldsymbol{u}_{abc}$ as the control input, the continuous-time state-space representation of the system in Fig. 1(a) is

$$\frac{\mathrm{d}\boldsymbol{x}(t)}{\mathrm{d}t} = \boldsymbol{F}\boldsymbol{x}(t) + \boldsymbol{G}\boldsymbol{u}(t) \tag{2a}$$

$$\boldsymbol{y}(t) = \boldsymbol{C}\boldsymbol{x}(t), \tag{2b}$$

where the voltage $\boldsymbol{v}_g$ models the grid source, $\boldsymbol{i}_g$ is the grid current, $\boldsymbol{i}_{\text{conv}}$ is the converter current, while $\boldsymbol{v}_c$ is the capacitor voltage. The reader is referred to [10] for the full derivation of the continuous-time differential equations which describe the system dynamics depicted in Fig. 1(b) as well as the definition of the matrices $\boldsymbol{F} \in \mathbb{R}^{8 \times 8}$, $\boldsymbol{G} \in \mathbb{R}^{8 \times 3}$, and $\boldsymbol{C} \in \mathbb{R}^{6 \times 8}$, respectively. Note that since an ideal grid is assumed (i.e., its angular frequency ω_g is constant), $\boldsymbol{F}$ is a time-invariant matrix. Likewise, due to the assumption of a constant dc-link voltage, $\boldsymbol{G}$ is also time invariant.

The discretized version of the continous-time system (2) is derived by using exact discretization with the sampling interval T_s, i.e.,

$$\boldsymbol{x}(k+1) = \boldsymbol{A}\boldsymbol{x}(k) + \boldsymbol{B}\boldsymbol{u}(k) \tag{3a}$$

$$\boldsymbol{y}(k) = \boldsymbol{C}\boldsymbol{x}(k), \tag{3b}$$

with $\boldsymbol{A} = \mathrm{e}^{\boldsymbol{F}T_s}$ and $\boldsymbol{B} = \int_0^{T_s} e^{\boldsymbol{F}\tau} \, \mathrm{d}\tau \, \boldsymbol{G} = -\boldsymbol{F}^{-1}\left(\boldsymbol{I}_8 - \boldsymbol{A}\right)\boldsymbol{G}$, while e is the matrix exponential, and $k \in \mathbb{N}$ denotes the discrete time step. Due to the absence of a modulator, T_s directly affects the switching frequency (and the *granularity of switching*), see [11].

Reformulation of the Constrained Direct MPC Problem in the Equivalent ILS Form

The block diagram of the presented MPC strategy is shown in Fig. 2. As a grid-connected power converter is considered in this work, the total demand distortion (TDD) of $\boldsymbol{i}_{g,abc}$ and $\boldsymbol{v}_{\text{pcc},abc}$ as well as the amplitude of the associated harmonics should meet the IEEE 519 [1] and IEC 61000-2-4 [2] grid standards. Given this, the main control objective is to regulate $\boldsymbol{i}_g$, $\boldsymbol{i}_{\text{conv}}$, and $\boldsymbol{v}_c$ along their sinusoidal references by directly manipulating the converter switches, i.e., without the use of a modulator. This needs to be achieved while operating the MV converter at a low switching frequency so that the switching power losses are kept low. The output references $\boldsymbol{y}_{\text{ref}}$, i.e. $\boldsymbol{y}_{\text{ref}} = [\boldsymbol{i}_{\text{conv,ref}}^T \, \boldsymbol{v}_{c,\text{ref}}^T \, \boldsymbol{i}_{g,\text{ref}}^T]^T$, are computed based on the real $P_{\text{in,ref}}$ and reactive $Q_{\text{in,ref}}$ power requirements at the transformer secondary side, with $Q_{\text{in,ref}} = 0$ at steady-state operation to achieve unity power factor $\mathrm{pf} = 1$. Given a prediction horizon of N_p time steps, these control objectives are mapped into a scalar by the quadratic objective function

$$J(k) = \sum_{\ell=k}^{k+N_p-1} \|\boldsymbol{y}_{\text{ref}}(\ell+1) - \boldsymbol{y}(\ell+1)\|_{\boldsymbol{Q}}^2 + \lambda_u \|\Delta\boldsymbol{u}(\ell)\|_2^2. \tag{4}$$

The first term in (4) denotes the deviation of the output variables y from their reference values y_{ref}. This error is weighted with the 6×6 positive semidefinite matrix $Q \succeq 0$, the diagonal entries of which prioritize the tracking accuracy among the different controlled variables. The second term in (4) penalizes the switching effort, i.e., $\Delta u(\ell) = u(\ell) - u(\ell-1)$, which directly relates to the definition of the average device switching frequency f_{sw} [11]. To set the trade-off between the tracking accuracy and the resulting f_{sw} at each time step $\ell = k, \ldots, k + N_p - 1$, the weighting factor $\lambda_u \in \mathbb{R}^{++}$ is introduced.

The aforementioned control goals have to be met while protecting the switching devices and LCL filter components from overcurrents and overvoltages, respectively. Therefore, during power transients, the amplitudes of i_{conv}, v_c, and i_g must be kept within given bounds. To this end, soft constraints are typically introduced in the optimization problem to include such physical limitations. Nevertheless, hard constraints are implemented in this work to facilitate the reformulation of the optimization problem, as shown in later sections. Hence, the constraints on the system output are designed in the $\alpha\beta$-plane and imposed at each time step $\ell + 1$, i.e.,

$$\|i_{\text{conv}}(\ell+1)\|_2 \le i_{\text{conv,max}}, \qquad \|v_c(\ell+1)\|_2 \le v_{c,\text{max}}, \qquad \|i_g(\ell+1)\|_2 \le i_{g,\text{max}}, \tag{5}$$

where the positive scalars $i_{\text{conv,max}}, v_{c,\text{max}}, i_{g,\text{max}} \in \mathbb{R}^+$ define the upper boundary values of the converter current, filter capacitor voltage, and grid current, respectively. As an example, considering i_{conv} bounded by $i_{\text{conv,max}}$, it follows that

$$i^2_{\text{conv},\alpha}(\ell+1) + i^2_{\text{conv},\beta}(\ell+1) \le i^2_{\text{conv,max}}, \tag{6}$$

which defines a circle in the $\alpha\beta$-plane of radius $i_{\text{conv,max}}$. The constraints on v_c and i_g are imposed in a similar manner by referring to $v_{c,\text{max}}$ and $i_{g,\text{max}}$, respectively.

Given the above, to compute the optimal switching sequence over the N_p prediction steps[4], i.e., $U^*(k) = \begin{bmatrix} u^{*T}(k) & u^{*T}(k+1) & \ldots & u^{*T}(k+N_p-1) \end{bmatrix}^T \in \mathbb{U} = \mathcal{U} \times \ldots \times \mathcal{U} = \mathcal{U}^{N_p} \subset \mathbb{Z}^{3N_p}$, that results in the best system performance—as quantified by (4)—while respecting the system dynamics and constraints—as expressed by (3) and (5), respectively—the constrained optimization problem underlying MPC needs to be solved. As shown in [8], to mitigate the computational burden of long-horizon direct MPC, while still guaranteeing optimality, function (4) can be written such that the associated optimization problem is a truncated ILS one. This problem can be subsequently solved in a computationally efficient manner with a dedicated branch-and-bound strategy named *sphere decoder* [8], [12].

By considering the unconstrained solution U_{unc}, i.e., the solution that minimizes (4) when relaxing the feasible set from $\mathbb{U}$ to $\mathbb{R}^{3N_p}$, the following equivalent constrained ILS problem can be defined

$$U^*(k) = \arg \min_{U(k) \in \mathbb{U}} \|VU(k) - \bar{U}_{\text{unc}}(k)\|_2^2 \tag{7a}$$

$$\text{subject to } U(k) \in \mathbb{U} = \mathcal{U}^{N_p}, \quad \|\Delta u(\ell)\|_\infty \le 1, \tag{7b}$$

$$\|i_{\text{conv}}(\ell+1)\|_2 \le i_{\text{conv,max}}, \quad \|v_c(\ell+1)\|_2 \le v_{c,\text{max}}, \tag{7c}$$

$$\|i_g(\ell+1)\|_2 \le i_{g,\text{max}}, \quad \forall \ell = k, \ldots, k+N_p-1, \tag{7d}$$

where $\bar{U}_{\text{unc}}(k) = VU_{\text{unc}}(k) \in \mathbb{R}^{3N_p}$, and $V \in \mathbb{R}^{3N_p \times 3N_p}$ is a nonsingular, upper triangular matrix, known as the *lattice generator* matrix. The latter generates the $3N_p$-dimensional discrete space (lattice), one point of which is the solution to problem (7), i.e., the lattice point with the shortest Euclidean distance to $\bar{U}_{\text{unc}}$. Note that, a further (hard) input constraint, i.e., $\|\Delta u(\ell)\|_\infty \le 1$, with $\Delta u(\ell) = u(\ell) - u(\ell-1)$, is imposed by (7b) to avoid a shoot-through in the converter due to the 3L-NPC topology [5]. Out of U^*, only the first element u^* is applied to the converter whereas the rest are discarded in line with the receding horizon (RH) control principle [13]. Following, the optimization procedure is repeated at $k+1$ based on a new $x(k)$ and a shifted prediction horizon N_p.

Note that the initial radius of the sphere ρ_{ini} affects the effectiveness of the search process as it defines the first upper bound of the branch-and-bound mechanism. Therefore, ρ_{ini} should be as small as possible

---

[4]Note that, the *feasible set* $\mathbb{U}$ is defined by the N_p-times Cartesian product of $\mathcal{U}$

to remove a priori as many candidate solutions as possible, while still ensuring feasibility of the optimization process, e.g., containing at least one lattice point. In [14], the initial radius ρ_{ini} is computed as

$$\rho_{\text{ini}}(k) = \min\left\{\rho_1(k), \rho_2(k)\right\}, \tag{8}$$

where the relative options are

$$\rho_1(k) = \left\|\bar{U}_{\text{unc}}(k) - VU_{\text{bab}}(k)\right\|_2 \quad \text{and} \quad \rho_2(k) = \left\|\bar{U}_{\text{unc}}(k) - VU_{\text{ed}}(k)\right\|_2. \tag{9}$$

Radius ρ_1 in (9) depends on the Babai estimate U_{bab}, which is the rounded unconstrained solution, i.e., $U_{\text{bab}}(k) = \lfloor U_{\text{unc}}(k) \rceil$. On the other hand, radius ρ_2 in (9) depends on an educated guess U_{ed}, which is the previously applied solution $U^*(k-1)$ shifted by one time step according to the RH policy, see [9].

Solving the Equivalent Constrained Integer Least-Squares Problem

The presence of constraint (7) affects the aforementioned computation of ρ_{ini}. Indeed, by adopting (8), there is a possibility that U_{unc} and/or some of the candidate solutions included in the sphere violate (7). To avoid this, the computation of the initial radius needs to be revised. This is done by mapping the output constraints (7) into (hard) input constraints, thus limiting the feasible set $\mathbb{U}$. The approach presented in this paper is based on [9] and extended to handle multiple constraints by simultaneously considering bounds on i_{conv}, v_c, and i_g.

By utilizing (3a) and (3b), the output dynamics at step $k+1$ are given by

$$\boldsymbol{y}(k+1) = \boldsymbol{C}\boldsymbol{A}\boldsymbol{x}(k) + \boldsymbol{C}\widetilde{\boldsymbol{B}}\boldsymbol{K}\boldsymbol{u}(k) \Rightarrow \boldsymbol{i}_{\text{conv}}(k+1) = \begin{bmatrix} \boldsymbol{I}_2 & \boldsymbol{0}_{2\times 6} \end{bmatrix}\boldsymbol{A}\boldsymbol{x}(k) + \gamma_{\text{conv}}\boldsymbol{I}_2\boldsymbol{u}_{\alpha\beta}(k) \tag{10a}$$

$$\boldsymbol{v}_c(k+1) = \begin{bmatrix} \boldsymbol{0}_{2\times 2} & \boldsymbol{I}_2 & \boldsymbol{0}_{2\times 4} \end{bmatrix}\boldsymbol{A}\boldsymbol{x}(k) + \gamma_c\boldsymbol{I}_2\boldsymbol{u}_{\alpha\beta}(k) \tag{10b}$$

$$\boldsymbol{i}_g(k+1) = \begin{bmatrix} \boldsymbol{0}_{2\times 4} & \boldsymbol{I}_2 & \boldsymbol{0}_{2\times 2} \end{bmatrix}\boldsymbol{A}\boldsymbol{x}(k) + \gamma_g\boldsymbol{I}_2\boldsymbol{u}_{\alpha\beta}(k), \tag{10c}$$

where $\widetilde{\boldsymbol{B}} = -\boldsymbol{F}^{-1}\left(\boldsymbol{I}_8 - \boldsymbol{A}\right)\widetilde{\boldsymbol{G}}$, with $\widetilde{\boldsymbol{G}} = (V_{\text{dc}}/2X_{fc})\begin{bmatrix} \boldsymbol{I}_2 & \boldsymbol{0}_{2\times 6} \end{bmatrix}^T$. Moreover, $\boldsymbol{u}_{\alpha\beta}(k) = \boldsymbol{K}\boldsymbol{u}(k)$, while $\boldsymbol{C}\widetilde{\boldsymbol{B}}\boldsymbol{K} = \boldsymbol{C}\boldsymbol{B}$ holds. In more detail, given $\boldsymbol{C} \in \mathbb{R}^{6\times 8}$ and $\widetilde{\boldsymbol{B}} \in \mathbb{R}^{8\times 2}$, (10a), (10b), and (10c), are derived by noticing the structure of the product $\boldsymbol{C}\widetilde{\boldsymbol{B}} \in \mathbb{R}^{6\times 2}$, i.e.,

$$\boldsymbol{C}\widetilde{\boldsymbol{B}} = \begin{bmatrix} \gamma_{\text{conv}} & 0 & \gamma_c & 0 & \gamma_g & 0 \\ 0 & \gamma_{\text{conv}} & 0 & \gamma_c & 0 & \gamma_g \end{bmatrix}^T = \begin{bmatrix} \gamma_{\text{conv}}\boldsymbol{I}_2 & \gamma_c\boldsymbol{I}_2 & \gamma_g\boldsymbol{I}_2 \end{bmatrix}^T. \tag{11}$$

As can be seen in (11), three 2×2 diagonal matrices appear, i.e., $\text{diag}(\gamma_{\text{conv}}) \succ 0 \in \mathbb{R}^{2\times 2}$, $\text{diag}(\gamma_c) \succ 0 \in \mathbb{R}^{2\times 2}$, and $\text{diag}(\gamma_g) \succ 0 \in \mathbb{R}^{2\times 2}$, with $\gamma_{\text{conv}} \neq \gamma_c \neq \gamma_g$. In particular, $\gamma_{\text{conv}}\boldsymbol{I}_2 = \begin{bmatrix} \boldsymbol{I}_2 & \boldsymbol{0}_{2\times 6} \end{bmatrix}\widetilde{\boldsymbol{B}}$, $\gamma_c\boldsymbol{I}_2 = \begin{bmatrix} \boldsymbol{0}_{2\times 2} & \boldsymbol{I}_2 & \boldsymbol{0}_{2\times 4} \end{bmatrix}\widetilde{\boldsymbol{B}}$ and $\gamma_g\boldsymbol{I}_2 = \begin{bmatrix} \boldsymbol{0}_{2\times 4} & \boldsymbol{I}_2 & \boldsymbol{0}_{2\times 2} \end{bmatrix}\widetilde{\boldsymbol{B}}$ can be easily computed. Given (10a), (10b), and (10c), the three one-dimensional constraints in (5) can be rewritten as

$$\|\boldsymbol{i}_{\text{conv}}(\ell+1)\|_2 \leq i_{\text{conv,max}} \Rightarrow \left\|\frac{\begin{bmatrix} \boldsymbol{I}_2 & \boldsymbol{0}_{2\times 6} \end{bmatrix}\boldsymbol{A}\boldsymbol{x}(k)}{\gamma_{\text{conv}}} + \boldsymbol{K}\boldsymbol{u}(k)\right\|_2 \leq \frac{i_{\text{conv,max}}}{\gamma_{\text{conv}}} \tag{12a}$$

$$\|\boldsymbol{v}_c(\ell+1)\|_2 \leq v_{c,\text{max}} \Rightarrow \left\|\frac{\begin{bmatrix} \boldsymbol{0}_{2\times 2} & \boldsymbol{I}_2 & \boldsymbol{0}_{2\times 4} \end{bmatrix}\boldsymbol{A}\boldsymbol{x}(k)}{\gamma_c} + \boldsymbol{K}\boldsymbol{u}(k)\right\|_2 \leq \frac{v_{c,\text{max}}}{\gamma_c} \tag{12b}$$

$$\|\boldsymbol{i}_g(\ell+1)\|_2 \leq i_{g,\text{max}} \Rightarrow \left\|\frac{\begin{bmatrix} \boldsymbol{0}_{2\times 4} & \boldsymbol{I}_2 & \boldsymbol{0}_{2\times 2} \end{bmatrix}\boldsymbol{A}\boldsymbol{x}(k)}{\gamma_g} + \boldsymbol{K}\boldsymbol{u}(k)\right\|_2 \leq \frac{i_{g,\text{max}}}{\gamma_g}. \tag{12c}$$

To simplify the derivation of the constrained ILS problem, let us first consider a generic one-dimensional constraint of the form

$$\left\|\frac{\widetilde{\boldsymbol{C}}\boldsymbol{A}\boldsymbol{x}(k)}{\gamma} + \boldsymbol{K}\boldsymbol{u}(k)\right\|_2 \leq \frac{y_{\text{max}}}{\gamma} \iff \|\boldsymbol{u}_{\text{constr}}(k) + \boldsymbol{K}\boldsymbol{u}(k)\|_2 \leq \rho_{\text{constr}}, \tag{13}$$

where the generic $\widetilde{\boldsymbol{C}}$, γ, and y_{max} are changed according to which constraint among (12a), (12b), and (12c) is taken into consideration. Note that (13) describes a feasible set in the form of a circle $\mathcal{C}$ in the $\alpha\beta$-plane centered at $\boldsymbol{u}_{\text{constr}}(k) = -\frac{\widetilde{\boldsymbol{C}}\boldsymbol{A}\boldsymbol{x}(k)}{\gamma}$ with radius $\rho_{\text{constr}} = \frac{y_{\text{max}}}{\gamma}$.

With (13), the constraints in (12) define $\mathcal{C}_{\text{conv}}$, $\mathcal{C}_c$, and $\mathcal{C}_g$ related to $\boldsymbol{i}_{\text{conv}}$, $\boldsymbol{v}_c$, and $\boldsymbol{i}_g$, respectively,

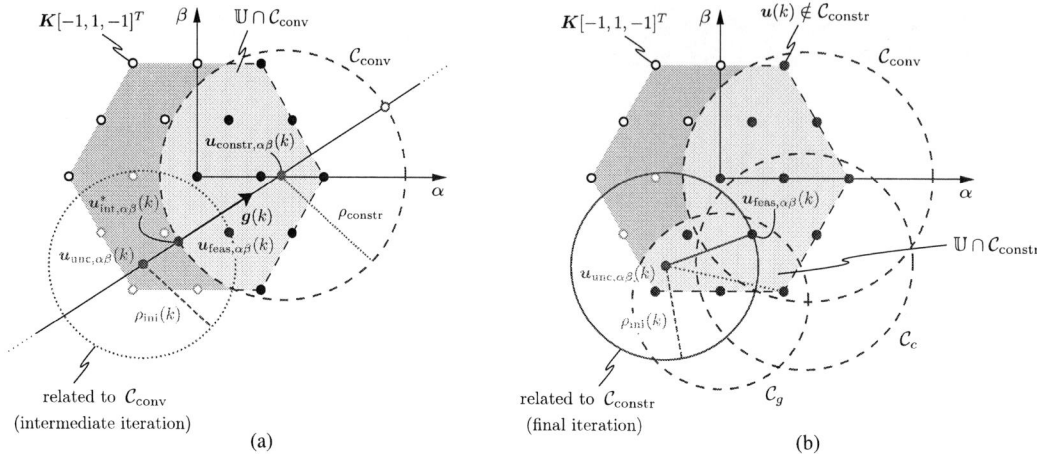

Fig. 3: Computation of the input constrained set $\mathcal{C}_{\text{constr}}$ and choice of the initial radius in the $\alpha\beta$-plane. (a) Computation of $\rho_{\text{ini}}(k)$ for a one-dimensional constraint (set $\mathcal{C}_{\text{conv}}$). (b) Computation of $\mathcal{C}_{\text{constr}}$ and $\rho_{\text{ini}}(k)$ based on all three constraints. Focusing on (a), the gradient, as depicted by the (scaled-down) vector $\boldsymbol{g}(k)$, shows the relative position of $\boldsymbol{u}_{\text{unc}}(k)$ with respect to the center of $\mathcal{C}_{\text{conv}}$, i.e., $\boldsymbol{u}_{\text{constr}}(k)$. Subsequently, the intersection point between $\boldsymbol{g}(k)$ and $\mathcal{C}_{\text{conv}}$, i.e., $\boldsymbol{u}_{\text{int}}^*(k)$ (indicated as green circle), that is closer to $\boldsymbol{u}_{\text{unc}}(k)$ (shown as red circle) is rounded up. Based on $\boldsymbol{u}_{\text{feas}}(k)$ (shown as blue circle) the radius $\rho_{\text{ini}}(k)$ (red dash-dotted line) of the hypersphere (shown as a red circle) is determined. The infeasible points enclosed in the hypersphere are indicated with light gray circles, whereas the black solid circles are feasible, but suboptimal, points.

which individually restrict the feasible set $\mathbb{U}$. This implies that to meet the output constraint (7), the optimal switching sequence $\boldsymbol{U}^*$ should lie within the intersection of all the constrained sets $\mathcal{C}_{\text{constr}} = \mathcal{C}_{\text{conv}} \cap \mathcal{C}_c \cap \mathcal{C}_g$ in the $\alpha\beta$-plane. Based on this, the *feasible* set of the integer-valued input vector $\boldsymbol{U}$ is defined as $\mathbb{U}_{\text{constr}} = \mathbb{U} \cap \mathcal{C}_{\text{constr}} = \boldsymbol{\mathcal{U}}_{\text{constr},1} \times \ldots \times \boldsymbol{\mathcal{U}}_{\text{constr},N_p}$, with

$$\boldsymbol{\mathcal{U}}_{\text{constr},i} = \{\boldsymbol{u}(\ell) \mid \boldsymbol{K}\boldsymbol{u}(\ell) \in \mathcal{C}_{\text{constr}},\, \boldsymbol{u}(\ell) \in \boldsymbol{\mathcal{U}}\} \qquad \forall \ell = k, \ldots, k+N_p-1 \,, \tag{14}$$

for $i \in \{1, \ldots, N_p\}$. A graphical example is depicted in Fig. 3(a).

Given (14), the initial radius ρ_{ini} of the hypersphere needs to be revised according to the restricted input feasible set $\mathbb{U}_{\text{constr}}$. While on the one hand the radius should meet the same objective as in the unconstrained ILS problem case, i.e., to be as small as possible, an additional goal is that the new hypersphere should include at least one *feasible* point. To satisfy both criteria, (8) is substituted by

$$\rho_{\text{ini}}(k) = \min\left\{ \widehat{\rho}_1(k), \widehat{\rho}_2(k) \right\}\,. \tag{15}$$

To compute ρ_{ini} in presence of the multiple output constraints, the following procedure is adopted.

Step 1: Radius $\widehat{\rho}_1$ is defined as

$$\widehat{\rho}_1(k) = \left\| \bar{\boldsymbol{U}}_{\text{unc}}(k) - \boldsymbol{V}\boldsymbol{U}_{\text{rnd}}(k) \right\|_2\,, \tag{16}$$

where the new initial guess $\boldsymbol{U}_{\text{rnd}}(k) = \left[\boldsymbol{u}_{\text{rnd}}^T(k) \ \ldots \ \boldsymbol{u}_{\text{rnd}}^T(k+N_p-1)\right]^T$ in the $\alpha\beta$-plane is equal to the Babai estimate $\boldsymbol{U}_{\text{bab}}$ from time step $k+1$ up to $k+N_p-1$, whereas the estimate at k results by guessing the feasible candidate solution $\boldsymbol{u}_{\text{feas}}(k) = \boldsymbol{K}\boldsymbol{u}_{\text{feas},abc}(k)$, with $\boldsymbol{u}_{\text{feas},abc} \in \boldsymbol{\mathcal{U}}_{\text{constr},1}$, closest to $\boldsymbol{u}_{\text{unc}}$ [9].

Step 2: Considering a single (one-dimensional) output constraint, the vector $\boldsymbol{g}(k) = \boldsymbol{u}_{\text{constr}}(k) - \boldsymbol{u}_{\text{unc}}(k)$ that spans the line passing through $\boldsymbol{u}_{\text{unc}}$ and the center of $\mathcal{C}$, i.e., $\boldsymbol{u}_{\text{constr}}$, is computed.

Step 3: The intersection points, $\boldsymbol{u}_{\text{int},1}$ and $\boldsymbol{u}_{\text{int},2}$, of the aforementioned line and $\mathcal{C}$ are calculated. The one closer to $\boldsymbol{u}_{\text{unc}}$ is computed with

$$\boldsymbol{u}_{\text{int}}^*(k) = \arg\operatorname{minimize} \left\| \boldsymbol{u}_{\text{int},i}(k) - \boldsymbol{u}_{\text{unc}}(k) \right\|_2 \quad \text{for } i = 1, 2\,.$$

Step 4: The feasible input closer to u_{int}^* is found. This is done by examining the sign of the elements of $g(k)$, i.e., the direction of the gradient in each dimension of the space. Following, u_{int}^* is *rounded* up or down (i.e., ceiled or floored, respectively) depending on the direction of the gradient, i.e.,

$$u_{\text{feas}}(k) = \begin{cases} \lceil u_{\text{int}}^*(k) \rceil & \text{if } g(k) \geq 0 \\ \lfloor u_{\text{int}}^*(k) \rfloor & \text{if } g(k) < 0 . \end{cases} \tag{17}$$

Step 5: The candidate solution u_{feas}—mapped onto the three-phase switch position by $u_{\text{feas},abc}(k) = K^{-1} u_{\text{feas}}(k)$—is the initial guess at time-step k. As for radius $\widehat{\rho}_2$, this depends on the educated guess U_{ed}. However, it is finite only if U_{ed} belongs to the constrained feasible set, i.e.,

$$\widehat{\rho}_2(k) = \begin{cases} \rho_2(k) & \text{if } U_{\text{ed}}(k) \in \mathbb{U}_{\text{constr}} \\ \infty & \text{if } U_{\text{ed}}(k) \notin \mathbb{U}_{\text{constr}} . \end{cases} \tag{18}$$

Then, the initial radius is found by solving $\rho_{\text{ini}}(k) = \min\{\widehat{\rho}_1(k), \widehat{\rho}_2(k)\}$.

At this point it should be mentioned that by adopting steps 1–5 for each one-dimensional constraint, three values of ρ_{ini}—related to three different values of u_{feas}—result. Hence, the new hypersphere should contain at least one point/solution within $\mathcal{C}_{\text{constr}} = \mathcal{C}_{\text{conv}} \cap \mathcal{C}_c \cap \mathcal{C}_g$. To achieve this, the following step is introduced.

Step 6: For each one-dimensional constraint $j \in \{1, 2, 3\}$, the outcome of steps 1–5 is $u_{\text{feas},j}$ (see (17)) that relates to $\rho_{\text{ini},j}$. Thus, $u_{\text{feas},j}$ which lies in $\mathcal{C}_{\text{constr}}$ is selected and the related $\rho_{\text{ini},j}$ is used to build the sphere. If multiple $u_{\text{feas},j}$ lie in $\mathcal{C}_{\text{constr}}$, the one with the minimal distance from u_{unc} is selected[5].

To elucidate the above-mentioned procedure, the example in Fig. 3(b) is provided. Therein, the three output constraints are visualized, with $u_{\text{feas},1} \leftarrow \mathcal{C}_{\text{conv}}$, $u_{\text{feas},2} \leftarrow \mathcal{C}_c$, and $u_{\text{feas},3} \leftarrow \mathcal{C}_g$, while only one candidate solution is assumed to lie in $\mathcal{C}_{\text{constr}}$ at time step k to simplify the demonstration. After computing the feasible input set and the refined initial radius, the sphere decoder is called with the difference that now the $3N_p$-dimensional candidate solution $U^*(k)$ enclosed in the hypersphere belongs to set $\mathbb{U}_{\text{constr}} = \mathbb{U} \cap \mathcal{C}_{\text{constr}}$. Nevertheless, in the worst-case scenario, if none of the resulting $u_{\text{feas},j}$ computed from steps 1–6 belongs to $\mathcal{C}_{\text{constr}}$, the set is first relaxed to $\mathcal{C}_{\text{constr}} = \mathcal{C}_{\text{conv}} \cap \mathcal{C}_c$, and then to $\mathcal{C}_{\text{constr}} = \mathcal{C}_{\text{conv}}$ in order to find the that best suboptimal solution that violates the smallest number of constraints.

Performance Assessment

The performance of the proposed direct MPC scheme is evaluated through MATLAB simulations. Considering $f_{\text{res}} = 304\,\text{Hz}$, the goal is to choose λ_u such that the switching frequency f_{sw} is close to f_{res}. To this aim, λ_u is chosen such that $f_{\text{sw}} \approx 400\,\text{Hz}$ results, i.e., $\lambda_u = 0.45$. A sampling interval $T_s = 150\,\mu\text{s}$ is chosen such that a high granularity of switching is achieved, as this improves the system performance [11]. Moreover, given that for a favorable steady-state and transient performance a long prediction interval in time is recommended [15], a ten-step ($N_p = 10$) horizon is implemented. Regarding the matrix Q, the tracking of $i_{g,abc}$ is prioritized over the reference tracking of the other variables to reduce the grid current TDD, $I_{g,\text{TDD}}$. Then, the converter current tracking is prioritized with respect to the voltage capacitor to (indirectly) avoid a deterioration in the tracking performance of $i_{g,abc}$. This yields $Q = \text{diag}(10, 10, 1, 1, 100, 100)$. The upper bounds $i_{\text{conv,max}} = 1.3$ per unit (p.u.), $v_{c,\text{max}} = 1.25\,\text{p.u.}$, and $i_{g,\text{max}} = 1.25\,\text{p.u.}$ are considered to limit overshoots during transients. Note that, given the chosen limits, the hard constraints do not affect the steady-state operation. All results in the sequel are in p.u.

Both steady-state and transient performances are depicted in Fig. 4 over two fundamental periods. More specifically, Fig. 4(a) shows the three-phase output waveforms produced by a direct MPC formulation without output bounds, i.e., when neglecting (7) and computing ρ_{ini} as in (8). On the other hand, Fig. 4(b) depicts the system response when the constrained MPC is taken into account, i.e., when considering (7) and the modified radius is computed according to (15).

---

[5]Note that, this procedure can be also adopted to evaluate a larger number of constraints n_c, i.e. $j = 1, ..., n_c$

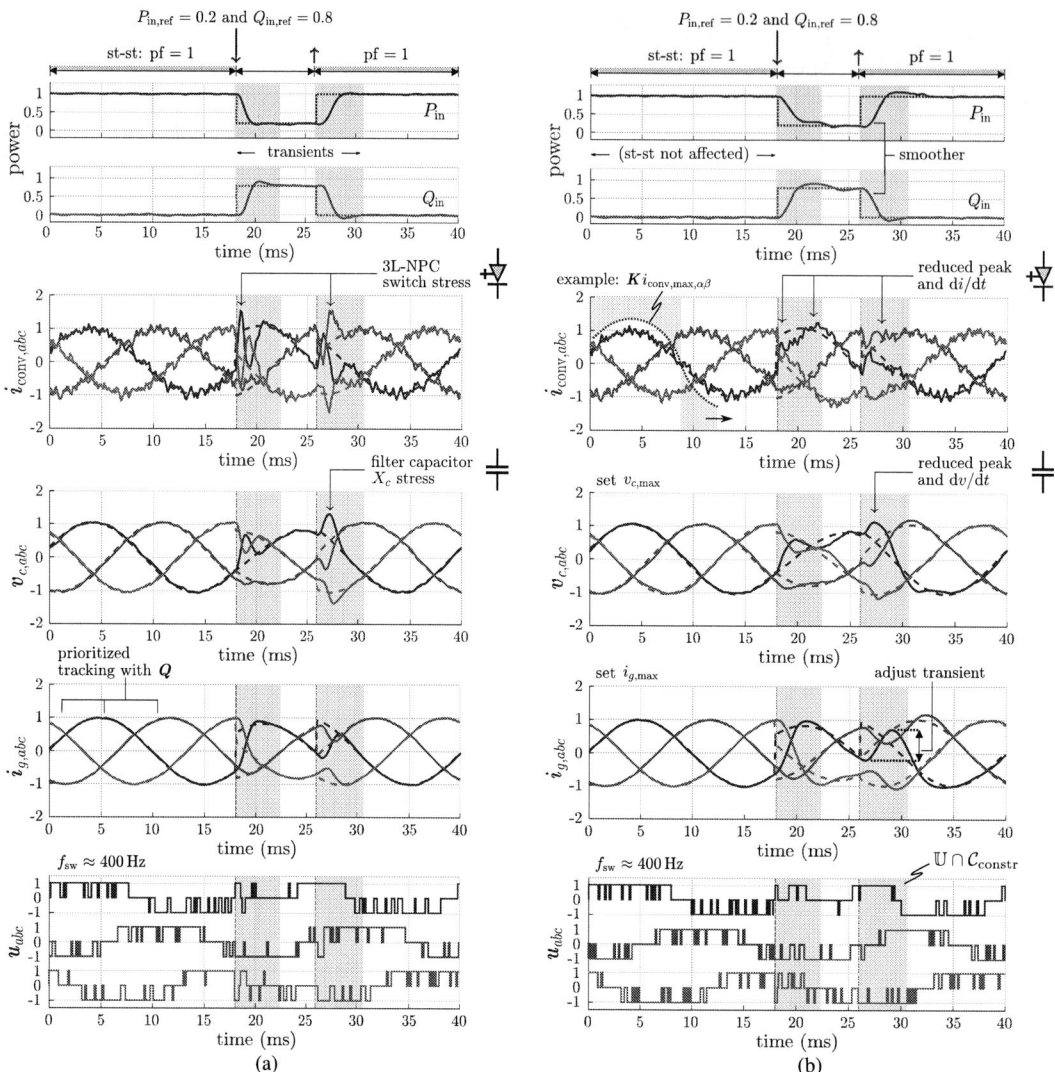

Fig. 4: Simulated waveforms produced by direct MPC when the output constraints are (a) not included, and (b) included. The results are shown over two fundamental periods $2T_g$, with $T_g = 20$ ms. From top to bottom: real P_{in} (blue line) and reactive power Q_{in} (green line) and their references (dashed lines); three-phase converter currents $\boldsymbol{i}_{\text{conv},abc}$ (with a, b and c shown with blue, red, and green lines, respectively) and the related references; three-phase capacitor voltage $\boldsymbol{v}_{c,abc}$; three-phase grid currents $\boldsymbol{i}_{g,abc}$; three-phase switch position $\boldsymbol{u}_{abc}$.

Large step-wise changes in the input power references are applied. At $t = 18$ ms, $P_{\text{in,ref}}$ is changed from 1 to 0.2 p.u. and back to 1 p.u. at $t = 26$ ms. Likewise, $Q_{\text{in,ref}}$ is changed from 0 to 0.8 p.u. and back to 0 p.u. at the same time instants. The power references are translated into the corresponding $\boldsymbol{y}_{\text{ref},abc}$ which are accurately followed by $\boldsymbol{y}_{abc}$. As shown in Fig. 4(a), when the output constraints are not considered, the variables $\boldsymbol{i}_{\text{conv},abc}$ and $\boldsymbol{v}_{c,abc}$ exhibit significant overshoots during transients, exceeding the associated trip levels defined by the translation of $i_{\text{conv,max}}$ and $v_{c,\text{max}}$ into the abc-plane via $\boldsymbol{K}$. To provide more insight into this point, and since the constraints are given in the $\alpha\beta$-plane, the dynamics in Fig. 4 are translated accordingly and shown in Fig. 5. In particular, given the time window from $t = 15$ ms to $t = 35$ ms, Fig. 5(a) shows and quantifies the violation of $\boldsymbol{y}_{abc}$. For example, as can be seen, $\boldsymbol{i}_{\text{conv},\alpha\beta}$ presents a peak of 1.70 p.u. which is 40% above $i_{\text{conv,max}} = 1.3$ p.u.

On the other hand, the effectiveness of the constrained MPC algorithm can be appreciated in Figs. 4(b) and 5(b), where $\boldsymbol{y}_{abc}$ (and $\boldsymbol{y}_{\alpha\beta}$) always remain within the imposed bounds. It is worth mentioning, however, that the constrained MPC is less aggressive since the output bounds restrict the input set to $\mathbb{U}_{\text{constr}} \subset \mathbb{U}$, i.e., the candidate solutions that could lead to faster settling times are limited as these can lead to violation of the constraints. This implies that the decisions the MPC algorithm makes when the

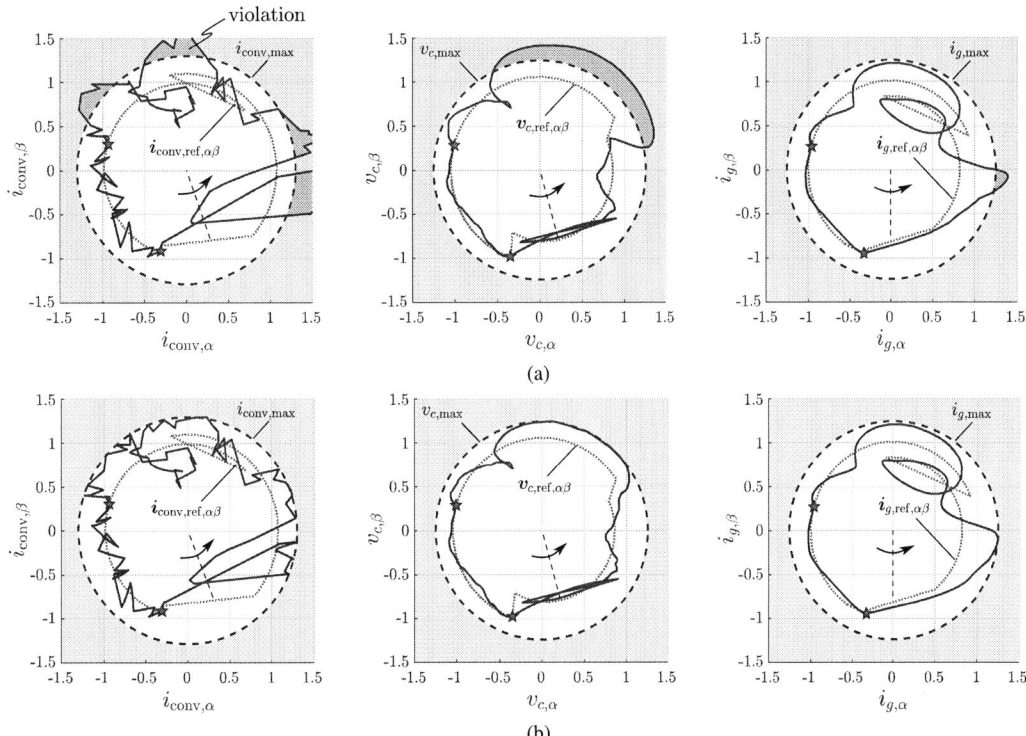

Fig. 5: Simulated waveforms of $i_{\mathrm{conv},\alpha\beta}$ (left), $v_{c,\alpha\beta}$ (center), and $i_{g,\alpha\beta}$ (right) produced by the proposed direct MPC algorithm when output constraints are (a) not included, and (b) included. The references and bounds are shown as (red) dotted and (black) dashed circles, respectively. The $\alpha\beta$-plots cover a time window from 15 to 35 ms. All variables in Fig. 4 are mapped from abc to $\alpha\beta$, i.e. $\boldsymbol{\xi}_{\alpha\beta} = \boldsymbol{K}\boldsymbol{\xi}_{abc}$.

output constraints are active would correspond to suboptimal solutions compared to the case where the constraints are not considered. As a result, the settling times of the power transients in Figs. 4(b) are slightly longer compared to Fig. 4(a).

Finally, it worth mentioning that when operating at steady-state—i.e., $P_{\mathrm{in,ref}} = 1$ and $Q_{\mathrm{in,ref}} = 0$ with pf $= 1$—the output constraints are not activated, hence both MPC algorithms perform the same.[6] In both cases, all output variables $\boldsymbol{y}_{abc}$ accurately track their sinusoidal reference values, despite operation at a switching frequency that is very close to the resonance one. The current and PCC voltage harmonic spectra produced by the proposed direct MPC are shown in Fig. 6. As can be seen, even though the harmonic energy for both $\boldsymbol{i}_{g,abc}$ and $\boldsymbol{v}_{\mathrm{pcc},abc}$ is spread over a wide range of frequencies due to the variable switching frequency, the grid codes, such as the IEEE 519 and IEC 61000-2-4 standards, are met with $I_{g,\mathrm{TDD}} \approx 1.25\%$ and $V_{\mathrm{pcc,TDD}} \approx 3.03\%$. For comparison purposes the spectra of conventional space vector modulation (SVM) are shown in the same figure, while considering the same f_{sw}. This is like having a simple closed-loop linear controller with a very low bandwidth [16]. As can be seen, the TDD values of both current and PCC voltage produced by SVM are higher than those obtained with MPC.

Conclusions

This paper presented a long-horizon direct MPC algorithm for an MV 3L-NPC converter connected to the grid via an LCL filter. The control problem formulation is augmented with output constraints that relate to physical limitations of the system components. The underlying optimization problem is formulated as an ILS problem where the output constraints are translated into input constraints that define a new feasible set. Subsequently, the employed sphere decoder finds the feasible point within the refined hypersphere in a computationally efficient manner. In doing so, operation within the safety operating region of the system is guaranteed without deteriorating its performance. As shown, the proposed approach

---

[6]Note that the constraint $\|\Delta\boldsymbol{u}(\ell)\|_\infty \le 1$ is activated and fully respected in both MPC formulations.

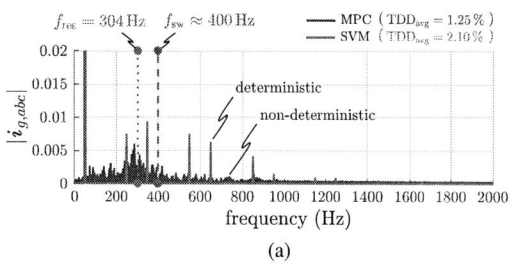
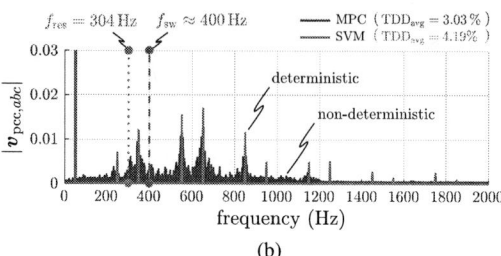

(a) (b)

Fig. 6: Harmonic spectra of (a) $i_{g,abc}$, and (b) $v_{\mathrm{pcc},abc}$ in p.u. For both cases, the harmonics do not violate their respective limits (in terms of TDD and amplitude) imposed by the IEEE 519 and IEC 61000-2-4 standards.

achieves excellent steady-state and transient performance, while meeting the relevant grid codes despite the low switching frequency which is very close to the resonance one. Finally, thanks to the adopted long prediction horizon and the full-state information, an additional active damping loop is unnecessary, resulting in a simpler control architecture compared to conventional control strategies.

References

[1] IEEE Std 519-2014 (Revision of IEEE Std 519-1992), "IEEE recommended practices and requirements for harmonic control in electrical power systems," pp. 1–29, Jun. 2014.

[2] IEC 61000-2-4, "Electromagnetic compatibility (EMC)—part 2-4: Environment—compatibility levels in industrial plants for low-frequency conducted disturbances," Sep. 2002.

[3] J. Dannehl, F. W. Fuchs, S. Hansen, and P. B. Thøgersen, "Investigation of active damping approaches for PI-based current control of grid-connected pulse width modulation converters with LCL filters," *IEEE Trans. Ind. Appl.*, vol. 46, no. 4, pp. 1509–1517, Jul./Aug. 2010.

[4] P. C. Loh and D. Holmes, "Analysis of multiloop control strategies for $LC/CL/LCL$-filtered voltage-source and current-source inverters," *IEEE Trans. Ind. Appl.*, vol. 41, no. 2, pp. 644–654, Mar./Apr. 2005.

[5] T. Geyer, *Model predictive control of high power converters and industrial drives.* Hoboken, NJ, USA: Wiley, 2016.

[6] P. Karamanakos, E. Liegmann, T. Geyer, and R. Kennel, "Model predictive control of power electronic systems: Methods, results, and challenges," *IEEE Open J. Ind. Appl.*, vol. 1, pp. 95–114, 2020.

[7] P. Cortés, M. P. Kazmierkowski, R. M. Kennel, D. E. Quevedo, and J. Rodriguez, "Predictive control in power electronics and drives," *IEEE Trans. Ind. Electron.*, vol. 55, no. 12, pp. 4312–4324, Dec. 2008.

[8] T. Geyer and D. E. Quevedo, "Multistep finite control set model predictive control for power electronics," *IEEE Trans. Power Electron.*, vol. 29, no. 12, pp. 6836–6846, Dec. 2014.

[9] P. Karamanakos., T. Geyer, and R. Kennel, "Constrained long-horizon direct model predictive control for power electronics," in *Proc. IEEE Energy Convers. Congr. Expo.*, Milwaukee, WI, USA, Sep. 2016, pp. 1–8.

[10] M. Rossi, P. Karamanakos, and F. Castelli-Dezza, "An indirect model predictive control method for grid-connected three-level neutral point clamped converters with LCL filters," *IEEE Trans. Ind. Appl.*, vol. 58, no. 3, pp. 3750–3768, May/Jun. 2022.

[11] P. Karamanakos and T. Geyer, "Guidelines for the design of finite control set model predictive controllers," *IEEE Trans. Power Electron.*, vol. 35, no. 7, pp. 7434–7450, Jul. 2020.

[12] P. Karamanakos, T. Geyer, and R. Kennel, "Reformulation of the long-horizon direct model predictive control problem to reduce the computational effort," in *Proc. IEEE Energy Convers. Congr. Expo.*, Pittsburgh, PA, USA, Sep. 2014, pp. 3512–3519.

[13] J. B. Rawlings and D. Q. Mayne, *Model predictive control: Theory and design.* Madison, WI: Nob Hill, 2009.

[14] P. Karamanakos and T. Geyer and R. Kennel, "Suboptimal search strategies with bounded computational complexity to solve long-horizon direct model predictive control problems," in *Proc. IEEE Energy Convers. Congr. Expo.*, Montreal, QC, Canada, Sep. 2015, pp. 334–341.

[15] T. Geyer, P. Karamanakos, and R. Kennel, "On the benefit of long-horizon direct model predictive control for drives with LC filters," in *Proc. IEEE Energy Convers. Congr. Expo.*, Pittsburgh, PA, USA, Sep. 2014, pp. 3520–3527.

[16] M. Rossi, E. Liegmann, P. Karamanakos, F. Castelli-Dezza, and R. Kennel, "Long-horizon direct model predictive control for a series-connected modular rectifier," in *Proc. Int. Exhib. and Conf. for Power Electron., Intell. Motion, Renew. Energy and Energy Manag.*, Nuremberg, Germany, 2020, pp. 342–349.

Performance Evaluation of SiC-based Isolated Bidirectional DC/DC Converters for Electric Vehicle Charging

Kaushik Naresh Kumar[*], Rafał Miśkiewicz[#], Przemysław Trochimiuk[#], Jacek Rąbkowski[#], Dimosthenis Peftitsis[*]

[*]NORWEGIAN UNIVERSITY OF SCIENCE AND TECHNOLOGY, Department of Electric Power Engineering, Trondheim, Norway

[#]WARSAW UNIVESITY OF TECHNOLOGY, Institute of Control and Industrial Electronics, Warsaw, Poland

E-Mail: kaushik.n.kumar@ntnu.no, rafal.miskiewicz@pw.edu.pl, przemyslaw.trochimiuk@pw.edu.pl, jacek.rabkowski@pw.edu.pl, dimosthenis.peftitsis@ntnu.no

Acknowledgements

This work is supported by the "Modularized, Reconfigurable and Bidirectional Charging Infrastructure for Electric Vehicles with Silicon Carbide Power Electronics" (MoReSiC) project, funded by the POLNOR programme in the Norway-Poland EEA Grants framework.

Keywords

«Dual Active Bridge (DAB)», «DC-DC converters», «Electric vehicle (EV) charging», «Silicon Carbide (SiC) MOSFET»

Abstract

In this paper, six 10 kW DC/DC isolated and bidirectional dual active bridge topologies, supplied by +750/0/-750V three-wire DC bus, are evaluated based on efficiency, loss distribution, volt-ampere semiconductors ratings and normalized cost of SiC MOSFETs for electric vehicle charging applications. The selected topologies are evaluated under the same voltage and power conditions, through electrothermal simulations and experiments. The simulation results were verified through experiments conducted on 5 kW prototypes of four of the considered topologies. The advantages and disadvantages of the topologies are discussed and analyzed based on the chosen performance metrics. It has been shown that series-resonant input-series output-parallel full-bridge DAB topology exhibits the highest efficiency while the active neutral point clamped (ANPC) DAB topology can be designed with the lowest cost. However, considering a fair trade-off between all the performance metrics, series-resonant ANPC DAB topology is shown to be the best design choice for the considered evaluation conditions.

1. Introduction

Electrification of the transportation sector plays a key role in the decarbonization of the environment. In this context, the installation of advanced electric vehicles (EVs) fast charging infrastructure with bidirectional capability is paramount considering challenges like EVs demand flexibility, power grid support, etc. [1]. The isolated bidirectional DC/DC converter (IBDC) is one of the key components of such an EV charging system and the first step for its design is the choice and performance evaluation of the utilized power electronics converter topology.

The advantages of dual active bridge (DAB) topology introduced in [2] are its soft-switching capability, higher power density, galvanic isolation, bidirectional power flow capability, and wide voltage gain range [3]. Silicon Carbide (SiC) metal oxide semiconductor field-effect transistors (MOSFETs) can switch at higher frequencies, block higher voltages for the same conduction loss performance, and operate at higher temperatures compared to Silicon insulated gate bipolar transistors (IGBTs) [4] [5]. It is inferred from other EV applications that employing SiC power devices may be highly beneficial in terms of efficiency and compactness [6] [7] [8]. Therefore, DAB topologies employing SiC MOSFETs are evaluated in this paper. The EV charging system considered in this evaluation is based on bipolar (three-wire) DC bus as depicted in Fig. 1. The advantages of using a bipolar DC bus include lower on-state losses in devices, higher power quality and enhanced flexibility [9] [10] [11].

Comparison of IBDC DAB converter topologies employing wide band gap (WBG) devices have been presented for applications such as EV fast chargers [12], more electric aircraft [13] and other power applications [14]. However, there is not much literature available for such a comparison considering an EV charging system using bipolar DC bus with appropriate experimental validation. In this point of view, the contribution of this work is on the performance evaluation of various IBDC DAB converter topologies through electrothermal modeling and simulation, as well as laboratory prototyping and experimental validation. The paper is organized as follows: Section 2 presents the topologies under investigation and their design considerations. In Section 3, the performance of the topologies based on volt-ampere ratings, normalized cost, efficiency, and loss distribution of SiC MOSFETs is evaluated using theoretical analysis and simulations. Section 4 provides a description of the experimental prototypes and the results from experiments are summarized and compared to simulation results in Section 5. Section 6 presents the conclusion.

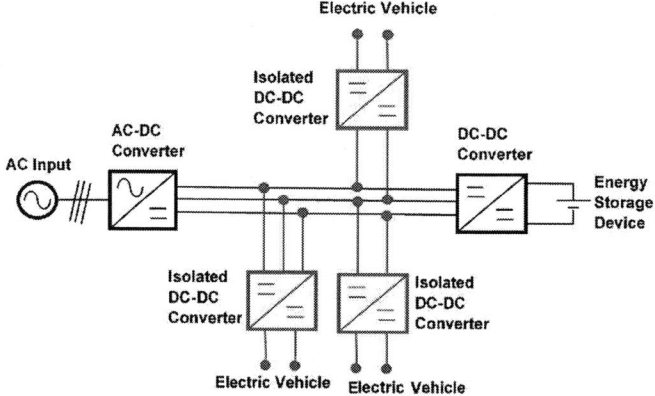

Fig. 1: Schematic of the EV Charging system based on bipolar DC bus.

2. Evaluated DAB-based topologies

DAB-based IBDC generally consists of two full-bridge (FB) circuits that are galvanically isolated by means of a high frequency transformer. For positive power flow, the primary FB operates as an inverter and the secondary FB functions as a rectifier. In the traditional single-phase-shift (SPS) modulation, the switches in both FBs operate at 50% fixed duty ratio and generate high frequency AC square-wave voltages across the transformer primary and secondary sides. The direction and magnitude of transferred power can be controlled by the phase difference between the FBs like conventional ac power systems.

Considering the bipolar (three-wire) nature of the DC bus and the two-terminal nature of EV battery load, two fundamental circuits for the isolated DAB converters have been chosen, namely, the full-bridge (FB) and the active neutral-point clamped (ANPC) circuits, as shown in Figs. 2(d) and (e), respectively. These circuits can be combined for designing either one-to-one isolated DAB converters comprising a single primary bridge and single secondary bridge or multiport counterparts. Fig. 2(c) depicts a one-to-one isolated DAB topology employing an ANPC circuit on the primary side and a FB circuit on the secondary side that are coupled though a high-frequency transformer.

To increase the flexibility in terms of voltage and electric power supplied, the FB and ANPC circuits can be connected in series or in parallel. The input-series output-parallel (ISOP) configuration is one such topology that provides voltage flexibility on the primary side through series-connected circuits and power scalability in the secondary side through parallel-connected circuits. Fig. 2(a) shows an ISOP FB DAB topology where the series-connected primary and parallel-connected secondary bridges consist of FB circuit. Fig. 2(b) depicts an ISOP ANPC DAB topology with series-connected ANPC circuit on the primary side and parallel-connected FB circuits on the secondary side. Both series-resonant (SR) and non-resonant variants of each of the three listed topologies are evaluated to study their performance, as well as advantages and disadvantages. Topologically, the SR DAB configurations require an additional capacitor to form a L_rC_r resonant tank (shown in blue in Figs. 2(a)-(c)) compared to the non-resonant DAB converters, which only have a series-inductance, L (shown in green in Figs. 2(a)-(c)). The series-inductance represents the sum of the transformer leakage inductance plus any external and parasitic inductance in the circuit.

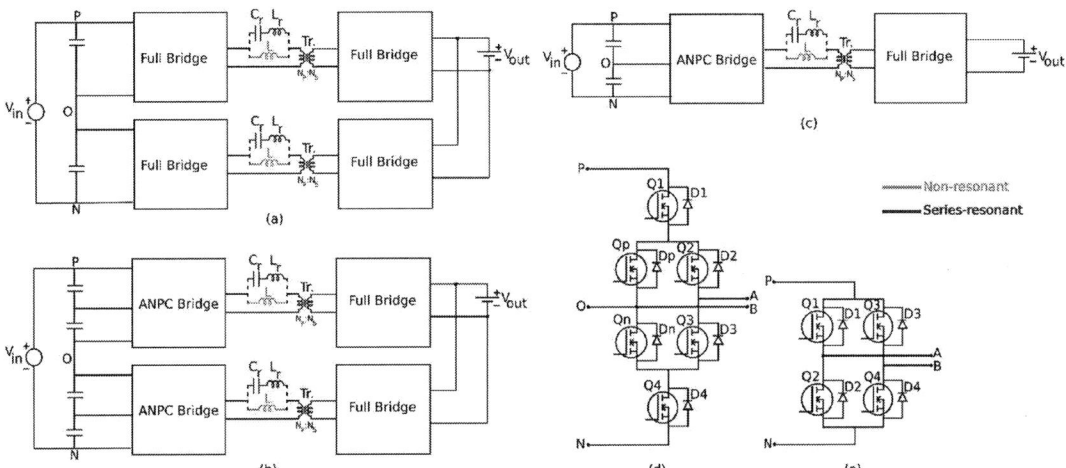

Fig. 2: Block diagrams of the evaluated topologies: (a) ISOP FB DAB, (b) ISOP ANPC DAB, (c) ANPC DAB, (d) ANPC bridge, and (e) Full bridge.

2.1. Design considerations

The selected topologies depicted in Fig. 2 are evaluated under the same voltage and power levels. The values considered for the dc input voltage (V_{in}), dc output voltage (V_{out}), output power (P_{out}), switching frequency (f_{sw}), non-resonant/resonant series-inductance (L/L_r), resonant capacitance (C_r), and transformer turns ratio (N) are summarized in Table I.

Table I: Design parameters for the converters.

DAB type	Non-resonant DAB	Series-resonant DAB
V_{in}	1500V	
V_{out}	400 V	
P_{out}	10 kW	
f_{sw}	100 kHz	
L/L_r	17µH – 70µH*	100 µH
C_r	-	28 nF
N	$V_{primary\_pk}/V_{secondary\_pk}$	

$V_{primary\_pk}$ - peak transformer primary voltage

$V_{secondary\_pk}$ - peak transformer secondary voltage

*Calculated using Eqn.2 depending on topology

Table II: SiC MOSFETs utilized in each topology

Topology	MOSFET used in primary bridge	MOSFET used in secondary bridge
ISOP FB DAB	NTH4L040N120SC1[1]	C3M0015065K[2]
ISOP ANPC DAB	IMZA65R039M1HXKSA1[3]	C3M0015065K[2]
ANPC DAB	NTH4L040N120SC1[1]	C3M0015065K[3]

[1]1200V, 40mΩ, 58A SiC MOSFET from ON semiconductor
[2]650V, 15mΩ,120A SiC MOSFET from CREE
[3]650V, 39mΩ, 50A SiC MOSFET from Infineon

The switching frequency, f_{sw} is chosen to be 100 kHz as a fair trade-off between the heat sink size, the size of magnetics and the switching losses. The selection of SiC MOSFETs was made considering the type of bridge circuit (i.e., ANPC or FB), considered voltage and power levels, better switching performance in terms of faster switching transients due to lower stray inductance in the gate loop (four-pin devices with Kelvin connection) and their availability in the market. Table II summarizes the chosen SiC MOSFETs for each topology and their ratings.

Since the performance evaluation based on efficiency and losses of SiC MOSFETs are carried out at a specific operating point, only the unidirectional power flow operation of the chosen converter topologies is considered. Therefore, the EV battery is modelled as a resistive load, R_{load} given by:

$$R_{load} = \frac{V_{out}^2}{P_{out}} \tag{1}$$

The power transfer equation for a non-resonant DAB converter operating with SPS modulation is given by:

$$P_{out} = N * V_{primary} * V_{secondary} * \frac{\left(D * (1-D)\right)}{2 * f_{sw} * L} \tag{2}$$

Where, D is the phase-shift ratio (0 < D < 1) between primary and secondary bridges. The value of inductor, L is calculated according to Eqn. 2 for a phase shift of D = 0.5 and for twice the nominal power level. The reason for this approach is to ensure lower reflow power at the nominal operating point, where reflow power refers to the power flow back to the source caused by circulating currents in the converter [15]. It is to be noted that the value of L is calculated at P_{out} = 10 kW per each module of the ISOP FB DAB and ISOP ANPC DAB configurations (each module is operated at 5 kW nominal power) and at P_{out} = 20 kW for the ANPC DAB configuration for a nominal operating power of P_{out} = 10 kW.

In the case of resonant DABs, the L_r and C_r values are chosen such that the resonance frequency, f_r is less than the switching frequency, f_{sw} of the converter to ensure continuous current mode operation [16]. The resonance frequency, f_r is given by:

$$f_r = \frac{1}{2 * \pi * \sqrt{L_r * C_r}} \tag{3}$$

The transformer turns ratio, N is chosen to be equal to the ratio of the transformer voltages as given in Table I to ensure zero voltage switching and to reduce the stress on transformer by reducing circulating currents.

3. Performance Evaluation

This section presents the performance evaluation of the topologies under study using theoretical analysis and simulations. In particular, the required volt-ampere ratings of power semiconductors, as well as the loss distribution among the primary and secondary FBs are evaluated based on simulations.

3.1. Evaluation based on Volt-Ampere (VA) rating and normalized cost of SiC MOSFETs

Table III summarizes the SiC MOSFETs' VA ratings and normalized costs for each of the evaluated topologies, where the VA ratings are calculated based on the total number of switches and ratings of the chosen SiC MOSFETs (see Table II).

From Table III, it is seen that the ANPC DAB is the most cost-effective configuration and requires the lowest VA ratings of SiC MOSFETs compared to the ISOP DAB topologies for the considered operating conditions. However, considering the reliability, the ISOP DABs are better compared to the ANPC DAB. This is because the two parallel DAB modules in a ISOP DAB topology share the total power transferred, which decreases the current stress on the switches. In terms of physical components, the series-resonant DABs require an additional resonant capacitance compared to the non-resonant variants which may contribute to a decrease in power density and an increase in losses due to its equivalent series resistance (ESR).

TABLE III: Summary of VA ratings and normalized costs

Topology	VA rating of SiC MOSFETs (kVA)			Normalized total cost of MOSFETs[*]
ISOP FB DAB	Primary bridge	Secondary bridge	Total	
	556.8 (8 * 1200V * 58A)	624 (8 * 650V * 120A)	1180.8	2.34
ISOP ANPC DAB	390 (12 * 650V * 50A)	624 (8 * 650V * 120A)	1014	2.32
ANPC DAB	417.6 (6 * 1200V * 58A)	312 (4 * 650V* 120A)	729.6	1.44

[*]Cost normalized per 200 USD

3.2. Evaluation based on efficiency and MOSFET loss distribution

As all the evaluated topologies are designed to operate at unity voltage transfer ratio and similar voltage and power levels, the major differentiating factor between them is the rating and number of SiC MOSFETs used. Therefore, the efficiency and loss distribution comparison have been performed based only on the SiC MOSFET losses. The calculation of efficiency and losses for the considered topologies is performed in PLECS simulation environment using the corresponding SiC MOSFET models.

Table II contains the information about the MOSFETs types and ratings used in simulations. The SiC MOSFET simulation model for C3M0015065K was provided on the manufacturer's (i.e., Wolfspeed) webpage. The PLECS models for the NTH4L040N120SC1 (from ON semiconductor) and IMZA65R039M1HXKSA1 (from Infineon) were developed using parameters extracted from their respective datasheets and by conducting simulations using their LT SPICE models. LT SPICE models were mainly used for estimation of switching energies. The modelling considerations for the evaluated topologies are as described in Subsection 2.1 and the simulation parameters correspond to Table I.

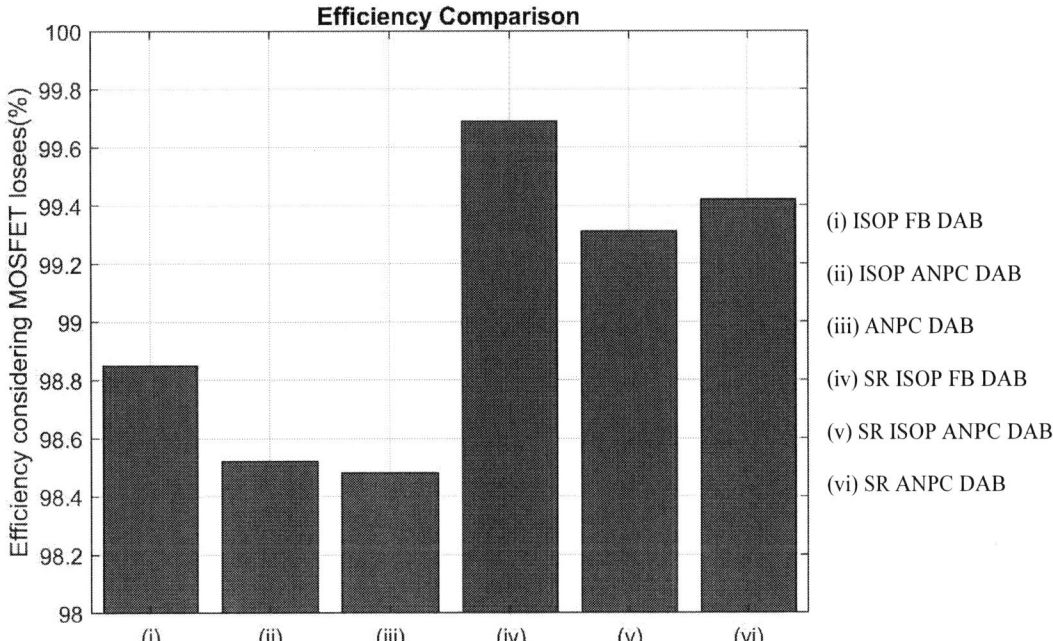

Fig. 3: Comparison of efficiency considering only the MOSFET losses based on simulations.

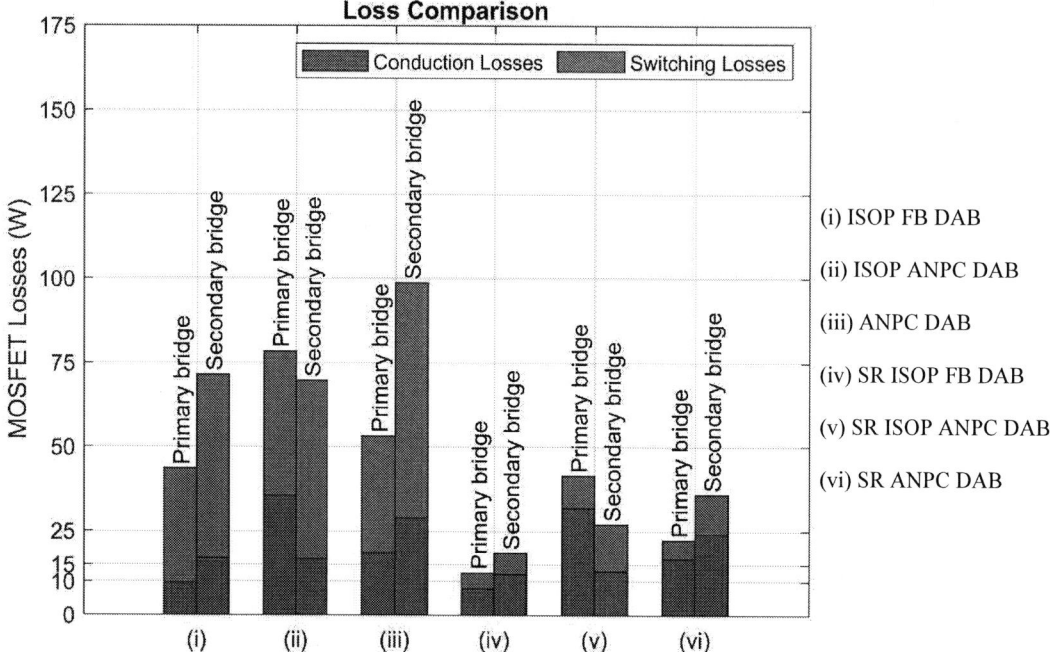

Fig. 4: Comparison of MOSFET loss distribution based on simulations.

In the comparison of efficiency based on MOSFET losses as shown in Fig. 3, the series-resonant DAB configurations outperform the non-resonant variants. One possible reason for the better efficiency of SR variants is due to the operation of SiC MOSFETs near soft switching region in both primary and secondary bridges. This fact is reflected in the MOSFET loss distribution shown in Fig. 4, where the switching losses in both primary and secondary bridges of the SR DABs (i.e., iv, v and vi) are far lower compared to the non-resonant DABs. An additional reason is the sinusoidal shape of the transformer currents in the SR DABs, which results in lower RMS currents and in turn lower conduction losses. This can be observed in Fig. 4, where the conduction losses of the SR DABs are lower in both primary and secondary bridges compared to the respective non-resonant variants.

Among ISOP FB DAB and ISOP ANPC DAB configurations, considering both non-resonant and SR variants, the main difference is the higher conduction losses in the primary bridge for the ISOP ANPC DAB as seen in Figs. 4(ii) and (v). The reason for this is that ISOP ANPC DABs carry almost double the current in the primary bridge compared to ISOP FB DABs for the same power transferred, as only half of the input voltage is reflected on the transformer primary. The higher secondary bridge current (almost double) in non-resonant ANPC DAB compared to the ISOP topologies causes higher switching losses, since the MOSFET switching energy is dependent on drain current. This can be observed in Fig. 4 (iii).

3.3. Evaluation Summary

The spider plot in Fig. 5 summarizes the evaluation of the considered topologies based on the performance metrics discussed in Subsections 3.1 and 3.2. The series-resonant topologies exhibit higher efficiencies compared to their non-resonant variants, which can be attributed to their significantly lower switching losses. The FB topologies exhibit lower conduction losses compared to the ANPC topologies. In terms of VA ratings and normalized cost of SiC MOSFETs, the ANPC DAB topology has a clear advantage compared to the ISOP FB and ISOP ANPC DAB configurations. The SR ANPC DAB topology combines the advantages of both the series-resonant configuration (i.e., lower switching losses and higher efficiency) and the ANPC topology (i.e., lower VA ratings and normalized cost of SiC MOSFETs). This makes it the best design choice for the evaluated application, considering all performance metrics.

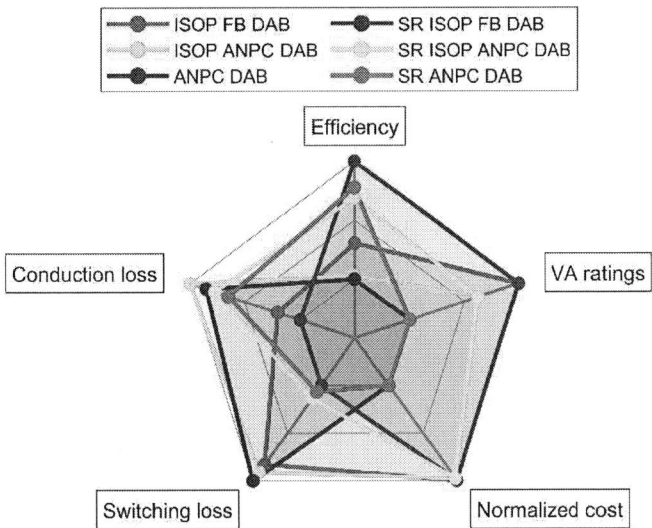

Fig. 5: Comparison of the evaluated topologies based on all the chosen performance metrics.

4. Laboratory prototype

Since the ISOP DAB configuration consists of two parallel-connected DAB modules sharing the total power transferred, validation of a single module is sufficient in terms of efficiency comparison. Therefore, 5 kW prototypes of single modules of non-resonant and series-resonant variants of ISOP FB and ANPC DABs were built-up for the experimental validation of the simulation results. Fig. 6 shows a photo of the non-resonant FB DAB laboratory prototype.

The experimental parameters summarized in Table IV correspond to half the dc input voltage and output power levels considered for simulations (See Table I), since they refer to a single module of non-resonant and SR versions of ISOP FB and ISOP ANPC DABs. The series-inductances values include the stray inductances of transformer windings measured using impedance analyzer. The actual transformer turns ratio for the ANPC configuration was 0.909 as opposed to the calculated value of 0.9375. In case of FB configuration, the calculated transformer turns ratio was achieved. The resonant capacitances have different values for the SR ANPC and FB DABs to keep the resonant frequency the same irrespective of the deviation in the transformer turns ratio and leakage inductances between the two configurations.

Fig. 6: 5 kW non-resonant FB DAB prototype: (1) Primary full bridge, (2) Secondary full bridge, (3) Series inductor, (4) Transformer, and (5) Control board.

Table IV: Experimental parameters

DAB type	Non-resonant DAB	Series-resonant DAB
V_{in}	750V	
V_{out}	400 V	
P_{out}	5 kW	
f_{sw}	100 kHz	
L/L_r	18.2[a] – 72.5μH[b]	100 μH
C_r	-	26.9[a] – 30.5[b] nF
N	0.909[a] – 1.875[b]	
R_{load}	30.5 Ohms	

[a] ANPC DAB
[b] FB DAB

The efficiency measurements were carried out using the YOKOGAWA WT5000 power analyzer and the transformer current and voltage waveforms were captured using Tektronix DPO 4034 mixed signal oscilloscope for all the DABs. Tektronix THDP0200 high-voltage differential probes were used for voltage measurements. Tektronix TCP0150 and TCP303 current probes were used for current measurements. The YOKOGAWA WT5000 power analyzer has an accuracy of ± (0.1% reading error + 0.1% measurement range error) for voltage, current and power measurements. The accuracy is also influenced by the temperature and aging of the equipment. The interesting observations from the simulations are related to comparison between the non-resonant and series-resonant DABs and the difference between the FB and ANPC configurations. Therefore, only four out of the six topologies considered were tested experimentally.

5. Experimental results

The experimental results consisting of converter efficiency measured in the DC circuit and the transformer voltage and current waveforms in the AC circuit are shown in Figs. 7(a)-(d) for the investigated topologies. In Figs. 7(a) and (b) the voltage per division is 500V and current per division is 10A. In Figs. 7(c) and (d) the voltage per division is 250V and current per division is 25A. It should be noted that the power analyzer data correspond to the overall converter efficiency whereas efficiency due to only MOSFET losses was considered in the simulations.

Most of the observations discussed in Subsection 3.2 can be validated from the experimental results. From Figs. 7(a) and (b) and Figs. 7(c) and (d), it can be observed that the SR DABs have a near sinusoidal transformer current waveform compared to their non-resonant counterparts, resulting in lower RMS bridge currents. From Figs. 7(a) and (c) and Figs. 7(b) and (d) it can be observed that the ANPC DABs carry almost double the current in the primary bridge and reflects half of the DC input voltage on transformer primary compared to FB DABs for the same power transferred.

Due to zero current switching and reduced RMS bridge currents in SR FB DAB as seen in Fig. 7(b), this topology exhibits the lowest losses. One of the reasons for high losses in the SR ANPC DAB can be attributed to hard switching at higher current values (nearly 10A) as seen in Fig. 7(d). Since the calculated value of transformer turns ratio did not match the actual value in the prototype, the zero current switching could not be achieved for the considered operating conditions. This shows the sensitivity of resonant circuits when operating them under practical conditions.

It should be noted that the value of resonant capacitor, C_r in both the SR DAB prototypes was achieved by series connection of two strings of 17 parallel-connected ceramic capacitors. The ESR of each ceramic capacitor at 100 kHz was found to be 2 ohms and the effective ESR of the C_r was estimated to be 0.235 ohms. Since the SR ANPC DAB has a primary RMS current of about 17A compared to 8A in SR FB DAB, the losses due to the ESR of C_r is about 68W and 15W, respectively. This is one of the major contributors to higher losses in the SR ANPC DAB.

Fig. 8 shows a comparison of the efficiency values obtained from the simulations with the experimental results for the investigated topologies. Simulation results are based on only MOSFET losses whereas experimental results correspond to the overall converter's efficiency. The difference between the experimental efficiencies and simulation efficiencies can be attributed to other losses like transformer core and copper losses, losses due to ESR of resonant capacitors, inductor losses, and other resistive losses. It can be observed that the general trend of efficiency data from the simulations agree with experimental results except for the SR ISOP ANPC DAB topology. The reasons for this deviation are as described above. In addition, it should be noted that the simulations were carried out based on calculated values of series-inductance, resonance capacitance and transformer turns-ratio without considering the actual experimental parameters like parasitic inductances, contact resistances, actual transformer turns-ratio and the ESR of resonant capacitor.

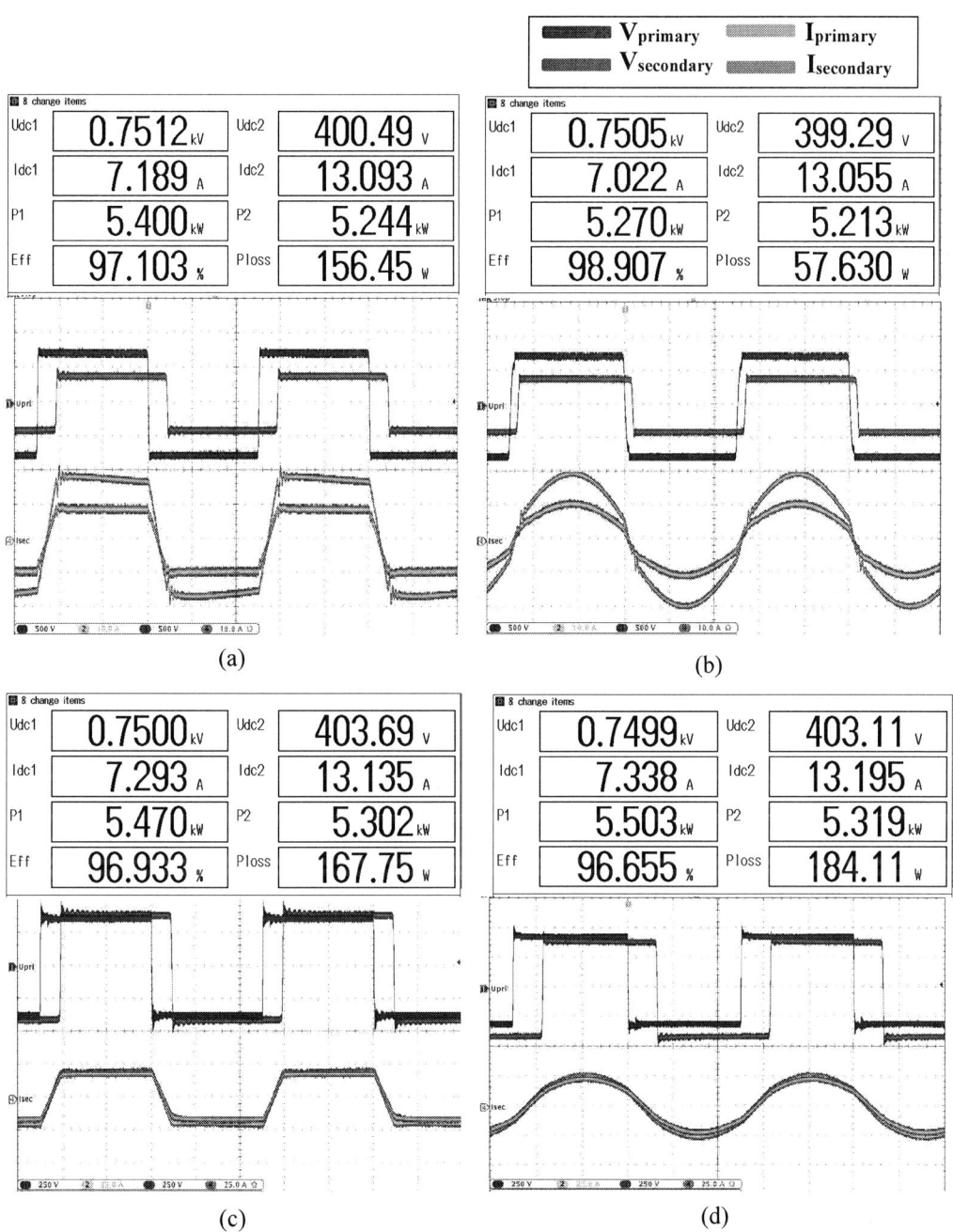

Fig. 7: Experimental results: Power analyzer data and transformer voltage and current waveforms for (a) non-resonant FB DAB, (b) series-resonant FB DAB, (c) non-resonant ANPC DAB and (d) series-resonant ANPC DAB

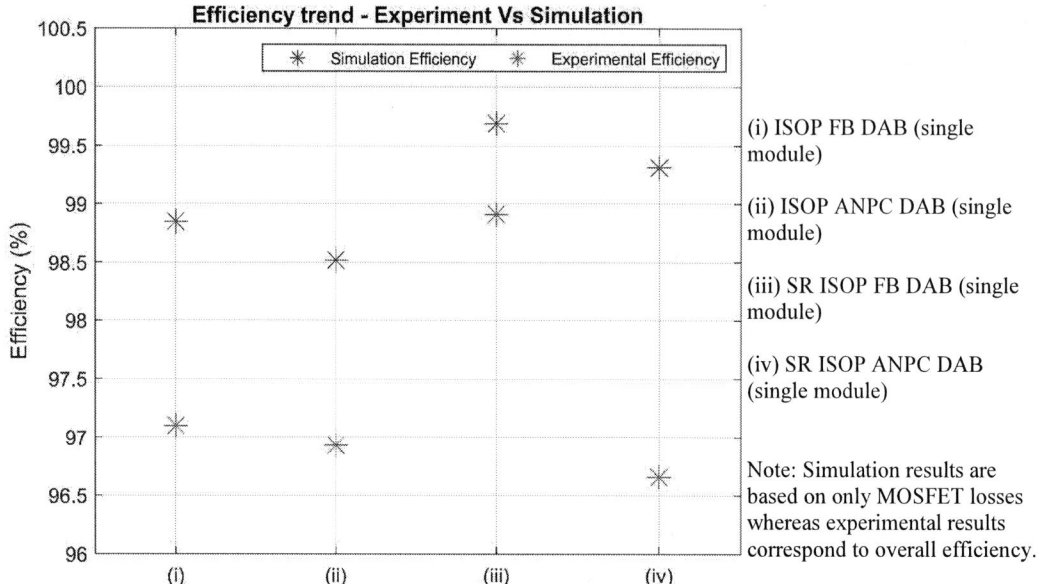

Fig. 8: Comparison of efficiency results from simulations and experiments

6. Conclusions

Six different configurations of the IBDC DAB topology have been evaluated for a 10-kW EV charging application. The topologies were selected considering the bipolar (three-wire) nature of input DC bus, two-terminal nature of EV battery load, flexibility of voltage as well as power scalability. In terms of additional passive component required, sensitivity of the circuit performance in the control scheme and control complexity the non-resonant DABs have an advantage over the SR DABs. However, SR DABs have a higher number of soft-switched (zero voltage and zero current switching) MOSFETs and near sinusoidal transformer currents, which result in lower switching and conduction losses. These positive performance impact factors of SR DABs outweigh their challenges for the considered application and operating conditions. In terms of efficiency and SiC MOSFET losses, the series-resonant variants outperform the non-resonant DABs with the SR ISOP FB DAB exhibiting the best performance. When the normalized cost and the VA ratings of SiC MOSFETs are considered, the ANPC DAB configuration seems to be the most promising one. Considering a fair trade-off between efficiency, loss distribution, VA ratings and the normalized cost, the SR ANPC DAB exhibits the best performance for this application.

7. References

[1] N. Matanov and A. Zahov, "Developments and Challenges for Electric Vehicle Charging Infrastructure," in *2020 12th Electrical Engineering Faculty Conference (BulEF)*, Sep. 2020, pp. 1–5. doi: 10.1109/BulEF51036.2020.9326080.

[2] R. W. A. A. De Doncker, D. M. Divan, and M. H. Kheraluwala, "A three-phase soft-switched high-power-density DC/DC converter for high-power applications," *IEEE Transactions on Industry Applications*, vol. 27, no. 1, pp. 63–73, Jan. 1991, doi: 10.1109/28.67533.

[3] B. Zhao, Q. Song, W. Liu, and Y. Sun, "Overview of Dual-Active-Bridge Isolated Bidirectional DC–DC Converter for High-Frequency-Link Power-Conversion System," *IEEE Transactions on Power Electronics*, vol. 29, no. 8, pp. 4091–4106, Aug. 2014, doi: 10.1109/TPEL.2013.2289913.

[4] J. Rabkowski, D. Peftitsis, and H.-P. Nee, "Silicon Carbide Power Transistors: A New Era in Power Electronics Is Initiated," *IEEE Industrial Electronics Magazine*, vol. 6, no. 2, pp. 17–26, Jun. 2012, doi: 10.1109/MIE.2012.2193291.

[5] J. Biela, M. Schweizer, S. Waffler, and J. W. Kolar, "SiC versus Si—Evaluation of Potentials for Performance Improvement of Inverter and DC–DC Converter Systems by SiC Power Semiconductors," *IEEE Transactions on Industrial Electronics*, vol. 58, no. 7, pp. 2872–2882, Jul. 2011, doi: 10.1109/TIE.2010.2072896.

[6] B. Whitaker *et al.*, "A High-Density, High-Efficiency, Isolated On-Board Vehicle Battery Charger Utilizing Silicon Carbide Power Devices," *IEEE Transactions on Power Electronics*, vol. 29, no. 5, pp. 2606–2617, May 2014, doi: 10.1109/TPEL.2013.2279950.

[7] D. Yıldırım, S. Ozturk, I. Cadirci, and M. Ermis, "All SiC PWM Rectifier Based Off-Board Ultrafast Charger for Heavy Electric Vehicles," *IET Power Electronics*, vol. 13, Aug. 2019, doi: 10.1049/iet-pel.2019.0583.

[8] X. Liang, S. Srdic, J. Won, E. Aponte, K. Booth, and S. Lukic, "A 12.47 kV Medium Voltage Input 350 kW EV Fast Charger using 10 kV SiC MOSFET," in *2019 IEEE Applied Power Electronics Conference and Exposition (APEC)*, Mar. 2019, pp. 581–587. doi: 10.1109/APEC.2019.8722239.

[9] Y. Du, X. Zhou, S. Bai, S. Lukic, and A. Huang, "Review of non-isolated bi-directional DC-DC converters for plug-in hybrid electric vehicle charge station application at municipal parking decks," in *2010 Twenty-Fifth Annual IEEE Applied Power Electronics Conference and Exposition (APEC)*, Feb. 2010, pp. 1145–1151. doi: 10.1109/APEC.2010.5433359.

[10] P. J. Grbović, P. Delarue, P. Le Moigne, and P. Bartholomeus, "A Bidirectional Three-Level DC–DC Converter for the Ultracapacitor Applications," *IEEE Transactions on Industrial Electronics*, vol. 57, no. 10, pp. 3415–3430, Oct. 2010, doi: 10.1109/TIE.2009.2038338.

[11] S. Rivera, B. Wu, S. Kouro, V. Yaramasu, and J. Wang, "Electric Vehicle Charging Station Using a Neutral Point Clamped Converter With Bipolar DC Bus," *IEEE Transactions on Industrial Electronics*, vol. 62, no. 4, pp. 1999–2009, Apr. 2015, doi: 10.1109/TIE.2014.2348937.

[12] M. Alharbi, M. Dahidah, V. Pickert, and J. Yu, "Comparison of SiC-based DC-DC modular converters for EV fast DC chargers," in *2019 IEEE International Conference on Industrial Technology (ICIT)*, Feb. 2019, pp. 1681–1688. doi: 10.1109/ICIT.2019.8843693.

[13] N. Keshmiri, M. I. Hassan, R. Rodriguez, and A. Emadi, "Comparison of Isolated Bidirectional DC/DC Converters Using WBG Devices for More Electric Aircraft," *IEEE Open Journal of the Industrial Electronics Society*, pp. 1–1, 2021, doi: 10.1109/OJIES.2021.3058196.

[14] M. Parvez, A. T. Pereira, N. Ertugrul, N. H. E. Weste, D. Abbott, and S. F. Al-Sarawi, "Wide Bandgap DC–DC Converter Topologies for Power Applications," *Proceedings of the IEEE*, vol. 109, no. 7, pp. 1253–1275, Jul. 2021, doi: 10.1109/JPROC.2021.3072170.

[15] Y. V. Pushpalatha and D. Peftitsis, "Design of Dual Active Bridge Converters with SiC MOSFETs for minimized reflow power operation," in *2021 IEEE 12th Energy Conversion Congress & Exposition - Asia (ECCE-Asia)*, Singapore, Singapore, May 2021, pp. 574–579. doi: 10.1109/ECCE-Asia49820.2021.9479059.

[16] Xiaodong Li and A. K. S. Bhat, "Analysis and Design of High-Frequency Isolated Dual-Bridge Series Resonant DC/DC Converter," *IEEE Trans. Power Electron.*, vol. 25, no. 4, pp. 850–862, Apr. 2010, doi: 10.1109/TPEL.2009.2034662.

Impact of threshold voltage shifting on junction temperature sensing in GaN HEMTs

Burhan Etoz[1], Jose Ortiz Gonzalez[1], Arkadeep Deb[1], Saeed Jahdi[2] and Olayiwola Alatise[1]

[1]SCHOOL OF ENGINEERING, UNIVERSITY OF WARWICK
Coventry, United Kingdom
[2]FACULTY OF ENGINEERING, UNIVERSITY OF BRISTOL
Bristol, United Kingdom
Tel.: +44(0)247 615 1437
E-Mail: burhan.etoz@warwick.ac.uk, J.A.Ortiz-Gonzalez@warwick.ac.uk,
Arkadeep.deb@warwick.ac.uk, saeed.jahdi@bristol.ac.uk, O.Alatise@warwick.ac.uk

Acknowledgements

This work was supported in part by the UK Engineering and Physical Sciences Research Council (EPSRC) through the grant reference EP/R004366/1.

Keywords

«Gallium Nitride (GaN)», «HEMT», «Leakage Current», «Condition Monitoring», «Junction Temperature»

Abstract

Junction temperature sensing in GaN HEMTs has been identified as a critical challenge for condition monitoring especially under power cycling conditions. The use of temperature sensitive electrical parameters has been widely studied. In GaN devices, the ON-state resistance and gate leakage currents have been identified as TSEPs as both are junction temperature sensitive. Circuits capable of measuring the gate leakage currents in commercially available GaN HEMTs have previously been presented, however, the impact of variability in the threshold voltage on junction temperature sensing requires further investigation. In this paper, junction temperature measurements are implemented using the gate current as a TSEP and are compared with the junction temperature inferred from the ON-state resistance. The measured junction temperatures were verified against electrothermal simulations using manufacturer provided thermal networks. Threshold shift from charge trapping in Schottky GaN HEMTs has been shown to impact the temperature dependence of the gate leakage currents and ON-state resistance. It is important to account for these changes when using them as temperature sensitive electric parameters for real time junction temperature estimation in GaN HEMTs.

1. Introduction

GaN HEMTs are capable of very high switching frequencies while maintaining high energy conversion efficiency [1]. This is due to the very low switching losses compared to comparatively rated SiC MOSFETs and IGBTs. Carrier confinement in the AlGaN/GaN hetero-interface means very high electron mobility since carriers are shielded from scattering mechanisms (like acoustic phonon, surface roughness or ionized dopants) that reduce carrier mobility. This high carrier mobility means low conduction losses and low specific ON-state resistance which can be traded for reduced parasitic capacitances by die shrinkage. Hence GaN devices have the lowest switching energy compared to all other comparatively rated technologies including SiC MOSFETs.

GaN HEMTs, due to spontaneous charge polarization and carrier confinement in quantum wells at the AlGaN/GaN interfaces, are normally ON. However, GaN e-HEMTs have been made normally OFF using advanced gate technologies. The two commercially available variants of GaN HEMTs are the Schottky gated GaN HEMTs (from GaN Systems) [2] and Ohmic gated GaN HEMTs (from Infineon) [3]. Both devices comprise of reverse biased PN junctions to deplete the 2DEG of carriers in the OFF state. Fig. 1 shows a simple schematic of the gate structure [4] of the Schottky gated GaN HEMT with the back-to-back diode arrangement that enables normally OFF operation. To turn the Schottky gated GaN HEMT ON, the breakdown voltage of the reverse bias Schottky diode must be exceeded and the GaN/AlGaN diode must be forward biased for hole injection into the AlGaN layer. For charge neutrality, electrons must diffuse from the AlGaN layer into the p-GaN gate. This means that unlike MOS gated devices, like MOSFETs and IGBTs, there is significant gate current (µA to mA depending on the gate technology and temperature) during steady state ON operation of GaN e-HEMTs. In MOSFETs and IGBTs, gate leakage currents are on the order of nanoamperes due to the fact that leakage currents are generated from carriers scaling the oxide interface as a result of thermal energy and high electric fields. Furthermore, at the nominal gate driving voltage the increase with temperature of the gate leakage currents [5] is very low in MOS devices, which result in measurement challenges.

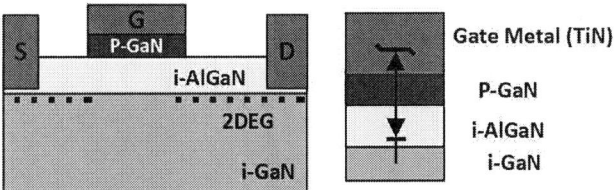

Fig. 1. Schottky gated GaN HEMT showing gate design

When power cycling GaN HEMTs, it is a requirement to have accurate measurements of the junction temperature [6]. This is to enable control of the power cycling system especially as the device packaging degrades and the thermal impedance of the package changes with the number of cycles. The gate leakage current in GaN HEMTs have previously been identified as TSEP for junction temperature estimation for both Schottky Gate and Ohmic Gate HEMTs [7, 8]. The gate leakage current in GaN e-HEMTs shows a very high temperature dependency, as shown in [9, 10] for Schottky gate GaN HEMTs. Modified gate driver circuits with diodes for sensing gate leakage currents have been developed for real-time junction temperature sensing in GaN devices and were presented in [8, 11]. The HEMT ON-state resistance has also been identified as a TSEP since it increases with temperature, with a temperature coefficient higher than SiC MOSFETs and silicon IGBTs [12]. However, the stability of TSEPs is important to investigate since unstable TSEPs can cause inaccurate junction temperature estimation, as shown in [13] for SiC MOSFETs. One of the device parameters that can influence the use of TSEPs in GaN is the threshold voltage. Threshold voltage (V_{TH}) shifting under gate voltage stress in GaN devices has been reported by various researchers, with both positive and negative shifts in V_{TH} reported depending on the magnitude of the gate voltage stress, the stress time and the stress temperature [2, 9, 14-16]. At low V_{GS} stress voltages, positive V_{TH} shifts have been reported due to negative charge trapping in the p-GaN gate. At higher V_{GS} stress, negative V_{TH} shifts have been reported to positive charge injection in the AlGaN layer. In this paper, V_{GS} stressing is combined with gate leakage current measurements to investigate the consistency of the gate leakage currents and ON-state resistance as TSEPs, similar to the studies done with SiC MOSFETs and the impact of V_{TH} shifts in TSEPs [13, 17]. Section 2 of the paper describes the experimental set-up for gate leakage current measurement and junction temperature estimation. Section 3 analysis the experimental measurements while section 4 discusses the impact of threshold voltage shifting.

2. Experimental Measurement of Gate Leakage as TSEP

a. Gate Driver for Junction Temperature Sensing

To evaluate the effectiveness of the gate leakage current as a TSEP, the gate driver circuit of the GaN HEMT has been modified to include the leakage current sensing diode. In this paper, commercially available normally-OFF 650V/30A GaN HEMTs from GaN Systems with datasheet references GS65508T have been evaluated. The gate driver has two isolated DC/DC converters: one with datasheet reference RP-0509S for providing the required voltage to an adjustable voltage regulator that defines the gate driver supply voltage and another DC/DC converter with datasheet reference RP-0512D, which provides the dual voltage required for powering an operational amplifier. The PCB prototype designed for testing GaN HEMTs, and sensing voltage measurement is shown in Fig. 2(a). The circuit schematic of the experimental set-up and further details of the temperature sensing circuit are shown in Fig. 2(b). When the device is ON, the gate leakage current forward biases the diode D_1 in series with the gate resistance R_{ON}, as shown in Fig. 2(b). Since the leakage current increases with the temperature, this causes an increase of the voltage across the diode. The diode type used for this study is pn rectifier diode with datasheet reference of 1N4007. A differential amplifier circuit is used to indicate the sensing voltage by amplifying the voltage difference between anode and cathode terminals of the pn rectifier diode, V_{AB}. The sensing voltage is proportional with the resistors used in the amplifier input. The relationship between V_{SENSE} and V_{AB} is given by (1).

$$V_{SENSE} = \frac{R_2}{R_1} V_{AB} \tag{1}$$

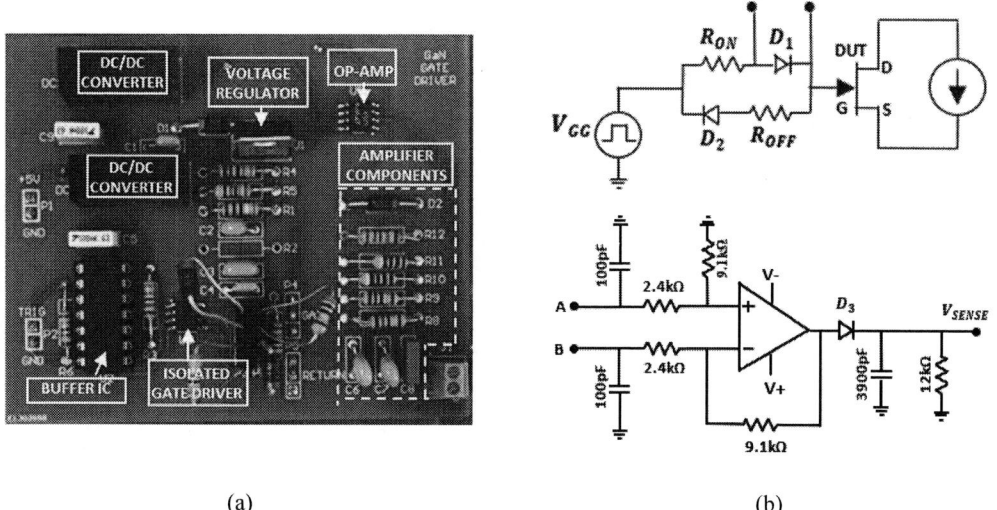

(a)   (b)

Fig. 2 (a) GaN Gate driver circuit and GaN HEMT with connections, (b) Circuit schematic of GaN HEMT and gate driver with current sensing diode and differential amplifier

In the diode voltage sensing circuit, R_1 and R_3 are 2.4 kΩ, and R_2 and R_4 are 9.1 kΩ. The differential amplifier's main component is a operational amplifier with the reference LT1253, and the op-amp works with -12V/+12V supplied by the RP-0512D. In addition, a passive low-pass filter was used at the output of the circuit to suppress unwanted frequencies and transmit signals at the desired frequencies. The cut-off frequency for the filter is selected as 3.4 kHz. The schematic of the gate driver with diode voltage sensing circuit is shown in Fig. 2(b). To measure the current across the diode in the turn-ON branch, the effective gate voltage (V_{GEFF}) of the Device Under Test (DUT) is measured after triggering the circuit with the voltage of 5 V. The gate leakage current causes a voltage drop on R_{ON} and D_1, influencing the effective gate voltage. The current across the diode depends on the applied gate voltage, voltage drop between gate and source, effective gate voltage and the value of turn-ON resistance. The value of the current can be calculated by using equation (2)

$$i_{leak} = \frac{V_{GG} - V_{D1} - V_{GEFF}}{R_{ON}} \tag{2}$$

b. Temperature Calibration Curves

For the calibration test, three GaN devices were characterized at different junction temperatures ranging from ambient temperature (T_{AMB}) to high temperature (150°C). The case temperature (T_C) was set using a small DC electric heater and sufficient time was allowed for the junction temperature (T_J) to reach thermal equilibrium with the case temperature. Table I shows the measured sensing voltage (V_{SENSE}) as a function of the case/junction temperature. The measurements of V_{SENSE} were performed using an oscilloscope (model TDS5054B from Tektronix). Fig. 3(a) shows the plot of the measured sensing voltage during a calibration pulse for one of the GaN HEMTs as a function of time. It can be seen from Fig. 3(a) that when the gate is triggered there is a short transient (less than 50 μs long) before V_{SENSE} reaches its steady state value. Fig. 3(b) shows the measured steady-state V_{SENSE} for 3 different GaN devices as a function of junction temperature set by the electric heater. It can be seen from Fig. 3(b), that the temperature dependence of the gate leakage current varies from device to device, hence, normalization techniques are required to make temperature extraction device invariant.

Table I: Sensing voltage at different temperatures

Case Temp (°C)	V_{SENSE} (V)		
	Device 1	Device 2	Device 3
T_{AMB}	1.680	1.668	1.669
50	1.690	1.671	1.681
75	1.700	1.689	1.715
100	1.720	1.709	1.763
125	1.770	1.760	1.839
150	1.825	1.815	1.927

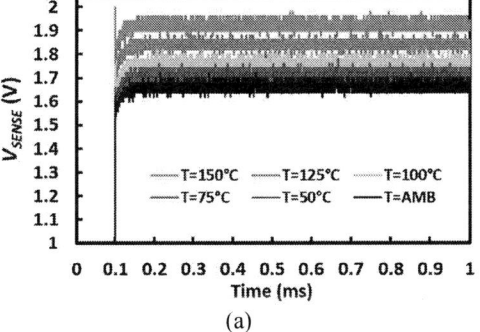 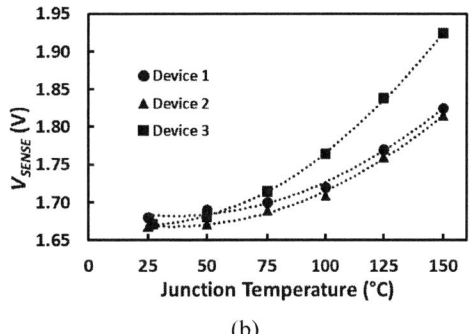

(a) (b)

Fig. 3(a) Calibration response of the sensing voltage at different temperatures for device 3, (b) TSEP calibration for the evaluated GaN HEMTs

It can be seen that the relationship between V_{SENSE} and temperature follows a quadratic formula. Hence, equations (3) and (4) can be used to extract the junction temperature from V_{SENSE} once the parameters A, B and C have been determined using curve fitting and normalization techniques.

$$V_{SENSE} = A \cdot T_j^2 + B \cdot T_j + C \tag{3}$$

$$T_J = \frac{-B + \sqrt{B^2 - 4 \cdot A \cdot (C - V_{SENSE})}}{2A} \tag{4}$$

3. Analysis of Experimental Measurements

To evaluate the use of the leakage current under power cycling conditions, the experimental set up shown in Fig. 4(a) was set up. This comprises of an IGBT switching in a constant current through the GaN HEMT which is the DUT. The heating and cooling of the DUT are controlled by the IGBT which is switched ON (for heating the DUT) and switched OFF (for cooling the DUT). The DUT is left ON during the test. Fig. 4(b) shows a typical heating/cooling sequence. A thermocouple is used for measuring the case temperature. The duration of the heating pulse as well as the magnitude of the current are used as parameters to control the junction temperature. Fig. 5(a) shows the measured V_{SENSE} for a DUT with 2 heating pulses (12 A and 15 A) at different pulse durations while Fig. 5(b) shows the measured ON-state resistance (R_{DS-ON}) for different heating pulses. The ON-state resistance has a known temperature dependency that can be read off from the datasheet. Fig. 6(a) compares the temperatures extracted using V_{SENSE} with those extracted using R_{DS-ON}, for a 15 A heating pulse. It can be seen from

Fig. 6(a) that for a 16 second pulse, the rise in case temperature (T_C) is under 5°C while the rise in junction temperature is over 55°C. It is also important to mention that V_{SENSE} (gate leakage current) allows to capture both the heating and cooling transient, as shown in Fig. 6(a). Fig. 6(b) shows good agreement between the peak junction temperatures predicted by the 2 TSEPs (V_{SENSE} and R_{DS-ON})

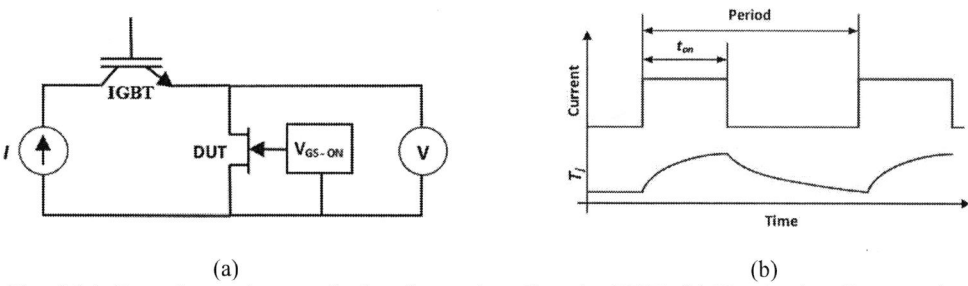

(a)   (b)

Fig. 4 (a). Experimental set-up for heating and cooling the DUT, (b) Heating/cooling transient

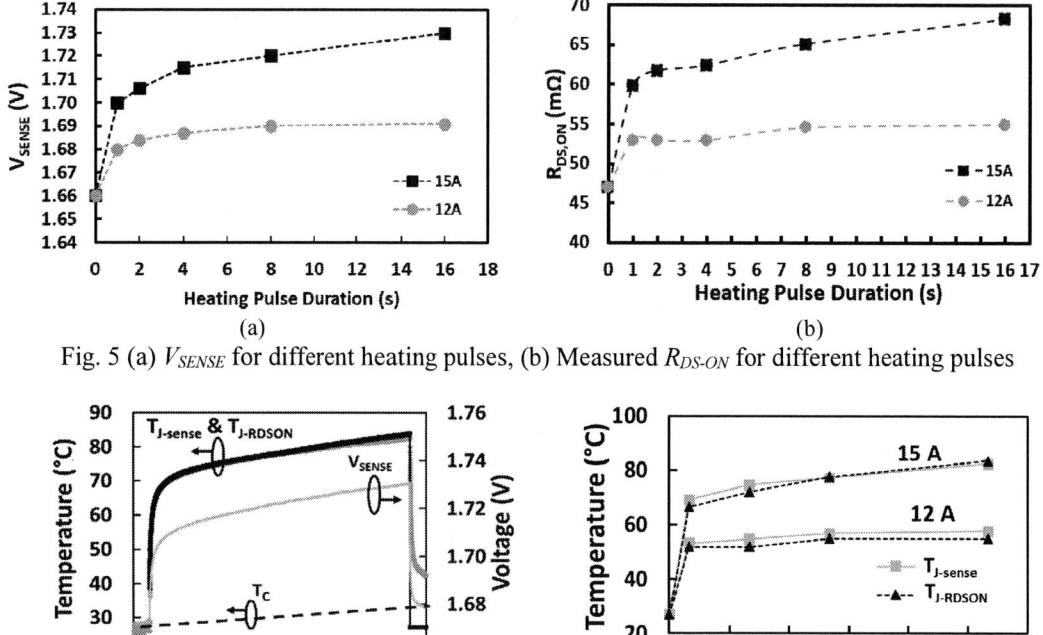

(a)   (b)

Fig. 5 (a) V_{SENSE} for different heating pulses, (b) Measured R_{DS-ON} for different heating pulses

(a)   (b)

Fig. 6 (a) Extracted temperature from V_{SENSE} and R_{DS-ON} for the GaN DUT during a heating transient (b) Comparison of the peak junction temperatures predicted using R_{DS-ON} and V_{SENSE}.

To check the validity of the junction temperature measured using the TSEPs, the junction temperature is simulated in Simulink, using the measured heating power (current and voltage) and the thermal network provided by the manufacturer. As the device is mounted on a heatsink, the thermal resistance and capacitance of the heatsink is also an input to the simulation. Fig. 7(a) shows the Cauer thermal network of the device while Fig. 7(b) shows the picture of the device on the heatsink, as well as the custom PCB for testing the GaN e-HEMT. The parameters of the thermal network used in the simulation are shown in Table II. The thermal capacitances and resistances are taken from the GaN e-HEMT datasheet while the parameters from the heatsink are calculated using the physical dimensions and material properties of the heatsink. Fig. 8(a) shows a comparison of the electrothermal simulations with the measured junction temperatures using the gate leakage current TSEP for a 1 second heating pulse while Fig 8(b) shows the same comparison for an 8 second pulse. Good matching of the measured and simulated junction temperatures demonstrates the accuracy of the electrothermal model.

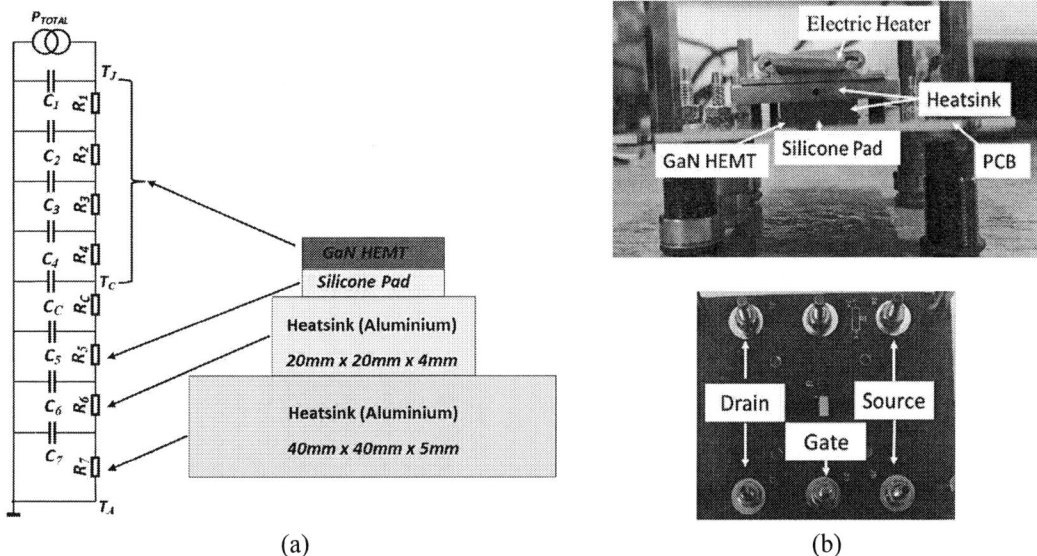

(a)   (b)

Fig. 7 (a). Schematic diagram showing thermal network of the GaN HEMTs and heatsink, (b) Pictures of the DUT on heatsink

Table II: Parameters of thermal network used in GaN Thermal simulations

Material	R (°C/W)		C (Ws/°C)	
GaN HEMT (GS66508T)	R_{TH1}	0.150	C_{TH1}	8×10^{-6}
	R_{TH2}	0.230	C_{TH2}	7.4×10^{-4}
	R_{TH3}	0.240	C_{TH3}	0.065
	R_{TH4}	0.015	C_{TH4}	0.002
Silicone Pad	R_{TH5}	2.100	C_{TH5}	0.060
Heat Sink (20 mm x 20 mm x 4 mm)	R_{TH6}	0.049	C_{TH6}	3.880
Heat Sink (40 mm x 40 mm x 5 mm)	R_{TH7}	0.015	C_{TH7}	19.440

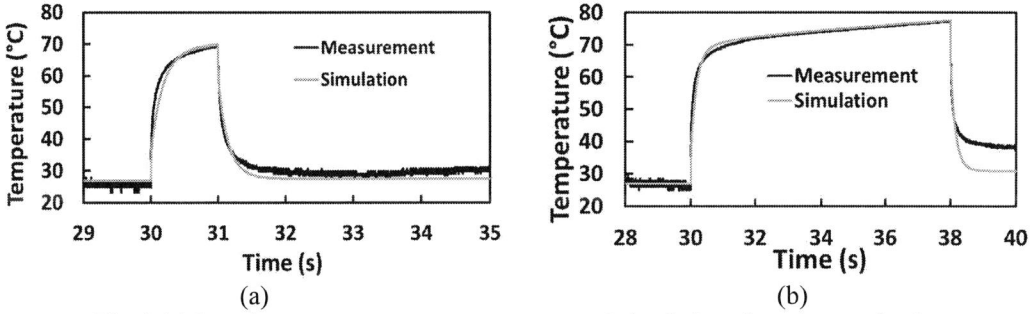

(a)   (b)

Fig. 8 (a) Junction temperature measurements and simulations for a 1 second pulse, (b) Simulated and measured junction temperature for an 8 second pulse

The measured and simulated peak junction temperatures are shown in Fig. 9(a), indicating good agreement between the model and the measurement. Additionally, one of the main benefits of this TSEP (V_{SENSE}) is that it allows to capture both the heating and cooling transient during power cycling. This is shown in Fig. 9(b) for a heating/cooling sequence of 40 pulses (2 s ON/ 2 s OFF) and a heating current of 15 A. The case temperature, measured with at thermocouple is also shown, indicating the effectiveness of the TSEP for monitoring the junction temperature during power cycling. The next section of the paper will perform some gate stress tests on the DUT and evaluate the impact of threshold voltage shift on the TSEPs. This is particularly relevant for power cycling, as it is the case of SiC MOSFETs [18].

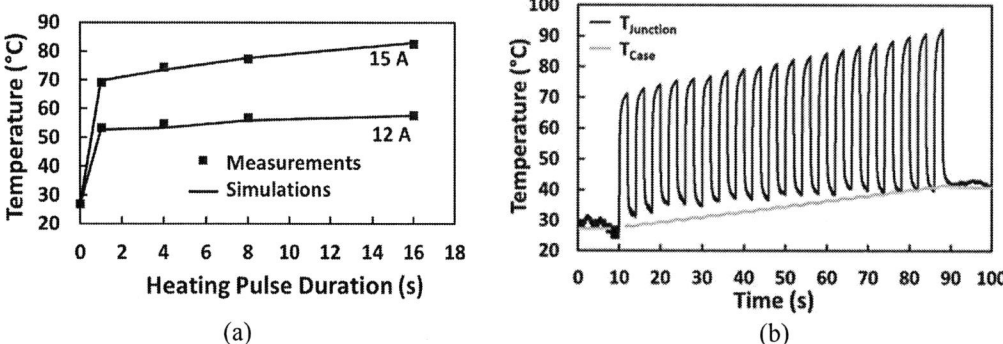

(a)   (b)

Fig. 9 (a). Simulated and measured peak junction temperature (using V_{SENSE} as TSEP)
(b) Repetitive heating/cooling pulses. 15 A heating current – 2 s ON – 2 s OFF

4. Impact of Gate Stress on Junction Temperature Measurement

At high gate stress voltages, V_{TH} in GaN HEMTs have been known to exhibit both upward and downward shifts [2, 19, 20] depending on which stress mechanism dominates i.e. if negative charge trapping in the p-GaN gate dominates, then V_{TH} shift upwards and if positive charge trapping in the AlGaN layer dominates, V_{TH} shifts downwards. The polarity of the V_{TH} shift is also recovery time dependent being initially positive (due to more electron trapping) and then becoming negative (due to faster electron release) [9]. Fig. 10(a) shows the results of positive gate voltage stress applied for 5 hours (25 Hz pulsed stress and 50% duty cycle) on the 650 V GaN e-HEMT with different stress voltages. It is clear that there is saturation of V_{TH} shift within the first hour. The recovery time (time between V_{GS} stress removal and V_{TH} measurement) in these measurements was over 600 seconds. The results show a negative shift in V_{TH} which increases with the magnitude of the V_{GS} stress. Fig. 10(b) shows the measured V_{TH} shift as a function of V_{GS} stress. This change in V_{TH} is likely to impact both the R_{DS-ON} (through the gate overdrive voltage: V_{GS} - V_{TH}) and V_{SENSE} (through the change in leakage current).

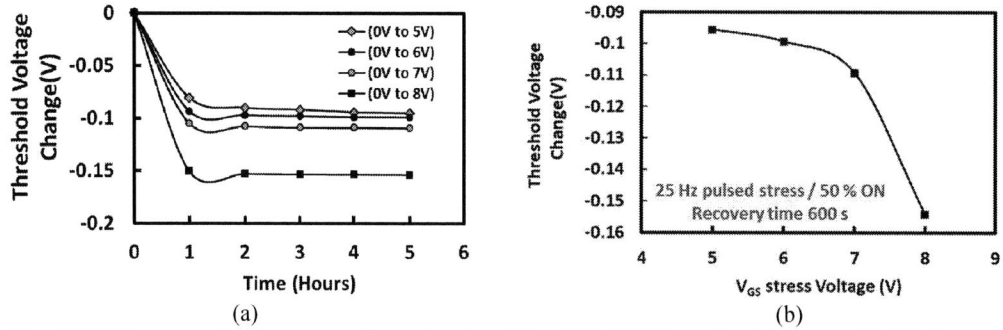

(a)   (b)

Fig. 10 (a) Impact of gate stress voltage level on V_{TH} shift (25 Hz pulsed stress and 50% duty cycle)
(b) Change in V_{TH} as a function of stress voltage

To investigate the impact of V_{TH} shift on the consistency of V_{SENSE} and R_{DS-ON} as TSEPs, accelerated V_{GS} stress tests were performed, by applying gate-source voltages higher than the rated value of the selected GaN HEMT. Fig. 11(a) shows the measured R_{DS-ON} for a GaN HEMT that has been subjected to cumulative gate stresses of 8 V and 8.5 V for 300 s at 150 °C. The effective $V_{GS-STRESS}$ values were 7.82 V and 8.28 V after accounting for the voltage drop across a 15 Ohm series resistance used for limiting the peak current during the gate stress sequence. These stress voltages were selected to cause a more permanent V_{TH} shift, enabling the study of the impact of the V_{TH} shift on the TSEPs. Fig. 11(b) shows the measurements of V_{SENSE} for the same device.

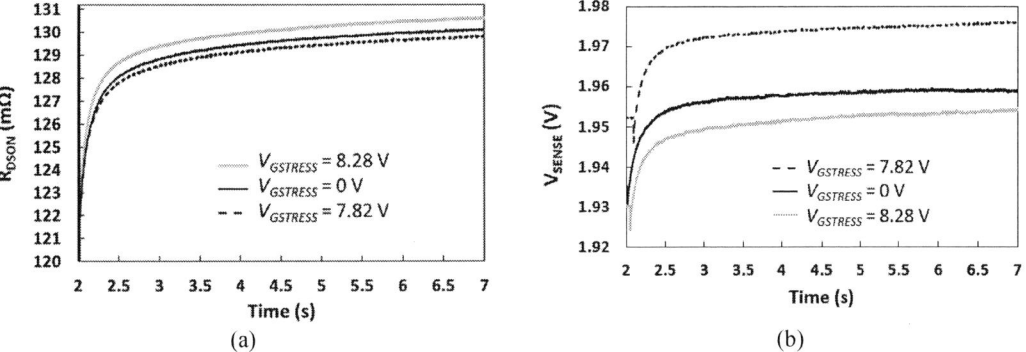

(a)   (b)

Fig. 11 (a) Impact of $V_{GSTRESS}$ on R_{DSON}, (b) Impact of $V_{GSTRESS}$ on V_{SENSE}

The characterization was performed using a 5 A and 5 second single pulse. In the absence of stress-induced V_{TH} shift, the characteristics in Fig. 11(a) and Fig. 11(b) should all be super-imposed with no apparent variation. However, it is clear from Fig. 11(a) that R_{DS-ON} shifts first downwards for $V_{GS-STRESS}$ = 7.82 V and upwards for $V_{GS-STRESS}$ = 8.28V. V_{SENSE} shifts in the opposite directions by first going up at $V_{GS-STRESS}$ = 7.82V and then going down after the 8.28V gate stress. This means both parameters will record different junction temperatures in contradiction to the plots shown in Fig. 6(b) where both TSEPs show a good agreement on junction temperature estimation. Evaluating both TSEPs after the stress sequences, it can be observed that the R_{DS-ON} is less affected than V_{SENSE}. The difference in R_{DS-ON} after the stresses is around ±0.5 mΩ and the impact on temperature estimation would be less than -1°C for the 7.82 V stress and around +1°C for the 8.28 V stress. The impact of the gate stress is more apparent in V_{SENSE}. For example, considering the 7.82 V stress, a difference of +17 mV is observed, which corresponds to a difference in temperature estimation around +4°C. For the 8.28 V stress, the difference is around -1°C.

5. Conclusion

The use of TSEPs in junction temperature estimation of GaN e-HEMT power devices is important not just for power cycling but potentially in condition monitoring systems where instantaneous junction temperature estimation is important. The ON-state resistance, measured from the forward voltage during ON-state and the gate leakage current have been cited as TSEPs in GaN. In this paper, a previously presented circuit used to measure the gate leakage current of the GaN device is used to estimate the junction temperature. The measured junction temperature is compared to the temperature measured using the ON-state resistance as TSEP and both are shown to have numerical agreement. Using electrothermal simulations, the accuracy of the peak junction temperature measured by the TSEPs (gate leakage current and ON-state resistance) was confirmed using the measured heating power and thermal network parameters taken from the device datasheet as well as heatsink parameters. Threshold voltage shifting from gate voltage stress impacts that ON-state resistance and gate leakage current. If the threshold voltage increases due to electron trapping in the p-gate of the GaN e-HEMT, the ON-state resistance increases due to lower carrier density while the sensing voltage (measured from the gate leakage current) decreases. Similarly, a decrease in the threshold voltage will cause a decrease in the ON-state resistance and an increase in the gate leakage current. Measurements show ON-state resistance is less sensitive to V_{TH} shifting compared to gate leakage current. However, the gate leakage current can provide junction temperature estimation during both the heating and cooling part of the thermal transients while the ON-state resistance gives junction temperature measurement only during the heating part of the transient.

References

[1] E. A. Jones, F. F. Wang, and D. Costinett, "Review of Commercial GaN Power Devices and GaN-Based Converter Design Challenges," *IEEE Journal of Emerging and Selected Topics in Power Electronics,* vol. 4, no. 3, pp. 707-719, 2016, doi: 10.1109/JESTPE.2016.2582685.

[2] J. He, G. Tang, and K. J. Chen, "V_{TH} Instability of p-GaN Gate HEMTs Under Static and Dynamic Gate Stress," *IEEE Electron Device Letters,* vol. 39, no. 10, pp. 1576-1579, 2018, doi: 10.1109/LED.2018.2867938.

[3] D. Varajao and B. Zojer, "Infineon Whitepaper 2021-11 - Gate drive solutions for CoolGaN™ GIT HEMTs," 2021.

[4] T. Detzel, "Reliability of GaN Power Devices from the Industrial Perspective - Tutorial - ESREF Conference," September 2018.

[5] J. O. Gonzalez, R. Wu, S. Jahdi, and O. Alatise, "Performance and Reliability Review of 650 V and 900 V Silicon and SiC Devices: MOSFETs, Cascode JFETs and IGBTs," *IEEE Transactions on Industrial Electronics,* vol. 67, no. 9, pp. 7375-7385, 2020, doi: 10.1109/TIE.2019.2945299.

[6] J. Franke, G. Zeng, T. Winkler, and J. Lutz, "Power cycling reliability results of GaN HEMT devices," in *2018 IEEE 30th International Symposium on Power Semiconductor Devices and ICs (ISPSD),* 13-17 May 2018, pp. 467-470, doi: 10.1109/ISPSD.2018.8393704.

[7] L. Zhang, P. Liu, S. Guo, and A. Q. Huang, "Comparative study of temperature sensitive electrical parameters (TSEP) of Si, SiC and GaN power devices," in *2016 IEEE 4th Workshop on Wide Bandgap Power Devices and Applications (WiPDA),* 7-9 Nov. 2016, pp. 302-307, doi: 10.1109/WiPDA.2016.7799957.

[8] A. Borghese, A. Di Costanzo, M. Riccio, L. Maresca, G. Breglio, and A. Irace, "Gate Current in p-GaN Gate HEMTs as a Channel Temperature Sensitive Parameter: A Comparative Study between Schottky- and Ohmic-Gate GaN HEMTs," *Energies,* vol. 14, no. 23, 2021, doi: 10.3390/en14238055.

[9] J. O. Gonzalez, B. Etoz, and O. Alatise, "Gate stresses and threshold voltage instability in normally-OFF GaN HEMTs," in *2020 22nd European Conference on Power Electronics and Applications (EPE'20 ECCE Europe),* 7-11 Sept. 020, pp. P.1-P.10, doi: 10.23919/EPE20ECCEEurope43536.2020.9215865.

[10] A. Borghese, M. Riccio, G. Longobardi, L. Maresca, G. Breglio, and A. Irace, "Gate leakage current sensing for in situ temperature monitoring of p-GaN gate HEMTs," *Microelectronics Reliability,* vol. 114, p. 113762, 2020/11/01/ 2020, doi: https://doi.org/10.1016/j.microrel.2020.113762.

[11] A. Borghese, M. Riccio, L. Maresca, G. Breglio, and A. Irace, "Gate Driver for p-GaN HEMTs with Real-Time Monitoring Capability of Channel Temperature," in *2021 33rd International Symposium on Power Semiconductor Devices and ICs (ISPSD),* 30 May-3 June 2021 2021, pp. 63-66, doi: 10.23919/ISPSD50666.2021.9452317.

[12] S. Zhu, A. Fayyaz, and A. Castellazzi, "Static and Dynamic TSEPs of SiC and GaN Transistors," *IET PEMD,* 2018.

[13] J. O. Gonzalez and O. Alatise, "Bias Temperature Instability and Junction Temperature Measurement Using Electrical Parameters in SiC Power MOSFETs," *IEEE Transactions on Industry Applications,* vol. 57, no. 2, pp. 1664-1676, 2021, doi: 10.1109/TIA.2020.3045120.

[14] M. Meneghini *et al.,* "Reliability and failure analysis in power GaN-HEMTs: An overview," in *2017 IEEE International Reliability Physics Symposium (IRPS),* 2-6 April 2017 2017, pp. 3B-2.1-3B-2.8, doi: 10.1109/IRPS.2017.7936282.

[15] M. Ruzzarin *et al.,* "Degradation Mechanisms of GaN HEMTs With p-Type Gate Under Forward Gate Bias Overstress," *IEEE Transactions on Electron Devices,* vol. 65, no. 7, pp. 2778-2783, 2018, doi: 10.1109/TED.2018.2836460.

[16] J. O. Gonzalez, B. Etoz, and O. Alatise, "Characterizing Threshold Voltage Shifts and Recovery in Schottky Gate and Ohmic Gate GaN HEMTs," in *2020 IEEE Energy Conversion Congress and Exposition (ECCE),* 11-15 Oct. 2020, pp. 217-224, doi: 10.1109/ECCE44975.2020.9235650.

[17] J. Ortiz Gonzalez and O. Alatise, "Bias temperature instability and condition monitoring in SiC power MOSFETs," *Microelectronics Reliability,* vol. 88-90, pp. 557-562, 2018, doi: https://doi.org/10.1016/j.microrel.2018.06.045.

[18] H. Luo, F. Iannuzzo, and M. Turnaturi, "Role of Threshold Voltage Shift in Highly Accelerated Power Cycling Tests for SiC MOSFET Modules," *IEEE Journal of Emerging and Selected Topics in Power Electronics,* pp. 1-1, 2019, doi: 10.1109/JESTPE.2019.2894717.

[19] Y. Shi *et al.,* "Bidirectional threshold voltage shift and gate leakage in 650 V p-GaN AlGaN/GaN HEMTs: The role of electron-trapping and hole-injection," in *2018 IEEE 30th International Symposium on Power Semiconductor Devices and ICs (ISPSD),* 13-17 May 2018, pp. 96-99, doi: 10.1109/ISPSD.2018.8393611.

[20] J. Wei *et al.,* "Dynamic Threshold Voltage in p-GaN Gate HEMT," in *2019 31st International Symposium on Power Semiconductor Devices and ICs (ISPSD),* 19-23 May 2019, pp. 291-294, doi: 10.1109/ISPSD.2019.8757602.

Comparison of Power Cycling Results of discrete GaN Cascodes for Automotive Power Electronics with high Temperature Swings

Florian Lippold, Philipp Hauenschild, Regine Mallwitz
INSTITUTE FOR ELECTRICAL MACHINES, TRACTION AND DRIVES (IMAB)
TU Braunschweig
Braunschweig, Germany
E-Mail: f.lippold@tu-braunschweig.de
URL: https://www.tu-braunschweig.de/imab

Keywords

«Lifetime», «Power cycling», «wide bandgap devices», «Gallium Nitride (GaN)», «Automotive component»

Abstract

High voltage rated GaN HEMTs are attractive for automotive applications such as on-board charger or auxiliary power supplies. Nowadays GaN is packaged as single chips in discrete housings. In this paper a power cycling set up and results for GaN components in discrete TO-247 package are presented.

I Introduction

Many automotive qualified semiconductors for power applications are offered in the TO-247 package. The TO-247 package, see Fig. 1a, is available for a long time. New semiconductors are also offered in this package. GaN components can increase the power density or reduce cooling requirements of an existing system. Some manufacturers offer normally-on GaN HEMTs which are combined with a low voltage Silicon MOSFET to reach normally-off behavior. For the new semiconductors different interconnection technologies are used for the upper and lower side of the chip. Nowadays it is common to use bond wires and solder connections. For investigating the lifetime of GaN cascodes in the TO-247 package with new interconnection technologies, power cycling tests are done.

In order to estimate the lifetime of power electronics, it is mandatory to create lifetime models for the used semiconductors. For generating a lifetime model, power cycling tests at different temperature swings and absolute temperatures have to be done. Wide Band Gap (WBG) components like Silicon Carbide (SiC) MOSFETS or Galliumnitride (GaN) also in cascodes are still exotic under lifetime aspects. Only a few results have been published [1, 2]. In [1] the focus was set on the aging effects, rather on lifetime model. In [2] a component in a different discrete package (TO-220) was tested, however the temperature swings were not as high as in this test and the measuring points are relativly close together. The tested GaN cascodes in [1] and [2] were compared with GaN HEMTs, not with different GaN cascodes. A general fitting model like LESIT known from Si-IGBTs still does not exist. The LESIT model [3] is a lifetime model which is the first destination when no information about the inner structure is available. However the new parameters have to be identified by curve fitting of the power cycling results.

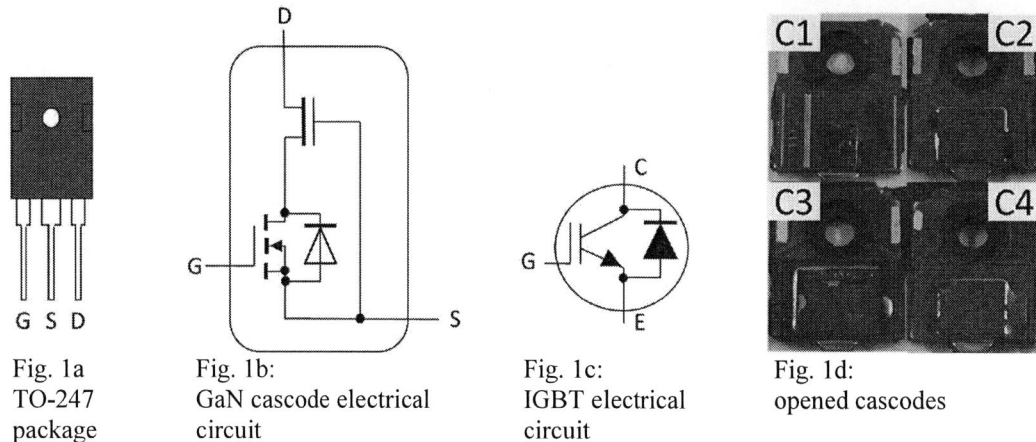

Fig. 1a:
TO-247
package

Fig. 1b:
GaN cascode electrical
circuit

Fig. 1c:
IGBT electrical
circuit

Fig. 1d:
opened cascodes

A GaN cascode, see Fig. 1b, consist of a normally-on GaN HEMT (top) and a normally-off silicon MOSFET (bottom). The MOSFET is used for switching the states of the HEMT, which blocks the high voltage. The load current flows through the HEMT and the MOSFET from Drain to Source. A body diode is included in the MOSFET. In comparison an IGBT, see Fig. 1c, which requires for the diode a separate chip. From this follows that the indirect measurement of the virtual junction temperature T_{vj} of the chips can be done in different ways and at different positions inside the component. The measurement of T_{vj} is described in chapter III "test criteria".

The structure of the components is shown in Fig. 1d. After removing the plastic housing the top view to the lead frame and the chip is possible. It can be seen that the position of the chips of the GaN cascodes within the package is different. While the chip size of component C2 and C4 (both Rds(on) = 50 mΩ) is very similar, the structure of the components C1 and C3 is very different.

II   Mechanical Setup for the Power Cycling Tests

During the power cycling test, the load current flows cyclically through the device under test (DUT), which heats up and cools down afterwards. There is a mechanical expansion in each material layer in each cycle. The virtual junction temperature of the chip rises to $T_{vj,max}$ and sinks to $T_{c/s,min}$. The temperature $T_{c/s}$ is measured below the DUT. Depending on the cooling system, also $T_{c/s}$ can change between $T_{c/s,min}$ and $T_{c/s,max}$. When the dimension of the cooling system is sufficient, the temperature $T_{c/s}$ should be $T_{c/s,max} = T_{c/s,min} = $ constant . The courses of the temperatures are shown in Fig. 2.

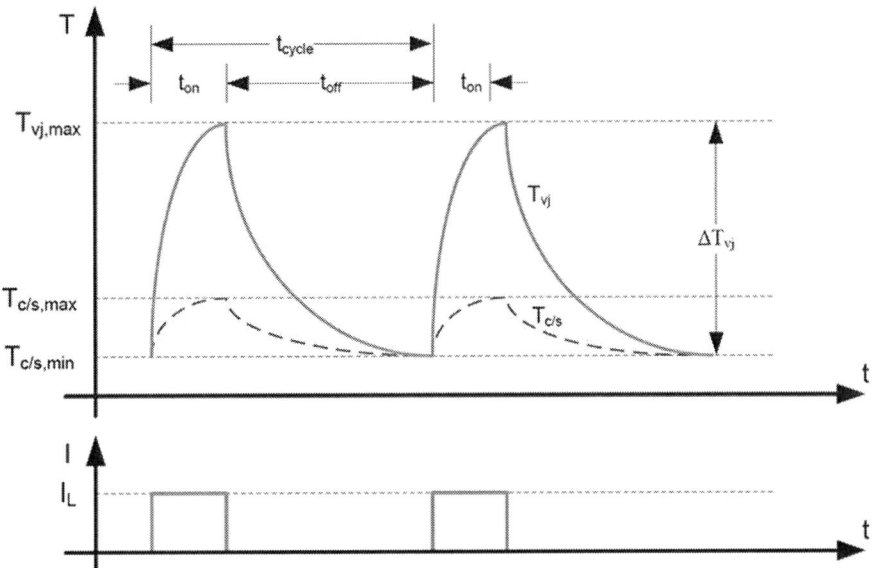

Fig. 2: Test parameters and power cycling definitions [4]

The mechanical setup for the measurements is based on the AQG324 [4] and includes a temperature sensor, a thermocouple type k, 2 mm below the DUT for $T_{c/s}$ (see Fig. 3). The mounting PCB is located at a 90° angle to the cooling plate. Mechanical stresses to the component resulting from temperature changes are compensated by a standard mounting clip. The mechanical construction is shown schematically and in real in Fig. 3a and 3b. The TO-247 housings are always positioned exactly the same using a mounting template.

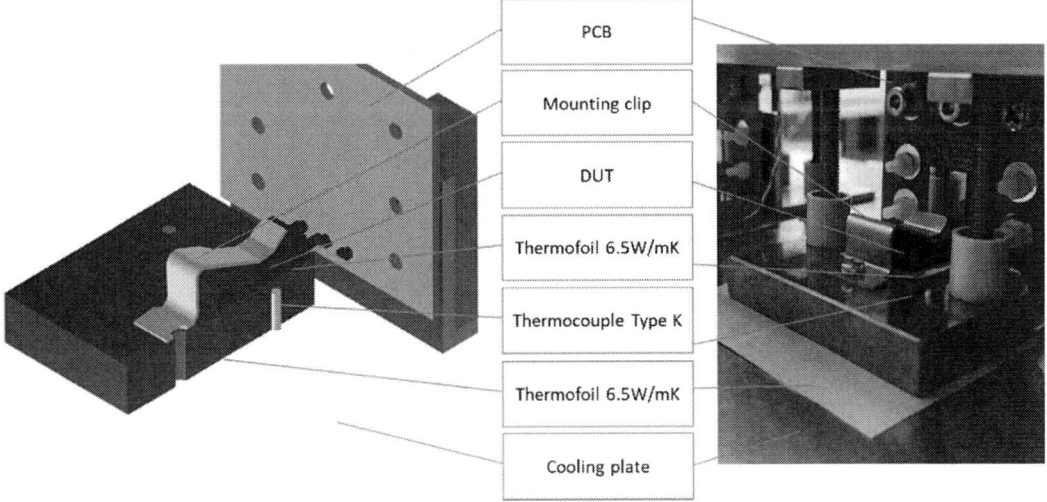

Fig. 3a: schematically setup   Fig. 3b: real setup

III   Test Criteria

Different failure modes are known from existing housing and interconnection technologies, such as chip delamination or bond wire defects. A short on-time t_{on} and a short off-time t_{off} are set in order to stress the chip-near interconnections. The mean junction temperature T_{jm}, which is defined as $(T_{vjmax} + T_{vjmin})/2$ should be the same for all tests. High temperature swings for T_{vj} are reducing the test time, because the stress for the component is higher. Two measuring points M1 and M2 with a large difference between the temperature swings $\Delta T_{vj}(M1)$ and $\Delta T_{vj}(M2)$ are used, see Table I.

For these tests measuring point M2 was chosen with a maximum junction Temperature T_{jmax} close to the allowed chip temperature of 175°C. The temperature swing $\Delta T_{vj}(M2) = 130$ K is relatively high. The on-time t_{on} and the off-time t_{off} are set to 4 seconds. A fine adjustment is done with the cooling temperature of the liquid cooling system, see Table II.

For the second measurement point M1 a smaller temperature swing $\Delta T_{vj}(M1)$ is set with $t_{on} = t_{off} = 0,5$s in order to achieve a difference between the two measuring points with about 60 K. The load current I_L for M1 and M2 is the same, see Table II. The planed test criteria are shown in Table I, however there are deviations for each DUT because of tolerances.

Table I: The planed test criteria

	M1	M2
ΔT_{vj}	70K	130K
T_{vjmax}	140°C	170°C
T_{vjmin}	70°C	40°C
T_{jm}	105°C	105°C
t_{on}	0.5s	4s
t_{off}	0.5s	4s

The junction temperature T_{vj} of the GaN cascodes is measured indirectly with the V_{sd}-Method [5]. The measurement current is -20mA from Source to Drain. The HEMT and the MOSFET are close together and heated up by the same load current.

The junction temperature T_{vj} of the IGBT is measured indirectly with a measurement current +20mA through the channel from Collector to Emitter, because the IGBT and the diode are two different chips. For each device a calibration from -25°C up to 125°C is done. The temperature is measured a short time (µs) after the load current is turned off. The tested devices are listed in Table II. For each test four devices were used. All devices have the TO-247 package.

Table II: The tested devices

Component	Manu-facturer	Technology	Voltage class	Rds(on)eff / VCEsat	Load Current I_L	$T_{cooling}$ M1	$T_{cooling}$ M2
C1	A	GaN cascode	650V	35 mΩ	35,5A	25°C	21°C
C2	A	GaN cascode	650V	50 mΩ	27,6A	25°C	22,4°C
C3	B	GaN cascode	650V	35 mΩ	36,2A	26,5°C	23,8°C
C4	B	GaN cascode	650V	50 mΩ	27,3A	26,5°C	25°C
C5	C	IGBT	650V	1,66V	49A	23,8°C	23,8°C

IV Test Results and Discussion

The aging and the following failure of the semiconductors can be observed by different criteria. Here the Drain-Source voltage $V_{DS,on}$ (see Fig. 4) and $T_{vj,max}$ (see Fig. 5) are recorded. The test was stopped when the junction temperature $T_{vj,max}$ has reached 250°C. Measurements of the thermal resistance from junction to sensor $R_{th,j-c/s}$ are automatically done in constant intervals by the test bench. These measurements are causing brief drops in the course of the temperatures and voltage $V_{DS,on}$ (see Fig. 4). The curves of the Thermal Resistance $R_{th,j-c/s}$ per device are presented in Fig. 6.

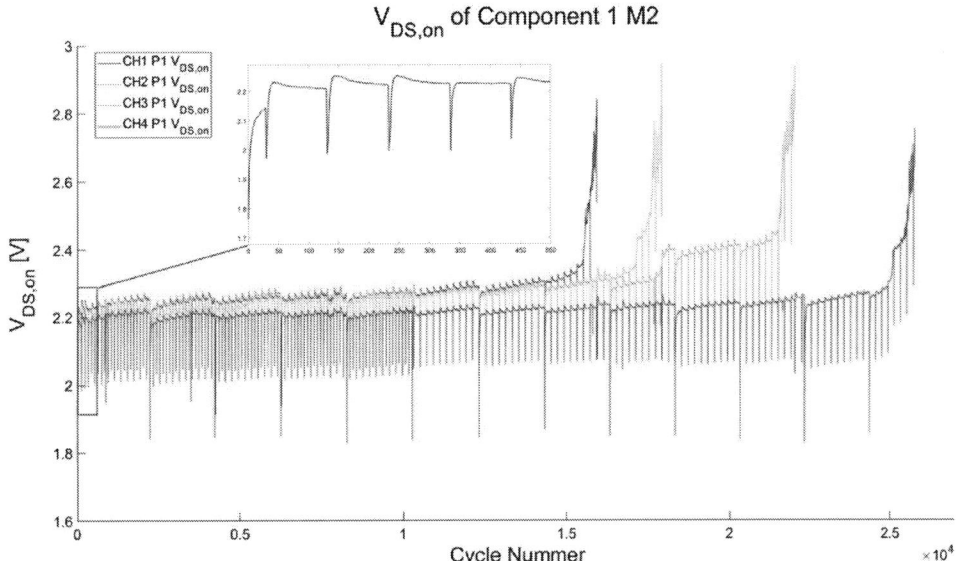

Fig. 4: Drain-Source voltage $V_{DS,on}$ and cycle number

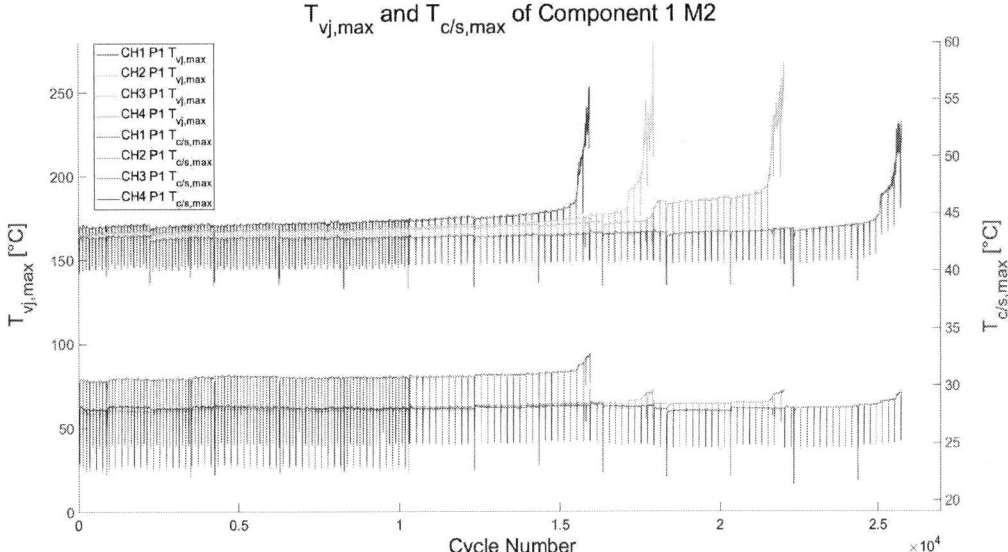

Fig. 5: Maximum junction temperature $T_{vj,max}$ and temperature $T_{c/s,max}$ and cycle number

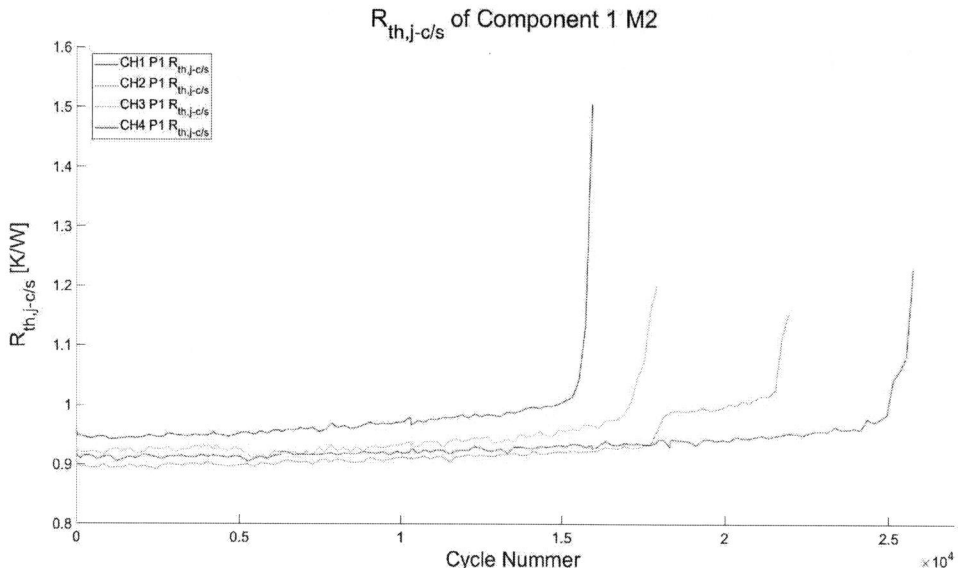

Fig. 6: Thermal resistance $R_{th,j\text{-}c/s}$ from junction to sensor $T_{c/s,max}$ and cycle number

The end of life criteria is defined as the starting value of the Drain-Source voltage $V_{DS,on}$ plus 5% and/or the thermal resistance $R_{th,j\text{-}c/s}$ starting value plus 20%, see [4]. The drops in the curves of the Drain-Source voltage $V_{DS,on}$ make the evaluation more complex. The end of life criteria of the thermal resistance $R_{th,j\text{-}c/s}$ plus 20% was reached when the test was stopped. For the lifetime models the total number of cycles are used. The results of the power cycling tests are shown in Fig. 7 (component 1: blue triangle, component 2: purple dot, component 3: red rectangle, component 4: orange rhombus, component 5: green dot).

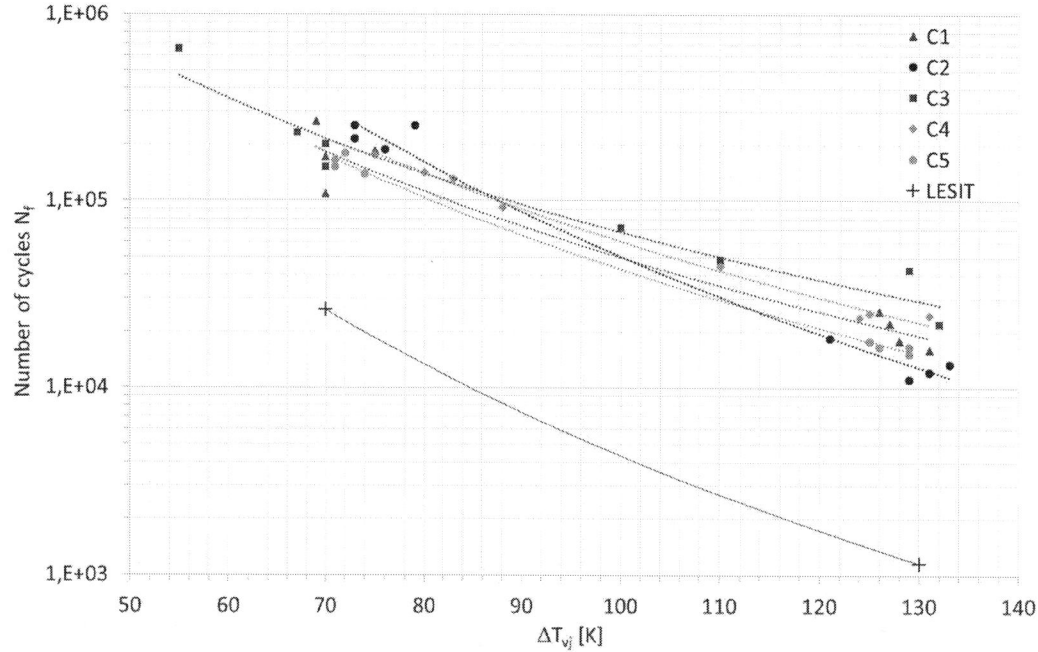

Fig. 7: Number of cycles until failure N_f for GaN cascodes (C1-C4) and IGBT (C5) in comparison to LESIT

The expected curve of the LESIT model for the two temperature swings from Table I and standard IGBT parameter [3] is depicted in Fig. 7. All tested devices reach higher number of cycles than the calculated LESIT values. For the lifetime modelling trend lines are integrated.

In Fig. 8 and 9 the results for GaN cascodes with 35mΩ and 50mΩ and the IGBT are shown in detail.

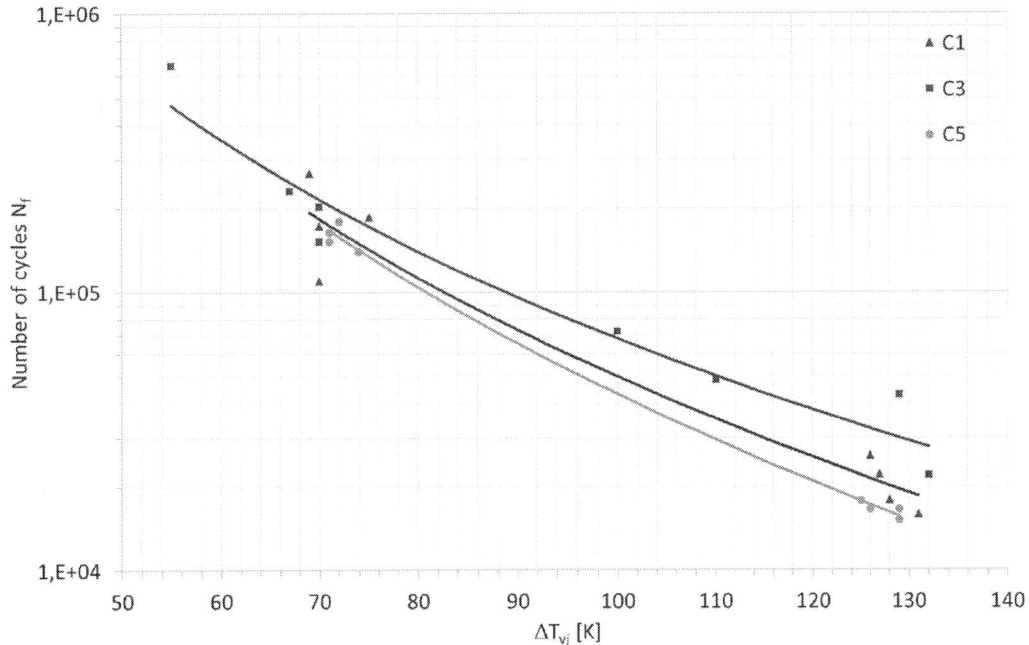

Fig. 8: Number of cycles until failure N_f for GaN cascodes 35mΩ (C1+C3) and IGBT (C5)

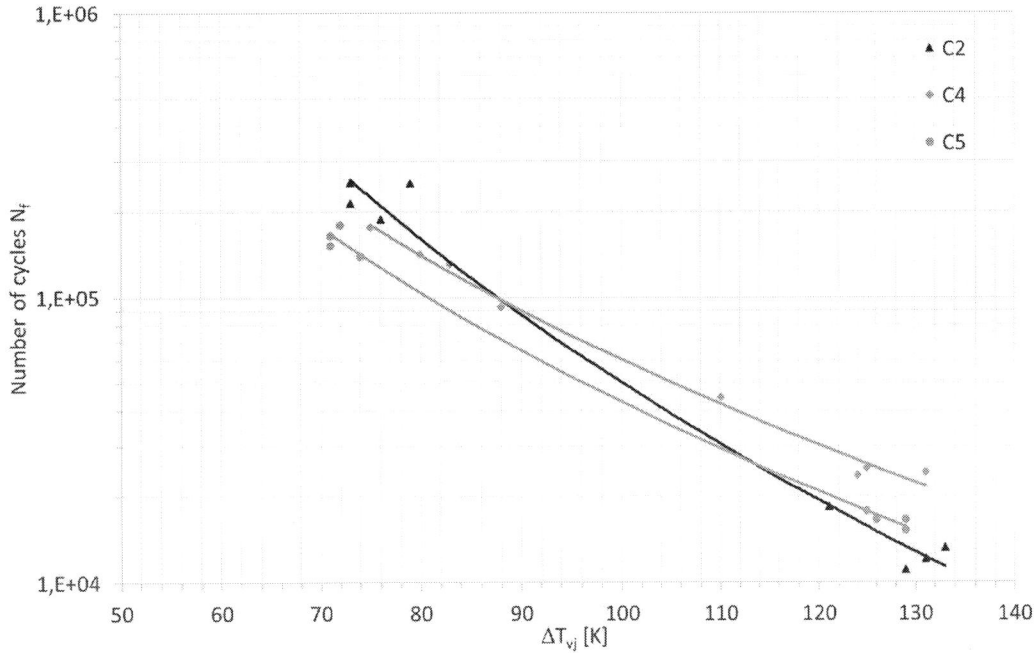

Fig. 9: Number of cycles until failure N_f for GaN cascodes 50mΩ (C2+C4) and IGBT (C5)

The interconnection technologies and inner structures of semiconductors in discrete packages are different than for semiconductors in modules. Failure modes for example bond wire lift off or cracks caused by temperature swings and plastic deformation are though comparable. The LESIT model takes this failure modes into account.

The LESIT formula [3]:

$$N_f = A \cdot \Delta T_j^{\alpha} \cdot e^{\frac{E_a}{k_b \cdot T_{jm}}}$$ (1)

For the LESIT model the exponents α and factors A must be extracted from the test results. The trend lines, see Fig. 7 – 9, are composed of potency functions. The exponents α_{new} are directly adopted, see Table III. The factors A_{new} are computed by rearranging the equation (1) of the LESIT model for each device and by using the mean value for the eight tested devices (M1: 4 devices and M2: 4 devices). For the calculations the activation energy $E_a = 9.891 \cdot 10^{-20}$ J and the Boltzmann-constant $k_b = 1.38065 \cdot 10^{-23}$ J/K are used. The modified formula is shown in equation (2).

The LESIT formula with modified exponent α_{new} and factor A_{new}:

$$N_f = A_{new} \cdot \Delta T_j^{\alpha_{new}} \cdot e^{\frac{E_a}{k_b \cdot T_{jm}}}$$ (2)

Table III: Trend line formulas, exponents and factors for the LESIT model

	Trend line formula	α_{new}	A_{new}
Component 1	$N_f = 10^{12} \cdot \Delta T_{vj}^{-3.661}$	-3.661	6,018
Component 2	$N_f = 10^{15} \cdot \Delta T_{vj}^{-5.221}$	-5.221	8,125,317
Component 3	$N_f = 20^{11} \cdot \Delta T_{vj}^{-3.231}$	-3.231	718
Component 4	$N_f = 20^{12} \cdot \Delta T_{vj}^{-3.786}$	-3.786	9,448
Component 5	$N_f = 40^{12} \cdot \Delta T_{vj}^{-3.963}$	-3.963	21,043
LESIT	$N_f = 50^{13} \cdot \Delta T_{vj}^{-5.039}$	-5.039	302,500

The exponents α_{new} of component C1 and C3 which have the same Rds(on) are very similar. Whereas component C2 with a higher Rds(on) also has a higher exponent α_{new}. The exponent value of component C4 lies between the values of component C1 and C2. The values of the factors A_{new} are far apart for all components. The power cycling tests show that for every component a different lifetime model must be made.

V   Conclusion

In this paper the results of power cycling tests of different, discrete 650V GaN cascodes with high temperature swings were presented. The inner structure of the GaN cascodes varies. The chips of MOSFET and HEMT are placed close together in the TO-247 package and the temperature of the chips was measured indirectly with the Vsd-method. The sample size of four devices per measurement point seems to be sufficient. As end of life criteria the $R_{th,j\text{-}c/s}$ was used, the course of the drain-source voltage $V_{DS,on}$ was not clear for the definition of limits.

The results of the power cycling tests show that differences of the lifetime between the GaN cascodes were not as big as expected. The results of [2] with a GaN cascode in TO-220 package were in the same range. For lifetime modelling the parameter of the LESIT model were adjusted. Compared to other lifetime models like CIPS08 no additional information about the inner structure are required. Generating one lifetime model for all tested GaN cascodes would not be useful due to big differences in the inner structure.

The lifetime of the tested GaN cascodes is comparable to the tested discrete, automotive qualified Si IGBTs, what makes the GaN cascodes interesting for automotive applications. In future work further tests with 650V SiC devices as well as semiconductors in other packages will be done.

References

[1] Xu C., Yang, F., Pu, s., Akin, B.: Performance Degradation of GaN HEMTs under accelerated power cycling tests, CPSS Transactions on power electronics and applications Vol 3 No 4 December 2018, pp. 269- 277

[2] Franke J., Zeng, G., Winkler T., Lutz J.: Power cycling reliability results of GaN HEMT devices, Proceedings of the 30th International Symposium on Power Semiconductor Devices & Ics, May 13-17, 2018 Chicago USA

[3] Held, M.; Jacob, P.; Nicoletti, G.; Scacco, P.; Poech, M.H.: Fast Power Cycling Test for IGBT Modules in Traction Application, Power Electronics and Drive Systems 1997, Conference Proceedings

[4] ECPE Guideline AQG 324: Qualification of power modules for use in power electronics converter in motor vehicles, Release no.: 02.1/2019, ECPE European Center of Power Electronics e.V., Nuremberg, p. 28

[5] Farkas, G., Sarkany, Z., Rencz, M.: Issues in Testing Advanced Power Semiconductor Devices, 32nd SEMI-THERM Symposium

Current distortion study for hybrid multi-level grid inverter with active neutral-point-clamped 4-Leg topology

Jonas Steffen[1], Matthias Klee[1], Fabian Schnabel[1], Axel Seibel[1] and Marco Jung[1,2]

[1] Fraunhofer Institute for Energy Economics and Energy System Technology (IEE)
Joseph-Beuys-Straße 8
Kassel, Germany
Phone: +49 561 7294 102
Email: jonas.steffen@iee.fraunhofer.de
URL: https://www.iee.fraunhofer.de

[2] Hochschule Bonn-Rhein-Sieg
University of Applied Sciences
Grantham-Allee 20
Sankt Augustin, Germany
Phone: +49 2241 865 316
Email: marco.jung@h-brs.de
URL: https://www.h-brs.de/

Acknowledgments

The authors would like to thank the German Federal Ministry for Economic Affairs and Climate Action as well as the Project Coordinator Jülich for funding the research project LEITNING (support code 03EI6030A) leading to this article.
Only the authors are responsible for the content of this publication.

Keywords

≪Multi-level inverters≫, ≪Dead-time≫, ≪Design optimization≫, ≪Silicon Carbide (SiC)≫, ≪Pulse Width Modulation (PWM)≫

Abstract

In multi-level inverters in particular, the switching blocking time (dead-time) can lead to undesired voltage errors, which have a significant influence on the current quality. When changing between levels, clamping must be applied to ensure the minimum switch-on and switch-off time of the power electronics. The trend towards higher switching frequencies and smaller grid filters increases the difficulty of feeding in a standard-compliant output current, especially in active neutral-point-clamped (ANPC) inverters. In this paper, the effects of clamping and a compensation method are simulated for ANPC inverters and verified on a 3-level grid forming demonstrator.

Introduction

In this paper, the behavior of active neutral-point-clamped (ANPC) inverters voltage zero-crossing is investigated. Depending on the topology and the grid coupling, different responses can be observed, ranging from DC-link oscillations to common-mode distortions. First, the causes of current distortions at voltage zero-crossing are investigated, then the proposed compensation method will be presented and analyzed.

In the last few years, inverters with multi-level topologies have gained popularity [1, 2]. In multi-level inverters, depending on the topology, n-voltage levels can be generated at the output. Classically, topologies with 2-levels, such as H4 or B6 inverters, are used in the low-voltage range [2, 3, 4] . The increase in possible output voltage levels represents a trade-off between of improved electromagnetic interference (EMI), lower voltage gradient (dV/dt), improved total harmonics distortion (THD), and on the other hand, more complex control and increased component count. The different voltage levels are derived from the intermediate link by splitting the DC-voltage. Different topologies are used to realize multi-level low voltage inverters, one of the most common being the (active) neutral-point-clamped inverter topology [1, 2, 3, 4, 5, 6]. Fig. 1 presents the investigated 3-level ANPC inverter topology. This particular topology features four inverter legs that allow fully unbalanced loads as well as advanced DC-link balancing techniques. The legs itself utilize a hybrid structure, using four classical silicon (Si) based IGBTs which will only toggle at twice the grid frequency and two fast switching silicon carbide (SiC) based MOSFETs [5]. With the further development of gallium nitride (GaN) semiconductors, multi-level topologies with GaN devices will also be increasingly used in the future [7]. The IGBTs will connect the MOSFETs to either the high or low side of the DC-link depending on the desired output voltage. While this topology does increase the overall efficiency of the inverter it comes at the costs of flexibility in the switching patterns. Advanced modulation schemes like space vector modulation (SVM) or selective harmonic elimination pulse width modulation (SHEPWM) suffer from limitations in their vector selection [8]. Furthermore, it is no longer possible to use the redundant vectors to control DC-link oscillations. Despite their benefits multi-level inverters also suffer from some drawbacks, mainly are

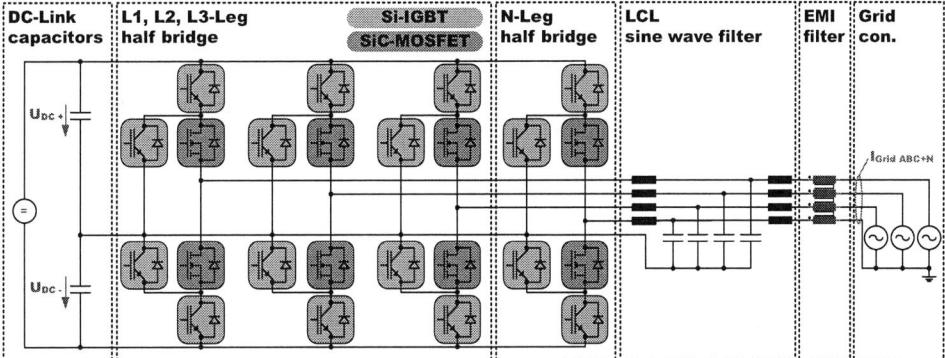

Fig. 1: Topology of a 3 Level 4 Leg ANPC-Inverter

DC-link balancing and current distortion during voltage zero crossing. Taking the ANPC 3-level topology as a reference, one DC-link capacitor is always charged and one discharged in each of the active inverter branches, depending on the current direction. This fluctuation leads to a constant 3rd-harmonic oscillation of the two DC-link halves. The impacts of DC-link oscillations and compensation strategies are already discussed in detail in references [9, 10] and will not be discussed further in this paper. The current distortions during voltage zero crossing on the other hand are still an interesting topic, that will gain even more relevance when moving to higher switching frequencies and smaller filters, especially for grid connected low voltage applications.

Cause of error

A negative influence on the current is caused by various superimposed effects; in addition to the already mentioned clamping, the voltage of the DC-link and the dead-time as well as the control of the inverter play a role. The clamping occurs especially at the zero crossing of the voltage and is explained in more detail below. A current distortion can be observed at the voltage zero crossing of a phase voltage of an inverter leg. In the reference inverter, the distortion occurs in the form of a current in the neutral leg due to the 4-leg design. Fig. 2 depicts the grid voltage and grid current for the three phases and the neutral conductor of the 4 wire 3 level reference inverter in simulation. For the exploration of the clamping error itself a simplified simulation is used, to highlight the clamping issues during zero crossing. It neglects dead-time effects, as well as the effects of the DC-link oscillation, to isolate and highlight the actual issue. The neglected effects will be explained in more detail in the simulation section. This current distortion seen in Fig. 2 can directly be traced back to an unbalance in the phase voltages during zero crossing. Unlike 2-level inverters, the inverter output current ripple in ANPC inverters reaches its minimum at zero voltage crossing as shown in Fig. 3 (a). This is a benefit when feeding active power in,

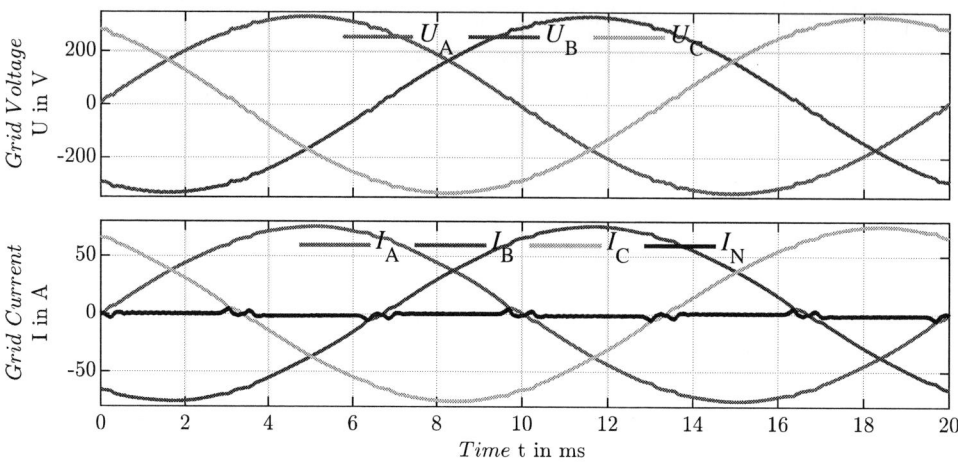

Fig. 2: Current distortion of the simulated reference inverter

as it reduces current distortions related to dead-time effects. However, there are still restrictions during the zero crossing, as for the reference system specifically during this time the slower IGBTs will switch over. Here it will be necessary to follow a specific order to avoid current distortions and always allow for a connection to the DC-link middle point. Another limitation is given by the SiC-MOSFET switches, which have a significant lower dead-time and allow for higher switching-frequencies compared to Si-IGBTs. A minimum on/off time is still required to ensure that the SiC-MOSFET switches have actually changed state completely. To guarantee a safe on-time of the MOSFETs, the minimum on-time t_{MIN} is defined as twice the dead-time. However, this restriction in the switching behavior also limits the achievable duty-cycle. And in turn limit the possible output voltages at low and high duty-cycles. Fig. 3 (a) shows this effects. As can be seen there will always be a small dead-band around the zero level. The dead-band Mod_{MIN} depends on the switching frequency f_{SW} and the minimum on-time t_{MIN} of the semiconductor, as defined by (1).

$$Mod_{\text{MIN}} = t_{\text{MIN}} \cdot f_{\text{SW}} \tag{1}$$

The inverter is not able to output any of the voltages within the dead-band, except for the zero level itself. The usual strategy, as depicted in Fig. 3 (a), is to set the output voltage directly to zero level, causing two voltage steps. An alternative strategy, as depicted in Fig. 3 (b), is to maintain the minimal voltage level for a period of time, then change to zero and finally back to the lowest possible negative output voltage level. The 3-step approach also results in distortions, but with a smaller error than the usual strategy. The behavior described is commonly known as "clamping" and can already be observed with 2-level inverters, but it only occurs with very large ($Mod \approx 1$) or very small ($Mod \approx 0$) duty cycles, as with over modulation. The voltage error caused by the clamping-mechanics during voltage zero crossing will result in a temporary occurrence of a zero-sequence in the output voltage of the clamping inverter leg. This creates a zero-sequence current in the connected neutral conductor, which also affects the DC-link voltages and leads to future phase current imbalances. In order to show the influence of different

Table I: Dead-time and minimum on-time assumptions

	Si IGBT	SiC MOSFET
Dead-time t_{DT}	1 µs	300 ns
Minimum on-time t_{MIN}	2 µs	600 ns

semiconductor technologies and switching frequency, the boundary conditions shown in Table I were defined for Si IGBTs and for SiC MOSFETs. To calculate the duration of the dead-band t_{C}, a line to line

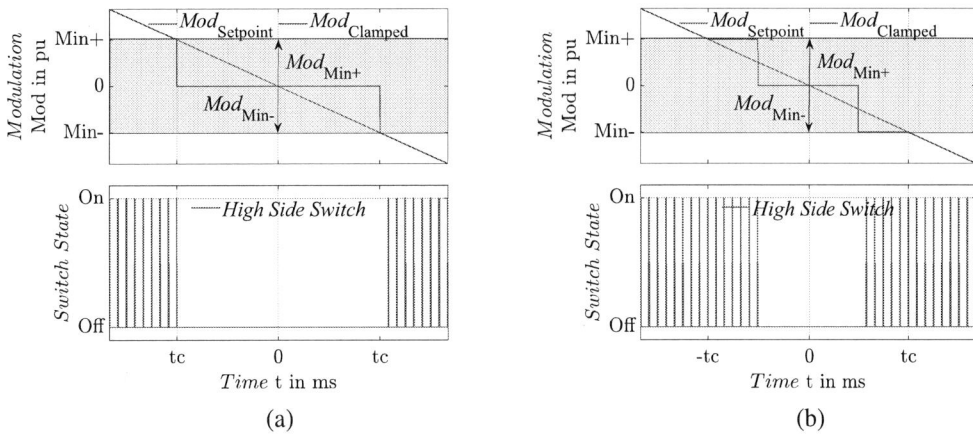

(a)   (b)

Fig. 3: Modulation and switching states in case of voltage zero crossing. (a) Direct clamping after reaching the boundary (b) 3-Step clamping for minimizing the voltage error

voltage of $U_{AC} = 400\,V$ and a DC-link voltage $U_{DC} = 800\,V$ were assumed. Using Eq. (2)

$$Mod_{MIN} = \frac{U_{AC}\sqrt{2} \cdot sin(\omega t)}{U_{DC}/2}, \qquad (2)$$

the duration of the dead-band t_C can be derived with Eq. (3)

$$t_C = \frac{1}{\omega}arcsin(\frac{Mod_{MIN} \cdot U_{DC}/2}{U_{AC}\sqrt{2}}). \qquad (3)$$

For an average value consideration of the voltage error applied to the inductance L, the line voltage is used for simplification (4). With this voltage the current in the inductance L can be calculated (5). By solving (5) and using the dead-band time t_C, the deviation of the inductor current Δi_L (6) can be determined. Using the boundary conditions from Table I, the dead-band for Si IGBTs with a switching

Table II: Comparison of Si IGBT und SiC MOSFET dead-band time and current error

f_{SW} in kHz	Si IGBT				SiC MOSFET			
	Mod_{MIN}	t_C in µs	L in µH	Δi_L in A	Mod_{MIN}	t_C in µs	L in µH	Δi_L in A
10	0,020	78	1200	0,261	0,006	23	1200	0,023
15	0,030	117	800	0,881	0,009	35	800	0,079
20	0,040	157	600	2,089	0,012	47	600	0,188
30	0,060	235	400	7,056	0,018	70	400	0,634
40	0,080	314	300	16,742	0,024	94	300	1,503
50	0,100	392	240	32,744	0,030	117	240	2,937
60					0,036	141	200	5,076
70					0,042	164	171	8,061
80					0,048	188	150	12,036
90					0,054	212	133	17,141
100					0,060	235	120	23,519

frequency of 20 kHz is comparable to SiC MOSFETs with a switching frequency of 70 kHz as listed in Table II. The value of the inductance L has been scaled according to the switching frequency f_{SW}. Due to the lower inductance value L with a comparable dead-band, there is a significantly higher current deviation Δi_L for SiC MOSFETs with 8 A compared to 2 A for Si IGBTs. A comparable current deviation

Δi_L is reached with SiC MOSFETs at approx. 45 kHz.

$$u_L = U_{AC}\sqrt{2} \cdot sin(\omega t) \tag{4}$$

$$i_L = \frac{1}{L} \int U_{AC}\sqrt{2} \cdot sin(\omega t)dt \tag{5}$$

$$\Delta i_L = \frac{U_{AC}\sqrt{2} \cdot (1 - cos(\omega t_C))}{\omega L} \tag{6}$$

Table II compares the different boundary conditions of Si IGBT and SiC MOSFET. The explanation points at $f_{sw} = 20$, 40 and 70 kHz are highlighted. Fig. 4 shows the current deviation Δi_L for the investigated Si and SiC power semiconductors over the switching frequency. It can be observed that the current deviation increases by the power of 2 with the switching frequency and has a stronger influence on the Si device due to the larger dead-time. Due to the higher switching frequencies when using modern SiC

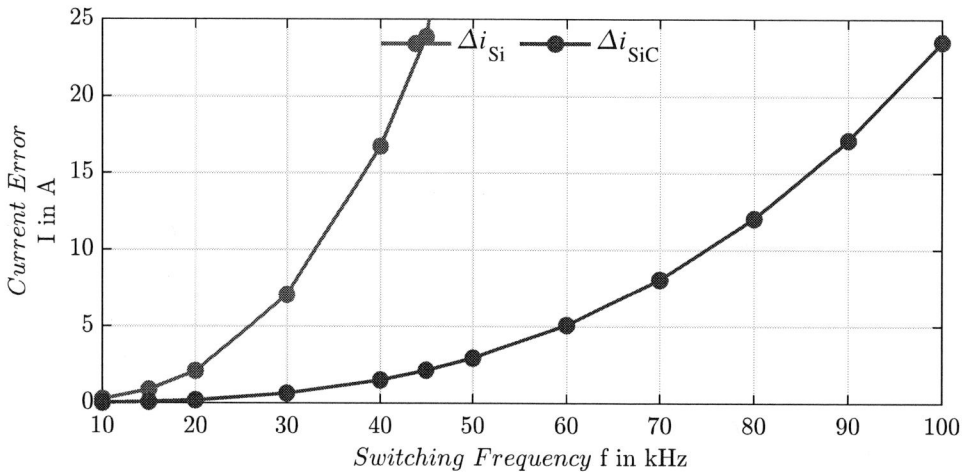

Fig. 4: Current error Δi_L depending on the switching frequency for Si IGBTs and SiC MOSFETs

and GaN switches allow to choose a small inductance which also increases the current ripple in contrast to Si devices, the resulting current error will increase. The effect will be shown in the simulation and laboratory measurements in the next chapters.

Effects on the current distortion in simulation

There are various effects caused by the inverter which influence the current quality. Besides the clamping, DC-link oscillations and the switch dead-time have an impact on the current quality. These effects will be explained in Fig. 5 by signal characteristics of the phase currents and the neutral current from the simulation in different operation conditions. The current distortions are observed, especially in the neutral leg. The blue signal curve shows the simplified simulation where the influences of the dead-time and the DC-link are deactivated. Only the clamping causes the current deviations in this simulation. Furthermore, shows the red signal curve in Fig. 5 the results when an extended model with a more realistic DC-link representation and balancing control for the two DC-link halves is used. In this model additional distortions in the neutral leg current can be observed, as well as small deformations of the phase currents. While the chosen modulation scheme compensates for most of the DC-link oscillations, the balancing controller will again introduce a small zero sequence in the inverters output current. However, the current distortion caused by the clamping is still present in the output currents. Looking at the peak-to-peak values of the disturbances at zero crossing of the voltage of the simplified simulation (blue curve) and the extended simulation (red curve), they remain unchanged and are only slightly shifted by the additional third harmonic introduced due to the DC-link oscillations and balancing control. In a similar manner the yellow signal curve on Fig. 5 showing the results of a model which introduces the dead-time effect and represents a nearly realistic simulation of an ANPC inverter. Additional current fluctuations in the neutral leg can also be observed in this case. This current oscillation relates to the effect of the dead-time in the output voltage of the inverter, it introduces a zero sequence voltage, which in turn influences the current in the neutral leg. This model does not feature a dead-time compensation, so the full effect will be visible. The dead-time also produces a third harmonic current which leads to a superposition with the already existing current distortion of the clamping. This is due to the fact that after taking into account the dead-time for MOSFET and IGBT during the voltage zero crossing with pure active power, the dead-time effect also occurs at the same time. A feasible method to compensate for the clamping error is presented in the next chapter.

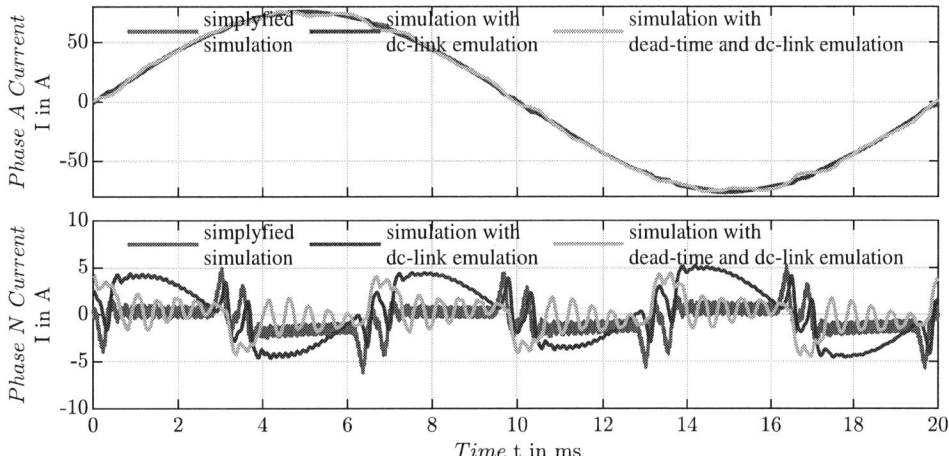

Fig. 5: Current distortion from phase A and N of the simulated 4-leg ANPC inverter

Compensation of the voltage clamping error

For Si inverters with lower switching frequencies and larger filters compared to recent SiC inverters, clamping caused minor distortion and no active compensation was required. However, current distortions increase with increasing switching frequency f_{SW}, as can be seen in (1). To compensate this problem the minimum on-time t_{MIN} of SiC semiconductors is also decreasing but not in the same extent as the increasing switching frequency f_{SW}. However, with highly optimized filters for the desired switching frequencies of SiC-inverters, the clamping mechanism plays a significant role in ensuring the inverters output current quality. In addition, the hybrid topology plays a decisive role whereby advanced modulation types cannot be used. The presented 4-leg-4-wire inverter, designed for high unbalanced load, allows simple compensation of the clamping effects. Since the voltage error occurs between the output voltages of the phases and the neutral is connected to the 4th active inverter branch, it is possible to completely eliminate the effects by adding a common mode voltage to all voltages once a phase voltage reaches its clamping level. To calculated the compensation $u_{PH_{CORR}}$, the minimal allowable positive and negative voltage $u_{MIN\pm}$ are calculated using (1), which leads to (7).

$$u_{MIN\pm} = Mod_{MIN} \cdot u_{DC\pm} \tag{7}$$

By interrogating the set voltage of the neutral conductor with (8), the output voltage of the neutral leg can be updated by the clamping. Finally, using $(u_N - u_{N_{CLAMPED}})$ as a common mode compensation voltage, the corrected phase voltages $u_{PH_{CORR}}$ can be calculated using (9). The modulation $Mod_{ABC,N}$ is derived from (10).

$$u_{\text{N}_{\text{CLAMPED}}} = \begin{cases} u_{\text{N}} & u_{\text{N}} \geq u_{\text{MIN}+} \\ u_{\text{MIN}+} & 0 < u_{\text{N}} < u_{\text{MIN}+} \\ 0 & u_{\text{N}} = 0 \\ u_{\text{MIN}-} & 0 > u_{\text{N}} > u_{\text{MIN}-} \\ u_{\text{N}} & u_{\text{N}} \leq u_{\text{MIN}-} \end{cases} \tag{8}$$

$$u_{\text{PH}_{\text{CORR}}} = u_{\text{PH}} - \left(u_{\text{N}} - u_{\text{N}_{\text{CLAMPED}}}\right) \tag{9}$$

$$Mod_{\text{PH}} = u_{\text{PH}_{\text{CORR}}}/u_{\text{DC}} \tag{10}$$

The standard modulation of the inverter is shown in Fig. 6 (a). The sinusoidal wave-forms are shown

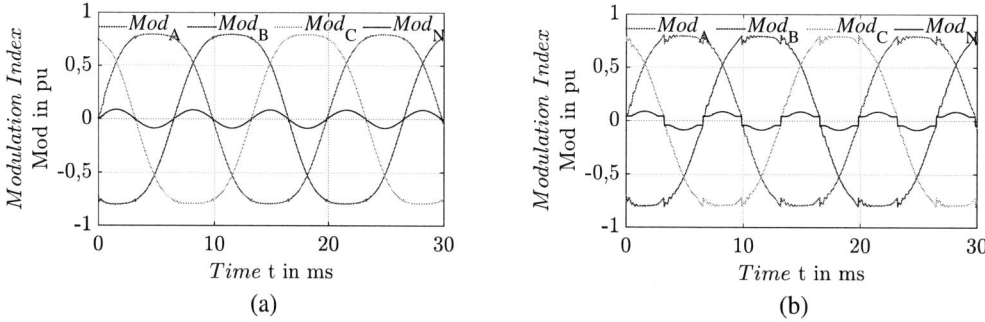

(a)   (b)

Fig. 6: Modulation index for a 4-leg-4-wire ANPC inverter. (a) Without clamping compensation (b) With clamping compensation

for phase A, B, C and N with injection of an third harmonic based oscillation. The modified modulation scheme is depicted in Fig. 6 (b). Due to the added common-mode voltage the inverter will always operate in a valid duty-cycle range. However this modification comes at the cost of additional common-mode distortions, which need to be considered during the inverters filter design.   Fig. 7 (a) shows

Table III: Result of the FFT analysis of the current in phase A and N without and with compensation in simulation

Har. Freq. [Hz]	Amplitude current of phase A I_A [A]		Amplitude current of phase N I_N [A]	
	Compensation off	Compensation on	Compensation off	Compensation on
50	75.06	75.05	0.06	0.01
150	0.47	0.11	1.47	0.60
450	0.46	0.12	1.43	0.35
750	0.32	0.12	0.94	0.37

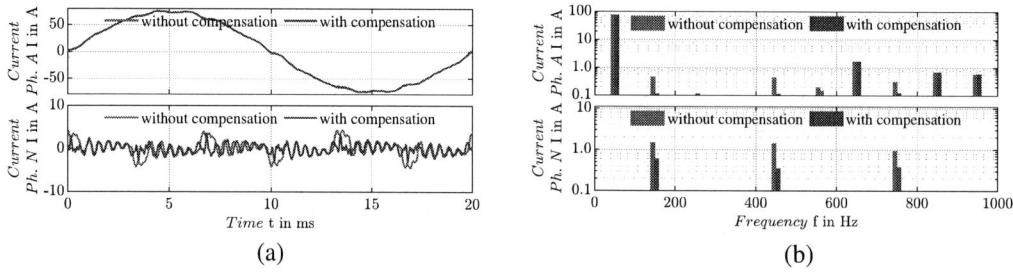

(a)   (b)

Fig. 7: Impact of the clamping compensation on the output currents of the 4-leg-4-wire ANPC inverter in simulation. (a) Currents of the phase A and N (b) FFT analysis of the currents

the comparison of the mains output currents without and with clamping compensation. The upper plot shows the current in phase A and the lower plot shows the current in the neutral conductor. A reduction in current disturbances can be seen in the curves. For accurate comparability of fundamental and harmonic current amplitudes, a fast fourier transform (FFT) was performed. In particular, the FFT on Fig. 7 (b) reveals a reduction of harmonics in phase A current and N. The results are summarized in Table III. The current amplitude of the fundamental wave corresponds with $I_A = 75\,\text{A}$ to the set value. In addition, the amplitude remains unaffected by the clamping compensation. Especially the amplitudes of the listed harmonics 3rd, 9th and 15th can be reduced by the compensation technique in phase and neutral current. The positive results of the compensation in the simulation are the motivation for the investigations on the laboratory demonstrator.

Validation using the laboratory demonstrator

A laboratory demonstrator was used for validation. The device is designed for use in the low voltage grid and has a power of $S = 43\,\text{kVA}$. The laboratory demonstrator has a topology based on Fig. 1 and allows through the configuration an on and off switching of the clamping compensation. The laboratory setup is shown on Fig. 8. The demonstrator inverter as well as the DC voltage supply and the data logger are depicted on Fig. 8 (a). The laboratory demonstrator is connected to a network emulation. The network emulator can be found on Fig. 8 (b). In the laboratory tests, the identical operating point as in the simulation is set, an active current with an amplitude of $I = 75\,\text{A}$. The measurement was carried out without and with compensation of the clamping. The exact structure of the Fraunhofer IEE power inverter can be seen in publication [11]. Fig. 9 (a) depicts the corresponding results as a measurement using the labora-

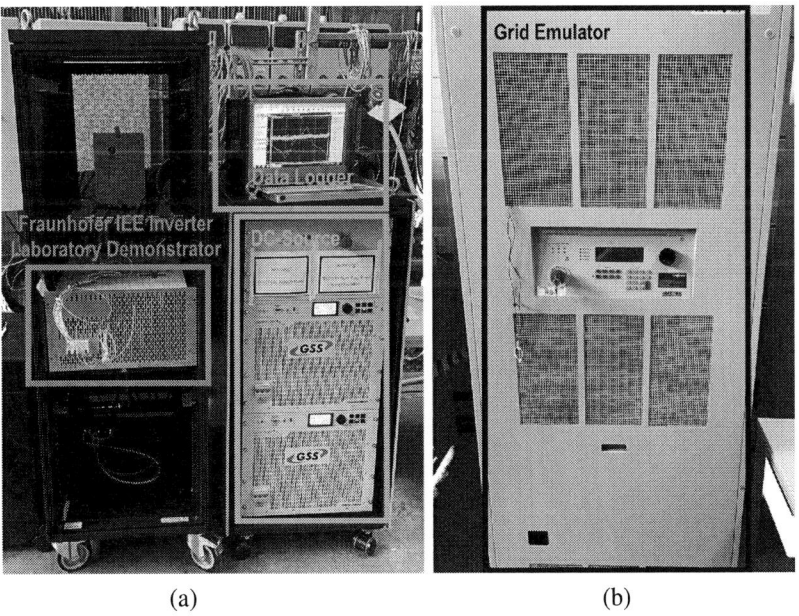

(a)   (b)

Fig. 8: Laboratory demonstrator of the 4-leg-4-wire ANPC inverter

tory demonstrator. It can be observed, that in the real system the proposed compensation strategy leads to an improvement in the differential output currents of the inverter. This is especially well highlighted by the FFT shown in Fig. 9 (b). It should be mentioned that the laboratory demonstrator, similar to the simulations, does not feature a dead-time compensation, hence the rather high additional harmonic currents. This decision was made to simplify the setup and allow for a more precise investigation of the presented effect and compensation. It can be seen, that the proposed compensation strategy reduces the neutral current caused by the clamping effect, as well as it reduces harmonic phase currents, not only for the simulation, but the measurements as well. The results from the simulation cannot be depicted completely with the laboratory measurements. The third harmonic in particular is not improved by the compensation, but improvements can be observed in the spectrum for the multiple harmonics. Due to the clamping, the 12th and 18th harmonics in phase N are generated in the laboratory device, and these could be completely compensated. In comparison with the simulation, it can be said that an improvement in the current quality also occurs on the laboratory demonstrator, but less well than in the simulation. The

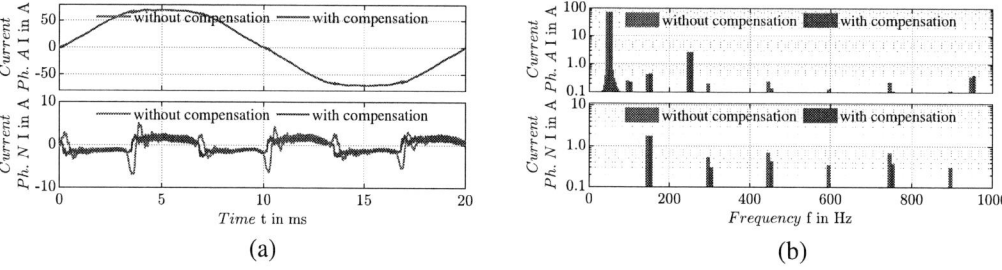

(a)   (b)

Fig. 9: Impact of the clamping compensation on the output currents of the 4-leg-4-wire ANPC inverter laboratory demonstrator. (a) Currents of the phase A and N (b) FFT analysis of the currents

Table IV: Result of the FFT analysis of the current in phase A and N without and with compensation from the measurements

Har. Freq. [Hz]	Amplitude current of phase A I_A [A]		Amplitude current of phase N I_N [A]	
	Compensation off	Compensation on	Compensation off	Compensation on
50	72.9	72.72	0.06	0.01
150	0.43	0.45	1.75	1.77
450	0.24	0.01	0.70	0.43
600	0.13	0.01	0.35	0.09
750	0.23	0.01	0.70	0.37
900	0.01	0.01	0.31	0.01

Table IV summarizes the results of the laboratory measurement. For the fundamental wave, the current in phase A has an amplitude of $I_A = 72.9$ A, which remains constant with compensation switched on. For the third harmonic, the amplitude of the current in phases A and N is $I_A = 0.45$ A and $I_N = 1.75$ A. In contrast to the simulation, a positive effect for the third harmonic cannot be achieved with the proposed compensation. For the harmonics of order 9th, 12th, 15th and 18th, the current distortions in the phase and neutral currents can be effectively reduced. Furthermore, compared to the values in Table III, it can be observed from the simulation that the harmonics are already weaker. However, the improvements with compensation for the typical harmonics 3rd, 5th and 7th are also lower than the simulation suggests.

Conclusion

The paper presents one of the challenges to ensure high current quality of next generation high efficient grid inverters. The hybrid ANPC topology in combination with optimized grid filter design leads to distortions at the zero voltage crossing. It was presented how the current of the inverter is affected by clamping. A simple simulation was carried out to explain the clamping effect. The simulation was extended by the typical influences of a power inverter such as DC-link oscillations and the dead-time. Subsequently, it was shown how the distortions in the inverter output current can be effectively reduced by slight adjustments in the preparation of the modulation. The evaluation by means of FFT in the simulation shows an improvement for the third-order harmonic and multiples, thereof in the current spectrum the amplitude could be reduced by 50% up to 70%. The simulations results presented are verified by measurements with a demonstrator inverter. In the laboratory, an improvement in power quality was also observed for the harmonics; in particular, the 12th and 18th harmonics were fully compensated. However, the third harmonic could not be influenced as in the simulation. Especially in phase N the 6th, 9th and 15th harmonics could be reduced by about 30%. In the future, with the trend of ever increasing switching frequencies, also with regard to GaN semiconductors, the negative effects of dead-times and clamping will continue to increase and effective compensation methods become even more important. As the investigations have shown, the proposed solution can improve the current quality, but further methods and improvement of compensation should be researched and developed to meet future requirements.

References

[1] L. Franquelo, J. Rodriguez, J. Leon, S. Kouro, R. Portillo and M. Prats: The age of multilevel converters arrives, IEEE Industrial Electronics Magazine, 2008, doi: 10.1109/MIE.2008.923519

[2] H. Akagi: Multilevel Converters: Fundamental Circuits and Systems, Proceedings of the IEEE, 2017, doi: 10.1109/JPROC.2017.2682105

[3] P. Barbosa, P. Steimer, J. Steinke, M. Winkelnkemper and N. Celanovic: Active-neutral-point-clamped (ANPC) multilevel converter technology,2005 European Conference on Power Electronics and Applications, 2005, doi: 10.1109/EPE.2005.219713

[4] H. Wang, L. Kou, Y. Liu and P. Sen: A Seven-Switch Five-Level Active-Neutral-Point-Clamped Converter and Its Optimal Modulation Strategy, IEEE Transactions on Power Electronics, 2017, doi 10.1109/TPEL.2016.2614265

[5] Q. Guan, C. Li, Y. Zhang, S. Wang, D. Xu, W. Li and H. Ma: An Extremely High Efficient Three-Level Active Neutral-Point-Clamped Converter Comprising SiC and Si Hybrid Power Stages, IEEE Transactions on Power Electronics, 2018, doi: 10.1109/TPEL.2017.2784821

[6] R. Teichmann and S. Bernet: A Comparison of Three-Level Converters Versus Two-Level Converters for Low-Voltage Drives, Traction, and Utility Applications, IEEE Transactions on Industry Applications, 2005, doi: 10.1109/TIA.2005.847285

[7] N. Keshmiri, D. Wang, B. Agrawal, R. Hou and A. Emadi: Current Status and Future Trends of GaN HEMTs in Electrified Transportation, IEEE Access, 2020, doi: 10.1109/ACCESS.2020.2986972

[8] K. Gnana Sambandam, A. K. Rathore, A. Edpuganti and D. Srinivasan: Current-fed multilevel converters: An overview of circuit topologies and modulation techniques, 2016 IEEE Industry Applications Society Annual Meeting, 2016, doi: 10.1109/IAS.2016.7731887

[9] W. Zhang, X. Li, C. Du, X. Wu, G. Shen and D. Xu: Study on neutral-point voltage balance of 3-level NPC inverter in 3-phase 4-wire system, The 2nd International Symposium on Power Electronics for Distributed Generation Systems, 2010, doi: 10.1109/PEDG.2010.5545910

[10] F. Rojas-Lobos, R. Kennel and R. Cárdenas-Dobson: 3D-SVM algorithm and capacitor voltage balancing in a 4-leg NPC converter operating under unbalanced and non-linear loads, 15th European Conference on Power Electronics and Applications (EPE), 2013, doi: 10.1109/EPE.2013.6634684

[11] T. Gühna, D. Stracke, M. Klee, F. Schnabel, A. Seibel and M. Jung: Hardware and Software Concept for Distributed Grid-Forming Inverters in Microgrids, 23rd European Conference on Power Electronics and Applications (EPE'21 ECCE Europe), 2021, doi: 10.23919/EPE21ECCEEurope50061.2021.9570688

Dynamic Maximum Power Point Tracking Method including Detection of Varying Partial Shading Conditions for Photovoltaic Systems

Rosalie Rouphael, Nezha Maamri, Jean-Paul Gaubert
Laboratoire d'Informatique et d'Automatique pour les Systèmes (LIAS), Université de Poitiers
2, rue Pierre Brousse
Poitiers, France
Phone: +33 (0) 549453509
Fax : +33 (0) 549454034
Email: rosalie.rouphael@univ-poitiers.fr,
nezha.trigeassou@univ-poitiers.fr,
jean.paul.gaubert@univ-poitiers.fr
URL: https://www.lias-lab.fr/

Keywords

≪Photovoltaic≫, ≪MPPT≫, ≪P&O MPPT≫, ≪Maximum Power Point Tracking Quadratic Converters≫, ≪Renewable energy systems≫.

Abstract

Photovoltaic (PV) systems, being one of the promising power generation systems, need performance optimization typically under partial shading conditions (PSCs). In such non-uniform irradiation conditions, not only conventional maximum power point tracking (MPPT) algorithms but also advanced techniques fail in operating at the power peak. So, global maximum power point tracking (GMPPT) methods are needed when several local maximum power points exist under PSCs. The major difficulty is not in detecting the GMPP but in finding the new one when the PSCs change. Consequently, the challenge becomes in detecting the variation of PSCs. The proposed method relies on monitoring the rate change on power at the output of the PV panel. Proving that this criterion alone is not accurate, a measurement Q is introduced and used as a watchdog for critical partially shaded cases. When in such case, a timer alternates between a scanning process and a dual-mode MPPT method based on perturb and observe (P&O) and constant voltage algorithms. Otherwise, the MPPT work in not interrupted unless a drastic power change takes place. The proposed algorithm proved its efficiency in all partially shaded irradiation conditions when implemented in a MATLAB/Simulink environment. Furthermore, it is compared to other partial shading detection (PSD) methods to verify the algorithm's performance in a generic case.

Introduction

Due to the increase of global environmental concerns, countries' intention and investment in finding solutions have grown over the past decade. As a result, the past 20 years have witnessed high deployment and integration of solar photovoltaic (PV) systems. Although PV energy remains one of the most promising eco-friendly sources of energy, several proceedings need to be treated in order to raise its efficiency. The performance of PV systems is greatly influenced by climatic conditions like sunlight, ambient temperature, airflow, dust, clouding, etc. Solar irradiation and temperature intervene explicitly in the model of the PV panel. Clouding or any shadow, creates non-uniform irradiation on the surface of the PV panel. Consequently, partial shading conditions (PSC) occur and produce a different form for the characteristic PV output [1].

To ensure operating at the maximum power point (MPP), many maximum power point tracking (MPPT) techniques were developed: conventional MPPT (P&O based [2, 3], Inc-Cond based [4, 5]) and advanced intelligent methods (fuzzy logic [6], neural networks [7]). Although these techniques show great performances under uniform irradiation, they fail to track the global maximum power point (GMPP) when PSCs occur [8]. In order to detect the global maximum, several algorithms suggest novel GMPPT methods. Most GMPPT algorithms propose a search method for the GMPP [9, 10].

Since some energy is lost during the search interval, the interest in reducing the duration or the occurrence of these searching mechanisms has increased. Therefore, the need for PSC detection methods has become important to avoid unnecessary searches. Partial shading detection (PSD) techniques proposed in literature used different criterion mainly relying on the variation of electrical measurements like power, voltage and current. For example a monitoring of current and voltage variation along with a fast tracking technique is proposed in [11]. Some PSD methods are based on the number of sign changes of voltage and power respectively [12, 13]. Also, adding to the power variation verification two equations based on voltages changes are suggested for instance in [14]. Other techniques propose examining the climatic conditions while monitoring only open-circuit and short-circuit measurements, such as [17, 18].

However, in some scenarios, most methods fail to detect any change in climatic conditions. These scenarios are presented later and critical PSC are discussed. Some methods eliminating these critical cases rely on hardware solutions [15] or computational solutions [16]. The proposed semi-dynamic GMPPT technique guarantees maximum power transfer in all varying climatic conditions and takes into consideration the critical cases without additional sensors or calculators. When an irradiation change occurs, a quick scan is launched to detect the new maximum power point whether there are many MPP or just one. When the GMPP is found, dual-mode algorithm, an adaptive P&O based technique, operates to stabilize the system with minimum steady-state oscillations. Expressly, the scanning mechanism starts if there is a sizable variation in power P_{PV} or if a possible critical case is detected. The transitions between subprograms are smoothed by varying the duty cycle which controls the DC/DC converter. The proposed aperiodic scan GMPPT method is tested in a MATLAB/Simulink environment under varying partial shading conditions and compared to a classical technique that does not consider critical cases.

System description

To form a PV plant, several modules are generally connected in series in order to attain a certain voltage, forming strings. The latter are connected in parallel to allow reaching a required current. The impact of shadow differs according to the electrical interconnection and the PV panel orientation. Several researches analyse the role of PV array size and configuration [19, 20]. Series-Parallel (SP) remain widely implemented due to its simplicity along with the Total Cross Tied (TCT) scheme that proved to offer the least mismatch losses.

During this work, the PV array studied regroups three PV modules in series creating a single PV string. The mismatches caused by PSCs can be generalized later when several strings are connected in parallel (SP configuration). A Conergy PowerPlus 214P PV module is chosen and used in simulations and for measurements. This choice is conditioned by the presence of these modules in the laboratory installation. A stand-alone PV system is considered in order to test the proposed control method (Fig. 1), so the PV array is only supplying a resistive load R. The impedance matching is adapted by a DC/DC quadratic boost converter whose role also includes voltage output regulation and maximum power extraction. The proposed converter structure presents two major advantages: a high voltage conversion ratio and having one controllable switch. The components used in the quadratic boost converter are presented in Table I. The control strategy calculates the value of the duty cycle commanding the MOSFET switch via a PWM block with a switching frequency of $10kHz$. The global maximum power point tracking algorithm provides the optimal value for the duty cycle in real-time. The strategy uses only the voltage and the current measured at the output of the PV array, hence minimizing the number of sensors in such configurations.

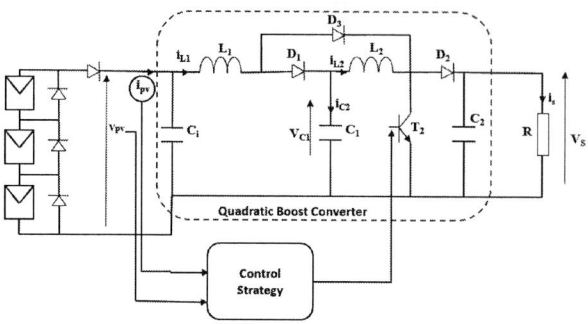

Fig. 1: Stand-alone PV system synoptic

Table I: Quadratic boost converter's parameters

C_i	L_1	r_{L1}	L_2	r_{L2}	C_1	C_2	R
$4.7\mu F$	$5mH$	$52m\Omega$	$10mH$	$134m\Omega$	$470\mu F$	$220\mu F$	150Ω

Proposed control strategy

Critical partial shading conditions

Various GMPPT methods suggest a curve tracing to locate the GMPP whether its a full scanning or a two points verification. During the search process, the GMPPT work is interrupted and the available energy is not all harvested. Consequently, it is more beneficial to avoid needless search when the PSC does not change. In order to detect the change in the climatic conditions, partial shading detection (PSD) methods are developed [11, 12, 13]. Several techniques monitor the variation of successive measurements of PV panel output variables: P_{PV}, I_{PV} and/or V_{PV}. Although many techniques proved good performance in varying PSC, they still fail to detect the change of irradiation in some cases. These cases are critical because P_{PV}, I_{PV} and V_{PV} do not vary. This happening occurs when the GMPP in the old irradiation profile is also a LMPP in the new irradiation pattern. Several irradiation profiles are investigated on a PV string formed by three PV modules in series, and each module is shunted by a by-bass diode. Some of the tested patterns are presented in Table II along with their P-V and I-V curves (Fig. 2 and Fig. 3 respectively).

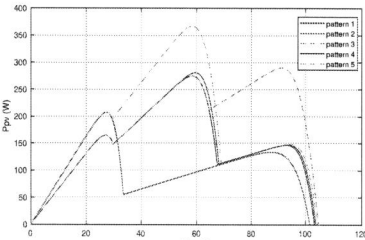

Fig. 2: P-V curve under different irradiation patterns

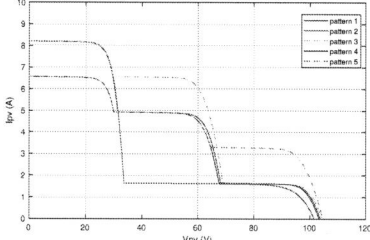

Fig. 3: I-V curve under different irradiation patterns

When the GMPP position is at the most left, the MPP current reaches its maximal possible value (as seen graphically for pattern 2). Expressly, the current delivered from the PV string is imposed by the PV module most exposed to sunlight since other modules are shunted. Therefore, even if one of the less illuminated modules encounters an increase of irradiation (for example a transition from pattern 2 to pattern 3 occurs), the corresponding diode remains conductive. The global current is intact and the change of irradiation is not perceived. Consequently, the PSD algorithm does not detect the need to scan

Table II: Irradiation values on the PV array in W/m^2

Pattern 1	Pattern 2	Pattern 3	Pattern 4	Pattern 5
1000	1000	1000	800	800
600	200	200	600	600
200	200	800	200	400

and notable power is lost because the operating MPP (still the most left) is not the GMPP of pattern 3. By analogy, this happening can occur in an all-parallel configuration so one module imposes its voltage and the increase of irradiation of the shaded module is imperceptible. In order to overcome these critical cases, the proposed method suggests a periodic scan when the GMPP is to the most left. One of the main advantages of this technique is being also suitable for systems with another configuration and having a larger number of modules. The challenge remains in finding whether or not the GMPP is the local maximum point most to the left.

Aperiodic scan-dual-mode

The proposed aperiodic scan technique combines two programs: a P-V curve scanning method and a MPPT algorithm for rapidly changing irradiation. The transition from one sub-program to another is subjected to different conditions. The flowchart of the proposed technique is shown in Fig. 4 where the scanning process, the MPPT dual-mode algorithm and the transition criteria are presented.

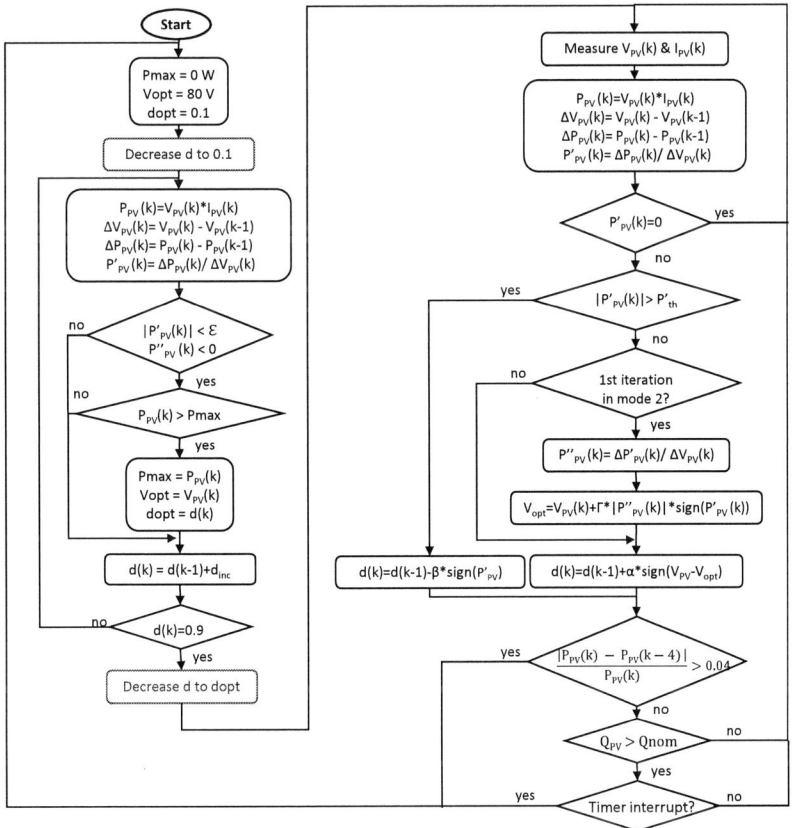

Fig. 4: The flowchart of the global algorithm

Dual-mode MPPT algorithm is a hybrid between the algorithms constant voltage and fixed step P&O. So, in mode 1, the operating point is far from the MPP, the P&O with a large step is activated for rapid convergence. The step's expression is based on the gradient method:

$$\delta d = -\beta * sign(P'_{PV}) \tag{1}$$

Where β is a constant whose value impacts the response time. However, in mode 2, the goal is to approach the MPP as close as possible so the algorithm brings the output voltage V_{PV} near its optimal value at MPP. Therefore, the oscillations in steady-state are significantly reduced. In this mode, the duty cycle step is constant:

$$\delta d = \alpha * sign(V_{PV} - V_{opt}) \tag{2}$$

Where α has a relatively small value. The optimal value of V_{PV} is estimated using the expression (3) found by comparing the curves P'_{PV} and P''_{PV} in terms of V_{PV} at different irradiation around the MPP.

$$V_{opt} = V_{PV} + \Gamma * |P''_{PV}| * sign(P'_{PV}) \tag{3}$$

Where Γ is a fixed constant. The value of V_{opt} can be verified later with the voltage value measured at the GMPP during the scan. The shifting from one mode to another is made by comparing the value of P'_{PV} to a threshold P'_{th}. The value of the power derivative with respect to the voltage is high in transient response while it approaches zero around the MPP. Expressly, when P'_{PV} is higher then P'_{th}, mode 1 is operating; otherwise, mode 2 is launched. The values of Γ and P'_{th} are chosen manually to approach the MPP. The chosen values for all the algorithm constants are: $\alpha = 0.001$, $\beta = 0.005$, $\Gamma = 0.5$ and $P'_{th} = 5$.

The proposed aperiodic scan-dual-mode method monitors continuously the variation of successive measures of P_{PV}. When the absolute difference between two consecutive PV output power values exceeds a threshold value, a scanning process is launched. The threshold value is a ratio of the current P_{PV} measure, it is determined to be superior to the steady-state power oscillations. Since the proposed dual-mode MPPT method presents the advantage of very small oscillation, the power threshold is fixed to 4%. The expression of the first scan-triggering criterion is:

$$\frac{|P_{PV}(k) - P_{PV}(k-4)|}{P_{PV}(k)} > 0.04 \tag{4}$$

However, in light of the above analysis, even if this criterion (4) is not met, partial shading variation might have occurred. This occurrence could be the result of critical partial shading conditions variation. Consequently, a critical case verification test is carried out, presenting the second scan-triggering criterion. A new indicator is introduced: the conductance $Q_{PV} = \frac{I_{PV}}{V_{PV}}$. The LMPP most to the left is characterized by a Q_{PV} relatively high because around left LMPP the current is maximal and the voltage is minimal. Therefore, the conductance of the current MPP is calculated. If Q_{MPP} exceeds the nominal value $Q_{nom} = \frac{I_{MPP,nom}}{V_{MPP,nom}}$, the current MPP is then positioned to the left. Hence, while the PV power has not changed, a scan is launched every T seconds. The periodic scan eliminates the critical cases without any considerable loss. In order to smooth the transitions from the dual-mode MPPT to the scanning process, the duty cycle is decreased from its current value to 0.1. After the scan, where the P-V curve is traced by varying the duty cycle from 0.1 to 0.8, the transition back to MPPT happen by decreasing the duty cycle down to its value at the GMPP. The smoothing blocks are presented in blue on the flowchart Fig. 4.

Validation and results

The proposed PV system (shown in Fig. 1) is evaluated in a Simulink/MATLAB environment for different shading patterns. In this paper, the PSCs scenario presented shows the occurrence of a critical case to verify the performance of the proposed approach. Then, the suggested method is compared to a similar approach only without the criterion Q which is the major inclusion in this algorithm. First, the scenario considered starts with the non-uniform irradiation conditions of pattern 1. Then at $t = 4s$, the change of PSCs occurs so the PV array becomes subjected to pattern 2. Finally at $t = 8s$, the shading pattern is set as pattern 3. All pattern curves and values are presented in Fig. 2, Fig. 3 and Table II respectively.

The performance of the PV system under the proposed scenario is shown in Fig. 5. From start to $t = 4s$, the PV array is subjected to pattern 1. The system is launched with a scan then dual-mode MPPT starts operating after finding the GMPP and its parameters. So, the algorithm operates at the GMPP with $P_{PV} = P_{GMPP} = 277W$. The proposed method guarantees minimum oscillations around the GMPP due to the choice of small step and the measurement accuracy of V_{PV}. Since there is no climatic change, the conductance is tested: $Q_{PV} = 0.076A/V$ is less than $Q_{nom} = 0.087A/V$, hence no periodic scan is needed.

At $t = 4s$, a change in irradiation to pattern 2 occurs which causes a high drop in power satisfying criterion (4). So, the scanning process is launched, and the new GMPP is found. Again, the proposed alternation between the scan and the dual-mode algorithm allows reaching the global maximum point quickly and staying around it with reduced fluctuations ($P_{PV} = 202W$). The shifting of modes of the dual-mode MPPT between 1 and 2 are shown in "mode" evolution figure, where $mode = 0$ indicates the curve tracing operation. However, the current GMPP conductance value is 0.28, consequently the periodic scanning process is activated. At $t = 6.5s$, the scan shows no irradiation variation. The PSCs vary from pattern 2 to pattern 3 at $t = 8s$. Since the criterion (4) is not met, this change can not be considered before the next scan. The search process at $t = 9s$ reveals the new PSC profile allowing the operation at the GMPP with power equal to $362W$.

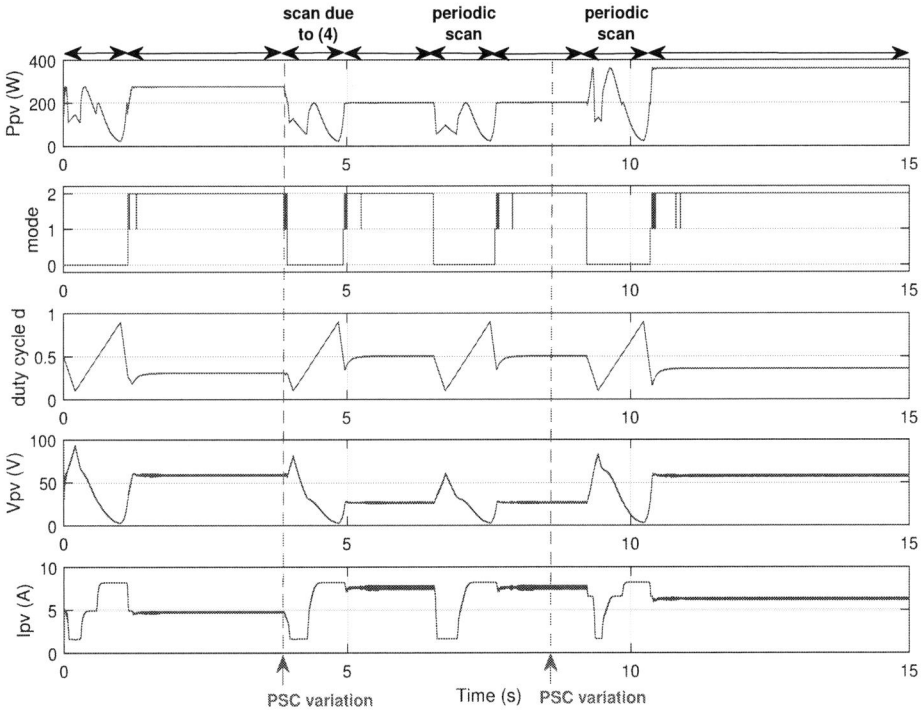

Fig. 5: The evolution of the system's outputs under varying PSC

The evolution of the duty cycle mainly controlling the converter also presented in Fig. 5 explicitly shows the variation of its value during the curve tracing and the smoothing transition phases between the two subprograms. Besides, the impact of this forced variation on the voltage and the current at the output of the PV array is displayed.

Moreover, in order to verify and quantify the performance of the proposed technique, it is compared to a similar method only eliminating the criterion Q originally used as a watchdog for the critical partial shading conditions. The two algorithms are tested under the scenario previously described highlighting a critical case where the variation from pattern 2 to pattern 3 can be undetected. Fig. 6 shows the variation of power of both methods. The proposed algorithm with Q criteria power evolution is the one detailed before along with the different measurements. Nevertheless, removing the critical PSCs watchdog, the variation at $t = 8s$ goes under the radar as expected. The algorithm continues to operate at the local maximum point positioned to the leftmost of the P-V curve with $P_{PV} = 202W$ although the new global maximum point is in the middle with $P_{PV} = 362W$.

This loss of power can be quantified, so the energy gathered during this simulation is presented in Fig. 7. From the start till $t = 6.6s$, the energy harvested is exactly the same while going through two P-V curve scans due to power variation. At $t = 6.6s$, an unnecessary scan causes a light drop in energy for the proposed method. Nonetheless, this energy loss is compensated from $t = 10.6s$ instant. The choice of irradiation variation instant and periodic scan duration surely impact the energy yield. This work presented an adapted time scale scenario similar to a lifelike irradiation variation profile. Furthermore, it quantified the power loss caused by an unnecessary scan and the one engendered by not tracking the global maximum power point. So depending on the application and PV placement, the user can decide which criteria to prioritize and adapt the parameters (timer's duration, Q_{nom} value, etc) accordingly.

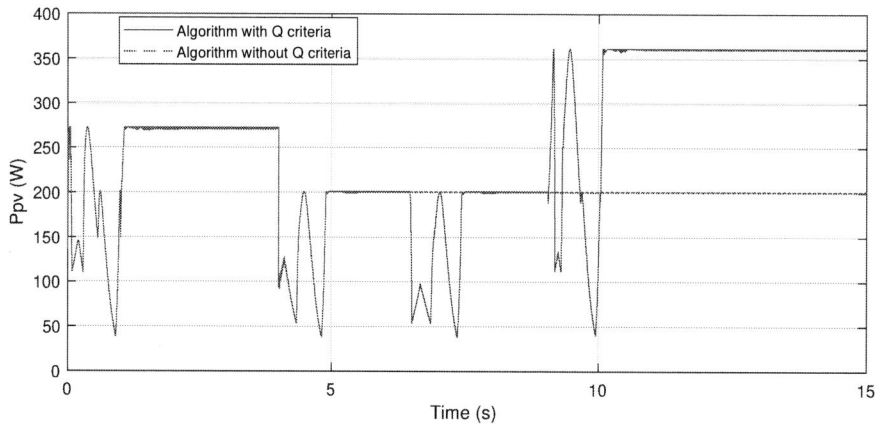

Fig. 6: The variation of power comparison for algorithms with/without Q criteria

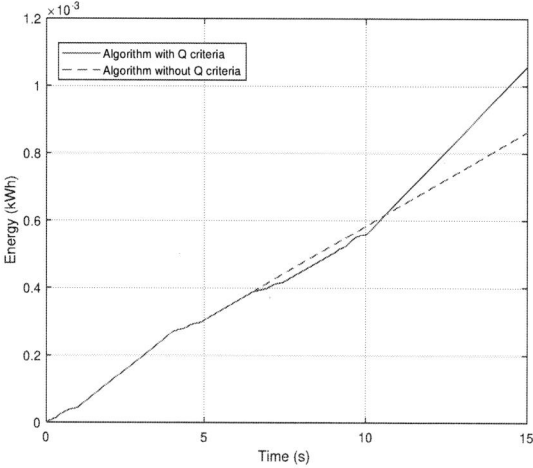

Fig. 7: The energy harvested comparison for algorithms with/without Q criteria

Conclusion

Optimizing the operational performance of GMPPT techniques during uniform or non-uniform irradiation variation has become a major challenge. PSC detection techniques are facing many obstacles. The proposed GMPPT method suggests a PSD while encountering the critical PSC variation cases. The critical PSCs are introduced in this paper and their occurrence is justified. Then, the proposed algorithm, controlling a DC/DC quadratic boost converter to extract maximum power from a PV array, is presented. It alternates non-regularly between two sub-programs: a novel MPPT algorithm and a rapid scanning procedure. The aperiodic scan-dual-mode method ensures operating at the GMPP even in critical PSC while reducing the transition to the scan as rarely as possible. The global PV system proved great performance in the newly introduced critical cases and notable superiority in all varying PSC, in a Simulink/MATLAB environment.

References

[1] Bai J, Cao Y, Hao Y, Zhang Z, Liu S, Cao F. Characteristic output of PV systems under partial shading or mismatch conditions. Solar Energy. 2015 Feb 1;112:41-54.

[2] Abdel-Salam M, El-Mohandes MT, El-Ghazaly M. An efficient tracking of MPP in PV systems using a newly-formulated P&O-MPPT method under varying irradiation levels. Journal of Electrical Engineering & Technology. 2020 Jan;15(1):501-13.

[3] Alik R, Jusoh A. An enhanced P&O checking algorithm MPPT for high tracking efficiency of partially shaded PV module. Solar Energy. 2018 Mar 15;163:570-80.

[4] Motahhir S, El Ghzizal A, Sebti S, Derouich A. Modeling of photovoltaic system with modified incremental conductance algorithm for fast changes of irradiance. International Journal of Photoenergy. 2018 Mar 13.

[5] Yatimi H, Ouberri Y, Aroudam E. Enhancement of power production of an autonomous PV system based on robust MPPT technique. Procedia Manufacturing. 2019 Jan 1;32:397-404.

[6] Farajdadian S, Hosseini SH. Design of an optimal fuzzy controller to obtain maximum power in solar power generation system. Solar Energy. 2019 Apr 1;182:161-78.

[7] Hamdi H, Regaya CB, Zaafouri A. Real-time study of a photovoltaic system with boost converter using the PSO-RBF neural network algorithms in a MyRio controller. Solar energy. 2019 May 1;183:1-6.

[8] Da Luz CM, Vicente EM, Tofoli FL. Experimental evaluation of global maximum power point techniques under partial shading conditions. Solar Energy. 2020 Jan 15;196:49-73.

[9] Belhachat F, Larbes C. Comprehensive review on global maximum power point tracking techniques for PV systems subjected to partial shading conditions. Solar Energy. 2019 May 1;183:476-500.

[10] Nadeem A, Hussain A. A comprehensive review of global maximum power point tracking algorithms for photovoltaic systems. Energy Systems. 2021 Aug 25:1-42.

[11] Pillai DS, Ram JP, Ghias AM, Mahmud MA, Rajasekar N. An accurate, shade detection-based hybrid maximum power point tracking approach for PV systems. IEEE Transactions on Power Electronics. 2019 Nov 13;35(6):6594-608.

[12] Al Abri W, Abri RA, Yousef H, Al-Hinai A. A Simple Method for Detecting Partial Shading in PV Systems. Energies. 2021 Jan;14(16):4938.

[13] Chandrasekaran K, Sankar S, Banumalar K. Partial shading detection for PV arrays in a maximum power tracking system using the sine-cosine algorithm. Energy for Sustainable Development. 2020 Apr 1;55:105-21.

[14] Ghasemi MA, Foroushani HM, Parniani M. Partial shading detection and smooth maximum power point tracking of PV arrays under PSC. IEEE Transactions on Power Electronics. 2015 Dec 1;31(9):6281-92.

[15] Ma J, Bi Z, Man KL, Yue Y, Smith JS. Automatic shading detection system for photovoltaic strings. In2018 International SoC Design Conference (ISOCC) 2018 Nov 12 (pp. 176-177). IEEE.

[16] Fadhel S, Diallo D, Delpha C, Migan A, Bahri I, Trabelsi M, Mimouni MF. Maximum power point analysis for partial shading detection and identification in photovoltaic systems. Energy Conversion and Management. 2020 Nov 15;224:113374.

[17] Ahmed J, Salam Z. An accurate method for MPPT to detect the partial shading occurrence in a PV system. IEEE transactions on industrial informatics. 2017 May 11;13(5):2151-61.

[18] Gosumbonggot J, Fujita G. Partial shading detection and global maximum power point tracking algorithm for photovoltaic with the variation of irradiation and temperature. Energies. 2019 Jan;12(2):202.

[19] Potnuru SR, Pattabiraman D, Ganesan SI, Chilakapati N. Positioning of PV panels for reduction in line losses and mismatch losses in PV array. Renewable Energy. 2015 Jun 1;78:264-75.

[20] Malathy S, Ramaprabha R. Comprehensive analysis on the role of array size and configuration on energy yield of photovoltaic systems under shaded conditions. Renewable and Sustainable Energy Reviews. 2015 Sep 1;49:672-9.

Novel Operation Mode of the Modular Multilevel Matrix Converter based on a Dimensioning Algorithm

Rebecca Dierks, and Axel Mertens
Leibniz University Hannover
Institute for Drive Systems and Power Electronics
Hannover, Germany
Email: rebecca.dierks@ial.uni-hannover.de
URL: http://www.ial.uni-hannover.de

Acknowledgments

This research was funded by the German Research Foundation (DFG) – Project 254417319.

Keywords

≪Modular Matrix Converter≫, ≪AC-to-AC Converter≫, ≪Converter Control≫

Abstract

The focus of the proposed operation mode of the modular multilevel matrix converter is to find the optimal trade-off between branch energy variation and branch currents. For this purpose, a weighting function is used to calculate the optimal circulating current parameters. This novel approach is compared to existing operation modes and validated on a low-voltage test bench.

Introduction

The modular multilevel matrix converter (M3C) shown in Fig. 1 (a) can directly convert a three-phase AC to another three-phase AC system without a DC voltage link. This type of converter with a modular branch design was first introduced by Glinka *et al.* [1] for a single phase. In later publications, this technology was also developed for two three-phase systems. Compared to conventional matrix converters, higher voltage classes can be achieved by the modular structure and output voltages with lower harmonics can be obtained due to smaller voltage steps.

A typical application of the M3C is the connection of a three-phase medium-voltage drive to the grid [2]. Furthermore, recent publications present the approach of connecting, for example, offshore wind farms via Low-Frequency-HVAC transmission (LF-HVAC transmission) [3]. In this case, the M3C is used as an onshore converter and thus large offshore converter platforms like those involved in HVDC transmission can be eliminated. In both the drive application and LF-HVAC transmission application, the M3C control must ensure that the system is also functional at low frequencies.

Similar to the DC-AC modular multilevel converters, a cascaded control structure with an outer branch energy control and an inner current control can be used [4] [5]. This has the advantage that the energies and currents with significantly different control time constants can be controlled decoupled from each other. So-called circulating currents can be used as degrees of freedom when controlling the branch energies, as these only circulate inside the M3C and do not influence the external variables. Beside the circulating current components that are responsible for controlling the M3C, which occur only in the transient state, additional circulating currents which exist also in the steady state can further be injected. In a simple operation (Normal Mode) of the M3C without steady-state circulating currents, high branch energy variations occur when one connected system is operated with frequencies close to zero or when both systems have the same frequency. High branch energy variation leads to a high installed module capacitance. For this reason, several operation modes have been introduced in the literature to reduce the branch energies. The operation modes of Korn *et. al* [6] introducing the Instantaneous Power Mode (IPM) and Kawamura *et. al* [7] introducing the mode Control III (Ctr3) deal with the injection of additional steady-state circulating currents to ensure the operation of the M3C even when one connected system is running at a low frequency. In addition, there are also several operation modes that deal with the situation where both connected systems have the same frequency and an additional injection of the star-point voltage can also be utilised as a degree of freedom [8] [9].

However, major disadvantages of these operation modes are that higher branch current losses occur due to the injection of additional circulating currents, and the operation modes do not consider a trade-off between branch energy variation and the resulting losses. This means that with a fixed module capacitor size, the branch energy variation often does not have to be reduced so drastically at lower frequencies and consequently not so high circulating currents have to be injected. For example in the publication of Engel et al. [10], the topology of the conventional Modular Multilevel Converter (M2C) is used to show how a trade-off between installed module capacitance and branch losses can be achieved when using a cost function containing branch energy variation and branch current at different modulation indices. In addition, as discussed in [11], the choice of the circulating current amplitudes should be made depending on the system frequencies. For example, the amplitudes of the circulating currents should be chosen relatively large if one of the connected systems operates at low frequencies in order to compensate the high branch energy variations. For increasing system frequencies, smaller circulating current amplitudes should be selected.

For this reason, the following operation mode is proposed, which is based on an analytical model determining the optimal weighting between installed module capacitances and occurring branch losses. When solving the optimisation problem, the optimal amplitudes of the circulating currents are calculated, which are dependent on the system frequencies. The new approach is derived with a generalised three-phase model with variable frequencies, so that the operation mode is suitable for both machine and LF-HVAC grid applications. Furthermore, the novel operation mode is validated on a low-voltage test bench.

Model Description of the Modular Multilevel Matrix Converter

This section gives a brief overview of the control structures used and the operation modes considered, which are derived from an analytical model.

Control Structure of the Modular Multilevel Matrix Converter

The M3C is controlled by the cascaded control system of Fig. 2, which is based on the approach of reference [4]. The branch energy control represents the outer control loop and is responsible for distributing the energy stored in the capacitors evenly between the branches. The current control setpoints resulting from the branch energy control are controlled in the inner current control loop. In the current control, the state space vector with

$$\mathbf{x}' = \begin{bmatrix} \mathbf{x} & v_{\text{st}} \end{bmatrix}^{\text{T}} = \begin{bmatrix} i_{\text{X}\alpha} & i_{\text{X}\beta} & i_{\text{Y}\alpha} & i_{\text{Y}\beta} & i_{\text{cir1}} & i_{\text{cir2}} & i_{\text{cir3}} & i_{\text{cir4}} & v_{\text{st}} \end{bmatrix}^{\text{T}} \tag{1}$$

represents the variables that need to be controlled. The vector $\mathbf{x}$ contains the system currents in alpha-beta coordinates of the system X with $i_{\text{X}\alpha}$, $i_{\text{X}\beta}$ and of the system Y with $i_{\text{Y}\alpha}$, $i_{\text{Y}\beta}$ and the four circulating currents $\mathbf{i}_{\text{cir}} = \begin{bmatrix} i_{\text{cir1}} & i_{\text{cir2}} & i_{\text{cir3}} & i_{\text{cir4}} \end{bmatrix}^{\text{T}}$. In addition, the state space vector $\mathbf{x}'$ is extended by the star-point voltage v_{st}, which also represents a degree of freedom in the control of the M3C. To implement different operation modes, additional circulating currents $\mathbf{i}_{\text{cir,add}}$ in

$$\mathbf{x}'_{\text{add}} = \begin{bmatrix} 0 & 0 & 0 & 0 & i_{\text{cir1,add}} & i_{\text{cir2,add}} & i_{\text{cir3,add}} & i_{\text{cir4,add}} & 0 \end{bmatrix}^{\text{T}} \tag{2}$$

are injected as offset circulating currents. The setpoints for the branch voltages $\mathbf{u}^* = \mathbf{v}_{\text{b}}^*$ with $v_{\text{b}ij}^*$ are calculated with the state space representation, as shown in Fig. 2. Index $i \in \{1, 2, 3\}$ stands for the phase of system X and $j \in \{1, 2, 3\}$ for the phase of system Y. A more detailed description of the control structure can be found in [4].

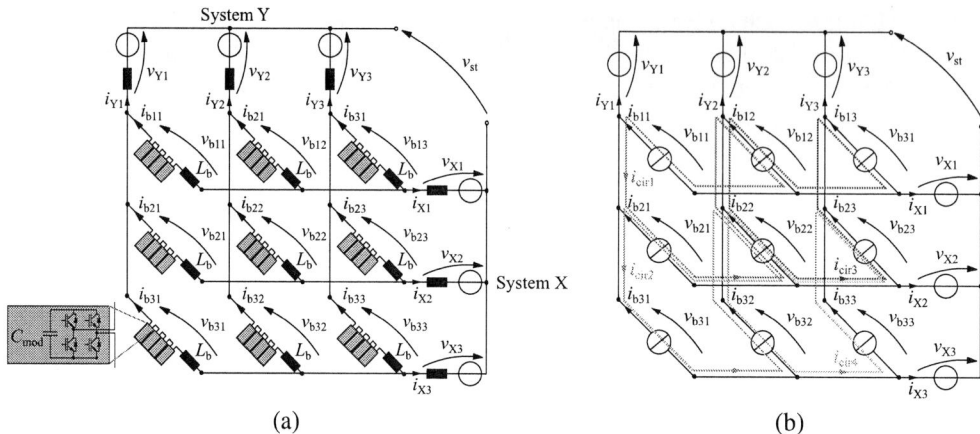

Fig. 1: Modular multilevel matrix converter: (a) configuration (b) simplified model

Analytical Derivation of the Operation Modes

For the derivation of the operation modes, an analytical model with a simplified equivalent circuit diagram (ECD) of the M3C from Fig. 1 (b) is considered. It is assumed that the current control reacts infinitely fast and thus the branch voltage source can be replaced by a branch current source and the branch inductance and branch losses can be neglected. Furthermore, system X and system Y are each represented by three voltage sources, which describe the voltages measured at the outputs of the converter. By using the node equations of the simplified ECD, the branch currents can be transformed with the matrix $\mathbf{T}_{ib}$ into

$$\mathbf{i}_b = \begin{bmatrix} i_{b11} & i_{b12} & i_{b13} & i_{b21} & i_{b22} & i_{b23} & i_{b31} & i_{b32} & i_{b33} \end{bmatrix}^T = \mathbf{T}_{ib} \cdot \begin{bmatrix} \mathbf{i}_x & \mathbf{i}_y & \mathbf{i}_{cir} \end{bmatrix}^T . \tag{3}$$

The used convention of the circulating currents $\mathbf{i}_{cir}$ can be taken from Fig. 1 (b). In addition, the three-phase system currents $\mathbf{i}_X = \begin{bmatrix} i_{X1} & i_{X2} & i_{X3} \end{bmatrix}^T$ and $\mathbf{i}_Y = \begin{bmatrix} i_{Y1} & i_{Y2} & i_{Y3} \end{bmatrix}^T$ are simplified as symmetrical currents

$$\mathbf{i}_X = \sqrt{2} \cdot I_X \cdot \begin{bmatrix} \cos(\theta_X - \varphi_X) \\ \cos(\theta_X - \frac{2\pi}{3} - \varphi_X) \\ \cos(\theta_X - \frac{4\pi}{3} - \varphi_X) \end{bmatrix} \text{ and } \mathbf{i}_Y = \sqrt{2} \cdot I_Y \cdot \begin{bmatrix} \cos(\theta_Y - \varphi_Y + \psi) \\ \cos(\theta_Y - \frac{2\pi}{3} - \varphi_Y + \psi) \\ \cos(\theta_Y - \frac{4\pi}{3} - \varphi_Y + \psi) \end{bmatrix} \tag{4}$$

with the phase differences φ_X and φ_Y between voltage and current and with the rotation angles $\theta_X = \omega_X \cdot t$ and $\theta_Y = \omega_Y \cdot t$. Furthermore, the angle ψ indicates the phase shift between system X and system Y. Similar to the description of the currents, the branch voltage vector

$$\mathbf{v}_b = \begin{bmatrix} v_{b11} & v_{b12} & v_{b13} & v_{b21} & v_{b22} & v_{b23} & v_{b31} & v_{b32} & v_{b33} \end{bmatrix}^T = \mathbf{T}_{vb} \cdot \begin{bmatrix} \mathbf{v}_x & \mathbf{v}_y & v_{st} \end{bmatrix}^T \tag{5}$$

can be represented with the help of the mesh equation of the simplified ECD as a function of the system voltages $\mathbf{v}_X = \begin{bmatrix} v_{X1} & v_{X2} & v_{X3} \end{bmatrix}^T$, $\mathbf{v}_Y = \begin{bmatrix} v_{Y1} & v_{Y2} & v_{Y3} \end{bmatrix}^T$ and the star-point voltage v_{st}. The matrices $\mathbf{T}_{ib}$ and $\mathbf{T}_{vb}$ are shown in (6).

$$\mathbf{T}_{ib} = \frac{1}{9} \cdot \begin{bmatrix} 3 & 0 & 0 & -3 & 0 & 0 & 16 & 8 & 9 & 4 \\ 3 & 0 & 0 & 0 & -3 & 0 & -8 & -4 & 8 & 4 \\ 3 & 0 & 0 & 0 & 0 & -3 & -8 & -4 & -16 & -8 \\ 0 & 3 & 0 & -3 & 0 & 0 & -8 & 8 & -4 & 4 \\ 0 & 3 & 0 & 0 & -3 & 0 & 4 & -4 & -4 & 4 \\ 0 & 3 & 0 & 0 & 0 & -3 & 4 & -4 & 8 & -8 \\ 0 & 0 & 3 & -3 & 0 & 0 & -8 & -16 & -4 & -8 \\ 0 & 0 & 3 & 0 & -3 & 0 & 4 & 8 & -4 & -8 \\ 0 & 0 & 3 & 0 & 0 & -3 & 4 & 8 & 8 & 16 \end{bmatrix}, \mathbf{T}_{vb} = \begin{bmatrix} 1 & 0 & 0 & -1 & 0 & 0 & 1 \\ 1 & 0 & 0 & 0 & -1 & 0 & 1 \\ 1 & 0 & 0 & 0 & 0 & -1 & 1 \\ 0 & 1 & 0 & -1 & 0 & 0 & 1 \\ 0 & 1 & 0 & 0 & -1 & 0 & 1 \\ 0 & 1 & 0 & 0 & 0 & -1 & 1 \\ 0 & 0 & 1 & -1 & 0 & 0 & 1 \\ 0 & 0 & 1 & 0 & -1 & 0 & 1 \\ 0 & 0 & 1 & 0 & 0 & -1 & 1 \end{bmatrix} \tag{6}$$

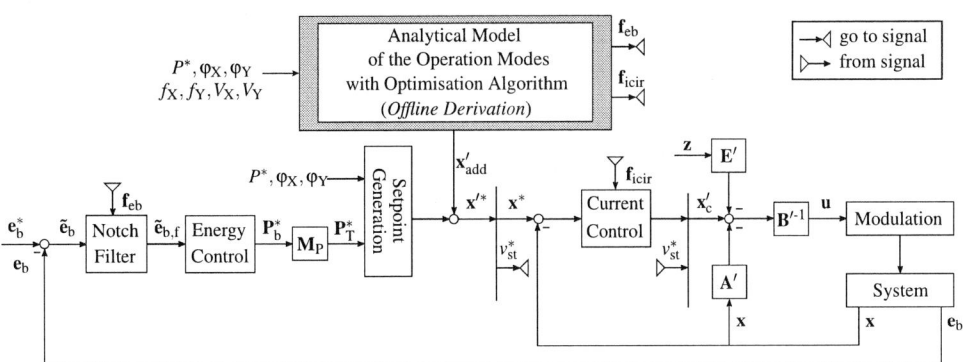

Fig. 2: Control structure of the M3C (see [4])

. For the system voltages

$$\mathbf{v}_X = \sqrt{2} \cdot V_X \cdot \begin{bmatrix} \cos(\theta_X) \\ \cos(\theta_X - \frac{2\pi}{3}) \\ \cos(\theta_X - \frac{4\pi}{3}) \end{bmatrix} \quad \text{and} \quad \mathbf{v}_Y = \sqrt{2} \cdot V_Y \cdot \begin{bmatrix} \cos(\theta_Y + \psi) \\ \cos(\theta_Y - \frac{2\pi}{3} + \psi) \\ \cos(\theta_Y - \frac{4\pi}{3} + \psi) \end{bmatrix} \quad (7)$$

symmetrical waveforms are also assumed. The branch powers $\mathbf{p}_b = \mathbf{i}_b \odot \mathbf{v}_b$ are then obtained by element-wise multiplication of branch currents and branch voltages. In the following, it should additionally be noted that the branch energies $\mathbf{e}_b$ are calculated by integrating the branch powers. This means that the frequencies of the harmonic power components appear in the denominator of the corresponding branch energy components. The following familiar operation modes with different definitions of the circulating currents $\mathbf{i}_{cir}$ are used for comparison purposes in this elaboration:

• Instantaneous Power Mode (IPM) [6]:

$$\mathbf{i}_{cir} = -\frac{1}{4P^*} \cdot \begin{bmatrix} (v_{Y1} \cdot i_{Y1} - v_{Y2} \cdot i_{Y2}) \cdot (i_{X1} - i_{X2}) \\ (v_{Y1} \cdot i_{Y1} - v_{Y2} \cdot i_{Y2}) \cdot (i_{X2} - i_{X3}) \\ (v_{Y2} \cdot i_{Y2} - v_{Y3} \cdot i_{Y3}) \cdot (i_{X1} - i_{X2}) \\ (v_{Y2} \cdot i_{Y2} - v_{Y3} \cdot i_{Y3}) \cdot (i_{X2} - i_{X3}) \end{bmatrix} \quad (8)$$

• Control III (Ctr3) [7]:

$$\mathbf{i}_{cir} = -\frac{1}{4P^*} \cdot \begin{bmatrix} \frac{3}{8} \cdot i_{aa} - \frac{\sqrt{3}}{8} \cdot i_{ab} - \frac{\sqrt{3}}{8} \cdot i_{ba} + \frac{1}{8} \cdot i_{bb} \\ \frac{\sqrt{3}}{4} \cdot i_{ab} - \frac{1}{4} \cdot i_{bb} \\ \frac{\sqrt{3}}{4} \cdot i_{ba} - \frac{1}{4} \cdot i_{bb} \\ \frac{1}{2} \cdot i_{bb} \end{bmatrix} \quad \text{with} \quad \begin{bmatrix} i_{aa} \\ i_{ab} \\ i_{ba} \\ i_{bb} \end{bmatrix} = \frac{I_Y \cdot V_Y}{\sqrt{2} \cdot V_X} \cdot \begin{bmatrix} \cos(\theta_3) \\ \sin(\theta_3) \\ -\sin(\theta_3) \\ \cos(\theta_3) \end{bmatrix} \quad (9)$$

with $\theta_3 = \theta_X + 2 \cdot (\theta_Y + \psi) - \phi_Y$.

It should be noted that the conventions of the circulating currents are different in the references and for the purposes of this paper have been transformed into the configuration shown in Fig. 1 (b). In addition, the star-point voltage of the operation modes considered is assumed to be zero, $v_{st} = 0$. For each operation mode, Table I shows the frequency components resulting from the analytical model for the circulating currents, branch currents and branch energies.

Table I: Frequency components of the operation modes

	NM	IPM	Ctr3
$\mathbf{i}_{cir}$	-	$f_X + 2f_Y, f_X - 2f_Y$	$f_X + 2f_Y$
$\mathbf{i}_b$	f_X, f_Y	$f_X, f_Y, f_X + 2f_Y, f_X - 2f_Y$	$f_X, f_Y, f_X + 2f_Y$
$\mathbf{e}_b$	$2f_X, 2f_Y, f_X \pm f_Y$	$2f_X, f_X \pm 3f_Y, f_X \pm f_Y, 2f_X \pm 2f_Y$	$2f_X, f_X + f_Y, 2f_X + 2f_Y, f_X + 3f_Y$

Table II: Parameters of the M3C analytical model ($k = 100$, only first row is used) and the MMC test bench prototype ($k = 1$)

System voltages	Power setpoint	Modules per branch	System frequencies	Reference values
$V_{X,s}; V_{Y,s}$	P_o^*	n_{mpb}	$f_X; f_Y$	$I_{b,ref}; \Delta e_{b,ref}$
$k \cdot \frac{80}{\sqrt{2}}$ V	$k^2 \cdot 1800\,\mathrm{W}$	exp.: 6; analyt.: 16	50 Hz, 10 Hz	$k \cdot \frac{15}{\sqrt{2}}$ A; $k^2 \cdot 7$ J

Branch inductance	Module capacitance	System X filter	System Y impedance	System Y filter
L_b	$C_{mod,max}$	$L_{X,f}$	$L_Y; R_Y$	$L_{X,f}$
350 µH	exp.: 324 µF; analyt.: 992 µF	1.7 mH	165 µH; 5 mΩ	3.13 mH

If it is assumed that f_X remains constant at 50 Hz and f_Y varies from 0 to 50 Hz, the branch energy components of the Normal Mode (NM), oscillating with $2f_Y$, become very large at frequencies close to zero. For this reason, different frequency components of the circulating currents are injected in the IPM and Ctr3, so that the frequency components of $2f_Y$ are eliminated in the branch energies. To investigate this for several operating points, the branch energy variation $\Delta e_b = \max(e_{bij}) - \min(e_{bij})$ and the RMS branch current $I_b = I_{bij}$ are shown in Fig. 3 (a) for the converter parameters ($k = 100$) of Table II, depending on the variable frequency f_Y. In this case, the analytical model is to represent a medium-voltage converter with 3300 V IGBT modules (Infineon - FZ1400R33HE4) and module capacitors with a capacitance of $C_{mod} = C_{mod,max} = 992$µF. Based on the components used, the limit values of the module voltages are set to $v_{mod,max} = 3150$ V and $v_{mod,min} = 1050$ V. It becomes obvious in the plot of Δe_b that the NM, in contrast to the IPM and Ctr3, is not suitable for low frequencies of f_Y. However, the IPM and Ctr3 have the disadvantage that the RMS value of the branch currents, as an indicator of the branch losses, is much larger due to the injection of the circulating currents. For this reason, it will be investigated in the following section which circulating currents have to be injected to achieve the best trade-off between branch energy variations and branch losses.

Novel Operation Mode with a Dimensioning Optimisation Algorithm

To describe the trade-off between the branch current and the branch energy variation, the following function ξ_{opt} is introduced:

$$\xi_{opt}(\kappa, \mathbf{f}_\kappa, f_Y) = \frac{1}{2} \cdot \left(w_{eb} \cdot \frac{\Delta e_b(\kappa, \mathbf{f}_\kappa, f_Y)}{\Delta e_{b,ref}} + w_{ib} \cdot \frac{I_b(\kappa, \mathbf{f}_\kappa, f_Y)}{I_{b,ref}} \right) \,. \tag{10}$$

with $\kappa = \begin{bmatrix} \kappa_1 & \kappa_2 & \cdots \end{bmatrix}^T$ and $\mathbf{f}_\kappa = \begin{bmatrix} f_{\kappa 1} & f_{\kappa 2} & \cdots \end{bmatrix}^T$. The dimension of the vectors depends on how many frequency components are used for the circulating currents. In the following, it is assumed that the circulating currents have two frequency components, like for Ctr3 and IPM, so that the calculation of the dimensioning algorithm with the analytical model does not require excessively long time. Both the branch energy variation $\Delta e_b(\kappa, \mathbf{f}_\kappa, f_Y)$ and the RMS value $I_b(\kappa, \mathbf{f}_\kappa, f_Y)$ in Equation (10) are dependent on the circulating currents, which are defined as follows

$$\mathbf{i}_{cir\kappa} = \sqrt{2} \cdot I_{cir\kappa} \cdot \begin{bmatrix} \kappa_1 \cdot \cos(\theta_{\kappa 1} + \varphi_{\kappa 1,1}) + \kappa_2 \cdot \cos(\theta_{\kappa 2} + \varphi_{\kappa 2,1}) \\ \kappa_1 \cdot \cos(\theta_{\kappa 1} + \varphi_{\kappa 1,2}) + \kappa_2 \cdot \cos(\theta_{\kappa 2} + \varphi_{\kappa 2,2}) \\ \kappa_1 \cdot \cos(\theta_{\kappa 1} + \varphi_{\kappa 1,3}) + \kappa_2 \cdot \cos(\theta_{\kappa 2} + \varphi_{\kappa 2,3}) \\ \kappa_1 \cdot \cos(\theta_{\kappa 1} + \varphi_{\kappa 1,4}) + \kappa_2 \cdot \cos(\theta_{\kappa 2} + \varphi_{\kappa 2,4}) \end{bmatrix}$$

$$\text{with} \quad \varphi_{\kappa 1} = \begin{bmatrix} \varphi_{\kappa 1,1} & \varphi_{\kappa 1,2} & \varphi_{\kappa 1,3} & \varphi_{\kappa 1,4} \end{bmatrix}^T \quad \text{and} \quad \varphi_{\kappa 2} = \begin{bmatrix} \varphi_{\kappa 2,1} & \varphi_{\kappa 2,2} & \varphi_{\kappa 2,3} & \varphi_{\kappa 2,4} \end{bmatrix}^T \tag{11}$$

with $\theta_{\kappa 1} = 2\pi \cdot f_{\kappa 1} \cdot t$ and $\theta_{\kappa 2} = 2\pi \cdot f_{\kappa 2} \cdot t$. Due to the fact that the branch energies and the branch currents have different units and scalings, they are normalised with the reference values $\Delta e_{b,ref}$ and $I_{b,ref}$, which represent the specified maximum values of the converter components. The maximum value $I_{b,ref} = I_{b,max}$ of the RMS branch current is determined with the help of the data sheet of the semiconductor used in the modules (mostly IGBTs) and should be below the defined maximum value of the continuous DC collector current. In order to limit the range of the circulating current RMS value, it is assumed that

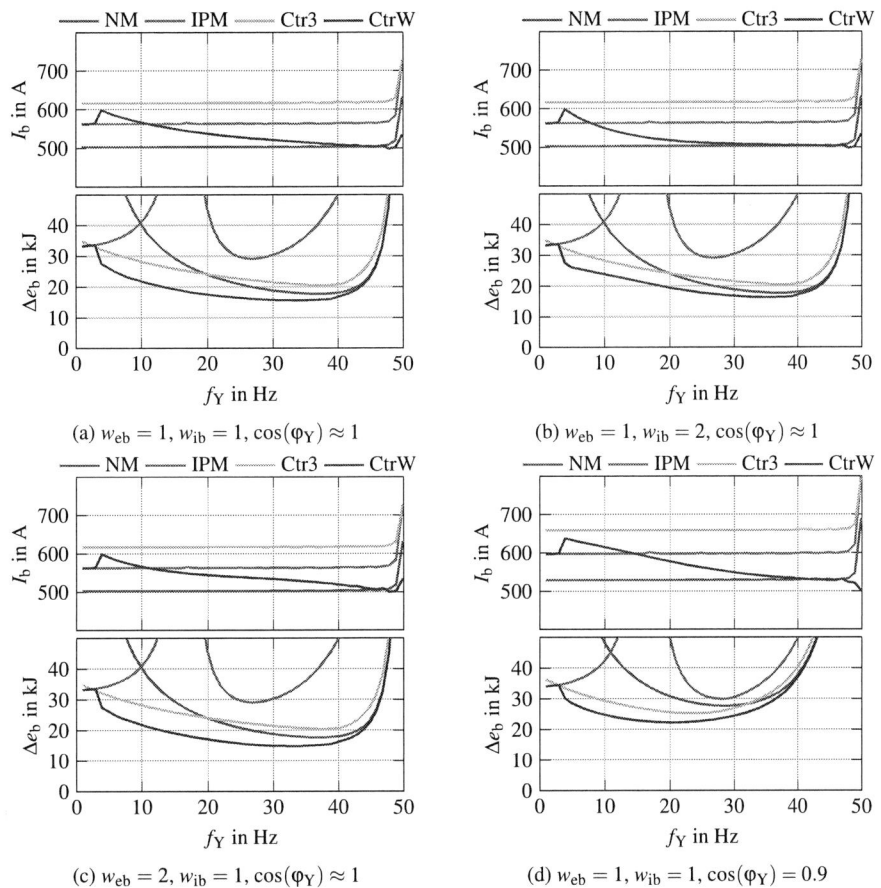

Fig. 3: Comparision of the operation modes regarding the branch energy variation and the branch current RMS value at different frequencies of f_Y with different weighting factors (a) - (c) and a different phase angle φ_Y (d) with $\psi = \varphi_X = 0$.

$\kappa_1 \cdot I_{cir\kappa}$ or $\kappa_2 \cdot I_{cir\kappa}$ does not exceed the reference value of the branch current $I_{b,ref}$. For this reason, $I_{cir\kappa}$ is chosen equal to the reference value of the branch currents $I_{cir\kappa} = I_{b,ref}$ and κ_1 and κ_2 vary from 0 to 1. The reference value of the branch energy variation

$$\Delta e_{b,ref} = \Delta e_{b,max} = \frac{1}{2} \cdot C_{mod,max} \cdot n_{mod} \cdot \left(v^2_{mod,max} - v^2_{mod,min}\right) \tag{12}$$

is calculated using the maximum desired module capacitance $C_{mod,max}$, the number of modules per branch n_{mod}, the maximum $v_{mod,max}$ and minimum value $v_{mod,min}$ of the module voltages. In addition, the weighting factors w_{eb} and w_{ib} are introduced so that some variability in the trade-off function ξ_{opt} is still possible. For each operating point of f_Y, the minimum is calculated from the weighting function for circulating currents with variable frequencies and amplitudes: $\min\left(\xi_{opt}(\kappa, \mathbf{f}_\kappa, f_Y)\right) \longrightarrow \kappa_{opt}, \mathbf{f}_{\kappa,opt}$. For each operating point, an optimal set of parameters $(\kappa_{opt}, \mathbf{f}_{\kappa,opt})|_{f_Y}$ is then obtained for the circulating currents.

As an example, in Fig. 4 (a), the weighting function ξ_{opt} is shown for an operating point of $f_Y = 10\,\text{Hz}$ when only one parameter κ_1 is active. In the zoomed-in plots shown alongside, it becomes visible that especially at the frequencies $f_{\kappa 1} = 30\,\text{Hz}$, $f_{\kappa 1} = 70\,\text{Hz}$, $f_{\kappa 1} = 90\,\text{Hz}$ the weighting function is comparatively small. The minimum of the weighting function is at the operating point for the operating point of $f_Y = 10\,\text{Hz}$ is located at the parameter set $(f_{\kappa 1,opt} = 70\,\text{Hz}, \kappa_{1,opt} = 0.6)|_{10\,\text{Hz}}$. In Fig. 4 (d), the calculated optimal parameters shown for several operating points. With the dotted plot of $\kappa_{1,opt}$ and $\kappa_{2,opt}$ the calculated optimal parameters at which ξ_{opt} reaches a minimum for different operating points are shown. To avoid having to examine an infinite number of operating points, the circulating current frequencies $f_{\kappa 1}$

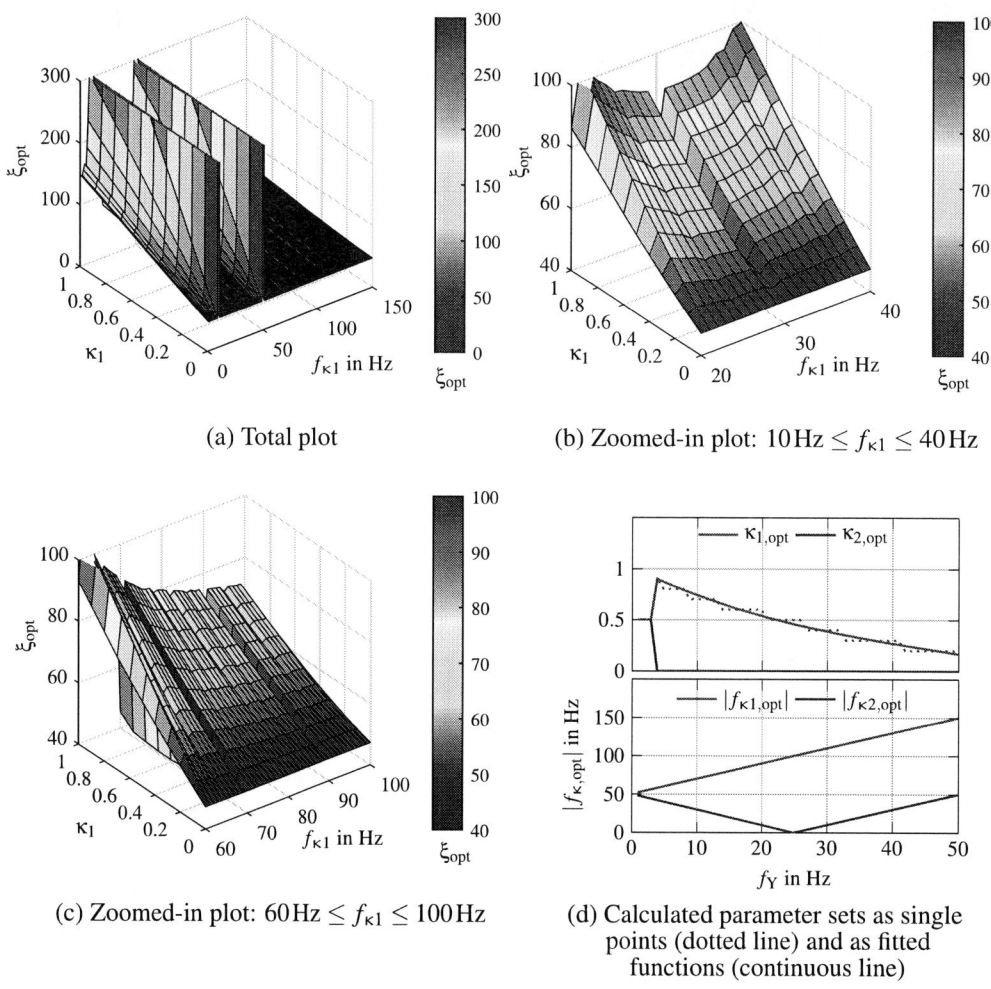

(a) Total plot

(b) Zoomed-in plot: $10\,\mathrm{Hz} \leq f_{\kappa1} \leq 40\,\mathrm{Hz}$

(c) Zoomed-in plot: $60\,\mathrm{Hz} \leq f_{\kappa1} \leq 100\,\mathrm{Hz}$

(d) Calculated parameter sets as single points (dotted line) and as fitted functions (continuous line)

Fig. 4: (a) - (c): Weighting function ξ_{opt} surface plots at $f_\mathrm{Y} = 10\,\mathrm{Hz}$ depending on κ and $\mathbf{f}_\kappa$ and (d) the optimal parameter sets for $f_\mathrm{Y} = 1\,\mathrm{Hz}$ to $50\,\mathrm{Hz}$. Parameters: $\psi = \varphi_\mathrm{X} = \varphi_\mathrm{Y} = 0$, $w_{\mathrm{ib}} = w_{\mathrm{eb}} = 1$. Minimum of ξ_{opt} for $f_\mathrm{Y} = 10\,\mathrm{Hz}$: $(f_{\kappa1,\mathrm{opt}} = 70\,\mathrm{Hz}, f_{\kappa2,\mathrm{opt}} = 30\,\mathrm{Hz}, \kappa_{1,\mathrm{opt}} = 0.6, \kappa_{2,\mathrm{opt}} = 0.0)|_{10\,\mathrm{Hz}}$.

and $f_{\kappa2}$ are analysed in the interval of $1\,\mathrm{Hz} : 1\,\mathrm{Hz} : 3 \cdot f_\mathrm{X}$ and amplitude factors κ_1 and κ_2 in the interval of $0 : 0.1 : 1$ for each operating point $1\,\mathrm{Hz} : 1\,\mathrm{Hz} : 50\,\mathrm{Hz}$ of the system frequency f_Y. To ensure that the points in between are also taken into account, the points are approximated by third-degree polynomials

$$\kappa_{1,\mathrm{fit}}(f_\mathrm{Y}) = q_{13} \cdot f_\mathrm{Y}^3 + q_{12} \cdot f_\mathrm{Y}^2 + q_{11} \cdot f_\mathrm{Y} + q_{10} \text{ with } \mathbf{q}_1 = \begin{bmatrix} q_{10} & q_{11} & q_{12} & q_{13} \end{bmatrix}^\mathrm{T}, \tag{13}$$

$$\kappa_{2,\mathrm{fit}}(f_\mathrm{Y}) = q_{23} \cdot f_\mathrm{Y}^3 + q_{22} \cdot f_\mathrm{Y}^2 + q_{21} \cdot f_\mathrm{Y} + q_{20} \text{ with } \mathbf{q}_2 = \begin{bmatrix} q_{20} & q_{21} & q_{22} & q_{23} \end{bmatrix}^\mathrm{T}. \tag{14}$$

In the designed optimisation algorithm, the polyfit() function of MATLAB is used to execute this regression and provide the coefficients $\mathbf{q}_1$ and $\mathbf{q}_2$. Instead of transferring an infinite number of points as a look-up table into the model, the fitted functions $\kappa_{1,\mathrm{fit}}$ are $\kappa_{2,\mathrm{fit}}$ are included in the calculation of the circulating currents in Equation (11). Furthermore, the minima of ξ_{opt} is always at $f_{\kappa1} = f_\mathrm{X} + 2f_\mathrm{Y}$ and $f_{\kappa2} = f_\mathrm{X} - 2f_\mathrm{Y}$ like shown in Fig. 4 (d). Although κ_2 and $f_{\kappa2}$ are only required at low frequencies. With this conclusions, the computational time of the optimisation algorithm can be reduced, as $f_{\kappa1}$ and $f_{\kappa2}$ can be determined by the system frequencies. Additionally, the phase shifts $\varphi_{\kappa1}$ and $\varphi_{\kappa2}$ are also a solution of the optimisation algorithm without dependence on the frequency f_Y:

- Case 1 with $\varphi_{\kappa1} = \left[-\varphi_X - \varphi_Y \quad -\frac{2\pi}{3} - \varphi_X - \varphi_Y \quad -\frac{4\pi}{3} - \varphi_X - \varphi_Y \quad -\varphi_X - \varphi_Y\right]^T$ and
  $\varphi_{\kappa2} = \left[\frac{\pi}{3} - \varphi_X + \varphi_Y \quad -\frac{\pi}{3} - \varphi_X + \varphi_Y \quad -\frac{\pi}{3} - \varphi_X + \varphi_Y \quad \pi - \varphi_X + \varphi_Y\right]^T$ if κ_1 and κ_2 are active,
- Case 2 with $\varphi_{\kappa1} = \left[-\varphi_Y \quad -\frac{2\pi}{3} - \varphi_Y \quad -\frac{4\pi}{3} - \varphi_Y \quad -\varphi_Y\right]^T$ if only κ_1 is active.

By including the adjusted circulating currents $\mathbf{i}_{cir\kappa}$ in the analytical equations from the previous section, the branch energy variation and RMS branch current of the novel operation mode with the weighting function, named Control W (CtrW), can be displayed for each operating point. In this publication, only the influences of the circulating currents as degrees of freedom are investigated and the star-point voltage ($v_{st} = 0$) is not considered, which as in reference [9] are used to compensate the significant high branch energy variation for $f_Y \approx f_X$. Fig. 3 shows in purple the plot of CtrW for the listed (converter) parameters of Table II. It can be seen that both the branch energy variation and the branch current can be reduced in contrast to the Ctr3 and thus a better trade-off can be found for frequencies below 25 Hz. For frequencies above 25 Hz, the CtrW approximates the NM, since the additional injection of circulating currents is no longer necessary. The analytical results show that the CtrW has advantages especially in the lower frequency ranges of f_Y, because a suitable trade-off between branch energy variation and branch currents can be found.

Validation of the Novel Operation Mode

For the validation of the novel operation modes, the control structure from Fig. 2 is used. The additional setpoints of the circulating currents are equal to the analytically calculated circulating currents from the last sections: $\mathbf{i}_{cir,add} = \mathbf{i}_{cir\kappa}$ of (11). This means that in steady state, the setpoints of the circulating currents $\mathbf{i}_{cir}^*$ correspond to the analytically calculated values $\mathbf{i}_{cir,add}$. In order to better compare the experimental results with the medium-voltage analytical model, both currents and voltages were scaled down by a factor of $k = 100$ for the measurements, so that the same parameter functions $\kappa_{1,fit}(f_Y)$ and $\kappa_{2,fit}(f_Y)$ of the CtrW can be implemented in the control system of the test bench for various operating points of f_Y (Table II). In Fig. 5 (a) the low-voltage test bench of the M3C is shown and in Fig. 5 (b) the implementation of the control structure is visualised. The voltage of system X is provided by the 400 V/50 Hz grid and transformers. On the other hand, system Y is supplied via a microgrid from the company Triphase, so that a variable frequency f_Y is possible. In the M3C, 120 V MOSFETs (Infineon - IPB036N12N3-G) are integrated in the 54 modules (nine branches, each six installed modules). The limitation of the module voltage is set to $v_{mod,max} = 90$ V, $v_{mod,min} = 30$ V and the maximum required module capacitance to $C_{mod,max} = 324 \mu F$. Here, the actual installed module capacitance is not taken into account, as this was designed for relatively large branch energy variation.

To verify that the control works in principle, Fig. 6 shows the active switching ($t = 150$ ms) for the operating point $f_Y = 10$ Hz from NM to CtrW. In addition to the system voltages $\mathbf{v}_X$, $\mathbf{v}_Y$ measured behind the filter inductances ($L_{X,f}$, $L_{X,f}$), the branch current i_{b11} and the branch energy control deviation $\tilde{e}_{b11}$ (defined in Fig. 2) are also displayed. It can be seen that the branch energy control is working and that, as shown in Fig. 3 (a) at $f_Y = 10$ Hz, the branch energy variation of the CtrW becomes smaller and the branch current larger compared to the NM. Moreover in Fig. 7, the measured circulating currents are shown. From time $t = 150$ ms onwards, additional circulating currents are injected by the CtrW, in order to reduce the branch energy variations. This results in increased branch currents and thus increased branch losses in contrast to the NM.

Finally, the analytical model is to be validated using the experimental steady state waveforms and compared with the analytical results. In Fig. 8 the branch currents and branch energies of the analytical model and of the test bench are shown. The parameter Δt_{steady} represents the time until the steady state of the measured values are reached. The phase shift between the systems ψ, which is not equal to zero for the measurements, is determined with the help of the time shift of both angles θ_X and θ_Y. In order to better compare the waveforms in Fig. 8, this resulting phase angle ψ was also included in the calculations of the analytical waveforms. Except for the switching harmonics and the parasitic influences on the test bench, the analytical and experimental waveforms match in amplitude and frequency components. In summary, the experimental results show that the dimensioning algorithm based on the analytical model also works for the test bench.

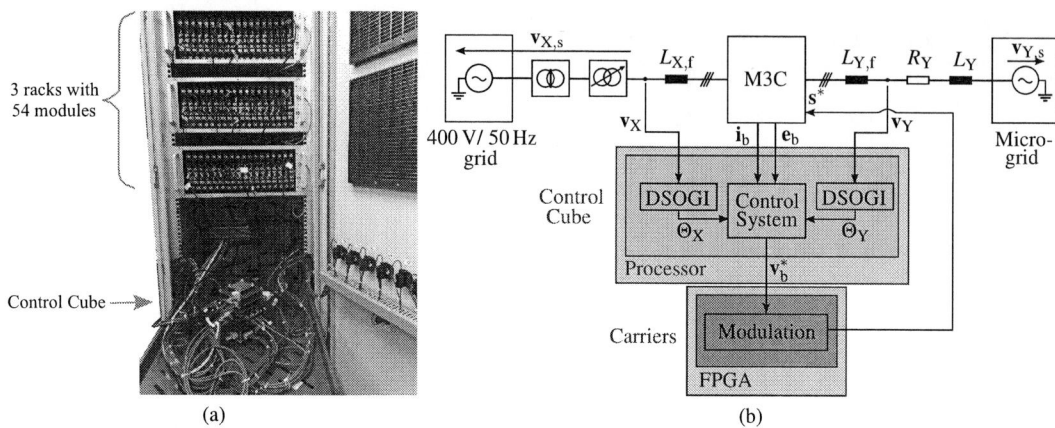

Fig. 5: (a) Low-voltage test bench of the M3C and (b) test bench set-up schematic.

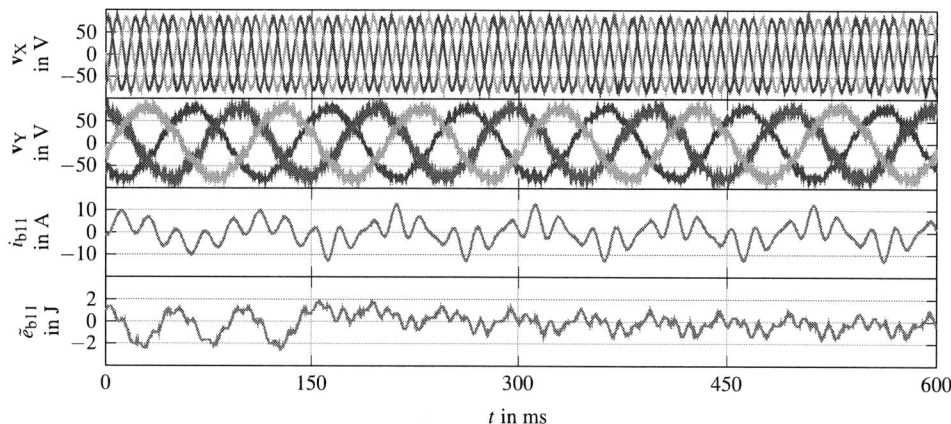

Fig. 6: Measured waveforms of the system voltages, branch current and branch energy during an active switchover from NM to CtrW at $t = 150\,\text{ms}$ for $f_Y = 10\,\text{Hz}$. Parameters: $\varphi_X = \varphi_Y = 0$, $w_{ib} = w_{eb} = 1$. Minimum of ξ_{opt} for $f_Y = 10\,\text{Hz}$: $(f_{\kappa1,opt} = 70\,\text{Hz}, f_{\kappa2,opt} = 30\,\text{Hz}, \kappa_{1,opt} = 0.6, \kappa_{2,opt} = 0.0)|_{10\text{Hz}}$.

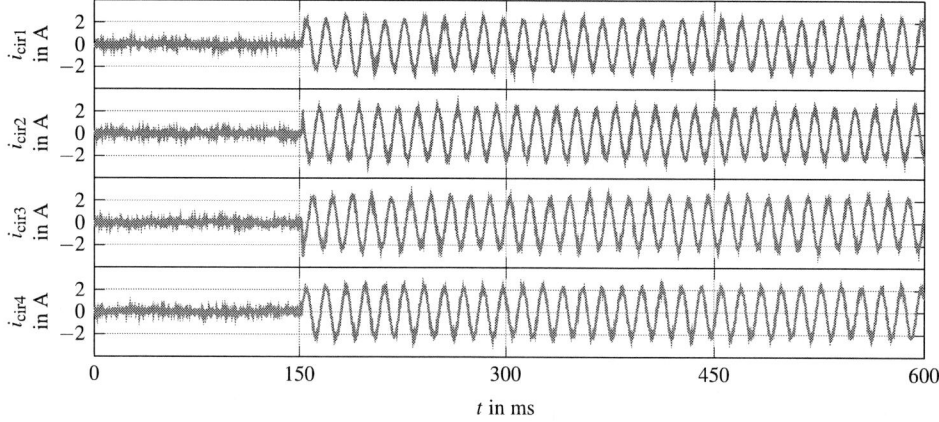

Fig. 7: Measured waveforms of the circulating currents during an active switchover from NM to CtrW at $t = 150\,\text{ms}$ for $f_Y = 10\,\text{Hz}$. Parameters: $\varphi_X = \varphi_Y = 0$, $w_{ib} = w_{eb} = 1$. Minimum of ξ_{opt} for $f_Y = 10\,\text{Hz}$: $(f_{\kappa1,opt} = 70\,\text{Hz}, f_{\kappa2,opt} = 30\,\text{Hz}, \kappa_{1,opt} = 0.6, \kappa_{2,opt} = 0.0)|_{10\text{Hz}}$.

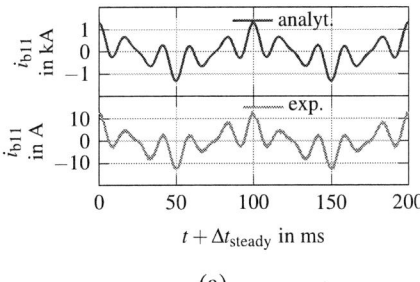
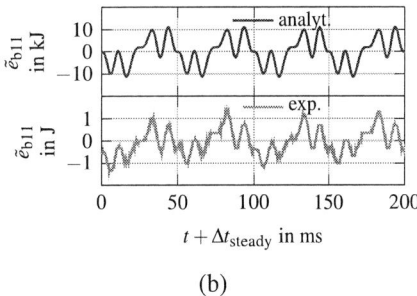

(a)   (b)

Fig. 8: Comparison of the calculated (analyt.) and measured waveforms (exp.) of (a) the branch currents and (b) branch energies of the CtrW at $f_Y = 10\,\text{Hz}$. Parameters: $\varphi_X = \varphi_Y = 0$, $w_{ib} = w_{eb} = 1$. Minimum of ξ_{opt} for $f_Y = 10\,\text{Hz}$: $(f_{\kappa1,opt} = 70\,\text{Hz}, f_{\kappa2,opt} = 30\,\text{Hz}, \kappa_{1,opt} = 0.6, \kappa_{2,opt} = 0.0)|_{10\,\text{Hz}}$.

Conclusion

Both the results of the analytical model and the experimental results have shown that an optimal weighting between installed module capacitances and losses in the branch can be found in contrast to the other operation modes considered, especially at lower frequencies, by applying the new operation mode (Control W). This is particularly interesting in machine applications or LF-HVAC where the components such as module capacitances and IGBT modules are predefined and the branch losses are to be reduced at lower frequencies. Therefore, it is necessary to have a more detailed knowledge of the converter's parameters and limitations in order to achieve the best optimisation result.

References

[1] M. Glinka and R. Marquardt, "A New AC/AC-Multilevel Converter Family Applied to a Single-Phase Converter," *The Fifth International Conference on Power Electronics and Drive Systems, 2003*, vol. 1, pp. 16–23, 2003.

[2] F. Kammerer et. al, "Operating Performance of the Modular Multilevel Matrix Converter in Drive Applications," *PCIM Europe 2015; International Exhibition and Conference for Power Electronics, Intelligent Motion, Renewable Energy and Energy Management; Proceedings of*, no. May, pp. 19–21, 2015.

[3] Y. Tang et. al, "Offshore low frequency AC transmission with back-to-back modular multilevel converter (MMC)," *IET Seminar Digest*, vol. 2015, no. CP654, pp. 1–8, 2015.

[4] D. Karwatzki et. al, "Generalized Control Approach for a Class of Modular Multilevel Converter Topologies," *IEEE Transactions on Power Electronics*, vol. 33, no. 4, pp. 2888–2900, 2018.

[5] F. Kammerer et. al, "A novel cascaded vector control scheme for the Modular Multilevel Matrix Converter," *IECON Proceedings (Industrial Electronics Conference)*, pp. 1097–1102, 2011.

[6] A. J. Korn et. al, "Direct Modular Multi-Level Converter for Gearless Low-Speed Drives," *Proceedings of the 2011 14th European Conference on Power Electronics and Applications, EPE 2011*, 2011.

[7] W. Kawamura et. al, "Control and Experiment of a Modular Multilevel Cascade Converter Based on Triple-Star Bridge Cells," *IEEE Transactions on Industry Applications*, vol. 50, no. 5, pp. 3536–3548, 2014.

[8] F. Kammerer et. al, "Energy Balancing of the Modular Multilevel Matrix Converter based on a New Transformed Arm Power Analysis," *2014 16th European Conference on Power Electronics and Applications, EPE-ECCE Europe 2014*, pp. 1–10, 2014.

[9] W. Kawamura et. al, "Experimental Verification of an Electrical Drive Fed by a Modular Multilevel TSBC Converter When the Motor Frequency Gets Closer or Equal to the Supply Frequency," *IEEE Transactions on Industry Applications*, vol. 53, no. 3, 2017.

[10] S. P. Engel et. al, "Control of the modular multi-level converter for minimized cell capacitance," *Proceedings of the 2011 14th European Conference on Power Electronics and Applications, EPE 2011*, pp. 1–10, 2011.

[11] B. Fan et. al, "An Optimal Full Frequency Control Strategy for the Modular Multilevel Matrix Converter Based on Predictive Control," *IEEE Transactions on Power Electronics*, vol. 33, no. 8, pp. 6608–6621, 2018.

On the Cosmic Ray Influence on the Electronics Design of a High Altitude Electric Aircraft

Philippe Morey, Mauro Carpita

HEIG-VD/IESE
Route de Cheseaux 1
Yverdon-les-Bains, Switzerland
Tel.: +41 / (0)24 557 73 74
E-Mail: philippe.morey@heig-vd.ch
URL : iese.heig-vd.ch

Acknowledgements

Part of this research is funded by the Swiss Federal Office of Energy (SFOE)

Keywords

Airplane, Aerospace, Silicon Carbide (SiC), Design optimization, Degradation, Failure modes

Abstract

Space in Earth's orbit is growing more and more scares. High altitude solar electric drones, known as High Altitude Pseudo-Satellites (HAPS), are becoming a viable solution to replace satellite functions. One of the issues that must be faced in their design are cosmic rays (CR). CR can cause failures in power electronics and their flux is significantly higher with altitude. Additionally, with lower pressures, arcing and thermal management become issues as well. This study aims to determine the effects of cosmic radiation at stratospheric altitudes and fix guidelines to be able to answer how to design a high-altitude electric aircraft power plant.

Introduction

Space junk is becoming a risk issue for the satellite services on which our modern society depends on and having an alternative is critical. One solution that is being investigated is in the form of high-altitude pseudo-satellite (HAPS) stratospheric drones. Currently, the stratosphere is a flight domain which is not or only very little exploited. The airspace is controlled up to FL660 (20.1 km), which is just above the Armstrong limit, altitude at which a short decompression event would be lethal for a human being. However, most of the traffic is below FL400 (12.2 km). In fact, commercial airliners are designed to fly close to the tropopause to avoid weather as well as for speed. The technology readiness for perpetual electric flight has been demonstrated with the Solar Impulse project [1] which was concluded in 2016. Private endeavors such as Airbus's Zephyr [2] are now exploring the field. The targeted altitudes require special attention due to the harsh stratospheric conditions, such as cosmic ray radiation levels, that are two to three orders of magnitude higher than at sea level, as well as the issues liked to the low-pressure environment. The goal here is to provide design guidelines to solve or mitigate the issues caused by the CR, to be able to design a reliable high-altitude electric power plant. Part of the work presented here is supported by the Swiss Federal Office of Energy in a demonstration project where the strategies and components for the future electrification of aeronautical power and propulsion systems are investigated [3]. The paper begins by presenting the structure of the power plant under investigation and then by explaining the methodology to be applied. Then, the preliminary results are presented, and design recommendations are discussed.

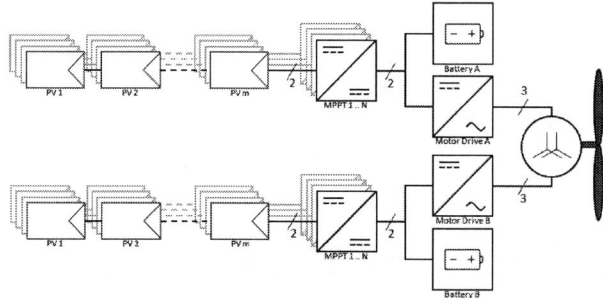

Fig. 1: Block diagram of the proposed power plant. Two independent power systems drive one motor with electrically insulated double windings.

General structure of the power plant

The general structure of the proposed power plant in the project is illustrated in Fig. 1. The chosen design calls for two fully independent power drive systems in parallel that both drive a single motor with electrically insulated double windings. Each system has its own photovoltaic (PV) energy supply chain and battery pack. In case of one side failing, the other can still provide half the power, allowing for safer return. The system voltage is based on a battery composed of 96 lithium-ion cells in series. This choice is motivated by the geometric flexibility, since 96 is a multiple of ten different integers. Furthermore, the resulting voltage level (290-400V) is a good compromise in terms of efficiency. It is high enough to improve efficiency, yet low enough for component technology and availability to enable implementation. It must be remarked that this also corresponds to the battery configuration of the Pipistrel VELIS Electro, the first and currently only type-certified electric aircraft in the world [4]. In this paper, examples we will be specially concentrated around the design of the motor drive. However, the same principles apply to the DC/DC MPPT converters.

Methodology

The effects of CR on electronics have been studied since they were suspected and confirmed in the 1980s [5]. In the beginning of the 1990s, experiments at ground level and in a salt mine 140 m below showed that the failure in time (FIT) of power devices such as GTO's and IGBT's was exponentially dependent on the applied voltage and could be attributed to CR [6]. Observation of MOSFET failures due to high energy neutrons soon followed [7]. CR are known to cause single event effects (SEE) [8]–[12] that, depending on the device can cause soft errors, such as bit-flips, or hard errors, such as single event burnouts (SEB). Neutrons, protons and pions are known to be the main source of SEEs at ground level [13], however at higher altitudes, heavy ions can also participate in the disturbance. Because of the high energy levels of CR, as well as the weight constraint on aircrafts, possible shielding has limited effect. CR must be delt with rather than attenuated. To be able to quantify the effects of CR on the electronic devices at high altitudes, it is imperative to know the nature of the radiation. The EXcel-based Program for calculating Atmospheric Cosmic-ray Spectrum (EXPACS) [14] tool, cited as a reference source by IEC 62396-1:2016 [13], allows to simulate the differential particle flux (Fig. 2) for different particles at different altitudes and different geomagnetic cutoff rigidities [15].

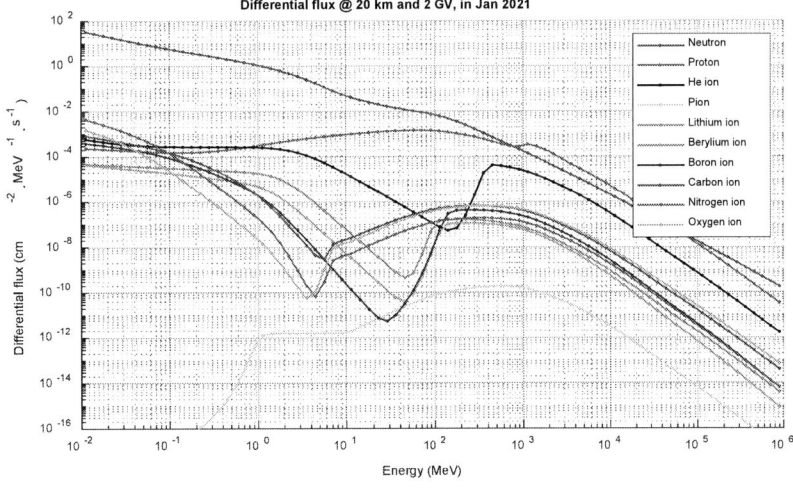

Fig. 2: Differential particle flux at an altitude of 20 km and a cutoff rigidity of 2 GV in January 2021.

Since the goal here is to design a power plant allowing perpetual solar electric flight, the latitude north and south of the equator is limited to a certain range because of the apparent height of the Sun. The subject of study here being CR, this limit can be described as a minimum vertical geomagnetic cutoff rigidity (as the vertical cutoff rigidity is maximum close to the equator and reduces to almost zero at the poles). In the present study, this minimum was chosen to be 2 GV.

Hereafter, there are three categories of issues that need to be addressed: i) Single Event Effects (SEE) on the information level (signal or memory) where errors can be addressed through redundancy and error detection, ii) Single Event Burnouts (SEB) and Single Event Gate Rupture (SEGR), which are hard errors that can have catastrophic consequences and need to be avoided through adequate sizing and derating, and iii) Total Ionizing Dose (TID) and displacement damage, which are the result of long term radiation exposure and determine a maximum lifespan base on experimental observations.

Single Event Effects

The effects of CR are different depending on the size and type of electronic component. Two energy thresholds are defined as significant in [13] depending on component feature size. For devices with feature sizes smaller than 150 nm, the threshold is set to 1 MeV, above it is set to 10 MeV. The particle fluxes with energies higher than the threshold is obtained by integrating the differential flux above the preset energy threshold. Data was extracted from EXPACS ranging from sea level up to an altitude of 50 km. The same was done over a whole time period ranging from 1950 to 2021 to observe the effect of the solar cycles. At solar minimum (low Sunspot count), the particle flux is at its highest, and vice versa. Fig. 3 shows the calculated fluxes with the two predefined thresholds, in January 2021, during a solar minimum. At solar maximum, depending on the intensity, the flux can be attenuated by 50%.

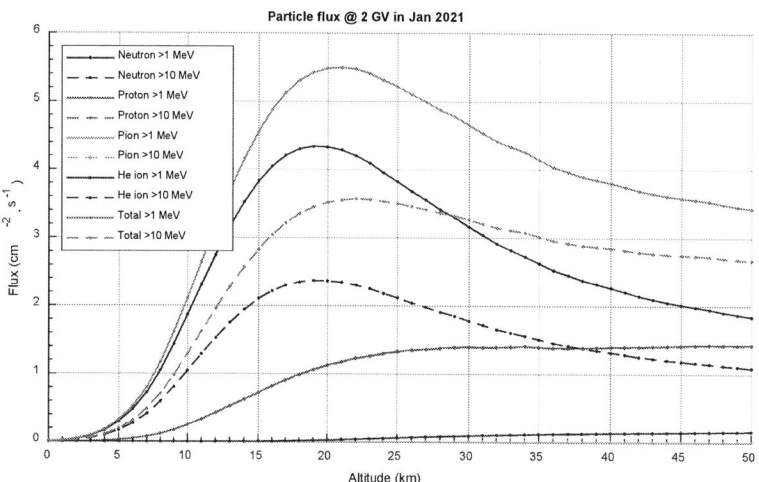

Fig. 3: Particle flux as a function of altitude, with energies greater than 1 MeV and 10 MeV for a cutoff rigidity of 2 GV in January 2021.

The total abundance of particles with energies higher than 1 MeV is significantly greater than the particles with energies greater than 10 MeV. However, as shown in Fig. 3, the difference is entirely in the abundance of neutrons[1]. The total flux at the peak is 700 to 800 times greater than at sea level. To calculate the upset rate (1), the cross section of a given device must be known[2]. This is determined experimentally for each component in a particle accelerator.

$$\text{Upset rate } [s^{-1}] = \text{Flux } [cm^{-2} \cdot s^{-1}] \cdot \text{Bit X-section } [cm^2] \qquad (1)$$

In [16], cross sections for different memory chips where measured as a function of proton energy as well as linear energy transfer (LET) for different heavy ions. What comes out is that cross section is strongly dependent on LET. What also appears, is that the cross sections for protons peaks at approximately the physical cross section of the memory bits and is otherwise three to four orders of magnitude lower. The cross section for heavy ions can be a thousand times more than the physical cross section, however they are also much less abundant.

In their experiments, the cross section for protons peaks around 1 MeV, which certainly corresponds to the specific energy for the proton's Bragg peak to occur in the sensitive area of the devices under test (DUT). Neutrons do not ionize directly since they do not experience the electronic stopping power (responsible for the Bragg peak). However, they produce secondary protons though nuclear collisions, which do. Simulations performed with the SPace ENVironment Information System (SPENVIS) [17] show that the interactions between neutrons and matter produce secondary protons are between two to four orders of magnitude less abundant than the neutrons. In any case, when using commercial off the shelf (COTS) components, few will have been tested to determine their cross sections and the information is usually not readily available. However, a worst-case scenario can be done by considering the total flux with energies above 1 MeV and the physical cross section. For example, considering a 50 KB RAM memory on a microcontroller and a 100 nm² cross section per bit, the total cross section is $4 \cdot 10^{-7}$ cm². With 5.5 particles cm²/s, there would be an average of 69 upsets per year. This shows that even

---

[1] Since neutrons are not stable alone and have a short lifespan (less than 15 minutes) [38], all neutrons present in the atmosphere are secondary particles resulting from collisions in the upper atmosphere.

[2] Here, the bit cross section is not the physical cross section, but rather the apparent one relative to a particle. It is therefore a function of the nature of the particles (proton, neutron, etc...) and their energies.

with a conservative approach, the upset rate is extremely low in regard of the microcontroller's clock frequency, however the measures presented further must be taken to make the system robust and error proof.

Single event burnout and single event gate rupture

SEEs, as presented above, are at a signal or information (memory) level and are in a low voltage environment. A SEB concerns power semiconductors which withstand large voltages and where the active part is subject to high electric fields. SEB occurs when a power semiconductor such as an IGBT or a MOSFET is off (blocking) and suffers catastrophic failure due to an ionizing particle (usually a heavy ion) causing it to switch on at an undesired moment. Concretely, the deposition of charge caused by the particle penetrating the drain-body-source region causes a partial forward bias in the junction. If the drain-source voltage (V_{DS}) is high enough, the remaining blocking part of the junction breaks down and current rushes through, leading to burnout. SEB occurs at drain-source voltages lower than the rated voltages of commercial components. The probability of destruction by cosmic radiation is reduced by the manufacturers of power semiconductors thanks to the dimensioning rules for the thickness and the distribution of the electric field in the components. On the ground, these rules specify a random failure rate of approximately 1–3 FIT per cm^2 of component surface, corresponding to 1 to 3 failures per billion hours of operation and cm^2 [18]. Protons and neutrons have also been shown to cause SEB and extensive research has been done to determine the safe derating levels at which no SEB has been observed [19]–[25]. These studies show that a derating from 20% to 50% reduces FIT by a factor 1000x [20]. Since at stratospheric altitudes the particle flux is 700 to 800 times higher than on the ground, derating should be further down at 66%. Furthermore, in support of this recommendation, it has been shown that SEB does not occur below 400V with 1200V SiC MOSFETs [24]. P-channel MOSFETS are almost immune to SEB [26], however their downside is that their internal resistance is approximately three times higher due to the difference in carrier mobility.

SEGR affects as much N-channel than P-channel power semiconductors. It occurs when a heavy ion strikes a MOSFET through the gate oxide region [27]. The ionized track leads to a localized increase in the oxide field which causes it to breakdown. Leakage starts to occur until rupture of the gate. Some events that appear to be SEB are actually caused by SEGR [28]. SEGR had been thought to be caused only by heavy ions, however it has be shown that high energy protons [29] as well as neutrons can also be the cause [30]. The only way to reduce the risk of SEGR is through derating [31].

Total Ionizing Dose

Total ionizing dose is the cumulative effect due to exposition to ionizing radiation. All particles participate in the total ionizing dose. The observable effect is a gradual degradation of the electrical properties of an electronic device. TID affects as much metal oxide semiconductors (MOS) as bipolar devices. The main effects are increased leakage currents, change of threshold voltages and degraded timing [32]. Based on more than two decades of experimental results, the lower boundary at which devices still perform within specifications has been found to be 10 Gy (J/kg) [13]. However, most components can withstand much more. IEC 62396-1:2016 [13] recommends setting the TID threshold for avionics at 5 Gy.

To calculate TID, a tool such as SPENVIS can be used, however the environment and the geometry around each component must be known. Since this is not feasible here, a specific case is studied to see what the TID levels are. Simulations were done considering a 2 mm thick aluminum sheet (airframe, and electronics casing) and a 0.2 mm thick silicon wafer. The differential flux shown in Fig. 2 was used as input data and was setup omnidirectionally. Table 1 summarizes the results.

Protons	2.176 µGy/h	Helium ions	0.356 µGy/h	Nitrogen ions	0.038 µGy/h
Neutrons	0.252 µGy/h	Lithium ions	0.005 µGy/h	Oxygen ions	0.179 µGy/h
Electrons	0.901 µGy/h	Beryllium ions	0.006 µGy/h	Fluorine ions	0.006 µGy/h
Positrons	0.734 µGy/h	Boron ions	0.033 µGy/h	Neon ions	0.026 µGy/h
Gamma rays	0.476 µGy/h	Carbon ions	0.096 µGy/h	Sodium ions	0.007 µGy/h

Table 1: Ionizing dose, by particle, absorbed by the 0.2 mm silicon wafer, calculated using SPENVIS, with the cosmic ray data extracted from EXPACS, at an altitude of 20 km and a vertical cutoff rigidity of 2 GV, in January 2021.

Table 1 shows the contributions of the heavy ions up to sodium, however contributions to TID were calculated up to aluminum. The TID calculated in the 0.2 mm silicon wafer sums up to 5.35 µGy/h. Rounding it up to 6 µGy/h, it would take 95 years to reach the 5 Gy threshold. Other simulations performed showed that changing the thickness of the silicon or the shielding changes only a little the TID. E.g., exposing directly a 1 mm thick silicon wafer without shielding results in a TID of 5.72 µGy/h. It is safe to say that TID is not an issue in the present study in normal operating conditions.

Displacement damage

Displacement damage refers to atoms being knocked out of their lattice by an energetic particle. It is a cumulative effect, as TID, however it is mainly due to heavy ions, protons and neutrons. Past a certain threshold, displacement damage causes electronic devices to no longer function normally. The threshold was defined experimentally on sensitive optocouplers with a 1 MeV equivalent neutron fluence at 10^{10} neutron/cm^2 [13]. MOS devices have a much higher threshold of 10^{15} neutron/cm^2 [26]. Two methods are used to calculate displacement damage.

The first method uses the non-ionizing energy loss (NIEL) function [33], [34], which is accepted as the best estimate of the potential displacement damage. NEIL, or D, is dependent on the nature of the incoming particle and its energy. According to ASTM E722-19 [35], the displacement damage cross section for a 1 MeV neutron, noted $D_n(1\ MeV)$, is 95 MeVmb. This value is defined as the normalizing value. NEIL scaling of any particle, with any energy, is then expressed in terms of hardness factor k, where:

$$k_{particle} = \frac{D_{particle}}{D_n(1\ MeV)} \tag{2}$$

Therefore, the hardness factor $k_{particle}$ is the displacement damage relative to a 1 MeV neutron. Using hardness data of neutrons and protons from [36] and applying it to their respective differential fluxes at 21 km, we obtain the $D_n(1\ MeV)$ equivalent differential flux (Fig. 4). These are then integrated over the full available energy spectrum to obtain the $D_n(1\ MeV)$ equivalent flux (Fig. 5).

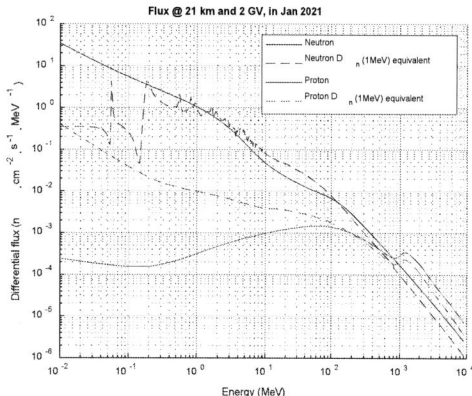

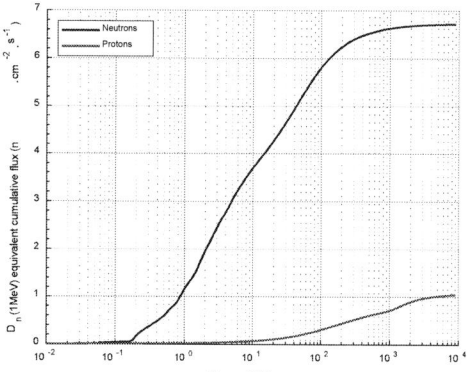

Fig. 4 : Neutron and proton differential fluxes and their respective resulting differential displacement damage fluxes.

Fig. 5: $D_n(1\ MeV)$ equivalent cumulative flux for neutrons and protons.

The total $D_n(1\ MeV)$ equivalent cumulative flux adds up to 7.75 $[n \cdot cm^{-2} \cdot s^{-1}]$. Rounding up to 10, to take into account heavy ions, results in a lifespan of over 277'000 hours (31 years).

The second method is a much simpler one and is the one proposed in IEC 62396-1:2016. This method only considers the neutron flux between 1 MeV and 1 GeV and applies a factor of three as a conservative measure. At an altitude of 21 km the neutron flux with energies between 1 MeV and 1 GeV is 4.3 neutron $\cdot\ cm^{-2} \cdot s^{-1}$. With the factor three, it takes more than 215'000 hours of flight at 21 km to reach the fluence threshold. This second method is indeed more conservative then the first. Nevertheless, even with the conservative method, displacement damage is yet not an issue before more than 24 years.

Summary and design recommendations

As shown, at the studied altitude SEEs are present and expected, but not at a high frequency. SEEs group several different effects. Among these are single event upsets (SEU), multiple bit upsets (MBU), multiple cell upsets (MCU), single event transients (SET), single event functional interrupt (SEFI) and single event latch up (SEL), which are all non-destructive and are reversible. SEEs also include single event burnout (SEB), single event gate rupture (SEGR) and single event induced hard error (SHE), which are all permanent damage.

SEU refers to a single bit being flipped in a memory, where the effect is reversible by rewriting the bit. This can be solved though simple error detection and correction algorithms. However, it will be shown further that more advanced error correction algorithms are necessary.

MBU and MCU are caused by a shower of secondary particles, originating from the same primary particle, which generate multiple simultaneous upsets. MBUs are when multiple bits in a same word are flipped, whereas MCU is when bits in different words are flipped during the same event. For critical functions, static RAM (SRAM) is preferable to dynamic RAM (DRAM) [13]. Because DRAM has space optimization logic, an upset in this logic can cause thousands of bits to be in error because of one event. Because of this, the memory should be physically distributed in different locations and more advanced error detection and correction algorithms should be used.

CR can also create transient disturbances at the signal level (in operational amplifiers, ADCs, etc.) in which case it is referred to as a SET. These can occur on analog or digital signals and lead to false readings or false commands. When data is being read or processed, repeating operations allows to detect SETs and eliminate them. Furthermore, more robust hardware can be used. To give an example, in the present case, the position sensing of the propeller shaft that is fed to the motor drives is done using a resolver rather than an encoder. Resolvers, though less accurate, are more robust than encoders.

When sending a command, e.g., to a power device's gate driver, an untimely turn on can have catastrophic consequences. To reduce this risk, gate drivers with both high side and low side signal inputs that reject the command if both signals are high can be used. Furthermore, a hardware short-circuit protection which pulls down the driver command signals can be implemented to further reduce the risk. Additionally, placing a pull-down resistor between the gate and source is always recommendable because it reduces the risk of an EM transient as well as charge deposition from CR to cause MOSFETs to auto-turn on during the high Z state of gate drivers when switching or powered down.

An SEU in the wrong place can cause a complex system such as a microprocessor to misfunction and is qualified as a SEFI. For example, a bit flip in a critical register or a program counter could cause the system to freeze. In some devices such as DRAMs and FPGAs, a SEFI can cause an increase in supply current [13]. A reset or a power cycle may need to be applied to resolve it. Detection and resolving of SEFI is done, depending on how critical the system is, through double or triple modular redundancy. In the current project, the chosen microcontroller is part of the Hercules family by Texas Instruments. This microcontroller has double modular redundancy built in. It is equipped with dual cores that execute the same instructions in parallel but offset by two clock cycles to ensure a single disturbance does not affect both executions the same way. A check is done to control that both results are coherent. Additionally, the two cores are not physically in the same location and are oriented differently to reduce the chances of the same bits being flipped simultaneously [37].

A SEL is a short circuit that can occur in an integrated circuit (IC). It is caused by the triggering of a parasitic structure that creates a low impedance path short circuiting the power supply. It can rapidly lead to the destruction of the IC. A power cycle is required to resolve the latch-up. A power supply monitoring system with an auto-power cycle must be implemented.

SEB and SEGR only occur on power devices in the off state with a high voltage present between drain and source. Both phenomena are due to a charge deposit that consequently overwhelms the dielectric strength of the device. As shown above, protection against SEB and SEGR can only be done thought appropriate derating. A derating of 66%, corresponding to a usage at 1/3 of the nominal voltage is recommended at stratospheric altitudes.

SHE is when a particle causes a memory bit to be permanently stuck. This is significant because it cancels simple error correction codes such as single error correct, double error detect (SECDED). If any other error occurs, it will not be corrected. For this reason, advanced error detection and correction algorithms are again necessary.

Total ionizing dose as well as displacement damage were shown not to be of concern, because even with a conservative calculation, the lifespan is expected to be more than 24 years. However, one of the effects of radiation on power devices is slow but progressive increase in leakage across insulating boundaries. The leakage between drain and gate can cause the gate to charge itself and the device to auto-turn on. To compensate for this leakage, in addition to the pull-down resistor between gate and source, a negative bias voltage should be applied as well.

Conclusion

CR can cause errors, disfunctions of complex electronic systems and can even lead to catastrophic failure of power devices. To quantify the risk posed by CR on avionics and power drives functioning at stratospheric altitudes, as well as to minimize it and compensate for the effects, the intensity of the CR was defined as well as their effects on electronic devices. Because of the high energies in play, the effectiveness of shielding to attenuate CR is limited. However, the effects are manageable and solutions exist to correct induced errors and minimize the risk of catastrophic failure. Single event effects are the main issue, whereas long term exposure is less. For each type of effect concrete recommendations were given allowing to define the design guidelines for a high-altitude power electronic system regarding cosmic rays.

References

[1] Solar Impulse SA, "Solar Impulse Clean Technologies to Fly Around the World," 2016. https://aroundtheworld.solarimpulse.com/%0Ahttp://www.solarimpulse.com/#airplane (accessed Dec. 17, 2021).

[2] Airbus, "Zephyr." https://www.airbus.com/en/products-services/defence/uas/uas-solutions/zephyr (accessed Dec. 17, 2021).

[3] SFOE, "Flexible Hochleistungskomponenten für die Elektrifizierung von zukünftigen aeronautischen Antriebssystemen basierend auf photoelektrischer Energieerzeugung (SPET) - Texts." https://www.aramis.admin.ch/Texte/?ProjectID=41460 (accessed Mar. 22, 2022).

[4] "Velis Electro – Pipistrel Aircraft." https://www.pipistrel-aircraft.com/aircraft/electric-flight/velis-electro-easa-tc/ (accessed Mar. 07, 2022).

[5] J. Ziegler and W. Lanford, "The effect of sea level cosmic rays on electronic devices," 1980, pp. 70–71, doi: 10.1109/isscc.1980.1156060.

[6] H. Kabza et al., "Cosmic radiation as a cause for power device failure and possible countermeasures," *IEEE Int. Symp. Power Semicond. Devices ICs*, pp. 9–12, 1994, doi: 10.1109/ISPSD.1994.583620.

[7] D. L. Oberg, L. L. Wert, E. Normand, P. P. Majewski, and S. A. Wender, "First observations of power MOSFET burnout with high energy neutrons," *IEEE Trans. Nucl. Sci.*, vol. 43, no. 6 PART 1, pp. 2913–2920, 1996, doi: 10.1109/23.556885.

[8] P. Raaby and J. Schultz, "Neutron-Induced Single Event Upsets in Static RAMS Observed at 10 KM Flight Altitude," *IEEE Trans. Nucl. Sci.*, vol. 40, no. 2, pp. 74–77, 1993, doi: 10.1109/23.212319.

[9] E. Normand and T. J. Baker, "Altitude and Latitude Variations in Avionics SEU and Atmospheric Neutron Flux," *IEEE Trans. Nucl. Sci.*, vol. 40, no. 6, pp. 1484–1490, 1993, doi: 10.1109/23.273514.

[10] E. Normand, "Single Event Upset in Avionics," *IEEE Trans. Nucl. Sci.*, vol. 40, no. 2, pp. 120–126, 1993, doi: 10.1109/23.212327.

[11] K. Johansson, P. Dyreklev, B. Granbom, M. Catherine Calvet, S. Fourtine, and O. Feuillatre, "In-flight and ground testing of single event upset sensitivity in static RAMs," *IEEE Trans. Nucl. Sci.*, vol. 45, no. 3 PART 3, pp. 1624–1627, 1998, doi: 10.1109/23.685251.

[12] R. D. Schrimpf and D. M. Fleetwood, *Radiation Effects And Soft Errors In Integrated Circuits And Electronic Devices*, vol. 34. WORLD SCIENTIFIC, 2004.

[13] International Electrotechnical Commission (IEC), "Process management for avionics – Atmospheric radiation effects – Part 1: Accommodation of atmospheric radiation effects via single event effects within avionics electronic equipment," *IEC 62396-1*, vol. 1, no. 1. IEC, Geneva, 2016.

[14] Japan Atomic Energy Agency, "EXPACS." https://phits.jaea.go.jp/expacs/ (accessed May 11, 2020).

[15] C. C. Finlay et al., "International Geomagnetic Reference Field: The eleventh generation," *Geophys. J. Int.*, vol. 183, no. 3, pp. 1216–1230, Dec. 2010, doi: 10.1111/j.1365-246X.2010.04804.x.

[16] H. Puchner, J. Tausch, and R. Koga, "Proton-induced single event upsets in 90nm technology high performance SRAM memories," *IEEE Radiat. Eff. Data Work.*, pp. 161–163, 2011, doi: 10.1109/REDW.2010.6062523.

[17] "SPENVIS - Space Environment, Effects, and Education System." https://www.spenvis.oma.be/ (accessed May 12, 2020).

[18] S. Linder, "Power semiconductors," *Power Semicond.*, no. ABB review 4, pp. 34–39, 2006, doi: 10.1201/b11535-4.

[19] A. Akturk, J. M. McGarrity, S. Potbhare, and N. Goldsman, "Radiation effects in commercial 1200 v 24 A silicon carbide power MOSFETs," *IEEE Trans. Nucl. Sci.*, vol. 59, no. 6, pp. 3258–3264, 2012, doi: 10.1109/TNS.2012.2223763.

[20] A. Akturk et al., "Predicting Cosmic Ray-Induced Failures in Silicon Carbide Power Devices," *IEEE Trans. Nucl. Sci.*, vol. 66, no. 7, pp. 1828–1832, 2019, doi: 10.1109/TNS.2019.2919334.

[21] A. Akturk, R. Wilkins, and J. McGarrity, "Terrestrial neutron induced failures in commercial SiC power MOSFETs at 27C and 150C," *IEEE Radiat. Eff. Data Work.*, vol. 2015-Novem, pp. 1–5, 2015, doi: 10.1109/REDW.2015.7336737.

[22] A. Akturk et al., "The effects of radiation on the terrestrial operation of SiC MOSFETs," *IEEE Int. Reliab. Phys. Symp. Proc.*, vol. 2018-March, pp. 2B.11-2B.15, 2018, doi: 10.1109/IRPS.2018.8353543.

[23] A. F. Witulski et al., "Single-Event Burnout Mechanisms in SiC Power MOSFETs," *IEEE Trans. Nucl. Sci.*, vol. 65, no. 8, pp. 1951–1955, Aug. 2018, doi: 10.1109/TNS.2018.2849405.

[24] K. F. Galloway et al., "Failure estimates for SiC power MOSFETs in space electronics," *Aerospace*, vol. 5, no. 3, pp. 1–7, 2018, doi: 10.3390/aerospace5030067.

[25] D. R. Ball et al., "Estimating Terrestrial Neutron-Induced SEB Cross Sections and FIT Rates for High-Voltage SiC Power MOSFETs," *IEEE Trans. Nucl. Sci.*, vol. 66, no. 1, pp. 337–343, 2019, doi: 10.1109/TNS.2018.2885734.

[26] A. Papadopoulou, "Single Event Effect Testing of Commercial Silicon Power MOSFETs," Democritus University of Thrace, 2020.

[27] M. Allenspach *et al.*, "SEGR: A unique failure mode for power MOSFETs in spacecraft," *Microelectron. Reliab.*, vol. 36, no. 11–12, pp. 1871–1874, Nov. 1996, doi: 10.1016/0026-2714(96)00218-1.

[28] R. Sheehy, J. Dekter, and N. Machin, "Sea level failures of power MOSFETs displaying characteristics of cosmic radiation effects," *PESC Rec. - IEEE Annu. Power Electron. Spec. Conf.*, vol. 4, pp. 1741–1746, 2002, doi: 10.1109/PSEC.2002.1023062.

[29] E. V. Mitin and V. G. Malinin, "Investigation of SEGR cross-section in power MOSFETs under proton irradiation," *Proc. Eur. Conf. Radiat. its Eff. Components Syst. RADECS*, vol. 2015-December, Dec. 2015, doi: 10.1109/RADECS.2015.7365624.

[30] A. Hands *et al.*, "Single event effects in power MOSFETs due to atmospheric and thermal neutrons," *IEEE Trans. Nucl. Sci.*, vol. 58, no. 6 PART 1, pp. 2687–2694, Dec. 2011, doi: 10.1109/TNS.2011.2168540.

[31] J.-M. Lauenstein, "Single-event gate rupture in power MOSFETS: A new radiation harness assurance approach," University of Maryland, 2011.

[32] F. B. McLean and T. R. Oldham, "Basic Mechanisms of Radiation Effects in Electronic Materials and Devices," *US Army Lab. Command*, pp. 0–11, Sep. 1987, Accessed: Jan. 04, 2022. [Online]. Available: https://apps.dtic.mil/sti/citations/ADA186936.

[33] M. A. Xapsos, E. A. Burke, F. F. Badavi, L. W. Townsend, J. W. Wilson, and I. Jun, "Nonionizing Energy Loss (NIEL) Calculations for High-Energy Heavy Ions."

[34] E. A. Burke, "Energy dependence of proton-induced displacement damage in silicon," *IEEE Trans. Nucl. Sci.*, vol. 33, no. 6, pp. 1276–1281, 1986, doi: 10.1109/TNS.1986.4334592.

[35] ASTM, *ASTM E722-19 : Standard Practice for Characterizing Neutron Fluence Spectra in Terms of an Equivalent Monoenergetic Neutron Fluence for Radiation-Hardness Testing of Electronics*. 2019.

[36] G. Lindstroem and A. Vasilescu, "Displacement Damage in Silicon," 2000. https://rd50.web.cern.ch/niel/ (accessed Jan. 21, 2022).

[37] Texas Instruments, "TMS570LS0914 16- and 32-Bit RISC Flash Microcontroller: Datasheet." 2016.

[38] "How Long Does a Neutron Live? | www.caltech.edu." https://www.caltech.edu/about/news/how-long-does-a-neutron-live (accessed May 27, 2022).

DC-Bus Control Considerations of Asymmetrical Multilevel Inverters with Embedded Buck-Boost Converter

Theodoros P. Mouselinos and Emmanuel C. Tatakis
UNIVERSITY OF PATRAS
Laboratory of Electromechanical Energy Conversion
Department of Electrical and Computer Engineering
26504, Rion-Patras, Greece
Tel.: +30 2610 996412.
E-Mail: t.mouselinos@ece.upatras.gr, e.c.tatakis@ece.upatras.gr
URL: http://www.lemec.ece.upatras.gr

Keywords

«DC-AC Converters», «DC-DC converter», «Multi-level Inverters», «Converter control», «DC voltage control»

Abstract

This paper focuses on the DC-bus control of asymmetrical Multilevel inverter family featuring a Buck-Boost converter to boost the input voltage. To investigate the dynamic performance of the system a thorough analysis is presented on the DC-bus dynamic behavior. It is shown that with higher system bandwidth the input capacitance requirements and the peak current through the boosting inductor are increased, compromising the power density and the efficiency of the whole DC/AC converter. To improve the transient performance of the system without increasing the volume of the passive components, a control scheme with a feedforward current estimator term is proposed. Finally, the correctness of the theoretical analysis is validated via experimental probing on a laboratory prototype.

Introduction

Multilevel inverters (MLIs) show many advantages compared to two-level inverter topologies such as the high-quality multi stepped output voltage waveform, the fault tolerant operation, the reduced blocking voltage of the semiconductor devices and the reduced common mode voltage. The state-of-the-art MLI topologies, named Cascaded H-Bridge MLI (CHB-MLI), Diode Clamped MLI (DC MLI) and Flying Capacitor MLI (FC MLI) are used in many renewable energy sources applications and industrial applications [1].

Even though the mentioned above topologies have been adopted in many applications, their main disadvantage is the increased number of semiconductors needed and the reduction of the switch count is a subject of research in recent years [2]. Also, as it can be seen from [3] the voltage boosting feature is often a must in renewable energy systems. Since, the reduction of the switching devices and the voltage boosting feature are required in many applications, a new research field is introduced and multilevel inverter topologies with embedded Buck-Boost DC-DC converters have been proposed in the literature [4]-[8].

In this paper the DC-bus voltage control of asymmetrical MLI family (AMLI) with embedded Buck-Boost converter is analyzed and a thorough analysis is presented on the dynamic performance of the system. DCM operation is selected over int-DCM and CCM as the magnetic component volume and the RMS value of the current of Buck-Boost main switch are decreased. Moreover, the influence of the controller bandwidth in double line frequency component of the input current is investigated and a full schematic control with a feedforward current estimator term is presented to improve the transient performance without adding extra current measurement circuitry.

Calculation of the Embedded Buck-Boost Converter Load Current

The topology of the five-level Buck-Boost Multilevel Inverter under investigation is depicted in Fig. 1. As presented in [8] the MLI topology is based on the simplified neutral point clamped topology with an active switch added (ANPC topology), since a proper commutation path must be provided to the load current in the case of resistive-inductive load.

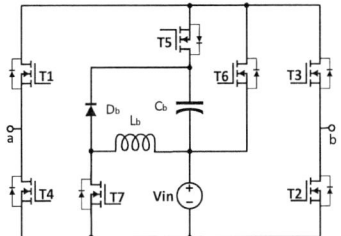

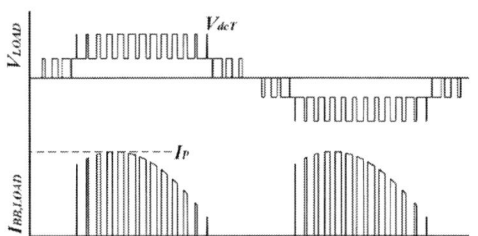

Fig. 1: The ANPC Buck-Boost embedded AMLI topology under investigation.

Fig. 2: Output voltage of the five-level MLI and load current waveform of the embedded Buck-Boost converter.

Since the average value of the output current of the Buck-Boost converter must be known for a proper small signal analysis, in this section its value is calculated. Considering the Buck-Boost output current waveform from Fig. 2 and a modulation technique similar to the one presented in [9] its mean value can be written as:

$$I_{BB,Load,AVG} = \frac{1}{T_h} \int_{arcsin(a)}^{\pi - arcsin(a)} i_{BB,Load}(t)dt \tag{1}$$

where T_h is the half period of the output voltage waveform and $a=1/[m_a \cdot (1+G)]$, where $G=V_{Cb}/V_{in}$ and m_a is the inverter modulation index. For the analytical calculation of the current value $I_{BB, Load, AVG}$ the following must be considered:

- m is the number of the switching cycles during the T_h time interval and it is equal to $m=round(T_h/T_s)$ T_s is the carrier switching frequency responsible for the modulation of the upper level
- the random angle ωt_i can be written as $\omega t_i=\pi \cdot i/m$ and
- the inverter output current follows a pure sine wave

Considering the above the mean value of the load current of the Buck-Boost converter is given from eq. (2).

$$I_{BB,Load,AVG} = I_p \cdot \cos\varphi \cdot \left[m_a \cdot \frac{(1+G')}{2} - m_a \cdot \frac{(1+G')}{\pi} \cdot arcsin(a) - \frac{G'}{\pi} \cdot \sqrt{1-a^2} \right], \tag{2}$$

where I_p is the peak amplitude of the output current waveform, φ is the load angle and G' is equal to $G'=1/G$.

In this point it must be noted that eq. (2) is valid for all the five-level Buck-Boost based AMLI topologies in which the generation of the lower voltage level is achieved only with the input voltage source and the generation of the upper level is achieved with the series connection of the input voltage source and the boosting capacitor C_b. As it can be also deduced from eq. (2) and [8] the load current of the Buck-Boost converter is relatively small, and hence the power processed by the embedded DC/DC converter is only a fraction of the total inverter output power.

In Buck-Boost based AMLI topologies the inductor is used to transfer energy from the input DC voltage source to the boosting capacitor. Due to the low power processed by the integrated Buck-Boost converter, the selection of an inductance for CCM operation will lead to a bulky and costly inductor. Also, for the selection of the operational mode of the embedded DC/DC converter two other indexes are also considered. The first index is the amplitude of the double line component of the input current, because its value determines the input capacitance value. The other index is the RMS value of the current through the *T7* switch which determines the conduction power losses of the transistor. In Fig. 3 and Fig. 4 comparative results are presented considering the critical inductance value for each mode of operation and DCM operation is chosen since it offers a good tradeoff between volume and power losses. The critical inductance for CCM is equal to 3.6 mH, for int-DCM it is equal to 120 uH, while for DCM operation the critical inductance is equal to 200 uH.

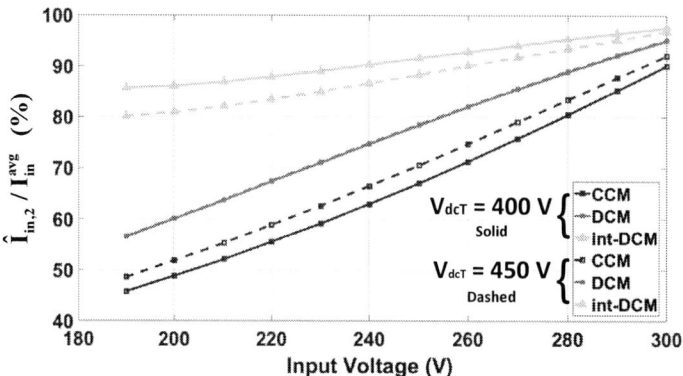

Fig. 3: Amplitude of the double line component of the input current over the average input current of the converter.

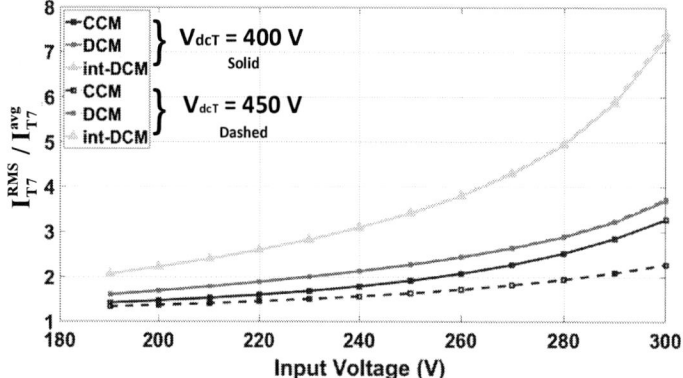

Fig. 4: RMS value over the average value of the current through transistor *T7*.

Small Signal Analysis and DC-bus Control Considerations

Since DCM operation is selected for the embedded Buck-Boost converter the DC-bus voltage should be controlled, since the output voltage of the DC/DC converter V_{Cb} is heavily depended on the output load. To begin with, the output load of the embedded Buck-Boost converter should be identified. This can be achieved considering an equivalent load resistance which it is given from (3) taking into account eq. (2).

$$R_{eq} = \frac{V_{Cb}}{I_p \cdot \cos\varphi \cdot \left[m_a \cdot \frac{(1+G')}{2} - m_a \cdot \frac{(1+G')}{\pi} \cdot arcsin(a) - \frac{G'}{\pi} \cdot \sqrt{1-a^2} \right]} \tag{3}$$

With the equivalent output resistance of the embedded converter been known the next step is the small signal modeling of the DC/DC converter to quantify its dynamic behavior. As it can be seen from the literature, reduced order models and full order models are presented [10]-[11]. In this paper full order average models with the correction proposed in [11] are used, since they offer increased accuracy in both magnitude and phase response in the high-frequency range.

When the switch $T7$ is ON (d_1T_s interval), energy is transferred from the input DC source to the inductor L_b while the output capacitor C_b supplies the load and hence (4) can be exported. When $T7$ is off, diode D_b conducts (d_2T_s interval) until the inductor current becomes zero. In d_2T_s interval the inductor transfers its stored energy to the capacitor C_b and the load R_{eq} and thus (5) is exported. The final time interval d_3T_s neither the diode D_b nor the $T7$ switch conducts, thus the inductor current is zero, the capacitor transfers energy to the load and hence (6) is exported. Deploying the averaging techniques, the modified average model can be extracted (7).

$$\frac{d}{dt}\begin{bmatrix} i_L \\ u_C \end{bmatrix} = \begin{bmatrix} 0 & 0 \\ 0 & -1/R_{eq}C_b \end{bmatrix} \cdot \begin{bmatrix} i_L \\ u_C \end{bmatrix} + \begin{bmatrix} 1/L_b \\ 0 \end{bmatrix} \cdot u_{in} \ , for \ d_1T_s \quad (4) \qquad \frac{d}{dt}\begin{bmatrix} i_L \\ u_C \end{bmatrix} = \begin{bmatrix} 0 & -1/L_b \\ 1/C_b & -1/R_{eq}C_b \end{bmatrix} \cdot \begin{bmatrix} i_L \\ u_C \end{bmatrix} + \begin{bmatrix} 0 \\ 0 \end{bmatrix} \cdot u_{in} \ , for \ d_2T_s \quad (5)$$

$$\frac{d}{dt}\begin{bmatrix} i_L \\ u_C \end{bmatrix} = \begin{bmatrix} 0 & 0 \\ 0 & -1/R_{eq}C_b \end{bmatrix} \cdot \begin{bmatrix} i_L \\ u_C \end{bmatrix} + \begin{bmatrix} 0 \\ 0 \end{bmatrix} \cdot u_{in} \ , for \ d_3T_s \quad (6) \qquad \frac{d}{dt}\begin{bmatrix} i_L \\ u_C \end{bmatrix} = \begin{bmatrix} 0 & \dfrac{-d_2}{L_b} \\ \dfrac{d_2}{(d_1+d_2)\cdot C_b} & \dfrac{-1}{R_{eq}C_b} \end{bmatrix} \cdot \begin{bmatrix} i_L \\ u_C \end{bmatrix} + \begin{bmatrix} \dfrac{d_1}{L_b} \\ 0 \end{bmatrix} \cdot u_{in} \quad (7)$$

After some mathematical manipulations the small signal model of the embedded Buck-Boost converter operating in DCM can be exported (8).

$$\frac{d}{dt}\begin{bmatrix} \tilde{i}_L \\ \tilde{u}_C \end{bmatrix} = \begin{bmatrix} -\dfrac{2\cdot G}{D_1\cdot T_s} & -\dfrac{D_1}{L_b\cdot G} \\ 1/C_b & -1/R_{eq}C_b \end{bmatrix} \cdot \begin{bmatrix} \tilde{i}_L \\ \tilde{u}_C \end{bmatrix} + \begin{bmatrix} \dfrac{D_1\cdot(2+G)}{L_b} & \dfrac{2\cdot V_{in}\cdot(1+G)}{L_b} \\ -\dfrac{D_1^2\cdot T_s}{2\cdot L_b\cdot C_b} & \dfrac{D_1\cdot T_S\cdot V_{in}}{L_b\cdot C_b} \end{bmatrix} \cdot \begin{bmatrix} \tilde{u}_{in} \\ \tilde{d}_1 \end{bmatrix} \quad (8)$$

With the small signal model of the converter been known, a *PI* controller can be properly tuned to regulate the voltage of the boosting capacitor C_b and hence the total DC-bus voltage V_{dcT} of the multilevel inverter. The main requirement of the control system is to be stable with a fast transient response. The stability and settling time of the overall system depends on the *PI* controller gains, from which the system bandwidth is determined. In Fig. 5 and Fig. 6 the key waveforms of the converter are depicted for the same step change of the load but with different gains for the *PI* controller. In Fig. 5 the *PI* controller gains are $k_p=0.1418$ and $k_i=1.3747$ leading to an 89.8° phase margin (PM) at 1010.25 rad/s and 52 dB gain margin (GM) at $410.389 \cdot 10^3$ rad/s. In Fig. 6 the *PI* controller gains are $k_p=0.002198$ and $k_i=0.02423$ with PM=90° (at 15.66 rad/s) and GM=88 dB (at $410.388 \cdot 10^3$ rad/s). As it can be deduced with a higher bandwidth the system features a faster transient response as it is known from the classic control theory, but in this case a double line frequency component is present in the inductor current waveform (Fig. 5). A part of this double line frequency component is drawn from the input DC source through the switch $T7$, leading to increased input capacitance requirement. In Fig. 6 the controller features narrower bandwidth compared to the controller used for the extraction of the results of Fig. 5. As can be seen, in this case a slower transient response is achieved but the double line frequency component of the inductor current is much reduced. In both cases a saturator is added to the controller output to avoid CCM operation.

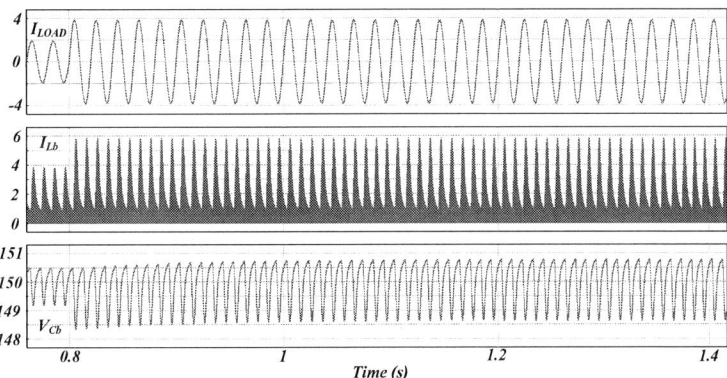

Fig. 5: Transient response of the C_b voltage for a step change of the load with $k_p=0.1418$ and $k_i=1.3747$.

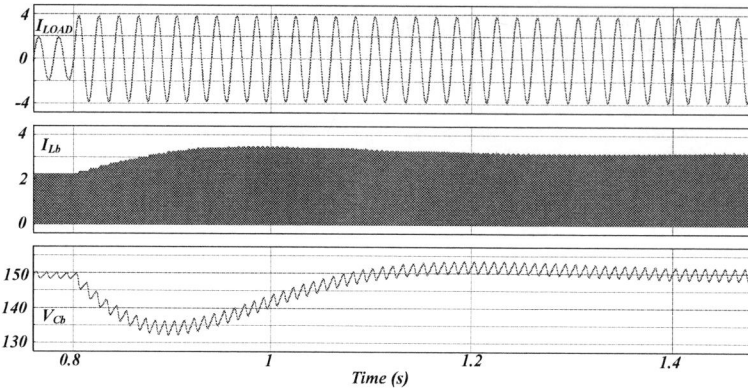

Fig. 6: Transient response of the C_b voltage for a step change of the load with $k_p=0.002198$ and $k_i=0.02423$.

To further investigate the effect of the controller bandwidth to the double line frequency component of the input current waveform multiple controllers are designed and simulated considering a 0.7 kW inverter load. Then, the double line frequency component is extracted along with the peak current flowing through the inductor L_b and the simulation results are depicted in Fig. 7 and Fig. 8. In Fig. 7 the minimum input capacitance requirement is also depicted given a 2% voltage ripple with the average value of the input voltage been equal to 250 V.

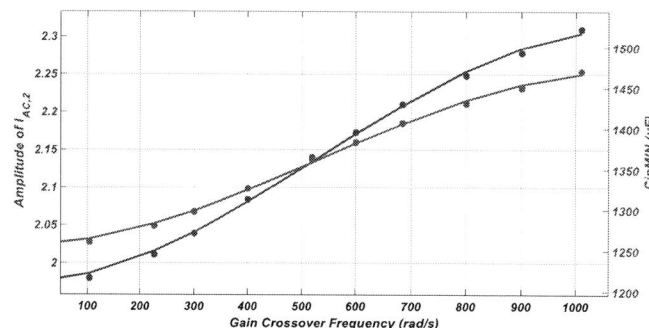

Fig. 7: The effect of the system bandwidth to the double line frequency component of the input current waveform.

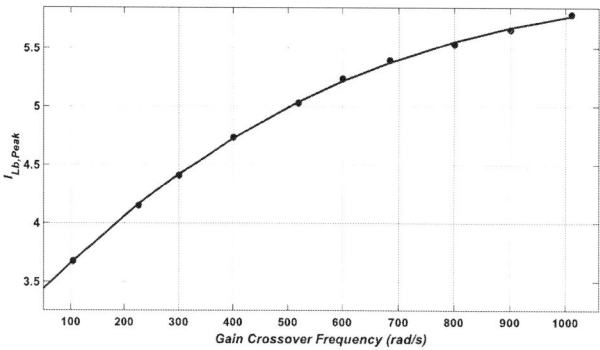

Fig. 8: The effect of the system bandwidth to the peak current flowing through L_b.

From Fig. 7 and Fig. 8 it is visible that the controller bandwidth plays a crucial role to the peak current through the boosting inductor and the double line frequency component of the input current. More specifically the double line frequency component of the input current waveform is increased by 17% and thus the minimum input capacitance requirements are also increased accordingly. Furthermore, if a designed converter for PV applications is considered with a specific capacitance installed on the input DC side, the increase double line frequency on the input current will cause an increased input voltage ripple, leading to decreased energy harvesting from the PV panels [13]. Moreover, with higher system bandwidth the peak current through the inductor is dramatically increased. Considering the case in which the system bandwidth is 1000 rad/s, the peak current through L_b is almost double compared to the case in which the bandwidth is 100 times smaller, which may lead to the saturation of the inductor if it is marginally designed according to the theoretical calculations. In conclusion, it can be deduced that the system bandwidth can affect the power density of the converter since with higher system bandwidth the input capacitance requirements are increased. Also, the Area Product requirement for the inductor L_b is increased leading to bulkier and more expensive inductor if the above results are taken into consideration in the inductor design procedure.

Closed Loop Control with Feedforward Current Estimator Term

As it is shown the fast transient response in the DC bus voltage control cannot be achieved without adding extra double line frequency component in the inductor current waveform. The design of higher order controllers in this case is a much more complicated procedure since the Buck-Boost converter features a right half plane zero and the output current waveform contains low order harmonics in this application. In this section a control scheme is proposed to meet the fast response requirement without the extra low order harmonic component in the inductor current waveform. This is achieved by estimating the output current of the embedded DC/DC converter to avoid the extra current measurement circuit. Then, the theoretical duty cycle of the embedded DC/DC converter can be estimated, and this value can be used as a feedforward term in the closed loop control system.

From eq. (2) it can be seen that the output current of the Buck-Boost converter depends on the amplitude and the phase of inverter output current. Moreover, the input voltage and the modulation index are also two parameters that must be known to properly estimate the output current of the embedded converter. Thereafter, it is determined that the inverter output current and the input voltage must be measured, which there are measured in almost all the cases for control and protection purposes. Hence, no extra measurement is deployed. The estimation of the inverter output current angle can be achieved using the SOGI PLL [12]. Deploying an orthogonal signal generator, the fictitious component β of the inverter output current can be obtained. Subsequently, the *Park* transformation of the inverter output current can be used to extract the quantity I_d which it is described by (9). Given all the above, the output current of the Buck-Boost converter can be estimated by (2) and the theoretical duty cycle of the embedded Buck-Boost DC/DC converter is given by eq. (10).

$$I_d = I_p \cdot \cos\varphi \qquad (9) \qquad\qquad d = \sqrt{\frac{2 \cdot I_{BB,Load,AVG} \cdot V_{Cb,AVG} \cdot L_b \cdot f_S}{V_{in,AVG}^2}} \qquad (10)$$

Consequenlty, the proposed control scheme is presented in Fig. 9. In this point it must be noted that the feedforward term cannot affect the stability of the feedback system as reported in [14], but in order to be applied correctly an accurate dynamic model of the converter must be available, pinpointing the importance of the analysis shown in the previous section.

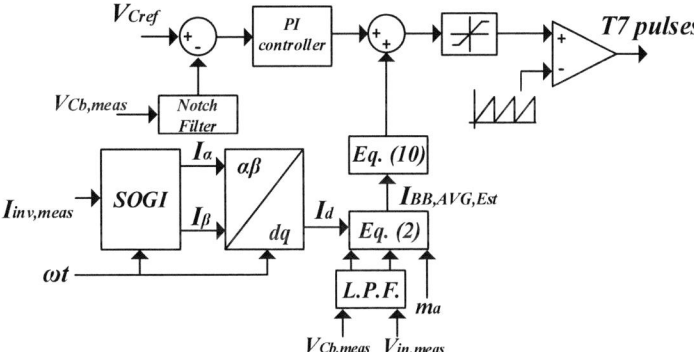

Fig. 9: Block diagram of the closed loop control with the feedforward current estimator term.

In Fig. 10 the performance is shown for the same step change of the load as shown in Fig. 5 and Fig. 6. The *PI* controller gains in this case are $k_p=0.002198$ and $k_i=0.024423$ leading to a narrower bandwidth and also a 100Hz notch filter is used on the feedback to completely remove the double line frequency component of inductor L_b. Notch filter can be deployed in this case, because the lower crossover frequency can be maintained since the notch filter attenuating frequency is much higher compared to the controller bandwidth.

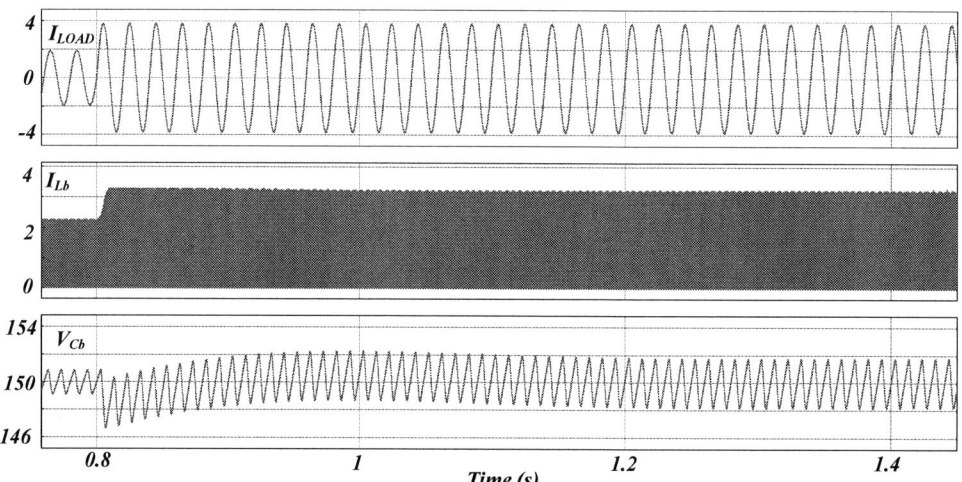

Fig. 10: Transient response of the C_b voltage for a step change of the load with the proposed closed loop control deploying the feedforward current estimator.

Experimental Results

The performance of the proposed control scheme is tested via a 1 kVA experimental prototype of the converter topology shown in Fig. 1. The main components and parameters of the laboratory prototype are summarized in Table I. Also, an LC filter is designed for the AC side of the inverter considering a 30% current ripple in the inductor current waveform and the filter capacitance is selected to obtain a resonant frequency about 10 times lower compared to the switching frequency of the inverter [15]. For the control of the converter the TMS320F28379D is selected. The sampling frequency for the control loop is selected equal to 25 kHz. The input voltage V_{in}, the capacitor voltage V_{Cb} and the output current of the inverter are measured, and then digital low pass filters are deployed featuring a bandwidth of 2 kHz to ensure the measurement quality. Moreover a 100 Hz notch filter is used for the filtering of the capacitor voltage to completely remove the unwanted double line frequency component.

Table I: Converter Parameters and Components

Parameters/ Components	Value/ Parameter
Input Voltage V_{in}	250 V
Output Voltage V_{ab}	230V/50Hz
Output Power P_o	1 kVA
PD- SPWM carrier frequency	25 kHz
Switching frequency of $T7$ switch	80 kHz
Power Switches	C3M0065090D
Diode D_b	IDW15G120C5B
Boosting Capacitor C_b	10x EEU-EB2V101 (100uF each) (898 uF measured)
Inductor L_b	190 uH (RM14 core, N87)
Output LC filter parameters	Lf: 2x350µH (E42/21/20 core, N27) Cf: 12x0.47µF (5.46µF measured)

In Fig. 11 a step change is taking place on the inverter output current, while the DC-bus voltage is regulated deploying a *PI* controller with $k_p=0.002198$ and $k_i=0.024423$ and the feedforward term in this case is disabled. As can be seen large overshoots and undershoots are visible in the capacitor voltage V_{Cb} and no double line frequency component exists in the inductor current waveform since the controller bandwidth is kept at lower levels.

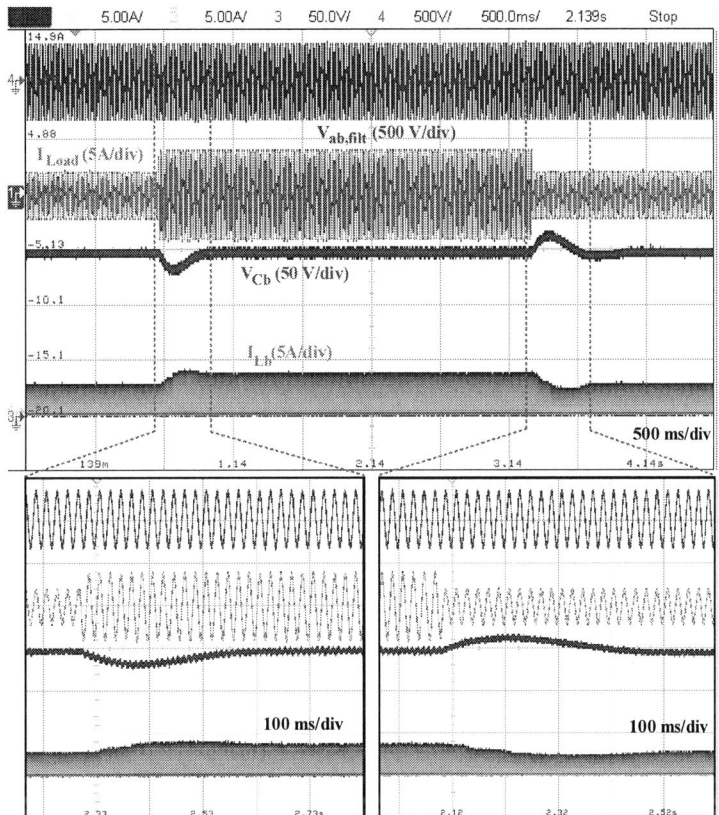

Fig. 11: Key waveforms of the converter for positive and negative changes of the load deploying a *PI* controller for the DC-bus voltage regulation.

In Fig. 12 the control scheme shown in Fig. 9 is deployed for the DC-bus voltage regulation with the feedforward term been activated. The gains of the *PI* controller in this case are the same as the ones used for the extraction of the results presented in Fig. 9 and the same step changes of the inverter load are taking place for comparison purposes. In both cases the input voltage is equal to 250 V, while the reference voltage for the boosting capacitor is equal to 150 V and the fundamental component of the output voltage waveform is kept equal to 230 V/ 50 Hz.

As it can be deduced from the experimental results, superior performance can be achieved using the proposed control scheme with the current estimator feedforward term as shown in Fig. 9, since almost no overshoot or undershoot is reported in the capacitor voltage waveform. Moreover, no double line frequency component is present in the inductor current waveform in this case also denoting the correctness of the analysis and the effectiveness of presented control method.

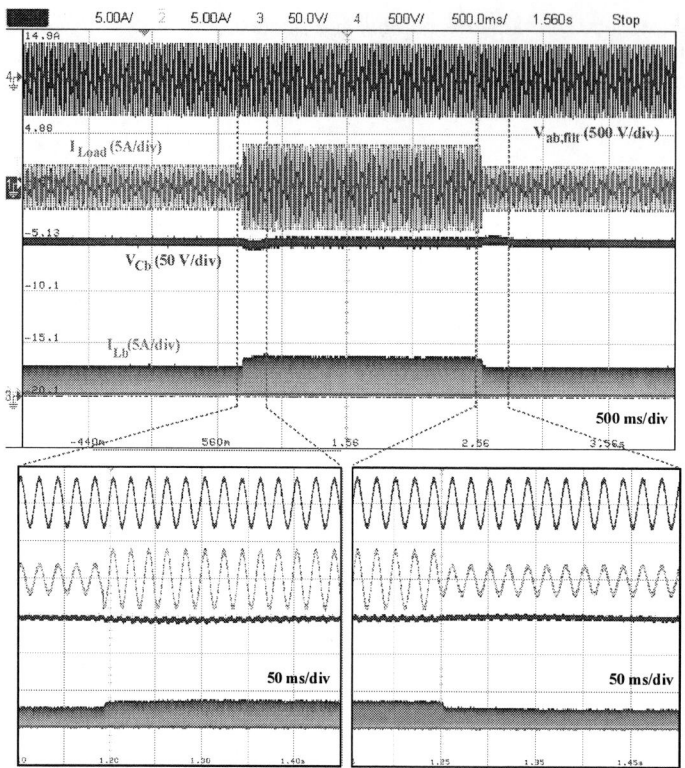

Fig. 12: Key waveforms of the converter for positive and negative changes of the load deploying the schematic control shown in Fig. 9 for the DC-bus voltage regulation with the current estimator feedforward term.

Conclusion

In this paper, the DC-bus control of AMLIs with an embedded Buck-Boost converter is investigated. It is shown that the controller bandwidth can be crucial for the proper operation of the topology despite that it offers the desired stability. The influence of the controller bandwidth on the double line frequency component of the inverter input current is studied via simulation and a full schematic control with feedforward current estimator term is shown to meet all the requirements. Concluding, the performance of the presented control method is also tested via experimental probing.

References

[1] S. Kouro, M. Malinowski, K. Gopakumar et al., "Recent advances and industrial applications of multilevel converters," IEEE Trans. Ind. Electron., vol. 57, no. 8, pp. 2553-2580, 2010.

[2] K.K. Gupta, A. Ranjan, P. Bhatnagar et al., "Multilevel inverter topologies with reduced device count: a review," IEEE Trans. Power Electron., vol. 31, no. 1, pp. 135-151, 2016.

[3] S. Strache, R. Wunderlich and S. Heinen, "A comprehensive quantitative comparison of inverter architectures for various PV systems PV cells and irradiance profiles," IEEE Trans. Sustain. Energy, vol. 5, no. 3, pp. 813-822, 2014.

[4] A. Anurag, N. Deshmukh, A. Maguluri and S. Anand, "Integrated DC– DC Converter Based Grid-Connected Transformerless Photovoltaic Inverter With Extended Input Voltage Range," IEEE Trans. Power Electron., vol. 33, no. 10, pp. 8322-8330, Oct. 2018.

[5] S. S. Lee, C. S. Lim, Y. P. Siwakoti and K. Lee, "Dual-T-Type Five- Level Cascaded Multilevel Inverter With Double Voltage Boosting Gain," IEEE Trans. Power Electron., vol. 35, no. 9, pp. 9522-9529, Sept. 2020.

[6] S. Dhara and V. T. Somasekhar, "An Integrated Semi-Double Stage- Based Multilevel Inverter With Voltage Boosting Scheme for Photovoltaic Systems," IEEE J. Emerg. Sel. Top. Power Electron, vol. 8, no. 3, pp. 2326-2339, Sept. 2020.

[7] S. Madhu Babu and B. L. Narasimharaju, "Single-phase boost DC-link integrated cascaded multilevel inverter for PV applications," IET Power Electronics, vol. 13, no. 10, pp. 2086-2095, 2020.

[8] T. P. Mouselinos and E. C. Tatakis, "A Buck-Boost Embedded Multilevel Inverter with Double Voltage Gain," 2021 23rd European Conference on Power Electronics and Applications (EPE'21 ECCE Europe), 2021, pp. 1-10.

[9] H. Wu, L. Zhu, F. Yang, T. Mu and H. Ge, "Dual-DC-Port Asymmetrical Multilevel Inverters With Reduced Conversion Stages and Enhanced Conversion Efficiency," in IEEE Trans. on Ind. Electron., vol. 64, no. 3, pp. 2081-2091, March 2017

[10] S. Cuk and R. D. Middlebrook, "A general unified approach to modeling switching DC-to-DC converters in discontinuous conduction mode," in Proc. IEEE PESC'77, 1977, pp. 36–57.

[11] Jian Sun, D. M. Mitchell, M. F. Greuel, P. T. Krein and R. M. Bass, "Averaged modeling of PWM converters operating in discontinuous conduction mode," in IEEE Transactions on Power Electronics, vol. 16, no. 4, pp. 482-492, July 2001.

[12] M. Ciobotaru, R. Teodorescu and F. Blaabjerg, "A new single-phase PLL structure based on second order generalized integrator," 2006 37th IEEE Power Electronics Specialists Conference, Jeju, 2006, pp. 1-6.

[13] Y. Xue, L. Chang, S. B. Kjaer, J. Bordonau, and T. Shimizu, "Topologies of single-phase inverters for small distributed power generators: an overview," IEEE Trans. Power Electron., vol. 19, no. 5, pp. 1305–1314, Sep. 2004.

[14] G. C. Goodwin, S. F. Graebe, and M. E. Salgado, Control System Design. New York: Prentice Hall, Sep. 2001.

[15] M. Liserre, F. Blaabjerg and S. Hansen, "Design and control of an LCL-filter-based three-phase active rectifier," IEEE Trans. Ind Appl., vol. 41, no. 5, pp. 1281-1291, Sept.-Oct. 2005.

A Seamless Modulation Strategy for Step-up/down Partial Power Processing Converter (SUD-P3C)

Chao Liu[1], Zhe Zhang[2], Ziwei Ouyang[1], Jiasheng Huang[1], Michael A. E. Andersen[1], Tiberiu Gabriel Zsurzsan[1].

TECHNICAL UNIVERSITY OF DENMARK

Anker Engelunds Vej 101

2800 Kongens Lyngby, Denmark

HEBEI UNIVERSITY OF TECHNOLOGY

No. 5340, Xiping Road,

Beichen District, Tianjin, China

E-Mail: chali@elektro.dtu.dk, zhezhangdtu@outlook.com, zo@elektro.dtu.dk, jiahuan@elektro.dtu.dk, ma@elektro.dtu.dk, tgzsur@elektro.dtu.dk

Keywords

« DC-DC converter», « Seamless transfer», «Efficiency», «DC voltage control», «Partial power processing».

Abstract

This paper proposes a seamless modulation strategy for a step-up/down partial power processing converter (SUD-P3C), to realize smooth mode transition and continuous output regulation. Due to the implementation of zero volt switching (ZVS) on the high voltage side, the proposed method effectively suppresses voltage spikes caused by hard switching of the high voltage side switches, resulting in improved system reliability. Moreover, according to the small-signal analysis, the two operating modes have the same output-to-control transfer function, so the converter can adopt only one unified controller rather than two dedicated ones and reduce controller design complexity. A 400 V-3 kW prototype has been built, and the experimental results verified the effectiveness of the proposed modulation strategy. Compared to the conventional modulation strategy, the power losses are reduced by up to 14 % due to ZVS switching.

Introduction

Partial Power Processing (PPP) technology has been widely used in different applications, such as solar photovoltaic systems [1]-[4], energy storage systems (ESSs) [5], and electric vehicle (EV) fast-charging stations [6],[7]. In PPP converters (P3C), only a small portion of the power being processed, i.e., the most considerable amount of energy flows directly from the source to the load without being processed, resulting in power converter downsizing and efficiency improvement.

Depending on the regulation target, P3C configurations can be divided into two main categories: Parallel-connected configurations (P-P3C) and Series-connected configurations (S-P3C) [8]. The P-P3Cs are usually employed in PV module strings, also widely called differential power processing, which maintains distributed local maximum power point operation by only processing a fraction of the total power. On the other hand, for S-P3C, the output voltage regulation is achieved by adjusting the difference between the input and output voltages. S-P3Cs can be further divided into three types: step-up P3C (SU-P3C), step-down P3C (SD-P3C), and step-up/down P3C (SUD-P3C). The experimental results in [9] reveal that the SUD-P3C processes the least active power over the same voltage variation range compared to the other type of P3Cs, resulting in higher efficiency and power density. Ref [10] proposed a unified modulation strategy for SUD-P3C, simplifying the pulse width modulation (PWM). However, the different operation modes in SUD-P3C pose new challenges, such as achieving reliable regulation between different modes and smooth mode transition.

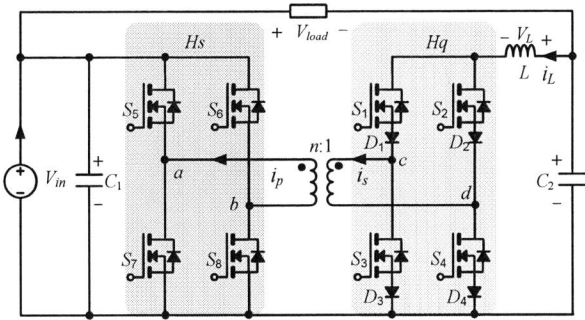

Fig. 1: Configuration of the SUD-P³C in [10].

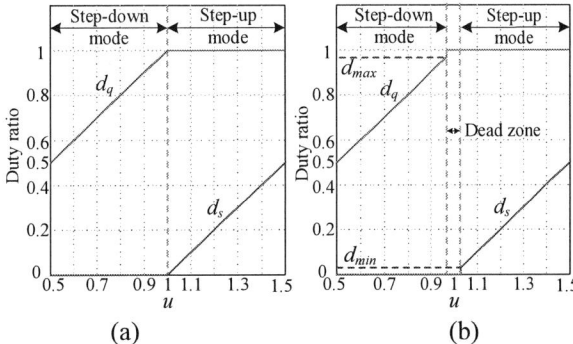

(a)   (b)

Fig.2: Duty ratios variation with u (a) without dead zone; (b) with dead zone

In addition, the converter suffers from mode transferring issues such as discontinuous current and power spikes due to parasitic parameters existing in the power stage when the transition goes through the region where the output voltage is very closed or equal to the input voltage. Therefore, this paper proposed a novel modulation strategy that achieves seamless mode transition, continuous output regulation, and ZVS operation of the high voltage side switches.

Topology and modulation

A. Mode Transfer Issue in Conventional Modulation Strategy

Fig.1 shows the topology of SUD-P³C in [10], which consists of two H-bridges, i.e., Hq and Hs, and a high-frequency transformer. The switches in Hq and Hs are modulated by the duty ratio d_q and d_s, respectively. In the conventional modulation strategy, the diagonal switches have the same duty ratio, while the duty ratios for upper and lower switches in the arm are phase-shifted by 180°. d_q and d_s in [10] are expressed by the unified modulation ratio u, as given in (1) and (2).

$$
d_q = \begin{cases} u & , 0.5 \leq u < 1 \\ 1 & , 1 \leq u \leq 1.5 \end{cases}
$$

(1)

$$
d_s = \begin{cases} 0 & , 0.5 \leq u < 1 \\ u - 1 & , 1 \leq u \leq 1.5 \end{cases}
$$

(2)

Obviously, S_{1-4} are switched in high frequency while S_{5-8} keep OFF when $0.5 \leq u < 1$, i.e., the converter operates as the traditionally isolated Boost converter. Similarly, it operates as an isolated Buck converter at $1 \leq u \leq 1.5$, when PWM signals drive S_{5-8} while S_{1-4} keeps ON. D_{1-4} avoids blocking the reverse voltage when C_2 voltage is negative. Neglecting power losses, the following function expresses the converter voltage gain.

$$
G(u) = \frac{V_{load}}{V_{in}} = \frac{n + 2u - 2}{n}, 0.5 \leq u \leq 1.5
$$

(3)

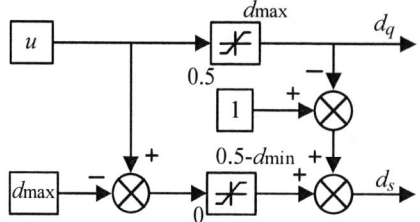

Fig.3: Diagram of the seamless modulation strategy.

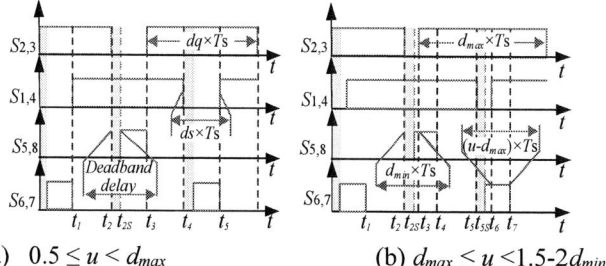

(a) $0.5 \leq u < d_{max}$ (b) $d_{max} \leq u < 1.5-2d_{min}$

Fig.4: PWM signals of the proposed modulation strategy.

Fig. 2 (a) indicates the operation area of two modes when the dead zone is not taken into consideration. Ideally, the mode transfer occurs at $u = 1$, i.e. $V_{load} < V_{in}$ when $u < 1$ and $V_{load} > V_{in}$ when $u > 1$.

When $u = 1$, the voltage over capacitor C_2 is zero, i.e. $V_{in} = V_{load}$, and thereby the source is directly connected to the load, and the converter processes zero power, thereby the system presents the peak efficiency. However, as u gradually approaches 1, the switching time will become very short, e.g., at a switching frequency (f_{sw}) of 100 kHz and $0.99 < u < 1.01$, the OFF time of S_{1-4} and the ON time of S_{5-8} are less than 100 ns, which is very challenging for switches and drivers to complete commutation. As presented in [10], incomplete charging and discharging of the switch's capacitor leads to a non-ideal voltage across the transformer.

Furthermore, the duty cycle limitation is necessary to recharge the bootstrap capacitor periodically when using half-bridge gate drivers. As a result, the duty cycle limitation is usually implemented to ensure the switching-time constraint, resulting in the dead zone in the duty cycle, as shown in Fig.2 (b). Define $d_{max} = 1 - d_{min}$, and there is a nonadjustable voltage range, i.e., $((1 \pm (2d_{min}/n)) \times V_{in}$. When the SUD-P³C is operated under the closed-loop control, the discontinuity leads to random jumps in the regulated voltage, i.e., the converter works at burst mode. It increases the controller design's complexity and leads to low-stability voltage regulation when u approaches 1.

B. The proposed Novel modulation strategy

In order to overcome such mode transferring issues, a novel modulation strategy is proposed and expressed in (4) and (5).

$$d_q = \begin{cases} u & , 0.5 \leq u < d_{max} \\ d_{max} & , d_{max} \leq u \leq 1.5 - 2 \cdot d_{min} \end{cases} \tag{4}$$

$$d_s = \begin{cases} 1 - u & , 0.5 \leq u < d_{max} \\ u - d_{max} + d_{min} & , d_{max} \leq u \leq 1.5 - 2 \cdot d_{min} \end{cases} \tag{5}$$

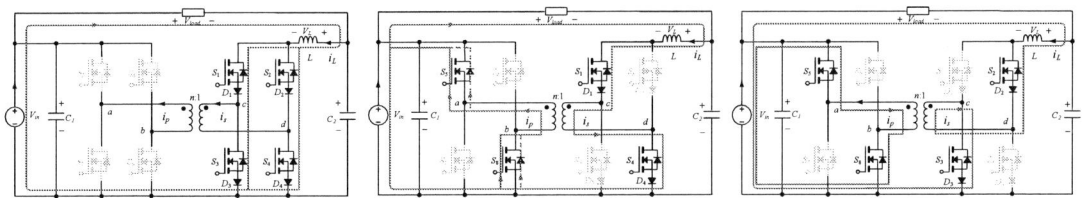

(a) State 1 in boost mode (b) State 2 in boost mode (c) State 3 in buck-boost mode

Fig. 5: Equivalent circuits for boost mode and buck-boost operating modes.

Fig. 3 shows the diagram of the proposed modulation strategy. Fig.4 (a) and (b) show the PWM signals when $0.5 \leq u < d_{max}$ and $d_{max} \leq u < 1.5-2d_{min}$, respectively. When $0.5 \leq u < d_{max}$, d_q and d_s are complementary. The operating mode is called boost mode as the converter works as an isolated boost converter, which has four symmetrical operating states. The first two states are analyzed below.

State 1 in boost mode: t_1 - t_2

From t_1 to t_2, S_{1-4} keeps ON while S_{5-8} keeps OFF. Power is transferred from source to load directly, as shown in Fig.5 (a). v_L can be expressed by (6).

$$v_L = L \cdot \frac{di_L}{dt} = v_{in} - v_{load} \tag{6}$$

State 2 in boost mode: $t_2 - t_3$

At t_2, S_2 and S_3 are turned off, i_L is forced to flow through $S_{1,4}$. In Hs, i_p flows through the antiparallel diodes of S_5 and S_8, dash line in Fig.5 (b), until S_5 and S_8 are turned on at t_{2S}, solid line in Fig.5 (b). Therefore, the dead-band delay between $S_{2,3}$ and $S_{5,8}$ as well as $S_{1,4}$ and $S_{6,7}$ can guarantee ZVS switching of S_{5-8}, as shown in the grey shaded area of Fig.4. v_L is expressed by (7).

$$v_L = L \cdot \frac{di_L}{dt} = v_{in} - v_{load} - \frac{v_{in}}{n} \tag{7}$$

Consequently, when $0.5 \leq u < d_{max}$, this novel modulation strategy's voltage gain can still be expressed by (3). When $u \geq d_{max}$, d_q is limited at d_{max} while $d_s = u - d_{max} + d_{min}$. To ensure d_s is smaller than 0.5, the maximum of u is $1.5-2d_{min}$. Compared to the boost mode, two symmetrical operating states are added in this operating mode. v_L for State1 and State 2 in this mode can also be expressed by (6) and (7), respectively. Fig.5 (c) shows the equivalent circuit for State 3 in this operating mode. At t_3, S_2 and S_3 are turned on again. i_L migrates from $S_{1,4}$ to $S_{2,3}$. As a result, the currents across the transformer (i_p and i_s) start to reduce. When currents pass through the $S_{1,4}$ and $S_{2,3}$ are equal, i_p and i_s are zero. Then i_p reversely increases. Due to the blocking of D_1 and D_4, the current pass through $S_{1,4}$ continues to reduce until it reaches 0. In contrast, the current pass through $S_{2,3}$ continues to increase. v_L in this period is expressed by (8).

$$v_L = L \cdot \frac{di_L}{dt} = v_{in} - v_{load} + \frac{v_{in}}{n} \tag{8}$$

Applying the volt-second balance principle, the output voltage function for $u \geq d_{max}$ is derived, which is the same as for $u < d_{max}$, thereby the $G(u)$ of the proposed strategy is expressed by (3), but the range of u is reduced to 0.5 to $1-2d_{min}$. As a result, the maximum output voltage is reduced to $(n + 1 - 4d_{min}) \times V_{in}/n$. Although the mode switching occurs at $u = d_{max}$, the output voltage equaling the input voltage still occurs when $u=1$. In other words, this operating mode achieves step-down voltage regulation when $u \geq d_{max}$ and step-up voltage regulation at $u \geq 1$, and is called buck-boost mode. Similar to the boost mode, the minimum switching-off time on S_{1-4} is ensured. On the one hand, the regulation dead zone when u approach 1 is avoided. On the other hand, the transformer is not short-circuited by the low-voltage side when $u = 1$. As a result, the proposed strategy addresses the transfer issue, and the only sacrifice is that the maximum regulation voltage is slightly reduced. Although compared to the step-up mode in the conventional modulation strategy, the proposed modulation strategy has a bit higher processing power due to the additional operating states.

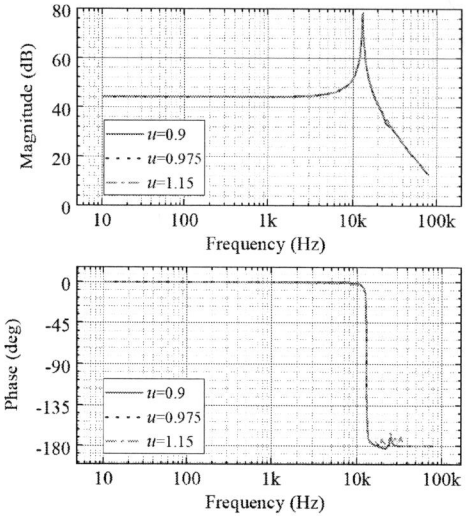

Fig. 6: Frequency response of the control to output voltage transfer function.

Thanks to the ZVS switching of S_{5-8}, the proposed strategy improves efficiency when $u < d_{max}$ and suppresses spikes caused by the hard switching of S_{5-8} when $u \geq d_{max}$.

C. Small Signal Analysis

Based on the operation principle, the averaged equations of inductor voltage and the capacitor current for the two modes are the same,

$$L\frac{d\langle i_L\rangle_{Ts}}{dt} = \frac{(n + 2u - 2)\langle v_{in}\rangle_{Ts}}{n} - \langle v_{load}\rangle_{Ts} \tag{9}$$

$$C_2\frac{d\langle v_{C2}\rangle_{Ts}}{dt} = \frac{\langle v_{load}\rangle_{Ts}}{R} - \langle i_L\rangle_{Ts} \tag{10}$$

Introducing perturbation to the state variables and other quantities and neglecting the second-order terms, the linearized small-signal equations are given by (11) and (12).

$$sL\,\hat{i}_L = \frac{\hat{v}_{in}}{n}(n + 2u - 2) + \frac{2V_{in}}{n}\hat{u} - \hat{v}_{load} \tag{11}$$

$$sC_2\left(\hat{v}_{in} - \hat{v}_{load}\right) = \frac{\hat{v}_{load}}{R} - \hat{i}_L \tag{12}$$

Arranging (12), $\hat{i}_L$ is derived.

$$\hat{i}_L = \frac{(1 + sRC_2)\,\hat{v}_{load}}{R} - sC_2\,\hat{v}_{in} \tag{13}$$

Introducing (13) to (11), the control-to-output transfer is derived by (14).

$$\left.\frac{\hat{v}_{load}}{\hat{u}}\right|_{\hat{v}_{in} = 0} = \frac{2RV_{in}}{s^2nLC_oR + snL + nR} \tag{14}$$

In order to verify the proposed small-signal analysis, the AC sweep simulation was carried out in the converter circuit model constructed by PLECS. The system specifications are listed in Table. I.

Table I: System specification of the simulation

Name	V_{in}	L	C_2	n	R	f_{sw}	d_{max}
Value	400 V	15 μH	10 μF	5:1	70 Ω	100 kHz	0.95

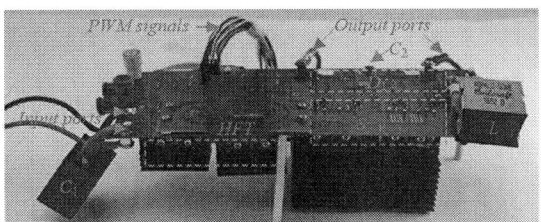

Fig. 7: Experimental prototype of the SUD-P³C.

TABLE II: Main Parameters of prototype

Parameter	Value	Notes
S_{1-4}	IRF200P223, 200 V/100 A	R_{dson}: 8.5 mΩ,
S_{5-8}	IMW65R072M1H, 650 V/26 A	R_{dson}: 72 mΩ, C_{oss}:100 pF
D_{1-4}	V30200C, 200 V/2×15 A	V_F @ I_F = 15 A: 0.648 V
L	AGP4233-153ME	Inductance: 15 μH, R_{dc}:2.85 mΩ
HFT	PCB transformer	20:4, Core: ELP58/11/38 N87

The frequency response curves in the case of $u = 0.9$ (Boost mode, $V_{load} = 384$ V) and $u = 0.975$ (Boost-buck mode, $V_{load} = 396$ V) and $u = 1.15$ (Boost-buck mode, $V_{load} = 424$ V) are plotted in Fig.6. The proposed strategy has the same frequency response for different operating modes and voltage regulation ranges. In the novel strategy, the transfer functions are almost identical in the frequency spectrum of 10 Hz - 80 kHz, so only one unified controller is required for the two operating modes.

Experimental results

A 3kW prototype, as shown in Fig.7, is designed, constructed and tested in order to verify the proposed modulation strategy. The specifications of the experimental system are listed in Table I. Table. II shows the parameters of the main components.

A 200 ns dead time is set to ensure ZVS switching of S_{5-8}. Fig. 8 shows the operating waveforms of the conventional modulation strategy in [10]. In Fig.8 (a), S_{5-8} can't be fully charged or discharged due to the too-short switching time, resulting in V_{ab} being only half of the theoretical value. The converter does not process power at $u = 1$ because S_{1-4} is ON while S_{5-8} is OF, as shown in Fig.8 (b). The hard switching of S_{5-8} results in a large voltage spike, 577 V, as shown in Fig.8 (c). As shown in Fig.9, the proposed modulation strategy ensures enough OFF time for S_{5-8} and ON time for S_{1-4} even at $u = 1$, and the voltage spike is suppressed. In order to observe the behavior during the mode transition, the control signal u was programmed to increase linearly from 0.7 to 1.2, as shown in Fig. 10. The variation of output is continuous and linear. Moreover, the transformer's voltage gap is avoided compared to the waveforms reported in [10]. Therefore, the proposed strategy effectively solves the switching issue between different modes. Fig. 11 shows the output voltage and system efficiency versus u for the modulation strategy in [10] and the proposed one.

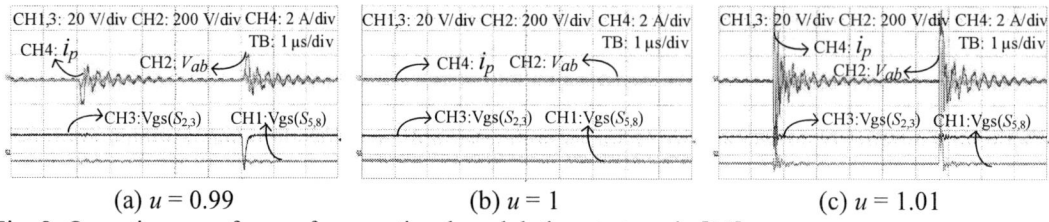

(a) $u = 0.99$ (b) $u = 1$ (c) $u = 1.01$

Fig. 8: Operating waveforms of conventional modulation strategy in [10].

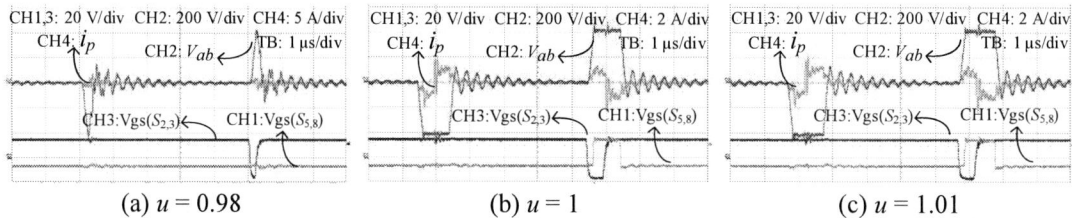

(a) $u = 0.98$ (b) $u = 1$ (c) $u = 1.01$

Fig. 9: Operating waveforms of proposed conventional modulation strategy.

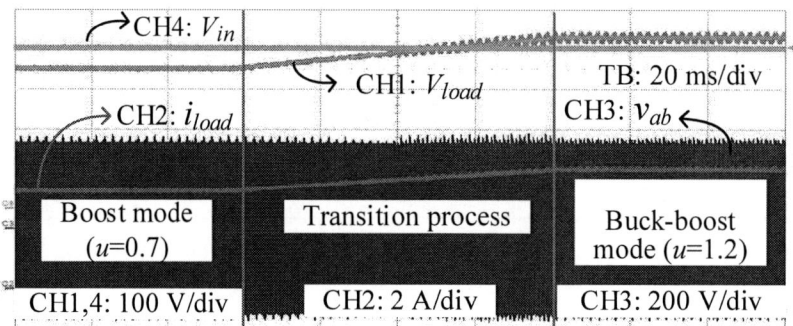

Fig. 10: Waveforms of operating mode transition.

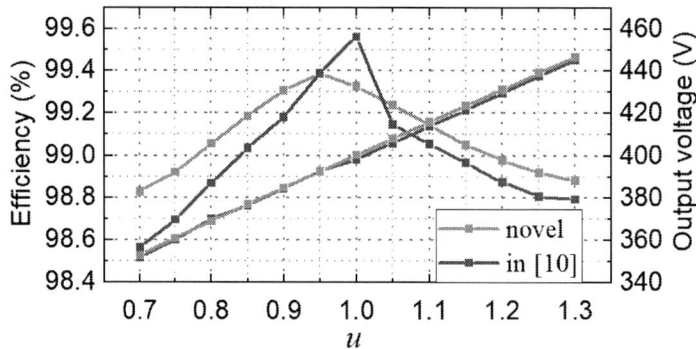

Fig. 11: Output voltage and system efficiency versus u.

The output voltage curve with the novel proposed strategy is highly linear, consistent with Fig.10. The novel approach presents a higher efficiency when $u < 0.95$ and a 14 % reduction in power losses at $u = 0.7$ due to ZVS switching.

CONCLUSION

This paper presents an improved modulation strategy with seamless mode switching, solving the stability problem when the input and output voltages are close to each other and achieving continuous regulation.

References

[1] B. Min, J. Lee, J. Kim, T. Kim, D. Yoo and E. Song, "A New Topology with High Efficiency Throughout All Load Range for Photovoltaic PCS," in *IEEE Transactions on Industrial Electronics*, vol. 56, no. 11, pp. 4427-4435, Nov. 2009, doi: 10.1109 /TIE. 2008.928098.

[2] H. Zhou, J. Zhao and Y. Han, "PV Balancers: Concept, Architectures, and Realization," in *IEEE Transactions on Power Electronics*, vol. 30, no. 7, pp. 3479-3487, July 2015, doi: 10.1109/TPEL.2014.2343615.

[3] K. A. Kim, P. S. Shenoy and P. T. Krein, "Converter Rating Analysis for Photovoltaic Differential Power Processing Systems," in *IEEE Transactions on Power Electronics*, vol. 30, no. 4, pp. 1987-1997, April 2015, doi: 10.1109/TPEL.2014.2326045.

[4] H. Jeong, H. Lee, Y. Liu and K. A. Kim, "Review of Differential Power Processing Converter Techniques for Photovoltaic Applications," in *IEEE Transactions on Energy Conversion*, vol. 34, no. 1, pp. 351-360, March 2019, doi: 10.1109/TEC.2018.2876176.

[5] M. C. Mira, Z. Zhang, K. L. Jørgensen and M. A. E. Andersen, "Fractional Charging Converter With High Efficiency and Low Cost for Electrochemical Energy Storage Devices," in *IEEE Transactions on Industry Applications*, vol. 55, no. 6, pp. 7461-7470, Nov.-Dec. 2019, doi: 10.1109/TIA.2019.2921295.

[6] T. Kanstad, M. B. Lillholm and Z. Zhang, "Highly Efficient EV Battery Charger Using Fractional Charging Concept with SiC Devices," *2019 IEEE Applied Power Electronics Conference and Exposition (APEC)*, 2019, pp. 1601-1608, doi: 10.1109/ APEC. 2019. 8722191.

[7] V. M. Iyer, S. Gulur, G. Gohil and S. Bhattacharya, "An Approach Towards Extreme Fast Charging Station Power Delivery for Electric Vehicles with Partial Power Processing," in *IEEE Transactions on Industrial Electronics*, vol. 67, no. 10, pp. 8076-8087, Oct. 2020, doi: 10.1109/TIE.2019.2945264.

[8] J. Anzola et al., "Review of Architectures Based on Partial Power Processing for DC-DC Applications," in *IEEE Access*, vol. 8, pp. 103405-103418, 2020, doi: 10.1109/ ACCESS.2020.2999062.

[9] J. R. R. Zientarski, M. L. d. S. Martins, J. R. Pinheiro and H. L. Hey, "Series-Connected Partial-Power Converters Applied to PV Systems: A Design Approach Based on Step-Up/Down Voltage Regulation Range," in *IEEE Transactions on Power Electronics*, vol. 33, no. 9, pp. 7622-7633, Sept. 2018, doi: 10.1109/TPEL.2017.2765928.

[10] Liu, C., Zhang, Z., & Andersen, M. A. E. (Accepted/In press). An Efficient Voltage Step-up/down Partial Power Converter (SUD-PPC) using Wide Bandgap devices. *In Proceedings of IEEE Workshop on Wide Bandgap Power Devices and Applications IEEE.*

Performances Analysis of Non-Model-Based
Speed Estimation Algorithms for Motor Drives

Gaetano Turrisi[1], Luigi Danilo Tornello[1], Giacomo Scelba[1],
Giulio De Donato[2], Giuseppe Scarcella[1]

University of Catania, Viale Andrea Doria 6, Catania, Italy [1]
University of Rome "La Sapienza", Via Eudossiana 18, Roma, Italy [2]
E-Mail: giacomo.scelba@unict.it, giulio.dedonato@uniroma1.it

Keywords

«Variable speed drives», «resolution», «position measurement», «speed estimation», «motion control», «speed control», «digital filters», «stability analysis».

Abstract

This paper investigates the performances of speed-controlled motor drives using non-model-based speed estimation algorithms. A suitable modelling of the speed estimation algorithms combined to the analytical representation of the instantaneous quantized speed of finite resolution position sensors are exploited to evaluate the filtering action of the estimation algorithms, and the stability and rejection to torque disturbances of speed-controlled drives at low rotational speeds; the last operating condition is very critical for motor drives, especially for that using low resolution position sensors. In this study, the theoretical analysis is experimentally validated on a 2kW PMSM drive.

Introduction

In several applications a closed loop speed control is required, as well as the rotor position for field orientation in AC motor drives [1]-[3] and for switching in different categories of synchronous drives, such as BLDC, SRM and wound field doubly salient generator drives [4]-[8]. This is accomplished using a single position sensor, and by estimating the speed through an appropriate speed estimation algorithm [1]. Alternatively, the position and speed can be carried out through model-based and high frequency injection-based sensorless algorithms [7]-[14].

Regardless of the technology employed in the position sensor (incremental/absolute encoders, resolvers, inductive position sensors or Hall effect sensors), the selection of it is influenced by several factors, such as accuracy, resolution, operating temperatures, sensibility to mechanical, magnetic and electrical disturbances, maximum speed rating, and cost [15]. In particular, the choice of the position sensor resolution can significantly impact on the accuracy and bandwidth of the estimated speed and cost of the drive, especially in low-cost applications where a considerable reduction of cost and size of the sensor is desirable [4], [5], [16], [17].

The rotor position measurement is characterized by a finite resolution that can be expressed in number of discrete states per revolution, N_{ds}. Besides the sensor, the speed estimation method plays a key role as it contributes to satisfy the desired performance requirements. Various speed estimation methods are available in the technical literature, which can be classified in two main categories: model-based and non-model-based speed estimation algorithms; the last category is still widely used in many industrial applications not only for their simple implementation but also because they do not need additional information on the properties of the mechanical load [18]-[19]. Some of non-model-based speed estimation algorithms are implemented according to the first-order Taylor approximation of rotor speed:

$$\omega_{re}(t) = \frac{d\theta_{re}(t)}{dt} \approx \hat{\omega}_{re} = \frac{\theta_{re}(t_1)-\theta_{re}(t_2)}{t_1-t_2} = \frac{\Delta\theta_{re}}{\Delta t} \tag{1}$$

where θ_{re} is the rotor angle, while ω_{re} and $\hat{\omega}_{re}$ are the actual and estimated rotor speeds. Two straightforward speed estimation implementations can be carried out starting from (1): the first, named as fixed time method (FTM) [18]-[27], is based on counting the number of position measurement updates in a set time interval, while the second method measures the time elapsed between two consecutive position measurement updates, and is known as fixed position method (FPM), or period-based speed estimation, [18]-[27]. While FTM is very accurate at high speeds, FPM is effective at low and medium speeds; motor drives can combine both methods to cover the full speed range, [18]. More advanced speed estimation algorithms have been also presented in the literatures, mainly based on higher-order Taylor approximations (TSE) [20], [23], polynomial interpolation [23]-[24], and least-squares fit (LSF) [20]-[24], where the main aim of all of these methods is to properly interpolate the quantized discontinuous position measurement. Most studies are mainly focused to the analysis of the filter characteristics of those algorithms, without providing an in-depth analysis of the dynamic response, stability and disturbance rejection capability associated to these estimation methods, which are also related to the concept of instantaneous quantized speed of finite resolution position sensors. With this paper, the authors strive to fill this gap, presenting a methodology for investigating the performances of FPM speed estimation algorithms, starting from their small signal modelling around an operating point. Although of general validity, the proposed approach has been applied to three different FPM algorithms, based on Taylor series expansion (TSE1) [28]-[29], and least-squares fit (LSF1/4 and LSF1/8) [20]-[22].

The rest of the paper is organized as follows: In the next section, a brief description of three different FPM algorithms considered in this study is presented; then, small signal models for each implementation are carried out. The performances of these estimation algorithms are thus analysed in the speed-controlled drive, according to their filtering, stability, and disturbance rejection capability. The theoretical study is experimentally evaluated on a 2.6kW PMSM drive, and conclusions are finally given in the last section.

Fixed Position-Based Speed Estimation Algorithms

The FPM implementation provides the estimated speed by measuring the time τ_d between two consecutive updates of the quantized angular position $\theta_{re}^{(q)}(t)$. The time τ_d is determined as in (2), by counting the number of clocks n of a high-frequency counter with frequency $f_{CLK}=1/T_{CLK}$. This count can be expressed as the reciprocal value of the product between the sensor's resolution N_{ds} and the frequency f_{re0} associated to a constant angular rotor speed ω_{re0}.

$$\tau_d = nT_{CLK} = \frac{1}{N_{ds}f_{re0}} \tag{2}$$

FPM can be implemented in different ways and three different methods have been considered in the following study.

(a). Taylor Series Expansion (TSE) Algorithm

In this approach, the estimated speed $\hat{\omega}_{re}(k)$ is computed according to the first order approximation of the Taylor series expansion, indicated in the following study as TSE1:

$$\hat{\omega}_{re}(k) \approx \frac{\Delta\theta_{re}(k)}{\Delta t(k)} \tag{3}$$

where k is the kth time elapsed between two consecutive position measurement updates.

(b). Least Square Fit (LSF) Algorithms

An alternative way to estimate the rotor speed is to construct a polynomial fitting of the input position/time data by using the least-squares fit (LSFs) method. In the LSFs technique an Nth-order polynomial can be fit through the M most recent data points $\theta_{re}(1), \theta_{re}(2), ..., \theta_{re}(M)$ acquired from the rotor position sensor, where typically $M > N+1$ [20], [22]. Generally, with LSFN/M is defined a least-squares fit method with a Nth-order polynomial to fit M points of the acquisition rotor position data. For example, with LSF1/4 is defined as a least-squares fit method with a 1st-order polynomial ($N=1$) (a straight line) to fit four rotor position data points ($M=4$), or LSF1/8 is a 1st-order polynomial ($N=1$) (a straight line) to fit eight rotor position data points ($M=8$), etc. If with $t(k)$ is indicated the generic instant of time for which a generic pulse $\theta_{re}(k)=2\pi k/N_{ds}$ of the position sensor occurs, the estimated time instant $\hat{t}(k)$ against which will occur next pulse is computed according to the polynomial fitting of $t(k)$, as:

$$\hat{t}(k) = a_0 + a_1\theta_{re}(k) + a_2\theta_{re}(k)^2 + \ldots + a_N\theta_{re}(k)^N \tag{4}$$

where k is the generic rotor position data point, with $k=1,\ldots,M$. Therefore, by considering M rotor position data points it is possible to define the linear system (5).

$$\begin{bmatrix} \hat{t}(1) \\ \hat{t}(2) \\ \vdots \\ \hat{t}(M) \end{bmatrix} = \begin{bmatrix} 1 & \theta_{re}(1) & \theta_{re}(1)^2 & \cdots & \theta_{re}(1)^N \\ 1 & \theta_{re}(2) & \theta_{re}(2)^2 & \cdots & \theta_{re}(2)^N \\ \vdots & \vdots & \vdots & \vdots\vdots\vdots & \vdots \\ 1 & \theta_{re}(M) & \theta_{re}(M)^2 & \cdots & \theta_{re}(M)^N \end{bmatrix} \begin{bmatrix} a_0 \\ a_1 \\ \vdots \\ a_N \end{bmatrix} \qquad \hat{t} = \boldsymbol{\Theta}\,\boldsymbol{a} \tag{5} \qquad \boldsymbol{\Theta} = \begin{bmatrix} 1 & 1 & 1^2 & \cdots & 1^N \\ 1 & 2 & 2^2 & \cdots & 2^N \\ \vdots & \vdots & \vdots & \vdots\vdots\vdots & \vdots \\ 1 & M & M^2 & \cdots & M^N \end{bmatrix} \tag{6}$$

where $\hat{t}$ is the estimated time for each rotor position data points acquired, $\boldsymbol{\Theta}$ is the coefficients matrix of the fitting polynomials, while $\boldsymbol{a}$ is the vector of the unknown coefficients of the fitting polynomials. The consider matrix $\boldsymbol{\Theta}$ depends on each rotor position data points and resolution N_{ds}. Generally, it is easier to rewrite the coefficients matrix $\boldsymbol{\Theta}$ by normalizing it with respect to $2\pi/N_{ds}$, and $\boldsymbol{\Theta}$ can be rewritten as shown in (6). The least squares method minimizes the sum of squared errors E, providing the coefficients of $\boldsymbol{a}$.

$$e = t - \hat{t} \tag{7} \qquad E = \sum e^2 = e^T e = (t - \boldsymbol{\Theta}\,a)^T (t - \boldsymbol{\Theta}\,a) \tag{8} \qquad a = (\boldsymbol{\Theta}^T \boldsymbol{\Theta})^{-1} \boldsymbol{\Theta}^T t \tag{9}$$

The rotor speed estimation can be evaluated by considering the derivative of (4):

$$\frac{d\hat{t}(k)}{d\theta_{re}(k)} = 0 + a_1 + 2a_2\theta_{re}(k) + \ldots + Na_N\theta_{re}(k)^{N-1} \tag{10} \qquad \frac{d\hat{t}}{d\boldsymbol{\Theta}} = q^T \bar{\boldsymbol{\Theta}} t \tag{11} \qquad g = (q^T \bar{\boldsymbol{\Theta}})^T \tag{12}$$

The previous polynomial coefficients $\boldsymbol{a}$ can be substitute in (10) obtaining a generalized form (11), where $q^T = [0\ 1\ 2M\ 3M^2 \ldots NM^{N-1}]$ is the vector of coefficients of (10), while $\bar{\boldsymbol{\Theta}} = (\boldsymbol{\Theta}^T \boldsymbol{\Theta})^{-1} \boldsymbol{\Theta}^T$. Finally, the coefficients to fit the rotor position data points are obtained in (12).

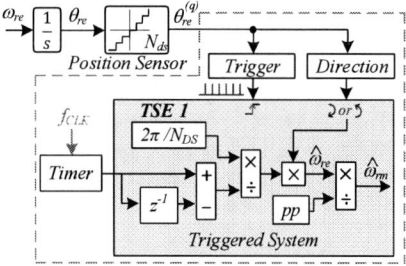

Fig. 1: FPM algorithm based on TSE1.

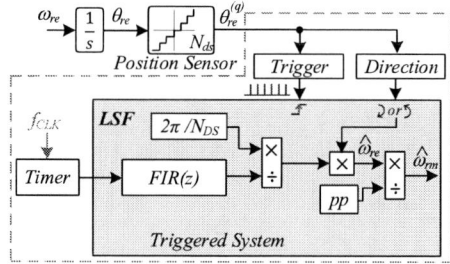

Fig. 2: FPM algorithm based on LSF.

It is shown in [2], [3] that the generic fitting polynomial $d\hat{t}(k)/d\Delta\theta_{re}(k)$ can be described through a finite impulsive response (FIR) digital filter. In this study two *LSFs* algorithms have been examined: LSF1/4 and LSF1/8. The first algorithm is based on a polynomial that fits four points on a straight-line, where the polynomial coefficients are $g_0 = 0.3$, $g_1 = 0.1$, $g_2 = -g_1$ and $g_3 = -g_0$ and the second algorithm LSF1/8 is implemented by considering a fitting polynomial of straight-line with eight points, where the polynomial coefficients are $g_0 = 0.0833$, $g_1 = 0.0595$, $g_2 = 0.0357$, $g_3 = 0.0119$, $g_4 = -g_3$, $g_5 = -g_2$, $g_6 = -g_1$ and $g_7 = -g_0$.

The discrete time function of the simplified FIR filter used in the LSF can be expressed as in (13), whose representation in the Laplace domain is given by (14).

$$FIR(z) = g_0 + g_1 z^{-1} + \ldots + g_{M-1} z^{-(M-1)} \tag{13} \qquad FIR(s) = g_0 + g_1 e^{-s\tau_d} + \ldots + g_{M-1} e^{-(M-1)s\tau_d} \tag{14}$$

The speed estimation algorithms described above inherently introduce a delay equal or multiple of τ_d. Moreover, the accuracy of the estimated speed is related to the ratio between f_{CLK} and $1/\tau_d = N_{ds}f_{re0}$, which is strongly related to the motor speed, [18]-[29]. This becomes particularly evident at low rotational speed, as explained in the following sections.

Small-Signal Modeling

To represent the dynamic behaviour of FPM algorithms, their transfer functions around an operating point are determined by assuming that the shaft rotates at a constant speed $\omega_{re0}=2\pi f_{re0}$. The following approach can be extended in a straightforward way to several FPM implementations.

(a). TSE1

As described in [28] and [29], the small-signal transfer function of the TSE1 method can be determined by considering the effects of a) the counter, represented with a moving average filter $H_{aveg}(s)$, b) the triggered system represented with a sample and hold $H_{S\&H}(s)$, and c) the control algorithm sampling time T_s, modelled with a further sample and hold $H_{Ts}(s)$. However, typical values of the sampling time T_s and clock time T_{CLK} are much smaller than τ_d, thus the effects of the control algorithm sampling time can be neglected, $H_{Ts}(s)\approx0$, and the moving average filter can be approximated to a sample and hold, $H_{aveg}(s)\approx H_{S\&H}(s)$, [28]-[29]. Therefore, the small-signal transfer function of TSE1, $H_{TSE1}(s)$, can be expressed as:

$$H_{TSE1}(s) = H_{aveg}(s)\, H_{S\&H}(s)\, H_{Ts}(s) \approx H^2_{S\&H}(s) = \frac{\left(1 - e^{-s\tau_d}\right)^2}{s^2\tau_d^2} \qquad (14)$$

$H_{TSE1}(s)$ can be expressed in terms of its magnitude, $|H_{TSE1}(j\omega)|$, which is a sinc-squared function (15), and its phase $\angle H_{TSE1}(j\omega)$ (16).

$$|H_{TSE1}(j\omega)| = \frac{4}{\omega^2\tau_d^2}\,sin^2\!\left(\frac{\omega\tau_d}{2}\right) \qquad (15) \qquad\qquad \angle H_{TSE1}(j\omega) = -\omega\tau_d \qquad (16)$$

(b). LSF

The small-signal transfer function representing the LSF algorithm around an operating point can be determined by using same approach applied for TSE1. As shown in Fig. 2, the small-signal transfer function of the generic LSF algorithm depends on these three elements: a) the digital filter $FIR(z)$, b) the triggered system represented with a sample and hold $H_{S\&H}(z)$, and c) the control algorithm sampling time T_s modelled by considering another sample and hold block $H_{Ts}(z)$. Even in this case the effect of $H_{Ts}(z)$ can be neglected compared to $H_{S\&H}(z)$. According to the definition of the sample and hold discrete time transfer function $H_{S\&H}(z) = 1-z^{-1}$, [30], the discrete time small-signal transfer function of the LSF can be defined as:

$$H_{LSF}(z) = FIR(z)H_{S\&H}(z)H_{Ts}(z) \approx FIR(z)\, H_{S\&H}(z) = \left(g_0 + g_1z^{-1} + ...+ g_{M-1}z^{-(M-1)}\right)\left(1 - z^{-1}\right) \qquad (17)$$

Which can be rewritten as:

$$H_{LSF}(z) \approx H_{FIR}(z)\, H_{S\&H}(z)^2 = \left[c_0 + c_1z^{-1}+...+c_{M-2}z^{-(M-2)}\right]\left(1 - z^{-1}\right)^2 \qquad (18)$$

where $H_{FIR}(z)$ is a modified digital filter FIR function of $FIR(z)$ with coefficients defined as $c_0 = g_0$, $c_1 = (g_1+g_0),...,c_{M-2} = (g_0+g_1+...+g_{M-2})$. Finally, the continuous time domain transfer function LSF around the operating point is obtained by considering the delay time τ_d as $z^{-1} = e^{-s\tau_d}$:

$$H_{LSF}(s) \approx H_{TSE1}(s)\, H_{FIR}(s) = \frac{\left(1 - e^{-s\tau_d}\right)^2}{s^2\tau_d^2}\left(c_0 + c_1\, e^{-s\tau_d}+...+c_{M-2}\, e^{-(M-2)s\tau_d}\right) \qquad (19)$$

From (19) it is possible to define the LSF1/4 small-signal transfer function:

$$H_{LSF1/4}(s) \approx H_{TSE1}(s)\, H_{FIR}(s) = \frac{\left(1 - e^{-s\tau_d}\right)^2}{s^2\tau_d^2}\left(0.3 + 0.4\, e^{-s\tau_d} + 0.3\, e^{-2s\tau_d}\right) \qquad (20)$$

where the magnitude and phase for this speed estimator are:

$$|H_{LSF1/4}(j\omega)| = |H_{TSE1}(j\omega)|\,|0.4+0.6cos(\omega\tau_d)| \qquad (21) \qquad \angle H_{LSF1/4}(j\omega) = -2\omega\tau_d \qquad (22)$$

The magnitude obtained in (21) behaves like a sinc-square function but with an additional term that increases the filtering action of the LSF1/4 algorithm with respect to TSE1. As regards the phase, this transfer function highlights a delay twice compared to TSE1.

Similar procedure has been used to carried out the transfer function of the LSF1/8 algorithm, considering the delay action given by the different FIR filter structure. Its magnitude, $|H_{LSF1/8}(j\omega)|$ includes an additional term furtherly increasing the filtering action compared to TSE1 and LSF1/4. As regards the phase, a delay four time higher than TSE1 is observed.

$$|H_{LSF1/8}(j\omega)| = |H_{TSE1}(j\omega)||0.1666cos(3\omega\tau_d)+0.2858cos(2\omega\tau_d)+0.3572cos(\omega\tau_d)+0.1905| \qquad (23)$$

$$\angle H_{LSF1/8}(j\omega) = -4\omega\tau_d \qquad (24)$$

Performance Analysis of FPM Algorithms

Starting from the small signal models, an in-depth analysis of FPMs performances can be carried out in terms of filtering action, closed loop stability analysis, torque load disturbance rejection. All these characteristics are strongly depending on the sensor finite resolution value N_{ds}.

(a). Filtering Action

The magnitude functions obtained in (15), (21) and (23) allow to evaluate the filtering action of each speed estimation algorithm on the quantization harmonics associated to the discontinuous position measurement. In particular, the study presented in [31] provided the analytical models of the quantized electrical rotor angle $\theta_{re}^{(q)}(t)$ and its time derivative, that is, the instantaneous quantized speed $\omega_{re}^{(q)}(t)$; these expressions are reported in (25) and (26) for a generic rotor position sensor of resolution N_{ds}, where it is assumed that the shaft rotates at ω_{re0}.

$$\theta_{re}^{(q)}(t) = \omega_{re0}t + \sum_{k=1}^{+\infty} \frac{2}{N_{ds}\,k} \sin(N_{ds}k\omega_{re0}t) \quad (25) \qquad \omega_{re}^{(q)}(t) = \omega_{re0} + \sum_{k=1}^{+\infty} 2\omega_{re0}\cos(N_{ds}k\omega_{re0}t) \quad (26)$$

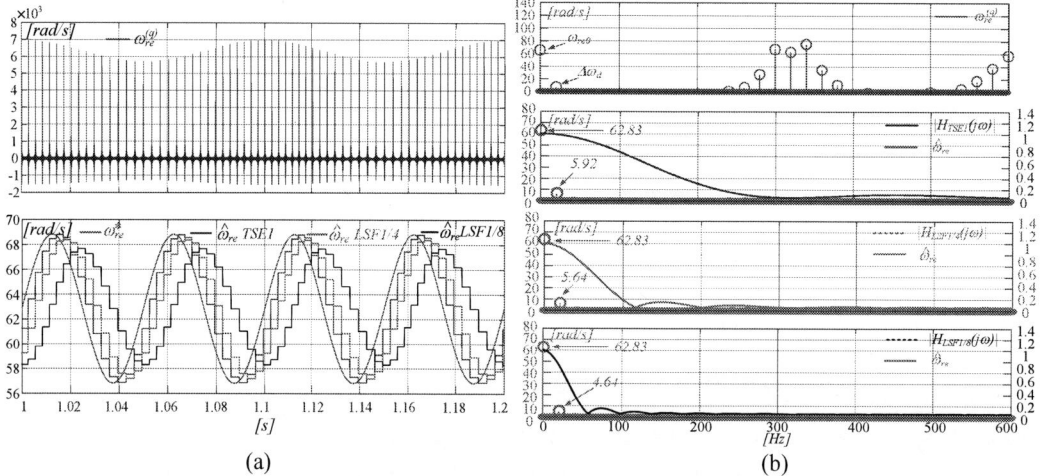

Fig. 3: Harmonic content of the instantaneous quantized $\omega_{re}^{(q)}$ and estimated speed $\hat{\omega}_{re}$ for the considered FPM algorithms: magnitudes $|H_{FPM}(j\omega)|$ (a); corresponding phases $\angle H_{FPM}(j\omega)$ (b).

Fig. 4 Time waveforms (a) and harmonic spectrum (b) of the quantized $\omega_{re}^{(q)}$ and estimated speeds $\hat{\omega}_{re}$, considering $\omega_{re0} = 2\pi 10$ rad/s and a superimposed sinusoidal disturbance with magnitude $\Delta\omega_d = 0.1\omega_{re0}$ and frequency $\omega_d = 2\pi 10$ rad/s.

The magnitude spectrum of $\omega_{re}^{(q)}(t)$ is displayed in Fig. 3a for $\omega_{re0}=2\pi10$ rad/s, and $N_{ds}=32$. The instantaneous quantized speed $\omega_{re}^{(q)}(t)$ features an infinite number of equally spaced quantization harmonics at $kN_{ds}\omega_{re0}$, each having an amplitude equal to $2\omega_{re0}$; the only physically meaningful harmonic is the dc component ω_{re0}. When the FPM methods are applied to $\omega_{re}^{(q)}$, the sinc-squared functions (15), (21) and (23) filter $\omega_{re}^{(q)}(t)$ by cancelling all quantization harmonics. It is worth noting that LSF1/4 and LSF1/8 feature a greater filtering action in the low frequency range.

With regards to the phase $\angle H_{FPM}(j\omega)$, delays equal to τ_d, $2\tau_d$ or $4\tau_d$ are included in the speed estimation loop with TSE1, LSF1/4, and LSF1/8 respectively. These delays have a significant impact on the stability of the speed closed loop, as will be outlined below.

Filtering action of FPMs can have a detrimental impact even on the estimation of external torque disturbances applied to the rotor shaft. Fig. 4 deals with this issue, displaying the waveforms and harmonic spectrums of the quantized $\omega_{re}^{(q)}$ and estimated speeds $\hat{\omega}_{re}$ carried out with the considered FPMs, when a speed sinusoidal disturbance is superimposed to ω_{re0}. An effective suppression of the quantization harmonics in the estimated speed are observed even in this operating condition for all FPM algorithms, Fig. 4b, even though TSE1 provides a more accurate estimation of the speed disturbance, Fig. 4a.

(b). Speed Controller Design

The stability analysis of the closed speed loop including the FPM algorithms has been performed by exploiting the modelling described in the previous section. In particular, the design of speed loop controller $C(s)$ is accomplished according to the speed loop bandwidth f_{BW} requested by the drive specifications. It is assumed that the actual shaft rotor speed is the feedback, and the mechanical system $P(s)$ is represented only with its inertia J, neglecting mechanical frictions, (27). The inner current control loop is also neglected as its bandwidth is significantly higher than external speed loop. A small signal closed loop control structure including the FPM algorithm is displayed in Fig. 5.

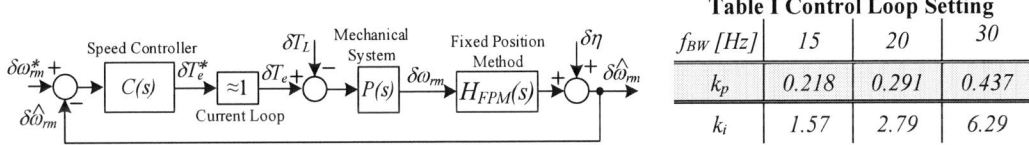

Table I Control Loop Setting

f_{BW} [Hz]	15	20	30
k_p	0.218	0.291	0.437
k_i	1.57	2.79	6.29

Fig. 5: Small signal closed loop control including FPM algorithms.

The proportional k_p and integral k_i gains of the PI controller $C(s)$ can be computed by imposing the two poles of the closed loop transfer function $W(s)$ (28):

$$P(s) = \frac{1}{sJ} \quad (27) \qquad C(s) = \frac{k_p s + k_i}{s} \qquad F(s) = C(s)P(s) = \frac{k_p s + k_i}{s^2 J} \qquad W(s) = \frac{F(s)}{1+F(s)} = \frac{k_p s + k_i}{s^2 J + k_p s + k_i} \quad (28)$$

In particular, $F(s)$ and $W(s)$ are represented in terms of the poles p_1 and p_2 as indicated in (29), allowing to express the crossover angular frequency $\omega_{ci} = 2\pi f_{ci}$ and the closed loop bandwidth $\omega_{BW} = 2\pi f_{BW}$ as indicated in (31) and (33) respectively, [30], [32].

$$F(s) = \frac{(p_1+p_2)s+p_1 p_2}{s^2} \qquad W(s) = \frac{(p_1+p_2)s+p_1 p_2}{s^2+(p_1+p_2)s+p_1 p_2} \qquad k_p=(p_1+p_2)J \quad k_i=p_1 p_2 J \quad (29)$$

$$|F(j\omega_{ci})| = 1 \quad (30) \qquad\qquad f_{ci} = \frac{1}{2\pi}\sqrt{\frac{(p_1+p_2)^2+\sqrt{(p_1+p_2)^4+4(p_1 p_2)^2}}{2}} \quad (31)$$

$$|W(j\omega_{BW})| = \frac{\sqrt{2}}{2}\,(32) \qquad f_{BW} = \frac{1}{2\pi}\sqrt{\frac{p_1^2+p_2^2+4p_1 p_2+\sqrt{(p_1^2+p_2^2+4p_1 p_2)^2+4(p_1 p_2)^2}}{2}} \quad (33)$$

A common approach in the selection of these two closed loop poles is to space them one decade apart from each other on the negative real axis, i.e. $p_2 = 0.1p_1$; in this way, it is possible from (33) to obtain an easy relationship linking the fastest pole p_1 and the closed loop bandwidth: $p_1 \approx 1.68\pi f_{BW}$. From the last expression, it is possible to set the speed controller gains k_p and k_i according to (29). Table I lists the gain values computed for different bandwidths f_{BW}.

(c). Stability Analysis

A metric providing the closed loop stability margin is given by the phase margin $m_{\varphi i}$ associated to the open loop transfer function $F(s)$:

$$m_{\varphi i} = \angle F(j\omega_{ci}) + \pi = atan\left(\frac{\omega_{ci}k_p}{k_i}\right) \tag{34}$$

In order to evaluate the FPMs speed estimator influence on the closed loop stability, it is necessary to include the transfer functions $H_{FPM}(s)$ of speed estimators in the above analysis, as displayed in Fig. 5. The corresponding open loop $F_{FPM}(s)$ and closed loop $W_{FPM}(s)$ transfer functions are respectively given by:

$$F_{FPM}(s) = C(s)\,P(s)\,H_{FPM}(s) \tag{35} \qquad W_{FPM}(s) = \frac{C(s)\,P(s)\,H_{FPM}(s)}{1 + C(s)\,P(s)\,H_{FPM}(s)} \tag{36}$$

The phase margin m_{φ} and crossover angular frequency ω_c are modified compared to the ideal speed closed loop, because of the delay time τ_d which impacts on the magnitude and phase of the open loop frequency responses. As an example, Fig. 6 displays magnitude and phase of $F(s)$ and $F_{FPM}(s)$ for two speed loop bandwidths f_{BW}, in case of $m_{\varphi}=0°$ and $m_{\varphi}=60°$; these values of m_{φ} correspond to very different values of τ_d. The figures display the crossover frequencies f_c when $H_{FPM}(s)$ are included in the open loop transfer function. The comparison between $F(s)$ and $F_{FPM}(s)$ highlights limited differences between the ideal crossover frequencies f_{ci} and f_c, especially at high phase margins m_{φ}.

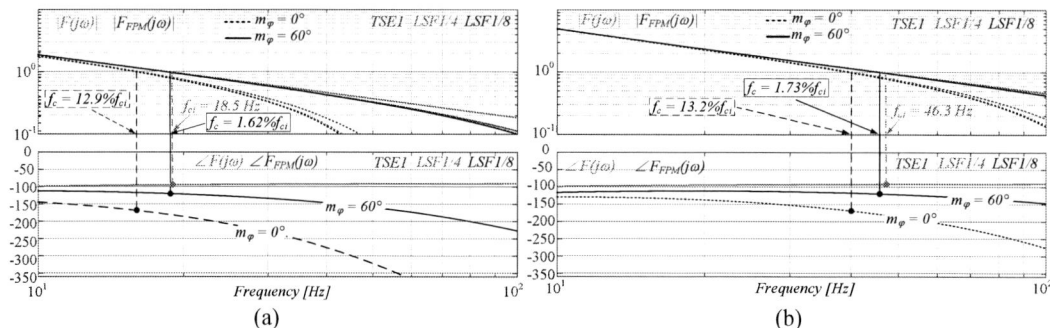

Fig. 6: Magnitude and phase comparisons between $F(s)$ and different $F_{FPM}(s)$ algorithms (TSE1, LSF1/4 and LSF1/8), (a) at $f_{BW}=20Hz$, for different m_{φ} and (b) at $f_{BW}=50Hz$ for different m_{φ}.

As a consequence, we can approximate $f_c \approx f_{ci}$ and thus compute the phase of $F_{FPM}(s)$ at f_{ci} and its corresponding phase margin:

$$\angle F_{FPM}(j\omega_{ci}) = atan\left(\frac{\omega_{ci}k_p}{k_i}\right) - \angle H_{FPM}(j\omega_{ci}) - \pi \qquad m_{\varphi} = \angle F_{FPM}(j\omega_{ci}) + \pi = m_{\varphi i} - \angle H_{FPM}(j\omega_{ci}) \tag{37}$$

Considering the phase delay defined for each FPM algorithms analysed in this paper, the corresponding phase margin evaluated at the ideal cross-over frequency f_{ci}, are:

$$TSE1:\ m_{\varphi} = m_{\varphi i} - \omega_{ci}\tau_d \qquad LSF1/4:\ m_{\varphi} = m_{\varphi i} - 2\omega_{ci}\tau_d \qquad LSF1/8:\ m_{\varphi} = m_{\varphi i} - 4\omega_{ci}\tau_d \tag{38}$$

The relationships (38) highlight the detrimental impact of τ_d on the stability margin and thus on the closed-loop damping ratio and rise time. For a given value of m_{φ} requested by the motor drive specifications at the lowest rotating speed ω_{rm0}, which is usually around $60°\div80°$, the max value of allowed delay time τ_{dmax} can be determined for each considered FPM as:

$$\tau_{dmaxTSE1} = \frac{atan\left(\frac{2\pi f_{ci}k_p}{k_i}\right) - m_{\varphi}}{2\pi f_{ci}}; \quad \tau_{dmaxLSF1/4} = \frac{\tau_{dmaxTSE1}}{2}; \quad \tau_{dmaxLSF1/8} = \frac{\tau_{dmaxTSE1}}{4} \tag{39} \quad \omega_{rm0} = \frac{2\pi}{pp\ \tau_{dmax}\ N_{ds}} \tag{40}$$

By exploiting (2), it is possible to link ω_{rm0} to sensor resolution N_{ds}, according to (40). Hence, for a specific set of N_{ds} and f_{BW}, the lowest operating speed ω_{rm0} will be different depending on the speed estimation algorithm.

(d). Torque Load Disturbance Rejection

Finally, the capability of the drive to mitigate the effects of load disturbances can be analysed by means of the dynamic stiffness transfer function. According to the control structure of Fig. 5, the dynamic stiffness is defined as:

$$DS_{FPM}(s) = \frac{\delta T_L(s)}{\delta \omega_{rm}(s)} = \frac{1+F_{FPM}(s)}{P(s)} = \frac{1+C(s)\,P(s)\,H_{FPM}(s)}{P(s)} \tag{41}$$

which is also affected by τ_d, as will be clearly underlined in the experimental tests.

Experimental Results

The above presented theoretical study has been validated through a wide campaign of experimental tests. In particular, a test bench has been arranged consisting of two PMSM drives whose shafts are mechanically coupled and sharing same DC bus. Motors specifications are summarized in Tables II and III. Both motor drives are fed by SiC inverters operated at 20kHz. A dSpace DS1006 have been used to control the IPM drive under test and implementing the FPMs algorithms with a f_{CLK} equal to 1MHz. Both electric drives are equipped with incremental encoders featuring $N_{ds} = 2048$, and different position sensor resolutions have been obtained by downsampling the encoder position measurement. Rotor field oriented control has been implemented in the drive under test, as displayed in Fig. 9, where the execution time of the current and speed loops are respectively $100\mu s$ and $200\mu s$. In all tests the current control loop bandwidth is fixed at 500Hz.

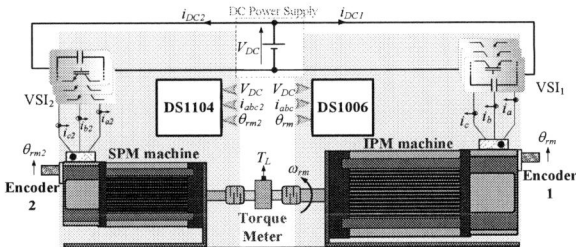

Fig. 7: Variable speed drive test bench.

Fig. 8: Test bench setup.

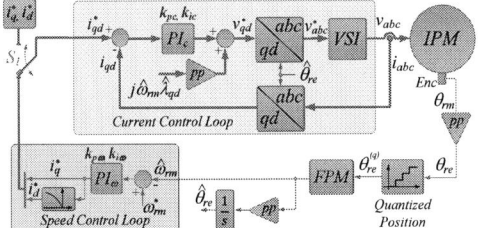

Fig. 9: Block diagram control of the drive under test.

Table II: SPM Motor Drive Data		Table III: IPM Motor Drive Data	
P_n	2 kW	P_n	3.6 kW
ω_n	6000 rpm	ω_n	2000 rpm
T_n	5 Nm	T_n	19.1 Nm
R_s	0.84 Ω	R_s	2.4 Ω
L_s	4.7 mH	L_q	12.189 mH
pp	3	L_d	9.947 mH
J	3.4 kgcm^2	pp	3
		J	21.7 kgcm^2

(a). Stability Limits of FPM Algorithms

Tests were undertaken to verify the effectiveness of the closed loop stability study. In particular, Fig. 10 depicts the stability limits ($m_\varphi =0°$) of the speed control loop implemented in the experimental setup (circle marks) for different rotor position sensor resolutions N_{ds} and by using different speed estimation algorithms; the same figure displays the stability limits determined by using the proposed analytical approach. The speed loop bandwidth is set at $f_{BW} = 20Hz$ for each N_{ds} configuration. It is noted that a good agreement between modelling and experimental tests is achieved for medium-low values of N_{ds}, while at higher N_{ds} the limited resolution of the encoder combined to the downsampling of the rotor position measurement and $f_{CLK}=1MHz$ yield to higher errors between simulations and experimental results. It has been verified via simulation that by increasing the f_{CLK} to $100MHz$ a significant reduction of the differences between simulations and experimental tests is achieved even for high N_{ds} (filled rhombus). It is worth noting that LSF-based methods underline greater instability region at low rotational speed compared to TSE1. Same figures display the minimum rotational speeds predicted by (40) guaranteeing specific value of m_φ.

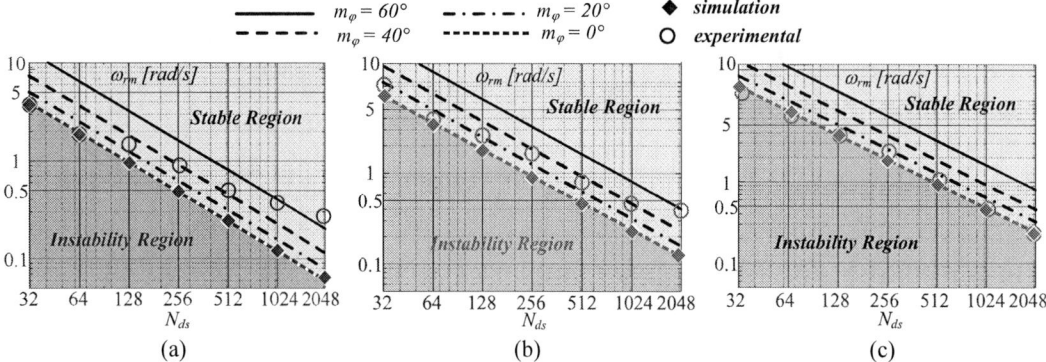

Fig. 10: Rotor speed stability vs instability regions at $f_{BW}=20Hz$: (a) TSE1, (b) LSF1/4 and (c) LSF1/8.

Further tests have been performed to validate the design of speed control loop including the FPM algorithms. In the following tests the drive is set according to the proposed approach to feature a closed speed loop bandwidth $f_{BW} = 20Hz$ ($k_p=0.29$, $k_i=2.79$), $N_{ds}=32$, and a phase margin $m_\varphi = 60°$ for the three considered speed loops. Based on the type of FPM, the fulfilment of these drive specifications is obtained by considering different reference speed ω^*_{rm} according to (40): *17rad/s, 34rad/s* and *69rad/s* for TSE1, LSF1/4 and LSF1/8 respectively. With these settings we expect to observe the same dynamic behaviour in all speed loops, even though they include different FPM algorithms. For each speed loop operating at ω^*_{rm}, an additional sinusoidal speed disturbance of magnitude $\Delta\omega_{rm} = 5rad/s$ and frequency variable from 1Hz to 100 Hz is superimposed to the reference speed to experimentally determine $|W_{FPM}(j\omega)|$.

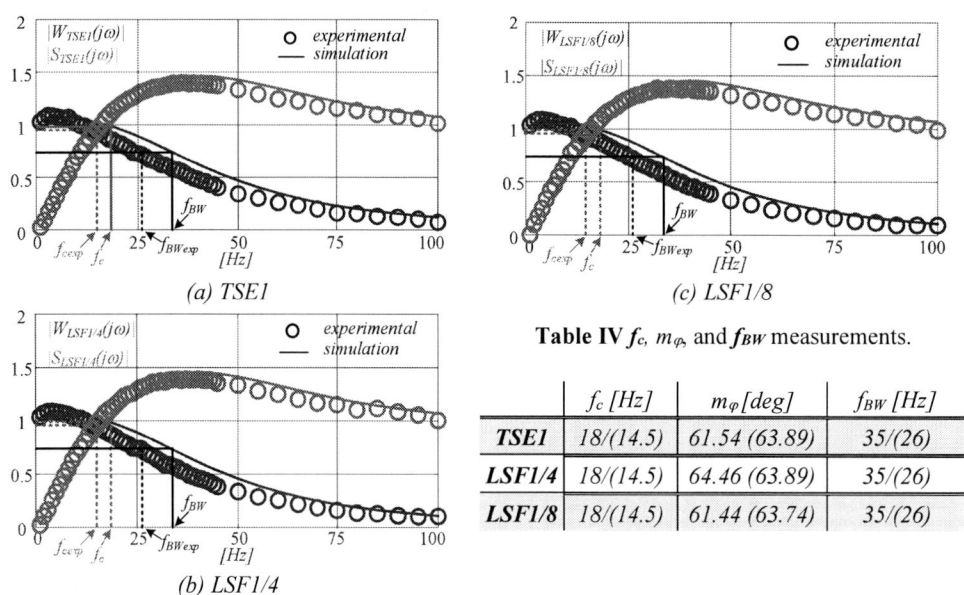

Table IV f_c, m_φ, and f_{BW} measurements.

	f_c [Hz]	m_φ [deg]	f_{BW} [Hz]
TSE1	18/(14.5)	61.54 (63.89)	35/(26)
LSF1/4	18/(14.5)	64.46 (63.89)	35/(26)
LSF1/8	18/(14.5)	61.44 (63.74)	35/(26)

Fig. 11 Comparison between the magnitude of the sensitivity $|S_{FPM}(s)|$ and closed loop $|W_{FPM}(s)|$ transfer functions carried out through the small signal models (within brackets) and experimental tests, with the following reference design specifications: $f_{BW}=20Hz$, $m_\varphi=60°$ and $N_{ds}=32$.

Then, a further campaign of measurement has been conducted for each speed loop by setting a constant ω^*_{rm}, and adding the same sinusoidal disturbance in the feedback speed $\hat{\omega}_{rm}$. In this way, the sensitivity transfer function $S_{FPM}(s)$ has been experimentally carried out $S_{FPM}(j\omega) = 1/(1 + F_{FPM}(j\omega))$. It is demonstrated by the control systems theory [32] that the cross point between the frequency responses

$|W_{FPM}(j\omega)|$ and $|S_{FPM}(j\omega)|$ provides the cross-over frequency f_c and the vector stability margin VM ($VM = 1/|W(j\omega_c)|$); the last quantity, in turns, can be related to m_φ by the following relationship: $m_\varphi =\arcsin(VM/2)$. Fig. 11 displays $|W_{FPM}(j\omega)|$ and $|S_{FPM}(j\omega)|$ for each FPM considered in this study. The cross-over frequency $f_{cexp} = 14.5Hz$ carried out with experimental tests is pretty much the same for all FPMs, confirming the theoretical analysis, whose deviation from the theoretical value $f_c=18Hz$ is lower than 20%. Looking at the phase margin determined from experimental test $m_{\varphi\_exp} = 63.8°$, the last is slightly higher compared the one used in (38)-(40). These tests confirm that TSE1 allows to achieve same dynamic performances compared to FPMs based on LSF, at lower operational speed. The f_{BW} can be experimentally evaluated by considering the frequency of $|W_{FPM}(j\omega)|$ when the condition $|W(j\omega_{BW})| = (\sqrt{2}/2)$ occurs. The tests underlined an increase of the speed loop bandwidth of the experimental test-bench f_{BWexp} higher than 30% compared to the requested f_{BW}, which can be considered fully satisfying the motor drive specifications.

(b). Torque Load Disturbance Rejection

Dynamic stiffness curves have been carried out by applying a sinusoidal torque disturbance featuring a magnitude of $\Delta T_L = 3Nm$ and a variable frequency range from 1Hz to 200 Hz with the drive controlled at a low and constant rotational speed $\omega^*_{rm}=20rad/s$ and $N_{ds}=32$. Fig. 12 summarizes the results of this analysis, where the theoretical curve (41) has been compared to the experimental tests. You can note that all methods feature a good disturbance rejection capability at medium high frequencies, while note a worsen behaviour in LSF-based algorithms around the medium frequencies because they feature a reduced phase margin m_φ.

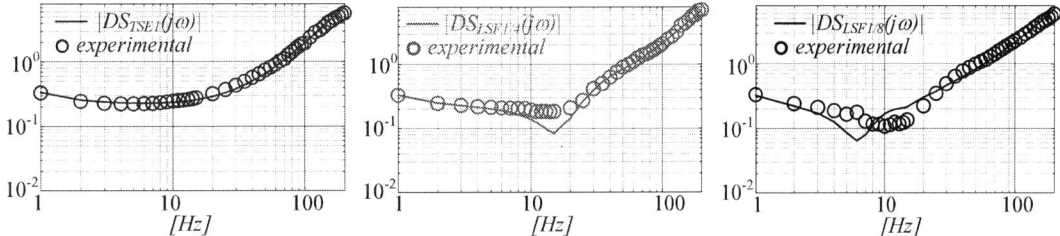

Fig. 12 Experimental vs theoretical Dynamic Stiffness magnitude for different FPM algorithms.

Conclusion

This paper has investigated the performances of speed-controlled motor drives using non-model-based speed estimation algorithms. Key observations include the following.

- Small signal modelling around an operating point of the speed closed loop allows to predict its filtering action, closed loop stability analysis, and torque load disturbance rejection, as a function of the requested motor drive specifications at the lowest rotating speed ω_{rm0} and position sensor resolution N_{ds}.
- The algorithm TSE1 features lower filtering action of the quantization speed harmonics compared to the LSF-based methods which limits its capability to mitigate the effects of measurement nonidealities and noise. On the other hand, its greater stability region allows the drive to operate at lower operating speeds.

Experimental results have confirmed the key results of this theoretical investigation.

References

[1] R.D. Lorenz and K. Van Patten, "High-resolution velocity estimation for all-digital, ac servo drives," IEEE Trans. Ind. Appl., vol. 27, no. 4, pp. 701 – 705, Jul./Aug. 1991.

[2] M. Cacciato, A. Consoli, G. Scarcella and G. Scelba, "Indirect Maximum Torque per Ampere control of induction motor drives," 2007 European Conference on Power Electronics and Applications, 2007, pp. 1-10.

[3] P. L. Jansen and R. D. Lorenz, "Transducerless position and velocity estimation in induction and salient AC machines," in IEEE Transactions on Industry Applications, vol. 31, no. 2, pp. 240-247, March-April 1995.

[4] Hyunbae Kim, Sungmo Yi, Namsu Kim and R. D. Lorenz, "Using low resolution position sensors in bumpless position/speed estimation methods for low cost PMSM drives," Fourtieth IAS Annual Meeting. Conference Record of the 2005 Industry Applications Conference, 2005., 2005, pp. 2518-2525 Vol. 4.

[5] B. Akin et al., "Low Speed Performance Operation of Induction Motors Drives Using Low-Resolution Speed Sensor," 2006 IEEE International Symposium on Industrial Electronics, 2006, pp. 2110-2115.

[6] M. Pulvirenti et al., "On-line stator resistance and permanent magnet flux linkage identification on open-end winding PMSM drives," IEEE Energy Conversion Congress and Exposition (ECCE), 2017, pp. 5869-5876.

[7] S. Morimoto et al., "Sensorless control strategy for salient-pole PMSM based on extended EMF in rotating reference frame," in IEEE Trans. on Industry Applications, vol. 38, no. 4, pp. 1054-1061, July-Aug. 2002.

[8] Jianrong Bu, Longya Xu, T. Sebastian and Buyun Liu, "Near-zero speed performance enhancement of PM synchronous machines assisted by low-cost Hall effect sensors," APEC '98 Thirteenth Annual Applied Power Electronics Conference and Exposition, 1998, pp. 64-68 vol.1.

[9] G. Scarcella, G. Scelba and A. Testa, "High performance sensorless controls based on HF excitation: A viable solution for future AC motor drives?," 2015 IEEE Workshop on Electrical Machines Design, Control and Diagnosis (WEMDCD), 2015, pp. 178-187.

[10] Y. -C. Kwon, J. Lee and S. -K. Sul, "Recent Advances in Sensorless Drive of Interior Permanent-Magnet Motor Based on Pulsating Signal Injection," in IEEE Journal of Emerging and Selected Topics in Power Electronics, vol. 9, no. 6, pp. 6577-6588, Dec. 2021.

[11] D. Raca et al., "Carrier-Signal Selection for Sensorless Control of PM Synchronous Machines at Zero and Very Low Speeds," in IEEE Trans. on Industry Applications, vol. 46, no. 1, pp. 167-178, Jan.-feb. 2010.

[12] J. Holtz, "Sensorless Control of Induction Machines—With or Without Signal Injection?," in IEEE Transactions on Industrial Electronics, vol. 53, no. 1, pp. 7-30, Feb. 2006.

[13] J. Holtz, "Developments in Sensorless AC Drive Technology," 2005 International Conference on Power Electronics and Drives Systems, 2005, pp. 9-16.

[14] L. D. Tornello et al., "Combined Rotor-Position Estimation and Temperature Monitoring in Sensorless, Synchronous Reluctance Motor Drives," in IEEE Trans. on Ind. Appl., vol. 55, pp. 3851-3862, 2019.

[15] Avago Technologies Motion Control Encoders in Electrical Motor Systems: Design Guide.

[16] G. Scelba, G. De Donato, G. Scarcella, F. Giulii Capponi and F. Bonaccorso, "Fault-Tolerant Rotor Position and Velocity Estimation Using Binary Hall-Effect Sensors for Low-Cost Vector Control Drives," in IEEE Transactions on Industry Applications, vol. 50, no. 5, pp. 3403-3413, Sept.-Oct. 2014.

[17] G. De Donato et al., "Low-Cost, High-Resolution, Fault-Robust Position and Speed Estimation for PMSM Drives Operating in Safety-Critical Systems," in IEEE Trans. on Pow. Electr., vol. 34, pp. 550-564, 2019.

[18] R. Petrella, M. Tursini, L. Peretti and M. Zigliotto, "Speed measurement algorithms for low-resolution incremental encoder equipped drives: a comparative analysis," 2007 International Aegean Conference on Electrical Machines and Power Electronics, 2007, pp. 780-787.

[19] L. Bascetta et al., "Velocity Estimation: Assessing the Performance of Non-Model-Based Techniques," in IEEE Transactions on Control Systems Technology, vol. 17, no. 2, pp. 424-433, March 2009.

[20] R. H. Brown, S. C. Schneider and M. G. Mulligan, "Analysis of algorithms for velocity estimation from discrete position versus time data," in IEEE Trans. on Ind. Electronics, vol. 39, no. 1, pp. 11-19, Feb. 1992.

[21] S. M. Phillips and M. S. Branicky, "Velocity estimation using quantized measurements," 42nd IEEE International Conference on Decision and Control (IEEE Cat. No.03CH37475), 2003, pp. 4847-4852 Vol.5.

[22] P. S. Carpenter et al., "On algorithms for velocity estimation using discrete position encoders," Proceedings of IECON '95 - 21st Annual Conference on IEEE Industrial Electronics, 1995, pp. 844-849 vol.2.

[23] Q. Ni et al., "A New Position and Speed Estimation Scheme for Position Control of PMSM Drives Using Low-Resolution Position Sensors," in IEEE Trans. on Ind. Appl., vol. 55, pp. 3747-3758, July-Aug. 2019.

[24] A. Anuchin, A. Dianov and F. Briz, "Synchronous Constant Elapsed Time Speed Estimation Using Incremental Encoders," in IEEE/ASME Trans. on Mechatronics, vol. 24, no. 4, pp. 1893-1901, Aug. 2019.

[25] A. Anuchin, V. Astakhova, D. Shpak, A. Zharkov and F. Briz, "Optimized method for speed estimation using incremental encoder," 2017 International Symposium on Power Electronics (Ee), 2017, pp. 1-5.

[26] Y. Vázquez-Gutiérrez et al., "Small-Signal Modeling of the Incremental Optical Encoder for Motor Control," in IEEE Trans. on Industrial Electronics, vol. 67, no. 5, pp. 3452-3461, May 2020.

[27] L. D. Tornello, G. Scelba, G. D. Donato, F. G. Capponi, G. Scarcella and M. Harbaugh, "Selection of Rotor Position Sensor Resolution for Variable Frequency Drives Utilizing Fixed-Position-Based Speed Estimation," 2021 IEEE Energy Conversion Congress and Exposition (ECCE), 2021, pp. 4846-4853.

[28] Gene F. Franklin, J. D. Powell, Michael Workman, "Digital Control of Dynamic Systems", Third Ed., 1998.

[29] G. Scelba et al., "Resolution of Rotor Position Measurement: Modeling and Impact on Speed Estimation," in IEEE JESTPE, vol. 10, no. 2, pp. 1992-2004, April 2022.

[30] Karl Johan Astrom, Richard M. Murray, "Feedback Systems: An Introduction for Scientists and Engineers", February, 2009.

A Method to Design Power Control System of Wayside Energy Storage System for Energy Saving in DC-electrified Railway

Kota Sato*, Keiichiro Kondo*, Hiroyasu Kobayashi**, Makoto Chida***

* Department of Electrical Engineering and Bioscience, Waseda University, Tokyo, Japan
**Department of Electrical and Electronic Engineering, Chiba University, Chiba, Japan
*** Innovation Department, West Japan Railway Company, Osaka, Japan
Tel.: +813 –5286 – 3184.
E-Mail: sugarballs@fuji.waseda.jp
URL: http://www.kondolab.eb.waseda.ac.jp/access/

Keywords

«Railway traction system», «DC railway power supply», «Energy storage», «Battery», «Regenerative power»

Abstract

The energy-saving effect of the use of a wayside energy storage system (WESS) power control method is improved by increasing the controller gain of the WESS for DC-electrified railways. However, excessive gain may cause instability. Therefore, this study proposes a method for designing a charge/discharge current controller.

1. Introduction

Regenerative braking enables to recover the kinetic energy at deceleration electrical energy and increase the energy efficiency of electric railways [1]. However, because DC-electrified railway systems generally use diode rectifiers at substations, regenerative energy cannot be returned to the AC grid. Thus, only a fraction of the regenerative energy is effectively used if powering trains absorbed all of the regenerative energy. One of the solutions to the problem of regenerative energy is installation of energy storage systems to utilize the excess regenerative energy. There are two types of energy storage systems: onboard energy storage systems [2] and wayside energy storage systems (WESSs) [3][4]. Onboard energy storage system can more efficiently utilize regenerative energy on the vehicle and is also effective for emergency power supply [5][6]. However, the amount of onboard batteries is greatly affected by the space and weight constraints of the railway vehicle. On the other hand, WESS, as shown in Fig.1, is more advantageous to avoid to increase the mass of the train and is more cost effective for the energy saving by recovering the waste regenerative energy. In the WESS, batteries are connected by bi-directional non-isolated DC/DC converter [7] to catenary line to boost up its voltage and for charge and discharge control. The charge and discharge control of the WESS is carried out by the current control based on the voltage at the overhead catenary line connection point, as shown in Fig. 2. The energy saving effect is improved by increasing the gain K_{ref}. However, if the gain K_{ref} becomes too high, the

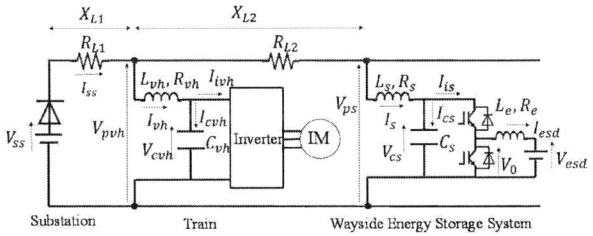

Fig. 1. WESS model with a powering train and a substation

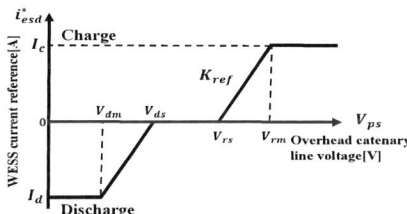

Fig. 2. Determination pattern of WESS current reference

phase margin of the control system decreases, which leads to unstable phenomena such as sustained oscillation or divergence [8].

Several methods have been proposed to suppress the oscillation of the system by connecting voltage-type power converters to a common DC link, similar to DC-electrified railway systems, utilizing the additional circuits [9]. However, in DC-electrified railways system, additional equipment is not desirable because of its higher cost. Another method involves the use of a control system model that suppresses oscillation by controlling the bidirectional power converter on the power-supply side [10] [11]. However, stabilization by controlling the power-supply side is impossible for DC-electrified railway systems because the diode rectifier is utilized as the power converter of the power supply substation, as previously mentioned. The value of the inductance of the filter reactor (FL) in a DC-electrified railway system is set to a large value to prevent the increase of the short-circuit current in the event of an accident. Therefore, when the discharge current on the low-voltage side of the WESS increases significantly, the current of the FL does not; hence, the current of the filter capacitor (FC) increases, while the FC voltage becomes large and oscillatory, leading to the instability phenomena.

Therefore, it is necessary to design the power control system of the WESS to maintain stability. However, a method for determining the gain K_{ref}, which causes instability, has not been established. In this study, a linearized model of a DC feeder circuit which includes WESS, is derived. Then, a design method for the WESS power control system based on the derived model is proposed to enhance the energy efficiency of the DC-electrified railway system. In the proposed linearization model, two patterns of the assumed circuit configurations are considered: (i) when the train is powering and the WESS is discharging, and (ii) when the train is regenerating and the WESS is charging. The proposed design method was verified experimentally utilizing a downscaled model. In addition, the energy efficiency of the proposed design method was investigated by the numerical simulation of a real-scaled DC-electrified railway system with a WESS. By establishing a method to determine the gain K_{ref}, it is possible to realize charge/discharge control with a high energy efficiency while maintaining the stability of the WESS.

2. Modeling and designing the WESS charge/discharge control system

In this section, a linearized model is derived for the charge/discharge power control system of a WESS. Based on this, the design method of the charge/discharge power control system using Bode diagrams is explained.

A. *Modeling the power feeding system and the controller for WESS*

In this section, a linearized model of the assumed system, as shown in Fig.1, is described. The two patterns described above are considered as the assumed system. In case (i), the overhead catenary line voltage of the train is lower than the voltage at the substation, and diodes of the substation are turned on. The circuit configuration that includes a substation is assumed. In case (ii), the overhead catenary

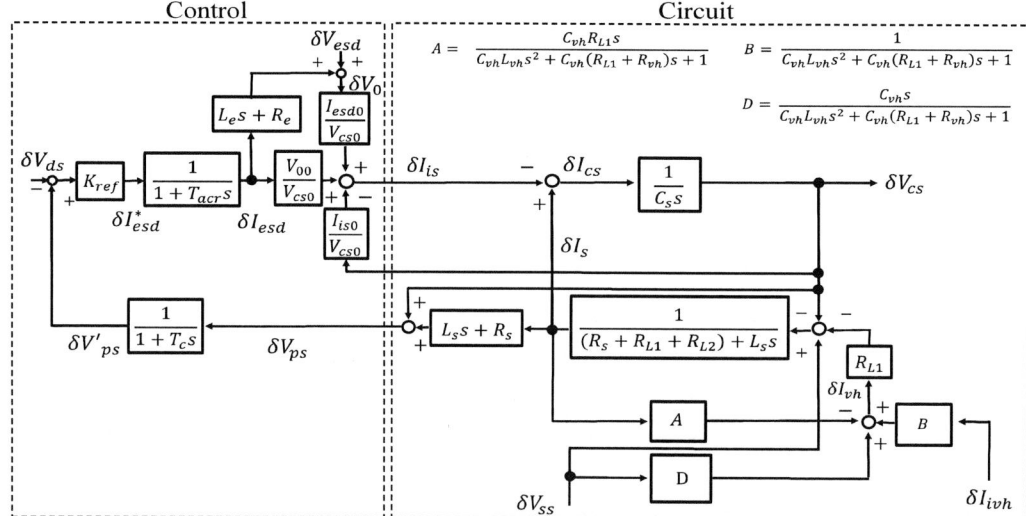

Fig. 3. A linearized model including powering-train, WESS, and substation

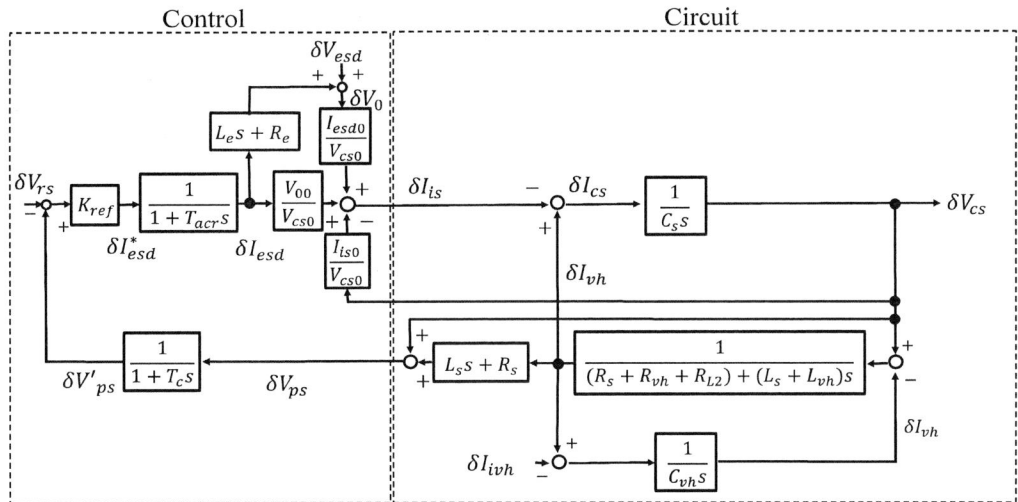

Fig. 4. A linearized model including regenerative train and WESS

line voltage of the train is higher than the voltage at the substation, and diodes of the substation are turned off. Therefore, a circuit configuration consisting of only a train and a WESS is assumed (Fig. 1). The control system and circuit equations in case (i) are explained.

The charge/discharge current I_{esd} of the WESS corresponds to the reference current I^*_{esd} with a first-order delay, and equation (1) holds.

$$I_{esd} = \frac{1}{1 + T_{acr}s} I^*_{esd} \tag{1}$$

For the two-quadrant chopper of the WESS, equation (2) holds, assuming that there are ~~zero~~ no losses in the power conversion.

$$V_{cs} \cdot I_{is} = V_0 \cdot I_{esd} \tag{2}$$

Equation (2) is a nonlinear equation; therefore, considering the small variation near the equilibrium point, it is linearized by equation (3) from the Taylor expansion. In equation (3), "0" represents the equilibrium point, and "δ" represents the small variation from the equilibrium point.

$$\delta I_{is} = \frac{V_{0(0)}}{V_{cs(0)}} \cdot \delta I_{esd} + \frac{I_{esd(0)}}{V_{cs(0)}} \cdot \delta V_0 - \frac{I_{is(0)}}{V_{cs(0)}} \cdot \delta V_{cs} \tag{3}$$

As equation (3) is a nonlinear equation, it can be expressed as a linear equation by the Taylor expansion around the equilibrium point.

The lower side voltage of the chopper V_0 is expressed by equation (4).

$$V_0 = V_{esd} - (R_e I_{esd} + L_e \frac{d}{dt} I_{esd}) \tag{4}$$

Equation (5) shows the relationship between the current and voltage of the DC-link capacitor of the WESS.

$$C_s \frac{d}{dt} V_{cs} = I_s - I_{is} \tag{5}$$

The overhead catenary line voltage of the WESS V_{ps} is expressed by equation (6).

$$V_{ps} = V_{cs} + (R_s I_s + L_s \frac{d}{dt} I_s) \tag{6}$$

Equation (7) holds for the DC-link voltage of the train V_{cvh}.

$$V_{cvh} = V_{ss} - R_{L1} I_{vh} - \left\{ (R_s + R_{L1} + R_{L2}) I_s + L_s \frac{d}{dt} I_s \right\} \tag{7}$$

Equation (8) shows the relationship between the current and voltage of the DC-link capacitor of the train.

$$C_{vh} \frac{d}{dt} V_{cvh} = I_{vh} - I_{ivh} \tag{8}$$

The voltage source inverter with constant current control on the AC side can be regarded as the current source from the DC side. Therefore, the train model consists of the FL, FC, and current source.

Next, the circuit equations for case (ii) are shown. In case (ii), equations (1) – (6) and (8) are the same as those in case (i); only equation (7) is different. As previously mentioned, diodes of the substation are turned off; hence, the circuit configuration in case (i) consists of the train and the WESS. Therefore, the train current I_{vh} is expressed by equation (9).

$$I_{vh} = -I_s \tag{9}$$

The voltage equation that includes the train DC-link voltage and the WESS DC-link voltage is expressed by equation (10).

$$V_{cs} = (R_s + R_{vh} + R_{L2})I_{vh} + (L_s + L_{vh})\frac{d}{dt}I_{vh} + V_{cvh} \tag{10}$$

The block diagrams for case (i) and (ii) are shown in Fig. 3 and Fig. 4, respectively.

B. *Calculation of theoretical stability limits*

The open-loop transfer function from the discharge start voltage δV_{ds} to the overhead catenary line voltage $\delta V'_{ps}$ in Fig. 3 is shown on the Bode diagram. The open-loop transfer function from the charge start voltage δV_{rs} to the overhead catenary line voltage $\delta V'_{ps}$ in Fig. 4 is also shown on the Bode diagram. From the phase margin, the stability limit gain K_{ref} is obtained. The parameters in Table I are obtained based on the parameters of the 1 kW class downscaled experimental system described in Section 3. The value of the feeder resistance R_{L1} is determined based on the case of $X_{L1} = 4$. The value of R_{L1} is set to 12.5 Ω to match the rate of voltage

Table I. Parameters of the assumed system

Symbol	Parameters	Value
C_{vh}	FC capacitance of the train	1320 µF
L_{vh}	FL inductance of the train	80 mH
R_{vh}	FL resistance of the train	0.7 Ω
C_s	FC capacitance of the wayside energy storage system	1320 µF
L_s	FL inductance of the wayside energy storage system	80 mH
R_s	FL resistance of the wayside energy storage system	0.7 Ω
X_{L1}	Distance between the train and the wayside energy storage system	4 km
X_{L2}	Distance between the train and substation	4 km
R_{L1}	Feeder resistance between the train and the wayside energy storage system	12.5 Ω
R_{L2}	Feeder resistance between the train and substation	12.5 Ω
L_e	WESS Low voltage side FL inductance	20 mH
R_e	WESS low voltage side FL resistance	0.25 Ω
T_{acr}	Time constant of current control	5 ms
T_c	Time constant of low-pass filter	50 ms

Table II. Equilibrium points and phase margin at each gain in case (i)

(a) Equilibrium points in case (i)

Parameters	K_{ref} $= 1$	K_{ref} $= 2$	K_{ref} $= 3$	K_{ref} $= 4$	K_{ref} $= 5$
Chopper inflow current I_{te0} [A]	3.88	3.84	3.83	3.83	3.82
WESS FC voltage V_{cs0}[V]	295.9	300.5	302.3	303.3	303.9
Chopper lower side voltage V_{00}[V]	49.4	49.1	49.0	48.9	48.8
Discharge current I_{esd0}[A]	10.4	11.6	12.1	12.3	12.5

(b) Phase margin in case (i)

Gain value	Phase margin[deg]
1	+45
2	+28
3	+15
4	+4
5	-5

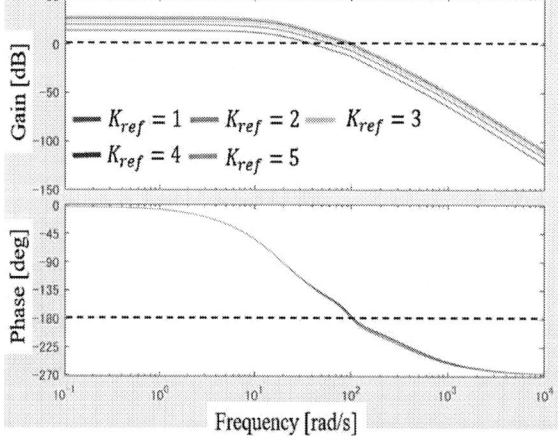

Fig. 5. Bode diagram of the open-loop transfer function from V_{ds} to V'_{ps} in case (i) (K_{ref}= 1, 2, 3,4, 5)

Table III. Equilibrium points and Phase margin at each gain in case (ii)

(a) Equilibrium points in case (ii)

Parameters	K_{ref} = 1	K_{ref} = 2	K_{ref} = 3	K_{ref} = 4	K_{ref} = 5
Chopper inflow current I_{is0} [A]	1.50	1.51	1.52	1.52	1.53
WESS FC voltage V_{cs0}[V]	313.6	309.3	307.8	307.1	306.7
Chopper lower side voltage V_{00}[V]	54.1	54.2	54.2	54.2	54.2
Discharge current I_{esd0}[A]	8.66	8.64	8.63	8.63	8.62

(b) Phase margin in case (ii)

Gain value	Phase margin[deg]
1	+36
2	+25
3	+10
4	-1
5	-9

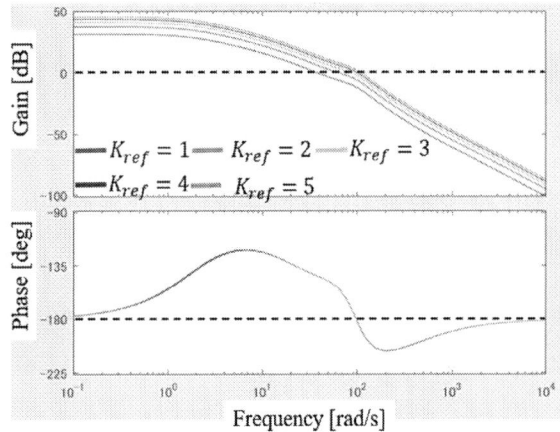

Fig. 6. Bode diagram of the open-loop transfer function from V_{ds} to V'_{ps} in case (ii) (K_{ref}= 1, 2, 3,4, 5)

drop at the real line section. Similarly, the feeder resistance R_{L2} is also set to 12.5 Ω. The current control time constant T_{acr} is set to 5 ms to ensure that the response is faster than the overall external feedback loop, and larger than the control calculation period.

Under the conditions listed in Table I, the Bode diagram of Fig. 3 with gain K_{ref} = 1–5 is shown in Fig. 5. The equilibrium points of the Bode diagram and the phase margin at each gain are listed in Table II. Fig. 5 and Table II show that the gain at the stability limit is K_{ref}= 4 in case (i). Similarly, a Bode diagram is described for case (ii). The Bode diagram in Fig. 4 with gain K_{ref} = 1–5 is shown in Fig. 6. The equilibrium points of the Bode diagram and the phase margin at each gain are listed in Table III. The Bode diagram exhibited the same tendency as that in Fig. 5 when the gain in the frequency band around the transition to the unstable side is considered. Focusing on this part, the stability limit gain ~~was~~ is considered as K_{ref} = 3 in case (ii).

3. Experimental verification utilizing downscaled model

The validity of the proposed gain design method in the section 2 is evaluated through an experiment utilizing downscaled model of the assumed system shown in Fig. 1.

A. Configurations of the experiment utilizing downscaled model

In this section, the results of the 1 kW class downscaled model experiments with different gains K_{ref} in models (i) and (ii) are presented. The experimental circuit configuration is shown in Fig. 7. In this experiment, a 200 V 50 Hz three-phase AC power supply is converted to 300 V DC using a transformer and diode rectifier to simulate the traction substation. The storage battery is modelled using a regenerative DC power source of 52 V. The parameters of the experimental setup are the same as those listed in Table I. The induction motor is driven by indirect field-oriented control.

B. Results of the experiment utilizing downscaled model

The value of the gain K_{ref} in the experiment is determined based on the results discussed in Section 2. In particular, for case (i), the experimental results for K_{ref} = 1 and 3 are shown in Fig. 8. For case (ii), the experimental results for K_{ref} = 2 and 4 are shown in Fig. 9. The difference between the overhead catenary line voltage and charge/discharge start voltage determines the value of the charging/discharging current. This current flows into the capacitor and determines the FC voltage of the WESS. The FC voltage of the

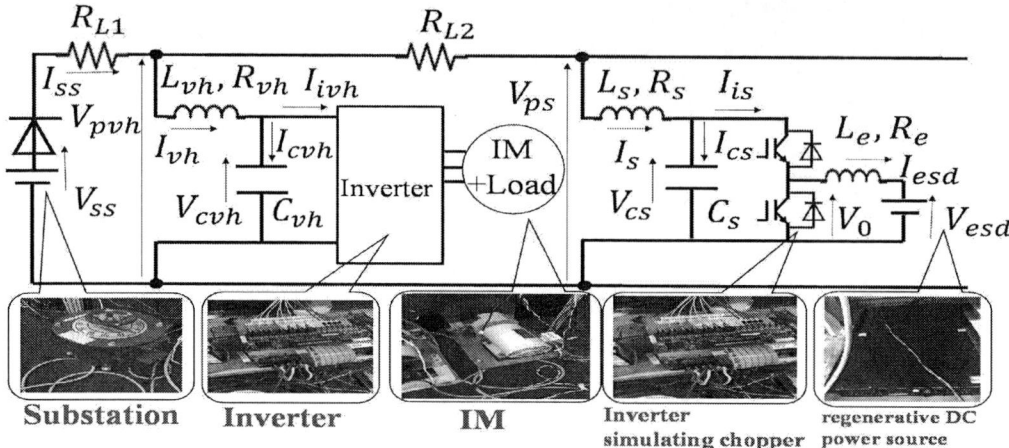

Fig. 7. Experimental circuit utilizing downscaled model

WESS, which is the output of the transfer function studied in Section 2, is used as the evaluation index in terms of stability. Fig. 6 shows that the oscillation increased as the gain K_{ref} increase, and divergence is observed at K_{ref} =3. The same tendency is observed in case (ii), where the oscillation of V_{cs} became more significant at K_{ref} = 4. Because there are parameter errors in the proposed model for the dynamic analysis in Section 2, it is necessary to design the K_{ref} by considering some phase margins. In other words, it is necessary to maintain a phase margin of about 25–30° from the unstable gain that is shown in the experimental results. Fig. 4 shows that for practical applications, it is better to design with a phase margin of +45° during discharge. Based on the same idea, it is preferable to have a phase margin of +25° during charging.

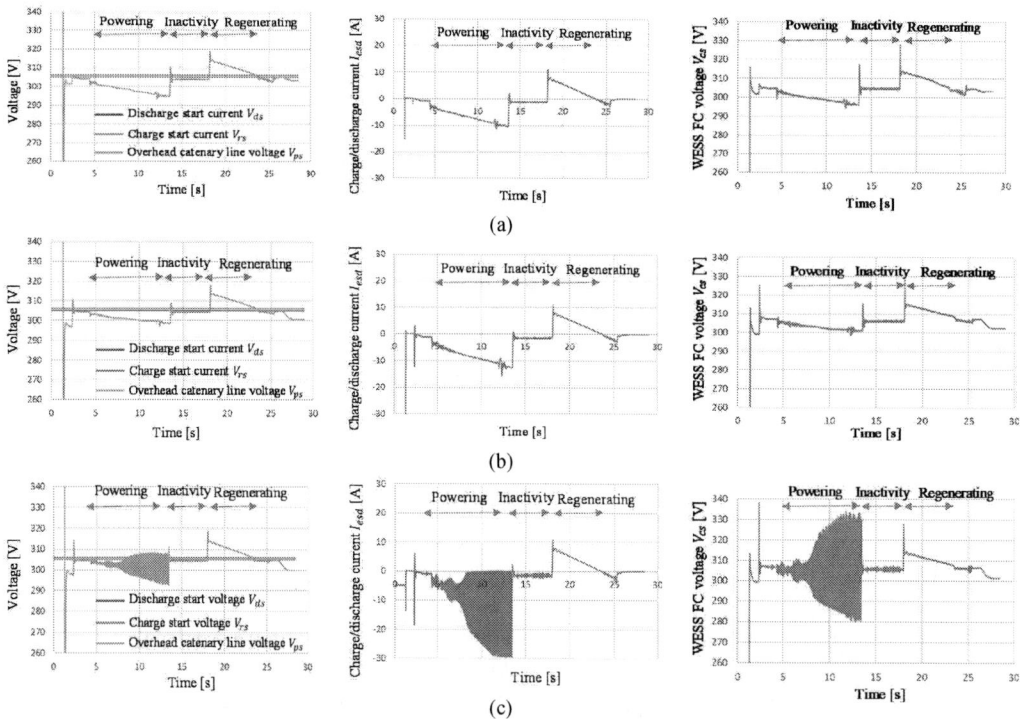

Fig. 8. Overhead catenary line voltage V_{ps}, charge/discharge current I_{esd}, and WESS FC voltage V_{cs} at each gain (a)K_{ref} =1, (b) K_{ref} =2, and (c)K_{ref} =3

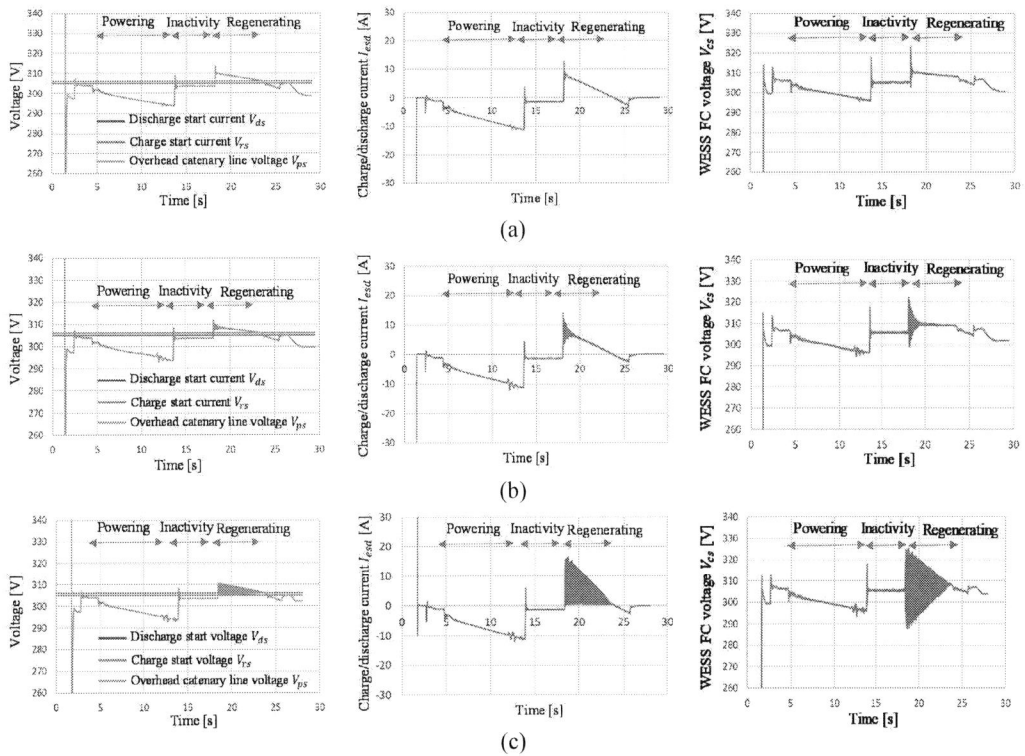

Fig. 9. Overhead catenary line voltage V_{ps}, charge/discharge current I_{esd}, and WESS FC voltage V_{cs} at each gain (a)K_{ref} =2, (b) K_{ref} =3, and (c)K_{ref} =4

4. Verification of the energy-saving effect by real-scale power feeder simulation

A. *Conditions for the evaluation of the energy saving by numerical simulation*

In this section, the gain is designed with a phase margin of +45° and +25° during discharging and charging, respectively, based on the results of the design in Sections 2 and 3. The parameters of the real scale system are listed in Table IV. The Bode diagram in Fig. 3 is shown in Fig. 10. The equilibrium points and the phase margin at each gain are shown in Table V. The Bode diagram for Fig. 4 is shown in Fig. 11. The equilibrium points and the phase margin at each gain are ~~is~~ shown in Table VI. The parameters are determined based on a real-scale system. The gains are determined by the proposed design method: the discharge gain $K_{ref} = 23$ and charge gain $K_{ref} = 18$. The energy-saving effect of the difference between the gain of the proposed design method and the small-gain K_{ref} is compared. The locations of the stations, substations, and the WESS are shown in Fig. 12. The train diagram is shown in Fig. 13. The train parameters are listed in Table VII.

B. *valuation of the energy saving effect*

The comparison of the energy-saving effect according to the difference between the gain K_{ref} of the proposed design method and the smaller gain $K_{ref} = 10$ is presented in Table VIII. $K_{ref} = 10$ is set as the actual conventional and conservative gain

Table IV. Real-scale parameters of the assumed system

Symbol	Parameters	Value
C_{vh}	FC capacitance of the train	37.5 mF
L_{vh}	FL inductance of the train	3.0 mH
R_{vh}	FL resistance of the train	0.1 Ω
C_s	FC capacitance of the wayside energy storage system	37.5 mF
L_s	FL inductance of the wayside energy storage system	3.0 mH
X_{L1}	Distance between the train and the wayside energy storage system	4 km
X_{L2}	Distance between the train and substation	4 km
R_s	FL resistance of the wayside energy storage system	0.1 Ω
R_{L1}	Feeder resistance between the train and the wayside energy storage system	0.142 Ω
R_{L2}	Feeder resistance between the train and substation	0.142 Ω
T_{acr}	Time constant of current control	2 ms
T_c	Time constant of low-pass filter	50 ms

Table V. Real-scale equilibrium points and phase margin at each gain in case(i)

(a) Equilibrium points in case (i) of the real-scale

Parameters	$K_{ref} = 10$	$K_{ref} = 23$	$K_{ref} = 50$
Chopper inflow current I_{is0} [A]	1890	1856	1834
WESS FC voltage V_{cs0}[V]	1405.5	1433.3	1451.9
Chopper lower side voltage V_{00}[V]	740.3	696.7	688.9
Discharge current I_{esd0}[A]	770.4	1040.4	1199.8

(b) Phase margin in case (i) of the real-scale

Gain value	Phase margin[deg]
10	+95
23	+45
50	+2

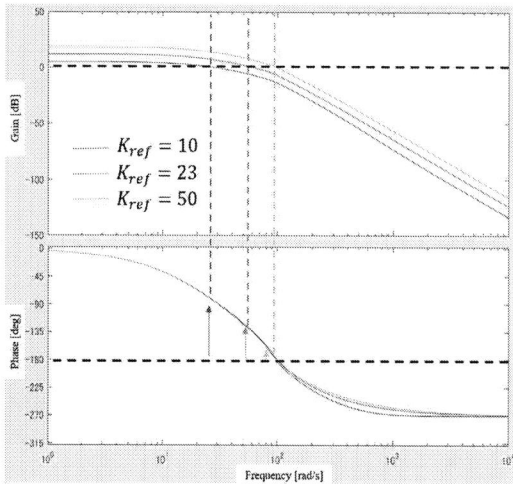

Fig. 10. Bode diagram of the open-loop transfer function from V_{ds} to V_{ps} in case (i) of the real-scale

Table VI. Real-scale equilibrium points and phase margin at each gain in case(ii)

(a) Equilibrium points in case (ii) of the real-scale

Parameters	$K_{ref} = 10$	$K_{ref} = 18$	$K_{ref} = 50$
Chopper inflow current I_{is0} [A]	444.3	556.1	686.5
WESS FC voltage V_{cs0}[V]	1663.5	1621.6	1572.9
Chopper lower side voltage V_{00}[V]	795.3	804.9	815.1
Discharge current I_{esd0}[A]	929.4	1120.5	1324.7

(b) Phase margin in case (ii) of the real-scale

Gain value	Phase margin[deg]
10	+29
18	+25
50	-3

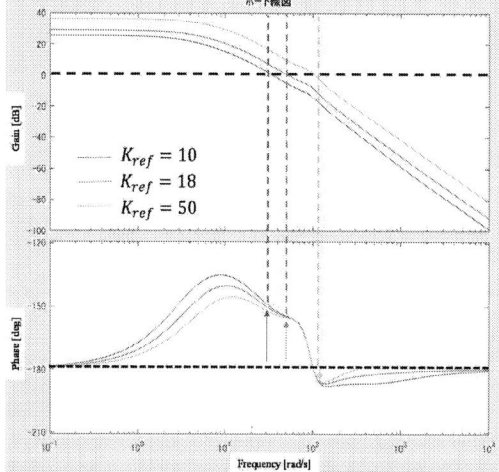

Fig. 11. Bode diagram of the open-loop transfer function from V_{ds} to V_{ps} in case (ii) of the real-scale

value. Table VIII shows that the output energy decrease only at the substations near the WESS. There are two reasons for this result.

Table VII Simulation parameters

	Parameters	Value
substation	No-load output voltage	1575 V
Train	Number of the trains	8
	Number of the motors	16
	Mass of units	306 t
	Maximum acceleration	3.5 km/h/s
	Deceleration	4.9 km/h/s
	Maximum speed	100 km/h
	Voltage to start to apply light-load regenerative brake control	1700 V
	Voltage to stop apply light-load regenerative brake control	1830 V
	SIV power consumption	10 kVA
WESS	Voltage to start to charge	1610 V
	Voltage to start to discharge	1500 V
	Power capacity of charging	1000 kW
	Power capacity of discharging	1000 kW
	Maximum input current	1330 A
	Maximum output current	1330 A
	Output voltage of the battery	748 V
Other	Feeder resistance	0.0356 Ω/km

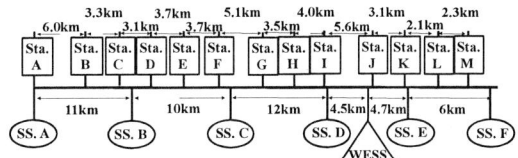

Fig. 12. Overview of simulation conditions

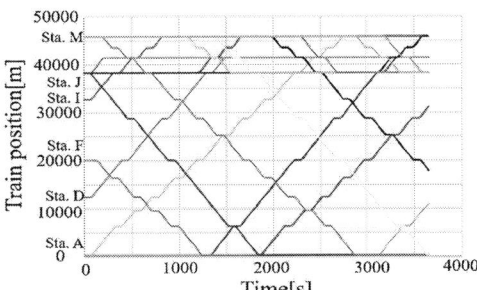

Fig. 13. Train diagram

Table VIII. Simulation results of substation energy consumption

Substation name	Discharging $K_{ref} = 23$ Charging $K_{ref} = 18$	Discharging $K_{ref} = 10$ Charging $K_{ref} = 10$
SS. A[kWh]	575.6	575.6
SS. B[kWh]	1077	1077
SS. C[kWh]	981.9	981.6
SS. D[kWh]	**1064**	**1071**
SS. E[kWh]	**1612**	**1619**
SS. F[kWh]	241.2	241.4

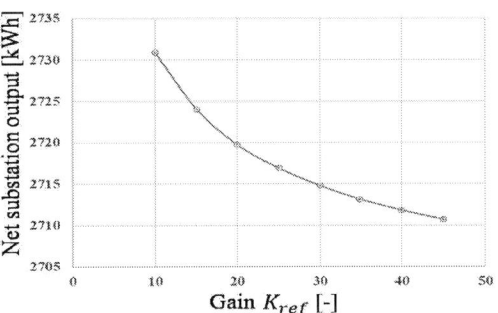

Fig. 14. Relationship between the gain K_{ref} and net output energy

1. A feeder circuit with a total length of approximately 50 km is assumed in the simulation. Therefore, the regenerative energy from the distant braking train cannot be charged to the WESS owing to the large voltage drop at the feeder resistance.

2. Because the effects of suppressing the overhead line voltage fluctuations are limited to the vicinity of the WESS, the effect of reducing the energy consumption of the powering-train is also limited to the vicinity of the WESS.

The slight increase in the output energy at substation C may be attributed to the change in power flow near the adjacent substation D as a result of the change in the output of the WESS. However, as seen in Table XII, the effect is small. For these reasons, the energy-saving effect is evaluated in substations D and E, which are located on both sides of the WESS. To make a comparison under fair conditions, the energy-saving effect is determined by calculating the net output energy from equation (11).

$$E_{out} = E_{SSout} + E_{WESSout} - E_{WESSin} \quad (11)$$

where E_{out} is the net output energy and E_{SSout} is the sum of the output energies of substations D and E in Table VIII. $E_{WESSout}$ and E_{WESSin} are the discharge energy and charge energy of the storage device, respectively. The relationship between the net output energy calculated from equation (11) and the gain K_{ref} is shown in Fig. 14. It can be seen that the larger the gain, the higher the energy-saving effect. When $K_{ref} = 45$, and the phase margin is zero, the energy can be reduced by 0.74%. When the phase margins +45 ° and +25 ° during discharging and charging, respectively, are assumed, an energy reduction of 0.37% is observed.

5. Conclusion

In this paper, a method for determining the gain K_{ref} was proposed, which was set to a conservative value by trial and error based on experience. Two feeder circuit models for the proposed design method were investigated: (i) when the train is powering and the WESS discharging; and (ii) when the train is regenerating and the WESS charging. From the Bode diagram of the open-loop transfer function of the proposed linearized models, the marginal gain K_{ref} from the perspective of stability can be obtained by focusing on the phase margin. The results of the analysis and downscaled model experiments showed good agreement, thus confirming the validity of the proposed design method. The energy-saving effect of applying the gain of the proposed design method to a real-scale system was also considered. By applying the gains determined from the proposed design method to the real-scale system, it was observed that the substation output energy decreased by up to 0.74%. In conclusion, the proposed method enables the design of the WESS power control system that achieves the maximum energy-saving effect while maintaining stability.

References

[1] T. Koseki: "Technical Trends of Rilway Traction in the World," IEEE IPEC, Sapporo, pp.2836-2841, (2010)

[2] H. Kobayashi, K.Kondo: "Control Method for Increasing Motor Power of DC-electrified Railway Vehicle with an Onboard Energy Storage System", IEEJ Journal of Industry Applications,Vol.10 No.5, pp.520-527, (2021)

[3] L. Alferi, L. Battistelli, and M. Pagano: "Impact on railway infrastructure of wayside energy storage systems for regenerative braking management: a case study on a real Italian railway infrastructure," IET Electr. Syst. Transp. Vol.9 lss.3, pp. 140–149, (2019)

[4] K. Pham, R. Eacker, M. Burnett, and M. Bardslkey, "A step forward or backward? Sound transit opts for 1500 VDC traction electrification," in Proc. ASME/IEEE Joint Railroad Conf., Newark, NJ, USA, pp. 67–72, (2000)

[5] K. Sato, H. Kato, and T. Fukushima: "Outstanding Technical Features of Traction System in N700S Shinkansen New Generation Standardized High Speed Train", IEEJ Journal of Industry Applications, Vol. 10 No.4, pp.402-410, (2021)

[6] K. Sato, H. Kato, and T. Fukushima: "Development of SiC Applied Traction System for Next-Generation Shinkansen High-Speed Trains", IEEJ Journal of Industry Applications, Vol. 9 No.4, pp.453-459, (2020)

[7] K. Tesaki, Y. Ishida, and M. Hagiwara: "Control and Experimental Verification of a Bidirectional Non-isolated DC-DC Converter Based on Three-level Flying-Capacitor Converters", IEEJ Journal of Industry Applications, Vol.10 No1, pp.114-123, (2021)

[8] H. Kobayashi, J. Asano, T. Saito, K. Kondo:"A Power Control method to Save Energy for Wayside Energy Storage Systems in DC-electrified Railway System", Electrical engineering in Japan, 2016-07, Vol.196, p.56-66.

[9] T. Funaki, N. Kimura, K. Matsu-ura:"Suppression of DC Line Current Oscillation of HV DC Transmission System Using Voltage Source Forced Commutation Converter", Electrical engineering in Japan, 1994, Vol.114, p.123-132

[10] A. Emadi, A. Khaligh, Claudio H. Rivetta, Geoffrey A. Williamson:"Constant power loads and negative impedance instability in automotive systems: definition, modeling, stability, and control of power electronic converters and motor drives", IEEE Transactions on Vehicular Technology, Vol.55, Issue 4, pp.1112-1125, (2006)

[11] Xuan Z, Lie X, Yongdong L, Zedong Z, Kui W: "Stabilization and assessment of interaction dynamics for More Electric Aircraft", Proceedings on the IEEE 8th International Power Electronics and Motion Control Conference (IPEMC-ECCE Asia) , DOI 10.1109/IPEMC.2016.7512401, 2016

A Reconfigurable Single-Stage Three-Phase Electric Vehicle DC Fast Charger Compatible With Both 400V and 800V Automotive Battery Packs

Mojtaba Forouzesh, Yan-Fei Liu, Paresh C. Sen
Queen's University
Department of Electrical and Computer Engineering
Kingston, ON, Canada
E-Mail: m.forouzesh@queensu.ca, yanfei.liu@queensu.ca, senp@queensu.ca
URL: https://www.ece.queensu.ca/research/labs/power-group.html

Keywords

«Electric vehicle (EV)», «Charging infrastructure for EVs», «AC-DC converter», «Power factor correction», «Resonant converter».

Abstract

A novel three-phase Electric Vehicle (EV) DC fast charger with a wide output voltage range is proposed in this paper. One of the main features of the proposed EV charger is that it can provide a wide output voltage range (i.e., 250 V to 850 V) using low voltage rating mainstream switches/diodes (i.e., 650 V). The proposed EV charger takes advantage of Inductor-Inductor-Capacitor (LLC) resonant tanks for each phase allowing soft-switching performance for all switches so the major power loss will be only conduction loss. Consequently, by operating at a high switching frequency the size of passive components can be reduced leading to a high-power density. Moreover, as the AC to DC conversion is being realized in an isolated single-stage approach, the total conversion efficiency for the proposed EV charger is higher than conventional two-stage EV chargers. The analysis of the power circuit design and control of the proposed EV fast charger is provided in the paper for both 400 V and 800 V automotive battery systems. Moreover, the performance is verified by computer simulation results and experimental results of a 1.5 kW laboratory prototype.

Introduction

In recent decades a lot of attention is paid to renewable energy technologies due to the increased greenhouse emissions and other environmental concerns [1]. One of the main contributors to the emission of CO_2 is transportation which is mainly consisted of internal combustion engine vehicles. To overcome this issue, it is widely accepted that more efficient and affordable Electric Vehicles (EVs) should be replaced with existing vehicles. Therefore, EV power train technologies have been the topic of a lot of research studies in the past decade [2] and [3]. Most recent EVs use a 400 V powertrain system which is well established and standardized through the years. From the advent of EVs, one of the early concerns about them was their limited driving range, which is being fulfilled by adding to the capacity of the long-range EV battery packs. The latter brings forward another challenge which is the long recharging time of a depleted high-capacity battery pack. To address the high power charging challenge, DC fast chargers that have been classified as the level 3 charging method for EV charging stations are meant to provide high power DC voltage directly to the EV battery [4].

The conventional method for DC fast charger is based on a two-stage approach to achieve the required specifications [5], with a direct three-phase AC-DC converter like Vienna rectifier at the grid side to achieve Power Factor Correction (PFC) [6]. In the second stage, soft-switching DC-DC converters like phase-shifted full-bridge converters or LLC resonant converters are preferred to provide voltage isolation and regulation while achieving high efficiency in DC-to-DC conversion [7] and [8]. To improve efficiency, interleaving multiphase DC-DC converters is a necessity for high-power chargers, which can be a challenging task for resonant converters [9]. Because of using two cascaded stages in the conventional method, the total AC-to-DC conversion efficiency is usually limited to below 96% and the power density is also suffering due to many active and passive components.

In recent years, single-stage EV chargers are becoming attractive due to their ability to improve both efficiency and power density [10]-[12]. In [10] a DAB-based direct three-phase AC-DC converter is proposed using a matrix converter at the primary of the transformer with back-to-back switches. In [11], a phase-modular three-phase AC-DC converter is proposed for EV battery charging applications using Cuk converter modules. This converter lacks soft-switching and suffers high current stress making it less attractive for high-power applications. In [12], a soft-switching phase-modular three-phase AC-DC converter is proposed for EV battery charging. The efficiency is low as a result of large current ripples due to using DCM boost inductors to obtain inherent PFC.

In the past decades, most of the literature only discussed EV chargers for a 400 V system while an 800 V system is beginning to take more attention in recent years [13]-[15]. In [16], a single-stage single-phase EV charger based on a bridgeless boost PFC rectifier and a three-level CLL resonant converter with three winding transformers is proposed. The secondary and tertiary windings of the transformer are connected in series with a coupled inductor to provide 800V at the battery side. The output voltage of this converter has a double line frequency ripple and hence needs a ripple cancellation method. In [17] a single-stage three-phase EV charger is proposed based on a phase-modular approach using a dual active bridge converter. Although it is mentioned that the proposed charger is suitable for a wide output voltage range, sufficient analysis and verification are not provided.

Since the 800 V EV battery that was first proposed by Porsche and now is being used by other automakers consists of two series-connected modules with equal cells [13], a Battery Selection Circuit (BSC) is proposed in [18] to be added at the output of exiting EV chargers allowing both 400 V and 800 V battery charging. The proposed BSC charges two 400 V battery modules interchangeably so the semiconductors are all rated for a 400 V system. However, this circuit introduces additional loss when charging a 400 V EV battery pack and it requires additional battery management systems at the charger side for safety reasons. In [19], a two-stage three-phase EV charger is proposed that is suitable for both 400 V and 800 V batteries. A direct three-phase Boost PFC is used at the input followed by a three-active bridge DC-DC converter with two transformers. The output of the transformers is connected in a reconfigurable fashion so both output charging voltage ranges can be met. The stress of components is large as a direct three-phase approach is used and the efficiency is not high enough due to the dual power processing stages.

In this paper, a novel single-stage three-phase EV charger is proposed with a wide output voltage range. Fig.1 illustrates a general block diagram of the proposed EV charger with reconfigurable output that can be used with both 400 V and 800 V EV battery voltages (V_B). Moreover, the proposed EV charger has a phase-modular structure using LLC PFC converter modules [20], which allows achieving soft-switching over a wide output voltage range. Furthermore, unlike most other power electronics converters suitable for the EV 800 V system, in the proposed EV charger there is no need for 1200 V SiC devices, and only 650 V switches/diodes are used. The proposed converter is introduced and analyzed in the next section. Then, simulation and experimental results are provided to verify the performance. Finally, the paper is concluded in the last section.

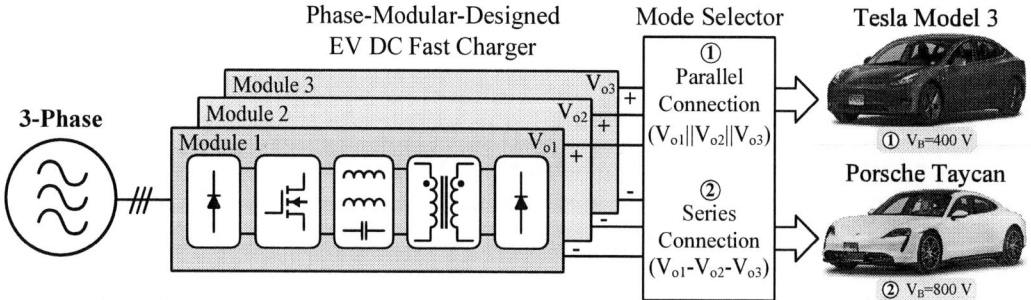

Fig. 1. Block diagram illustration of the proposed EV charger with reconfigurable output.

The proposed single-stage three-phase EV charger

Fig. 2 illustrates the proposed single-stage three-phase EV charger with reconfigurable output. When two double pole relays are in position 1 the output is set for 400 V battery systems ($V_o^{400} = V_{o1} = V_{o2} = V_{o3}$) and when the relays are in position 2 the output is set for 800 V battery systems ($V_o^{800} = V_{o1} + V_{o2} + V_{o3}$). The proposed three-phase charger has a phase-modular structure, so the maximum stress is distributed among components leading to high reliability, high efficiency, and low cost. Each phase is consisted of an LLC converter module to take advantage of Zero Voltage Switching (ZVS) and Zero Current Switching (ZCS) for all the switching devices. The latter allows high switching frequency implementation in the proposed three-phase rectifier leading to a high power density. The PFC is achieved by changing the switching frequency of each phase with respect to the line AC voltages. In this way, using a low-quality factor (Q) resonant tank design allows for achieving high voltage gains around the parallel resonant frequency (f_p), which is required around the line Voltage Zero Crossing (VZC) area. Moreover, the voltage gain requirement around the peak line voltage is minimum and usually is set to be around the series resonant frequency (f_s) keeping the operation in the below resonant area to take advantage of ZCS for the diodes.

Considering that the switching frequency is much higher than the AC line frequency, there is negligible high-frequency current flowing into the output capacitors. Then, the rectified output current of each phase ($i_{o1,2,3}$) can be represented by its fluctuating average value, which can be written based on the instantaneous input voltages ($v_a(t), v_b(t), v_c(t)$) and input currents ($i_a(t), i_b(t), i_c(t)$) of a balanced three-phase system with 120° phase displacement connected to a unity power factor correction lossless three-phase circuit. Then in a balanced three-phase system, the fluctuating part of the voltage and current get canceled in the output. Hence, the battery voltage and current for both output modes can be written based on the average voltages and currents and the Root Mean Square (RMS) value of the AC voltages (V_{ac}) and currents (I_{ac}) in the following forms.

$$V_B^{400} = V_{o1} = V_{o2} = V_{o3} \, , I_B^{400} = I_{o1} + I_{o2} + I_{o3} = 3 \times \frac{V_{ac}I_{ac}}{V_B^{400}} \tag{1}$$

$$V_B^{800} = V_{o1} + V_{o2} + V_{o3} = 3 \times \frac{V_{ac}I_{ac}}{I_B^{800}} \, , I_B^{800} = I_{o1} = I_{o2} = I_{o3} \tag{2}$$

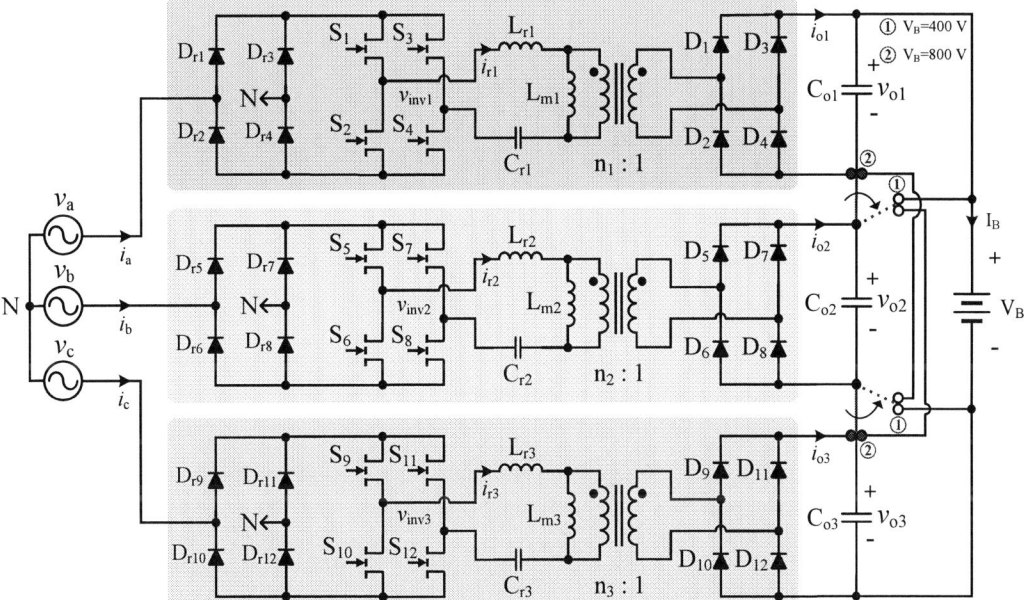

Fig. 2. The proposed single-stage three-phase EV charger with reconfigurable output.

The equivalent model of the three-phase AC-DC LLC converter in both output configuration modes is shown in Fig. 3. The input of each LLC resonant tank is a square wave voltage coming from the inverter bridge that is dependent on the line angle and its initial value, θ and θ_0, respectively. In a balanced three-phase system with 120° phase displacement, θ_0 is equal to zero for phase 1, it is equal to -120° for Phase 2, and it is equal to +120° for phase 3. The equivalent load resistance transferred to the primary side of the transformer of each phase can be written in terms of fluctuating output power over the line cycle that is equal to the input power for a lossless circuit (i.e., $p_{in}(\theta + \theta_0) = p_o(\theta + \theta_0)$).

$$p_{o(1,2,3)}(\theta + \theta_0) = 2 \times \sin^2(\theta + \theta_0) \times P_{o(1,2,3)} = 2 \times \sin^2(\theta + \theta_0) \times \frac{v_{oe(1,2,3)}^2}{R_{oe(1,2,3)\_FL}} \tag{3}$$

$$R_{oe(1,2,3)\_FL} = \frac{8 \times n_{(1,2,3)}^2}{\pi^2} \times R_{L(1,2,3)\_FL} = \frac{8 \times n_{(1,2,3)}^2}{\pi^2} \times \frac{V_{o(1,2,3)}}{I_{o(1,2,3)\_FL}} \tag{4}$$

$$R_{o(1,2,3)}(\theta + \theta_0) = \frac{v_{oe(1,2,3)}^2}{p_{o(1,2,3)}(\theta + \theta_0)} = \frac{R_{oe(1,2,3)\_FL}}{2\sin^2(\theta + \theta_0)} = \frac{4 \times n_{(1,2,3)}^2}{\pi^2 \sin^2(\theta + \theta_0)} \times R_{L(1,2,3)\_FL}$$

$$= \frac{4 \times n_{(1,2,3)}^2}{\pi^2 \sin^2(\theta + \theta_0)} \times \frac{V_{o(1,2,3)}}{I_{o(1,2,3)\_FL}} \tag{5}$$

where $R_{oe(1,2,3)\_FL}$ is the equivalent load resistance transferred to the primary side of the transformers, $R_{L(1,2,3)\_FL}$ is the output load resistance of each phase, and $I_{o(1,2,3)\_FL}$ is the output load current of each phase when operating at full output power ($P_{o(1,2,3)}$).

Then with a balanced power distribution between the three-phase modules ($P_{o(1,2,3)} = P_o/3$), the total equivalent load resistance at the primary of the transformers for the three-phase system is not dependent on the line cycle angle as the AC line fluctuation gets canceled in the output in a balanced system.

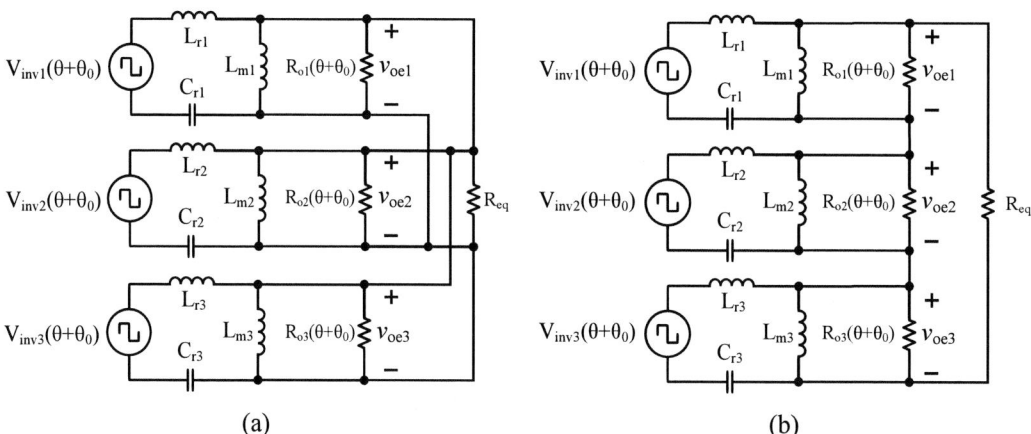

Fig. 3. The equivalent model of the three-phase AC-DC LLC converter, (a) Mode1, and (b) Mode 2.

Design considerations and control scheme

The first step in the design procedure of the proposed EV charger is to find the turn ratio of the transformer. It is important to make sure all the boundary output voltage conditions can be met. The detailed design of the single-stage PFC operation of three-phase soft-switching LLC-based rectifiers has been discussed in the literature [21] and [22]. The voltage range of the 400 V battery systems is considered to be from 250 V to 420 V and the voltage range of the 800 V battery systems is considered to be from 550 V to 850 V. To find out the transformer turn ratio, the minimum required gain by the rectifier at θ=90° is equal to the minimum available gain by the LLC converter. Please note that the maximum switching frequency swing will happen at the lowest output voltage condition of Mode 2 (i.e., a series connection of phase modules), which is for 550 V battery voltage. Hence, the output voltage range for each phase of the proposed EV charger should be designed from 550/3 V to 420 V. As the lower voltage limit for battery charging has a derating condition due to constant current charging, it is important to minimize circulating conduction losses for the voltages close to the rated battery voltage (e.g., 400 V and 800 V). Therefore, only for the 550 V condition, the switching frequency goes into an above series resonant region ($f_s < f_{sw}$) where a minimum required LLC voltage gain of 0.8 is realized. Hence, the turn ratio of the transformer ($n_{1,2,3} = N_{p1,2,3}/N_{s1,2,3}$) can be found in the following equations.

$$G_{req}{}^{min}(\theta) = \frac{V_{o1,2,3}{}^{min}}{V_{ac}} \rightarrow G_{req}{}^{min}(90°) = \frac{550/3}{V_{ac(pk)}} \tag{6}$$

$$G_{LLC(total)}{}^{min}(f_{sw}) = G_{LLC}{}^{min}(f_{sw}) \times \frac{1}{n_{1,2,3}} \rightarrow G_{LLC(total)}{}^{min}(f_{sw}) = \frac{0.8}{n_{1,2,3}} \tag{7}$$

$$G_{req}{}^{min}(\theta) = G_{LLC(total)}{}^{min}(f_{sw}) \rightarrow \frac{550/3}{V_{ac(pk)}} = \frac{0.8}{n_{1,2,3}} \rightarrow n_{1,2,3} = \frac{0.8 \times V_{ac(pk)}}{550/3} \tag{8}$$

Fig. 4 illustrates the design gain curves for the maximum and minimum switching frequency swing conditions, which is for V_o=550 V and V_o=420 V, respectively. It should be mentioned that this design is for a 1.5 kW prototype that was implemented in the laboratory. These design gain curves show the available gain curves for different line angles corresponding to different output power levels. The resonant tank parameters are designed such that the required minimum gain can be achieved for both boundary voltage conditions of each phase while minimizing the circulating current at the primary side of the LLC modules.

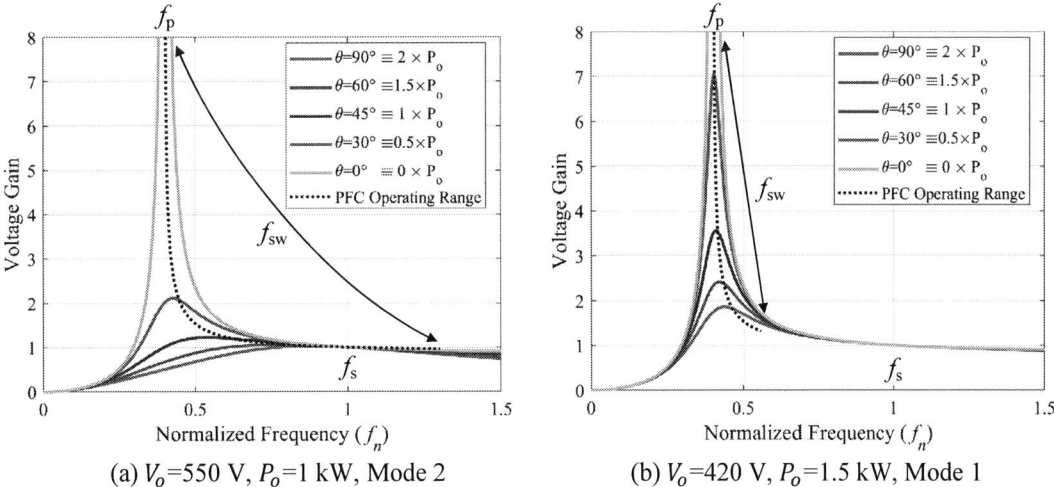

(a) V_o=550 V, P_o=1 kW, Mode 2   (b) V_o=420 V, P_o=1.5 kW, Mode 1

Fig. 4. LLC module design gain curves for (a) maximum switching frequency swing and (b) minimum switching frequency swing.

Fig. 5 illustrates a typical charging profile of both 400 V and 800 V battery systems with similar charging capacity being connected to the same power level EV charger. In the proposed EV charger, the maximum charging current is limited to I_{ch} for 400 V battery and it is limited to $I_{ch}/2$ for 800 V battery. Commonly, the first charging mode in battery charging is Constant Current (CC) followed by a Constant Voltage (CV) mode to maximize the battery State of Charge (SOC). The charging current at the end of a charging cycle is usually around 10 % of the CC charging current. It should be mentioned that in EV DC fast charging only the maximum deliverable power in CC mode is critical in order to increase the battery SOC from 10-20 % to 60-80 %. It should be mentioned that the DC fast charging profile can be variable for different EVs to meet the temperature stress of the battery cells which is also dependent on the available battery cooling capacity.

The overall control scheme of the proposed EV charger is demonstrated in Fig. 6. In the control unit, the input three-phase voltages and currents are sensed to realize PFC, and the output voltage and current are sensed to realize the desired battery charging profile. Based on the pre-charging communication between the battery and the EV charger the proper signal for CC/CV charging and output mode selection of the proposed charger is decided to reflect the type, capacity, and SOC of the installed battery on the EV. Furthermore, three current reference signals are generated based on the sensed AC voltages that form three fast inner current loops for PFC. In the outer loop that is slower than the inner loops, the voltage/current feedback signals are compared with the reference signals and a control signal (v_c) is generated based on the desired charging mode that is then used in the inner current loops.

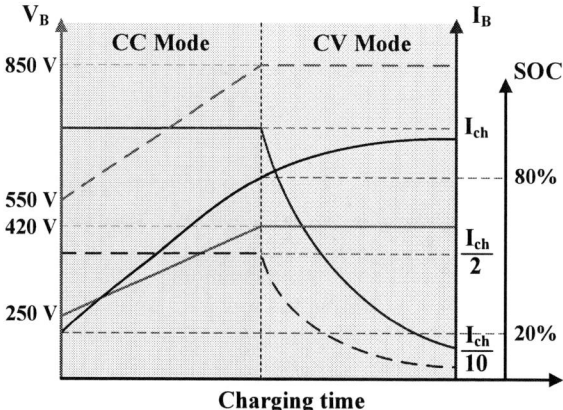

Fig. 5. The charging strategy of the proposed EV charger for both 400 V and 800 V batteries with similar capacity using the same power rating charger with a maximum current of I_{sh}.

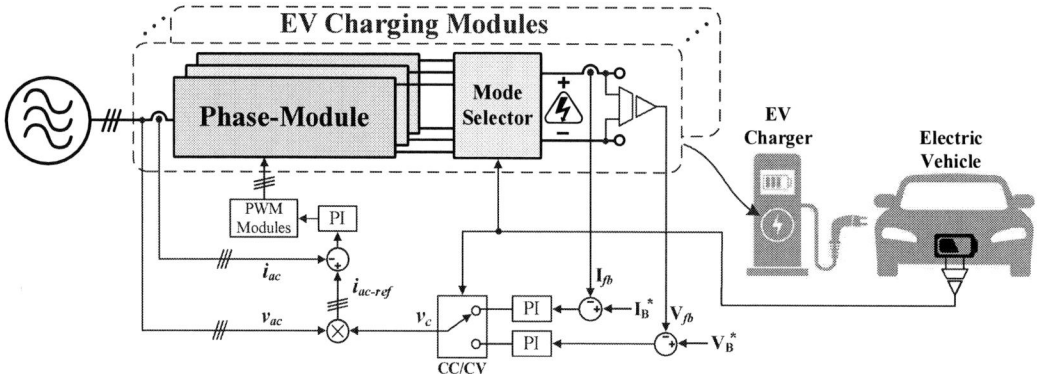

Fig. 6. The overall control scheme of the proposed wide output voltage range EV charger.

Simulation results

The proposed EV charger is simulated in the PSIM environment to verify its performance. Table I shows the design parameters used for both computer simulation and experiment. The resonant tank components are selected based on the design curves shown in Fig. 3 so all the operating conditions can be met. It should be mentioned that a small LC filter is used at the input of each module to filter the switching frequency ripple at the grid side. In the practical EV charging profile, around 30 % derating is considered for the depleted battery conditions, i.e., for 250 V in the 400 V battery and 550 V in the 800 V battery, and the full power should be available from V_o=320 V for 400 V battery system and from V_o=650 V for 800 V battery systems.

Table I: The parameters used in both simulation and experiment.

Parameters/Description		Values
Output Power Range (P_o)		1 kW - 1.5 kW
Input Voltages (V_a, V_b, V_c)		220 V_{ac}
Output Voltage Range (V_o)		250 V_{dc} - 850 V_{dc}
Switching Frequency Range ($f_{sw[1,2,3]}$)		200 kHz - 600 kHz
Parallel Resonant Inductor ($L_{m[1,2,3]}$)		120 µH
Series Resonant Inductor ($L_{r[1,2,3]}$)		22.5 µH
Series Resonant Capacitor ($C_{r[1,2,3]}$)		4.8 nF
Parallel Resonant Frequency (f_p)		191 kHz
Series Resonant Frequency (f_s)		484 kHz
Transformer Turns Ratio ($n_{[1,2,3]} : 1$)		1.35
Input LC Filter	Inductor ($L_{f[1,2,3]}$)	25 µH
	Capacitor ($C_{f[1,2,3]}$)	0.5 µF
Output Capacitor (C_{o1}, C_{o2}, C_{o3})		120 µF

Fig. 7. shows the steady-state line cycle simulation results for 400 V battery systems. For this condition, the output of the three-phase modules is connected in parallel. Fig. 7 (a) shows the simulation results for V_o=250 V. A scaled-down waveform of phase one's voltage is demonstrated to show a near unity power factor along with three-phase sinusoidal currents. As intended, the switching frequency of the three modules is close to the parallel resonant frequency around VZC points over the line cycle. Fig. 7 (b) shows the simulation results for V_o=420 V at rated output power. It can be observed that for the higher battery voltage condition also a unity power factor is achieved with near sinusoidal currents and low THD. Moreover, the output voltage ripple in both boundary conditions is below 100 mV.

The simulation results for the 800 V battery charging condition with series output connection of the three-phase modules are shown in Fig. 8. As can be observed from Fig. 8 (a) the switching frequency swing is from 200 kHz to 600 kHz for V_o=550 V which is in accordance with the design criteria mentioned in the previous section. Fig. 8 (b) illustrates the simulation results for V_o=850 V at rated output power. In both boundary conditions of the 800 V battery charging a unity power factor is achieved with a near sinusoidal current shape at the AC side. Furthermore, it can be observed that the output voltage of the proposed EV charger does not carry any low-frequency ripple in both parallel connection and series connection of the three-phase modules.

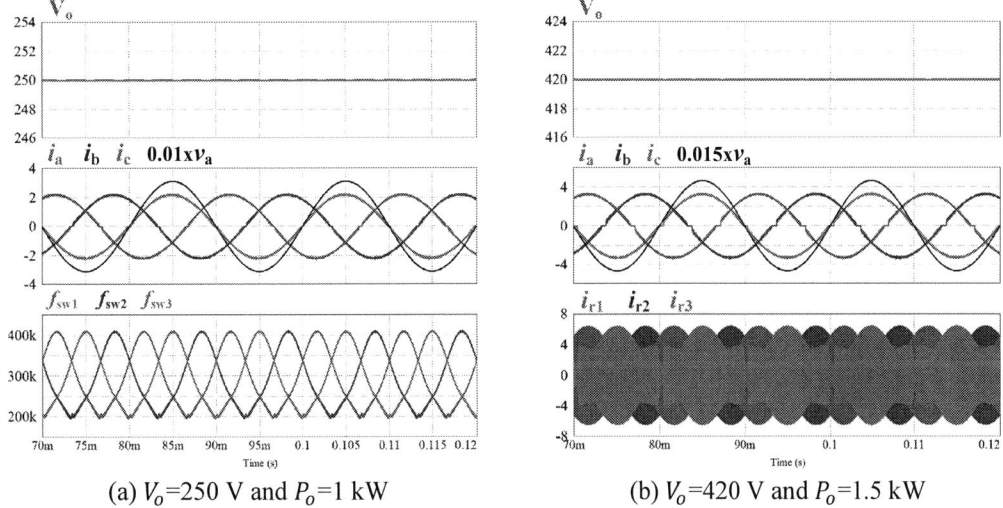

(a) V_o=250 V and P_o=1 kW　　　　　　(b) V_o=420 V and P_o=1.5 kW

Fig. 7. Line cycle simulation results for the 400 V battery charging profile.

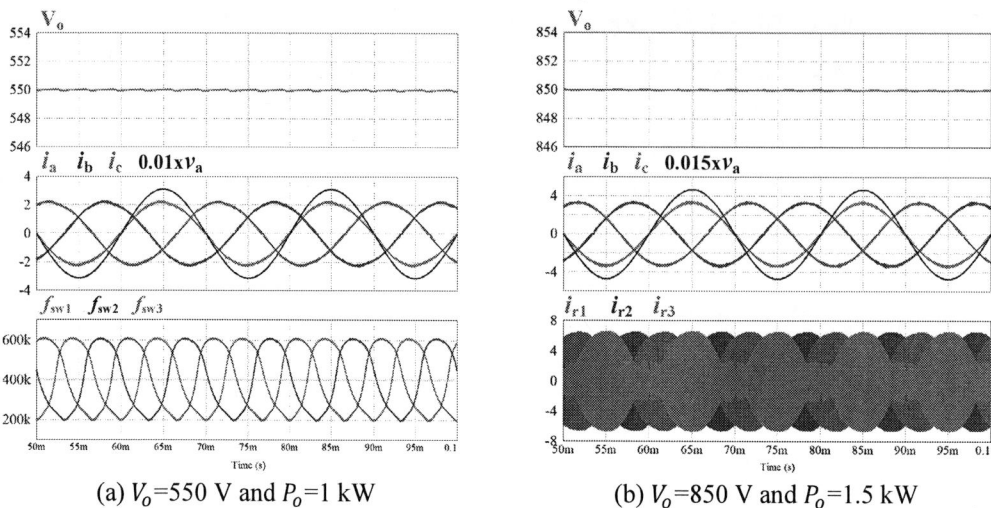

(a) V_o=550 V and P_o=1 kW   (b) V_o=850 V and P_o=1.5 kW

Fig. 8. Line cycle simulation results for the 800 V battery charging profile.

Fig. 9 illustrates the simulation results of resonant current and magnetizing current for the peak power delivery condition in the positive line cycle (i.e., at θ=90°) for both minimum and maximum voltage levels of the 400 V and 800 V battery charging profiles. As can be observed from Fig. 9 (a), the operation of the LLC tank for 250 V output voltage is close to the series resonant frequency. The circulating current at the primary side of the transformer is 35 % in this condition. Fig. 8 (b) shows that the switching frequency for 420 V output voltage is far away from the series resonant frequency with the highest circulating current of 55 %. As expected from the design criteria shown in Fig. 4 (b), the switching frequency range is narrow for the 420 V condition. Fig. 9 (c) illustrates the minimum charging voltage for the series output connection in the 800 V battery charging profile with the maximum switching frequency operation and the circulating current of 18 % that is happening in the above resonant region. As demonstrated in Fig. 4 (a) the maximum switching frequency swing happens for the 550 V output voltage condition to reduce the RMS current and improve the circulating current for the 420 V output voltage condition. Fig. 9. (d) illustrates the maximum charging voltage condition with the switching frequency of 339 kHz and a 32 % circulating current.

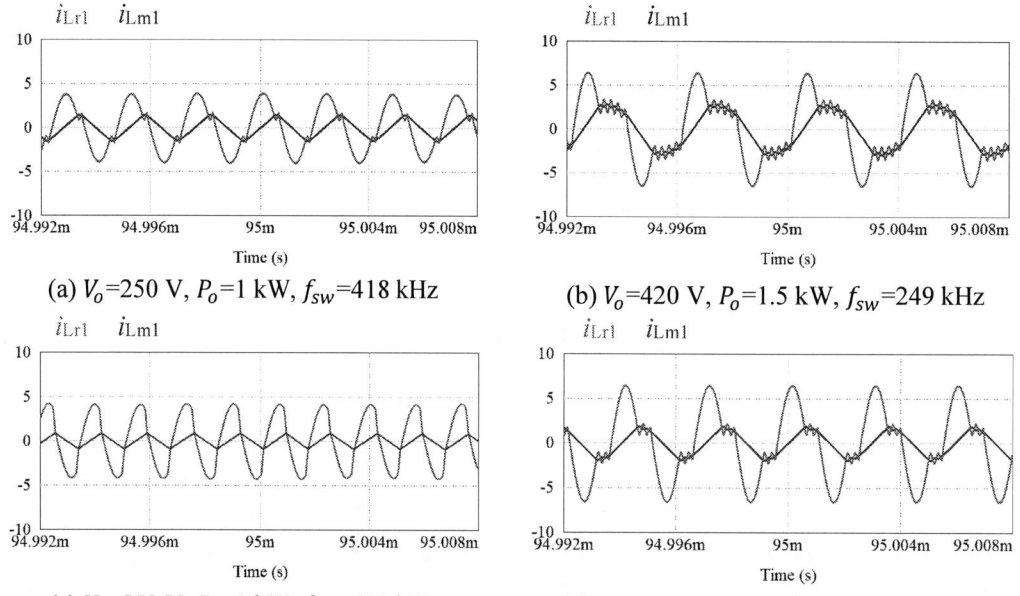

(a) V_o=250 V, P_o=1 kW, f_{sw}=418 kHz   (b) V_o=420 V, P_o=1.5 kW, f_{sw}=249 kHz

(c) V_o=550 V, P_o=1 kW, f_{sw}=600 kHz   (d) V_o=850 V, P_o=1.5 kW, f_{sw}=339 kHz

Fig. 9. Zoomed-in waveforms of peak power delivery at θ=90° for all boundary conditions of the 400 V and 800 V charging profiles.

Experimental results

A 1.5 kW prototype is built in the laboratory to verify the performance of the proposed EV DC fast charger. A single Microchip dsPIC microcontroller is used to perform digital control implementation of the proposed three-phase charger. In the MCU, a look-up table is used to implement the reference signals for the three-phase input current to realize a proper PFC. It should be mentioned that in the prototype the input diode rectifiers are 650 V Silicon diodes and 650 V GaN HEMTs are used for the primary switching bridges while 650 V SiC diodes are used for the output rectifiers of each module. Hence, all the switching devices of the proposed EV DC fast charger are rated for 650 V, which increases the reliability while keeping the implementation costs down.

Fig. 10 (a) shows the experimental result for three-phase currents over the line frequency for the 420 V output voltage condition. In this case, double pole relays in Fig. 2 are in position 1 so the output of the three-phase modules are connected in parallel. As can be observed, a proper PFC is achieved in all phases and the output voltage does not consist of any low-frequency ripple. Fig. 10 (b) shows the line frequency experimental results for the 850 V output voltage condition. In this case, the double pole relays are set in position 2 so the output of the three-phase modules is connected in series.

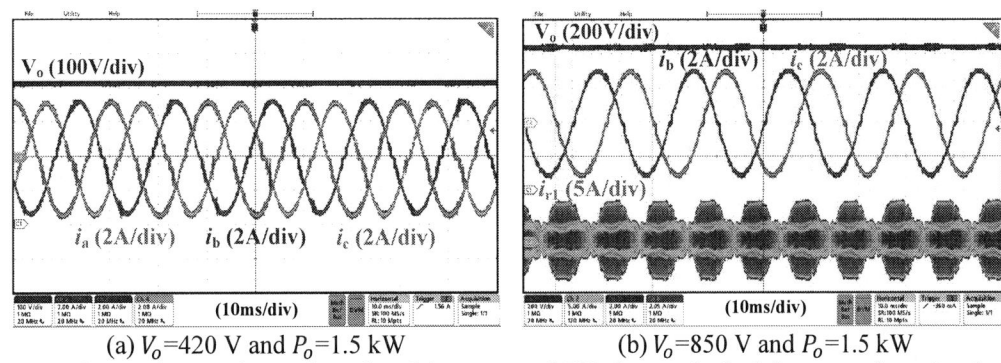

(a) V_o=420 V and P_o=1.5 kW   (b) V_o=850 V and P_o=1.5 kW

Fig. 10. Line cycle experimental results of the proposed EV charger for both battery voltage levels.

The zoomed-in resonant current waveforms are shown in Fig. 11 for both 420 V and 850 V output voltage conditions. The maximum switching frequency that happens at θ=90° is 240 kHz for V_o=420 V and the maximum switching frequency for V_o=850 V is equal to 337 kHz. Both conditions are as expected and according to the simulation results shown in Fig. 7 to Fig. 9.

The efficiency measurements for V_o=420 V - P_o=1.5 kW shows around 96.5 % efficiency and for V_o=850 V - P_o=1.5 kW shows around 96.2 % efficiency. It should be mentioned that based on power loss estimations the efficiency can be improved by around 1 % after changing the input rectifiers with Si MOSFETs and using SiC MOSFETs for the synchronous rectifiers.

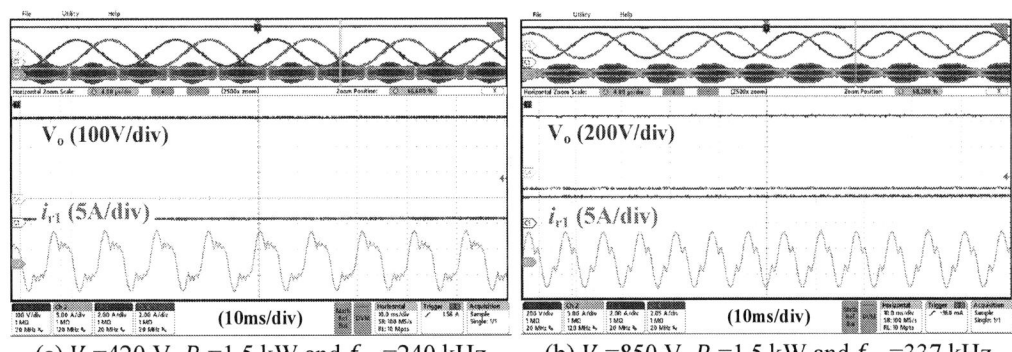

(a) V_o=420 V, P_o=1.5 kW and f_{sw}=240 kHz   (b) V_o=850 V, P_o=1.5 kW and f_{sw}=337 kHz

Fig. 11. Zoomed-in resonant current waveforms of different battery pack voltage levels.

Conclusion

A new single-stage three-phase EV charger with reconfigurable output has been proposed in this paper. The main characteristics of the proposed EV charger include soft-switching, no dc-link capacitor requirement, and suitability for both 400 V and 800 V automotive battery systems. The analysis demonstrated a suitable resonant tank design for a wide output voltage operation in the proposed three-phase EV DC fast charger. Computer simulation results verified the proper PFC performance and small output voltage ripple of the proposed single-stage three-phase EV charger over a wide output voltage range from 250 V to 850 V. Moreover, a 1.5 kW laboratory prototype has been built to verify the practicability of the proposed EV DC fast charger for both 400 V and 800 V battery charging applications.

References

[1] O. Ellabban, H. Abu-Rub and F. Blaabjerg. "Renewable energy resources: Current status, future prospects and their enabling technology." *Renewable Sustain. Energy Rev.,* vol 39, pp. 748-764, 2014.

[2] B. Wang, M. Xu and L. Yang. "Study on the economic and environmental benefits of different EV powertrain topologies." *Energy Convers. Manage.,* vol 86, pp. 916-926, 2014.

[3] M. Pahlevani and P. K. Jain, "Soft-Switching Power Electronics Technology for Electric Vehicles: A Technology Review," *IEEE J. Emerg. Sel. Topics Ind. Electron.*, vol. 1, no. 1, pp. 80-90, July 2020.

[4] S. Chon, M. Bhardwaj and H. Nene. "Maximizing power for Level 3 EV charging stations." *Texas Instrument Article*, pp. 1-33, 2018.

[5] Y. Li, J. Schäfer, D. Bortis, J. W. Kolar and G. Deboy, "Optimal Synergetic Control of a Three-Phase Two-Stage Ultra-Wide Output Voltage Range EV Battery Charger Employing a Novel Hybrid Quantum Series Resonant DC/DC Converter," in *Proc. IEEE 21st Workshop on Control and Modeling for Power Electronics (COMPEL)*, pp. 1-11, 2020.

[6] M. Leibl, J. W. Kolar and J. Deuringer, "Sinusoidal Input Current Discontinuous Conduction Mode Control of the VIENNA Rectifier," *IEEE Trans. Power Electron.*, vol. 32, no. 11, pp. 8800-8812, Nov. 2017.

[7] C. Lim, Y. Jeong, M. Lee, K. Yi and G. Moon, "Half-Bridge Integrated Phase-Shifted Full-Bridge Converter With High Efficiency Using Center-Tapped Clamp Circuit for Battery Charging Systems in Electric Vehicles," *IEEE Trans. Power Electron.*, vol. 35, no. 5, pp. 4934-4945, May 2020.

[8] C.-C. Hua, Y.-H. Fang and C.-W. Lin. "LLC resonant converter for electric vehicle battery chargers." *IET Power Electron.,* vol. 9. no. 12, pp. 2369-2376, 2016.

[9] M. Forouzesh and Y. -F. Liu, "Interleaved LCLC Resonant Converter With Precise Current Balancing Over a Wide Input Voltage Range," *IEEE Trans. Power Electron.*, vol. 36, no. 9, pp. 10330-10342, Sept. 2021.

[10] D. Das, N. Weise, K. Basu, R. Baranwal and N. Mohan, "A Bidirectional Soft-Switched DAB-Based Single-Stage Three-Phase AC–DC Converter for V2G Application," *IEEE Trans. Transport. Electrific.*, vol. 5, no. 1, pp. 186-199, March 2019.

[11] N. Rathore, S. Gangavarapu, A. K. Rathore and D. Fulwani, "Emulation of Loss Free Resistor for Single-Stage Three-Phase PFC Converter in Electric Vehicle Charging Application," *IEEE Trans. Transport. Electrific.*, vol. 6, no. 1, pp. 334-345, March 2020.

[12] T. Mishima and S. Mitsui, "A Single-Stage High Frequency-link Modular Three-Phase Soft-Switching AC-DC Converter for EV Battery Charger," in *Proc. IEEE Energy Conversion Congress and Exposition (ECCE)*, 2019, pp. 2141-2147.

[13] V. Reber. (2016). *E-Power: New Possibilities With 800-Volt Charging.* Porsche Eng. Mag. Accessed: Dec. 2021. [Online]. Available: https://www.porscheengineering.com/peg/en/about/magazine/

[14] C. Jung, "Power Up with 800-V Systems: The benefits of upgrading voltage power for battery-electric passenger vehicles," *IEEE Electrific. Mag.*, vol. 5, no. 1, pp. 53-58, March 2017.

[15] I. Aghabali, J. Bauman, P. J. Kollmeyer, Y. Wang, B. Bilgin and A. Emadi, "800-V Electric Vehicle Powertrains: Review and Analysis of Benefits, Challenges, and Future Trends," *IEEE Trans. Transport. Electrific.*, vol. 7, no. 3, pp. 927-948, Sept. 2021.

[16] M. Abbasi and J. Lam, "An SiC-Based AC/DC CCM Bridgeless Onboard EV Charger With Coupled Active Voltage Doubler Rectifiers for 800-V Battery Systems," in *Proc. IEEE Applied Power Electronics Conference and Exposition (APEC)*, 2020, pp. 905-910.

[17] S. Sen, L. Zhang, T. Chen, J. Zhang and A. Q. Huang, "Three-phase Medium Voltage DC Fast Charger based on Single-stage Soft-switching Topology," in *Proc. IEEE Transportation Electrification Conference and Expo (ITEC)*, 2018, pp. 1123-1128.

[18] J. -Y. Kim, B. -S. Lee, D. -H. Kwon, D. -W. Lee and J. -K. Kim, "Low Voltage Charging Technique for Electric Vehicles With 800 V Battery," *IEEE Trans. Ind. Electron.*, doi: 10.1109/TIE.2021.3109526. (Early Access)

[19] G. V. Bharath, S. Kiran Voruganti, V. T. Nguyen, V. Uttam Pawaskar and G. Gohil, "Performance Evaluation of 10kV SiC-based Extreme Fast Charger for Electric Vehicles with Direct MV AC Grid Interconnection," in *Proc. IEEE Applied Power Electronics Conference and Exposition (APEC)*, 2020, pp. 3547-3554.

[20] W. Liu, A. Yurek, B. Sheng, Y. Chen, Y. -F. Liu and P. C. Sen, "A Single Stage 1.65kW AC-DC LLC Converter with Power Factor Correction (PFC) for On-Board Charger (OBC) Application," in *Proc. IEEE Energy Conversion Congress and Exposition (ECCE)*, 2020, pp. 4594-4601.

[21] M. Forouzesh, Y. -F. Liu and P. C. Sen, "Implementation of an Isolated Phase-Modular-Designed Three-Phase PFC Rectifier Based on Single-Stage LLC Converter," in *Proc. IEEE Energy Conversion Congress and Exposition (ECCE)*, 2021, pp. 2266-2273.

[22] M. Forouzesh, Y. -F. Liu and P. C. Sen, "A Novel Soft-Switched Three-Phase Three-Wire Isolated AC-DC Converter With Power Factor Correction," in *Proc. IEEE Applied Power Electronics Conference and Exposition (APEC)*, 2022, pp. 887-894.

Efficiency Improvement of Single-Stage AC-DC LLC Converter Using a Line Cycle Synchronous Rectifier (SR) Driving Strategy

Mojtaba Forouzesh, Yan-Fei Liu, Paresh C. Sen
Queen's University
Department of Electrical and Computer Engineering
Kingston, ON, Canada
E-Mail: m.forouzesh@queensu.ca, yanfei.liu@queensu.ca, senp@queensu.ca
URL: https://www.ece.queensu.ca/research/labs/power-group.html

Keywords

«AC-DC converter», «DAB-LLC converter», «Synchronous rectifier (SR)», «Silicon Carbide (SiC) MOSFET», «Gallium Nitride (GaN) HEMT», «Wide bandgap devices».

Abstract

Synchronous rectification that is being widely used in high-power Inductor-Inductor-Capacitor (LLC) DC-DC converters to improve efficiency can be challenging in single-stage AC-DC LLC converters with high output voltage levels (i.e., >100V) where synchronous driving ICs cannot be used. In this paper, a simple line cycle SR driving strategy with direct MCU control is proposed for single-stage AC-DC LLC converters. The principles of operation and methodology behind the proposed line cycle SR driving strategy are discussed. Simulation and experimental results validated the performance of the proposed SR driving strategy for a 250 V to 400 V output voltage range AC-DC LLC converter. A full-load efficiency of 98.1 % was achieved for the 250 V output voltage condition and a full-load efficiency of 97.3 % was achieved for the 400 V output voltage condition. Compared with a fixed ON time method, around 0.5 %, and 0.8 % efficiency improvements were observed in 250 V and 400 V output voltage conditions, respectively. In addition, it is observed that the proposed simple SR driving method obtains the same efficiency levels as more complex adaptive SR driving approaches.

Introduction

Resonant converters have been the focus of many research studies over the past decades. Some studies discussed the optimal design of resonant converters using various computer-aided methods [1] and [2]. Some research studies focused on magnetics improvement for resonant converters [3] and [4]. Some studies discussed the impact of wide bandgap semiconductors such as Gallium Nitride (GaN) HEMTs and Silicon Carbide (SiC) MOSFETs in improving the efficiency and power density of resonant converters [5] and [6]. In high-power DC-DC resonant converters diode conduction loss is a major contributor to the total power conversion loss. Hence, using synchronous rectification is a necessity for low voltage and high current applications to achieve high conversion efficiency at kW levels. Therefore, a lot of research has been done on Synchronous Rectifier (SR) driving strategies for LLC DC-DC converters [7]-[12].

One of the common practices for SR driving in resonant converters is the drain-source voltage sensing using SR driving ICs [7]. This method is widely used in many low output voltage high current applications [8] and [9]. The performance of driving ICs is promising; however, it can be highly dependent on the effect of parasitic components of the circuit that reduces the reliability of the converter. On the other hand, the operating voltage of most of these driving ICs is limited to below 100V and hence they are not suitable for SR driving at higher output voltage levels. In [10], some additional components are added to the famous NCP4306 SR driving IC to make it work for high output voltage applications, however, the disadvantages of low voltage sensing still exist.

Another common practice for SR driving is a model-based approach that generates an appropriate gate pulse in the MCU to drive the SR. In [11] the SR driving signal is generated based on the primary bridge signals using a simulation model to find out the turn ON and OFF delay that was inserted into a look-up table for all the operating conditions. In [12] a rather fixed conduction time is used with specific turn ON and -OFF delays for the LLC converter of an EV battery charger. In this approach, the output voltage and current should be considered to correctly reflect the output power and tune proper driving signal delays accordingly to avoid inaccurate gate signals. In [13], an adaptive sensor-less model-based digital driving scheme is developed for the LLC converter in DC-DC applications. In this method, the output voltage and current are sensed to find out the load conditions, then based on the switching frequency the conduction time of the SR is calculated and tuned online.

Most of the literature for SR driving of resonant converters only discussed the DC-DC application so far. In recent years there is a high interest in the implementation of the LLC converter in single-stage AC-DC conversion with PFC capability which is due to its soft-switching performance [14]-[17]. Near the line Voltage Zero Crossing (VZC) area of AC to DC conversion, the input power delivered to the load is much smaller than the rated power, and in such light load conditions, both SR late turn ON and early turn OFF can happen which makes the SR driving of LLC converter more challenging. For low output voltage conditions, it is still possible to use the same dedicated SR controller ICs as in DC-DC converters. However, this method is not transparent to the designer, is not immune to noise, and there is no full control over the driving signals. Therefore, a proper SR driving strategy for LLC converter in AC-DC conversion is missing to achieve enhanced power conversion efficiency in high-power applications and high output voltage levels (i.e., >100V).

In this paper, a simple line cycle SR driving strategy is proposed for single-stage AC-DC LLC converters. The proposed method is a sensor-less approach with digital driving signals coming directly from the MCU. The principles of the proposed SR driving method are described in the next section followed by simulation results and experimental results from a 500 W wide bandgap-based prototype to verify the performance of the proposed line cycle SR driving strategy for the single-stage AC-DC LLC converters. This paper is concluded in the last section.

The Proposed Line Cycle SR Driving Strategy

Fig. 1 illustrates the structure of a single-stage AC-DC LLC converter with active switches for input and output rectifiers. Some of the main characteristics of this converter are demonstrated in Fig. 1. The input rectifier bridge switches (S_1-S_4) operate at line frequency (f_{line}) with Zero Current Switching (ZCS) performance, which can be implemented with Si MOSFETs. It is mentioned in the literature that to achieve PFC while maintaining soft-switching the operating switching frequency (f_{sw}) of the primary bridge of the LLC converter should vary over the half-line cycle between the parallel resonant frequency (f_p) and series resonant frequency (f_r) [14]. In this way, primary side bridge switches (Q_1-Q_4) can take advantage of Zero Voltage Switching (ZVS) and the secondary side bridge switches (SR_1-SR_4) can take

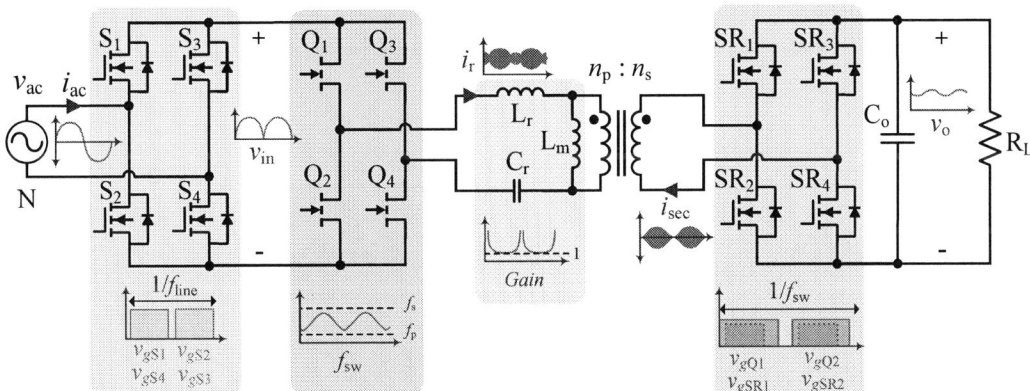

Fig. 1. The single-stage AC-DC LLC converter with synchronous rectifier bridges.

advantage of ZCS. The latter allows high switching frequency implementation leading to high power density designs. Since the output rectifier switches should be controlled based on the synchronized pulses with the respective gate pulses of the primary bridge switches, this method is called synchronous rectification.

Details of different operating conditions over the half-line cycle with primary and secondary bridge driving signals are illustrated in Fig. 2. Over the half-line cycle, the LLC tank goes through a combination of various modes such as O, P, and N modes. Based on the desired operating condition of the LLC converter in PFC mode, the majority part of the line cycle operation is around the peak line voltage while the LLC tank is in PO mode delivering the highest power to the load. It is followed by OPO modes at the sides and O modes near the line VZC where no power is delivered to the load hence the SRs should be turned OFF to avoid reverse power flow. It should be noted that since the PN and PON modes are more likely to lose ZVS, they are avoided in the design procedure, and hence the N mode which normally happens in a very short time is neglected in the analysis. Moreover, the NP, NOP, and OP modes that happen in the above resonant area are avoided in the LLC resonant tank design as it is desirable to operate below the series resonant frequency to take advantage of ZCS for the SRs.

From Fig. 2 it can be observed that in PO mode the turn ON instant of the SR is synchronized with the respective primary switch with a small turn ON delay (1-2% of the switching cycle) considering the propagation delay of the driver ICs. In OPO mode the turn ON delay should be adjusted based on the switching frequency and load condition to avoid reverse power flow. The conduction time of the SR is related to the load and switching frequency that can be calculated from

$$t_{ON}(\theta) = \frac{1}{2f_{sw}(\theta)} \sqrt{-\frac{\pi}{4}\frac{\omega_{sw}(\theta)L_m}{n^2 R_L(\theta)} + \sqrt{\frac{\pi^2}{16}\left(\frac{\omega_{sw}(\theta)L_m}{n^2 R_L(\theta)}\right)^2 + \frac{\pi}{2}\frac{\omega_{sw}(\theta)L_m}{n^2 R_L(\theta)}\left(\frac{f_{sw}(\theta)}{f_r}\right)^2}} \quad (1)$$

where f_r is the series resonant frequency, $\omega_{sw}(\theta)$ is the line phase angle dependent angular switching frequency and $R_L(\theta)$ is the line phase angle dependent load that can be calculated as follows.

$$f_r = \frac{1}{2\pi\sqrt{L_r \times C_r}} \quad (2)$$

$$\omega_{sw}(\theta) = 2\pi f_{sw}(\theta) \quad (3)$$

$$R_L(\theta) = \frac{4 \times \left(n_p/n_s\right)^2}{\pi^2 \sin^2(\theta)} \times R_{L\_FL} \quad (4)$$

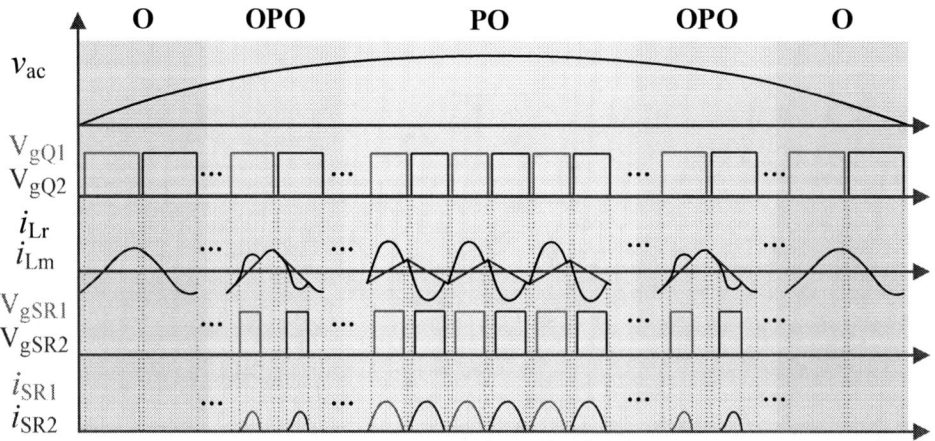

Fig. 2. Different operating conditions and respective switching operations over the half-line cycle.

In (4) R_{L_FL} is the load resistance corresponding to the rated output power (i.e., P_{o_FL}). Due to brevity, the design of LLC tank parameters for AC-DC conversion is not discussed here as the detailed analysis and design are provided in [15]. The SR turn ON time for two different output voltage and load conditions are plotted over a quarter line cycle in Fig. 3. It can be observed that only after $\theta=45°$ a noticeable reduction can be observed in the calculated turn ON time. The latter suggests using a fixed turn ON for the PO mode where the highest power is being delivered to the load. Moreover, to perform a proper SR driving strategy it is crucial to find out the boundary condition between the OPO and PO modes.

In the proposed SR driving strategy, the SRs are turned ON with a fixed calculated ON time for the PO mode and they are turned OFF as soon as the SR turn ON delay becomes large in OPO mode. In this way, the majority of AC power (~70%) is delivered to the load when SRs are operating. This method improves efficiency over fixed turn ON of the SRs by avoiding reverse power flow in O mode. Moreover, it achieves the same efficiency as implementing a complicated online calculation or an adaptive method over the line cycle that requires expensive fast microprocessors and/or additional voltage/current sensing.

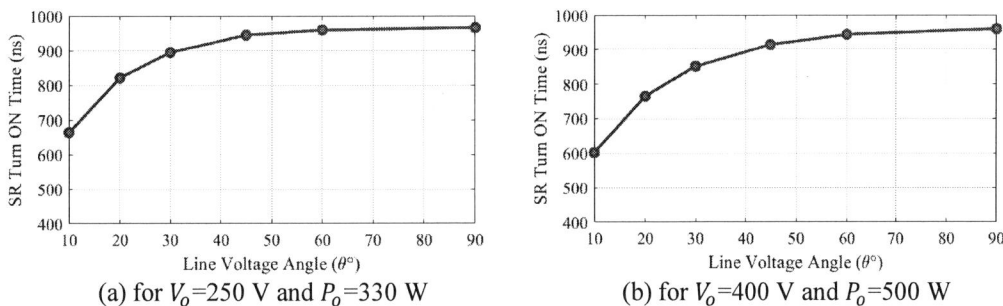

(a) for V_o=250 V and P_o=330 W (b) for V_o=400 V and P_o=500 W

Fig. 3. Calculated SR conduction time over a quarter line cycle at different line voltage angles (θ).

Simulation and experimental results

A simulation model is built in the PSIM environment to verify different operating modes of the LLC converter over the AC line cycle. Table 1 listed the design parameters used for computer simulations as well as the components used in the laboratory prototype.

Fig. 4 illustrates the simulation results over the line cycle for V_o=400 V, P_o=500 W condition to show different operating modes of the single-stage LLC converter in AC-DC conversion. It should be mentioned that in all the simulation results a green background pattern is used to show power delivery P mode and a dark blue background pattern is used to show O mode where there is no power delivery to the output. Both Fig. 4 (a) and Fig. 4 (b) show the PO mode at $\theta=90°$ and $\theta=45°$. As can be observed the current of SR₁ starts to rise at the instant Q₁ turn ON and it ends after a certain conduction time. Fig. 4 (c) shows the OPO mode where the power delivery to the output is reduced and a phase shift between the primary gate pulse and the SR current starts to develop that demands a turn ON delay for the SR₁ driving. Fig. 4 (d) illustrates the O mode where there is no power delivery to the output and hence the SRs should be turned OFF in this mode to avoid reverse current flowing.

Fig. 5 shows the simulation results over the line cycle for V_o=250 V, P_o=330 W condition. Operating modes are mostly similar to Fig. 4 for V_o=400 V, P_o=500 W condition, however, the P mode is mostly longer in PO and OPO modes that is because the switching frequency is closer to the series resonant frequency for V_o=250 V, P_o=330 W condition.

Table 1: The parameters used in both simulation and experiment.

Parameters/Descriptions		Values
Output Power Range (P_o)		333 W - 500 W
AC Voltage (V_{ac})		220 V_{RMS}
Output Voltage Range (V_o)		250V - 400V
Line Frequency (f_{line})		50 Hz
Switching Frequency Range (f_{sw})		200 kHz - 450 kHz
Parallel Resonant Inductance (L_m)		120 µH (PQ3230 - 3F36)
Series Resonant Inductance (L_r)		23 µH (PQ2620 - 3F36)
Series Resonant Capacitance (C_r)		4.8 nF
Integrated Transformer Turns Ratio ($n_p : n_s$)		24 : 19
Input LC Filter	Inductance (L_f)	15 µH
	Capacitance (C_f)	1 µF
Output Capacitance (C_o)		200 µF
Input Rectifier Bridge Si MOSFET		IPP60R099P7 (650V - 31A - 99mΩ)
Inverter Bridge GaN E-HEMT		GS66504B (650V - 15A - 130mΩ)
Output Rectifier Bridge SiC MOSFET		C3M0120065J (650V - 21A - 157mΩ)

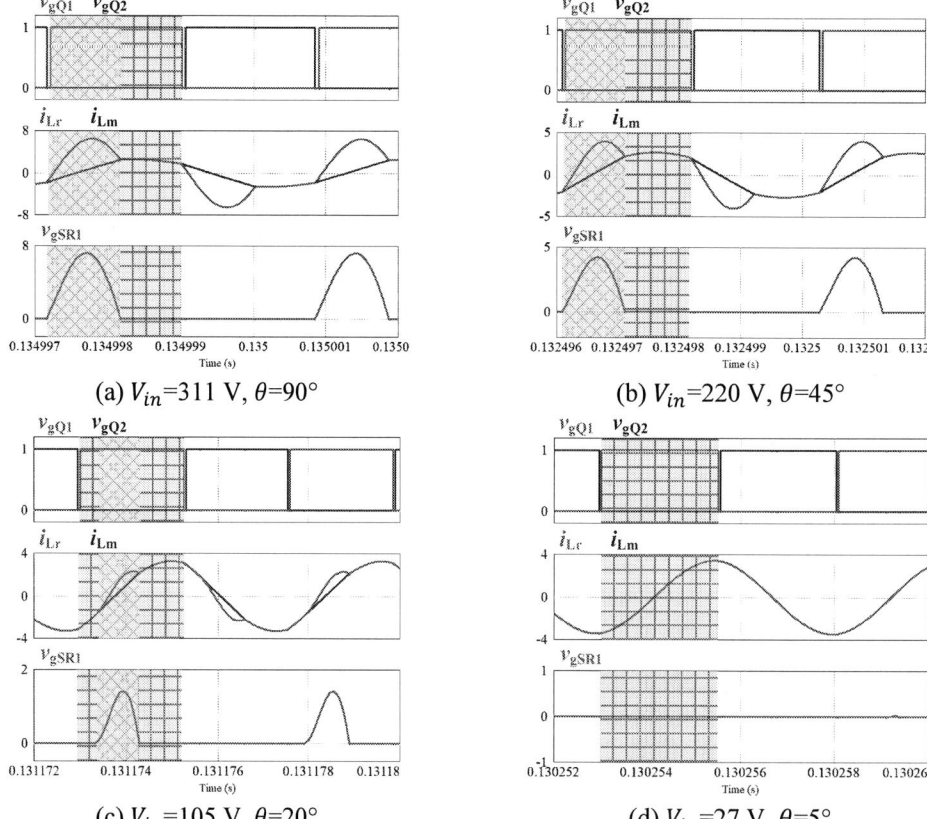

(a) V_{in}=311 V, θ=90°

(b) V_{in}=220 V, θ=45°

(c) V_{in}=105 V, θ=20°

(d) V_{in}=27 V, θ=5°

Fig. 4. Simulation results over the line cycle for V_o=400 V, P_o=500 W condition showing PO, OPO, and O operating modes.

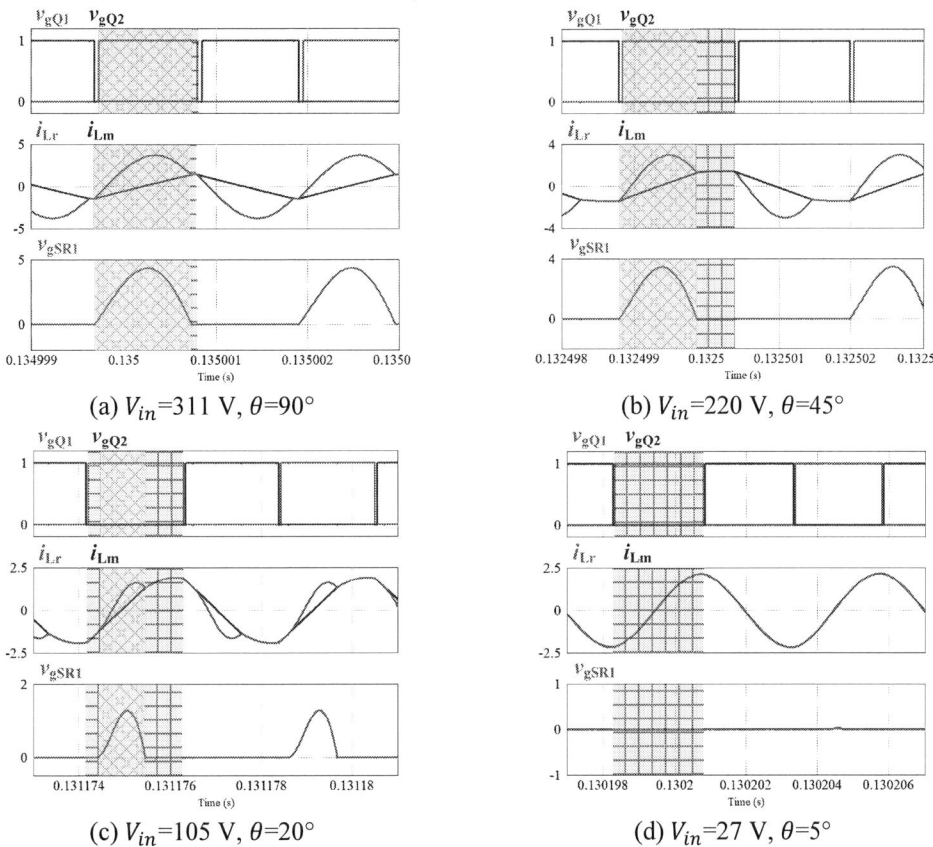

(a) V_{in}=311 V, θ=90°   (b) V_{in}=220 V, θ=45°

(c) V_{in}=105 V, θ=20°   (d) V_{in}=27 V, θ=5°

Fig. 5. Simulation results over the line cycle for V_o=250 V, P_o=330 W condition showing PO, OPO, and O operating modes.

A wide bandgap-based (GaN and SiC) laboratory prototype is built to verify the performance of the proposed SR driving strategy for single-stage AC-DC LLC converters. A picture of the laboratory prototype is shown in Fig. 6. A low-cost Microchip dsPIC microcontroller is used to implement pulse frequency modulated control and perform PFC by providing proper driving signals for the switching bridges of the single-stage AC-DC LLC converter over the line cycle. It should be mentioned that input rectifiers are 650V Silicon MOSFETs and 650 V GaN HEMTs are used for the primary switching bridge while 650 V SiC MOSFETS are used for the synchronous rectifiers. Different scenarios are implemented in practice for the sake of comparison. First, the SR is turned on with a fixed ON time found from the calculation. Second, the proposed method is implemented by turning OFF the SRs over the line cycle after a certain voltage level that is found from the calculation and fine-tuned based on simulation. Last, an adaptive SR driving is implemented using a look-up table mapped with the delay and conduction time data extracted from the simulation to compare the efficiency with the proposed SR driving method.

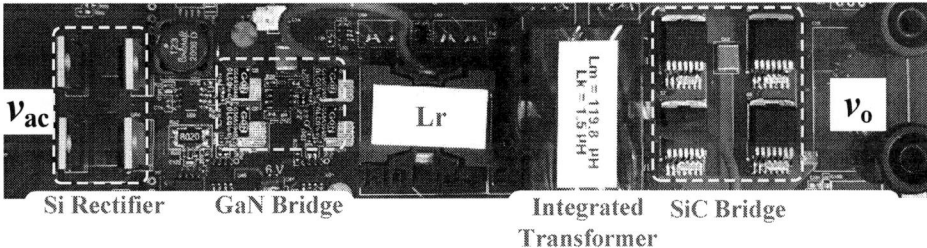

Fig. 6. A picture of the 500W LLC AC-DC converter laboratory prototype.

Fig. 7 (a) shows the experimental results of the single-stage AC-DC LLC converter with a fixed SR ON time over the line cycle. As mentioned before, around the line VZC points the LLC converter operates in O mode, and hence by modulating SRs in that area some reverse power flow is introduced which decreases the power conversion average efficiency over the line cycle. Zoomed-in waveforms for $\theta=24°$, $\theta=20°$, and $\theta=10°$ are shown in Fig. 5 (b) to Fig. 5 (d), respectively. The reverse current due to early turn ON of the SRs can be observed in the secondary current waveforms showing the OPO mode operation of the LLC tank. The reverse current problem gets worse when the O stage at the beginning of the OPO mode gets larger at smaller line angles.

The proposed SR driving strategy for V_o=400V is shown in Fig. 8 where the SRs are turned ON after $\theta=26°$ to avoid reverse power transfer in OPO and O modes. From Fig. 8 (b) to Fig. 8 (d) it can be observed that a fixed conduction time of 940 ns with a 30 ns turn ON delay is providing a proper turn ON and OFF instants for the SRs. Below $\theta=26°$ the turn ON delay starts to become larger and hence the SRs are turned OFF after that angle to avoid reverse current flow. The exact border angle at which the SRs need to be turned OFF should be checked with the simulation results as the theoretical values calculated in Fig. 3 are based on First Harmonic Approximation (FHA) and hence are not so accurate for frequencies far away from the series resonant frequency. In Fig. 8 (e) and Fig. 8 (f), the operating condition at $\theta=20°$ and $\theta=10°$ are shown, respectively. It can be observed that there is no reverse current as the body diode of the SiC MOSFETs are fast enough and hence operating properly.

Furthermore, to emulate online tuning SR driving over the line cycle an adaptive SR driving is implemented via a look-up table using the parameters found from the simulation to compare the efficiency with the proposed SR driving method. The latter resulted in no reverse current in OPO mode, and the SRs were turned ON and OFF at the right instants from $\theta=90°$ to VZC points and vice versa. Moreover, it is found that the efficiency with the complicated online tuning SR driving will either be similar to or less than 0.1 % higher than the proposed simple SR driving method, which does not justify the complex implementation via a powerful microcontroller.

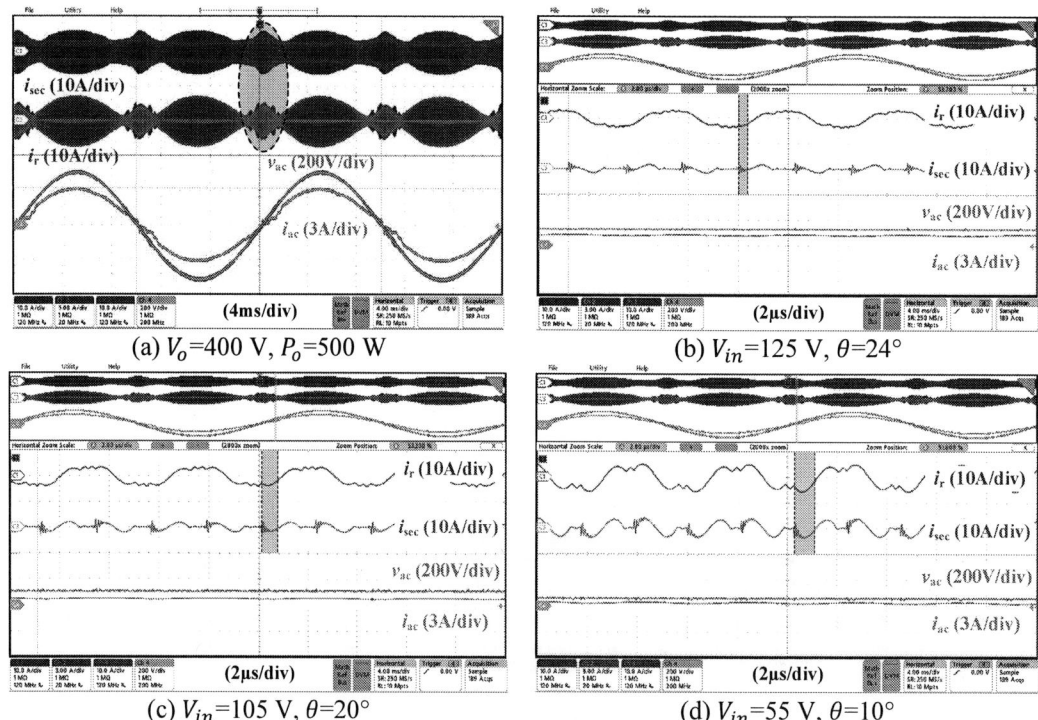

Fig. 7. Experimental results using a fixed SR ON time over the line cycle.

Fig. 9 (a) illustrates the full-load efficiency of 97.3% that was achieved for V_o=400 V at P_o=500 W condition and is 0.8 % higher than using a fixed SR conduction time over the line cycle. The full-load efficiency for V_o=250 V at P_o=330 W condition was measured to be 98.1 % which is shown in Fig. 9 (b). For the same output level condition, the efficiency with a fixed SR conduction time was 97.6 %. It should be mentioned that all the efficiency measurements were carried out using an LMG671 Zimmer precision power analyzer.

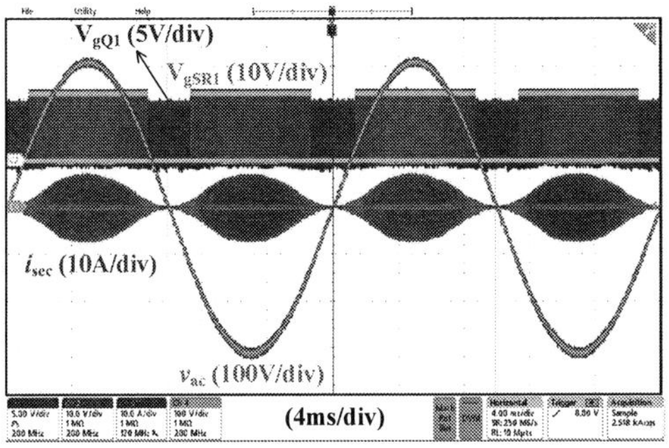

(a) V_o=400 V, P_o=500 W

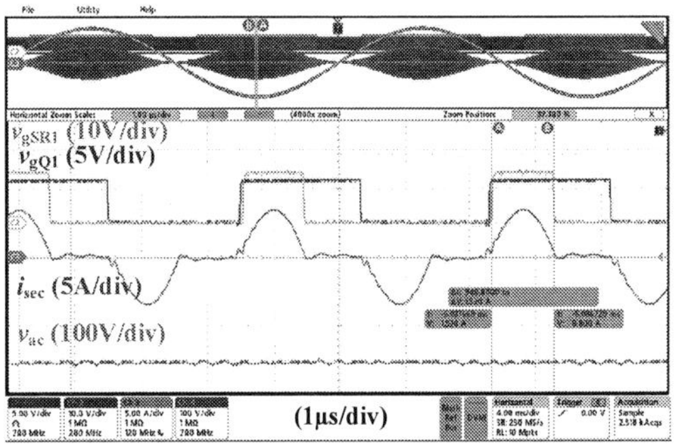

(b) V_{in}=311 V, θ=90°, SR_{ON}=940 ns

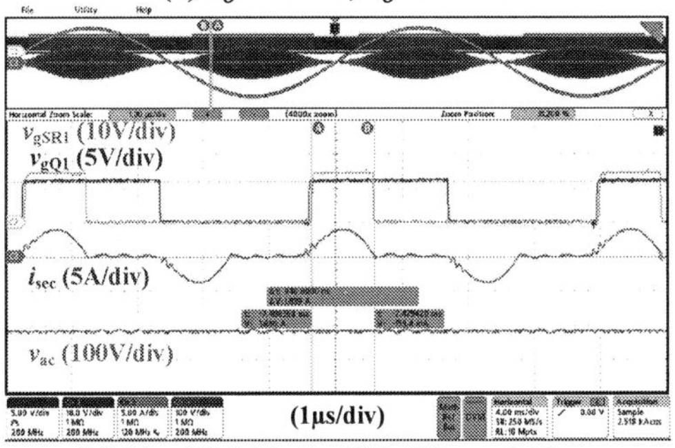

(c) V_{in}=220 V, θ=45°, SR_{ON}=940 ns

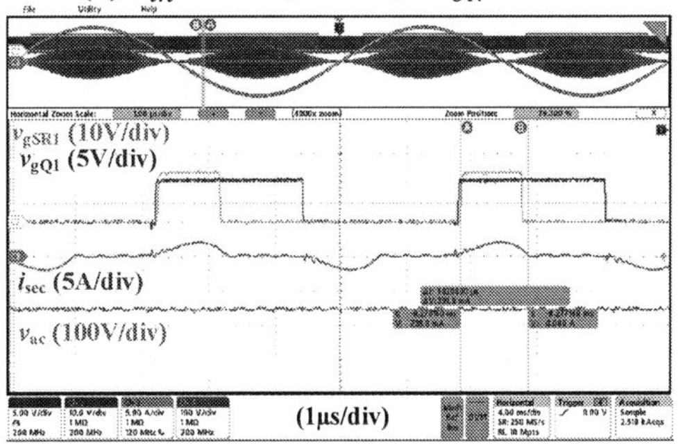

(d) V_{in}=155 V, θ=30°, SR_{ON}=940 ns

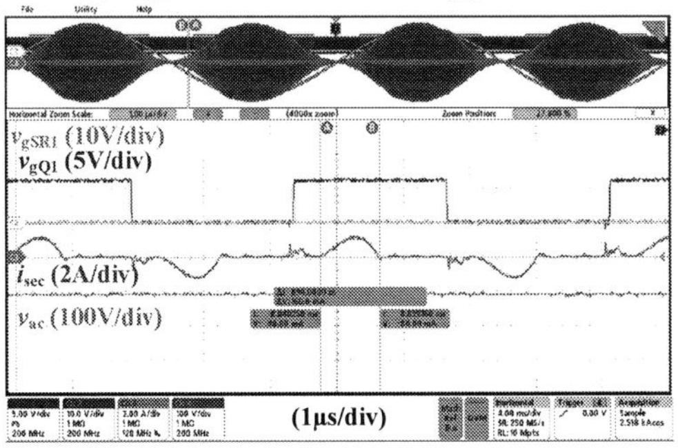

(e) V_{in}=105 V, θ=20°, SR_{ON}=890 ns

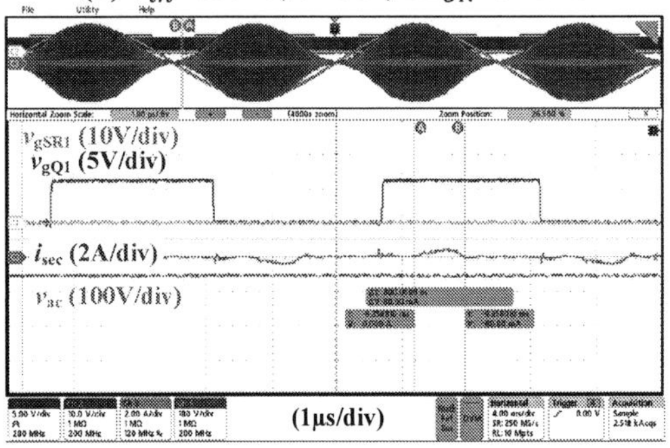

(f) V_{in}=55 V, θ=10°, SR_{ON}=800 ns

Fig. 8. Experimental results of the proposed line cycle SR driving strategy for V_o=400 V, P_o=500 W.

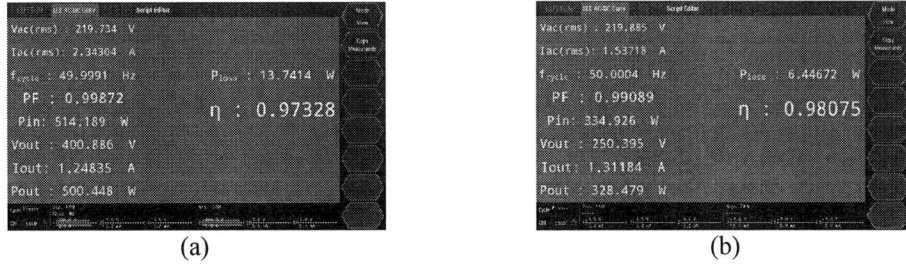

(a)

(b)

Fig. 9. Efficiency measurement results with the proposed synchronous rectification strategy for (a) V_o=400 V, P_o=500 W, and (b) V_o=250 V, P_o=330 W.

Conclusion

This paper proposed a new SR driving strategy for single-stage AC-DC LLC converters. The simplicity and effectiveness of the proposed method is a key feature that allows reliable efficiency improvement through proper SR operation of AC-DC LLC converters in high output voltage applications above 100 V. The principles of operation and analysis of the proposed SR driving strategy are discussed. Moreover, simulation and experimental results verified the analysis and the efficacy of the proposed direct MCU SR driving strategy for two different output voltage and load conditions. Furthermore, the experimentally measured efficiency comparison with a more complex alternative method verified that the proposed simple method can effectively improve the conversion efficiency.

References

[1] X. Fang, H. Hu, F. Chen, U. Somani, E. Auadisian, J. Shen, and I. Batarseh, "Efficiency-Oriented Optimal Design of the LLC Resonant Converter Based on Peak Gain Placement," *IEEE Trans. Power Electron.*, vol. 28, no. 5, pp. 2285-2296, May 2013.

[2] Y. Wei, Q. Luo, Z. Wang and H. A. Mantooth, "A Complete Step-by-Step Optimal Design for LLC Resonant Converter," *IEEE Trans. Power Electron.*, vol. 36, no. 4, pp. 3674-3691, April 2021.

[3] J. Zhang, W. G. Hurley, W. H. Wolfle and M. C. Duffy, "Optimized design of LLC resonant converters incorporating planar magnetics," in *Proc. IEEE Applied Power Electronics Conference and Exposition (APEC)*, 2013, pp. 1683-1688.

[4] M. A. Saket, N. Shafiei and M. Ordonez, "LLC Converters With Planar Transformers: Issues and Mitigation," *IEEE Trans. Power Electron.*, vol. 32, no. 6, pp. 4524-4542, June 2017.

[5] H. Zhou, W. Liu and E. Persson, "Evaluation of GaN, SiC and Superjunction in 1 MHz LLC converter," in *Proc. International Exhibition and Conference for Power Electronics, Intelligent Motion, Renewable Energy and Energy Management (PCIM Europe 2015)*, 2015, pp. 1-6.

[6] M. Forouzesh, B. Sheng and Y. Liu, "Interleaved SCC-LCLC Converter with TO-220 GaN HEMTs and Accurate Current Sharing for Wide Operating Range in Data Center Application," in *Proc. IEEE Applied Power Electronics Conference and Exposition (APEC)*, 2020, pp. 482-489.

[7] "Secondary Side Synchronous Rectification Driver for High Efficiency SMPS Topologies NCP4306," ON Semiconductor, Phoenix, AZ, USA, Aug. 2021. [Online]. Available: https://www.onsemi.com/pdf/datasheet/ncp4306-d.pdf

[8] C. Fei, Q. Li, and F. C. Lee, "Digital implementation of adaptive synchronous rectifier (SR) driving scheme for high-frequency LLC converters with microcontroller," *IEEE Trans. Power Electron.*, vol. 33, no. 6, pp. 5351–5361, Jun. 2018.

[9] X. Zhou, B. Sheng, W. Liu, Y. Chen, A. Yurek, Y.F. Liu, and P.C. Sen, "Analysis and Design of SR Driver Circuit for LLC DC-DC Converter Under High Load Current Application," in *Proc. IEEE Energy Conversion Congress & Expo (ECCE) 2019*, Baltimore, MD, USA, Sept 2019, pp. 1375-1381.

[10] "GaN Based Ultra-high Power Density Adapter 300W," ON Semiconductor, Phoenix, AZ, USA, Nov. 2019. [Online]. Available: https://www.onsemi.com/pub/Collateral/EVBUM2684-D.PDF

[11] C. Duan, H. Bai, W. Guo, and Z. Nie, "Design of a 2.5-kW 400/12-V high-efficiency DC/DC converter using a novel synchronous rectification control for electric vehicles," *IEEE Trans. Transport. Electrific.*, vol. 1, no. 1, pp. 106–114, Jun. 2015.

[12] H. Li, Z. Zhang, S. Wang, J. Tang, X. Ren, and Q. Chen, "A 300-kHz 6.6-kW SiC Bidirectional LLC Onboard Charger," *IEEE Trans. Ind. Electron.*, vol. 67, no. 2, pp. 1435-1445, Feb. 2020.

[13] X. Zhu, H. Li, Z. Zhang, Y. Yang, Z. Gu, S. Wang, X. Ren, and Q. Chen, "A Sensorless Model-Based Digital Driving Scheme for Synchronous Rectification in 1-kV Input 1-MHz GaN LLC Converters," *IEEE Trans. Power Electron.*, vol. 36, no. 7, pp. 8359-8369, July 2021.

[14] Y. Qiu, W. Liu, P. Fang, Y. Liu and P. C. Sen, "A mathematical guideline for designing an AC-DC LLC converter with PFC," in *Proc. IEEE Applied Power Electronics Conference and Exposition (APEC)*, 2018, pp. 2001-2008.

[15] M. Forouzesh, Y. -F. Liu, and P. C. Sen, "Implementation of an Isolated Phase-Modular-Designed Three-Phase PFC Rectifier Based on Single-Stage LLC Converter," in *Proc. IEEE Energy Conversion Congress and Exposition (ECCE)*, 2021, pp. 2266-2273.

[16] X. Li, L. Guo, T. Lang, D. Lu, K. Alluhaybi, and H. Hu, "Steady-State Characterization of LLC-Based Single-Stage AC/DC Converter Based on Numerical Analysis," *IEEE Trans. Power Electron.*, vol. 36, no. 9, pp. 9970-9983, Sept. 2021.

[17] M. Forouzesh, Y. -F. Liu and P. C. Sen, "A Novel Soft-Switched Three-Phase Three-Wire Isolated AC-DC Converter With Power Factor Correction," in *Proc. IEEE Applied Power Electronics Conference and Exposition (APEC)*, 2022, pp. 887-894.

Influence of DC Supply Voltage Unbalances on the Performance of ARCP Inverters

Gholamreza Tabrizi[*], Sebastian Sprunck[*], Marco Jung[*†]

[*] Fraunhofer Institute for Energy Economics and Energy System Technology IEE
Joseph-Beuys-Straße 8
Kassel, Germany
Tel: +49/(0)561 7294-1590
Gholamreza.Najfi.Tabrizi@iee.fraunhofer.de
https://iee.fraunhofer.de/

[†] Hochschule Bonn-Rhein-Sieg University of Applied Sciences
Grantham-Allee 20
Sankt Augustin, Germany
Tel: +49/(0)-2241 865 316
Marco.Jung@h-brs.de
https://www.h-brs.de/

Acknowledgements

The authors gratefully acknowledge the valuable supports from their colleagues at Fraunhofer IEE. We acknowledge the support of our work by the German Federal Ministry for Economic Affairs and Energy within the project "Methoden, Verfahren und Komponenten zur Reduzierung von Schaltverlusten in schnell taktenden PV-Stromrichtern für zielgerichtete Gewichts- und Kostenreduktion" (FKZ 03EE1011B). Only the authors are responsible for the content of this paper.

Keywords

«DC-AC converter», «Voltage Source Inverter (VSI)», «Soft switching», «ZCZVS converters», «Calculation method»

Abstract

The auxiliary resonant commutated pole inverter (ARCPI) is an attractive soft switching topology due to its small Electromagnetic Interference (EMI), voltage and current stresses. This topology has previously been investigated for balanced DC input voltages, which not always occur in practical applications. This paper therefore presents an analysis of an ARCPI with unbalanced DC supply voltage to determine necessary conditions to also achieve soft switching for reduced losses in the load current switches under such conditions. First, the necessary timings of the switches are calculated and validated through simulation. Then, possibilities to optimize the behavior of this topology are discussed.

Introduction

Increasing the switching frequencies of power electronic converters is a proven technique to enable smaller passive components within converters and, consequently, to reach a higher power density. However, switching frequencies cannot be increased to arbitrarily high values due to the switching losses of power semiconductors and EMI radiation that occur in hard-switching topologies. One approach to handle these limitations for power electronic inverters is the Auxiliary Resonant Commutated Pole Inverter (ARCPI), introduced in [1, 2]. The ARCPI (Figure 1) permits zero voltage switching (ZVS) conditions and provides significantly reduced switching losses, dv/dt and di/dt for its main switches [3]. To ensure soft switching for real-world applications, the influence of unbalanced input dc voltage should be investigated carefully. In this paper, this effect on the ARCPI operation is analytically analyzed and confirmed through simulation.

For this investigation, the split dc input capacitors are modeled through voltage sources V_{S1} and V_{S2}, respectively (see Figure 1). Furthermore, we assume ideal properties for all sources, passive components and switches. A constant output load current I_{Load} is assumed over the investigated commutation cycle, flowing in the direction marked in Figure 1. The resonance frequency ω_r, the effective resonant capacitance C_r, the characteristic impedance of the auxiliary resonant circuit Z_r, and the input DC voltage V_{dc} are defined through (1) − (4).

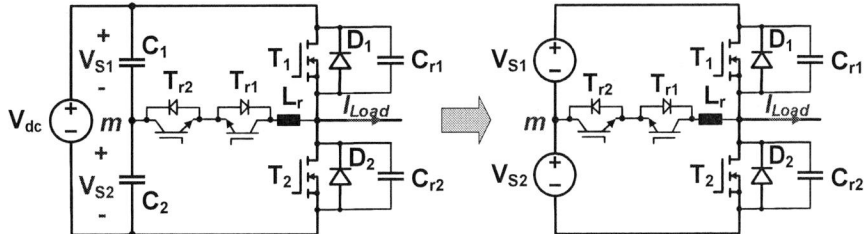

Figure 1: Circuit scheme and equivalent circuit of the auxiliary resonant commutated pole inverter.

$$C_r = 2\,C_{r1} = 2\,C_{r2} \tag{1}$$

$$\omega_r = \frac{1}{\sqrt{L_r\,C_r}} \tag{2}$$

$$Z_r = \sqrt{L_r/C_r} \tag{3}$$

$$V_{dc} = V_{S1} + V_{S2} \tag{4}$$

A variable timing method that ensures soft switching with minimized losses over a wide DC input voltage range has been presented in [3]. The timings of the gate-signals of the auxiliary switches T_{r1} and T_{r2} vary depending on the load current and dc-link voltage. The length of the overlap between the two gate signals of one of the main switches T_1 and T_2 and the corresponding auxiliary switch is called overlapping time. The overlapping time for a commutation from the high-side switch to the low-side switch is called t_{ovp1} while the overlapping time for the reverse commutation is called t_{ovp2}. This paper will focus on the commutation from D_2 through T_2 and T_{r2} to T_1, illustrated in separate phases of the commutation in Figure 2, and therefore the calculation of t_{ovp2}. The pulse width of the auxiliary switch has to be calculated and applied to variable timing control for each commutation process since the output current changes between each switching event, thus altering the necessary timing values [3].

Operational modes and calculation of the resonant current

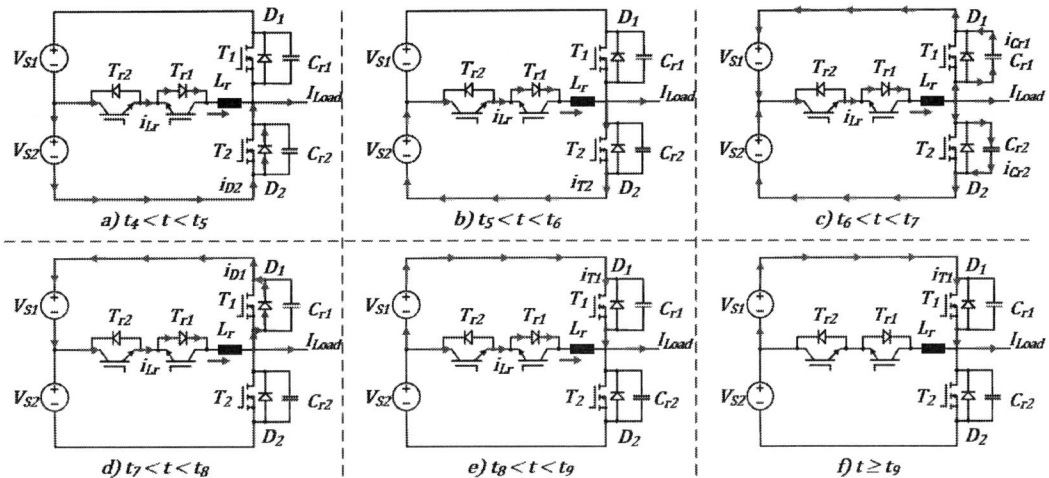

Figure 2: Transition states and time intervals of the commutation from a conducting lower diode D_2 to the upper main switch T_1.

The circuit can operate both with balanced or unbalanced dc input voltages, if the voltage imbalance is not too large. The following investigation is performed for a positive output current I_{Load}, starting with both main switches T_1 and T_2 turned off while D_2 conducts the load current (Fig. 2a) (5).

$$i_{D2}(t_0) = I_{Load} \qquad (5)$$

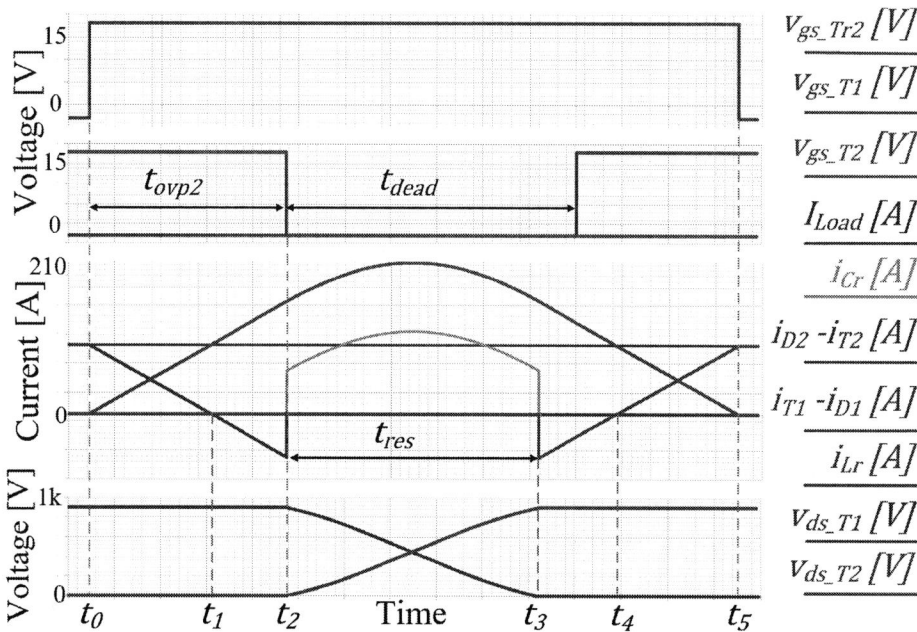

Figure 3: Voltage and current waveforms for commutation from D_2 to T_1 for $V_{S1} = V_{S2}$ and $I_{Load} > 0$.

A. Balanced operating mode: $V_{S1} = V_{S2} = V_{dc}/2$

The balanced operation of the ARCPI was previously analyzed in [3]. Figure 3 shows the ideal commutation from D_2 to T_1 and subsequently ZVS and ZCS operation of the main switch T_1. Referring to [3], the timing values for balanced dc voltages can be calculated as (6) – (8), where I_{off_T2} is the cutoff current of T_2 at $t = t_2$.

$$t_{ovp2} = t_2 - t_0 = t_5 - t_3 = \frac{2\,L_r}{V_{dc}}\left(I_{Load} + I_{off\_T2}\right) \qquad (6)$$

$$t_{diode\_cond} = t_4 - t_3 = t_2 - t_1 = \frac{2\,L_r}{V_{dc}}\,I_{off\_T2} \qquad (7)$$

$$t_{res} = 2\sqrt{L_r\,C_r}\,tan^{-1}\left(\frac{V_{dc}}{2\,Z_r\,I_{off\_T2}}\right) \qquad (8)$$

These results will now be derived for and compared to the operation under unbalanced dc input voltages.

B. Unbalanced operating mode: $V_{S1} \neq V_{S2}$

The following calculations correspond to the commutation phases shown in Figure 2; the resulting calculated waveforms for unbalanced operation are shown in Figure 5.

Phase a and b: $t_0 \leq t \leq t_2$

T_{r2} is switched on and the current of the auxiliary inductance L_r rises linearly (9). Consequently, the diode current I_{D2} decreases and reaches zero at $t = t_1$ (10). The gate of T_2 is activated at or before $t = t_0$ and has to be kept active until $t = t_2$ so that T_2 starts conducting in the opposite direction as D_2 (11) and reaches I_{off_T2} (12).

$$i_{Lr}(t) = \frac{V_{S2}}{L_r}(t - t_0) \qquad (9)$$

$$i_{D2}(t) = I_{Load} - \frac{V_{S2}}{L_r}(t - t_0) \tag{10}$$

$$i_{T2}(t) = \frac{V_{S2}}{L_r}(t - t_1) \tag{11}$$

$$I_{off\_T2} = i_{T2}(t = t_2) = \frac{V_{S2}}{L_r} t_{ovp2} - I_{Load} \tag{12}$$

The overlapping time t_{ovp2} can be expressed in terms of the cutoff current I_{off_T2} of T_2 (13) and the conduction time of T_2 can be calculated as (14) by inserting $t = t_2$ into (11).

$$t_{ovp2} = \frac{I_{Load} + I_{off\_T2}}{V_{S2}} L_r \tag{13}$$

$$t_2 - t_1 = \frac{I_{off\_T2}}{V_{S2}} L_r \tag{14}$$

Phase c: $t_2 < t \leq t_3$

T_{r2} is still on and T_2 is switched off, thus begins the resonant period. The time interval $[t_3 - t_2]$ is called resonance time t_{res}. During this period, the resonant capacitor C_{r1} will discharge from V_{dc} to almost zero, while C_{r2} will charge vice versa. The current of the resonant inductor i_{Lr} and the voltages of the resonant capacitors v_{Cr1} and v_{Cr2} can be obtained through (15) − (17).

$$i_{Lr}(t) = I_{Load} + I_{off\_T2} \cos(\omega_r(t - t_2)) + \frac{V_{S2}}{Z_r} \sin(\omega_r(t - t_2)) \tag{15}$$

$$v_{Cr1}(t) = V_{S1} + V_{S2} \cos(\omega_r(t - t_2)) - I_{off\_T2} Z_r \sin(\omega_r(t - t_2)) \tag{16}$$

$$v_{Cr2}(t) = V_{S2} - V_{S2} \cos(\omega_r(t - t_2)) + I_{off\_T2} Z_r \sin(\omega_r(t - t_2)) \tag{17}$$

At the end of the resonant period, i.e. at $t = t_3$, the resonant capacitor C_{r1} is almost discharged. Therefore, the resonant time can be calculated from the equations above, through evaluation at $t = t_3$, as (18).

$$t_{res} = t_3 - t_2 = 2\sqrt{L_r C_r} \tan^{-1}\left(\frac{-I_{off\_T2} Z_r + \sqrt{(I_{off\_T2} Z_r)^2 + V_{S2}^2 - V_{S1}^2}}{V_{S2} - V_{S1}}\right) \tag{18}$$

To ensure resonant transition under the unbalanced operating conditions, the radical expression of (18) must be zero or positive. If $V_{S1} \leq V_{S2}$ this requirement is automatically fulfilled. However, if $V_{S1} > V_{S2}$, the necessary condition (19) can be obtained. Substituting (12) into (19), the requirement for the overlapping time can be given as (20).

$$I_{off\_T2} > \sqrt{\frac{C_r}{L_r} \frac{V_{dc}}{2} (V_{S1} - V_{S2})} \tag{19}$$

$$t_{ovp2} > \sqrt{L_r C_r} \sqrt{\left(\frac{V_{S1}}{V_{S2}}\right)^2 - 1} + \frac{I_{Load}}{V_{S2}} L_r \tag{20}$$

Equation (20) is a necessary requirement to achieve ZVS if the load current I_{Load} is positive and if $V_{S1} > V_{S2}$. Similarly, this requirement can be expressed for negative load currents and for $V_{S1} < V_{S2}$ as (21).

$$t_{ovp2} > \sqrt{L_r C_r} \sqrt{\left(\frac{V_{S2}}{V_{S1}}\right)^2 - 1} + \frac{|I_{Load}|}{V_{S1}} L_r \tag{21}$$

Phase d: $t_3 < t \le t_4$

At $t = t_3$, the resonant period of the auxiliary circuit is finished. Substituting (18) into (15) yields (22) and the current of the resonant inductor for $t > t_3$ can be expressed as (23). At $t = t_3$, a negative linear ramp current starts to flow through D_1 at the same rate as i_{Lr} (24). After the gate of T_1 is activated, this current commutates to the channel of T_1. At $t = t_4$, the current of D_1 reaches zero, hence the diode conduction time can be calculated by rearranging (23) as (25).

$$I_{Lr}(t_{res}) = |i_{Lr}(t = t_3)| \tag{22}$$

$$i_{Lr}(t) = I_{Lr}(t_{res}) - \frac{V_{S1}}{L_r}(t - t_3) \tag{23}$$

$$i_{D1}(t) = I_{Lr}(t_{res}) - I_{Load} - \frac{V_{S1}}{L_r}(t - t_3) \tag{24}$$

$$t_{diode\_cond} = t_4 - t_3 = \frac{I_{Lr}(t_{res}) - I_{Load}}{V_{S1}} L_r \ne t_2 - t_1 \tag{25}$$

Comparing the result of the balanced operation (14) with the result of the unbalanced operation (25) reveals that the diode conduction times are not identical. To achieve ZVS turn-on, T_1 must be switched on during this time interval. In other words, the diode conduction time must be sufficiently long so that T_1 can be switched on.

Phase e: $t_4 < t \le t_5$

When the gate signal of the upper switch T_1 is applied during the diode conduction time, T_1 starts conducting under ZVS conditions and a linear ramp current with a positive slope starts flowing through T_1 (26). At $t = t_5$, the resonant current i_{Lr} reaches zero and consequently, the current of the upper switch T_1 reaches the load current. Therefore, the commutation is completed and since the current slope in L_r is constant between not only t_4 and t_5, but also between t_3 and t_5, the time interval $[t_5 - t_3]$ can be derived as (27).

$$i_{T1}(t) = \frac{V_{S1}}{L_r}(t - t_4) \tag{26}$$

$$t_5 - t_3 = \frac{I_{Lr}(t_{res})}{V_{S1}} L_r \tag{27}$$

Again comparing the balanced operation result (13) to (27), it can be concluded that the length of the last linear ramp in the time interval $t_3 \le t \le t_5$ is not equal to the overlapping time (28). Through these equations, the waveforms for the possible voltage imbalances $V_{S1} < V_{S2}$ and $V_{S1} > V_{S2}$ can be illustrated as shown in Figure 5.

$$t_{ovp2} = t_2 - t_0 \ne t_5 - t_3 \tag{28}$$

Simulation results

To check the accuracy of the presented equations, a single phase ARCPI model was created in Matlab/Simulink using the "Specialized Power Systems Blockset", investigating its operation at $I_{Load} = 95\ A$ and $V_{dc} = 900\ V$. In order to maintain a low current stress in the auxiliary circuit, the design strategy presented in [3] was adopted. The resonant tank parameters were then calculated to $L_r = 625\ nH$ and $C_r = 29\ nF$.

Figure 4 shows the schematic of the simulated circuit. This schematic is derived from the one shown in Fig. 1.

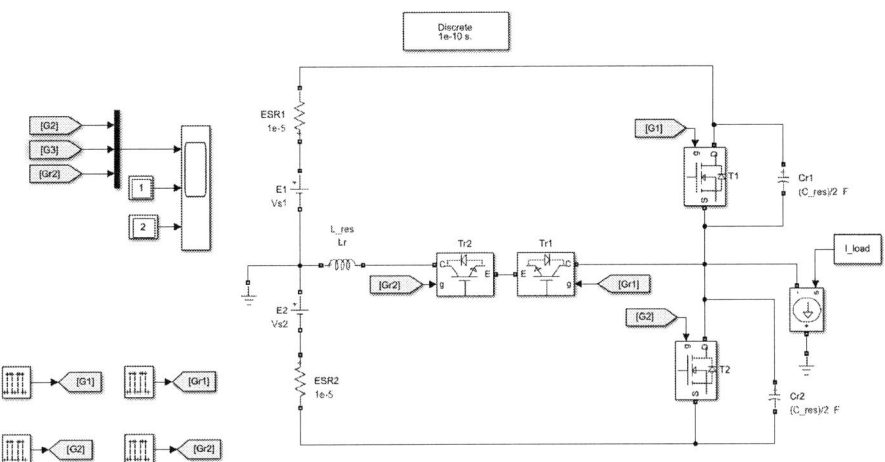

Figure 4: The simulated circuit

Simulink solver is able to work with a variable- or fixed-step sample time. Fixed-step discrete solver computes the time of the next simulation step by adding a fixed step size to the current time, and the accuracy of the resulting simulation depends on the size of the steps taken by the simulation. The smaller the step size, the more accurate the results are but the longer the simulation takes. To achieve a high accuracy, a fixed- step discrete solver with a step size of $\Delta t = 100\ ps$ was selected.

Three different simulations were conducted.

1. $V_{S1} = 300\ V$ and $V_{S2} = 600\ V$
2. $V_{S1} = 450\ V$ and $V_{S2} = 450\ V$
3. $V_{S1} = 600\ V$ and $V_{S2} = 300\ V$

The semiconductors and the passive components are modeled using the "Specialized Power Systems" device models. The two halves of the DC-link are modeled using ideal voltage sources and no parasitic properties are taken into consideration.

Table 1: Input parameters and the resulting calculated and simulated values.

Input Parameters				Results			
Case	V_{S1} [V]	V_{S2} [V]	t_{ovp2} [ns]		$|\hat{I}_{Lr}|$	t_{res2}	t_{diode_cond}
1	300	600	160	Calculated	236,91 A	217,82 ns	263,21 ns
				Simulated	236,52 A	217,3 ns	262,1 ns
				Difference	-0,1 %	-0,2 %	-0,4 %
2	450	450	215	Calculated	208,9 A	274,11 ns	83,06 ns
				Simulated	208,6 A	271,9 ns	84,2 ns
				Difference	-0,1 %	-0,8 %	+1,4 %
3	600	300	460	Calculated	236,43 A	219,07 ns	59,82 ns
				Simulated	236,36 A	217,8 ns	60,3 ns
				Difference	-0,03 %	- 0,06 %	+0,8 %

Table 1 summarizes important input parameters and the simulation results. Cases 1 and 3 represent extreme imbalances in the input voltages, while case 2 investigates the balanced operation. For the critical case 3 with $V_{S1} > V_{S2}$, the minimum value of t_{ovp2} can be obtained using (20):

$$t_{ovp2\_min} = 431\ ns \tag{29}$$

If this minimum time was selected, then the time interval to turn on T_2 under ZVS condition would be close to zero. Therefore, the overlapping time for the simulation is selected to $t_{ovp2} = 460\ ns$, which also provides a safety margin to account for component tolerances in physical setups. Table 1 indicates a high precision for the proposed approach with less than 1.5 % deviation between the calculated and simulated data, confirming the derived formula.

Figures 5 und 6 shows the simulated voltage and current waveforms of the ARCP inverter under the unbalanced input voltage cases. Figure 6 shows the resonant currents $i_{Lr}(t)$ for these cases and compares them to the operation with balanced input voltage.

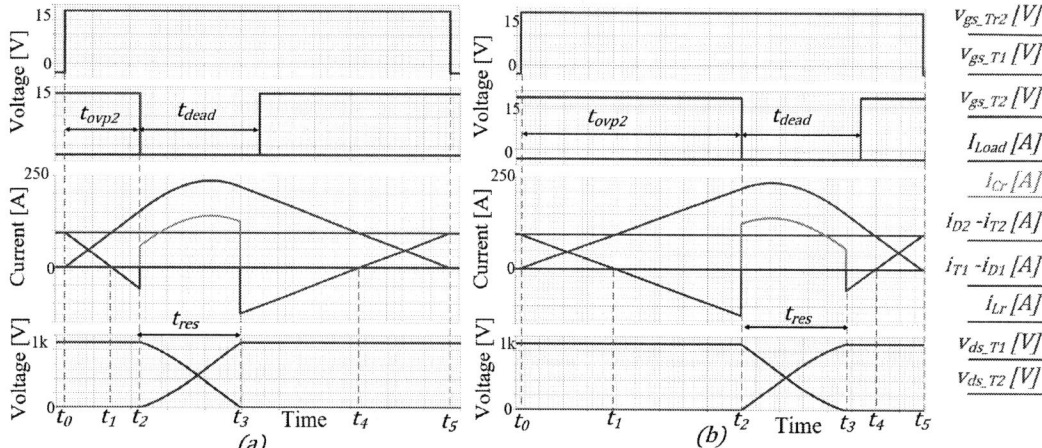

Figure 5: Voltage and current waveforms for positive load current commutation from D_2 to T_1 (a) for $V_{S1} < V_{S2}$, (b) for $V_{S1} > V_{S2}$.

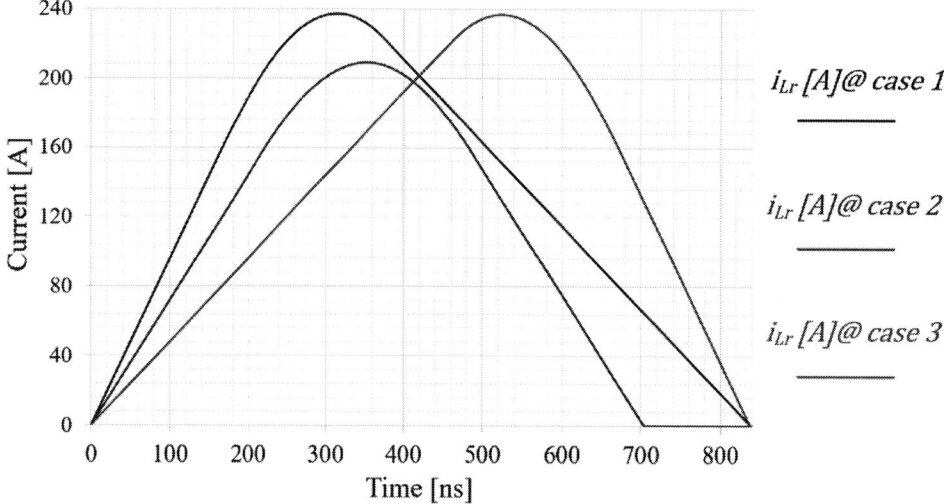

Figure 6: Comparison of the simulated resonant currents in the investigated input voltage scenarios for $L_r = 625\ nH$, $C_r = 29\ nF$ and $I_{Load} = 95\ A$.

As shown in Figure 6, the resonant currents differ considerably from each other. The minimum commutation time occurs under the balanced operating condition (blue curve). From Figures 5 and 6, it can be seen that wide variations of commutation time can be expected when input supply voltage unbalances occur.

Comparing results for different overlapping times

As previously mentioned, to achieve ZVS, the overlapping time must meet the requirement (20). If the selected overlapping time is less than the minimum value (20), then the resonant capacitor C_{r1} cannot fully discharge during the resonant period because the radical expression of (18) would be negative. Thus the diode D_1 cannot be biased in forward direction and when the main switch T_1 turns on, it quickly discharges C_{r1}, i.e. the capacitor voltage is forced to zero. This leads to an inrush current through the main switch T_1 and subsequently causes additional switching losses in the switch T_1, since the voltage across T_1 has not yet fully dissipated at $t = t_3$, as shown in Figure 7 (b).

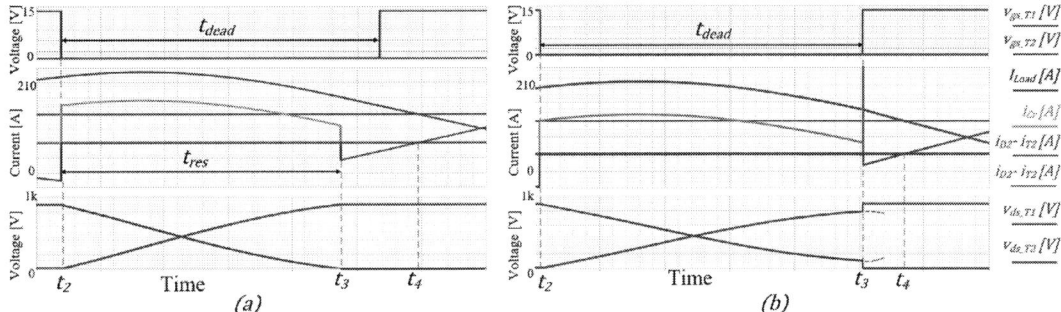

Figure 7 Extended waveforms for positive load current commutation and unbalanced input voltages $V_{S1} > V_{S2}$. (a) $t_{ovp2} = 460\,ns > t_{ovp2\_min}$, (b) $t_{ovp2} = 420\,ns < t_{ovp2\_min}$

From Figure 7 (b), it can be seen that no full ZVS is possible when the overlapping time is selected as $t_{ovp2} < t_{ovp2\_min}$. If the selected overlapping time remains below the minimum overlapping time (29), additional switching losses occur, which remarkably decrease the efficiency of the ARCP inverter and limit the applicable switching frequency. To prevent this from occurring, two other important design parameters can be considered:

(a) Increasing only the dead time t_{dead}
Looking at the voltage waveform of the main switches in Figure 7(b), it could be concluded that a longer dead time could further decrease the voltage across the main switch T_1. However, since L_r and C_r form a resonant network, this voltage would not further decrease, but continue to oscillate if the resonant period is not interrupted. Therefore, the resonant capacitor C_{r1} would begin to charge in the opposite direction (dashed line), leading to even higher switching losses than in the investigated case where the resonant period is interrupted at the minimum possible voltage at the end of the dead time. This approach therefore cannot overcome the incomplete ZVS problem.

(b) Decreasing the resonant inductance L_r
As previously analyzed in [3], choosing an appropriate value of L_r can minimize the total conduction losses of the auxiliary circuit. Equation (20) shows that a reduction in the resonant inductance causes the minimum allowed overlapping time to decrease further, but simultaneously increases conduction losses of the auxiliary circuit. Consequently, this requires more robust components for the auxiliary circuit, likely increasing their size.
In other words, the advantages of ZVS can only be fully realized for unbalanced input voltage operation when the resonant inductance properly decreases, at the cost of increased conduction losses.

From these investigations, it can be concluded that the best solution to achieve true ZVS commutation is to increase the overlapping time t_{ovp2} to values larger than t_{ovp2_min}, with a suitable safety margin to account for component tolerances etc.

In addition, it can be clearly seen that the imbalances in the input supply voltages have a significant influence on the commutation timing and consequently on the inverter performance in terms of efficiency. Therefore, both input supply voltages should be monitored for each switching cycle in addition to the load current. Depending on these acquired data, the timing of the output signals have to be calculated accordingly to achieve true ZVS commutation.

Conclusion

Effects of imbalances in the supply voltage of the ARCP inverter based on the variable timing control have been explored in this paper. The major conclusion is that the imbalances in the DC input voltage can significantly affect the performance of an ARCP inverter and have a large influence on the timing control of the ARCPI to achieve ZVS.

Therefore, the DC input voltage sources should be monitored, and applied to the variable timing control of the ARCPI on a cycle-by-cycle basis. Due to these imbalances, a minimum overlapping time must be observed to ensure soft switching. It was shown that simulation results have good agreement with the results of the theoretical analysis and they reconfirm the presented theoretical analysis.

An optimized design method of the ARCPI should be investigated in the future, which takes into account the imbalances in input supply voltage and its effect on the selection and dimensioning of the relevant component.

References

[1] R. W. De Doncker, J. P. Lyons: "The Auxiliary Resonant Commutated Pole Converter" 1990 IEEE Transactions on Industry Applications

[2] J. G. Cho, J. W. Baek, D. W. Yoo, C. Y. Won: "Three Level ARCPI for high power Application, 27th Annual IEEE Power Electronics Specialist Conference, 23-27 June 1996

[3] G. Tabrizi, F. Schnabel, M. Jung: "Optimized Design Method and Control of an ARCPI for two level photovoltaic inverters ", Conference PCIM Europe 2021, pp. 1662- 1669

[4] R. Teichmann, S. Bernet: "Investigation and comparison of auxiliary resonant commutated pole converter topologies ", 29th Annual IEEE Power Electronics Specialist Conference, 1998

[5] M. J. Oberley; J. G. Cizeki: The operation and interaction of the Auxiliary Resonant Commutated Pole Converter in a shipboard DC power distribution network, December 1996 Monterey, California Naval Postgraduate School

[6] J. Voss, R. W. De Doncker: "Modified ARCP Applied in a Three-Phase Dual-Active Bridge DC/DC Converter" IEEE Transactions on Power Electronics, Volume: 35, Issue: 2, Feb. 2020

Grid-Forming Control for Enhanced Microgrid Interconnection

Tobias Erckrath[1], Christian Bendfeld[1], Peter Unruh[1], Axel Seibel[1], Marco Jung[1,2]

[1]Fraunhofer Institute for Energy Economics
and Energy System Technology (IEE)
Joseph-Beuys-Str. 8
Kassel, Germany
Tel. +49 561 7294 - 1569
Tobias.Erckrath@iee.fraunhofer.de
https://www.iee.fraunhofer.de

[2]Hochschule Bonn-Rhein-Sieg
University of Applied Sciences
Grantham-Allee 20
Sankt Augustin, Germany
Tel. +49 2241 865 316
Marco.Jung@h-brs.de
https://www.h-brs.de

Acknowledgements

The authors acknowledge the support of the presented work by the German Federal Ministry for Economic Affairs and Climate Action within the project "RuBICon: Rule-Based Initialisation of Converter Dominated Grids" (ID number 03EI4003A). Only the authors are responsible for the content of this publication.

Keywords

«Microgrid», «Grid-forming converter», «Plug and Play control», «Synchronization», «Droop Control»

Abstract

This paper demonstrates the potential of an enhanced grid-forming control for grid-forming converters, to optimize the interconnection process of black started microgrids (MG) after a global blackout. A point of common coupling switch (PCC switch) is proposed, which does not require any higher-level control entities (e.g. MG-controller) since the switching condition only considers phase angle differences, which enables the interconnection of unintended MGs. The verification of the grid-forming control for enhanced interconnection process with overlaid grid is performed by the presented simulation study.

1 Introduction

The increasing integration of distributed generation (DG) is leading to a progressive decentralization of the entire power grid structure. To ensure a secure power supply and stable grid operation in future grids, a trend towards increasing grid-forming capabilities of distributed inverters can be expected [1]. As grid-forming capabilities are increasingly used, inverters will be characterized as independent voltage phasors, contributing to the formation of the grid voltage for other grid participants. In addition to improved secure power supply, this also opens up potential for upcoming challenges such as new black start concepts after a blackout (e.g. [2, 3]). One approach for repowering future power grids is the build-down strategy [2], which aims to restore the power grid by successively synchronizing isolated and black-started low voltage MGs. To achieve this, connecting the local and individual black-started MGs to the medium voltage level (MV) is an essential step to successfully restore the entire power supply [4] (see Fig. 1). Without a supervised control instance, the MG voltage $\underline{U}_{MG}$ at the PCC is expected to deviate significantly from the nominal values, since considered droop controllers without secondary control instances do not force the frequency f_{MG} to its nominal value [5]. Following this reasoning, it is expected that the voltage amplitude of MG U_{MG} at the PCC (see Fig. 1) also have a significant deviation from the nominal amplitude, as without overlaid control, PCC voltage is not forced as a controlled variable. A typical way for MG synchronization is based on centralized microgrid-controllers (MG-C) located at the MV/LV substation [5]. With MG-C, a seamless interconnection can be achieved as the

MG voltage $\underline{U}_{MG}$ (amplitude, frequency, phase angle) is adjusted according to the utility voltage $\underline{U}_{util}$ in terms of synchronization [6]. After minimizing the MG voltage deviation to meet the switching condition of the PCC switch, the interconnection process is performed [4].

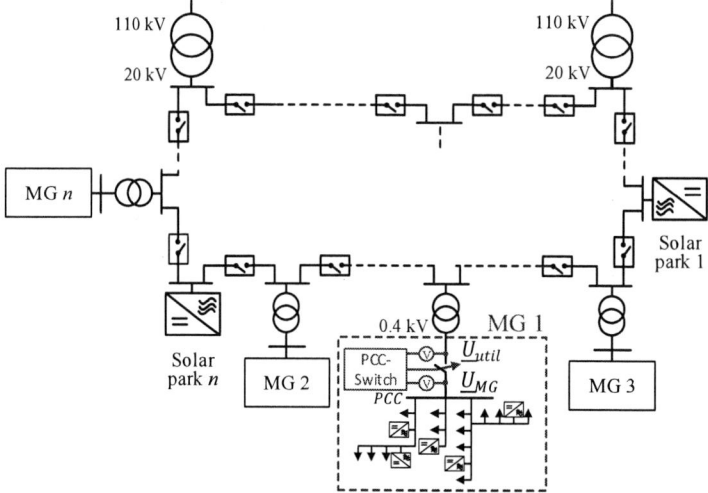

Fig. 1: Build-down strategy based on locally black-started MGs to restore the utility grid.

On the other hand, numerous methods for seamless MG synchronization based on additional or modified control loops within the inverter control can be found in [7, 8].

The paper investigates the connection of a MG to the utility grid with minimum requirements, based on the capabilities offered by grid-forming inverters with presented grid-forming control to handle the resulting dynamics during transients. For the interconnection process considered, only the phase angle deviation of the two grids is taken into account to satisfy the switching condition where all required voltages can measured locally at PCC switch. The synchronization approach does not require MG-C or specific modifications within the inverter control in the form of additional loops specifically designed for the synchronization process. Therefore, the proposed synchronization method can improve the interconnection of undedicated MGs after grid faults, since it does not require individual communication with specific grid participants.

A simulation model of a MG characterized with distributed generation is used to validate the grid-forming control in case of interconnection of MGs with two different grids. First, the interconnection of two MGs connected via MV grid is considered, which represents the first step for grid interconnection after a global blackout in a bottom up approach. Afterwards, the MG connection with stiff MV grid model shows the interconnection in an advanced state of grid restoration.

1.1 Role of Grid-Forming Control for Interconnection

Various grid-forming inverter control schemes have been recently reviewed and discussed [9–11]. The concept of grid-forming inverter control boils down to how the provided voltage phasor of the voltage source converter (VSC) is driven. In grid-forming inverter control, the voltage phasor is supposed to be autonomous in a sense, but still capable of synchronizing with other voltage sources. By imposing inertia on the voltage phasor, meaning that the voltage phasor reacts retarded to grid-side excitations, the provision of a momentary power reserve is inherent [12]. This so-called persistency can be considered in terms of both voltage amplitude and angle. In inductively coupled power grids, the angle persistency leads to an inherent active power provision and the amplitude persistency to an inherent reactive power flow [1].

This relation and the behavior of a grid-forming inverter can be deduced from a very simplified system (Fig. 2, Fig. 3): An inverter voltage $\underline{U}$ is connected to a grid voltage source $\underline{E}$ via an impedance $\underline{Z}$. Thereby the grid voltage source is a composition of several voltage sources, which together set up a more or less stiff voltage. The power flow from node 1 to 2 depending on the angle and amplitude difference of both voltage phasors forms the controlled system. The control task consists in guiding a voltage phasor with retarded behavior, which is likewise well-damped and experiences a synchronizing torque.

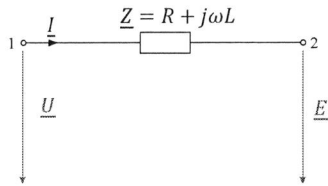

Fig. 2: Simplified equivalent circuit.

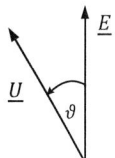

Fig. 3: Phasor diagram of two voltage sources.

The following section presents the study of grid-forming inverter model of MG interconnection.

2 Grid-Forming Inverter Model

Fig. 4 shows the model of a grid-forming inverter with LCL filter and the control instance. For the inverter, the following measurements are assumed: the inverter current I, the voltage at the output filter capacitor U_c as well as the measurement of the output voltage at the output terminal of the inverter U_o.

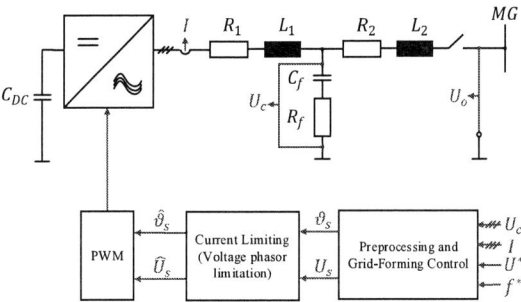

Fig. 4: Overview of grid-forming inverter model.

The grid-forming inverter control is introduced in the following section. For current limiting, a voltage phasor based current limiting is used based on the restriction of the amplitude $\widehat{U}_s$ and the phase angle $\hat{\vartheta}_s$ of the provided voltage phasor of the grid-forming control for implicit current limiting. The current limiting will be published in [13].

2.1 Used Grid-Forming Control

The grid-forming inverter control used here, with provided voltage amplitude U_s as well as provided angle ϑ_s is illustrated in Fig. 5. Based on power droops (k_p, k_q) [1], it utilizes angle feedforward loops (k'_q, k'_p) for improved damping and synchronization [14].

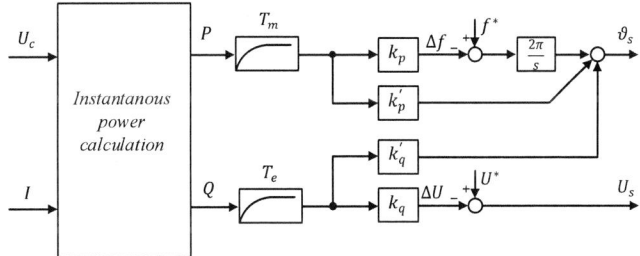

Fig. 5: Schema of the grid-forming inverter control [14].

This assists to align the interconnected grids during transient and reduce overshoot. Notable is the cross-coupling between the Q/U-loop and the phase angle ϑ_s. In this way, synchronization between several voltage sources in low-inductive grid applications can be improved in combination with reduced output filters with respect of high-frequency switching [15]. Furthermore, the grid-forming inverter is equipped with first-order delays with time constants T_m and T_e. With first-order delays, the grid-forming inverter is able to emulate virtual inertia and can therefore be dimensioned according to the time constants of a synchronous machine, for example (see [12]). In the following, a basic simulation model for validating the grid-forming control for the interconnection of MG and overlaid grid models is introduced.

3 System Scenario and Simulation Model

The following investigations are considering bottom-up approach to grid restoration of the overall grid after system blackout. The grid restoration is performed by interconnecting already black-started low voltage MGs (LV-MGs) (see Fig. 1) [2, 3], with appropriate concepts for individual black-start capability (see e.g. [3]). The process for interconnecting LV-MGs with the overlaid grid uses a simplified switching condition for the LV-PCC switch, which is introduced in section 3.2. The presented switching condition enables the interconnection of unintended MGs without the need for additional superior communication or MG-C.

3.1 Basic MG-Simulation Model

A LV-MG connected to an overlaid grid model through MV (see Fig. 6) is used to validate the interconnection process with simplified switching condition for the synchronization switch. Two different overlaid grid models (Grid model 1 and Grid model 2) are used to simulate different stages of the grid restoration process.

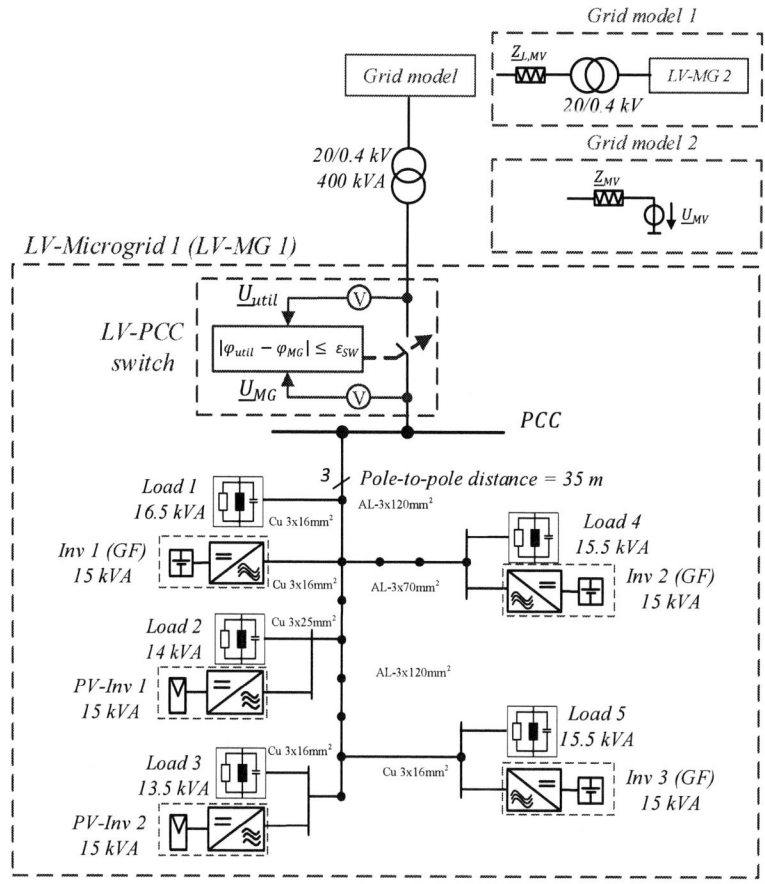

Fig. 6: Basic simulation model for interconnection of LV-MG 1 with overlaid grid-models, with LV-MG 1 defined based on [16].

The interconnection of LV-MG 1 and "Grid model 1" represents the earliest stage in grid restoration process, in which two individual black-started MGs were interconnected via MV-level. In contrast, the connection of LV-MG 1 with "Grid model 2" represents a more advanced state in which various black-started and interconnected MGs form a stiff grid voltage $\underline{U}_{MV}$.

The simulation model of LV-MG 1 was defined in terms of a modified version of "Benchmark LV-Microgrid" presented in [16]. LV-MG 1 is mainly characterized by decentralized grid-forming battery inverters (Inv1 (Gf) - Inv3 (Gf)), which enable both islanded and grid-tied operation [17]. Furthermore, two grid-following PV inverters (PV-Inv 1- PV-Inv 2), one constant power load (Load 1) and four constant impedance loads (Load 2-5) are included. The two overlaid grid models for the interconnection process are introduced in section 4 and section 5.1. In the following, the model of the synchronization switch (LV-PCC switch) with the corresponding switching condition is presented.

3.2 Concept and Model of LV-PCC Switch

For the following considerations, MG voltage $\underline{U}_{MG}$ and utility grid voltage $\underline{U}_{util}$ are defined as follows: $\underline{U}_{MG} = U_{MG} \cdot e^{j(2\pi f_{MG}+\varphi_{MG})}$, $\underline{U}_{util} = U_{util} \cdot e^{j(2\pi f_{util}+\varphi_{util})}$. Conventional switching conditions aim at equal grid values of both grids (in terms of voltage amplitude, frequency and phase angle), which enables a seamless transition during interconnection process. However, seamless interconnection with the equal grid values requires overlaid control instances of the grid participants to adjust the grid voltage at the PCC.

This publication aims for interconnecting two grids with optimized switching condition to reduce the need for an overlaid control instances such as MG-C, which allows interconnection of unintended MGs with overlaid utility grid. Thus, refurbishing existing LV cells with additional MG-Cs for the ability to interconnect with the utility grid is not required.

The presented approach for PCC switches reduces the switching condition to equal phase angles of both grids, regardless of voltage and frequency deviations (see (1) for $\varepsilon_{SW} = 0$).

$$|\varphi_{util} - \varphi_{MG}| \leq \varepsilon_{SW} \tag{1}$$

The condition of equal phase angles can be repeatedly achieved without an overlaid control instance as long as frequencies of both grids are unequal, where the opportunity to satisfy the switching condition can be achieved every $1/(|f_{util} - f_{MG}|)$ seconds. In the following, a simulation model of the PCC switch with corresponding voltage measurements is introduced (see Fig. 7)

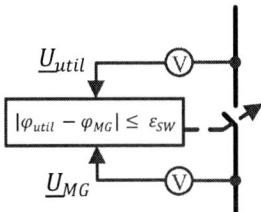

Fig. 7: Schema of synchronization switch.

Due to various delays, the phase angle difference $\Delta\varphi_{SW} = \varphi_{util} - \varphi_{MG}$ may not be equal to zero after the processor unit triggers the closing command. Fig. 8 shows the process from measuring the voltages to the closing process of the load switches. The voltage measurement usually delays the signal by about 10 µs. The time needed for the signal processing depends on the frequency of the processor and might be very small (~1 µs) when using a FPGA. The processor output is amplified using a bipolar transistor, for example, with a likewise negligible delay. However, the load contactor causes by far the largest time delay. The closing process is divided into the coil energization and the closing of contacts itself. The time for closing the load contacts varies in a wide range of several tens of milliseconds. Even if the delays are taken into account and compensated for earlier triggering of the load contactor, the tolerances for the closing time is in the range of milliseconds.

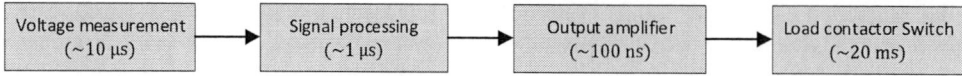

Fig. 8: Chain of delays.

In case the switch is triggered with equal phase angles of both grids according to (1), t_{delay} describes the resulting time for closing the physical load conductors at the switch. For unequal frequencies of both grids, the delay causes a phase angle difference according to (2) at the time of closing the load contactor.

$$\Delta \varphi_{SW} = \left| t_{\text{delay}} (f_{\text{util}} - f_{\text{MG}}) \cdot 360° \right| \tag{2}$$

In the following, a frequency difference of 2.5 Hz is assumed, which is justified in section 4. For the delay time t_{delay}, various data sheets of circuit breakers were evaluated. The closing time varies in a wide range depending on the power category and nominal voltage. In the following, the automatic circuit breaker specified in [18] is assumed. The maximum closing time is specified as 80 ms. This results at 2.5 Hz with (2) with a phase difference $\Delta\varphi_{SW} = 72°$. Furthermore, a smaller delay time of 20 ms ($\Delta\varphi_{SW} = 18°$) is considered, since the closing time can be compensated, so that this only represents the component tolerances. The following section investigates the interconnection process of LV-MG 1 with Grid-model 1 based on Fig. 6.

4 Simulation Analysis 1 – Interconnection of two MG's

The first simulation analysis focuses on the first instance of grid restoration, where two individual black started MGs were interconnected (see Fig. 10 for interconnection of LV-MG 1 and LV-MG 2).
Since the grid-forming inverters used are equipped with active and reactive power droops (see Fig. 5) without superior control instance, the power flow of grid-forming inverters is linked to frequency and voltage deviation (see Fig. 9) [1].

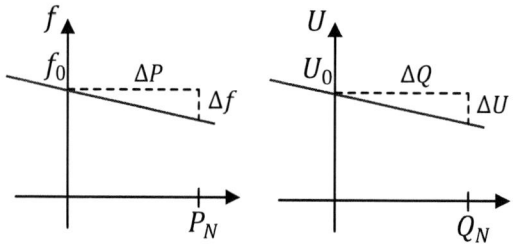

Fig. 9: Active and reactive power droops.

The droop coefficients k_p and k_q are dimensioned with a slope of $k_p = \Delta f / P_N$ and $k_p = \Delta U / Q_N$ which leads to maximum active power flow of P_N at a frequency deviation of $\Delta f = \pm 2.5 \ Hz$ and a maximum reactive power flow Q_N at a voltage difference of $\Delta U = \pm 0.2 \cdot U_N$.
The operation points of both voltages $\underline{U}_{MG,1}$ and $\underline{U}_{util,1}$ before interconnection are chosen according to Table I. The specific frequency and voltage deviation between both grids can be realized by proper dimensioning of the loads and set points for the PV inverters in both MGs.

Table I: Operation points for MG interconnection

LV-MG 1 Voltage	Utility Voltage at PCC Switch 1	Differences
$U_{MG,1} = 0.85 \ pu$	$U_{util,1} = 1.05 \ pu$	$\Delta U = 0.2 \ pu$
$f_{MG,1} = 48 \ Hz$	$f_{util,1} = 50.5 \ Hz$	$\Delta f = 2.5 \ Hz$

With the definition of the operation points according to Table I, MG 1 is in a state where grid-forming inverters supply loads together with PV inverters, resulting in frequency and voltage deviation at the PCC that are smaller than nominal values ($f_{MG,1} < f_N, U_{MG,1} < U_N$). In contrast, LV-MG 2 represents a scenario where PV generation exceeds load demand, resulting in battery charging of the grid-forming inverters and voltage and frequency difference at LV-PCC switch 1 that are higher than the nominal values ($f_{util,1} < f_N, U_{util,1} < U_N$). The used grid-forming inverter models are equipped with grid-forming control (see Fig. 5) and current limiting based on [13].

Before interconnecting LV MG 1 and LV MG 2 at LV-PCC switch 1, it is assumed that MV has already been energized by LV-MG 2 (see closed LV-PCC switch 2 in Fig. 10).

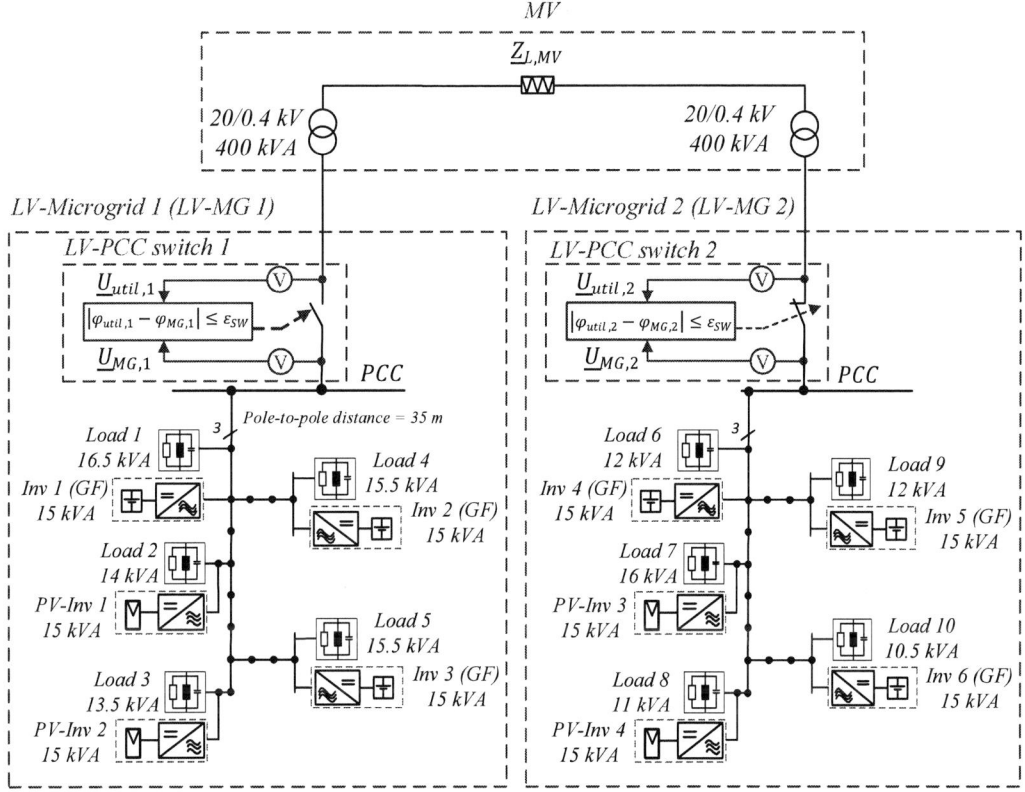

Fig. 10: Interconnection of two MGs (based on Fig. 6) as earliest stage of black start sequence.

With the definition of operation points according to Table I, both grids are interconnected with significant amplitude and frequency deviations before switching.

Fig. 11-Fig. 13 shows the inverter current of Inv 1 (GF) of LV-MG 1 at interconnection for different phase angle differences $\Delta\varphi_{SW}$ at LV-PCC switch 1, according to section 3.2. For the evaluation of the synchronization behavior, the voltage amplitude $U_{util,1}$ of the utility grid 1 is compared with the output voltage $U_{inv,1}$ (referred as U_o in Fig. 4) of Inv 1 (GF). In addition, considering the phase angle difference $\Delta\varphi = \varphi_{util,1} - \hat{\vartheta}_s$ of the internal phase angle set point $\hat{\vartheta}_s$ of inverter 1 (see Fig. 5) and the phase angle $\varphi_{util,1}$ of the three phase voltage $\underline{U}_{util,1}$ at LV-PCC switch 1 (see Fig. 10) serves as an indicator of the synchronization behavior after a dynamic event.

Before synchronization ($t < 50\ ms$), the grid-forming inverters supply the maximum current to the loads (see $I_{Inv1}(GF)$ in Fig. 11, Fig. 12 and Fig. 13), resulting in an overall frequency and amplitude deviation of the voltages, as shown in Table I. Overall, the presented grid-forming control and in case of idealized switching[1] ($\Delta\varphi_{SW} = 0$) as well as switching with common phase angle difference ($\Delta\varphi_{SW} = $

---

[1] Phase angle difference $\Delta\varphi = \varphi_{util,1} - \hat{\vartheta}_s$ shown in Fig. 11 is an indicator for synchronization between the grid-forming inverters of both MGs. The switching condition of the PCC switch can not be seen in Fig. 11, as it is defined according to (1), based on phase angle of utility grid $\varphi_{util,1}$ and phase angle of LV-MG 1 $\varphi_{MG,1}$ at LV-PCC switch 1.

18°), fast synchronization with well damped behavior after interconnection at $t = 50\,ms$ can be achieved. In addition, over current operation can be prevented due to the used current limiting method. After synchronization, both grids approach the new operation point according to the dynamic of the grid-forming control (See Fig. 11 and Fig. 12). Fig. 13 shows an interconnection with a maximum phase angle difference of $\Delta\varphi_{SW} = 72°$, where the maximum overshoot of the inverter current is limited to 1.3 pu by an additional fast reacting pulse break (see current peak of $I_{Inv1}(GF)$ at 50 ms). In addition, the switching operation causes a voltage drop in both grids to 0.5 pu.

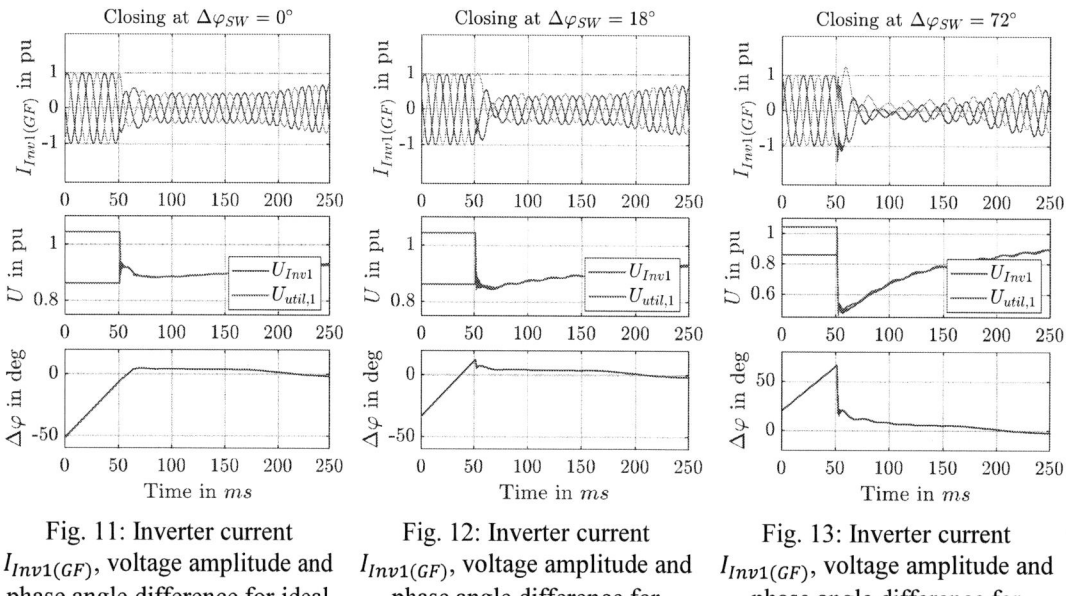

Fig. 11: Inverter current $I_{Inv1(GF)}$, voltage amplitude and phase angle difference for ideal interconnection $(\Delta\varphi_{SW} = 0°)$.

Fig. 12: Inverter current $I_{Inv1(GF)}$, voltage amplitude and phase angle difference for interconnection with $\Delta\varphi_{SW} = 18°$.

Fig. 13: Inverter current $I_{Inv1(GF)}$, voltage amplitude and phase angle difference for interconnection with $\Delta\varphi_{SW} = 72°$.

Depending on the operation time and the depth of the voltage drop, a fault condition (e.g. FRT-state - Fault ride through) of the inverters may occur during the interconnection for relatively high phase angle delays (see e.g. [19]). After interconnection, synchronization of the two grids depending on the characteristic of the grid-forming control takes place.

Overall, with the grid-forming control and the phase angle difference of LV-PCC switch with $\varphi_{SW} = 18°$ at switching, the interconnection of two MGs can be achieved without violating existing current and voltage limits for normal operation of grid-forming inverters. For switches with higher time delays (see e.g. $\Delta\varphi_{SW} = 72°$), delay compensation methods are proposed to avoid undesirable grid states for voltage amplitude and frequency. The next analysis focuses on interconnection of MG with stiff utility grid.

5   Simulation Analysis 2 – Interconnection of MG and Utility Grid Model

The second analysis focuses on the interconnection of LV-MG 1 with a simplified MV model (see "Grid model 2" in Fig. 6). The consideration of the MV model represents an advanced state in grid restoration process for the bottom-up approach, where the stiff utility voltage is formed as a composition of several black-started and interconnected MGs.

5.1 Equivalent MV Model

Fig. 14 shows an equivalent grid model consisting of an equivalent grid impedance $\underline{Z}_{MV}$ as well as an ideal MV voltage source with $\underline{U}_{MV}$.

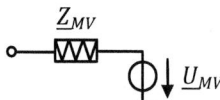

Fig. 14: Thévenin equivalent MV model.

The equivalent grid impedance can be calculated as in [20], depending on short circuit ratio $SCR = SCC/S_N$ [21], with the short circuit capacity SCC and the nominal power at the MV transformer S_N as shown in (3)

$$Z_{MV} = \frac{U_{MV}^2}{SCR \cdot S_N} = \frac{U_{MV}^2}{SCC} \tag{3}$$

SCC is chosen from [22] for MV grids with $U_{MV} = 20\ kV$ between 30 MVA for weak grids and 500 MVA for strong grids. As the strong grid model represents worst case consideration of the interconnection process due to the stiff voltage phasor characteristic of utility grid voltage, the equivalent grid impedance is calculated based on (3) to $Z_{MV} = (20\ kV)^2/500\ MVA = 0.8\ \Omega$. Finally, the equivalent resistance R_{MV} and the equivalent reactance X_{MV} can be calculated as in (4) and (5) [20], with a corresponding short circuit angle of $\Psi_{SC} = 85°$ for a strong MV grid based on [22].

$$R_{MV} = Z_{MV} \cdot \cos\Psi_{SC} = 69.7\ m\Omega \tag{4}$$

$$X_{MV} = Z_{MV} \cdot \sin\Psi_{SC} = 797\ m\Omega \tag{5}$$

5.2 Simulation Analysis

In the following, the interconnection of LV MG 1 (see Fig. 6) and the equivalent MV model (Fig. 14) is analyzed. The operation points of both grids were chosen in the same way as in the previous simulation study according to Table I.

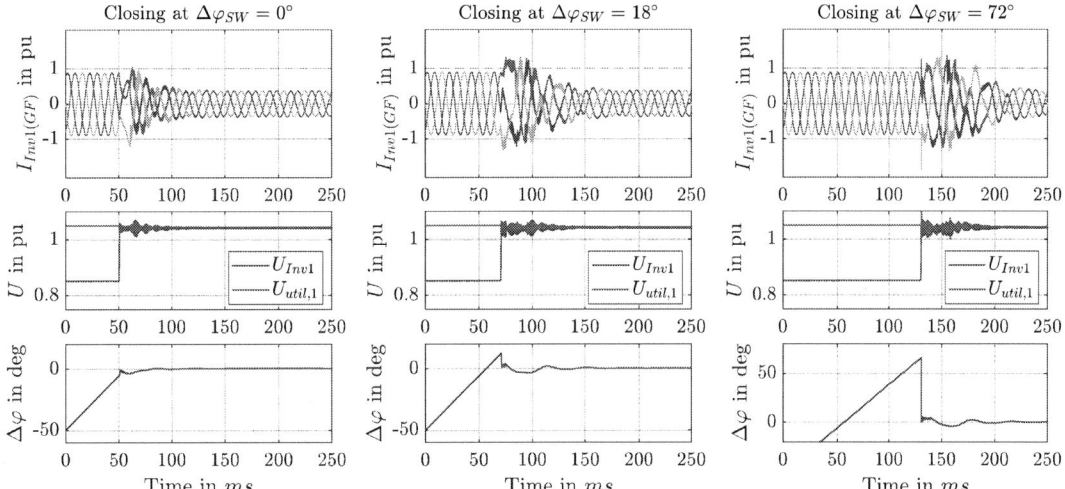

Fig. 15: Inverter current $I_{Inv1(GF)}$, voltage amplitude and phase angle difference for ideal interconnection ($\Delta\varphi_{SW} = 0°$).

Fig. 16: Inverter current $I_{Inv1(GF)}$, voltage amplitude and phase angle difference for interconnection with $\Delta\varphi_{SW} = 18°$.

Fig. 17: Inverter current $I_{Inv1(GF)}$, voltage amplitude and phase angle difference for interconnection with $\Delta\varphi_{SW} = 72°$.

Again, the inverter current of $I_{inv1}(GF)$ serves as an indicator of the overall supply to the loads in LV-MG 1 before interconnection, as all grid-forming inverters are equipped with the same droop slope for equivalent active power sharing. For switching with idealized phase angle difference $\Delta\varphi_{SW} = 0$ at LV-PCC switch (see Fig. 15), the interconnection can be achieved with slight noncritical overcurrent, followed by synchronization process of the both grids within around 50 ms (see $\Delta\varphi$ in Fig. 15). Furthermore, the transient response of the inverter leads in high frequency oscillations due to excitation of output filter resonances.

Fig. 16 shows the influence of $\Delta\varphi_{SW} = 18°$ phase angle difference leading to noncritical short-term maximum overcurrent of 1.05 pu as well as a settling time of around 50 ms with additive harmonic oscillations due to filter resonances.

In case of a phase angle difference of $\Delta\varphi_{SW} = 72°$ (see Fig. 17), an additional pulse break [23] of the inverter is activated (see Fig. 17 at $t = 125\ ms$) to avoid undesired overcurrent operation. Furthermore, the settling time of system response increases to 125 ms after the dynamic event.

Overall, in the scope of this simulation analysis, interconnection with simplified switching conditions, combined with appropriate control and current limiting for grid-forming inverters, allows enhanced interconnection of MGs with overlaid grids under consideration of realistic phase angle difference.

6 Conclusion

The potential of the presented grid-forming control and current limitation for successful MG interconnection with minimized switching condition for the PCC switch was demonstrated. The proposed switching condition is based only on the phase angle deviation of both grids. With the presented grid-forming control and the example phase angle deviation of 18°, successful resynchronization could be achieved without violating voltage and frequency limits for interconnection of two MGs as well as interconnection of MG in an advanced grid-restoration process. It is proposed to use delay compensation methods for synchronization switches with significant switching delays (e.g. 72°). Further investigations to verify the presented interconnection approach in a laboratory MG test setup according to [14] are planned.

7 References

[1] J. Rocabert, A. Luna, F. Blaabjerg, and P. Rodríguez, "Control of Power Converters in AC Microgrids," *IEEE Trans. Power Electron.*, vol. 27, no. 11, pp. 4734–4749, 2012, doi: 10.1109/TPEL.2012.2199334.

[2] F. O. Resende, N. J. Gil, and J. A. P. Lopes, "Service restoration on distribution systems using Multi-MicroGrids," *Euro. Trans. Electr. Power*, vol. 21, no. 2, pp. 1327–1342, 2011, doi: 10.1002/etep.404.

[3] M. Mirzadeh, R. Strunk, T. Erckrath, A. Mertens, "Power Hardware-in-the-Loop Verification of a Cold Load Pickup Scenario for a Bottom-up Black Start of an Inverter-dominated Microgrid," *24rd European Conference on Power Electronics and Applications (EPE'22 ECCE Europe)*, 2022.

[4] J. Wang, A. Pratt, M. Baggu, "Integrated Synchronization Control of Grid-Forming Inverters for Smooth Microgrid Transition," *2019 IEEE Power & Energy Society General Meeting (PESGM)*, 2019, doi: 10.1109/PESGM40551.2019.

[5] T. L. Vandoorn, B. Meersman, de Kooning, and L. Vandevelde, "Transition From Islanded to Grid-Connected Mode of Microgrids With Voltage-Based Droop Control," *IEEE Trans. Power Syst.*, vol. 28, no. 3, pp. 2545–2553, 2013, doi: 10.1109/TPWRS.2012.2226481.

[6] J. Lopes, C. L. Moreira, and A. G. Madureira, "Defining Control Strategies for MicroGrids Islanded Operation," *IEEE Trans. Power Syst.*, vol. 21, no. 2, pp. 916–924, 2006.

[7] A. Vukojevic and S. Lukic, "Microgrid Protection and Control Schemes for Seamless Transition to Island and Grid Synchronization," *IEEE Trans. Smart Grid*, vol. 11, no. 4, pp. 2845–2855, 2020.

[8] G. G. Talapur, H. M. Suryawanshi, L. Xu, and A. B. Shitole, "A Reliable Microgrid With Seamless Transition Between Grid Connected and Islanded Mode for Residential Community With Enhanced Power Quality," *IEEE Trans. on Ind. Applicat.*, vol. 54, no. 5, pp. 5246–5255, 2018, doi: 10.1109/TIA.2018.2808482.

[9] D. B. Rathnayake *et al.*, "Grid Forming Inverter Modeling, Control, and Applications," *IEEE Access*, vol. 9, pp. 114781–114807, 2021, doi: 10.1109/ACCESS.2021.3104617.

[10] R. Rosso, X. Wang, M. Liserre, X. Lu, and S. Engelken, "Grid-Forming Converters: Control Approaches, Grid-Synchronization, and Future Trends—A Review," *IEEE Open Journal of Industry Applications*, vol. 2, pp. 93–109, 2021, doi: 10.1109/OJIA.2021.3074028.

[11] P. Unruh, M. Nuschke, P. Strauß, and F. Welck, "Overview on Grid-Forming Inverter Control Methods," *Energies*, vol. 13, no. 10, p. 2589, 2020, doi: 10.3390/en13102589.

[12] M. Guan, W. Pan, J. Zhang, Q. Hao, J. Cheng, and X. Zheng, "Synchronous Generator Emulation Control Strategy for Voltage Source Converter (VSC) Stations," *IEEE Trans. Power Syst.*, vol. 30, no. 6, pp. 3093–3101, 2015, doi: 10.1109/TPWRS.2014.2384498.

[13] T. Erckrath, P. Unruh, M. Jung, "Voltage Phasor Based Current Limiting for Grid-Forming Converters," *2022 IEEE Energy Conversion Congress & Exposition, Detroit, Michigan (USA)*, Oct. 2022 (unpublished).

[14] P. Unruh, R. Brandl, A. Seibel, M. Jung, "Enhanced Grid-Forming Inverters in Future Power Grids," in *2018 20th European Conference on Power Electronics and Applications (EPE'18 ECCE Europe)*, Institute of Electrical and Electronics Engineers (IEEE), Ed.

[15] T. Gühna, D. Stracke, M. Klee, F. Schnabel, A. Seibel, M. Jung, "Hardware and Software Concept for Distributed Grid-Forming Inverters in Microgrids," *23rd European Conference on Power Electronics and Applications (EPE'21 ECCE Europe)*, 2021.

[16] S. Papathanassiou, N. Hatziargyriou, K. Strunz, "A Benchmark Low Voltage Microgrid Network," *Proceedings of the CIGRE Symposium: Power Systems with Dispersed Generation*, Apr. 2005.

[17] H. Deng, J. Fang, Y. Qi, Y. Tang, and V. Debusschere, "A Generic Voltage Control for Grid-Forming Converters with Improved Power Loop Dynamics," *IEEE Trans. Ind. Electron.*, pp. 7–8, 2022, doi: 10.1109/TIE.2022.3176308.

[18] ABB, "SACE Emax: Low voltage air circuit-breakers," 2013. [Online]. Available: https://library.e.abb.com /public/53b6b91b30694e35c1257d7900380633/1SDC200006D0209_EMAX%20EN.pdf

[19] R. Rosso, S. Engelken, and M. Liserre, "On The Implementation of an FRT Strategy for Grid-Forming Converters Under Symmetrical and Asymmetrical Grid Faults," *IEEE Trans. on Ind. Applicat.*, vol. 57, no. 5, pp. 4385–4397, 2021, doi: 10.1109/TIA.2021.3095025.

[20] R. Yin, Y. Sun, S. Wang, and L. Zhang, "Stability Analysis of the Grid-Tied VSC Considering the Influence of Short Circuit Ratio and X/R," *IEEE Trans. Circuits Syst. II*, vol. 69, no. 1, pp. 129–133, 2022, doi: 10.1109/TCSII.2021.3076058.

[21] L. Yu, H. Su, S. Xu, B. Zhao, J. Zhang, "Critical system strength evaluation of the power system with high penetration of renewable energy generations," *CSEE JPES*, 2021, doi: 10.17775/CSEEJPES.2021.03020.

[22] G. Arnold, M. Braun, T. Reimann, T. Stetz, B. Valov, "Optimal Reactive Power Supply in Distribution Networks - Technological and Economic Assessment for PV Systems," *in Proc. 24th European Photovoltaic Solar Energy Conference and Exhibition*, 2009.

[23] Y. Yinfu, L. Jingbo, Z. Dangsheng, "Pulse by pulse current limiting technique for SPWM inverters," *Proceedings of the IEEE 1999 International Conference on Power Electronics and Drive Systems. PEDS'99*, 1999.

Low Phase Shift Filter for Current Sensing based on the Difference between AC Machine Models with and without Iron Losses

Niklas Himker, Marcel Krümpelmann and Axel Mertens
Leibniz University Hannover
Institute for Drive Systems and Power Electronics
Welfengarten 1
Hannover, Germany
Email: niklas.himker@ial.uni-hannover.de
URL: https://www.ial.uni-hannover.de

Acknowledgments

This work was funded by the Deutsche Forschungsgemeinschaft (DFG, German Research Foundation) – project identification number 424944120.

Keywords

≪AC machine≫, ≪Vector control≫, ≪Current sensor≫, ≪Field Programmable Gate Array≫

Abstract

For a current measurement with the goals of a high bandwidth and a high sampling rate, a filter with minimal phase shift is proposed. The filter is analytically derived using the differential equation of an induction machine (IM) model and is based on the difference in the stator current behaviour of machines with and without iron losses. To use this filter, only two parameters need to be set.

Introduction

To operate an electrical machine, current measurement and control are essential in most cases. Typically, the current for the control is measured once or twice per pulse-width-modulation (PWM) period of the converter. In a digital controller, analogue-to-digital (A/D) conversion is required, which can be performed by a sampling A/D converter [1, 2] or an averaging A/D converter [3]. Both systems introduce a phase shift of the measured quantity as a result of using the averaging filter, an analogue filter, or both of these.

In recent years, the use of the field programmable gate array (FPGA) in the field of self-sensing control has been investigated in order to implement oversampling of the A/D conversion with a sampling rate of $f_{AD} \geq 1\,MS/s$ [4, 5]. By using oversampling, the bandwidth of the analogue filter is set higher than in a conventional design, introducing a lower phase shift. In this field of research, the effect of oversampling has not yet been investigated in an encoder-based field-oriented control (FOC).

Of course, the use of an FPGA with oversampling permits the implementation of a moving average filter, which leads to an averaging A/D converter. If only a few samples are used to implement the moving average filter, a lower phase shift is introduced. With regard to the high bandwidth of the analogue filter, this can lead to a measurement error, due to the influence of iron losses on the measured current. The iron losses introduce a superimposed step response of a first-order low-pass filter on the stator current of the electrical machine. For a converter-fed machine, this becomes apparent for each change in switching state. The time constant of the superimposed step response is mainly dependent on the leakage inductance and the iron loss resistance of the machine, leading to a small time constant (here:

$T_{\text{Fe}} < T_{\text{PWM}}$). This makes the measured current dependent on the time elapsed between the last switching instant of the converter and the sampling instant of the controller. In literature, iron losses have only been discussed with regard to the fundamental frequency-efficiency behaviour of an AC machine [6, 7].

The idea of this paper is to use the difference between machine models with and without iron losses as a current filter. The time constant T_{Fe} does not exist in the machine model without iron losses. The proposed filter estimates and subtracts the superimposed step response from the stator current of the machine. In this way, stator current measurement noise is reduced, as this is identified as a component of the superimposed step response of the iron losses. This can be beneficial for high dynamic current control as it reduces measurement noise and can increase the quality of state estimation for electrical machines. The proposed filter is analytically derived using the differential equation of the IM, and simplified to a grey box model with only two parameters, which can be easily tuned. The resulting transfer function of the filter with a small phase shift is discussed and it is validated in both simulations and an experimental investigation.

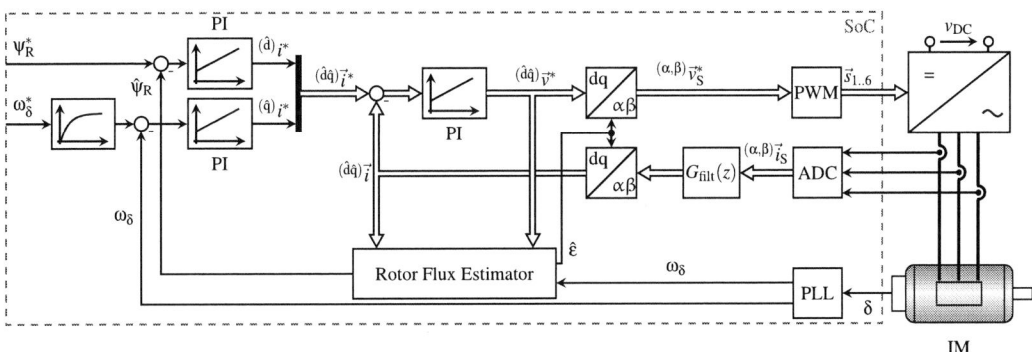

Fig. 1: FOC of the IM (feed-forward not shown)

Model of the Induction Machine

In this chapter, the models used to represent the IM are described. Initially, the fundamental model is presented, which is defined by the following differential equations in the (α, β) reference frame

$$^{(\alpha,\beta)}\vec{v}_S = R_S\,^{(\alpha,\beta)}\vec{i}_{S,f} + \frac{\mathrm{d}}{\mathrm{d}t}\,^{(\alpha,\beta)}\vec{\psi}_{S,f} \tag{1}$$

$$0 = R_R\,^{(\alpha,\beta)}\vec{i}_{R,f} + \frac{\mathrm{d}}{\mathrm{d}t}\,^{(\alpha,\beta)}\vec{\psi}_{R,f} - \mathbf{J}\omega_\delta\,^{(\alpha,\beta)}\vec{\psi}_{R,f} \text{ with } \mathbf{J} = \begin{pmatrix} 0 & -1 \\ 1 & 0 \end{pmatrix} \text{ and } \omega_\delta = \frac{\mathrm{d}}{\mathrm{d}t}\delta \ . \tag{2}$$

The stator voltage is $^{(\alpha,\beta)}\vec{v}_S$, R_S and R_R are the resistances, and $^{(\alpha,\beta)}\vec{i}_{S,f}$ and $^{(\alpha,\beta)}\vec{i}_{R,f}$ are the currents of the stator and rotor, respectively. The rotor angular position is δ. To transform from the rotor-oriented (x,y) reference frame to the stator-oriented (α, β) reference frame, the matrix

$$^{(\alpha,\beta)}\mathbf{T}_{(x,y)} = \begin{pmatrix} \cos(\delta) & \sin(\delta) \\ -\sin(\delta) & \cos(\delta) \end{pmatrix} \tag{3}$$

is used. The stator flux linkage $^{(\alpha,\beta)}\vec{\psi}_{S,f}$ and rotor flux linkage $^{(\alpha,\beta)}\vec{\psi}_{R,f}$ are defined by

$$^{(\alpha,\beta)}\vec{\psi}_{S,f} = L_S\,^{(\alpha,\beta)}\vec{i}_{S,f} + L_{SR}\,^{(\alpha,\beta)}\vec{i}_{R,f} \text{ with } L_S = L_{SR} + L_{S\sigma} \tag{4}$$

$$^{(\alpha,\beta)}\vec{\psi}_{R,f} = L_R\,^{(\alpha,\beta)}\vec{i}_{R,f} + L_{SR}\,^{(\alpha,\beta)}\vec{i}_{S,f} \text{ with } L_R = L_{SR} + L_{R\sigma} \ .$$

where L_S and L_R are the inductances and $L_{S\sigma}$ and $L_{R\sigma}$ are the leakage inductances of the stator and rotor, respectively. The magnetisation inductance is L_{SR}. The equivalent circuit of the fundamental IM is shown in Fig. 2(a). Additionally, the magnetisation current

$$^{(\alpha,\beta)}\vec{i}_{M,f} = {}^{(\alpha,\beta)}\vec{i}_{R,f} + {}^{(\alpha,\beta)}\vec{i}_{S,f} \tag{5}$$

is depicted. In the next step, the model of the IM with iron losses is introduced. In the equivalent circuit, the iron loss resistance R_{Fe} and iron loss current $^{(\alpha,\beta)}\vec{i}_{S,Fe}$ are added in parallel to the magnetisation inductance, leading to the circuit of Fig. 2(b).

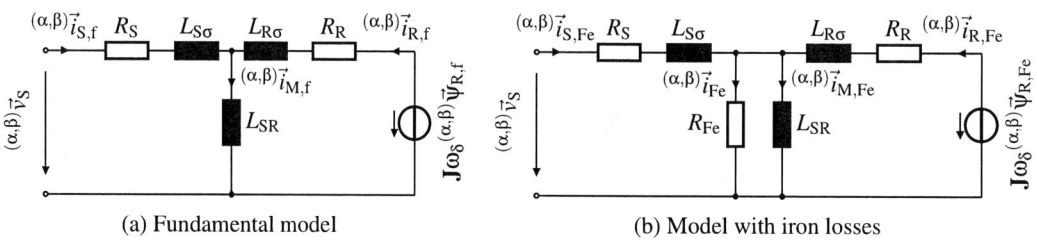

(a) Fundamental model   (b) Model with iron losses

Fig. 2: Model of the IM used

The differential equation for the IM with iron losses is derived from its equivalent circuit [6] and results in

$$\frac{d}{dt}{}^{(\alpha,\beta)}\vec{i}_{S,Fe} = \frac{1}{L_{S\sigma}}\left[{}^{(\alpha,\beta)}\vec{v}_S - (R_S + R_{Fe}){}^{(\alpha,\beta)}\vec{i}_{S,Fe} - R_{Fe}{}^{(\alpha,\beta)}\vec{i}_{R,Fe} + R_{Fe}{}^{(\alpha,\beta)}\vec{i}_{M,Fe}\right] \tag{6}$$

$$\frac{d}{dt}{}^{(\alpha,\beta)}\vec{\psi}_{R,Fe} = {}^{(\alpha,\beta)}\vec{\psi}_{M,Fe}\left(-\frac{R_{Fe}}{L_{SR}L_{R\sigma}} + \frac{R_R}{L_{R\sigma}} + \frac{R_{Fe}}{L_{R\sigma}} + \frac{R_{Fe}}{L_{SR}}\right) \tag{7}$$

$$+ {}^{(\alpha,\beta)}\vec{\psi}_{R,Fe}\left(\frac{R_R}{L_{R\sigma}} + \frac{R_{Fe}}{L_{R\sigma}}\right) + \mathbf{J}\omega_\delta{}^{(\alpha,\beta)}\vec{\psi}_{R,Fe}$$

$$\frac{d}{dt}{}^{(\alpha,\beta)}\vec{\psi}_{M,Fe} = R_{Fe}{}^{(\alpha,\beta)}\vec{i}_{S,Fe} + \frac{R_{Fe}}{L_{R\sigma}}{}^{(\alpha,\beta)}\vec{\psi}_{R,Fe} - \frac{R_{Fe}L_R}{L_{SR}L_{R\sigma}}{}^{(\alpha,\beta)}\vec{\psi}_{M,Fe} \quad . \tag{8}$$

The variables deviating from the fundamental-frequency model are marked with the subscript *Fe*. Additionally, the magnetisation flux linkage $^{(\alpha,\beta)}\vec{\psi}_{M,Fe}$ is introduced. The flux linkages of the stator and rotor are defined by

$$^{(\alpha,\beta)}\vec{\psi}_{S,Fe} = L_{S\sigma}{}^{(\alpha,\beta)}\vec{i}_{S,Fe} + {}^{(\alpha,\beta)}\vec{\psi}_{M,Fe} \quad \text{and} \quad {}^{(\alpha,\beta)}\vec{\psi}_{R,Fe} = L_{R\sigma}{}^{(\alpha,\beta)}\vec{i}_{R,Fe} + {}^{(\alpha,\beta)}\vec{\psi}_{M,Fe} \quad . \tag{9}$$

Deriving the Grey Box Model

The grey box model is derived in this section. The model is based on the equivalent circuit of an IM with iron losses. The effect of iron losses can be seen as a superimposed step response of a first-order low-pass filter on the stator current of the IM, if the machine is driven by a converter with a DC-voltage bus. This is shown in Fig. 5 by the current $^{(\alpha,\beta)}\vec{i}_{S,sim}$. The idea of the grey box model is to filter the stator current in such a way that it does not contain the effect of the iron losses. In order to do this, the grey box model is derived in three steps:

1. Estimating the voltage $^{(\alpha,\beta)}\vec{v}_{M,Fe}$ across the iron loss resistance R_{Fe}
2. Calculating the influence of the iron loss current $^{(\alpha,\beta)}\vec{i}_{Fe}$ on the stator current $^{(\alpha,\beta)}\vec{i}_{S,Fe}$
3. Subtracting the estimated current $^{(\alpha,\beta)}\vec{i}_{S,grey}$ from the stator current $^{(\alpha,\beta)}\vec{i}_{S,Fe}$

Estimating the Voltage Across the Iron Loss Resistance

The voltage $^{(\alpha,\beta)}\vec{v}_{M,Fe}$ is defined by

$$^{(\alpha,\beta)}\vec{v}_{M,Fe} = \frac{d}{dt}{}^{(\alpha,\beta)}\vec{\psi}_{M,Fe} = R_{Fe}{}^{(\alpha,\beta)}\vec{i}_{S,Fe} + \frac{R_{Fe}}{L_{R\sigma}}{}^{(\alpha,\beta)}\vec{\psi}_{R,Fe} - \frac{R_{Fe}L_R}{L_{SR}L_{R\sigma}}{}^{(\alpha,\beta)}\vec{\psi}_{M,Fe} \quad . \tag{10}$$

By using the derivative of (10), the voltage can be described with the differential equation

$$\frac{L_{SR}L_{R\sigma}}{R_{Fe}L_R}\frac{\mathrm{d}}{\mathrm{d}t}{}^{(\alpha,\beta)}\vec{v}_{M,Fe} + {}^{(\alpha,\beta)}\vec{v}_{M,Fe} = \frac{L_{SR}L_{R\sigma}}{L_R}\frac{\mathrm{d}}{\mathrm{d}t}{}^{(\alpha,\beta)}\vec{i}_{S,Fe} + \frac{L_{SR}}{L_R}\frac{\mathrm{d}}{\mathrm{d}t}{}^{(\alpha,\beta)}\vec{\psi}_{R,Fe} \ . \tag{11}$$

The goal of the derivation is to achieve an easy-to-implement filter. Equation (11) has two inputs ${}^{(\alpha,\beta)}\vec{i}_{S,Fe}$ and ${}^{(\alpha,\beta)}\vec{\psi}_{R,Fe}$. Both inputs are differentiated, which makes the voltage dependent on higher frequencies. The assumption is made that the derivative of the stator current is higher than the derivative of the rotor flux linkage, leading to $L_{R\sigma}\frac{\mathrm{d}}{\mathrm{d}t}{}^{(\alpha,\beta)}\vec{i}_{S,Fe} \gg \frac{\mathrm{d}}{\mathrm{d}t}{}^{(\alpha,\beta)}\vec{\psi}_{R,Fe}$. This is valid due to the derivative of the current ${}^{(\alpha,\beta)}\vec{i}_{S,Fe}$ being proportionally dependent on the voltage ${}^{(\alpha,\beta)}\vec{v}_S$, which is driven by the PWM and thus contributes the largest component at the switching frequency f_{sw} of the model. Making the assumption of $\frac{\mathrm{d}}{\mathrm{d}t}{}^{(\alpha,\beta)}\vec{\psi}_{R,Fe} \approx 0$, the equation is simplified to

$$\frac{L_{SR}L_{R\sigma}}{R_{Fe}L_R}\frac{\mathrm{d}}{\mathrm{d}t}{}^{(\alpha,\beta)}\vec{v}_{M,Fe} + {}^{(\alpha,\beta)}\vec{v}_{M,Fe} = \frac{L_{SR}L_{R\sigma}}{L_R}\frac{\mathrm{d}}{\mathrm{d}t}{}^{(\alpha,\beta)}\vec{i}_{S,Fe} \ . \tag{12}$$

By simply using ${}^{(\alpha,\beta)}\vec{v}_{M,Fe} = {}^{(\alpha,\beta)}\vec{i}_{Fe}R_{Fe}$, the iron loss current can be estimated with

$$\frac{L_{SR}L_{R\sigma}}{R_{Fe}L_R}\frac{\mathrm{d}}{\mathrm{d}t}{}^{(\alpha,\beta)}\vec{i}_{Fe} + {}^{(\alpha,\beta)}\vec{i}_{Fe} = \frac{L_{SR}L_{R\sigma}}{R_{Fe}L_R}\frac{\mathrm{d}}{\mathrm{d}t}{}^{(\alpha,\beta)}\vec{i}_{S,Fe} \ . \tag{13}$$

Calculating the Influence of the Iron Loss Current on the Stator Current

Using (13), the iron loss current ${}^{(\alpha,\beta)}\vec{i}_{Fe}$ can be calculated applying a first-order, low-pass filter to the measured stator current. This current is divided between the stator, rotor and magnetising paths. Its influence on the stator current can be written in the frequency domain as

$$^{(\alpha,\beta)}\vec{i}_{S,grey} = {}^{(\alpha,\beta)}\vec{i}_{Fe}\frac{1}{j\omega L_{S\sigma} + R_S}\left(\frac{1}{j\omega L_{S\sigma} + R_S} + \frac{1}{j\omega L_{SR}} + \frac{1}{j\omega L_{R\sigma} + R_R}\right)^{-1} \ . \tag{14}$$

The goal is to find the higher frequency component of this current, and then subtract it from the measured stator current. This makes the assumptions $j\omega L_{S\sigma} \gg R_S$ and $j\omega L_{R\sigma} \gg R_R$ reasonable, leading to $R_S = R_R \approx 0$ and resulting in

$$^{(\alpha,\beta)}\vec{i}_{S,grey} = {}^{(\alpha,\beta)}\vec{i}_{Fe}\frac{1}{L_{S\sigma}}\left(\frac{1}{L_{S\sigma}} + \frac{1}{L_{SR}} + \frac{1}{L_{R\sigma}}\right)^{-1} \ . \tag{15}$$

Subtracting the Estimated Current from the Stator Current

In the last step, the estimated current ${}^{(\alpha,\beta)}\vec{i}_{S,grey}$ is subtracted from the stator current ${}^{(\alpha,\beta)}\vec{i}_{S,Fe}$ to yield the estimated stator current without iron losses

$$^{(\alpha,\beta)}\vec{i}_{S,filt} = {}^{(\alpha,\beta)}\vec{i}_{S,Fe} - {}^{(\alpha,\beta)}\vec{i}_{S,grey} \tag{16}$$

Combining all three steps leads to the block diagram in Fig. 3, with $K_1 = \frac{L_{SR}L_{R\sigma}}{L_R}$, $K_2 = \frac{1}{R_{Fe}}$, $K_3 = \frac{1}{L_{S\sigma}}\left(\frac{1}{L_{S\sigma}} + \frac{1}{L_{SR}} + \frac{1}{L_{R\sigma}}\right)^{-1}$ and $T_2 = \frac{L_{SR}L_{R\sigma}}{L_R R_{Fe}}$. In this block diagram, a rate limiter with a maximal slope of $2v_{DC}/L_{S\sigma}f_{AD}$ is added, in order to filter out any ringing effects from the current measurements.

The transfer function of the filter is given by

$$G_{grey}(s) = \frac{{}^{(\alpha,\beta)}\widehat{I}_{S,grey}(s)}{{}^{(\alpha,\beta)}\widehat{I}_{S,Fe}(s)} = \frac{sT_1}{1 + sT_2} \quad \text{and} \quad G_{filt}(s) = \frac{{}^{(\alpha,\beta)}\widehat{I}_{S,filt}(s)}{{}^{(\alpha,\beta)}\widehat{I}_{S,Fe}(s)} = \frac{1 + s(T_2 - T_1)}{1 + sT_2} \ , \tag{17}$$

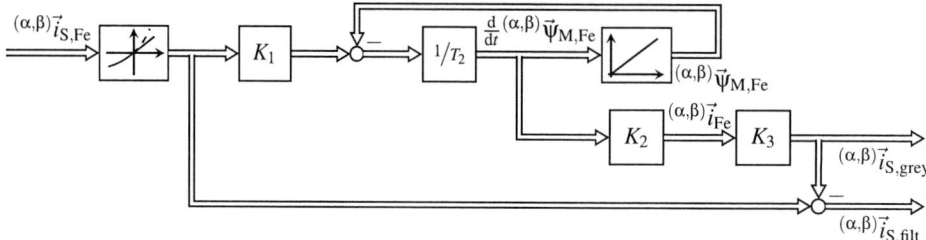

Fig. 3: Resulting filter of the IM to eliminate the superimposed step response

using $T_1 = K_1 K_2 K_3 = \frac{L_{SR} L_{R\sigma}}{L_R L_{S\sigma} R_{Fe}} \left(\frac{1}{L_{S\sigma}} + \frac{1}{L_{SR}} + \frac{1}{L_{R\sigma}} \right)^{-1}$. To determine the parameters of the grey box model, a time constant and a proportional gain (T_2 and T_1) need to be determined. In the transfer function of the filter, two time constants (T_2 and $T_2 - T_1$) must be tuned. Thus, only two parameters have to be analytically or empirically set, which makes this filter easy to implement even if the parameters of the machine are not known. The Bode plot of the filter is shown in Fig. 4. On the left-hand side, the plot for estimating $^{(\alpha,\beta)}\vec{i}_{S,grey}$ is depicted, and on the right-hand side, the Bode plot for $^{(\alpha,\beta)}\vec{i}_{S,filt}$ is shown.

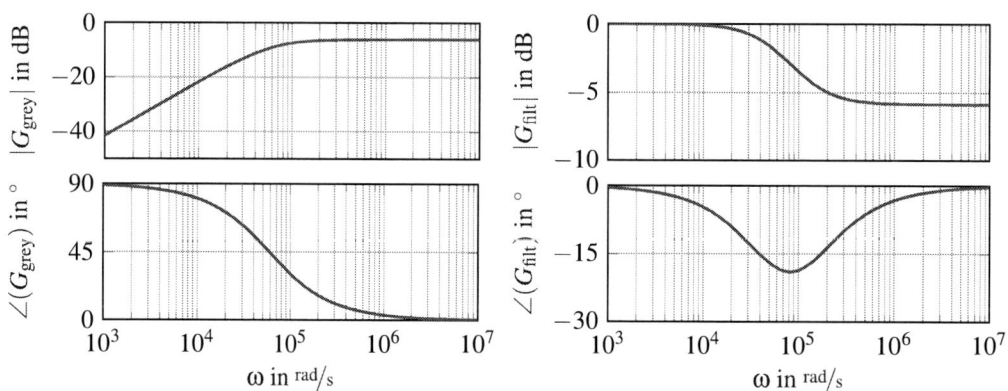

Fig. 4: Bode plot of the grey box model and the resulting filter

Discussion of the Characteristics of the Filter

The transfer function of the filter $G_{filt}(s)$ represents a general first-order transfer function with $T_2 > (T_2 - T_1) > 0$, leading to $G_{filt}(s \to \infty) < 1$. Thus, a low-pass filter with a minimal phase shift is introduced, which derives the IM current that would flow without iron losses from the measured current including iron losses. This can be used in a highly dynamic current control to reduce measurement noise. Another application can be found in the field of state estimation (e.g., rotor position) of machines. The filter can be used to improve the quality of the estimated parameters, as the measured current is less distorted by the high-frequency behaviour of the machine and the noise from the current sensor.

Simulation results using an FOC [8] of the IM and the filter are presented in Fig. 5. The control scheme used is shown in Fig. 1. In this paper, a switching frequency f_{sw} of 4 kHz and a current measurement sampling rate f_{AD} of 1 MS/s are used. In the simulation results, two cases are presented. In Fig. 5(a), an operating point with no load and high speed is shown. An operating point with rated torque and zero speed is shown in Fig. 5(b). On both graphs, the (d)-axis of the controller aligns with the (α)-axis of the machine. When comparing the currents at the two operating points, one can observe the superimposed step response of the first-order, low-pass component in the current $^{(\alpha,\beta)}\vec{i}_{S,sim}$. The filtered current $^{(\alpha,\beta)}\vec{i}_{S,filt}$ shows a constant slope during each switching state of the converter, meaning that it is independent of the iron loss current $^{(\alpha,\beta)}\vec{i}_{Fe}$.

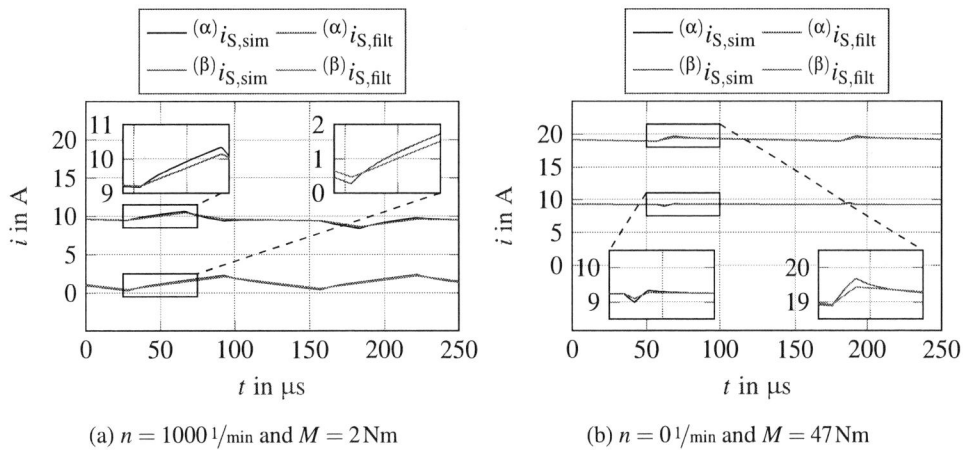

(a) $n = 1000\,1/\mathrm{min}$ and $M = 2\,\mathrm{Nm}$ (b) $n = 0\,1/\mathrm{min}$ and $M = 47\,\mathrm{Nm}$

Fig. 5: Simulation results of the proposed filter

Experimental Results

To validate the simulation results, the filter is implemented on a test bench with an FPGA-based control system. The data concerning the used machine are given in Table I and a picture of the test bench is shown in Fig. 6. In order to validate the filter presented, two different current measurements are used. One measurement is based on a relatively low bandwidth Hall-effect current sensor ($f_{\mathrm{BW},1} = 100\,\mathrm{kHz}$) and the other measurement uses a shunt sensor with a higher bandwidth of $f_{\mathrm{BW},2} = 300\,\mathrm{kHz}$.

Table I: Machine data of the IM

Rated phase voltage (RMS)	V_r	230 V
Rated phase current (RMS)	I_r	15.2 A
Rated power	P_r	7.5 kW
Rated rotational speed	n_r	$1{,}480\,\mathrm{min}^{-1}$
Stator resistance	R_S	0.52 Ω
Rotor resistance	R_R	0.51 Ω
Iron loss resistance	R_Fe	250 Ω
Magnetisation inductance	L_SR	100 mH
Stator leakage inductance	$L_{\mathrm{S}\sigma}$	3.5 mH
Rotor leakage inductance	$L_{\mathrm{R}\sigma}$	3.4 mH
Number of pole pairs	p	2

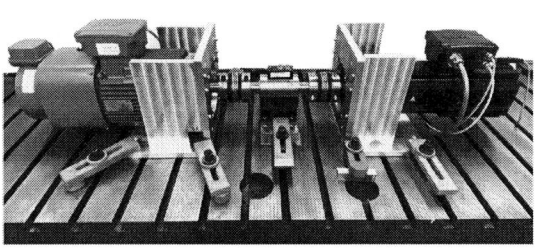

Fig. 6: Experimental test bench

Setting of Grey Box Model Parameters

Within simulations, the parameters of the filter are set by using the parameters of the IM. These parameters vary across the operating range of the IM, making the setting via analytical equation difficult. For the experimental validation, the parameters T_1 and T_2 are identified offline by parameter optimisation. The proposed method is presented, using the results of the shunt sensor. To determine the parameters, measurements of the machine are taken across the operating range. In this case, the setpoints of the rotor flux ψ_R^* and the current $^{(\hat{\mathrm{q}})}i^*$ are varied throughout ranges of $\psi_\mathrm{R}^* = [0.1, 0.15...1]\,\mathrm{Vs}$ and $^{(\hat{\mathrm{q}})}i^*[0, 2...20]\,\mathrm{A}$. To achieve long switching states of the converter, a square-wave signal with an amplitude of $100\,\mathrm{V}$ is used [9].

For each resulting operating point, the parameters of the filter are optimised in a two-step method. The first step is shown in the upper plot of Fig. 7(a). The current $^{(\alpha)}i_{\mathrm{S,f,ref}}$ is fitted to the current $^{(\alpha)}i_{\mathrm{S,shunt}}$ while the converter is in the passive switching state marked with ①. This is achieved using the model of the IM based on the fundamental behaviour ((1) and (2)). As inputs for the model, the actual switching state of the converter, the DC-link voltage, the current of the machine and the estimated flux are used.

Now the sum of leakage inductance of the stator $L_{S\sigma}$ and the rotor $L_{R\sigma}$ is varied until the sum of the square of the error of the first step $e_1 = {}^{(\alpha)}i_{S,f,ref} - {}^{(\alpha)}i_{S,shunt}$ is minimised. In the next step, the parameters T_1 and T_2 are optimised. The result of this optimisation is shown in the bottom plot of Fig. 7(a), marked with ②. In this case, each current sample is used to reduce the error of the second step $e_2 = {}^{(\alpha)}i_{S,filt,fit} - {}^{(\alpha)}i_{S,f,ref}$.

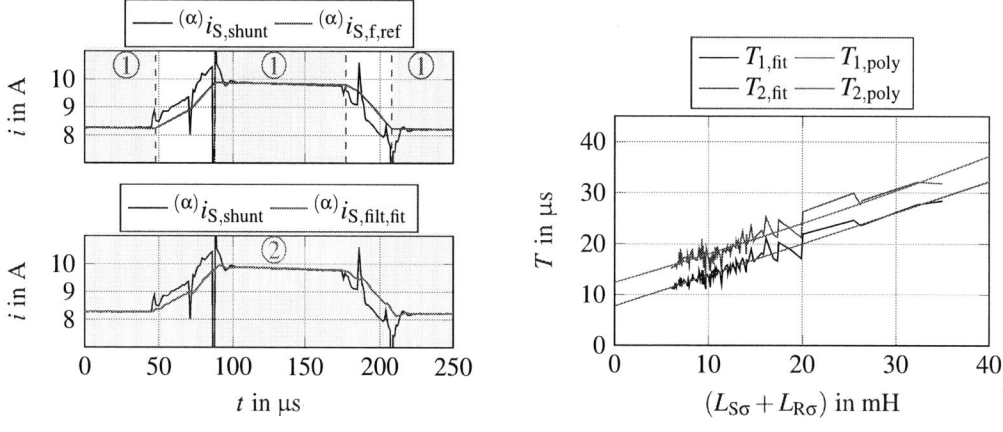

(a) Two-step method used for the parameter optimsation (b) Results of the parameter optimisation

Fig. 7: Parameter optimisation of the proposed filter

The result of the parameter optimisation is shown in Fig. 7(b). Here, the parameters $T_{1,fit}$ and $T_{2,fit}$ are plotted against the sum of the leakage inductances $(L_{S\sigma} + L_{R\sigma})$. One can see that the results for this machine can be approximated with a first-order polynomial function of $T_{1,poly}$ and $T_{2,poly}$ and that the difference between the parameters $T_{1,poly}$, and $T_{2,poly}$ is approximately constant.

The results of the polynomial function are stored in a look-up table of the FPGA and are adapted online, using the sum of the leakage inductances $(L_{S\sigma} + L_{R\sigma})$ as an input. The sum of the leakage inductances $(L_{S\sigma} + L_{R\sigma})$ is estimated using the operating point of the machine described by the setpoints of the rotor flux ψ_R^* and the current $^{(\hat{q})}i^*$. For each operating point, the identified sum of the leakage inductances $(L_{S\sigma} + L_{R\sigma})$ is also stored in a look-up table of the FPGA. The same method is applied to the measurement using the Hall-effect current sensor.

Closed-Loop Performance of the Filter

The experimental results for no load at a speed of $n = 1000\,{}^1/\text{min}$ are shown in Fig. 8. The (d)-axis aligns with the (α)-axis of the machine. In each case, the offline optimised parameters $T_{1,poly}$ and $T_{2,poly}$ are used.

The results show reduced noise in the filtered current $^{(\alpha,\beta)}\vec{i}_{S,filt}$ for both sensors. In both measured currents $^{(\alpha,\beta)}\vec{i}_{S,Hall}$ and $^{(\alpha,\beta)}\vec{i}_{S,shunt}$, ringing appears at the switching instant of the converter leading to one or more distorted samples. The higher bandwidth of the shunt-based sensor leads to the recognisable superimposed step response of the first-order, low-pass filter in the current $^{(\alpha,\beta)}\vec{i}_{S,shunt}$. Further results for an operating point at rated torque and standstill are shown in Fig. 9. Due to the shorter active switching states of the converter, the results show a higher impact of distorted current samples due to ringing. Still, the noise of the current samples is reduced for both sensors. Again, the results from the shunt-based sensor show the higher bandwidth of the sensor.

In the next results, the effect on the output voltage $^{(\hat{d}\hat{q})}\vec{v}^*$ of the current controller is investigated. In Fig. 10, the spectra of the voltage $^{(\hat{d}\hat{q})}\vec{v}^*$ of the current controller are shown for steady-state operation at standstill with no-load. By comparing the amplitudes of the spectra without and with using the the proposed filter, one can see less noise when using the filter. The difference between the two spectra in the frequency range from $f = [1...2]$kHz is defined as the damping D.

Results at rated speed and no load are shown in Fig. 11. The results with the Hall-effect-based current sensor show only small differences between using the filter and not using the filter. At standstill, an

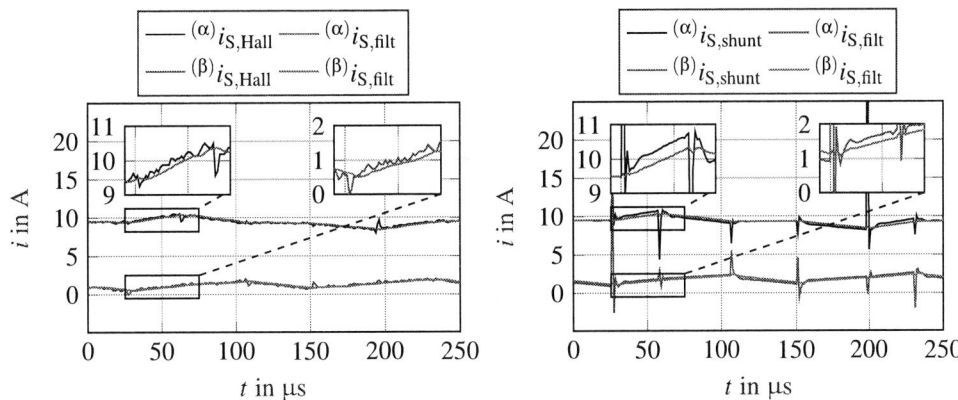

(a) Results with a Hall effect-based current sensor   (b) Results with a shunt-based current sensor

Fig. 8: Experimental results at the operating point of $n = 1000\,^1/\mathrm{min}$ and $M = 2\,\mathrm{Nm}$

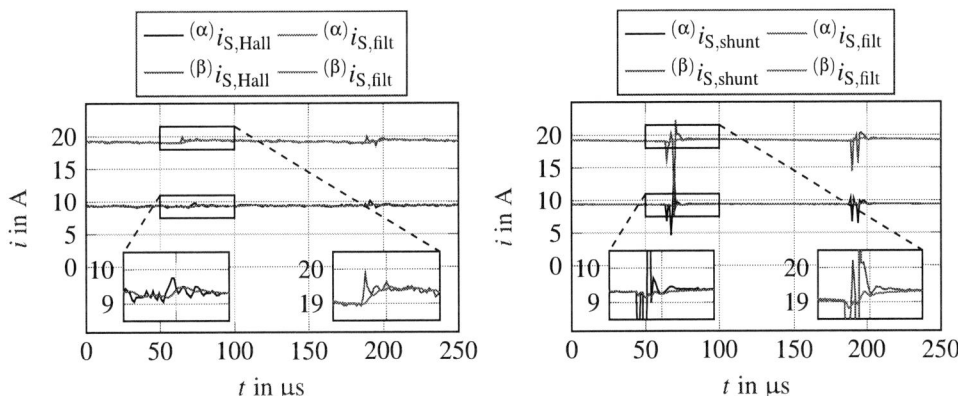

(a) Results with a Hall effect-based current sensor   (b) Results with a shunt-based current sensor

Fig. 9: Experimental results at the operating point of $n = 0\,^1/\mathrm{min}$ and $M = 47\,\mathrm{Nm}$

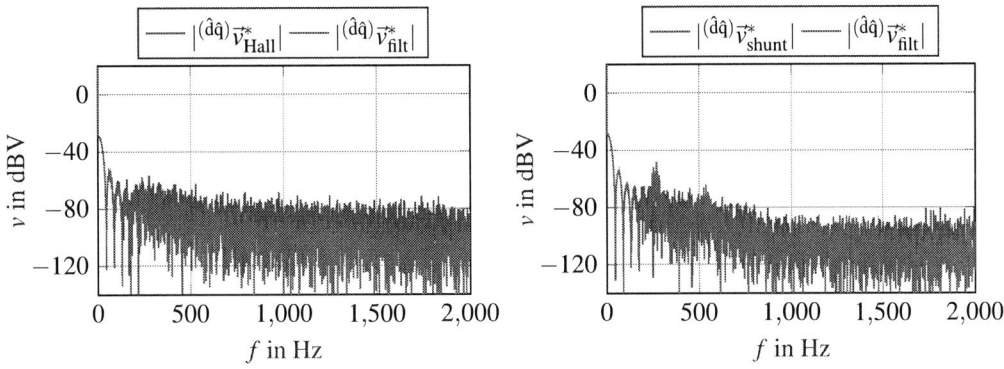

(a) Results with a Hall effect-based current sensor   (b) Results with a shunt-based current sensor

Fig. 10: Experimental results showing the spectra of the voltage $^{(\hat{d}\hat{q})}\vec{v}^*$ at the operating point of $n = 0\,^1/\mathrm{min}$ and $M = 0\,\mathrm{Nm}$

improvement is achieved, because the white noise of the measurement is reduced. Now, at a higher speed, the noise is generated systematically by the experimental setup. Thus, the signal-to-noise ratio is not improved. But for the shunt-based current sensor, an improvement becomes evident. This can be explained by the high output level of the PWM. Due to this, the effect of ringing leads to a distorted current sample when the filter is not being used.

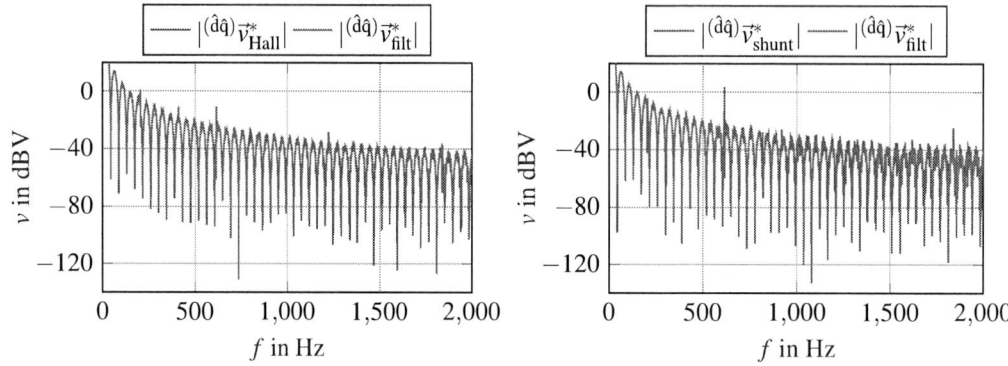

(a) Results with a Hall-effect-based current sensor

(b) Results with a shunt-based current sensor

Fig. 11: Experimental results showing the spectra of the voltage $^{(\hat{d}\hat{q})}\vec{v}^*$ at the operating point $n = 1500\,^{1}/\mathrm{min}$ and $M = 0\,\mathrm{Nm}$

A comparison of the damping of the voltage noise $^{(\hat{d}\hat{q})}\vec{v}^*$ for the whole operating range is shown in Fig. 12. For the results with the Hall-effect- and shunt-based current sensors, a positive effect can be seen at standstill with damping ranging up to 16 dB for the Hall-effect-based sensor and up to 11 dB for the shunt-based sensor. In the area of mid-range speed only, small differences are apparent. For the shunt-based sensor, the reduction of ringing effects at high speed leads to a damping of up to 19 dB, while there is no difference at high speed for the Hall-effect-based sensor.

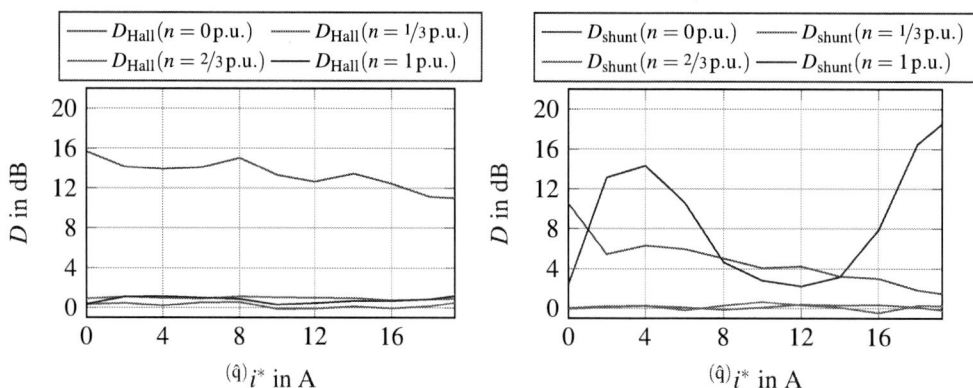

(a) Results with a Hall-effect-based current sensor

(b) Results with a shunt-based current sensor

Fig. 12: Experimental results showing damping across the operating range of the machine

Conclusion

The proposed filter for high-bandwidth, high-sampling-rate current measurement meets the requirement of a low phase shift. Through the use of this filter, the signal-to-noise ratio of the investigated current measurements is improved, leading to less noisy current control of the electrical machine investigated. In future work, this filter can be experimentally validated with synchronous machines.

References

[1] S.-H. Song, J.-W. Choi, and S.-K. Sul, "Current measurements in digitally controlled AC drives," *IEEE Industry Applications Magazine*, vol. 6, no. 4, pp. 51–62, Jul. 2000.

[2] F. Briz, D. Díaz-Reigosa, M. W. Degner, P. García, and J. M. Guerrero, "Current sampling and measurement in PWM operated AC drives and power converters," in *The 2010 International Power Electronics Conference*, Jun. 2010, pp. 2753–2760.

[3] A. Mertens and D. Eckardt, "Voltage and current sensing in power electronic converters using sigma-delta A/D conversion," *IEEE Transactions on Industry Applications*, vol. 34, no. 5, pp. 1139–1146, Sep. 1998.

[4] P. Landsmann, D. Paulus, A. Dötlinger, and R. Kennel, "Silent injection for saliency based sensorless control by means of current oversampling," in *2013 IEEE International Conference on Industrial Technology (ICIT)*, Feb. 2013, pp. 398–403.

[5] B. Weber, G. Lindemann, and A. Mertens, "Reduced observer for anisotropy-based position estimation of PM synchronous machines using current oversampling," in *2017 IEEE International Symposium on Sensorless Control for Electrical Drives (SLED)*, Sep. 2017, pp. 121–126.

[6] G. Garcia, J. Santisteban, and S. Brignone, "Iron losses influence on a field-oriented controller," in *Proceedings of IECON'94 - 20th Annual Conference of IEEE Industrial Electronics*, vol. 1, Sep. 1994, pp. 633–638 vol.1.

[7] D. S. Ding, X. J. Wang, and X. P. Luo, "A Study of Vector Control of Induction Machine Considering Iron Loss," *Advanced Materials Research*, vol. 268-270, pp. 129–137, Jul. 2011.

[8] R. D. Lorenz and D. B. Lawson, "Performance of Feedforward Current Regulators for Field-Oriented Induction Machine Controllers," *IEEE Transactions on Industry Applications*, vol. IA-23, no. 4, pp. 597–602, Jul. 1987.

[9] S. Kim, J.-I. Ha, and S.-K. Sul, "PWM Switching Frequency Signal Injection Sensorless Method in IPMSM," *IEEE Transactions on Industry Applications*, vol. 48, no. 5, pp. 1576–1587, Sep. 2012.

Design and analysis of a voltage clamping active delay control method for series connected SiC MOSFETs

Rui Wang, Asger Bjørn Jørgensen, Hongbo Zhao, Stig Munk-Nielsen
AAU Energy, Aalborg University
Pontoppidanstræde 101
9220 Aalborg, Denmark
*E-Mail: rwa@energy.aau.dk
URL: https://www.energy.aau.dk

Keywords

«Medium voltage», «MOSFET», «Silicon Carbide (SiC)», «Small signal stability», «Controllers»

Abstract

Series connection of power devices is an attractive approach to overcome the obstacle of the blocking voltage limitation of a single power device. However, voltage balancing measures should be taken to assure the anticipated performance of series connected power devices. In this paper, based on emerging silicon carbide (SiC) metal-oxide semiconductor field-effect transistors (MOSFETs), a clamping resistor-capacitor-diode circuit-based voltage clamping active delay control method is proposed to improve their voltage balancing performance. Compared with existing active delay control methods which sample the drain-source voltages of SiC MOSFETs as feedbacks, this proposed method utilizes the voltages of clamping capacitors as control criteria, which exhibits two prominent advantages: (1) an accurate model of the system is easier to attain (2) the feedback loop is simpler to design. After detailed demonstration of this method, the corresponding model is established to help determine appropriate control parameters, and experiments finally validate the effectiveness of the proposed method.

Introduction

The development of wide band gap devices has prompted power electronics converters towards a higher frequency and a higher power density. Among them, emerging silicon carbide (SiC) metal-oxide semiconductor field-effect transistor (MOSFET) has drawn much attention particularly in medium voltage applications. However, the blocking voltage of a commercial single SiC MOSFET is still limited, maximum to 3.3kV according to the data from GeneSiC [1]. Therefore, in order to accommodate a higher voltage rating, series connection of SiC MOSFETs is called for [2 - 5].

To solve the voltage unbalancing problem of series connected SiC MOSFETs, active voltage balancing methods are more prevalent compared with passive ones in published literature, owing to the advantage of lower induced loss [6]. Adding compensating circuits to the gate loop of SiC MOSFET is the general idea of active voltage balancing methods, by considering the inevitable difference of gate loop parameters during the switching transient [7]. In addition, the delay deviation of switching signals between series connected SiC MOSFETs is also one major factor that causes the voltage unbalancing. On the other hand, it can be utilized as a measure to control the voltage balancing by properly adjusting this deviation, and that is the basic principle of active delay control concept [8 - 12].

With additional feedback units and delay executive units, the active delay control methods for voltage balancing feature limited penalty of switching performance of series connected SiC MOSFETs. In general, the drain-source voltages of series connected SiC MOSFETs are divided by resistor network and then sampled by analog-to-digital convertors (ADCs) for comparison, which is used as the control criteria. After calculation by the controller, the additional delay executive units are acting to subtly give the turn-off delay compensations for the voltage balancing purpose. During this process, a potential stability problem exists, as mentioned in [10], in other words, the parameters in the control algorithm should be accurately designed by modelling the entire active delay control system. However, the relationship between the derived delay and the voltage unbalancing degree of series connected SiC

MOSFETs is uncertain, and it is usually identified by an experimental way [10, 11], which increases the design burden and reduces the flexibility.

Consequently, in this paper, a clamping Resistor-Capacitor-Diode (RCD) circuit is combined with the active delay control method to overcome the above drawback since a certain system model can be attained. Besides, the clamping RCD circuit not only helps to improve the voltage balancing performance, but also makes both the feedback unit and the delay executive unit easier to design as: (1) the voltage across the capacitor in the clamping RCD circuit is sampled and converted to frequency through the optic fiber, which provides galvanic isolation; (2) Digital signal processer (DSP) controller detects the feedback frequency by its integrated Enhanced Capture (eCAP) module, then gives the delay compensation through its inside High Resolution Pulse-Width-Modulation (HRPWM) module directly. The demonstration of the proposed hybrid voltage balancing method, also named as the voltage clamping active delay control method, is presented, followed by the detailed modeling and stability analysis. Afterwards, the validity is confirmed by a buck chopper circuit using two series connected SiC MOSFETs.

The demonstration of the proposed voltage clamping active delay control method

The overall hardware structure of the proposed voltage clamping active delay control method is shown in Fig. 1. Two SiC MOSFETs T_1 and T_2 are connected in series for meeting a higher blocking voltage requirement, and the same clamping RCD circuit is equipped with each SiC MOSFET for two benefits: (1) during the static state, the static voltage balancing of T_1 and T_2 is realized by forming a current loop as: diode $D_1 \rightarrow$ resistor $R_1 \rightarrow$ diode $D_2 \rightarrow$ resistor R_2. (2) during the dynamic state, the voltage overshoot of T_1 (T_2) is clamped once the voltage exceeds the clamping value of capacitor C_1 (C_2) in the clamping RCD circuit.

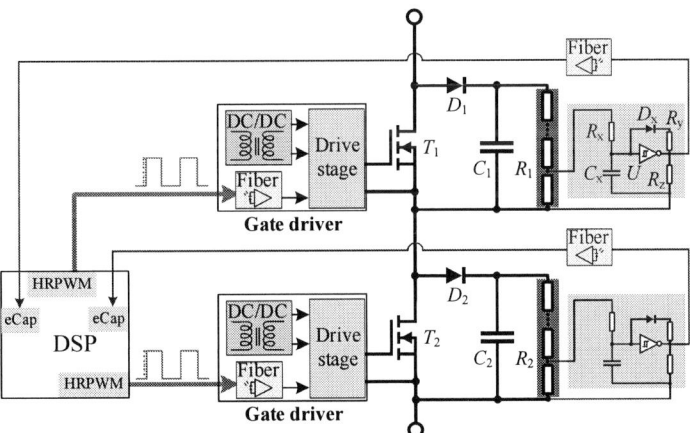

Fig. 1: The overall hardware structure of the proposed voltage clamping active delay control method

General active delay control methods attempt to control the voltage balancing of series connected SiC MOSFETs by sampling the drain-source voltage v_{ds1} (v_{ds2}) of T_1 (T_2), which requires expensive high-speed ADCs. In this paper, instead, the voltage v_{c1} (v_{c2}) across the capacitor of the clamping RCD circuit is sampled as the feedback as follows:

R_1 (R_2) consists of N series connected resistors, and the voltage v_{c1} (v_{c2}) is scaled down by N as the input of the voltage-to-frequency circuit. The voltage-to-frequency circuit, consisting of resistor R_i ($i=$ x, y, z), capacitor C_x, diode D_x and Schmidt-input comparator U, can output a frequency f_1 (f_2) proportional to the input voltage, which will be illustrated in the next section. Then, an optic fiber is applied to provide galvanic isolation, and it is captured by the eCAP module inside the DSP controller as the control feedback. After the calculation with a preset algorithm, the DSP controller directly gives the switching signal delay compensation by using its inside HRPWM module. Then the gate drivers output appropriate timing-compensated driving voltages after a short-time interval, and the voltage balancing of v_{ds1} and v_{ds2} is achieved. During this interval, the clamping RCD circuit also contributes to tolerating the short-time voltage unbalancing.

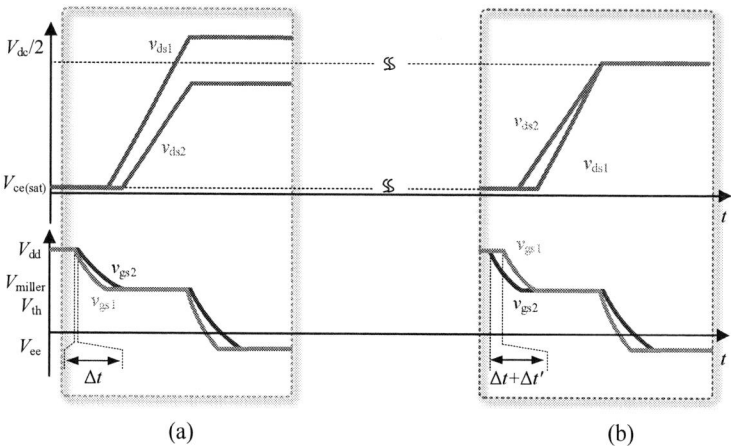

(a)   (b)

Fig. 2: The voltage sharing of v_{ds1} and v_{ds2} (a) without delay compensation (b) with delay compensation

Without the delay compensation, taking the turn-off process as an example, the voltage sharing of v_{ds1} and v_{ds2} is shown in Fig. 2(a). The DSP controller outputs two identical switching signals for T_1 and T_2. However, due to the total delay deviation of gate driver and optic fiber network, the gate source voltage v_{gs1} of T_1 starts to decrease earlier than v_{gs2} of T_2. In addition, since parasitic parameter variation of the devices exist, the decreasing of v_{gs1} is faster than that of v_{gs2}, and the increasing of v_{ds1} is faster than that of v_{ds2} as well in this assumed case. Consequently, a large voltage unbalancing of v_{ds1} and v_{ds2} occurs. It is pointed out that, the clamping RCD circuit helps to reduce the voltage unbalancing in the initial switching cycles, however, the clamping voltage of T_1 gets much higher as more energy is accumulated in the capacitor C_1, and a stable voltage unbalancing is formed gradually.

Consequently, a closed-loop compensation is required to balance v_{ds1} and v_{ds2}. As presented in Fig. 1, v_{c1} (v_{c2}) is converted into frequency f_1 (f_2), then it is captured when the preset time interrupt arrives and a control algorithm (PI control) in the DSP is applied to generate a compensation delay $\Delta t'$ for the next switching cycle. To make it clear, the detailed system flowchart is presented in Fig. 3.

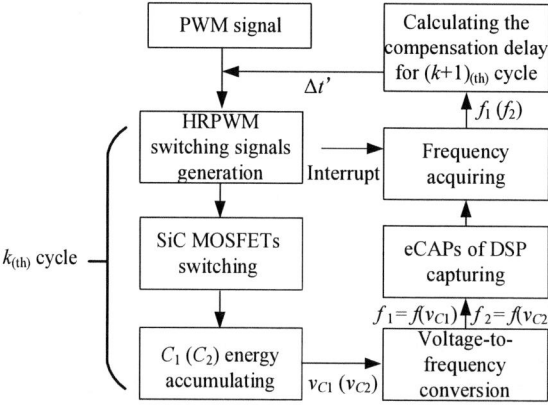

Fig. 3: The detailed system flowchart of the proposed voltage clamping active delay control method

With the delay compensation, the voltage balancing of v_{ds1} and v_{ds2} is shown in Fig. 2(b). $\Delta t'$ makes v_{gs1} start to decrease later than v_{gs2}, and eventually v_{ds1} and v_{ds2} reach to $V_{dc}/2$ synchronously due to the closed-loop control. Therefore, with a dynamic balance between charging and discharging of C_1 (C_2) in the clamping RCD circuit, the clamping values of T_1 and T_2 are nearly the same, and v_{ds1} and v_{ds2} are well-balanced.

Modeling and analysis of the proposed voltage clamping active delay control method

As mentioned, PI control is applied as the algorithm in the DSP, and how to properly choose the K_p and K_i parameters remains. Instead of using a method of trial and error, modeling the whole closed-loop system is a more effective way to help choose parameters while maintaining the system stability. Therefore, every separate section presented in Fig. 3 should be modeled accordingly.

In published literatures, the relationship between the derived delay and the voltage unbalancing degree of series connected SiC MOSFETs is generally attained by an experimental way, which limits the design flexibility. Instead, in the proposed circuit, v_{C1} and v_{C2} are sampled as the control feedback, and the relationship between $\Delta t'$ and the difference Δv_C of v_{C1} and v_{C2} is obtained as:

$$\Delta t' = \frac{C_1 \cdot (v_{C1} - v_{C2})}{I_d} = \frac{C_1 \cdot \Delta v_C}{I_d} \tag{1}$$

where I_d is the power loop current, and C_1 is equal to C_2.

In addition, the voltage to frequency conversion part together with optic fiber connection provides galvanic isolation between the power circuit and the DSP control circuit. Here, a simple Schmitt-trigger-input comparator-based circuit is designed to realize the goal of voltage to frequency conversion, while commercial voltage-to-frequency chips could also be an option. The relationship between f_1 (f_2) and v_{C1} (v_{C2}) can be described as:

$$f_i \approx \frac{v_{Ci}}{N \cdot R_x \cdot C_x \cdot (V_{th}^+ - V_{th}^-)}, \qquad i = 1, 2 \tag{2}$$

where V_{th}^+ is the positive threshold voltage of U, and V_{th}^- is the negative threshold voltage of U.

Since the DSP is applied as the controller, the closed-loop system is a discrete system. Defining the sample period as T_s, based on (1) and (2), the control block diagram is shown in Fig. 4.

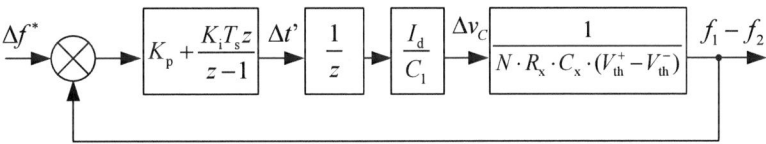

Fig. 4: The control block diagram of the proposed voltage clamping active delay control method

According to Fig. 4, the closed-loop z-transfer function can be solved as:

$$G(z) = \frac{\dfrac{I_d \cdot K_p + I_d \cdot K_i \cdot T_s}{C_1 \cdot N \cdot R_x \cdot C_x \cdot (V_{th}^+ - V_{th}^-)} \cdot z - \dfrac{I_d \cdot K_p}{C_1 \cdot N \cdot R_x \cdot C_x \cdot (V_{th}^+ - V_{th}^-)}}{z^2 + \left(\dfrac{I_d \cdot K_p + I_d \cdot K_i \cdot T_s}{C_1 \cdot N \cdot R_x \cdot C_x \cdot (V_{th}^+ - V_{th}^-)} - 1\right) \cdot z - \dfrac{I_d \cdot K_p}{C_1 \cdot N \cdot R_x \cdot C_x \cdot (V_{th}^+ - V_{th}^-)}} \tag{3}$$

Consequently, by applying Routh–Hurwitz stability criterion, the stability conditions of the system can be obtained as:

$$\begin{cases} \dfrac{C_1 \cdot N \cdot R_x \cdot C_x \cdot (V_{th}^+ - V_{th}^-)}{I_d} > K_p > -\dfrac{C_1 \cdot N \cdot R_x \cdot C_x \cdot (V_{th}^+ - V_{th}^-)}{I_d} \\[4mm] \dfrac{2 \cdot C_1 \cdot N \cdot R_x \cdot C_x \cdot (V_{th}^+ - V_{th}^-) - 2 \cdot I_d \cdot K_p}{I_d \cdot T_s} > K_i > 0 \end{cases} \tag{4}$$

Based on the above, according to (4), K_p and K_i can be properly chosen.

Experimental verification

Voltage-to-frequency conversion response

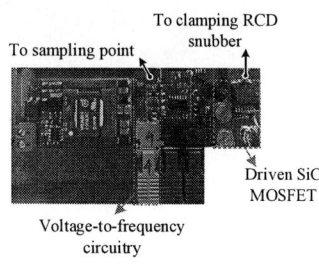

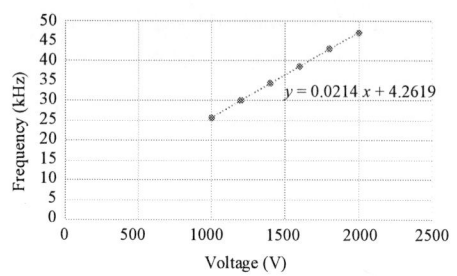

Fig. 5: Photograph of the individual gate driver part of SiC MOSFET

Fig. 6: The measured relationship between input voltage (v_{C1}, v_{C2}) and output frequency

The voltage-to-frequency circuitry plays an important role in the feedback loop, therefore, it is important to evaluate its response firstly. As shown in Fig. 5, this circuitry is integrated in the gate driver part of SiC MOSFET with a fiber connector to send the feedback to DSP controller, and the key circuit parameters relevant to the calculation result is listed in Table I. Based on that, the relationship between the input voltage (v_{C1}, v_{C2}) and the measured output frequency is drawn in Fig. 6. It is observed that their relationship is linear, which is consistent with the derived formula (2).

Effectiveness of the voltage clamping active delay control method

Further, based on the above analysis, the experimental validation is based on a buck chopper circuit, where the circuit diagram is shown in Fig. 7 and the corresponding parameters are given in Table I.

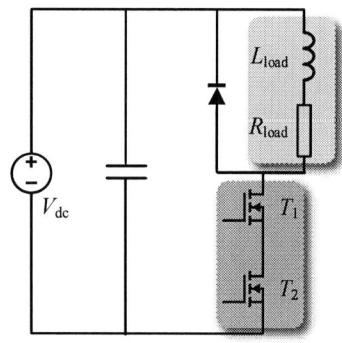

Fig. 7: The buck chopper circuit for testing

Table I: Key Parameters

	Name	Parameters
Gate driver side	N	10
	U	74HC132D
	$C_1(C_2)$, C_x	100 nF, 680 pF
	R_1 (R_2), R_x	2 MΩ, 47 Ω,
Power side	T_1, (T_2)	G2R120MT33J
	L_{load}, R_{load}	70 μH, 30 Ω

When the delay compensation is not applied as shown in Fig. 8, it is observed that a voltage unbalancing of two series connected SiC MOSFETs occurs, and the voltage unbalancing degree is measured to be 10% (maximum (v_{ds1}-v_{ds2})/ (v_{ds1}+v_{ds2})) when the switching frequency is 10 kHz.

Therefore, the delay compensation is applied by using the HRPWM module of DSP, and the time resolution can reach 150 ps, which can satisfy the requirement of active delay control. In this case, T_s is chosen as 0.2 s to ease the controller burden, thanks to the clamping RCD circuit which helps tolerate the short-time voltage unbalancing. Moreover, I_d is set as 20 A, and the calculation could be made according to (4) since the other parameters regarding of the real circuit are known. Therefore, it is concluded that K_i should be smaller than $5.8*10^{-7}$ and larger than 0 when K_p is 0. To verify that, when K_p and K_i are not chosen properly ($K_i = 10^{-6}$, $K_p = 0$), an unstable voltage balancing process is observed as shown in Fig. 9. Since T_s is 0.2 s, during this 0.2 s interval, the unbalancing degree is continuously increased to strike a balance with the clamping RCD circuit. Then in the next 0.2 s interval, the excessive compensation causes reverse unbalancing of v_{ds1} and v_{ds2}, and repeating over time, which is not allowed in series connection.

EPE'22 ECCE Europe

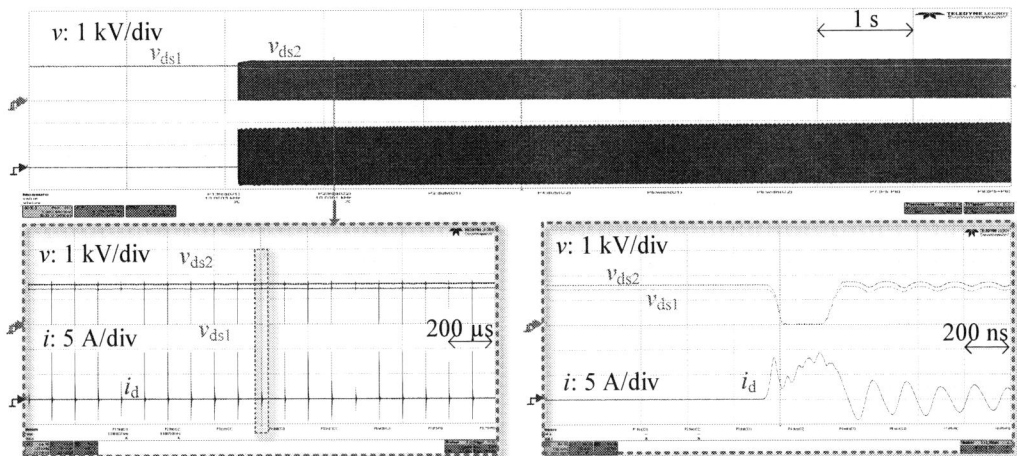

Fig. 8: Voltage sharing of SiC MOSFETs without delay compensation

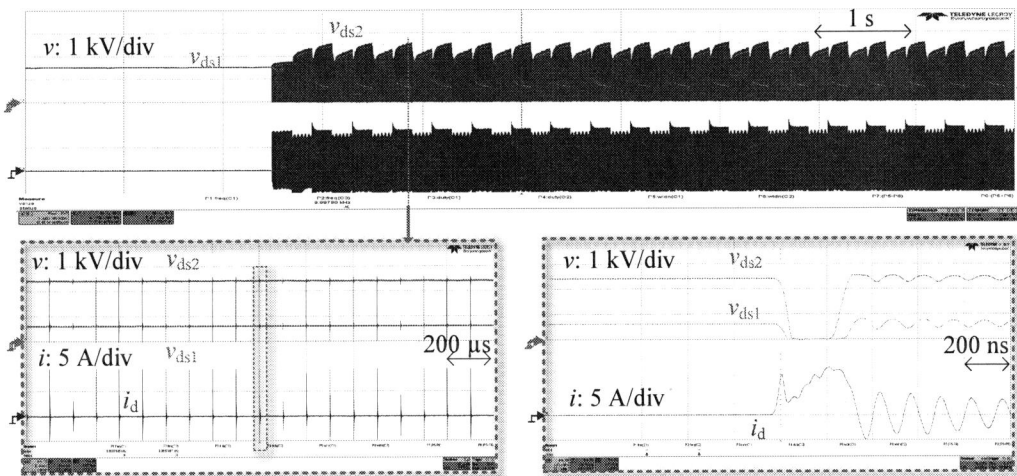

Fig. 9: Voltage sharing of SiC MOSFETs with unstable delay compensation ($K_i = 10^{-6}$)

By contrast, when the delay compensation is applied and K_p and K_i are well chosen ($K_i = 10^{-8}$, $K_p = 0$), a stable voltage balancing is observed as shown in Fig. 10, and the voltage unbalancing degree is measured to be within 1%.

It is worthy of noticing that, despite that the condition $K_i = 10^{-7}$, $K_p = 0$ can satisfy the requirement to be stable as well and it can accelerate the voltage balancing process by choosing a large K_i, but it will cause some undesired oscillation in the meantime, as shown in Fig. 11(a). In the contrary, a smaller value with $K_i = 0.000000001$ can make the voltage balancing process smoother, as shown in Fig. 11(b), but the voltage balancing process becomes slow. Therefore, if a fast response speed is not required, a smaller K_i could be chosen.

Conclusion

In this paper, a voltage clamping active delay control method is proposed and analyzed. Compared with existing active delay control methods, the entire closed-loop system model is simpler to obtain since a clamping RCD circuit is combined and the voltages of clamping capacitors are utilized as control criteria. The feedback unit is designed by using a voltage-to-frequency conversion, and galvanic isolation is provided through an optic fiber. Based on the derived model, the principles to choose the control parameters are provided explicitly. Finally, the experimental results validate the effectiveness of the proposed voltage clamping active delay control method, also, the accuracy of the established model gets verified as well.

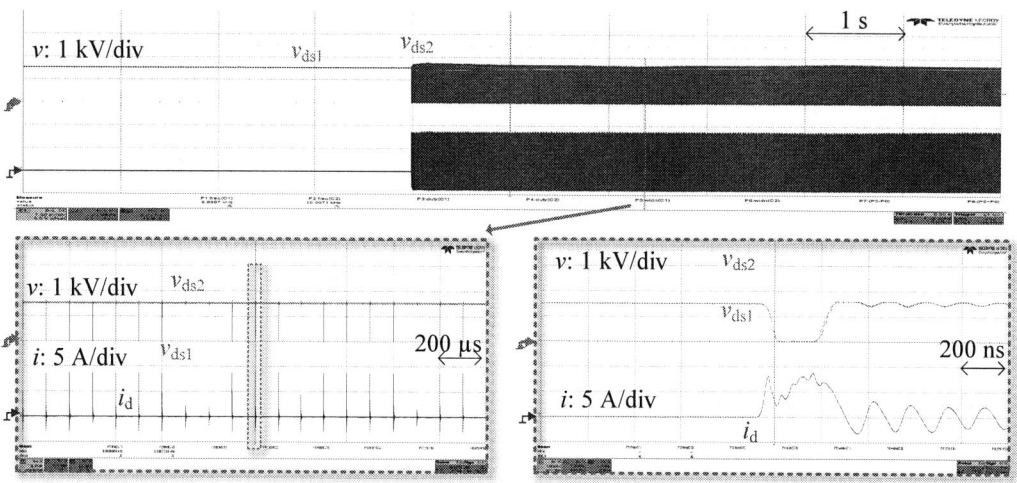

Fig. 10: Voltage sharing of SiC MOSFETs with stable delay compensation ($K_i = 10^{-8}$)

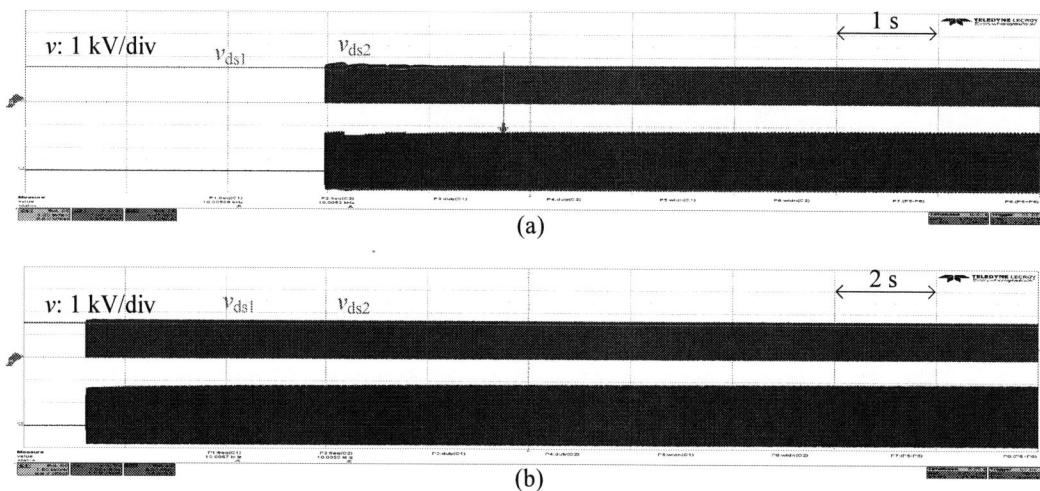

Fig. 11: Voltage sharing cases with stable delay compensation (a) $K_i = 10^{-7}$ (b) $K_i = 10^{-9}$

References

[1] GeneSiC Semiconductor, Accessed 20-12-2021, Online: https://www.genesicsemi.com/sic-mosfet/

[2] R. Wang, A.B. Jørgensen, S. Munk-Nielsen, "An enhanced single gate driven voltage-balanced SiC MOSFET stack topology suitable for high-voltage low-power applications," in IET Power Electronics, vol. 15, no. 3, pp. 251-262, Feb. 2022.

[3] Z. Lu et al., "Medium Voltage Soft-Switching DC/DC Converter With Series-Connected SiC MOSFETs," in IEEE Transactions on Power Electronics, vol. 36, no. 2, pp. 1451-1462, Feb. 2021.

[4] X. Lin, L. Ravi, Y. Zhang, R. Burgos and D. Dong, "Analysis of Voltage Sharing of Series-Connected SiC MOSFETs and Body-Diodes," in IEEE Transactions on Power Electronics, vol. 36, no. 7, pp. 7612-7624, Jul. 2021.

[5] R. Wang, L. Liang, Y. Chen and Y. Kang, "A Single Voltage-Balancing Gate Driver Combined With Limiting Snubber Circuits for Series-Connected SiC MOSFETs," in IEEE Journal of Emerging and Selected Topics in Power Electronics, vol. 8, no. 1, pp. 465-474, Mar. 2020.

[6] V. U. Pawaskar, G. Gohil and P. T. Balsara, "Study of Voltage Balancing Techniques for Series-Connected Insulated Gate Power Devices," in IEEE Journal of Emerging and Selected Topics in Power Electronics, vol. 10, no. 2, pp. 2380-2394, Apr. 2022.

[7] A. Marzoughi, R. Burgos and D. Boroyevich, "Active Gate-Driver With dv/dt Controller for Dynamic Voltage Balancing in Series-Connected SiC MOSFETs," in IEEE Transactions on Industrial Electronics, vol. 66, no. 4, pp. 2488-2498, Apr. 2019.

[8] P. Trochimiuk, R. Kopacz, G. Wrona and J. Rąbkowski, "Active Voltage Balancing of Series-Connected 1.7 kV/325 A SiC MOSFETs Enabling Continuous Operation at Medium Voltage," in IEEE Access, vol. 9, pp. 8604-8614, Jan. 2021.

[9] P. Wang et al. "An Integrated Gate Driver with Active Delay Control Method for Series Connected SiC MOSFETs," in 2018 IEEE 19th Workshop on Control and Modeling for Power Electronics (COMPEL), 2018, pp. 1-6.

[10] M. Zhao, H. Lin and T. Wang, "An Adaptive Driving Signals Delay Control for Voltage Balancing of Series-Connected SiC MOSFETs," in 2020 IEEE Energy Conversion Congress and Exposition (ECCE), 2020, pp. 2814-2818.

[11] K. Wada and K. Shingu, "Voltage Balancing Control for Series Connected MOSFETs Based on Time Delay Adjustment Under Start-Up and Steady-State Operations," in 2018 IEEE Energy Conversion Congress and Exposition (ECCE), 2018, pp. 5495-5499.

[12] C. Li, R. Chen, S. Chen, C. Li, H. Luo, W. Li, X. He, "Analytical Model and Design of Voltage Balancing Parameters of Series-Connected SiC MOSFETs Considering Non-Flat Miller Plateau of Gate Voltage," in Energies, 2022, 15(5), 1722.

Practical Implementation of a Concept for In-Situ Detection of Humidity-Related Degradation of IGBT Modules

Benedikt Kostka, Axel Mertens
Institute for Drive Systems and Power Electronics
Welfengarten 1A
30167 Hanover, Germany
Phone: +49 (0) 511-762 18833
Email: Benedikt.Kostka@ial.uni-hannover.de
URL: https://www.ial.uni-hannover.de

July 14, 2022

Acknowledgments

This work is part of the project ReCoWind and was funded by the Federal Ministry for Economic Affairs and Climate Action on the basis of a decision by the German Bundestag. Funding number: 0324336E. The authors are responsible for the content of this publication.

Special thanks go to CONVERTERTEC GMBH for providing the back-to-back converter investigated.

Keywords

≪Breakdown≫, ≪Condition monitoring≫, ≪IGBT≫, ≪Power semiconductor device≫, ≪Reliability≫.

Abstract

The high voltage, high humidity, high temperature, reverse bias (HV-H^{3}TRB) test reveals that humidity affects the blocking characteristic of an IGBT module, resulting in increased leakage current, decreased voltage blocking capability, or some combination of these. This paper presents the practical implementation of a previously proposed measurement concept for in-situ detection of moisture-induced degradation of IGBT modules in a full-scale back-to-back converter for doubly fed electrical machines typically used in wind turbines.

Introduction

In recent years, various field data analyses have found a correlation between relative ambient humidity and the failure rate of wind turbines [1], [2]. It has been shown that the phase module, which consists of the semiconductor modules and the DC-link capacitor, is responsible for a large proportion of the failures. Furthermore, the failures are more likely to occur during the start-up process after the wind turbines have been subjected to a prolonged standstill period [2]. Indeed, [3] shows that IGBT modules can be damaged by moisture by exposing the devices to the HV-H^{3}TRB test. Consequences can include reduced voltage-blocking capability and increased leakage current.

Therefore, further investigations into the condition monitoring of the system, especially the power semiconductors, are necessary. There are a lot of publication covering the online determination of the junction temperature of an IGBT using temperature-sensitive parameters (TSEPs) and/or online detection of temperature-related degradation, such as bond wire liftoff and solder joint fatigue [4], [5]. However, to date, there have been only a few publications that deal with online condition monitoring with regard to moisture-related degradation or moisture in the inverter itself, with the aim of estimating its electro-chemical condition on the basis of service life models (e.g., Peck's Model) [6]. A first approach for monitoring humidity-related degradation of semiconductors is presented in [7]; a concept is detailed that enables the in-situ measurement of a device state inside an inverter by measuring the blocking charac-teristics of IGBTs in a half-bridge configuration. However, the results have only been obtained on a laboratory scale. Based on the findings in [7], this paper presents the practical implementation of the concept in a full-scale, back-to-back converter typically used for doubly fed electrical machines in wind turbines.

The structure of the paper is as follows. First, the functional principle of the concept under investigation is explained and a functional description of the hardware developed is given. Then the implementation of the measurement board in the converter is described. Next, the measurements, which are initially carried out with non-degraded modules, are presented. Afterwards, the measurements are also conducted with previously degraded modules connected to the inverter under test. Finally, a conclusion to this work is given.

Measurement Concept

In this section, the developed measurement concept is presented. After the general function is explained, the structure of the developed hardware is described.

Principle of Operation

The general idea of the condition-monitoring concept is shown in Fig. 1 (left). The concept is designed to conduct the measurement during the standstill phases of the wind turbine, as restarting the wind turbine after downtime increases the risk of a phase module failure [8]. In this state, the switches (e.g., IGBT modules) are actively switched off and a DC-link voltage V_{dc} must be applied. To detect module degradation before it becomes critical, the DC link must be precharged above its nominal voltage $V_{dc,N}$. Ideally, the DC-link voltage should be at least as high as the highest voltage occurring during normal operation (e.g., the overvoltage during IGBT switch-off). This can be realised, for example, by using a precharging transformer, which is usually used in wind turbines and must be adapted for a higher voltage. The load machine connected to the phase terminals acts like a DC short circuit between all phases during the measurement.

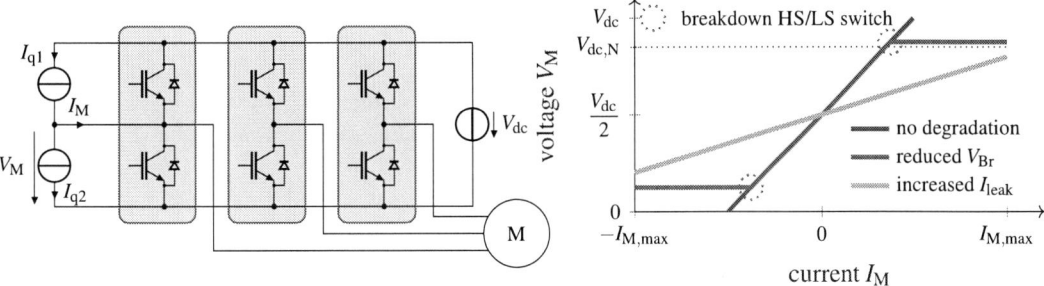

Fig. 1: Measurement concept for detecting moisture-related damage to semiconductor chips in a three-phase inverter (left); Possible I_M-V_M waveforms for the same states of degradation of the high-side and low-side switches (right)

The concept is based on two additional DC current sources (I_{q1} and I_{q2}). These inject a measuring current I_M into the phase terminals which corresponds to the difference between I_{q1} and I_{q2}. Since the voltage V_M depends on the leakage currents of the switched-off high-side and low-side switches, the voltage V_M is adjusted using the current I_M. Safe operation of the concept is ensured so long as the combined breakdown voltage of the high-side and low-side switches is greater than the applied DC-link voltage V_{dc}.

The voltage across the low-side switches ($V_{LS} = V_M$) increases with increasing positive current I_M. It continues to rise until either the applied DC-link voltage V_{dc} is reached, a premature avalanche of the low-side switches is detected, or the leakage current of the low-side switches at V_{dc} is higher than the positive maximum current of I_M, which is limited by the additional current sources. As would be expected, the voltage V_M decreases when the current I_M decreases, because then the voltage across the high-side switches increases and the voltage across the low-side switches decreases ($V_{HS} = V_{dc} - V_M$). The voltage V_M will decrease until it reaches $0\,V$, a premature avalanche of the high-side switches is detected, or the leakage current of the high-side switches at V_{dc} is higher than the negative maximum current of I_M, which is limited by the additional current sources. The maximum current limits $I_{M,max}$ and $-I_{M,max}$ must be chosen carefully so that detection of degradation is possible, but also so that the current does not destroy the semiconductor components in the inverter.

Fig. 1 (right) outlines three possible I_M-V_M characteristics, which assume that the state of degradation of the high-side and low-side switches is the same. Thus, for a given voltage, the same leakage current flows through both the high-side and low-side switches, and all I_M-V_M characteristics cross at the point $V_M = \dfrac{V_{dc}}{2}$ and $I_M = 0\,A$.

The blue line shows the case in which no degradation of the modules is observed. In this case, the voltage V_M increases with increasing current I_M. Consequently, as the current decreases, the voltage also decreases. As V_M reaches the value of the applied DC-link voltage V_{dc} or $0\,V$, it can be stated that the high-side and low-side switches have a blocking capability of at least V_{dc}. A more detailed characterisation is not possible at this point.

In contrast, the red line shows the case in which the maximum breakdown voltage of the high-side and low-side switches is reduced, such as might result from degradation caused by humidity. Since only a reduction in breakdown voltage and no increase in leakage current is assumed, the red line coincides with the blue line in the area where no breakdown is provoked. As soon as the voltage V_M reaches the breakdown voltage of the low-side switches, the current I_M immediately increases to the maximum current value $I_{M,max}$ without any significant change in the voltage V_M. Here, the exact value of the breakdown voltage of the weakest low-side switch can be determined directly from the measurement data. The procedure is similar for the high-side switches, except that now the voltage V_M decreases as the current I_M decreases. This means that the voltage across the high-side switches increases. Thus, as the voltage V_M decreases further, the breakdown of the weakest high-side switch is triggered and the current I_M falls to its maximum negative value $-I_{M,max}$ without the voltage V_M changing significantly.

The yellow line shows the I_M-V_M characteristic for an increased leakage current through the high-side and low-side switches without a drop in breakdown voltage. As can be seen in Fig. 1 (right), an increase in the leakage current through the modules from DC+ to DC- reduces the influence of the injected current I_M on the voltage V_M. If the leakage current through the low-side switches at V_{dc} is greater than $I_{M,max}$, then V_M does not reach the applied DC-link voltage V_{dc}. The same applies to the high-side switches, with the difference that V_M does not then reach $0\,V$. The voltage value V_M at maximum negative current and maximum positive current is therefore an indicator of whether or not degradation of the blocking characteristics has occurred, which could be caused by humidity.

Hardware

Fig. 2 (left) shows the printed circuit board (PCB) designed for this concept. The PCB has the size of $130\,mm$ x $70\,mm$ with a voltage rating of $1400\,V$. Furthermore, Fig. 2 (right) shows the schematic and the main components used on the PCB.

In general, only two additional current sources need to be implemented, and these are attached to the DC link from DC+ to DC-. To inject the current I_M, the current sources I_{q1} and I_{q2} have to be connected to one of the phase terminals. Given the low power consumption of the concept, it can be fed directly from the power supply of the digital control unit (DCU) of the inverter and the injected current I_M is provided by the DC-link capacitor.

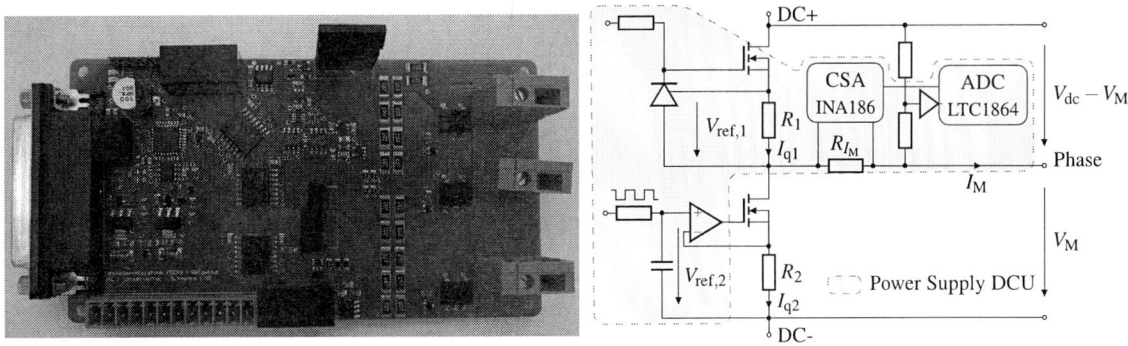

Fig. 2: Hardware design used in this work; Designed PCB (left); Simplified schematic (right)

I_{q1} is implemented as a constant current source. It is built out of a high voltage blocking MOSFET, a resistor R_1 and a shunt regulator. If the shunt regulator is fed with a sufficient current (approx. 200 µA), it regulates the voltage $V_{ref,1}$ occurring across the resistor R_1 to a constant value. The MOSFET must be capable of blocking the whole applied DC-link voltage V_{dc}. The current I_{q1} can then be adjusted by choosing an appropriate value of R_1.

The principle of operation is similar for I_{q2}. But instead of a constant current, now an adjustable current is needed to adjust the current I_M, which is equal to the difference between I_{q1} and I_{q2} ($I_M = I_{q1} - I_{q2}$). Therefore, no shunt regulator is used. Instead, the reference voltage $V_{ref,2}$ is created by a PWM signal, which is filtered by a low-pass RC filter, so that $V_{ref,2}$ corresponds to the product of the amplitude $\hat{V}_{PWM}$ and the duty cycle d of the PWM signal ($V_{ref,2} = d \cdot \hat{V}_{PWM}$). Since the subsequent operational amplifier (OPA) is installed with its inverting input directly connected to the source of the MOSFET and its output connected to the gate, the OPA regulates the output voltage so that the voltage at the inverting input corresponds to the voltage present at the non-inverting input $V_{ref,2}$. In this way, the current I_{q2} can be controlled by the duty cycle of the PWM signal provided by a microcontroller (µC).

To monitor the condition of the switches in the inverter, the current I_M and voltage V_M needs to be captured. The current measurement is performed with a shunt resistor in parallel to a current-sense amplifier (CSA). The current measurement range of the custom-designed PCB in Fig. 2 (left) is from $I_{M,min} = -3.5$ mA to $I_{M,max} = 5.8$ mA. The output voltage of the CSA is then recorded by an analog-to-digital converter (ADC). The recorded values are sent via the serial peripheral interface (SPI) to the µC, which sends the data to the processing unit (here: a computer) over a controller area network (CAN) bus. In contrast, the voltage V_M is captured with a high-impedance voltage divider. An identical ADC is used to record and send the measured data for V_M to the µC.

The measurement is started when the inverter is in an off state (i.e., a standstill phase of a wind turbine). All switches are then actively turned off and a DC-link voltage V_{dc}, which is higher than the nominal DC-link voltage $V_{dc,N}$, is applied. Afterwards, the PCB gets an signal to start the measurement. At the beginning, the duty cycle of the PWM signal is zero, which means that the reference voltage $V_{ref,2}$ present at the resistor R_2 is 0 V. So no current I_{q2} flows. In this case, I_M is equal to the constant current I_{q1}. In the following, the duty cycle is increased, which subsequently results in a reduction in the current I_M until the minimum current of $I_M = I_{q1} - I_{q2,max}$ is reached.

Experimental Implementation

The concept proposed in Fig. 1 (left) is implemented in a back-to-back converter typically used for doubly fed electrical machines in wind turbines. Fig. 3 shows the complete circuit diagram of the inverter including the additionally implemented measurement concept.

The converter consists of a line-side converter (LSC) connected to the grid via its three phases. The LSC converts the three-phase AC voltage into a DC voltage and feeds it into the DC link, which contains a

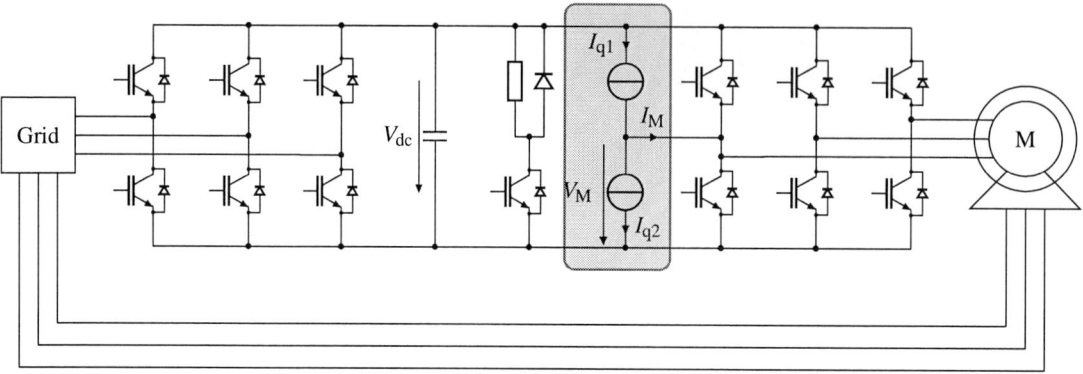

Fig. 3: Schematic of a back-to-back converter for doubly fed electrical machines with the integrated measurement concept at the machine-side converter

DC-link capacitor C_{dc} and a chopper. If the DC-link voltage exceeds a predefined value, the chopper switches on to limit the overvoltage and thus protect the components connected to the DC link. The resistor in series with the chopper switch is then connected directly to the DC link and dissipates the excess energy stored in C_{dc}. As a result, the DC-link voltage falls until the chopper switch is turned off again at a desired value of V_{dc}. As mentioned in [7], the chopper switch cannot be monitored using the proposed concept without further effort.

Therefore, the condition of the chopper switch needs to be investigated in advance in order to safely apply the proposed concept in a back-to-back converter for wind turbines. One option might be to measure the (very small) voltage drop across the chopper resistance to monitor the leakage current through the chopper. In this case, only an increase in the leakage current is detectable. Otherwise the chopper resistance must be disconnected, for example, by an additional contactor or relay. Then the condition of the chopper switch and diode might be monitored using the proposed concept in Fig. 1, in the same way as discussed in the previous section.

In addition to the LSC, a back-to-back converter also includes a machine-side converter (MSC) that shares the DC link with the LSC. The MSC converts the DC-link voltage into a three-phase AC voltage in order to drive the electrical machine. The three phases leaving the MSC are connected to the rotor of the electrical machine. In contrast, the stator of the doubly fed electrical machine is connected directly to the grid. Thus, power transfer between the grid and stator is also possible.

As shown in Fig. 3, the concept is integrated into the conventional back-to-back inverter. For the experiments conducted in this paper, the concept is connected to the second phase of the MSC. Furthermore, as mentioned in the previous section, the concept must be connected to the DC link. In the configuration shown in Fig. 3, only the condition of the switches in the MSC is monitored.

Since it is assumed that the wind turbine is in a standstill phase during the measurement, the converters (LSC and MSC) are actively switched off and are in a blocking state. Normally, no DC-link voltage is present at C_{dc} in this state. To carry out the measurement, the DC-link voltage V_{dc} must therefore be supplied externally. The applied DC-link voltage V_{dc} should be higher than the nominal DC-link voltage $V_{dc,N}$, but must be lower than the value at which the chopper switch is activated, otherwise the DC-link voltage would be discharged immediately. In the experiments conducted in this paper, the chopper switch was deactivated.

The DC-link capacitor C_{dc} can be precharged with the help of a precharge transformer, which wind turbines are often equipped with. In the converter investigated, the transformer is usually designed to precharge the DC link to a voltage of only 990 V, which is below the nominal DC-link voltage $V_{dc,N}$ of the converter used. In this case, it is only possible to monitor modules that show an increase in leakage current as a result of humidity-related degradation. To detect a reduced blocking capability of the semiconductor components in the converter, the secondary voltage of the precharging transformer must be adjusted, which is easily possible. The adjusted precharge transformer allows the precharging of the DC-link capacitor up to 1200 V. This means that the measurement concept may be applied to both converters (LSC and MSC) without any concerns regarding EMI, since no steep current or voltage slopes occur due to switching.

Experimental Results

First, the designed measurement board in Fig. 2 (left) is mounted in a full scale back-to-back converter as shown in Fig. 3. Normally, the DC link can be precharged to a voltage of $V_{dc} = 990\,V$ by the use of the precharge circuit. Since a higher DC-link voltage than the nominal DC-link voltage is needed to exploit the presented concept to its full extend, the precharge circuit is modified for the measurements presented in this paper. Now the DC link can be precharged to a voltage of $V_{dc} = 1200\,V$. The measured I_M-V_M characteristics are recorded and plotted in Fig. 4 (left).

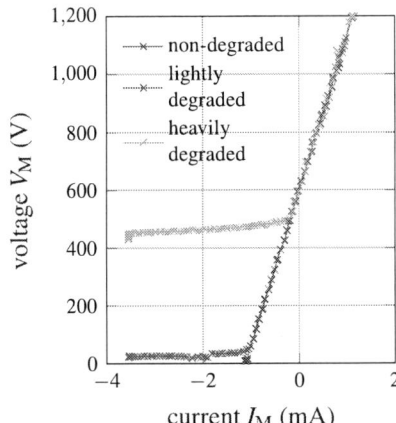 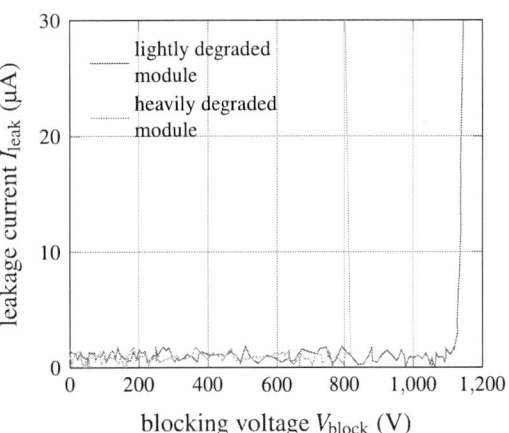

Fig. 4: Measured I_M-V_M characteristic of the MSC when the DC link is precharged to a value of $V_{dc} = 1200\,V$ using a modified precharge circuit (left); blocking characteristics of the modules previously degraded in the HV-H³TRB test (right)

During the measurements, the chopper is deactivated, since it would be triggered at a voltage of $V_{dc} = 1170\,V$ and would discharge the DC-link capacitor immediately.

Results without Degraded Modules

First of all, the measurement is carried out in the whole converter system without any emulation of humidity-related degradation. The I_M-V_M characteristic obtained is shown in Fig. 4 (left) by the blue line. It reaches the maximum current of $I_M \approx 1.1\,mA$ at a measured voltage of V_M, which is equal to the applied DC-link voltage of $V_{dc} = 1200\,V$. This means that the voltage across the high-side switches V_{HS} is equal to $0\,V$ and the full DC-link voltage V_{dc} is applied across the low-side switches of the MSC ($V_{LS} = V_{dc}$). Therefore, the measured current I_M here corresponds to the total leakage current through all the low-side switches connected in parallel, since the connected machine acts like a DC short between the three phases. However, the measured current of $I_M \approx 1.1\,mA$ is quite high to be caused only by the semiconductors alone. The linear shape of the I_M-V_M characteristic curve indicates resistive behaviour. This means that the characteristic of the intact modules is dominated by additional resistors and not by the semiconductors themselves. In the converter under investigation, the gate driver (GD) of the second phase has an additional measuring circuit for the collector-emitter voltage v_{ce}, which is provided with the help of a high-impedance voltage divider with a total resistance of $1.62\,M\Omega$. Additionally, it should be noted that the GD of the first and third phases differs from the GD of the second phase, where the desaturation detection circuit is implemented using diodes. Therefore the leakage current of these GDs can be neglected in comparison to the leakage current of the GD of the second phase. Anyway, the measured equivalent resistance $R_{M,LS}$ of the low side can easily be calculated as the ratio between the applied DC-link voltage V_{dc} and the maximum of the measured current $I_{M,max}$, according to Ohm's law, since the voltage above the low side V_{LS} equals the measured voltage V_M ($V_{LS} = V_M$).

$$R_{M,LS} = \frac{V_{dc}}{I_{M,max}} \quad . \tag{1}$$

Thus, this resistance is $R_{M,LS} \approx 1.1\,M\Omega$. In other words, $R_{M,LS}$ is of the same order of magnitude as the resistance of the collector-sense circuit at the GD. As expected, the measured resistance is lower than the $1.62\,M\Omega$ of the second phase's GD, since the other components (e.g. IGBT modules) connected from the phase to DC- contribute an additional leakage current which is also captured by the concept utilised. Since all phases can be considered parallel, the leakage current of all other components on the low side, except the $1.62\,M\Omega$ resistance of the second phase's desaturation detection circuit, add up to $360\,\mu A$. Obviously, the high leakage current is dominated by the resistance of the collector-sense circuit and not by the modules themselves.

Therefore, a very small change in leakage current cannot be detected with this setup. However, the datasheet states the maximum collector-emitter cut-off current to be as high as $I_{CES} = 5\,mA$ at $25\,°C$ and the maximum applied collector-emitter voltage to be $V_{ce} = 1700\,V$. So it is possible to monitor a substantial increase in the leakage current of the modules in the range of mA as a sign of humidity-related degradation, even though this is still within the specification of the modules. This is sufficient for monitoring the condition of the modules.

Since the measured I_M-V_M characteristic is linear and reaches $V_M = \dfrac{V_{dc}}{2} = 600\,V$ at $I_M = 0\,A$, it can be stated that the measured resistance of the high side is equal to the measured low-side resistance ($R_{M,HS} = R_{M,LS}$). Otherwise $R_{M,HS}$ can be calculated in the same way as $R_{M,LS}$ in (1). The voltage above the high side V_{HS} is equal to the difference between the DC-link voltage V_{dc} and the measured voltage V_M.

$$V_{HS} = V_{dc} - V_M \quad . \tag{2}$$

According to (2), the whole DC-link voltage is applied to the high side when V_M is zero. Since no voltage is then applied to the low side, the whole (negative) current ($-I_{M,max}$) is provided by the components of the high side at the applied DC-link voltage V_{dc}.

Results with Degraded Modules

To emulate humidity-related degradation of an IGBT switch on the high side in the following, single-switch modules are connected in parallel to the high-side switches of the second phase of the MSC, which were previously degraded in the HV-H[3]TRB test. Fig. 4 (right) shows the blocking characteristics of the degraded modules. For more information about the test condition and underlying degradation mechanism, refer to [7]. In this work a lightly degraded module and a heavily degraded module are used. The lightly degraded module was kept for 168 hours under HV-H[3]TRB test conditions, while the heavily degraded module was kept for 1168 hours under the same conditions. As can be seen in Fig. 4 (right), both modules show a reduction in breakdown voltage but no increase in leakage current as signs of humidity-related degradation. For the lightly degraded module, the remaining maximum blocking voltage is approximately $1140\,V$ and for the heavily degraded module it is approximately $800\,V$.

The obtained I_M-V_M characteristics for the MSC with the degraded modules connected to the high side of the second phase are shown in Fig. 4 (left). In the figure, the red line represents the measurement results with the lightly degraded module connected and the yellow line shows the results with the heavily degraded module connected. Both measured characteristics show a linear section which fits well to the measured I_M-V_M characteristic without any degraded module connected to the MSC. They both start at the point $V_M = V_{dc} = 1200\,V$ / $I_M = 1.1\,mA$. Obviously, no change to the low-side switches is detected, since no adjustments are done on the low side.

By reducing the current I_M, the voltage V_M decreases. Consequently, the voltage above the high side V_{HS} increases according to (2). For the red I_M-V_M characteristic, the voltage V_M decreases linearly with the current I_M only until V_M is approximately $42\,V$. Then, the voltage V_M does not decrease significantly even though the current I_M is further reduced. This indicates that a breakdown on the high side is taking place. According to (2), the breakdown voltage of the high side for the I_M-V_M characteristic with the lightly degraded module connected to the high side (see red line in Fig. 4 (left)) is determined to be approximately $1160\,V$. This fits well to the previously captured breakdown voltage of $V_{BR} \approx 1140\,V$ of the lightly degraded module (red line in Fig. 4 (right)).

In contrast to this, for the yellow I_M-V_M characteristic shown in Fig. 4 (left) when the heavily degraded module is connected to the high side, the breakdown is detected at a voltage of $V_M = 490\,V$. This corresponds to a determined breakdown voltage $V_{BR} \approx 710\,V$ on the high side and is in good agreement with the captured blocking voltage of the heavily degraded module of $800\,V$ in Fig. 4 (right). The small changes in the determined breakdown voltage in Fig. 4 (left) and (right) might be caused by differences in the ambient temperature, since changes in room temperature cannot be excluded between the measurements. However, the results show that the degradation of the high-side switch and even the severeness of the degradation are detected successfully.

Conclusion

This paper presents a first approach to the practical implementation of a concept for the in-situ detection of humidity-related degradation of IGBT modules in a full-scale, back-to-back converter usually used in wind turbines, which was previously tested only on a laboratory scale in [7].

The measured current I_M reveals that the leakage current is dominated by an additional resistance, which is identified as the resistance of the collector-sense circuit attached to the gate driver. The influence of this external resistance could be calculated and subtracted from the measured data in order to determine the leakage current through the switches. However, other influences such as temperature, tolerances, etc., on the resistance must be taken into account. A substantial increase in the leakage current is detectable, but might still meet the specifications given on the module datasheet. The results also show that not only the modules but also the gate driver is monitored. Therefore, a change in the I_M-V_M characteristic cannot be clearly assigned to the gate driver, the module or other components. Nevertheless, this would suggest a notable change in the system compared to the original state, which would indicate some kind of degradation or damage to the system. Ultimately, the entire system must be observed from the beginning or characterised in advance in order to obtain a lookup table for the initial system under different environmental conditions, such as temperatures. Only then can a statement about the degradation of the system be made.

To exploit the concept to the fullest extent, the applied DC-link voltage V_{dc} during the measurement should be higher than the nominal DC-link voltage during operation $V_{dc,N}$. The DC-link voltage in an active blocking state of both converters (LSC and MSC) can be set by the additional precharge circuit. Since the voltage supplied by the precharge circuit is normally chosen to be below the nominal DC-link voltage $V_{dc,N}$, only modules which exhibit an increase in leakage current as the result of humidity-related degradation could be monitored in this case. Otherwise, the precharge circuit needs to be adjusted to provide a higher DC-link voltage V_{dc}. This would enable modules which show a reduction in the breakdown voltage to be monitored as well.

In this work, the precharge circuit has been adjusted so that the DC link could be charged up to $1200\,V$. To avoid the chopper turning on, it was deactivated for the time of the measurement. This work shows that it is possible to conduct the measurement safely inside the full-scale, back-to-back converter during the standstill phases of wind turbines, even though the DC-link voltage V_{dc} is above the nominal DC-link voltage $V_{dc,N}$. Moreover, it has been shown that the detection of humidity-related degradation is possible and even its severeness may be distinguished.

References

[1] K. Fischer, M. Steffes, K. Pelka, B. Tegtmeier, and M. Dörenkämper, "Humidity in power converters of wind turbines—field conditions and their relation with failures," *Energies*, vol. 14, no. 7, p. 1919, 2021.

[2] K. Fischer, K. Pelka, A. Bartschat, B. Tegtmeier, D. Coronado, C. Broer, and J. Wenske, "Reliability of power converters in wind turbines: Exploratory analysis of failure and operating data from a worldwide turbine fleet," *IEEE Transactions on Power Electronics*, vol. 34, no. 7, pp. 6332–6344, 2019.

[3] C. Zorn and N. Kaminski, "Acceleration of temperature humidity bias (thb) testing on igbt modules by high bias levels," in *2015 IEEE 27th International Symposium on Power Semiconductor Devices & IC's (ISPSD)*. IEEE, 10.05.2015 - 14.05.2015, pp. 385–388.

[4] D. Herwig, T. Brockhage, and A. Mertens, "Combining multiple temperature-sensitive electrical parameters using artificial neural networks," in *2020 22nd European Conference on Power Electronics and Applications (EPE'20 ECCE Europe)*, 2020, pp. 1–10.

[5] M. A. Eleffendi and C. M. Johnson, "In-service diagnostics for wire-bond lift-off and solder fatigue of power semiconductor packages," *IEEE Transactions on Power Electronics*, vol. 32, no. 9, pp. 7187–7198, 2017.

[6] W. Holzke, A. Brunko, H. Groke, N. Kaminski, and B. Orlik, "A condition monitoring system for power semiconductors in wind energy plants," in *PCIM Europe 2018; International Exhibition and Conference for Power Electronics, Intelligent Motion, Renewable Energy and Energy Management*, 2018, pp. 1–7.

[7] B. Kostka, D. Herwig, M. Hanf, C. Zorn, and A. Mertens, "A concept for detection of humidity-driven degradation of igbt modules," *IEEE Transactions on Power Electronics*, vol. 36, no. 12, pp. 13 355–13 359, 2021.

[8] K. Fischer, T. Stalin, H. Ramberg, J. Wenske, G. Wetter, R. Karlsson, and T. Thiringer, "Field-experience based root-cause analysis of power-converter failure in wind turbines," *IEEE Transactions on Power Electronics*, vol. 30, no. 5, pp. 2481–2492, 2015.

Design For Enhanced Noise Immunity of PCB Coils used for Sensing Current through Power Devices

Aamir Rafiq
Indian Institute of Technology Delhi
Hauz Khas
New Delhi, India
Email: aamirrzargar@ieee.org

Sumit Pramanick
Indian Institute of Technology Delhi
Hauz Khas
New Delhi, India
Email: sumit.pramanick@ieee.org

Acknowledgments

This work was supported by the Ministry of Electronics and Information Technology, Government of India under the NaMPET Phase 3 programme (Project ID: NaMPET-III/SP-01/NH-CON-01).

Keywords

≪Current Sensor≫, ≪Silicon Carbide (SiC)≫, ≪Half bridge≫, ≪High frequency power converter≫, ≪Measurements≫.

Abstract

PCB coil, when placed near a current carrying trace, senses the rate of change of current by virtue of its flux linkages through the coil. Thereafter, the coil output is integrated to obtain the sensor output which is proportional to the trace current. Proximity of the PCB coil to external current carrying conductors can make the coil susceptible to undesired flux linkages which can interfere with the operation of the sensor. Furthermore, presence of a switching node near the coil, with high dv/dt, can influence the sensor output through parasitic coupling with the coil. This paper investigates the influence of external magnetic fields as well as on-board dv/dt on the performance of PCB coil based current sensors. Based on theoretical analysis followed by experimental investigations on two PCB coil prototypes on a half-bridge board, design guidelines for enhancing noise immunity of PCB coils are presented.

Introduction

Current sensing in power electronic devices requires high sensing bandwidth owing to the wide frequency spectrum of the pulsed switching current. Moreover, the sensor should introduce very low inductance inside the power-loop to minimize peak voltage overshoot in the drain-source voltage of the power device. For that purpose, the sensing element should posses a small size. Commonly available high bandwidth sensors like CTs are bulky and are difficult to integrate inside switching power-loops. Rogowksi coils have been demonstrated to be effective for sensing the current in power modules with high bandwidth [1]. Rogowksi coils are simple air cored coils which pickup the rate of change of flux linkage through the coil. The flux linkage through the coil is due to the magnetic field generated by the device current under measurement.

Similar to Rogowksi coils, PCB pick-up coils have been effectively utilized for sensing current on PCB traces with a high bandwidth [2–5]. Unlike circular Rogowski coils, these PCB coils are not fabricated with a symmetrical distribution of its turns around the current carrying conductor at the center of the coil. Consequently, the PCB coils lack the external magnetic field rejection capability which is inherent with circular Rogowski coils. The factors which influence the noise immunity of PCB coils towards external magnetic fields are identified in this paper. Further, based on these factors, methodology for the design of PCB coils with enhanced noise immunity is presented.

In addition to noise coupled by external magnetic fields, another source of noise which gets coupled to the PCB coil output is through the high dv/dt switching node, which is inherent with the power electronic circuit. While faster switching wide-bandgap (WBG) devices like silicon carbide (SiC) MOSFETs and gallium nitride (GaN) HEMTs are helpful for improving the efficiency of power converter, the high switching node slew rate (dv/dt) is responsible for higher noise being coupled to sensitive control and sensor circuitry. When this voltage node with high dv/dt gets coupled with the PCB coil through parasitic capacitances, a displacement current flows through the PCB coil which produces noise in the sensor output. An experimental scheme is presented to investigate the influence of dv/dt induced noise due to the structure of the PCB coil as well as the power-board layout.

PCB Coil Based Current Sensing

Fig. 1(a) shows a half-bridge circuit schematic. The hardware implementation of this circuit featuring embedded PCB coils is shown in Fig. 1(b). The PCB coil is fabricated around the trace which connects the power MOSFET to the DC busbars. The trace (T_2) is shown zoomed in Fig. 1(b), and it serves to connect source of M_2 with the negative DC bus. PCB layout of T_2 along with the PCB coil is depicted in Fig. 2. T_2 is routed on the bottom copper layer as shown in Fig. 2(a). Placement of the PCB coil above T_2 is shown in Fig. 2(b). Further, on a four-layered PCB, the PCB coil can be fabricated either between layer 2 and the top copper layer (Coil A) or between layer 1 and layer 2 (Coil B). For both these coil structures, the output voltage of the coil (V_C) is expressed as

$$V_C(t) = M \frac{dI_M(t)}{dt} \tag{1}$$

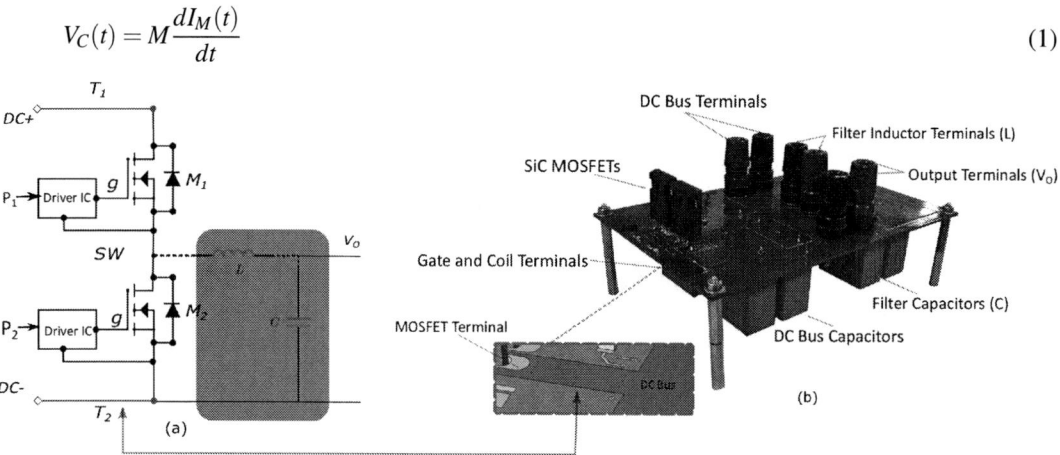

Fig. 1: (a) Half-bridge circuit. (b) PCB design of the half-bridge circuit with embedded PCB coils for trace T_1 and T_2.

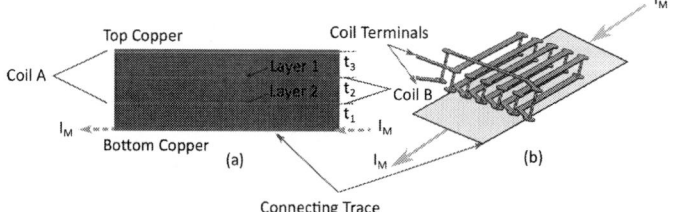

Fig. 2: (a) Possible coil designs on a four-layered PCB. (b) Embedded PCB coil above the trace.

where M is the mutual inductance between the power trace and the PCB coil. The coil voltage is then integrated using an analog integrator circuit with an ideal gain of $1/R_i C_i$ to produce the sensor output voltage (V_S), expressed as

$$V_S(t) = \frac{MI_M(t)}{R_iC_i}.$$ (2)

M is influenced by the size of the PCB coil as well as its proximity (t_1) with the power-loop trace. Both the coils, A and B, are fabricated and tested separately and the sensor voltage (V_S) is measured for both the coils as shown in Fig. 3(a) and Fig. 3(b), respectively. Current measurements are also compared with measurements from a commercial Tektronix current probe (TRCP0300), indicated as I_M. The drain-source voltage (V_{DS}) and the gate-source voltage (V_{GS}) of the MOSFETs are shown for both the tests. With the same integrator circuit gain ($R_i = 470\,\Omega$ and $C_i = 0.1$ nF) current measurement with coil A exhibits a sensitivity of 80 mV/A while the measurement with coil B exhibits a sensitivity of 65 mV/A.

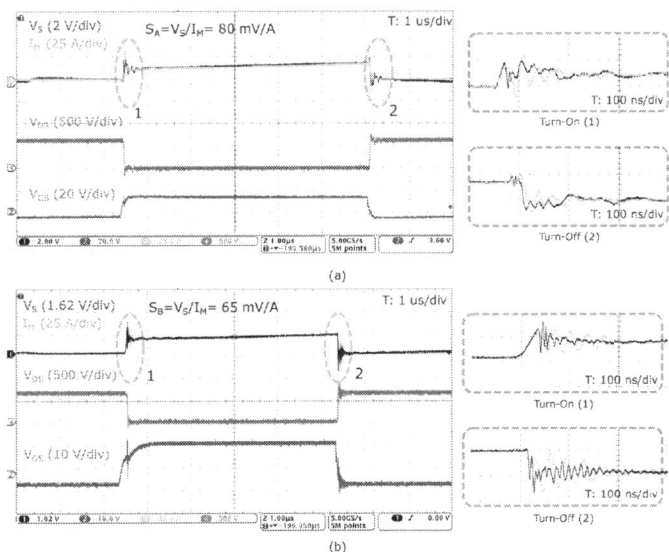

Fig. 3: (a) Pulse test results with coil A. (b) Pulse test results with coil B.

Therefore, coil A exhibits a higher mutual inductance than coil B with the power trace, owing to its larger size. Higher mutual inductance between the PCB coil and the power trace enhances the sensitivity of the sensor and helps reduce error in the sensor output due to op-amp input offset voltage [6]. However, a larger coil size can also make it susceptible to external magnetic fields. For both the coil A and coil B, an external current carrying conductor is placed 1.5 mm above the center of the PCB coil and the output voltage of the coil is measured, as shown in Fig. 4. Coil A shows a higher induced voltage due to the time-varying external current (I_L) on account of the larger coil size. An alternate approach to enhance M is by reducing the separation between the coil with the power trace. The separation (t_1) of both these coil designs to the current carrying trace is fixed by the number of layers in the PCB and the total finished PCB thickness. To quantify these factors, an analytical approach towards determining the influence of external magnetic fields on the performance of PCB coil based current sensor is now given.

Noise Due to External Magnetic Fields

Consider that the PCB coil lies in a region where an external magnetic field (B) is present. Let α denote the angle with which B penetrates the PCB coil, as shown in Fig. 5(a). Due to this magnetic field, the flux linking the PCB coil is expressed as

$$\lambda = (NA)B\sin\alpha = kB\sin\alpha$$ (3)

where N is the number of turns of the PCB coil and A is the cross-sectional area of each turn. The factor

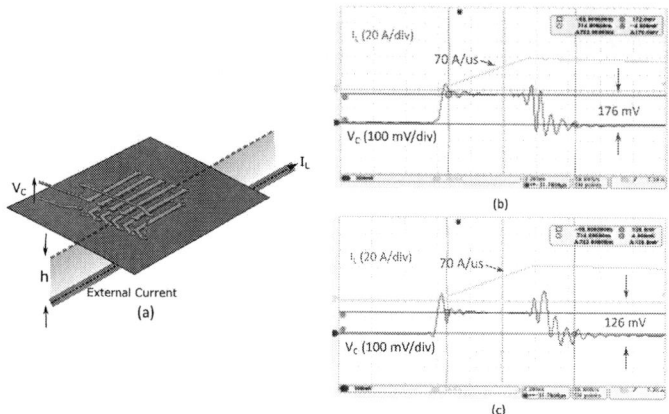

(b)

(c)

Fig. 4: (a) PCB coil in the presence of an external conductor carrying a current I_L. (b) Output voltage of coil A due to I_L. (b) Output voltage of coil B due to I_L.

$k = NA$ is, therefore, determined by the dimensions of the PCB coil. If B is time-varying, the voltage induced across the PCB coil is expressed as

$$V_c(t) = k \sin \alpha (dB/dt). \tag{4}$$

Therefore, this induced coil voltage can be a source of error in the sensor output. To estimate the error, B, as an illustration, can be considered as a homogeneous sinusoidal varying field as

$$B = B_m \sin \omega t. \tag{5}$$

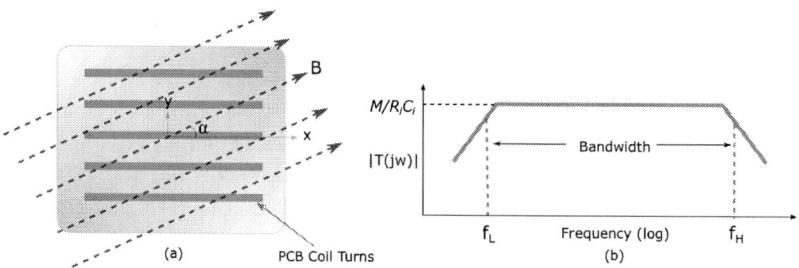

Fig. 5: (a) PCB coil in the presence of a penetrating magnetic field (B). (b) Frequency response of the PCB coil based current sensor.

If the frequency (ω) of the externally varying field (B) lies within the bandwidth of the sensor, as depicted in Fig. 5(b), the sensor output when measuring a current $I_M(t)$ is expressed as

$$V_s(t) = \frac{M I_M(t)}{R_i C_i} + \frac{k B_m \sin \omega t \sin \alpha}{R_i C_i}. \tag{6}$$

To analyze the noise immunity performance of the PCB coil based current sensor, an error can be defined in terms of the peak value of interference in the sensor output and the sensor output corresponding to the steady state value of device current ($I_{M_{SS}}$) as

$$E = \frac{kB_m/R_iC_i}{MI_{M_{SS}}/R_iC_i} = \frac{kB_m}{MI_{M_{SS}}}. \tag{7}$$

Therefore, immunity of the coil towards external fields can be enhanced by increasing M. However, increasing M by increasing the coil size will also lead to an increased k, and therefore not enhance the immunity of the coil towards external magnetic fields. Consequently, the factor k/M needs to be reduced for enhancing the noise-immunity of coils towards external magnetic fields. One approach to reduce k/M is by reducing the separation (t_1) between the PCB coil and the power trace. While t_1 can be reduced by employing a PCB with higher number of layers, clearances needed for track isolation at high voltages need to be respected. An illustration to demonstrate this concept is provided below.

Consider the case of two PCB stack-ups with four (A) and eight (B) PCB layers with different thickness, specified in Table 1 and Table 2, as provided by a PCB manufacturer. As shown in Fig. 6, if similar sized PCB coils are designed for both these stack-ups, stack-up B results in a coil with a mutual inductance of 3.5 nH between the coil and the power-trace and stack-up A results in a coil with a mutual inductance of 2.9 nH between the coil and the power-trace. Therefore, noise immunity can be enhanced by 20 % with a different stack-up configuration. Therefore, depending on the intended application, PCB stack-up can play an important factor in determining the noise immunity of PCB coils against external magnetic fields.

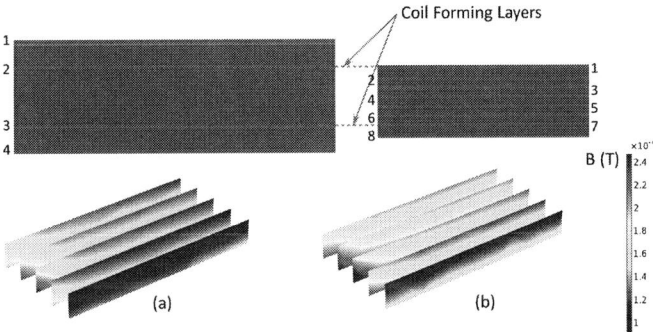

Fig. 6: PCB coil design and the magnetic field distribution through the coil due to a 1 A current flow through the power-loop for (a) PCB coil designed for a four-layered PCB of 1.6 mm thickness (Stack A) (b) PCB coil designed for an eight-layered PCB of 1.2 mm thickness (Stack B).

Finished Thickness	Tolerance	Prepreg 1-2	Core 2-3	Prepreg 3-4
1.6	±10%	0.31	0.73	0.31

Table I: Stackup thickness of a 4-Layer PCB (mm).

Finished Thickness	Tolerance	Prepreg 1-2	Core 2-3	Prepreg 3-4
1.2	±10%	0.08	0.13	0.14
	Core 4-5	Prepreg 5-6	Core 6-7	Prepreg 7-8
	0.13	0.14	0.13	0.08

Table II: Stackup thickness of an 8-Layer PCB (mm).

dv/dt coupled noise

Fig. 7 shows the lumped model of the PCB coil interfaced with a non-inverting integrator circuit [6]. Noise is injected from the switching node of the power device to the PCB coil through parasitic capacitance (C_P). Even though this capacitive coupling is distributed over the entire length of the PCB coil, the coupling is depicted with a lumped capacitance (C_P) in Fig. 7(a) for simplicity.

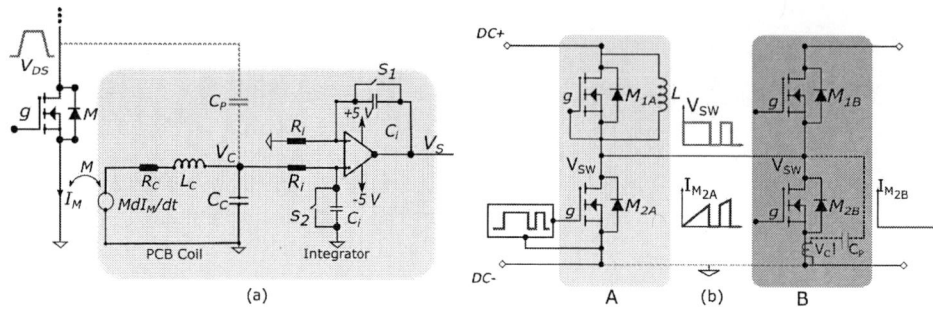

(a)

Fig. 7: (a) Coupling of the switch node with the PCB coil through a lumped parasitic capacitance C_P. (b) An experimental scheme to investigate the noise in the coil output due to parasitic coupling with the switch node.

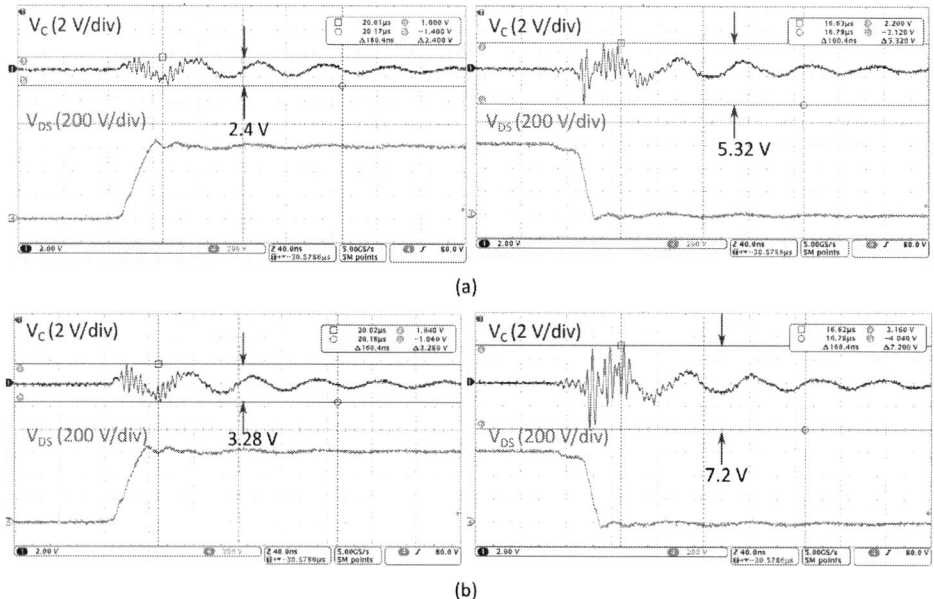

Fig. 8: Coil voltage due to signal injected by the switch node due to parasitic capacitive coupling with the PCB coil for (a) Coil B and (b) Coil A.

Due to to complexity of the parasitic capacitance owing to its distributed nature, dv/dt induced noise on the sensor output is difficult to model analytically [7]. To show the influence of coil structure on the dv/dt noise immunity of the PCB coils, an experimental scheme is shown in Fig. 7(b). The setup consists of two half bridge boards A and B. A pulse test is carried with M_{2A} on board A which causes causes high dv/dt on the switching node of the bridge. Further, the node is connected to the switching node of bare half-bridge board B. Consequently, board B only has a high dv/dt on its switching node and without any switching current through the devices. Since the only coupling the switching node will have with the PCB coils on board B are due to parasitic capacitance between the switching node and the coil, the output voltage of the PCB coil will be due to dv/dt induced noise.

The coil output voltage is measured for both the A and B coil configurations demonstrated earlier in Fig. 2. The experimental results are shown in Fig. 8. Coil A exhibits a higher coupling of the coil with the switch node traces owing to its larger size than coil B. Therefore, not only does a larger coil size increase susceptibility towards external magnetic fields, there is also an increase in the amount of noise coupled on account of parasitic coupling with the switch node traces on the PCB.

Conclusion

Noise immunity of PCB coils used for measuring fast transitioning switch current was investigated in this paper. Based on the theoretical analysis and the experimental investigations, it is inferred that the noise immunity of the coil is dependent on the coil size as well as the proximity of the coil with the current carrying power-trace. Consequently, layout of the coil on the power-board PCB is especially critical towards determining the noise immunity of the PCB pickup coils. Therefore, a proper PCB stack-up specification can be selected for adjusting the proximity of the coils with the power-trace, based on the desired noise immunity of the PCB coil. Moreover, coils with small size and high mutual inductance have better immunity to dv/dt generated noise due to lesser parasitic coupling between the switching node and the PCB coil.

References

[1] D. Gerber, T. Guillod, R. Leutwyler, and J. Biela, "Gate unit with improved short-circuit detection and turn-off capability for 4.5-kv press-pack IGBTs operated at 4-ka pulse current," *IEEE Trans. Plasma Sci.*, vol. 41, no. 10, pp. 2641–2648, Oct 2013.

[2] Y. Kuwabara, K. Wada, J. Guichon, J. Schanen, and J. Roudet, "Implementation and performance of a current sensor for a laminated bus bar," *IEEE Trans. Ind. Appl.*, vol. 54, no. 3, pp. 2579–2587, May 2018.

[3] U.-J. Kim, "Design of a rectangular pickup coil fabricated on a pcb using wbg power semiconductor in discrete package," *Applied Sciences*, vol. 11, no. 5, 2021.

[4] K. Wang, X. Yang, H. Li, L. Wang, and P. Jain, "A high-bandwidth integrated current measurement for detecting switching current of fast GaN devices," *IEEE Trans. Power Electron.*, vol. 33, no. 7, pp. 6199–6210, July 2018.

[5] A. Rafiq, S. Pramanick, and R. Maheshwari, "Design of pcb coil based high bandwidth current sensor with power-loop stray inductance characterization," *IEEE Trans. Ind. Electron.*, pp. 1–1, 2020.

[6] A. Radun, "An alternative low-cost current-sensing scheme for high-current power electronics circuits," *IEEE Trans. Ind. Electron.*, vol. 42, no. 1, pp. 78–84, Feb 1995.

[7] C. Hewson and W. Ray, "The effect of electrostatic screening of rogowski coils designed for wide-bandwidth current measurement in power electronic applications," in *2004 IEEE 35th Annual Power Electronics Specialists Conference (IEEE Cat. No.04CH37551)*, vol. 2, 2004, pp. 1143–1148 Vol.2.

Measurement Principle for Measuring High Frequency Bearing Currents in Electric Machines and Drive Systems

Benjamin Knebusch, Lennart Jünemann, Pauline Höltje, Axel Mertens and Bernd Ponick
Leibniz University Hannover
Institute of Drive Systems and Power Electronics
Welfengarten 1
30167 Hannover, Germany
Phone: +49 511-762 2408
Fax: +49 511 762 3040
Email: benjamin.knebusch@ial.uni-hannover.de
URL: https://www.ial.uni-hannover.de

Keywords

≪Bearing currents≫, ≪Component for measurements≫, ≪Impedance measurement≫

Abstract

In this paper, a new principle for measuring bearing currents will be presented and validated using a replica of a drive system. The idea is to insulate the bearing and create a new current path on a PCB for the bearing current, together with a defined impedance in that path.

Introduction

Electric motors powered through frequency converters are increasingly being used in modern drive systems. Modern frequency inverters - especially with wide-bandgap (WBG) power semiconductors - use high voltage gradients, which enable dynamic operation with both low switching losses and smaller passive components. This leads to electrical voltages across bearings and other machine elements, which can result in bearing currents. Machine elements that are subjected to critical electrical stress can be massively damaged. This can lead to the failure of the affected component, the subcomponent or the entire system.

Various studies exist regarding the analytical prediction of the electrical stress on bearings and other machine elements [1, 2, 3, 4, 5, 6]. However, for the validation of these approaches, it is necessary to compare the calculations to measurements. The challenge with this measurement is that the electrical structure of the overall system should not be changed significantly by the insertion of the measuring tools.

In this paper, a methodology for measuring the electrical current through bearings is presented and validated. After a brief introduction to the origin of bearing currents in section *High Frequency bearing currents*, the general measuring principle for bearing currents is presented in section *Explanation of the measuring principle*. The rest of this section provides a description of the structure of a measurement board. For the validation of this measurement methodology, the construction of the circuit board developed and its installation in a bearing replica are explained in section *Design of the test bench*. The results of the measurement are given in section *Results* and are compared with two other measurements.

High frequency bearing currents

WBG power semiconductors offer a number of advantages in comparison to conventional silicon power semiconductors. The use of silicon carbide (SiC) semiconductors increases the possible voltage gradient du/dt by about one order of magnitude. This reduces losses, but causes critical high frequency excitations and increases the common mode currents through parasitic capacitances.

A distinction is made between circular and electric discharge machining (EDM) bearing currents. Circular bearing currents are due to capacity current flow across the parasitic capacitance between the stator lamination and the winding. This creates a circular magnetic field in the stator yoke, which induces a voltage in the conductor loop of the rotor shaft, bearing, end shields and housing, causing a current to flow.

EDM bearing currents are the result of the galvanic separation of winding, stator and rotor. The common mode voltage U_{CM} ingressed to the stator winding excites a capacitive voltage divider across the entire drive, as shown in Figure 1. The resulting electric voltage across the bearings is calculated using the bearing voltage ratio

$$f_{BVR} = \frac{C_{WR}}{C_{WR} + C_{DE} + C_{NDE} + C_{RS}} \tag{1}$$

which results in a bearing voltage of

$$U_B = U_{CM} \cdot f_{BVR}. \tag{2}$$

A third high frequency bearing current phenomenon is called rotor-ground currents. On systems where the ground impedance between motor and inverter has its lowest path through the system attached to the shaft, the common-mode (CM) current will most likely take its path through the shaft and therefore directly through one or more bearings.

In order to measure the bearing current, a defined path for the bearing current is necessary to apply a valid measurement probe. Common practice, the bearings stationary part is insulated from the bearing shield and a wire or litz-wire is used to close the loop and attach a current sensor [7]. This type of application will force the current to take a completely different current path and will have an unknown impact on the overall current path impedance. Another method is the application of one or several rogowski coils around the shaft as shown in [8], the downside of this being the limited bandwith of the rogowski coils.

The presented system will overcome both downsides: On the one hand, the current path will stay symmetrical, and the known input impedance will help to validate simulations with measurements, and, on the other hand, the target bandwidth will be higher than 40 MHz.

Explanation of the measuring principle

This section deals with the general procedure for measuring bearing currents and with the implementation of the new concept using an integrated measuring PCB.

General Measurement of Bearing Currents

To measure a bearing current, it must flow along a defined path through a probe. To do this, all potential paths for the bearing current must be identified and then prevented by insulation. Figure 1 shows a motor with a housing, two bearings and the rotor shaft. The bearings need to be insulated by a nonconducting sleeve as the first step. It must be ensured that the additional capacitance does not - relatively to the other capacitances in the machine - become significantly large, in order to prevent an influence on f_{BVR}. However, a sufficiently thick insulation layer must be employed to prevent an insulation breakdown. With these modifications, the circuit is interrupted. The current is now conducted over a defined measuring path from the bearing outer ring to the housing. In order to be able to investigate the electrical stress of each bearing individually, it must be possible to detach or switch off the measuring path. The EDM bearing stress is determined by measuring the bearing current via the measuring path. This is performed individually for each bearing. To clearly identify the circular current in the system, the insulation of all bearings or at least two which encircle the circular flux have to be bridged. This also includes EDM bearing currents if they occur. Depending on the condition of each bearing, one or the other type of bearing current will dominate.

Design of the measuring PCB

In this section, the measuring board, which functions as a measuring path, is presented. The measuring principle is based on acquiring the voltage over a shunt resistor [9]. Using the measured voltage and a

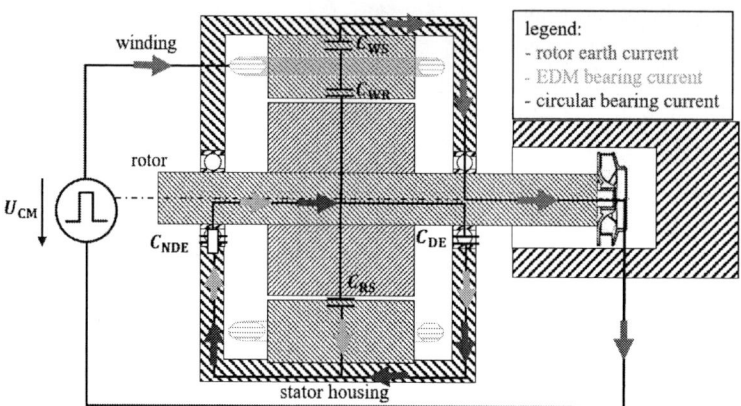

Fig. 1: Distinction of Different Kinds of Bearing Current

known impedance value, the current can be calculated. The current is picked up on the bearing outer race via contact pins. Contact probe tips from the manufacturer Ingun (GKS-913305230A2502Z) are used. The shunt is placed on the measuring board, with the electric voltage being measured at both sides of the shunt. The current is then conducted back to the housing through relays [10]. The relays are responsible for manually switching off each circuit board in order to measure the bearings in the drive individually.

Care must be taken to ensure that all components are distributed as evenly as possible radially around the circumference to ensure an equipotential surface and thus reduce measurement inaccuracies. This direct measurement concept at the bearing offers the advantage that the current does not have to be conducted to the outside via a high impedance path, which often leads to severely compromised results. Furthermore, the effort of rebuilding the machine is also much lower. In [11], a large part of the machine is milled off to get to the bearing, which also leads to a modification in the system. The novel measuring board is barely larger than the bearing itself and takes up only a few millimeters of axial length in the machine. With this concept, even in machines close to series production, only minor changes are necessary for the implementation of the measuring PCB. Even if it is possible to use this methode in close to series production drives, a measurement of bearing currents in a series drive is not meaningful. In addition, the measurement methodology is much less expensive than measuring with high-precision probes. It can also be used without the probe tips directly mounted between two isolated surfaces in a machine.

Selection of components

Due to the high frequencies of the bearing currents, it is indispensable to produce a structure that is as clean as possible. Therefore it is important to select the components accordingly. However, there must be a trade-off between high frequency capability and component size, since there is not much installation space for additional components in most e-machines. The shunt also needs to be able to withstand a certain load, since the bearing current can reach several amperes.

The shunt selected here is a $40.2\,\Omega$ high frequency thin film chip resistor in a 0402 package from Vishay [9]. This exhibits purely resistive behavior into the gigahertz range. There are 60 resistors evenly connected in parallel around the circumference. Thus, the effective resistance is $670\,\text{m}\Omega$. It is necessary to achieve a low resistance value, so that there is no significant impact on the value of the bearing current.

To pick up the bearing current from the bearing outer race, pins contact the outer race as evenly as possible. The contact pins are made of beryllium copper with a gold coating. Again, the focus here is on the current carrying capability and a low impedance that is as independent of frequency as possible. Depending on the length, the contact pins have an internal resistance of $100\,\text{m}\Omega$. The relays are Panasonic AQV255GS [10]. These relays, switching via optocouplers, have a small size and a low parasitic capacitance. A parallel connection ensures an even current distribution. U.FL connectors are used to control the relays and to measure the voltage across the shunt. These also have a low physical volume.

Measurement Methodology

The voltage right before and after the shunt is tapped relative to a separate measuring ground. This offers the advantage that the measurement is independent of the housing's local potential, which normally acts as the measuring ground. However, due to currents in the housing, a potential shift can occur, which compromises the measurement results.

The control for the relays is also located on a separate PCB layer to keep the control as decoupled as possible from the measurement to avoid electromagnetic interference. Figure 2 shows the top layer of the PCB. On this layer, the contact pins pick up the current from the bearing. These are mounted in retaining bushes. Furthermore, the relays are visible on this side. The shunts and U.FL connections are located on the bottom layer.

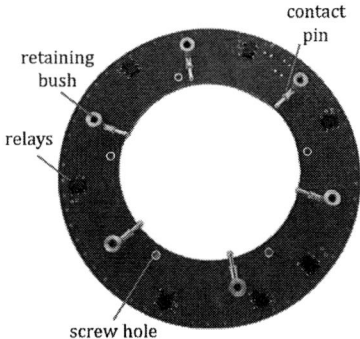

Fig. 2: Measuring circuit board

Design of the test bench

This section deals with the design of the test bench. First, the replica of the e-machine is explained. Subsequently, further components for carrying out the test are presented.

Replica of the e-machine

In order to validate the measurement concept, the most important components of the an e-machine are reproduced to simulate the assembly and installation work. The system setup for this purpose is shown in Figure 3. The replica consists of a shaft onto which a 6008 ball bearing is pressed. As described above, this must be isolated from the rest of the machine, so that an insulating sleeve is fitted over the bearing outer race. This is followed by the housing, which has recesses in which the contact pins of the axially mounted PCB are located.

To validate the measurements, a passive current sensor is integrated around the shaft to determine the current flowing in various subsequent measurements. This is completed by the end shield and a BNC connector through which the current is supplied to the e-machine replica.

The shaft is also isolated from the housing. So the resulting current path begins on the signal line of the BNC connector, flows through the shaft, through the bearing to the PCB and through the screw back to the ground line of the BNC connector on the housing. This path is now very simular to the path of the real bearing current in an e-machine.

Experimental procedure

The test bench is shown in Figure 4. The test bench consists of the bearing replica, the control system belonging to the measuring board - consisting of microcontroller and control board - a power supply, a signal generator, three different measuring sensors and two oscilloscopes. The second oscilloscope, a Keysight Technologies InfiniiVision MSO7104A [12], is used to display the signal from the high frequency sensor. The voltage source is used to switch on the relays by supplying a voltage to the optocoupler. The circuit is then enabled by the microcontroller which connects the power circuit through

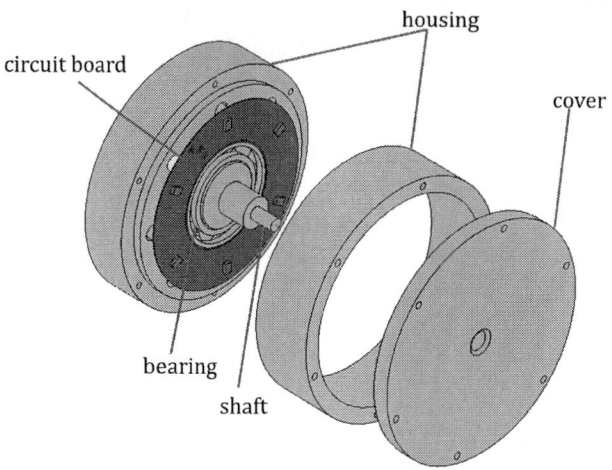

Fig. 3: Assembly of the e-machine replica

the MOSFET. The signal generator - AimTTi TGA12100 - is directly connected to the BNC terminal of the replica. A sinusoidal voltage with an amplitude of 10 V at various frequencies up to a maximum of 40 MHz is applied.

The first current measurement is performed on the shaft of the replica with a high precision HF Current Transformer CT-D1.0-B from AMS technologies (passive sensor). There is a small axial preload on the bearing to ensure that the bearing is electrically conductive at standstill, the voltage drops across shaft, bearing, contact pins, PCB and finally the screw connection back to the housing.

The high frequency differential voltage probe 1134A 7 GHz InfiniiMax from Keysight Technologies measures the voltage difference across the shunts on the PCB. In addition, two U.FL connectors measure the electrical potential before and behind the shunts relative to a defined measurement ground on the other side of the board. The cables of the three sensors are fed through a hole in the housing to the oscilloscope, which is a Teledyne LeCroy HDO8108A. This determines the voltage difference between the two U.FL connectors. The second oscilloscope - which is pictured in the figure - serves to display the HF sensor.

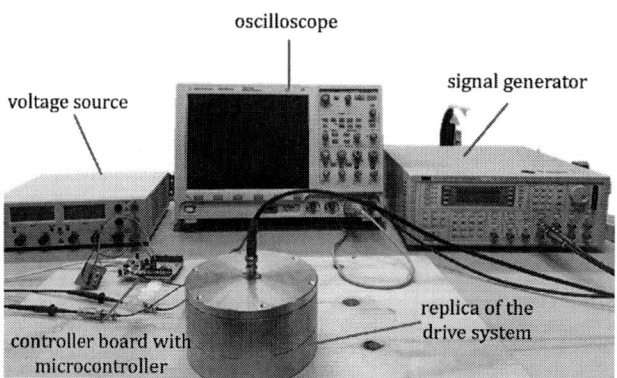

Fig. 4: Test bench

Results

In this section, the results of the impedance measurement are presented first, and these can be used to determine the currents corresponding to the measured voltages. Then, the results of the current measurements at different frequencies are shown.

Measurement of the impedance

The impedance of the measurement board was recorded with a Wayne Kerr 65120B. Since only U.FL connections are integrated onto the board, an adapter cable from U.FL to BNC connection was also used; therefore, the impedance of the cable was determined in a second measurement. Thus, the results of the board measurement including the adapter cable could be corrected for the influences of the cable. In Figure 5, the measured resistance of the board is shown against the frequency. Up to a frequency of 20 MHz, the resistance of the board is constant at 690 mΩ. Beyond this point, the resistance increases with increasing frequency. At a frequency of 96 MHz, the resistance is 2.6 Ω. Above this frequency, the resistance decreases with increasing frequency due to parasitic capacitive couplings. This drop is most probably to be due to resonance in the cable.

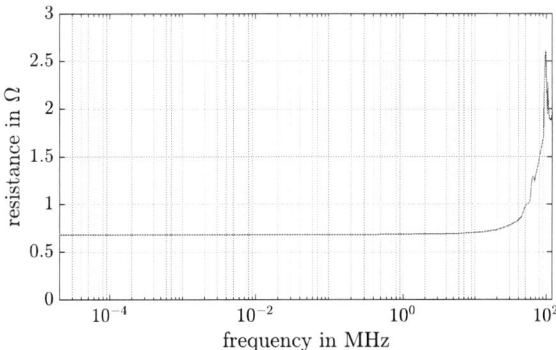

Fig. 5: Frequency-dependent shunt resistance

For the validation of the system and measuring at constant frequency, the impedance curve can be interpolated and the corresponding impedance can be used to calculate the bearing current from the measured voltages using

$$i_{\text{bearing}} = \frac{u_{\text{shunt}}}{|\underline{Z}_{\text{shunt}}(f)|}. \tag{3}$$

When measuring transient currents in real drive systems a post processing is necessary. Therefore the measured shunt voltage is transformed in the frequency domain

$$u(t) \xleftrightarrow{\mathcal{F}} \underline{U}(f). \tag{4}$$

The impedance curve is also interpolated to the sampling points of the FFT in magnitude and phase, and the corresponding current components for each frequency are calculated using

$$\underline{I}_{\text{bearing}}(f) = \frac{\underline{U}_{\text{shunt}}(f)}{\underline{Z}_{\text{shunt}}(f)}. \tag{5}$$

The result is then back-transformed into time domain

$$\underline{I}_{\text{bearing}}(f) \xleftrightarrow{\mathcal{F}^{-1}} i_{\text{bearing}}(t) \tag{6}$$

to get the corresponding transient signal.

Current measurement

The results of the three different measurement methods with the passive current sensor (current sensor), the HF voltage differential probe (HF sensor) and the differential measurement using the U.FL connectors at shunts (shunt) will now be compared to each other. To determine the currents, the measured voltage differences from the HF sensor and the U.FL connectors are divided by the frequency-dependent board resistance. Figure 6 shows the measured currents as a function of time for a sinusoidal voltage with a frequency of 1 MHz, provided by the signal generator. All three measurement methods identify the

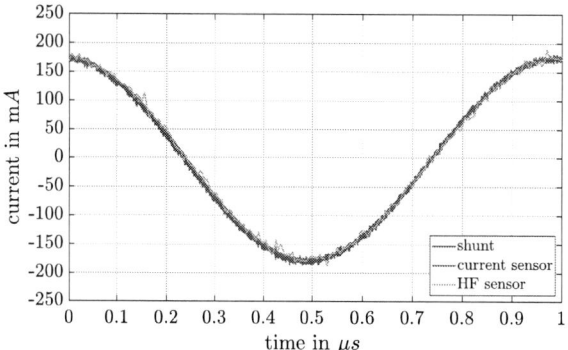

Fig. 6: Measured currents at a frequency of 1 MHz

same current with an amplitude of approximately 170 mA. Even at a frequency of 10 MHz, as shown in Figure 7, the currents measured with the three methodologies correspond well to each other. Some deviations exist between the three curves, which can be traced back to parasitic effects. The amplitudes are slightly different, because of different voltage supplies.

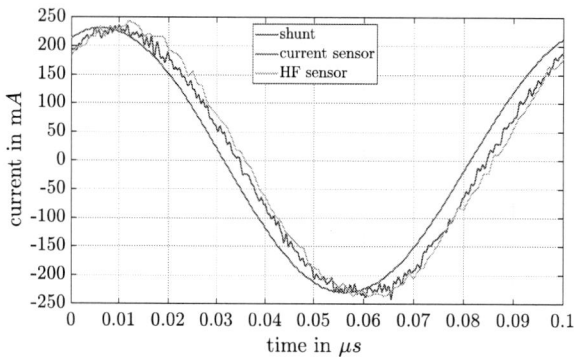

Fig. 7: Measured currents at a frequency of 10 MHz

Figure 8 compares the results for a supply with a frequency of 40 MHz. The current sensor on the shaft measures the highest current with an amplitude of almost 197 mA. In comparison, the shunt measurement records an amplitude of 186 mA and the HF sensor only 151 mA. The current sensor measures the total current flowing through the shaft, but the shunts and the HF sensor only capture the current on the measuring PCB.

Due to parasitic capacitances within the experimental setup, it can be assumed that not the total current flows across the PCB, as some of the current is diverted towards the housing. The differences between the shunt and the HF sensor measurements are due to the different connecting leads. In order to connect the HF sensor to the PCB, short pieces of wire were soldered on which the sensor sticks to. Even if these connecting wires are kept as short as possible, they create a conductor loop which slightly distorts the measurement.

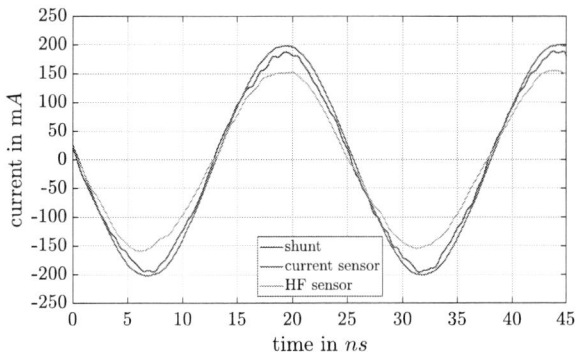

Fig. 8: Measured currents at a frequency of 40 MHz

Conclusion

In this paper, a novel PCB for measuring bearing currents, intended to allow measuring bearing currents in real drive systems with much lower modification effort compared to the state of art, is presented and validated with the help of a simplified example measurement. For the measurement, it is necessary to electrically isolate the bearing from the housing and only allow a current flow along a defined path via a measuring PCB. On the measuring PCB, the current is determined from the voltage drop across shunts. The measuring PCB can be switched on and off using relays. The measurement principle designed here was tested on a simplified e-machine replica and validated against two other measurement methods for frequencies up to 40 MHz. In a future study, the principle will also be tested at higher frequencies. Furthermore, the PCB will be integrated and tested in a real e-machine.

Acknowledgement

The basic concept for measuring high-frequency bearing currents with circumferentially arranged shunt resistors was proposed by Urs Obernolte of Lenze SE and is part of the project UmSiCht, funded by the Federal Ministry for Education and Research on the basis of a decision by the German Bundestag. Funding number: 16EMO0252.

References

[1] B. J. Hamrock, D. Dowson, and T. Tallian, *Ball bearing lubrication: The elastohydrodynamics of elliptical contacts.* NASA National Aeronautics and Space Administration, 1982.

[2] A. Furtmann, H. Tischmacher, and G. Poll, "Extended hf equivalent model of a drive train," in *2016 XXII International Conference on Electrical Machines (ICEM).* IEEE, 2016, pp. 2244–2250.

[3] E. Wittek, M. Kriese, H. Tischmacher, S. Gattermann, B. Ponick, and G. Poll, "Capacitances and lubricant film thicknesses of motor bearings under different operating conditions," in *The XIX International Conference on Electrical Machines-ICEM 2010.* IEEE, 2010, pp. 1–6.

[4] Y. Gemeinder, M. Schuster, B. Radnai, B. Sauer, and A. Binder, "Calculation and validation of a bearing impedance model for ball bearings and the influence on edm-currents," in *2014 International Conference on Electrical Machines (ICEM).* IEEE, 2014, pp. 1804–1810.

[5] M. Schuster and A. Binder, "Comparison of different inverter-fed ac motor types regarding common-mode bearing currents," in *2015 IEEE Energy Conversion Congress and Exposition (ECCE).* IEEE, 2015, pp. 2762–2768.

[6] H. Tischmacher, *Systemanalysen zur elektrischen Belastung von Wälzlagern bei umrichtergespeisten Elektromotoren.* Hannover: Gottfried Wilhelm Leibniz Universität Hannover, 2017.

[7] A. Muetze and A. Binder, "Techniques for measurement of parameters related to inverter-induced bearing currents," *IEEE transactions on industry applications*, vol. 43, no. 5, pp. 1274–1283, 2007.

[8] A. E. B. P. D. Heide, B. Knebusch, , "Measurement of Parasitic High-Frequency Currents in Inverter-Fed Low-Speed Electrical Machines Using Rogowski Current Sensors.* 23rd European Conference on Power Electronics and Applications (EPE'21 ECCE Europe), 2021.

[9] *High Frequency 50 GHz Thin Film Chip Resistor*, Vishay, 2020, revised 10.2021.

[10] *PhotoMOS - Miniature SOP6-pin type with high capacity of 3A load current*, Panasonic Industry, 2020, revised 10.2021.

[11] A. M. Bubert, *Optimierung des elektrischen Antriebsstrangs von Elektrofahrzeugen mit Betrachtung parasitärer Ströme innerhalb der elektrischen Maschine.* Aachen: RWTH Aachen Universität, Aachener Beiträge der ISEA, Band 141, 2020.

[12] *Infiniium Oscilloscope Probes and Accessories*, Keysight Technlogies, 2021, revised 10.2021.

Climatically Induced Insulation Degradation in Power Semiconductor Modules of Wind Turbines

Timo Lichtenstein[1], Sören Fröhling[1,2], Bernd Tegtmeier[1], and Katharina Fischer[1]

[1]Fraunhofer Institute	[2]Leibniz University Hannover
for Wind Energy Systems IWES	Institute for Drive Systems and Power
Postkamp 12	Electronics, Welfengarten 1
30159 Hannover, Germany	30167 Hannover, Germany
Email: timo.lichtenstein@iwes.fraunhofer.de	Email: soeren.froehling@ial.uni-hannover.de
URL: https://www.iwes.fraunhofer.de	URL: https://www.ial.uni-hannover.de

Acknowledgments

This work is part of the project ReCoWind and was funded by the Federal Ministry for Economic Affairs and Climate Action on the basis of a decision by the German Bundestag, funding number: 0324336A. We would like to thank Bender GmbH & Co. KG for the provision of the measuring instrumentation and the climate chamber.

Keywords

≪Lifetime≫, ≪Condition Monitoring≫, ≪Degradation≫, ≪Humidity≫, ≪Wind energy≫

Abstract

Power converters are among the most frequently failing subsystems in wind turbines. Humidity-induced degradation plays a key role, especially in locations with geographically high absolute humidity. In a laboratory experiment, a failure of a power module induced by humidity and condensation could be replicated under extreme climatic conditions.

Introduction

Power converters in wind turbines (WT) stand out with high failure rates [1]. The SPARTA initiative reported an average of 1.32 repairs per turbine and year from the converter subsystems of 1045 offshore wind turbines in the waters of the UK [2]. A study of Fraunhofer IWES based on a worldwide fleet of 2734 onshore and offshore turbines revealed an annual average of 0.48 converter system failures per turbine, with a third of it associated with the core components including the IGBT semiconductor modules, their gate-driver units, DC link capacitors, and busbars [3]. In view of the fact that the frequent and typically unexpected failures cause substantial repair costs and downtime-related revenue losses, enhancing converter reliability is an important task to further reduce the cost of wind energy.

Previous research has shown that environmental influences, in particular humidity, play a key role in the emergence of converter failures while well-understood thermo-mechanical fatigue is relegated to the background as a life-limiting mechanism in wind turbines [4, 5]. Failures are not limited to power semiconductors but spread over the whole converter subsystem. Among the issues reported from the field, degradation of the insulation foils between the DC terminals of power modules as well as in the area of DC busbars or laminates under the influence of moisture or condensation is a repeatedly observed phenomenon: The insulating materials degrade until the insulation strength drops below a critical limit and short-circuit failure occurs between DC+ and DC− [4]. In addition to improvements in the design and protection of the converter hardware, condition monitoring is an important measure to prevent impending failures and reduce downtime. In order to be effective, it must cover the failure modes and mechanisms prevailing in the field. As a consequence, condition monitoring approaches should be capable of detecting not only faults developing at the semiconductor chip or packaging level, but also insulation degradation in the converter system.

The investigations of this paper serve as a first step toward potential monitoring approaches of these insulation properties. A power module from a wind turbine is examined in a simplified electrical load setup in a climatic chamber: The test specimen is subjected to a typical DC link voltage level used in low-voltage converters in the field but without a continuous pulse width modulation (PWM). The applied climate profile is based on field-measurement data from India and replicates the highly fluctuating climatic conditions to which power converters in wind turbines are exposed [5].

Measurement Configuration

The following chapter describes the experimental configuration and discusses the selection of the operating parameters. Important aspects of the test setup are detailed here and the operational control is presented.

Experimental Setup

The test setup as shown in Fig. 1 consists of two devices under test (DUT). The first module is a Semikron SKiiP 513GB 172CT (DUT 1) consisting of a total of four individual half-bridges of which three are connected on the AC side in this test setup. The module has an integrated gate driver unit and a heat sink for liquid cooling that was not used during the test. Snubber capacitors are mounted directly to the DC terminals. The second DUT is a submodule (DUT 2) from a Semikron SKiiP 603GB 123CT but without the DCB (direct copper bond) substrate and, thus, without power semiconductor components. Also, the gate driver unit is removed and no snubber capacitors are mounted in this case. As the two DUTs are from different module generations, the materials of their insulation foils used at the DC terminals differ. Though the material thickness is identical with both types measuring 0.25 mm, differences can be identified mainly in color: the foil of DUT 2 is transparent, whereas on DUT 1 it has an additional orange tint.

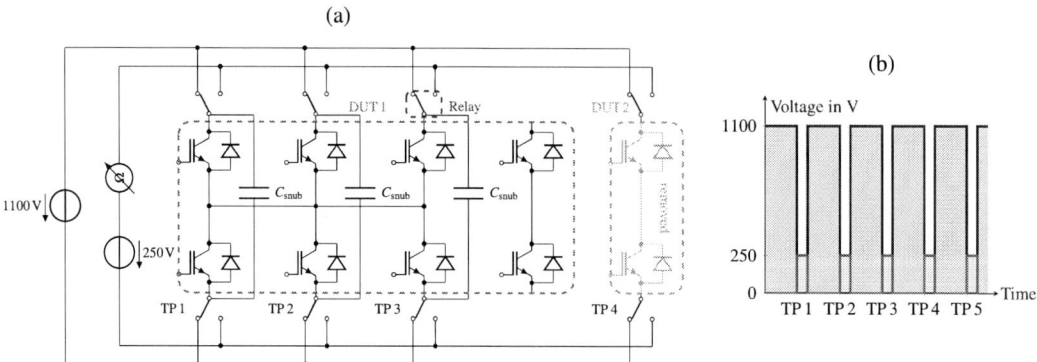

Fig. 1: Schematics of the experimental setup: Simplified circuit diagram of the test setup consisting of a high-voltage source (a): an insulation resistance meter with a laboratory source for the offset voltage, the two DUTs, and the relays for switching the test configuration (TP 1 to TP 4). Schematic of the applied voltage, with cycling between static load states and the individual test points (b): Here, TP 5 is a parallel connection of TP 1-3.

A high voltage source is used to apply a voltage of 1.1 kV to the DUTs, which corresponds to the DC link voltage commonly used in low voltage wind turbine converters. The switch-on time of the high voltage is about 150 ms, which means that a significant influence of overvoltage peaks due to parasitic inductances in the setup can be excluded here. After 15 min in load state, the setup automatically switches to measurement state and measures one of the test points (TP) for 6 min after which it returns to load state, see the right part of Fig. 1 for a schematic. To be able to carry out the corresponding insulation resistance measurements, the DC link voltage is switched off during the measuring process and the voltage source is disconnected by contactors. The individual measuring sections are switched to the measuring circuit via relays. TP 1 to TP 3 are associated with the respective half-bridges in DUT 1, see Fig. 1, TP 4 is the submodule without semiconductors (DUT 2), and TP 5 switches to a parallel connection of the three half-bridges TP 1-3 (DUT 1). During measurement, the insulation resistance meter is connected in series with a 250 V voltage source to force the diodes into a reverse-biased operating point. This is necessary to match the measurement procedure of the insulation resistance meter to the requirements of the experiment. The IGBTs remain switched off throughout the whole experiment.

Climatic Conditions during the Experiment

With the aim of exposing the power modules to extreme climatic stress, we have chosen parameters based on field data from India [5]. Fig. 2 shows the temperature and humidity conditions measured inside the cabinet air of the power converter with a temporal resolution of 1 min. The 10-day excerpt of the time series in the left part of the figure illustrates the pronounced day-night cycles to which power converters in wind turbines are typically exposed, cf. [5]. The climatogram in the right part of the figure summarizes the climatic conditions encountered during the 2-year measurement campaign and indicates the climatic conditions selected for the present experiment.

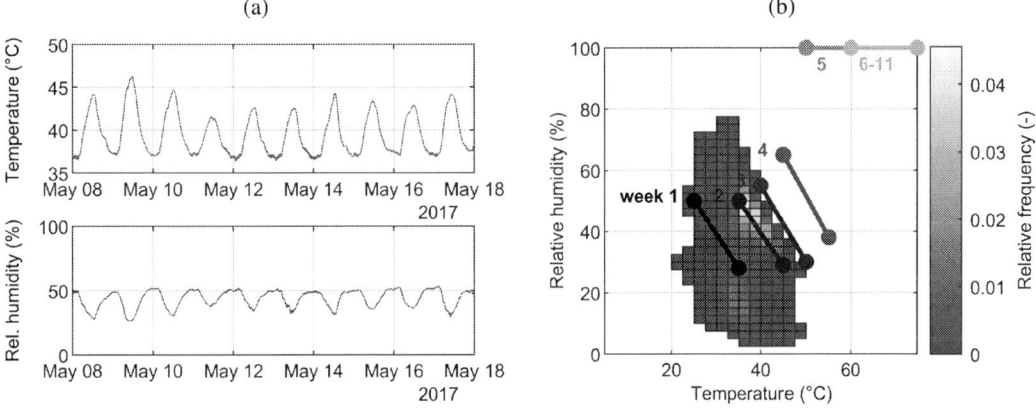

Fig. 2: Climatic conditions measured inside the converter cabinet of a WT in India; time series excerpt with typical day-night cycles of temperature and relative humidity (a), climatogram based on the entire 2-year measurement campaign with climatic conditions chosen for the experiment (b).

The test can be divided into two parts: During the so-called pre-stress phase of the first four weeks, a weekly increase of humidity and temperature is carried out. Within each week, day-night cycles occurring in the field are replicated in a simplified manner by toggling every 12 h between a state with higher temperature and lower relative humidity and a state with lower temperature and higher humidity. As Fig. 2 indicates, the climatic conditions applied during the first three weeks of the experiment correspond to conditions that are representative of those inside the considered wind turbine: The profile applied in week 1 replicates a level of absolute humidity encountered during spring and autumn; week 2 is typical of the Indian summer, while week 3 represents the strongest climatic stress exposure of converter components possible during summer. From week 4, the climatic stress is increased beyond conditions measured in the field. At the end of each week, an additional 12 h cycle with a temperature of 35 °C and 20 % relative humidity is conducted to allow for drying of the DUTs. After four weeks, i.e. 28 days, the stress phase begins: The DUT is subjected to extreme climatic stress by setting the humidity to 100 % and further increasing the level of the temperature cycles. First, the temperature is toggled between 50 and 60 °C, thereafter between 60 and 75 °C. The drying cycles at the end of every week are also implemented in the stress phase.

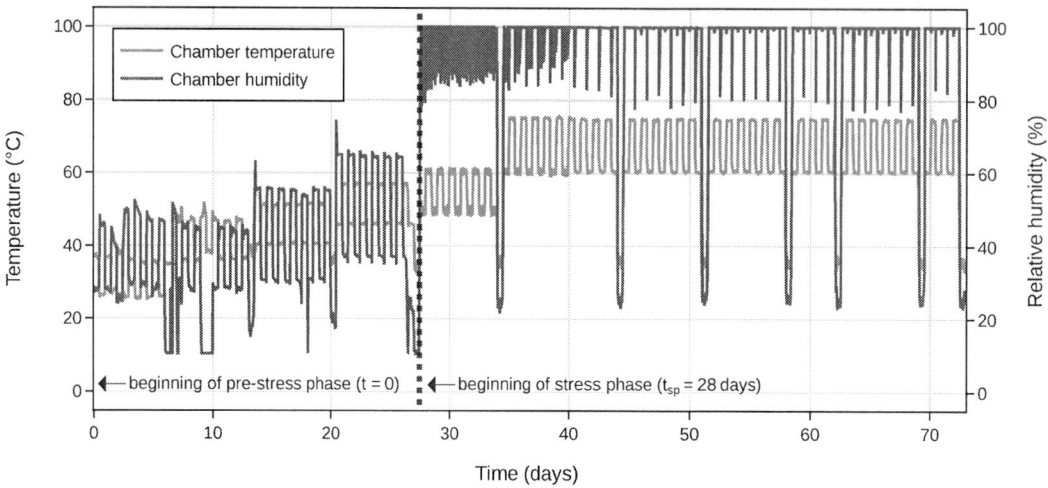

Fig. 3: Measured temperature and humidity inside the climate chamber over time. The dotted line indicates the transition between the two phases of the experiment.

It is important to note that the climatic conditions of the stress phase in the second part of the experiment include condensation. These conditions are far outside the specification of the power modules. While the level of absolute humidity used to enforce the precipitation of liquid water is excessive, the occurrence of condensation in converters of wind turbines in the field is a matter of fact. As the DUTs are subjected to voltage only and no continuous PWM operation with load currents is carried out, there are only negligible leakage currents through the semiconductors such that the modules experience no considerable self-heating. Therefore, the temperatures of the modules follow the climatic conditions inside the chamber with a time delay caused by their thermal capacity. The climatic conditions inside the test chamber during our experiment are shown in Fig. 3.

While the levels of the applied climatic cycles are intended to remain identical during each week of the test period, Fig. 3 shows that the climatic conditions recorded inside the chamber during the experiment deviate from the intended time series to some extent: During the pre-stress phase, the humidity control failed several times, causing an unplanned reduction to 10 % relative humidity. Furthermore, at the beginning of the stress phase, the climate chamber had issues regulating to 100 % relative humidity, identifiable by the noisy data between days 28 and 35. Additionally, due to a failure incident—see results—on TP 4 after 61 days we had to stop the chamber and restart it, beginning with a drying cycle. For that reason, there is a period shorter than 7 days between two drying cycles between days 58 and 62.

Results and Discussion

In the following chapter, we present and discuss the measured resistances, a post-mortem analysis of a failed module, and an investigation of parasitic conductive paths at the power modules.

Measurement Results

The recorded resistance data for the four test points (TP 1, TP 2, TP 3, and TP 4) and the parallel connection of TP 1-3 (TP 5) over the course of the experiment are displayed in Fig. 4.

During the pre-stress phase, a 12 h toggle of the resistance can be observed for TP 1-3. The mean value of this toggle decreases from 300 MΩ to 70 MΩ with increasing temperature. Due to the parallel connection of TP 1-3, a lower resistance value of $\approx \frac{1}{3}$ of that of the single TPs is expected for TP 5. The measured resistance time series decreasing from 100 MΩ to 25 MΩ agrees with this expectation. All variations observed in this phase show a typical strong temperature dependence of the semiconductor components

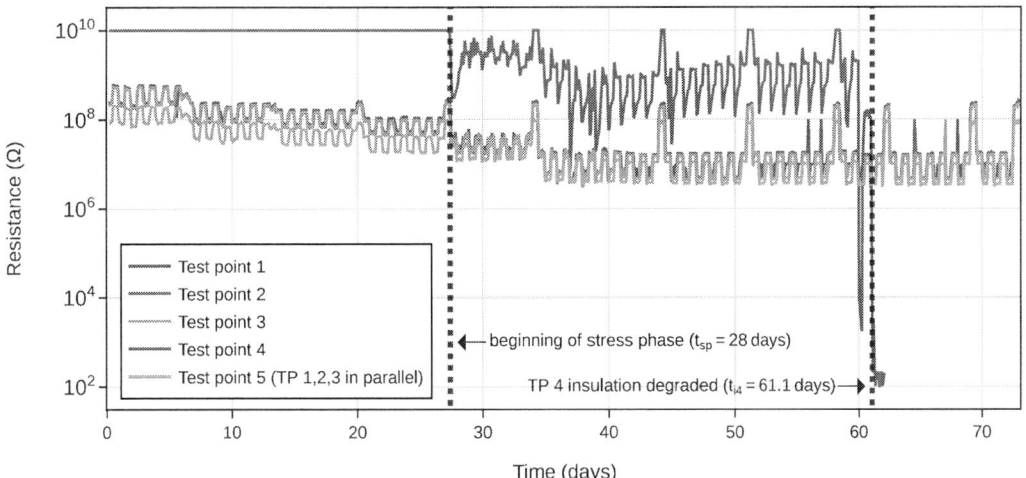

Fig. 4: Raw measured resistances from the resistance meter for each test point.

and are, therefore, attributed to this behavior, see also Eq. (1) in the chapter after next. Furthermore, during the drying cycles the resistance always stays at $250\,\mathrm{M\Omega}$ ($90\,\mathrm{M\Omega}$ for TP 5), thus, we see no signs of irreversible degradation in the pre-stress phase. No change in the resistance is observed for TP 4 as its insulation resistance remains above the detection limit of $10\,\mathrm{G\Omega}$.

With the beginning of the climatic stress phase at $t_{\mathrm{sp}} = 28\,\mathrm{d}$ and its harsh ambient conditions, a further decrease of the toggling behavior of the resistances at TP 1-3 and a significant drop in the insulation resistance of TP 4 is observed. After this initial drop, a $12\,\mathrm{h}$ toggle of the resistance is also visible for TP 4, although not as regular as for TP 1-3 or TP 5. Since the humidity is constant at $\approx 100\,\%$ in this phase, changes in the insulation resistance are mainly attributed to a temperature dependence of the power semiconductor path, dominating at TP 1-3 and 5, as well as condensation, dominating at TP 4.

Compared to the pre-stress phase, the slopes of the resistances of TP 1-3 keep their regular toggle also in the stress phase, as expected by a semiconductor's temperature dependence. We observe values of around $100\,\mathrm{M\Omega}$ in the first week and $10\,\mathrm{M\Omega}$ thereafter. During the drying cycles, the resistances at TP 1-3 always recover to values around $250\,\mathrm{M\Omega}$, i.e. no evidence of an irreversible degradation of the insulation in DUT 1 during the time of the experiment is found in the data.

On TP 4 during the first week of the stress test, i.e. $t < 35\,\mathrm{d}$, the mean value remains around $2.5\,\mathrm{G\Omega}$, after half of the second week it reaches approximately $350\,\mathrm{M\Omega}$. The drying periods also show a recovery of the previously reduced insulation resistance. As the recorded data exceeds the measuring range of the insulation resistance meter, it remains unclear if during these periods the insulation recovers to its original value at the start of the measurement campaign.

After 60 days, during ambient conditions of a temperature of $75\,°\mathrm{C}$ and $100\,\%$ relative humidity, the resistance at TP 4 suddenly drops to a value of merely $1.7\,\mathrm{k\Omega}$. It recovers to $150\,\mathrm{M\Omega}$ during the subsequent cycle with a lower temperature of $60\,°\mathrm{C}$, but breaks down completely to the measuring range limit of $100\,\Omega$ during the next cycle on day 61 showing the collapse of the insulation of TP 4. This event is highlighted by a dashed line in Fig. 4. The incident's timestamp is denoted by $t_{\mathrm{i}4}$. At this point in our experiment, a low resistance path has formed triggering one of the protection measures of the setup: The overcurrent protection (OCP) of the voltage source intervenes, not allowing to apply high voltages during the load part. Due to this feature, the affected insulating area at the DC terminals is not completely destroyed yet at $t_{\mathrm{i}4}$. This also means that the plotted data does not show the final resistance.

To investigate the condition of the insulation, we reduced the target voltage to determine the residual voltage strength of the section. After approximately $200\,\mathrm{V}$ of residual voltage strength, the OCP shut down the power supply. This value could not be clearly reproduced after a recurrent break-in. Another attempt to set the voltage source back into operation was made by resetting the OCP and subsequently ramping up the DC link voltage, albeit, without success. A fault in the measuring system could be ruled out. After several endeavors we recognized the complete destruction of the insulating foil, also displayed in Fig. 6 (b), that we wanted to enforce within the experiment with its harsh climatic condition. After this discovery, the connections to TP 4 were detached from the setup to be able to continue the experiment under voltage load at the remaining test points.

Post-Mortem Analysis

After the removal of the damaged submodule from the test bench, significant, visually recognizable damage can be observed, see also Fig. 5 and Fig. 6. Residues from repetitive water condensation are found on the surface of the entire submodule. Metallic surfaces have undergone corrosive processes, particularly visible on screw connections and load terminals. The insulation foil which is folded between the DC terminals of the module has lost its insulation properties due to partial disintegration, see Fig. 6 (b). It is likely that the presence of condensed water in combination with the voltage between the terminals has driven the degradation process. We suspect that the low resistance path that formed due to the degradation of the film resulted in an increased temperature at the spot and that the temperature impact finally became severe enough for carbonization to occur.

Fig. 5: Close-up image of the damaged insulation foil between the DC contacts.

The disintegrated area on the foil is located at the folding zone of the foil between the two busbars. There is also massive damage to the housing of the power module in the form of melted plastic and black stain at this point, see Fig. 6 (a).

Apart from the mentioned damage in the vicinity of the DC connectors and stain on the housing, the tin-plated busbars inside the module remain in relatively speckless condition. Humidity-induced degradation due to moisture penetration into the silicone of the module is discussed in the literature [6, 7]. In the present case, however, also no change in structure or color of the insulating foil are visible on the inside of the submodule. Thus, we assume that there has only been faint, if any, condensation or capillary motion of water to this part of the module.

Apart from traces of corrosion and deposits, the other four half-bridge modules—with different insulating foil—show no externally visible damage, which is consistent with the measurement results.

Investigation of Parasitic Conductivity

As shown in the post-mortem analysis, the insulation foil at TP 4 has been destroyed during the experiment. As the module DUT 2 at TP 4 contains no DCB with semiconductors, its changes in resistance must be solely attributed to the insulation foil between the DC contacts. Consequently, we also analyzed the changes in the resistances at TP 1-3 as well as TP 5 (the parallel connection of TP 1-3) for a

(a)   (b)

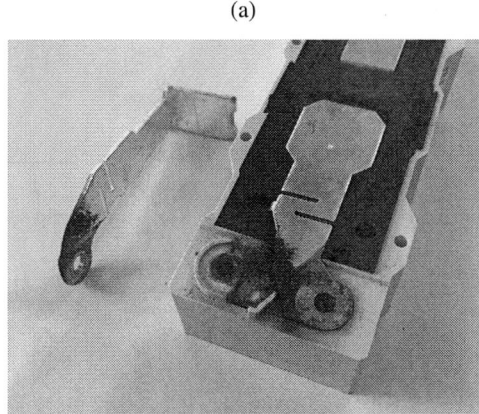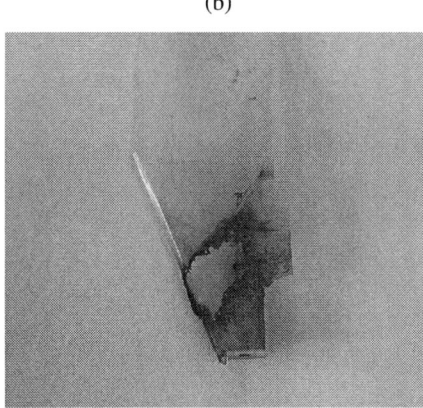

Fig. 6: Close-up image of the defective area with the insulation foil removed and the upper part of the housing disassembled (a). Damaged insulation foil (b).

possible emerging parasitic conductive path. Due to the nature of semiconductors, the resistance in the blocking state of the DUT itself is highly dependent on the semiconductor temperature. In the present case without a load, the DCB temperature signal available from the gate driver unit recorded during the experiment provides a good approximation of the semiconductor temperature and is, therefore, used for the following analysis.

Assuming only minor changes in resistance caused by parasitic paths during the pre-stress phase, data recorded within the first 28 days of our experiment are used to fit an empirical semiconductor resistance model:

$$R(T) = R_0 \cdot \exp\left(\frac{E}{2k_\mathrm{B}T}\right) \tag{1}$$

This model comprises only two fit parameters R_0 and E, whereas k_B is the Boltzmann constant. This simple model is used under the assumption that all conductive paths of the DUT that are connected in parallel to our measurement device undergo a similar temperature dependence, i.e. their band gaps and induced charge carrier densities closely resemble each other. As the main resistance should be dominated by the properties of the IGBTs in their blocking state, we see this as a valid approach. Furthermore, it provides a straightforward solution to model an otherwise complex system.

The parameters are estimated separately for each test point based on the corresponding data set using a Levenberg-Marquardt algorithm. The resulting regression functions are shown as dotted curves on top of the underlying data in Fig. 7 (a), the corresponding parameter values are listed in Table I.

Table I: Resulting optimized fit parameters of the empiric semiconductor resistance model (1) with standard deviation directly obtained from the optimizer used as uncertainty.

Test point	R_0 (mΩ)	E (eV)
1	$0.67 \pm 14\%$	$1.40 \pm 0.5\%$
2	$0.28 \pm 7\%$	$1.44 \pm 0.25\%$
3	$0.38 \pm 7\%$	$1.43 \pm 0.27\%$
5	$0.20 \pm 12\%$	$1.41 \pm 0.5\%$

Fig. 7 (b) exemplarily summarizes the procedure for TP 1 on a logarithmic scale with both the underlying data of the pre-stress phase $t < t_\mathrm{sp}$ as purple circles as well as the data of the stress phase $t \geq t_\mathrm{sp}$ as red crosses. The regression function is depicted by the dashed black line accompanied by thin purple dashed lines that show the extent of the algorithm's uncertainty region. Although this uncertainty region is slightly smaller for the data of the other test points, the overall behaviors strongly resemble each other.

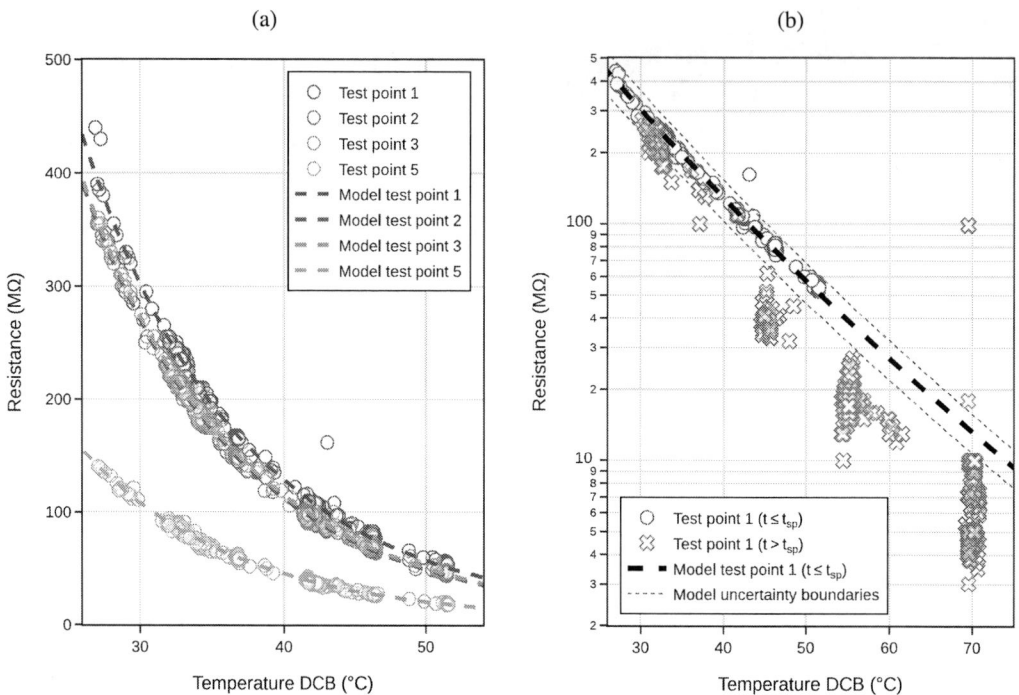

Fig. 7: Visualizations of model (1) to the data of $t < t_{sp}$ (a). Detailed logarithmic plot of the regression function with uncertainty region resulting from the standard deviations of the parameters and corresponding data $t < t_{sp}$ in purple circles as well as data of stress phase $t \geq t_{sp}$ for the calculation of the parasitic resistance in (2) in red crosses (b).

As the next step, all measured resistance data is subtracted from the model. Under the assumption that the temperature dependence of the semiconductor part of the total resistance does not change during the stress phase, i.e. the resulting parameters of (1) are also valid for $t > t_{sp}$, the main reason for any deviation must be a parasitic conductive path parallel to the DC connectors. The resistance of this parasite is described by

$$R_{parasitic} = \frac{R_{model} \cdot R_{measure}}{R_{model} - R_{measure}}. \tag{2}$$

As the deviation from the model $R_{deviation} = R_{model} - R_{measure}$ can become very small or even negative due to scattering of data around the model, the resulting parasitic resistances are capped at a maximum of 50 GΩ and also set to 50 GΩ for negative values. The latter results when the scatter of data is above the regression function, cf. Fig. 7. The resulting data is displayed in Fig. 8 for each test point: TP 1, 2, and 3 depicted by purple, red, and green circles in the top pane, TP 5 by yellow circles in the bottom pane. To show a general trend, rolling mean values with a time window of 24 h over the data for TP 1-3 in semi-opaque black and for TP 5 in semi-opaque brown are added to the plot.

During the first phase of the measurement, most data of TP 1-3 up to t_{sp} remains above 10 GΩ. Nevertheless, the scattering goes down to values as low as 1 GΩ. Only one outlier reaches 170 MΩ. With the beginning of the stress phase at $t > t_{sp}$ with its 100 % relative humidity, the parasitic resistances drop to a mean of 50 GΩ, decreasing to a mean of 28 MΩ when the temperature is further increased during the stress test. This value keeps decreasing to values of 18 to 20 MΩ at the end of the experiment. During the drying phases, the resistance recovers to higher values of 1 to 10 GΩ. Albeit, the values do not reach the pre-stress phase values but remain two orders of magnitude lower. After the first drying phase of the stress phase, two domains of parasitic resistances can be observed in the raw data of TP 1-3: One around

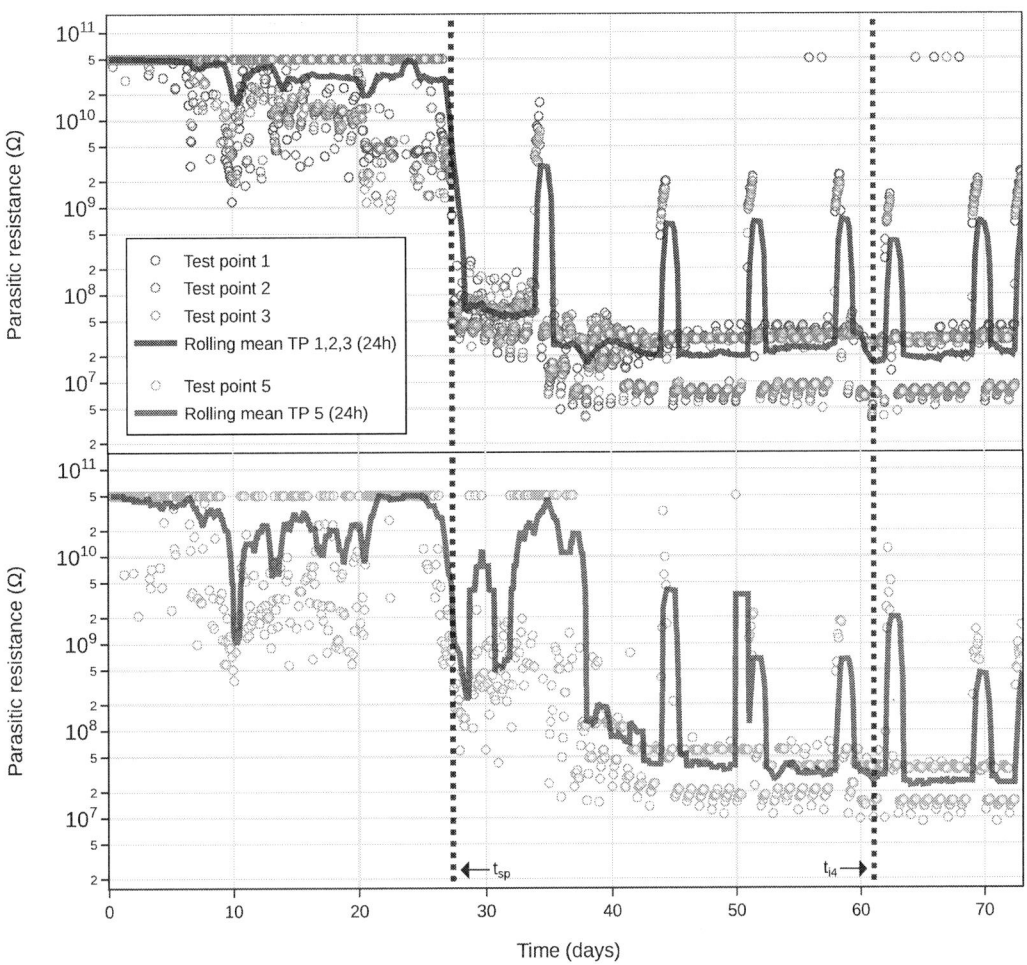

Fig. 8: Deduced parasitic resistances parallel to the DC connector of each test point and a 24 h rolling mean over all data points highlighting the general trend.

30 and the other one around 8 MΩ. They are directly related to the toggle between the two temperatures in this phase. Nevertheless, these temperatures are outside the scope of the training data for the model (1). Therefore, the uncertainty of the model might also play a role here.

When considering the data at TP 5, the general trend is similar. For $t < t_{sp}$ also most data is above 50 GΩ, although the scattering is stronger leading to values as low as 200 MΩ. After the beginning of the stress phase, the decrease in parasitic resistance is not as strong as for TP 1-3 although a parasitic resistance of $\approx \frac{1}{3}$ of the parasitic resistances of TP 1-3 due to their parallel connection might be expected. However, the value range of the recorded data of TP 5 is much lower, leading to an already higher uncertainty that might be amplified in the calculation of the parasitic resistance. The values further decrease during the run time of the experiment, also undergoing recovery to former values when drying phases take place. After the fifth drying phase in the stress phase on day 63, the values of the parasitic resistance show a similar range of values as for TP 1-3.

Conclusion

Using a climate chamber, we exposed IGBT power modules of type SKiiP3 under voltage load to increasingly harsh climatic conditions in a laboratory experiment. We have been able to reproduce the failure of an insulating foil as previously observed in the field in the form of partial disintegration. The failure pattern on the empty module can clearly be attributed to climatic and electrical stress. Nevertheless, some uncertainties in clarifying the actual failure mechanism remain. One important question is how exactly the conductive path is formed. Due to the very harsh climate in the chamber, condensation has taken place on the devices under test, forming a water film on the whole surface of the housing. This has also led to various residual depositions on the housing of the modules and visible corrosion on all metallic connections.

Besides the observation of a complete disintegration of an insulating foil on a power module with removed semiconductors, an increasing parasitic conductive path in parallel to the DC connectors could also be observed on an unmodified power module in the presence of condensed water during the measurement's stress phase. Although the calculated resistances do not qualify for quantitative analysis, a general trend of an increasing parasitic conductivity with increasing temperature and humidity can be seen. These resistances recover during drying cycles, but only occasionally reach their original values after prolonged exposure to 100 % relative humidity at high temperatures. One reason might be an insufficient drying time, leaving residual humidity or even liquid water on or inside the foil. Furthermore, this observation is solely attributed to water condensation: An optical inspection during the post-mortem analysis showed no signs of irreversible degradation and the resistances returned to their original values on the unmodified power modules after the experiment.

While some questions remain unanswered at this stage, a key result of the present work is the successful identification of a precursor of insulation failure in IGBT power modules. In view of the urgent need for condition monitoring methods capable of detecting also humidity-induced degradation and faults in power converters, future work will be dedicated to adapting and testing the insulation-monitoring approach on larger units such as power stacks including DC link components in order to assess the suitability for application in the field.

References

[1] S. Pfaffel, S. Faulstich, and K. Rohrig, Performance and Reliability of Wind Turbines: A Review, Energies, vol. 10, no. 11, p. 1904, 2017, doi: 10.3390/en10111904.

[2] Portfolio Review 2016: System Performance, Availability and Reliability Trend Analysis. [Online]. Available: https://ore.catapult.org.uk/wp-content/uploads/2018/02/SPARTAbrochure_20March-1.pdf

[3] K. Fischer et al., Reliability of Power Converters in Wind Turbines: Exploratory Analysis of Failure and Operating Data From a Worldwide Turbine Fleet, IEEE Trans. Power Electron., vol. 34, no. 7, pp. 63326344, 2019, doi: 10.1109/TPEL.2018.2875005

[4] K. Fischer et al., Exploring the Causes of Power-Converter Failure in Wind Turbines based on Comprehensive Field-Data and Damage Analysis, Energies, vol. 12, no. 4, p. 593, 2019, doi: 10.3390/en12040593.

[5] K. Fischer, M. Steffes, K. Pelka, B. Tegtmeier, and M. Dörenkämper, Humidity in Power Converters of Wind Turbines-Field Conditions and Their Relation with Failures, Energies, vol. 14, no. 7, p. 1919, 2021, doi: 10.3390/en14071919.

[6] C. Zorn and N. Kaminski, Acceleration of temperature humidity bias (THB) testing on IGBT modules by high bias levels, in 2015 IEEE 27th International Symposium on Power Semiconductor Devices IC's (ISPSD), Hong Kong, China, May. 2015 - May. 2015, pp. 385388.

[7] K. Zhang, G. Schlottig, E. Mengotti, O. Quittard, and F. Iannuzzo, Study of moisture transport in silicone gel for IGBT modules, Microelectronics Reliability, vol. 114, p. 113773, 2020, doi: 10.1016/j.microrel.2020.113773.

Comparison of Magnetic Noise Compensation Techniques for Dual Three-Phase Electrically Excited Synchronous Machines

Jonas Henkenjohann, Jan Andresen,[*] Axel Mertens
Leibniz University Hannover
Institute for Drive Systems and Power Electronics
Welfengarten 1
Hannover, Germany
Phone: +49 (0) 511 762 5619
Fax: + 49 (0) 511 762 3040
Email: jonas.henkenjohann@ial.uni-hannover.de
URL: https://www.ial.uni-hannover.de/

Acknowledgments

This work is part of the project *DampedWEA* and was funded by the Federal Ministry for Economic Affairs and Climate Action on the basis of a decision by the German Bundestag. Funding number: 03EE2008D. The authors would like to thank the industrial project partner Enercon for the inspiration to compare HFI and FFI when applied to an EESM, and the continuous support during the process.

Keywords

≪Force Control≫, ≪Acoustic noise≫, ≪Multiphase drive≫, ≪Vibration suppression≫

Abstract

This paper presents a detailed comparison of two commonly known magnetic noise compensation methods. One is based on forces generated by harmonic currents interacting with the fundamental field, while the other uses a harmonic field. Both methods are applied to dual three-phase electrically excited synchronous machines. The required compensation currents and parasitic effects, such as the induced voltage in the field winding, are compared. Furthermore, known precalculation methods are adapted to dual three-phase electrically excited synchronous machines and validated with FEM calculations.

Introduction

The ever-increasing number of electrical machines permeating more and more aspects of human life is leading to increased attention regarding the noise which they generate. Particularly due to the transformation of the energy system and the associated increase in the use of synchronous electrical machines in applications with a high power density (e.g., wind energy and traction drives), multiphase machines are now a strong focus of noise-reduction research. As [1] describes, electrically excited synchronous machines (EESM) offer various advantages over permanent magnet synchronous machines (PMSM). For example, the implementation of field weakening is easier, rare-earth magnets are not required and irreversible demagnetisation of an EESM cannot occur.

The noise emissions of electrical machines is often dominated by the radial mode 0 [2]. For this reason, there are already a large number of approaches in the literature for influencing the noise behaviour of electrical drives by injecting harmonic currents to compensate for the electromagnetic force excitation

---

[*]Now with: KEB Automation KG, Südstraße 38, Barntrup, Germany

of mode 0. In [3–7], optimal harmonic currents are determined with the help of parameter variations and numerical simulations. However, these methods are computationally intensive.

To overcome this issue, precalculation methods based on the rotating field theory have also been developed in the literature [8–10]. However, these methods neglect the tangential field component and for this reason are not able to predict the tangential tensile stresses within the machine. In [11], the analytical relationships for a PMSM are derived to enable the calculation of both the radial and tangential tensile stresses that result from injecting an additional spatial fundamental harmonic field. In this work, this is referred to as fundamental field injection (FFI). In a second step, the analytical relationships can be used to calculate the harmonic currents needed for compensation. In [12], the knowledge gained is validated on a test bench. Similar predictions of valid noise compensating currents for EESMs have not yet been documented in the literature.

This paper extends the analytical methods from [11] to include dual three-phase EESMs and evaluates the additional degrees of freedom that arise when applying for the FFI method. As a second approach, the investigations of dual three-phase PMSMs in [3], which have shown that compensation of the radial mode 0 can also be achieved by injecting spatial harmonic fields, are also examined within the scope of this work. This procedure is called harmonic field injection (HFI) in the following text. This paper derives an analytical equation relating tensile stress to harmonic current, so that computation intensive parameter variations as reported in [3] can be avoided for the HFI method.

Noise compensation in three-phase machines

This paper builds on the work of [11], uses the same notation, and adopts the consideration of the air-gap field as a sum of individual rotating waves. Since the machine under investigation is a synchronous machine, the dependence on time is replaced by the angular rotor position in this paper. The air-gap field can thus be described by [13, 14] as

$$B = \hat{i}_1 \cdot \sum_{\nu'} G_{\nu'} \cos\left(\nu'\gamma' - \varepsilon' - \varphi_0\right) \tag{1}$$

where $\gamma' = p\gamma$ is the circumferential angle, $\varepsilon' = p\varepsilon$ the mechanical angle between the d-axis and α-axis and ν' the circumferential order. The number of pole pairs is given as p. The proportionality factor $G_{\nu'}$, which describes the relationship between harmonic current and generated flux density, can be derived from [15]. The harmonic current used in [11] for compensation is an alternating fundamental current that can be described by three degrees of freedom, namely the amplitude $\hat{i}_k$, the phase φ_k, and the injection direction α_k. The air-gap field caused by injecting this alternating current in the dq-reference frame can thus be described by

$$B = \hat{i}_k \cos\left(k\varepsilon' + \varphi_k\right) \cdot \sum_{\nu'} G_{\nu'} \cos\left(\nu'\gamma' - \varepsilon' - \alpha_k\right). \tag{2}$$

The equations for calculating the resulting tensile stress waves are well known in the literature [8–10]. For this reason, their derivation is omitted here. The following Equation (3) describes the relevant radial tensile stresses resulting from Equation (2), and is correspondent with [14]:

$$\sigma_r = \cdots + \frac{\hat{B}_{1,1,k}\hat{B}_{1,1}}{2\mu_0} \cos\left(k\varepsilon' + \varphi_k\right) \cdot \left(\cos\left(-\alpha_k + \varphi_0\right) + \cos\left(2\gamma' - 2\varepsilon' - \alpha_k - \varphi_0\right)\right). \tag{3}$$

Here, the only relevant terms are those which are independent of γ' and thus represent a force pulsation with spatial order 0.

Equation (3) has therefore shown that by injecting a harmonic current in the dq-reference frame, a force of spatial order 0 with an adjustable frequency can be generated.

Noise compensation of dual three-phase EESMs

In high power applications like energy generation in wind turbines, it is common to use multiphase electric machines. Interest in using multiphase electric machines as traction drives is also growing in the automotive industry [16]. The dual three-phase winding, often also called asymmetrical six-phase winding, is used most frequently [3]. The advantage is that limited operation is still possible despite a failure of one system or inverter. Furthermore, the use of two unconnected neutral points suppresses the zero component in the current, which in turn significantly simplifies the control of the machine.

Harmonic Currents

In addition to the advantages mentioned above, the increased number of phases also results in new degrees of freedom (DOF) that can be exploited when injecting harmonic currents. In this way, the currents in each of the two three-phase systems can be controlled independently. This makes it possible to inject an alternating fundamental current as performed in FFI, as well as to impose a phase shift of 180 between the harmonic currents of the two systems. The advantage of this strategy is that the influences of these two harmonic currents on the fundamental field in the machine cancel each other out. As a result, exclusively spatial harmonic fields are injected (HFI). This method can be illustrated with the equivalent circuit diagram of an EESM in the dq-reference frame (Fig. 1a and b). The definitions of the given parameters can be found in [17]. Fig. 1c and d also show a simplified waveform of the currents in the dq-reference frame. Here, an enlarged amplitude and an arbitrary phase angle of the harmonic current were chosen for the purposes of clarity.

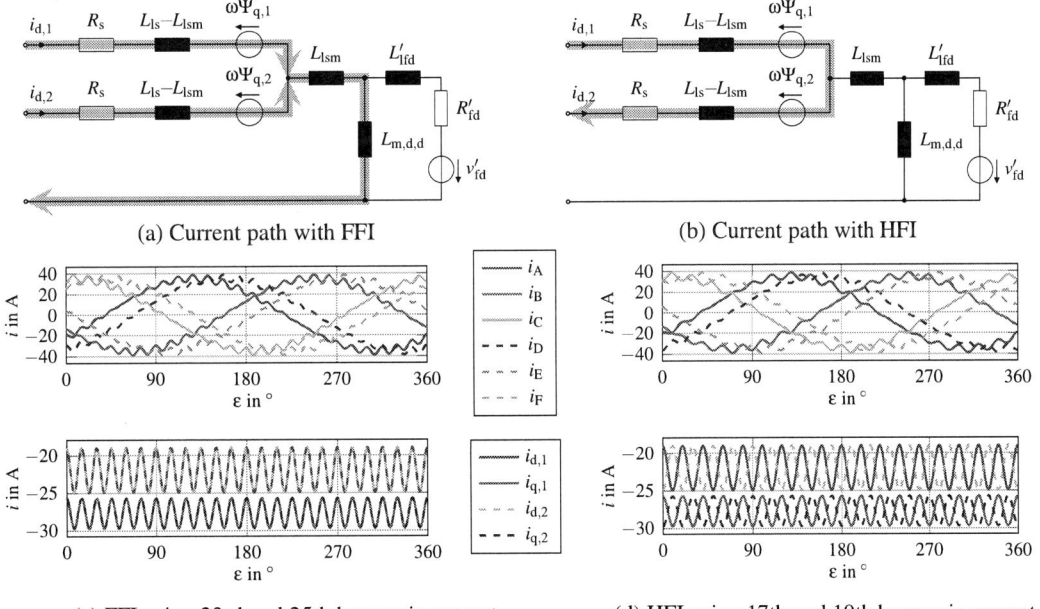

(a) Current path with FFI

(b) Current path with HFI

(c) FFI using 23rd and 25th harmonic current

(d) HFI using 17th and 19th harmonic current

Fig. 1: (a) and (b): Equivalent circuit diagram of an EESM in the dq-reference frame [17–19]. Only the d-axis is shown. (c) and (d): Waveforms of the currents in the dq-reference frame.

Stator Flux Density Harmonics

As shown in [20], dual three-phase machines do not produce the same harmonics as conventional three-phase machines. Due to the magnetic coupling between the two three-phase systems, spatial harmonics with an order which is an even multiple of six do not occur (marked as red fields in Table I). The time and spatial harmonics which do appear are marked as green fields in Table I.

The harmonics used in [11] are marked with FFI. The choice of the harmonic order was set to $k = 24$, since the machine investigated here has a mechanical resonance frequency with spatial order 0 in the tangential direction in this frequency range.

Table I: Time and spatial harmonics for dual-three phase electrical machines [3, 20].

Spatial harmonic ν'	Time harmonic $1 \pm k$								
	1	5	7	11	13	17	19	23	25
± 1								FFI	FFI
± 5							HFI		
± 7						HFI			
± 11									
± 13									

Due to the phase-opposed injection of the harmonic currents with the abbreviation HFI, there is no alternating fundamental field in the machine. Because of this, the force from the FFI method (cf.: Equation (3)) is not generated. In order to still generate a force with spatial order 0 and time order 24, the HFI method considers the interaction of the harmonics $(1 \pm k, \nu') = (-17, 7)$ and $(19, -5)$, which are marked with the abbreviation HFI, and the harmonics $(5, 5)$ and $(7, 7)$ generated by the rotor. In order to generate the desired spatial orders, two rotating currents are considered this time. The air-gap field caused by injecting the two rotating currents -17 and 19 can be described by

$$B = \hat{i}_{-17} \sum_{\nu'} G_{\nu'} \cos\left(\nu'\gamma - (-17\varepsilon') - \alpha_{-17}\right) + \hat{i}_{19} \sum_{\nu'} G_{\nu'} \cos\left(\nu'\gamma - 19\varepsilon' - \alpha_{19}\right). \tag{4}$$

A detailed description of the resulting rotating waves can be taken from [21]. Another way of visually representing the resulting harmonics is illustrated in Fig. 2. Here, the generated flux within the machine resulting from a sinusoidal current is shown in Fig. 2a. In Fig. 2b, it can be seen that, by injecting the harmonic currents with FFI, a field with spatial order $\nu' = p$ is created, which is also coupled to the rotor winding. In contrast, Fig. 2c indicates that the resulting fields with HFI are spatial harmonics that are not coupled or barely coupled to the rotor winding.

 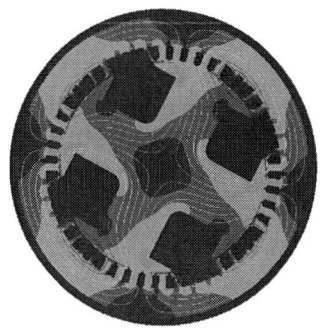 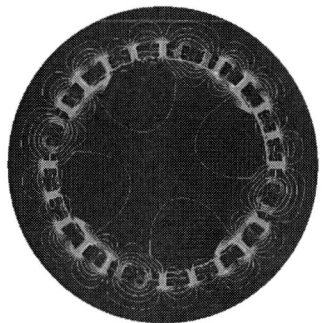

(a) Only fundamental current

(b) Only 23rd and 25th harmonic currents (FFI)

(c) Only 17th and 19th harmonic currents (HFI)

Fig. 2: Flux distribution with $I_{\text{fd}} = 0\,\text{A}$. Equivalent images like Fig. 2a and Fig. 2c for a PMSM can be found in [3].

Force Waves

If only the field waves with $\nu' = -5$ and $\nu' = 7$ are considered, the following spatial harmonics result from (4)

$$B = \cdots + \hat{i}_{-17} G_{7p} \cos\left(7\gamma - (-17)\varepsilon' - \alpha_{-17}\right) + \hat{i}_{19} G_{-5p} \cos\left(-5\gamma - 19\varepsilon' - \alpha_{19}\right). \tag{5}$$

Together with the rotor hamonics

$$B_{5,5} = \hat{B}_{5,5} \cos\left(5\gamma - 5\varepsilon' - \varphi_{5,5}\right) \text{ and } B_{7,7} = \hat{B}_{7,7} \cos\left(7\gamma - 7\varepsilon' - \varphi_{7,7}\right) \tag{6}$$

Table II: Operating point investigated

S_{N}	$17.3\,\mathrm{kVA}$
$I_{\mathrm{d,N}}$	$-21.93\,\mathrm{A}$
$I_{\mathrm{q,N}}$	$-27.68\,\mathrm{A}$
I_{fd}	$22.08\,\mathrm{A}$
p	2
$|\underline{\sigma}_{\mathrm{n},0}|$	$226.2\,\mathrm{N\,m}^{-2}$
$|\underline{\sigma}_{\mathrm{t},0}|$	$214.5\,\mathrm{N\,m}^{-2}$

Table III: Currents required to compensate for the radial and tangential modes 0

FFI		HFI	
$\hat{\imath}_{-23}$	$0.425\,\mathrm{A}$	$\hat{\imath}_{-17}$	$40.7\,\mathrm{A}$
$\hat{\imath}_{25}$	$0.302\,\mathrm{A}$	$\hat{\imath}_{19}$	$13.12\,\mathrm{A}$
$|\underline{\sigma}_{\mathrm{n},0}^{*}|$	$0.887\,\mathrm{N\,m}^{-2}$	$|\underline{\sigma}_{\mathrm{n},0}^{*}|$	$0.526\,\mathrm{N\,m}^{-2}$
$|\underline{\sigma}_{\mathrm{t},0}^{*}|$	$0.464\,\mathrm{N\,m}^{-2}$	$|\underline{\sigma}_{\mathrm{t},0}^{*}|$	$0.028\,\mathrm{N\,m}^{-2}$

the relevant terms of the tensile stress are:

$$\sigma_{\mathrm{r}} = \cdots + \frac{1}{2\mu_0}\left(\hat{\imath}_{19}G_{-5\mathrm{p}}\hat{B}_{5,5}\cos\left(-24\varepsilon' - \varphi_{5,5} - \alpha_{19}\right) + \hat{\imath}_{-17}G_{7\mathrm{p}}\hat{B}_{7,7}\cos\left(24\varepsilon' + \varphi_{7,7} - \alpha_{-17}\right)\right). \quad (7)$$

The same approach can be applied to the tangential force. However, the analytical calculation of the tangential force is outside the focus of the paper. The difference between FFI and HFI can also be demonstrated with the help of the force equation. If (7) is compared with (2), it can be seen that, in contrast to [11], two tensile stresses with the same frequency arise. The amplitude of the tensile stress can be influenced by the amplitude of the harmonic currents and their phase by the choice of the injection direction.

Comparison of the currents required by FFI and HFI

This section presents the calculation of the compensation currents needed to produce a change in the tensile stress by the value $\Delta\underline{\sigma}$. These calculations were carried out with the help of FEM simulations. From (7), it is clear that the relationship between current and tensile stress can be described by a complex factor. With reference to [11], this factor will be called $\underline{\eta}$. The change in the tensile stress in the radial direction $\Delta\underline{\sigma}_{\mathrm{n}}$ caused by the complex harmonic current from [3] can consequently be described by

$$\Delta\underline{\sigma}_{\mathrm{n}} = \underline{\eta}_{\mathrm{n},-17} \cdot \underline{i}_{-17} + \underline{\eta}_{\mathrm{n},19} \cdot \underline{i}_{19}. \quad (8)$$

Considering the tangential forces, this equation expands into the matrix notation

$$\begin{bmatrix} \Delta\underline{\sigma}_{\mathrm{n}} \\ \Delta\underline{\sigma}_{\mathrm{t}} \end{bmatrix} = \begin{bmatrix} \underline{\eta}_{\mathrm{n},-17} & \underline{\eta}_{\mathrm{n},19} \\ \underline{\eta}_{\mathrm{t},-17} & \underline{\eta}_{\mathrm{t},19} \end{bmatrix} \begin{bmatrix} \underline{i}_{-17} \\ \underline{i}_{19} \end{bmatrix}. \quad (9)$$

With the condition that the determinant of the matrix is not equal to zero, the matrix can be inverted. This leads to the relationship between the complex harmonic currents and the tensile stress, which can be described by

$$\begin{bmatrix} \underline{i}_{-17} \\ \underline{i}_{19} \end{bmatrix} = \begin{bmatrix} \underline{\eta}_{\mathrm{n},-17} & \underline{\eta}_{\mathrm{n},19} \\ \underline{\eta}_{\mathrm{t},-17} & \underline{\eta}_{\mathrm{t},19} \end{bmatrix}^{-1} \begin{bmatrix} \Delta\underline{\sigma}_{\mathrm{n}} \\ \Delta\underline{\sigma}_{\mathrm{t}} \end{bmatrix}. \quad (10)$$

A determinant of nonzero also shows that the radial and tangential spatial directions are linearly independent and can therefore be compensated independently. With an algorithm similar to the one developed in [14], the complex parameters $\underline{\eta}$ from (10) can be identified using FEM simulations. This makes it possible to calculate the right harmonic currents to compensate for a given tensile stress. The calculations are performed iteratively as described in [14]. This gives the possibility to compensate for nonlinearities and inaccuracies of the determination of the complex parameters.

The calculation was carried out for a salient-pole EESM. Data concerning the machine and the tensile stresses to be compensated can be found in Table II. The results of the compensation current calculation are shown in Table III. These show that harmonic currents of $\hat{\imath}_{-23} = 0.435\,\mathrm{A}$ and $\hat{\imath}_{25} = 0.302\,\mathrm{A}$ are required to reduce the radial and tangential tensile stresses to below 1 % using the FFI method. These figures correspond to 1.2 % and 0.9 % of the fundamental amplitude, respectively, so no significant effects on the winding losses are expected.

Compensation employing the HFI method means that significantly higher harmonic amplitudes are required to similarly reduce the radial and tangential tensile stresses to around 1 %. This requires $\hat{i}_{-17} = 40.7\,\text{A}$ and $\hat{i}_{19} = 13.12\,\text{A}$. These figures correspond to 115 % and 37.2 % of the fundamental amplitude, respectively, and are thus of the same order of magnitude as the fundamental amplitude. The high compensation currents can be explained, for instance, by the chorded winding in the machine. Since the fields with a spatial order of -5 and 7 are suppressed, significantly higher compensation currents are required for the HFI method.

In order to evaluate the compensation methods, their influence on the losses within the machine is of crucial importance. The iron losses within the machine increase by 2 % with the FFI method. The HFI method increases the iron losses by 4 %. These numbers were determined by means of an FEM simulation. There is a more significant effect on the winding losses in the stator. Here, the losses increase by 200 % with the HFI method. With the FFI method, the winding losses increase by just 0.2 %. This leads to the conclusion that the FFI method is to be preferred for efficiency-optimized operation of the EESM.

Parasitic Effects

In addition to the desired change in the radial and tangential tensile stresses with spatial order 0, a number of unwanted parasitic effects arise from injecting the harmonic currents. Due to the alternating fundamental field of the FFI method (see Fig. 3), a voltage with k times the fundamental frequency is induced into the field winding (see Fig. 4). Since there is no alternating fundamental field with HFI, the induced voltage is considerably lower than with the FFI method. Still, regular operation with sinusoidal currents exhibits even smaller induced voltages.

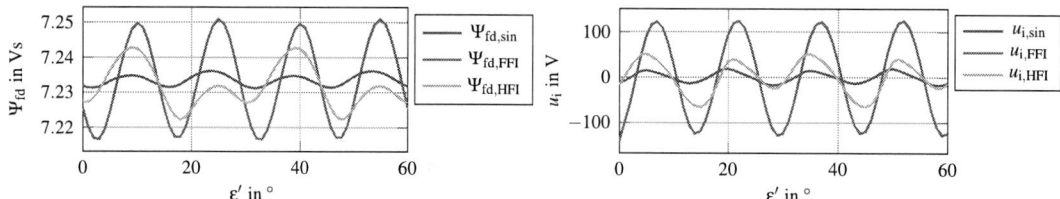

Fig. 3: Rotor flux linkage   Fig. 4: Induced voltage in the field winding

Typical voltage supplies for a field winding are not able to compensate for the induced voltage. For this reason, a harmonic current will flow in the rotor, which, according to Lenz's law, will counteract the alternating fundamental field. To investigate the influence of this effect on the calculation of the harmonic currents, FEM simulations with voltage injection into the field winding were also performed. It can be seen in Fig. 6 that a harmonic current of $\hat{i}_{\text{fd},24} = 0.25\,\text{A}$ with a frequency of $f_{\text{fd},24} = 24 \cdot f_1$ flows through the field winding and thus changes the rotor flux linkage (see Fig. 5). This increases the harmonic currents required for FFI compensation to $\hat{i}_{-23} = 0.624\,\text{A}$ and $\hat{i}_{25} = 0.449\,\text{A}$. These are increases of 46.8 % and 48.7 %, respectively. Fig. 5 and Fig. 6 also show that the influence of the HFI method on the rotor is lower than that of the FFI method. The required compensation currents change by 23 % and −38 % to become $\hat{i}_{-17} = 50.1\,\text{A}$ and $\hat{i}_{19} = 8.1\,\text{A}$, respectively. The effects of an pulsating excitation current on the force excitation and also on the induced voltage in the stator has also been demonstrated in [5].

In addition, the harmonic currents in the stator cause new tensile stresses which were not considered in the derivation. For FFI, this includes the spatial order $2p$ with the harmonic orders 2 ± 24 (see Fig. 7). A reduction in the tensile stress waves is shown in green and an increase in red. Fig. 8 shows the force changes resulting from the HFI method. The changes in the parasitic forces are significantly larger for HFI than for FFI, due to the significantly higher current amplitude needed for compensation. The parasitic forces primarily arise from the interaction between the additional spatial harmonic fields and the fundamental field.

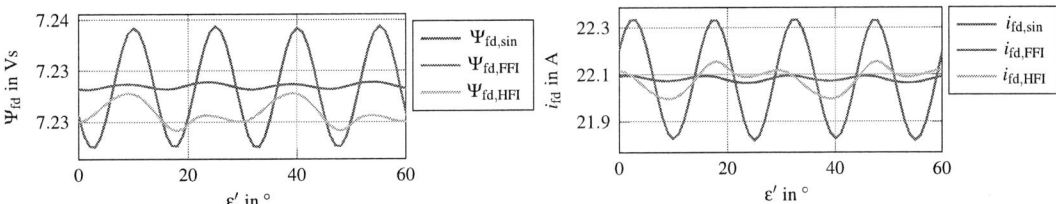

Fig. 5: Rotor flux linkage with voltage injection

Fig. 6: Current in the field winding with voltage injection

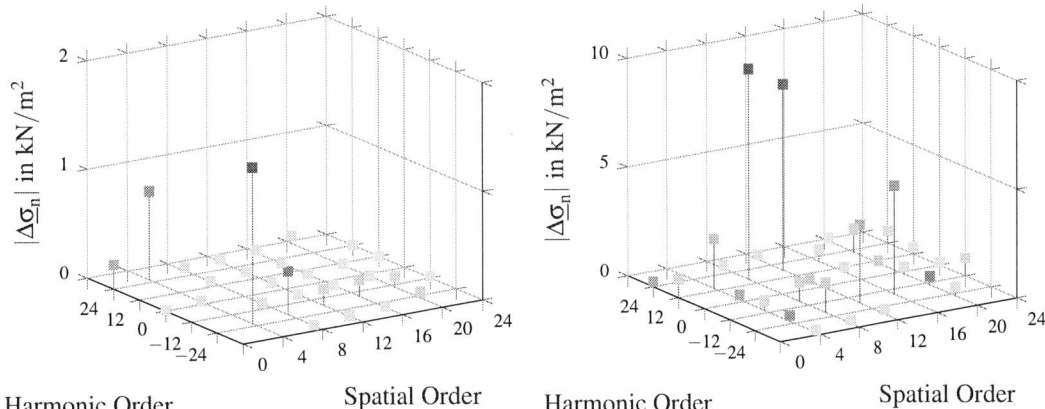

Fig. 7: Changes in the tensile stress with FFI

Fig. 8: Changes in the tensile stress with HFI

Conclusion

This paper provides an extensive comparison between two methods known from the literature for noise suppression in dual three-phase EESMs: fundamental field injection (FFI) and harmonic field injection (HFI). For HFI, the analytical prediction of harmonic currents, which has already been demonstrated for FFI, was adapted. Using FEM calculations, it is shown that compensation of the radial and tangential tensile stresses of mode 0 can be achieved with both FFI and HFI. FFI compensation requires significantly lower harmonic current amplitudes and thus also causes considerably lower winding losses than HFI compensation. In addition, it can be seen that with FFI, a voltage is induced into the field winding by the alternating fundamental field. Therefore, a harmonic field current controller with sufficient voltage reserve is essential to prevent harmonic currents from flowing in the field winding. If suitable harmonic current controllers are not available, it is shown in this paper that the required harmonic currents in the stator become significantly larger. Future research will analyze whether the selection of the compensation method depending on the operating point is advisable.

References

[1] J. Redlich, J. Juergens, K. Brune, and B. Ponick, "Synchronous machines with very high torque density for automotive traction applications," in *2017 IEEE International Electric Machines and Drives Conference (IEMDC)*, 2017, pp. 1–8.

[2] A. Hofmann, F. Qi, T. Lange, and R. W. De Doncker, "The breathing mode-shape 0: Is it the main acoustic issue in the pmsms of today's electric vehicles?" in *2014 17th International Conference on Electrical Machines and Systems (ICEMS)*, 2014, pp. 3067–3073.

[3] P. Hollstegge, A. Wanke, and R. W. De Doncker, "Noise mitigation in dual three-phase internal permanent magnet machines by injection of current harmonics," in *The Journal of Engineering*, vol. 2019, no. 17. IET, 2019, pp. 4273–4277.

[4] M. Harries, M. Hensgens, and R. W. De Doncker, "Noise reduction via harmonic current injection for concentrated-winding permanent magnet synchronous machines," in *2018 21st International Conference on Electrical Machines and Systems (ICEMS)*, 2018, pp. 1157–1162.

[5] F. Evestedt, J. J. Pérez-Loya, C. J. D. Abrahamsson, and U. Lundin, "Controlling airgap magnetic flux density harmonics in synchronous machines using field current injection," *Electrical Engineering*, vol. 103, no. 1, pp. 195–203, 2021.

[6] J. Nägelkrämer, A. Heitmann, and N. Parspour, "Application of dynamic programming for active noise reduction of pmsm by reducing torque ripple and radial force harmonics," in *2018 AEIT International Annual Conference*, 2018, pp. 1–6.

[7] S. Ciceo, F. Chauvicourt, J. Gyselinck, and C. Martis, "PMSM current shaping for minimum joule losses while reducing torque ripple and vibrations," *IEEE Access*, vol. 9, pp. 114 705–114 714, 2021.

[8] B. Cassoret, R. Corton, D. Roger, and J.-F. Brudny, "Magnetic noise reduction of induction machines," *IEEE Transactions on Power Electronics*, vol. 18, no. 2, pp. 570–579, 2003.

[9] D. Belkhayat, D. Roger, and J. F. Brudny, "Active reduction of magnetic noise in asynchronous machine controlled by stator current harmonics," *IET Conference Proceedings*, pp. 400–405(5), 1997, publisher: Institution of Engineering and Technology.

[10] D. Franck, M. van der Giet, and K. Hameyer, "Active reduction of audible noise exciting radial force-density waves in induction motors," in *2011 IEEE International Electric Machines Drives Conference (IEMDC)*, 2011, pp. 1213–1218.

[11] J. Andresen, S. Vip, A. Mertens, and S. Paulus, "Theory of influencing the breathing mode and torque pulsations of permanent magnet electric machines with harmonic currents," in *2020 22nd European Conference on Power Electronics and Applications (EPE'20 ECCE Europe)*, 2020, pp. P.1–P.9.

[12] J. Andresen, S. Vip, A. Mertens, and S. Paulus, "Compensation of the radial and circumferential mode 0 vibration of a permanent magnet electric machine based on an experimental characterisation," in *2020 22nd European Conference on Power Electronics and Applications (EPE'20 ECCE Europe)*, 2020, pp. P.1–P.9.

[13] G. Müller and B. Ponick, *Theorie elektrischer Maschinen*, 6th ed., ser. Elektrische Maschinen. Weinheim: Wiley-VCH, 2009, no. 3.

[14] J. Andresen, "Aktive Geräuschunterdrückung in einer permanentmagneterregten Synchronmaschine mit Hilfe von Stromoberschwingungen," Dissertation, Faculty of Electrical Engineering and Computer Science, Leibniz University Hannover, Germany, 2021.

[15] J. F. Gieras, J. C. Lai, and C. Wang, *Noise of polyphase electric motors*, ser. Electrical and computer engineering. Boca Raton, FL: CRC/Taylor & Francis, 2006, no. 129.

[16] A. Salem and M. Narimani, "A review on multiphase drives for automotive traction applications," *IEEE Transactions on Transportation Electrification*, vol. 5, no. 4, pp. 1329–1348, 2019.

[17] R. F. Schiferl and C. M. Ong, "Six phase synchronous machine with ac and dc stator connections, part i: Equivalent circuit representation and steady-state analysis," *IEEE Transactions on Power Apparatus and Systems*, vol. PAS-102, no. 8, pp. 2685–2693, 1983.

[18] S. Gradev, J. Reuss, and H.-G. Herzog, "A general voltage-behind-reactance formulation of a multivoltage n 3-phase hybrid-excited synchronous machine," *IEEE Transactions on Energy Conversion*, vol. 31, no. 4, pp. 1452–1461, 2016.

[19] J. Heseding, "Baukastensysteme für den Antriebsstrang von Elektrofahrzeugen," Dissertation, Faculty of Electrical Engineering and Computer Science, Leibniz University Hannover, Germany, 2018.

[20] D. G. Dorrell, C. Y. Leong, and R. A. McMahon, "Analysis and performance assessment of six-pulse inverter-fed three-phase and six-phase induction machines," *IEEE Transactions on Industry Applications*, vol. 42, no. 6, pp. 1487–1495, 2006.

[21] P. Hollstegge, "Injektion raumzeigerzerlegter Stromharmonischer zur Minderung tonaler Geräuschanteile in asymmetrisch sechsphasigen Permanentmagnetsynchronmaschinen," Dissertation, Faculty of Electrical Engineering and Information Technology, RWTH Aachen University, Germany, 2021.

PCB Technology Comparison Enabling a 900V SiC MOSFET Half Bridge Design for Automotive Traction Inverters

Matthias Spieler[1], Che-Wei Chang[1], Ayman EL-Refaie[2],
Muhammad H Alvi[3], Dong Dong[1], Rolando Burgos[1]

[1]Center for Power Electronics Systems
Virginia Tech
1185 Perry St, Blacksburg, VA 24061
United States
mspieler@vt.edu

[2]Opus College of Engineering
Marquette University
1637 W. Wisconsin Ave, Milwaukee, WI 53233
United States
ayman.el-refaie@marquette.edu

[3]General Motors Research & Development
30565 William Durant Blvd. Warren, Michigan 48092-2031
United States
muhammad.alvi@gm.com

Acknowledgments

This material is based upon work supported by the U.S. Department of Energy's Office of Energy Efficiency and Renewable Energy (EERE) under the Advanced Vehicle Technologies Office Award Number DE-EE0009190.

Keywords

≪Silicon Carbide (SiC)≫, ≪Electric Vehicle≫, ≪Traction application≫, ≪Inverter design≫, ≪Parasitic inductance≫

Abstract

The design of automotive traction inverters for an $800\,V$ dc-bus typically utilize $1.2\,kV$ silicon carbide (SiC) metal oxide semiconductor field effect transistors (MOSFET). The $1.2\,kV$ power devices allow for high overshoot voltages during switching transients but experience a high on-state resistance due to the die's thick drift layer region. This paper proposes the usage of $900\,V$ SiC MOSFETs for $800\,V$ automotive traction inverter applications. The proposed half-bridge design combines discrete power semiconductors and the dc-link capacitor on one printed circuit board (PCB). This design approach enables a small current commutation loop and thus a small overshoot voltage. Three different PCB technologies are compared based on their suitability for the traction inverter application. An $800\,V$ half-bridge prototype is designed, simulated, and tested. Measurement results are provided and show an overshoot voltage of $49.6\,V$ at a switching speed of $69\,V/ns$ under maximum load conditions.

Introduction

The transition to electrified automobiles accelerates the push to greater power densities and efficiency targets. Thus, in academia and industry a shift towards higher dc-bus voltages of $800\,V$ and the usage of high voltage SiC MOSFETs is observed [1–4]. SiC MOSFETs offer lower losses and faster switching speeds compared to insulated-gate bipolar transistors (IGBTs). The increase in dc-bus voltage reduces the current stress on the drive train while maintaining the power output constant. These changes help to increase the efficiency and power density [4]. The US Department of Energy defined a research and development road map for automotive traction inverter that pursue an inverter efficiency increase to more than 98%, and power densities exceeding $100\,kW/l$ by 2025 [5].

EPE'22 ECCE Europe

One step towards achieving the efficiency and power density goals is the transition from 1.2 kV SiC MOSFETs to 900 V SiC MOSFETs. The lower breakdown voltage allows for a thinner die epitaxial layer and thus reduces the MOSFET's on-state resistance [6]. As the dc-bus voltage can reach 850 V under normal operating conditions [2], special attention needs to be paid on a low inductance current commutation loop design to minimize the drain-source overshoot voltage during switching transients.

Many studies focus on reducing the current commutation loop stray inductance [7–14]. These studies can be generally divided into stray inductance reduction achieved by: 1) dc-link capacitor selection, 2) laminated busbars, 3) the placement of decoupling capacitors, 4) and low inductance power module integration.

Moorthy et al. [15] described the usage of a PCB as the dc-busbar for an automotive traction inverter. Their design connects the dc-link capacitors, power modules, and adjacent decoupling capacitors through the PCB. This results in a low current commutation loop stray inductance of 7.56 nH. Their inverter has an overshoot voltage of 163 V at a drain current of 440 A. Although no information on the di/dt is provided, it is unlikely that the stray inductance reduction resulting from the sole usage of a heavy copper PCB as a dc-busbar is sufficient to enable the usage of 900 V devices. Yet, the technology shows promise and is adapted for our half-bridge module design.

The current commutation loop stray inductance previously mentioned originates largely from the high-power module layout. Thus, to shrink the stray inductance further, the traditional half-bridge packaging design needs to be rethought. Plenty of research investigated different power module designs, such as a vertical current commutation loop integration [7, 11], or reducing the commutation loop inductance by inductance cancellation [13]. Yet, the module's spacers and screw connections increase the overall stray inductance and cannot be omitted while using power modules. One approach to reduce the inductance is to use discrete devices instead of power modules for the inverter build-up. The authors in [16] opted for a discrete SiC MOSFET based 15 kW Matrix converter design with PCB embedded ceramic inlays for better heat transfer from junction to heatsink. By choosing this design, the authors were able to integrate a vertical current commutation loop design and thus reduce the overall current commutation loop stray inductance. Another promising approach is to embed SiC MOSFET dies directly into the PCB. Marczok et al. [17] designed a six phase 100 kW traction inverter using embedded dies and achieving a design with a current commutation loop stray inductance of 1.69 nH resulting in an approximate overshoot voltage of 50 V.

This paper investigates which PCB technology is most suitable for a 900 V SiC MOSFET integrated half-bridge design. By combining the dc-busbar, dc-link capacitor, and power stage into one PCB, the current commutation loop is minimized and thus enables the usage of 900 V SiC MOSFETs in traction inverter applications. In section one, a brief overview of the inverter structure is provided, explaining the its build-up and its impact on the overshoot voltage in detail. In section two, a high-level view of the 900 V device selection and the benefits of 900 V SiC MOSFETs over 1.2 kV FETs is presented. The FETs associated losses play a key role in the thermal requirements and current density requirements which the PCB technologies have to fulfill. Section four provides a detailed approach on the PCB technology selection based on thermal, dc-bus current stress criteria, and loop inductance criteria. Section five includes switching transient measurements. Lastly, conclusions are drawn on the efficacy of this study.

Traction inverter build-up

The usage of 900 V SiC MOSFETs implies challenging constraints on the inverter design. The fast turn-off switching transient can induce a high overshoot voltage, causing an increased voltage strain on the low side SiC MOSFET. The maximum drain-source voltage during switching transient can be expressed as follows:

$$V_{DS;max} = L_{stray,sum} \cdot \frac{di_D}{dt} + V_{DC}, \tag{1}$$

where $L_{stray,sum}$ is the pooled current commutation loop stray inductance, di_D/dt is the drain current slope during the turn-off switching transient, and V_{DC} is the applied dc-bus voltage. Fig. 1 depicts the general equivalent electric circuit for one phase leg and the prevalent stray inductances. The total current commutation loop stray inductance can be described as follows:

$$L_{stray,sum} = ESL + L_{busbar} + 2 \cdot (L_D + L_S), \qquad (2)$$

where ESL is the dc-link capacitor's parasitic stray inductance, L_{busbar} is the busbar's intrinsic inductance, and L_D and L_S are the MOSFET package's intrinsic inductance.

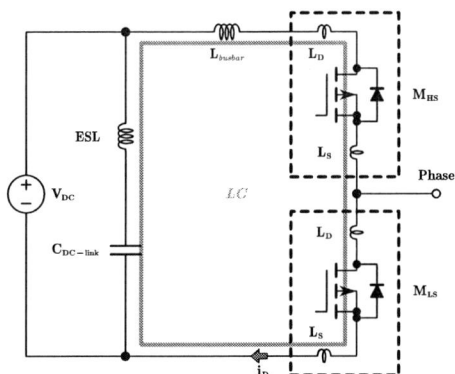

Fig. 1: Equivalent eclectic circuit of current commutation loop with intrinsic inductance of components.

The dc-bus voltage of an electric vehicle fluctuates under normal operating conditions between 650 V and 850 V depending on the battery package's state of charge [2]. This results in an overshoot voltage limit of maximum 50 V, if 900 V SiC FETs are used. As the switching time should be minimized to reduce the MOSFET's switching losses, a small current commutation loop stray inductance must be achieved. The inverter is designed for a switching transient voltage slope of less than 30 V/ns.

The proposed inverter design uses a heavy copper PCB as a base. The dc-link capacitor is directly soldered onto the PCB, a planar busbar can be achieved with multiple PCB layers, and discrete SiC MOSFETs are placed near the dc-link capacitor resulting in a small current commutation loop. Ceramic capacitors were chosen as the dc-link capacitor, as their parasitic inductance is smaller than the traditionally used film capacitors.

The discrete SiC MOSFET's package must be selected with care to reduce the intrinsic stray inductance. A current commutation loop implemented with the standard TO-247 package has a stray inductance of 8.18 nH [18]. Solely this induced stray inductance would result in an overshoot voltage exceeding the MOSFET's break down voltage. Thus, a through-hole mounted device package is not applicable for an 800 V dc-bus traction inverter with 900 V semiconductor devices. More suitable for the inverter application are SMD packages that create a current commutation loop stray inductance of 2.93 nH [18], or similarly the use of bare dies.

The device package constraints and dc-link capacitor design establish the PCB design boundaries. The PCB must combine the three following characteristics: 1) small current commutation loop stray inductance, 2) high thermal conductivity from junction to heatsink to allow reducing the number of paralleled SiC MOSFETs, 3) and an acceptable trace current density for the high current carrying PCB traces. To derive the cooling requirements, the device losses are simulated in the next section.

MOSFET selection consideration

PLECS simulations were conducted to compare losses of different SiC dies. Based on the simulation results, a conclusion can be made about the die type and die quantity needed for the phase leg. Criteria for the minimum number of parallel devices is dependent on the MOSFET's losses at a selected switching frequency. The PLECS simulation uses the switching loss and die's conduction loss to determine the overall MOSFET loss. It needs to be noted that this loss approximation neglects any reverse-recovery losses, input capacitance charge loss, and gate losses. A three-phase inverter with a discontinuous pulse width modulation control (DPWM) is simulated. The inverter's specifications are a nominal dc-bus voltage V_{nom} of 800 V, maximum output power $P_{out;max}$ of 200 kW, a switching frequency f_{sw} within the range of 5 kHz to 50 kHz, and a die junction temperature T_j of 150 °C. The MOSFET's total loss is the sum of conduction and switching loss.

$$P_{tot} = P_{con} + P_{sw} \tag{3}$$

$$P_{con} = i_D^2 \cdot R_{DS;on} \tag{4}$$

$$P_{sw}(i_D) = [E_{sw;on}(i_D) + E_{sw;off}(i_D)] \cdot f_{sw} \tag{5}$$

Where i_D is the drain current, $R_{DS;on}$ is the MOSFET's on-state series resistance, $E_{sw;on}(i_D)$ and $E_{sw;off}(i_D)$ is the turn-on switching energy loss and turn-off switching energy loss respectively as stated in the datasheet as a function of the drain current. By increasing the number of paralleled MOSFETs per phase, the current stress per device is reduced. Thus, the conduction loss decreases. The switching losses per device decrease as the drain current per device is reduced. With increasing switching frequency, the switching loss increases approximately linearly.

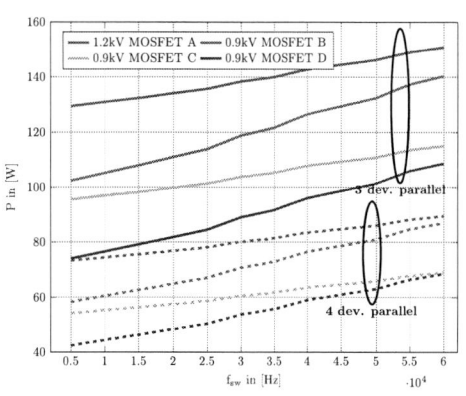

Fig. 2: PLECS simulation based MOSFET die loss for different MOSFETs as a function of switching frequency and number of paralleled devices.

In total four devices were compared with respect to their total losses. Three devices have a breakdown voltage of 900 V, whereas one device has a breakdown voltage of 1.2 kV. Despite choosing a high efficiency 1.2 kV MOSFET, it is worth noting that 900 V devices lead to a significant reduction of losses highlighting the loss savings by using 900 V FETs instead of 1.2 kV FETs for automotive traction inverter applications.. Fig. 2 depicts the total MOSFET die loss at different switching frequencies, and different number of paralleled devices. MOSFET D has the lowest overall losses within the selected frequency range while paralleling three or four devices. MOSFET D is selected for the inverter build-up. The cooling requirements are derived based on MOSFET D's total losses at a switching frequency of 20 kHz.

PCB technology selection

To enable the usage of 900 V devices the current commutation loop stray inductance has to be minimized. As the dc-busbar is integrated within the PCB, the layer cross-sectional area must be sufficient to achieve a current density of less than $5 A/mm^2$ [19]. A PCB trace-width calculator is used to determine the minimum trace-width [20]. To ensure no PCB overheating due to trace losses, the thermal conductivity from the device junction to the coolant must be high enough to allow for sufficient cooling. A higher thermal conductivity enables fewer numbers of paralleled SiC MOSFETs and hence reduces the bill-of-material list and increases the power density. All three requirements must be fulfilled by the PCB technologies selected for closer comparison. The PCB technologies are compared based on analytical equations and finite-element simulations.

The three PCB technologies, 1) copper inlay PCB, 2) ceramic inlay PCB, and 3) embedded dies within the PCB were considered for this study. A cross section of the PCB technologies and their thermal resistance path are displayed in Fig. 3.

The copper inlay PCB is a heavy copper PCB with small copper coins press fitted into PCB cavities. The maximum diameter of the inlay is 8 mm. By placing the inlay beneath the SiC MOSFET SMD footprint a high thermal conductive path from the SMD package to the heatsink placed at the PCB bottom is created. The bottom side copper polygon has a footprint of 115 mm^2 to allow for better heat spreading

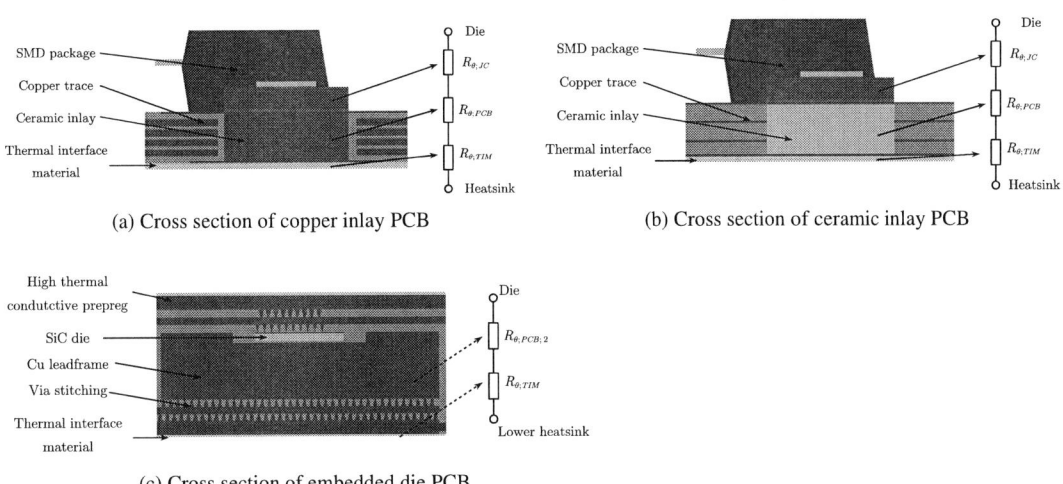

(a) Cross section of copper inlay PCB

(b) Cross section of ceramic inlay PCB

(c) Cross section of embedded die PCB

Fig. 3: PCB technology cross section with resulting thermal resistance path from junction to heatsink. The TIM material is highlighted in yellow. The isolating prepreg material is highlighted in blue.

and a larger thermal interface material area. Due to the manufacturing process, the bottom side is not electrically insulated from the MOSFET's drain pad. Thus, a non-conductive thermal interface material is required. A PCB with a maximum of 6 layers with an outer layer copper thickness of 70 μm and inner layer copper thickness of 500 μm can be manufactured. This allows for a low inductance design, with low trace losses. The copper inlay PCB cross section is displayed in Fig. 3a.

The operating principle of the ceramic inlay is similar to the copper inlay PCB. The inlay provides a high thermal conductive path for the heat transfer from the device junction to the heatsink. A PCB with $10x10\ mm^2$ ceramic inlays is considered for the inverter application. The ceramic inlay's advantage over the copper inlay is its electrical insulating properties. This enables the usage of a non-electrical insulating thermal interface material and thus increases the overall thermal conductivity. A detailed comparison is given in the following sub sections. The ceramic inlay PCB can be manufactured with a layer stack of four 70 μm copper layers. Due to the thin copper layers, it is expected that the dc-busbar current density is the highest of all three PCB technologies. Fig. 3b displays the cross section of a ceramic inlay PCB.

The embedded die technology enables the placement of bare SiC dies into the PCB. Fig. 3c shows the embedded die cross section. The lack of any SMD package omits the intrinsic package stray inductance. While die embedding allows for a compact trace design resulting in a low stray inductance, and a good thermal conductivity to the heatsink. The die is sintered onto a $11x11\ mm^2$ copper lead frame. The lead frame is then embedded into the PCB and electrically connected with buried vias. The bottom side prepreg material (highlighted in blue) electrically isolates the embedded inlay to the heatsink. Thus, a high thermal conductive and electrically non-insulating interface material can be used.

Thermal modeling

Equation based thermal resistance

The PCB design and technology must enable sufficient heat transfer from the power semiconductor to the heatsink. Although the cooling requirements are increased for fewer paralleled SiC-MOSFETs, a lower number of paralleled devices is preferred, as the power density is increased, and system cost is reduced. Therefore a phase leg design with three devices in parallel at a switching frequency of 20 kHz is chosen. As depicted in Fig. 2, the simulated die loss is 82 W at the selected maximum load operating point. To handle the high-power loss, the thermal resistance between the MOSFET junction and the coolant must be small to keep the junction temperature at a predefined value. The general equation for

Table I: Calculated thermal resistances of each material used for the PCB layer stack up

Parameter	Copper	Ceramic	Embedded
$R_{\theta,j-c}$ in $°C/W$	0.31	0.31	0
$R_{\theta,PCB}$ in $°C/W$	0.13	0.06	0.177
$R_{\theta,TIM}$ in $°C/W$	0.66	0.057	0.057
$R_{\theta,tot}$ in $°C/W$	1.1	0.427	0.234

the total thermal resistance is given below.

$$R_{\theta;tot} = \frac{T_{j;FET} - T_{coolant}}{P_{FET}},\qquad(6)$$

where $T_{j;FET}$ is the MOSFET junction temperature, $T_{coolant}$ is the coolant temperature, and P_{FET} is the discrete MOSFET total loss. The predefined maximum junction temperature is set to 150 $°C$. The coolant temperature is set to constant 70 $°C$ resulting in a maximum thermal resistance for MOSFET D of 0.98 $°C/W$ if the devices are switched at 20 kHz.

The previously mentioned PCB technologies need to fulfill the requirement of a junction-to-heatsink thermal resistance not greater than 0.58 $°C/W$ under the assumption that the heatsink's thermal resistance is not greater than 0.4 $°C/W$ per MOSFET. If a PCB technology doesn't meet the thermal requirements, it will not be considered further for the inverter application.

First, the thermal resistance is derived based on analytical equations. If the analytical based thermal resistance does not fulfill the requirement, the PCB technology is no longer considered. All PCB technologies with an analytical based thermal resistance lower than given maximum thermal resistance will be simulated in ANSYS Icepak. This step is necessary, as the analytical based thermal resistance tends to be lower than the actual thermal resistance, due to the assumed equal heat spreading within the thermal path stack up.

The junction-to-heatsink thermal resistance $R_{\theta;j-h}$ is the sum of thermal resistances introduced by each material within the heat transfer path. The thermal resistance paths for each PCB layout are shown in Fig. 3. Each material thermal resistance is calculated based on the following equation:

$$R_{\theta} = \frac{\delta x}{A \cdot K},\qquad(7)$$

where δx represents the material thickness, A is the material area, and K equals the material's specific thermal conductivity. The two thermal interface materials Thermflow T777 Material from Parker and TM-TIFX200 from Futura Italia are used for the PCB comparison [21,22]. The resulting thermal resistances for each material in the junction to heatsink thermal path are listed in Table I.

The copper inlay PCB junction-to-heatsink thermal resistance is 1.1 $°C/W$. Largest contributor to the thermal resistance is the thermal interface material which electrically isolates the heatsink from the PCB bottom. The high total thermal resistance does not allow for sufficient cooling if only three SiC-MOSFETs are paralleled. Thus, the copper inlay PCB technology is not applicable for a high-power density traction inverter with 900 V SiC devices.

The ceramic inlay PCB has the advantage of an electric isolating inlay over the copper inlay PCB technology. Thus, allowing the usage of an electric conducting thermal interface material. This results in a junction-case thermal resistance of 0.427 $°C/W$. As the thermal resistance fulfills the requirement, the thermal heat path will be simulated in the following subsection.

By embedding the die into the PCB, the SMD package is removed from the heat transfer path resulting in a smaller total thermal resistance. The die is sintered to a copper lead frame which is connected to the below layers through via stitching. The electric isolation is achieved via a high thermal conductive

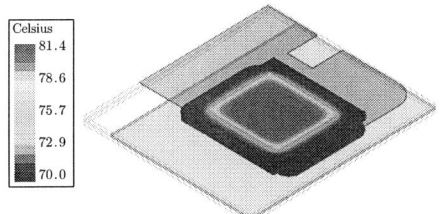 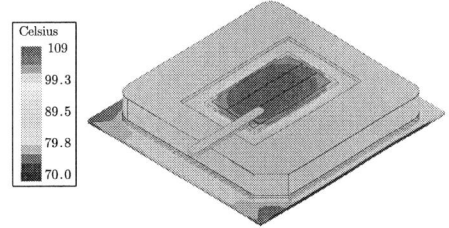

(a) Ceramic inlay PCB ANSYS ICEPAK simulation. The temperature ranges from blue 70 °C to red 81.43 °C.

(b) Embedded die PCB ANSYS ICEPAK simulation. The temperature ranges from blue 70 °C to red 109.03 °C.

Fig. 4: Simulated die temperature for the ceramic inlay PCB design and double-sided-cooling embedded die PCB design.

prepreg layer without any vias. This results in an analytical-based junction-to-heatsink thermal resistance of 0.234 °C/W. The combined embedded PCB thermal resistance is below 0.5 °C/W. Thus, the thermal resistance will be investigated closer in the following subsection

Simulation based thermal resistance

To evaluate the thermal behaviour of the ceramic inlay PCB and the embedded die PCB technology, ANSYS Icepak simulations are conducted. The PCB and its layer stack up are designed in Altium designer and exported to ANSYS Icepak.

To further evaluate the PCB thermal resistance, the die temperature or the SMD footprint were assigned with a power loss of 82 W and the PCB's bottom side copper layer is set to a constant temperature of 70 °C. The thermal interface material's impact on the junction-to-heatsink thermal resistance is added post simulation. The via stitching is approximated by changing the area specific prepreg material's thermal conductivity to the combined averaged thermal conductivity value of the buried vias and the FR4. The temperature of the die or the SMD footprint is simulated, and its temperature is used to calculate the thermal resistance based on eq. 6. The simulation results are depicted in Fig. 4a and Fig. 4b.

The finite element simulations based thermal resistance for the ceramic inlay PCB stack up and embedded die PCB are 0.44 °C/W and 0.45 °C/W respectively. Both thermal resistances are below the set maximum. Thus, both PCB technologies are considered for further trace resistance analysis.

Trace loss modeling

The dc-busbar is integrated into the PCB. The thinner PCB layers increase the overall conduction loss compared to an external thick copper busbar. Therefore, it is essential to investigate the trace losses created by the PCB design. Too high losses increase the trace temperature. The trace-width must be sufficient to limit the PCB's temperature increase to 25 °C. Allocco states that a good rule of thumb is to limit the current density to 5 A/mm^2 for passively cooled busbar designs [19]. Additionally, the designed PCB dc-bus trace-width is compared to the minimum trace-width derived by an online trace-width calculator based on the IPC-2221 standard [20].

The ceramic capacitors are soldered to the top and bottom of the PCB with the dc-busbar connecting all components. The capacitor footprint is adapted for a low impedance design. Through-hole vias connect the PCB layers. The required clearance between the high voltage potentials, defined by the standard IPC-2221B [23], creates the layer layout of one half-bridge dc-link capacitor as shown in Fig. 5.

The high current traces narrow down to 2.4 mm in areas with the through hole vias. It is expected to have the highest current density in these areas. To analyze the current density, finite element analysis were conducted. The trace design is exported to ANSYS Icepak and the current flow path is defined. As the trace layout of $dc+$ and $dc-$ traces are similar, only the $DC+$ trace current density is simulated. The $dc+$

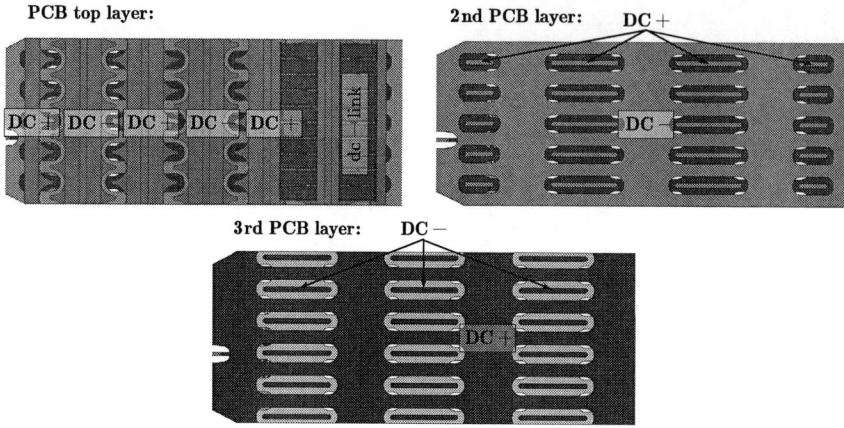

Fig. 5: Top view of layer stack up of first three half-bridge PCB layers. Layer layout of layer two and three alternate for all following four layers

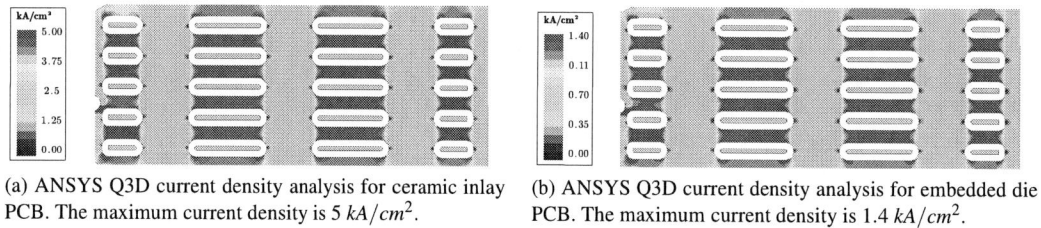

(a) ANSYS Q3D current density analysis for ceramic inlay PCB. The maximum current density is 5 kA/cm^2.

(b) ANSYS Q3D current density analysis for embedded die PCB. The maximum current density is 1.4 kA/cm^2.

Fig. 6: Equivalent electric circuit diagram and simulated die losses of 1.2 kV MOSFET and 900 V MOSFETs

trace-width is 14.7 mm. To minimize simulation time, the busbar of one half-bridge was simulated. The average current stress is derived from the PLECS simulations conducted in Section 2 and is set to 134 A per phase. If three half-bridges are paralleled per phase, the current stress per half-bridge is 44.67 A.

Current density simulation

As previously mentioned the ceramic inlay PCB has four layers with a layer thickness of 70 μm. The top and bottom layers are used to mount the dc-link capacitors restricting the high current flow to the two inner layers. This results in a simulated current density of up to 43.3 A/mm^2 as shown in Fig. 6a. Additionally, the design does not meet the minimum trace-width derived using the trace-width calculator. Hence, the high dc-bus current density of the ceramic PCB technology is not suitable for the inverter application in question.

The eight-layer PCB used for the embedded PCB design has a combined trace thickness of 255 μm for each dc-trace. The current density is reduced to a maximum of 11 A/mm^2, which is above the recommended 5 A/mm^2. Yet, the total dc-bus trace-width of 14.7 mm per half bridge is greater than the PCB trace-width calculator's minimum trace-width limit of 11.7 mm [20]. Future research will include testing the dc-busbar heating to investigate these two contradicting results. Accordingly, the embedded PCB technology is further analyzed.

Current commutation loop stray inductance

The eight-layer PCB used for the embedded build-up allows for a vertical current commutation loop integration. The vertical integration achieves good mutual inductance cancellation resulting in a small current commutation loop stray inductance. The loop design is depicted in Fig. 7. The $dc+$ trace is routed on the third layer and is connected with vias to the high-side lead frame and subsequently to the drain pad. The AC trace is also routed on the third layer connecting the high-side die source pad with the low-side MOSFET drain. The $dc-$ trace is routed from the low-side die source pad on the second

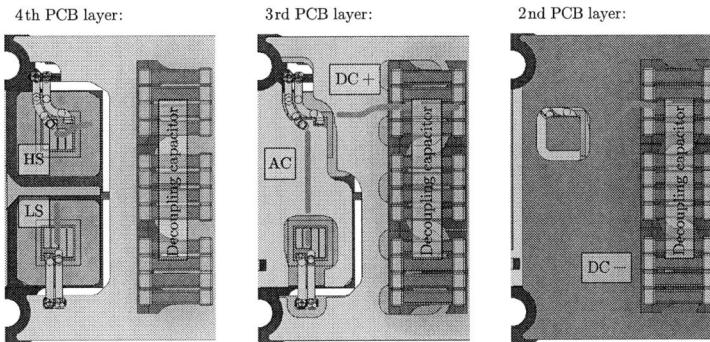

Fig. 7: Current commutation loop design. The current path is highlighted by the pink line on each PCB layer. The abbreviations describe the trace potential.

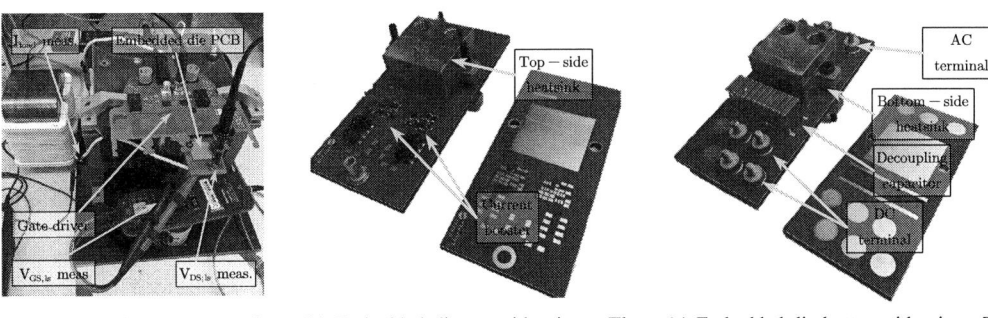

(a) Double pulse test setup for testing one embedded die half-bridge PCB

(b) Embedded die top side view. The embedded die half-bridge is located beneath the heatsink pad.

(c) Embedded die bottom side view. The embedded die half-bridge is located beneath the heatsink pad.

Fig. 8: Double-pulse setup and turn-off switching transient

PCB layer back to the dc-link capacitor. A cut-out on the second layer above the high side MOSFET allows for a better high thermal conductivity path from the die to a top side heatsink. Future work will show the thermal performance of a double-sided-cooling embedded die PCB design. The PCB design is analyzed in ANSYS Q3D at a frequency of 10 MHz. The design current commutation loop inductance and increase in drain-source capacitance for the high-side and low-side MOSFET are 744 pH, 32.9 pF, and 59.01 pF respectively. To analyze the MOSFET's overshoot voltage during switching transients, double pulse tests are carried out in the next section.

Embedded die PCB double pulse measurement

To evaluate the embedded die half-bridge PCB, a prototype was manufactured. As a safety measure, a SiC MOSFET with a breakdown voltage of 1.2 kV was used instead of a 900 V SiC MOSFET. Also, a modular phase design was chosen to allow for easy replacement of defective half-bridges. The embedded die PCB is shown in Fig. 8b and Fig. 8c. The half-bridge prototype assembly is depicted in Fig. 8a. The following measurement probes were used: passive high voltage probe (TPP0850) for measuring the MOSFET drain-source voltage, two differential probes (THDP0200) to measure the dc-bus voltage and the low-side gate signal, and one Rogowski coil (CWTUM/3/B) to measure the load inductor current.

The double pulse tests were conducted at a dc-bus voltage of 855 V, a drain current of 142 A per half-bridge, and a switching transient speed of 69 V/ns. Due to the lack of a current measurement method with minimal impact on the current commutation loop stray inductance, the switching transient current cannot be measured. Thus, the overshoot voltage is evaluated based on the turn-off voltage switching transient and the drain current. Fig. 9 depicts the turn off switching transient. The resonant frequency between the MOSFET input capacitance and the current commutation loop stray inductance is 273 MHz. The

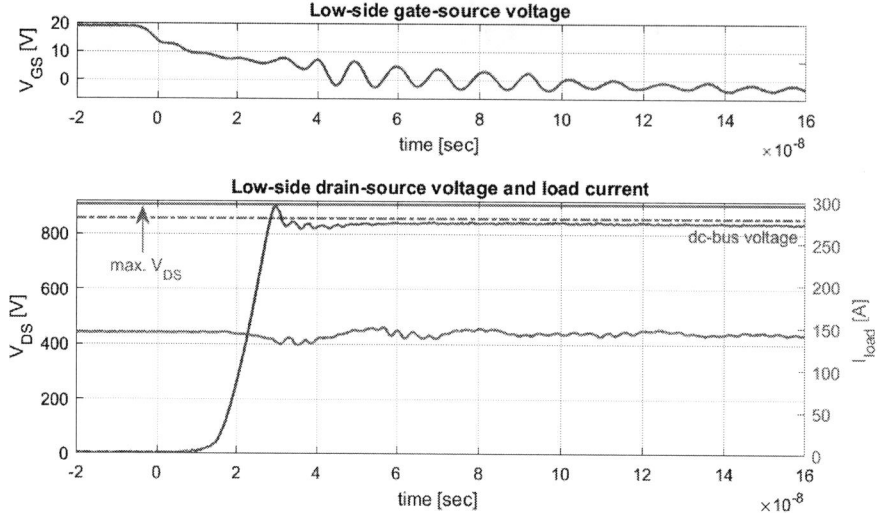

Fig. 9: Experimental test results of low side MOSFET's drain-source voltage during turn-off switching transient at $V_{DS} = 855\ V$, $I_{load} = 142\ A$, $dv/dt = 69.9V/ns$. The maximum drain-source voltage peak is 904.6 V.

overshoot voltage reaches 49.6 V indicating that the embedded die PCB design is capable of operating at a dc-bus voltage of 850 V, a turn-off voltage transient speed of 69 V/ns, and a drain current of 142 A without reaching the breakdown voltage of a 900 V SiC MOSFET.

Conclusions

A comprehensive PCB technology comparison enabling 900 V SiC MOSFETs for automotive traction inverter applications with a dc-bus voltage of 800 V was presented in this work. The copper inlay PCB technology, ceramic inlay PCB technology, and embedded die PCB technology were compared in terms of the PCB's thermal conductivity, the dc-bus current stress, and the overshoot voltage during switching transients. The copper inlay PCB technology was shown to have an insufficient thermal conductivity form junction-to-heatsink due to the electrically isolated thermal interface material. The ceramic inlay on the other hand led to thin copper traces making it unsuitable for the inverter application, as the dc-bus current density exceeds the recommended of 5 A/mm^2. The embedded die PCB technology also had a current density above the threshold value, but it exceeded the minimum trace-width derived by the trace-width calculator used. Lastly, the switching transient overshoot voltage caused by the current commutation loop stray inductance was evaluated, where the measurements conducted showed a maximum drain-source overshoot voltage of 49.6 V, at a switching speed of 60 V/ns, and a drain current per die of 142 A for the embedded die PCB technology. The embedded die technology consequently showed to be a viable alternative for the use of 900 V SiC MOSFET devices in for automotive traction inverter applications with nominal dc-bus voltages of 800 V.

Future work will include testing multiple paralleled embedded die PCB half-bridges as well as analyzing the thermal resistance based on the standard JEDEC JESD51-14 [24], and conducting continuous maximum load tests to investigate the dc-busbar heating.

References

[1] M. Su, C. Chen, S. Sharma, and J. Kikuchi, "Performance and cost considerations for sic-based hev traction inverter systems," in *2015 IEEE 3rd Workshop on Wide Bandgap Power Devices and Applications (WiPDA)*, Conference Proceedings, pp. 347–350.
[2] C. Jung, "Power up with 800-v systems: The benefits of upgrading voltage power for battery-electric passenger vehicles," *IEEE Electrification Magazine*, vol. 5, no. 1, pp. 53–58, 2017.

[3] W. Taha, B. Nahid-Mobarakeh, and J. Bauman, "Efficiency evaluation of 2l and 3l sic-based traction inverters for 400v and 800v electric vehicle powertrains," in *2021 IEEE Transportation Electrification Conference Expo (ITEC)*, Conference Proceedings, pp. 625–632.

[4] J. Zhu, H. Kim, H. Chen, R. Erickson, and D. Maksimović, "High efficiency sic traction inverter for electric vehicle applications," in *2018 IEEE Applied Power Electronics Conference and Exposition (APEC)*, Conference Proceedings, pp. 1428–1433.

[5] U. D. Partnership, "Electrical and electronics technical team roadmap," p. https://bit.ly/3o1egLl, 2017.

[6] R. W. Erickson and D. Maksimovic, *Fundamentals of power electronics*. Springer Science Business Media, 2007.

[7] C. Marczok, E. Hoene, T. Thomas, A. Meyer, and K. Schmidt, "Low inductive sic mold module with direct cooling," in *PCIM Europe 2019; International Exhibition and Conference for Power Electronics, Intelligent Motion, Renewable Energy and Energy Management*, Conference Proceedings, pp. 1–6.

[8] F. Hou, W. Wang, R. Ma, Y. Li, Z. Han, M. Su, J. Li, Z. Yu, Y. Song, Q. Wang, M. Chen, L. Cao, G. Zhang, and B. Ferreira, "Fan-out panel-level pcb-embedded sic power mosfets packaging," *IEEE Journal of Emerging and Selected Topics in Power Electronics*, vol. 8, no. 1, pp. 367–380, 2020.

[9] D. J. Kearney, S. Kicin, E. Bianda, and A. Krivda, "Pcb embedded semiconductors for low-voltage power electronic applications," *IEEE Transactions on Components, Packaging and Manufacturing Technology*, vol. 7, no. 3, pp. 387–395, 2017.

[10] A. Tablati, N. Alayli, T. Youssef, O. Belnoue, L. Theolier, and E. Woirgard, "New power module concept in pcb-embedded technology with silver sintering die attach," *Microelectronics Reliability*, vol. 114, p. 113891, 2020. [Online]. Available: https://www.sciencedirect.com/science/article/pii/S0026271420305102

[11] L. Zhang, P. Liu, A. Q. Huang, S. Guo, and R. Yu, "An improved sic mosfet-gate driver integrated power module with ultra low stray inductances," in *2017 IEEE 5th Workshop on Wide Bandgap Power Devices and Applications (WiPDA)*, Conference Proceedings, pp. 342–345.

[12] G. Regnat, P. O. Jeannin, J. Ewanchuk, D. Frey, S. Mollov, and J. P. Ferrieux, "Optimized power modules for silicon carbide mosfet," in *2016 IEEE Energy Conversion Congress and Exposition (ECCE)*, Conference Proceedings, pp. 1–8.

[13] K. Takao and S. Kyogoku, "Ultra low inductance power module for fast switching sic power devices," in *2015 IEEE 27th International Symposium on Power Semiconductor Devices IC's (ISPSD)*, Conference Proceedings, pp. 313–316.

[14] J. Schnack, J. P. Goerdes, S. Stahl, U. Schuemann, R. Mallwitz, and H. P. Tiedemann, "Design of bypass network for fast switching inverters using an innovative hybrid dc-link," in *2021 IEEE 15th International Conference on Compatibility, Power Electronics and Power Engineering (CPE-POWERENG)*, Conference Proceedings, pp. 1–8.

[15] R. S. K. Moorthy, B. Aberg, M. Olimmah, L. Yang, D. Rahman, A. N. Lemmon, W. Yu, and I. Husain, "Estimation, minimization, and validation of commutation loop inductance for a 135-kw sic ev traction inverter," *IEEE Journal of Emerging and Selected Topics in Power Electronics*, vol. 8, no. 1, pp. 286–297, 2020.

[16] V. Baker, B. Fan, R. Burgos, V. Blasko, and W. Chen, "3d commutation-loop design methodology for a silicon-carbide based 15 kw, 380:480 v matrix converter with pcb aluminum nitride cooling inlay," in *2020 IEEE Energy Conversion Congress and Exposition (ECCE)*, Conference Proceedings, pp. 233–238.

[17] C. Marczok, M. Martina, M. Laumen, S. Richter, A. Birkhold, B. Flieger, O. Wendt, and T. Paesler, "Sicmodul - modular high-temperature sic power electronics for fail-safe power control in electrical drive engineering," in *CIPS 2020; 11th International Conference on Integrated Power Electronics Systems*, Conference Proceedings, pp. 1–6.

[18] F. Denk, K. Haehre, S. Eizaguirre Cabrera, C. Simon, M. Heidinger, R. Kling, and W. Heering, "Rds(on) vs. inductance: comparison of sic mosfets in 7pin d2pak and 4pin to-247 and their benefits for high-power mhz inverters," *IET Power Electronics*, vol. 12, no. 6, pp. 1349–1356, 2019, https://doi.org/10.1049/iet-pel.2018.5838. [Online]. Available: https://doi.org/10.1049/iet-pel.2018.5838

[19] J. M. Allocco, "Laminated bus bars for power system interconnects," in *Proceedings of APEC 97 - Applied Power Electronics Conference*, vol. 2, Conference Proceedings, pp. 585–589 vol.2.

[20] A. Circuits, "Pcb trace width calculator." [Online]. Available: https://www.4pcb.com/trace-width-calculator.html

[21] P. Chomerics, "Thermflow® non-silicone, phase-change thermal interface pads." [Online]. Available: prker.co/3xJCugk

[22] F. Italia, "Thermal interface materials," 2017. [Online]. Available: https://bit.ly/futura-italia

[23] "Generic standard on printed board design," 2012. [Online]. Available: https://bit.ly/IPC-2221B

[24] "Transient dual interface test method for the measurement of the thermal resistance junction-to-case of semiconductor devices with heat flow through a single path," 2010. [Online]. Available: https://bit.ly/jedec-standard

Desaturated turn-off of low-saturation IGBTs with clamping method to reduce turn-off energy losses

Vishwas Acharya Nayampalli, Hans-Günter Eckel

UNIVERSITY OF ROSTOCK
Albert-Einstein-Str. 2
18059, Rostock, Germany
Tel.: +49 (0) 381-498 7134
Fax: +49 (0) 381-498 7102
E-Mail: vishwas.nayampalli@uni-rostock.de

Acknowledgements

This work was funded by the German Federal Ministry for Economic Affairs and Climate Action under grant 0350055B.

The authors would like to thank Mr. Gurunath Vishwamitra Yoganath of the University of Rostock for a set of IGBT models used in the simulations presented in this paper.

Keywords

«IGBT», «Smart Gate Drivers», «Driver Concepts», «Switching Devices»

Abstract

Investigations into the desaturated turn-off of low-saturation IGBTs with a novel hardware-based clamping approach are presented in this paper. Effects of the clamping voltage level and duration of desaturation on the reduction in turn-off energy losses are investigated. Turn-off measurement results with experimental low-saturation IGBT chips with the implemented clamping method demonstrate significant reduction in turn-off losses relative to the intrinsic turn-off of the device. Furthermore, TCAD based turn-off simulations with the clamping approach applied to different IGBT device models are presented with the objective of determining the potential of this method in reducing turn-off losses, as well as to obtain an insight into the necessary device on-state carrier concentration profile to obtain large reductions in turn-off energy losses for the advantageous application of this turn-off method.

Introduction

The Insulated Gate Bipolar Transistor (IGBT) finds itself in an increasingly large number of applications, such as in wind energy converters, VSC-HVDC applications, medium voltage drives among others. The rapid move towards green energy and energy efficient applications has propelled it to the status of being an indispensable device in medium to high voltage applications. With the goal of increasing the efficiency and decreasing the size of converters, recent research has focused on increasing the carrier concentration inside the device in an effort to reduce the on-state forward conduction voltage drop ($V_{CE,sat}$) [1] and thereby the conduction losses in the device. This is accomplished by increasing the front-end emitter-sided plasma concentration in the device during conduction by increasing the emitter efficiency [2]. Several methods have been proposed in literature and implemented in practice to increase the carrier concentration in the front-end emitter side of the device. The 'Injection-Enhancement effect' to increase the electron injection by suppressing the flow of hole current at the emitter side by reducing the trench-trench distance, also known as the mesa-width, was proposed in [3], which led to the device known as the IEGT. A further reduction of the mesa to sub-micron levels towards attaining the 'Silicon limit' for IGBTs was proposed in [4]. A n-doped layer below the p-base of the device between the trenches, which led to the device CSTBT, has been proposed [5] to further limit the hole current and lead to enhanced electron current in the MOS region. Recently, an attempt to further reduce the $V_{CE,sat}$ with carrier confinement and increased channel width has led to the MPT-IGBT concept with wide channels and low mesa-widths [6][7].

The result of increasing the front-end emitter sided-hence the device carrier concentration is a decreased $V_{CE,sat}$ but has the consequence of increased amount of carriers to be extracted from the device during turn-off, which could lead to higher turn-off energy losses necessitating more sophisticated turn-off methods. A desaturated turn-off of trench-fieldstop IGBTs was first shown to reduce the turn-off energy losses by about 17% in such devices

[8]. It has been further shown [9] through simulations, that such desaturated switching is especially beneficial in low-saturation IGBTs with increased front-end emitter sided carrier concentration compared to conventional IGBTs, which feature a higher back-end emitter sided plasma concentration.

Recently, turn-off measurements have been performed on a low-saturation 3.3kV IGBT chip [10] with a feed-forward based desaturation pulse based on the methodology introduced in [9]. This approach showed an optimistic 21% reduction in E_{OFF} compared to normal turn-off. However, investigations also revealed that the purely feed-forward-based approach also suffers from significant disadvantages in practical implementation. The outcome, it has been shown, is highly sensitive to control parameters such as the switching instants and also to the current carried by IGBT when it is switched off. It was also shown through TCAD simulations that a clamping-based approach to desaturated turn-off could potentially result in large reduction in turn-off energy losses and inherently being a feedback-based method, also helps avoid the disadvantages associated with the feed-forward-based approach.

The implementation of this method would require a sophisticated intelligent driver to time the activation of the clamping circuit based on feedback signals proportional to the collector to emitter voltage V_{CE} during turn-off. There has been a recent trend in research towards customized drivers equipped with fast FPGAs or microcontrollers to achieve particular objectives during device switching, for instance, to implement a gate voltage behavior-based distinction during turn-off of a reverse conducting IGBT [11] and in the implementation of a sophisticated turn-off of SiC mosfets [12]. Although not standard at the moment, it could be expected that application-specific drivers would eventually find commercial applications.

In the work presented in this paper, a previously implemented FPGA-based gate driver [10] is now supplemented with an additional active clamping circuit, the switching of which is also controlled by the FPGA as required to include the collector-gate clamping action during turn-off to implement the desaturation pulse. Scaled single chip turn-off measurement results with the desaturation pulse are presented with two different experimental low-saturation 3.3kV class IGBT chips to show the advantageous application of this method in terms of reduction of E_{OFF}. Investigations to the effect of desaturation voltage level and duration on the reduction of E_{OFF} are presented. TCAD simulation results with different low-sat IGBT models are presented to obtain certain qualitative insight into the required device on-state plasma profile to gain maximum benefit from this turn-off method in terms of reduced E_{OFF}.

Collector-gate clamping based desaturation pulse – driver implementation and control

The relevant functional sections of the FPGA-based high voltage gate driver used to implement the collector-gate clamping (CG-Clamping) desaturation pulse are shown in Fig. 1. The push-pull actuator stage of the driver comprises of three different turn-off mosfet-$R_{G,off}$ combinations to enable the tuning of the turn-off dynamic during the initial fast intrinsic turn-off phase, the desaturation pulse phase and the final commutation phase. The logic signals required to switch these actuator mosfets are provided by the FPGA, which is programmed with a state machine algorithm to implement the turn-off method. This figure also shows the active clamping circuit used to implement the desaturation pulse. This circuit comprises of a TVS diode and a high voltage mosfet. The breakdown voltage of the TVS diode determines the clamping voltage level between the collector and gate and therefore between the collector and emitter ($V_{CE,clamp}$).

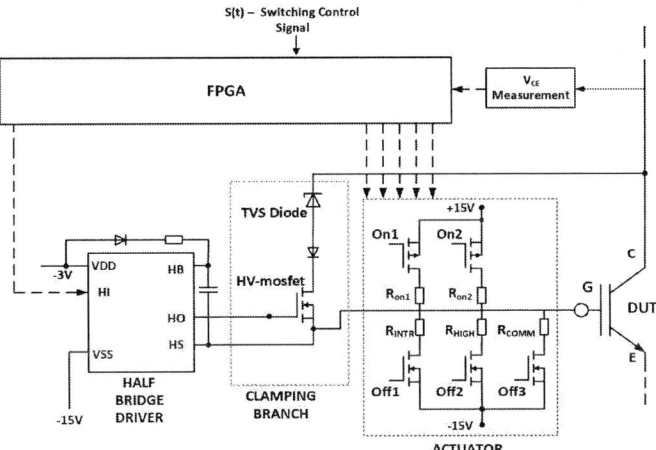

Fig. 1: Schematic representation of the FPGA-based HV gate driver with active clamping circuit

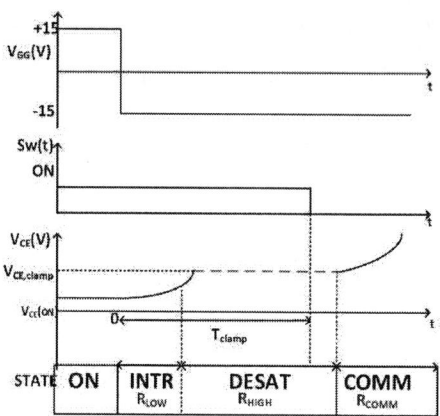

Fig. 2: Switching control diagram to implement the collector-gate clamping desaturation pulse-based turn-off

The high voltage mosfet, rated to block the full rated voltage of the DUT IGBT, is switched by the half bridge driver IC. The switching of this mosfet determines the clamping duration (t_{clamp}) during the desaturation pulse phase of the turn-off.

The control timing diagram to implement the desaturation pulse is shown in Fig. 2. The clamping HV- mosfet is turned-on (shown as Sw(t)) a few micro seconds after the turn-on of the DUT-IGBT. Following the turn-off signal at t=0, the IGBT turn-off initially progresses intrinsically with a turn-off gate resistance of R_{LOW} until shortly before the clamping voltage is attained. At this point, the gate is switched to a high resistance R_{HIGH} as the clamping action begins, to limit the current through the clamping circuit. The clamping action continues until the clamping mosfet is turned-off. The gate is now switched to R_{COMM} to lead to the completion of the turn-off process. The switching of the clamping mosfet is carried out in a feedforward manner by programming the FPGA.

Measurement Results with the Collector-gate clamping desaturation pulse-based turn-off

Turn-off measurement results with the desaturation pulse, performed on experimental low-saturation IGBT chips, are presented in this section. These experimental devices have a rated blocking voltage of 3.3kV and a nominal current ($I_{C,nom}$) of 106A. The stray inductance in the commutation circuit is scaled to correspond to the stray inductance seen in comparable IGBTs in an XHP module package. Normal turn-off is carried out with a turn-off resistance of 26Ω for $R_{G,COMM}$. For the implementation of the desaturation pulse, a R_{HIGH} of 5kΩ is used, and R_{COMM} of 23Ω is used during the final commutation phase of the turn-off.

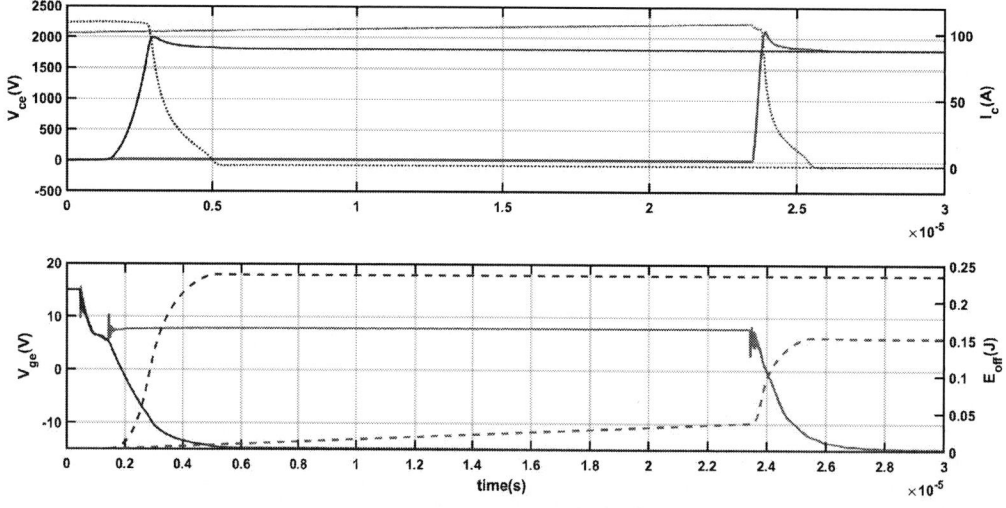

Fig. 3: Turn-off measurements for IGBT type W1
Top: V_{CE} and I_C, bottom V_{GE} and E_{OFF}
Legend: Blue→ Normal Turn-off ($R_{G,off}$ = 25Ω), Red→ Desat pulse-based turn-off (5V TVS diode)

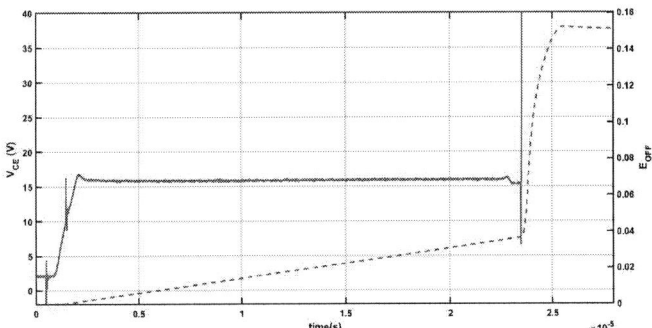

Fig. 4: V_{CE} during desaturation pulse corresponding to Fig. 3

Fig. 3 shows the comparison between normal and desaturation pulse-based turn-off measurements for one of the experimental low-saturation IGBT devices (W1), which has a $V_{CE,sat}$ of 1.75V at $I_{C,nom}$ and 125°C. Normal turn-off (shown in blue) results in a E_{OFF} of 0.237J. Desaturation pulse-based turn-off with a 5V TVS diode to set the CG-clamping voltage level, as described earlier, results in a E_{OFF} of 0.15J for a clamping duration (t_{clamp}) of 22µs, which is a significant reduction of 37% compared to the normal turn-off case. The clamped V_{CE} during the turn-off is shown magnified in Fig. 4. The V_{CE} remains more or less constant during the desaturation pulse. This indicates the adequacy of the driver implementation applied to drive the clamping mosfet.

Additional desaturation pulse-based turn-off measurements are performed on this IGBT chip to determine the optimal combination of collector-emitter clamping voltage $V_{CE,clamp}$ and duration t_{clamp} to obtain the minimum value for E_{OFF}. Fig. 5 shows the summary of these measurements. Here, the E_{OFF} normalized with respect to the intrinsic E_{OFF}, are plotted for different values of t_{clamp} and $V_{CE,clamp}$. A noticeable aspect from these results is that the E_{OFF} attains a local minimum for each value of clamped voltage and then starts increasing again. For this IGBT chip, it can be seen that a $V_{CE,clamp}$ of 16V and t_{clamp} of 22µs leads to the best outcome regarding reduction of E_{OFF}. It is to be noted that the turn-off energy losses during the turn-off with the desaturation pulse comprise of the additional on-state losses during the desaturation pulse E_{clamp} and the energy losses during final commutation E_{comm}. To further investigate the effect of the parameters $V_{CE,clamp}$ and t_{clamp}, the normalized E_{clamp} and E_{comm} corresponding to the turn-off measurement summary of Fig. 5 are shown in Fig. 6a and Fig. 6b respectively. It is seen from Fig. 6a that the E_{comm} decreases faster with respect to t_{clamp} at higher values of $V_{CE,clamp}$. It is also seen that the E_{comm} initially decreases rapidly with respect to t_{clamp} and rather sluggishly later. From Fig. 6b, it is immediately discernible that the losses during the desaturation pulse, E_{clamp} increases rather linearly with respect to t_{clamp} as expected, and is proportional to the $V_{CE,clamp}$. Therefore, although the E_{comm} decreases rapidly at larger values of $V_{CE,clamp}$, higher clamping voltages do not present an advantage because of the much larger amount of additional losses incurred in the device during the desaturation pulse phase of the turn-off due to larger on-state voltage across the device. It is noted from Fig. 6a however, that the normalized commutation losses for the chip W1 could potentially be decreased to 38% compared to the value 100% without the desaturation pulse, but at larger clamping voltages.

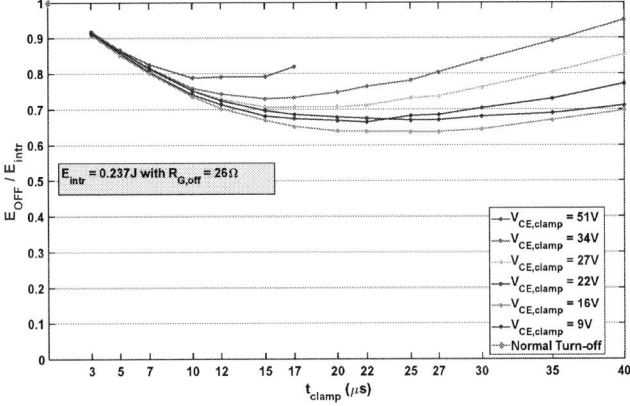

Fig. 5: Summary of Turn-off Measurements with desaturation pulse for low-sat IGBT chip W1

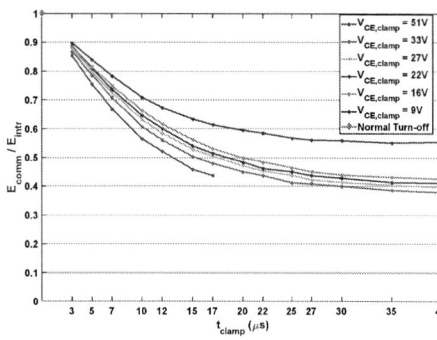

Fig. 6a: Normalized E_{comm} for W1

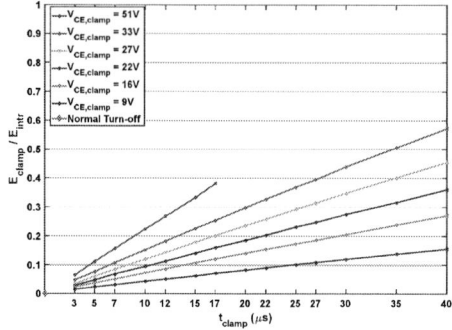

Fig. 6b: Normalized E_{clamp} for W1

Turn-off measurements have also been performed on a different experimental low-sat IGBT chip (W2), which has a $V_{CE,sat}$ of 2.4V at $I_{C,nom}$ and 125°C. This chip is known to have a much lower front-end emitter sided plasma concentration compared to chip W1. A summary of these measurements is shown in Fig. 7a. For this chip, it is seen that a lower $V_{CE,clamp}$ of 12V with a t_{clamp} of 22μs yields a maximum reduction of about 27% in E_{OFF}. The plot of normalized commutation losses corresponding to these measurements is shown if Fig. 7b. It is seen from these results that for comparable values of $V_{CE,clamp}$ and t_{clamp}, the chip W2 shows a smaller reduction in E_{OFF} and also in E_{comm} compared to the chip W1. A larger amount of carrier concentration in chip W1 in the on-state implies that not only the E_{OFF} is higher for this chip but also the potential to reduce the E_{OFF} due to removal of a larger amount of carriers during the desaturation pulse, which is an advantage for chip W1.

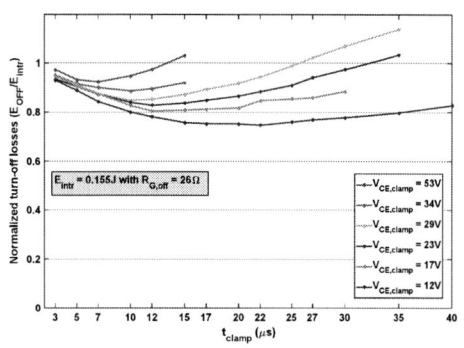

Fig. 7a: Normalized E_{OFF} for chip W2

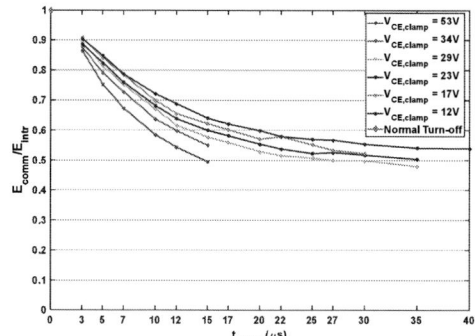

Fig. 7b: Normalized E_{comm} for chip W2

An advantageous application of this turn-off method requires a good performance at collector current values other than the nominal current. Turn-off measurements have been performed for the optimal combination of $V_{CE,clamp}$ and t_{clamp} determined for the nominal current case. Table I shows these results for chip W1. It can be seen that for a range of currents between 20% and 150% of the nominal current, this turn-off method yields a consistent reduction in E_{OFF} similar to the reduction seen in case of the nominal current. This table also shows a small but rather insignificant variation in V_{CE} as a function of the current through the device. This is in contrast to the large dependence of V_{CE} on I_C and various other control parameters from the results reported previously [10] with the feed-forward implementation of the desaturation pulse. Similar measurements for chip W2 have been made at

Table I: Desat-pulse based turn-off measurement results at different I_C for chip W1

I_C (A)	E_{OFF} (intr) (J)	E_{OFF} with desat pulse (J) (5V TVS diode, t_{clamp} 22μs)	% Reduction in E_{OFF}	V_{CE} during desaturation pulse (V)
159	0.316	0.22	30	17.3
120	0.256	0.166	35	16.6
90	0.204	0.132	35	15.8
75	0.175	0.113	36.5	15.3
53	0.133	0.085	36	14.8
21	0.059	0.039	33	14

Table II: Desat-pulse based turn-off measurement results at different I_C for chip W2

I_C (A)	E_{OFF} (intr) (J)	E_{OFF} with desat pulse (J) (No TVS diode, t_{clamp} 22µs)	% Reduction in E_{OFF}	V_{CE} during desaturation pulse (V)
159	0.158	0.116	21	12.8
120	0.175	0.134	23	12.2
90	0.143	0.103	28	10.9
75	0.122	0.09	26	10.3
53	0.095	0.068	29	9.5
21	0.043	0.032	25	8.8

its respective optimal combination of $V_{CE,clamp}$ and t_{clamp} determined for the nominal current case. These results are shown in table II. It can be seen again that this turn-off method shows consistent performance in terms of reduction in E_{OFF} in the entire range of currents considered in the measurement.

The measurements on W1 and W2 could be viewed comparatively in a trade-off diagram where the on-state voltage drop $V_{CE,sat}$ is plotted along the abscissa and the absolute turn-off energy loss E_{OFF} plotted along the ordinate. For each device, the normal turn-off is indicated in black whereas the result from the turn-off with desaturation pulse is shown in red. As could be expected, the result of a lesser $V_{CE,sat}$ for device W1 is an increased E_{OFF} compared to device W2. However, this diagram now brings out the advantage of the desaturation pulse-based turn-off. It can be seen that as a result of the desaturation pulse-based turn-off, the absolute reduction in E_{OFF} for device W1 is much larger than the corresponding reduction in device W2. Therefore, the E_{OFF} values for the devices are now closer to each other compared to the normal turn-off case. Obviously, the best result would have been if the E_{OFF} after the desaturation pulse were to be almost the same while W1 featured a much lower $V_{CE,sat}$ compared to W2.

Investigations in this direction are conveniently pursued with the TCAD simulations approach, which will be presented in the following section.

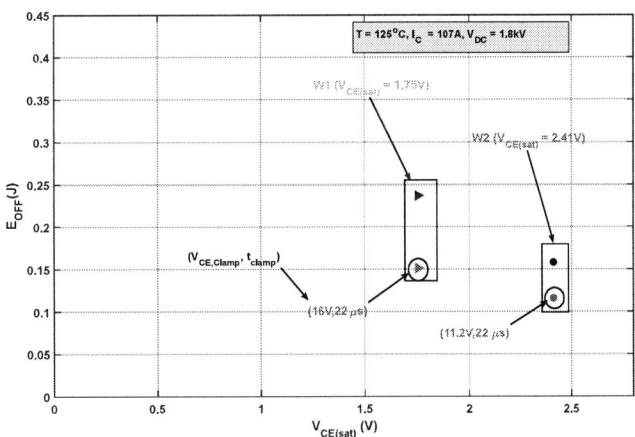

Fig. 8: Trade-off diagram comparing the devices W1 and W2
Legend: Black: Normal turn-off, Red: Turn-off with desaturation pulse

Collector-gate clamping based desaturation pulse – TCAD Simulation Results

TCAD turn-off simulations of the clamping-based desaturation pulse are carried out to further investigate the potential of this turn-off method to determine the extent of reduction in E_{OFF} as well as to obtain an insight into the required profile of the on-state carrier concentration in the device to achieve such a result. A trade-off between $V_{CE,sat}$ and E_{OFF} (normal turn-off) results when the plasma concentration is increased in the device. It will be investigated if along with a lower $V_{CE,sat}$ in such IGBTs, a lower E_{OFF} could also be achieved with the implementation of the desaturation pulse, which is desirable from the point of view of reduction of overall inverter losses. Previous work [10] presented simulation results with the clamping method with a low-saturation IGBT model, which promisingly showed up to 52% reduction in turn-off energy losses. In this section, further simulation results with different low-sat IGBT models are presented.

The turn-off simulation circuit is shown in Fig. 9. The breakdown voltage of diode D determines the collector-gate clamping voltage level. The duration for which the switch Sw is turned on determines the duration of the clamping desaturation pulse during the turn-off of the IGBT.

Fig. 10 shows the on-state plasma profiles of the simulated 3.3kV IGBT models at a nominal current of 450A corresponding to a XHP module. The model M1 features a moderate slope between the front-end emitter and the collector of the device, and resulted in 52% reduction in turn-off energy losses with the clamping-desaturation pulse-based turn-off method [10]. Model M2 has the same back-end emitter plasma concentration but features a higher front-end concentration compared to the model M1 whereas in model M5, the back-end concentration is increased while keeping the front-end concentration unchanged. Model M3 features a reduced back-end concentration compared to, and same front-end concentration as M2 and model M4 has a much higher front-end concentration and same back-end concentration as M3.

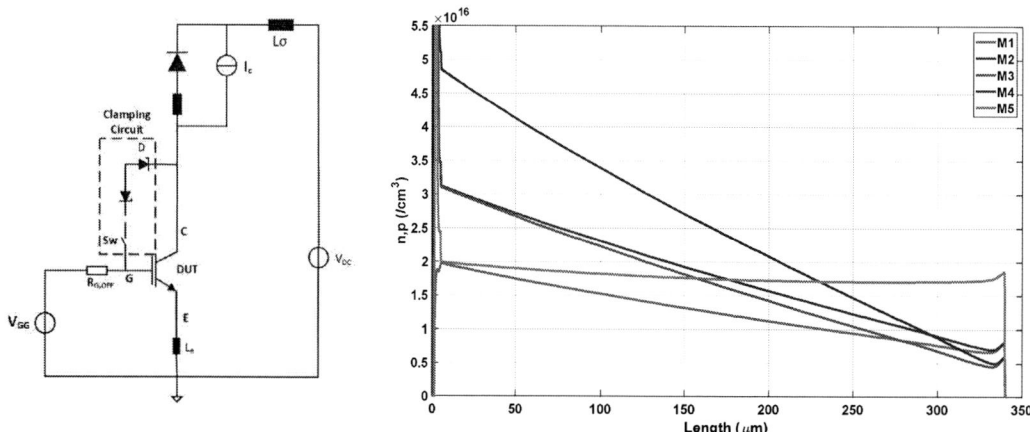

Fig. 9: Simulation Circuit for the clamping-based desaturation pulse

Fig. 10: On-state carrier concentration profiles of simulated models

Fig. 11 shows the summary of desaturation pulse-based turn-off simulations in a trade-off diagram similar to Fig. 9, where the $V_{CE,sat}$ is plotted along the x-axis and the E_{OFF} along the y-axis. The black markers show the points corresponding to normal turn-off whereas the red markers show optimal results from the desaturation pulse-based turn-off. As is evident, models M2, M3, M4 and M5 as expected, have a lower $V_{CE,sat}$ compared to model M1 as a result of higher plasma concentration in the device. Model M4, with the steepest plasma concentration profile between the emitter and collector has a $V_{CE,sat}$ of 1.29V at I_{nom} compared to $V_{CE,sat}$ of 1.96V seen in M1. With the desaturation pulse-based turn-off, the E_{OFF} in case of M4 is reduced from the intrinsic E_{OFF} of 1.031J to 0.42J, which is a significant reduction of about 60%. Therefore, the model M4 not only has a 35% reduced $V_{CE,sat}$ compared to model M1, but also an almost same value of E_{OFF} after the desaturation pulse. The device M4 therefore, offers both a large reduction in on-state conduction losses and with the desaturation pulse-based turn-off, also a higher absolute reduction in E_{OFF} bringing it to a value almost at par with the E_{OFF} of model M1

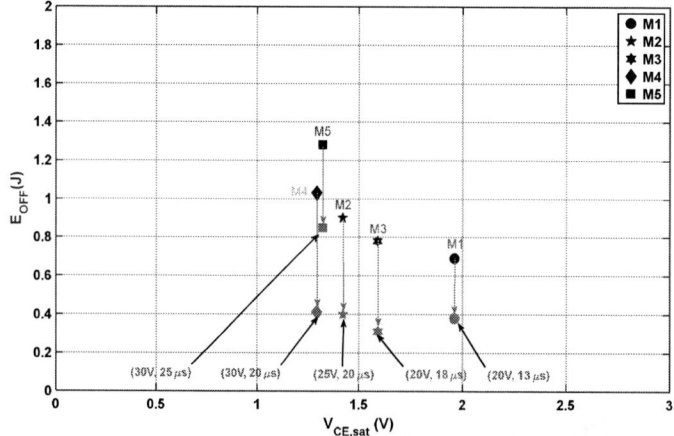

Fig. 11: Trade-off diagram from the desaturation-pulse based turn-off simulations
Legend: Black: Normal turn-off, Red: Turn-off with the desaturation pulse

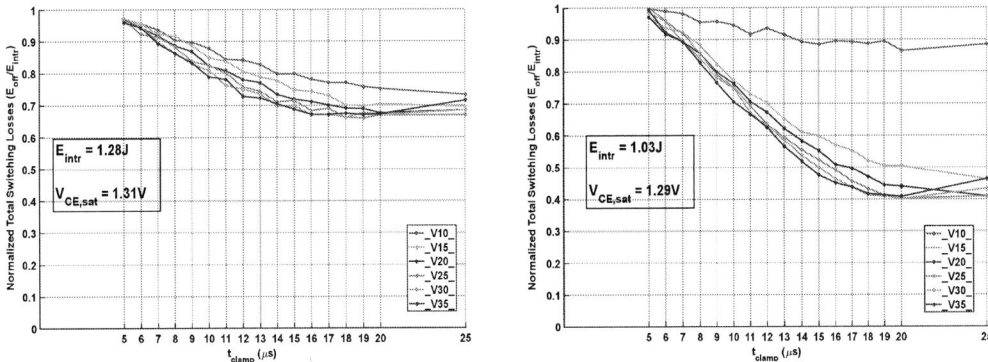

Fig. 12: Normalized E_{OFF} – Left: Model M5 and Right: Model M4

The turn-off energy losses with the desaturation pulse can be split into the conduction losses during the clamping duration and the losses during the final commutation. An analysis of the effect of the plasma profile on the switching and commutation losses provide insight into the desirable plasma profile from the point of view of possible further reduction in turn-off energy losses. The summary of turn-off simulation results for models M4 and M5 are presented in Fig.12, with the normalized E_{OFF} plotted for different collector-gate clamping voltages during the desaturation pulse ($V_{CG,clamp}$) and different values of t_{clamp}. It can be seen that the model M5, with higher collector-sided plasma concentration shows a maximum reduction of about 34% in E_{OFF} at the optimal combination of $V_{CG,clamp}$ = 25V and t_{clamp} = 20µs. However, the model M4 with a much higher front end plasma concentration and much lower back-end doping shows a significantly higher reduction of about 60% in E_{OFF} at the optimal combination of $V_{CG,clamp}$ = 30V and t_{clamp} = 20µs. It is noted, that both these models have almost the same $V_{CE,sat}$. The corresponding normalized commutation losses E_{comm} for both models are plotted in Fig. 13. The reason for a much larger reduction in E_{OFF} with desaturation pulse in case of model M4 is immediately evident from this figure. Model M4 shows a large reduction of about 85% in commutation losses at the determined minimum loss point whereas model M5 shows only about 58% at its respective minimum loss point. Therefore, it could be inferred that a steep on-state plasma profile is desirable from the loss reduction point of view after the desaturation pulse.

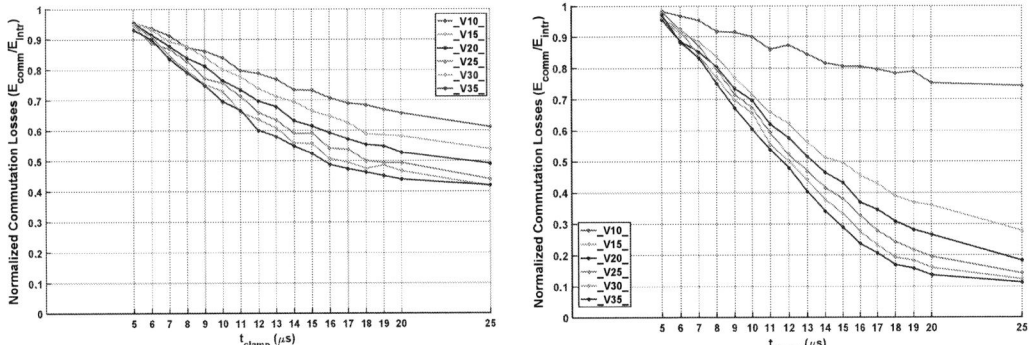

Fig. 13: Normalized E_{comm} – Left: Model M5 and Right: Model M4

The effect of the collector-sided plasma concentration on reduction in E_{OFF} and the necessary t_{clamp} are further investigated by plotting the normalized E_{OFF} and E_{comm} as a function of t_{clamp} at a particular level of $V_{CG,clamp}$ of 20V for the considered simulation models. From Fig. 14a, it is discernable that the model M1 shows the fastest reduction in E_{OFF} with respect to t_{clamp}, whereas the models with the higher device plasma concentration require increasingly larger values of t_{clamp} for a substantial reduction in E_{OFF}. Fig. 14b shows that the corresponding reductions in E_{comm} during the desaturation pulse appear to depend directly on the collector-sided plasma concentration in the device. This is evidenced by the fact that while model M5 shows the lowest reduction in E_{comm}, the models M3 and M4 show the largest reductions. An increase in the emitter sided concentration from M3 to M4 means that M4 requires a longer t_{clamp} to achieve this reduction but results in a better tradeoff between $V_{CE,sat}$ and E_{OFF} after the desaturation pulse as seen from the tradeoff diagram in Fig. 11.

The evolution of plasma inside the device during the progress of the desaturation pulse is shown in Fig. 15 for both the models, for a $V_{CG,clamp}$ of 20V. It can be seen from the figures that at the end of the desaturation pulse, model M5 still contains a large amount of plasma at the collector side as compared to model M4, which has desaturated substantially throughout the drift region of the device. This implies that a large reduction in losses

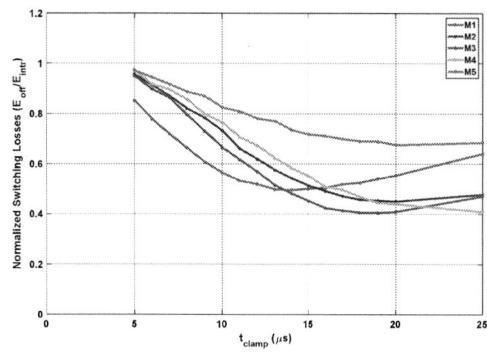

Fig. 14a: Normalized E_{OFF} for all models at $V_{CG,clamp}= 20V$

Fig. 14b: Normalized E_{comm} for all models at $V_{CG,clamp}= 20V$

during the final commutation is possible in case of model M4, thereby leading to a much better result from the desaturation pulse.

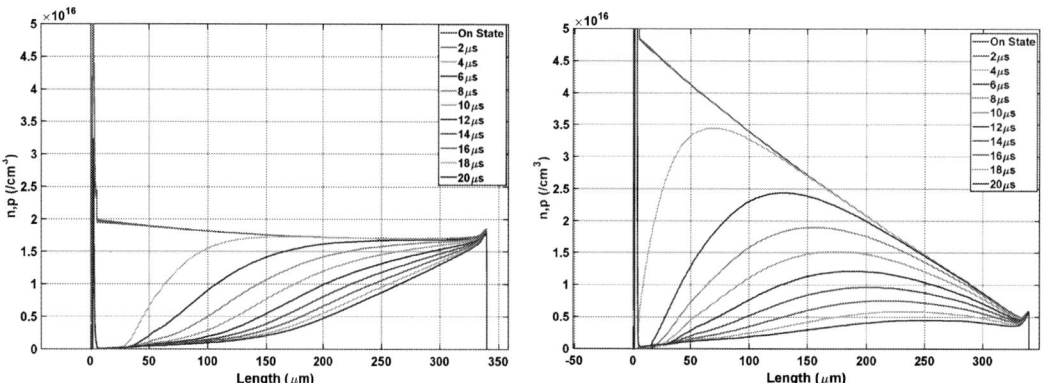

Fig. 15: Evolution of device plasma concentration during the desaturation pulse - Left: Model M5 and Right: Model M4

Conclusion

In this paper, further investigations into the turn-off of modern low-saturation type 3.3kV IGBT devices with the desaturation pulse have been presented. Turn-off measurements with the desaturation pulse based on a novel active-clamping based method applied on two different experimental 3.3kV IGBT chips have been presented. These measurements have shown up to 37% reduction in turn-off energy losses for one of the chips. Further measurements with that this method show a consistent reduction in turn-off energy losses over a range of currents carried by the device. TCAD based turn-off simulation results with desaturation pulse have been presented for different low-saturation IGBT models. These simulations have indicated the possibility of a device having not only a very low on-state collector-emitter voltage drop but also a potentially a much larger reduction of up to 60% in turn-off energy losses for a device with a steep gradient of the on-state carrier concentration profile.

References

[1]   N. Iwamuro and T. Laska, "IGBT History, State-of-the-Art, and Future Prospects", in IEEE Transactions on Electron Devices, vol 64, no. 3, pp. 741-752, March 2017

[2]   J. Lutz et. al, "Semiconductor Power Devices", Springer International Publishing AG, 2018

[3]   N.Kitagawa et al., "A 4500V injection enhanced insulated gate bipolar Transistor (IEGT)", IEEE IEDM Tech., Dig., p.679-682, 1993

[4]   A. Nakagawa, "Theoretical Investigation of Silicon Limit Characteristics of IGBT", Proc. International Symposium on Power Semiconductor Devices and ICs, Naples, 2006

[5]   H.Takahashi et al., "Carrier stored trench- gate bipolar transistor – a novel power device for high voltage application", Proc. of ISPSD, p.349-352, 1996

[6]   C. Jaeger et. al., "A new sub-micron trench cell concept in ultrathin wafer technology for next generation 1200V IGBTs", Proc. International Symposium on Power Semiconductor Devices and ICs, Sapporo, Japan, 2017

[7]   F.-J. Niedernostheide et. al, "Progress in IGBT Development", IET Power Electronics, vol 11, no. 4, pp 646-653, 2018

[8]   M. Bohllaender et. al, "Desaturated switching of Trench-Fieldstop IGBTs", Proc. PCIM Europe 2006, Int. Exhibition and Conf. for Power Electronics, Intelligent Motion, Renewable Energy and Energy Management, Nuremberg, Germany, May/June 2006

[9]   D. Lexow, Q. TienTran, and H.-G. Eckel, "Concept of an IGBT Desaturation Pulse to Reduce Turn-off Losses", 2018 20th European Conference on Power Electronics and Applications (EPE'18 ECCE Europe)

[10] V. Acharya Nayampalli, J. da Cunha, H.-G. Eckel, "Turn-off strategies for low-saturation IGBTs to reduce turn-off losses", 2021 23rd European Conference on Power Electronics and Applications (EPE'21 ECCE Europe)

[11] D. Lexow and H. -G. Eckel, "Performance Improvement for Plug-In Reverse Conducting IGBTs through Gate-Voltage Observation", 2020 22nd European Conference on Power Electronics and Applications (EPE'20 ECCE Europe), 2020

[12] Z. Li, R. W. Maier, M. -M. Bakran, D. Domes and F. -J. Niedernostheide, "How to Turn off SiC MOSFET with Low Losses and Low EMI Across the Full Operating Range", PCIM Europe 2021; International Exhibition and Conference for Power Electronics, Intelligent Motion, Renewable Energy and Energy Management, 2021

Impact of Bond Wire Configuration on the Power Cycling Capability of Discrete SiC-MOSFET Devices

Patrick Heimler, Nick Thönelt, Josef Lutz, Thomas Basler
CHEMNITZ UNIVERSITY OF TECHNOLOGY
Reichenhainer Straße 70
Chemnitz, Germany
Tel.: +49 / (0) 371–531 37907
Fax: +49 / (0) 371–531 837907
E-Mail: patrick.heimler@etit.tu-chemnitz.de
URL: https://www.tu-chemnitz.de/etit/le/

Keywords

«Power Cycling », «Discrete Power Device», «Reliability», «Silicon Carbide (SiC)», «MOSFET»

Abstract

This work investigates the power cycling capability of SiC MOSFETs (60 mΩ/1200 V) in TO-247-packages with two different bonding configurations. As a result, a difference in lifetime by a factor of 1.5 to 2 is determined. The failure mode was an increase in forward voltage drop by degradation of the bond wire connection. The ANSYS simulations (thermal-electrical and mechanical) confirm that several thinner bond wires have a higher power cycling capability compared to a few bond wires with larger diameters.

Introduction

In a power converter, SiC MOSFET devices achieve a higher efficiency compared to their Si equivalents [1]. Furthermore, the temperature dependency of $R_{DS(ON)}$ plays a major role in the application. It was found that SiC MOSFET devices exhibit a smaller increase in resistance compared to their Si counterparts for a temperature rise from 25 °C to 100 °C as an example [2]. Beside these advantages, reliability plays a decisive role. The power cycling test is an important method for accelerated lifetime investigations in order to evaluate the durability of interconnect technologies. This test is used for power modules as well as discrete devices. Therefore, 1.2 kV SiC MOSFETs with the same $R_{DS(ON)}$ and different bond wire configuration in discrete packages were investigated to evaluate their power cycling capability.

Test specimens

SiC MOSFETs from one manufacturer with two different numbers of bond wires and correspondingly different diameters in TO-247-package (see Figure 1) have been selected as test specimens. They have an identical blocking voltage of 1200 V, an $R_{DS(ON)}$ of 60 mΩ and a nominal current of 13 A, respectively. The configuration of the bond wires for test series A consists of 2*375 μm and for B of 3*250 μm bond wires.

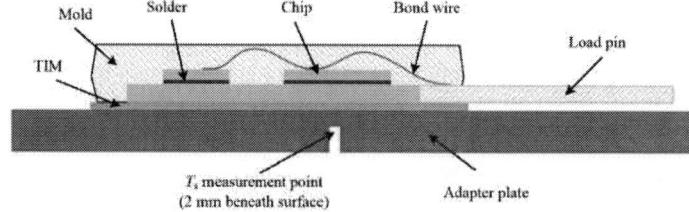

Figure 1: Schematic picture of a cross-section of TO-247 package. Figure from [3]

Results and Discussion

Junction temperature determination

The precise determination of the virtual junction temperature (T_{vj}) is an essential part of the power cycling test to ensure that statements on the determined lifetime are sufficiently accurate. This measurement is done during the test via the body diode with fully closed n-channel by using the $V_{CE}(T)$-method adapted to MOSFETs [4, 5, 6]. An insufficient low off-state gate voltage leads to shifts in the calibration curves during the test, depending on the test mode (Diode or MOSFET) and accordingly to incorrect temperature measurements [7]. Therefore, forward measurements of the body diode at different gate voltages ($V_{GS} = 0 \ldots -11$ V) and temperatures (60 °C and 150 °C) have been carried out before the test with the Keysight B2901A SMU. These results are shown in Figure 2 for a device of test series A. The measurements allow to conclude that a gate voltage of $V_{GS} = -8$ V is sufficient to close the n-channel for these devices. Therefore, it is of major importance that a verification of the closed channel in the test range has been carried out, especially with the temperatures of T_{jmin} and T_{jmax}. However, the required gate voltage may differ from manufacturer to manufacturer [8] and the minimal allowed values specified in the respective data sheets should be taken into account.

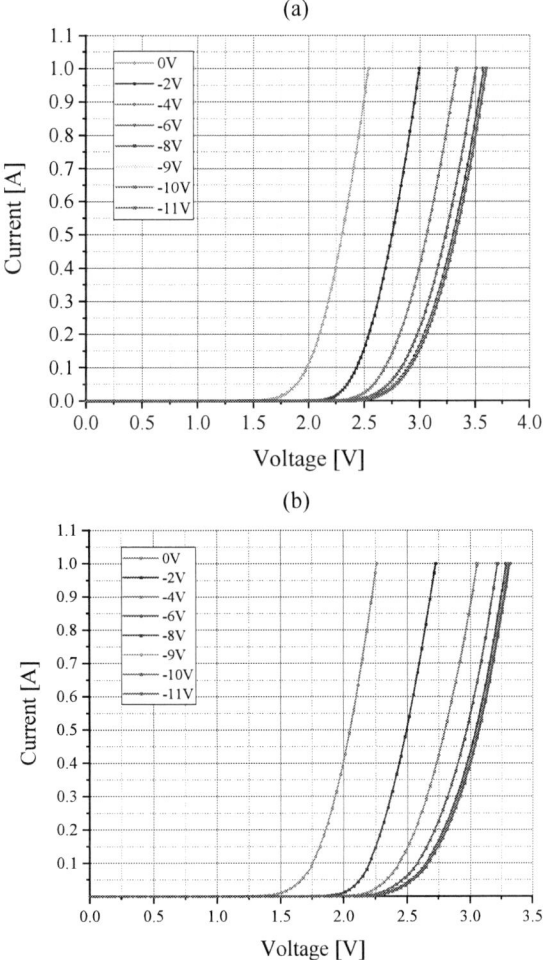

Figure 2: Forward characteristic of the body diode at different gate voltages ($V_{GS} = 0 \ldots -11$ V) at a temperature of 60 °C (a) and 150 °C (b) of test series A

Before and after the power cycling test, temperature calibration curves for the adapted $V_{CE}(T)$ method have been measured in the range from about 50°C to 160°C at constant gate voltage of -8 V. An example for a device of test series A is given in Figure 3. The maximum deviation between the two curves is approx. 1.2 K, but the measurement accuracy of the thermocouple should be considered. It can be stated that there were no changes in the calibration curves and therefore, no incorrect measurements in the temperature determination.

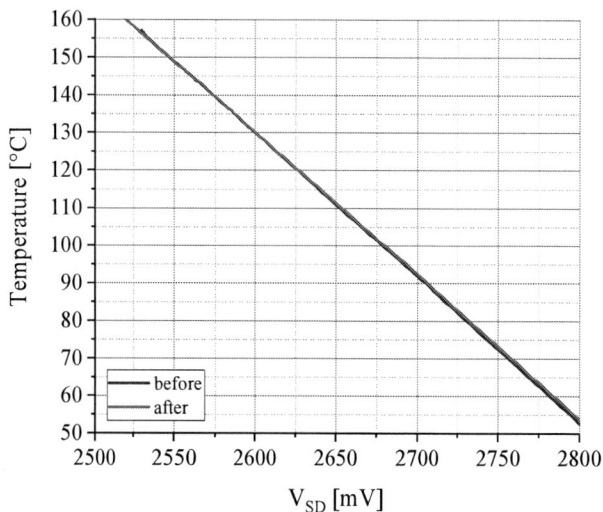

Figure 3: Temperature calibration curves before and after the power cycling test at a gate voltage of V_{GS} = - 8 V for a device from test series A with measurement current I_{mess} = 100 mA

Power cycling capability

The power cycling test was performed for the two test series with the same on-time of t_{on} = 2 s, temperature swing ΔT_j of about 90 K and a maximum junction temperature T_{jmax} of approximately 150 °C. The load current I_{Load} was 21.1 A for test series A and 23.2 A for B, respectively. The gate voltage was set in the range of 14-18 V in order to achieve similar test conditions, such as temperature swing. This means also that the power cycling tests were executed above the temperature compensation point (TCP) and there was a positive temperature coefficient of the $R_{DS(ON)}$. The maximum junction temperature T_{jmax} has been measured 160 µs after switching off the load current and this measurement delay results in an offset of T_{jmax} of approx. 7.5 K, which has been calculated with the square-root-t-method. [9, 10, 11]. The test specimens were attached to the heat sink adapter plates with standard mounting clips. An overview of the test conditions without delay time correction is given in Table 1.

Table 1: Overview of the test conditions for the two test series A and B in the power cycling test

	A	B
T_{jmax} [°C]	146-150	148-151
ΔT_j [K]	87-90	89-92
I_{load} [A]	21.1	23.2
I_{load} per bond or I_{load} per bond foot [A]	10.55	7.73
bond config.	2*375µm	3*250µm
t_{on} [s]	2	
t_{off} [s]	4	
T_{inlet} [°C]	55	
$V_{GS(on)}$ [V]	14-15	15-18
$V_{GS(off)}$ [V]	-8	

Between the two test series, a difference of 100% in power cycling capability can be observed. The development of V_{DS} and R_{thjhs} (junction to heatsink) is shown for one device of each test series in Figure 4 as example. All devices failed with an increase of V_{DS} of + 5 %. The decrease in $R_{th,jhs}$ can be attributed to the run-in behavior of the used thermal conducting foil (Kerafol 86/60), which also serves as an electrical isolation between devices and adapter plates.

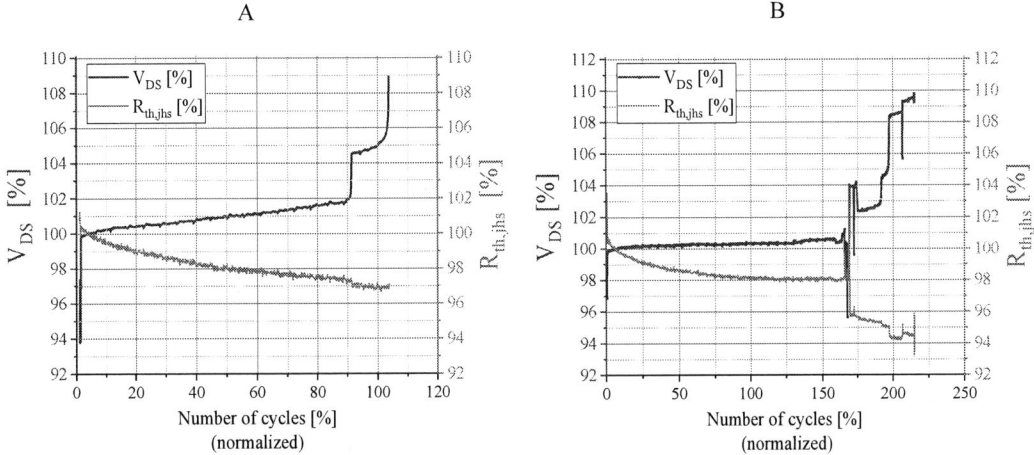

Figure 4: Power cycling test results for the test series A and B

The cycles achieved (raw data) are presented in Figure 5(a). A difference in lifetime by a factor of 1.5 to 2 between the two variants can be observed although the current density per bond of B is higher. For a clearer comparison, the test results are normalized to a temperature swing of 90 K, mean temperature of 105 °C and an application-near nominal current of 13 A according to the data sheet with a lifetime model for discrete devices [3]. This nominal current results in a current per bond or current per bond foot of 6.5 A for A and 4.33 A for B. According to this values, a normalization of the measured current per bond foot was done. The results of this normalization are shown in Figure 5(b) indicating that lifetime of the test series B is two to three times higher than that of the test series A. Furthermore, there is no difference in lifetime between DC power cycling tested SiC MOSFET devices as well as those with additional switching losses [12].

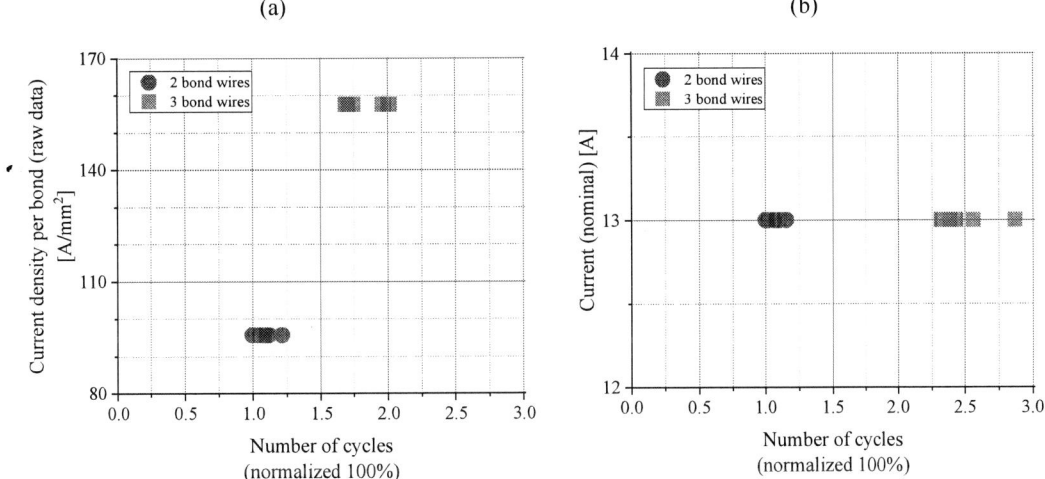

Figure 5: Cycles until EoL: (a) raw data and (b) normalized to the same temperature swing of 90 K, mean temperature of 105 °C and nominal current of 13 A

The normalization was performed according to the following equation with the specific parameters for discrete Si IGBT devices from [3]:

$$N_{\text{f, normalized } (\Delta T_{\text{j}},\, T_{\text{jm}},\, I_{\text{b}})} = N_{\text{f, measured}} \cdot \frac{\Delta T_{\text{j, desired}}^{\alpha}}{\Delta T_{\text{j, measured}}^{\alpha}} \cdot \frac{e^{\frac{E_{\text{A}}}{k \cdot T_{\text{jm, desired}}}}}{e^{\frac{E_{\text{A}}}{k \cdot T_{\text{jm, measured}}}}} \cdot \frac{I_{\text{b, desired}}^{\gamma}}{I_{\text{b, measured}}^{\gamma}} \tag{1}$$

Failure analysis

After the power cycling test, a failure analysis has been performed to determine the cause of failure. In the first step, Scanning Acoustic Microscopy (SAM) images with focus on the bonding wires and the solder layer, respectively, were generated for both test series (see Figure 6). No solder deterioration is observed, but both test series show bond feet degradation.

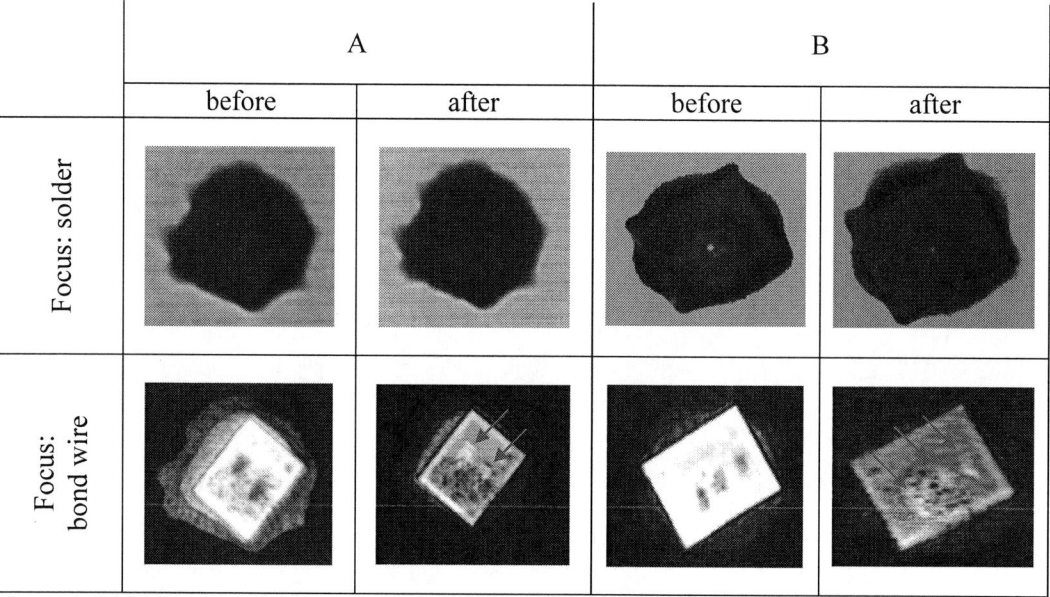

Figure 6: SAM images of both test series with focus on the bond wires and the solder layer

To be able to determine whether the specimen was affected by a bond wire lift off or a heel crack, the failed parts were opened. In test series A, only bond wire lift offs occurred, whereas in test series B both, bond wire lift offs and heel cracks were detected. Such failure modes are also observed with Si IGBTs or diodes in discrete packages [3, 13, 14].

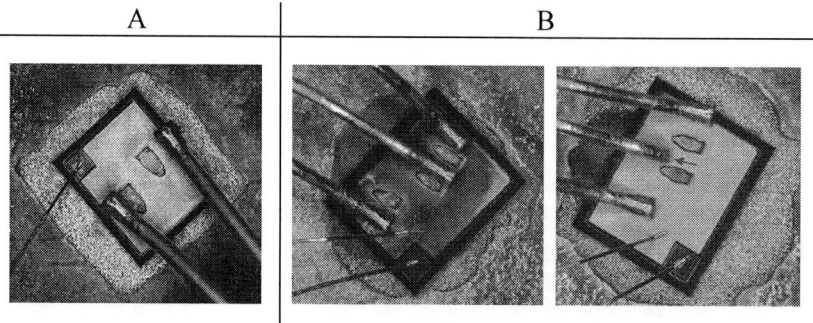

Figure 7: Bond wire lift off and heel crack (red arrow)

Simulation results

In order to gain further insight into the behavior of the test specimens, simulation models were created in ANSYS for both configurations (see Figure 8). The material properties of the individual layers are identical for both models. Only the position of the chip and the bond wire setup have been adapted to the test specimens. 10 cycles were simulated until the system reached a steady state.

A   B

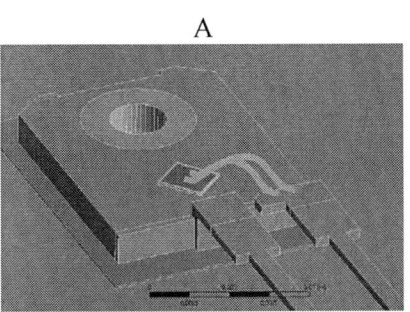 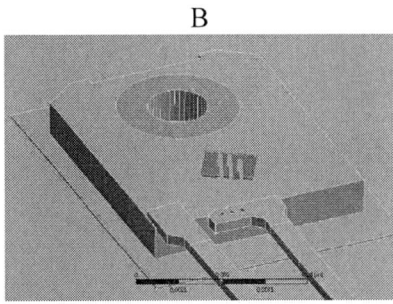

Figure 8. Simulation model for both bond wire configurations (A and B)

The cross-sections along one bond wire are shown in Figure 9. A more homogeneous temperature distribution for the configuration with three bond wires can be observed. The two bond wire configuration results in a more concentrated heat transfer and consequently in a heavily inhomogenious temperature distribution. Therefore, the lower thermal stress due to the temperature distribution of the configuration B correlates with the higher lifetime found in the experiments.

A   B

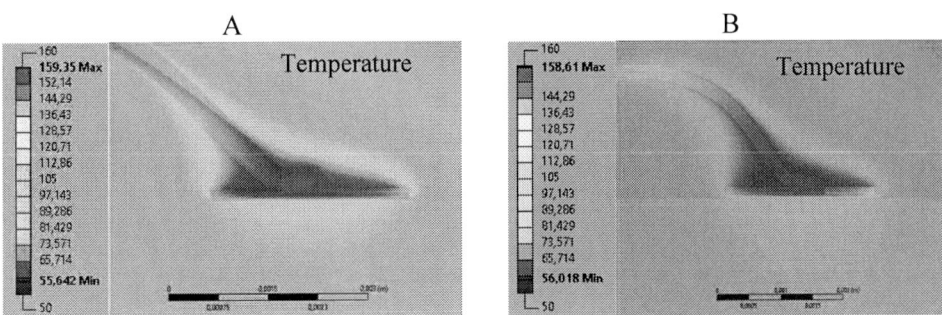

Figure 9: Cross section after the tenth heating cycle (56 s) reaching a steady state along the bond wire for both bond wire configurations (A and B)

The temperature distribution with focus on the chip surface and bond feet for both configurations is presented in Figure 10. At a similar maximum junction temperature of 150 °C, the maximum and minimum temperatures on the chip surface are almost identical for both configurations. The temperature gradient along the corresponding bond feet at A is larger than at B. In addition, the temperature gradient over the bond foot at B is more depending on the position on the chip.

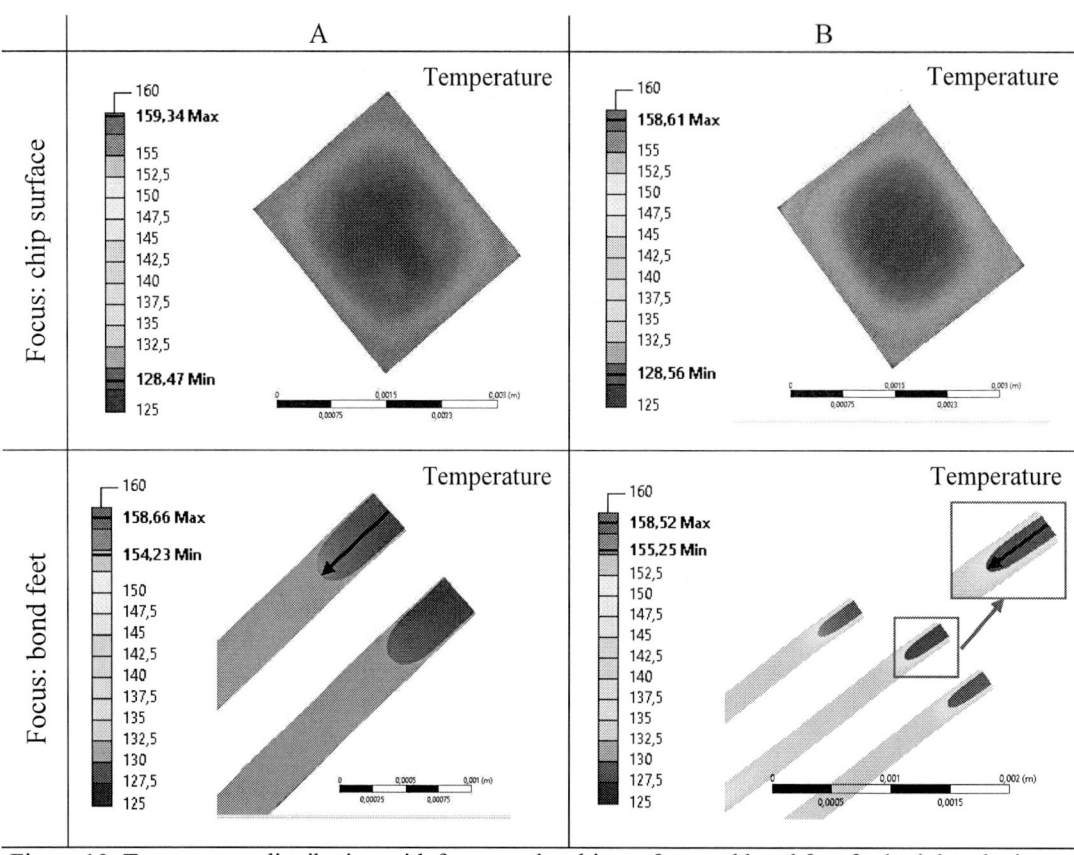

Figure 10: Temperature distribution with focus on the chip surface and bond feet for both bond wire configurations (A and B) after the tenth heating cycle (56 s) reaching steady state

Figure 11 provides a more detailed analysis of the temperature gradient across a selected bond foot for both bond wire configurations. The temperature is plotted along a path according to the black arrows in Figure 10. It is found that the temperature difference for A is approx. 3 K and for B approx. 0.5 K.

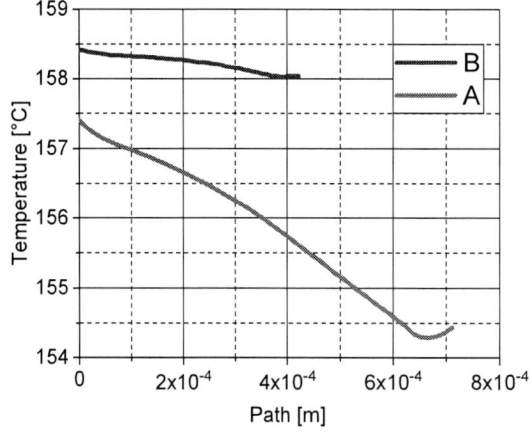

Figure 11: Temperature along the path (according to the black arrows in Figure 10) for both bond wire configurations (A and B) after the tenth heating cycle (56 s) reaching steady state

The total strain per cycle with focus on the metallization is presented in Figure 12 for both configurations (A and B) from the results of [12]. An important factor in these static structural (mechanical) simulation is that the Anand model was used for the solder layer [15] and a model with nonlinear stress-strain-temperature behavior for the metallization and bond wires [16]. Configuration A results in a maximum total strain per power cycling swing of $\varepsilon = 3,88$ m/mm and B of $\varepsilon = 1,96$ mm/m [12]. This means that the maximum strain at B is about two times lower than at A [12]. Therefore, these mechanical simulation findings are being added on top of the more homogenous temperature gradient in the thinner bond wire and confirm the experimental measurement results. Consequently, it follows that three bond wires with a smaller diameter have a higher power cycling capability than two bond wires with a larger diameter.

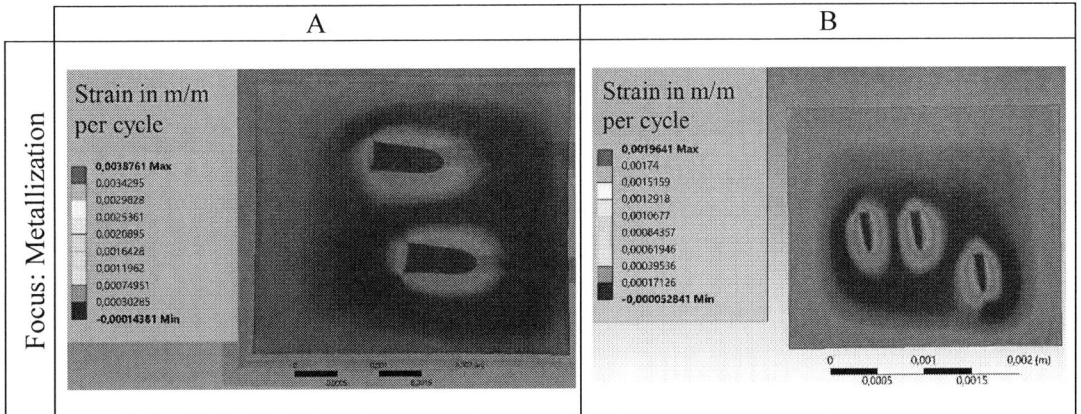

Figure 12: Total strain analysis per power cycling swing for A and B with focus on the metallization. Figures from [12]. Note: The scale is not equal

Conclusion

An influence of the bond wire configuration on the power cycling capability of SiC MOSFETs in TO-247 was found. The lifetime of the test series B with three thinner bond wires is two to three times higher than that of the test series A with two thicker bond wires although the current density per bond of A is smaller after the normalization. This corresponds to the findings for MOSFETs in modules [17]. As cause of failure, an increase of $V_{DS} + 5$ % has occurred for all test specimens and bond wire fatigue has been verified as the dominant mechanism by failure analysis. The ANSYS simulation shows a more homogeneous temperature distribution for the three bond wire configuration than for the two bond wire configuration. Also the mechanical simulation shows a higher total strain per power cycling swing for the 2 bond wire configuration. Therefore, the number and the associated different diameter of the bond wires have a significant influence on the power cycling capability and should be considered during the chip and package design.

References

[1] R. Siemieniec, R. Mente, W. Jantscher, D. Kammerlander and U. Wenzel, "600 V power device technologies for highly efficient power supplies," 2021 23rd European Conference on Power Electronics and Applications (EPE'21 ECCE Europe), 2021, pp. P.1-P.10, doi: 10.23919/EPE21ECCEEurope50061.2021.9570498.

[2] R. Siemieniec, R. Mente, W. Jantscher, D. Kammerlander, U. Wenzel and T. Aichinger, "650 V SiC Trench MOSFET for high-efficiency power supplies," 2019 21st European Conference on Power Electronics and Applications (EPE '19 ECCE Europe), 2019, pp. P.1-P.16, doi: 10.23919/EPE.2019.8915154.

[3] G. Zeng, L. Borucki, O. Wenzel, O. Schilling and J. Lutz, "First Results of Development of a Lifetime Model for Transfer Molded Discrete Power Devices," PCIM Europe 2018; International Exhibition and Conference for Power Electronics, Intelligent Motion, Renewable Energy and Energy Management, 2018, pp. 1-8.

[4] U. Scheuermann and R. Schmidt, "Investigations on the VCE(T)-Method to Determine the Junction Temperature by Using the Chip Itself as Sensor," in PCIM Europe, 2009

[5] F. Hoffmann and N. Kaminski, "Evaluation of the VSD-method for temperature estimation during power cycling of SiC-MOSFETs," IET Power Electronics, 2019, 12. Jg., Nr. 15, S. 3903-3909

[6] C. Herold, J. Sun, P. Seidel, L. Tinschert and J. Lutz, "Power cycling methods for SiC MOSFETs," 2017 29th International Symposium on Power Semiconductor Devices and IC's (ISPSD), 2017, pp. 367-370, doi: 10.23919/ISPSD.2017.7988994.

[7] M. Gerlach, R. Boldyriew-Mast, F. Bruchhold, J. Lutz, T. Basler and H. Schwarzmann, "Influence of different test strategies on the power cycling test results of 6.5 kV SiC MOSFETs," in Microelectronics Reliability vol. 126, Elsevier, 2021

[8] C. Kempiak and A. Lindemann, "Impact of Threshold Voltage Instabilities of SiC MOSFETs on the Methodology of Power Cycling Tests," PCIM Europe digital days 2021; International Exhibition and Conference for Power Electronics, Intelligent Motion, Renewable Energy and Energy Management, 2021, pp. 1-9.

[9] D. L. Blackburn and F. F. Oettinger, "Transient thermal response measurements of power transistors." IEEE Transactions on Industrial Electronics and Control Instrumentation 2 (1975): 134-141

[10] C. Herold, M. Beier, J. Lutz and A. Hensler, "Improving the accuracy of junction temperature measurement with the square-root-t method," 19th International Workshop on Thermal Investigations of ICs and Systems (THERMINIC), 2013, pp. 92-94, doi: 10.1109/THERMINIC.2013.6675204.

[11] E. Deng and J. Lutz, "Measurement Error Caused by the Square Root t Method Applied to IGBT Devices during Power Cycling Test," 2020 32nd International Symposium on Power Semiconductor Devices and ICs (ISPSD), 2020, pp. 545-548, doi: 10.1109/ISPSD46842.2020.9170083.

[12] C. Schwabe, N. Thönelt and T. Basler, "Reliability investigation of SiC MOSFETs under switching operation in various packages," Proceedings CIPS2022.

[13] G. Zeng, "Some aspects in lifetime prediction of power semiconductor devices," PhD thesis, Universitätsverlag Chemnitz 2019

[14] R. Amro, J. Lutz and A. Lindemann, "Power cycling with high temperature swing of discrete components based on different technologies," 2004 IEEE 35th Annual Power Electronics Specialists Conference (IEEE Cat. No.04CH37551), 2004, pp. 2593-2598 Vol.4, doi: 10.1109/PESC.2004.1355239.

[15] Z. Johnson, "A Compilation of Anand Parameters for Selected SnPb and Pb-free Solder Alloys," in DOI: 10.13140/RG.2.2.10895.00166, 2012.

[16] Y. Yogo, M. Sawamura, R. Harada, K. Miyata, N. Iwata and T. Ishikawa, "Stress-strain curve of pure aluminum in a super large strain range with strain rate and temperature dependency," in Procedia Engineering Bd. 207, 2017

[17] C. Ehrhardt, U. Geissler, J. Hoefer, M. Broll, M. Schneider-Ramelow, K.-D. Lang and S. Schmitz, "Influence of wire material and diameter on the reliability of Al-H11, Al-CR, Al-R and AlX heavy wire bonds during power cycling," CIPS 2016; 9th International Conference on Integrated Power Electronics Systems, 2016, pp. 1-6.

A Low-Leakage, Low-Loss Magnetic Transformer Structure for High-Frequency Applications

Allen Nguyen, Ajinkya Phanse, Michael Solomentsev, Alex J. Hanson
University of Texas at Austin
2501 Speedway
Austin, USA
Phone: +1 (512) 232-8110
Email: {allentn372}, {ajinkya.phanse}, {mys432}, {ajhanson}   @utexas.edu
URL: https://www.utexas.edu/

Acknowledgments

This material is based upon work supported by Enphase Energy.

Keywords

≪Transformer≫, ≪Magnetic device≫, ≪Conduction losses≫, ≪Core loss≫

Abstract

Energy-storage transformers, such as transformers for flyback or LLC converters, have different design constraints than typical transformers. Since primary and secondary currents are not in phase, interleaving does not necessarily reduce high-frequency losses. Such transformers often must be designed with low leakage as well. In this work, we propose design guidelines for a transformer structure that uses field shaping to achieve current conduction along most of the skin of the conductors (double-sided conduction), equal current sharing between paralleled turns, even for out-of-phase currents, and near zero MMF drop across the leakage reluctance paths. The transformer therefore has low leakage inductance and low conduction loss without the use of litz wire and can be used effectively at frequencies beyond a few megahertz. Step-by-step design guidelines are proposed and a prototype transformer is built which achieves a leakage to magnetizing ratio of 1.12%, a power loss $14 - 17\%$ of a traditional lumped-gap transformer, and current sharing variation less than 1.5% between paralleled turns.

Introduction

Energy-storage transformers must be designed for low conduction loss; but high-frequency effects pose several challenges. Skin depth limitations already lower the effective cross sectional area in which current flows. Then when unbalanced H-fields are present (typically seen with lumped gaps), additional eddy currents within conductors and poor current sharing between parallel conductors can result in orders-of-magnitude higher losses. These adverse high-frequency conduction patterns cannot be mitigated by the use of litz wire above a few MHz where the skin depth becomes thinner than 48 AWG wire [1]. Thus, effective strategies are needed to achieve low conduction loss at MHz frequencies without relying on litz wire.

Energy stored in the leakage inductance of a transformer can also significantly deteriorate the efficiency of a power converter if it is not recovered in each switching cycle. The flyback converter, for example, first stores energy in the magnetizing inductance of the transformer (and inadvertently in the primary leakage) before transferring the magnetizing energy to the output. The energy stored in the leakage is lost if no additional circuitry for recovering the energy is included [2]. For typical leakage-to-magnetizing inductance ratios (2-5%), this efficiency loss is often unacceptable. Even when the leakage problem is directly addressed through added circuitry, the design of auxiliary leakage-recovery circuits become more constrained for larger leakage inductances. Minimizing leakage inductance is therefore a continuing challenge for many power converter designs.

Leakage is typically reduced by tightly packing ([3, 4]) and/or interleaving the windings. These strategies pose challenges at high frequencies, as tight packing will increase parasitic capacitance [5] and interleaving may not produce good current distributions in energy-storage transformers (such as for flyback or LLC converters) where the primary and secondary current are not in phase.

In this work we propose a low-leakage energy-storage transformer structure for high-frequency applications. The proposed structure uses distributed gaps to shape the H-fields around the conducting wires, resulting in near-zero MMF drops across the core window. This feature forces the flux across the window (leakage) to be near zero, independent of the inter-turn spacing. In addition, the distributed gap balances the H-fields adjacent to two sides of the conductors, even for out-of-phase currents, resulting in a more even distribution of current densities (double-sided conduction) [6]. We use Finite Element Analysis (FEA) simulations to evaluate the transformer and to identify design guidelines. We then test a hardware prototype that achieves a low leakage to magnetizing inductance ratio alongside low power losses and successful current sharing among parallel-connected turns.

Geometry Overview

Fig. 1 shows a cross-section of the proposed transformer structure, which resembles a pot core with a center post, an outer shell, and top and bottom core pieces. Instead of a single air gap, the proposed structure has multiple smaller gaps in both the center post and outer shell that form a quasi-distributed gap [7]. The number of gaps is the same as the number of turns in the transformer, which enables the field-shaping approach proposed in the following sections. The primary and secondary windings of the transformer are interleaved, and each turn of the primary winding is paired with a secondary turn and each such pair is aligned with one core section (for ideal circular turns – we explore the implications of helical turns in the following sections).

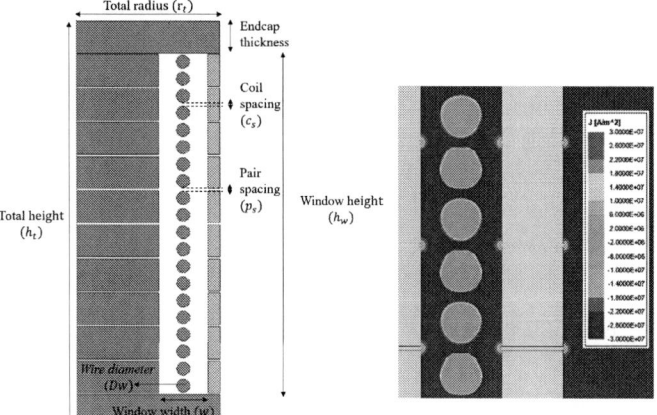

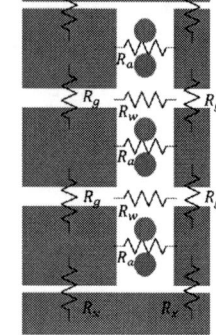

Fig. 1: Radial cross-section of the proposed transformer

Fig. 2: FEA simulation of balanced H-fields and double-sided conduction

Fig. 3: Transformer showing the various reluctances in the magnetic flux paths

Balanced H-fields for Low Conduction Loss

The introduction of smaller, distributed gaps allows us to strategically manipulate the H-fields around the conductors. The first use of this is to balance the H-fields around each individual conductors. High-frequency current crowds near the conductor edge adjacent to the highest H-field region, increasing copper loss. By balancing the MMF drops across the gaps of the center post and the outer shell we can achieve balanced H-fields on two sides of the conductor, causing current to evenly distribute on those two sides (known as double-sided conduction [6]). This more even distribution of the current reduces copper losses. An example of balanced H-fields achieving double sided conduction is shown in Fig. 2. This balancing is achieved by making the total reluctance on the center-post equal to the total reluctance in the outer-shell. Fig. 4 shows the lumped reluctance model of the structure. Here, Rc_{post} is the total reluctance of the core pieces in the center post and Rg_{post} is the total reluctance of the air gaps in the center post. Similarly, Rc_{shell} and Rg_{shell} are the reluctances of the core and the air gaps of the outer shell and R_f is the reluctance faced by fields that fringe outside of the structure. Balancing is achieved by enforcing

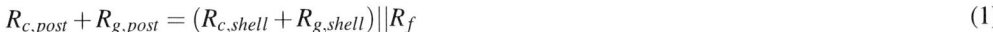

$$R_{c,post} + R_{g,post} = (R_{c,shell} + R_{g,shell}) \| R_f \qquad (1)$$

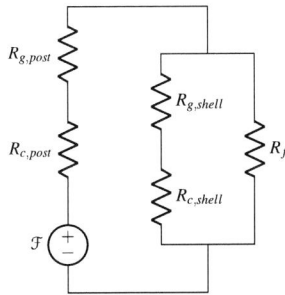

Fig. 4: Magnetic circuit model used to balance the H-fields in the proposed structure

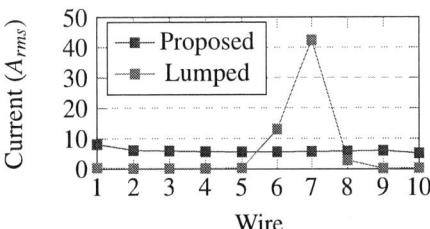

Fig. 5: Current distribution as seen in FEA simulation of the proposed transformer vs. a lumped-gap structure

Double-sided conduction has been achieved in transformers [8], but only for in-phase currents. In this case, double sided conduction is achieved when the currents in the two windings are out of phase (for example, in a flyback converter where only one winding carries current at a time). For high-current or high-turns-ratio transformers, it may also be desirable to arrange multiple turns in parallel. On its own, this can be beneficial for leakage inductance by permitting current to distribute itself to minimize stored energy [5]. However, in the presence of a lumped gap, the current distribution in the paralleled turns can be very uneven, with the turns closest to the region of high H-field carrying the most current.

In the proposed transformer geometry, the turns can be paralleled while still achieving approximately the same net current in each turn. This is a consequence of each conductor pair being in a magnetically similar environment as achieved through the quasi-distributed gap. Fig. 5 shows an FEA comparison between the current distribution in the paralleled turns of the proposed transformer against an identical transformer with a lumped gap (that spanned the center post and outer shell) of the same net reluctance. The proposed approach achieves much better current sharing than the conventional approach, decreasing conduction loss by a very substantial amount. As an example, when the primary is excited with 5 A and the secondary port is open, the conduction loss in the proposed structure was 1.44 W compared to the lumped gap structure with 44.1 W.

MMF Cancellation for Low Leakage Inductance

The quasi-distributed gap structure grants a great deal of design freedom, which we use here to achieve low leakage. We propose to use the many available gaps to balance the MMF drops around the main magnetic path such that the MMF drop across the window (the path of leakage flux through R_w) is zero. Fig. 3 highlights the non-negligible reluctances in different magnetic paths in the proposed structure for a three-turn transformer, with the magnetic circuit equivalent shown in Fig. 6. Each turn of the primary is paired with a turn of the secondary; this will be known as a conducting pair, the space between the primary and secondary turn in a conducting pair will be denoted as conductor spacing (c_s), and the space between conducting pairs will be denoted as pair spacing (p_s). We use the constraint of double-sided conduction to enforce that the reluctance of the center post and the outer shell be equal. For initial

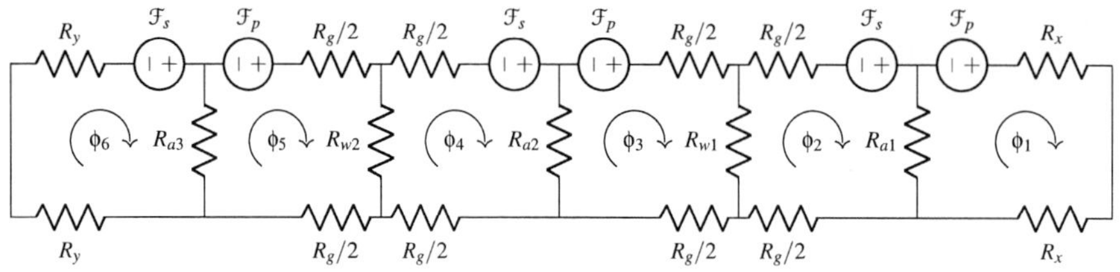

Fig. 6: Magnetic circuit model of the structure shown in Fig. 3

calculations we will assume that there are no large fringing fields outside the structure (i.e assume R_f is negligible) and that only the leakage path associated with the gap between two conducting pairs (R_w) is significant. We will introduce the leakage path associated with the gap between the primary and secondary wire in a conducting pair (R_a) afterwards.

Consider the structure of Fig. 3 excited only by a single winding with the vast majority of flux flowing through the core-and-gap path (not crossing the window). It is plain to see, then, that there will be MMF *gains* traveling up the center post and MMF *drops* traveling down the outer shell. The total MMF gains and drops must be equal; it is therefore conceivable that the sum of the MMF gains and drops around each primary-secondary conductor pair could be engineered to *individually* be zero. This would imply that the MMF drops across the window would also be zero. For this balanced condition, the inclusion of the R_w paths would not change the result because the flux through each R_w would be equal to zero (not just approximately zero). Finally, because the flux across the window (the only flux that does not couple all of the primary and secondary turns) is zero, the leakage inductance must also be zero. In the Appendix, we derive that this such MMF cancellation is possible under the condition that $R_x = R_g/2$.

It is worth dwelling on this conclusion. While most approaches seek to minimize leakage by maximizing *reluctances* associated with leakage (e.g. by tightly packing turns), this approach recognizes the possibility of minimizing the *MMF drop* across such reluctances through judicious balancing of reluctances around the main flux path. In principle, it is possible to reduce the main contributors to leakage inductance to zero. In practice, some flux will be generated locally around each wire and in between the primary/secondary wires, so we expect low but non-zero leakage leakage.

Finite reluctance between the paired primary/secondary turns (R_a in Fig. 3) permits some leakage flux to flow. We examine the effect of R_a numerically by calculating the elements of the inductance matrix through the flux linkage equation, $\underline{\lambda} = \underline{L}\,\underline{i}$. When the primary is driven and the secondary is left open, the flux flowing through the V_p MMF sources correspond to the total flux that is linking with the primary winding, for which we adopt the notation (λ_{1po}) [flux linkage in the (1)st (primary) winding with the (p)rimary driven and the other winding (o)pen]. A summary of the open-circuit test equations are shown below.

$$L_{11} = \frac{\lambda_{1po}}{I_1} = \phi_1 + \phi_3 + \phi_5 \qquad L_{22} = \frac{\lambda_{2so}}{I_2} = \phi_2 + \phi_4 + \phi_6 \tag{2}$$

$$L_{12} = \frac{\lambda_{2po}}{I_1} = \phi_2 + \phi_4 + \phi_6 \qquad L_{21} = \frac{\lambda_{1so}}{I_2} = \phi_1 + \phi_3 + \phi_5 \tag{3}$$

$$L_{l,p} = L_{11} - L_{12} \qquad\qquad\qquad L_{l,s} = L_{22} - L_{21} \tag{4}$$

The previous conclusion that leakage will be zero when $R_x = R_g/2$ is not true anymore and the optimization of gap lengths and expected leakage must be explored. A MATLAB script was written to perform circuit analysis of the magnetic circuit of Fig. 3, with leakages calculated according to the equations above. Fig. 7 shows the variation in the leakage inductance when the gap reluctances that border the end caps (R_x and R_y) are changed while holding constant the reluctances of flux paths through the window (R_a's and R_w's) and the total reluctance of the gaps ($2R_x + 2R_y + 4R_g$). Thus the variable $R_x/(R_x + R_y)$ becomes a proxy for how symmetric the gaps at the end caps are and $(R_x + R_y)/(R_g/2)$ becomes a proxy for how distributed the air gaps are. Fig. 7 then shows that a symmetric structure, i.e. $R_x = R_y$, is desirable to minimize leakage, though an unbalanced structure does not deviate far from optimal. Fig. 7 also shows that the sum of the primary and secondary port leakages (L_{lkg}) decreases as $(R_x + R_y)/(R_g/2)$ increases. This means that if we are designing a transformer with leakage inductance as a prime design parameter, then a symmetric structure with lumped gaps (and no other gaps) at the endcaps will achieve a lower leakage inductance than the proposed structure. However, the graph is quite flat and rarely are transformers designed considering only the leakage inductance. Instead, the proposed design with $R_x = R_y = R_g/2$ has some important advantages (low leakage achieved alongside better current distributions in the conductors and between paralleled conductors) compared to a structure with lumped gaps near the endcaps. Fig. 8 shows the dependence of L_{lkg} on the window reluctance. Because the

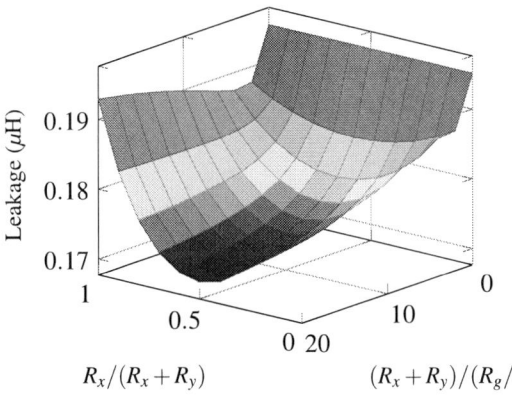 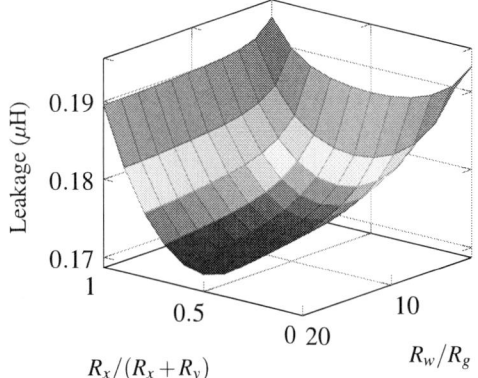

Fig. 7: Leakage inductance as a function of $R_x/(R_x+R_y)$ and $(R_x+R_y)/(R_g/2)$

Fig. 8: Leakage inductance as a function of $R_x/(R_x+R_y)$ and R_w/R_g

MMF drop across R_w is close to zero, it can be seen that the leakage path reluctance has little influence on leakage inductance as long as R_w is much greater than R_g. This is expected to be the case in most designs.

Lastly, we introduce simulations to discuss the effect helical turns rather than circular turns such that conductor pairs will not always be centered on their corresponding core piece, as analyzed. To investigate this issue, we run 2-D simulations with windings misaligned from the center of the core (displaced vertically). The leakage inductance is shown to not be sensitive to this misalignment (Fig. 9). Overall, careful placement of the air gaps reduce the leakage inductance of the structure and reduces parameter sensitivities (such as window reluctances and wire placement) on this leakage.

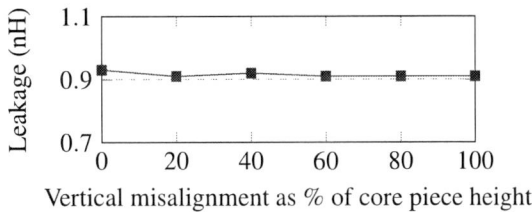

Fig. 9: Total leakage inductance as seen in FEA simulations vs. misalignment of the winding.

Design Guidelines for an Optimized Design

Core Material: Core material selection is application dependent. High-frequency applications are likely to be core loss limited rather than saturation limited and core material is selected based on its performance factor in the desired frequency range as compared in [9].

Number of Turns: The turns ratio and magnetizing inductance are assumed to be set by the application. The total number of turns (and corresponding total core reluctance) is usually chosen by trading off copper loss and considerations relating to the core (saturation and core loss). When the application is saturation limited, the minimum number of turns consistent with avoiding saturation is used. When the application is core loss limited, an optimal balance between core loss and copper loss is found. High-frequency (MHz) applications are more likely to be core loss limited than saturation limited

The proposed structure uses the same number of conductors for the primary and the secondary. Therefore, for non-unity-turns-ratio transformers, it may be necessary to connect some turns in parallel on the low-turns side. The optimal way to do this may require additional FEA investigation, and may be difficult to optimize for turns ratios in which one side is not a multiple of the other. For example, a 6:3 transformer will have 6 secondary conductors with pairs connected in parallel to form 3 turns (there is still some flexibility to determine which pairs are connected together). By contrast, a 7:3 transformer may have a variety of sensible ways to form 3 turns from 7 conductors on the secondary. For high-current transformers, it is possible to use parallel turns on both the primary and secondary. A 1:1 transformer could have N conductors on both the primary and secondary, all connected in parallel on each side. Nevertheless, for many designs it may be more advantageous to simply use as few parallel-connected wires as possible.

Vertical and Horizontal Fill: For a given window height, large wire diameter restricts the available spacing between adjacent conductors but provides larger conduction area (even in skin-depth-limited designs) leading to lower copper losses. Small space between conductors in a conducting pair causes the little flux that does flow across the window in the conductor spacing region (c_s) to be concentrated with high H-fields, causing current to concentrate in that narrow area. Thus it is important to choose an appropriate wire diameter for a given window height. Fig. 10 and Fig. 11 show the results of simulations

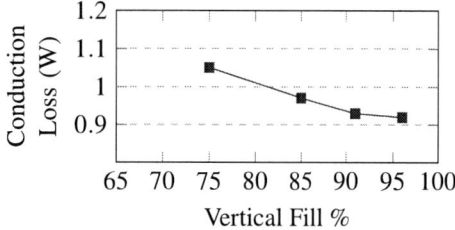

Fig. 10: Conduction losses plotted against vertical fill when c_s is constant p_s is decreased

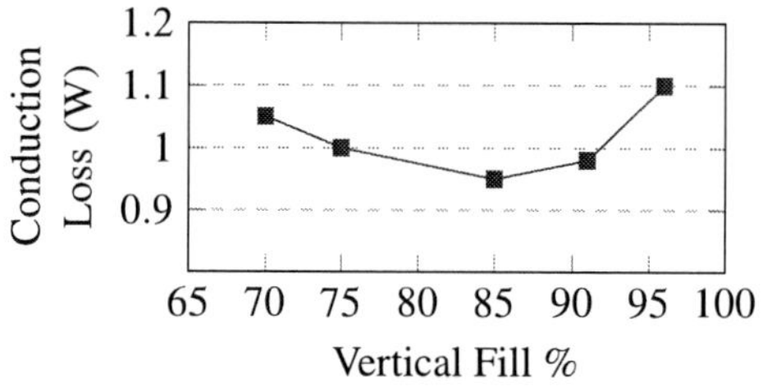

Fig. 11: Conduction losses plotted against vertical fill when p_s is constant c_s is decreased

performed for a 10:10 transformer with an approximately 1:1 aspect ratio (diameter to height). In Fig. 10 the spacing between the conducting wires in a conducting pair (c_s) constant while varying the wire size which is accommodated by varying the spacing between two conducting pairs (p_s). In Fig. 11 we do the opposite, and held p_s constant while varying c_s. While performing these simulations, the width of the window was kept three times the diameter of the biggest wire that would fit (wire corresponding to a vertical fill of 100%) so that the fielding effects due to the changing wire diameter (i.e. R_f, the fields that fridge outside the structure) are negligible. It can be seen that the conduction losses do not increase at higher vertical fill when p_s is decreased (simulated through increased wire diameter). This result is because of the near zero H-field in that region due to the near zero MMF drop across R_w in the window which does not result in any change in current crowding. However, decreasing c_s leads to increased losses. Thus, for a unity turns ratio, it is recommended to have very low pair spacing (p_s) while selecting wire diameter to achieve a vertical fill of $75 - 90\%$. It may be possible for this spacing to be automatically applied using the insulation of the wires.

Once the wire diameter has been chosen to keep the conduction losses and leakage inductance low, the window width should be chosen such that the total losses in the transformer are minimized. A small window gives more core area and reduces core loss but may expose the wires to fringing fields from the small gaps and increases the leakage flux that flows in the conductor spacing region. Fig. 12 shows the core and copper losses for different horizontal fills (while holding vertical fill at 85% and simulating with a square aspect ratio).

It can be seen that the optima are shallow and maintaining a horizontal fill between $30 - 60\%$ keeps the total losses low. Larger values of horizontal fill (narrower windows) lead to a larger leakage inductance, incentivizing lower horizontal fill factors for leakage constrained designs. However, this is the conclusion with a square aspect ratio, as shown in the next section, different aspect ratios will require different horizontal fills to reach power loss minimums. Thus we recommend to select the aspect ratio early in the design process.

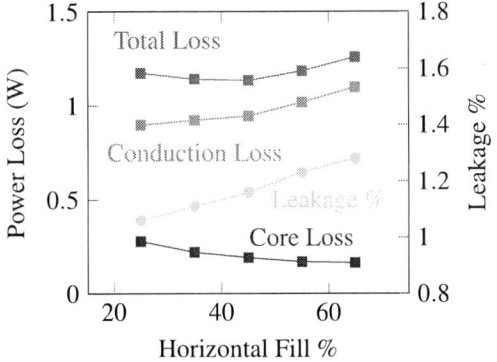 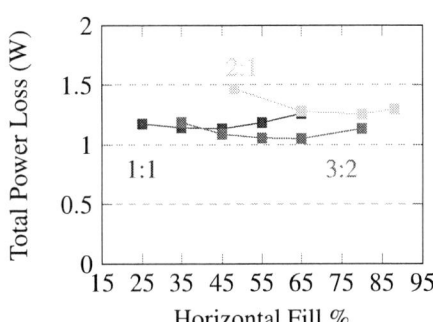

Fig. 12: Power losses observed in FEA simulations for different horizontal fill (all structures had a vertical fill of $\approx 85\%$ and a square aspect ratio)

Fig. 13: Total power loss observed in FEA simulations for different Aspect Ratios (all structures had a vertical fill of $\approx 85\%$)

Aspect Ratio: For a given volume, the transformer can have an aspect ratio (height:diameter) that is square (1:1) or it can be tall and skinny or short and wide. This design choice has several implications for other aspects of the design, such as wire length, wire spacing, gap length, etc. For example, a taller structure would allow the use of wires with larger diameters for the same vertical fill resulting in lower copper losses, but would result in smaller core cross-sectional area, increasing core losses. To observe the effect of the aspect ratio, we ran several simulations of transformers with the same volume and different aspect ratio while optimizing the horizontal fill (window width) under the same vertical fill ($\approx 85\%$). As observed by Fig. 13, relatively square aspect ratios are preferred, with the optimal closer to 1.5:1.

Gap Lengths: Since magnetizing inductance and number of turns have been fixed, the number of gaps and the gap size are determined from the center post area and the outer shell area (with an optional correction factor to account for fringing fields).

End Caps: The choice of end cap size is a trade off between cross-sectional area (height) and transformer volume. The trade off is not linear, however, as flux tends to crowd near the edges of the window. It is recommended to start the design process with a reasonable end cap height, then tune the height of the end caps after the choice of the center-post radius, outer-shell thickness and wire diameter have been made. Fig. 14 below shows the comparison between the end cap volume to the end cap core loss contribution.

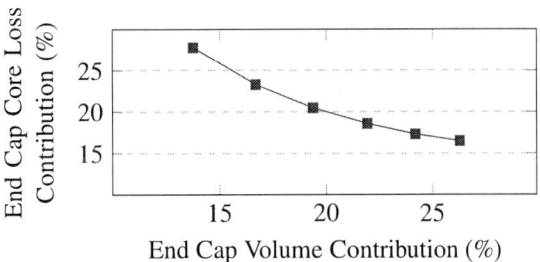

Fig. 14: Increasing end cap thickness and its effects on lowering core loss

This figure shows as the percentage of end cap volume to the total volume ($volume_{endcap}/volume_{total}$) is increased past $\approx 25\%$, the reduction in core loss is not substantial. For near-square aspect ratio transformers, end cap volumes between $20-25\%$ are likely to keep the core loss contribution of the end caps low without adding too much volume.

Prototype Design and Test Results

In order to validate the design procedure outlined above, a 1:1 and a 1:10 transformer with parallel turns were built at a target magnetizing inductance of 11 μH with a desired operating frequency of 2 MHz. The transformer was designed under a volume constraint of $< 1in^3$ with a target 3:2 aspect ratio, a $\sim 88.5\%$ vertical fill, and a $\sim 55\%$ horizontal fill.

Design Parameters: First, due to its high performance factor at 2-3 MHz [9, 10], Fair-Rite 80 was chosen for all core pieces (center post discs, end caps, and outer shell pieces) of the transformer.

As for geometric considerations, the volume and aspect ratio selection gave us an initial transformer diameter of 23 mm and a height of 34.5 mm. We then initially set the end cap lengths to be 4.25 mm each. Thus the window height is 26 mm. Then, because of this window height and the selected vertical fill, 17 AWG magnet wire was selected as the conducting material. Then, to achieve the selected horizontal fill given the 17 AWG wire, the widow width was selected to be 2.1 mm.

Now, in order to ensure that the H-fields on each side of the conductor are balanced and that we minimize leakage inductance, the end cap gap lengths were set to be equal to each other ($R_x = R_y$) and set to be half the length of the center post and shell gap lengths ($R_x = R_g/2$). This, alongside the target magnetizing inductance of ≈ 11 μH, the total of ten conducting pairs, and the area selection, we required that the end cap gap lengths (R_x, R_y) are 50 μm, and the center post and shell gaps (R_g) are 100 μm. This selection alongside the window height, required that each core section of the center post and the outer shell to be 2.5 mm.

Lastly, after simulating this transformer, the end cap length's were decreased to 3.5 mm to reduce volume, without significantly impacting simulated core loss values. Table I below summarizes all of the design parameters.

Table I: Geometry and specifications of the simulated transformer

Parameter	Value	Unit
Magnetizing Inductance	11	μH
Total Diameter	23	mm
Total Height	33	mm
Center Post Diameter	15.6	mm
Window Height	26	mm
Window Width	2.1	mm
Core Section Height	2.5	mm
End Cap Height	3.5	mm
End Cap Gap Lengths	50	μm
Center Core Gap Lengths	100	μm
Wire Selection	17	AWG

Fig. 15: (left) Shell construction on a half-cylindrical jig. (middle) Disassembled modules of the transformer. (right) Final built transformer.

Prototype Construction: The transformer core consists of several different segments: the center post, the outer shells, the end caps, and the windings. The center post was constructed by stacking center post disks with laser-cut shim stock spacers at the optimized gap height in between them. The outer shell was constructed in the same manner. Due to the C shape of the shells, a half cylindrical jig was used to assist in the construction as shown in Fig. 15.

During construction, the center post and shell height were increased in order to account for the helical nature of the windings, this change adjusted both the aspect ratio ($\sim 3:2 \rightarrow \sim 5:3$) and the vertical fill value (88.5% → 75%) and was accounted for in simulated results. Also for ease of construction, we kept the pair and coil spacing constant to allow the insulation of the wires to provide the spacing between conductors.

Inductance Measurements: The 1:1 transformer's inductances were measured through a E5061B Network Analyzer using open- and short-circuit one-port measurements, with results in Table II. These values then allow us to calculate the ratio of the leakage inductance to the magnetizing inductance, with results in Table III.

Table II: Constituent inductances for the proposed transformer, experimental and simulated

	Measured (μH)	Simulation (μH)
Lm	9.324	11.386
L11	9.389	11.456
L22	9.429	11.446

Table III: Leakage to magnetizing inductance ratios for the proposed transformer

	Measured	Simulation
Ll1/Lmag	0.70%	0.62%
Ll2/Lmag	1.12%	0.53%

Port inductance measurements do vary ($\sim 18\%$) from simulation, which could be due to structural changes and defects associated with prototyping (such as non-ideal, gap spacing due to the curled edges of the laser cut plastic spacers). However, the prototype transformer does achieves a low leakage to magnetizing ratio of 0.70% on the primary port and a ratio of 1.12% on the secondary. Also, through the network analyzer, we measured a resonant frequency of the structure around 20 MHz, indicating that the structure also achieves a low parasitic capacitance.

Power Loss Measurements: The power losses of the transformer were measured on the primary and the secondary windings through a series resonant based approach [11, 12]. This technique allows for accurate loss measurements at high-frequency by measuring only sinusoidal voltage amplitudes. These results are found in Table IV and Table VI, both showing good agreement with simulation across different current drives (validating this method as a form of loss measurement).

Table IV: Primary side power losses for the prototype 1:1 transformer, values measured at 1.918 MHz

Current (A)	Measured (W)	Simulation (W)	Error
1.59	0.254	0.273	6.95%
2.04	0.429	0.475	9.63%
2.38	0.600	0.662	9.29%

Table V: Primary side power losses for a lumped-gap equivalent, values measured at 1.965 MHz

Current (A)	Measured (W)
1.61	1.602
2.03	2.502
2.44	3.574

Table VI: Secondary side power losses for the prototype 1:1 transformer, values measured at 1.916 MHz

Current (A)	Measured (W)	Simulation (W)	Error
1.62	0.253	0.265	4.71%
1.99	0.385	0.412	6.57%
2.37	0.558	0.618	9.85%

Table VII: Secondary side power losses for a lumped-gap equivalent, values measured at 1.947 MHz

Current (A)	Measured (W)
1.59	1.689
1.98	2.614
2.37	3.810

We also used the same power loss measurements on an identical transformer with a lumped gap in the center post and the outer shell as opposed to the distributed gap of the proposed structure. These results are found in Table V and Table VII. Overall, the lumped gap transformer had $6-7$ times the amount of measured loss as compared to our prototype design.

Current Sharing: Lastly, to verify that each wire was in a magnetically similar environment and could share current as predicted, we tested a 10:1 structure where the paralleled one-turn port (secondary) turns were interleaved with the ten-turn port (primary) turns. This winding procedure resulted in ten separate paralleled current paths spaced evenly across the length of the transformer. Driving the primary of this transformer allowed us to measure the current from each of the secondary paths resulting in Fig. 16 below.

Experimentally, the percentage of the net current in each parallel turn ranged from 8.5% to 10.9%, indicating successful current sharing.

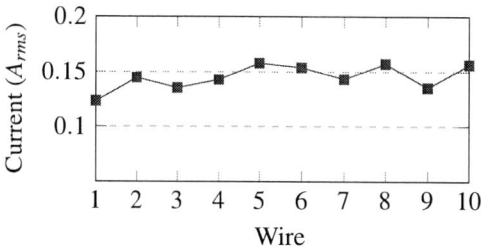

Fig. 16: Current measured from each paralleled wire in the prototype transformer

Conclusion

In this work we proposed a low loss, low leakage transformer that is suitable for application in high frequency power converters. The transformer achieves low conduction losses by balancing the H-fields on the two sides of the windings so that the current flows along nearly the entire skin of the conductor. Distributed gaps shape the H-fields such that each turn of the transformer windings are in a magnetically similar environment. This allows paralleling of the turns of the transformer winding such that each turn carries almost equal current. These properties make the transformer especially suitable for applications with large currents and requiring high turns ratios. The proposed structure also achieves almost zero MMF drop across the window. Because of this, the transformer achieves a very small leakage inductance. This low leakage inductance enables the use of this transformer structure in power converters like coupled-inductor boost or flyback converters which would otherwise require additional circuitry to recover the energy stored in the leakage inductance. The design guidelines provided in this work were confirmed experimentally with a prototype transformer of which achieved both low losses and a low leakage to magnetizing inductance ratio.

Appendices

Derivation to reduce MMF drop across R_w to zero: For the magnetic circuit model, $\mathcal{F}_p$ is the MMF introduced by each primary turn, $\mathcal{F}_s$ is the MMF introduced by each primary turn, and ϕ_i are fluxes in the magnetic circuit meshes (in Fig. 6, six meshes are shown for illustration). We first analyze the magnetic circuit without the parasitic leakage path in between the paired primary/secondary turn, R_a. For an n-stage network, the MMF drop across the ith window reluctance R_w can be given by the sum of the MMF contributions by the sources on the right of the ith R_w ($\mathcal{F}\{R_{w_i}\}_{right}$) and those on its left ($\mathcal{F}\{R_{w_i}\}_{left}$).

$$\mathcal{F}\{R_{w_i}\}_{right} = \frac{((i-1)(2R_g)+R_z)}{2R_z+(n-2)(2R_g)}(n-i)\mathcal{F}_p \tag{5}$$

$$\mathcal{F}\{R_{w_i}\}_{left} = \frac{(((n-2)-(i-1))(2R_g)+R_z)}{2R_z+(n-2)(2R_g)}(i)(-\mathcal{F}_p) \tag{6}$$

Here, R_z is equal to $2R_x + R_g$. Then by setting $|\mathcal{F}\{R_{w_i}\}_{right}| = |(\mathcal{F}\{R_{w_i}\}_{left}|$, we find that the MMF drop across R_{w_i} becomes zero when $R_x = R_g/2$. Thus, if the length of air gap at the endcaps is half of the length of air gap in the center post, we get zero flux through the window reluctances.

References

[1] C. R. Sullivan, "Prospects for advances in power magnetics," Proc. 9th Int. Conf. Integr. Power Electron. Syst., 2016

[2] Q. Zhao and F. Lee, "High-efficiency, high step-up dc-dc converters," IEEE Transactions on Power Electronics, vol. 18, no. 1, pp. 65–73, 2003.

[3] B. Tamyurek and D. A. Torrey, "A three-phase unity power factor single-stage ac–dc converter based on an interleaved flyback topology," IEEE Transactions on Power Electronics, vol. 26, no. 1, pp. 308–318, 2011.

[4] B. Tamyurek and B. Kirimer, "An interleaved high-power flyback inverter for photovoltaic applications," IEEE Transactions on Power Electronics, vol. 30, no. 6, pp. 3228–3241, 2015.

[5] Z. Ouyang, O. C. Thomsen, and M. A. E. Andersen, "The analysis and comparison of leakage inductance in different winding arrangements for planar transformer," in 2009 International Conference on Power Electronics and Drive Systems (PEDS), 2009, pp. 1143–1148.

[6] R. S. Yang, A. J. Hanson, D. J. Perreault, and C. R. Sullivan, "A low-loss inductor structure and design guidelines for high-frequency applications," in 2018 IEEE Applied Power Electronics Conference and Exposition (APEC), 2018, pp. 579–586.

[7] J. Hu and C. Sullivan, "The quasi-distributed gap technique for planar inductors: design guidelines," in IAS '97. Conference Record of the 1997 IEEE Industry Applications Conference Thirty-Second IAS Annual Meeting, vol. 2, 1997, pp. 1147–1152 vol.2.

[8] O. Okeke, M. Solomentsev and A. J. Hanson, "Double-Sided Conduction: A Loss-Reduction Technique for High Frequency Transformers," 2022 IEEE Applied Power Electronics Conference and Exposition (APEC), 2022.

[9] A. J. Hanson, J. A. Belk, S. Lim, C. R. Sullivan, and D. J. Perreault, "Measurements and performance factor comparisons of magnetic materials at high frequency," IEEE Transactions on Power Electronics, vol. 31, no. 11, pp. 7909–7925, 2016.

[10] High Frequency Power Materials, Fair-Rite Products Corp., 2 2018.

[11] Y. Han, G. Cheung, A. Li, C. R. Sullivan, and D. J. Perreault, "Evaluation of magnetic materials for very high frequency power applications," IEEE Transactions on Power Electronics, vol. 27, no. 1, pp. 425–435, 2012.

[12] M. Solomentsev, O. Okeke and A. J. Hanson, "A Resonant Approach to Transformer Loss Characterization," 2022 IEEE Applied Power Electronics Conference and Exposition (APEC), 2022

Temperature Distribution of an IGBT Chip during Repetitive Switching Events under Consideration of Front-Side Ageing

Christian Bäumler, Bo Zhang, Maximilian Goller, Xing Liu, Thomas Basler
CHEMNITZ UNIVERSITY OF TECHNOLOGY
Reichenhainer Str. 70
09126 Chemnitz, Germany
Tel.: +49 / (0) – 371 531-31726
E-Mail: christian.baeumler@etit.tu-chemnitz.de
URL: www.tu-chemnitz.de/etit/le

Acknowledgements

The authors would like to thank Mr. Marco Müller from Würth Elektronik eiSos GmbH & Co. KG for the support of components for assembly and contacting.

Keywords

«Module temperature measurement», «IGBT», «Reliability»

Abstract

Ageing effects are considered to provoke an inhomogeneous current distribution within power devices. The resulting temperature distribution at the junction and the surface of an IGBT chip was investigated in detail at different ageing states and for different switching frequencies during repetitive hard-switching events. Furthermore, the limitations of utilized methods for temperature determination were discussed. The observations were judged with respect to reliability issues.

1. Introduction

In application, power semiconductor devices such as IGBTs and diodes are exposed to repetitive switching events with a certain setting of frequency, duty cycle, temperature and current level. As a result of the simultaneous presence of current and voltage during switching, conduction, and blocking phase, the power losses are generated in the semiconductor device. Depending on the package structure, the power loss is mainly dissipated to the environment via the front- or backside of the semiconductor. A temperature distribution across the device, including the packaging, is the result. This temperature distribution is expected to reach a steady state, when the generated power losses meet the cooling capabilities.

In order to determine the temperature distribution within and on the surface of a power semiconductor chip, detailed information about the device composition like thicknesses, dimensions and material properties are necessary. However, this distribution is only valid for a virgin device and is expected to vary with respect to the ageing state. In addition, the quantitative results are highly dependent on detailed information about the packaging technology and front-side interconnection technology, like the amount of bond-feet, the back-side interconnection technology, and the used cooling.

In this investigation, the development of the temperature during repetitive switching events across the chip surface, at the case of the module package, and at the die junction is shown, with respect to a certain state-of-life (SoL) up to and beyond end-of-life (EoL). According to AQG324 [1] the EoL state in power cycling is defined as 5 % increase of forward voltage drop $V_{CE,cold}$, which is the result of a degraded front-side interconnection. Or an increase of 20 % in thermal resistance between junction and case $R_{th,jhs}$ which suggests a degradation in the thermal path. The transient response of the temperature was monitored for two different switching frequencies at fixed duty cycle of 50 %.

2. Test setup

The converter-like setup in Fig. 1 was used to perform repetitive hard-switching events for a desired device under test (DUT). The semiconductor devices of the high side (HS) and low side (LS) have been part of one module. This investigation focuses on the behavior of LS-IGBT, which is further referenced as DUT. The investigated module was an IGBT in an Econo package with a nominal current rating of 100 A and a nominal voltage of 650 V. As protection IGBT (PIGBT), a 1500 A and 4500 V IGBT was used with V_{CE} fault detection in order to turn-off right after a destructive device failure.

The clamping circuit, as introduced by [2], was utilized to measure the voltage $V_{CE(on)}$ during the device conduction state with high resolution.

The current I_C was measured by a 30 MHz Rogowski coil close to the DC-Link capacitor in order to avoid voltage slope interferences [3].The voltages V_{GE} and $V_{CE(off)}$ were measured with passive probes of 500 and 400 MHz. Owing to the package without kelvin connections that are directly connected to the emitter surface, the measured voltages did always include package components like bond-wires and copper traces.

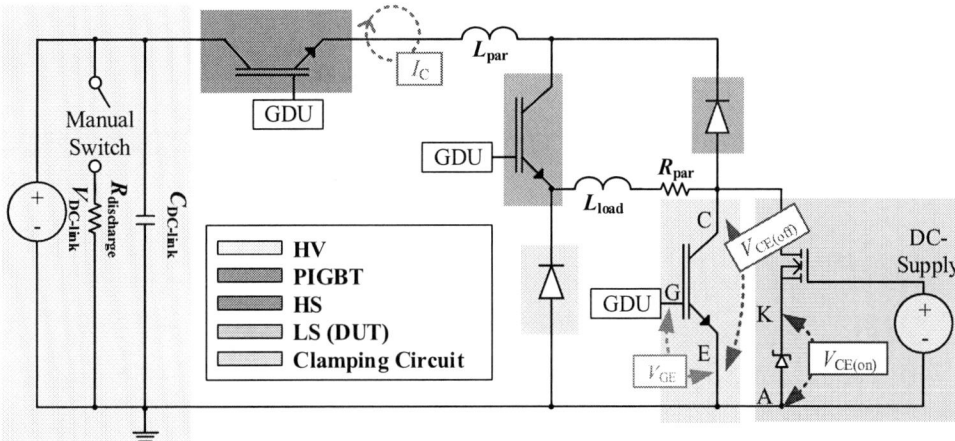

Fig. 1: Converter-like setup for repetitive switching events including relevant measurement points, clamping and protection circuit

3. Pulse pattern and switching principle

By utilizing the pulse pattern in Fig. 2, a converter like application is possible, where only conduction and switching losses have to be compensated by the power source.

During t_1, the HS and LS IGBT are turned on in addition to the permanent-on PIGBT. The current I_C ramps up according the following equation (1):

$$\left(V_{DC-Link} - V_{CE(on)\_PIGBT} - V_{CE(on)\_HS-IGBT} - V_{CE(on)\_LS-IGBT} - i_C \cdot R_{par}\right) = L_{load} \cdot \frac{di_C}{dt_1} \quad (1)$$

R_{par} is thereby the parasitic resistance of all conducting paths within the test bench, the power modules, and the inherent resistance of the used load inductance of about 10 mΩ.

The duration of t_1 is chosen in order to obtain a slight increase of I_C at the beginning of the subsequent pulse. In Fig. 2, a difference between $I_{C(DC+)}$ and $I_{C(rep)}$ is highlighted. During t_1 and t_3, the current $I_{C(DC+)}$ is flowing from the DC-link capacitor through the LS-IGBT and is recorded by the Rogowski coil. The current $I_{C(rep)}$ is present during the free-wheeling phase t_2 and t_4. The voltage $V_{CE(on)}$ is recorded during the on-time of the DUT and is increasing according to the current level.

During t_2, HS-IGBT is off and the current drops during the free-wheeling phase across LS-IGBT and LS-Diode. During t_3, HS IGBT is on and the current is ramped up again according to equation (1) to compensate the drop within t_2. In section t_4, the LS-IGBT is off and the current commutates to the HS free-wheeling circuit.

A detailed depiction of the initial pulse is given in Fig. 3

Period	1/2f			1/2f
GDU	t_1	t_2	t_3	t_4
PIGBT	on	on	on	on
HS	on	off	on	on
LS (DUT)	on	on	on	off

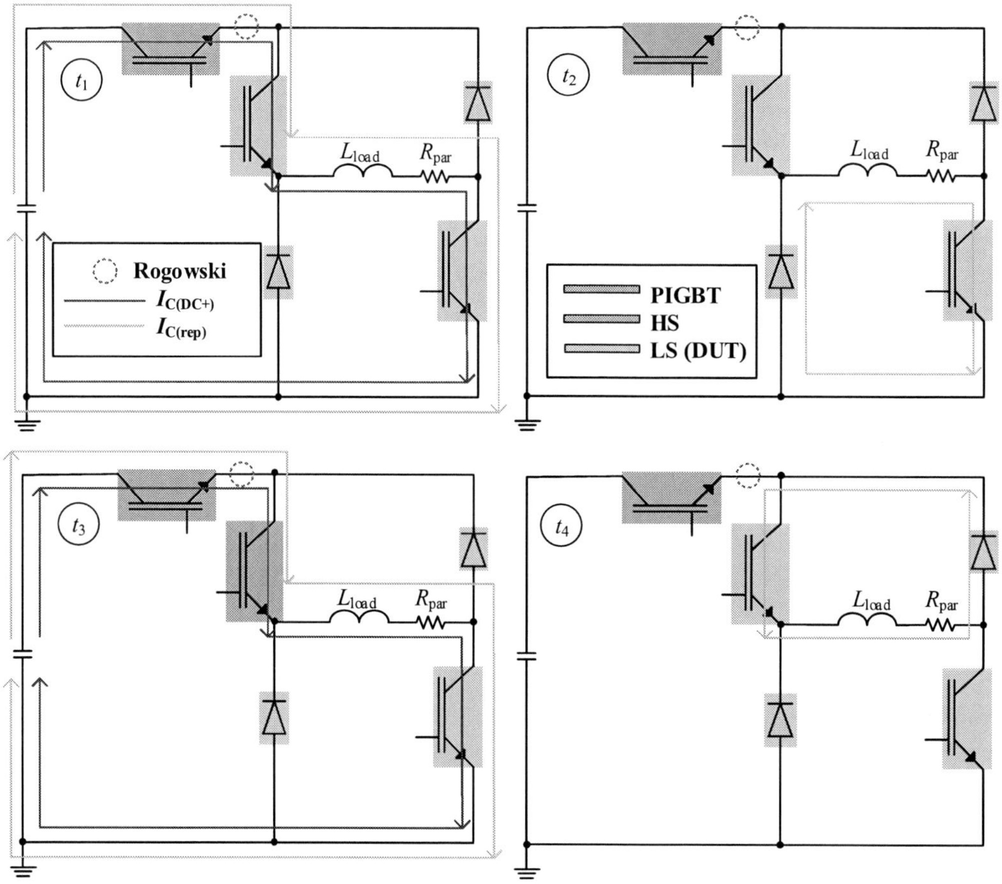

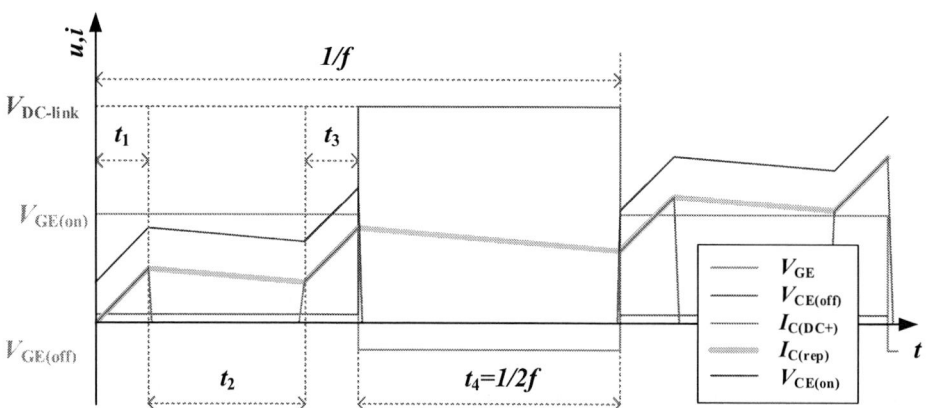

Fig. 2: Pulse pattern and corresponding course of load current for one cycle of the repetitive switching operation. The difference between $I_{C(DC+)}$ and $I_{C(rep)}$ is highlighted for each phase (L_{par} not shown).

Fig. 3: Initial pulse of the repetitive switching investigation according to Fig. 2

Performing repetitive pulses with fixed periods of t_1 to t_4 results into different phases for the repetitive operation as shown in Fig. 4. The displayed voltages and the current $I_{C(DC+)}$ belong to the DUT.

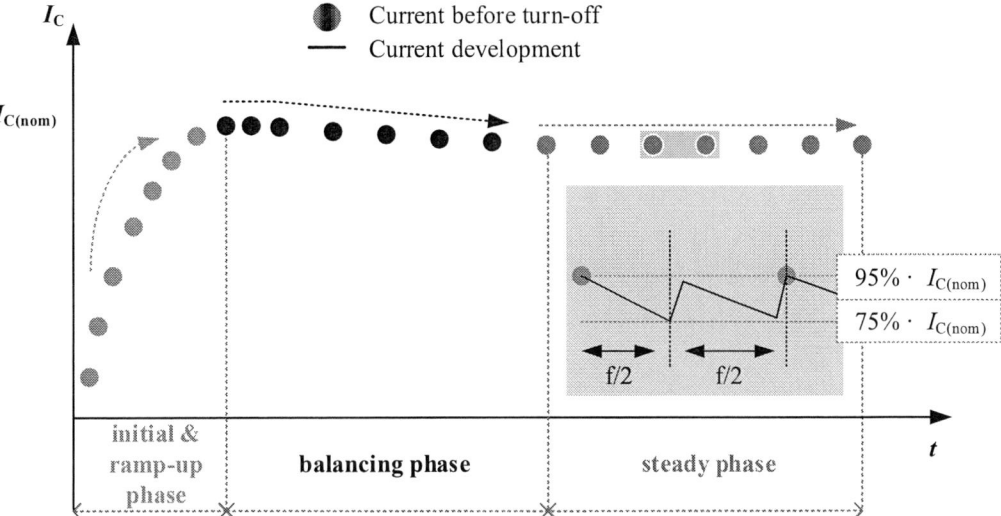

Fig. 4: Different phases of the repetitive switching operation caused by static pulse pattern and self-heating effects. $I_{C(nom)}$ corresponds to the desired nominal current level of 100 A for this investigation.

During the ramp-up phase, the current increases steadily until the desired current level is reached. During the balancing phase, the current level decreases slightly, according to joule heating of conducting elements and self-heating of the active devices. During the steady phase, an equilibrium between static pulse pattern, temperature induced effects and power losses is obtained.

The repetitive switching operation was performed for 20 min with a switching frequency of 500 and 1000 Hz. Within the first 200 pulses the balancing phase was already reached. The balancing phase was unaffected by the ageing state and switching frequency and took about 15 minutes for the tested system. The remaining time was performed within the steady phase. Taking the obtained current development into consideration, as sketched in Fig. 4, an average value of 86 % of $I_{C(nom)}$ was applied during the steady phase, with a minimum of 75 % of $I_{C(nom)}$ and a maximum of 95 % $I_{C(nom)}$. A higher load inductance L_{load} and a dynamically self-adjusting period t_1 would enable higher average values, hence less ripple.

4. Temperature determination

Three different techniques were used to monitor the temperature development during the repetitive switching operation, as depicted in Fig. 5.

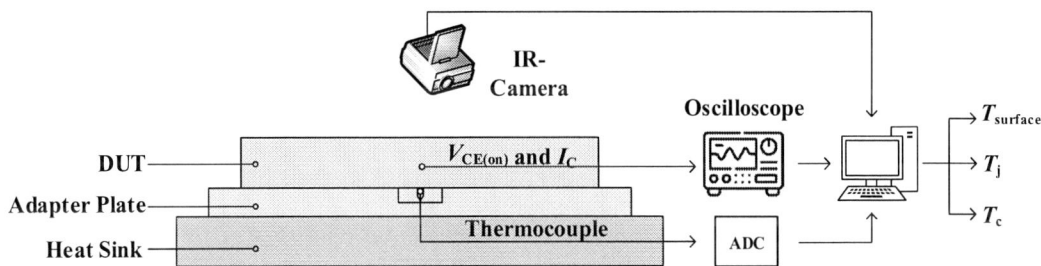

Fig. 5: Different methods of temperature tracking during repetitive switching operation

The case temperature T_c was measured with a thermocouple inserted into an adapter-plate below the power module. Due to the limited accessibility of the adapter plate, a certain distance between thermocouple and DUT had to be taken into account.

The surface temperature $T_{surface}$ was measured by an infrared (IR) camera, after the silicon gel of the DUT was removed and the device was dyed by thermal lacquer to obtain a homogeneous emission coefficient. After the gel removal, the leakage current at nominal blocking voltage was not increased.

For the determination of the junction temperature T_j, different temperature sensitive electrical parameters (TSEPs) were considered according to Table 1. The voltage $V_{CE(on\text{-}sense)}$ corresponds to the on-state voltage drop of the DUT, while a 100 mA sense current is applied. In contrast, the voltage $V_{CE(on\text{-}load)}$ drops across the DUT at load current levels of about $I_{C(nom)} = 100$ A (shortly before turn-off).

Table 1: Applicability of different TSEPs during repetitive switching operation. Green – beneficial | Light red – adverse | Dark red - disqualifier

TSEP	Self-heating within calibration?	Interruption during repetitive operation?	Additional measurement circuit necessary?	Further drawbacks
$V_{CE(on\text{-}sense)}$	No	Yes	Yes	Measurement delay due to settling times and tail current phase of IGBT [4]
$V_{CE(on\text{-}load)}$	Yes	No	Yes	Re-calibration after (artificial) ageing mandatory
$V_{GE(th)}$	No	No	Yes	Difficult to distinguish due to plasma dominated turn-off process
di/dt_{max}	Yes	No	No	Additional non-linear load current dependency
I_{tail}	Yes	No	No	Current level difficult to distinguish

From this, the following considerations can be derived for the determination of the junction temperature under repetitive conditions:

- $V_{CE(on\text{-}sense)}$: This method could not be used, according to the requirement to not interrupt the repetitive operation. However, this is the method typically applied to measure the T_{vj} during power cycling.

- $V_{CE(on\text{-}load)}$: Owing to the device package, the voltage drop during on-state included bond-wires, copper traces and the semiconductor chip itself. Hence, re-calibration after each artificial ageing step was mandatory. In order to determine the voltage drop during on-state with sufficiently high resolution and to not damage the oscilloscope during blocking state, a clamping circuit according to [2] could be used.

- $V_{GE(th)}$: It was a requirement to determine the temperature at the maximum current level within one switching cycle, hence during turn-off or right before. Limited by plasma-induced effects of an IGBT, a clear link between current zero-crossing and measured gate-voltage during turn-off is not given. As a consequence, this TSEP was not considered.

- di/dt_{max}: For different temperatures and current levels, the maximum current slope was recorded. This investigation revealed a non-linear current dependency. Hence, due to the enormous additional calibration effort for different currents, as present during the repetitive operation, this TSEP was not considered.

- I_{tail}: In [5], the tail current was presented as TSEP and verified by simulation and measurement. However, the determination of the amplitude of the tail current is difficult and unprecise and strongly depends on switching condition and the chip technology.

Taking all requirements into account, the voltage drop during on-state $V_{CE(on\text{-}load)}$ at load current was utilized as TSEP. A strong dependency of $V_{CE(on\text{-}load)}$ on the packaging condition and hence ageing state, as mentioned in [6], was omitted by performing the calibration again for each SoL. The calibration was performed as follows and is depicted in Fig. 6: The DUT was heated up to different temperature levels between room temperature and 150 °C by a heat plate underneath the power module. At different temperatures, a pulse of about 130 µs with a linear current slope was applied. From this pulse, a calibration matrix could be obtained that links $V_{CE(on\text{-}load)}$ voltage drops to certain current levels at a pre-adjusted temperature. This procedure was repeated for each ageing state. Exemplary calibration curves for different ageing and current levels are depicted in Fig. 7.The junction temperature of the DUT was determined according to Fig. 3 right before turn-off at the end of period t_3.

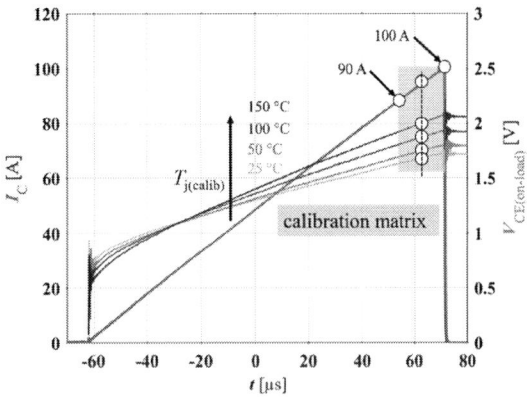

Fig. 6: Calibration pulse at one ageing state for different temperatures.

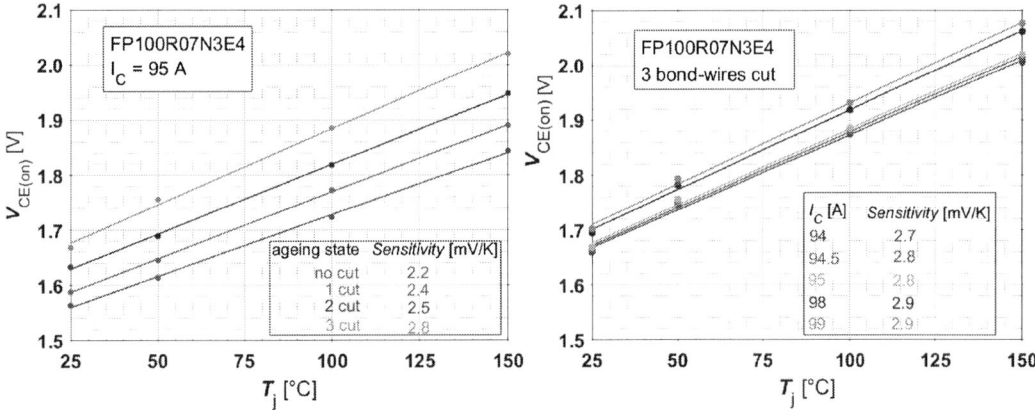

Fig. 7: $V_{CE(on\text{-}load)}$ calibration curves for different ageing states (left) and different current levels (right). Measurement points were fitted by a linear approximation of 1st degree.

5. Measurement results with respect to different ageing states and switching frequencies

A fixed pulse pattern was used to reach a steady state of the switched current level through the load inductance and hence the LS (DUT). Due to the applied current, the device heats up and the forward voltage drop $V_{(CE)on}$ increases. According to equation (1), the di_C/dt decreases slightly, because the remaining voltage across the load inductance decreases. As a consequence, the current level does not remain constant and declines over the whole period of repetitive switching events to about 95 % of the initial current level, which was reached in the balancing phase shown in Fig. 4. This trend is shown in Fig. 8 for 3 different turn-off pulses of the overall 20 min repetitive operation.

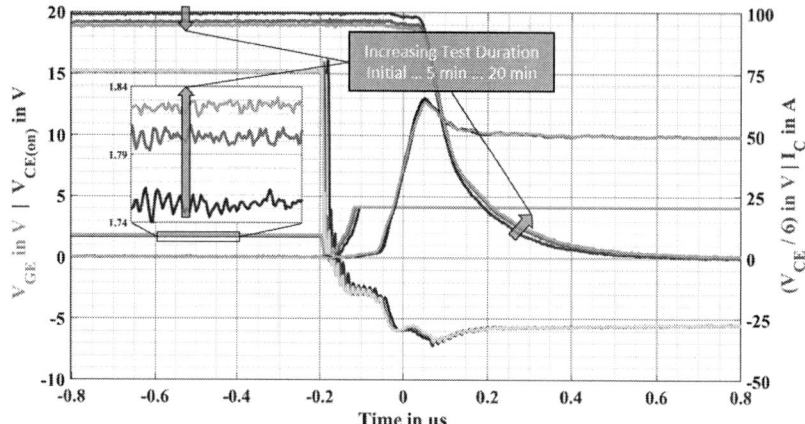

Fig. 8: Recorded turn-off waveforms at the end of t_3 for different test durations within one repetitive switching investigation of overall 20 minutes.
Repetitive Switching at 500 Hz, $I_C = 100$ A, $V_{CE} = 300$ V, $V_{GE(on)} = 15$ V, $V_{GE(off)} = -5$ V, $R_{G(ext)} = 0$ Ω

Every two seconds, a turn-off event according to Fig. 8 was recorded and the junction temperature was derived from $V_{CE(on)}$ at a corresponding current level. The courses of the junction temperature is depicted for different SoL in Fig. 9. From this, different slopes of temperature development over the test duration can be seen. The slope increases from ageing state SoL-0 to SoL-3 within the first 3 seconds. After about 3 seconds, the temperature development reveals a higher slope in case of SoL-2. A higher stationary junction temperature T_j follows. This surprising result is addressed within the evaluation in section 6.1.

Starting from the initial state SoL-0, different SoL were obtained by artificially ageing which should simulate a bond-wire lift of during a power-cycling test. In fact, bond-wires were cut intentionally within the loop. The cut location was selected with respect to experiences from active power cycled bond-wire lift-offs [7] and the hot-spot analysis by IR-camera. The chronology of the cuts is shown in Fig. 10b). Due to the consecutive cuts, the $V_{CE(on)}$ at room temperature increased as listed in Fig. 9.

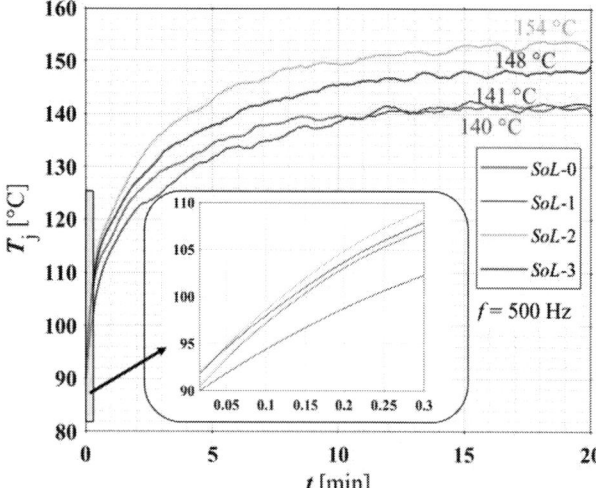

SoL	$V_{CE(on)}$ (@ RT)
0	100.0 %
1	101.6 %
2	104.6 %
3	108.7 %

Fig. 9: Junction temperature development during repetitive switching events at 500 Hz with respect to different SoL. The temperature T_j does not start at room temperature (RT), because the first recorded turn-off pulse was captured after two seconds repetitive operation. A zoomed plot for the first 18 seconds is shown.

The temperature profile in Fig. 10a) was extracted along the 8 different segments and could reveal the position of the bond wires, as well as the point of maximum temperature.

The development of surface temperature was further evaluated as shown in Fig. 11. The displayed temperature T_{surf} was obtained by averaging 2 points close to the bond foot as shown in Fig. 10a) for segment 1.

From Fig. 11 it can be evaluated, that segments with cut bond-wire become thermally relieved, whereas surrounding segments become hotter. For each ageing state, a maximum $T_{surf(max)}$, mean $T_{surf(mean)}$, and minimum $T_{surf(min)}$ surface temperature can be determined. The mean temperature was obtained by averaging all segment temperatures within one SoL.

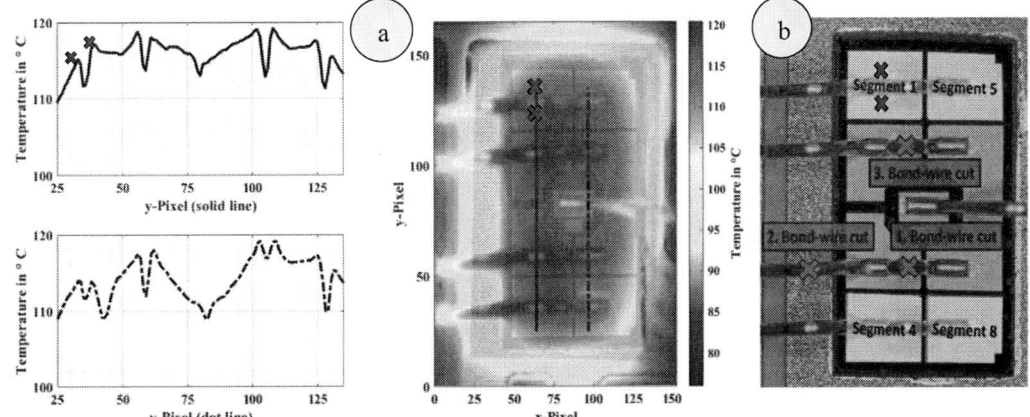

Fig. 10: a) Exemplary surface temperature analysis with IR camera for initial state SoL-0. Image was taken after 20 min repetitive switching operation at 500 Hz. In addition to two temperature profiles along the surface, measurement points (x) next to the bond foot were inserted and evaluated in order to obtain the temperature for each segment. b) Further, the chronology of bond-wire cuts (x) is indicated.

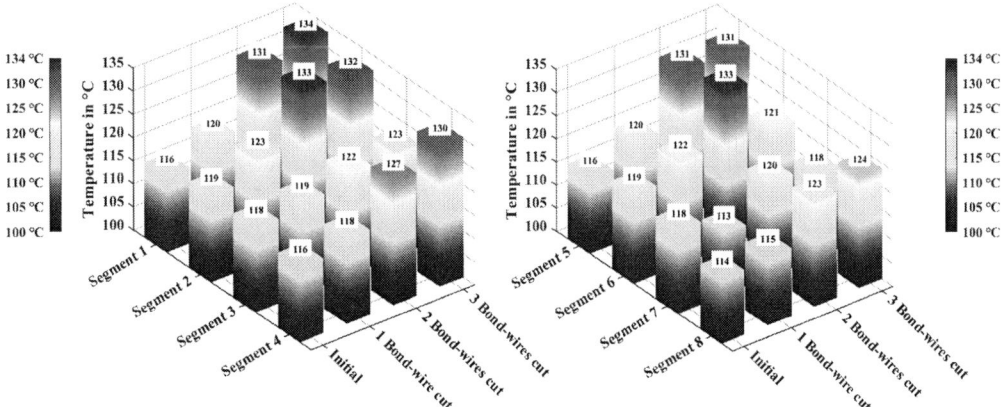

Fig. 11: Evaluation of the surface temperature for each segment of the emitter metallization at different SoL after 20 min repetitive switching events at 500 Hz.

6. Development of temperatures and their distribution with respect to SoL and switching frequency

How the junction, case, and surface temperature have changed with respect to the SoL of the DUT and the applied switching frequency, is described in detail within the following subsections.

6.1. Junction temperature

As shown in Fig. 12, the junction temperature T_j increases more, with respect to the SoL, when a higher switching frequency of 1000 Hz was applied during the test. Already at SoL-1 with 1.6 % $V_{CE(on)}$ increase (the DUT did not reach EoL (PCT 5% limit)) the evaluation reveals an exceedance of the recommended junction temperature T_{vjopt} of 150 °C, when switched with 1000 Hz. At SoL-2 and an increase of $V_{\text{CE(on)}}$ by 4.6 %, the junction temperature is exceeding 150 °C with 500 Hz, too.

An unexpected trend was revealed in case of a switching frequency with 500 Hz. In fact, a further increased ageing state from SoL-2 to SoL-3 revealed a junction temperature drop from 154 °C down to 148 °C. This development for the repetitive operation at 500 Hz contrasts with the investigation at 1000 Hz.

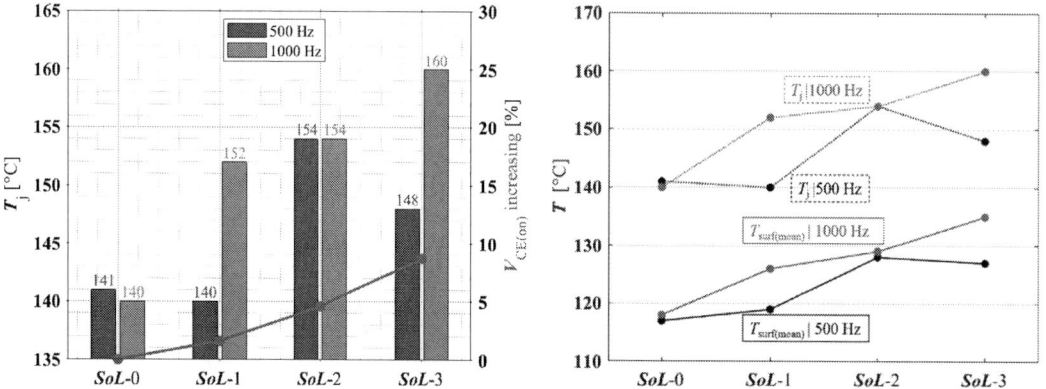

Fig. 12: (Left) Junction temperature T_j in steady state with respect to $V_{\text{CE(on)}}$ increase. (Right) Trend of the mean surface temperature $T_{\text{surf(mean)}}$ compared with the junction temperature. Both for different frequencies and ageing states.

The trend of the junction temperature T_j for 500 and 1000 Hz was also revealed for the mean surface temperature $T_{\text{surf(mean)}}$, as shown in Fig. 12 (right). The hypothesis, that self-balancing mechanisms become effective after a certain ageing state and can only take place up to a certain frequency, has to be proven by FEM simulations.

6.2. Surface temperature

The maximum surface temperature evaluation is shown Fig. 13 (left). A deviation of about 5 K occurs at the initial SoL. The highest temperature within all segments increases stronger in case of 1000 Hz. The deviation between hottest and coldest segment increases for both investigated frequencies with increasing degradation. The deviation can be directly linked to the degree of inhomogeneous temperature distribution at the surface. Within the investigated ageing range, the deviation goes up to 16 K and is almost independent from the switching frequency.

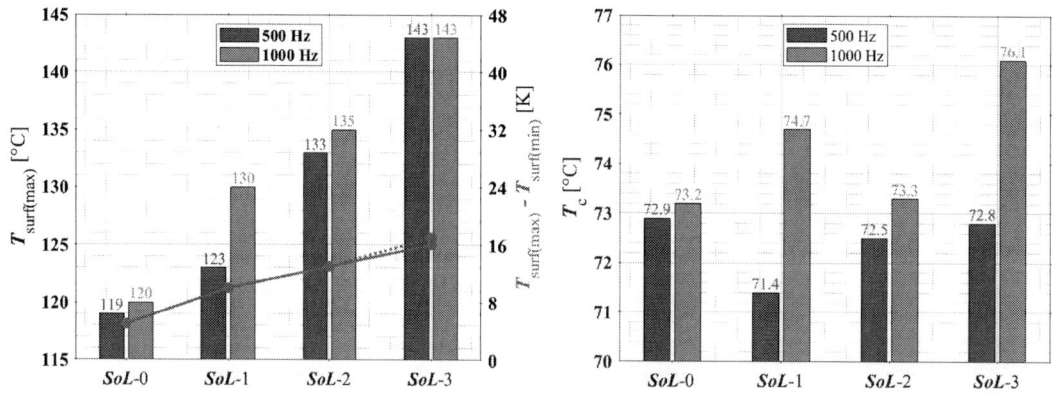

Fig. 13: (Left) Maximum surface temperature in steady state for different switching frequencies and ageing states. The temperature deviation represents the relative difference between hottest and coldest segment temperature with respect to all segments and accounts for the inhomogeneous temperature distribution. (Right) Case temperature in steady state for different switching frequencies and ageing states.

6.3. Case temperature

The case temperature development is presented in Fig. 13 (right). At 1000 Hz, the case temperature reveals an increasing trend with respect to a higher amount of cut bond-wires. At 500 Hz, the case temperature remains almost stable with a small change rate in comparison to the surface and junction temperature. Therefore, the benefit of the TC-measurement for this study is low.

7. Challenges of current adjustment and temperature determination

The presented results are part of the first attempt to evaluate the temperature development during converter-like repetitive hard switching events. Some limitations were found, which must be considered.

7.1. Junction temperature determination

Limited by the package of the DUT without a direct Kelvin connection, $V_{CE(on-load)}$ consists of the voltage drop across the chip $V_{CE(on-chip)}$ and a packaging share $V_{CE(on-package)}$ due to bond-wires and a common conduction path for load and sense current. During the initial and re-calibration after each artificially ageing step, a homogeneous temperature distribution is assumed. However, during repetitive switching operation, the package and the chip face an inhomogeneous temperature distribution, as could be evaluated from IR-camera pictures. Further analysis by FEM simulations and a DUT with kelvin connection have to be conducted in order to estimate the error in junction temperature determination. An additional junction temperature reference by $V_{CE(on-sense)} = f(T)$ method [4] could be useful.

7.2. Pulse pattern

Again, it has to be pointed out, that due to the static pulse pattern a repetitive turn-off current of 95 % $I_{C(nom)}$ was reached. However, the average current during the repetitive conduction phase, accounted for only 85 % $I_{C(nom)}$. A higher load inductance along with a dynamically adjusted pulse pattern could be used to operate closer to the permitted limits of the datasheet.

7.3. Mismatch between surface and junction temperature

The junction temperature was evaluated right before turn-off, where the heat development within a pulse reached the maximum. Limited by IR camera technology, the image could not be recorded at the same time. A not distinguishable difference in timing makes a quantitative comparison between determined junction temperature and recorded surface temperature difficult.

7.4. Artificial ageing

Front-side ageing mechanisms, as described and reported in detail for accelerated lifetime investigation by [7], are not covered by bond-wire cuts, only. In addition, the modification of the top-side metallization has to be taken into account. In dependency of the applied temperature swing, an increased sheet resistance according to [8] is expected, which could lead to an even higher inhomogeneous temperature distribution between segments [7].

8. Conclusion

In this paper, the development of temperatures at different positions of an IGBT chip and its package was monitored and analyzed with respect to different state-of-life and frequencies during repetitive hard-switching events. The junction temperature was determined by using the forward voltage drop $V_{CE(on-load)}$ at load current. Different ageing states were obtained by artificial cuts within bond-wire loops. It can be concluded that all investigated temperatures are affected by ageing and the switching frequency. With a higher degree of degradation and higher switching frequency, the found junction temperature and the inhomogeneous temperature distribution on the surface became larger. Following the presented results, before reaching the power-cycling end-of-life criterion of 5 % - $V_{CE(on)}$ increase, the tested device faced a higher peak temperature as it is recommended in the datasheet. The reliability of the results was discussed and further analysis is necessary to clear uncertainties, especially in case of the junction temperature determination.

References

[1] T. Harder, „ECPE Guideline AQG 324 - Qualification of Power Modules for Use in Power Electronics Converter Units in Motor Vehicles," 2019.

[2] M. Goller, M. A. Thiem, J. Song, J. Kowalsky, J. Franke und J. Lutz, „Investigation of the current collapse behaviour in GaN power HEMTs with highly adjustable pulse and measurement concept," in *In31st European Symposium on Reliability of Electron Devices, Failure Physics and Analysis, ESREF*, Microelectronics Reliability, Athen, 2020.

[3] I. Application Note, „High Speed SW Current measurement by Rogowski Coil Current Probe," [Online]. Available: https://www.pmk.de/web/editor/files/High%20Speed%20SW%20Current%20measurement%20 by%20Rogowski%20Coil%20Current%20Probe_Iwatsu%20AppNote.pdf. [Zugriff am 18 May 2022].

[4] C. Herold, J. Franke, R. Bhojani, A. Schleicher und J. Lutz, „Requirements in power cycling for precise lifetime estimation," in *Microelectronics Reliability, 58:82-89*, 2016.

[5] C. Chen, V. Pickert, B. Ji, C. Jia, A. C. Knoll und C. Ng, „Comparison of TSEP Performances Operating at Homogeneuous and Inhomogeneous Temperature Distribution in Multichip IGBT Power Modules," IEEE JOURNAL OF EMERGING AND SELECTED TOPICS IN POWER ELECTRONICS, VOL. 9, NO. 5 October, 2021.

[6] N. Degrenne und S. Mollov, „Robust On-line Junction Temperature Estimation of IGBT Power Modules based on Von during PWM Power Cycling," 2019 IEEE International Workshop on Integrated Power Packaging (IWIPP), Toulouse, France, 2019.

[7] C. Bäumler, M. Hernes, J. Kowalsky und J. Lutz, „Short Circuit Robustness of an Aged High Power IGBT-Module," in *EPE*, Genoa, 2019.

[8] J. Lutz, T. Herrmann, M. Feller, R. Bayerer und T. Licht, „Power cycling induced failure mechanisms in the viewpoint of rough temperature environment," Proceedings of the 5th International Conference on Integrated Power Electronic Systems, 2008.

Boosting Pilot-Diode Reverse-Conducting IGBTs Turn-ON and Reverse-Recovery Losses with a Simple Gate-Control Technique

Daniel Lexow[1,2], Hans-Günter Eckel[1]

[1]UNIVERSITY OF ROSTOCK
Albert-Einstein Str. 2
18059 Rostock, Germany
Phone: +49 (0) 381-498 7112
Fax: +49 (0) 381-498 7102
Email: daniel.lexow@uni-rostock.de
URL: http://www.iee.uni-rostock.de

[2]NORDEX SE
Langenhorner Chaussee 600
22419 Hamburg, Germany
Phone: +49 (0) 40-300030-1000
Fax: +49 (0) 40-30030-1101
Email: dlexow@nordex-online.com
URL: https://www.nordex-online.com/de/

Keywords

«Power density optimization», «IGBT», «Diode», «control methods for electrical systems», «Driver Concepts»

Abstract

PD-RC-IGBTs reverse-recovery and turn-ON losses are significantly improvable by reducing the inverter interlock time in diode conduction mode. Device Measurements with 1200 V PD-RC-IGBTs reveal a loss reduction potential of up to 27,5 % for reverse-recovery and 26,2 % for turn-ON energy losses, compared to state-of-the-art inverter interlock times. A gate control realizing this selective inverter interlock time adaption for the diode mode, as well as the corresponding measurements and loss calculations, are presented.

Introduction

Today, the concept of RC-IGBTs has been widely known for well over two decades [1], [2]. But in 2004, Mitsubishi established a 1200 V RC-IGBT module [3] which can be seen as the initial starting point for consumer awareness and the introduction of the RC-IGBT into the field of high-power semiconductors. [4]. In order to achieve a reverse conducting functionality, the implementation of n-shorts into the IGBTs collector-sided p-layer was accomplished. In this way, the diode merged with the IGBT and an RC-IGBT, based on only one chip – the so-called: *"first-generation" (1. Gen.)* RC-IGBTs (see figure 1) was created. Related to a full 6.5 kV *High-Voltage (HV)* IGBT module, which usually consists of 24 IGBT and 12 diode chips, the same HV-RC-IGBT module could now be equipped with 36 RC-IGBT chips. This not only results in a significantly higher electrical power output but also increases the device lifetime due to the new chip properties. Since the current can now be conducted bidirectionally in one single chip, temperature ripple and resulting thermal-mechanical stress are significantly reduced [5], [6].

Despite all the positive features mentioned above, the 1. Gen. RC-IGBT comes with a challenging issue. Contrary to conventional IGBTs, the electron-hole plasma concentration in the RC-IGBT is highly dependent on the applied *gate-emitter voltage (V_{GE})* as long as the current is driven in the diode direction. A positive V_{GE} ($V_{GE} > V_{THRESHOLD}$, e.g., $V_{GE} = +15$ V) creates an electron channel under the gate, allowing electrons to bypass and causing a unipolar current flow which drastically reduces the emitter efficiency in diode mode. Therefore, it is inevitable to turn OFF V_{GE} (e.g., $V_{GE} = -15$ V) in diode mode in order to ensure high emitter efficiency and low diode *forward losses (V_F)*. To guarantee proper device behaviour, the control state, either the *Gate Drive Unit (GDU)* or a higher-level control unit, relies on some sort of current direction detection that utilizes a static MOS-Control as stated in [7].

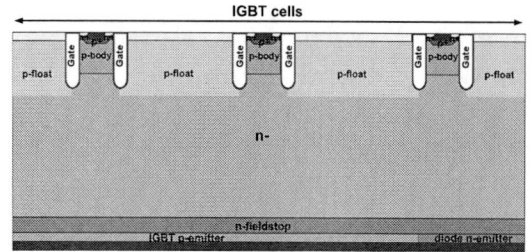 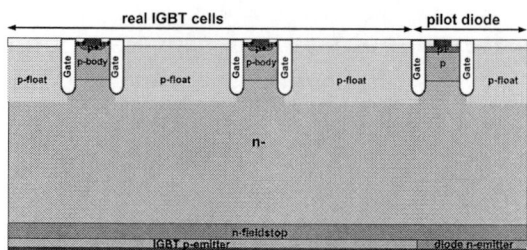

Fig. 1: Cross-section of a conventional RC-IGBT (1. Gen.) with homogeneous frontside and collector-sided n+-shorts

Fig. 2: Cross-section of a PD-RC-IGBT (2. Gen.) with inhomogeneous frontside due to pilot diode areas

Unfortunately, not only the need for a static MOS-Control is a disadvantage of the 1. Gen. RC-IGBT. The significantly increased diode area within an entire module (36 RC-IGBT chips) leads to a considerably higher reverse-recovery charge and, therefore, greater reverse-recovery as well as turn-ON losses. This issue needs to be addressed by implementing a Dynamic MOS-Control [7] into the gate control scheme. Herby, the emitter efficiency in diode mode is deliberately reduced by turning on V_{GE} for a short time ($t_{DESAT} \sim 5...20$ µs) right before reverse-recovery. It is possible to build a functional RC-IGBT control that is able to include both MOS-Control strategies, but it is complicated and demanding, as shown in [8], [9]. Furthermore, high-frequency parasitic oscillations on the load current need to be addressed to protect the GDU from overheating due to multi-switching, as explained in [10]. In this respect, customers are very cautious about using these components within their applications.

To minimize the effort of control and thus make the RC-IGBT more user-friendly, the *Pilot-Diode (PD) RC-IGBT* was developed. This *"second generation" (2.Gen.)* RC-IGBT is characterized by additionally implemented pilot diode areas guarantying low V_F in diode mode without the necessity of special gate control mechanisms [11]. The pilot diode areas ensure sufficient electron-hole plasma in the device even at $V_{GE} = +15$ V in diode mode. Here, the electron channel reduces the electron-hole plasma concentration in all IGBT cells, but this does not affect the pilot diode areas. Due to the modified front side of the PD-RC-IGBT, no electron channel is created in PD areas, and therefore, sufficient electron-hole plasma to obtain low diode ON-state losses is supplied.

Hence, PD-RC-IGBT can be used with conventional IGBT GDUs. They simplify the applicability and allow easy access of the components into the commercial market. Nevertheless, producers are aware of the fact that they still have an adverse issue caused by the *inverter interlock time* (t_{INT}), leading to high reverse-recovery and turn-ON losses [11]. The problem itself and its proposed solution, supported by device measurements with real PD-RC-IGBTs, are described in the ongoing chapters of this paper.

The latest scientific research shows different forms of RC-IGBTs, which unit both exceptional electrical and thermal-mechanical performance as well as independence from special gate control mechanisms [12]. Those RC-IGBTs are not "hardware-based" at the moment. They are only available for device simulations and might be produced and used in the future. They are referred to as *"third-generation" (3.Gen.)* RC-IGBTs and mentioned here for the sake of completeness.

Inverter interlock time (t_{INT}) and its negative impact on PD-RC-IGBTs

In 2015 Rahimo et al. introduced a novel 6.5 kV/1000 A PD-RC-IGBT referred to as "6500 V Plug-In BIGT" [11]. In their publication, they not only described the internal structure of the component but also made detailed reflections on the switching losses. In this context, they choose an interlock time of $t_{INT} = 10$ µs, typical for the voltage class of 6500 V, to present their devices' switching losses. The working group further shared a key finding: concerning interlock time reduction. They stated: *"By reducing locking time moderately to 5 usec, a further 5 % reduction in turn-ON losses and a 10 % reduction in reverse recovery losses can be obtained"*[11].

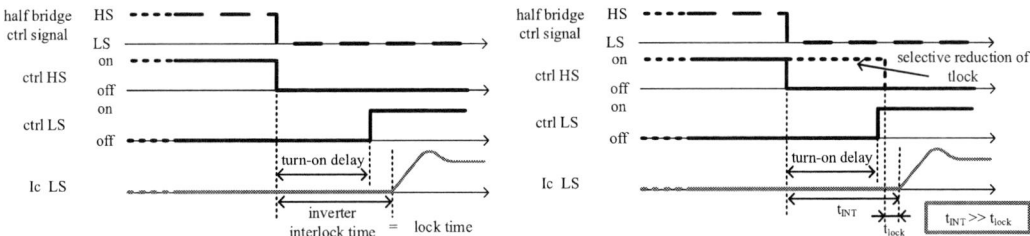

Fig. 3: Definition of the inverter interlock time (t_{INT}) [device point of view]

Fig. 4: Definition of the lock time (t_{LOCK})

The reason for this device behaviour is that the conventional RC-IGBT frontside (see figure 2) of the PD-RC-IGBT is still V_{GE} dependent in diode conduction mode. As long as the gate-emitter voltage is turned on to $V_{GE} = + 15$ V, electron-hole plasma in those IGBT areas is reduced. The low-loss diode on-state properties (low V_F) are, in this case, ensured by the pilot diode regions, but at the same time, a minimum electron-hole plasma is present in the device. Therefore, it would be desirable to start the reverse-recovery process instantly as soon as the turn-OFF signal from the inverter control is received. Unfortunately, this is prohibited due to the inverter interlock time by several µs (figure 3, HS RC-IGBT in diode mode). Within the t_{INT}, the PD-RC-IGBT keeps conducting current in diode direction. Still, V_{GE} is now turned OFF to $V_{GE} = -15$ V. This results in an electron-hole plasma increase in the conventional RC-IGBT areas, leading to unnecessarily high reverse-recovery and turn-ON losses at the time of the actual reverse-recovery process.

If it is assumed that the HS IGBT is in diode mode, then for the conventional diode turn-OFF process, the inverter interlock time (t_{INT}) is equal to the lock time (t_{LOCK}). Figure 4 shows the behaviour of t_{LOCK} for an applied selective lock time reduction method. In this case, the lock time (t_{LOCK}) becomes significantly smaller than the inverter interlock time (t_{INT}). Deriving from this, in diode conduction mode, it is preferable to minimize the inverter interlock time as much as possible (see figure 4) in order to achieve the lowest reverse-recovery losses. For IGBT conduction mode, this is not possible because t_{INT} prevents half-bridge short circuits. Therefore, there has to be some sort of distinguishing process which ensures that a reduction of t_{INT} is only applied while conducting current in diode direction.

Implementation of a selective lock time reduction in diode conduction mode for PD-RC-IGBTs GDU

As discussed and introduced in [13], realizing a selective lock time reduction for the PD-RC-IGBTs diode mode relies on a current direction detection method. Hereby, three different methods can be used:

Firstly, the direct current measurement via a current sensor (e.g., LEM-Converter). This method impresses with its accuracy and precision but is not recommended for the present application due to its high costs. In this respect, this variant will not be further considered.

Secondly, the indirect current measurement via high-voltage desaturation diodes. Here, V_{CE} is monitored by the HV desaturation diode circuit. The decision of whether a positive or negative current is conducted through the device is based on either the positive or negative decoupling voltage. This method impresses with its simplicity and cost-efficiency. Since the desaturation diodes need a specific creepage distance that must be obeyed if installed directly on the GDU, it is rather recommended for lower voltage classes (< 1,7 kV).

Lastly, another indirect current measurement method can be utilized as fully described in [12], [13]. The "gate observation method" distinguishes between IGBT and diode mode by keeping track of the miller plateau in the turn-OFF process. If a miller plateau is detected by a time-dependent gate voltage threshold, the GDU performs a conventional IGBT turn-OFF. If no miller plateau is detected by the time-dependent gate voltage threshold, a special diode mode turn-OFF is performed.

As a result, the gate voltage is instantly turned back ON again and remains "ON" until lock time is sufficiently reduced. This method is easy to implement into the GDU (see GDU in [14]) and very cost-efficient. Unfortunately, it is only suitable for higher voltage classes (>1,7 kV) since a pronounced miller plateau, especially at low currents, is needed to determine the correct turn-OFF mode. RC-IGBTs in voltage classes below 1,7 kV do not show a sufficiently pronounced miller plateau, as shown in figure 9.

Measurement Results

All measurement results were obtained using 1200 V; 1000 A PD-RC-IGBTs with nominal gate resistors @ a temperature of T= 125 °C. The used module is a half-bridge module, where each RC-IGBT (low side and high side switch) can be controlled individually by external GDUs. The GDU responsible for achieving the t_{LOCK} reduction utilizes HV desaturation diodes and an FPGA (see [14]).

Double pulse tests for measurements (A and B) were performed using a test bench (Fig. 5) equipped with a DC-link capacitor (C = 2,82 mF) and a load inductance (L = 543 µH) to adjust the current. The parasitic stray inductance in the commutation path was Lσ = 39 nH.

The oscillating load current measurements (C) were performed in a different test-bench (Fig. 6) with the following characteristics: DC-link capacitor (C = 2,82 mF), a load inductance (L = 130 µH) and a load capacitor (C = 175uF). Due to different test-bench conditions and mechanical superstructures, parasitic stray inductance in the commutation path increased to Lσ = 44 nH.

A) Reverse-recovery and turn-ON losses with successively reduced interlock time

Within the performed double pulse measurements, the inverter interlock time for the high-side PD-RC-IGBT was successively reduced. As shown in figures 7 and 8, the lock time reduction results in significantly decreased collector peak currents, turn-ON losses as well as reverse-recovery peak currents, and reverse-recovery losses. All results are shown in Table I. For the value of the "reduction" columns, the row value (e.g. t_{LOCK} = 0,3 µs) is set in relation to the reference value (t_{INT} = t_{LOCK} = 5 µs [state-of-the-art inverter interlock time], bold). The output value (in %) thus indicates how much the test value is below the reference value. Measurements were also accomplished with a maximal t_{LOCK} = 10 µs (first row of Table I), showing that plasma barely increases with longer lock times.

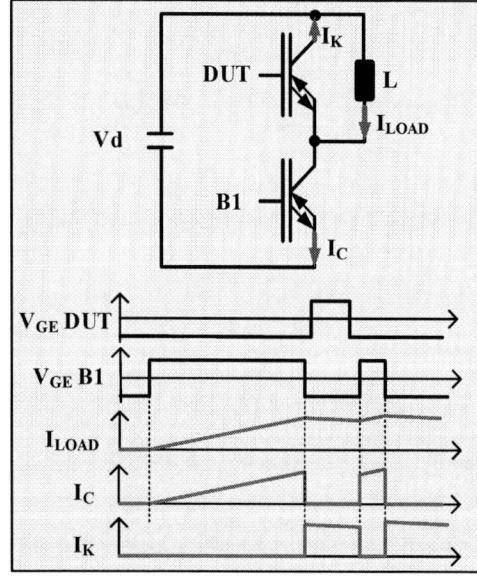

Fig. 5: Double pulse test with a corresponding pulse pattern

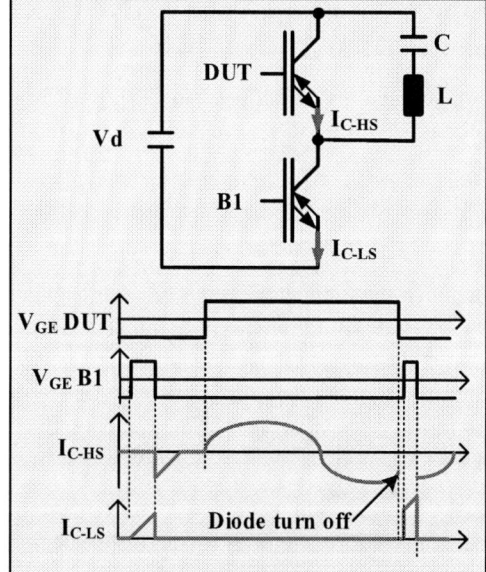

Fig. 6: Oscillating load current test with a corresponding pulse pattern

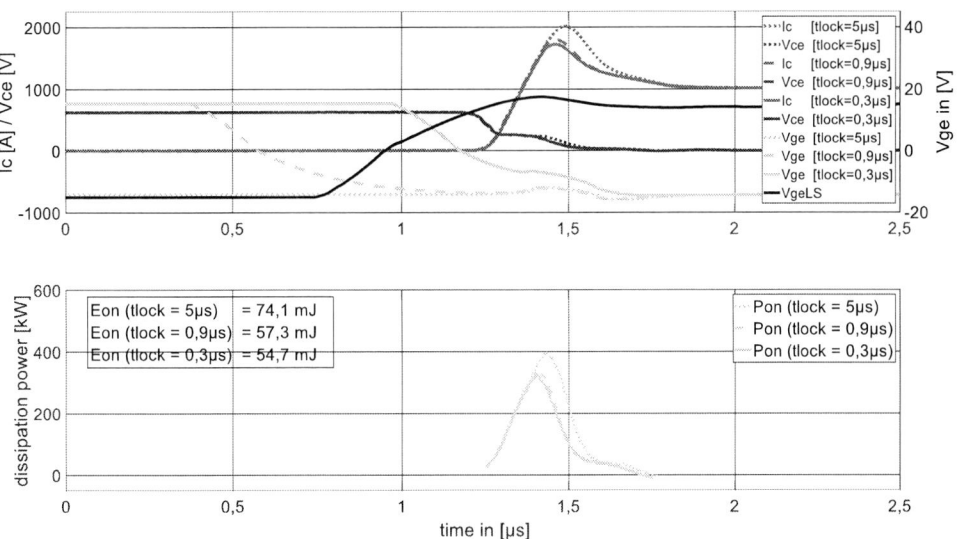

Fig. 7: Turn-ON curves and turn-ON losses for different t_{LOCK}

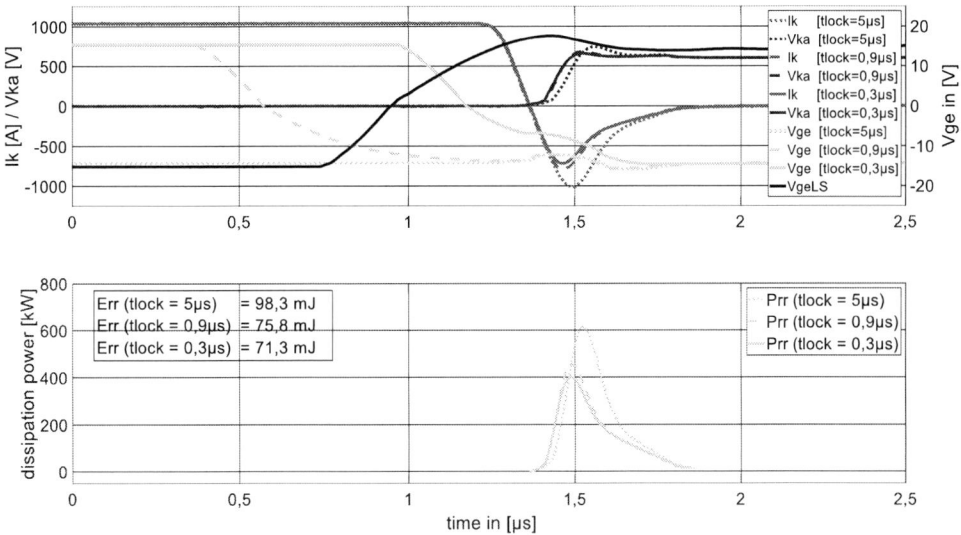

Fig. 8: Reverse-recovery curves and reverse-recovery losses for different t_{LOCK}

Table I: Overview of turn-ON and reverse-recovery losses as a function of t_{LOCK}

t_{LOCK} [μs]	I_{C_max} [A]	reduction [%]	E_{ON} [mJ]	reduction [%]	E_{RR} [mJ]	reduction [%]	E_{TOTAL} [mJ]	reduction [%]
10	2038	+1,6	74,7	+0,8	102,0	+3,8	176,7	+2,5
5	**2005**	**0**	**74,1**	**0**	**98,3**	**0**	**172,4**	**0**
2,5	1910	-4,7	65,5	-11,6	87,8	-10,7	153,3	-11,1
1,5	1839	-8,3	61,1	-17,5	80,3	-18,3	141,4	-18,0
0,9	1787	-10,9	57,3	-22,7	75,8	-22,9	133,1	-22,8
0,6	1750	-12,7	55,6	25,0	73	-25,7	128,6	-25,4
0,4	1720	-14,2	55,1	-25,6	71,5	-27,3	126,6	-26,6
0,3	1716	-14,4	54,7	-26,2	71,3	-27,5	126	-26,9

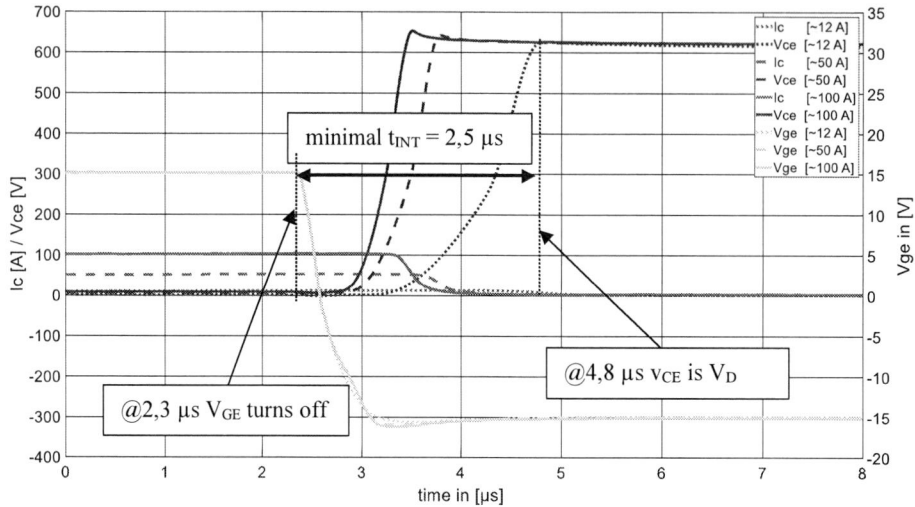

Fig. 9: Turn-OFF times dependent on the collector current

B) IGBT mode turn-OFF with the nominal turn-OFF gate resistor

This series of tests aims to determine the maximum IGBT mode turn-OFF times. Since the turn-OFF process is strongly dependent on the load current and takes longer with decreasing collector currents, an attempt was made to determine the maximum turn-OFF time by switching off a minimum current (see figure 9). Knowing the maximum turn-off time at minimum current makes it possible to define a fair and comparable inverter interlock time for the individual component (locking time = 5 μs with safety margin). The results obtained from the previous tests can be related to this defined interlock time to specify the objectively achievable loss savings.

The obtained results allow two different procedures: Firstly, the locking time is adopted almost without a time safety buffer. This means that the obtained turn-OFF time of 2,5 μs (figure 9) is rounded up to a 3 μs inverter locking time. However, this would be relatively "tight" and would leave no room for any delays. Accordingly, the choice of the second method is recommended, which entails the introduction of a time safety buffer that can counteract possible delays. Here, the locking time could be increased "moderately" to 4 μs or "conservatively" to 5 μs. The results for those cases are shown in Table II.

Table II: Overview of turn-ON and reverse-recovery losses for three different t_{INT}

	inverter locking time [μs]	E_{RR} + E_{ON} [mJ]	reduced lock time [μs]	E_{RR} + E_{ON} [mJ]	reduction [%]	reduced lock time [μs]	E_{RR} + E_{ON} [mJ]	reduction [%]
"tight"	3	159,6	0,9	133,1	16,6	0,3	126	21,1
"moderate"	**4**	**167,3**	**0,9**	**133,1**	**20,4**	**0,3**	**126**	**24,7**
"conservative	5	172,4	0,9	133,1	22,8	0,3	126	26,9

C) GDUs functionality validation with the help of an L-C oscillating circuit

This last series of tests aims to validate the GDUs functionality in a close to real application. Hereby, an oscillating (L-C) circuit was utilized to provide the DUT with a low-frequency oscillating load current (figure 6). Based on that provided current, the GDU needs to self-adapt its switching behaviour. Within the switching experiment, the resonant circuit capacitor is first charged by switching ON the lower RC-IGBT. Subsequently, the oscillation of the load current can start as soon as the DUT receives its turn-ON signal from the higher-level control. Hence, two experiments were performed:

1. The turn-OFF signal for the DUT was given within the positive load current half-wave

The GDU must recognize that a positive collector current is conducted and that an IGBT switch-OFF process has to be carried out accordingly. This must be initiated immediately in order to have reliably switched off the DUT before the opposite RC IGBT is switched on after the interlock time of t_{INT} = 5 μs. This behaviour can be traced in Figure 10. It is important to realise that the driver switches OFF correctly. The gate-emitter signal depicted in green shows that the CTRL signal from the higher-level controller, which is displayed in black, is followed directly. The time offset (~250 ns) between the two signals arises from the processing times during signal processing in the FPGA and during forwarding on the driver. These can vary depending on the driver (components) and FPGA (clock frequency) but can be regarded as constant variables and are therefore not critical in the design for timing aspects such as inverter locking times. For the PD-RC-IGBT GDU, the resulting t_{LOCK} = 5 μs can be seen between the upper (green, solid) and lower (green, dashed) RC-IGBT.

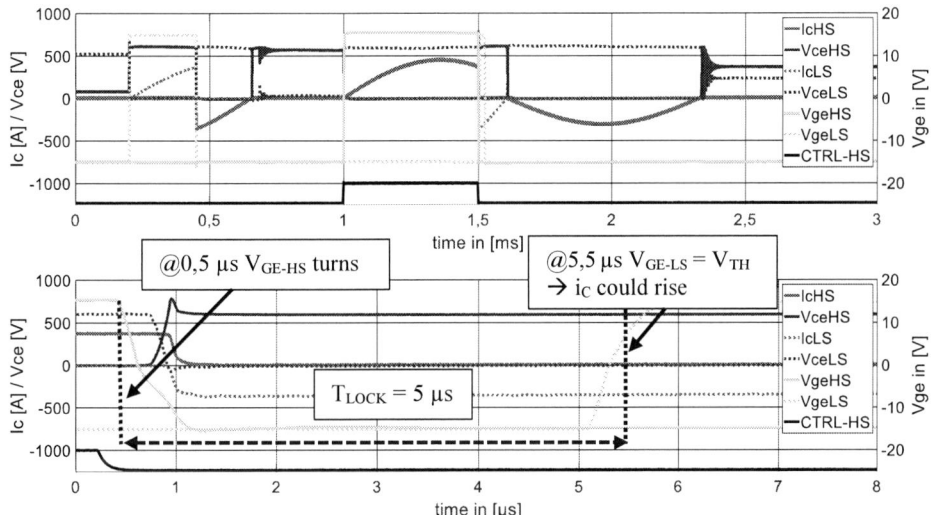

Fig. 10: Top (macroscopic view): IGBT turn-OFF in positive load current half-wave
Bottom (microscopic view): lock time determination based on V_{GE-HS} and V_{GE-LS}

2. The turn-OFF signal for the DUT was given within the negative load current half-wave

The driver must recognize that a negative collector current is conducted and that a modified diode turn-OFF process must be carried out accordingly. In contrast to a conventional IGBT driver, the higher-level turn-OFF signal is not to be followed directly. Instead, the driver must delay the turn-OFF. The delay time can be defined individually for each voltage class and application. In the present measurement example, it was set to 4 μs and thus led to a moderate lock time (t_{LOCK}) of 1 μs.

Figure 11, like Figure 10 for the IGBT turn-OFF process, now shows the corresponding curve for the diode turn-OFF process. It can be seen that the interlock time of t_{INT}= 5 μs (approx. CRTL_HS (black) "off" to Vge_LS "on" (dashed green)) specified by the higher-level controller is not executed. The driver detects the diode mode with the help of its current direction detection. Accordingly, the FPGA modifies the turn-OFF process, which results in a reduction of the effective lock time. The lock time set in the present example is a moderate 1 μs and is calculated as seen in the bottom part of figure 11. The gate driver's (hardware) functionality, combined with the proposed control principle (software), is thus proven. The effects of the presented control principle in terms of loss balance (switching losses) are considered below.

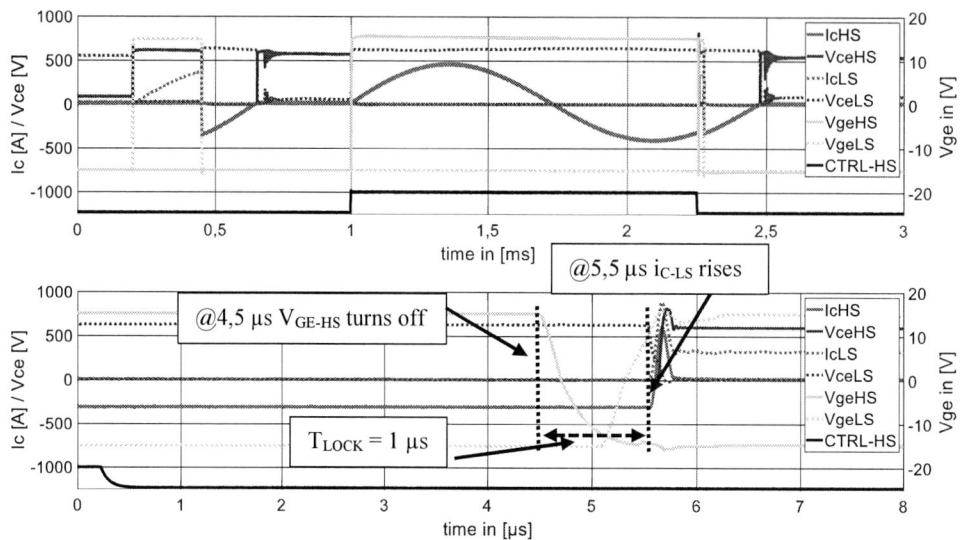

Fig. 11: Top (macroscopic view): Diode turn-OFF in negative load current half-wave
Bottom (microscopic view): lock time determination based on $V_{GE\text{-}HS}$ and $i_{C\text{-}LS}$

Figure 12 shows the course of the diode turn-OFF, and Figure 13 shows the turn-ON of the opposite RC-IGBT. The gate emitter voltage (green, dotted) shows the influence of the driver on the turn-OFF process. To better illustrate the positive effect on the E_{RR} and E_{ON}, these were contrasted with the resulting reverse recovery and turn-ON processes with a conventional IGBT gate control ($t_{INT} = t_{LOCK} = 5$ µs) in a second measurement experiment. For this purpose, a blind gate driver (green, solid) was used, which exclusively follows the incoming signals of the higher-level control. Within the measurements, it was found that there is a significant reduction in switching losses, as already shown in Tables I and II. The introduced PD-control is able to reduce switching losses [mean reduction value = (reduction Eon + reduction Err)/2] for the given turn-ON current by 25,9 %. The comparison, especially with regard to current and voltage characteristics as well as the significant loss reduction between those two GDU variants, makes it evident that the use of the novel PD-RC-IGBT control is unrestrictedly recommended.

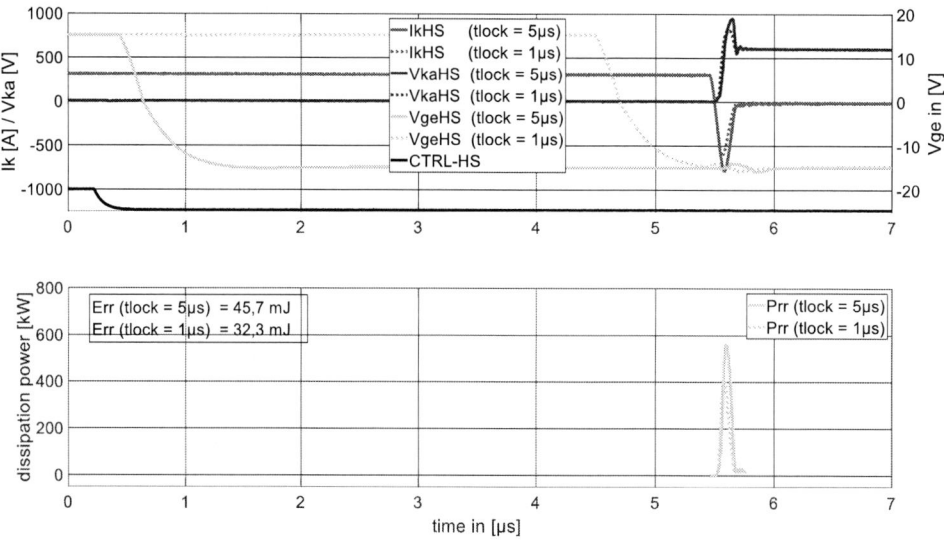

Fig. 12: Diode reverse-recovery behaviour with corresponding loss calculation

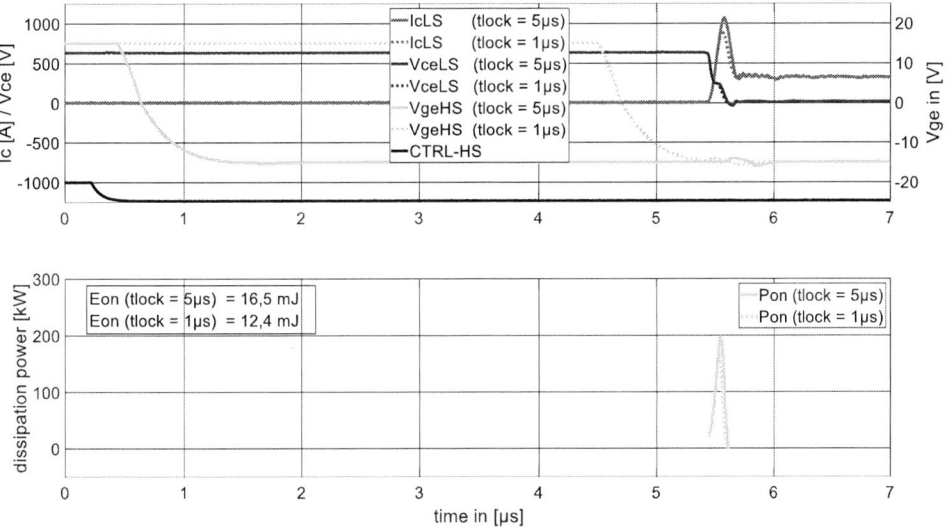

Fig. 13: IGBT turn-ON behaviour with corresponding loss calculation

Conclusion

The results of this paper are, due to the inclusion of high voltage measurements on real PD-RC-IGBTs, able to prove that an adapted gate control (selective locking time reduction) can significantly reduce the reverse-recovery and the turn-ON energy losses of PD-RC-IGBTs. To put the obtained results into a comparative perspective and guarantee an appropriate classification, Table III provides a comparison to a conventional IGBT module. Please note that the RC-IGBT switching losses are optimized to meet the results of the conventional module. Therefore, V_F is increased. The Fuji – 2MBI800XNE120-50 exhibits equivalent chip technology and blocking voltage and is scaled up to 1000 A to allow proper comparison with the Fuji 1200 V 1000 A PD-RC-IGBT (2MBI1000XRNE120-50). Latter is presented with data sheet values (column 2) and own measurements (columns 3 and 4). Results indicate a huge switching loss (E_{ON} and E_{RR}) advantage (18,1 %) over the conventional module while simultaneously introducing all positive features (higher current density, greater diode surge current capability and reduced thermo-mechanical stress and therefore longer life-time expectancy) of the RC-technology into the application.

Table III: Performance comparison of 1200 V (RC) IGBT [Fuji]

	data sheet values		*own measurements*	
Product	1200 V 800 A IGBT	1200 V 1000 A RC-IGBT	1200 V 1000 A RC-IGBT	1200 V 1000 A RC-IGBT
Product name	2MBI800 XNE120-50	2MBI1000 XRNE120-50	2MBI1000 XRNE120-50	2MBI1000 XRNE120-50
MOS-Control	no	no	no	t_{LOCK} reduction to 0,3 μs
V_F @ V_{GE} = 0 V	1,65	1,9	1,9	1,9
E_{RR} @ T = 125 °C in [mJ]	52,9	72,8	91	71,3
E_{ON} @ T = 125 °C in [mJ]	70,2	85,2	68,6	54,7
E_{ON} + E_{RR} in [mJ]	123,1	158	159,6	126
E_{ON} + E_{RR} scaled up to 1000 A in [mJ]	153,9	/	/	/
Losses (E_{ON} + E_{RR}) relatively to conventional IGBT	100%	102,70%	103,70%	***81,90%***

At this point, it should be emphasised once again that the general control effort is drastically reduced compared to the RC-IGBTs of the first generation. The PD-RC-IGBT does not require any current direction-dependent control for normal operation and can therefore be operated at $V_{GE} = +15$ V in diode mode. The implemented current direction detection ensures the selective determination of the operating mode exclusively when receiving the turn-off signal from the higher-level inverter control. In the case of a diode mode, this results in the desired lock time (t_{LOCK})reduction. Further, the adaptation effort for the GDU is rather low or almost non-existent if the used GDU already provides an FPGA or V_{CE} monitoring.

Conclusively, it would be desirable that the results obtained in this paper contribute to making the PD-RC-IGBT more user-friendly and pushing it into technical high-power applications.

References

[1] Y. Seki, Y. Takahashi, T. Koga, M. Ichijyou, and H. Kirihata, "(Power pack IGBT: High power 2.5 kV 1 kA) RC-IGBT with highly reliable flat package," 1995.

[2] M. Hiyoshi, S. Yanagisawa, K. Nishitani, K. Kotaka, H. Matsuda, and S. Teramae, "A 1000 A 2500 V pressure mount RC-IGBT," in *EPE'95*, 1995, pp. 1.051-1.055.

[3] Takahashi, Yamamoto, Aono, and Minato, "1200V reverse conducting IGBT," in *2004 Proceedings of the 16th International Symposium on Power Semiconductor Devices and ICs*, 2004, pp. 133–136, doi: 10.1109/WCT.2004.239844.

[4] G. Majumdar, "Future of power semiconductors," in *2004 IEEE 35th Annual Power Electronics Specialists Conference (IEEE Cat. No.04CH37551)*, 2004, vol. 1, pp. 10-15 Vol.1, doi: 10.1109/PESC.2004.1355704.

[5] M. Rahimo, U. Schlapbach, R. Schnell, A. Kopta, J. Vobecky, and A. Baschnagel, "Realization of higher output power capability with the Bi-mode Insulated Gate Transistor (BIGT)," in *2009 13th European Conference on Power Electronics and Applications*, 2009, pp. 1–10.

[6] D. Werber et al., "6.5kV RCDC: For increased power density in IGBT-modules," *Proc. Int. Symp. Power Semicond. Devices ICs*, pp. 35–38, 2014, doi: 10.1109/ISPSD.2014.6855969.

[7] R. Hermann, E. U. Krafft, and A. Marz, "Reverse-conducting-IGBTs - A new IGBT technology setting new benchmarks in traction converters," 2013.

[8] D. Domes, "Control Method for a Reverse Conducting IGBT," in *2015 International Exhibition and Conference for Power Electronics, Intelligent Motion, Renewable Energy and Energy Management (PCIM Europe)*, 2015, no. May, pp. 147–154.

[9] D. Lexow, H. Wiencke, D. Domes, K. Fleisch, and H.-G. Eckel, "Optimized Control Method for Reverse Conduction IGBTs," in *2017 19th European Conference on Power Electronics and Applications, EPE 2017*, 2017, pp. 1–9.

[10] D. Lexow, H. Wiencke, and H.-G. Eckel, "Improved Gate-Drive Unit for RC-IGBT to Overcome Load Current Oscillations," in *2018 International Exhibition and Conference for Power Electronics, Intelligent Motion, Renewable Energy and Energy Management (PCIM Europe)*, 2018, pp. 1–9.

[11] M. Rahimo, C. Papadopoulos, C. Corvasce, and A. Kopta, "An optimized plug-in BIGT with no requirements for gate control adaptations," *PCIM Eur. 2017 - Int. Exhib. Conf. Power Electron. Intell. Motion, Renew. Energy Energy Manag.*, no. May, pp. 16–18, 2017.

[12] Q. T. Tran, F.-J. Niedernostheide, F. Pfirsch, A. Mauder, R. Baburske, and H.-G. Eckel, "RC-GID IGBT – A novel reverse-conducting IGBT with a gate voltage independent diode characteristic and low power losses," in *The 33rd International Symposium on Power Semiconductor Devices and ICs (ISPSD)*, 2021, pp. 347–350, doi: 10.23919/ispsd50666.2021.9452199.

[13] D. Lexow, Q. T. Tran, and H. Eckel, "Outcome Improvement for Pilot Diode Reverse Conducting IGBTs through selective locking time reduction in diode mode," in *2021 23nd European Conference on Power Electronics and Applications (EPE'21 ECCE Europe)*, 2021, pp. 1–10.

[14] D. Lexow and H. G. Eckel, "Performance Improvement for Plug-In Reverse Conducting IGBTs through Gate-Voltage Observation," in *2020 22nd European Conference on Power Electronics and Applications, EPE 2020 ECCE Europe*, 2020, p. P.1-P.7, doi: 10.23919/EPE20ECCEEurope43536.2020.9215764.

Modeling of an Interleaved DC-DC Boost Converter for a Direct Model Predictive Control Strategy

Thomas Effenberger*, Hannes Börngen*, Eyke Liegmann*, Michael Hoerner*[†],
Petros Karamanakos[‡], and Ralph Kennel*

*Chair of High-Power Converter Systems, Technical University of Munich
Arcisstr. 21, 80333 Munich, Germany
[†]Institute ELSYS, Technische Hochschule Nuremberg
Kesslerplatz 12, 90489 Nuremberg, Germany
[‡]Faculty of Information Technology and Communication Sciences, Tampere University
FI-33101 Tampere, Finland
Email: eyke.liegmann@tum.de

Keywords

≪Interleaved converters≫, ≪Modelling≫, ≪MPC (Model-based Predictive Control)≫.

Abstract

This paper presents a model predictive control (MPC) algorithm for interleaved dc-dc boost converters with coupled inductors. The prediction model covers the switching nature of the converter and all possible operating states. The MPC algorithm is realized in MATLAB and designed such to facilitate its real-time implementation on a field programmable gate array (FPGA) using the MATLAB HDL Coder. Open-loop measurement results demonstrate the accuracy of the system model, while the effectiveness of the controller is validated in simulation.

Introduction

The interleaved dc-dc boost converter is a multi-branch converter topology consisting of multiple parallel boost stages, as depicted in Fig. 1. The branches share the input voltage, the output capacitance and—in the case of coupled inductors—the magnetic core, resulting in a compact design. This enables the reduction of the input current ripple and an increase of the output current, as shown in [1], [2]. However, as there is no inherent mechanism to guarantee proper current sharing among the parallel paths a suitable controller is required to address this challenging issue.

Another challenge that a controller needs to deal with is that the output voltage of the boost converter has a non-minimum-phase behavior with respect to the control input, i.e., the switching action. To mask this,

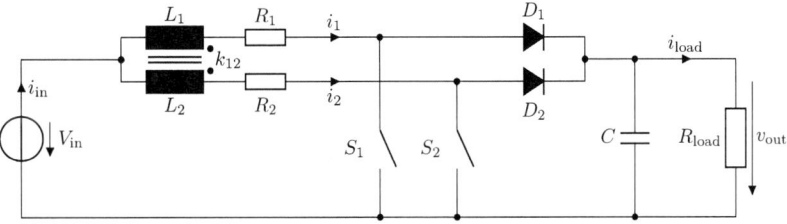

Fig. 1: Interleaved dc-dc boost converter with coupled inductors.

the standard control approach is to design the controller as a current controller, average the continuous-time dynamics associated with the different modes of operation, and to linearize them at the operating point [3], [4]. This, however, complicates the controller design and can potentially deteriorate the dynamic performance.

Direct model predictive control (MPC) with reference tracking—also referred to as finite control set MPC (FCS-MPC)—is a control alternative that can effectively deal with the aforementioned challenges as it allows one to directly include constraints in the design phase and to account for the switching or hybrid nature of dc-dc converters [5]. According to this method, the control inputs are modeled as integers that directly relate to the switch positions of the converter, thus bypassing any modulation stage. Examples of FCS-MPC for dc-dc converters can be found in [6]–[9]. In [6] and [7], FCS-MPC for a dc-dc boost converter is proposed for the direct control of the inductor current and output voltage, respectively. In a similar direction, [8] focuses on FCS-MPC of a single-phase boost converter feeding a constant power load. Finally, an FCS-MPC strategy for an interleaved buck converter with four phases, but without magnetically coupled inductors is presented in [9].

These works clearly demonstrate the advantages of FCS-MPC, such as the great design flexibility and fast dynamic response. Nevertheless, direct manipulation of the converter switches also implies that the optimization problem underlying FCS-MPC is an integer program. A straightforward approach to solve such problems is to use exhaustive enumeration, i.e., to test all candidate solutions before concluding to the optimal one. As integer problems are typically computationally demanding, such a brute-force solution method may not be a realistic option when the number of candidate solutions is not small. To reduce the computational complexity of the associated optimization problem, some methods have been proposed that limit the feasible set [10] or propose nontrivial horizons [11]. Alas, they have been mostly tested on a simulation level [10], [12].

Moreover, considering that the computing time is limited to values smaller than $10\,\mu s$ to achieve a high sampling frequency, and thus enable a high switching granularity [13], a central processing unit (CPU) based control hardware would not be fast enough to test all options in real time within that time interval. A solution is to implement the algorithm on an FPGA, which not only allows for a computationally efficient implementation of the FCS-MPC algorithm, but also facilitates the utilization of a complex model for the interleaved boost converter that fully describes its dynamics.

The above motivates the design of a long-horizon FCS-MPC for the interleaved dc-dc boost converter with coupled inductors that will fully exploit the advantages of FCS-MPC, while facilitating its real-time implementation with a system model covering all physically feasible operating states. To this aim, a detailed model of the converter of interest is first derived that is suitable for all physically feasible operating states. Specifically, the discrete-time model of the converter is designed such that it accurately predicts the plant behavior when operating both in continuous (CCM) and discontinuous conduction mode (DCM). Subsequently, an FCS-MPC strategy is designed as a voltage-mode controller with the main control objectives of regulating the output voltage, balancing the phase leg utilization and limiting the phase currents. Such a control method offers design simplicity as it directly addresses the voltage control problem—and thus non-minimum-phase nature of the converter—without requiring additional control loops, while simultaneously meeting additional control objectives. The presented experimental results based on a low-voltage test bench demonstrate the validity of the proposed model, while simulation results highlight the advantages of the presented control strategy.

System Model

The interleaved boost converter with coupled inductors, shown in Fig. 1, is a dc-dc converter that boosts the input voltage to a higher output voltage. Each converter branch $n \in \{1,2\}$ consists of a diode D_n and an active switch S_n. Each switch S_n can be controlled actively, whereas diode D_n conducts current depending on the applied voltage. In doing so, the current can be stored in the coupled inductors and, subsequently, deliver energy to the output, thus boosting its voltage value. These coupled inductors L_1 and L_2 and their mutual coupling k_{12} can be modeled by utilizing the equivalent Y-model with mutual

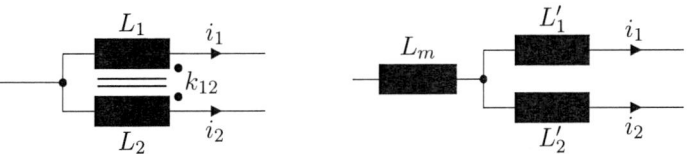

Fig. 2: Coupled inductors and equivalent Y-model.

inductance L_m and uncoupled inductances $L_1' = L_2' = L'$, as shown in Fig. 2. In doing so, the inductances can be mathematically described as follows

$$L_m = k_{12}\sqrt{L_1 L_2}, \tag{1a}$$
$$L_1' = L_1 - L_m, \tag{1b}$$
$$L_2' = L_2 - L_m. \tag{1c}$$

The system model of the interleaved boost converter with coupled inductors differs from the system model of a standard boost converter in two major ways. Firstly, by interleaving, the system state should account for the branch currents, e.g., in the case of two branches, the state vector is chosen to be $\boldsymbol{x} = [i_1\, i_2\, v_{\mathrm{out}}]^T$, with v_{out} being the output voltage. Secondly, by coupling the inductors, a change of the inductor current i_1 induces a voltage in the second winding, thus affecting branch current i_2, and vice versa.

The switch positions of S_1 and S_2 are modeled by the binary variables $u_1, u_2 \in \mathcal{U} = \{0,1\}$, respectively, together forming the input vector $\boldsymbol{u} = [u_1\, u_2]$. In total, four distinct operating modes can be identified for each branch individually. The state machine in Fig. 3 visualizes the system states and the transitions for a single branch.

Within every discrete time interval T_s each converter branch is in one of the four modes, resulting in a total of 16 possible operating modes. In the following, the operating modes are denoted with a symbolic (XY), where X and Y indicate the operating mode of the first and second converter branch, respectively. With switch S_n on, i.e., $u_n = 1$, the respective branch operates in mode ① with current i_n increasing and always being larger than 0. For all remaining modes it holds that $u_n = 0$. In mode ② the corresponding current is decreasing, still being larger than zero, i.e., $i_n(k+1) > 0$. Mode ③ is an intermediate state of modes ② and ④, in which the branch current falls to 0 A, i.e., $i_n(k+1) = 0$. Finally, in mode ④ the current remains 0. An illustrative example of combined modes for the two-branched system is shown in Fig. 4. In addition, all modes (except mode ③) with the corresponding current paths (▬) are depicted in detail in Fig. 5.

The system matrices for mode ③ are calculated by linearly averaging modes ② and ④, weighted with

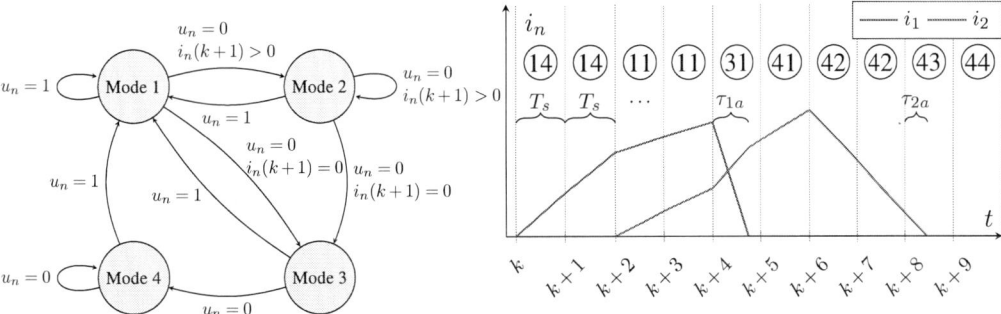

Fig. 3: State machine of branch n [7].

Fig. 4: Illustrative subset of modes of interleaved boost converter.

Modeling of an Interleaved DC-DC Boost Converter for a Direct Model Predictive Control Strategy

EFFENBERGER Thomas

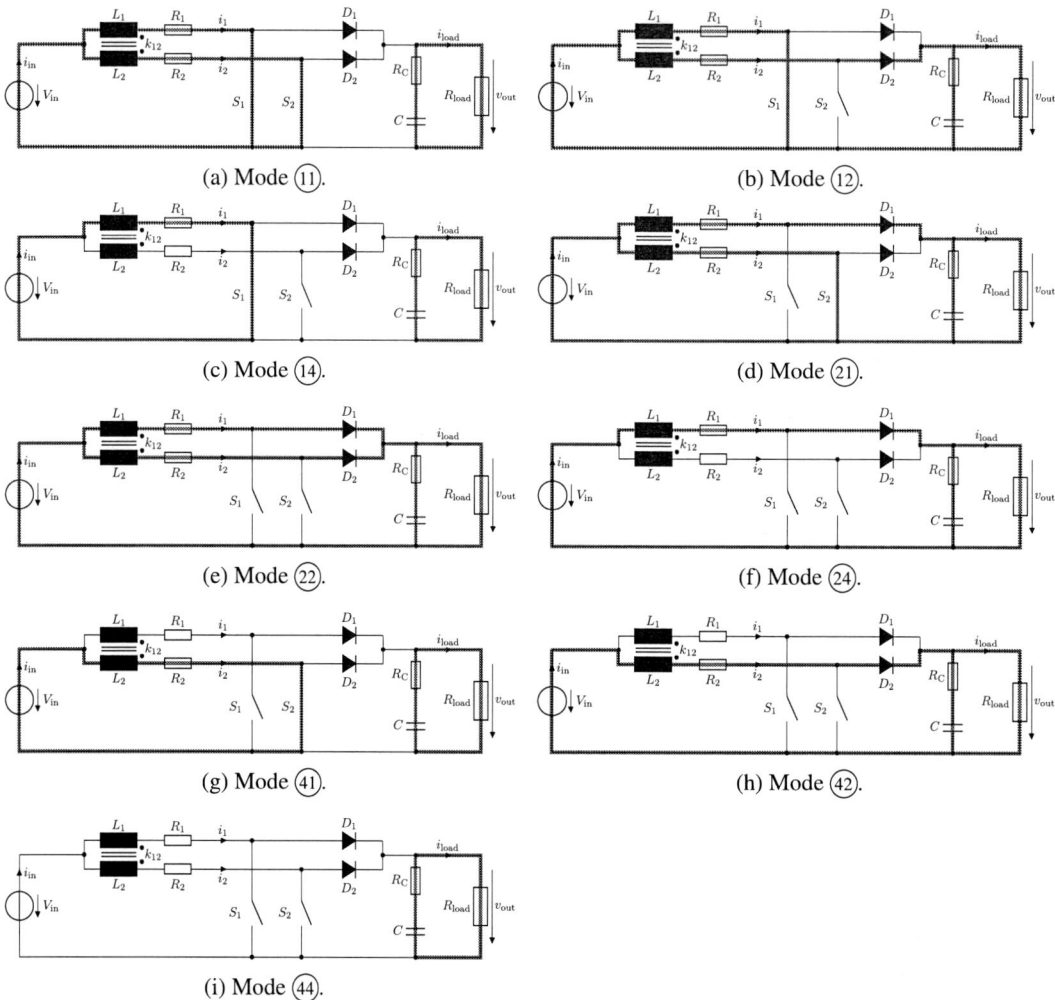

Fig. 5: Active current paths (———) of the interleaved boost converter in different operating modes.

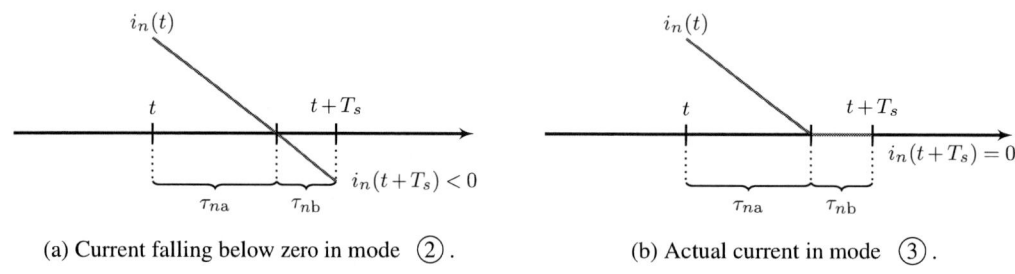

(a) Current falling below zero in mode ②.

(b) Actual current in mode ③.

Fig. 6: Current in mode ③.

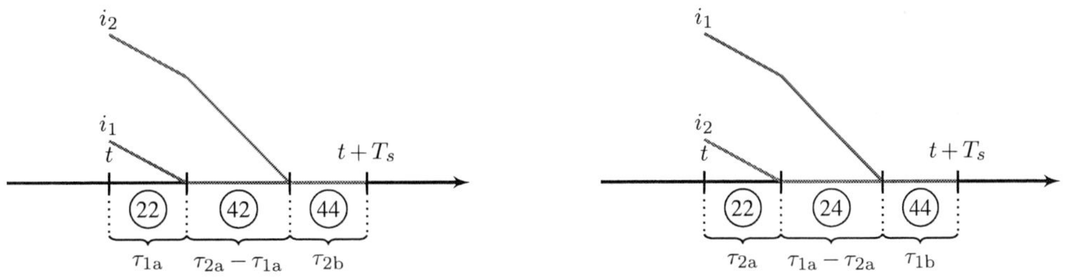

(a) Submodes when i_1 reaches 0 A first. (b) Submodes when i_2 reaches 0 A first.

Fig. 7: Submodes of mode (33).

the duration τ_{na} given by

$$\tau_{na} = \frac{i_n(k)}{i_n(k) - i_n(k+1)} T_s . \tag{2}$$

This interval is the time in which the respective branch current is greater than 0 A, as shown in Fig. 6. Based on that, τ_{nb} can be obtained as $\tau_{nb} = T_s - \tau_{na}$. This is calculated online, assuming operation in mode (2). If a zero-crossing of the branch current is detected, the operation mode is changed to mode (3).

To further clarify this point, consider as an example mode (3Y). For this mode it holds that

$$\boldsymbol{A}_{(3Y)} = \left(\boldsymbol{A}_{(4Y)} + \tau_{1a}/T_s \left(\boldsymbol{A}_{(2Y)} - \boldsymbol{A}_{(4Y)}\right)\right) . \tag{3}$$

In mode (33), both currents fall to 0 A within the same interval. This leads to three distinct intervals within one sampling period, as illustrated in Fig. 7. It can be seen that there are two variants of mode (33), depending on which branch current falls to 0 A first. For example, for the case depicted in Fig. 7a, it holds that

$$\boldsymbol{A}_{(33)} = \frac{1}{T_s}\left(\tau_{1a}\boldsymbol{A}_{(22)} + (\tau_{2a} - \tau_{1a})\boldsymbol{A}_{(42)} + \tau_{2b}\boldsymbol{A}_{(44)}\right) . \tag{4}$$

Given all the above, the system model is derived as a bi-linear state space model of the form

$$\frac{\mathrm{d}\boldsymbol{x}}{\mathrm{d}t} = \boldsymbol{A}(\boldsymbol{x},\boldsymbol{u})\,\boldsymbol{x}(t) + \boldsymbol{B}(\boldsymbol{u}) . \tag{5}$$

The state matrix $\boldsymbol{A}$ is given in (6). The individual entries, which depend on the switching state u_n and the branch current i_n are provided in Table I.

$$\boldsymbol{A} = \begin{bmatrix} A_{11} & A_{12} & A_{13} \\ A_{21} & A_{22} & A_{23} \\ A_{31} & A_{32} & A_{33} \end{bmatrix} \tag{6}$$

As for the entries of the input matrix $\boldsymbol{B}$, these are

$$\boldsymbol{B}_{(11)(22)(12)(21)} = \begin{bmatrix} \frac{L'}{D} \\ \frac{L'}{D} \\ 0 \end{bmatrix} V_{\text{in}}, \quad \boldsymbol{B}_{(14)(24)} = \begin{bmatrix} \frac{1}{L_m+L_1'} \\ 0 \\ 0 \end{bmatrix} V_{\text{in}}, \quad \boldsymbol{B}_{(42)(41)} = \begin{bmatrix} 0 \\ \frac{1}{L_m+L_1'} \\ 0 \end{bmatrix} V_{\text{in}}, \quad \boldsymbol{B}_{(44)} = \boldsymbol{0} \tag{7}$$

Finally, as the controller is designed in the discrete-time domain, in a subsequent step, the continuous-

Table I: Entries of matrix $\boldsymbol{A}$ depending on operating mode, with $D = L'\left(2L_m + L'\right) = \left(L' + L_m\right)^2 - L_m^2$.

Mode	A_{11}	A_{12}	A_{13}	A_{21}	A_{22}	A_{23}	A_{31}	A_{32}	A_{33}
⑪	$-\frac{(L_m+L')R}{D}$	$\frac{L_mR}{D}$	0	$\frac{L_mR}{D}$	$-\frac{(L_m+L')R}{D}$	0	0	0	$-\frac{1}{CR_{\text{load}}}$
㉒	$\vdots$	$\vdots$	$-\frac{L'}{D}$	$\vdots$	$\vdots$	$-\frac{L'}{D}$	$\frac{1}{C}$	$\frac{1}{C}$	$\vdots$
⑫	$\vdots$	$\vdots$	$\frac{L_m}{D}$	$\vdots$	$\vdots$	$-\frac{L_m+L'}{D}$	0	$\frac{1}{C}$	$\vdots$
㉑	$-\frac{(L_m+L')R}{D}$	$\frac{L_mR}{D}$	$-\frac{L_m+L'}{D}$	$\frac{L_mR}{D}$	$-\frac{(L_m+L')R}{D}$	$\frac{L_m}{D}$	$\frac{1}{C}$	0	$\vdots$
⑭	$-\frac{R}{L_m+L'}$	0	0	0	0	0	0	0	$\vdots$
㊶	0	0	0	0	$-\frac{R}{L_m+L'}$	0	0	0	$\vdots$
㉔	$-\frac{R}{L_m+L'}$	0	$-\frac{1}{L_m+L'}$	0	0	0	$\frac{1}{C}$	0	$\vdots$
㊷	0	0	0	0	$-\frac{R}{L_m+L'}$	$-\frac{1}{L_m+L'}$	0	$\frac{1}{C}$	$\vdots$
㊹	0	0	0	0	0	0	0	0	$-\frac{1}{CR_{\text{load}}}$

time system matrices are discretized using forward Euler method. This yields

$$\boldsymbol{x}(k+1) = \left(\boldsymbol{I} + \boldsymbol{A}T_s\right)\boldsymbol{x}(k) + \boldsymbol{B}T_s \tag{8}$$

with T_s being the sampling interval.

In the next section, the proposed converter model is verified experimentally.

Model Verification

The proposed model is tested using the FPGA-based open-source control platform UltraZohm [14], [15], which is depicted in Fig. 8. The platform enables rapid control prototyping of power electronic systems. The calculation unit is a Xilinx Zynq UltraScale+ 9EG MPSoC that consists of several ARM processors and an FPGA on the same silicon chip. The platform is expandable with adapter cards that offer digital and analog interfaces to the converter and sensor. For FPGA hardware development Xilinx Vivado Design Suite is used. The software for the generated hardware processor is programmed with Xilinx Vitis.

Fig. 9 shows the prototype of the coupled inductors, while Table II presents the parameters of the interleaved boost converter. A power inverter board equipped with gallium nitride (GaN) transistors, shown in Figs. 10 and 11, provides the switching cells. The GaN inverter adapter card offers three half-bridges

Fig. 8: Experimental setup with UltraZohm.

Fig. 9: Coupled inductors prototype.

Fig. 10: GaN adapter card top side.

Fig. 11: GaN adapter card bottom side.

with a shared dc-link, originally designed for drive applications. It is compatible with the digital adapter slots of the UltraZohm system, easing the initial setup effort. The power switches are fully integrated with logic level input drivers and protection against over-current. The availability of phase current, phase voltage, and dc-link voltage measurements allows for use with various control schemes.

In the preliminary experimental setup, push-pull stages are used, i.e., D_1 and D_2 are replaced with active switches, limiting the set of operating modes to only ⑪, ⑫, ㉑, and ㉒, leading to a forced CCM type of operation. Here, a more reasonable approach to not drive the top-side switches in each half-bridge, and thus yield a diode-like behavior, was not recommended by the manufacturer of the power switches [16]. As such, the top-side switches are driven with the inverse signal of the respective bottom-side switches, leading to the aforementioned forced CCM.

With this setup, the presented mathematical model is verified by applying a pre-calculated switching sequence directly to the switches, resulting in an open-loop start-up operation. The parameters of the open-loop experiment are provided in Table III. The sampling frequency is chosen to be $50\,\mathrm{kHz}$, resulting in an average switching frequency of $18.625\,\mathrm{kHz}$. A comparison of the simulated start-up with the corresponding measurements is depicted in Fig. 12. As can be seen, there is a close matching between the simulated and experimental output voltage v_{out}, shown in Figs. 12a and 12b, respectively. Moreover, the simulated and experimental input current i_{in} depicted in Figs. 12c and 12d, respectively, shows a comparable pattern that corroborates the validity of the derived model.

The differences in the branch currents i_1 and i_2, predicted by the simulation compared to the actual current in the test setup is briefly explained in the following. For this, the branch currents, in addition to the input current, are presented over a shorter time interval to allow for insightful observations, see Fig. 13. In simulation (see Fig. 13a) both branch currents match exactly, whereas Fig. 13b depicts drastically different waveforms. This is an effect caused by the peculiarities of the push-pull topology, in which the timing of the "rectifying" switches becomes critical—in comparison to a topology using conventional diodes. Here, even a small delay of one half-bridge—with respect to the other—will cause a steep change in the currents through both inductances L_1' and L_2', which are, with the coupling factor chosen, in the one-digit µH range. For example, at time $3.22\,\mathrm{ms}$, the sequence of commanded modes is ⑪→㉒, where an additional intermediate mode, i.e., mode ㉑, can be identified that stems from an unwanted delay in the gate driving. During the short time mode ㉑ is active, the output voltage is directly applied to L_1' and L_2', in the negative and in the positive direction respectively, forcing a change of the

Table II: Parameters of interleaved boost prototype. Table III: Parameters for open-loop measurements.

Parameter	Value
L_1, L_2	$190\,\mu\mathrm{H}$
k_{12}	0.993
R_1, R_2	$150\,\mathrm{m}\Omega$
C	$146\,\mu\mathrm{F}$

Parameter	Value
V_{in}	$10\,\mathrm{V}$
v_{out}^*	$20\,\mathrm{V}$
R_{load}	$20\,\Omega$

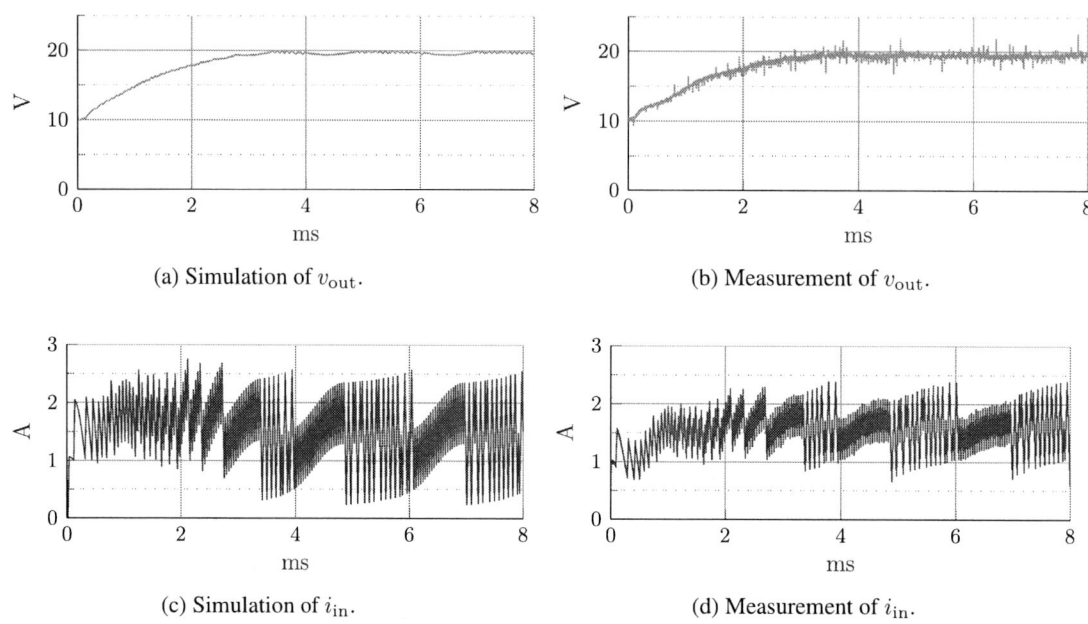

(a) Simulation of v_{out}.

(b) Measurement of v_{out}.

(c) Simulation of i_{in}.

(d) Measurement of i_{in}.

Fig. 12: Comparison between simulation and experimental measurements.

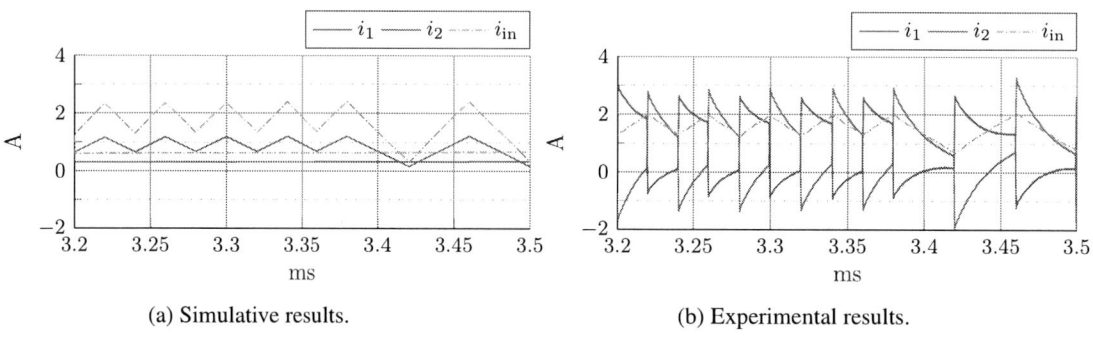

(a) Simulative results.

(b) Experimental results.

Fig. 13: Branch currents i_1 ——, i_2 ——— and input current i_{in} ⋯⋯.

branch currents according to the voltages applied.

Despite the aforementioned shortcomings of the setup, it can be concluded that the presented results from the *feasible* subset of operating modes clearly demonstrate the validity of the developed model of the converter.

Optimal Control Algorithm

In this section, the previously derived discrete-time model of the converter is leveraged in a direct MPC scheme. In MPC, the control action is obtained online by solving the underlying optimization problem. At each time step, the switching commands are optimized for N time steps ahead in the future, i.e., the prediction horizon, with respect to a given objective function. The formulated optimization problem, besides accounting for the discrete-time model of the system, can account for explicit constraints. The optimal sequence of control inputs is the one that minimizes the objective function. To introduce feedback, and thus deal with model uncertainties and disturbances, only the first element of the optimal sequence of control actions is applied to the converter. The described approach is known in literature as *receding horizon policy* [17]. At the next time step, the optimization problem is solved again with updated measurements.

The objectives of the proposed MPC strategy are the minimization of the difference between the predicted and the reference output voltage v_{out}^*, i.e., the tracking error, and the minimization of excess branch currents, i.e., currents larger than an upper current bound i_{bnd}. These objectives are mapped into a scalar with the following objective function

$$J(k) = \sum_{\ell=k}^{k+N-1} (v_{\mathrm{out}}^*(\ell) - v_{\mathrm{out}}(\ell))^2 + \lambda_i J_i(\ell) + \lambda_b \left| \sum_{m=1}^{k+N-1} i_1(m) - i_2(m) \right| \quad \text{with} \tag{9a}$$

$$J_i(\ell) = \sum_{n=1}^{2} \begin{cases} 0 & i_n(\ell) < i_{\mathrm{bnd}} \\ (i_n(\ell) - i_{\mathrm{bnd}})^2 & \text{otherwise.} \end{cases} \tag{9b}$$

It is worth mentioning that in (9a) a third term is added to achieve a balanced utilization of the two converter branches. This term minimizes the difference of the branch currents, $i_1 - i_2$ over the whole operation of the converter, starting from the first time step $m = 1$. Moreover, the weighting factors $\lambda_i, \lambda_b > 0$ are introduced to prioritize among the control goals.

To find the optimal switching sequence that results in the smallest cost $J_{\mathrm{opt}}(k)$ of (9), an exhaustive search is performed at every time step. Algorithm 1 shows the basic structure of the proposed controller with an outer for-loop iterating over all possible switching combinations. An inner loop, that iterates over N, performs the prediction and the cost calculation for each candidate solution. In a final step, the calculated cost is compared with the tentative minimal cost value. If a candidate with a lower associated cost is found, the upper bound of J_{opt} is updated accordingly. Once all candidates have been explored, the switching pattern resulting in J_{opt} is identified as the optimal solution.

Algorithm 1 Pseudocode of proposed MPC algorithm.

1: **for** each $U(k) = \begin{bmatrix} u(k) & u(k+1) & \dots & u(k+N-1) \end{bmatrix}^T \in \mathcal{U}^{2N}$ **do**
2:      **for** each step $\ell \in \{1, \dots, N\}$ **do**
3:          Predict next state based on the operating mode of the converter.
4:          Calculate intermediate cost.
5:      **end for**
6:      **if** cost of candidate $\leq J_{\mathrm{opt}}$ **then**
7:          Update J_{opt}
8:      **end if**
9: **end for**
10: **return** $u^*(k)$

Simulation Results

In this section, the proposed control scheme is scrutinized in a simulation environment. The input voltage of $50\,\mathrm{V}$ is boosted to a reference output voltage of $100\,\mathrm{V}$ at a load of $20\,\Omega$. The sampling and control frequency is $100\,\mathrm{kHz}$, while the switching frequency is variable due to the direct nature of the control scheme. The weighting factors are $\lambda_i = 0.05, \lambda_b = 0.01$, with the soft constraint on the branch current being activated above $i_{\mathrm{bnd}} = 5$. Fig. 14 displays simulation results for a prediction horizon of $N = 3$ and $N = 6$, as well as for two different coupling factors k_{12}. For three exemplary parameter sets, the input current i_{in}, the branch currents i_1 and i_2, and the output voltage v_{out} are plotted. Below, the gate signals u_1 and u_2 are depicted.

It is apparent that the reference output voltage can be tracked accurately in all depicted cases. For $N = 3$, (Figs. 14a and 14b) the algorithm exhibits similar behavior for both coupling factors in steady-state operation. Nevertheless, significant differences can be found when focusing on the output voltage ramp-up section. For $k_{12} = 0.5$, the algorithm chooses operating points that are in CCM, particularly for i_2. This enables a higher input current (with comparable branch currents) and thus speeds up the initial ramp up. For $k_{12} = 0.9$, however, the uncoupled inductances are significantly smaller. As a result, the changes

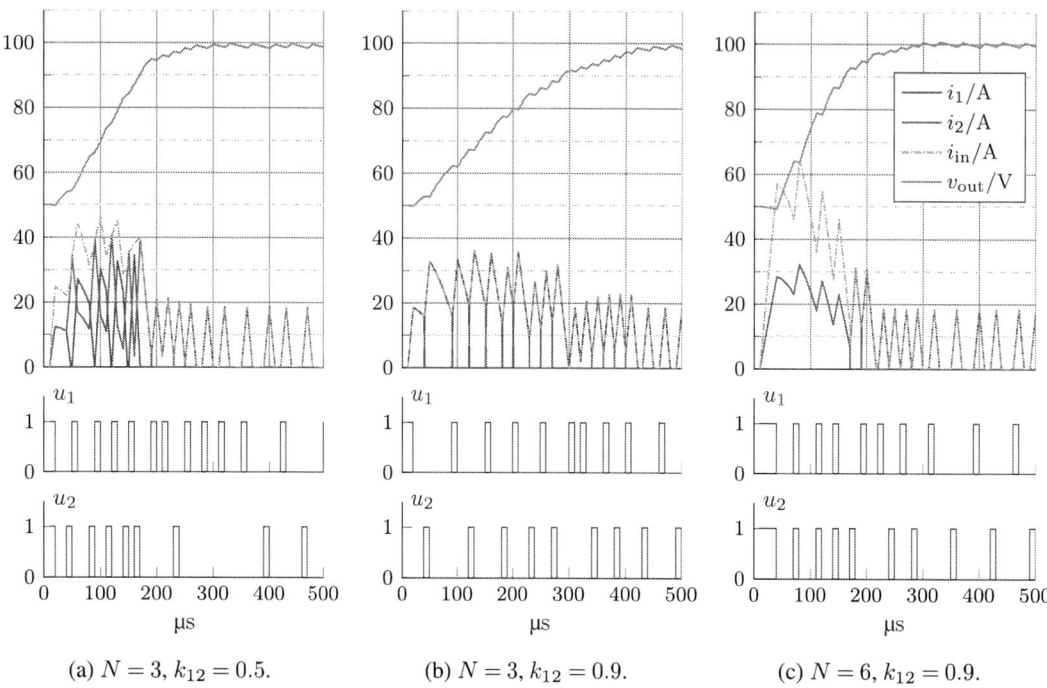

(a) $N = 3$, $k_{12} = 0.5$. (b) $N = 3$, $k_{12} = 0.9$. (c) $N = 6$, $k_{12} = 0.9$.

Fig. 14: Start-up behavior of the interleaved dc-dc boost converter for different N and k_{12}.

in the current, and thus its ripple, are bigger, leading to a DCM operation. This gives rise to a slower start-up scenario with a slower transient time by a factor of 2. It is noteworthy that the shorter prediction horizon hinders the algorithm to fully align the cost of branch current balancing while achieving output tracking error minimization.

Comparing this behavior with the case where the same coupling factor of 0.9 is used along with a longer prediction horizon of $N = 6$ steps, it can be seen that the proposed direct MPC strategy is able to provide a faster dynamic response. Specifically, the algorithm outputs a symmetric start-up pattern to minimize the cumulative tracking error by means of a faster initial ramp-up, while maintaining perfectly balanced branch currents.

Finally, it is noteworthy that the non-minimum-phase nature of the boost converter is visible in the first simulation steps in Fig.14, where the output voltage drops while the coupled inductors are being magnetized. In simulation, a minimum prediction horizon length of $N = 2$ steps is required to achieve a stable system behavior for the specific problem setting. One possible explanation is that the controller needs to be able to predict past the initial voltage dip caused by the non-minimum-phase behavior [7].

Conclusion

This paper presented a switched model of an interleaved boost converter with coupled inductors that covers all possible operating modes of the two-branched system. To derive an as accurate model as possible, the influence of the coupling of the inductors was included in the model. Based on this model, an FCS-MPC scheme was developed and its effectiveness was validated in a MATLAB simulation. The presented results demonstrate that a favorable operation can be achieved as the horizon length increases, even when the magnetic coupling of the inductors is strong.

In a next step, the experimental setup will be changed with respect to the power switches used to avoid operating limitations and thus enable operation in DCM. Additionally, the controller will be implemented on the FPGA of the open-source control platform UltraZohm. In doing so, the whole control scheme, consisting of the derived model and the presented control algorithm, will be validated in real time.

References

[1] P.-W. Lee, Y.-S. Lee, D. Cheng, and X.-C. Liu, "Steady-state analysis of an interleaved boost converter with coupled inductors," *IEEE Trans. Ind. Electron.*, vol. 47, no. 4, pp. 787–795, Aug. 2000.

[2] H. Kosai, S. McNeal, B. Jordan, *et al.*, "Coupled inductor characterization for a high performance interleaved boost converter," *IEEE Trans. Magn.*, vol. 45, no. 10, pp. 4812–4815, Oct. 2009.

[3] P. Athalye, R. W. Erickson, and D. Maksimović, "Variable-frequency predictive digital current mode control," *IEEE Power Electron. Lett.*, vol. 2, no. 2, pp. 113–116, Dec. 2004.

[4] T. Gomez, W. Hernández, W. Martínez, and C. A. Cortes, "Control techniques for interleaved dc/dc converters with magnetic coupling," in *IEEE Works. on Power Electr. and Power Qual. Appl.*, Bogota, Colombia, May 2017, pp. 1–6.

[5] P. Karamanakos, E. Liegmann, T. Geyer, and R. Kennel, "Model predictive control of power electronic systems: Methods, results, and challenges," *IEEE Open J. Ind. Appl.*, vol. 1, pp. 95–114, 2020.

[6] P. Karamanakos, T. Geyer, and S. Manias, "Direct model predictive current control strategy of dc-dc boost converters," *IEEE J. Emerg. Sel. Topics Power Electron.*, vol. 1, no. 4, pp. 337–346, Dec. 2013.

[7] P. Karamanakos, T. Geyer, and S. Manias, "Direct voltage control of dc-dc boost converters using enumeration-based model predictive control," *IEEE Trans. Power Electron.*, vol. 29, no. 2, pp. 968–978, Feb. 2014.

[8] Z. Karami, Q. Shafiee, S. Sahoo, *et al.*, "Hybrid model predictive control of dc–dc boost converters with constant power load," *IEEE Trans. Energy Convers.*, vol. 36, no. 2, pp. 1347–1356, Jun. 2021.

[9] T. Hausberger, A. Kugi, A. Eder, and W. Kemmetmüller, "High-speed nonlinear model predictive control of an interleaved switching dc/dc-converter," *Control Eng. Pract.*, vol. 103, pp. 1–13, Oct. 2020.

[10] P. Karamanakos, T. Geyer, and S. Manias, "Model predictive control of the interleaved dc-dc boost converter with coupled inductors," in *Proc. Eur. Power Electron. Conf.*, Lille, France, Sep. 2013, pp. 1–10.

[11] P. Karamanakos, T. Geyer, N. Oikonomou, F. D. Kieferndorf, and S. Manias, "Direct model predictive control: A review of strategies that achieve long prediction intervals for power electronics," *IEEE Ind. Electron. Mag.*, vol. 8, no. 1, pp. 32–43, Mar. 2014.

[12] Y. Liang, Z. Liang, D. Zhao, Y. Huangfu, and L. Guo, "Model predictive control for interleaved dc-dc boost converter based on Kalman compensation," in *Int. Power Electr. and Appl. Conf. and Expo.*, Shenzhen, China, Nov. 2018, pp. 1–5.

[13] P. Karamanakos and T. Geyer, "Guidelines for the design of finite control set model predictive controllers," *IEEE Trans. Power Electron.*, vol. 35, no. 7, pp. 7434–7450, Jul. 2020.

[14] S. Wendel, A. Geiger, E. Liegmann, *et al.*, "UltraZohm—A powerful real-time computation platform for MPC and multi-level inverters," in *Proc. IEEE Int. Symp. Pred. Control of Elect. Drives and Power Electron.*, Quanzhou, China, May 2019, pp. 1–6.

[15] E. Liegmann, T. Schindler, P. Karamanakos, A. Dietz, and R. Kennel, "UltraZohm——An open-source rapid control prototyping platform for power electronic systems," in *Int. Aegean Conf. on Elect. Mach. and Power Electron. and Int. Conf. on Optim. of Elect. and Electron. Equip.*, Brasov, Romania, Sep. 2021, pp. 1544–1551.

[16] *LMG342xR030 600-V 30-mOhm GaN FET with integrated driver, protection, and temperature reporting*, SNOSDA7D, Revised March 2022, Texas Instruments, Sep. 2020. [Online]. Available: https://www.ti.com/lit/gpn/LMG3425R030.

[17] T. Geyer, *Model predictive control of high power converters and industrial drives*. Hoboken, NJ: Wiley, 2016.

Static analysis and control strategies of the Single Active Bridge Converter

Alexis A. Gómez[1], Alberto Rodríguez[1], Marta M. Hernando[1], Diego G. Lamar[1], Javier Sebastián[1], Ibán Ayarzaguena[2], Jose Manuel Bermejo[2], Igor Larrazabal[2], David Ortega[2], Francisco Vázquez[3]

[1] Power Supply Systems Group. University of Oviedo, [2] Ingeteam Power Technology S.A., [3] Ingeteam R&D Europe.

[1] Gijón, Spain; [2] Zamudio, Spain; [3] Zamudio, Spain

E-Mail: gomezalexis@uniovi.es

URL: https://sea.grupos.uniovi.es/

Acknowledgements

This work was financed by the European project UE-18-POWER2POWER-826417, by the Principado de Asturias through project SV-PA-21-AYUD/2021/51931, and by the Spanish Ministry of Science, Innovation and Universities through projects MCI-21-PDC2021-121242-I00 and MCI-20-PID2019-110483RB-I00.

Keywords

«DC-DC converter», «Isolated converter», «DC-DC power converter control», «Switching frequency control», «Single active bridge»

Abstract

In this paper, a brief static analysis of the Single Active Bridge, a unidirectional isolated DC-DC converter, is done. In this initial analysis the SAB is operating at a fixed frequency and controlled by varying the duty cycle. Additionally, two variable frequency control strategies are proposed that extends the soft-switching capabilities. This paper focuses on one of them that fixes the duty cycle and uses the switching frequency as the sole control variable. The conduction modes of the converter using the proposed control are analyzed and expressions for the voltage conversion ratio are obtained, which will be used in a design guide that provides ZVS over the full range of operation. A comparative current analysis is performed in which average, rms and switching currents of both control methods are depicted. Finally theoretical, simulation and experimental results are compared with good agreement.

Introduction

The Dual Active Bridge (DAB) is one of the reference converters when it comes to bidirectional and high-power density isolated DC-DC converters [1], [2]. In the case of applications where the power flows always in the same direction, it is possible to replace the secondary transistor bridge by a diode H-bridge; this substitution can potentially decrease the cost of the converter while improving its reliability and power density. The resulting converter has been called the Single Active Bridge (SAB), and several publications have examined the static behavior of this converter operating at a fixed frequency and controlled by varying the duty cycle [3]–[7].

Isolated DC-DC converters are subjects of development in different areas such as, battery chargers for electro mobility, renewable energies or second stages of grid tied converters. Several converter topologies are adequate for these applications, including the DAB and its variations. This document briefly introduces the static study of the SAB converter operating at a fixed frequency and controlled by varying the duty cycle and its conduction modes. Finally, two variable frequency control strategies are proposed, and one is analyzed. A design guide and simulation and experimental validation are provided. A general scheme of the SAB converter is shown in Fig. 1.

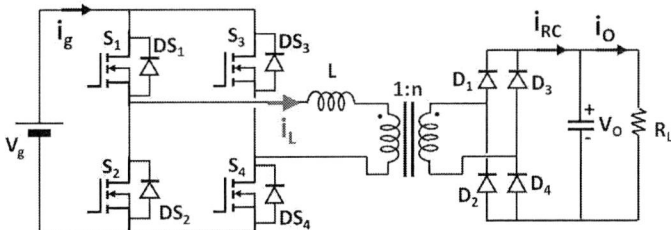

Fig. 1: Single Active Bridge (SAB) descriptive schematic.

Static analysis of the Single Active Bridge

A complete static analysis of the SAB operating at a fixed frequency and controlled by varying the duty cycle is presented in [6]. A brief explanation is introduced in this section, as a reference, to propose the variable switching frequency control, which is the main contribution of this paper.

The SAB can operate in two distinct conduction modes, Continuous Conduction Mode (CCM) and Discontinuous Conduction Modes (DCM). The operating mode of the converter is defined according to the current through the inductor L.

Continuous Conduction Mode (CCM)

This mode is characterized by an inductor current waveform that does not remain at zero, but only crosses zero on the transitions from the positive to the negative current stages and vice versa.

The analysis of the converter in this conduction mode starts by splitting a full switching period into 6 different stages and analyzing the current through the inductor (i_L) to find the voltage conversion ratio. This is done by averaging the transferred charge to the RC network over the full period. The i_L waveform for this mode is graphed in Fig. 2 and can be calculated by means of Faraday's Law.

Now analyzing each stage depicted in Fig. 2 a), from t_1 to t_2, S1 and S4 are closed, and the inductor withstands the input voltage minus the output voltage as seen from the primary side of the transformer, both voltages are supposed with negligible ripple. When the current reaches i_{L2} and from t_2 to t_3, S4 and DS2 conduct, the inductor withstands negative output voltage seen from the primary side. To finalize the positive current stages, from t_3 to t_4, DS2 and DS3 conduct and the voltage on the inductor is the negative sum of the input and output voltages, the latter as seen from the primary side. This analysis can be repeated in an analogue manner for the negative i_L stages.

Considering the current injected into the RC output network (i_{RC}), depicted in Fig. 2 b), it is possible to calculate the output voltage and the voltage conversion ratio for the SAB converter when working in CCM.

$$V_{o\_CCM} = R_L\, i_{RC\ average\ CCM} = \frac{R_L}{2\pi L T_s}\left[V_g t_c (T_s - t_c) - \frac{T_s^2 \left(\frac{V_o}{n}\right)^2}{4V_g}\right] \tag{1}$$

$$N_{CCM} = \frac{V_o}{n\,V_g} = \frac{4\,(1-d)d}{k + \sqrt{k^2 + 4\,(1-d)\,d}} \tag{2}$$

$$t_c = t_2 - t_0 \tag{3}$$

$$d = \frac{t_c}{T_s} \tag{4}$$

$$k = \frac{2Ln^2}{R_L \frac{T_s}{2}} \tag{5}$$

Where, n is the transformer relation, d is the duty cycle, T_s is the switching period and, t_c is the time in which current flows through the input on a semi period, V_o and V_g are the output and input voltages, respectively. These equations show that the voltage conversion ratio when the converter operates in

CCM is heavily dependent on the connected load (R_L), contrary to other step-down converters when operating in this mode such as, the Buck or the Phase-Shifted Full Bridge (PSFB).

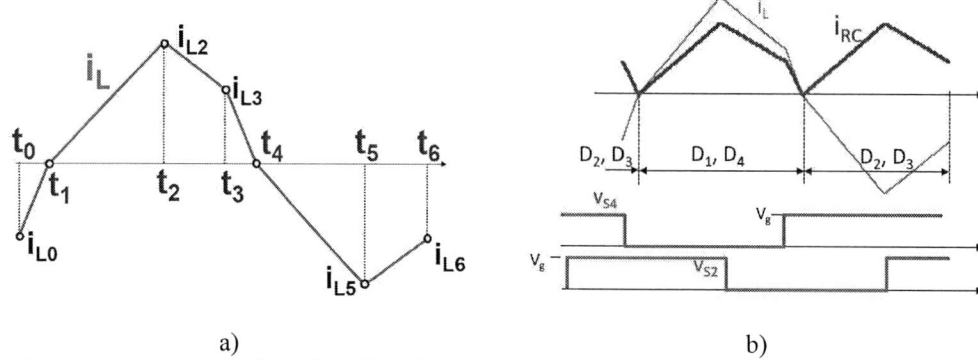

a)   b)

Fig. 2: a) Inductor current in CCM. b) Inductor current and injected current to the RC network for CCM and voltage on the semiconductors.

Discontinuous Conduction Mode (DCM)

When the converter operates in DCM, the inductor and the RC network currents, graphed in Fig. 3, remains at zero from t_3 to t_4 and from t_0 to t_1, and no current flows through any semiconductor. The analysis methodology is analogue to that employed for CCM, resulting in:

$$V_{o\_DCM} = R_L\, i_{RC\ average\ DCM} = \frac{V_g\, R_L}{V_o L T_s}\left[V_g - \frac{V_o}{n}\right] t_c^2 \tag{6}$$

$$N_{DCM} = \frac{V_o}{n\, V_g} = \frac{2d}{d + \sqrt{d^2 + k}} \tag{7}$$

In this mode, the output voltage and the voltage conversion ratio depend again on the connected load, as for other step-down converters.

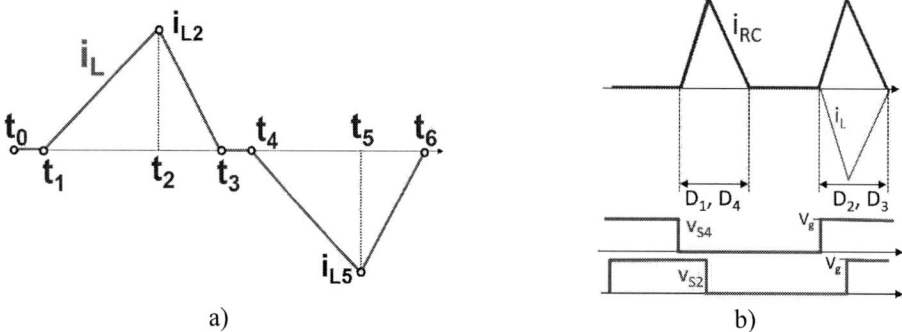

a)   b)

Fig. 3: a) Inductor current in DCM. b) Inductor current and injected current to the RC network for DCM and voltage on the semiconductors.

Boundary between CCM and DCM

The converter operates in the boundary between both conduction modes when t_3 and t_4; t_0 and t_1 are the same instant, respectively; and the inductor current is zero. As in the boundary mode the converter does not operate strictly in neither conduction mode, both voltage conversion relations must be correct, therefore equations (2) and (7) must be valid for this operating point, resulting in:

$$k_{crit} = 1 - 2d_{crit} \tag{8}$$

$$k_{crit} = \frac{2Ln^2}{R_{L\,crit}\,\dfrac{T_s}{2}} \tag{9}$$

$$N_{crit} = \left(\frac{V_o}{n\,V_g}\right)_{crit} = 2d_{crit} \tag{10}$$

As with other similar converters, such as the Buck or the PSFB, the converter enters DCM from CCM when the connected load is greater than a specific value, function of the parameters of the converter and control variables values.

Variable frequency control

Up to this point the converter was considered operating at a fixed frequency and controlled by varying the duty cycle (d), as defined in (4). This strategy allows for a regulated output current or voltage by changing the duty cycle as a function of input voltages and power levels. As the switching period appears on the voltage conversion ratio equation for DCM and CCM, the switching frequency can also be used as a control variable. To maximize the range of operation, a duty cycle controlled SAB must operate in both CCM and DCM. On the one hand, the SAB converter while operating in CCM can achieve Zero Voltage Switching (ZVS) [8], [9]; however, when working in DCM this is not possible. On the other hand, conduction losses are generally higher for CCM than DCM, this is due to a greater semiconductor RMS current value, greater average currents, and the presence of excess recirculating currents, necessary for soft-switching.

To operate the SAB in CCM over a wide range while maintaining a minimum amount of recirculating current, the SAB can be controlled by varying the frequency. Two possible control possibilities arise from this additional control variable, a control with fixed duty cycle and a variable switching frequency; and a control strategy where both are variable. For both control methods, the main idea is to operate in CCM at any operating point, guaranteeing ZVS. For the first strategy the necessary recirculating current to achieve soft-switching must be guaranteed, therefore this level of current is stablished at the maximum operating frequency and minimum load; as the duty cycle is fixed, the recirculating current at greater power levels, lower operating frequencies, will be greater than necessary. If the second alternative is chosen, as there are two control variables, the recirculating currents can be kept at the minimum level over the entire range of operation, at the expense of an extended switching frequency range. This paper will focus on the first approach. Ideal current waveforms for both approaches are graphed in Fig. 4.

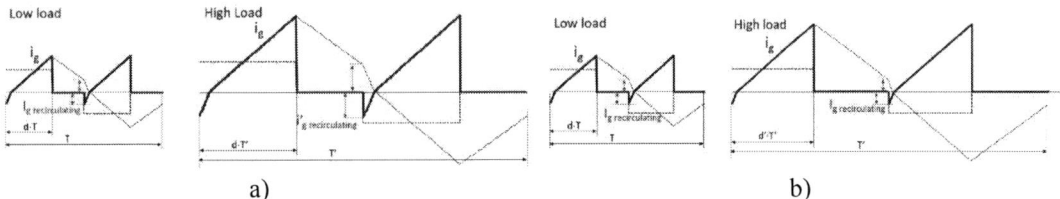

a)   b)

Fig. 4: Ideal inductor current waveforms with minimum recirculating current to achieve ZVS at different loads. a) Fixed duty cycle and variable switching frequency. b) Variable duty cycle and switching frequency.

Static analysis for variable frequency

By observing (2) and (7), if the duty cycle remains constant, to maintain a constant voltage conversion ratio the product $T\,R_L$ must remain constant, this implies a linear relation between the switching frequency and the connected load. Reformulating (2) and (7), the linear relation previously mentioned is clearly appreciated in the following equations.

$$f_{CCM} = \left(\frac{(1-d)d}{2N} - \frac{N}{8}\right)\frac{R_L}{n^2 L} \tag{11}$$

$$f_{DCM} = \frac{(1-N)d^2}{N^2}\frac{R_L}{Ln^2} \tag{12}$$

In Fig. 5 a) and b) the "open loop" behavior of the converter is shown, understood as the converter response to a fixed control variable as a function of the connected load and for different duty cycles and switching frequencies. In Fig. 5 c) and d) the "closed loop" behavior is depicted, understood as the control variable value required to obtain a specific response; this is done again as a function of load and for several duty cycles and normalized voltage conversion ratios. Fig. 5 and (10) show that the boundary between conduction modes is set by the value of the duty cycle, therefore if it is greater than half of the normalized voltage conversion ratio, CCM operation is guaranteed over all the operating range.

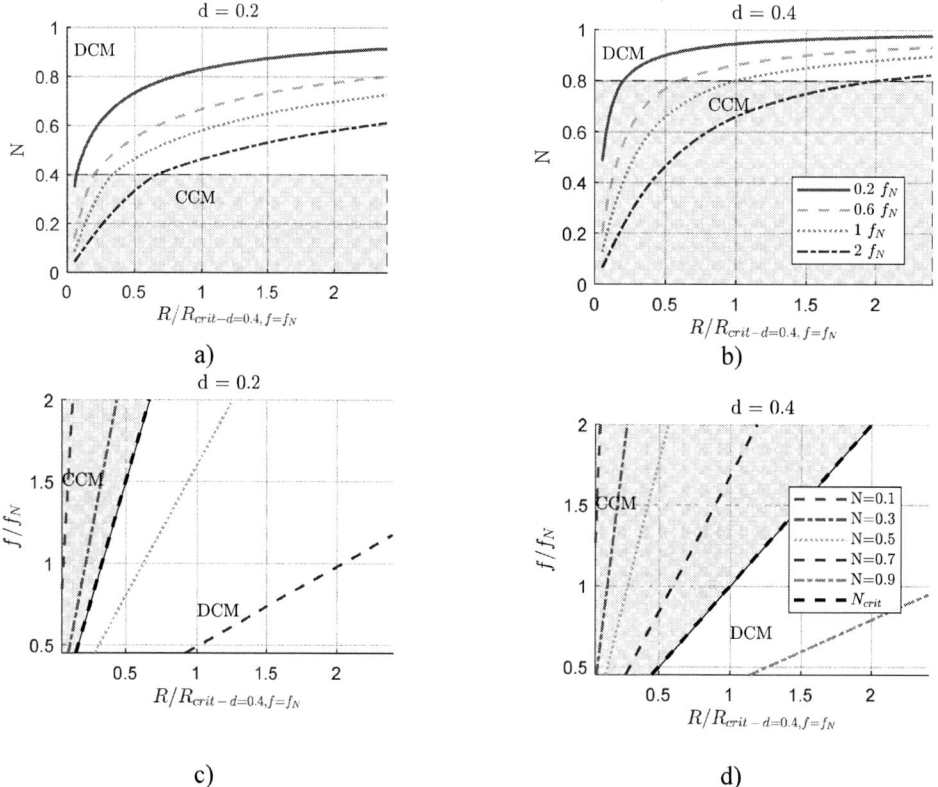

Fig. 5: a) and b) Normalized voltage conversion ratio (N) as a function of normalized load, for different switching frequencies. c) and d) Normalized switching frequency as a function of normalized load, for different values of N.

For the current analysis, input and recirculating average currents are compared for variable duty cycle and variable frequency control strategies. Input and output voltages are supposed ripple free, and all converter components are considered ideal. For both conduction modes, the average input current is described by:

$$\overline{I_g} = 2/T \left[\int_{t_1}^{t_2} i_g(t) \ dt + \int_{t_0}^{t_1} i_g(t) \ dt \right] \tag{13}$$

It can be appreciated in (13) that current only flows through the input in two intervals (from t_1 to t_2 and from t_0 to t_1). The current on the first interval is positive and appears on both conduction modes, the second interval is characterized by a negative current in CCM and null in DCM, this is the recirculating current. Equations (14) and (15) express the average input current for CCM and DCM respectively. To obtain the average recirculating current expression (18), an analogue method is followed.

$$\overline{I_{g\,CCM}} = \frac{1}{TL} \left[\left(V_g - \frac{V_o}{n} \right) (t_2 - t_1)^2_{MCC} - \left(V_g + \frac{V_o}{n} \right) (t_1 - t_0)^2_{CCM} \right] \tag{14}$$

$$\overline{I_{g\,DCM}} = \frac{1}{TL}\left[\left(V_g - \frac{V_o}{n}\right)(t_2 - t_1)^2_{DCM}\right] \tag{15}$$

$$(t_2 - t_1)_{CCM} = \frac{T}{2}\left(d + \frac{V_o}{2V_g n}\right) \tag{16}$$

$$(t_2 - t_1)_{DCM} = dT \tag{17}$$

$$\overline{I_{recirculating}} = \frac{1}{TL}\left(V_g + \frac{V_o}{n}\right)(t_1 - t_0)^2_{CCM} \tag{18}$$

$$(t_1 - t_0)_{CCM} = \frac{T}{2}\left(d - \frac{V_o}{2V_g n}\right) \tag{19}$$

By relating (18) and (14) equation (20) is obtained. It shows that the relation of average recirculating current to the average input current is independent of the switching frequency; and only depends on the converter design, input and output voltages and duty cycle.

$$\frac{\overline{I_{recirculating}}}{\overline{I_{g\,CCM}}} = \frac{\left(V_g + \frac{V_o}{n}\right)\left(d + \frac{V_o}{2nV_g}\right)^2}{\left(V_g - \frac{V_o}{n}\right)\left(d - \frac{V_o}{2nV_g}\right)^2 - \left(V_g + \frac{V_o}{n}\right)\left(d + \frac{V_o}{2nV_g}\right)^2} \tag{20}$$

Design guide

This design process aims to design a SAB converter that operates in CCM for all operating points, therefore achieving ZVS. With this purpose, the maximum critical duty cycle (d_{crit}) is set, and this value will serve as the inferior limit of the available duty cycle range (21). The initial specifications necessary for the design process are:
- Input and output voltage ranges.
- Output current range.
- Switching frequency range.

As mentioned, the first step is to set an adequate maximum critical duty cycle. The actual duty cycle should be superior to this value; the closer they both are, the less recirculating current.

$$d_{crit} \leq d \leq 0.5 \tag{21}$$

The duty cycle is used in (22) to find the transformation relation of the transformer. Once calculated, the inductor value that achieves the minimum voltage conversion ratio (23) at the maximum frequency specified is obtained. This is the minimum value of L. By doing this with (24), the minimum power operating point is fixed at the top of the specified frequency range. Finally, the minimum frequency is checked to be within the specified frequency range with (26).

$$n_{min} = \frac{V_{o\,max}}{2d_{crit}V_{g\,min}} \tag{22}$$

$$N_{min} = \frac{V_{o\,min}}{n\,V_{g\,max}} \tag{23}$$

$$L = \left(\frac{(1-d)d}{2N_{min}} - \frac{N_{min}}{8}\right)\frac{\frac{V_{o\,min}}{I_{o\,min}}}{n^2 f_{max}} \tag{24}$$

$$N_{max} = \frac{V_{o\,max}}{n\,V_{g\,min}} \tag{25}$$

$$f_{min} = \left(\frac{(1-d)d}{2N_{max}} - \frac{N_{max}}{8}\right)\frac{\frac{V_{o\,max}}{I_{o\,max}}}{n^2 L} \tag{26}$$

Alternatively, the operating point to fix during the design process can be the one corresponding to the minimum frequency; to achieve this, an analogue process should be followed. In this occasion, the L value calculated corresponds to the maximum, and it is necessary to check the maximum frequency to be below the specified maximum. This is done by using expressions (27) and (28).

$$L = \left(\frac{(1-d)d}{2N_{max}} - \frac{N_{max}}{8} \right) \frac{\frac{V_{o\,max}}{I_{o\,max}}}{n^2 f_{min}} \tag{27}$$

$$f_{max} = \left(\frac{(1-d)d}{2N_{min}} - \frac{N_{min}}{8} \right) \frac{\frac{V_{o\,min}}{I_{o\,min}}}{n^2 L} \tag{28}$$

A comparative design example between this design guide and the one exposed in [6] is done with the following specifications:
- Input voltage range: [800 V, 850 V].
- Output voltage range: [350 V, 400 V].
- Output current range: [0.5 A, 5.5 A].
- Acceptable switching frequency range: [22 kHz, 300 kHz].

The picked maximum critical duty (d_{crit}) cycle for both design guides is 0.25, and the variable frequency design has a fixed duty cycle of 0.275. The results of both design procedures are in Table I. Control variable values are graphed in Fig. 6 as a function of the output current for different input and output voltages. Fig. 6 a) corresponds to the design of variable duty cycle and can be appreciated that the converter operates in both conduction modes; whereas Fig. 6 b), corresponds to the variable frequency design and only operates in CCM, as this was one of the criteria used for the design guide. The switching frequency of the first design is 33 kHz.

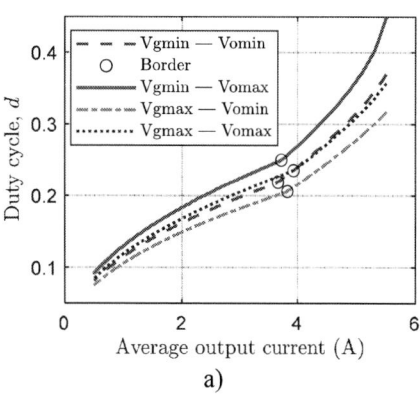
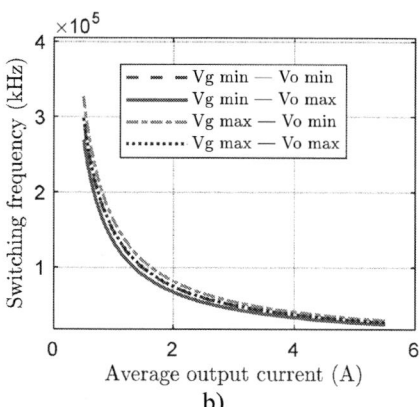

a)   b)

Fig. 6: Control variable values as a function of the average output current for different input and output voltages. a) Converter controlled by a variable duty cycle. b) Converter controlled by a variable switching frequency.

Table I: Resulting designs.

Design	Transformer relation (n)	Inductor value (L)
Variable duty cycle	1	408 µH
Variable switching frequency	1	444 µH

Current through the semiconductors

To compare both designs, rms and average current though the semiconductors have been calculated (these values provide a comparison of conduction losses). All currents are calculated for minimum input and maximum output voltages.

Fig. 7 a) shows the rms currents through a pair of semiconductors on the input side as a function of the average output current. With equal transformer relations, in general the variable frequency design operates with lower levels of rms current, due to a better inductor current waveform. Because of the design guide criteria, the variable frequency converter suffers the presence of a small recirculating current at all operating points that increases proportionally with the average output current, as predicted by (20). On the contrary, the variable duty cycle design only has recirculating current at CCM operating points.

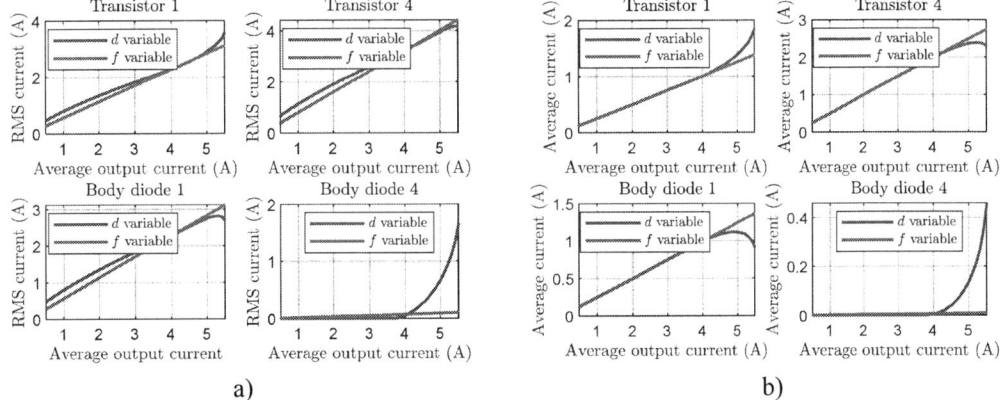

a)   b)

Fig. 7: Semiconductor currents for both designs. a) RMS currents. b) Average currents.

For the average currents, represented in Fig. 7 b), in the regions where the variable duty cycle design operates in DCM, the average currents are similar, this changes once both alternatives operate in CCM. Finally, Fig. 9 shows the inductor's current, i_L, at the switching events, this is of interest to compare switching losses. All the simulations were done using the software PSIM considering all components ideal; accurate correlation between analytical predictions and simulation results were obtained.

Experimental results

In this section experimental results of a duty cycle controlled SAB and a switching frequency controlled SAB are provided. The prototype used for the experimental validation was designed with the design guide of [6] resulting in the specifications of Table . The inductor is integrated in the leakage inductance of the transformer. With this prototype all previous equations are validated, and results are provided for both control strategies. The input and output voltages for all tests are 800 V and 400 V, respectively.

Table II: Experimental prototype specifications.

Transformer relation (n)	Inductor value (L)
1	408 µH

Equations (14), (15), (18) and (20) are graphed in Fig. 8. Fig. 8 a) and c) correspond to a converter controlled by a variable duty cycle and designed with the design guide from [6]. Fig. 8 b) and d) correspond to a converter designed to be controlled by a variable frequency. Fig. 8 a) and b) represent the average input current as a function of the control variable. Fig. 8 c) and d) represent the average normalized recirculating current as a function of the control variable for each converter design. From Fig. 8 it is possible to understand that high duty cycles result in a higher power transfer, but more relative recirculating current, which increases conduction losses. For this reason, to maximize the use of the converter the duty cycle must remain high, so the transferred power is high without reaching for excessively low switching frequencies; at the same time if the preferred operating mode is CCM, the maximum critical duty cycle should be high as well to minimize the recirculating current. Fig. 8 d) shows a relative recirculating current constant over the complete range, which results in excess recirculating current at higher power levels than the minimum.

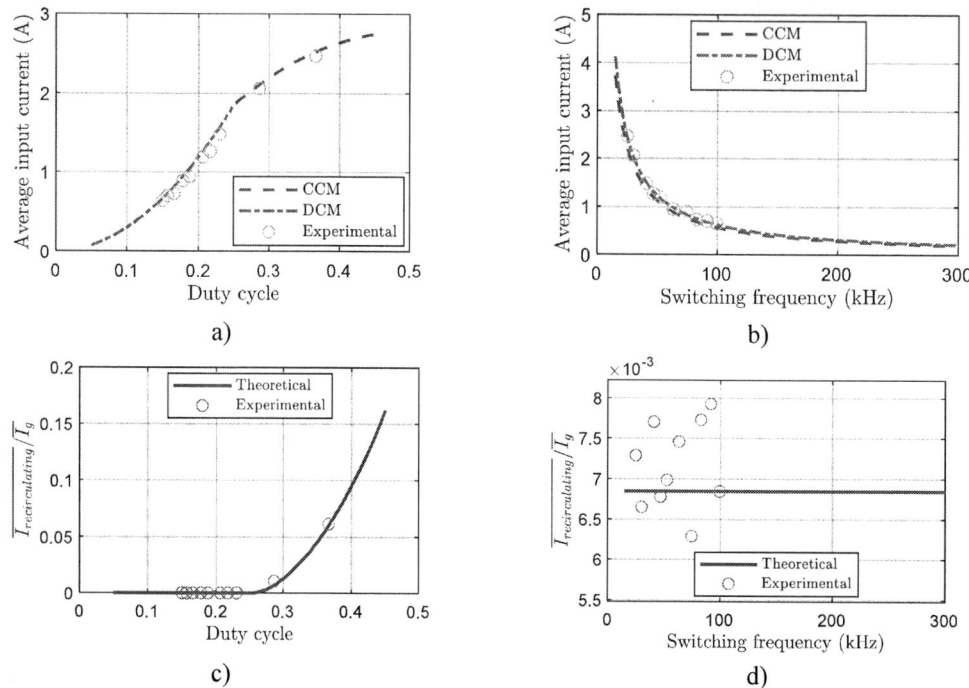

Fig. 8: a) and b) Average input current as a function of the control variable; duty cycle and switching frequency, respectively. c) and d) Relation of average recirculating and input currents as a function of the control variable; duty cycle and switching frequency, respectively.

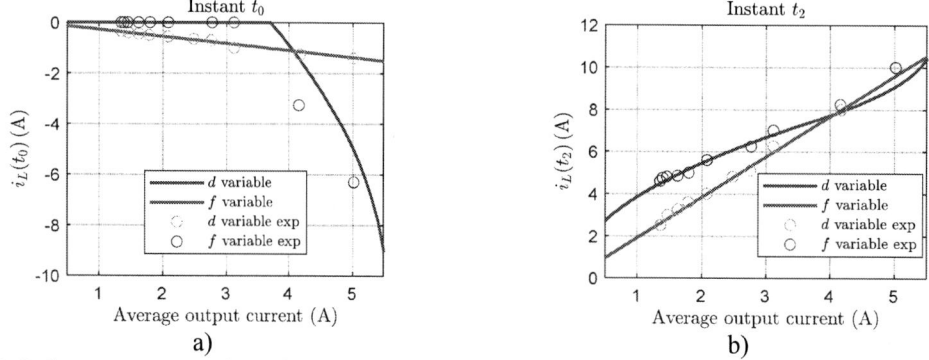

Fig. 9: Inductor current at the switching instants; theoretical and experimental, for both designs.

Fig. 10 a) Shows the theoretical, simulated, and experimental value of the switching frequency for the fixed duty cycle design when the input and output voltages are minimum and maximum respectively. Fig. 10 b) and c) are oscilloscopes captures when the converter operates at 91 % and 29 % of the nominal load. Yellow and green [5 V/div] are gate signals corresponding to channels 1 and 2 respectively; purple [5 A/div] is channel 3 and measures the inductor's current; magenta [200 V/div] is the drain voltage of one of the switches. [10 µs/div].

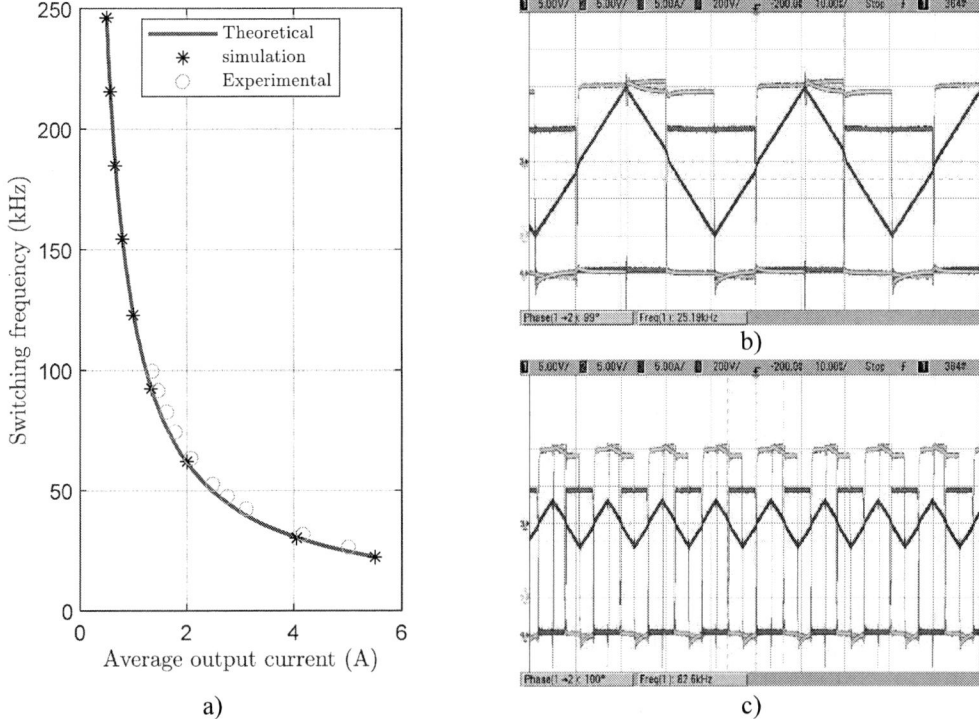

Fig. 10: a) Switching frequency as a function of the average output current; theoretical, simulated, and experimental results. b) and c) Oscilloscope captures of two operating points.

Finally, an efficiency comparison between the two control methods is done and graphed in Fig. 11. In this figure the efficiency of both control strategies as a function of the average output current is shown. In general, the proposed control strategy yields better efficiency, specially at high loads. Only at lower power levels the variable duty cycle achieves better performance, due in part to the absent recirculating currents.

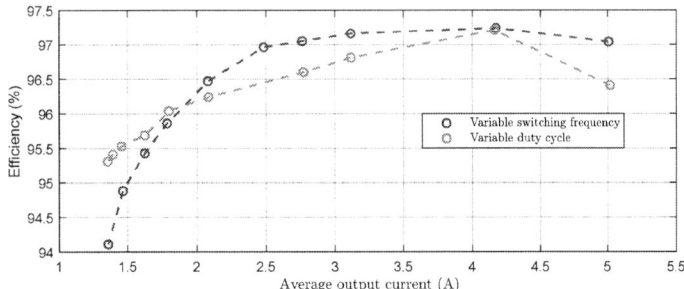

Fig. 11: Efficiency comparison.

Conclusion

A brief static analysis of the SAB converter is done, it shows a directly proportional relation between the switching frequency and the connected load, in CCM and DCM. Recirculating current proportion to average input current is independent of switching frequency, which makes a variable frequency control adequate to maintain ZVS over a wider range than a duty cycle based control. Due to the linearity between load and frequency, the frequency range can be excessively high, which makes for difficult and decreased performance magnetic design. From the current analysis, conduction losses seem similar in both cases, and efficiency is in general slightly improved except at low power. A control strategy with both, the switching frequency and the duty cycle, as control variables could reduce the excessive frequency range.

References

[1] A. R. Rodríguez Alonso, J. Sebastian, D. G. Lamar, M. M. Hernando, and A. Vazquez, "An overall study of a Dual Active Bridge for bidirectional DC/DC conversion," in *2010 IEEE Energy Conversion Congress and Exposition*, Sep. 2010, pp. 1129–1135. doi: 10.1109/ECCE.2010.5617847.

[2] A. Rodríguez, A. Vázquez, D. G. Lamar, M. M. Hernando, and J. Sebastián, "Different Purpose Design Strategies and Techniques to Improve the Performance of a Dual Active Bridge With Phase-Shift Control," *IEEE Transactions on Power Electronics*, vol. 30, no. 2, pp. 790–804, Feb. 2015, doi: 10.1109/TPEL.2014.2309853.

[3] C. Fontana, M. Forato, M. Bertoluzzo, and G. Buja, "Design characteristics of SAB and DAB converters," in *2015 Intl Aegean Conference on Electrical Machines & Power Electronics (ACEMP), 2015 Intl Conference on Optimization of Electrical & Electronic Equipment (OPTIM) & 2015 Intl Symposium on Advanced Electromechanical Motion Systems (ELECTROMOTION)*, Side, Turkey, Sep. 2015, pp. 661–668. doi: 10.1109/OPTIM.2015.7427025.

[4] A. Averberg and A. Mertens, "Characteristics of the single active bridge converter with voltage doubler," in *2008 13th International Power Electronics and Motion Control Conference*, Poznan, Poland, Sep. 2008, pp. 213–220. doi: 10.1109/EPEPEMC.2008.4635269.

[5] G. D. Demetriades, "On small-signal analysis and control of the single- and dual-active bridge topologies," Stockholm, 2005.

[6] A. Rodriguez *et al.*, "An Overall Analysis of the Static Characteristics of the Single Active Bridge Converter," *Electronics*, vol. 11, no. 4, Art. no. 4, Jan. 2022, doi: 10.3390/electronics11040601.

[7] C. A. Tuan, H. Naoki, and T. Takeshita, "Unidirectional Isolated High-Frequency-Link DC-DC Converter Using Soft-Switching Technique," in *2019 IEEE 4th International Future Energy Electronics Conference (IFEEC)*, Singapore, Singapore, Nov. 2019, pp. 1–7. doi: 10.1109/IFEEC47410.2019.9015117.

[8] G. D. Demetriades and H. P. Nee, "Characterisation of the Soft-switched Single-Active Bridge Topology Employing a Novel Control Scheme for High-power DC-DC Applications," in *IEEE 36th Conference on Power Electronics Specialists, 2005.*, Aachen, Germany, 2005, pp. 1947–1951. doi: 10.1109/PESC.2005.1581898.

[9] C. Fontana, M. Forato, K. Kumar, M. T. Outeiro, M. Bertoluzzo, and G. Buja, "Soft-switching capabilities of SAB vs. DAB converters," in *IECON 2015 - 41st Annual Conference of the IEEE Industrial Electronics Society*, Yokohama, Nov. 2015, pp. 003485–003490. doi: 10.1109/IECON.2015.7392640.

Multi-port Inductive Power Transfer System Considering Charging Auxiliary Battery in EVs

Zhuoqi ZHANG[1*], Ryosuke OTA*, Ryohei OKADA*, Nobukazu HOSHI*
* Tokyo University of Science
2641 Yamazaki, Noda-city
Chiba, 278-8510, Japan
Email: [1]7321553@ed.tus.ac.jp
URL: https://www.rs.tus.ac.jp/nhoshi/

Keywords

≪Electric vehicle≫, ≪Wireless power transmission≫, ≪Power converters for EV≫, ≪Battery charger≫, ≪On-board auxiliary power supply system≫.

Abstract

In electric vehicles (EVs), the main battery used for the traction system and an auxiliary battery for the low-voltage system are equipped. When inductive power transfer (IPT) is used to charge the main battery, it requires high-voltage ratings and high-cost switching devices for the onboard AC/DC converter. In addition, the onboard auxiliary battery is generally charged by an isolated DC/DC converter connected to the main battery. Due to this configuration, the high-voltage rating switching devices are needed for the isolated DC/DC converter in addition to IPT. For this, the utilization rate of the high-voltage rating switching devices for IPT is low, and that increases cost. A previous work proposed a multi-port system to integrate the transformer for the isolated DC/DC converter and the transmission coil for IPT to reduce the cost. In the previous system, though the high-voltage rating devices could be reduced, the number of the switching device was the same as in the general system. Therefore, to reduce the cost, this paper proposes a multi-port onboard converter that integrates an IPT system and an auxiliary battery charger system by using high-voltage rating switching devices for IPT. This integration reduces the number of switching devices by four compared to the general system. In this paper, the operation and characteristics of the proposed system are clarified, and its validity is verified by the experiment. Regarding the experimental result, it is confirmed the characteristics of the transmission power agreed well with the theoretical result.

Introduction

In recent years, inductive power transfer (IPT) has gathered attention as a new charging technology for electric vehicles (EVs) [1]–[10]. This IPT technology is assumed to charge the main battery used for the traction system, which has high voltage. However, in EVs, an auxiliary battery is also used for low-voltage systems such as lighting, ECUs, and car navigation system. Here, Fig. 1 shows a conventional onboard power electronic system to which an IPT system is applied. In the system shown in Fig. 1, the power is transmitted from the power supply system on/in the ground to the onboard system, and then the main battery is temporally charged. Afterward, the auxiliary battery is charged via an isolated DC/DC converter [11, 12]. However, the system configuration requires high-voltage rating switching devices depending on the main battery voltage. In addition, the power conversion stage in the onboard system increases. The literature [13] proposes a multi-port system with two receiver coils on EVs to address this issue. The system can charge the main and auxiliary batteries with the two coils. In addition, the system does not need high-voltage rating switching devices for the isolated DC/DC converter. However, the number of the switching device is the same as in the conventional system.

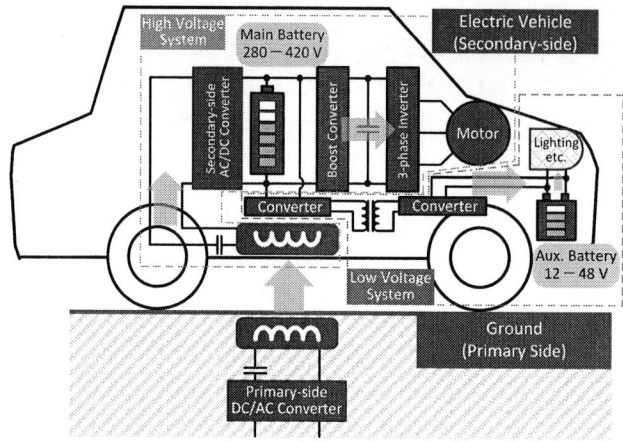

Fig. 1: Onboard power electronics system with an IPT system [13].

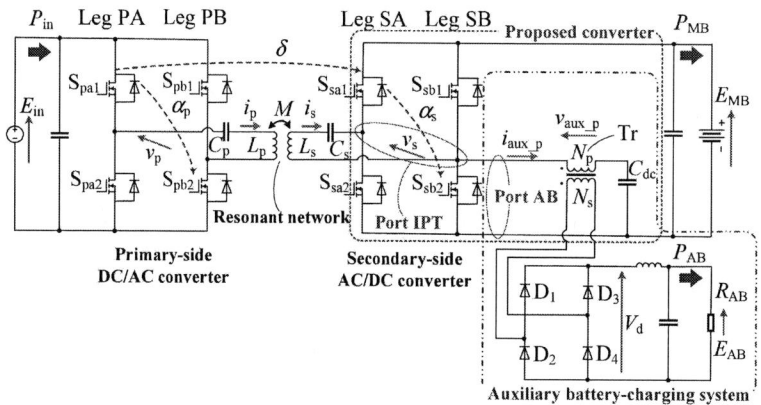

Fig. 2: IPT system with the proposed multi-port converter.

Therefore, this paper proposes a novel multi-port converter shown in Fig. 2. In the proposed system, the one-side leg in the AC/DC converter for IPT is also used for the isolated DC/DC converter, i.e., these two functions are integrated. This integration reduces the number of switching devices by four compared to the conventional systems [11]–[13]. This is expected to contribute the cost reduction. This paper shows the basic operation of the proposed system and verifies its operation with the experiment.

Overview of Proposed System

In this chapter, the proposed system shown in Fig. 2 is explained. In addition, the equations regarding the transmission power of the proposed system are shown.

Configuration of Proposed System

Here, Table I shows the definitions of the variables in this paper. In Fig. 2, the main battery is simulated by a constant voltage source E_{MB}, and a load resistor R_{AB} is used instead of the auxiliary battery. In addition, it is only assumed to transmit power from the ground-side system to the EV. The proposed system comprises a DC/AC converter on the ground (primary) side, a resonant network, and the proposed multi-port AC/DC converter on the vehicle (secondary) side. The SS topology is applied to the resonant network. The primary/secondary-side voltages $\dot{V}_p$, $\dot{V}_s$ on the resonant network are controlled by the phase-shift angle between the legs in each converter and the duty ratio of the switches. In addition, the

Table I: Definitions of variables in this paper.

Variables	Definitions	Variables	Definitions
P_{in} [W]	Transmission power from the primary side	P_{con} [W]	Total conduction losses in the primary/secondary-side converters
P_{MB} [W]	Transmission power to the main battery	δ [rad]	Phase-shift angle between the gate signals for S_{pa1} and S_{sa1}
P_{AB} [W]	Transmission power to R_{AB}	v_p, v_s [V]	Voltages on the primary/secondary-sides of the resonant network
P_{rn} [W]	Total conduction losses on the resonant network	i_p, i_s [A]	Currents flowing in the primary/secondary-sides of the resonant network
P_{rec} [W]	Total conduction losses in the rectifier	$\dot{V}_p, \dot{V}_s$ [V]	Phase notation for the fundamental component of v_p, v_s
E_{in} [V]	DC supply voltage	$\dot{I}_p, \dot{I}_s$ [A]	Phase notation for the fundamental component of i_p, i_s
E_{MB} [V]	Main battery voltage	r_p, r_s [Ω]	Total ESRs on the primary/secondary-sides of the resonant network
R_{AB} [Ω]	Equivalent resistance for the auxiliary battery	r_{sw_on} [Ω]	On resistance of the switching device
f [Hz]	Switching frequency in the converters	L_p, L_s [H]	Inductance of the transmission coils
M [H]	Mutual inductance expressed as $M = k\sqrt{L_p L_s}$	N_p, N_s	Turn numbers of Tr in the primary/secondary sides
k	Coupling coefficient between L_p and L_s	L_r [H]	Leakage inductance in Tr
d_{PA}, d_{PB}	Duty ratios of S_{pa1} and S_{pb1}	v_{aux_p} [V]	Voltage on the primary-side of Tr
d_{SA}, d_{SB}	Duty ratios of S_{sa1} and S_{sb1}	i_{aux_p} [A]	Current flowing in the primary-side of Tr
α_p [rad]	Phase-shift angle between Leg PA and Leg PB	$\dot{V}_{aux\_p}$ [V]	Phase notation for the fundamental component of v_{aux_p}
α_s [rad]	Phase-shift angle between Leg SA and Leg SB	$\dot{I}_{aux\_p}$ [A]	Phase notation for the fundamental component of i_{aux_p}
β_p, β_s	Phase angles of $\dot{V}_p, \dot{V}_s$	V_{rec} [V]	Threshold voltage in a rectifier diode based
$\cos\varphi$	Displacement factor in Port IPT	V_{d_av}	Average value of rectified voltage V_d
η [%]	Total efficiency of the whole system		

converters are operated at 85 kHz constantly.

Moreover, the transformer Tr for the auxiliary battery charging system and the capacitor C_{dc} to block direct current is connected to the Leg SB. The power transmitted to the auxiliary battery charging system is regulated by the duty ratio d_{SB} of the switch S_{sb1} in the Leg SB. With the above configuration, the auxiliary battery charging system can be constructed without additional switching devices to the general IPT system.

Transmission Power via Resonant Network

The transmission power of the resonant network can be derived from the fundamental components of the voltages and currents because the resonant network with the SS topology works as a bandpass filter of the resonant angular frequency ω [rad/s] [14, 15]. Here, $\dot{V}_p, \dot{V}_s$ are respectively expressed as

$$\dot{V}_p = \frac{E_{in}}{\sqrt{2}\pi}\sqrt{6 + 2\cos\{2\pi(d_{PA}+d_{PB})\} - 4\cos(2\pi d_{PA}) - 4\cos(2\pi d_{PB})}\,\varepsilon^{j(\beta_p+\alpha_p-2\pi)} \quad \text{and} \tag{1}$$

$$\dot{V}_s = \frac{E_{MB}}{\sqrt{2}\pi}\sqrt{6 + 2\cos\{2\pi(d_{SA}+d_{SB})\} - 4\cos(2\pi d_{SA}) - 4\cos(2\pi d_{SB})}\,\varepsilon^{j(\delta+\beta_s+\alpha_s-2\pi)}, \tag{2}$$

where the phase shift angle between the primary/secondary-sides δ is fixed at $\pi/2$ in this paper. In addition, the primary/secondary-side currents $\dot{I}_p, \dot{I}_s$ are expressed as

$$\dot{I}_p = -\frac{r_s \dot{V}_p - j\omega M \dot{V}_s}{(j\omega M)^2 - r_p r_s} \quad \text{and} \tag{3}$$

$$\dot{I}_s = \frac{r_p \dot{V}_s - j\omega M \dot{V}_p}{(j\omega M)^2 - r_p r_s}. \tag{4}$$

When considering the power losses in the whole system, the transmission power P_{in} from the primary side can be expressed by

$$P_{\text{in}} = P_{\text{MB}} + P_{\text{AB}} + P_{\text{rn}} + P_{\text{rec}} + P_{\text{con}} = |\dot{V}_{\text{p}}||\dot{I}_{\text{p}}|\cos\varphi, \tag{5}$$

where $\cos\varphi$ represents the displacement factor of the resonant network. The details of the control method for $\dot{V}_{\text{p}}$, $\dot{V}_{\text{s}}$, P_{MB}, and P_{AB} are described in the next chapter.

Operation Mode of Proposed System

When $P_{\text{in}} \geq 0$ is assumed, the operation of the proposed system is roughly divided into the following three modes.

(i)   Simultaneous Charging (SC) mode
(ii) Auxiliary Battery Charging (ABC) mode
(iii) Main Battery Charging (MBC) mode

Here, Fig. 3 shows the operation modes of the proposed converter. In Fig. 3, α_{s} is set to 0 in the SC mode for simplicity. Fig. 4 shows the representative waveforms of the gate signals and the port voltages/currents in the proposed converter. In the SC mode, the main and auxiliary batteries are simultaneously charged from the primary side. In the ABC mode, the auxiliary battery is charged from the main battery without

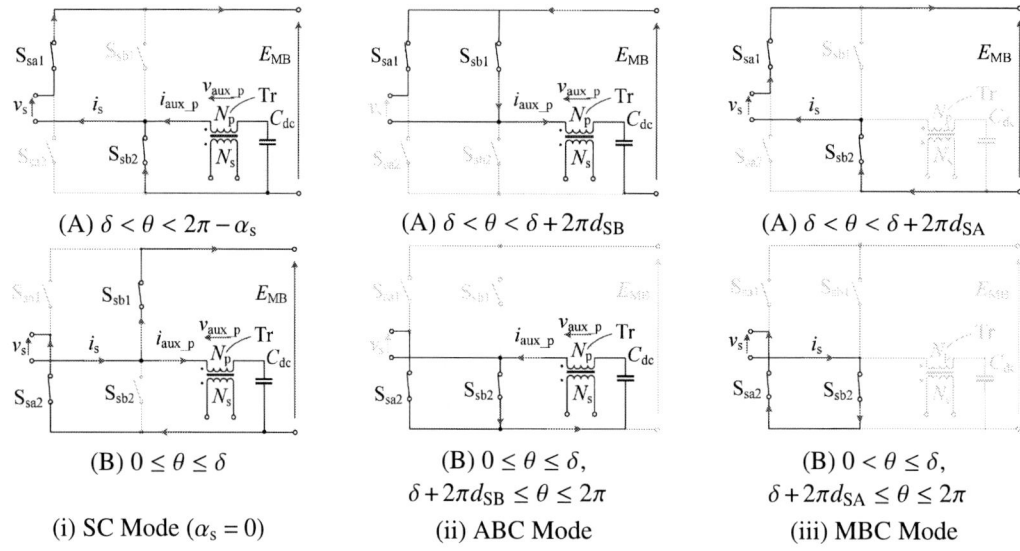

Fig. 3: Operation modes of the proposed converter.

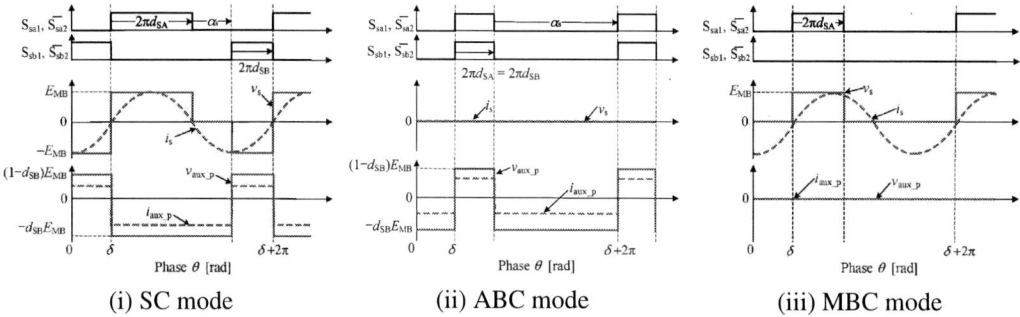

Fig. 4: Representative waveforms of the gate signals and the port voltages/currents of the proposed converter.

Table II: Relationship between the operation modes and the control parameters in the proposed converter.

Variables	SC mode (a)	SC mode (b)	SC mode (c)	ABC mode	MBC mode
α_s [rad]	$2\pi(1-d_{SA}-d_{SB})$	$2\pi(1-d_{SA}-d_{SB})$	$2\pi(1-d_{SA}-d_{SB})$	$2\pi(1-d_{SB})$	–
d_{SA}	0 to 1	0 to 1	0 to 1	d_{SB}	0 to 1
d_{SB}	d_{SA}	$1-d_{SA}$	$\neq d_{SA}, \neq 1-d_{SA}$	0 to 1	0

Definition of the parameters

SC mode: Simultaneous Charging mode

ABC mode: Auxiliary Battery Charging mode

MBC mode: Main Battery Charging mode

Table III: Theoretical formulas of the control parameters in each mode.

	Variables	SC mode	ABC mode	MBC mode
Port IPT	$\dot{V}_s$	Equation (2)	0	$\dot{V}_s/2$ in SC mode
	P_{rn}	$\|\dot{I}_p\|^2 r_p + \|\dot{I}_s\|^2 r_s$	0	P_{rn} in SC mode
	P_{con}	$2(\|\dot{I}_p\|^2 + \|\dot{I}_s\|^2)r_{sw\_on}$	0	P_{con} in SC mode
	P_{MB}	$P_{in} - P_{AB} - P_{rn} - P_{rec} - P_{con}$	$-P_{AB} - P_{rec}$	$P_{in} - P_{rn} - P_{con}$
Port AB	$\dot{V}_{aux\_p}$	$(E_{MB}/\pi)\sqrt{1-\cos(2\pi d_{SB})}\varepsilon^{j\alpha_s}$		0
	P_{AB}	$V_{d\_av}^2/R_{AB}$		0
Efficiency definition	η	$(P_{MB}+P_{AB})/P_{in}$	P_{AB}/P_{MB}	P_{MB}/P_{in}

Definition of the parameters

SC mode: Simultaneous Charging mode

ABC mode: Auxiliary Battery Charging mode

MBC mode: Main Battery Charging mode

IPT on the primary side. It is assumed that the EV is not driving/parking on an IPT charging lane in this situation. On the other hand, in the MBC mode, only the main battery is charged through IPT without charging the auxiliary battery. Here, Table II shows the relationship between the operation modes and the control parameters in the proposed converter.

SC mode

In Table II, the SC mode is moreover divided into the three modes (a), (b), and (c) according to the relationship between d_{SA} and d_{SB}. In the SC mode shown in Fig. 4 (i), $\dot{V}_s$ is controlled with d_{SA}, d_{SB}, and α_s. In this paper, the gate signals of S_{sa1} and S_{sb2} are controlled to turn them on simultaneously. According to this assumption, there is a relationship where $\alpha_s = 2\pi(1-d_{SA}-d_{SB})$ [rad]. Then, $\dot{V}_s$ is represented in Table III. Additionally, in Fig. 4 (i), $\dot{V}_{aux\_p}$ is controlled with d_{SB}, which is represented in Table III.

In the SC mode, the transmission power to the auxiliary battery P_{AB} can be controlled with $\dot{V}_{aux\_p}$ as shown in Table III. Moreover, P_{MB} is represented in Table III. The above relationship between the transmission powers can be visually represented in Fig. 5. Here, Table IV shows the circuit parameters used in the analysis. The areas (a), (b), and (c) shown in Fig. 5 represent the operation areas of SC mode (a), (b), and (c), respectively. These operation areas are overlapped each other as shown in Fig. 5. In the SC mode, P_{MB} gets up to 3.2 kW, and P_{AB} gets up to 362 W theoretically.

ABC mode

During the ABC mode shown in Fig. 4 (ii), both legs of the proposed converter are operated in-phase. In the case, there is the relationship where $\alpha_s = 2\pi(1-d_{SB})$ [rad] and $d_{SA} = d_{SB}$. For this, $\dot{V}_s$ becomes 0

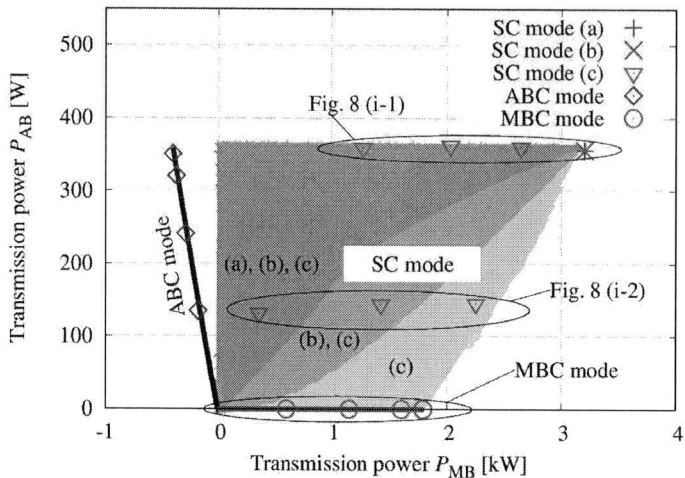

Fig. 5: Possible output area for each mode of the proposed system.

as shown in Table III, and then the voltage is not applied to the secondary side of the resonant network. On the one hand, $\dot{V}_{\text{aux\_p}}$ can be controlled with d_{SB} as in Table III. Then, P_{AB} is represented in Table III. Moreover, in the ABC mode, the auxiliary battery is charged by the main battery, and then P_{MB} equals $-P_{\text{AB}} - P_{\text{rec}}$ as shown in Table III.

The above relationship between the transmission powers can be visually represented in Fig. 5. In the ABC mode, only the auxiliary battery is charged, and there is the relationship of $P_{\text{AB}} = -P_{\text{MB}} - P_{\text{rec}}$, where P_{AB} becomes 362 W at the maximum.

MBC mode

During the MBC mode shown in Fig. 4 (iii), the proposed converter works as a half-bridge converter, and $\dot{V}_{\text{s}}$ is controlled by only d_{SA}, as represented by Table III. Moreover, in the MBC mode, $\dot{V}_{\text{aux\_p}}$ becomes 0 as shown in Table III, because d_{SB} is set to 0. In other words, the output voltage from the Leg SB is not applied to the transformer Tr. For this, P_{AB} becomes 0 as shown in Table III. Then, the transmission power of the main battery P_{MB} equals $P_{\text{in}} - P_{\text{rn}} - P_{\text{con}}$ as shown in Table III.

The above relationship between the transmission powers can be visually represented in Fig. 5. In the MBC mode, the maximum power of P_{MB} is 1.8 kW.

In each operation mode, the relationship between the transmission powers P_{MB} and P_{AB} is theoretically clarified. In the next chapter, the theoretical results are verified by an experiment.

Experiment

Experimental Condition

In the experiment, the behavior of the proposed converter shown in the previous chapter is confirmed. In addition, the theoretical results based on the previous chapter are verified. Fig. 6 shows the appearance of the experimental equipment of the proposed system. Here, Table IV shows the circuit parameters used in the experimental equipment. In the proposed system, for simplicity, the primary/secondary-side converters are synchronized and controlled using a single controller. The main battery is simulated using a DC power supply unit with 200 V. The auxiliary battery is simulated using a variable resistor.

Experimental Result

Fig. 7 shows the output waveforms of the proposed system in the experiment. Here, the current/voltage waveforms in Fig. 7 are measured when P_{MB} and P_{AB} are maximum. In the SC mode, it is confirmed

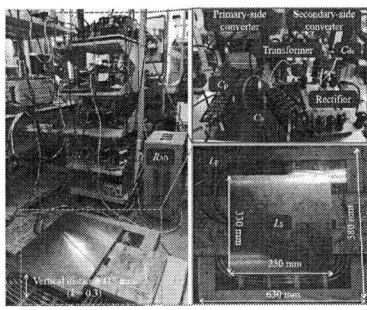

Fig. 6: Appearance of the experimental equipment.

Table IV : Circuit parameters in the experimental equipment.

Coupling coefficient k between L_p and L_s	0.3
Switching frequency f in the converters	85 kHz
DC supply voltage E_{in}	250 V
Main battery voltage E_{MB}	200 V
Inductances L_p, L_s of the transmission coils	63.3 μH, 75.0 μH
Turn numbers $N_p : N_s$ of Tr in the primary/secondary sides	25 : 12
Leakage inductance L_r in Tr	4.7 μH
Load R_{AB} instead of auxiliary battery	5.0 Ω
Threshold voltage V_{rec} in a rectifier diode based on the datasheet	1.1 V
Total ESRs r_p, r_s in L_p and C_p, L_s and C_s	155.0 mΩ, 152.0 mΩ
Turn-on ESR r_{sw_on} in the primary/secondary-sides converters	18.0 mΩ
Diodes in the rectifier	*Microsemi* APT2X41DC60J
Switching devices in the primary-side converter	*ROHM* BSM080D12P2C008
Switching devices in the proposed multi-port AC/DC converter	*ROHM* BSM120D12P2C005

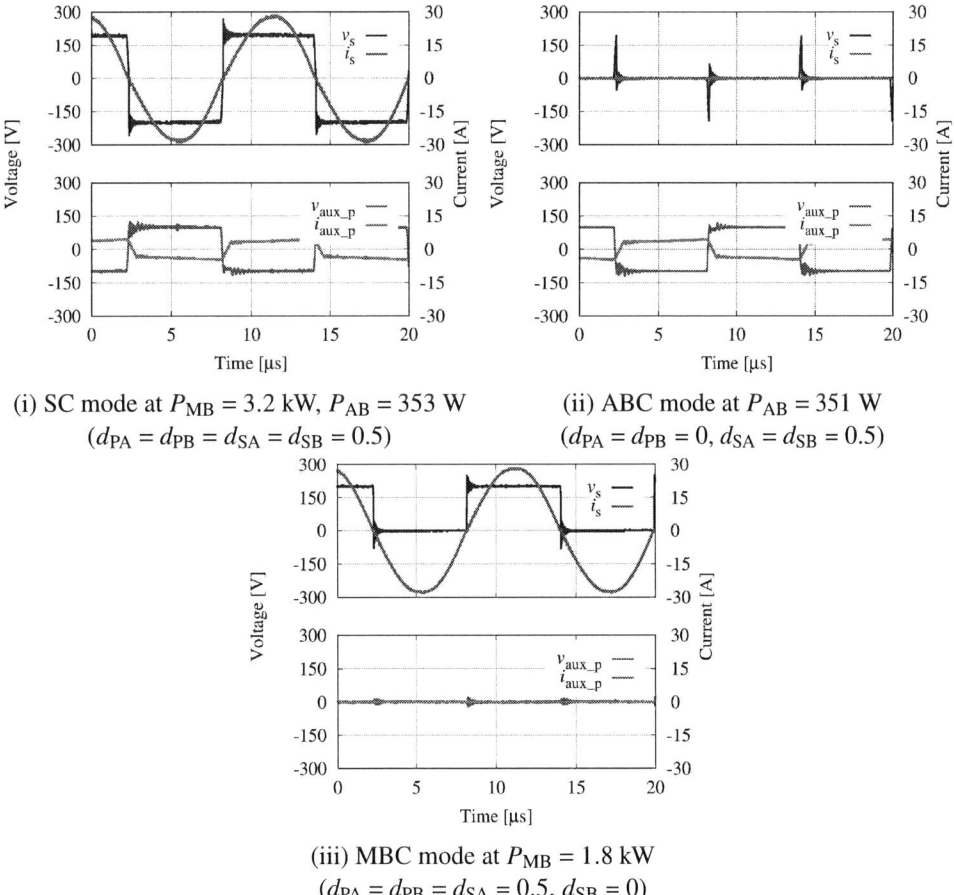

(i) SC mode at $P_{MB} = 3.2$ kW, $P_{AB} = 353$ W
$(d_{PA} = d_{PB} = d_{SA} = d_{SB} = 0.5)$

(ii) ABC mode at $P_{AB} = 351$ W
$(d_{PA} = d_{PB} = 0, d_{SA} = d_{SB} = 0.5)$

(iii) MBC mode at $P_{MB} = 1.8$ kW
$(d_{PA} = d_{PB} = d_{SA} = 0.5, d_{SB} = 0)$

Fig. 7: Output waveforms of the proposed system in the experiment.

that v_s and v_{aux_p} are simultaneously output, and the power can also be transmitted to each battery side simultaneously. In the ABC mode, the main battery is not charged because of $v_s = 0$ and $i_s = 0$. However, there are surge voltages in the waveforms of v_s at the timing of the switching. During the deadtime, the secondary-side converter outputs E_{MB} in a short time, and then the surge voltages occur. On the other hand, the power can be transmitted to the auxiliary battery by Leg SB. Additionally, in the MBC mode,

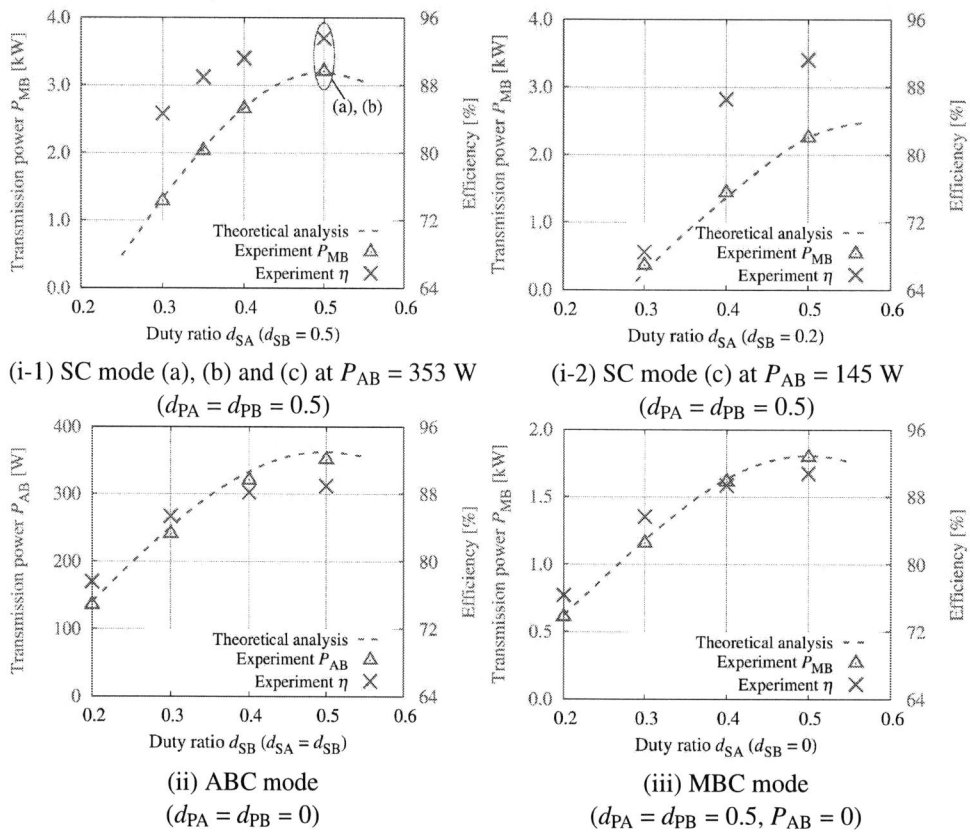

(i-1) SC mode (a), (b) and (c) at $P_{AB} = 353$ W
$(d_{PA} = d_{PB} = 0.5)$

(i-2) SC mode (c) at $P_{AB} = 145$ W
$(d_{PA} = d_{PB} = 0.5)$

(ii) ABC mode
$(d_{PA} = d_{PB} = 0)$

(iii) MBC mode
$(d_{PA} = d_{PB} = 0.5, P_{AB} = 0)$

Fig. 8: Transmission power and the total efficiency of the proposed system in the experiment.

it can be confirmed that only Leg SA works and receives power from the primary side. However, the power is not transmitted to the auxiliary battery. From the above, it can be confirmed the proposed system operates as shown in Figs. 3 and 4.

Next, Fig. 8 shows the transmission power and the total efficiency of the proposed system in the experiment. These characteristics are measured according to the operation points shown in Fig. 5. Regarding the SC mode, there are two characteristics in Fig. 8. In (i-1), the transmission power P_{AB} is regulated to 353 W constantly. In (i-2), P_{AB} is regulated to 145 W constantly. However, in (i-1), the proposed converter operates with the SC modes (a) and (b) at the maximum transmission power point because the maximum power cannot be transmitted in the SC mode (c). Except for this point, the proposed converter operates with the SC mode (c). From (i-1) and (i-2) in Fig. 8, it can be confirmed that the experimental result agrees well with the theoretical result for P_{MB}. Concerning the efficiency, the definition for each operation mode is shown in Table III. In the high power range, it is confirmed that the proposed converter operates with high efficiency. On the other hand, in the low power range, the efficiency also becomes low because of low power factor and switching loss.

Regarding P_{AB} in the ABC mode, it can also be confirmed that the experimental result agrees well with the theoretical result. However, the maximum efficiency is quite low compared to the other operation modes. This efficiency drop is caused by the power loss in the diode rectifier. In order to improve efficiency, the device of the diode needs to be reconsidered.

Regarding P_{AB} in the MBC mode, it can also be confirmed that the experimental result agrees well with the theoretical result. Though the maximum efficiency in the MBC mode is not so high, the efficiency corresponding to transmission power is higher than that in the other operation modes. By using this half-bridge operation that uses the Leg SA, the efficiency in the low power range of the SC mode could be improved.

The above results confirmed the operation of the proposed converter shown in the previous chapter. In addition, the theoretical result regarding transmission power based on the previous chapter was verified.

Conclusion

This paper proposed a multi-port converter that integrates the auxiliary battery-charging system and the IPT system for EVs. In this paper, three typical operation modes of the proposed converter, and the power transfer methods in each mode were also shown.

In addition, the characteristic of the transmission power of the proposed system was theoretically analyzed, and the experiment verified the theoretical result. As a result, it was confirmed that the experimental result agreed well with the theoretical result. However, there is still room for improving the efficiency of the proposed system. For future work, an efficient operation mode and switching devices need to be considered.

From the above, a portion of the effectiveness of the proposed system was clarified, and it can be said that onboard systems with an IPT system could be lower cost by using the proposed system.

Acknowledgement

This work is partially supported by JSPS KAKENHI Grant Number 20K14723.

References

[1] D. Patil, M. K. Mcdonough, J. M. Miller, B. Fahimi and P. T. Balsara, "Wireless Power Transfer for Vehicular Applications: Overview and Challenges," *IEEE Trans. Ind. Electron.*, vol. 62, no. 12, pp. 7436-7447, Dec. 2015.

[2] G. Buja, M. Bertoluzzo and K. N. Mude, "Design and Experimentation of WPT Charger for Electric City Car," *IEEE Trans. Transp. Electrification*, vol. 4, no. 1, pp. 3–37, Mar. 2018.

[3] D. H. Tran, V. B. Vu and W. Choi, "Design of a High-Efficiency Wireless Power Transfer System With Intermediate Coils for the On–Board Chargers of Electric Vehicles," in *IEEE Trans. Power Electron.*, vol. 33, no. 1, pp. 175–187, Jan. 2018.

[4] A. Ahmad, M. S. Alam and A. A. S. Mohamed, "Design and Interoperability Analysis of Quadruple Pad Structure for Electric Vehicle Wireless Charging Application," *IEEE Trans. Transp. Electrification*, vol. 5, no. 4, pp. 934–945, Dec. 2019.

[5] H. S. Wang, K. W. E. Cheng and J. F. Hu, "An Investigation of Compensation Networks for Three-coil Wireless Power Transfer," in *Proc. of 2020 8th Int. Conf. on Power Electron. Syst. Appl.*, pp. 1–6, Dec. 2020.

[6] S. Ann and B. K. Lee, "Analysis of Impedance Tuning Control and Synchronous Switching Technique for a Semibridgeless Active Rectifier in Inductive Power Transfer Systems for Electric Vehicles," in *IEEE Trans. Power Electron.*, vol. 36, no. 8, pp. 8786–8798, Aug. 2021.

[7] T. Imura, K. Suzuki, K. Hata and Y. Hori, "Comparison of Four Resonant Topologies Based on Unified Design Procedure for Capacitive Power Transfer," in *IEEJ J. Ind. Appl.*, vol. 10, no. 3, pp. 339-347, May. 2021.

[8] V. Yenil and S. Cetin, "Load Independent Constant Current and Constant Voltage Control of LCC-Series Compensated Wireless EV Charger," in *IEEE Trans. Power Electron.*, vol. 37, no. 7, pp. 8701–8712, Jan. 2022.

[9] H. Ishida, T. Kyoden and H. Furukawa, "Application of Parity-Time Symmetry to Low-Frequency Wireless Power Transfer System," in *IEEJ J. Ind. Appl.*, vol. 11, no. 1, pp. 59-68, Jan. 2022.

[10] R. Okada, R. Ota and N. Hoshi, "Novel Soft-Switching Active-Bridge Converter for Bi-directional Inductive Power Transfer System," in *IEEJ J. Ind. Appl.*, vol. 11, no. 1, pp. 97-107, Jan. 2022.

[11] M. K. Yang and W. Y. Choi, "Design of High-Efficiency Power Conversion System for Low-Voltage Electric Vehicle Battery Charging," in *Proc. of 2014 IEEE Int. Conf. on Ind. Technol.*, pp. 289–294, Mar. 2014.

[12] D. Mishra, B. Singh and B. K. Panigrahi, "Improved Dual Battery Charging System for Grid Connected Bi-directional EV Charger," in *Proc. of 2018 IEEE India Int. Conf. on Power Electron.*, Dec. 2018.

[13] D. S. Nugroho, R. Ota and N. Hoshi, "A Novel Multi-port Bi-directional Inductive Power Transfer System with Simultaneous Main and Auxiliary Battery Charging Capability," in *Proc. of 2019 IEEE 4th Int. Future Energy Electron. Conf.*, pp. 122–129, Nov. 2019.

[14] I. Awai, "BPF Theory-Based Design Method for Wireless Power Transfer System by Use of Magnetically Coupled Resonators," in *IEEJ Trans. EIS*, vol. 130, no. 12, pp. 2192-2197, Dec. 2010.

[15] B. Luo, S. Wu and N. Zhou, "Flexible Design Method for Multi-Repeater Wireless Power Transfer System Based on Coupled Resonator Bandpass Filter Model," in *IEEE Trans Circuits Syst I Regul Pap*, vol. 61, no. 11, pp. 3288-3297, Nov. 2014.

Influence of IGBT and Diode Parameters on the Current Sharing and Switching-Waveform Characteristics of Parallel-Connected Power Modules

Y. Ando*, J. Sakai, K. Hatori
Mitsubishi Electric Corporation
1-1-1, Imajukuhigashi Nishi-ku
Fukuoka 819-0192, Japan
Tel.: +81 – 92 – 805 – 4251
Fax: +81 – 92 – 805 – 3676
*E-Mail: Ando.Yu@bx.MitsubishiElectric.co.jp
URL: http://www.MitsubishiElectric.com

N. Soltau, E. Wiesner*
Mitsubishi Electric Europe B.V.
Mitsubishi-Electric-Platz 1,
40882 Ratingen, Germany
Tel.: +49 2102 486 0
Fax: +49 2102 486 7220
*E-Mail: Eugen.Wiesner@meg.mee.com
URL: http://www.Mitsubishichips.eu

Keywords

«Power semiconductor device», «Paralleling», «Parallel operation», «Device characterization», «Current balancing»

Abstract

The parallel connection of IGBT power modules allows flexible power converter designs. However, differences in parameters like forward and threshold voltage (ΔV_{CEsat}, $\Delta V_{GE(th)}$, and ΔV_{EC}) between parallel-connected power modules may particularly influence static and dynamic current sharing, and switching characteristics, among several factors. Therefore, when designing converters with parallel-connected power modules, current sharing and switching characteristics of individual power modules must be considered to ensure operation within safe operating area and temperature limits. Derating is to be considered if necessary. Identifying the influential parameters in a parallel-connection, and quantifying their influence on current sharing and switching characteristics, are essential to understand and optimize the amount of derating. This paper first describes the measurement results of ten parallel-connected pairs of 3.3 kV IGBT power modules. Afterwards, the imbalance of switching characteristics (such as switching energy or current amount) is related to power-module parameters, by multiple linear regression. Fanally a methodology for defining the derating ratio for each switching characteristic is described.

Introduction

In many applications, the parallel connection of IGBT power modules allows more flexible power converter designs due to the scalability of converter output power. In the railway application, for example, this topic is discussed in the Horizon 2020 Project "Roll2Rail" [1].
The main challenge in parallel connecting IGBT power modules is homogeneous current sharing among parallel connected devices. The current sharing must be considered during the conduction and switching of the devices respectively [2][3]. For this, special power modules for simplified parallel connection have become available, as well as setup for the characterization [4][5][6].
However, homogenous current sharing remains challenging as it is strongly influenced by power-module parameters and the converter design. Besides dedicated and deliberate converter design, proper selection of power modules by its parameters is required.
This paper analyses the influence of parameter differences of parallel-connected power modules on current sharing and switching energies. Firstly, the test method, devices under test, and test setup are introduced briefly. After that, multiple regression analysis is performed on the measurement results to identify highly influential parameters. Finally, these parameters are used to calculate the derating ratio for each characteristic and define the selection criteria for devices used for parallel connection.

Methodology

This paper analyzes the influence of different IGBT power module parameters, i.e. collector-emitter saturation voltage V_{CEsat}, gate-emitter threshold voltage $V_{GE(th)}$, and diode forward voltage V_{EC}, on switching-waveform characteristics of parallel-connected power modules. For this, IGBT turn-off, IGBT turn-on, and diode turn-off (reverse recovery) have been measured with ten parallel-connected pairs of power modules, respectively. Each pair has a unique value in terms of the differences of power module parameters (ΔV_{CEsat}, $\Delta V_{GE(th)}$, and ΔV_{EC}). Switching characteristics, i.e. switching current or switching energies, are derived from each measurement and related to the differences in parameters of each pair. Then, via multiple regression analysis, the highly influential parameters are identified for each switching characteristic. Afterwards, formulas are defined for the imbalance ratio of each characteristic and its prediction interval.

Devices under Test and Test Setup

The devices under test are MITSUBISHI ELECTRIC 3.3 kV IGBT power modules with a current rating of 450 A, as shown in Figure 1. Power modules use the LV100 package [6]. Ten pairs of these devices are measured in parallel connections of two modules. The devices have been selected to get a wide distribution of parameter differences between pairs. Table I shows the parameters of individual power modules and the parameter differences between them when paired. The parameter differences are calculated for power module A-power module B.

Table I: Parameters of Devices under Test and Parameter Difference of Pairs

Parameter	Power Module A			Power Module B			Pair A-B		
	V_{CEsat}	$V_{GE(th)}$	V_{EC}	V_{CEsat}	$V_{GE(th)}$	V_{EC}	ΔV_{CEsat}	$\Delta V_{GE(th)}$	ΔV_{EC}
Conditions / Pair No.	I_C=450A, T_j=150°C	I_C=45mA, V_{GE}=10V, T_j=25°C	I_E=450A, V_{GE}=0V, T_j=150°C	I_C=450A, T_j=150°C	I_C=45mA, V_{GE}=10V, T_j=25°C	I_E=450A, V_{GE}=0V, T_j=150°C	I_C=450A, T_j=150°C	I_C=45mA, V_{CE}=10V, T_j=25°C	I_E=450A, V_{GE}=0V, T_j=150°C
1	2.70 V	6.57 V	2.20 V	2.72 V	6.57 V	2.22 V	-0.02 V	0.00 V	-0.02 V
2	2.70 V	6.57 V	2.20 V	2.63 V	7.70 V	2.34 V	0.07 V	-0.51 V	-0.14 V
3	2.61 V	6.98 V	2.43 V	2.81 V	7.53 V	2.37 V	-0.20 V	-0.54 V	0.06 V
4	2.61 V	6.98 V	2.43 V	2.70 V	6.57 V	2.20 V	-0.09 V	0.42 V	0.23 V
5	2.70 V	6.57 V	2.20 V	2.72 V	7.25 V	2.45 V	-0.02 V	-0.68 V	-0.26 V
6	2.64 V	6.97 V	2.45 V	2.64 V	6.94 V	2.25 V	0.00 V	0.03 V	0.20 V
7	2.64 V	6.97 V	2.45 V	2.67 V	6.97 V	2.35 V	-0.03 V	0.00 V	0.10 V
8	2.81 V	7.53 V	2.37 V	2.67 V	6.56 V	2.35 V	0.14 V	0.97 V	0.02 V
9	2.63 V	6.95 V	2.45 V	2.64 V	6.97 V	2.45 V	-0.01 V	-0.02 V	0.00 V
10	2.81 V	7.53 V	2.37 V	2.63 V	7.07 V	2.34 V	0.17 V	0.45 V	0.03 V

As mentioned in the introduction, the key requirement from the railway market for these new packages is scalability. It allows converter manufacturers to have flexibility in different projects to design the required output power. For example, by using the one, two or six CM450DA-66X modules, the current ratings of 450 A, 900 A or 2700 A can be achieved respectively.

The power modules in parallel connection must be operated properly, but since different converter manufacturers have different designs, railway manufacturers agreed on the reference test setup for the evaluation of parallel connections [5]. This reference setup, as depicted in Figure 2, allows evaluation of up to six paralleled modules. The reference test setup was designed to minimize the influence of external factors on the parallel connection. Each power module is connected to a DC-link capacitor with symmetrical and equivalent stray inductance. Furthermore, the gate driver connection should have a symmetrical connection from the gate driver to each parallel-connected power module. The benefit of using this reference test setup is that the user can reproduce the results presented here because the components used, such as gate driver, DC-link capacitors, current probes, and their geometries are standardized. The reference test setup shown in Figure 2 was used to evaluate the ten pairs of power modules shown in Table I. In this evaluation, only two module positions No. 3 and No. 4 with two DC-Link capacitors were used.

Fig. 1: MITSUBISHI ELECTRIC 3.3 kV IGBT power module using the LV100 package [6]

Fig. 2: Reference test setup for LV100 device evaluation [5]

Results

For each pair, IGBT turn-off, IGBT turn-on and diode reverse recovery have been measured under the following conditions, respectively: DC-link voltage $V_{CC} = 1800$ V, target collector current $I_C = 450$ A/module, junction temperature $T_j = 150°C$, DC-link inductance $L_s = 65$ nH/module, gate voltage $V_{GE} = +15$ V / -10 V, turn-on and turn-off gate resistance $R_{G(on)} = 3.0$ Ω/module and $R_{G(off)} = 51$ Ω/module. Examples of waveforms for each measurement are shown in Figure 3, Figure 4 and Figure 5.

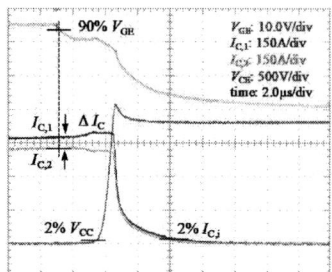

Fig. 3: Exemplary turn-off waveform with indication of characteristic values

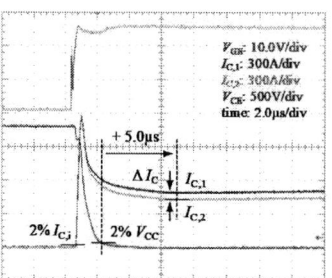

Fig. 4: Exemplary turn-on waveform with indication of characteristic values

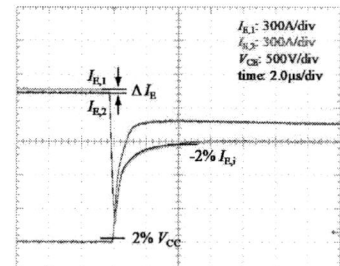

Fig. 5: Exemplary reverse recovery waveform with indication of characteristic values

Regarding IGBT turn-off switching

The turn-off collector currents $I_{C,i}$ (cf. Figure 3) and turn-off energy

$$E_{\text{off},i} = \int_{t(v_{CE}=2\%V_{CC})}^{t(i_{C,i}=2\%I_{C,i})} i_{C,i} \cdot v_{CE} \, dt \tag{1}$$

had been evaluated. The index $i \in \{1,2\}$ refers to the individual power modules in the parallel connection.

Table II shows the switching characteristics of individual power modules obtained from the measurements and the pair imbalance ratio calculated based on them. ΔI_C and ΔE_{off} are calculated for power module A-power module B, respectively.

Table II: Switching Characteristics of Each Power Module and Pair Imbalance Ratio in Turn-off Measurements

Pair No.	Power Module A		Power Module B		Imbalance Ratio	
	$I_{C,1}$	$E_{off,1}$	$I_{C,2}$	$E_{off,2}$	$x_{I_C} = 0.5\ \Delta I_C/I_{C,avg}$	$x_{E_{off}} = 0.5\ \Delta E_{off}/E_{off,avg}$
1	447 A	447 A	0.86 J	0.86 J	0.0 %	0.2 %
2	438 A	462 A	0.89 J	0.94 J	-2.7 %	-2.6 %
3	474 A	424 A	1.02 J	0.82 J	5.6 %	10.8 %
4	463 A	437 A	0.95 J	0.90 J	2.9 %	2.9 %
5	451 A	448 A	0.91 J	0.87 J	0.4 %	2.2 %
6	446 A	452 A	0.94 J	0.98 J	-0.7 %	-2.1 %
7	453 A	446 A	0.96 J	0.94 J	0.7 %	1.1 %
8	434 A	463 A	0.81 J	1.00 J	-3.3 %	-10.6 %
9	449 A	455 A	0.96 J	0.98 J	-0.7 %	-1.1 %
10	429 A	471 A	0.82 J	1.02 J	-4.7 %	-10.9 %

Multiple regression analysis has been performed on the power module parameter differences in Table I for each pair and the imbalance ratio in Table II. The power module parameters affecting each switching characteristic were identified.

Figure 6 shows the turn-off collector current imbalance ratio of $0.5\ \Delta I_C/I_{C,avg} = (I_{C,1}-I_{C,2})/(I_{C,1}+I_{C,2})$, which depends on the collector-emitter saturation voltage differences $\Delta V_{CEsat} = V_{CEsat,1}-V_{CEsat,2}$ only. By linear regression analysis [7], the measured results can be defined as regression equation (2).

$$x_{I_C\_typ} = 0.5\ \Delta I_C/I_{C,avg} \approx -0.28\ \text{V}^{-1} \cdot \Delta V_{CEsat} \qquad \text{with R}^2 = 0.969 \qquad (2)$$

R^2 is the correlation coefficient, and Figure 7 shows the correlation of the imbalance ratio between the measured results and the calculated results using the regression equation (2). The closer the correlation coefficient R^2 is to 1, the regression equation (2) is the more reliable.

The green line in Figure 7 is the 99% prediction interval (PI=99%). It is given to quantify the accuracy of the regression line, and defined by the following equation.

$$x_{I_C\_PI=99\%} = x_{I_C\_typ} \pm 3.69 \times \sqrt{2.76 \times 10^{-5} \times \left(1 + \frac{1}{10} + \frac{(\Delta V_{CEsat} - 0.002\ \text{V})^2}{0.103\ \text{V}^2}\right)} \qquad (3)$$

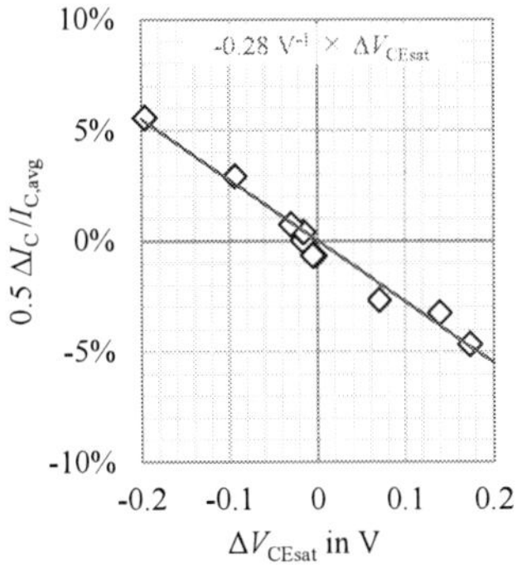

Fig. 6: Imbalance ratio of turn-off current x_{I_C} in dependence of ΔV_{CEsat}

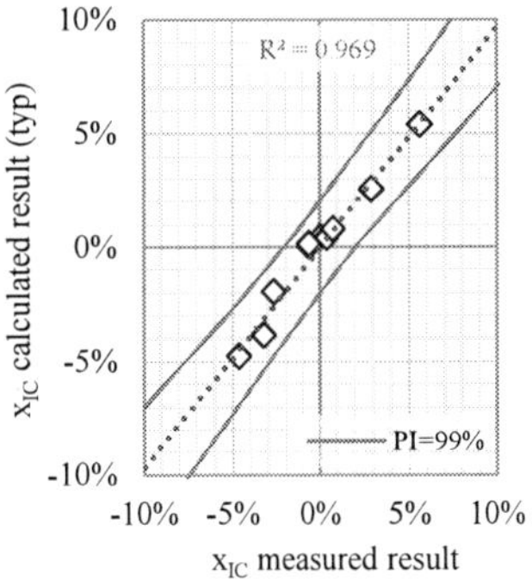

Fig. 7: Correlation of x_{I_C} between calculated and measured results

Differently, the measured results of the imbalance ratios of the turn-off energy $0.5\ \Delta E_{off}/E_{off,avg} = (E_{off,1}-E_{off,2})/(E_{off,1}+E_{off,2})$ cannot be described satisfactorily in dependence on one parameter only. Instead, it is defined as in regression equation (4) considering both ΔV_{CEsat} and $\Delta V_{GE(th)}$.

$$x_{E_{off}\_typ} = 0.5\ \frac{\Delta E_{off}}{E_{off,avg}} \approx -0.50\ \text{V}^{-1} \cdot \Delta V_{CEsat} - 0.03\ \text{V}^{-1} \cdot \Delta V_{GE(th)} - 0.009 \qquad \text{with R}^2 = 0.993 \qquad (4)$$

Figures 8 and 9 show the imbalance ratios of the turn-off energy $0.5\ \Delta E_{off}/E_{off,avg}=(E_{off,1}-E_{off,2})/(E_{off,1}+E_{off,2})$ that depend on the collector-emitter saturation voltage differences $\Delta V_{CEsat}=V_{CEsat,1}-V_{CEsat,2}$ and the gate-emitter threshold voltage differences $\Delta V_{GE(th)}=V_{GE(th),1}-V_{GE(th),2}$, respectively. Figure 8 includes the intercept of a regression equation (4).

The equation for the 99% prediction interval shown in Figure 10 is defined as (5).

$$x_{E_{off}\_PI=99\%} = x_{E_{off}\_typ} \pm 4.03 \times \sqrt{3.57 \times 10^{-5} \times \left(1 + \frac{1}{10} + \frac{(\Delta V_{CEsat} - 0.002\ V)^2}{0.103\ V^2} + \frac{(\Delta V_{GE(th)} - 0.012\ V)^2}{2.331\ V^2}\right)} \tag{5}$$

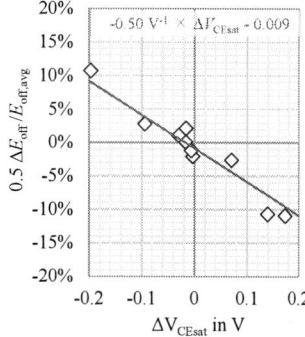

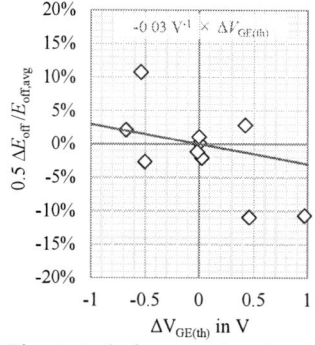

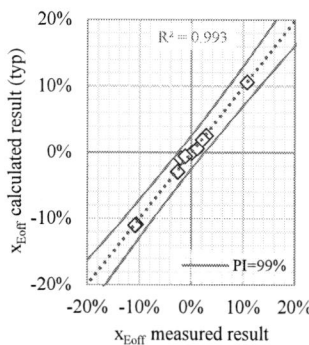

Fig. 8: Imbalance ratio of turn-off energy ($x_{E_{off}}$) in dependence of ΔV_{CEsat}

Fig. 9: Imbalance ratio of turn-off energy ($x_{E_{off}}$) in dependence of $\Delta V_{GE(th)}$

Fig. 10: Correlation of $x_{E_{off}}$ between calculated and measured results

Regarding IGBT Turn-on Switching

The turn-on collector current $I_{C,i}$ (cf. Figure 4) (6) and turn-on energy $E_{on,i}$ (7) have been evaluated.

$$I_{C,i} = i_{C,i}(t(v_{CE} = 2\%V_{CC}) + 5\mu s) \tag{6}$$

$$E_{on,i} = \int_{t(i_{C,i}=2\%I_{C,i})}^{t(v_{CE}=2\%V_{CC})} i_{C,i} \cdot v_{CE}\ dt \tag{7}$$

Table III shows the switching characteristics of individual power modules obtained from the measurements and the pair imbalance ratio calculated based on them. ΔI_C and ΔE_{on} are calculated for power module A-power module B, respectively.

Table III: Switching Characteristics of Each Power Module and Pair Imbalance Ratio in Turn-on Measurements

Pair No.	Power Module A		Power Module B		Imbalance Ratio	
	$I_{C,1}$	$E_{on,1}$	$I_{C,2}$	$E_{on,2}$	$x_{I_C} = 0.5\ \Delta I_C/I_{C,avg}$	$x_{E_{on}} = 0.5\ \Delta E_{on}/E_{on,avg}$
1	564 A	0.697 J	556 A	0.682 J	0.7 %	1.1 %
2	586 A	0.774 J	517 A	0.708 J	6.3 %	4.5 %
3	570 A	0.849 J	514 A	0.822 J	5.2 %	1.6 %
4	527 A	0.693 J	590 A	0.751 J	-5.7 %	-4.0 %
5	606 A	0.769 J	511 A	0.680 J	8.5 %	6.1 %
6	531 A	0.759 J	559 A	0.785 J	-2.6 %	-1.7 %
7	548 A	0.770 J	549 A	0.770 J	-0.1 %	0.0 %
8	499 A	0.752 J	593 A	0.800 J	-8.7 %	-3.1 %
9	546 A	0.780 J	543 A	0.772 J	0.3 %	0.5 %
10	519 A	0.844 J	574 A	0.872 J	-5.0 %	-1.6 %

As previously, multiple linear regression analysis has been performed on the measured switching results. For the turn-on collector current imbalance, correlation on ΔV_{EC} and $\Delta V_{GE(th)}$ was evident. The regression equation can be defined as (8).

$$x_{I_{C\_typ}} = 0.5\,\frac{\Delta I_C}{I_{Cavg}} \approx -0.09\ \text{V}^{-1} \cdot \Delta V_{EC} - 0.09\ \text{V}^{-1} \cdot \Delta V_{GE(th)} \qquad \text{with } R^2 = 0.992 \qquad (8)$$

Figures 11 and 12 shows the imbalance ratios of the turn-on collector current $0.5\,\Delta I_C/I_{C,avg}=(I_{C,1}-I_{C,2})/(I_{C,1}+I_{C,2})$ that depend on the emitter-collector forward voltage differences $\Delta V_{EC}=V_{EC,1}-V_{EC,2}$ and the gate-emitter threshold voltage differences $\Delta V_{GE(th)}=V_{GE(th),1}-V_{GE(th),2}$, respectively.
The equation for the 99% prediction interval shown in Figure 13 is defined as (9).

$$x_{I_{C\_PI=99\%}} = x_{I_{C\_typ}} \pm 3.83 \times \sqrt{2.93 \times 10^{-5} \times \left(1 + \frac{1}{10} + \frac{(\Delta V_{EC} - 0.022\ V)^2}{0.187\ V^2} + \frac{\left(\Delta V_{GE(th)} - 0.012\ V\right)^2}{2.331\ V^2}\right)} \qquad (9)$$

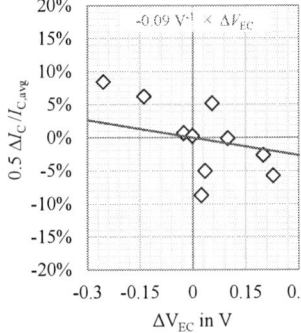
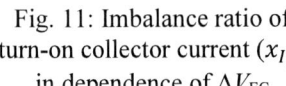
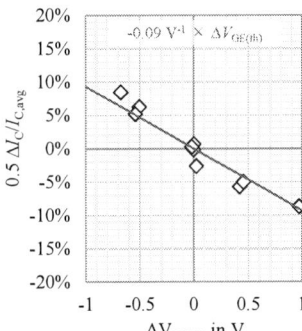
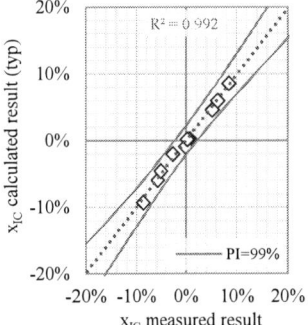

Fig. 11: Imbalance ratio of turn-on collector current (x_{I_C}) in dependence of ΔV_{EC}

Fig. 12: Imbalance ratio of turn-on collector current (x_{I_C}) in dependence of $\Delta V_{GE(th)}$

Fig. 13: Correlation of x_{I_C} between calculated and measured results

The imbalance ratio of turn-on energy is also correlated to ΔV_{EC} and $\Delta V_{GE(th)}$, and the regression equation is defined as (10).

$$x_{E_{on\_typ}} = 0.5\,\frac{\Delta E_{on}}{E_{on,avg}} \approx -0.12\ \text{V}^{-1} \cdot \Delta V_{EC} - 0.04\ \text{V}^{-1} \cdot \Delta V_{GE(th)} + 0.007 \qquad \text{with } R^2 = 0.991 \qquad (10)$$

Figures 14 and 15 shows the imbalance ratios of the turn-on energy $0.5\,\Delta E_{on}/E_{on,avg}=(E_{on,1}-E_{on,2})/(E_{on,1}+E_{on,2})$ that depend on the emitter-collector forward voltage differences $\Delta V_{EC}=V_{EC,1}-V_{EC,2}$ and the gate-emitter threshold voltage differences $\Delta V_{GE(th)}=V_{GE(th),1}-V_{GE(th),2}$, respectively. Figure 14 includes the intercept of regression equation (10).
The equation for the 99% prediction interval shown in Figure 16 is defined as (11).

$$x_{E_{on\_PI=99\%}} = x_{E_{on\_typ}} \pm 4.03 \times \sqrt{1.19 \times 10^{-5} \times \left(1 + \frac{1}{10} + \frac{(\Delta V_{EC} - 0.022\ V)^2}{0.187\ V^2} + \frac{\left(\Delta V_{GE(th)} - 0.012\ V\right)^2}{2.331\ V^2}\right)} \qquad (11)$$

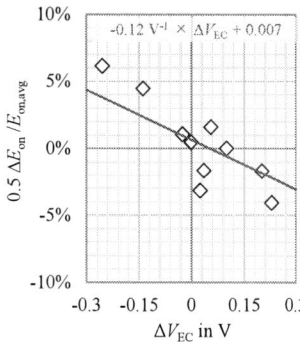
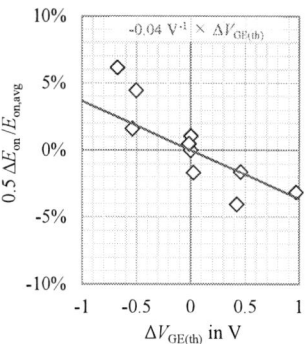
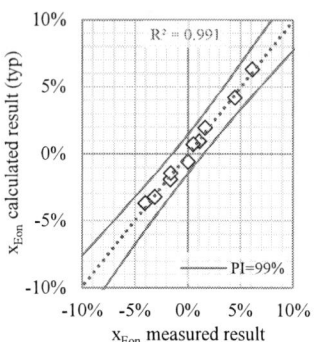

Fig. 14: Imbalance ratio of turn-on switching energy $(x_{E_{on}})$ in dependence of ΔV_{EC}

Fig. 15: Imbalance ratio of turn-on switching energy $(x_{E_{on}})$ in dependence of $\Delta V_{GE(th)}$

Fig. 16: Correlation of $x_{E_{on}}$ between calculated and measured results

Regarding Diode Reverse-recovery Switching

The reverse-recovery emitter current $I_{E,i}$ (cf. Figure 5) and reverse-recovery energy $E_{rec,i}$ (12) has been evaluated.

$$E_{rec,i} = -\int_{t(v_{EC}=2\%V_{cc})}^{t(i_{E,i}=-2\%I_{E,i})} i_{E,i} \cdot v_{CE} \, dt \tag{12}$$

Table IV shows the switching characteristics of individual power modules obtained from the measurements and the pair imbalance ratio calculated based on them. ΔI_E and ΔE_{rec} are calculated for power module A-power module B, respectively.

Table IV: Switching Characteristics of Each Power Module and Pair Imbalance Ratio in Reverse-recovery Measurements

Pair No.	Power Module A		Power Module B		Imbalance Ratio	
	$I_{E,1}$	$E_{rec,1}$	$I_{E,2}$	$E_{rec,2}$	$x_{I_E} = 0.5 \, \Delta I_E/I_{E,avg}$	$x_{E_{rec}} = 0.5 \, \Delta E_{rec}/E_{rec,avg}$
1	453 A	0.913 J	450 A	0.943 J	0.2 %	-1.6 %
2	472 A	0.906 J	417 A	0.835 J	6.2 %	4.1 %
3	432 A	0.740 J	461 A	0.841 J	-3.3 %	-6.4 %
4	404 A	0.767 J	486 A	0.983 J	-9.2 %	-12.4 %
5	502 A	0.945 J	393 A	0.760 J	12.2 %	10.8 %
6	418 A	0.733 J	467 A	0.930 J	-5.6 %	-11.9 %
7	435 A	0.735 J	449 A	0.851 J	-1.6 %	-7.3 %
8	448 A	0.797 J	436 A	0.835 J	1.4 %	-2.3 %
9	452 A	0.782 J	460 A	0.823 J	-0.9 %	-2.6 %
10	441 A	0.780 J	459 A	0.852 J	-2.0 %	-4.4 %

As so far, multiple linear regression analysis has been performed on the measured switching results. For the reverse-recovery emitter current imbalance, correlation on only ΔV_{EC} is evident. The regression equation can be defined as (13).

$$x_{I_E\_typ} = 0.5 \frac{\Delta I_E}{I_{Eavg}} \approx -0.39 \text{ V}^{-1} \cdot \Delta V_{EC} \qquad \text{with } R^2 = 0.923 \tag{13}$$

Figures 17 shows the imbalance ratios of the reverse-recovery emitter current $0.5 \, \Delta I_E/I_{E,avg} = (I_{E,1}-I_{E,2})/(I_{E,1}+I_{E,2})$ that depend on the emitter-collector forward voltage differences $\Delta V_{EC} = V_{EC,1}-V_{EC,2}$. The equation for the 99% prediction interval shown in Figure 18 is defined as (14).

$$x_{I_E\_PI=99\%} = x_{I_E\_typ} \pm 3.69 \times \sqrt{2.76 \times 10^{-4} \times \left(1 + \frac{1}{10} + \frac{(\Delta V_{EC} - 0.022\ V)^2}{0.187\ V^2}\right)} \tag{14}$$

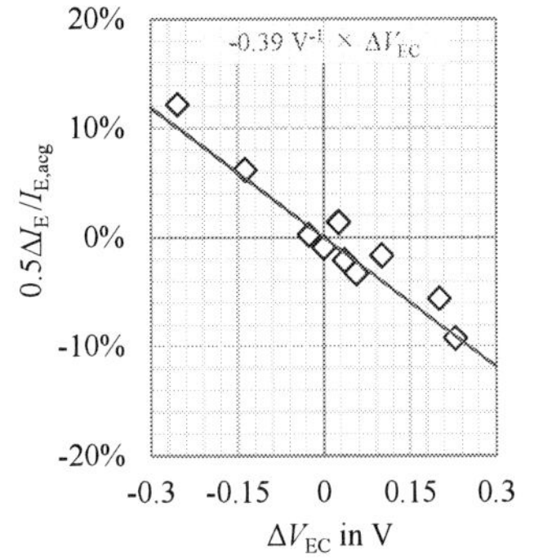

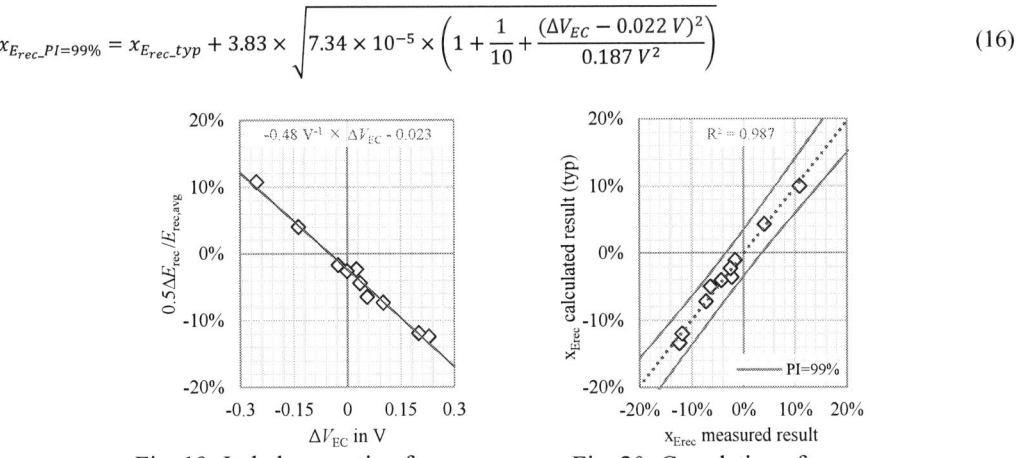

Fig. 17: Imbalance ratio of reverse-recovery emitter current (x_{I_E}) in dependence of ΔV_{EC}

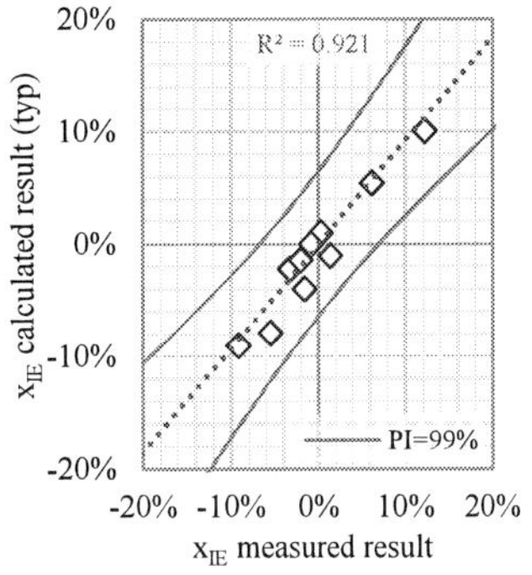

Fig. 18: Correlation of x_{I_E} between calculated and measured results

The imbalance ratio of reverse-recovery energy is also correlated to only ΔV_{EC}, and the regression equation is defined as (15).

$$x_{E_{rec}\_typ} = 0.5 \frac{\Delta E_{rec}}{E_{rec,avg}} \approx -0.48\ V^{-1} \cdot \Delta V_{EC} - 0.023 \qquad \text{with } R^2 = 0.987 \tag{15}$$

Figures 19 shows the imbalance ratios of the reverse-recovery energy $0.5\ \Delta E_{rec}/E_{rec,avg} = (E_{rec,1} - E_{rec,2})/(E_{rec,1} + E_{rec,2})$ that depend on the emitter-collector forward voltage differences $\Delta V_{EC} = V_{EC,1} - V_{EC,2}$. Figure 19 includes the intercept of regression equation (15).
The equation for the 99% prediction interval shown in Figure 20 is defined as (16).

$$x_{E_{rec}\_PI=99\%} = x_{E_{rec}\_typ} + 3.83 \times \sqrt{7.34 \times 10^{-5} \times \left(1 + \frac{1}{10} + \frac{(\Delta V_{EC} - 0.022\ V)^2}{0.187\ V^2}\right)} \tag{16}$$

Fig. 19: Imbalance ratio of reverse-recovery switching energy ($x_{E_{rec}}$) in dependence of ΔV_{EC}

Fig. 20: Correlation of $x_{E_{rec}}$ between calculated and measured results

Analysis of Required Parameter Derating Based on the Module Parameters

The regression equations defined in the previous chapters to define the current and energies imbalance ratios are used to calculate the derating ratio required in case of parallel connection of more than two modules. For this, it will be assumed that one of the paralleled modules has a minimum characteristic (resulting in maximum switching energy or current) while all other modules have the maximum characteristics (leading to minimum switching energy or current). As an example, the formula for the turn-off collector current derating ratio is shown as (17).

$$\frac{I_{C,max}}{I_{C,avg}} - 1 = \frac{I_{C,max}}{\left((n-1)I_{C,min}+I_{C,max}\right)/n} - 1 = \frac{n \cdot I_{C,max}}{\left((n-1)\frac{1-x}{1+x}I_{C,max}+I_{C,max}\right)} - 1 = \frac{n}{\left((n-1)\frac{1-x}{1+x}+1\right)} - 1 \tag{17}$$

The parameter n is in this formula the number of paralleled modules. The parameter x is the identified imbalance ratio from this measurement. As a result, the derating dependency on the power-module parameters can be defined as shown in Figure 21. The figure shows that the prediction intervals, as determined by the regression analysis are also very helpful with respect to derating ratios in case of parallel connection of more than two modules.

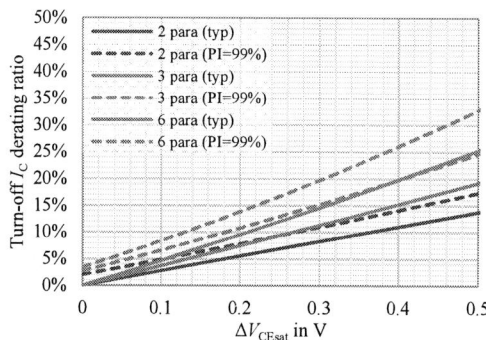

Fig. 21: Turn-off I_C derating ratio vs. ΔV_{CEsat}

As above, the derating ratios for other switching characteristics have been also calculated and defined as shown in Figure 22 ~ Figure 29 for each correlated power module parameter.

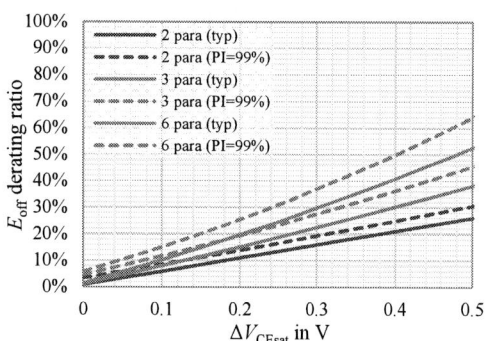

Fig. 22: E_{off} derating ratio vs. ΔV_{CEsat}

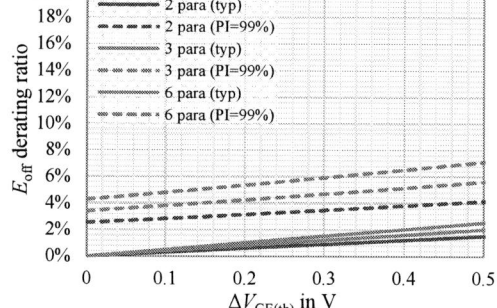

Fig. 23: E_{off} derating ratio vs. $\Delta V_{GE(th)}$

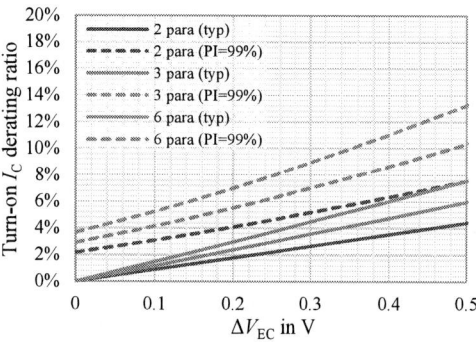

Fig. 24: Turn-on I_C derating ratio vs. ΔV_{EC}

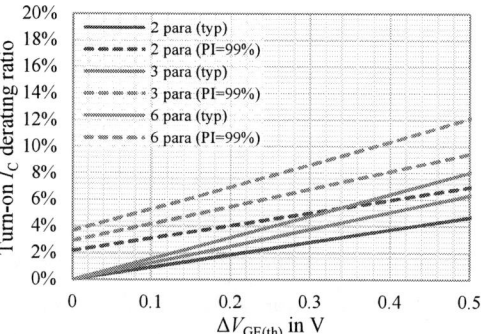

Fig. 25: Turn-on I_C derating ratio vs. $\Delta V_{GE(th)}$

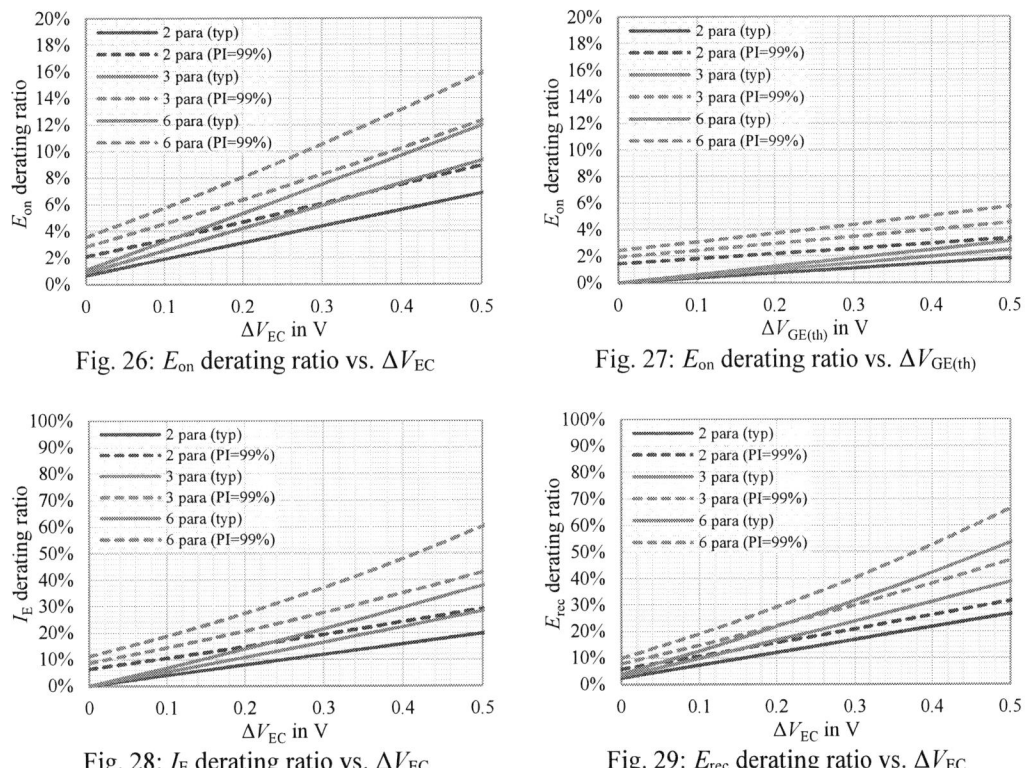

Fig. 26: E_{on} derating ratio vs. ΔV_{EC}

Fig. 27: E_{on} derating ratio vs. $\Delta V_{GE(th)}$

Fig. 28: I_E derating ratio vs. ΔV_{EC}

Fig. 29: E_{rec} derating ratio vs. ΔV_{EC}

Conclusion

This paper has shown a methodology to evaluate the influence of IGBT power module parameters on switching characteristics of power modules connected in parallel. For this purpose, the switching waveforms of ten different pairs of IGBT power modules connected in parallel have been analyzed. Using multiple linear regression on the measured data identifies correlation between power-module parameters and switching characteristics as current imbalance and inhomogeneous switching energies. The methodology allows the definition of pairing criteria for IGBT power modules in parallel connection. Moreover, the results can also be used to calculate the derating for two or more power modules in parallel.

In actual applications, the imbalance in current sharing causes the temperature of devices with high current flow to rise. However, if forward voltage characteristics of IGBTs and diodes have a positive temperature dependence, the current sharing imbalance will decrease. Hence, actual derating, including switching losses, might be lower compared to the results calculated here.

References

[1] T. Wiik, „D1.2, New generation power semiconductor, Common specification for traction and market analysis, technology roadmap, and value cost prediction,“ Roll2Rail, H2020 - 636032, 2016.

[2] J. Weigel, J. Boehmer, A. Nagel und R. Kleffel, „Paralleling high power dual modules: A challenge for application engineers and power device manufacturers,“ in 2017 19th European Conference on Power Electronics and Applications (EPE'17 ECCE Europe), Warsaw, Poland, 2017.

[3] N. Chen, F. Chimento und e. al, „Dynamic Characterization of Parallel-Connected High-Power IGBT Modules,“ IEEE Trans. on Ind. Appl., Bd. 51, Nr. 1, pp. 539-546, 2015.

[4] J. Weigel, J. Boehme und e. al, „Paralleling of High Power Dual Modules: Standard Building Block Design for Evaluation of Module Related Current Mismatch,“ in 2018 20th European Conference on Power Electronics and Applications (EPE'18 ECCE Europe), Riga, Latvia, 2018.

[5] A. Nagel, J. Weigel, et. al., „Paralleling reference setup," Shift2Rail, Pinta, H2020 - 730668, 2019.

[6] Mitsubishi Electric Corporation, Press Release No. 3104, Mitsubishi Electric to Launch LV100-type X-Series HVIGBT Modules, Tokyo, 2017.

[7] N. Inagaki, Mathematical Statistics, (in Japanese), Japan: Shokabo Co. Ltd., 2009, pp. 196-205.

Innovative driving scheme for electrical generators in more electric aircrafts employing series active filtering

Nena Apostolidou, Nick Papanikolaou
DEMOCRITUS UNIVERSITY OF THRACE
DEPARTMENT OF ELECTRICAL AND COMPUTER ENGINEERING
ELECTRIC MACHINES LABORATORY
ECE Campus, Building B', Office 012
Kimmeria Xanthi, Greece, 67132
Tel.: +30.2541079739, +30.2541079921
E-Mail: paposto@ee.duth.gr, npapanik@ee.duth.gr

Acknowledgements

 This research was supported by the Hellenic Foundation for Research and Innovation (HFRI) under the HFRI PhD Fellowship grant (Fellowship Number: 118)

Keywords

«Aerospace», «Switched reluctance drive», «Microgrid», «Generation of electrical energy», «Active filter».

Abstract

As more electrification of the future aircrafts becomes a trend, more electric functions, such as the Starter/Generator (S/G) that are incorporated in the More Electric Aircrafts (MEA) concept, increase the need for electrical power systems with high power supply capabilities, facilitating both motor (starter) and generator operation. Under this light, an innovative driving scheme for electrical generators in MEA that employs series active filtering (SAF) is proposed in this paper, to impose the desired output current in the dc microgrid (DCMG) of the MEA, thus facilitating the multi-operating-point range capability of the machine drive. The proposed SAF-based system that facilitates the incorporation of multiple energy storage units (ESUs) constitutes a robust S/G solution that serves this double-mode operation requirement (set by the specific application), along with the wide set point range necessity of the power generation mode. Moreover, an appropriately developed Design Tool allows for the selection of the optimal SAF parameters, according to the DCMG and the machine drive specifications, facilitating high performance and reliability of the proposed drive unit. The developed S/G driving scheme is verified through MATLAB/Simulink simulations and, additionally, the performance of the power generation control of the system under study is evaluated via real-time control hardware in-the-loop (CHIL) tests, with the use of a dedicated microcontroller (dsPIC30f4011) and dSPACE 1202 (MicroLabBox) platform.

Introduction

The innovative driving scheme deals with the electrical energy production of the electric generators that are incorporated in the MEA [1], under the necessity of the efficient operation of these machines over a wide operating point range, that is imposed by the vast load changes of the MEA power system; under this light, the key-feature of the innovative driving scheme lies in the direct power control over the electrical generator, through the incorporation of a SAF in its control unit, as depicted in Fig. 1(a). In addition, an appropriate vector control scheme is implemented on the SAF (incorporating various ESUs), which is suitable for the S/G drive unit, facilitating both the acceleration of the machine during starting mode (enhancing the machine's starting electromagnetic torque production, T_e) and the efficient power control at various operational points during generator mode.

The developed active power control scheme is extendable, thus applicable to any converter-fed, multi-phase machine (i.e., N_{Ph} number of phases can be considered), which is advantageous in terms of power quality improvement, by mitigating the current/torque pulsation impact at the point of common coupling (PCC); at the same time, the SAF is compatible with any type of electrical machine drive, thus allowing for the incorporation of various types of alternative SAF solutions, comprising of any n-level DC/AC converter (n, number of levels of the multilevel converter) and various alternative voltage source units (such as capacitors and/or batteries) that are suitable for each machine drive case, as Fig. 1(a) illustrates.

In addition, an appropriately developed Design Tool (Fig. 2) allows for the selection of the optimal SAF parameters (i.e., the Design Tool outputs, according to Fig. 2a), thus constituting a general tool that is totally adaptable to the specifications of any candidate electric drive system (i.e., the Design Tool inputs, according to Fig. 2a) to be integrated in the MEA DCMG, as well as in other similar systems that use electrical machine drives under this frame (which can be modeled in an appropriate dedicated software environment).

Furthermore, the innovative driving scheme allows for the minimization of the electrical variables' real time feedback, while the SAF incorporation inherently serves the fault-ride through capability (FRTC) of the drive system (i.e., in terms of power generation loss avoidance in cases of under/over voltage conditions) and facilitates the incorporation of energy management/storage electric apparatus.

On the other hand, the developed Design Tool facilitates the enhancement of the reliability of the total drive system through the optimum design and integration of electrical apparatus/units of high reliability, such as film capacitors, which can be used in the SAF (replacing the bulky electrolyte capacitors which are used in relevant applications). These features are of outmost importance for the MEA DCMG operation.

In this paper, for the implementation of the proposed SAF-based active power control concept, the switched reluctance machine (SRM) is used as the S/G unit, being among the most attractive candidates to perform the electric starting of the main engine, due to its unique physical characteristics that favor its direct coupling to the propulsive unit. Thus, the design of the system parameters as well as the presented results, i.e., the MATLAB/Simulink simulations (regarding the S/G concept, i.e., motor and generator operation) along with the real-time CHIL tests – verifying the effectiveness of the control scheme implementation on the switched reluctance generator (SRG), are based on the torque production mechanism of the specific machine type (i.e., the SRM's current and torque profile). However, these results verify in an excellent manner the proposed vector-controlled SAF concept (including the circuit topology and the Design Tool), which can be used in various machine drives (with the appropriate adjustment), as well.

The proposed innovate driving scheme along with the developed Design Tool have been patented under the Hellenic Industrial Property Organization (Patent Nr.1010204).

Case study: Implementation of the innovative driving scheme in the SRM

When it comes to the potential MEA electrical generators, the SRM has been recognized as one of the most promising candidates to be incorporated in the energy generation/distribution system of the MEA [2]. In light of this, the innovative driving scheme has been implemented for the control of the generated power of an SRG that is incorporated in the MEA high-voltage DCMG; in this case study, the Design Tool implementation regards the determination of the optimum SAF parameters that facilitate the specific machine's operational features, as it will be presented.

In the relative literature, there are numerous publications regarding the incorporation of the SRG in high power systems, such as those which are used in electrified aircrafts [3], that focus on the control upon the optimal operational point of the machine, either under motor or generator mode, especially for the S/G function [4]; however, these studies do not conclude to a reliable solution that facilitates the MEA multi-operational control feature that is discussed in the Introduction. More specifically, although in recent literature there are many studies regarding several SRM control schemes [5], [6], these studies do not focus on high power systems, such as the DCMG of the electrified aircraft, where the necessity of multi-operational generated power levels calls for the uninterrupted operation of the SRG under various operational points. What is more, the active front-end converter filtering solutions that have been recently proposed in relative literature [7], [8] (attempting to enhance the output power range and efficiency of the SRG in a wider speed operating range), face specific limitations, which are analyzed in depth in [9].

Under this light, the innovative driving scheme implementation (with the use of the developed Design Tool) in the SRM's case has been verified by both simulations (in MATLAB software environment) and real time implementation, with the use of an appropriate CHIL set-up (dSPACE 1202 platform) along with an external dedicated signal processor (dsPIC30f4011). It is noted that the presented simulation results regarding the S/G concept, that are oriented upon the high-voltage (270V) and high-power MEA application (which present both the steady state operation and the control response under transient conditions, highlighting the FRTC and dispatchability features of the innovative driving scheme), proved that the innovative active power control scheme is capable of enhancing the reliability and reducing the computational burden of the electrified aircrafts' application under study. On the other hand, the real time performance results that are presented in this paper are oriented upon the verification of the effectiveness of the real time control implementation during generator operation, under reduced voltage and speed conditions (resulting from the hardware limitations, as it will be explained).

Description of the Innovative Driving Scheme

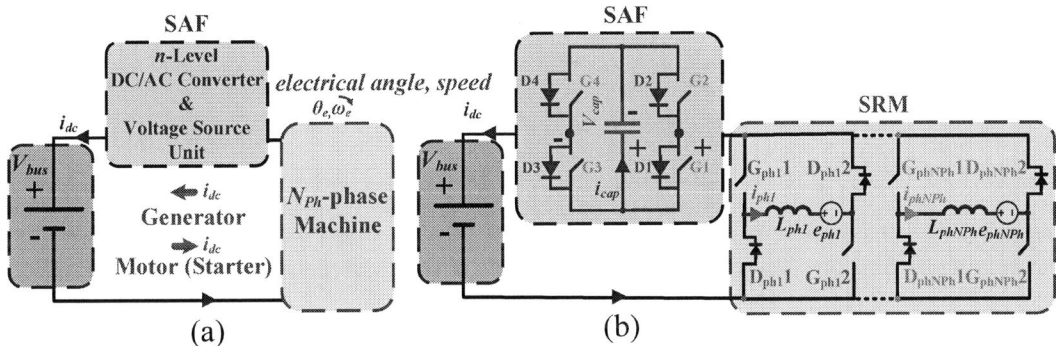

Fig. 1: DCMG voltage, V_{bus}, and instantaneous DCMG current, i_{dc}, (a) general-type machine drive case: SAF with a n-level DC/AC converter and a voltage source unit; N_{Ph}-phase machine (b) SRM drive case: SAF with a single-phase HBC and a capacitive voltage source (average SAF capacitor voltage, V_{cap}); N_{Ph}-phase SRM, with integrated asymmetric H-bridge converter (AHBC) – i_{ph}, e_{ph} and L_{ph} are the phase current, back-electromotive force (BEMF) – voltage – and inductance, respectively.

Regarding Fig. 1, the innovation of the control scheme as far as a N_{Ph}-phase SRM is considered – Fig.1(b), lies on the imposition of the desired SRM's output current (mean value), I_{dc} (in the specific application the reference current regards the high voltage DCMG side, V_{bus}), which is achieved through the independent control of the phases' magnetization current, I_{mag}, and the DCMG current, i_{dc}, with the application of the proper operational states of the switches in a 2-level H-bridge converter – HBC (these states are referred as voltage vectors (v_x), according to Tables I and II) – acting as the SAF.

In generator mode the goal is to provide constant current to the DCMG, thus the vector control is oriented upon i_{dc} feedback signal, according to the following ordinary differential equations (ODEs), corresponding to Fig. 1 (generator mode) and Table I:

$$\left.\frac{di_{dc}}{dt}\right|_{x=1,4,6} = \frac{-v_x - \omega_e \overbrace{\frac{dL_{phNPh}}{d\theta_e} i_{phNPh}}^{e_{phNPh}}}{L_{phNPh}} + \dots - \frac{v_x - \omega_e \overbrace{\frac{dL_{ph1}}{d\theta_e} i_{ph1}}^{e_{ph1}}}{L_{ph1}}, \tag{1}$$

$$\left.\frac{di_{dc}}{dt}\right|_{x=2,3,5,7} = \frac{-v_x - \omega_e \overbrace{\dfrac{dL_{ph1}}{d\theta_e} i_{ph1}}^{e_{ph1}}}{L_{ph1}} + \ldots + \frac{-v_x - \omega_e \overbrace{\dfrac{dL_{phNPh}}{d\theta_e} i_{phNPh}}^{e_{phNPh}}}{L_{phNPh}} \qquad (2)$$

The impact of the application of each voltage vector on i_{dc} and v_{cap} (instantaneous SAF capacitor voltage) during generator mode is presented in Table I, regarding the use of a capacitive ESU in the SAF input, which is properly charged and discharged (according to the applied voltage vector), so that it preserves the SAF capacitor voltage level ($V_{cap,ref}$).

Table I: Possible operational States of SAF circuit during generator mode, with capacitive ESU (v_{cap}): ($\uparrow$) Increase, ($\downarrow$) Decrease, (==) No change.

Voltage Vector	SAF Active Switch(es), Diode(s)	Generator Mode Active Voltage	i_{dc}	v_{cap}
1/4	G1, D3/G4, D2	$v_{1,4}=V_{bus}$ (Over-magnetization[1])	$\downarrow$	==
2/3	G2, D4/G3, D1	$v_{2,3}=V_{bus}$ (Generation)	$\uparrow$	==
5	G2, G3	$v_5=V_{bus}-v_{cap}$ (Generation)	$\uparrow$	$\downarrow$
6	G1, G4	$v_6=V_{bus}+v_{cap}$ (Over-magnetization[1])	$\downarrow$	$\downarrow$
6	D1, D4	$v_6=V_{bus}+v_{cap}$ (Under-magnetization[2])	$\downarrow$	$\uparrow$
7	D1, D4	$v_7=V_{bus}+v_{cap}$ (Generation)	$\downarrow$	$\uparrow$

[1] $I_{mag}<0$ (with respect to the current flow in Fig. 1, generator mode), over-magnetization [9].
[2] $I_{mag}>0$ (with respect to the current flow in Fig.1, generator mode), under-magnetization [9].

On the other hand, the SAF operation during starter mode is based upon the control of the SRM's phases' currents sum, $i_{Ph}=i_{ph1}+\ldots+i_{phNPh}$, with respect to Fig. 1(b), according to the following ODEs, corresponding to Fig. 1 (motor mode) and Table II:

$$\left.\frac{di_{Ph}}{dt}\right|_{x=1\div 7} = \frac{v_x - \omega_e \overbrace{\dfrac{dL_{ph1}}{d\theta_e} i_{ph1}}^{e_{ph1}}}{L_{ph1}} + \ldots + \frac{v_x - \omega_e \overbrace{\dfrac{dL_{phNPh}}{d\theta_e} i_{phNPh}}^{e_{phNPh}}}{L_{phNPh}} \qquad (3)$$

$$\left.\frac{di_{Ph}}{dt}\right|_{x=1\div 7} = \frac{v_x - \omega_e \overbrace{\dfrac{dL_{ph1}}{d\theta_e} i_{ph1}}^{e_{ph1}}}{L_{ph1}} + \ldots + \frac{-v_x - \omega_e \overbrace{\dfrac{dL_{phNPh}}{d\theta_e} i_{phNPh}}^{e_{phNPh}}}{L_{phNPh}} \qquad (4)$$

Table II presents the voltage vectors that are used in the motor (starter) mode, with the incorporation of a various type ESU (implemented as a constant voltage source, V_{ESU}), providing the necessary power to the SRM for the starting torque production.

The proposed control scheme is general/adaptable in various conditions/parameters of the system, such as the speed, the amounts of the produced energy/power, current and voltage level restrictions (nominal values of the SRM, the SAF and the DCMG) and load conditions (which determine the over/under magnetization operation of the SRG), resulting in the capability of the machine drive in a wide operating point range (dispatchability feature), as well as in a smooth transition from generator to motor operation and vice versa, under any speed. In any case, the power flow control is implemented without the need for real time processing of the electrical machine parameters (the machine's power control is independent from the conventional angle power control of the SRMs), which leads to the reduction of the computational burden of the control system and enhances the total reliability of the system.

Table II: Possible operational States of SAF circuit during motor mode, with any type ESU (constant voltage source, V_{ESU}): ($\uparrow$) Increase, ($\downarrow$) Decrease, (==) No change.

Voltage Vector	SAF Active Switch(es), Diode(s)	Motor Mode Active Voltage	i_{Ph}	ESU
1/4	G1, D3/G4, D2	$v_{1,4}=V_{bus}$	$\uparrow/\downarrow$ [2]	==
2/3	-[1]	$v_{2,3}=0$	$\uparrow$	==
5	G2, G3	$v_5=V_{bus}-V_{ESU}$	$\uparrow$ [3]	discharge
6	G1, G4	$v_6=V_{bus}+V_{ESU}$	$\uparrow$	discharge
7	-[1]	$v_7=0$	$\downarrow$	==

[1] i_{ph} flows through G_{ph} and D_{ph} in Fig. 1(b) [9].
[2] dictated by BEMF's magnitude, which is proportional to ω_e and the phase current [9].
[3] refers only to the phases' commutation interval [9].

Development of the SAF Design Tool

The developed Design Tool of the proposed SAF – general case, Fig. 2(a) – facilitates the adaptation of the control design to the requirements of the system/microgrid under study – general case, Fig. 2(a) – and the characteristics of the incorporated machine type. In the SRG case under study (the analytical description of the control scheme is given in [9]), the selection of the optimal values of the total system's parameters of the circuitry of Fig. 1(b) is based upon the system's specifications/restrictions which regard the maximum produced power level ($I_{dc,max}$), the maximum SAF capacitor voltage level ($V_{cap,max}$), the maximum SAF capacitor voltage deviation ($\Delta V_{cap,max}$) and the maximum SAF switching losses ($P_{sw,loss,,max}$); these values are the inputs of the Design Tool and, with the implementation of the appropriate control algorithm that interacts with the developed simulation model – Fig. 2(b), they determine the outputs of the Design Tool (the analytical description of the developed Design Tool is given in [9]); in the present SRG case study, where a capacitive ESU is considered and the hysteresis current control is implemented, the outputs of the Design Tool regard the maximum i_{dc} reference value ($I_{ref,max}$) that is used for the SRG's control implementation pattern, as well as the optimal parameters of the SAF, regarding the minimum capacitance value C_{min} (so that film capacitors can be used) and the minimum hysteresis zone (H_{min}), so that good i_{dc} profile (more constant) is achieved. Under this frame, the system is designed upon the maximum operating point (MOP), which is defined as MOP($I_{dc,max}$ (A), $V_{cap,max}$ (V)).

The advantages of the innovative active power control scheme regarding the SRM case study are summarized in the following features:
- Robust driving of the machine in the double-mode operation of the S/G application; reliable starting/acceleration of the SRM and smooth transition from motor to generator mode, though the implementation of the same control pattern (SAF vector control).
- Controllability over a wide operating point range for the electrical energy production, in generator mode. This is achieved through the independent control of I_{mag} and i_{dc}, thanks to the integration of the SAF; the magnetization current control also facilitates the control/definition of the desired/acceptable SAF capacitor voltage level ($V_{cap,ref}$).
- Limitation of the switching frequency of the SRM's phases converter (AHBC) to the basic electrical frequency (out of the SRM's rotational speed, RPM), which is a requirement for the production of high energy amounts, with high efficiency (limitation of the AHBC's switching losses).
- Capability of integrating an appropriate electrical unit for the energy management and storage, with the use of highly reliable electrical apparatus (e.g., film capacitors).
- High reliability of the machine under both stand-alone and dc-microgrid connected operation (e.g., aircrafts' systems).

- Elimination of additional passive elements, such as inductors and capacitors, which reduce the reliability and the gravimetric/volumetric power density of the system.
- Low complexity of the control scheme; the AHBC operates only in two basic modes (that is the phases' magnetization when the switches are on and the phases' generation through the free-wheeling diodes), while the total control scheme is implemented by the SAF.
- Inherent FRTC; during impermanent faulty conditions (e.g., DCMG voltage overshoots/dips), the SAF facilitates the uninterruptable SRM's energy production.

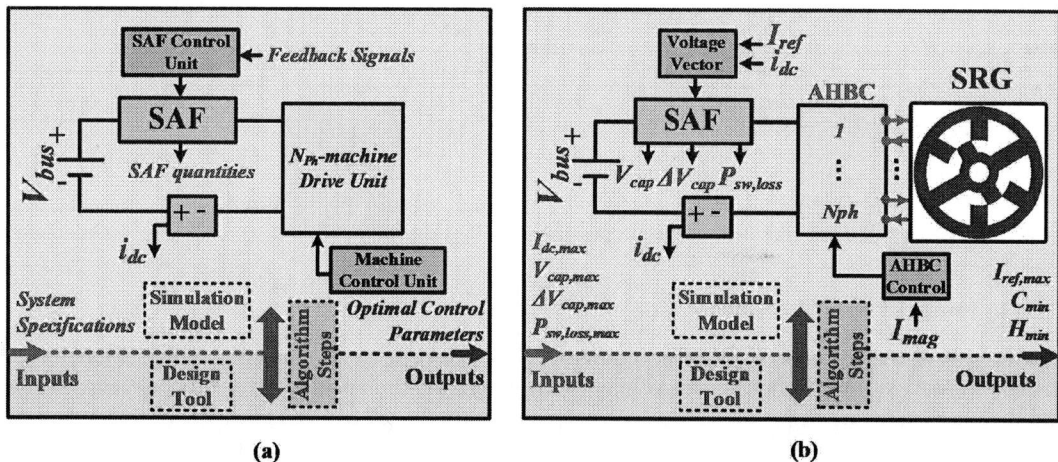

Fig. 2: Design Tool for the selection of the optimal values of the total system's parameters; (a) General electrical machine drive system and SAF, according to Fig. 1(a), (b) SAF and SRG drive system according to Fig. 1(b).

CHIL Verification results regarding the generator operation

The real-time performance of the developed control scheme in the 6/4, 3-phase SRG case study [9] has been verified through an appropriate CHIL set-up – Fig. 3(d), comprising of a dSPACE MicroLabBox 1202 platform (combined with MatlabR2014b software), running at a 10 kHz sampling frequency, in cooperation with an external microcontroller hardware, i.e., the dsPIC30f4011 DSP, running at 7.37 MHz; more specifically, dsPIC30f4011 receives real time feedback signals (as inputs) from the simulation model which is built in MicroLabBox 1202, implements the proposed control algorithm and sends the appropriate AHBC and SAF control signals (as outputs) back into MicroLabBox 1202. dSPACE Control Desk software performs data acquisition and visualization of the real time signals. It is noted that due to the MicroLabBox 1202 hardware limitation of 10 kHz real time simulation, the SRM model is evaluated to the limit of RPM=800 rpm and V_{bus}=48 V.

Fig. 3(a) and (b) present the CHIL implementation results regarding the real time response of the control (through the appropriate response of I_{mag}), under step changes of i_{dc} reference value, I_{ref}. In Fig. 3(a), I_{mag} is decreased from -49.1 A to -50.7 A when I_{ref} is increased from 60 A to 125 A (at 4 s time spot), while I_{mag} is further decreased from -50.7 A to -74.3 A when I_{ref} is further increased from 125 A to 200 A (at 24 s time spot). In any case, the SAF capacitor charges and discharges (according to the pattern of Table I) to preserve the reference voltage value, $V_{cap,ref}$, which in this case has been selected to 48V.

Fig. 3(c) illustrates the response of the control scheme under a V_{bus} drop, from 48 V to 0 V, from 3.525 s until 3.672 s time spot (the 147 ms abnormal voltage drop duration complies with MIL-STD-704F disturbances limits). During this interval, the SAF capacitor absorbs the SRG's energy, highlighting the FRTC capability of the developed control scheme under real time implementation.

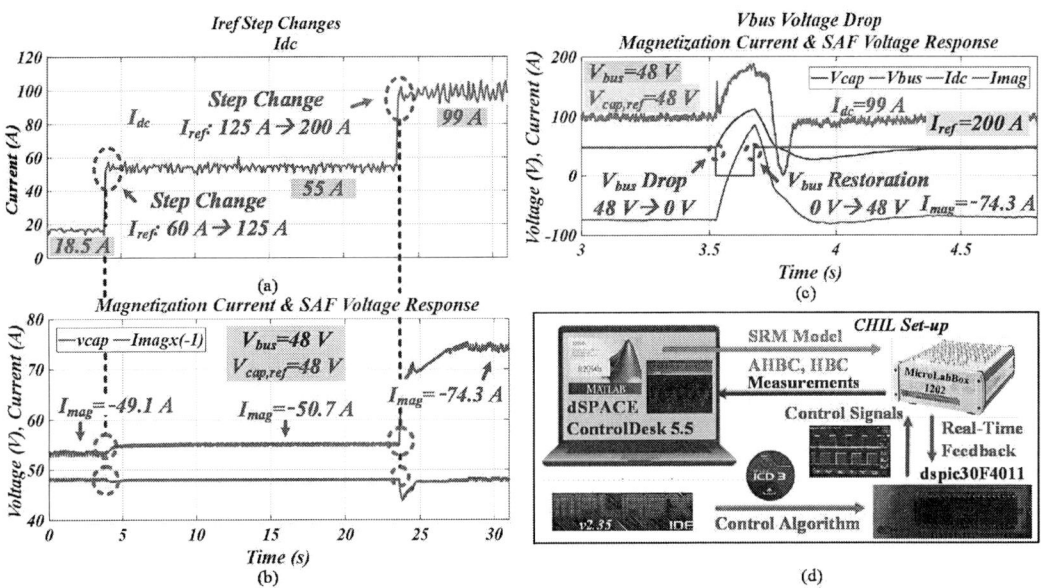

Fig. 3: (a), (b) i_{dc}, I_{mag} (the sign of I_{mag} is according to the current flow of Fig. 1, generator mode) and v_{cap} response, under I_{ref} step changes, from 60 A to 125 A at 4 s time spot and from 125 A to 200 A at 24 s time spot, (c) I_{mag} and V_{cap} response, under V_{bus} voltage drop at 3.525 s time spot, MOP(100 A, 48 V), V_{bus} restoration at 3.672 s time spot, (d) Block diagram of the CHIL set-up for the real time validation of the proposed control scheme [9].

MATLAB simulation results regarding the S/G concept

The performance of the developed S/G control scheme (regarding the 6/4, 3-phase SRG case study [9] of the CHIL tests) is modeled in MATLAB R2018b software, according to the circuitry of Fig. 1(b) and 2(b), as depicted in the results depicted in Fig. 4.

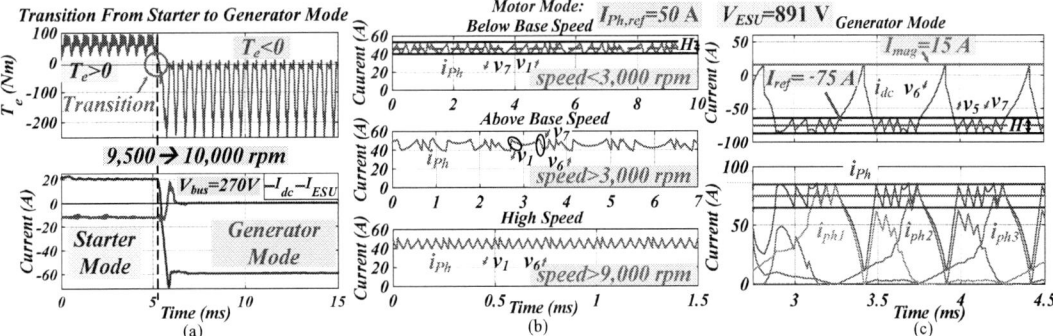

Fig. 4: (a) Transition from starter mode (T_e, I_{dc}, i_{dc}>0, with respect to the direction of the current flow in motor mode in Fig. 1) to generator mode (T_e, I_{dc}, i_{dc}<0, with respect to the direction of the current flow in motor mode in Fig. 1), generator , (b) vector control implementation during starter mode (Table II), under i_{Ph} reference value, $I_{Ph,ref}$=50 A and V_{ESU}=891 V, below/above base speed and at high speed, (c) vector control implementation during generator mode (according to Table I, where $v_{cap}=V_{ESU}$), with I_{ref}=-75 A (i_{dc}<0, with respect to Fig. 1, motor mode) and V_{ESU}=891 V – i_{ph1}, i_{ph2}, i_{ph3} are the phase currents.

The SRM predetermined model (incorporating a detailed mapping over the SRM technical characteristics based on FEA) that is used in the developed simulation model, along with the SAF control implementation parameters have been selected with the use of the developed Design Tool, which has been presented in [9], allowing for the determination of the optimum energy balance point (EBP) of the system under study (Fig. 1 and 2). The EBP refers to the condition that the ESU neither provides nor absorbs energy during a control pattern implementation period (regarding the generator operation), thus the average ESU current, I_{ESU}, is zero.

Fig. 4(a) verifies the successful transition of the SRM from starter to generator mode (with the aid of the vector control of Tables I and II); during starter mode ($I_{dc}>0$) the ESU provides energy to the SRM (average current of the ESU, $I_{ESU}>0$), while during generator mode ($I_{dc}<0$) the EBP ($I_{dc}=-60A$) is reached ($I_{ESU}=0$). Furthermore, Fig. 4(b) illustrates the vector control implementation during motor (starter) mode, accelerating the SRM from standstill to the speed that the machine enters the generator mode (in this case 10,000 rpm [9]); the latter mode of operation is depicted in Fig. 4(c), which regards the power generation ($i_{dc}<0$, with respect to Fig. 1, motor mode) at $I_{ref}=-75A$ and $I_{mag}=15A$ (over-magnetization).

Conclusion

The innovative driving scheme facilitates the direct power control of the electrical generators, through the imposition of the desired machine's output current value, thus facilitating the capability of the multi-operating point range machine mode, in line with the dispatchability requirements of MEA electric power systems. In addition, the incorporation of an ESU in the SAF circuitry facilitates the reliable starting torque production during motor (starter) mode of the S/G application; the developed Design Tool facilitates the adaptability of the power control scheme implementation under various specifications and restrictions of the system under study. The innovative control scheme has been verified via the developed simulation models both in software and in real time processing hardware.

References

[1] W. Cao, B. C. Mecrow, G. J. Atkinson, J. W. Bennett, and D. J. Atkinson, "Overview of electric motor technologies used for more electric aircraft (MEA)," IEEE Trans. Ind. Electron., vol. 59, no. 9, pp. 3523–3531, Sept. 2012.

[2] V. Madonna, P. Giangrande and M. Galea, "Electrical power generation in aircraft: review, challenges, and opportunities," IEEE Trans. Transp. Electrif., vol. 4, no. 3, pp. 646–659, Sept. 2018.

[3] S. Li, S. Zhang, T. G. Habetler, and R. G. Harley, "Modeling, design optimization and applications of switched reluctance machines – a review," IEEE Trans. Ind. Appl., vol. 55, no. 3, pp. 2660–2681, May/Jun. 2019.

[4] J. K. Nøland, M. Leandro, J. A. Suul, and M. Molinas, "High-power machines and starter-generator topologies for more electric aircraft: a technology outlook, IEEE access, vol. 8, pp. 130104–130123, Jul. 2020.

[5] S. Song, R. Hei, R. Ma and, W. Liu, "Model predictive control of switched reluctance starter/generator with torque sharing and compensation," IEEE Trans. Transp. Electrif., vol. 6, no. 4, pp. 1519–1527, Dec. 2020.

[6] R. Rocca, F. G. Capponi, S. Papadopoulos, G. D. Donato, M. Rashed, and M. Galea, "Optimal advance angle for aided maximum-speed-node design of switched reluctance machines," IEEE Trans. Energy Conv., vol 35, no. 2, pp. 775–785, Jun. 2020.

[7] A. Klein-Hessling, B. Burkhart and R. W. De Doncker, "Active source current filtering to minimize the DC-link capacitor in switched reluctance drives," IEEE Trans. Power Electron. Appl., vol. 4, no. 1, pp. 62–71, Mar. 2019.

[8] Q. Wang, H. Chen, H. Cheng, S. Yan and S. Abbas, "An active boost power converter for improving the performance of switched reluctance generators in dc generating systems," IEEE Trans. Power Electron., vol. 35, no. 5, pp. 4741–4754, May 2020.

[9] N. Apostolidou and N. Papanikolaou, "Active Power Control of Switched Reluctance Generator in More Electric Aircraft," IEEE Trans. Veh. Techn., Vol. 70, Issue 12, pp. 12604-12616, Dec. 2021.

Field-measurement based hygrothermal modelling of the converter-cabinet climate in wind turbines

Katharina Fischer [1], Katherina Göhler [1,2]

[1] Fraunhofer Institute for Wind Energy Systems IWES
Postkamp 12, Hannover, Germany
E-Mail: Katharina.Fischer@iwes.fraunhofer.de
URL: https://www.iwes.fraunhofer.de

[2] Leibniz University Hannover
Welfengarten 1,
Hannover, Germany
URL: https://www.uni-hannover.de

Acknowledgements

This work is part of the project ReCoWind, which was funded by the German Federal Ministry for Economic Affairs and Climate Action under grant 0324336A. The authors would like to thank the project partners for supporting the research by data-logger installation and re-collection in the field and the provision of corresponding wind-turbine operating data. ERA5 reanalysis data were obtained from the Copernicus Climate Change and Atmosphere Monitoring Services.

Keywords

«Wind Energy», «Reliability», «Humidity», «Measurements», «Modelling»

Abstract

Power converters in wind turbines suffer from frequent failures. Root-cause analyses point to environmental influences as important drivers of converter failure. Based on comprehensive field measurements in wind turbines, we derive hygrothermal models describing the dependence of temperature and humidity in the converter cabinet on the ambient climatic conditions and turbine operation. The results show that lumped-parameters models of minimal complexity are suitable for describing the conditions with reasonable accuracy and that publicly available ERA5 reanalysis data may be used to consider the site-specific climatic conditions outside of the wind turbine. In addition, we demonstrate that the hygrothermal model derived for a turbine type can successfully be transferred to identical turbines operating in other countries. The models can therefore serve as a basis for refined requirement specifications as well as for the derivation of application-specific test procedures for power converters and their components.

Introduction

Power converters in wind turbines (WT) show high failure rates. This causes considerable repair cost and downtime and ranks them among the most frequently failing subsystems of WT [1]–[3]. According to field-data analyses based on a worldwide and heterogeneous wind-turbine fleet reported in [4], the overall converter system failed with an average rate of 0.48 a^{-1} per WT, with a third of the failure events being attributed to the core components of the converter, including the IGBT modules, driver boards, DC-link capacitors and busbars.
While power- and thermal-cycling induced fatigue effects in the power electronics have been found not to contribute significantly to the field failures, seasonal patterns and regional differences in the failure behavior indicate that climatic influences, in particular humidity, play a central role in the emergence of failures [5], [6].

This brings the climatic conditions into the focus that power converters in wind turbines are exposed to. Fraunhofer IWES has collected and analyzed field-measurement data from the converter cabinets of more than 30 wind turbines, including turbines at onshore and offshore sites, with fully or partially rated converters, with power converters located in the tower base, the nacelle or distributed over both places, and including both liquid-cooled and water-cooled converters. Air temperature and humidity inside the power cabinet have been found to be influenced strongly by the climatic conditions around the WT as well as by its operation [7].

The present work seeks to identify WT-type specific hygrothermal models from the abovementioned field-measurement data that are suitable for estimating the dynamic temperature and humidity conditions in the power cabinet based on the ambient conditions around the turbine and the active power fed into the grid. Such models offer the potential to simulate the temperature and humidity timeseries in turbines of the considered types at any possible location in the world based on the ambient environmental conditions at this site. This is valuable not only for deepened field-data based root-cause analysis, which is so far limited to the use of WT-external climatic conditions; it is also a powerful basis for requirement specification as well as for the derivation of application- and even region-specific test profiles for the reliability qualification of WT converter systems and their components.

Modelling approach

In view of the available field data and information about the WT design, a grey-box modelling approach is chosen: In a first step, based on general knowledge about the turbine design and on the physical basic laws for heat and mass transport, a model structure is defined. In a second step, the parameters of this model are identified from the field data to achieve an optimal agreement between the measured and the simulated climatic conditions in the power-converter cabinet. The model is implemented as a lumped-parameter model (R-C network, see e.g. [8], [9]), making use of electrical analogies for the transport and storage processes of heat and moisture, respectively. The complexity of the model structure is kept as low as necessary to still achieve a sufficiently accurate representation of the climatic conditions in the power converter. As illustrated in Fig. 1, the hygrothermal model consists of a thermal model and a hygric model.

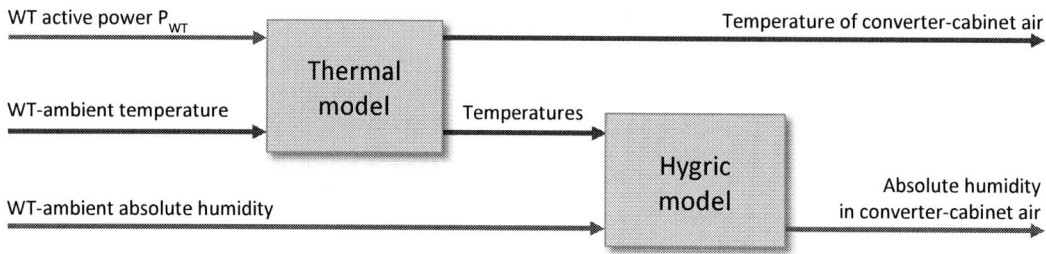

Fig. 1: Hygrothermal model for heat and moisture transport between the ambient air of the wind turbine and the converter-cabinet air

Data underlying the model development

The hygrothermal models are derived using the following data:
- Temperature and humidity timeseries measured in the WT converter-cabinet air during the measurement campaigns as described in [7] (temporal resolution: 1 min, 15 min or 30 min)
- Timeseries of the active power fed into the grid from the same turbines, typically from the Supervisory Control and Data Acquisition (SCADA) system (temporal resolution: 10 min)
- Site-specific environmental data: ambient temperature and humidity timeseries (a) collected outside of the WT as part of the measurement campaigns, or (b) from the publicly available ERA5 reanalysis data [10], which provide hourly estimates of many atmospheric and oceanographic variables based on global modal data, covering the earth on a grid of approximately 30 km x 30 km. Please refer to [11] for further information on the ERA5 reanalysis data and to [7] for information on how these data have been processed for the present work.

We use different parts of the available timeseries for parameter identification and for model validation, respectively. The model accuracy is assessed by means of the root mean squared error (RMSE) between simulated and measured temperature and humidity timeseries.

Thermal model

Following the approach presented in [8], the analogy of heat and charge transfer summarized in Table I is used to model the heat transfer between the WT environment and the power cabinet by means of a simple R-C network.

Table I: Analogy of thermal and electrical quantities (adopted from [8])

Thermal quantity	Symbol	Unit	Electrical quantity	Symbol	Unit
Temperature	T	K	Voltage	U	V
Heat flow	$\dot{Q}$	W	Current	I	A
Thermal resistance	R	K/W	Resistance	R	Ω
Heat capacitance	C	J/K	Capacitance	C	F

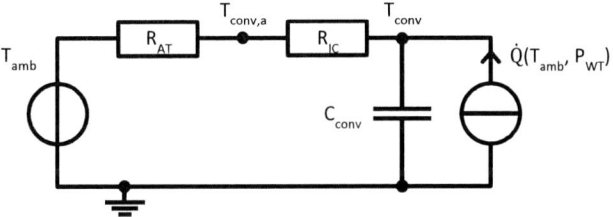

Fig. 2: Thermal R-C network for heat transport between WT-ambient air and converter-cabinet air

Fig. 2 shows the thermal network resulting in case of a wind turbine with the power converter located in the tower base: The WT-ambient temperature T_{amb} corresponds to an ideal voltage source. The thermal resistances of the WT tower and the cabinet enclosure are lumped in the resistor R_{AT}, including the convective heat transfer between air volumes and solids as well as conductive heat transfer. (Note that a first model structure also contained a thermal capacitance in this part representing the thermal mass of the tower wall; however, the simplified model structure without this element shown in Fig. 2 provides equally accurate results and is therefore given preference.) $T_{conv,a}$ denotes the air temperature inside the converter cabinet, for which measured timeseries are available. The thermal mass of the converter equipment inside the power cabinet is lumped into the heat capacitance C_{conv}.

The overall heat flow released into the converter components equals the losses generated inside the converter P_{loss} reduced by the heat flow withdrawn by means of the cooling system $\dot{Q}_{cool}$. In the R-C network, it is represented by an ideal current source applying the heat flow $\dot{Q}$.

Operating-point dependent heat release into the converter

The overall heat released into the cabinet-internal converter components varies with the electrical load, i.e. with the operating point of the turbine. Modelling it poses a particular challenge: While the power losses generated inside the converter can be calculated from the converter design and the electrical load condition, the heat flow transferred to the WT-ambient air by means of the cooling system is unknown.

In the present work, we therefore derive the resulting operating-point dependent heat release into the converter directly from the field data. For this purpose, we undertake an intermediate step in which we limit the consideration to time periods with quasi-stationary conditions. These are characterized by variations of T_{amb} and $T_{conv,a}$ by no more than ± 0.5 K and variation of the WT active power by up to $\pm 2.5\%$ of the turbine's rated power. In a stationary state, the R-C network reduces to the equivalent circuit shown in Fig. 3.

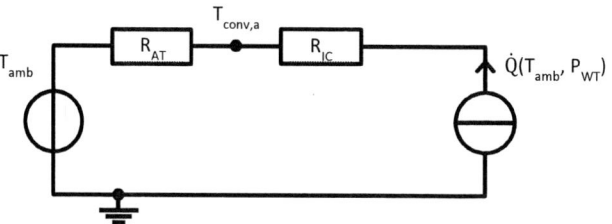

Fig. 3: Thermal R-C network for stationary conditions

In this case, the heat release into the converter is related to the temperature difference between cabinet air and ambient air through:

$$\dot{Q} = P_{loss} - \dot{Q}_{cool} = \frac{T_{conv,a} - T_{amb}}{R_{AT}} \tag{1}$$

Fig. 4 illustrates how the field data from periods with quasi-stationary conditions can be used to derive a regression function that expresses the temperature difference ($T_{conv,a} - T_{amb}$) as a function of the WT active power P_{WT} and the ambient temperature T_{amb}.

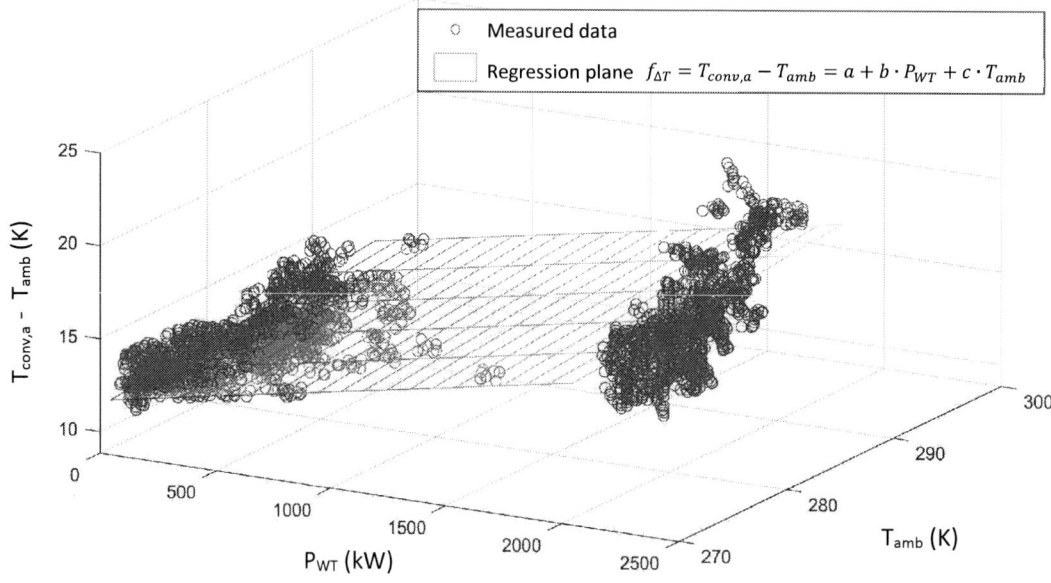

Fig. 4: Derivation of the operating-point dependent heat release into the converter from field data by means of regression, based on periods with quasi-stationary conditions

For the sake of enhanced model quality, two different regression functions $f_{\Delta T}$ are determined for every turbine: one for periods in which the turbine is in operation ($P_{WT} > 0$ kW) and one for standstill periods in which the turbine does not feed active power to the grid ($P_{WT} \leq 0$ kW) due to a lack of wind, faults, or maintenance.

Equations representing the thermal model

The resulting set of equations underlying the thermal model is:

$$\frac{dT_{conv}}{dt} = -\frac{1}{C_{conv}\cdot(R_{AT}+R_{IC})}\cdot T_{conv} + \frac{1}{C_{conv}\cdot(R_{AT}+R_{IC})}\cdot T_{amb} + \frac{1}{C_{conv}\cdot R_{AT}}\cdot f_{\Delta T} \tag{2}$$

$$T_{conv,a} = \frac{R_{AT}}{(R_{AT}+R_{IC})}\cdot T_{conv} + \frac{R_{IC}}{(R_{AT}+R_{IC})}\cdot T_{amb} \tag{3}$$

The model parameters to be estimated by means of parameter identification are the thermal resistances R_{AT} and R_{IC} and the thermal capacitance C_{conv}. T_{amb} and $f_{\Delta T} = f(T_{amb}, P_{WT})$ are the time-dependent input signals, while the air temperature inside the converter cabinet $T_{conv,a}$ is the output signal of the model.

Hygric model

Using the analogy of mass and charge transport summarized in Table II, also the moisture transfer between the WT environment and the power cabinet can be modelled by means of an R-C network. At the current stage, the model does not cover the occurrence of condensation. The structure of the R-C network for the moisture balance is shown in Fig. 5.

Table II: Analogy of hygric and electrical quantities (adopted from [8])

Hygric quantity	Symbol	Unit	Electrical quantity	Symbol	Unit
Moisture concentration	c	kg/m³	Voltage	U	V
Mass flow	ṁ	kg/s	Current	I	A
Diffusion resistance	R_D	s/m³	Resistance	R	Ω
Volume	V	m³	Capacitance	C	F

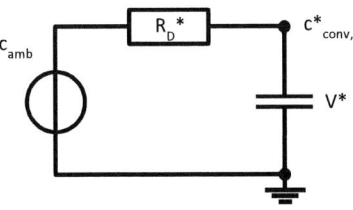

Fig. 5: R-C network for moisture transport between WT-ambient air and converter-cabinet air with equivalent quantities and parameters marked by (*)

The diffusion paths between the ambient air and the cabinet-internal air through gaps are represented by a lumped diffusion resistance R_D^*. The moisture storage in the cabinet air is modelled by means of a lumped equivalent volume V^*. It is important to note that, as explained and derived in detail in [8], transformation to equivalent quantities and parameters is necessary whenever the system-internal temperatures differ from the ambient temperature or different materials are involved. This takes account of the fact that, in the balanced state, there is no equilibrium of the water vapor concentrations but of the water vapor partial pressures.

Equations representing the hygric model

While the thermal model consists of a set of linear equations, the temperature dependence of diffusion and solubility coefficients as well as the transformation to equivalent quantities introduces nonlinearities into the equations of the hygric model. After expressing V^* and R_D^* according to the theory and equations in [8] and [12], the resulting set of equations representing the hygric model is:

$$\frac{dc^*_{conv,a}}{dt} = \frac{1}{k \cdot (273\ K)^{1.8}} \cdot T_{conv,a} \cdot T_{amb}^{0,8} \cdot \left(c_{amb} - c^*_{conv,a}\right) \tag{4}$$

$$c_{conv,a} = c^*_{conv,a} \cdot \frac{T_{amb}}{T_{conv,a}} \tag{5}$$

In these equations, k is a constant parameter to be identified from the measured data, c_{amb} and T_{amb} are input signals, $T_{conv,a}$ is an input signal provided by the thermal model and the moisture concentration (i.e. absolute humidity) in the converter-cabinet air $c_{conv,a}$ is the output signal of the hygric model.

Results and Discussion

The grey-box model identification, simulation and visualization has been implemented using MATLAB (Release 2022a). The results are presented and discussed in the following.

Measured and simulated climatic conditions in a wind-turbine converter

In Fig. 6 to 8, we present the results for the case of an onshore WT in the UK with doubly fed induction generator and liquid-cooled partially rated converter, the latter located in the tower base. The WT has a rated power of approx. 2 MW. Fig. 6 shows the measured temperature timeseries through the entire measurement period together with the simulated temperature timeseries obtained from the identified thermal model and the varying operating point of the turbine. Note that the model parameters have been identified using the timeseries data until end of December only, to save the last month for model validation.

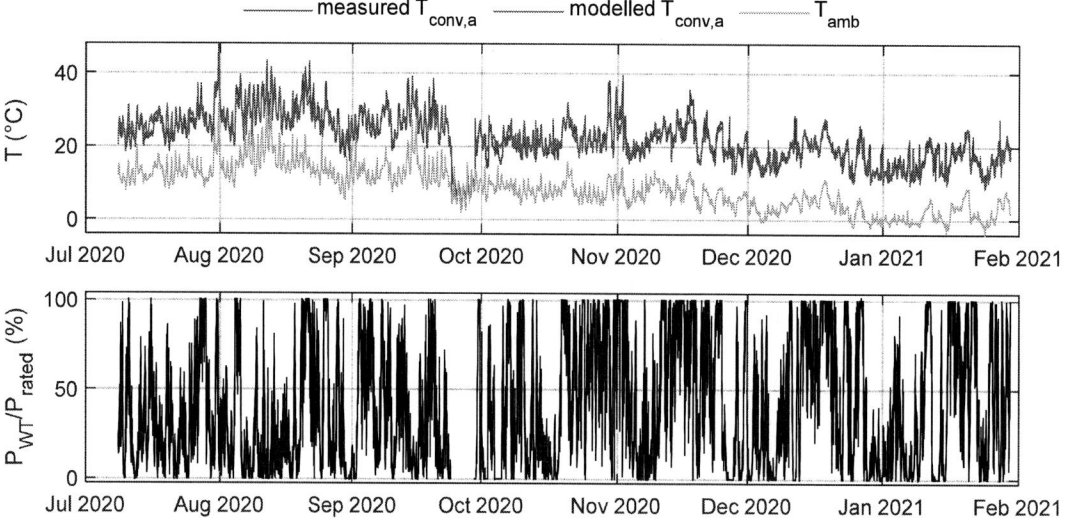

Fig. 6: Results of the identified thermal model of a WT in UK with liquid-cooled, partially rated converter located in the tower base; overall view of the measured and simulated temperature timeseries and the normalized WT active power fed to grid

Excerpts from the full timeseries are presented in Fig. 7, showing how day-night cycles and the WT load conditions influence the WT-internal temperatures. The excerpts cover periods of full-load and part-load operation as well as standstill periods of the turbine.

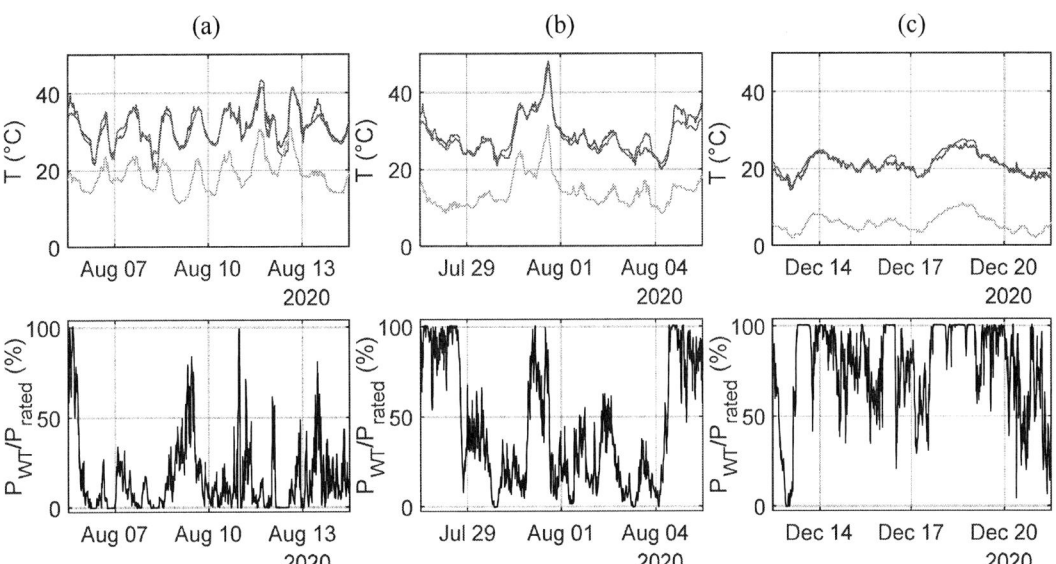

Fig. 7: Results of the identified thermal model of a WT in UK with liquid-cooled, partially rated converter located in the tower base (modelled cabinet-temperature timeseries: red, corresponding measured timeseries: blue; WT-ambient temperature: yellow); periods with (a) part-load operation and pronounced day-night cycles, (b) particularly dynamic variation of temperatures and electrical load, (c) operation close to and at rated power

Figures 6 and 7 show that the derived thermal model can estimate the air temperature inside the converter cabinet with reasonable accuracy. Both the seasonal variation, the day-night cycles and influence of the turbine operation are successfully replicated by the model. Also during a 6-day standstill period with disconnection from the power grid during end of September 2020 (cf. Fig. 6), simulated and measured timeseries are in good agreement. For the data period of the first 5.5 months used for parameter identification, the RMSE of the converter-cabinet temperatures is $RMSE_{T,tot} = 1.95$ K. For the validation period of the last month, it is even lower with $RMSE_{T,tot} = 1.43$ K. In case of the wind turbine considered here, the ambient temperature is found to have a stronger influence on the converter temperature than the operating point, i.e. electrical load of this turbine: At constant ambient temperature, the converter-cabinet temperature varies by no more than approx. 5 K between low part-load and full-load operation.

Fig. 8 presents the results of the hygric model, which are obtained with $T_{conv,a}$ from the thermal model, and their comparison with the measured moisture concentrations. It can be observed that the cabinet-internal and the WT-external absolute humidity are almost identical. This possibly surprising observation is in agreement with our findings from the evaluation of measurement campaigns in >30 WT reported in [7] and appears to be typical of wind turbines without active dehumidification. While at first sight, the internal moisture concentration seems to fully match that in the WT environment, a closer observation makes clear that the temperature difference between the WT-ambient and the converter-cabinet air gives rise to a small shift, i.e. leads to slightly reduced moisture concentrations inside the WT (cf. Equation (5)).

Also in case of the hygric model, the timeseries estimated with the derived model are in good agreement with the measured timeseries most of the time – both during the data period used for parameter identification and during the validation period in the last month. The corresponding RMSE values are $RMSE_{H,tot} = 0.45$ g/m³ in the period used for parameter estimation and $RMSE_{H,tot} = 0.26$ g/m³ in the period used for model validation. An interesting effect, which requires further investigation and is not yet reproduced by the present model, is the short-term peak in moisture concentration observed during the pre-heating routine after the 6-day standstill period end of September 2020 (cf. Fig. 6 and 8), which is likely related to moisture storage in e.g. polymers in the power cabinet.

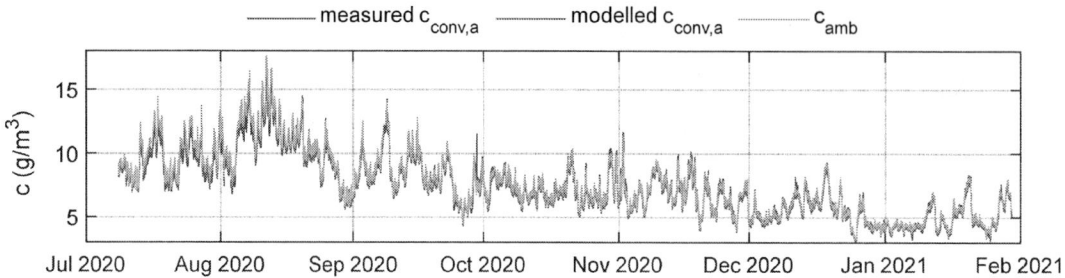

Fig. 8: Results for the moisture concentration (absolute humidity) obtained with the identified hygrothermal model of the same WT (modelled timeseries: red, corresponding measured timeseries: blue; WT-ambient conditions: yellow)

Using the modelling and parameter-identification approach described above does not provide the exact value of each single model parameter, but of their products (R·C) and (R$_D$*·V*), respectively. These can be used to determine the time constants of the heat and the moisture transfer between the WT-ambient air and the converter. For the turbine considered above, the overall thermal time constant (R$_{AT}$ + R$_{IC}$)·C$_{conv}$ is approx. 1.5 h. The hygric time constant (R$_D$*·V*) is influenced by the temperature levels. For the temperature conditions encountered inside and around the above WT, the hygric time constant is approx. 0.3 h and thus considerably shorter than the thermal time constant.

Dependence of model accuracy on environmental input data

In the case presented above, measured temperature and humidity timeseries are available from both the converter cabinet and from outside the wind turbine. In many other cases, measured data are available only from inside the turbine. In that case, ERA5 reanalysis data are used instead to include the site-specific ambient conditions of the WT. An interesting question is to which extent this influences the accuracy of the derived hygrothermal models, i.e. of the WT-internal timeseries calculated with these models. This is investigated for the same wind turbine as above.

The temperature and humidity timeseries obtained based on ERA5 data are shown in Fig. 9. The RMSE values for the entire measurement period in case of measured vs. ERA5 data for the environmental conditions are summarized in Table III.

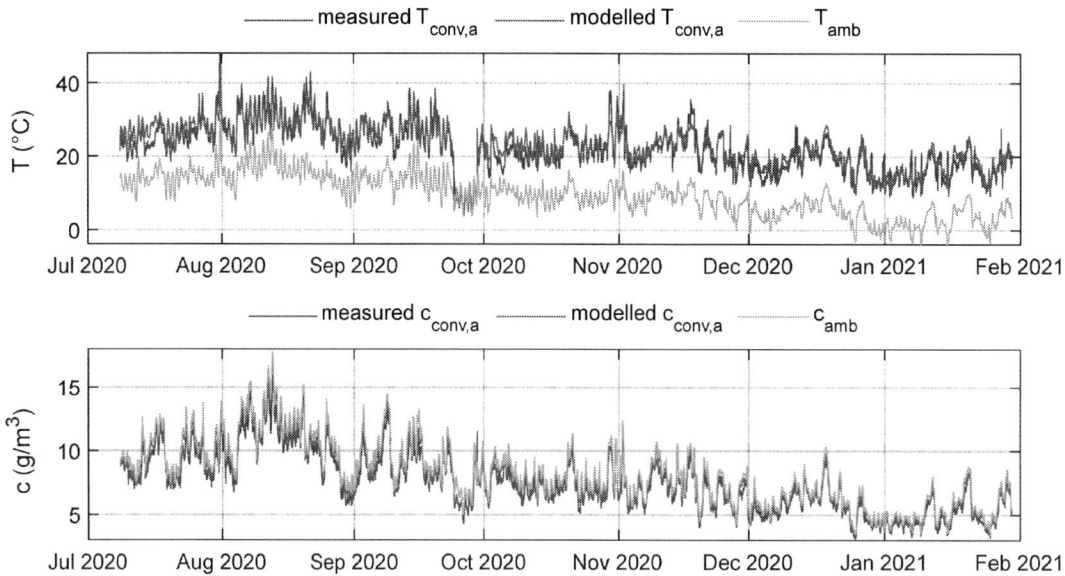

Fig. 9: Temperature and moisture-concentration timeseries based on site-specific ERA5 data for the ambient conditions

A comparison of the timeseries in Fig. 6, 8 and 9 makes clear that the differences are small: The ERA5 reanalysis data are in good agreement with the data measured on site. In consequence, also the modelled timeseries show little differences between the scenario using measured ambient data (Fig. 6 and 8) and the one using ERA5 ambient data (Fig. 9).

Table III: Comparison of model qualities by means of root-mean squared errors

Scenario	Thermal model			Hygric model		
	$RMSE_{T,op}$	$RMSE_{T,st}$	$RMSE_{T,tot}$	$RMSE_{H,op}$	$RMSE_{H,st}$	$RMSE_{H,tot}$
Model based on measured T_{amb} and c_{amb}	1.80 K	2.36 K	1.87 K	0.42 g/m³	0.50 g/m³	0.43 g/m³
Model based on T_{amb} and c_{amb} from ERA5 reanalysis	2.31 K	2.56 K	2.34 K	0.56 g/m³	0.80 g/m³	0.60 g/m³

This is confirmed by the RMSE values in Table III: While the model accuracy is – expectedly – higher (i.e. RMSE values are lower) when measured timeseries are used, the RMSE values are increased by no more than 0.51 K and 0.3 g/m³ in the case based on ERA5 data. Taking into consideration that the accuracy of the temperature and humidity loggers used in the measurement campaign is 0.3 K and 2% relative humidity [7], this is considered a very limited loss in model accuracy – a result that encourages the use of ERA5 data for the purpose of modelling, simulating, and investigating the site-specific climatic conditions in wind turbines.

Model transfer to identical turbines at different sites

Another important question is if the model parameters derived from a measurement campaign at a certain site may be transferred to a turbine of the same type operating in another region. We investigate this for the above case of DFIG-based turbine type for which measurement and operating data are available not only from a site in the UK but also from a site in Germany.

Fig. 10 shows the temperature and humidity timeseries calculated using the transferred model parameters of the previous WT in UK in combination with environmental data (here: ERA5 data) and the active-power timeseries of the turbine in Germany. The visual impression that the transferred model can describe the climatic conditions with similar accuracy as those in the original WT for which its model parameters were identified is confirmed by the corresponding RMSE values: With a value of $RMSE_{T,tot}$ = 2.53 K for the temperature timeseries in the WT in Germany, the measured temperatures and those simulated with the transferred thermal model show only a slightly higher deviation compared with the original WT in the UK (with $RMSE_{T,tot}$ = 2.34 K, see Table III). In case of the humidity timeseries, the value of $RMSE_H$ = 0.32 g/m³ indicates that the timeseries simulated with the transferred hygric model match the humidity data measured on this turbine even better than on the original WT located in the UK (with $RMSE_H$ = 0.60 g/m³, see Table III).

Similarly successful transfers of hygrothermal models between WT of identical type could also be observed in the cases of further turbines not presented here. This is an important finding as it confirms that the derivation of application-typical requirement specifications and test procedures is not limited to sites at which field-measurement campaigns have been carried out. Instead, the hygrothermal models derived from the field-measurement data can be combined with ambient climatic data from any site in the world (including e.g. challenging climates in tropical regions) to obtain region-specific test profiles or requirement specifications.

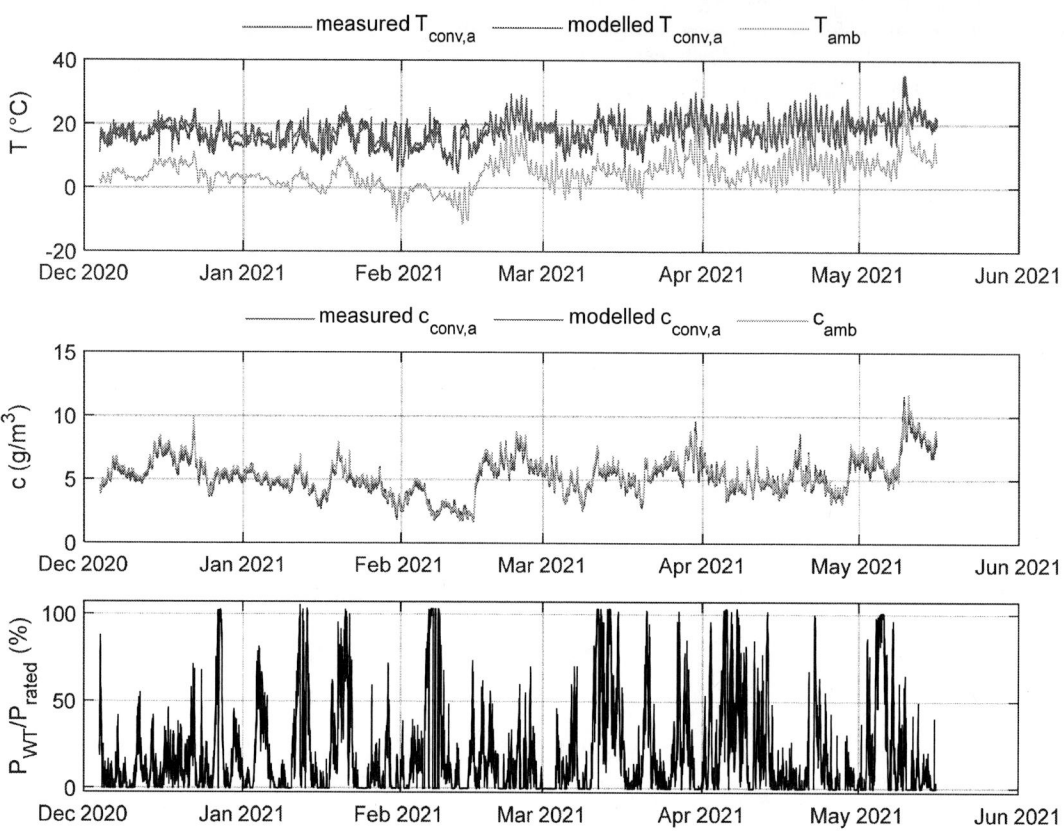

Fig. 10: Results of model transfer from the WT in UK to a WT of identical type in Northern Germany (Note that the simulated strong temperature drops in mid of December 2020, are caused by gaps in the active-power timeseries and therefore not attributable to the model quality.)

Conclusion

Using a lumped-parameter modelling approach, we have been able to derive hygrothermal models of low complexity that are capable of describing the climatic conditions in power converters of wind turbines. The field-data based models successfully replicate the seasonal variation of the climatic conditions experienced by the power converter throughout the year as well as the variations on a shorter timescale related to day-night cycles and load fluctuations. For the wind turbine with water-cooled power converter in the tower base considered in the present work, model qualities with an average root mean squared error of 1.4-2.5 K for the cabinet-air temperature and 0.3-0.6 g/m³ for the cabinet-internal moisture concentration have been achieved. The models describe periods with turbine operation slightly more accurately than standstill periods. For the considered turbine, the identified time constants for heat and moisture transport are in the range of 1.5 h and 0.3 h, respectively, indicating thermal balancing to be much slower than the hygric balancing processes.

We have demonstrated that it is not necessary to have measured temperature and humidity timeseries from the environment of the wind turbine. Global ERA5 reanalysis data for the site of interest can be used instead with only a slight reduction in model accuracy. In addition, we could show that it is possible to transfer the model parameters identified from measurement data from a certain site to turbines of identical type located at other sites.

In contrast to Schuster et al. [13] who investigated and modeled the climatic conditions in railway traction converters and reported massively increased moisture concentration inside the converter cabinets, we found the absolute humidity in the converter cabinets of wind turbines to be mostly very similar to the ambient conditions of the turbine. However, an interesting effect in wind turbines requiring further investigation and model refinement are short-term peaks in absolute humidity during the pre-heating routines applied after longer standstill periods of wind turbines, which could be related to moisture storage in polymers.

Field-data based hygrothermal models make it possible to simulate the temperature and humidity conditions in wind turbines at any possible location in the world based on the ambient environmental conditions at this site. Thus, they play an important role for the adjustment of requirement specifications as well as for the derivation of application-specific test procedures for power converters and their components.

References

[1] J. B. Gayo, "Final Publishable Summary of Results of Project RELIAWIND," 2011.

[2] Y. Lin, L. Tu, H. Liu, and W. Li, "Fault analysis of wind turbines in China," *Renewable and Sustainable Energy Reviews*, vol. 55, pp. 482–490, Mar. 2016, doi: 10.1016/j.rser.2015.10.149.

[3] "System Performance, Availability and Reliability Trend Analysis - Portfolio Review 2016," 2017. [Online]. Available: https://ore.catapult.org.uk/wp-content/uploads/2018/02/SPARTAbrochure_20March-1.pdf

[4] K. Fischer *et al.*, "Reliability of Power Converters in Wind Turbines: Exploratory Analysis of Failure and Operating Data From a Worldwide Turbine Fleet," *IEEE Trans. Power Electron.*, vol. 34, no. 7, pp. 6332–6344, Jul. 2019, doi: 10.1109/TPEL.2018.2875005.

[5] K. Fischer *et al.*, "Exploring the Causes of Power-Converter Failure in Wind Turbines based on Comprehensive Field-Data and Damage Analysis," *Energies*, vol. 12, no. 4, p. 593, Feb. 2019, doi: 10.3390/en12040593.

[6] K. Pelka and K. Fischer, "Field-data based Reliability Analysis of Power Converters in Wind Turbines: Assessing the Effect of Explanatory Variables," presented at the Wind Energy Science Conference WESC 2021, Hannover, May 27, 2021.

[7] K. Fischer, M. Steffes, K. Pelka, B. Tegtmeier, and M. Dörenkämper, "Humidity in Power Converters of Wind Turbines—Field Conditions and Their Relation with Failures," *Energies*, vol. 14, no. 7, p. 1919, Mar. 2021, doi: 10.3390/en14071919.

[8] R. Bayerer, M. Lassmann, and S. Kremp, "Transient Hygrothermal-Response of Power Modules in Inverters—The Basis for Mission Profiling Under Climate and Power Loading," *IEEE Trans. Power Electron.*, vol. 31, no. 1, pp. 613–620, Jan. 2016, doi: 10.1109/TPEL.2015.2408117.

[9] Z. Staliulionis, S. Mohanty, and Hattel, J.H., "Resistor-Capacitor Approach for Modelling of Temperature and Humidity Response Inside Electronic Enclosures," presented at the 2019 20th International Conference on Thermal, Mechanical and Multi-Physics Simulation and Experiments in Microelectronics and Microsystems (EuroSimE). doi: 10.1109/EuroSimE.2019.8724538.

[10] ECMWF, "Climate reanalysis | Copernicus." https://climate.copernicus.eu/climate-reanalysis (accessed Jul. 01, 2022).

[11] H. Hersbach *et al.*, "The ERA5 global reanalysis," *Q.J.R. Meteorol. Soc.*, vol. 146, no. 730, pp. 1999–2049, Jul. 2020, doi: 10.1002/qj.3803.

[12] M. Lassmann, "Moisture modeling in complex systems," presented at the ESREF 2018, Oct. 2018.

[13] O. Schuster, A. Nagel, and B. Laska, "Observation and simulation of dynamic humidity in power converters for railway applications due to moisture diffusion in plastics," presented at the 23rd European Conference on Power Electronics and Applications (EPE'21 ECCE Europe) 2021, 2021. doi: 10.23919/EPE21ECCEEurope50061.2021.9570624.

A Multi-Mode Control Based Asymmetrical Dual-Active-Bridge Series-Resonant DC-DC Converter (DABSRC)

M. Yaqoob, Grover Torrico, Wang Shuqin
Huawei Digital Power R&D Center
Stockholm, Sweden
Email: yaqoob.muhammad@huawei.com
URL: https://digitalpower.huawei.com/en/

Keywords

≪Bi-directional converters≫, ≪Dual-Active-Bridge (DAB) DC-DC converter≫, ≪Resonant converter≫, ≪Switching and conduction losses≫, ≪High power density systems≫.

Abstract

A multi-mode control for asymmetrical dual-active-bridge series-resonant DC-DC converter (DABSRC) based on half-bridge and full-bridge switching configurations is proposed. The proposed control is configured to eliminate burst-mode operation while supporting an efficient wide-voltage range voltage variation handling and bidirectional power-flow capabilities. Under heavy and medium load conditions, the proposed control method regulates converter's output by varying the switching frequency. When switching frequency exceeds its predetermined maximum value under light load conditions, a second control mode is employed (instead of opting for burst mode) where the switching frequency is fixed to maximum value and converter output is regulated by varying both the phase shift and duty cycle. The effectiveness of the proposed method is validated by experimental results with the peak efficiency of 98.95 % and power density of approximately 218 W/in^3 or 13.3 kW/L (including output filter and auxiliary-power circuit).

Introduction

A control method for a bidirectional dual-active-bridge series-resonant DC-DC converter (DABSRC) involving variable input and output DC voltage sources is proposed. A typical application of such a control method could be in DABSRC based DC-DC power converter stage of an AC-DC rectifier system used in power telecom equipment and DC-AC inverter system for photovoltaic (PV) applications. Generally, for an efficient and reliable performance, the DC-DC converter stage should be able to operate efficiently under wide-range voltage variations without burst-mode operation. Over the past several years, various topological and control solutions for the DC-DC conversion stage and its control were proposed to strive for aforementioned traits. These solutions are summarized below:

- An internal phase-shift or duty-cycle variation based control methods were proposed in [1] and [2]. Both of these control methods are applicable for two inductors and one capacitor (LLC) based DC-DC converter topology. However, this converter topology lacks the wide-range voltages variation handling capability along with inability of enabling bi-directional power flow.
- Reference [3] proposed a control method for half-bridge and full-bridge type switching networks based dual-active-bridge (DAB) converter. The proposed method is unable to handle wide-range voltage variations efficiently, and it lacks the burst mode elimination which could lead to high-voltage stress when required power is more than the rated power.
- A conventional LC-type DABSRC operating with single-phase-shift (SPS) modulation was proposed in [4]. With SPS modulation, power is modulated with one degree of freedom using the external phase shift between the primary and secondary bridge voltages but this method can result in

substantial conduction and switching loss, particularly under light-load and non-unity voltage-gain conditions. An efficient control method to overcome drawbacks associated with [4] was proposed in [5] for symmetrical full-bridge type DABSRC. This method utilized two degrees of freedom i.e., using SPS modulations and frequency variations. However, both aforementioned control methods suffer from back-flow power and burst-mode is required for light-load conditions.

- The authors of [6] proposed an efficient four-degrees-of-freedom control method for symmetrical full-bridge type DABSRC. However, similar to [1]-[5], burst mode is required to achieve low power levels. Furthermore, the burst mode requirement may move up to near to the medium power levels, when the value inductor and capacitor is chosen to handle the power more than rated power (overload condition).

To overcome the associated drawbacks of prior literature, a new control method is proposed to handle the wide-range voltage variations efficiently and eliminate the burst mode to allow high power density (by opting for smaller value the inductor). The proposed control method is based on two modes: mode 1 utilizes phase-shift modulation and frequency variation typically at medium-to-high power levels, while mode 2 is employed to regulate the low power levels by making use of both duty cycle and phase shift with frequency being fixed at its maximum value. A detailed analysis, derivation, control structure and its validation are illustrated in next sections of the paper.

Analysis of the Proposed Control Method

A circuit diagram of asymmetrical half-bridge and full-bridge based DABSRC and typical operating waveforms of the proposed control method are shown in Fig 1a-c. The choice of half-bridge (i.e., two switches $S_1' - S_2'$) switching network for high-voltage (HV) side and full-bridge switching network forming low-voltage (LV) side (i.e., four switches $S_3' - S_6'$) allow lower number of semiconductor switches (and hence the high-power density) compared to the most of the prior literature solutions. For the mode 1, the voltages v_1 and v_2 generates the near-sinusoidal currents i_1 and i_2 by exciting the series-resonant impedance Z. This property of currents i_1 and i_2 being near sinusoidal leads to lower current stress and reduced high-order harmonics, and hence lower conduction and core losses in transformer and inductor. The use of aforementioned switching-network configurations under mode 1 operate with near 50% ON and OFF time i.e., $D=0.5$ for the all semiconductor switches $S_1' - S_6'$ create a v_1 with the levels of 0 and $+V_1$ and v_2 with the levels of $-V_2$ and $+V_2$. As for the mode 2, the switching network at the HV operates with $D<0.5$ and the currents i_1 and i_2 opt to be piece-wise linear. Due the presence of capacitor C in series with inductor L and transformer T_r, (i.e., the choice of series-resonant impedance Z) gives inherent DC-bias removal generated by HV-side switching network. Both Mode 1 and Mode 2 operations are further explained below:

- **Mode 1:** By considering the power flow from HV→LV and fundamental component analysis, the average power and current at LV side can be calculated using the HV or LV bridge voltage (i.e., $v_1(t)$ or $v_1(t)$) and tank current $i_1(t)$ or $i_2(t)$; both approaches lead to the same expressions given by (1) and (2) [4].

$$
\begin{aligned}
P_{o,2} &= \frac{1}{2\pi} \int_0^{2\pi} v_1(t) i_1(t) dt \\
&= \frac{4NV_1V_2 \sin\theta}{\pi^2 Z} = \frac{2MV_1^2 \sin\theta}{\pi^2 Z}
\end{aligned}
\tag{1}
$$

$$
I_2 = \frac{4NV_1 \sin\theta}{\pi^2 Z}
\tag{2}
$$

where N is transformer T_r turn ratio and $-\pi/2 \leq \theta \leq +\pi/2$ is the phase shift between v_1 and v_2 determining the direction and the magnitude of the power/current flow between HV and LV sides. The positive value of $+\theta$ depicts the power/current from HV→LV side while negative value of $-\theta$ leads to the power/current flow from LV→HV side. Furthermore, voltage gain M (across

A Multi-Mode Control Based Asymmetrical Dual-Active-Bridge Series-Resonant DC-DC Converter (DABSRC)   YAQOOB Muhammad

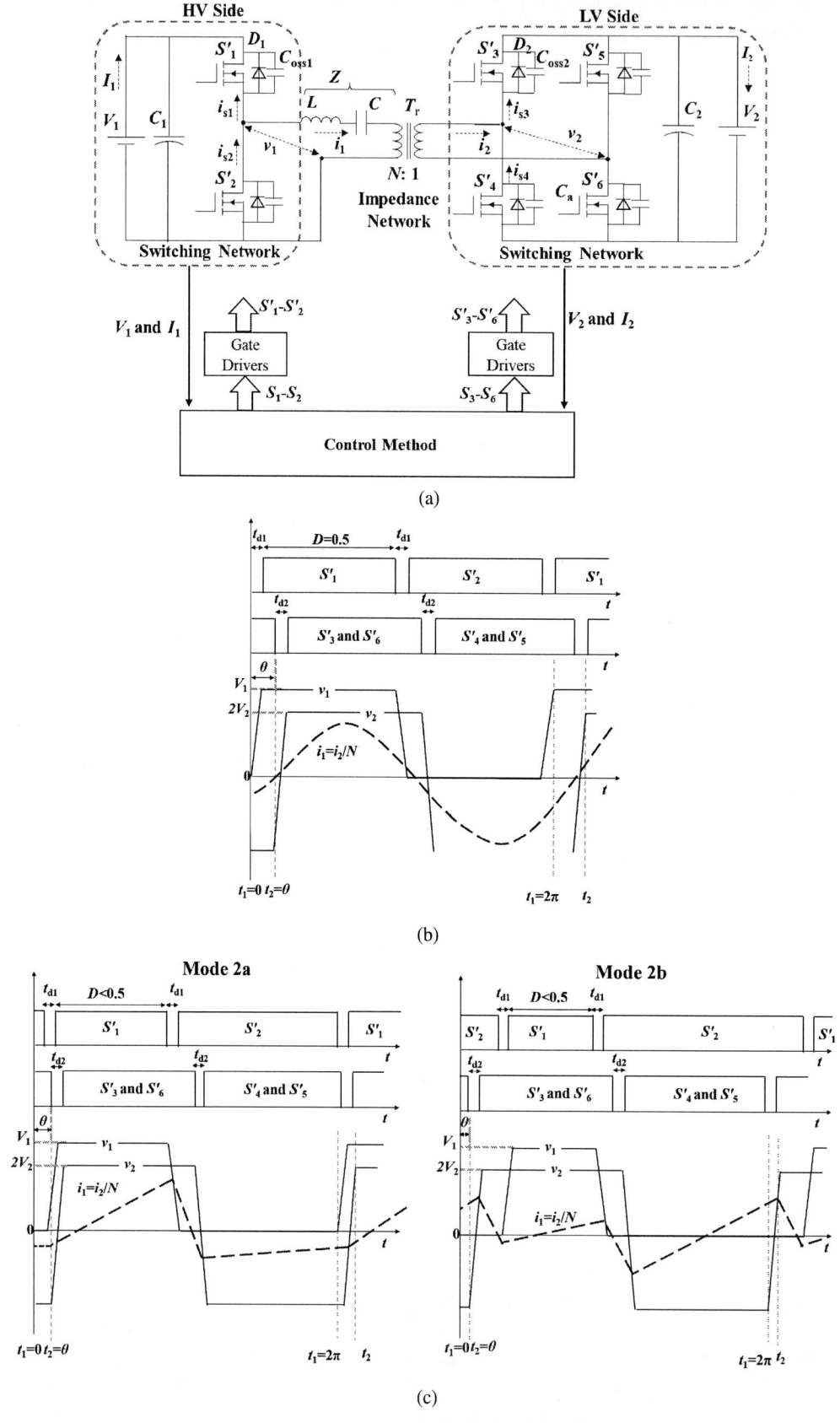

(a)

(b)

(c)

Fig. 1: (a) Circuit diagram of the half-bridge and full-bridge switching networks based DABSRC, (b)

impedance Z) and impedance Z are given by (3) and (4) with f_s as switching frequency,

$$M = \frac{NV_2}{0.5V_1},$$ (3)

$$Z = 2\pi f_s L - \frac{1}{2\pi f_s C}.$$ (4)

The conduction loss in DABSRC is mainly dependent on rms values of $i_1 = i_2/N$ which is given by (5),

$$I_{1,\text{rms}} = \frac{\pi I_2}{4N \sin \theta} \times \sqrt{2M^2 - 4M \cos \theta + 2}.$$ (5)

In order to find the minimum rms current $I_{1,\text{rms,min}}$ through Z for the given value of I_2, the first derivative of $I_{1,\text{rms}}/I_2$ with respect to phase shift θ should be kept zero and solve it for the required values of phase shift θ. The final result is presented as (6)[6],

$$\frac{\partial}{\partial \theta} \left(\frac{I_{1,\text{rms}}}{I_2} \right) = 0 \Rightarrow \theta = \begin{cases} \cos^{-1}(M) & \text{for} \quad M \leq 1 \\ \cos^{-1}(1/M) & \text{for} \quad M > 1 \end{cases}$$ (6)

Substituting (6) into (5) leads to the relationship of $I_{1,\text{rms,min}}$ and I_2 which is given by (7),

$$I_{1,\text{rms,min}} = \frac{I_{2,\text{rms,min}}}{N} = \begin{cases} \frac{\pi \times \sqrt{2}}{4N} I_2 & \text{for} \quad M \leq 1 \\ \frac{\pi M \times \sqrt{2}}{4N} I_2 & \text{for} \quad M > 1 \end{cases}$$ (7)

From (7), it can be seen that for the $M > 1$, the current increases proportionally to an increase in M. Hence, by choosing the value of N using (3) to keep $M \leq 1$ would be an appropriate design decision. Furthermore, by substituting (6) into (2), I_2 can be rewritten as (8),

$$I_2 = \begin{cases} \frac{4NV_1 \sqrt{1-M^2}}{\pi^2 Z} & \text{for} \quad M \leq 1 \\ \frac{4NV_1 \sqrt{1-\frac{1}{M^2}}}{\pi^2 Z} & \text{for} \quad M > 1 \end{cases}$$ (8)

At $M = 1$, from (6) phase shift $\theta = 0$ and from (8) output current $I_2 = 0$. To avoid such situation, there is a need to limit the values of M around 1 i.e., $M < 1$. Equation (6) is plotted in Fig 2 for $M < 1$ along with the maximum limit of M i.e., M_{max}, and therefore, the minimum phase-shift θ_{min} at M_{max} can be given as (9):

$$\theta_{\text{min}} = \cos^{-1}(M).$$ (9)

Equations (6) and (9) can be written together as (10) and plotted in Fig 2.

$$\theta = \cos^{-1}(M) \quad \text{for} \quad M \leq M_{\text{max1}}$$ (10)

The zero-voltage-switching (ZVS) operation at the HV-side switches $S_1' - S_2'$ depends on the minimum value of phase shift θ_{min} (or inductor current i_1 within dead time t_{d1}) which consequently depends on the choice of M_{max} given by (11),

$$M_{\text{max}} = \frac{\sqrt{(\pi I_{2,\text{min,zvs}} t_{\text{d1}})^2 - (4NQ_{1,\text{max}})^2}}{\pi I_{2,\text{min,zvs}} t_{\text{d1}}}$$ (11)

where $I_{2,\text{min,zvs}}$ is the minimum output current at LV side above which ZVS at HV is guaranteed and below this value the ZVS at HV side is not possible, t_{d1} is the dead time between $S'_1 - S'_2$, and $Q_{1,max} = $ max of $(V_1 C_{oss1})$ is the maximum charge stored on S'_1 with C_{oss1} as the parasitic output capacitance of the HV-side switches. The ZVS operation for the LV-side switches $S'_3 - S'_6$ is always guaranteed following (10) under mode 1 [5].

From (8), it can be seen that to change the I_2, Z (or f_s from (4)) can be varied because θ is already bounded by M i.e., variations in input and output voltages. Hence, the minimum required impedance Z_{min} to deliver given maximum $I_{2,max}$ can be calculated for M_{max} and $V_{1,min}$ and the result can be given by (12),

$$Z_{min} = \frac{4NV_{1,min}\sqrt{1-M_{max}^2}}{\pi^2 I_{2,max}} = 2\pi f_{s,min}L - \frac{1}{2\pi f_{s,min}C}, \tag{12}$$

where $V_{1,min}$, $f_{s,min}$ are the given minimum input voltage and frequency. Furthermore, the given resonance frequency $f_r < f_s$ (to behave inductive) of the impedance Z can be given by (13),

$$f_r = \frac{1}{2\pi\sqrt{LC}}. \tag{13}$$

Equations (12) and (13) can be solved simultaneously to determine the values of L and C to form the impedance Z.

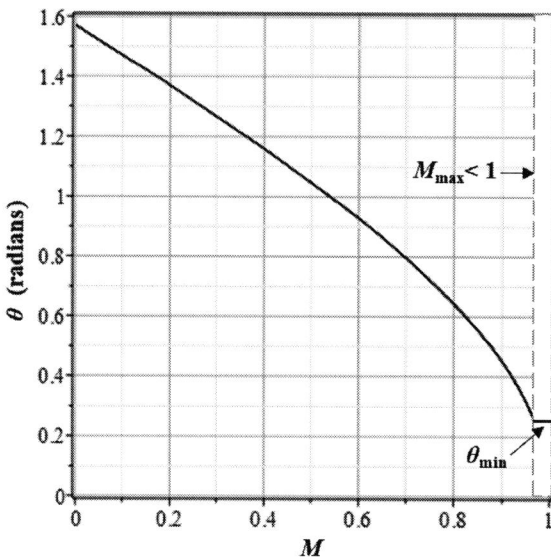

Fig. 2: Variation of voltage gain M vs. phase shift θ to achieve minimum rms current and ZVS operation at HV and LV sides.

- **Mode 2:** From (2) and (4), it can be seen that at $f_{s,min}$, the I_2 is of maximum value and at the $f_{s,max}$ (user defined based on hardware specifications) the I_2 would be of minimum value. In order to achieve the value of I_2 further lower than what $f_{s,max}$ can provide, it is a normal practice to opt for burst mode where all of the switching devices are periodically turned ON and OFF. However, when the M (i.e., input and output voltages) is varied under the wide-range, converter is pushed to deliver power more than its rated power and inductor is required to be reduced for increased power density, the start of burst mode is shifted from lower-to-medium values of I_2. Under this specific operation where burst mode starts at higher values of I_2, the reliability of the switching devices decrease and could lead to converter's failure. Hence, it is desired to eliminate the burst mode, and from mode 1 it is proposed to go to mode 2 (instead of starting burst-mode) after reaching $f_{s,max}$. Mode 2 is initiated when $f_s = f_{s,max}$, and instead of varying f_s to reduce I_2, duty-cycle at HV-side is reduced from 0.5 i.e., $D < 0.5$ and the remaining value of θ is reduced to modulate I_2 by stop following (10). By varying both D and θ, the currents waveforms i_1 and i_2 become piece-wise linear leading to two sub-modes i.e., mode 2a and mode 2b depicted in Fig 1c. For the power flow of HV→LV side i.e., $+\theta$, mode 2a appear when the rising edge of the v_1 is leading the rising edge of v_2, while mode 2b starts when rising edge of v_2 is leading the v_1 (c.f., Fig 1c). As for the power flow of LV→HV side i.e., $-\theta$, rising edge of v_2 leads v_1 for the mode 2a, vice versa. In this mode 2, minimum rms current operation cannot be maintained due to not following (10). As for ZVS operation, it remains same for HV-side switches (as in mode 1) with $M \leq M_{\max}$. For the LV-side switches, under the mode 2a, the two switches undergo hard-switching while for mode 2b, all switches at LV undergo ZVS operation. This aforementioned mode 2 of varying D and θ generally occurs at the medium-to-lower (depending on M) values of the I_2, and hence it doesn't have a big impact on the converter's overall performance.

- **Overall Control Structure:** Fig 3a-b, represents an overall control implementation of aforementioned modes of operations. It can be seen from Figure 3a, that for the power flow from HV→LV, a desired reference voltage V_{2ref} is compared with sensed voltage V_2, and the resulting error e is fed to the proportional-integral (PI) controller. Two multiplexers and a comparator are used to choose between two modes of operations. The output value of PI is passed to frequency (F_m) and duty-cycle and phase-shift (D_m-θ_m) modulators. For the mode 1 (i.e., when $x=0$ is given by the comparator), the value of PI is converted into frequency by F_m i.e., $f_s= f_{s,PI}$ to control and achieve the desired value of voltage V_2, while phase shift $\theta'= \theta$ from (10), and $D=0.5$ is used to maintain minimum rms current and ZVS operations. When f_s hits the maximum allowed limit with $f_s= f_{s,max}$ i.e., mode 2, PI value is converted into duty-cycle and phase-shift by D_m-θ_m modulator, which is then used to control $\theta =\theta_{PI}$ and $D=D_{PI}$ (i.e., when $x=1$ is given by the comparator) to deliver the remaining power flow with frequency limited to $f_s= f_{s,max}$. Furthermore, the PI gain K_m is changed based on the x i.e., based on mode of operation. The value of the K_m is dependent the converter's model and behavior in response to the variations in f_s, θ and D. Control parameters f_s, θ and D are fed to the PWM generator which comprises of a voltage-controlled oscillator, dead-time, phase-shift and duty-cycle generator. For the case, where power flows from LV→HV (c.f., Fig 3b), V_{1ref} is compared to the sensed value of V_1 to generate the error e. The rest of the control method behave similarly as in the case of HV→LV power flow, except that the output θ of the multiplexer 2 is multiplied with -1 to generate a negative phase shift -θ (i.e., LV side voltage v_2 is leading the HV side v_1). Fig 3 is shown to explain the control at block levels and the aforementioned control method can be implemented using both analog and digital (using microcontroller unit (MCU)) approaches.

Experimental Results

In order to validate the proposed control method, a hardware prototype with specifications given in Table I is built to imitate bidirectional DC-DC converter with maximum power flow of 3 kW (4 kW in overload conditions) for voltage sources of V_1 (380-430 V) and V_2 (42-58 V). The waveforms for a case of power from $V_2 \rightarrow V_1$ i.e., LV→HV are depicted in Fig 4a-c. The optimal combinations of the control parameters are used according to Fig 3a-b with objectives of power transfer with high efficiency, burst-mode elimination and high-power density.

Furthermore, the measured efficiency curves of the converter's operation for the power flow of HV→LV under various voltage variations are depicted in Fig 5a with a peak efficiency of 98.95% (including output filter and auxiliary power supply). Relatively, lower efficiency values were observed for maximum input voltage $V_{1,max} = 430$V and minimum output voltage $V_{2,min} = 42$V, representing the case of $M = 0.68$. This can be explained by observing Fig 2 showing higher values of θ for lower M, which consequently leads to higher switching (due to high current turn-off at HV-side MOSFETs) and conduction losses (due to circulating current at HV-side). Fig 5b represents the comparison of measured efficiency values for the power flow from both HV→LV and LV→HV. The difference in efficiencies at light load is due to the different values of D and θ provided by PI in mode 2 operation.

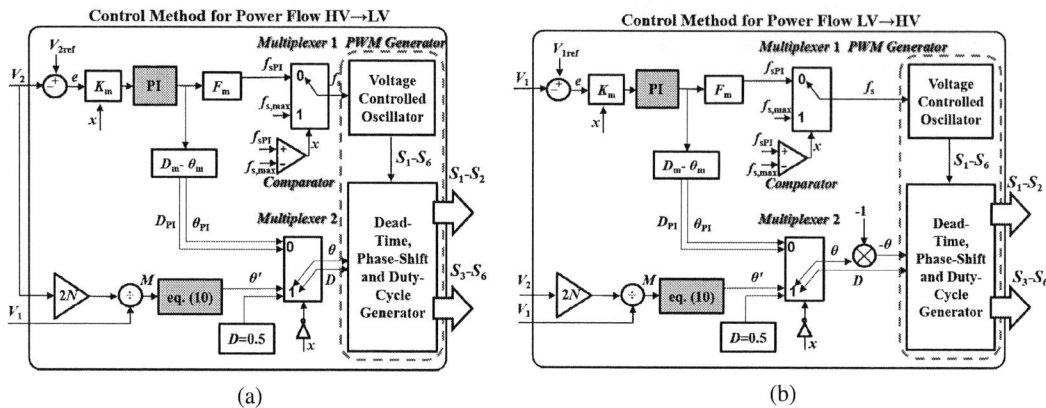

Fig. 3: Proposed control methods, (a) for the power flow from HV→LV, and (b) for the power flow from LV→HV.

Table I: Specifications of the Converter Prototype

Input Voltage Range V_i	380-430 V
Output Voltage Range V_o	42-58 V
Variation in Switching Frequency f_s	100-220 kHz
Maximum Rated Output Power P_o	3 kW
Primary Side MOSFETs	GaN
Secondary Side MOSFETs	Si
Control Implementation	Digital using STM32G4 MCU

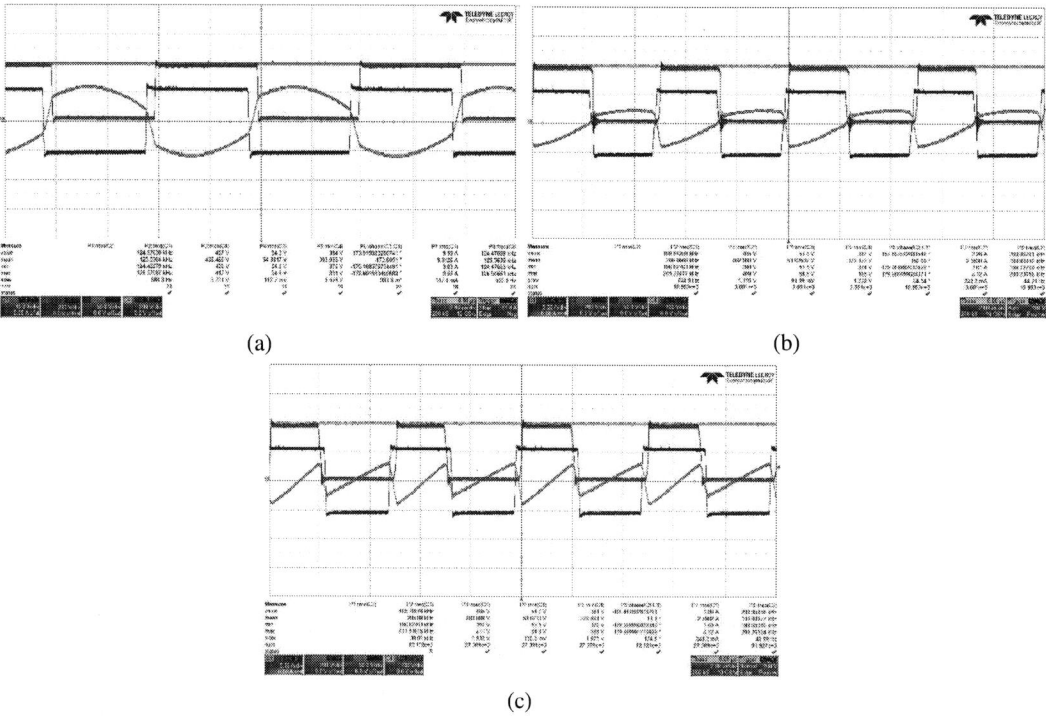

Fig. 4: Power transfer from LV→HV with $V_1 = 380$ V and $V_2 = 54$ V, with green V_1, pink v_1, blue v_2 and yellow i_1 (a) mode 1 with frequency variation, (b) mode 2a, and (c) mode 2b with phase-shift θ and duty-cycle D variations when frequency hits the maximum allowed limit of $f_{s,max}$.

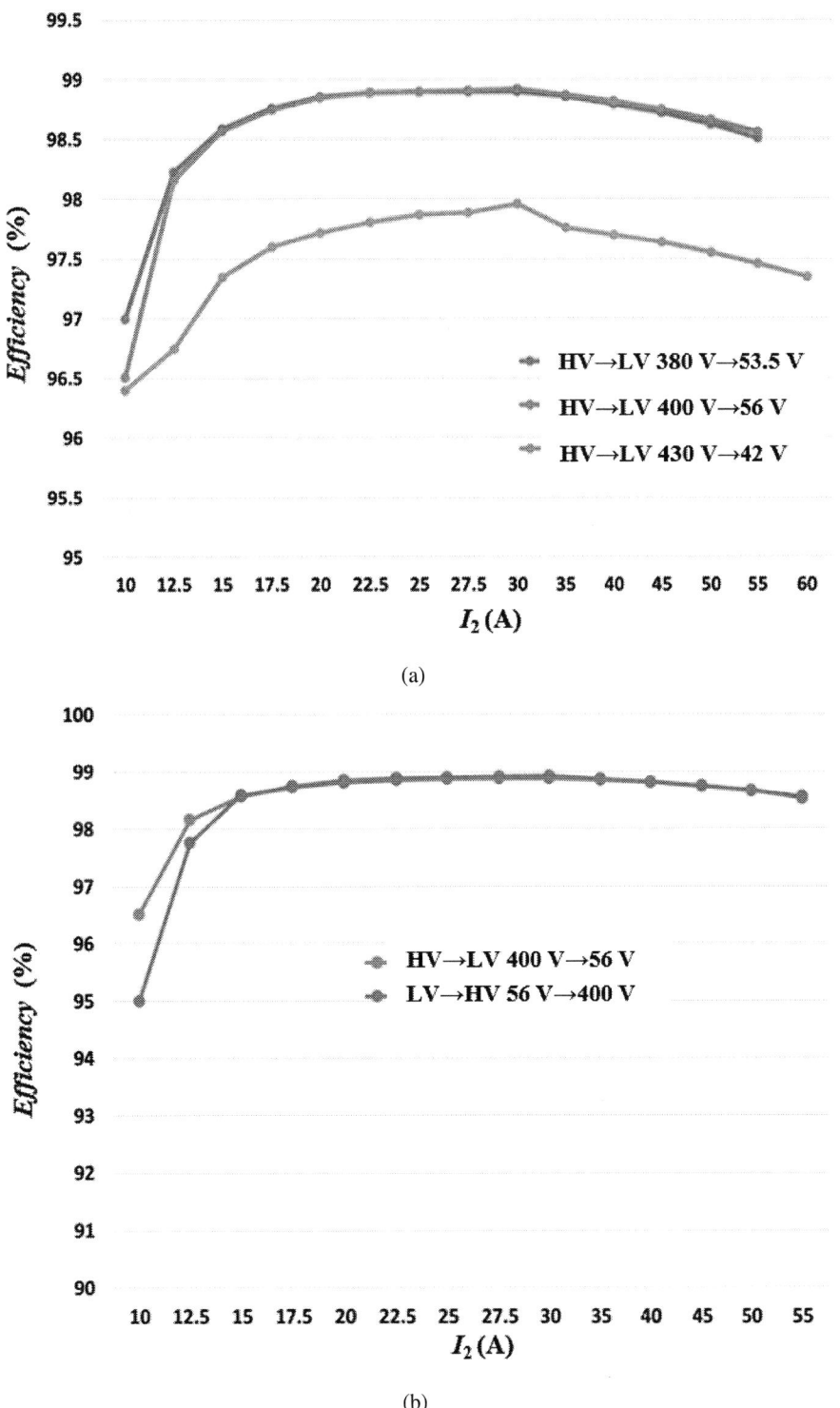

(a)

(b)

Fig. 5: Measured efficiency, (a) for various input and output voltage levels with power flow from HV→LV, (b) comparing both power flows i.e., HV→LV and LV→HV at typical operating point.

Conclusion

A control method is proposed for an asymmetrical half-bridge (HV-side) and full-bridge (LV-side) switching networks based DABSRC to address the drawbacks of the past literature. The proposed method opts for two modes of operation. The first mode uses the frequency variation to control the power flow while soft-switching and minimum rms current operations are achieved by varying phase shift. The second mode starts when the pre-assigned maximum value of frequency is achieved and the converter utilizes the HV-side's duty cycle and phase shift (between HV and LV sides) to control the power. The aforementioned modes of operations allows burst-mode elimination along with efficient wide-range voltage handling and high-power density capabilities.

References

[1] A. Awasthi, S. Bagawade and P. K. Jain, "Analysis of a Hybrid Variable-Frequency-Duty-Cycle-Modulated Low-Q LLC Resonant Converter for Improving the Light-Load Efficiency for a Wide Input Voltage Range," in IEEE Transactions on Power Electronics, vol. 36, no. 7, pp. 8476-8493, July 2021.

[2] N. Shafiei, M. Ordonez, M. Craciun, C. Botting and M. Edington, "Burst Mode Elimination in High-Power LLC Resonant Battery Charger for Electric Vehicles," in IEEE Transactions on Power Electronics, vol. 31, no. 2, pp. 1173-1188, Feb. 2016.

[3] J. Hiltunen, V. Väisänen, R. Juntunen and P. Silventoinen, "Variable-Frequency Phase Shift Modulation of a Dual Active Bridge Converter," in IEEE Transactions on Power Electronics, vol. 30, no. 12, pp. 7138-7148, Dec. 2015.

[4] Xiaodong Li and A. K. S. Bhat, "Analysis and Design of HighFrequency Isolated Dual-Bridge Series Resonant DC/DC Converter," IEEE Transactions on Power Electronics, vol. 25, no. 4, pp. 850–862, apr 2010.

[5] M. Yaqoob, K. H. Loo and Y. M. Lai, "Modeling the effect of dead-time on the soft-switching characteristic of variable-frequency modulated series-resonant DAB converter," 2017 IEEE 18th Workshop on Control and Modeling for Power Electronics (COMPEL), 2017.

[6] M. Yaqoob, K. H. Loo and Y. M. Lai, "A Four-Degrees-of-Freedom Modulation Strategy for Dual-Active-Bridge Series-Resonant Converter Designed for Total Loss Minimization," in IEEE Transactions on Power Electronics, vol. 34, no. 2, pp. 1065-1081, Feb. 2019,

Extended Balancing and Dimensioning of Capacitors in MMC Double Submodules

Ali Sharaf Addin, Christopher Dahmen, and Thomas Brückner
Universität der Bundeswehr München
Institute for Electrical Energy Systems
Werner-Heisenberg-Weg 39
85577 Neubiberg, Germany
e-mail: ali.sharaf@unibw.de

Abstract—**The Double-Zero Submodule in Double Connection (DZDCSM) is one of the most promising submodule topologies for future Modular Multilevel Converter (MMC) applications. Internal paralleling of semiconductors and capacitors in the DZDCSM during the time spans of high load currents results in low losses and reduced energy pulsation. Furthermore, full controllability of the capacitors enables their critical balancing in a passive and lossless manner. This paper presents an analytical investigation of the internal balancing process as well as the energy pulsation reduction in the full operating range of the MMC utilizing the DZDCSM. A simulation, applying the nearest level modulation and a basic sorting algorithm, verifies the general energy pulsation reduction in comparison to a conventional Full-Bridge Submodule (FBSM).**

I. INTRODUCTION

The Modular Multilevel Converter (MMC) has proven itself as the most reliable Voltage-Source Converter (VSC) for high-voltage applications and provides technical freedom to fulfill requirements of future meshed HVDC and MVDC grids [1]. On this account, advanced submodule topologies have been introduced that aim at reducing power losses and capacitor size while maintaining the DC fault blocking capability, as well as enhancing the submodule reliability [2]–[5]. One promising topology is that of the Double-Zero Submodule (DZSM), which has a full-bridge functionality and a controllable capacitor, offering a reduction of power losses, especially when employing SiC semiconductors [6]–[8]. In addition, the switchable capacitor offers an improved electronic protection against failures within the submodule.

Further reduction of semiconductor losses as well as reduction of capacitor energy pulsation can be achieved when applying the principle of a double connection of two DZSM resulting in the Double-Zero Submodule in Double Connection (DZDCSM). This topology is shown in Fig. 1 and features four terminal voltage levels (state (2): $+2V_C$, state (1): $+V_C$, state (0): 0 V, state (-1): $-V_C$) [5], [7]. The switching states are illustrated in Fig. 2 for both arm current directions. The switching state (2), where both capacitors C_1 and C_2 are connected in series, can cause a voltage imbalance between the capacitors due to capacitance tolerances. However, C_1 and C_2 can be passively balanced in the following switching state (1). Considering the initial condition $V_{C1} > V_{C2}$ at the beginning of state (1), the capacitor C_2 will be charged during state (1)

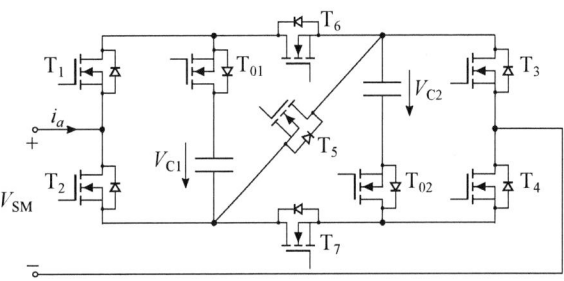

Fig. 1. Topology of the Double-Zero Submodule in Double-Connection equipped with SiC-MOSFET

(blue current path in Fig. 2), if the arm current is positive ($T_{6,7} = 1$, $T_{01,02} = 0$). On the other hand, the capacitor C_1 will be discharged during state (1) (green current path in Fig. 2), if the arm current is negative ($T_{6,7} = 0$, $T_{01,02} = 1$). The soft paralleling ($T_{6,7,01,02} = 1$) can be initiated at almost zero voltage difference, resulting in neglectable commutation and equalization losses [8]. The capacitor voltages can also be balanced in state (-1) but only for a negative arm current. If there is a remaining voltage difference between C_1 and C_2 when switching to state (-1) at a positive arm current, parallel connection of the capacitors can only be achieved by hard paralleling. This would result in high equalization currents that should be avoided in order to keep the semiconductor devices within the safe operating area.

In this paper, a detailed study of the balancing capability of the DZDCSM is carried out, covering the complete area of practically relevant operating points. Furthermore, different mechanisms are proposed to avoid hard paralleling in case state (1) ends with unbalanced voltages. Finally, a general analytical investigation of the energy pulsation is presented and put in perspective with a simulation of an MMC arm. In this manner, a universal comparison regarding the capacitor size of the DZDCSM and a conventional Full-Bridge Submodule (FBSM) will be extracted.

II. BALANCING OF CAPACITOR VOLTAGES

Assuming capacitors with a nominal capacitance C_0 and a relative tolerance $\pm\epsilon$, the capacitor voltages and the voltage

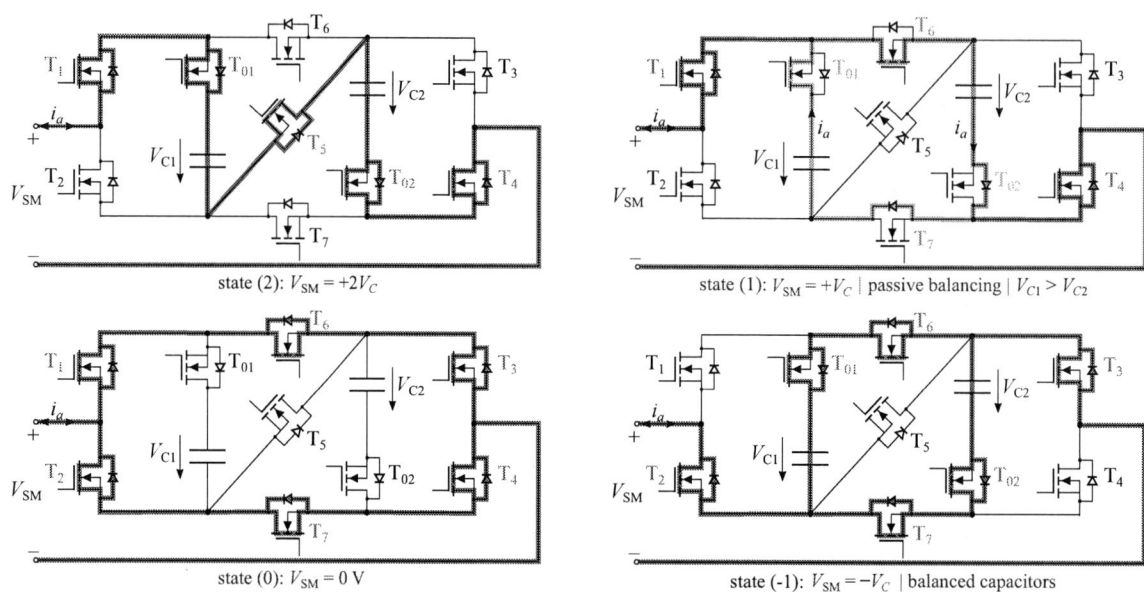

state (2): $V_{SM} = +2V_C$

state (1): $V_{SM} = +V_C$ | passive balancing | $V_{C1} > V_{C2}$

state (0): $V_{SM} = 0$ V

state (-1): $V_{SM} = -V_C$ | balanced capacitors

Fig. 2. Switching states of the DZDCSM

imbalance at the end of state (2) can be expressed as

$$\frac{V_{C1}|_{t_{x1}}}{\hat{i}_{AC}/C_0} = (1+\epsilon) \int_{t_{x0}}^{t_{x1}} i_a'(t)\, dt + \frac{v_C|_{t_{x0}}}{\hat{i}_{AC}/C_0}, \qquad (1)$$

$$\frac{V_{C2}|_{t_{x1}}}{\hat{i}_{AC}/C_0} = (1-\epsilon) \int_{t_{x0}}^{t_{x1}} i_a'(t)\, dt + \frac{v_C|_{t_{x0}}}{\hat{i}_{AC}/C_0}, \qquad (2)$$

$$\frac{\Delta V_C|_{t_{x1}}}{\hat{i}_{AC}/C_0} = 2\epsilon \int_{t_{x0}}^{t_{x1}} i_a'(t)\, dt, \qquad (3)$$

where i_a' is the arm current (example: upper arm)

$$i_a'(t) = \frac{i_a(t)}{\hat{i}_{AC}} = \frac{1}{4}k\cos(\varphi_i) + \frac{1}{2}\sin(\omega t - \varphi_i) \qquad (4)$$

normalized to the peak value of the AC current, neglecting the circulating currents. The modulation factor k is defined as the ratio of the fundamental AC-voltage peak value ($\hat{v}_{SM,AC,1}$) to the DC voltage ($V_{SM,DC}$) generated by the submodule and the arm [7]

$$k = \frac{\hat{v}_{SM,AC,1}}{V_{SM,DC}} = \frac{\hat{v}_{a,AC,1}}{V_{a,DC}}. \qquad (5)$$

Assuming a basic 4-level pulse pattern (see Fig. 4a), the generated voltage difference ΔV_C during state (2) can be nullified in the following state (1). This can be achieved if a charge ΔQ_{bal} is fed into the capacitor with the smaller voltage or extracted from the one with the higher voltage for the balancing. Since their capacitances are not equal, ΔQ_{bal} differs depending on the current direction

$$\text{sgn}(\Delta Q_{bal}) = \begin{cases} +1 & , i_a' > 0 \\ -1 & , i_a' < 0. \end{cases} \qquad (6)$$

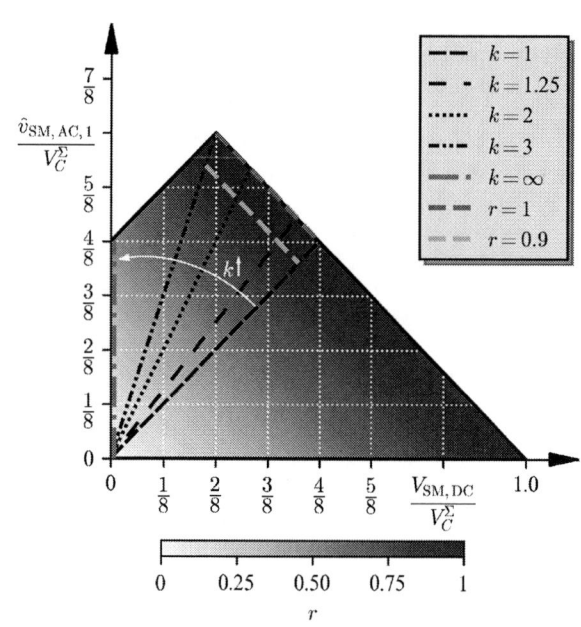

Fig. 3. Operation range of the DZDCSM applying a linear modulation scheme, the color map denotes the utilization factor of the DZDCSM

ΔQ_{bal} can be normalized to the net charge available in state (1)

$$|\Delta Q_{bal}'| = \frac{|\Delta Q_{bal}|}{|\Delta Q_{state(1)}|}. \qquad (7)$$

For a better clarity, the time spans of $|\Delta Q_{bal}|$ and $|\Delta Q_{state(1)}|$

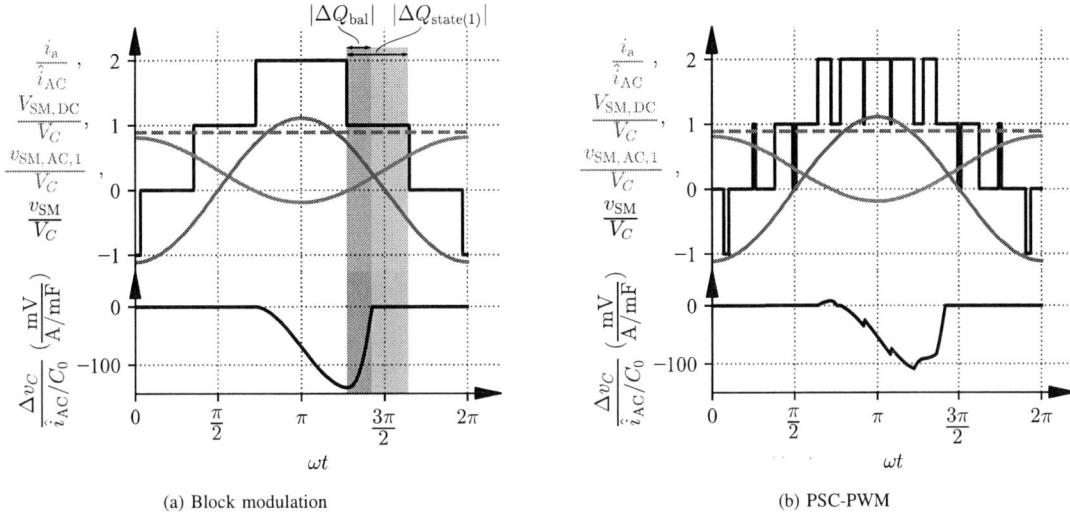

(a) Block modulation   (b) PSC-PWM

Fig. 4. Example of a balancing process applying different modulation schemes with the same utilization factor | $k = 1.25$, $r = 1$, $\cos\varphi_i = 1$, $\epsilon = 10\%$

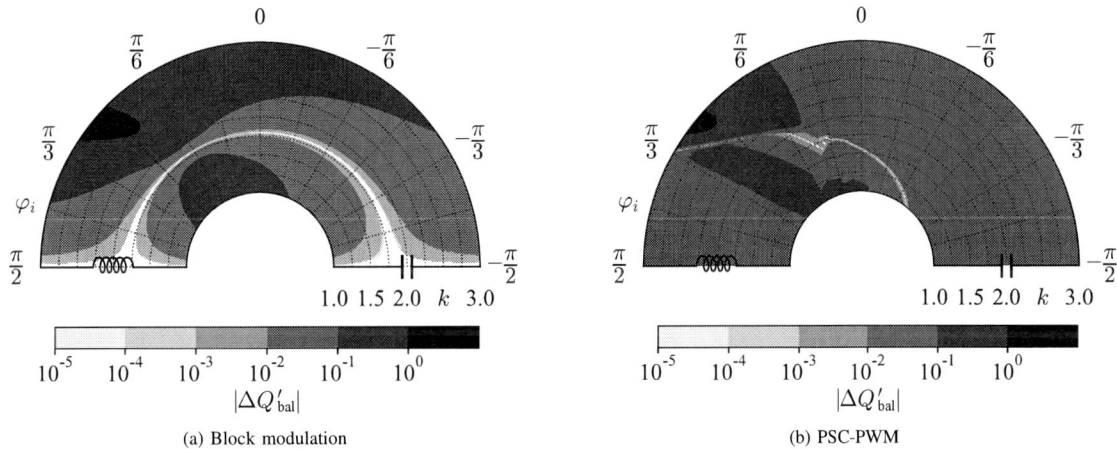

(a) Block modulation   (b) PSC-PWM

Fig. 5. Charge amount required to balance the capacitors in state (1) normalized to the net charge available in state (1) (see Eq. 7) | $\epsilon = 10\%$, operating points according to dashed red line in Fig. 2

are indicated exemplary in Fig. 4a. It can be noted, that in this operating point the ratio $|\Delta Q'_{\text{bal}}|$ is less than unity and therefore a complete balancing is achieved in state (1).

A. Evaluation of the balancing capability

In order to quantify the balancing capability of the DZ-DCSM, operating points have to be defined. Fig. 3 illustrates the operating region of the submodule when using a linear modulation scheme such as phase-shifted carrier modulation (PSC-PWM). The operating points along the red dashed line are chosen for this investigation, where the generated AC and DC voltages for a given k and the voltage imbalance generated in state (2) are maximized (worst case for the balancing process). The blue color gradient denotes the utilization factor

of the submodule

$$r = \frac{\hat{v}_{\text{SM,AC,1}} + V_{\text{SM,DC}}}{V_C^{\Sigma}} = \frac{\hat{v}_{\text{a,AC,1}} + V_{\text{a,DC}}}{n_{\text{SM}} \cdot V_C^{\Sigma}}, \quad (8)$$

where V_C^{Σ} describes the sum of submodule capacitor voltages within one submodule and n_{SM} the number of submodules within an arm.

Fig. 4a and Fig. 4b show normalized waveforms of submodule terminal voltages and arm currents, calculated for a typical operating point applying block modulation and PSC-PWM, respectively. Note, that both schemes generate the same DC and AC voltage levels. As can be seen, the voltage difference generated in state (2) is balanced out during state (1) before coming to the switching state (-1).

In Fig. 5 $|Q'_{\text{bal}}|$ is plotted as a function of k and φ_i for a wide range of operating points applying block modulation

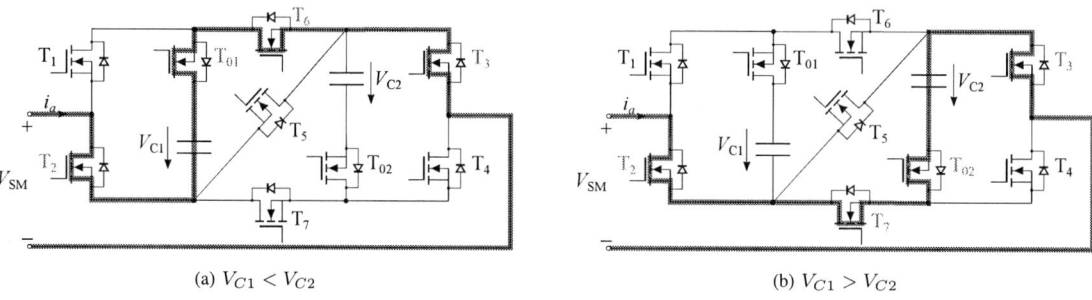

(a) $V_{C1} < V_{C2}$ (b) $V_{C1} > V_{C2}$

Fig. 6. Additional switching states for inserting the submodule in state (-1) with unbalanced voltages and $i_a > 0$

and PSC-PWM. The blue color gradient denotes the success of the balancing process within state (1). Solely in the black colored areas of the operating region, the balancing process does not finish during state (1). Note, that a good design of an MMC, equipped with submodules that have full-bridge functionality, is typically in the range $1.0 < k < 1.5$ where the arm energy pulsation, crucial for the capacitor size, and the semiconductor losses, are low. In this region the balancing process ends successfully and efficiently. However, the "black" region is present at $k > 2.3$ and $\frac{\pi}{6} > \varphi_i > \frac{\pi}{3}$ for both schemes investigated. These are not typical operating points but might be encountered during transients. Also during DC side failures, where the MMC can work as a STATCOM while maintaining the AC voltage and driving the DC voltage to zero (e.g. this corresponds to a horizontal line in Fig. 3 drawn from $k = 1.25 \mid r = 0.9$ to $k = \infty$), the normal passive balancing performs well. However, in order to deal with the problem of unbalanced capacitors in the "black" region, mechanisms to force the balancing need to be considered. This is important for enhancing the submodule availability over all operating points.

B. Mechanisms to avoid hard paralleling of C_1 and C_2 in state (-1)

As already mentioned, state (1) can be utilized to balance the capacitor voltages in a wide area of MMC operating points. Nonetheless, procedures to deal with incompletely balanced voltages have to be considered to avoid a hard paralleling of C_1 and C_2 in state (-1). Therefore, three possible approaches will be investigated in the following.

1) Inserting only one capacitor in state (-1): A straightforward approach to avoid a hard paralleling in state (-1) due to unbalanced capacitor voltages and $i_a > 0$ would be to insert only one capacitor instead of the parallel connection of C_1 and C_2. Considering that, there exist two possible switching states, illustrated in Fig. 6, for $V_{C1} < V_{C2}$ and $V_{C1} > V_{C2}$. As can be seen, only the capacitor with the smaller voltage can be inserted and, as a result, the remaining ΔV_C from state (1) will be increased instead of decreased, unfortunately.

Fig. 7 illustrates this issue for an operating point from the "black" region of Fig. 5a, where the lower diagram indicates the voltage imbalance in the initial period as a solid line and in the following period as a dashed line, i.e. its initial value

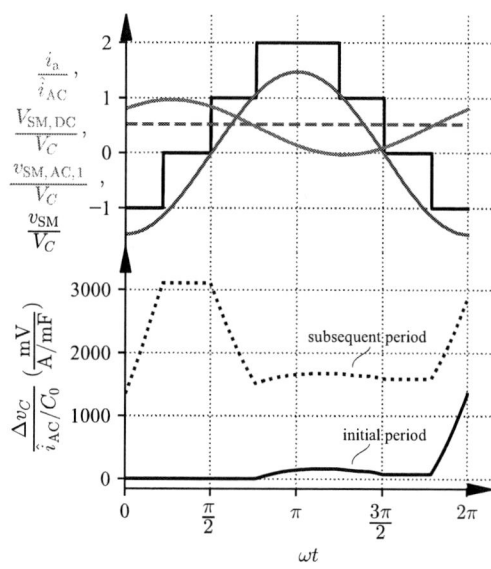

Fig. 7. Inserting only one capacitor of the DZDCSM in state (-1) with existing voltage imbalance | operating point from the "black" region of Fig. 5a according to Tab. I

is given from the previous period. Since the current-time area of the arm current is larger during state (-1) in contrast to state (1), the voltage imbalance would continue to increase over the periods, e.g. $\Delta V_C / V_{C0}$ would be 20 % at the end of the initial period for $\hat{i}_{AC} = 1.5$ kA and $V_{C0} = 2$ kV. Hence, this approach is not suitable for the voltage balancing. Note, that inserting the capacitor with the higher voltage (activating $T_{7,02}$ instead of $T_{6,01}$ in Fig. 6a) leaves no remaining diode in the blocking direction and thus leads to a high equalization current, which is not an option here.

2) Applying an extended balancing control: Considering an MMC arm equipped with DZDCSM, it is possible to extend the voltage-balancing algorithm used for inserting the submodules with a control scheme that ensures the passive balancing of each submodule over a complete period. For explanation, an MMC arm will be investigated in an operating point from the "black" region of Fig. 5a, see Tab. I.

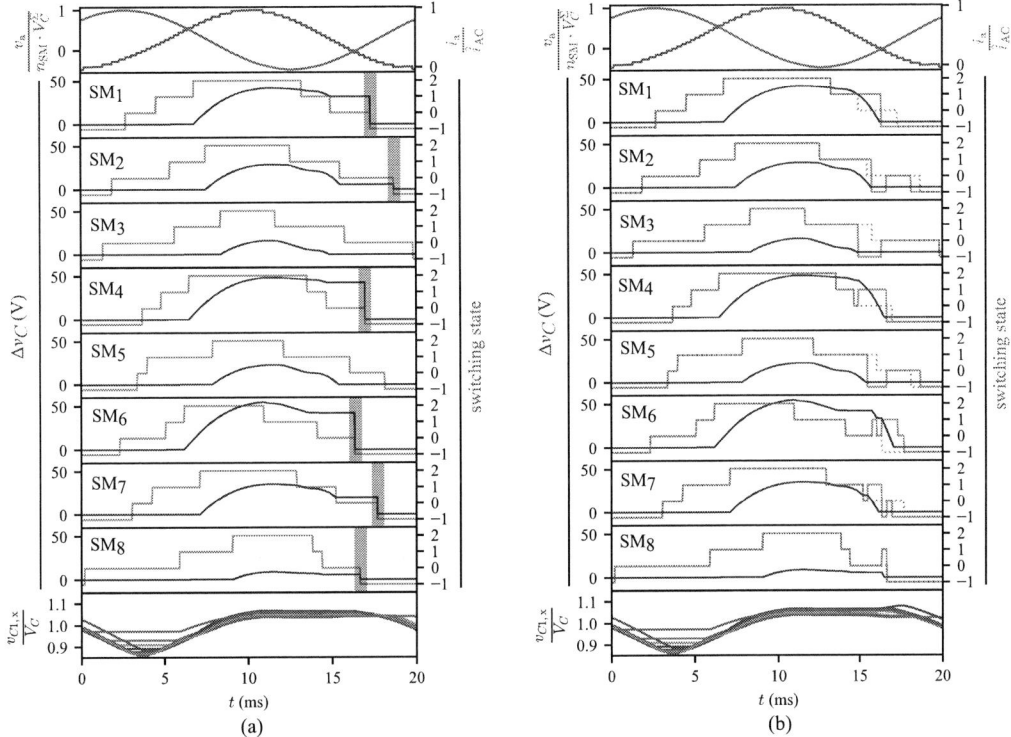

Fig. 8. (a): Operation of an MMC arm in the operating point specified in Tab. I, where the normal passive balancing does not finish successfully, (b): extending the voltage-balancing algorithm with a control scheme to enable the passive balancing (dashed line: original pulse pattern, solid line: modified pulse pattern)

TABLE I
OPERATING POINT FROM THE "BLACK" REGION OF FIG. 5A FOR INVESTIGATING ALTERNATIVE BALANCING STRATEGIES FOR THE DZDCSM

Parameter	Value
k	2.8
φ_i	48°
n_{SM}	8
ϵ	10 %
C_0	5 mF
C_1	4.54 mF
C_2	5.5 mF

The upper, middle and lower sections of Fig. 8a illustrate waveforms of the arm current and modulated voltage, the switching states of each submodule with the corresponding capacitor voltage imbalance and the submodule voltages, respectively. The nearest-level modulation (NLM) is used to realize a minimal submodule switching frequency of 150 Hz. A voltage-balancing algorithm within the arm for sorting the capacitor voltages is applied to keep the average of submodule capacitor voltages at a constant value. As can be seen from Fig. 8a, six submodules are switched to the state (0), while ΔV_C is not completely nullified. In the following state (-1) the capacitors are inserted in parallel resulting in high equalization currents due to the remaining ΔV_C that should be avoided. The

voltage-balancing algorithm should be adjusted in such a way, that the submodules are enabled to complete the balancing of their capacitors before entering the switching state (-1).

To achieve this objective, when the modulation triggers a reduction of one level of the modulated voltage, then only balanced submodules should be first switched from state (1) to (0) or to (-1) to let unbalanced submodules complete balancing their capacitors by remaining in state (1) or switching from state (0) back to (1). As can be seen from Fig. 8b, the balanced SM_3 is switched from state (1) to (-1) while SM_1 is retained in state (1) for the balancing together with SM_4, which is switched back from state (0) to (1). In the same manner, many switching operations can be performed for the purpose of balancing, when the modulation triggers. However, the net level reduction should be always equal to unity. Following this approach, a complete passive and lossless balancing can be achieved by the end of the period. The strategy leads to an increase of the average submodule switching frequency (in this example from 150 Hz to 200 Hz). However, as mentioned above, this is not a typical operating point of the MMC but could only occurs in infrequent situations such as transient events.

3) Equipping the DZDCSM with discharging resistors: The DZDCSM can be extended with two switchable resistors (see Fig. 9), that can be used

- to enable a voltage balancing of the capacitors by dis-

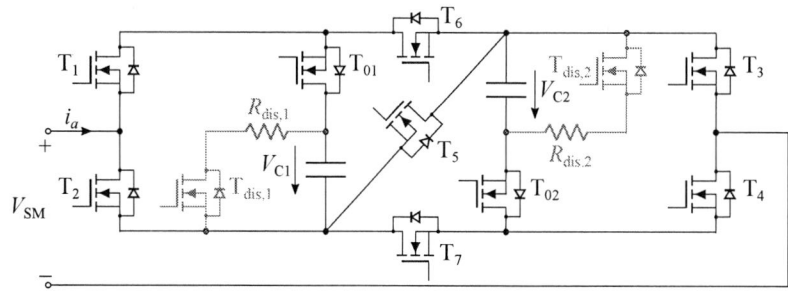

Fig. 9. The Double-Zero Submodule in Double-Connection extended with two switchable discharging resistors highlighted in red

charging the capacitor with the higher voltage to the level of the other one and

- for discharging the capacitors after a failure within the submodule for enhancing the submodule reliability.

The capacitor balancing using the discharging resistors should take place in state (0) in case state (1) ends with unbalanced capacitor voltages. Defining a time span t_{dis} for the balancing process, the required resistance can be calculated by

$$R_{\text{dis},i} = \frac{t_{\text{dis}}}{C_i \cdot \ln\left(\frac{V_{C_{i0}}}{V_{C_{i0}} - \Delta V_C}\right)}, \tag{9}$$

where $i = 1, 2$. In this time an energy amount of

$$E_{\text{dis},i} = \frac{1}{2} \cdot C_i \cdot (V_{Ci}^2(t_0) - V_{Ci}^2(t_0 + t_{\text{dis}})) \tag{10}$$

would be dissipated over the resistor. If designed to dissipate the total capacitor energy after a failure within the submodule, the resistors should be able to handle the balancing power during transients for a sufficient amount of time.

To give an estimation of the energy loss, the example of the balancing shown in Fig. 8 will be used. Thereby, the submodules switched to state (0) with existing voltage imbalance between their capacitors will use the discharging resistors for the balancing. The sum of the discharging losses can be normalized to the total arm capacitor energy

$$E'_{\text{dis}} = \frac{\sum E_{\text{dis}}}{n_{\text{SM}} \cdot \frac{1}{1 - \epsilon^2} \cdot C_0 \cdot V_{C_0}^2} = 0.88\ \%. \tag{11}$$

This value depends also on $\hat{i}_{\text{AC}}$, that drives the magnitude of ΔV_C.

After introducing the three strategies for the balancing, it can be concluded that only the strategies of *2)* and *3)* can be applied. Both strategies *2)* and *3)* cause little extra losses. Introducing discharging resistors is generally valuable in terms of enhancing the submodule reliability and solving the balancing on submodule level. For pure balancing purposes, the extended balancing algorithm also performs well.

III. INVESTIGATION OF SUBMODULE CAPACITOR ENERGY PULSATION

As already mentioned in the introduction of this paper, the DZDCSM offers many benefits in comparison to conventional FBSM such as the reduction of power losses, enhancing

the submodule reliability against failures through electronic protection as well as reducing the capacitor size. In [9] an analysis was carried out for comparing the energy pulsation – which has a direct impact on determining the capacitor size [10]–[13] – of one DZDCSM with two FBSM connected in series based on predefined pulse patterns for switching the submodules. It was assumed that all submodules within the arm are switched in the same manner as the investigated submodules. It was shown, that the capacitor size of the DZDCSM can be reduced by 50% compared to FBSM at specific operating points. However, in practical applications the switching states of the submodules for synthesizing the arm voltage are set by the voltage-balancing algorithm. The main task of this procedure is to ensure the equal share of power between all submodules while maintaining a constant switching frequency of the submodules. In this section an analysis for calculating the energy pulsation is introduced, considering all submodules within one MMC arm and taking the working principle of the voltage-balancing algorithm into account. The analysis results will be compared with simulation results.

A. Calculation of normalized arm power

In order to generalize the analysis, calculating the power of an MMC arm in a normalized form is fundamental, viz.

$$p'_a(t) = v'_a(t) \cdot i'_a(t). \tag{12}$$

The normalized arm current i'_a was already given in (4). The arm voltage v_a consists of an AC und DC component

$$v_a(t) = V_{\text{a,DC}} - \hat{v}_{\text{a,AC,1}} \sin(\omega t). \tag{13}$$

The DC component and AC-peak value can be described as functions of the peak value of the arm voltage $\hat{v}_a$ and the modulation factor k

$$V_{\text{a,DC}} = \frac{1}{k+1} \hat{v}_a, \tag{14}$$

$$\hat{v}_{\text{a,AC,1}} = \frac{k}{k+1} \hat{v}_a. \tag{15}$$

Substituting (14) and (15) into (13) yields

$$v_a(t) = \frac{1}{k+1} \hat{v}_a (1 - k \cdot \sin(\omega t)). \tag{16}$$

$v_a(t)$ can be normalized to the sum of installed arm capacitor voltages

$$v'_a(t) = \frac{1}{(k+1)} \frac{\hat{v}_a}{n_{\mathrm{SM}} \cdot V_C^\Sigma} (1 - k \cdot \sin(\omega t)). \quad (17)$$

Applying (8) in (17) simplifies the equation of the normalized arm voltage further,

$$v'_a(t) = \frac{r}{k+1} (1 - k \cdot \sin(\omega t)), \quad (18)$$

which will be used in (12) to calculate the normalized arm power

$$p'_a(t) = \frac{r}{k+1} \left[-\frac{1}{4} k^2 \cos(\varphi_i) \sin(\omega t) + \frac{1}{4} k \cos(2\omega t - \varphi_i) \right.$$
$$\left. + \frac{1}{2} \sin(\omega t - \varphi_i) \right]. \quad (19)$$

Fig. 10a illustrates the waveform of p'_a for a typical operating point of the MMC.

B. Definition of the insertion index

The insertion index a defines the ratio of inserted submodules within the arm ($V_{\mathrm{SM}} \neq 0$ V) to the number of installed submodules. For an arm equipped with FBSM the insertion index is defined as

$$a_{\mathrm{FBSM}}(t) = \frac{|v_a(t)|}{n_{\mathrm{SM}} \cdot V_C^\Sigma}. \quad (20)$$

V_C^Σ describes the sum of submodule capacitor voltages within one submodule. Note, that the DZDCSM has to be compared to a series connection of two FBSM and therefore a differentiation of V_C^Σ has to be done, viz.

$$V_C^\Sigma = \begin{cases} V_C & \text{for FBSM,} \\ 2V_C & \text{for DZDCSM.} \end{cases} \quad (21)$$

The blue waveform in Fig. 10b illustrates a_{FBSM} according to (20).

Defining the insertion index for an MMC arm equipped with DZDCSM cannot be done straightforward as with FBSM, because there are two groups of DZDCSM that need to be considered when calculating the insertion index:

- DZDCSM$_{\mathrm{parallel}}$: submodules whose capacitors are inserted in parallel (state (-1) and (1))
- DZDCSM$_{\mathrm{series}}$: submodules whose capacitors are inserted in series (state (2))

Hence, two insertion indices will be defined, one for each group. The insertion index of the first group is defined as

$$a_{\mathrm{DZDCSM}}^{\mathrm{parallel}}(t) = \begin{cases} 2\frac{|v_a(t)|}{n_{\mathrm{SM}} V_C^\Sigma} & , v_a(t) \leqslant \frac{1}{2} n_{\mathrm{SM}} V_C^\Sigma, \\ 2 - 2\frac{|v_a(t)|}{n_{\mathrm{SM}} V_C^\Sigma} & , v_a(t) > \frac{1}{2} n_{\mathrm{SM}} V_C^\Sigma, \end{cases} \quad (22)$$

and its waveform can be seen in Fig. 10b (green waveform). At the instant $v_a(t) = \frac{1}{2} \cdot n_{\mathrm{SM}} \cdot V_C^\Sigma$ all submodules are inserted with parallel capacitors. When exceeding this point, some of those submodules will also be inserted with series-connected

capacitors. Thus, the group DZDCSM$_{\mathrm{series}}$ exists only when the modulated voltage is greater than the half of the total installed arm voltage

$$a_{\mathrm{DZDCSM}}^{\mathrm{series}}(t) = \begin{cases} 0 & , v_a(t) \leqslant \frac{1}{2} n_{\mathrm{SM}} V_C^\Sigma, \\ 2\frac{|v_a(t)|}{n_{\mathrm{SM}} V_C^\Sigma} - 1 & , v_a(t) > \frac{1}{2} n_{\mathrm{SM}} V_C^\Sigma, \end{cases} \quad (23)$$

as can be seen from the orange waveform in Fig. 10b. Finally, the hybrid power share, illustrated in Fig. 10c, of both groups can be defined

$$p'(a_{\mathrm{DZDCSM}}^{\mathrm{series}}, t) = \frac{2a_{\mathrm{DZDCSM}}^{\mathrm{series}}(t)}{2a_{\mathrm{DZDCSM}}^{\mathrm{series}}(t) + a_{\mathrm{DZDCSM}}^{\mathrm{parallel}}(t)} p'_a(t), \quad (24)$$

$$p'(a_{\mathrm{DZDCSM}}^{\mathrm{parallel}}, t) = \frac{a_{\mathrm{DZDCSM}}^{\mathrm{parallel}}(t)}{2a_{\mathrm{DZDCSM}}^{\mathrm{series}}(t) + a_{\mathrm{DZDCSM}}^{\mathrm{parallel}}(t)} p'_a(t). \quad (25)$$

C. Analytical approach for calculating the energy pulsation

Considering an MMC arm, the task of the voltage modulation is to synthesize the arm voltage from the available voltage levels. The dynamic insertion of the submodules is carried out by the voltage-balancing algorithm that aims at keeping the mean voltage of the submodule capacitors at a constant value over the time. When applying a high switching frequency per submodule, the ripple of capacitor voltages can be reduced to considerably low values. However, if the submodule switching frequency is decreased towards the minimum, the ripple of capacitor voltages would increase as well as the deviation between the submodule voltages.

For a reasonable capacitor design, two insertion strategies will be introduced to estimate a potential range of the capacitor energy pulsation. The analytical approach will be first explained on the basis of a FBSM equipped MMC arm. The boundary conditions of this analysis are the following:

1) The MMC arm consists of a high number of submodules, so that the modulated voltage can be assumed to be continuous.
2) The submodules are switched with the minimal switching frequency, i.e. 50 Hz or 100 Hz for FBSM.

FBSM: In order to estimate the energy pulsation of the capacitors, the instantaneous power share of the inserted submodules within the arm has to be considered. This can be calculated by dividing the arm power by the associated insertion index

$$p'_{\mathrm{FBSM}}(t) = \frac{p'_a(t)}{a_{\mathrm{FBSM}}(t)}. \quad (26)$$

The inserted submodules from time t_β to time t_γ (compare Fig. 10) would encounter a positive or negative alteration of their stored energy

$$\Delta \tilde{W}'_{\mathrm{FBSM}} = \frac{1}{T_1} \int_{t_\beta}^{t_\gamma} p'_{\mathrm{FBSM}}(t) dt \quad ,$$
$$= \tilde{w}'_{\mathrm{FBSM}}(t_\gamma) - \tilde{w}'_{\mathrm{FBSM}}(t_\beta) \quad (27)$$

where T_1 is the fundamental period. The next step is to determine the dwell time of the inserted submodules (t_β and

t_γ). Since the minimum submodule switching frequency is presumed here, inserting the submodules with positive or negative terminal voltages happens only when the absolute arm voltage $|v_a|$ increases. The inserted submodules can be bypassed only when the absolute arm voltage starts decreasing. Based upon the definition of t_β and t_γ, two cases (best case and worst case of the encountered energy pulsation) will be investigated.

The best case scenario will be explained using Fig. 10d. Assume that the first submodule is inserted at $v_a(t_\beta) = 0$ when $|v_a|$ starts increasing. This submodule can then be bypassed only when $|v_a|$ reaches its maximum at t_γ. The maximum energy alteration ($\Delta \tilde{W}_{\mathrm{FBSM}}^{\prime min}$) encountered by this submodule during this period in the positive ($\Delta \tilde{W}_{\mathrm{FBSM}}^{\prime min+}$) or negative ($\Delta \tilde{W}_{\mathrm{FBSM}}^{\prime min-}$) range of v_a is the largest single energy rise, which is unavoidable. All other submodules inserted and bypassed later during the investigated fundamental period experience smaller energy pulsations. However, since the net energy exchange over the fundamental period per submodule is not necessarily zero — as it is for the complete converter arm during steady state — the energy levels of the individual submodules are slowly shifted and the voltage-balancing algorithm cannot limit the energy pulsation of the submodules to $\Delta \tilde{W}_{\mathrm{FBSM}}^{\prime min}$. Hence, the considered case is a boundary condition that cannot be avoided.

In the same manner, a maximum energy pulsation ($\Delta \tilde{W}_{\mathrm{FBSM}}^{\prime max}$) can be determined, that might not be exceeded during operation. This corresponds to inserting a submodule for the complete dwell time between two zero crossing points of the arm voltage allowing a high energy pulsation, see Fig. 10e. At this operating point $\Delta \tilde{W}_{\mathrm{FBSM}}^{\prime max}$ is located in the negative voltage range ($\Delta \tilde{W}_{\mathrm{FBSM}}^{\prime max-}$), Finally, the real energy pulsation of a FBSM within an MMC arm equipped with an integer number of submodules is expected to be between $\Delta \tilde{W}_{\mathrm{FBSM}}^{\prime min}$ and $\Delta \tilde{W}_{\mathrm{FBSM}}^{\prime max}$.

DZDCSM: Determining a maximum and a minimum band for the energy pulsation ($\Delta \tilde{W}_{\mathrm{DZDCSM}}^{\prime min}$ and $\Delta \tilde{W}_{\mathrm{DZDCSM}}^{\prime max}$) of the DZDCSM is far more complex. When speaking about the energy pulsation of the DZDCSM, only one of its capacitors is considered.

In Section III-B the insertion indices of the submodules inserted with parallel capacitors (group DZDCSM$_{\mathrm{parallel}}$) and with series capacitors (group DZDCSM$_{\mathrm{series}}$) as well as their power share within the arm were introduced. For determining $\Delta \tilde{W}_{\mathrm{DZDCSM}}^{\prime min}$ and $\Delta \tilde{W}_{\mathrm{DZDCSM}}^{\prime max}$ not only the inserting time of the submodule t_β as well as the bypassing time t_γ are important to know, but also the selection of the instantaneous group should be taken into account. Therefore, the time t_θ for switching the submodule from group DZDCSM$_{\mathrm{parallel}}$ to group DZDCSM$_{\mathrm{series}}$ as well as the time t_σ for switching it back to group DZDCSM$_{\mathrm{parallel}}$ should be defined.

For the best case scenario ($\Delta \tilde{W}_{\mathrm{DZDCSM}}^{\prime min}$) it is assumed that a submodule is switched on at $v_a(t_\beta) = 0$ in the positive range of v_a. At the instant t_θ where the group DZDCSM$_{\mathrm{series}}$ starts to exist, the submodule is switched to the series connection until t_σ where it is switched back to group DZDCSM$_{\mathrm{parallel}}$. t_σ is

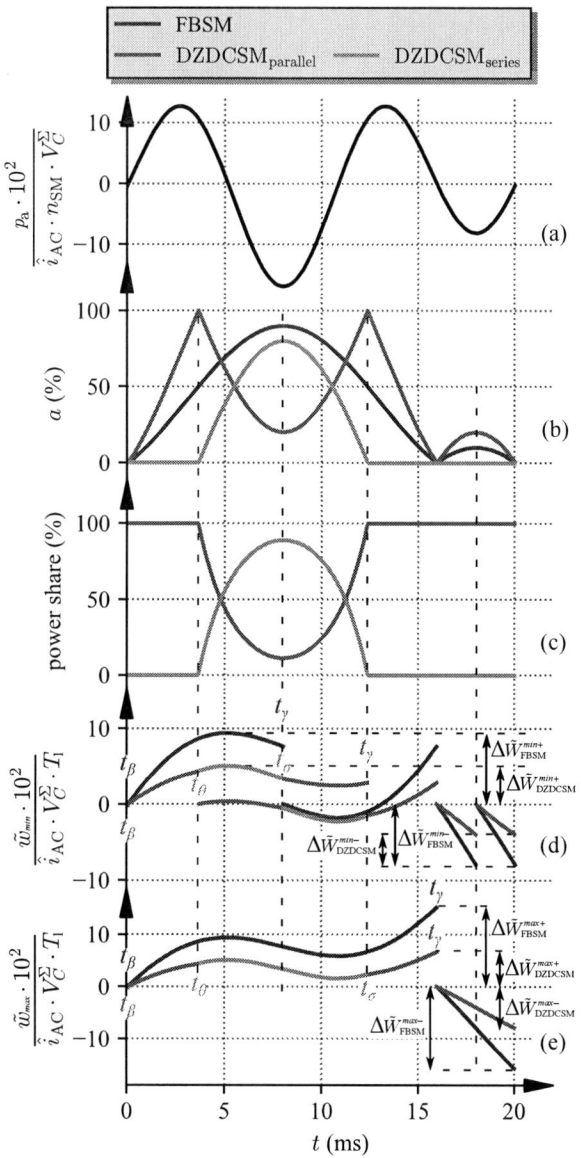

Fig. 10. (a): normalized arm power | (b): insertion indices of FBSM and DZDCSM | (c): percentage power share of the DZDCSM inserted with parallel and series connected capacitors within an arm | (d): minimum energy pulsation | (e): maximum energy pulsation | operating point: $k = 1.25, r = 0.9, \cos(\varphi) = 1$

the instant, where the group DZDCSM$_{\mathrm{series}}$ starts decreasing. The next time, where the submodule can be bypassed, is the time t_γ, where all submodules in group DZDCSM$_{\mathrm{series}}$ are switched back to group DZDCSM$_{\mathrm{parallel}}$ and the last one starts decreasing, see Fig. 10d. Thus, the submodule that sees the largest unavoidable energy pulsation during parallel insertion is selected, with the largest unavoidable pulsation during series insertion added on top. This is close, but not precisely the best case. The variety of possible combinations during different operating conditions makes an analytical and general identification of the actual best case nearly impossible. The energy pulsation in the negative range of v_a can be calculated similar to FBSM, since in the negative range only the group DZDCSM$_\text{parallel}$ exists. In Fig. 10d can be seen, that $\Delta \tilde{W}'^{min}_\text{DZDCSM}$ exists in the positive range of v_a at the current operating point.

For defining the maximum energy pulsation $\Delta \tilde{W}'^{max}_\text{DZDCSM}$ it is assumed that a submodule is inserted for the complete dwell time between two zero crossing points of the arm voltage. Furthermore, it will be switched to the group DZDCSM$_\text{series}$, as long as this group exists allowing a high energy pulsation, see Fig. 10e. At this operating point $\Delta \tilde{W}'^{max}_\text{DZDCSM}$ is located in the negative voltage range ($\Delta \tilde{W}'^{max-}_\text{DZDCSM}$). Similar to the FBSM, it is expected for the energy pulsation of a DZDCSM within an MMC arm to be located between $\Delta \tilde{W}'^{min}_\text{DZDCSM}$ and $\Delta \tilde{W}'^{max}_\text{DZDCSM}$.

D. Comparison with the simulation

In this section the capacitor energy pulsation calculated using the analytical approach introduced above will be compared to simulation results. Fig. 11 illustrates the calculated upper and lower limits of the energy pulsation for the FBSM (━━) and for the DZDCSM (━ ━). As already mentioned, it is expected for the simulation results of every topology to be included in between the associated limits. The results of the simulation, which is performed using 16 FBSM (■) and 8 DZDCSM (■) applying NLM and a voltage-balancing algorithm are also illustrated in Fig. 11. The simulation results are generally placed between the proposed upper and lower limits of the energy pulsation – which verifies the consideration of the proposed approach – except in one operating point at $k = 1$ for the FBSM. This deviation is due to the fact, that in the simulation the arm voltage is modulated using 16 submodules and thus contains harmonics. These harmonics could be useful in terms of reducing the energy pulsation in some operating points like this one. Injecting harmonic components to the arm voltage and its impact on reducing the energy pulsation was not included in the proposed analytical approach.

For comparison, Fig. 11 also shows the energy pulsation of the DZDCSM according to the analytical approach of [9] (green curve), where a pulse pattern for switching one DZDCSM with 150 Hz is used to calculate the energy pulsation. The black curve in Fig. 11 indicates the ideal energy pulsation, assuming that all submodules take their power share synchronously at every instant (dividing the energy pulsation of the arm by the number of installed capacitors) [9]. This could theoretically be achieved, if the submodule switching frequency were infinite. The limits for the energy pulsation calculated with the proposed method are higher than the ideal energy pulsation. This is logical, because the submodules cannot share the power of the arm equally at every moment, if the switching frequency of the submodules is limited. The energy pulsations calculated following [9] are also similar to

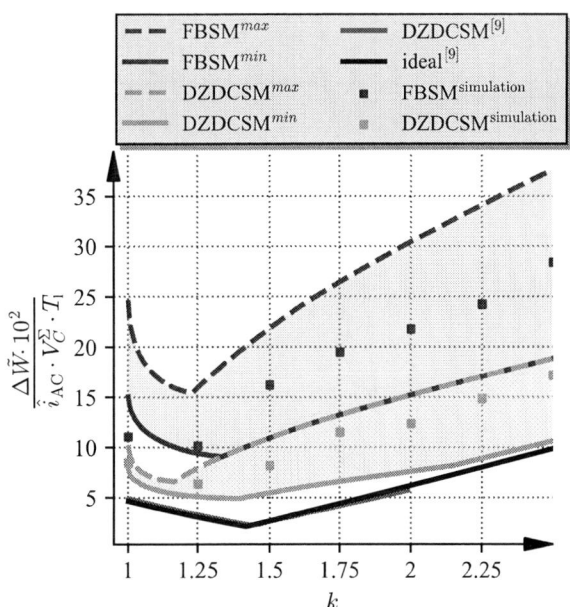

Fig. 11. Comparison of the introduced analytical approach for calculating an upper and lower limit for the energy pulsation with simulation results | $\cos(\varphi_i) = 1$ | $r = 0.9$

the ideal values, because all submodules within the arm are assumed to be switched with the same pattern.

Finally, the ratio of the energy pulsation of the DZDCSM to this of the FBSM is compared in Fig. 12a using the lower limits of the energy pulsation and in Fig. 12b using the upper limits. The ratio of the simulation results is also plotted in triangles using the same color scaling. It is evident, that the energy pulsation of the DZDCSM is generally smaller than the one of the FBSM. It can be noted, that the ratio is generally higher in the range where the modulation factor is also high. However, this is not a typical operating range for designing the MMC. The ratios calculated from the simulation are also higher than those from the analytical approach. This deviation is due to the fact, that only a limited number of submodules were used in the simulation and the impact of harmonics in the arm voltage on the energy pulsation was not considered in the analysis. Still, at a typical operating point, e.g. k = 1.25 | r = 0.9, the energy pulsation, and thus the capacitor size, could be reduced by 63% in the simulation and up to 50% according to the analytical approach compared to FBSM, if the arm consists of high number of submodules.

IV. CONCLUSION

The DZDCSM has shown an efficient voltage balancing capability of its capacitors in a wide area of MMC operating points. Still, there are operating points, where the arm voltage and current are phase shifted in such a way, that the voltage imbalance generated in state (2) cannot be passively and completely nullified in the following state (1). In order to avoid a hard paralleling of the capacitors, solutions to enable

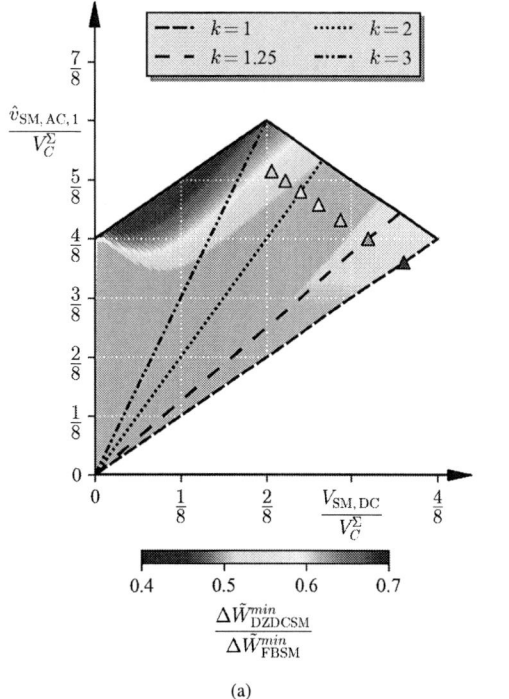
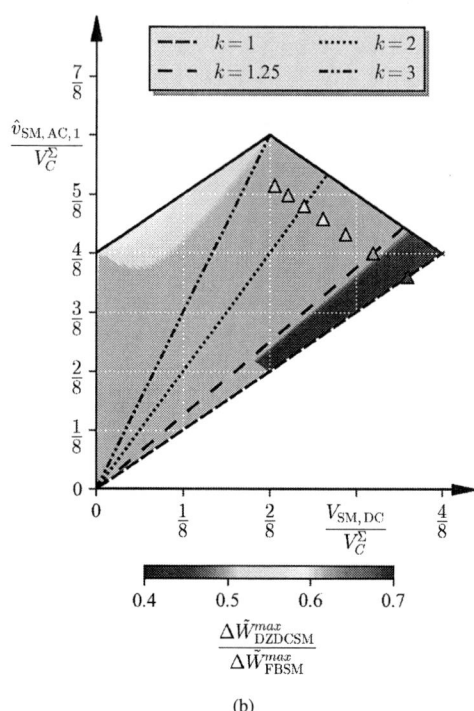

(a)

(b)

Fig. 12. Ratio of the energy pulsation of the DZDCSM and the FBSM calculated with the proposed approach | triangles denote the ratio calculated from the simulation results illustrated in Fig. 11 | $\cos(\varphi_i) = 1$

a complete balancing were introduced. Discharging resistors for dissipating the capacitor energies after a failure within the submodule can be also utilized for the voltage balancing. The introduced extended balancing algorithm is capable of achieving a passive balancing and ensuring a safe operation without the existence of equalization currents.

The capacitor energy pulsation of the DZDCSM was investigated and compared to this of the FBSM following an analytical approach, which defines an upper and lower limit for the energy pulsation. The results show that the energy pulsation of one capacitor of the DZDCSM is always smaller compared to this of the FBSM. In typical operating range of the MMC, the energy pulsation and thus the capacitor size reduction can be up to 50 %.

Acknowledgment

This research work has been carried out within the project DEFINE and is funded by dtec.bw – Digitalization and Technology Research Center of the Bundeswehr, which we gratefully acknowledge.

References

[1] R. Marquardt, "Modular multilevel converters: State of the art and future progress," *IEEE Power Electron. Mag.*, vol. 5, no. 4, 2018.

[2] K. Ilves, L. Bessegato, L. Harnefors, S. Norrga, and H.-P. Nee, "Semi-full-bridge submodule for modular multilevel converters," *Proc. 9th Int. Conf. Power Electron. ECCE Asia*, 2015.

[3] S. Heinig, K. Jacobs, K. Ilves, L. Bessegato, P. Bakas, S. Norrga, and H.-P. Nee, "Implications of capacitor voltage imbalance on the operation of the semi-full-bridge submodule," *IEEE Trans. Power Electron.*, vol. 34, no. 10, 2019.

[4] K. Jacobs, S. Heinig, D. Johannesson, S. Norrga, and H.-P. Nee, "Comparative evaluation of voltage source converters with silicon carbide semiconductor devices for high-voltage direct current transmission," *IEEE Trans. Power Electron.*, vol. 36, no. 8, 2021.

[5] C. Dahmen and R. Marquardt, "Power losses of advanced mmc submodule topologies using si- and sic-semiconductors," *Proc. 19th Eur. Conf. Power Electron. Appl.*, 2019.

[6] R. Marquardt, "Modular multilevel converter: Impact on future applications and semiconductors," *VDE, 7. ETG-Fachtagung, Bad Nauheim*, 2017.

[7] C. Dahmen, F. Kapaun, and R. Marquardt, "Analytical investigation of efficiency and operating range of different modular multilevel converters," *Proc. IEEE 12th Int. Conf. Power Electron. Drive Syst.*, 2017.

[8] C. Dahmen and R. Marquardt, "Charge balancing for advanced mmc-double-submodules with ultra-low loss," *CPE-POWERENG*, 2019.

[9] ——, "Reduced capacitor size and on-state losses in advanced mmc submodule topologies," *Proc. 20th Eur. Conf. Power Electron. Appl.*, 2020.

[10] K. Ilves, S. Norrga, L. Harnefors, and H.-P. Nee, "On energy storage requirements in modular multilevel converters," *IEEE Trans. Power Electron.*, vol. 29, no. 1, 2014.

[11] C. Zhao, Y. Li, Z. Li, P. Wang, X. Ma, and Y. Luo, "Optimized design of full-bridge modular multilevel converter with low energy storage requirements for hvdc transmission system," *IEEE Trans. Power Electron.*, vol. 33, no. 1, 2018.

[12] Q. Song, W. Yang, B. Zhao, S. Xu, H. Rao, and Z. Zhu, "Energy storage requirement reduction using negative-voltage states of a full-bridge modular multilevel converter," *IEEE Trans. Power Electron.*, vol. 34, no. 6, 2019.

[13] S. Fuchs, M. Jeong, and J. Biela, "Reducing the energy storage requirements of modular multilevel converters with optimal capacitor voltage trajectory shaping," *Proc. 20th Eur. Conf. Power Electron. Appl.*, 2020.

Saliency Extraction and Torque Sharing Estimation of Dual Motor Drive using Special Current Sensor Configuration

E. Rodriguez Montero[1], M. Vogelsberger[2], T. Wolbank[1]

TU WIEN[1]
Gusshausstrasse 27/29 E370-2, 1040
Vienna, Austria.
Tel.: +43 (1) 58801 – 370226[1]
thomas.wolbank@tuwien.ac.at[1]
ALSTOM TRANSPORT AUSTRIA GMBH[2]
Rolling Stock & Components
Bogies & DRIVES – Product Family
Hermann Gebauer Straße 5, 1220
Vienna, Austria.
Tel.: +43 (1) 25110 – 599[2]
markus.vogelsberger@alstomgroup.com[2]

Acknowledgements

The authors want to thank ALSTOM, especially Mr. H. Mannsbarth (head of Components-Bogies/Drives department in global Rolling Stock Platform/Components division) and Mr. Cedric Zanutti (head of R&D/Technology in RSC-Bogies/Drives), for their generous support, research/development funding and project supervision. Furthermore, the authors want to thank colleagues from ALSTOM group (Mr. Newesely, Mr. Cepak, Mr. Harasleben, Mr. Ganster and Mr. E. Moser), as well as Prof. Dr. H. Ertl (TU-Wien) for all their feedback and great support.

The authors are very indebted to LEM Company (especially Mr. A. Hürlimann/Chairman of Board of Directors, Dr. W. Teppan & Mr. J. Burk) for cooperation and generous support.

Keywords

«Current sensor», «Induction motors», «Parallel operation», «Sensorless control», «Signal processing»

Abstract

Dual motor drives refer to the control of two motors connected in parallel and fed by a single inverter. This type of drives is usual in some industry fields such as railway applications or conveyor drives. The control performance as well as the feasibility for saliency extraction are highly impacted by the arrangement and amount of current sensors equipped in the drive. Commonly, four current sensors in total are attached to two phases of each motor, thus permitting torque sharing calculation and individual spatial saliency extraction.

In this work, two different current sensor configurations are applied to an experimental test stand consisting of two induction motors connected in parallel and fed by a single inverter. The two current sensor configurations are investigated and discussed for encoderless saliency extraction, control performance and torque sharing estimation capability.

Introduction

Voltage step excitation methods [1-4] are a robust way to extract motor saliencies. Using the rotor information that saliencies contain, these methods permit encoderless speed control. They are based on the estimation of the transient leakage inductance, since, within it, motor saliencies are present.

According to the stator voltage equation, the resulting time current derivative relates to the transient leakage inductance after the inverter performs a voltage step. Yet, it also relates to the back EMF (electro-motive force) and stator resistance. Therefore, signal processing is required.

The vector combination of certain phase current slopes resulting from specific voltage steps leads to a saliency-offset vector, composed of an offset and a saliency-modulated term. In multiple ways, signal offset can be compensated. Preferably, offset can be accurately estimated using specific vector combinations [3]. In case it is mathematically not feasible, feedforward compensation can be done [1]. In the field of dual motors, voltage step excitation was applied in [1]. A current sensor configuration was proposed, where two current sensors were attached to the first motor (M1), and one to the second motor (M2). Using specific voltage step excitations and signal processing, a saliency vector was obtained for each motor separately. While M1 saliency vector achieved offset elimination by vector combinations of phase current slopes, M2 offset was eliminated via feedforward compensation. In addition, in [2] load sharing calculation was possible using a M2 stator current phasor estimator. According to literature [5-7], awareness of load sharing knowledge can improve performance.

In this work, two different current sensor configurations are applied to an experimental test stand consisting of two induction motors connected in parallel and fed by a single inverter. The two configurations are investigated and discussed for encoderless saliency extraction, control performance and torque sharing estimation capability. A FOC (field-oriented control) strategy is proposed, based on a single torque and flux estimator is implemented. The torque and flux estimator receives the average rotor angle and the average stator current as input.

In the first current sensor configuration (conf1), two current sensors are attached to two inverter output phases, thus measuring sum current of both machines. A third sensor is attached to an arbitrary motor phase (e.g. phase C M2). It will be shown that, using this configuration, both M1 and M2 require feedforward offset compensation in order to access motor saliencies. Besides, load sharing calculation is possible using a stator current estimator for each motor.

In the second current sensor configuration (conf2), two current sensors are attached to the A and B phases of M1. A third sensor is attached to M2 phase C. As it will be shown, conf2 allows for acquisition of an offset-less saliency vector for M1. However, for M2, a saliency-offset vector is obtained, thus requiring elimination. Besides, load sharing calculation is possible using a single stator current estimator for motor M2, at which only one current sensor is attached. Because the average stator current is not measured but estimated, performance might decrease.

Experimental results prove the functionality of both current sensor configurations for encoderless vector control and torque-sharing calculation, and discuss the benefits and drawbacks of each current sensor configuration.

Current sensor configurations

For vector control of two parallel motors, a standard practice is to measure the stator current vector of each motor by attaching two current sensors at each motor [5-10]. An advantage of this current sensor configuration is that the torque and flux of each motor can be calculated individually. Awareness of torque sharing can improve drive operation in terms of stability and performance [5-7].
As mentioned before, this paper proposes using only three current sensors, arranged in two different ways and compares them regarding saliency extraction, vector control and load-sharing estimation capability. For the analysis, three-phase balanced currents are assumed.

Both current sensor configurations are shown in Fig. 1. As observed, sum stator current vector is measured in conf1, together with phase C current of M2. In contrast, in conf2, stator current vector of M1 together with phase C current of M2.

In Fig. 1, M1 and M2 are represented by their transient leakage inductance L, back EMF E, and stator resistance R_S. ABC denote the three motor phases, and N an arbitrary phase. Within the inverter, SVPWM (space vector pulse width modulation) inverter output states are depicted.

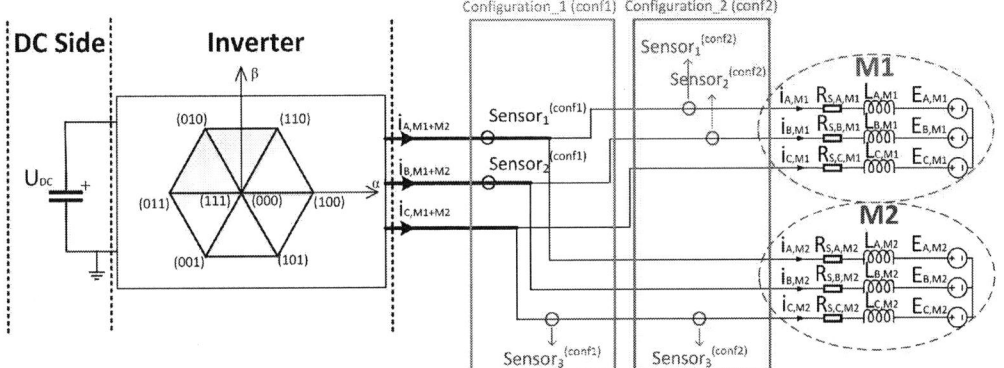

Fig. 1: Schematic representation of the DC-link, inverter and dual motors. The two current sensor configurations are shown (conf1 and conf2).

Saliency extraction using voltage step excitation

Voltage step excitation methods [1-4] rely on the extraction of the machine saliencies observed at the leakage inductance. The leakage inductance is visible in the phase current response that results from inverter active switching. According to the M1 and M2 models of Fig. 1, (1)-(3) hold true:

$$U_{AB} = R_{S,A} \cdot i_A + L_A \cdot di_A/dt + E_A - (R_{S,B} \cdot i_B + L_B \cdot di_B/dt + E_B) \tag{1}$$

$$U_{CA} = R_{S,C} \cdot i_C + L_C \cdot di_C/dt + E_C - (R_{S,A} \cdot i_A + L_A \cdot di_A/dt + E_A) \tag{2}$$

$$U_{BC} = R_{S,B} \cdot i_B + L_B \cdot di_B/dt + E_B - (R_{S,C} \cdot i_C + L_C \cdot di_C/dt + E_C) \tag{3}$$

As observed in (1-3), the phase current slopes after inverter switching transition are linked to the transient leakage inductances L_A, L_B, L_C. Moreover, U_{AB}, U_{AC}, U_{BC} vary according to Table I depending on the switching state.

Table I: Inverter output voltages for 2-level 3-phase inverter

	100	110	010	011	001	101	000&111
U_{AB}	U_{DC}	0	$-U_{DC}$	$-U_{DC}$	0	U_{DC}	0
U_{BC}	0	U_{DC}	U_{DC}	0	$-U_{DC}$	$-U_{DC}$	0
U_{CA}	$-U_{DC}$	$-U_{DC}$	0	U_{DC}	U_{DC}	0	0

Due to the presence of saliencies in the induction motor, the transient leakage can be expressed as:

$$L_A = L_0 + L_m \cdot sin(2 \cdot \pi \cdot f_{sal} \cdot t) \tag{4}$$

$$L_B = L_0 + L_m \cdot sin(2 \cdot \pi \cdot f_{sal} \cdot t + 2 \cdot \pi/3) \tag{5}$$

$$L_C = L_0 + L_m \cdot sin(2 \cdot \pi \cdot f_{sal} \cdot t + 4 \cdot \pi/3) \tag{6}$$

Where L_0 denotes symmetrical transient inductance or transient inductance offset, and $L_m \cdot sin(2 \cdot \pi \cdot f_{sal} \cdot t)$ the modulation of the transient inductance due to one saliency that revolves with f_{sal} frequency and L_m amplitude.

Saliency extraction for conf1

The extraction of the saliencies of M1 and M2 for conf1 is explained in this subsection. The excitation sequence chosen in this work is the three sets of antiparallel inverter states (100&011, 010&101 and 001&110). The resulting phase current slopes, depending on the current sensor configuration, are combined to form a saliency-offset vector and are further processed to extract rotor-slotting saliency.

For M1, the calculated current slopes are vector combined as in (7) for the 100-011 excitation sequence. Applying Kirchhoff's law, M1 phase C current can be obtained as in (8).

$$\Delta i_{M1,100}^{(conf1)} - \Delta i_{M1,011}^{(conf1)} = \left(\frac{di_{C,M1,100}^{(conf1)}}{dt} - \frac{di_{C,M1,011}^{(conf1)}}{dt}\right) \tag{7}$$

$$\frac{di_{C,M1,100}^{(conf1)}}{dt} - \frac{di_{C,M1,011}^{(conf1)}}{dt} =$$

$$= -\left(\frac{di_{A,M1+M2,100}^{(conf1)}}{dt} - \frac{di_{A,M1+M2,011}^{(conf1)}}{dt}\right) - \left(\frac{di_{B,M1+M2,100}^{(conf1)}}{dt} - \frac{di_{B,M1+M2,011}^{(conf1)}}{dt}\right) - \left(\frac{di_{C,M2,100}^{(conf1)}}{dt} - \frac{di_{C,M2,011}^{(conf1)}}{dt}\right) \tag{8}$$

For M2, (9) is applied.

$$\Delta i_{M2,100}^{(conf1)} - \Delta i_{M2,011}^{(conf1)} = \left(\frac{di_{C,M2,100}^{(conf1)}}{dt} - \frac{di_{C,M2,011}^{(conf1)}}{dt}\right) \tag{9}$$

After computing (7-9) for the other sets of excitation sequence, the saliency-offset vectors for M1 and M2 are formed using (10) and (11).

$$\bar{v}_{sal,M1}^{(conf1)} = \left(\Delta i_{M1,100}^{(conf1)} - \Delta i_{M1,011}^{(conf1)}\right) \cdot e^{i \cdot 5 \cdot \frac{\pi}{3}} + \left(\Delta i_{M1,010}^{(conf1)} - \Delta i_{M1,101}^{(conf1)}\right) \cdot e^{i \cdot 9 \cdot \frac{\pi}{3}} + \left(\Delta i_{M1,001}^{(conf1)} - \Delta i_{M1,110}^{(conf1)}\right) \cdot e^{i \cdot \frac{\pi}{3}} \tag{10}$$

$$\bar{v}_{sal,M2}^{(conf1)} = \left(\Delta i_{M2,100}^{(conf1)} - \Delta i_{M2,011}^{(conf1)}\right) \cdot e^{i \cdot 5 \cdot \frac{\pi}{3}} + \left(\Delta i_{M2,010}^{(conf1)} - \Delta i_{M2,101}^{(conf1)}\right) \cdot e^{i \cdot 9 \cdot \frac{\pi}{3}} + \left(\Delta i_{M2,001}^{(conf1)} - \Delta i_{M2,110}^{(conf1)}\right) \cdot e^{i \cdot \frac{\pi}{3}} \tag{11}$$

The result of (10) and (11) is a saliency-offset vector for both M1 and M2 as shown in (12) and (13), with $L_{0,M2} \cdot e^{i\left(\frac{\pi}{6}\right)}$ as offset and $\frac{1}{2} \cdot L_{m,M2} \cdot e^{i \cdot (2 \cdot \pi \cdot f_{sal,M2} \cdot t)}$ as saliency-modulated term.

$$\bar{v}_{sal,M1}^{(conf1)} = \left(L_{0,M2} \cdot e^{i\left(\frac{\pi}{6}\right)} - \frac{1}{2} \cdot L_{m,M2} \cdot e^{i \cdot (2 \cdot \pi \cdot f_{sal,M2} \cdot t)}\right) \cdot cst_{M1} \tag{12}$$

$$\bar{v}_{sal,M2}^{(conf1)} = \left(L_{0,M2} \cdot e^{i\left(\frac{\pi}{6}\right)} - \frac{1}{2} \cdot L_{m,M2} \cdot e^{i \cdot (2 \cdot \pi \cdot f_{sal,M2} \cdot t)}\right) \cdot cst_{M2} \tag{13}$$

Where cst is constant at constant torque and constant DC voltage.

$$cst_{M1} = U_{DC} \cdot \left(L_{0,M1}^2 - 1/4 \cdot L_{m,M1}^2\right)^{-1} \tag{14}$$

$$cst_{M2} = U_{DC} \cdot \left(L_{0,M2}^2 - 1/4 \cdot L_{m,M2}^2\right)^{-1} \tag{15}$$

Saliency extraction for conf2

The extraction of the saliencies of M1 and M2 for conf2 is explained in this subsection.

The calculated M1 current slopes, using the available current sensors, are vector combined as in (16-17) for the 100-011 excitation sequence.

$$\Delta i_{M1,100}^{(conf2)} - \Delta i_{M1,011}^{(conf2)} =$$

$$= \frac{di_{A,M1,100}^{(conf)}}{dt} - \frac{di_{A,M1,011}^{(conf2)}}{dt} + \left(\frac{di_{B,M1,100}^{(conf2)}}{dt} - \frac{di_{B,M1,011}^{(conf2)}}{dt}\right) \cdot e^{i\left(2 \cdot \frac{\pi}{3}\right)} + \left(\frac{di_{C,M1,100}^{(co\ )}}{dt} - \frac{di_{C,M1,011}^{(conf)}}{dt}\right) \cdot e^{i\left(4 \cdot \frac{\pi}{3}\right)} \tag{16}$$

$$\frac{di_{C,M1,100}^{(conf2)}}{dt} - \frac{di_{C,M1,011}^{(conf2)}}{dt} = -\left(\frac{di_{A,M1,100}^{(conf2)}}{dt} - \frac{di_{A,M1,011}^{(conf2)}}{dt}\right) - \left(\frac{di_{B,M1,100}^{(conf2)}}{dt} - \frac{di_{B,M1,011}^{(conf2)}}{dt}\right) \tag{17}$$

For M2, (18) is applied.

$$\Delta i_{M2,100}^{(conf2)} - \Delta i_{M2,011}^{(conf2)} = \left(\frac{di_{C,M2,100}^{(conf2)}}{dt} - \frac{di_{C,M2,011}^{(conf2)}}{dt}\right) \tag{18}$$

After computing (16-18) for the other sets of excitations, the saliency-offset vectors for M1 and M2 are formed using (19) and (20).

$$\bar{v}_{sal,M1}^{(conf2)} = \Delta i_{M1,100}^{(conf2)} - \Delta i_{M1,011}^{(conf2)} + \left(\Delta i_{M1,010}^{(conf2)} - \Delta i_{M1,101}^{(conf2)}\right) \cdot e^{i \cdot 2 \cdot \frac{\pi}{3}} + \left(\Delta i_{M1,001}^{(conf2)} - \Delta i_{M1,110}^{(conf2)}\right) \cdot e^{i \cdot 4 \cdot \frac{\pi}{3}} \tag{19}$$

$$\bar{v}_{sal,M2}^{(conf2)} = \left(\Delta i_{M2,100}^{(conf2)} - \Delta i_{M2,011}^{(conf2)}\right) \cdot e^{i \cdot 5 \cdot \frac{\pi}{3}} + \left(\Delta i_{M2,010}^{(conf2)} - \Delta i_{M2,101}^{(conf2)}\right) \cdot e^{i \cdot 9 \cdot \frac{\pi}{3}} + \left(\Delta i_{M2,001}^{(conf2)} - \Delta i_{M2,110}^{(conf2)}\right) \cdot e^{i \cdot \frac{\pi}{3}} \tag{20}$$

The result of (19) and (20) is a saliency vector for M1 as shown in (21) and a saliency-offset vector for M2 as shown in (22).

$$\bar{v}_{sal,M1}^{(conf2)} = -3/2 \cdot L_{m,M1} \cdot e^{i \cdot (2 \cdot \pi \cdot f_{sal,M1} \cdot t)} \cdot cst_{M1} \tag{21}$$

$$\bar{v}_{sal,M2}^{(conf2)} = \left(L_{0,M2} \cdot e^{i\left(\frac{\pi}{6}\right)} - \frac{1}{2} \cdot L_{m,M2} \cdot e^{i \cdot (2 \cdot \pi \cdot f_{sal,M2} \cdot t)}\right) \cdot cst_{M2} \tag{22}$$

Stator current vector estimation

According to literature, awareness on the torque sharing can improve dual motor drive performance [5-7]. In order to compute torque sharing, the stator current as well as rotor position of M1 and M2 must be available. Rotor angle is in this work calculated from the slotting saliency as explained in previous section. When it comes to stator current phasor, it is proposed to calculate M1 current phasor $\bar{\imath}_{S,M1}$ using Sensor1(conf2) and Sensor2(conf2), and estimate M2 current phasor $\bar{\imath}_{S,M2}$ using signal processing, summarized in (23-26).

$$\bar{\imath}_{S,M1}^{(conf2)} = i_{A,M1}^{(conf2)} + i_{B,M1}^{(conf2)} \cdot a + \left(-i_{A,M1}^{(conf2)} - i_{B,M1}^{(conf2)}\right) \cdot a^2 \tag{23}$$

Phase C currents of M1 and M2 are represented as in (24), with I_C amplitude and δ_C phase shift.

$$i_C = I_C \cdot \sin(w_e \cdot t + \delta_C) \tag{24}$$

Since both motors have the same electrical frequency, $\bar{\imath}_{S,M2}$ can be estimated using (25-26).

$$\bar{k}_{M2}^{(conf2)} = \frac{I_{C,M2} \cdot e^{j \cdot \delta_{C,M2}}}{I_{C,M1} \cdot e^{j \cdot \delta_{C,M1}}} = \frac{I_{C,M2}}{I_{C,M1}} \cdot e^{j \cdot (\delta_{C,M2} - \delta_{C,M1})} \tag{25}$$

$$\bar{\imath}_{S,M2}^{(conf2)} = \bar{\imath}_{S,M1}^{(conf2)} \cdot \bar{k}_{M2}^{(conf2)} \tag{26}$$

Where the amplitude division and phase shift difference of (25) is computed using FFT over a half electrical period.

Regarding conf1, the sum current phasor $\bar{\imath}_{S,M1+M2}^{(conf1)}$ is computable as in (27). Since the electrical frequency is equal for M1 and M2, $\bar{\imath}_{S,M1}^{(conf1)}$ and $\bar{\imath}_{S,M2}^{(conf1)}$ can be estimated using (28-31).

$$\bar{\imath}_{S,M1+M2}^{(conf1)} = i_{A,M1+M2}^{(conf1)} + i_{B,M1+M2}^{(conf1)} \cdot a + \left(-i_{A,M1+M2}^{(conf1)} - i_{B,M2}^{(conf1)}\right) \cdot a^2 \tag{27}$$

$$\bar{k}_{M1}^{(conf1)} = \frac{I_{C,M1} \cdot e^{j \cdot \delta_{C,M1}}}{I_{C,M1+M2} \cdot e^{j \cdot \delta_{C,M1+M2}}} = \frac{I_{C,M1}}{I_{C,M1+M2}} \cdot e^{j \cdot (\delta_{C,M1} - \delta_{C,M1+M2})} \tag{28}$$

$$\bar{\iota}_{S,M1}^{(conf1)} = \bar{\iota}_{S,M1+M2}^{(conf1)} \cdot \bar{k}_{M1}^{(conf1)} \tag{29}$$

$$\bar{k}_{M2}^{(conf1)} = \frac{I_{C,M2} \cdot e^{j \cdot \delta_{C,M2}}}{I_{C,M1+M2} \cdot e^{j \cdot \delta_{C,M1+M2}}} = \frac{I_{C,M2}}{I_{C,M1+M2}} \cdot e^{j \cdot (\delta_{C,M2} - \delta_{C,M1+M2})} \tag{30}$$

$$\bar{\iota}_{S,M2}^{(conf1)} = \bar{\iota}_{S,M1+M2}^{(conf1)} \cdot \bar{k}_{M2}^{(conf1)} \tag{31}$$

Where the amplitude divisions and phase shift differences of (28) and (30) are computed using FFT over a half electrical period.

Control strategy

This section shows and compares the control scheme used for conf1 and conf2. The control strategy used for both configurations in this work is based on average FOC using a single flux and torque estimator. The input to the average FOC scheme is the average stator current phasor, and the average rotor angle. Note that, for conf1, average stator current is available given the sensor arrangement, while for conf2, average stator current must be calculated using (27) and (31). The block diagram of the control scheme for conf1 and conf2 are shown in Fig. 2 and Fig. 3, respectively.

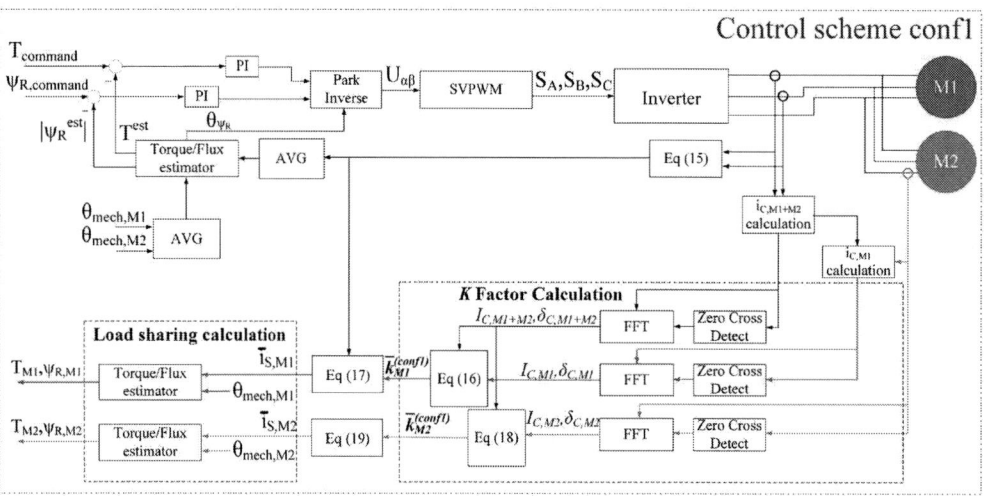

Fig. 2: control scheme for conf1.

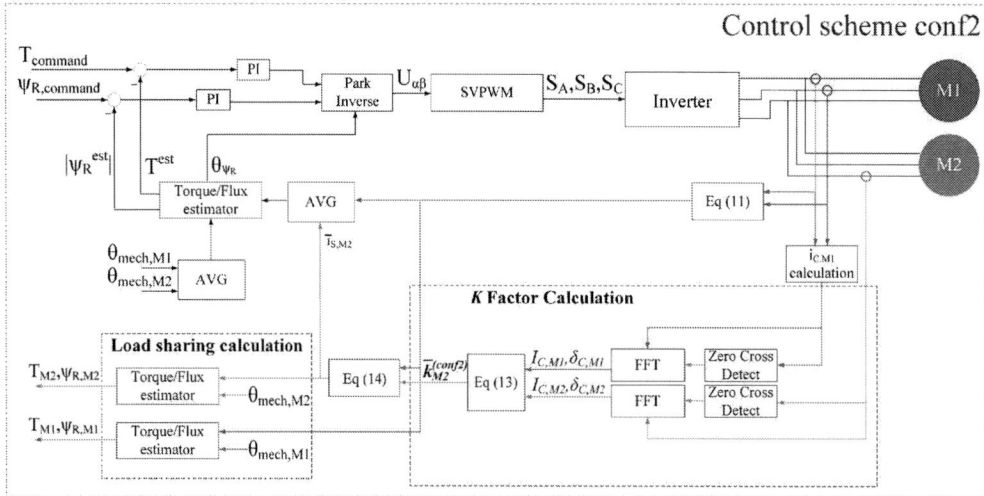

Fig. 3: control scheme for conf2.

Experimental results

The proposed conf1 and conf2 sensor arrangements are analyzed on a single-inverter dual induction motor test stand. Both motors are mechanically coupled by a toothed belt drive that ensures synchronous rotation. Both motors have 13.8kW rated power.

Torque/flux estimation for conf1 and conf2

During following measurements, average rotor flux and average torque were controlled using the single torque and flux estimator of Fig. 2 and Fig. 3. The speed of both M1 and M2 is kept constant by a coupled machine. The speed is set to 30rpm.

In order to evaluate the control performance and torque/flux estimation, average rated flux is commanded. Average torque is varied stepwise. The results are shown in Fig. 4.

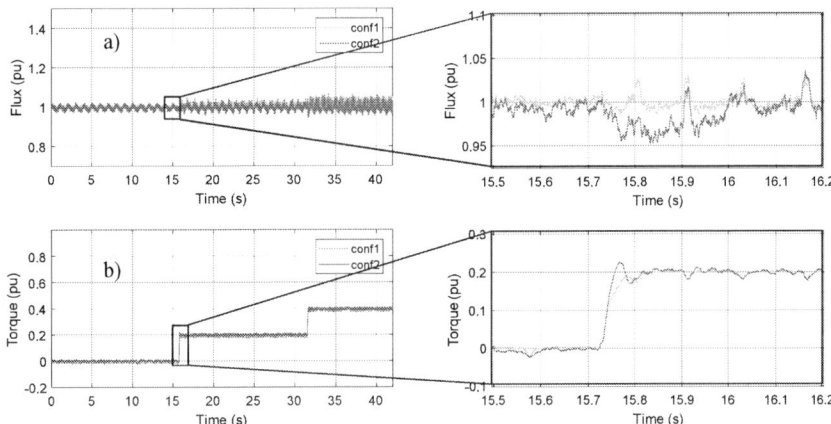

Fig. 4: conf1(green) and conf2 (magenta) flux and torque. a): average flux and zoom of it at 15.5s-16.2s. b): average torque and zoom of it at 15.5s-16.2s.

Regarding Fig. 4, it is worth noting that the torque and flux estimator in conf1 receives the stator sum current $\bar{\iota}_{S,M1+M2}^{(conf1)}$, which is actually being measured. However, conf2 needs to estimate $\bar{\iota}_{S,M2}^{(conf2)}$ in order to obtain the sum current. This explains the deviation from conf1 and conf2 torque and flux during transients (see Fig. 4). This implies that conf1 is more robust during transient operation than conf2. Yet, the transient characteristics of average flux and torque using conf2 strongly depend on the employed method to estimate $\bar{\iota}_{S,M2}^{(conf2)}$.

The M1 and M2 flux and torque are depicted in Fig. 5 using both configurations. Note that M1 conf2 torque and flux are magnitudes computed using the measured stator current vector of M1 (Fig. 5 a) and c)). For M2, both conf1 and conf2 stator currents are estimated using (26) and (31). Therefore, M2 "actual" torque and flux (additionally calculated using 2 current sensors on M2 A and B phases) are plotted together with the estimated conf1 and conf2 stator current vectors of M2.

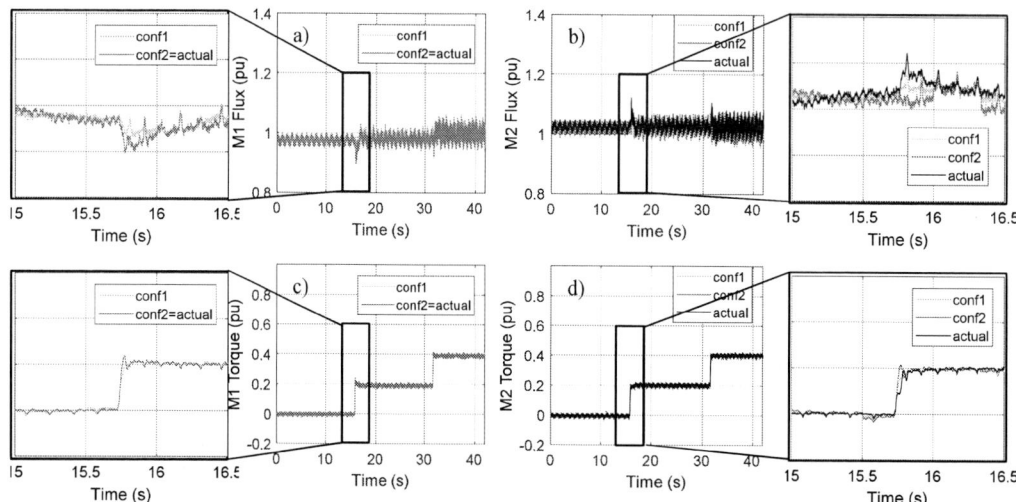

Fig. 5: conf1(green) and conf2 (magenta) current components. a): M1 flux. b): M2 flux. c): M1 torque. d): M2 torque.

As observed in Fig.5 b) and d), transient deviations in the M2 torque and flux are observed. This is due to the need for estimating the M2 stator current vector in both conf1 and conf2. Regarding M1, Fig.5 a) and c) show transient deviations in the M2 torque and flux for conf1. The M2 torque and flux estimator of conf2 uses the measured M2 stator current vector, thus representing the "actual" magnitudes.

Saliency amplitudes

For the two induction motors under test, multiple saliencies are present in the saliency vector $\bar{v}_{sal}$. In order to access rotor slotting, saturation and intermodulation (non-control saliencies) are identified using FFT. The offset present at $\bar{v}_{sal,M2}^{(conf2)}$, $\bar{v}_{sal,M1}^{(conf1)}$ and $\bar{v}_{sal,M1}^{(conf1)}$ is also identified.

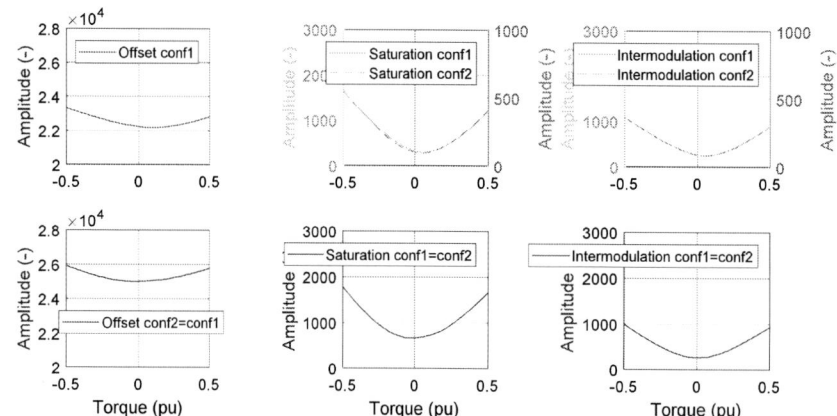

Fig. 6: Identified offset and non-control saliencies using conf1 and conf2. 1st row: M1. 2nd row: M2. 1st column: offset amplitude. 2nd column: saturation amplitude. 3rd column: intermodulation amplitude.

In order to access rotor slotting, offset and non-control saliencies are feed-forward compensated.

Encoderless operation

This subsection shows the saliency-offset vectors and encoderless operation for both motors and for both conf1 and conf2. During following measurement, the speed of both M1 and M2 is kept constant by a coupled machine at 20rpm. The dual motors are controlled at rated rotor flux and variable torque, varied from 20-50%. The resulting saliency-offset vector of each motor for each configuration is processed to deliver the slotting angle. It is achieved by means of feedforward compensating the offset (for the configurations where it is present) as well as the non-control saliencies. The results of the encoderless operation are shown in Fig. 7.

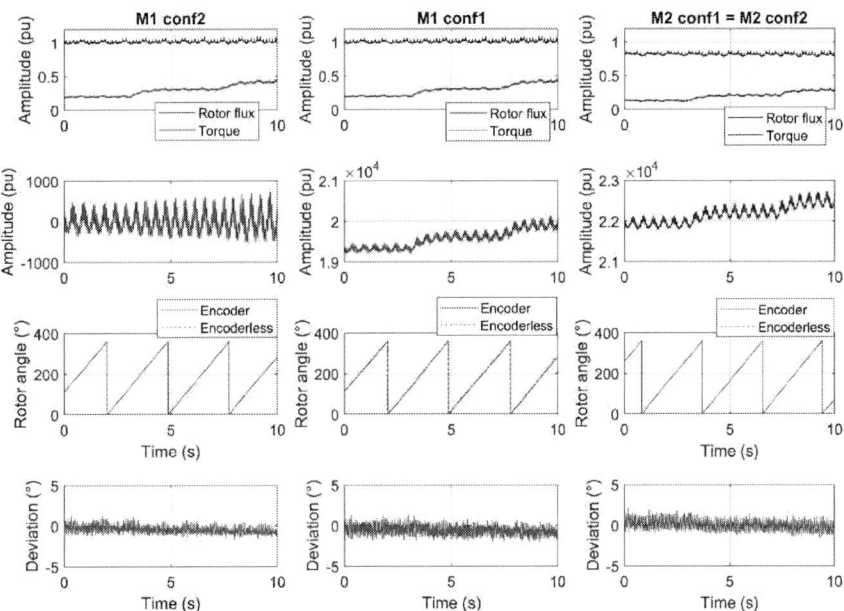

Fig. 7: Encoderless operation results for M1 conf2 (1st column), M1 conf1 (2nd column) and M2 (both conf1 and conf2, 3rd column). 1st row: rotor flux and torque. 2nd row: real part of saliency-offset vector. 3rd row: encoderless and shaft encoder angle. 4th row: deviation between encoderless and encoder angles.

As observed in Fig. 7, the saliency-offset vector of M1 for conf2 (2nd row 1st column) has zero offset due to the mathematical compensation as it was derived in (21). However, the saliency-offset vector of M1 for conf1 as well as that from M2 (for both conf1 and conf2) presents an offset. Both offset and non-control saliencies are feedforward compensated, thus only rotor slotting remains in the saliency vector. The angle of the slotting saliency is transformed into a rotor angle using the rotor slot number. The resulting encoderless angles are shown in Fig. 7, 3rd row. As the 4th row shows, very accurate rotor angles are obtained, having a maximum deviation of ±1.5° compared to the rotor angle measured by a shaft encoder. The angle deviation of M1 when conf2 is used is slightly smaller than when using conf1. This is due to the increased amount of current sensors in M1 for conf2.

Conclusion

Single-inverter dual induction motor drives commonly rely on four current sensors in total that are attached to two phases of each motor, thus permitting torque sharing calculation and individual spatial saliency extraction.

This work has investigated the use of two configurations (conf1, conf2) with reduced number of current sensors applied to a drive with two induction motors (M1, M2) connected in parallel and fed by a single inverter. Both configurations consist of only three current sensors in total. In conf1, two current sensors are attached to two inverter output phases, and the remaining current sensor is attached to one phase of M2. In conf2, two current sensors are attached to two phases of M1, and the remaining current sensor is attached to one phase of M2. In order to extract motor saliencies, voltage step excitation is used. Besides, the control strategy is field-oriented control, where a single torque and flux estimator is implemented with the average stator current and average rotor angle as input.

Both current sensor configurations have been discussed and compared in terms of torque sharing capability, control performance and spatial saliency extraction. An experimental test stand equipped with two similar induction motors connected in parallel has been presented. The experimental measurements proved that a saliency-offset vector can be obtained when a single current sensor is attached to the motor, such as for M1 (conf1) and M2 (conf1 and conf2). To access saliencies, the offset was feedforward compensated. If two sensors are attached (M1, conf2), the offset can be mathematically eliminated and the saliency signal to noise ratio is increased, which leads to the acquisition of a slightly better rotor angle. Regarding torque sharing calculation, transient deviations in the rotor flux and torque were measured at M1 (conf1) and M2 (conf1 and conf2). This is because the torque and flux estimator employed for field-oriented control received an estimated averaged stator current phasor. In this sense, conf1 is more prone to transient deviations in torque sharing calculation. Experimental results on average torque and flux further suggested that conf1 has increased stability and robustness since the sum stator current is directly measured.

References

[1] E. Rodriguez Montero, M. Vogelsberger, M. Bazant, T. Wolbank, "Saliency-based Speed Sensorless Control of Single-Inverter Dual Induction Machines using Reduced Amount of Current Sensors", *IEEE Energy Conversion Congress and Exposition (ECCE)*, pp. 5098-5103, 2020.

[2] E. Rodriguez Montero, M. Vogelsberger, M. Bazant, T. Wolbank, "A New Approach to Detect Load Sharing of Dual-Motors Driven and Controlled by a Single Converter using Only Three Current Sensors", *IEEE International Conference on Electrical Machines (ICEM)*, pp.1040-1045, 2020.

[3] Y. Sua, M. Sumner, G. Asher, Q. Gao, K, Saleh, "Improved sensorless control of a permanent magnet machine using fundamental pulse width modulation excitation." *IET Electric Power Application*, Vol. 5, pp. 359-370, Apr. 2011.

[4] M. X. Bui, D. Guan, D. Xiao, and M. Faz Rahman, "A modified sensorless control scheme for interior permanent magnet synchronous motor over zero to rated speed range using current derivative measurements," *IEEE Trans. Ind. Electron., vol. 66, no. 1*, pp. 102–113, Jan. 2019.

[5] L. Guo, Z. Yang, F. Lin, and X. Tu, "Weighted torque current control for high speed train with dual induction motors fed by a single inverter," in *Annual Conference of the IEEE Industrial Electronics Society*, pp. 2723–2728, Nov 2015.

[6] F. Xu, L. Shi, "Unbalanced thrust control of multiple induction motors for traction system", in *IEEE Industrial Electronics and Applications Conference*, pp. 2752-2757, June 2011.

[7] F. Xu, L. Shi, and Y. Li, "The weighted vector control of speed-irrelevant dual induction motors fed by the single inverter," *IEEE Trans. Power Electron., vol. 28, no. 12*, pp. 5665–5672, Dec. 2013.

[8] T. I. Yeam, D. C. Lee, "Design of Sliding-Mode Speed Controller With Active Damping Control for Single-Inverter Dual-PMSM Drive Systems", in *IEEE Trans. on Power. Elect.*, vol. 36, iss. 5, Mai 2021.

[9] Š. Janouš, J. Talla, V. Šmídl, Z. Peroutka, "Constrained LQR Control of Dual Induction Motor Single Inverter Drive", in *IEEE Trans. on Ind. Elect.*, vol. 68, no. 7, pp. 5548-5558, July 2021.

[10] Q. Geng, Z. Li, M. Zhang, Z. Zhou, H. Wang, T. Shi, "Sensorless Control Method for Dual Permanent Magnet Synchronous Motors Driven by Five-Leg Voltage Source Inverter", in *IEEE Journal of Emerging and Selected Topics in Power Electronics*, June 2021.

Soft-switching Converter for Inductive Power Transfer System with Double-sided LCC Resonant Network

Ryohei Okada[1*], Ryosuke Ota[*], Nobukazu Hoshi[*]
[*]Tokyo University of Science
2641 Yamazaki Noda-city
Chiba, Japan
Email: [1]r_okada@alumni.tus.ac.jp

Keywords

≪Wireless power transmission≫, ≪Electric vehicle≫, ≪Soft switching≫, ≪ZVS converters≫, ≪Bi-directional≫

Abstract

In a general soft switching method for inductive power transfer (IPT) systems, a reactive current is intentionally generated in the resonant network and then utilized to achieve soft switching. However, the whole-system efficiency worsens because the reactive current circulates in the system. Against this problem, the soft-switching active bridge (SAB) converter is proposed. The SAB converter comprises the full-bridge active converter and the LC circuit, which can generate the reactive current in place of the resonant network. As a result, the reactive current circulating in the system is reduced, and high efficiency can be achieved. This paper shows how to apply the SAB converter to IPT systems with the double-sided LCC topology. The theoretical analysis clarifies the effect of the design of the LC circuit on the whole-system efficiency, and the design guideline for the LC circuit is discussed. In the experiment, the validities of the theoretical analysis result and the design guideline are shown. The SAB-IPT system can get higher efficiency over wide operation range as well as the theoretical analysis result. Especially, the efficiency at 2.5 kW-output is improved by 0.69 points, and the efficiency at 1.5 kW-output is improved by 0.88 points.

Introduction

Recently, as a means to dramatically extend the driving range of electric vehicles (EVs), inductive power transfer (IPT) technology is expected [1–4]. Currently, the IPT system is required to increase power for rapid charging. However, we have challenges regarding the increase in power loss in the system and switching noise from the semiconductor devices.

Soft switching is a technique to reduce the electromagnetic noise and the loss caused by switching. In a general soft switching method for IPT systems, a reactive current is intentionally generated in the resonant network and then utilized to achieve soft switching [5–7]. However, it has been indicated that the whole-system efficiency is significantly worsened in some operation points [5–9]. The reason is that the reactive current generated in the resonant network circulates in the system. As a result, its conduction loss significantly influences efficiency, especially in the operation points where the reactive current is large. In other words, there is a trade-off relationship between switching-noise reduction and efficiency improvement. To solve this problem, a method where the reactive current is alternatively generated by the LC circuit added to the full-bridge active (FBA) converter has been proposed [10]. The reactive current generated by the LC circuit does not circulate in the whole system. In addition, the resonant network, which has large ESRs, does not need to generate a large reactive current due to the LC circuit. As a result, the proposed method can largely reduce the total conduction loss in the whole system.

In the literature [10], the effectiveness of the FBA converter with the LC circuit, which is called the soft switching active bridge (SAB) converter, is clarified for IPT systems with the series/series (SS) compensation resonant network. However, there are various compensation topologies for the resonant network. These input-impedance characteristics and behaviors of the coupling change between the transmission coils are different. Especially in charging in-motion, such as dynamic IPT systems for EVs, the double-sided LCC topology shown in Fig. 1 could be a better choice than the SS topology [5, 11, 12]. Therefore, it is necessary to clarify how to apply the SAB converter to the double-sided LCC resonant network and its effectiveness.

Against this background, this paper clarifies the operation method, the design guideline, and the effectiveness of the SAB-IPT system shown in Fig. 1. This paper discusses these with the theoretical analysis and the experiment compared with the conventional FBA-IPT system. In this paper, the IPT system comprising the FBA converters is called "FBA-IPT system", and the IPT system comprising the SAB converters is called "SAB-IPT system".

System Overview

System Configuration

The system shown in Fig. 1 comprises a double-sided LCC resonant network and the SAB converters on the primary/secondary sides. In a SAB converter, one LC circuit is connected to one-side leg of the general FBA converter. Moreover, the LC circuits comprise the inductors L_{LCp}, L_{LCs} to generate the reactive current and the capacitors C_{LCp}, C_{LCs}, which absorb the DC components of the voltages on S_{p4}, S_{s2}. The DC-bus voltages on both SAB converters are constant, and the AC-side voltages on the SAB converters are regulated with phase-shift control between the legs. Then, the duty ratio of each switch is fixed to 0.5, and the switching frequency is fixed to 85 kHz. In addition, the amount of the reactive current in the resonant network can be regulated with phase-shift control between the primary and secondary sides.

Power Transfer in SAB-IPT System with Double-sided LCC Resonant Network

Fig. 2 shows the operation waveforms in the primary-side SAB converter. Since the operation waveforms in the secondary side can be drawn as well as the primary side, these descriptions are omitted. From Fig. 2, the phasor notation for fundamental components $\dot{V}_{\mathrm{p1}}$, $\dot{V}_{\mathrm{s1}}$ [V] of the

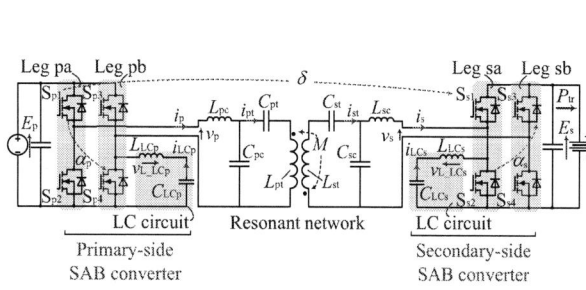

Fig. 1: SAB-IPT system with the double-sided LCC resonant network.

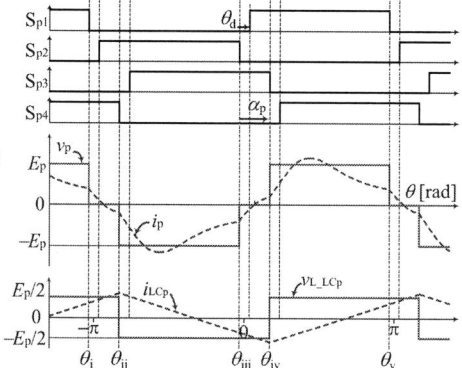

Fig. 2: Operation waveforms in the primary-side SAB converter.

output/input voltages on the primary/secondary-sides SAB converters are expressed as

$$\dot{V}_{p1} = \frac{2\sqrt{2}E_p}{\pi}\cos\frac{\alpha_p}{2}\varepsilon^{j(-\alpha_p/2+\theta_d/2)}, \quad \dot{V}_{s1} = \frac{2\sqrt{2}E_s}{\pi}\cos\frac{\alpha_s}{2}\varepsilon^{j(\delta-\alpha_s/2+\theta_d/2)}, \tag{1}$$

where E_p [V] represents the primary-side DC voltage, and E_s [V] represents the battery voltage. α_p and α_s [rad] represent the phase-shift angles between the legs in the primary/secondary-side converters, respectively. Additionally, δ [rad] denotes the phase-shift angle between the primary/secondary-side SAB converters, and θ_d [rad] represents the dead-time converted to the angle. For simplicity, this chapter does not consider the harmonic components of each voltage. When ESRs of each element are ignored and the switching angular frequency of both converters matches the resonant angular frequency ω in the resonant network, the phasor notation for fundamental components $\dot{I}_{p1}$ and $\dot{I}_{s1}$ [A] of the output/input currents on the primary/secondary SAB converters are respectively expressed as [5, 11]

$$\dot{I}_{p1} = \frac{M\dot{V}_{s1}}{j\omega L_{pc}L_{sc}}, \quad \dot{I}_{s1} = \frac{M\dot{V}_{p1}}{j\omega L_{pc}L_{sc}}, \tag{2}$$

where L_{pc} and L_{sc} [H] represent the compensation inductors, and M [H] represents the mutual inductance between the transmission coils L_{pt} and L_{st} [H]. Moreover, the phasor notation for fundamental components $\dot{I}_{pt1}$ and $\dot{I}_{st1}$ [A] of the currents flowing in the transmission coils L_{pt} and L_{st} can be respectively expressed as

$$\dot{I}_{pt1} = \frac{\dot{V}_{p1}}{j\omega L_{pc}}, \quad \dot{I}_{st1} = \frac{\dot{V}_{s1}}{j\omega L_{sc}}. \tag{3}$$

Then, the transmission power P_{tr} [W] of the resonant network is given by [11]

$$P_{tr} = \frac{M|\dot{V}_{p1}||\dot{V}_{s1}|}{\omega L_{pc}L_{sc}}\cos\phi \quad \left(\phi = \frac{\pi}{2} - \theta_{vps}\right), \tag{4}$$

where $\cos\phi$ represents the displacement factor in the resonant network, and ϕ can be controlled with the phase-shift angle θ_{vps} [rad] between $\dot{V}_{p1}$ and $\dot{V}_{s1}$. In this paper, the resonant-network efficiency η_{rn} [%] is defined as [5]

$$\eta_{rn} = \frac{100P_{tr}}{P_{tr} + r_{Lpc}|\dot{I}_{p1}|^2 + r_{pt}|\dot{I}_{pt1}|^2 + r_{st}|\dot{I}_{st1}|^2 + r_{Lsc}|\dot{I}_{s1}|^2}, \tag{5}$$

where r_{Lpc} and r_{Lsc} [Ω] represent the ESRs in L_{pc} and L_{sc}, respectively. In addition, r_{pt}, r_{st} [Ω] represent the total ESRs in L_{pt} and C_{pt}, L_{st} and C_{st}, respectively. The equation (5) does not consider the conduction losses in the compensation capacitors C_{pc} and C_{sc} [F].

Here, by substituting the equations (2)–(4) into (5), η_{rn} can be expressed as

$$\eta_{rn} = \frac{100}{1 + \dfrac{\left|\frac{\dot{V}_{s1}}{\dot{V}_{p1}}\right|\left(r_{Lpc}+r_{st}\frac{L_{pc}^2}{M^2}\right) + \left|\frac{\dot{V}_{p1}}{\dot{V}_{s1}}\right|\left(r_{Lsc}+r_{pt}\frac{L_{sc}^2}{M^2}\right)}{\frac{M}{\omega L_{pc}L_{sc}}\cos\phi}}. \tag{6}$$

From the denominator of (6), the requirements for the maximum efficiency of the resonant network can be derived by using the relationship of the arithmetic-geometric mean [6, 7]. As a result, these requirements can be expressed as

$$\begin{cases} \dfrac{|\dot{V}_{s1}|^2}{|\dot{V}_{p1}|^2} = \dfrac{|\dot{I}_{p1}|^2}{|\dot{I}_{s1}|^2} = \dfrac{r_{pt}L_{sc}^2 + r_{Lsc}M^2}{r_{st}L_{pc}^2 + r_{Lpc}M^2}, & (7) \\ \phi = 0. & (8) \end{cases}$$

Moreover, (7) can be transformed to

$$r_{\text{Lsc}}|\dot{I}_{\text{s1}}|^2 + r_{\text{pt}}|\dot{I}_{\text{pt1}}|^2 = r_{\text{Lpc}}|\dot{I}_{\text{p1}}|^2 + r_{\text{st}}|\dot{I}_{\text{st1}}|^2, \tag{9}$$

where the left hand side represents the losses regarding $\dot{V}_{\text{p1}}$ and the right hand side represents the losses regarding $\dot{V}_{\text{s1}}$. In other words, the requirement means that the power losses regarding $\dot{V}_{\text{p1}}$ and $\dot{V}_{\text{s1}}$ need to be balanced. On the other hand, (8) means the requirement where the displacement factor of the resonant network becomes a unity. In other words, the fundamental component of the reactive current in the resonant network is minimized with (8). Thus, when (7) and (8) are satisfied, the resonant network gets high efficiency.

Operation Modes of SAB Converter

This chapter explains the overview of the operation modes of the SAB converter based on the literature [10]. This chapter omits the detailed operation modes because it is explained in [10]. Fig. 3 shows the operation modes of the primary-side SAB converter whose waveforms correspond to Fig. 2. The dead-time period is omitted for simplicity. The operation of the secondary-side SAB converter is not explained because we can explain it as well as that of the primary side.

The operation of the SAB converter can be explained by superimposing the operations of the LC circuit and the resonant network. Therefore, each operation is explained.

Operation Mode of LC circuit

The operation mode of the LC circuit depends on the operation of the Leg pb in Fig. 3. When the duty ratios of the switches are set to 0.5, the waveforms of the voltage v_{LCp} [V] and the current i_{LCp} [A] are represented as shown in Fig. 2. In this case, the currents i_{LCp}, i_{LCs} [A] flowing in the LC circuits are expressed as

$$i_{\text{LCp}} \simeq \sum_{n=1}^{m} \frac{1}{n^2 \pi \omega L_{\text{LCp}}} 2E_{\text{p}} \sin \frac{n\pi}{2} \sin\left\{ n\left(\theta - \alpha_{\text{p}} + \frac{\theta_{\text{d}}}{2} - \frac{\pi}{2} \right) \right\}, \tag{10}$$

$$i_{\text{LCs}} \simeq \sum_{n=1}^{m} \frac{1}{n^2 \pi \omega L_{\text{LCs}}} 2E_{\text{s}} \sin \frac{n\pi}{2} \sin\left\{ n\left(\theta - \frac{\alpha_{\text{p}} - \alpha_{\text{s}} - 2\theta_{\text{vps}}}{2} + \frac{\theta_{\text{d}}}{2} + \frac{\pi}{2} \right) \right\}. \tag{11}$$

In this paper, $m = 5$ is assumed.

Operation Mode of Resonant Network

The resonant network is driven by the four operation modes shown in Fig. 3 as well as when the general FBA-IPT system is connected to it. In this paper, the only phase-shift angle between the legs is controlled, and the duty ratios of all switches are set to 0.5 constantly.

The resonant-network currents i_{p} and i_{s}, which consider harmonics, are derived by

$$i_{\text{p}} \simeq \sum_{n=1}^{l} \sqrt{2} I_{\text{pn\_rms}} \sin\left(n\theta + \theta_{\text{ipn}} \right), \tag{12}$$

$$i_{\text{s}} \simeq \sum_{n=1}^{l} \sqrt{2} I_{\text{sn\_rms}} \sin\left(n\theta + \theta_{\text{isn}} \right), \tag{13}$$

where the RMS values $I_{\text{pn\_rms}}$ and $I_{\text{sn\_rms}}$ [A] of harmonic components of these currents are shown in Table I. In addition, θ_{ipn} and θ_{isn} [rad] represent the phase angles of $\dot{I}_{\text{pn}}$ and $\dot{I}_{\text{sn}}$. In this paper, $l = 7$ is assumed.

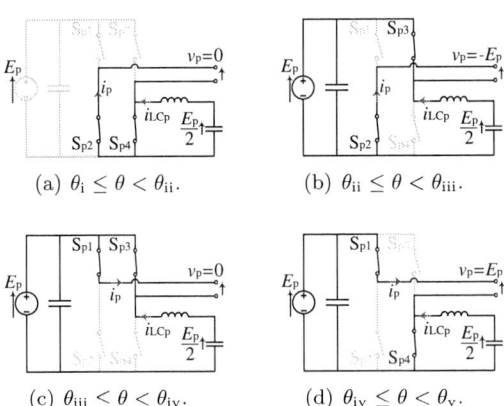

(a) $\theta_{\mathrm{i}} \leq \theta < \theta_{\mathrm{ii}}$.   (b) $\theta_{\mathrm{ii}} \leq \theta < \theta_{\mathrm{iii}}$.

(c) $\theta_{\mathrm{iii}} \leq \theta < \theta_{\mathrm{iv}}$.   (d) $\theta_{\mathrm{iv}} \leq \theta < \theta_{\mathrm{v}}$.

Fig. 3: Operation modes of the primary-side SAB converter.

Table I: Detail of variables related to $I_{\mathrm{pn\_rms}}$ and $I_{\mathrm{sn\_rms}}$

$I_{\mathrm{pn\_rms}}$	$\left\| \dfrac{(\dot{x}_{\mathrm{b}}\dot{x}_{\mathrm{c}}\dot{x}_{\mathrm{d}} - \dot{x}_{\mathrm{b}}\dot{x}_{\mathrm{g}}^2 - \dot{x}_{\mathrm{d}}\dot{x}_{\mathrm{f}}^2)\dot{V}_{\mathrm{pn}} + \dot{x}_{\mathrm{e}}\dot{x}_{\mathrm{f}}\dot{x}_{\mathrm{g}}\dot{V}_{\mathrm{sn}}}{\dot{x}_{\mathrm{a}}(\dot{x}_{\mathrm{b}}\dot{x}_{\mathrm{c}}\dot{x}_{\mathrm{d}} - \dot{x}_{\mathrm{b}}\dot{x}_{\mathrm{g}}^2 - \dot{x}_{\mathrm{d}}\dot{x}_{\mathrm{f}}^2) - \dot{x}_{\mathrm{e}}^2(\dot{x}_{\mathrm{c}}\dot{x}_{\mathrm{d}} - \dot{x}_{\mathrm{g}}^2)} \right\|$
$I_{\mathrm{sn\_rms}}$	$\left\| \dfrac{-\dot{x}_{\mathrm{e}}\dot{x}_{\mathrm{f}}\dot{x}_{\mathrm{g}}\dot{V}_{\mathrm{pn}} + (\dot{x}_{\mathrm{c}}\dot{x}_{\mathrm{e}}^2 + \dot{x}_{\mathrm{a}}\dot{x}_{\mathrm{f}}^2 - \dot{x}_{\mathrm{a}}\dot{x}_{\mathrm{b}}\dot{x}_{\mathrm{c}})\dot{V}_{\mathrm{sn}}}{\dot{x}_{\mathrm{a}}(\dot{x}_{\mathrm{b}}\dot{x}_{\mathrm{c}}\dot{x}_{\mathrm{d}} - \dot{x}_{\mathrm{b}}\dot{x}_{\mathrm{g}}^2 - \dot{x}_{\mathrm{d}}\dot{x}_{\mathrm{f}}^2) - \dot{x}_{\mathrm{e}}^2(\dot{x}_{\mathrm{c}}\dot{x}_{\mathrm{d}} - \dot{x}_{\mathrm{g}}^2)} \right\|$
$\dot{V}_{\mathrm{pn}}$	$\dfrac{2\sqrt{2}E_{\mathrm{p}}}{n\pi}\cos\dfrac{n\alpha_{\mathrm{p}}}{2}\varepsilon^{jn\left(-\frac{\alpha_{\mathrm{p}}}{2}+\frac{\theta_{\mathrm{d}}}{2}\right)}$
$\dot{V}_{\mathrm{sn}}$	$\dfrac{2\sqrt{2}E_{\mathrm{s}}}{n\pi}\cos\dfrac{n\alpha_{\mathrm{s}}}{2}\varepsilon^{jn\left(\delta-\frac{\alpha_{\mathrm{s}}}{2}+\frac{\theta_{\mathrm{d}}}{2}\right)}$
$\dot{x}_{\mathrm{a}}$	$jn\omega L_{\mathrm{pc}} + \dfrac{1}{jn\omega C_{\mathrm{pc}}}$
$\dot{x}_{\mathrm{b}}$	$jn\omega L_{\mathrm{pt}} + \dfrac{1}{jn\omega C_{\mathrm{pt}}} + \dfrac{1}{jn\omega C_{\mathrm{pc}}}$
$\dot{x}_{\mathrm{c}}$	$jn\omega L_{\mathrm{st}} + \dfrac{1}{jn\omega C_{\mathrm{st}}} + \dfrac{1}{jn\omega C_{\mathrm{sc}}}$
$\dot{x}_{\mathrm{d}}$	$jn\omega L_{\mathrm{sc}} + \dfrac{1}{jn\omega C_{\mathrm{sc}}}$
$\dot{x}_{\mathrm{e}}$	$\dfrac{-1}{jn\omega C_{\mathrm{pc}}}$
$\dot{x}_{\mathrm{f}}$	$-jn\omega M$
$\dot{x}_{\mathrm{g}}$	$\dfrac{-1}{jn\omega C_{\mathrm{sc}}}$

Soft-switching Requirement

This chapter describes the soft-switching requirement for SAB converter and the method to achieve soft switching. The type of soft switching is divided into soft-switching turn-off and -on. However, the detail of the requirement for the soft switching turn-off is not explained. This reason is that the parasitic D–S capacitance works so as to prevent the rapid voltage rising when the switching device turns off [13]. Therefore, this paper assumes that the soft-switching turn-off is always achieved.

On the other hand, a switch can turn on with soft switching by discharging the charge in the D–S parasitic capacitor during the dead-time period before the turn-on. Then, the requirements for soft switching in the primary-side SAB converter are shown in the following.

$$\begin{cases} i_{\mathrm{p}}\left(-\theta_{\mathrm{d}}/2\right) \leq 0 \\ i_{\mathrm{p}}\left(\theta_{\mathrm{d}}/2\right) \leq 0, \end{cases} \quad \begin{cases} i_{\mathrm{p}}\left(\alpha_{\mathrm{p}} - \theta_{\mathrm{d}}/2\right) + i_{\mathrm{LCp}}\left(\alpha_{\mathrm{p}} - \theta_{\mathrm{d}}/2\right) \leq 0 \\ i_{\mathrm{p}}\left(\alpha_{\mathrm{p}} + \theta_{\mathrm{d}}/2\right) + i_{\mathrm{LCp}}\left(\alpha_{\mathrm{p}} + \theta_{\mathrm{d}}/2\right) \leq 0, \end{cases} \tag{14}$$

where the left-side condition represents the soft-switching requirement for Leg pa, and the right-side condition represents the soft-switching requirement for Leg pb. In addition, the requirements for soft switching in the secondary-side SAB converter are shown in the following.

$$\begin{cases} i_{\mathrm{s}}\left(-\delta - \theta_{\mathrm{d}}/2\right) + i_{\mathrm{LCs}}\left(-\delta - \theta_{\mathrm{d}}/2\right) \geq 0 \\ i_{\mathrm{s}}\left(-\delta + \theta_{\mathrm{d}}/2\right) + i_{\mathrm{LCs}}\left(-\delta + \theta_{\mathrm{d}}/2\right) \geq 0, \end{cases} \quad \begin{cases} i_{\mathrm{s}}\left(-\delta + \alpha_{\mathrm{s}} - \theta_{\mathrm{d}}/2\right) \geq 0 \\ i_{\mathrm{s}}\left(-\delta + \alpha_{\mathrm{s}} + \theta_{\mathrm{d}}/2\right) \geq 0, \end{cases} \tag{15}$$

where the left-side condition represents the soft-switching requirement for Leg sa, and the right-side condition represents the soft-switching requirement for Leg sb. In the requirement for the legs, the term regarding the currents i_{LCp} and i_{LCs} in the LC circuits is contained. Thus, the SAB converter can also utilize the LC circuit current for soft switching. Therefore, the degree of freedom of i_{p} and i_{s} increases while achieving soft switching, and the resonant network can be driven with a near unity displacement factor. As a result, the high efficient operation area can be expanded.

However, the conduction loss by i_{LCp} and i_{LCs} cannot be ignored when i_{LCp} and i_{LCs} are large. Therefore, we need to design the LC circuit considering the balance between the degree of freedom and the conduction losses. In the next chapter, the effect of the LC circuit on the whole-system efficiency is analyzed. In addition, a design guideline for the LC circuit is discussed.

Efficiency Analysis of SAB-IPT System with Double-sided LCC Resonant Network

In this chapter, the effect of the LC circuit design on the whole-system efficiency is clarified based on the theoretical analysis. In this analysis, the circuit parameters shown in Table II are used. This paper analyzes under the condition $L_{\text{LCp}} = L_{\text{LCs}} = L_{\text{LC}}$ for simplicity.

Definition of Whole System Efficiency

Here, the whole-system efficiency η [%] in the soft-switching operation is defined as

$$\eta = \frac{100 P_{\text{tr}}}{P_{\text{tr}} + P_{\text{rn}} + P_{\text{sw\_con}} + P_{\text{sw\_turn off}} + P_{\text{LC}}}, \tag{16}$$

where P_{rn} [W] represents the total conduction loss in the resonant network, and P_{LC} [W] represents the total conduction loss of the LC circuits. In addition, $P_{\text{sw\_con}}$ and $P_{\text{sw\_turn off}}$ [W] represent the total conduction loss and the total turn-off loss in all switches, respectively. These losses are calculated based on the calculation method shown in [10].

Analysis of Relationship between L_{LC} and Whole-System Efficiency

Fig. 4 shows the theoretical maximum-efficiency characteristics of the whole system at $P_{\text{tr}} = 2.5$ kW when L_{LC} is varied so as to satisfy the soft switching requirement. The characteristics of $\eta_{\text{max}}(L_{\text{LC}})$ are exhaustively explored from sets of α_{p}, α_{s} and ϕ with satisfying the following equation.

$$\cos\frac{\alpha_{\text{p}}}{2}\cos\frac{\alpha_{\text{s}}}{2}\cos\phi = \frac{\pi^2 \omega L_{\text{pc}} L_{\text{sc}}}{8 M E_{\text{p}} E_{\text{s}}} P_{\text{tr}}{}^*, \tag{17}$$

where P_{tr}^* represents the rated power. In Fig. 4, when $L_{\text{LC}} > L_{\text{LC\_max}}$, i_{LCp} and i_{LCs} are small. In other words, in order to satisfy (14) and (15), we need to use the help of the reactive current from the resonant network. However, when $L_{\text{LC}} \leq L_{\text{LC\_max}}$, i_{LCp} and i_{LCs} are large. As a result, (14) and (15) are satisfied without the reactive current of the resonant network. As a result, the high-efficiency condition (8) of the resonant network can be satisfied. In addition, the whole-system efficiency is maximized when $L_{\text{LC}} = L_{\text{LC\_max}}$, i_{LCp} and i_{LCs} are minimized. However, it is confirmed that $L_{\text{LC\_max}}$ varies depending on the coupling coefficient from Fig. 4.

Table II: Circuit parameters in the experimental system

Coupling coefficient k between L_{pt} and L_{st}	0.2, 0.3
Primary/Secondary-sides DC voltages E_{p}, E_{s}	240 V
Switching frequency in each converter f	85 kHz
Transmission coils L_{pt}, L_{st}	62.1, 69.0 µH
Compensation inductor L_{pc}, L_{sc}	17.1, 22.9 µH
Inductors in the LC circuits L_{LCp}, L_{LCs}	20.0, 34.0 µH
Switching devices	*ROHM* BSM080D12P2C008

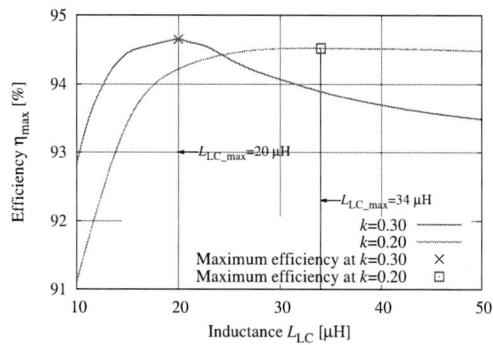

Fig. 4: Theoretical maximum-efficiency characteristics of the whole system at $P_{\text{tr}} = 2.5$ kW when L_{LC} is varied so as to satisfy the soft-switching requirement.

This is due to the following reasons. From (4), larger coupling coefficient k results in larger transmission power P_{tr}. Then, in order to keep P_{tr} constant, v_{p} and v_{s} need to be controlled to smaller values by increasing α_{p} and α_{s}. When α_{p} and α_{s} are large, the current values of i_{p} and i_{s} at switching are also large in the one-side leg. In other words, a larger current is needed for soft switching in this case. From the above, it can be said the larger k brings the smaller $L_{\text{LC\_max}}$.

From the above characteristics, it is preferable to make L_{LC} small for a system where it is not difficult to maintain close coupling of the coils. On the other hand, it is preferable to make L_{LC} large for a system where it is difficult to maintain close coupling of the coils such as a dynamic IPT system. In the next section, the validity of the theoretical analysis result is evaluated by using two inductors L_{LC} (20.0, 34.0 μH). A detailed discussion of the design method for the LC circuit will be reported in the future.

Comparison between FBA-IPT System and SAB-IPT System

This chapter compares the power losses and the efficiencies of the FBA-IPT/SAB-IPT systems in a theoretical analysis. In addition, its validity is shown by experimental results. The analysis and the experiment are conducted based on the parameters shown in Table II. Through the comparison, it is confirmed that the effectiveness of the SAB-IPT system to which the double-sided LCC is connected.

Theoretical Analysis

Fig. 5 shows the whole-system efficiency based on the theoretical analysis ($k = 0.3$). The soft-switching operation area is only drawn in Fig. 5. The line on the rim of the operation represents the maximum efficiency points for each transmission power. From Fig. 5, it can be confirmed that the soft-switching operation area of the SAB-IPT system expands compared with that of the FBA-IPT system. In other words, as mentioned in the previous chapter, the degree of freedom of θ_{vps} increases, and then the SAB-IPT system can be operated with $\theta_{\text{vps}} = \pi/2$ ($\cos\phi = 1$) at $L_{\text{LC}} = 20.0$ μH. Additionally, the SAB-IPT system can be operated with a higher displacement factor than the FBA-IPT system even at $L_{\text{LC}} = 34.0$ μH. As a result, the whole-system efficiency is also improved.

Here, Fig. 6 shows the maximum efficiency characteristics of FBA/SAB-IPT systems controlled according to Fig. 5. From Fig. 6, it can be confirmed that the SAB-IPT system achieves higher efficiency than the FBA-IPT system over wide operation range. Here, Fig. 7 shows the power-loss breakdown based on the theoretical analysis at $k = 0.2$. From Fig. 7, it can be confirmed that each power loss except for P_{LC} in the SAB-IPT system is smaller than that in the FBA-IPT

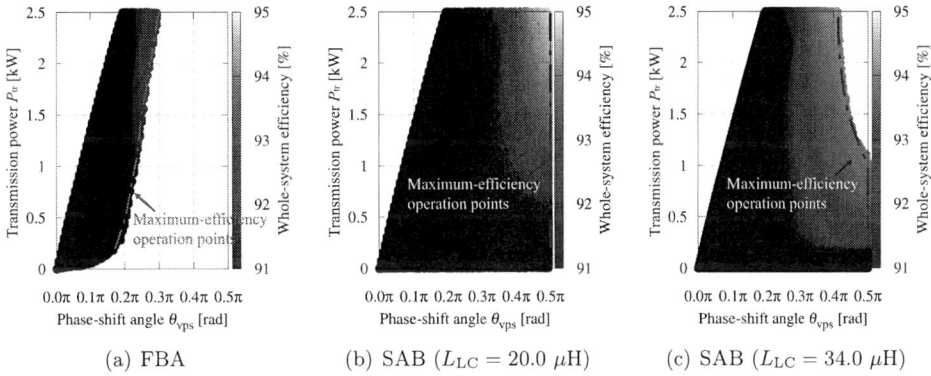

Fig. 5: Whole-system efficiency based on the theoretical analysis ($k = 0.3$).

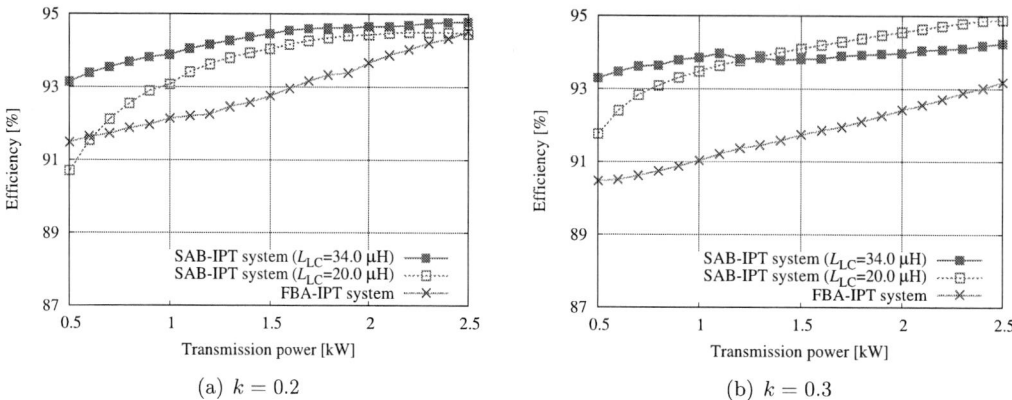

(a) $k = 0.2$ (b) $k = 0.3$

Fig. 6: Maximum efficiency characteristics of FBA/SAB-IPT systems controlled according to Fig. 5.

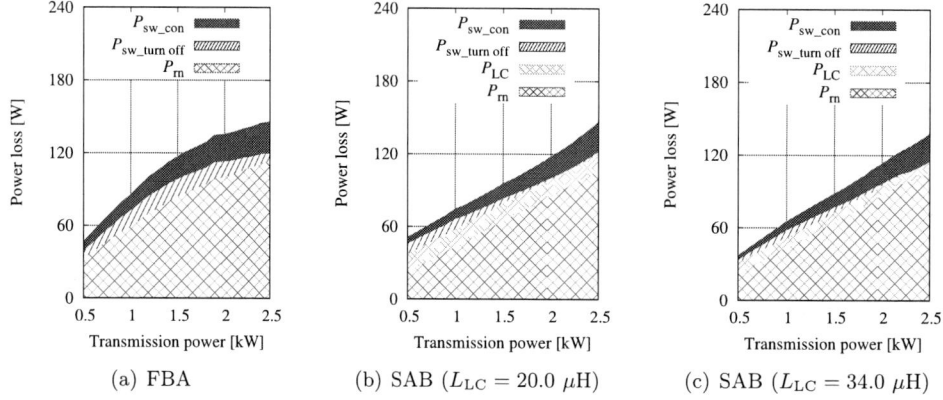

(a) FBA   (b) SAB ($L_{\mathrm{LC}} = 20.0~\mu\mathrm{H}$)   (c) SAB ($L_{\mathrm{LC}} = 34.0~\mu\mathrm{H}$)

Fig. 7: Power-loss breakdown based on the theoretical analysis at $k = 0.2$.

system over a wide operation range. In addition, it can also be confirmed that the power loss P_{LC} in the LC circuit is small, especially with $L_{\mathrm{LC}} = 34.0~\mu\mathrm{H}$.

However, in Fig. 6(a), there are some points where the efficiency of the SAB-IPT system is slightly lower than that of the FBA-IPT system. When $P_{\mathrm{tr}} = 2.5$ kW, α_{p} and α_{s} in the FBA-IPT system are small, i.e., the reactive current to achieve soft switching is also small. In this case, the FBA-IPT system operates with relatively high efficiency. On the other hand, in the SAB-IPT system, a larger current is generated from the LC circuit than the necessity for soft switching. As a result, the efficiency of the SAB-IPT system (with $L_{\mathrm{LC}} = 20.0~\mu\mathrm{H}$) that uses a larger LC current is lower than that of the FBA-IPT system.

Here, the power loss in the LC circuit is constant regardless of the transmission power because the duty ratios of the legs and the DC source voltages are constant. Due to this, in the low power range, the ratio of P_{LC} to the whole system becomes high. As a result, the efficiency of the SAB-IPT system (with $L_{\mathrm{LC}} = 20.0~\mu\mathrm{H}$) that uses a larger LC current is lower than that of the FBA-IPT system. The above theoretical analysis shows that the SAB-IPT system is more efficient than the FBA-IPT system when L_{LC} can be properly designed.

Experiment

In the experiment, the switches in each system are operated with soft switching and controlled with open-loop based on the maximum-efficiency operation points shown in Fig. 5. Fig. 8 shows the output waveforms of the SAB converters at $k = 0.2$ and $P_{\mathrm{tr}} = 2.5$ kW in the experiment.

From Fig. 8, it can be confirmed that these waveforms satisfy the soft-switching requirements in (14) and (15), and the soft-switching operation is achieved.

Next, Fig. 9 shows the maximum efficiency characteristics of FBA/SAB-IPT systems in the experiment. From Fig. 9, we can confirm that the SAB-IPT system is more efficient than the FBA-IPT system over wide operation range when $L_{\mathrm{LC}} = 34.0$ μH. Especially, the efficiency at $k = 0.3$ and 2.5 kW is improved by 0.69 points, and the efficiency at $k = 0.2$ and 1.5 kW is improved by 0.88 points. Here, Fig. 10 shows the power-loss breakdown at $k = 0.2$ based on the measured current values. In Fig. 10, P_{rn} and P_{LC} are calculated using the measured current values and circuit parameters in Table II. In addition, "Other losses" represent the power loss subtracting P_{rn} and P_{LC} from the total power loss in the system. However, the power losses in the converters are dominant in the other losses. From Fig. 10, it is confirmed that the SAB-IPT system with $L_{\mathrm{LC}} = 34.0$ μH can reduce the whole-system loss over wide operation range.

On the other hand, the efficiency of the SAB-IPT system, even with $L_{\mathrm{LC}} = 34.0$ μH, is lower than that of the FBA-IPT system by 0.24 points when $k = 0.2$, $P_{\mathrm{tr}} = 2.5$ kW. In a comparison of Figs. 7 and 10, it is confirmed that P_{rn} and P_{LC} in the experiment agree with these in the analysis. However, there is a large deviation in the power loss regarding the converter. This deviating causes different results from the theoretical analysis, and the SAB-IPT system is not more efficient than FBA-IPT system at the point. This problem can be improved by improving the power loss model of the converter and the design method for L_{LC}.

However, it can be confirmed that the tendency in Fig. 10 is close to the theoretical result in

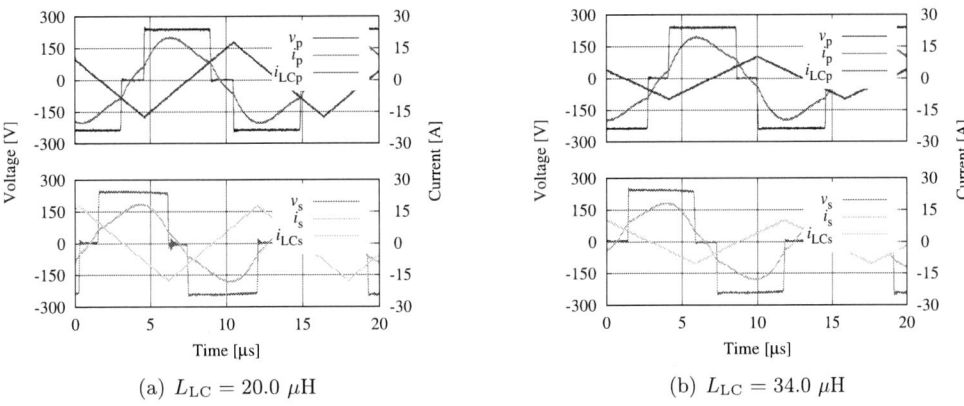

Fig. 8: Output waveforms of the SAB converters at $k = 0.2$ and $P_{\mathrm{tr}} = 2.5$ kW.

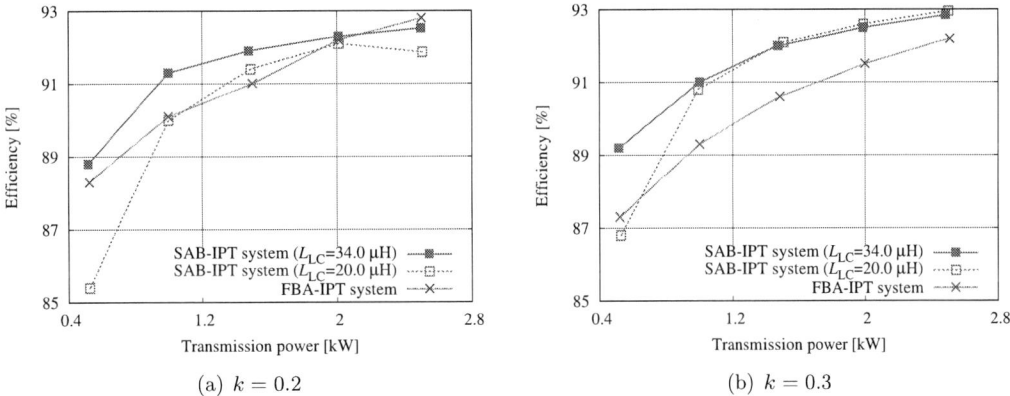

Fig. 9: Maximum efficiency characteristics of FBA/SAB-IPT systems in the experiment.

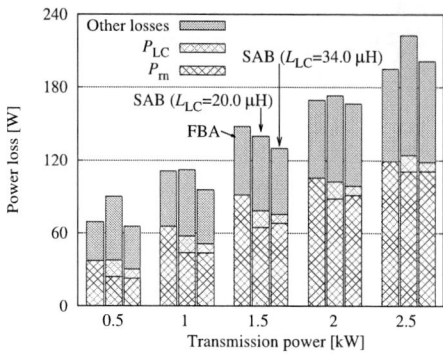

Fig. 10: Power-loss breakdown at $k = 0.2$ based on the measured current values.

Fig. 7, and it can be said that the theoretical analysis is valid. It can also be said that SAB-IPT system is also efficient when the double-sided LCC topology is applied.

Conclusion

This paper showed how to apply the SAB converter to IPT systems with the double-sided LCC topology. The theoretical analysis clarified the effect of the design of the LC circuit on the whole-system efficiency. As a result, regarding systems that are easy to maintain close coupling of the transmission coils, making L_{LC} small is an effective design for system efficiency. On the other hand, regarding the other system, making L_{LC} large is an effective design for system efficiency over a wide operation range.

In the experiment, the validities of the theoretical analysis result and the design guideline were evaluated by using two inductors L_{LC} (20.0, 34.0 μH). The SAB-IPT system could get higher efficiency over wide operation range as well as the theoretical analysis result. Especially, when $L_{\mathrm{LC}} = 34.0$ μH, the efficiency at $k = 0.3$ and $P_{\mathrm{tr}} = 2.5$ kW was improved by 0.69 points, and the efficiency at $k = 0.2$ and $P_{\mathrm{tr}} = 1.5$ kW was improved by 0.88 points. However, in contrast to the result of the theoretical analysis, the efficiency of the SAB-IPT system at $k = 0.2$ and $P_{\mathrm{tr}} = 2.5$ kW worsened by 0.24 points. To improve this problem, some improvements regarding the power loss model of the converter and the design method for L_{LC} are needed.

However, it was confirmed that the characteristics of the experimental result had a close tendency to that of the theoretical result, and it can be said that the theoretical analysis is valid. From the above, it can also be said that the SAB-IPT system has an effect when the double-sided LCC topology is applied. In future work, the power loss models and the design method for the LC circuit will be improved to achieve even higher efficiency.

Acknowledgment

The proposal for the SAB converter and the construction of the FBA converter are partially supported by JSPS KAKENHI Grant Number 20K14723. The proposal for the design guideline for the LC circuit to apply to the SAB converter with the double-sided LCC is partially supported by TEPCO Memorial Foundation.

References

[1] H. Sumiya, E. Takahashi, N. Yamaguchi, K. Tani, S. Nagai, T. Fujita and H. Fujimoto, "Coil Scaling Law of Wireless Power Transfer Systems for Electromagnetic Field Leakage Evaluation for Electric Vehicles," *IEEJ Journal of Industry Applications*, vol. 10, no. 5, pp. 589–597, 2021.

[2] H. Matsumoto, T. Zaitsu, R. Noborikawa, Y. Shibako and Y. Neba, "Control for Maximizing Efficiency of Three-Phase Wireless Power Transfer Systems At Misalignments," *IEEJ Journal of Industry Applications*, vol. 9, no. 4, pp. 401–407, 2020.

[3] K. Nara, N. Madoiwa, K. Maeshiro and Y. Kaneko, "Wireless Power Transfer System with Ideal Transformer Characteristics Determined Solely by Coil Turns Ratio," *IEEJ Journal of Industry Applications*, vol. 9, no. 6, pp. 656–662, 2020.

[4] B. Ji, K. Hata, T. Imura, Y. Hori, S. Shimada and O. Kawasaki, "Wireless Power Transfer System Design with Power Management Strategy Control for Lunar Rover," *IEEJ Journal of Industry Applications*, vol. 9, no. 4, pp. 392–400, 2020.

[5] X. Zhang, T. Cai, S. Duan, H. Feng, H. Hu, J. Niu and C. Chen, "A Control Strategy for Efficiency Optimization and Wide ZVS Operation Range in Bidirectional Inductive Power Transfer System," *IEEE Transactions on Industrial Electronics*, vol. 66, no. 8, pp. 5958–5969, 2019.

[6] R. Ota, D. J. Thrimawithana, U. K. Madawala and G. A. Covic, "Boundary of Soft-switching for Efficient Operation of Bi-directional IPT Systems," in *Proc. of 2020 IEEE PELS Work-shop on Emerging Technologies: Wireless Power Transfer (WoW)*, pp. 164–169, 2020.

[7] R. Ota, D. S. Nugroho and N. Hoshi, "A consideration on maximum efficiency of resonant circuit of inductive power transfer system with soft-switching operation," *World Electr. Veh. J.*, vol. 10, no. 3, 2019.

[8] G. R. Kalra, B. S. Riar, and D. J. Thrimawithana, "An integrated boost active bridge based secondary inductive power transfer converter," *IEEE Transactions on Power Electronics*, vol. 35, no. 12, pp. 12716–12727, 2020.

[9] N. Fu, J. Deng, Z. Wang, W. Wang, and S. Wang, "A hybrid mode control strategy for lcc-lcc-compensated wpt system with wide zvs operation," *IEEE Transactions on Power Electronics*, vol. 37, no. 2, pp. 2449–2460, 2022.

[10] R. Okada, R. Ota and N. Hoshi, "Novel Soft-switching Active-Bridge Converter for Bi-directional Inductive Power Transfer System," *IEEJ Journal of Industry Applications*, vol. 11, no. 1, pp. 97–107, 2022.

[11] R. Ota and N. Hoshi, "Basic Study of Integrated On-board Converter for Dynamic WPT EV," in *Proc. of 5th International Electric Vehicle Technology Conference 2021*, 2021.

[12] J. Zhang, Z. He, A. Luo, Y. Liu, G. Hu, X. Feng and L. Wang, "Total Harmonic Distortion and Output Current Optimization Method of Inductive Power Transfer System for Power Loss Reduction," *IEEE Access*, vol. 8, pp. 4724–4736, 2020.

[13] R. Ota, D. S. Nugroho, and N. Hoshi, "A capacitance design guideline of snubber capacitors for soft switching in bi-directional inductive power transfer system considering battery charging cycle," in *Proc. of 2018 7th International Conference on Renewable Energy Research and Applications (ICRERA)*, 2018, pp. 1080–1085.

Ultra Low Loss – MMC Submodules favorable for SiC-FET enabling High Functional Safety

Christopher Dahmen, Rainer Marquardt
Universität der Bundeswehr München
Institute for Electrical Energy Systems
Werner-Heisenberg-Weg 39
85577 Neubiberg, Germany
Email: christopher.dahmen@unibw.de

Abstract—**Further progress of Modular Multilevel Converters (MMC) is mainly related to advanced submodule topologies (SM), semiconductors (SiC) and improved control concepts. Fully electronic failure management for external failures (i. e. DC-short circuit) and protection against internal failures of the converter are key issues. A new submodule for MMC is presented, which meets these requirements. In addition, a novel SiC-JFET super-cascode is investigated, suitable for these and other applications.**

I. INTRODUCTION

Modular Multilevel Converters have revolutionized high power voltage source converters (VSC) and have become key components for future HV- and MV-DC-grids, the integration of wind and solar power into the grid and many other important future applications [1]–[4]. The elimination of bulky AC-filters and – more importantly – the elimination of large DC-capacitors from the DC-bus enables fully electronic failure management and protection. These important issues are twofold:

a) Management of external failures, especially electronic current limitation at the AC- <u>and</u> DC-side including overvoltage clamping

b) Management of internal failures, especially protection against explosion of semiconductors, arcing and mechanical damage at submodule level.

A submodule topology, which meets both requirements is investigated in this paper. Ultra low power loss and reduction of capacitor size is enabled, additionally [5]–[7]. The topology is suitable for combining Si- and SiC-semiconductors in a favorable manner. In the following paper, a realization using reverse conducting IGBT and one SiC-cascode switch will be investigated. For a fully SiC-implementation, which may be possible in the near future, even greater advantages – compared to Full-Bridge (FB)-SiC – are achieved [7].

II. DOUBLE-ZERO SUBMODULE: MODES OF OPERATION

Fig. 1 shows the investigated Double-Zero submodule (DZ-SM). It uses four reverse conducting IGBT ($T_1/D_1 \ldots T_4/D_4$) and one reverse conducting SiC-switch (T_0/D_0). This switch can be implemented in several favorable manners. The most reasonable options are:

- SiC-MOSFET [8]
- SiC-JFET or SiC-SIT [9]
- SiC-JFET cascode switch [10]

- SiC-JFET super cascode switch [11]–[13].

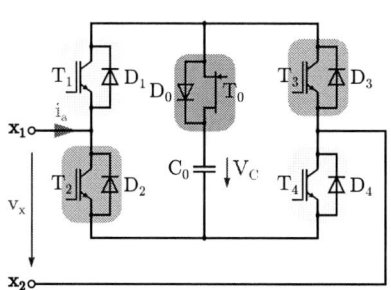

Fig. 1. Double-Zero submodule (DZ-SM)

With respect to development status of high voltage SiC-devices, mainly the first and the last option mentioned have to be considered. The superior current limiting characteristic of JFET (or SIT) combined with high temperature sustainability makes JFET a good choice under these conditions, here. Note that the decision between "normally on" and "normally off" behavior is not an essential issue for this topology.

The DZ-SM (see Fig. 1) meets essential requirements, which seem to be in contradiction, at first glance:

- Full-Bridge functionality with essentially reduced power-losses (see requirement a))
- Three independent semiconductor switches are in the critical "shoot through" path of the SM (see requirement b)).

For further explanation, Fig. 2 shows a typical arm current $i_a(t)$ and corresponding voltage modulation $d(t)$ of an MMC in inverter mode. Because there is no DC-power supply of the submodules in MMC, the energy and charge balancing enforces, that high arm current amplitudes always occur at low voltage modulations. Two intervals with charging the DC-capacitor (I and III) and two intervals with discharging (II and IV) can be distinguished for further analysis (see Tab. I).

A typical switching sequence in interval I is shown in Fig. 3: At zero terminal voltage ($v_x = 0\,\text{V}$), the SiC-switch is kept in off-state and the silicon devices of the high- and low-side are sharing the arm current – reducing the on-state losses especially at high arm currents. In order to switch to the charging state ($v_x = +V_C$), the IGBT (T_2 and T_3) must be turned off (at approx. half the arm current). When switching back to $v_x = 0\,\mathrm{V}$, these IGBT are turned on, again. This can be done very fast and with very low losses, because no compromise regarding Si-diode recovery is necessary. Evidently, the SiC-switch can be turned on in parallel to the SiC-diode structure to reduce the on-state losses (see right picture in Fig. 3).

loss reduction of minus 30%), when implementing in a fully SiC-version (compare 2nd and 4th bar in Fig. 5). Applying SiC-devices in a FB-SM is evidently not a favorable choice since no conduction loss improvement is achievable with this topology.

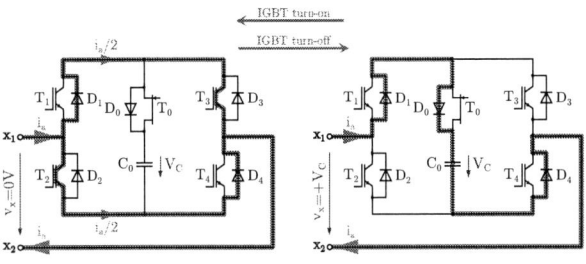

Fig. 3.   Switching sequence in Interval I (see Tab. I)

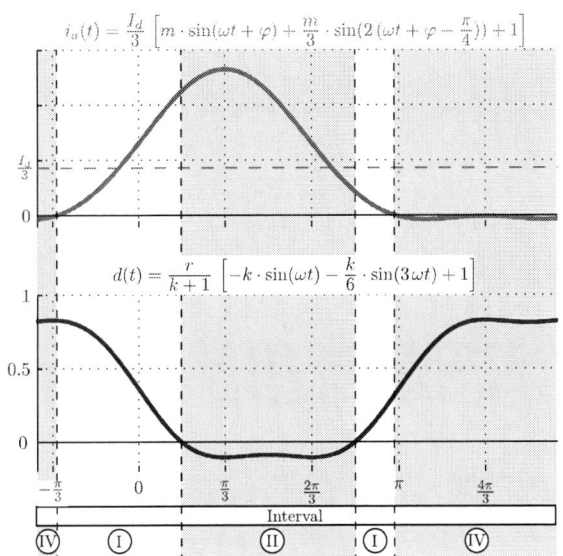

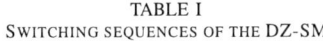

Fig. 2.   Arm current $i_o(t)$ and corresponding voltage modulation $d(t)$ of an MMC in inverter mode ($\varphi = 30°$; $k = 1.5$; $r = 0.9$)

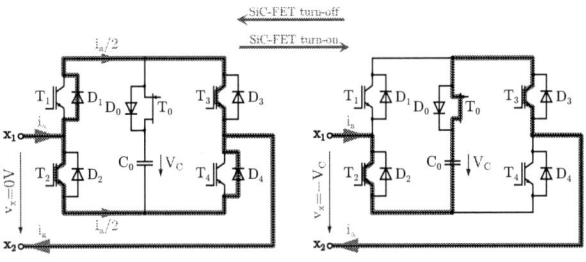

Fig. 4.   Switching sequence in Interval II (see Tab. I)

TABLE I
SWITCHING SEQUENCES OF THE DZ-SM

Interval	sgn(i_a)	sgn(v_x)	sgn(i_C)	sgn(P_a)	Capacitor
I	$\oplus$	$\oplus$	$\oplus$	$\oplus$	charging
II	$\oplus$	$\ominus$	$\ominus$	$\ominus$	discharging
III	$\ominus$	$\ominus$	$\oplus$	$\oplus$	charging
IV	$\ominus$	$\oplus$	$\ominus$	$\ominus$	discharging

Fig. 4 shows the switching procedure in interval II, where only the SiC-device is switching. Hence, the switching losses are very low, too. Only at SiC-turn-on, reduced speed (with respect to recovery of the Si-diode structures) has to be accepted. The switching conditions in intervals III and IV are similar to I and II, respectively.

In summary, reduced power losses of the DZ-SM can be achieved in the order of minus 20% – compared to a full silicon FB-SM utilizing $3.3\,\mathrm{kV}$ semiconductor devices (compare 1st and 3rd bar in Fig. 5). The resulting on-state loss of the SiC-switch is low, caused by the low RMS-current at a typical voltage modulation of an MMC. Therefore, the required chip area of the SiC-switch is very small in comparison to the four main semiconductors [6], [7], [14]. Nevertheless, the improvement of the DZ-SM is even more essential (power

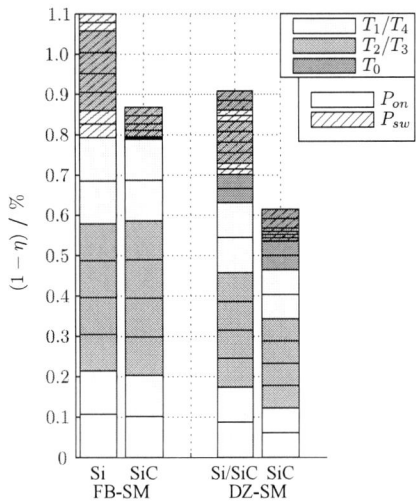

Fig. 5.   Normalized submodule power losses (considering only DC- and fundamental components of current and voltage modulation[1]) utilizing $3.3\,\mathrm{kV}$ semiconductor devices ($I_d = 2.6\,\mathrm{kA}$; $k = 1.5$; $r = 0.9$; $\varphi = 30°$; IGBT: $U_{T0} = 1.2\,\mathrm{V}$; $R_T = 0.75\,\mathrm{m\Omega}$; FET: $R_{DSon} = 1.5\,\mathrm{m\Omega}$; $f_p = 150\,\mathrm{Hz}$). The colors in the bar graphs are indicating the power losses of the corresponding semiconductor devices in Fig. 1.

---

[1] m: current modulation factor, k: voltage modulation factor, r: utilization factor of the MMC arm, f_p: semiconductor switching frequency [7]

III. CASCODE SWITCH

Cascode switches are well known to be favorable for high voltage applications. However, the realization of the low voltage switch was considered as an essential drawback in the past. Tremendous development progress of low voltage silicon FET has changed this situation, completely. The choice between a "standard" gate driver (a) and a "cascode type" gate driver (b) can be explained according to Fig. 6. Both choices have one specific main drawback:

a) The Miller capacitance (C_{GD}) slows down switching speed and aggravates problems with parasitic high frequency oscillations.

b) The n-channel FET of the gate driver circuit ($T_{1,GU}$ in Fig. 6) has to carry the full load current.

Solely, if the small p-channel FET of the gate driver is omitted, additional differences in switching performance occur. On the other hand, these issues are usually not mandatory.

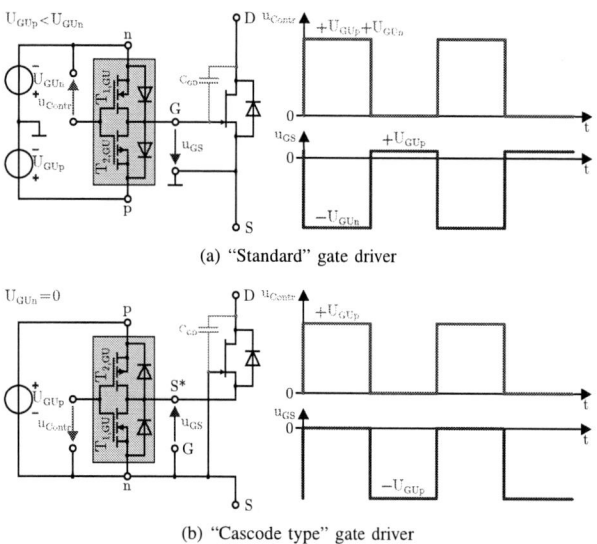

(a) "Standard" gate driver

(b) "Cascode type" gate driver

Fig. 6. Conventional gate drivers of JFET semiconductors

IV. STATE OF THE ART SUPER-CASCODES

Generally, super-cascodes – as stacked cascodes – are favorable topologies to achieve higher blocking voltages by internal series connection of semiconductors. In this way, conduction losses can be reduced compared to single semiconductor chips with a high blocking voltage capability. On the contrary, these advantages are coming with a higher complexity of the circuit as well as general challenges for the symmetrical distribution of the semiconductor voltages.

Fig. 7 illustrates an overview of the state of the art SiC-JFET super-cascodes – shown in a three voltage stage implementation. The indices of the devices in the pictures clarify the affiliation to the voltage stages. The main terminals of the structure are indicated as gate (G), main-drain (MD) and main-source (MS).

The first version of the super-cascode uses solely avalanche diodes between the gates of the JFET to balance the drain-source voltages (see Fig. 7(a)). The switching process of the different stages is generally sequential. To have a closer look at the super-cascode turn-on process, it can be recognized that the switching is initiated by turning on the low voltage MOSFET T_0. The decaying of its drain-source voltage leads to the exceeding of the (negative) pinch-off voltage of J_1 and therefore also to its turn-on. Since the gate potential of J_2 is initially kept constant, a decay of the drain-source voltage of J_1 leads to the conducting of J_2 as well. This mechanism continues until all transistors in the main path are turned on. The turn-off sequence advances from the bottom to the top stage as well. It can be understood from Fig. 7(a) that the sequential turn-off is mainly dependent on the breakdown voltage of the avalanche diodes. This and the absence of a defined reverse current path through the whole main circuit – since J_3 does not own an avalanche diode – is a fundamental drawback of this circuit.

The super-cascode concept from Fig. 7(b) is an improvement to the previous variant. On the one hand, utilizing RC-elements (R_1, C_1 and R_2, C_2) enables besides the static voltage distribution between the stages also a dynamic voltage distribution. On the other hand, the gate-source resistors R_{Di} counteract the aforementioned reverse current. The absence of a defined reverse current path through the whole main circuit is still problematic. Regarding the stacked structure – the reverse currents of the JFET decrease towards the lower voltage stages as the currents of the avalanche diodes increase – an asymmetrical design of the RC-elements is therefore resulting.

The super-cascode concept from Fig. 7(c) shows a variant with a distributed arrangement of the RC-elements. Due to the independent supply of the gate charge, this design is simpler as the previous version. The defined reverse current path through the whole main circuit represents another advantage compared to the latter version. On the contrary, the capacitor voltages are not identical – the voltage ratio of C_1 is $\frac{1}{3}$ and the voltage ratio of C_2 is $\frac{2}{3}$ to the total circuit voltage. An asymmetrical design is also the result of this super-cascode concept and limits the modularity.

The super-cascode concept from Fig. 7(d) illustrates a combination of the latter two variants – the resistors R_1 and R_2 are solely shifted into the gate path of the SiC-JFET. Series resistors R_{Di} increase the ruggedness of the avalanche diodes. The resistor R_3 ensures a defined reverse current path through the whole main circuit.

In Fig. 7(e) a super-cascode concept is shown, which looks different to the previous versions, at first glance. The bottom voltage stage is controlled by a HV-SiC-MOSFET with similar ratings to the main SiC-JFET, here. Nonetheless, T_1 could be replaced by a switching unit made of a HV-SiC-JFET and a LV-MOSFET, once more. A general improvement of this structure can therefore not be expected. The main advantage over the previous concepts is the completion of the RC-divider through the whole circuit. High resistor values R_{Si} ensure an improved voltage distribution in static state.

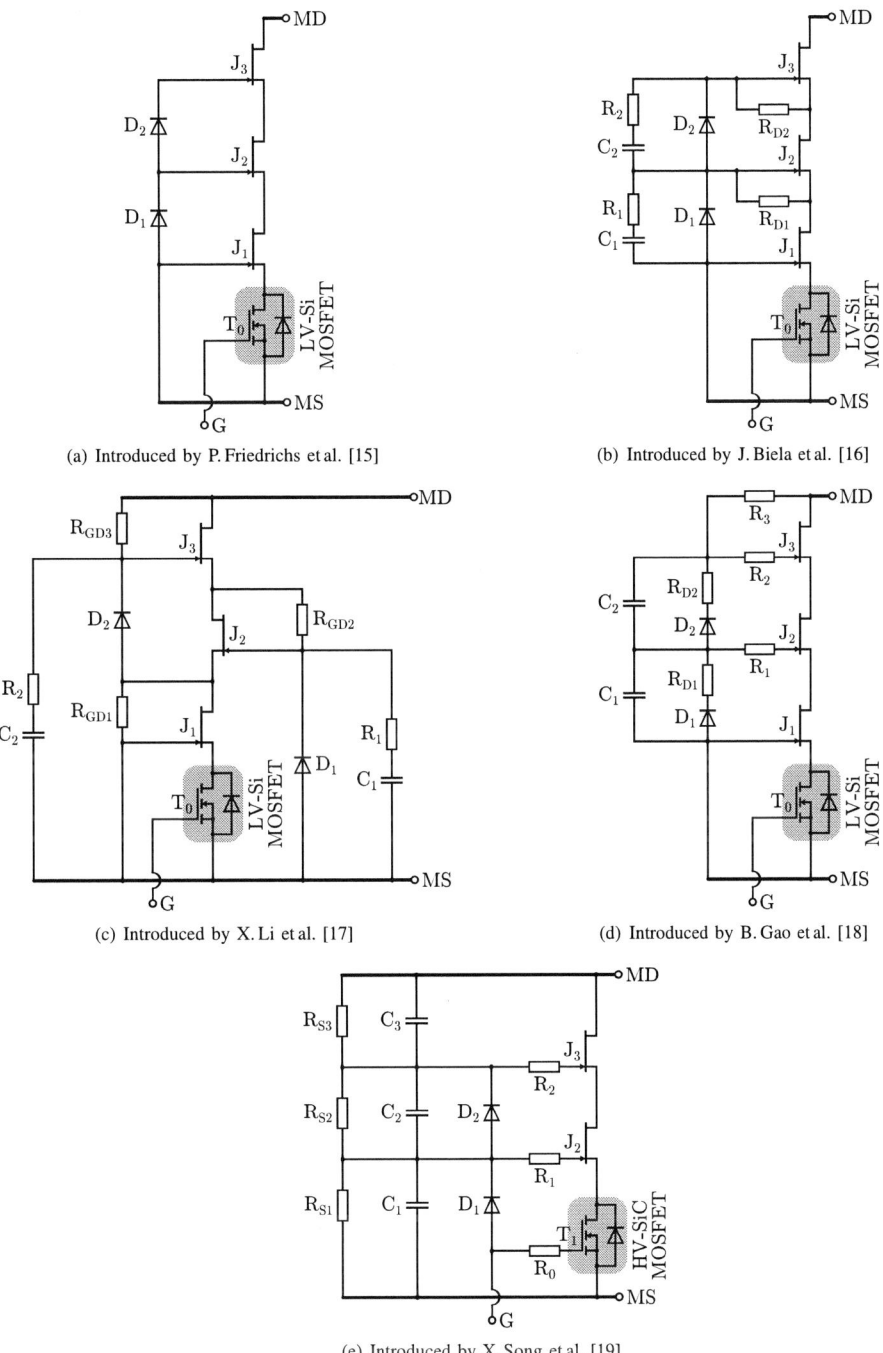

(a) Introduced by P. Friedrichs et al. [15]

(b) Introduced by J. Biela et al. [16]

(c) Introduced by X. Li et al. [17]

(d) Introduced by B. Gao et al. [18]

(e) Introduced by X. Song et al. [19]

Fig. 7. State of the art SiC-JFET super-cascodes (three voltage stage implementation)

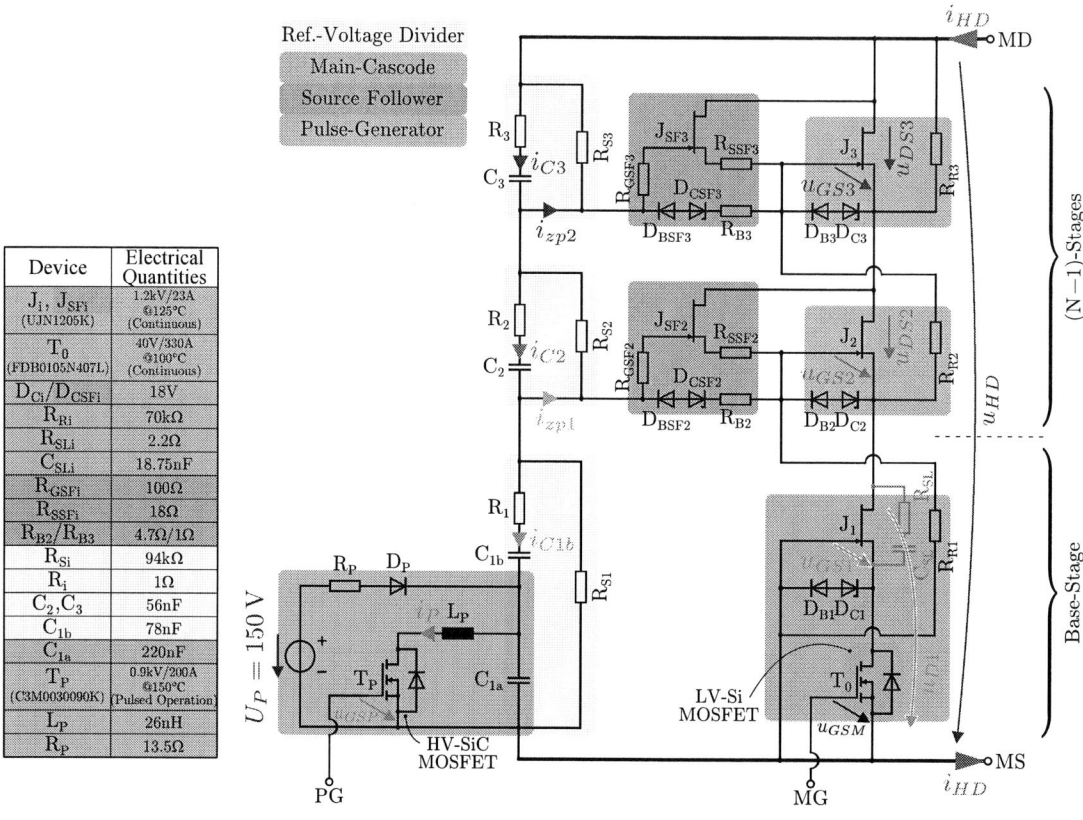

Device	Electrical Quantities
J_i, J_{SFi} (UJN1205K)	1.2kV/23A @125°C (Continuous)
T_0 (FDB0105N407L)	40V/330A @100°C (Continuous)
D_{Ci}/D_{CSFi}	18V
R_{Ri}	70kΩ
R_{SLi}	2.2Ω
C_{SLi}	18.75nF
R_{GSFi}	100Ω
R_{SSFi}	18Ω
R_{B2}/R_{B3}	4.7Ω/1Ω
R_{Si}	94kΩ
R_i	1Ω
C_2,C_3	56nF
C_{1b}	78nF
C_{1a}	220nF
T_P (C3M0030090K)	0.9kV/200A @150°C (Pulsed Operation)
L_P	26nH
R_P	13.5Ω

Fig. 8. Novel SiC-JFET super-cascode (three voltage stage implementation)

Regarding the state of the art SiC-JFET super-cascodes, general conclusions can be extracted:

a) The use of avalanche diodes to maintain symmetrical distribution between the voltage stages is disadvantageous. A reliable dimensioning of the current rating and defined reverse currents for the blocking state are critical with respect to semiconductor parameter scattering.

b) The "individual tuning" of the RC-elements to correct the switching sequence of the voltage stages is disadvantageous for industrial series production.

c) The switching sequence of the voltage stages at turn-off – the lower main JFET blocks first – is unfavorable, in principal. The resulting sequence is in contradiction to the balancing requirement of the main JFET – protective turn-on in order to limit individual overvoltage.

V. NOVEL SiC-JFET SUPER-CASCODE

A. General structure and function

Fig. 8 illustrates the novel SiC-FET super-cascode which has been developed to eliminate the described disadvantages. An improved scalability – in order to enable high currents and voltages – and a uniform design with low sensitivity to component tolerances are achieved, too. An additional current path in the novel super-cascode enables the access to the

upper main JFET to allow a favorable turn-off sequence of the voltage stages – the lower main JFET will be turned off after the blocking of the upper JFET is initiated. In addition to a high circuit efficiency, special focus was laid on a high degree of modularity of the super-cascode. High blocking voltages as well as high current rating can be achieved including overcurrent capability.

The novel super-cascode in Fig. 8 can generally be divided into a base-stage (including the driving of the LV-MOSFET) and the $(N-1)$-stages, which build up the super-cascode structure (these are all equipped identically). The circuit can be split into four specific function blocks. The reference-voltage divider and the main-cascode are also implemented in a similar form to the state of the art super-cascode topologies (see Fig. 7(b) to 7(e)). Additional function blocks in the novel topology are a source follower and a pulse-generator. The function blocks can generally be described as follows:

- Reference-voltage divider: The voltage divider is equipped with a dynamic component (R_i, C_i) as well as a static component (R_{Si}). Regarding the use of a source follower, the decoupling of the impedance between the main-cascode and the voltage divider can be achieved. In this manner, the components of the voltage divider can be designed identically for every voltage stage. Due to the absence of a current path including an avalanche diode, the static component R_{Si} maintains a symmetrical voltage distribution in the blocking state of the super-cascode – resulting in low static power losses. It can be noted that the capacitor C_1 of the base-stage is split into C_{1a} and C_{1b} – the capacitance C_{1a} is providing the interface for the coupling of the pulse-generator.

- **Main-cascode**: The main-cascode consists generally of the SiC-JFET which conduct the load current and block the super-cascode voltage (the capacitor voltage V_C for application in the DZ-SM). The diode D_{Ci} limits the gate-source voltages to approx. 20 V ensuring safe operation of the main JFET. D_{Bi} blocks this path for positive currents (this will be explained in the segment of the source follower). R_{Ri} provides a quiescent current (from source-to gate-electrode) for the main JFET, which flows through the whole circuit. That way, the (slight) conducting of the main JFET can be avoided.

- **Source follower**: The function block of the source follower consists of two current paths – one for charging and one for discharging the upper SiC-JFET gates of the main-cascode. The positive path has an additional HV-SiC-JFET (J_{SFi}) and corresponding resistors. The gate charge for the transistor turn-on is provided through the drain-source path of J_{SFi}. Additionally, J_{SFi} raises the gate potential in the on-state, reducing the R_{DSon} of the upper main JFET. For discharging J_2 and J_3, the devices D_{BSFi} and R_{Bi} on the other hand maintain a connection to the RC-divider to evacuate their gates at turn-off.

- **Pulse-generator**: For an improved balancing between the different voltage stages at turn-off, a switching sequence has to be considered in such way, that the $(N-1)$-stages start to block before the base-stage (T_0 and J_1). The pulse-generator is located at the bottom of the voltage divider (coupled through the capacitor C_{1a}) and has to be initiated (PG) slightly before the main-gate (MG) is turned off. The capacitor C_{1a} is charged from an external (low current) voltage source – this will happen passively, if the super-cascode is in the conducting-state.

B. Switching sequence

To get a deeper understanding of the function of the novel super-cascode, the switching sequence shall be analyzed in this paragraph. Since the turn-on switching sequence of the main JFET – from bottom up – is generally not critical in super-cascodes, the focus will be set on the turn-off sequence, exclusively. Fig. 9 shows SPICE simulation results of the novel super-cascode at turn-off corresponding to Fig. 8 performing a hard commutation in a half bridge configuration with freewheeling diode under inductive load ($I_L = 100$ A ; $U_{HD} = 1.4$ kV). The main JFET (J_1, J_2, J_3) are utilized with a parallel connection of 8 TO-247 semiconductor devices in every voltage stage – rated for 23 A per device (see table in Fig. 8). The instances of the turn-off sequence can be described as follows:

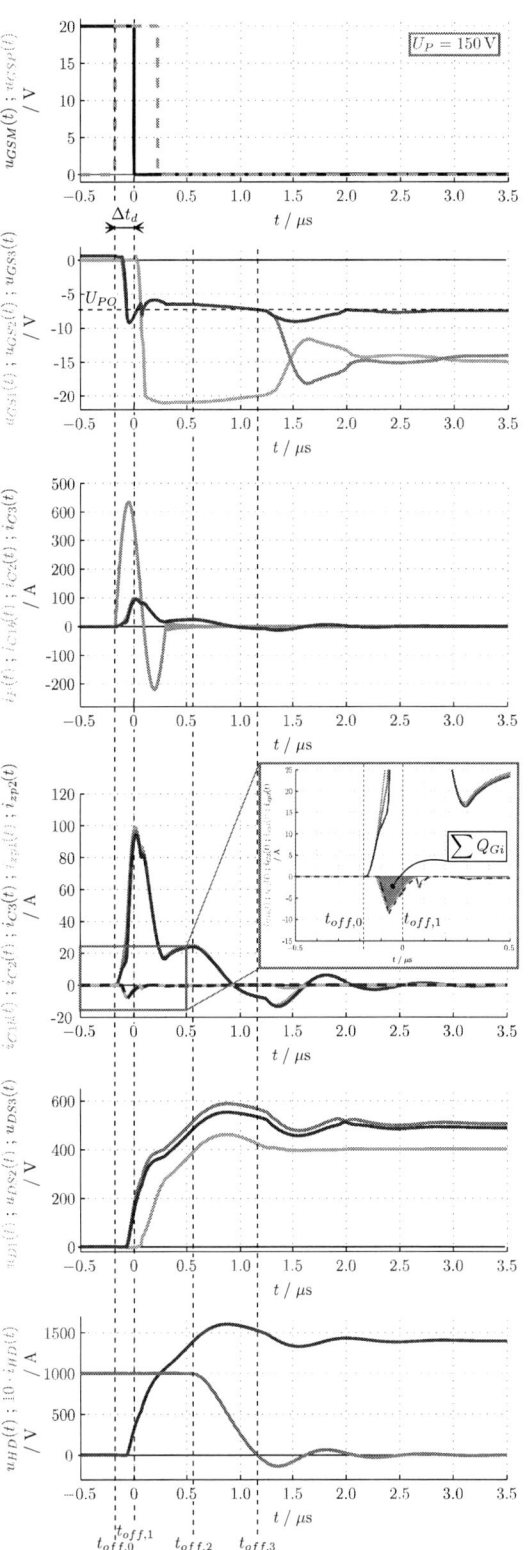

Fig. 9. Simulation of the main electrical quantities of the novel SiC-JFET super-cascode at turn-off ($I_L = 100$ A ; $U_{HD} = 1.4$ kV)

- $t < t_{off,0}$: Before the initiation of the turn-off, the super-cascode is in conducting-state. It can be noted, that the gate voltages of the upper main JFET (u_{GS2} and u_{GS3}) show significantly positive values – provided by the source followers – advantageously.

- $t_{off,0} < t < t_{off,1}$: The turn-off sequence is initiated with a positive pulse of the pulse generator (u_{GSM}). The capacitor C_{1a} is discharged via the (small) loop inductance L_P and generates an impulse current i_P. A partial component of the current is extracted to the reference-voltage divider (i_{C1b}, i_{C2}, i_{C3}) resulting in a negative current of the upper gate-paths of the main JFET (J_2 and J_3) – clearing out their gate charge. The gate-source voltages u_{GS2} and u_{GS3} decay. When reaching the negative maximum of the currents i_{zp1} and i_{zp2}, the voltages u_{GS2} and u_{GS3} fall below their pinch-off voltage resulting in an increase of the drain-source impedance of J_2 and J_3. The voltages u_{DS2} and u_{DS3} increase therefore as well. It can be recognized, that the magnitudes of i_{C1b}, i_{C2} and i_{C3} are primary affected by the super-cascode current i_{HD}, which commutates in this time span to the reference-voltage divider – due to the raising of u_{DS2} and u_{DS3}. The simulation was done with an increased gate-source charge of $\sum C_{GSi} = 45.6\,\mathrm{nF}$ (instead of $8\,\mathrm{nF}$) to illustrate the suitability of the circuit for a high number of paralleled main JFET.

- $t_{off,1} < t < t_{off,2}$: Since J_2 and J_3 already blocking voltage, the main-gate will be turned off (u_{GSM}) – blocking also the JFET J_1 of the base-stage. It can be noted that u_{GS2} and u_{GS3} will rise above the pinch-off voltage once again resulting in a reduced drain-source-voltage slope due to the voltage increase of the voltage divider. Here, these voltages adjust themselves to maintain a quasi-conducting state enabling a uniform dynamic voltage distribution between the different voltage stages (u_{D1}, u_{DS2} and u_{DS3}) near the pinch-off threshold. For a practical realization, it can be favorable to limit the voltage slope of the base-stage (u_{DS1}) applying an RC-element (R_{SL} and C_{SL}, grayed out in Fig. 8) to the circuit. For a better comparison of the later shown experimental results, the optional RC-element was also implemented in the simulation.

- $t_{off,2} < t < t_{off,3}$: The super-cascode voltage u_{HD} reaches the blocking voltage of $1.4\,\mathrm{kV}$. At this instance, no voltage drop occurs at the freewheeling diode, initiating its conducting. The load current i_{HD} trough the circuit is now descending. Regarding the parasitic inductances in the commutation loop and the resulting negative current slope $\frac{di_{HD}}{dt}$ an overshoot (of approx. $200\,\mathrm{V}$) can be noted. The transistors J_2 and J_3 still operate in a quasi-conducting state, enabling a controlled voltage distribution between u_{D1}, u_{DS2} and u_{DS3}.

- $t_{off,3} < t$: The current i_{HD} through the super-cascode is crossing zero – all the main JFET are blocking. A shifting of the upper gate-source voltages is stimulated by the turn-off voltage overshoot. After the decay of the oscillation, the gate-source voltages are pulled towards the static blocking state – via the current path through the resistors R_{Ri}.

Since the source followers are not completely blocking in the time span $t > t_{off,3}$, they supply a quiescent current what slows the clearing of the gate-charge (of J_2 and J_3) down. The opposing operation of the source followers (charging of the upper main JFET-gates) and the path through R_{Ri} (discharging of the upper main JFET-gates) was on the other hand not restricted in the circuit design. A compromise can be found between high switching frequency and steady state power losses. For the utilization of the novel SiC-JFET super-cascode in the capacitor switch of the DZ-SM (see T_0 in Fig. 1, 3 and 4) not more than two switching operations take place within a fundamental period of $20\,\mathrm{ms}$, typically.

C. Dimensioning

The dimensioning of the novel super-cascode will be briefly discussed in this paragraph. A specific focus is laid on the components of the reference-voltage divider and the pulse generator. The selection of the essential components of the main-cascode and the source follower is straightforward and is generally based on the current rating and voltage blocking rating of the super-cascode circuit application:

- C_i, R_i, R_{Si}: A basic approach for designing the voltage divider is to define an acceptable limit of energy loss in the capacitors C_i at super-cascode turn-off. The condition for one voltage stage (and $N = 3$) is:

$$C_i \leq \frac{I_L \cdot t_f}{\frac{U_{HD}}{N}}. \tag{1}$$

For the investigated circuit, a fall time of $t_f = 500\,\mathrm{ns}$ was assumed resulting in a capacitance of approx. $100\,\mathrm{nF}$ – rounded down to $56\,\mathrm{nF}$ for higher ruggedness.

The selection of the resistance R_i requires a compromise. On the one hand, a low-resistive design of the divider is necessary for minimal impairment of the pulse generator. On the other hand, oscillation at switching has to be dampened – stimulated by the parasitic loop inductances. A value of $R_i = 1\,\Omega$ complies with the aforementioned conditions and also keeps the initial voltage drop well below the per stage voltage:

$$R_i \cdot I_L \ll \frac{U_{HD}}{N}. \tag{2}$$

The design of the resistance R_{Si} follows a compromise to assure accurate voltage distribution in static state of the super-cascode as well as low static power losses – a value of about $100\,\mathrm{k}\Omega$ is generally adequate.

- C_{1b}, C_{1a}: The most critical issue for designing the pulse generator is – for a sufficient embedding of the pulse generator into the voltage divider – the selection of the capacitances C_{1a} and C_{1b}. Here, it is constructive to separate the design into two steps (see Fig. 10):

  - Step A: Fig. 10(a) shows a simplified equivalent circuit of the novel super-cascode of Fig. 8 and

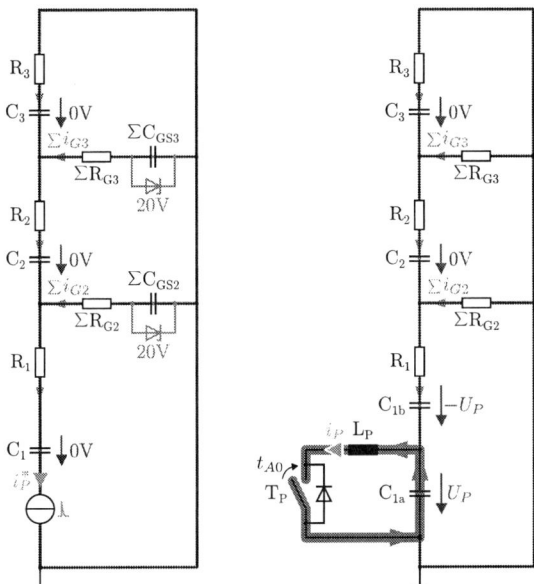

(a) Step A: Idealization of the pulse generator

(b) Step B: Neglecting of the gate-source capacitances

Fig. 10. Equivalent circuit of the novel SiC-JFET super-cascode for the stepwise dimensioning of the pulse generator capacitances

illustrates the pulse generator as an ideal current source with the function of clearing the gate charge of the upper main JFET (J_2 and J_3) prior to the turn-off of the base-stage semiconductor (J_1). The pulse generator needs to provide the negative gate-currents $\sum i_{G2}$ and $\sum i_{G3}$ (of the paralleled JFET) to discharge $\sum C_{GS2}$ and $\sum C_{GS3}$ from $+2\,V$ to $-18\,V$ – indicated by the zener diodes. If the voltage at the capacitances C_i as well as the voltage drop at the main JFET are neglected, it can be assumed that the gate-source capacitances will be charged before a voltage raise at C_i is noted. For this matter, C_i need to have significantly larger values than the summarized gate-source capacitances. For a number of 8 paralleled JFET per stage, their ratio can be determined as follows:

$$\frac{C_i}{\sum C_{GSi}} = \frac{56\,\text{nF}}{8\,\text{nF}} \gg 1. \tag{3}$$

– Step B: Fig. 10(b) shows the integration of the pulse generator into the voltage divider. For an adequate decoupling of pulse generator and voltage divider, C_{1a} has to be greater than the series connection of C_{1b}, C_2 and C_3 (forming C_{ers}). Considering that the series connection of C_{1a} and C_{1b} has to be identical to C_i, the capacitors can be found to $C_{1a} = 220\,\text{nF}$ and $C_{1b} = 78\,\text{nF}$ leading to a ratio of:

$$\frac{C_{1a}}{C_{ers}} = \frac{220\,\text{nF}}{20\,\text{nF}} \gg 1. \tag{4}$$

D. Prototype and measurement results

The general structure of a mechanical prototype of the novel SiC-JFET super-cascode is depicted in Fig. 11. The side view (see Fig. 11(a)) illustrates that the cascode-current flows vertically through the (per stage) assembly of:

1) Heatsink
2) Drain-contact of the main JFET (rear contact of the TO-247-package)
3) Source-contact of the main JFET (soldered to the PCB)
4) PCB (its back plane is fully contacted and mechanically pressed to the heatsink of the next voltage stage).

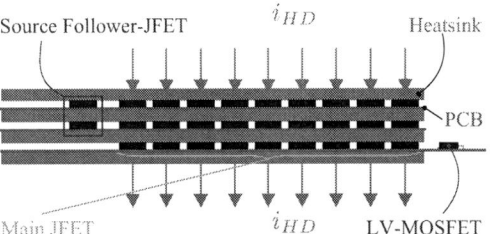

(a) Side view of the three voltage stages

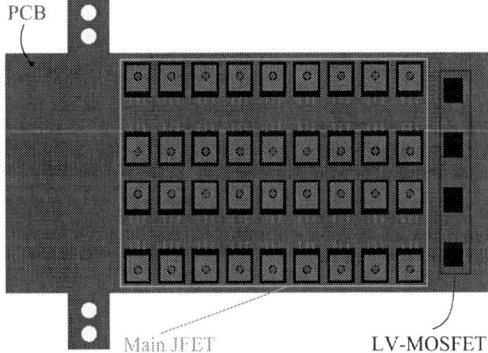

(b) Top view of the (fully assembled) base-stage

Fig. 11. General structure of the mechanical prototype of the novel SiC-JFET super-cascode

The results of a double pulse test applying three voltage stages and 8 paralleled SiC-JFET per stage (connected to a freewheeling diode of a CM400HB-90H IGBT-module) are shown in Fig. 12, 13 and 14.

The oscilloscope graphs are clearly demonstrating the excellent behavior of the super-cascode at turn-off – maintaining uniform voltage distribution between the different voltage stages. The similarity of simulation and experiment verifies the result of the basic dimensioning as an adequate design indicator.

VI. CONCLUSION

The Double-Zero submodule enables improved functional safety in combination with essentially reduced power losses – compared to Full-Bridge submodules. It is well suitable for SiC-devices or a hybrid Si/SiC-realization. Fully electronic

Ultra Low Loss - MMC Submodules favorable for SiC-FET enabling High Functional Safety

DAHMEN Christopher

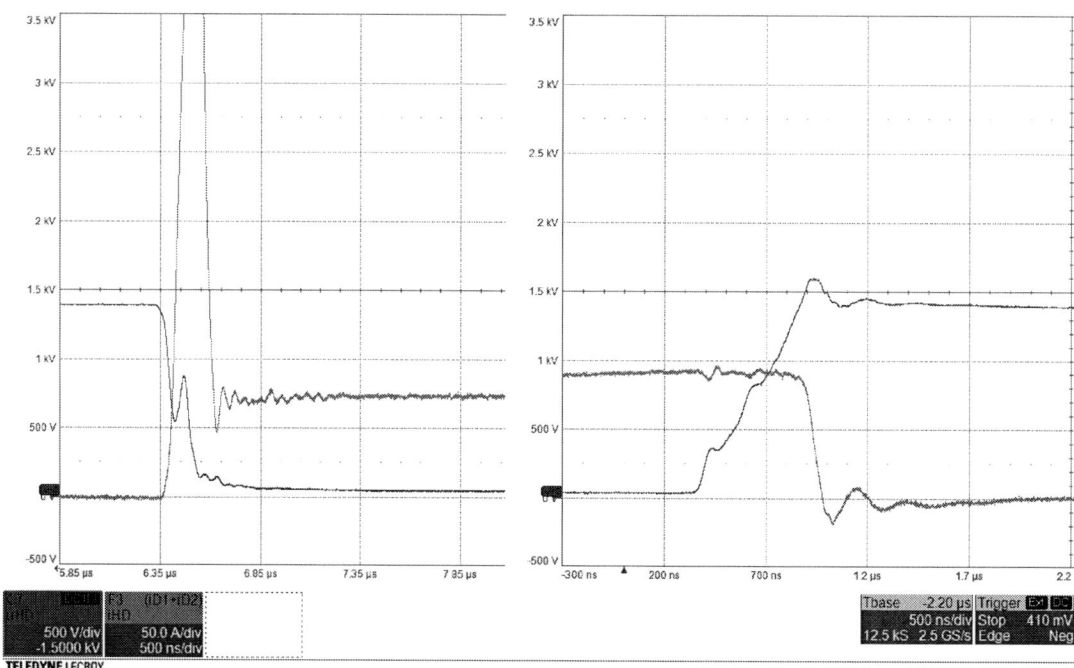

Fig. 12. Measurement of the turn-on (left picture) and turn-off (right picture) of the novel SiC-JFET super-cascode (u_{HD}, i_{HD})

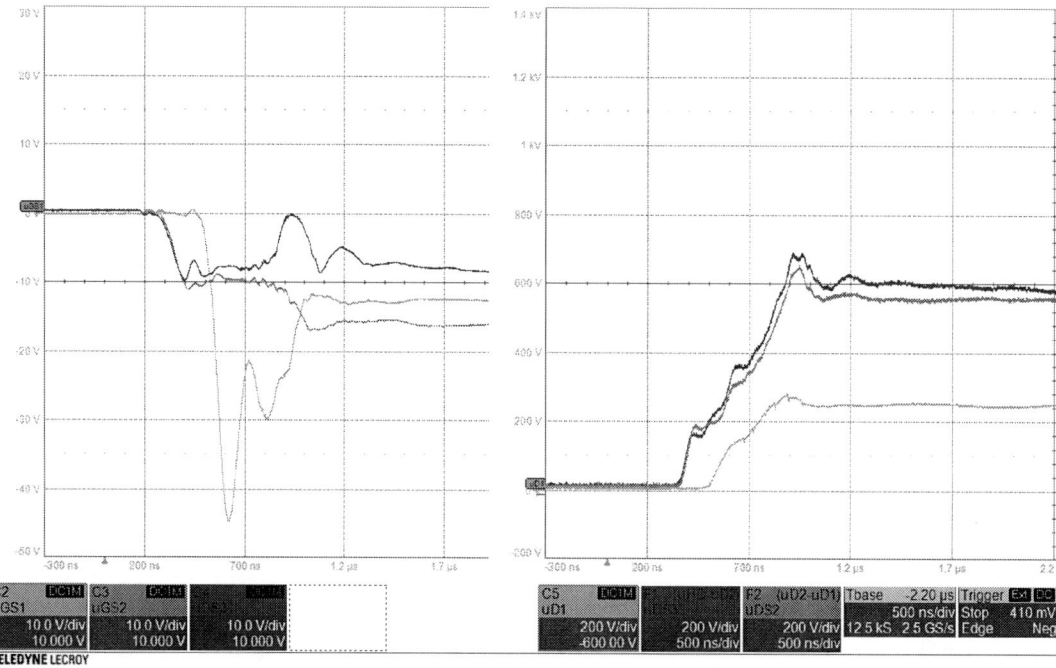

Fig. 13. Measurement of the gate-source voltages (left picture: u_{GS1}, u_{GS2}, u_{GS3}) and drain-source voltages (right picture: u_{D1}, u_{DS2}, u_{DS3}) at turn-off of the novel SiC-JFET super-cascode

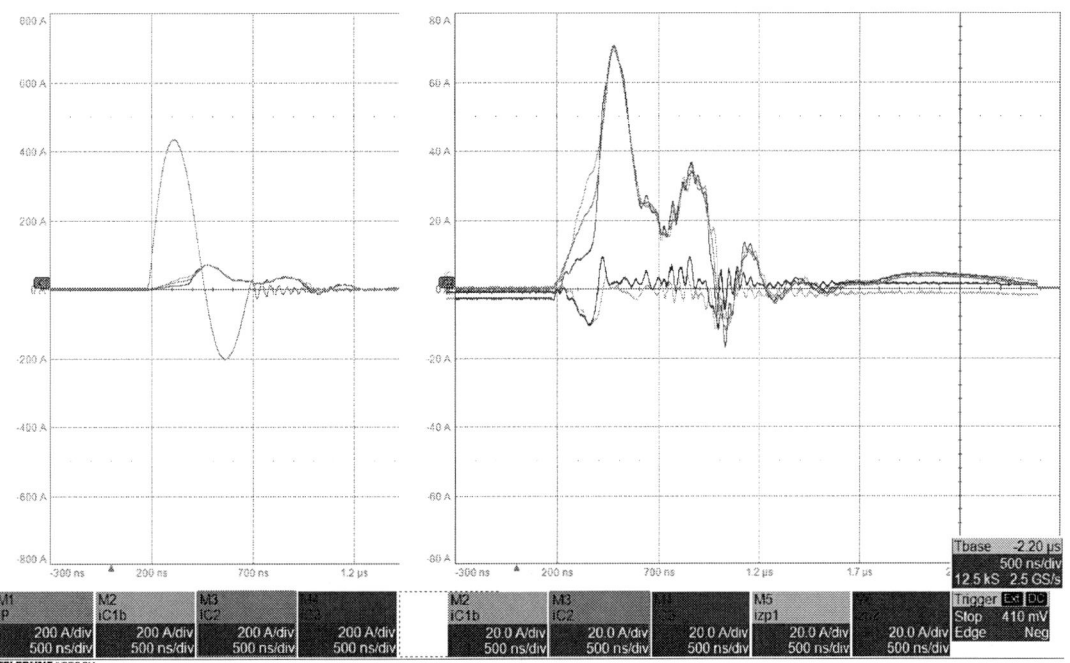

Fig. 14. Measurement of the currents inside the reference-voltage divider (i_{C1b}, i_{C2}, i_{C3}, i_{zp1}, i_{zp2}) of the novel SiC-JFET super-cascode (left picture: including the pulse generator current ; right picture: including the upper main JFET gate-paths currents)

failure management (incl. intrinsic protection against explosion of the submodule) in an MMC is achieved, too. Additionally, a novel SiC-JFET super-cascode is presented and investigated enabling the following improvements with respect to state of the art super-cascode topologies:

- A modular structure allows uncritical scalability of the blocking voltage as well as the current rating of the super-cascode enabling a suitability for highest cascode currents.
- A favorable turn-off sequence is enforced with an additional signal path (implemented with a pulse generator) to maintain a uniform, symmetrical voltage distribution of the drain-source-voltages.

REFERENCES

[1] R. Li, L. Xu, D. Holliday, F. Page, S. J. Finney, and B. W. Williams, "Continuous operation of radial multiterminal hvdc systems under dc fault," *IEEE Transactions on Power Delivery (Vol. 31, Issue: 1)*, 2016.

[2] Y. Chen, Z. Li, S. Zhao, X. Wei, and Y. Kang, "Design and implementation of a modular multilevel converter with hierarchical redundancy ability for electric ship mvdc system," *IEEE Journal of Emerging and Selected Topics in Power Electronics (Vol. 5, Issue: 1)*, 2016.

[3] S. Debnath and M. Saeedifard, "A new hybrid modular multilevel converter for grid connection of large wind turbines," *IEEE Transactions on Sustainable Energy (Vol. 4, Issue: 4)*, 2013.

[4] H. Nademi, A. Das, R. Burgos, and L. E. Norum, "A new circuit performance of modular multilevel inverter suitable for photovoltaic conversion plants," *IEEE Journal of Emerging and Selected Topics in Power Electronics (Vol. 4, Issue: 2)*, 2016.

[5] C. Dahmen and R. Marquardt, "Charge balancing for advanced mmc-double-submodules with ultra-low loss," *CPE-POWERENG*, 2019.

[6] ——, "Power losses of advanced mmc submodule topologies using si- and sic-semiconductors," *EPE*, 2019.

[7] ——, "Reduced capacitor size and on-state losses in advanced mmc submodule topologies," *EPE*, 2020.

[8] L. Fursin, X. Li, Z. Li, M. O'Grady, W. Simon, and A. Bhalla, "Reliability aspects of 1200v and 3300v silicon carbide mosfets," *IEEE 5th Workshop on Wide Bandgap Power Devices and Applications (WiPDA)*, 2017.

[9] K. Yano, Y. Tanaka, and M. Yamamoto, "Extremely low on-resistance sic cascode configuration using buried-gate static induction transistor," *IEEE Electron Device Letters (Vol. 39 , Issue: 12)*, 2018.

[10] K. Zhu, M. O'Grady, J. Dodge, J. Bendel, and J. Hostetler, "1.5 kw single phase ccm totem-pole pfc using 650v sic cascodes," *IEEE 4th Workshop on Wide Bandgap Power Devices and Applications (WiPDA)*, 2016.

[11] J. Biela, D. Aggeler, D. Bortis, and J. W. Kolar, "5kv/200ns pulsed power switch based on a sic-jfet super cascode," *IPMHVC*, 2008.

[12] X. Song, A. Q. Huang, L. Zhang, P. Liu, and X. Ni, "15kv/40a freedm super-cascode: A cost effective sic high voltage and high frequency power switch," *IEEE Energy Conversion Congress and Exposition (ECCE)*, 2016.

[13] X. Li, H. Zhang, and A. Bhalla, "Medium voltage power module based on sic jfets," *IEEE Applied Power Electronics Conference and Exposition (APEC)*, 2017.

[14] C. Dahmen, F. Kapaun, and R. Marquardt, "Analytical investigation of efficiency and operating range of different modular multilevel converters," *PEDS*, 2017.

[15] P. Friedrichs, H. Mitlehner, R. Schorner, K.-O. Dohnke, R. Elpelt, and D. Stephani, "Stacked high voltage switch based on sic vjfets," *ISPSD*, 2003.

[16] J. Biela, D. Aggeler, D. Bortis, and J. W. Kolar, "Balancing circuit for a 5kv/50ns pulsed-power switch based on sic-jfet super cascode," *IEEE Transactions on Plasma Science (Vol.: 40, Issue: 10)*, 2012.

[17] X. Li, A. Bhalla, P. Alexandrov, J. Hostetler, and L. Fursin, "Series-connection of sic normally-on jfets," *ISPSD*, 2015.

[18] B. Gao, "Scalable medium voltage and high voltage super cascode power modules," Ph.D. dissertation, North Carolina State University, 2018.

[19] X. Song, A. Q. Huang, X. Ni, and L. Zhang, "Comparative evaluation of 6kv si and sic power devices for medium voltage power electronics applications," *IEEE 3rd Workshop on Wide Bandgap Power Devices and Applications (WiPDA)*, 2015.

Control of an Active Gate Driver for an Electric Vehicle Traction Inverter Using Artificial Neural Networks

Julius Wiesemann, Jacob Dumtzlaff, Axel Mertens
LEIBNIZ UNIVERSITY HANNOVER
Welfengarten 1
Hannover, Germany
Phone: +49 (0) 511 762-5224
Email: julius.wiesemann@ial.uni-hannover.de
URL: http://www.ial.uni-hannover.de

Keywords

≪Smart Gate Drivers≫, ≪Neural network≫, ≪EMC/EMI≫, ≪Silicon Carbide (SiC)≫, ≪Power converters for EV≫.

Abstract

Electric vehicle drivetrains using wide-bandgap semiconductors face challenges regarding EMI and accelerated machine aging due to the fast switching transients. This paper presents a method of controlling a variable-resistance active gate driver with the help of a neural network in order to reduce the drawbacks of fast switching while increasing efficiency. Measurements covering the whole MOSFET operating range and a sinusoidal inverter output current prove that the proposed method effectively reduces losses while also reducing switching speed and, in this way, reduces EMI issues and machine damage.

Introduction

The efficiency of drive systems in electric vehicles (EV) can be increased through the use of fast-switching power semiconductors like SiC MOSFETs. However, the steep gradients of the voltage pulses at the inverter outputs lead to increased electromagnetic interference (EMI) issues [1] in the vehicle and increased aging of the electrical machine, especially the winding insulation and machine bearings [2]. For this reason, extensive research has been performed with the objective of reducing these effects. The use of input and output filters is the most common method employed [3]. However, filters are bulky, heavy and increase the system losses. Active gate drivers (AGDs) offer another solution to the problem by influencing the switching behavior of MOSFETs or IGBTs through variation of the current that flows into the gate of the semiconductor [4, 5]. In this way, active gate drivers can reduce the switching speed of the MOSFET without substantially increasing the switching losses. In [6], a method to estimate the switching behavior of an IGBT driven by an AGD through a neural network is presented. Afterwards, an optimization of the switching behavior is performed using the neural network. However, this approach requires additional algorithms to find the optimum driver settings after the neural network has been fitted. Additionally, the method is limited to estimating only one operating point through the neural network. Another approach is presented in [7] where the switching behavior of a MOSFET is altered to minimize the switching losses while tolerating higher dv/dt values. The output current range is divided into four stages, and in each stage a different driver setting is used. However, the variability of both the DC link voltage and the junction temperature are not accounted for and the division of the output current into just four ranges leaves room for improvement.

In this paper, a method to determine the optimal settings for an active gate driver using an artificial neural network (NN) in an electric vehicle drivetrain is presented. The network is trained to automatically determine the optimum driver settings across a wide range of operating conditions. In the first section, the operating conditions in an EV are analyzed. Afterwards, the structure and basic control of the AGD used are explained and the neural-network-based control is introduced. In the following section, a measurement setup to derive training data for a neural network is presented. After introducing the training of the neural network, the validation of the control method in the operating range of the MOSFET and for sinusoidal inverter load currents is provided. Lastly, a conclusion is drawn and an outlook of future work is given.

Influence of Changing Operating Conditions in Electric Vehicles on the Switching Behavior of a SiC MOSFET

Drive inverters in electric vehicles face varying operating conditions. The load current vary across a wide range depending on the vehicle speed and acceleration, while the battery voltage, and thus the DC link voltage, varies with the load current and the battery's state of charge [8]. The junction temperature of a SiC MOSFET also mainly depends on the load current [9] and cooling conditions.

When designing an EV drivetrain, the switching behavior must be taken into account. The slope of the drain-source voltage v_{ds} at turn-on (dv/dt_{on}) and turn-off (dv/dt_{off}) and the switching energies E_{on} and E_{off} mainly influence the electromagnetic compatibility (EMC) and the efficiency of the inverter as well as the negative impacts on the electrical machine. For the design of the cooling system, the assessment of EMC, and the design of EMI filters, the worst-case values of switching energies and switching times have to be considered, which, however, leads to oversizing for other working conditions.

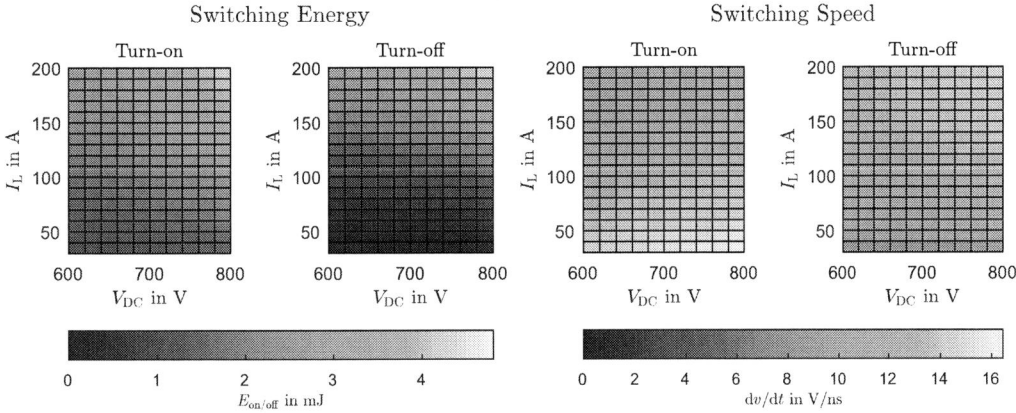

Fig. 1: Experimental results showing the variation of the switching behavior of an Infineon FF6MR12KM1 SiC MOSFET measured at $T_j = 90\,°C$ with $R_{g,on} = 2\,\Omega$ and $R_{g,off} = 4\,\Omega$ at different operating points of DC link voltage and load current.

Fig. 1 shows how variations in operating conditions affect the switching performance of a SiC MOSFET. A changing load current especially leads to a significant variation in switching times and switching energies. An active gate driver can modulate the switching behavior of a SiC MOSFET and thus lead to a higher utilization of the maximum permissible dv/dt, keeping the losses low.

Active Gate Driver

The AGD used in this paper influences the switching behavior of the SiC MOSFET by using a variable gate resistance. It consists of four parallel gate resistors for turn-on and four parallel gate resistors for turn-off. Each gate resistor can be activated and deactivated with a timing accuracy of 2 ns through small-signal MOSFETs, driven by two four-channel auxiliary gate drivers and signal isolators. A detailed description of the gate driver can be found in [5]. A picture of the AGD can be seen in Fig. 4. To drive the MOSFET, the active gate driver uses the gate resistance vector $\mathbf{R_g} = [R_{g,1}; R_{g,2}]^T$. Following the initial activation of $R_{g,1}$, the gate resistor $R_{g,2}$ is activated after the variable time period t_{sw} has passed. By using separate vectors for turn-on and turn-off, a small gate resistor is initially active during turn-on to reduce the rise time of the drain current i_D. Next, a large gate resistor is activated to increase the voltage fall time. The control during the turn-off works similarly, but in the opposite order: first, a large gate resistor is activated to slow down the voltage rise and this is then followed by a small gate resistor to reduce the current fall time.

Neural-Network-Based Control

The neural network is supposed to always choose the optimal t_{sw} settings for the AGD, based on load current, DC link voltage and MOSFET junction temperature. In order to fit the optimal driving data, a neural network with the structure shown in Fig. 2 is used.

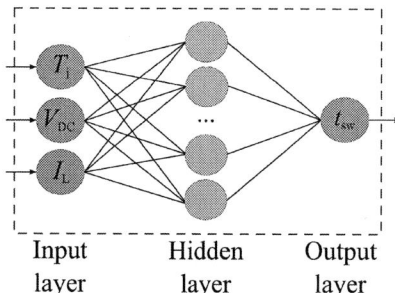

Fig. 2: Graphical representation of the neural network used to determine the optimal driver settings.

Two networks are trained, one for the turn-on and another for the turn-off process. The networks consist of three input neurons, a hidden layer of ten neurons, and a single output neuron to generate the variable switching time t_{sw}. The use of ten neurons in the hidden layer has proven to be a good compromise between fitting accuracy and complexity. The sigmoid function is used as the activation function. The network is created off-line for one specific MOSFET, which, due to the effort of training the network, makes it especially attractive for inverters that are produced in large numbers like automotive inverters. An adaption of the network during the operation of the inverter is not foreseen because it would require additional sensors and more computational power in the inverter, which reduces robustness and cost efficiency.

The training process for the neural network is shown in Fig. 3.

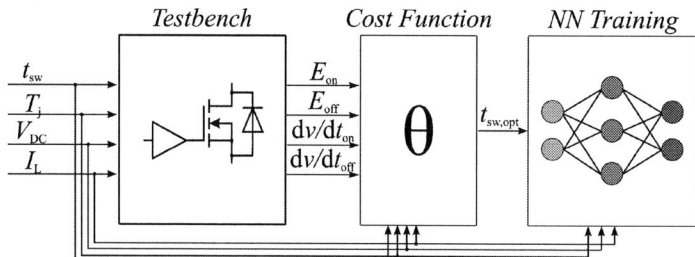

Fig. 3: Flow diagram for the neural network training process.

Test data is recorded across a wide range of operating conditions and t_{sw} settings and the cost function (1), similar to the one introduced in [5], is employed to find the optimal driving vector for each operating point. It is parameterized to determine the gate driver settings at which the $dv/dt_{on/off}$ is limited to a

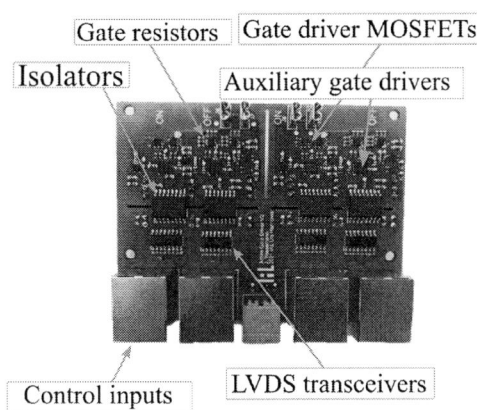 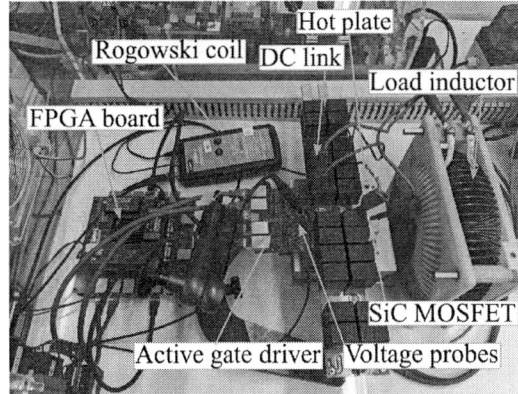

Fig. 4: Active gate driver and test bench with which the training and validation measurements were done.

setpoint $\mathrm{d}v/\mathrm{d}t_{\mathrm{set}}$:

$$
\theta = \begin{cases} \left(1 + \dfrac{\sqrt{(\frac{\mathrm{d}v}{\mathrm{d}t_{\mathrm{set}}} - \frac{\mathrm{d}v}{\mathrm{d}t_{\mathrm{on}}})^2}}{\frac{\mathrm{d}v}{\mathrm{d}t_{\mathrm{set}}}}\right)^3 & \text{for turn-on} \\[2em] \left(1 + \dfrac{\sqrt{(\frac{\mathrm{d}v}{\mathrm{d}t_{\mathrm{set}}} - \frac{\mathrm{d}v}{\mathrm{d}t_{\mathrm{off}}})^2}}{\frac{\mathrm{d}v}{\mathrm{d}t_{\mathrm{set}}}}\right)^3 & \text{for turn-off} \end{cases}
\tag{1}
$$

This way, the MOSFET will switch at the $\mathrm{d}v/\mathrm{d}t$ setpoint throughout the whole operating range. The value for $\mathrm{d}v/\mathrm{d}t_{\mathrm{set}}$ can be set according to the limits imposed by EMC and the electrical machine. After the optimal settings have been derived from the measured data using the cost function, the neural network is trained.

Derivation of Neural Network Training Data

The training data for the neural network was acquired in double-pulse tests using the variable-resistance active gate driver, an Infineon FF6MR12KM1 1200 V SiC MOSFET, and the test bench shown in Fig. 4. Different DC link voltages, load currents, baseplate temperatures and driving vectors for the active gate driver were tested. Table I shows the measurement conditions.

Table I: Measurement Conditions

Item	Symbol	Range
Load current	I_{L}	$[20\,\mathrm{A}, 30\,\mathrm{A}, \dots 200\,\mathrm{A}]$
DC link voltage	V_{DC}	$[600\,\mathrm{V}, 640\,\mathrm{V}, \dots 800\,\mathrm{V}]$
Junction temperature	T_{j}	$[25\,^\circ\mathrm{C}, 60\,^\circ\mathrm{C}, \dots 130\,^\circ\mathrm{C}]$
Turn-on gate resistor vector	$\mathbf{R_{g,on}}$	$[0.44\,\Omega; 4\,\Omega]^T$
Turn-off gate resistor vector	$\mathbf{R_{g,off}}$	$[8\,\Omega; 0.73\,\Omega]^T$
Switching time for turn-on	$t_{\mathrm{sw,on}}$	$[2\,\mathrm{ns}, 4\,\mathrm{ns}, \dots 102\,\mathrm{ns}]$
Switching time for turn-off	$t_{\mathrm{sw,off}}$	$[94\,\mathrm{ns}, 96\,\mathrm{ns}, \dots 192\,\mathrm{ns}]$

The drain current i_{D} was measured with a 30 MHz CWT UltraMini Rogowski coil. A Testec TT-HV 150 voltage probe with a bandwidth of 300 MHz was used to measure the drain-source voltage v_{DS}. All data were recorded using a 1 GHz LeCroy MDA810A oscilloscope. An example of the recorded waveforms for turn-on and turn-off can be seen in Fig. 5. Only selected variations of t_{sw} are shown for better clarity. The waveforms at the optimal value of t_{sw} are marked in red.

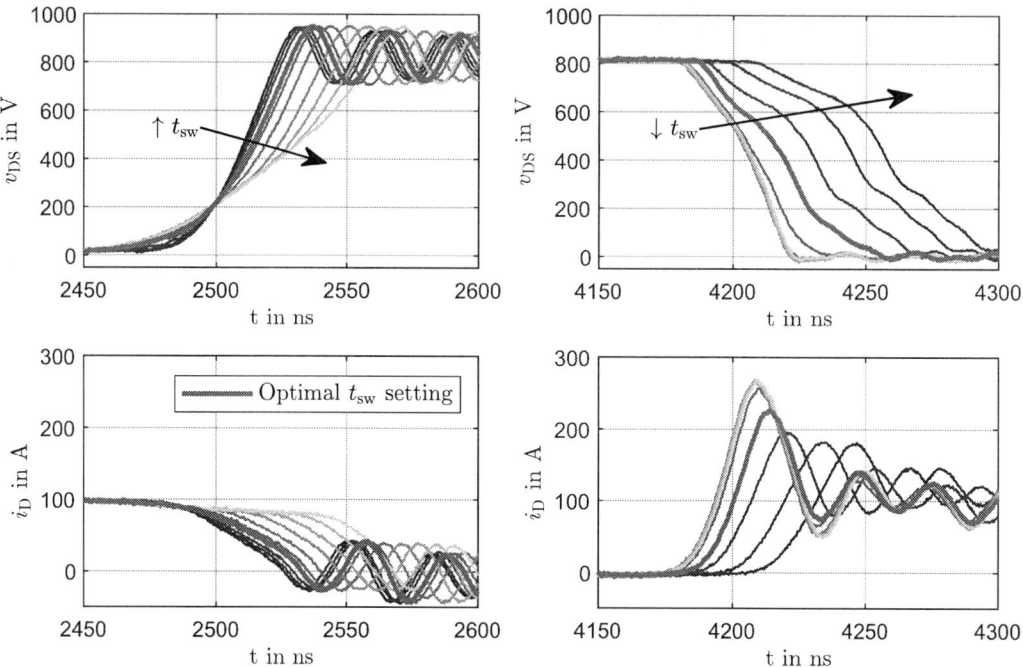

Fig. 5: Measured waveforms of v_{DS} and i_D at $V_{DC} = 800$ V, $I_L = 100$ A, $T_j = 60$ °C and at different values of t_{sw}. The optimal t_{sw} setting is marked in red. For a better overview, only every fifth variation of t_{sw} is shown.

Neural Network Training

After all the data points have been measured and evaluated by the cost function, only one value of both $t_{sw,on}$ and $t_{sw,off}$ is selected for each combination of V_{DC}, I_L, T_j, reducing the number of samples from its original 23256 to 456. These data are fed into a Matlab-based neural network training program that uses the Bayesian regularization algorithm to fit the neural network to the training data. 80 % of the data are used for training, while 10 % are used for validation and 10 % for testing.

Fig. 6 shows the input value of the neural network training process for turn-on and turn-off at two different temperatures. A gradual change in the optimal switching time between the different voltage, current and temperature values can be seen. For the turn-on, the optimal switching time rises with the current and decreases with the DC link voltage. The reason for this can be found in the switching behavior of

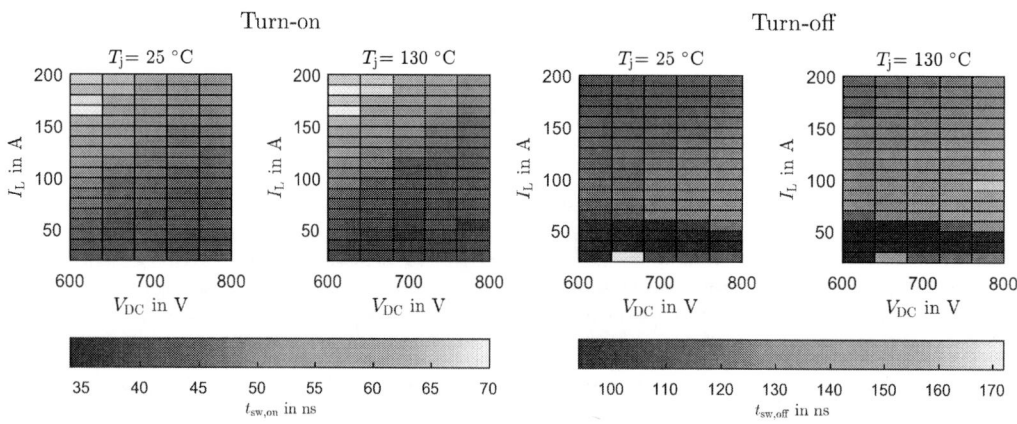

Fig. 6: Neural Network Training Data at $T_j = 25$ °C and $T_j = 130$ °C.

the SiC MOSFET: first, the drain current rises before the voltage across the MOSFET drops. At high currents, the current rise phase takes longer, which requires $R_{g,1}$ to be active longer in order to speed up the current rise phase. The desired $\mathrm{d}v/\mathrm{d}t$ is set in the following voltage fall phase. Considering that the voltage slope at higher voltages is less steep (see Fig. 1), a lower setting for t_{sw} is sufficient to reach the desired $\mathrm{d}v/\mathrm{d}t$ at high DC link voltages. The optimal value of t_{sw} for the turn-off process, however, increases with the DC link voltage and shows a maximum at currents in the middle of the current range. At very low currents, the switching speed of the MOSFET cannot be influenced much by the AGD since it is limited by the charging rate of the semiconductor output capacitance [10]. As the voltage fall and current rise phase overlap more at turn-off than at turn-on (see Fig. 5), the optimal switching time is also influenced by the drain current charging the output capacitance.

Neural Network Validation

To test the neural-network-based control, double-pulse measurements were performed. Comparative measurements were also done with a conventional gate driver with different gate resistor values. Since the gate loop design has an influence on the switching behavior [11], using a different gate driver would lead to results which could not be compared. For this reason, the AGD was used with only one gate resistor activated during a double-pulse test to replicate the behavior of a CGD.

The neural network was exported as Matlab code from the training software and implemented on the Xilinx Zynq-700 SoC that controls the AGD using the Matlab Embedded Coder. The driver for the AGD is implemented in the FPGA part, while the neural network runs on the CPU. The main interrupt of the CPU is triggered at a frequency of 10 kHz. An analysis of the code runtime on the CPU shows that the calculation of one neural network takes 2.5 µs which means that the calculation of the six neural networks needed for a three-phase inverter requires 15 % of the total interrupt time. A deeper investigation shows that the majority (2.2 µs) of the time required to calculate the neural network output goes into the evaluation of the sigmoid activation function. If timing becomes critical, a replacement of the activation function through a lookup table could increase the calculation speed.

Constant Current Validation

Double-pulse tests at various voltages and currents spread throughout the whole operating range are conducted. The resulting $\mathrm{d}v/\mathrm{d}t$ can be seen in Fig. 7.

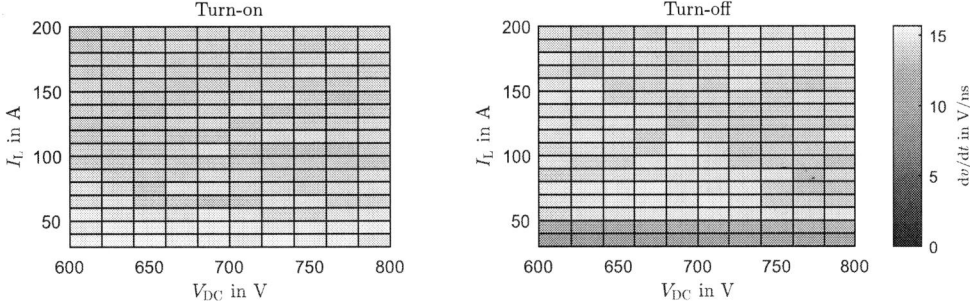

Fig. 7: Double-pulse test result with the AGD controlled by the neural network at different voltages and currents at $T_j = 90\,°C$ and a $\mathrm{d}v/\mathrm{d}t$ setting of 15 V/ns.

In comparison to Fig. 1, the turn-on process shows an almost constant $\mathrm{d}v/\mathrm{d}t$ throughout the whole operating range. At low currents during turn-off, the $\mathrm{d}v/\mathrm{d}t$ is reduced due to the slow charging of the output capacitance by the low drain current.

To evaluate the performance of the neural-network-driven AGD, it is compared to a CGD with $R_{G,on} = 2\,\Omega$ and $R_{G,off} = 4\,\Omega$. The differences in switching energies $\Delta E = E_{AGD} - E_{CGD}$ and switching speed $\Delta\mathrm{d}v/\mathrm{d}t = \mathrm{d}v/\mathrm{d}t_{AGD} - \mathrm{d}v/\mathrm{d}t_{CGD}$ are shown in Fig. 8.

The figure shows that especially the turn-on energies at high currents are reduced by the AGD. This is mostly achieved by increasing the $\mathrm{d}v/\mathrm{d}t$ at these currents, since the MOSFET's turn-on process at high

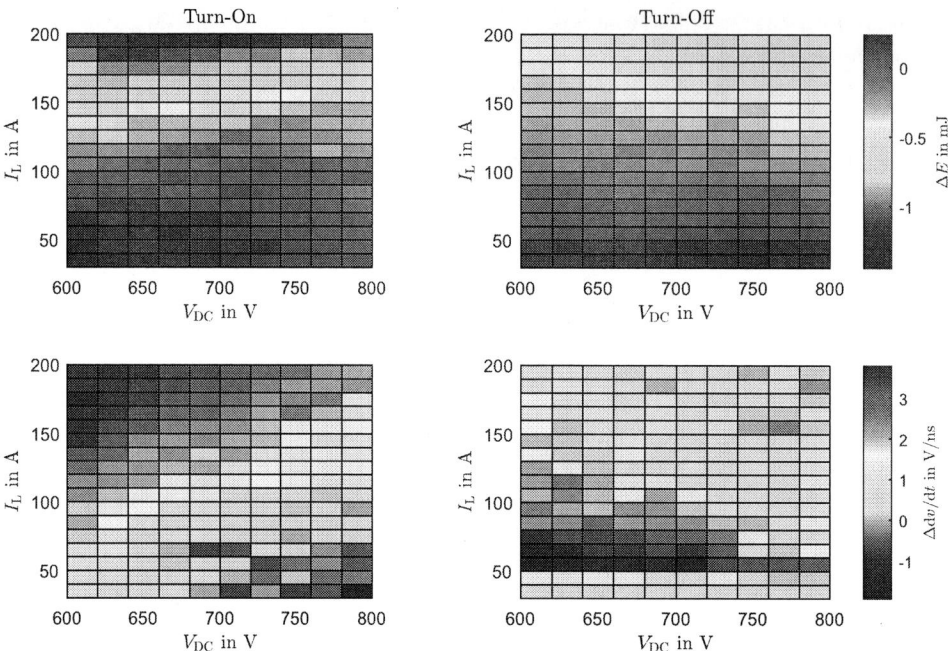

Fig. 8: Difference between the AGD controlled by the neural network and the CGD at different voltages and currents with $T_j = 90\,°C$, a dv/dt setting of $15\,V/ns$, $R_{G,on} = 2\,\Omega$ and $R_{G,off} = 4\,\Omega$.

currents is slower than the desired switching speed. At high DC link voltages and low currents, the CGD turns the MOSFET on faster than the desired rate of $15\,V/ns$. In this area, the AGD reduces the switching speed so as to avoid exceeding the maximum dv/dt.

The turn-off process behaves differently from the turn-on process, where the highest dv/dt occurs at high currents. However, the switching speed does not exceed the maximum dv/dt, which is why the AGD control does not influence the dv/dt. Even though the dv/dt is left almost unchanged, the switching energies are reduced through the use of the AGD. Although the switching speed is increased at low currents, since the switching energies at low currents are small, the increased switching speed does not influence the switching energy much.

Sinusoidal Current Validation

To evaluate the performance of the neural-network-based control in an EV drive inverter, double-pulse measurements were done to reproduce the behavior of the inverter. A virtual sinusoidal current defined by its RMS value I_{RMS}, a switching frequency f_s and an output current base frequency f_b was used to determine the currents at which the MOSFET in an inverter would switch during half of an output current period. Double-pulse experiments with the AGD and CGD were carried out at the pre-calculated currents. The gate resistor values of the CGD of $R_{G,on} = 2\,\Omega$ and $R_{G,off} = 4\,\Omega$ were chosen to limit the maximum dv/dt to around $15\,V/ns$ in the whole operating range.

The result of the validation measurements at a virtual load current of $I_{RMS} = 142\,A$ are shown in Fig. 9.

The switching times of the AGD show less dependence on the load current compared to the CGD. The switching energies are reduced, with a larger effect on the turn-on energy visible. The switching losses in a half bridge are calculated by adding up the switching losses at the predefined currents for the duration of one fundamental frequency period:

$$P_{sw} = f_g \cdot \sum_{0}^{f_g^{-1}} [E_{on} + E_{off}] \tag{2}$$

Since the reverse recovery losses in the complementary body diode are negligibly small compared to

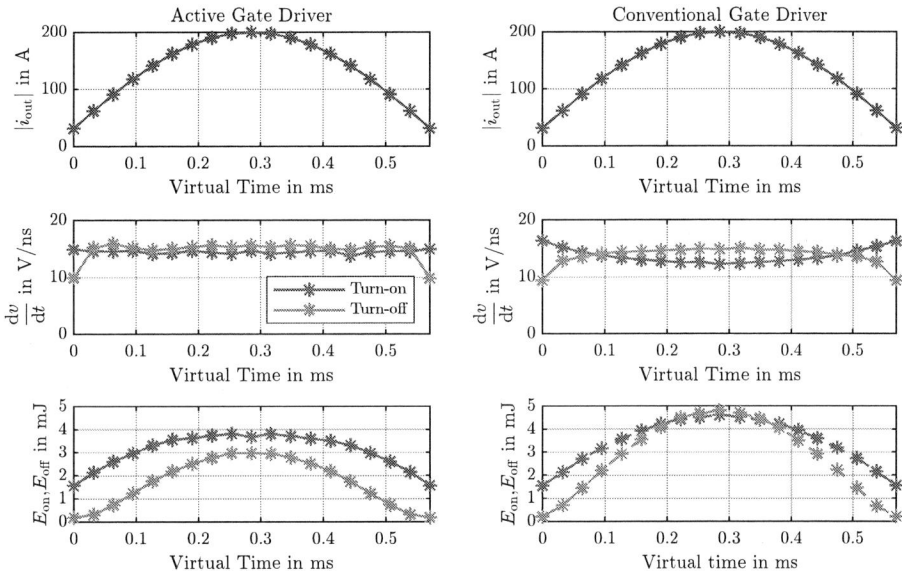

Fig. 9: Performance of the AGD in a virtual inverter with a sinusoidal output current with $T_j = 80\,°C$, $f_s = 24\,kHz$, $f_b = 600\,Hz$, and a DC link voltage of 800 V.

the switching losses, and hard to measure accurately, they are neglected [10]. The MOSFET and diode conduction losses are not influenced by the AGD and can thus be excluded from the comparison. The switching losses of the MOSFET driven by the AGD are 114 W, while the MOSFET driven by a CGD generates average switching losses of 142 W. The average dv/dt is 14.3 V/ns for the AGD and 13.7 V/ns for the CGD. The maximum dv/dt mainly responsible for EMI issues [1] is 15.2 V/ns for the AGD and 16.4 V/ns for the CGD.

Further measurements at DC link voltages of 600 V and 800 V and a sinusoidal load current of 142 A are carried out. The CGD measurements are carried out with different gate resistor combinations. Two AGD settings for dv/dt setpoints of 10 V/ns and 15 V/ns are used. The results can be seen in Fig. 10. The data show that at $V_{DC} = 800\,V$, the CGD version with $R_{on} = 1.3\,\Omega$ and $R_{on} = 2.6\,\Omega$ has losses similar to the AGD version with $dv/dt_{set} = 15\,V/ns$. However, the average and peak values of dv/dt are considerably higher for the CGD version. At $V_{DC} = 600\,V$, the differences in P_{loss} and $\overline{dv/dt}$ are reduced and only the dv/dt_{max} still shows a considerable difference, which means that there is still an influence on the EMC behavior of the inverter. If AGD and CGD at similar $\overline{dv/dt}$ are compared, the CGD with $R_{on} = 2\,\Omega$ and $R_{on} = 4\,\Omega$ shows a similar $\overline{dv/dt}$ compared to the 15 V/ns AGD. In this case, the CGD losses are higher than the AGD losses at $V_{DC} = 800\,V$. At $V_{DC} = 600\,V$, the $\overline{dv/dt}$ of the CGD decreases below the desired value of 15 V/ns. For this reason, the losses of the CGD are higher than the AGD losses since the CGD does not utilize the maximum permissible dv/dt.

Conclusion

In this paper, a method of controlling an AGD through an artificial neural network for use in an EV drive inverter is presented. The influence of changes in the working conditions of electric vehicle inverters, such as changing load currents, battery voltages or MOSFET junction temperatures, on the switching behavior of a SiC MOSFET is shown. After an introduction into the AGD used in this paper, a method of compensating for these changes using a neural network is introduced, and after deriving the optimal driver settings for all working conditions from data gathered in a double-pulse experiment, neural networks to control the AGD are trained. The network performance is tested in a double-pulse experiment throughout the whole operating range of the MOSFET and at current values matching the behavior of an EV inverter. Compared to a CGD, the switching losses are reduced, while almost constant switching times are achieved, meaning that throughout the operating area of the inverter, the SiC MOSFETs are

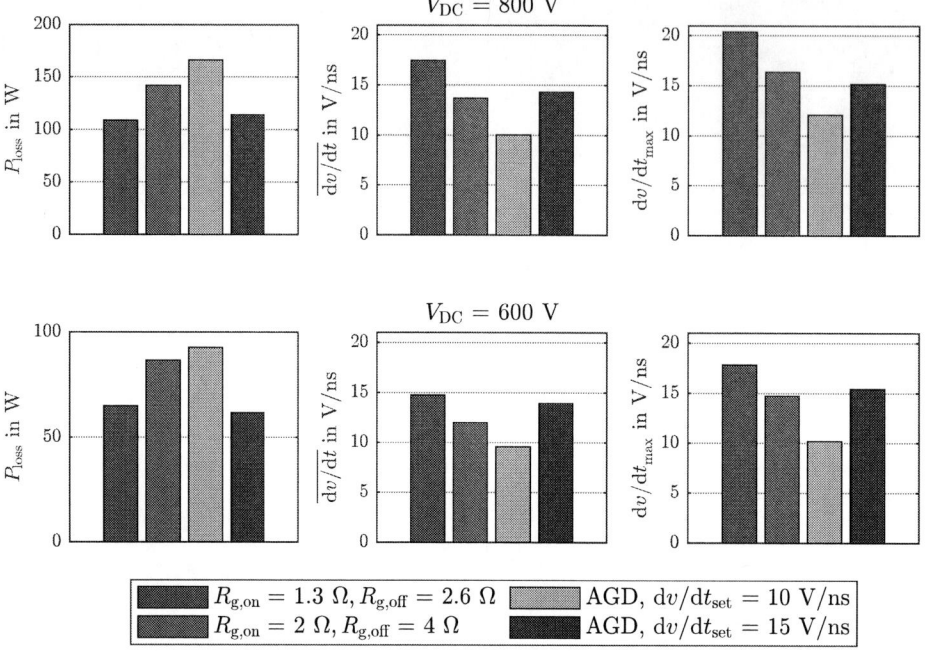

Fig. 10: Comparison of the performance of an AGD with different dv/dt settings and a CGD with different gate resistors at various DC link voltages.

switching at the maximum dv/dt imposed by the EMC requirements and machine characteristics. In future work, the effect of the AGD on the CM currents, EMC behavior and losses in an actual inverter will be investigated.

References

[1] R. Shirai, K. Wada, T. Shimizu, and D. Xu, "Suppressing emi noise to can communication by using active gate driver," in *2020 IEEE 9th International Power Electronics and Motion Control Conference (IPEMC2020-ECCE Asia)*, 2020, pp. 2648–2651.

[2] Y. Xu, X. Yuan, F. Ye, Z. Wang, Y. Zhang, M. Diab, and W. Zhou, "Impact of high switching speed and high switching frequency of wide-bandgap motor drives on electric machines," *IEEE Access*, vol. 9, pp. 82 866–82 880, 2021.

[3] H. Movagharnejad and A. Mertens, "Design methodology for dimensioning emi filters for traction drives with sic inverters," in *2021 23rd European Conference on Power Electronics and Applications (EPE'21 ECCE Europe)*, 2021, pp. 1–10.

[4] S. Zhao, X. Zhao, Y. Wei, Y. Zhao, and H. A. Mantooth, "A review of switching slew rate control for silicon carbide devices using active gate drivers," *IEEE Journal of Emerging and Selected Topics in Power Electronics*, vol. 9, no. 4, pp. 4096–4114, 2021.

[5] J. Wiesemann and A. Mertens, "An isolated variable-resistance active gate driver for use in sic-driven inverters," in *IECON 2021 47th Annual Conference of the IEEE Industrial Electronics Society*, 2021, pp. 1–6.

[6] Z. Meng, Y. Yang, Y. Gao, S. Ai, Y. Zhang, Y. Lv, Z. Zhang, Y. Wen, L. Wu, P. Zhang, and J. D. Thomson, "Prediction method of driving strategy of high-power igbt module based on mea-bp neural network," *IEEE Access*, vol. 8, pp. 94 731–94 747, 2020.

[7] J. D. Kagerbauer and T. M. Jahns, "Development of an active dv/dt control algorithm for reducing inverter conducted emi with minimal impact on switching losses," in *2007 IEEE Power Electronics Specialists Conference*, 2007, pp. 894–900.

[8] R. C. Kroeze and P. T. Krein, "Electrical battery model for use in dynamic electric vehicle simulations," in *2008 IEEE Power Electronics Specialists Conference*, 2008, pp. 1336–1342.

[9] E. Trancho, E. Ibarra, P. Prieto, A. Arias, A. Lis, and A. P. Pai, "Novel thermal management strategy for improved inverter reliability in electric vehicles," *Applied Sciences*, vol. 10, no. 22, 2020. [Online]. Available: https://www.mdpi.com/2076-3417/10/22/8024

[10] J. Ebersberger, J.-K. Mller, and A. Mertens, "Dynamic characterization of a sic-mosfet half bridge in hard- and soft-switching and investigation of current sensing technologies," in *2020 22nd European Conference on Power Electronics and Applications (EPE'20 ECCE Europe)*, 2020, pp. P.1–P.8.

[11] D. Han and B. Sarlioglu, "Study of the switching performance and emi signature of sic mosfets under the influence of parasitic inductance in an automotive dc-dc converter," in *2015 IEEE Transportation Electrification Conference and Expo (ITEC)*, 2015, pp. 1–8.

Cascaded H-Bridge Converter Designs for Future Short-Range All-Electric Aircraft Propulsion

Maximilian Hagedorn[1,2], Malte Lorenz[1,2], and Axel Mertens[1,2]
[1]Institute for Drive Systems and Power Electronics
Leibniz University Hannover, Germany
[2]Cluster of Excellence SEA Sustainable and Energy-Efficient Aviation
Technische Universität Braunschweig, Germany
Email: max.hagedorn@ial.uni-hannover.de
URL: http://www.ial.uni-hannover.de

Acknowledgments

We would like to acknowledge the funding by the Deutsche Forschungsgemeinschaft (DFG, German Research Foundation) under Germany's Excellence Strategy EXC 2163/1 - Sustainable and Energy Efficient Aviation Project-ID 390881007. Furthermore, we would like to thank Walter Cistjakov (Technische Universität Braunschweig, Germany) for the useful discussions on battery technologies that contributed to this work.

Keywords

≪Cascaded H-Bridge≫, ≪DC-AC converter≫, ≪Modular converter≫, ≪Green aviation≫

Abstract

The cascaded H-bridge converter topology makes a high voltage power supply structure with distributed energy storage for all-electric aircraft possible. In this paper, different cascaded H-bridge converter topologies are compared regarding mass and efficiency, with a focus on the requirements for the energy storage in a battery-powered short-range aircraft.

Introduction

The ongoing electrification of mobility has reached the aviation sector, where multiple research projects are evaluating the possibilities for electrifying aircraft propulsion [1]. The document *Flightpath 2050* expects to see short-range all-electric aircraft (SAEA) as an integral part of the aviation market by 2050 [2]. While for mid-range applications, hybrid concepts are suggested, fully battery-powered all-electric aircraft are being considered for short-range applications [3]. There are different concepts concerning how the onboard power supply can be realized, but all have a common design goal, which is the mass optimization of the whole propulsion system while maintaining a high efficiency. While converter topologies for a single DC backbone have been studied in [4], this paper focuses on a decentralized approach, where the power conversion system does not rely on a centralized battery system, but instead uses multiple battery packs which are distributed across separate modules. One feasible inverter topology is the cascaded H-bridge Converter (CHB), which is examined in this paper in different module variants. The CHB supplies each phase of the machine independently, which results in an H-bridge input current that consists of superimposed DC and AC components. The AC component of the current increases the root-mean-square (RMS) value, which results in higher losses in the internal battery resistance compared to the pure DC current, leading to higher battery temperatures which may negatively affect the battery lifetime [5]. The increased losses furthermore reduce the system efficiency, which results in a higher battery mass.

For this reason, it is considered impractical to directly connect the module battery to the H-bridge and instead additional circuitry is needed to filter the battery current.

To give an example of how a filtered battery current influences the battery mass, a factor $r_I = \frac{I_{bat}}{\bar{i}_{bat}}$ is introduced describing the relationship between the RMS current I_{bat} and the DC component $\bar{i}_{bat}$. Two different values for r_I are compared in a small calculation example, which is based on the reference mission profile from [4], where a battery energy of about 10 MWh is needed. Assuming a constant discharge, the battery output power is 2.5 MW. Without any filter circuitry, the factor r_I would be 1.31 at a power factor of $\cos\varphi = 0.84$. The battery is assumed to have an energy density of 700 Wh kg^{-1}, which is proposed in [6] as a feasible value for future battery technologies. To consider the losses at the internal battery resistance caused by I_{bat}, the battery resistance is estimated for the resulting parallel and serial cell stack at an assumed battery voltage of 1 kV. Due to the lack of data on the internal resistance in ongoing research on new cell technologies, the internal battery resistance is estimated based on a state-of-the-art automotive lithium-ion battery cell [7] which leads to a battery resistance of 11.7 mΩ. With these parameters, an r_I of 1.31 would lead to a battery that is 286 kg heavier than that necessary with an r_I of 1.01, which can be achieved with appropriate filter circuitry.

Since an SAEA is expected to be realized by 2050, higher blocking voltages in the upcoming generations of SiC MOSFETs can be expected. This is accounted for in a scalable semiconductor model which is presented in [4] and used in this work.

The CHB variants considered in this paper either contain a single-stage LC filter (CHBSLC), a dual-stage LC filter (CHBDLC), a DC-DC converter (CHBDCDC) or an interleaved DC-DC converter (CHBIL). As the design parameters for the battery and machine are not yet fully specified, the module variants are compared in two parameter studies to give an idea of the inverter designs that can be expected. Regarding the battery, the cell voltage factor $r_{uc} = \frac{v_{c,max}}{v_{c,min}}$, resulting from the different cell voltages at the highest and lowest states of charge (SoC), and the factor r_I are key parameters in the inverter design. The influence of these two battery-related parameters on the inverter design is investigated in the first parameter study. Moreover, the design of the passive components of the inverter is strongly affected by the fundamental machine frequency f_1, and the switching frequency of the DC-DC converter f_{dcdc}. Therefore these two parameters are varied in the second parameter study.

Converter Topology

Fig. 1 shows the topology of the propulsion system for a propeller that is assumed to be driven by a permanent magnet synchronous machine (PMSM). The inverter supplying the PMSM is a CHB, whose phase legs contain a number of n_{mod} modules connected in series, where i_{ph} is the phase current and v_{mod} is the module output voltage. The first module of the first phase is depicted in detail with its H-bridge (blue box), a filter circuit dependent on the module design (gray boxes), and a battery. For simplicity, the other two phases are represented by ideal voltage sources. To modulate the phase voltage v_{ph}, a phase-shifted pulse-width modulation is assumed, whereby each module receives the same sinusoidal reference signal and a shifted carrier signal, resulting in a switching frequency of $f_{s,eff}$ in v_{ph}. Neglecting the switching components, the H-bridge input current

$$i_d = \frac{M\hat{i}_{ph,1}}{2}\left(\cos(\varphi_1) - \cos(2\omega_1 t + \varphi_1)\right) \tag{1}$$

has a second harmonic component that oscillates at twice the fundamental angular frequency $2\omega_1$ and furthermore depends on the modulation index $M = \frac{\hat{v}_{mod,set}}{v_d}$ and the load angle φ_1 [8]. Here, $\hat{v}_{mod,set}$ denotes the module's setpoint voltage and v_d the module's input voltage.

The design of the CHB modules follows the calculations in [9], where the sum of all H-bridge input voltages in one phase is defined to be greater than $v_{ph,max} = a_{res}\sqrt{2}V_{ph,N}$, where $V_{ph,N}$ is the RMS phase voltage of the machine and a_{res} is a safety factor which accounts for deviations in the control system. The change of the H-bridge's input voltage due to the SoC variation in the battery needs to be accounted for, since it determines the maximum operating voltage of the MOSFETs. The semiconductor model

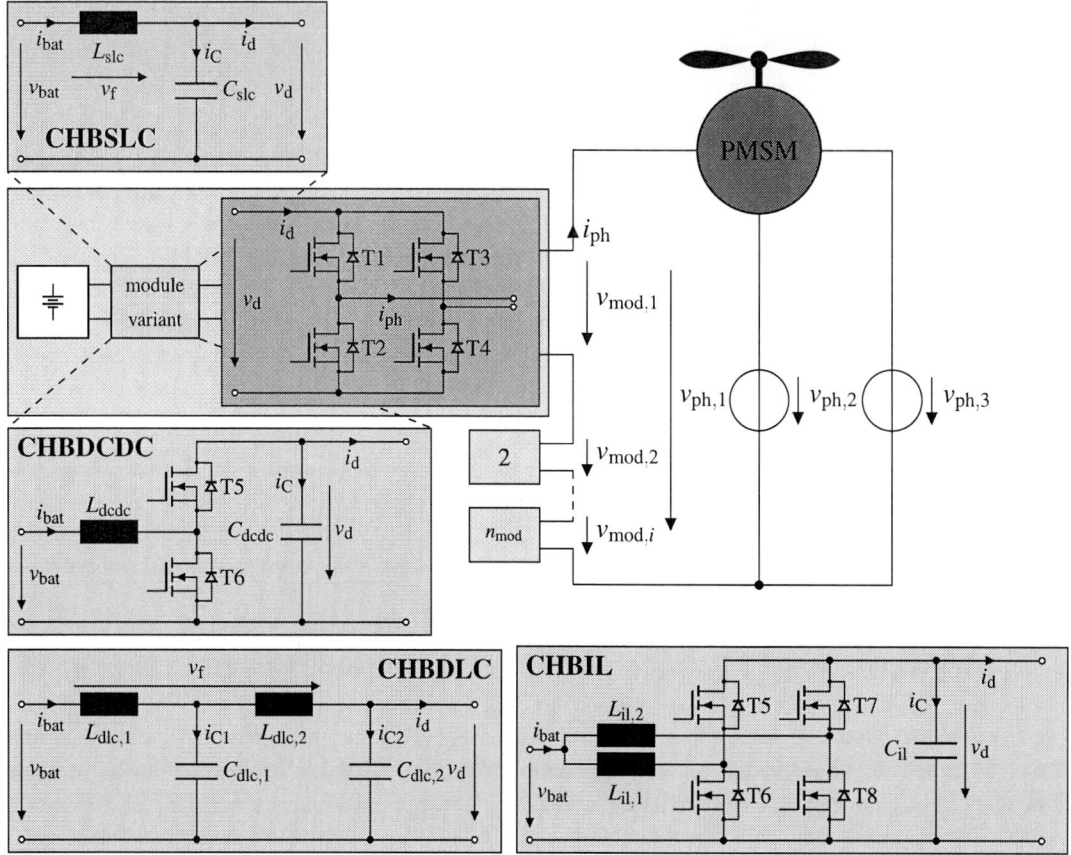

Fig. 1: Topology overview of the propulsion system with a three-phase PMSM supplied by a CHB inverter. The first phase of the CHB containing n_{mod} modules is shown in detail with the H-bridge (blue box) and the CHBSLC, CHBDLC, CHBDCDC, and CHBIL ($n_{IL} = 2$) module variants in the gray boxes.

implemented allows for an arbitrary breakdown voltage, which is why n_{mod} does not depend on v_{ph} or the battery voltage. Due to cosmic radiation, which can lead to critical failures, the maximum blocking voltage of the MOSFETs is reduced by a derating factor k_{der} [10].

Module Variants

Different module variants can be used to compensate for the power variation in i_d. In this section, the modules of the different CHB variants CHBSLC, CHBDLC, CHBDCDC, and CHBIL are described. To account for the highest possible currents, the module designs are carried out for the lowest SoC of the battery.

LC Filter Variants

The LC filter variants considered in this paper are a single-stage LC filter (SLC) and a dual-stage LC filter (DLC). In both these variants, the H-bridge input voltage is $v_d = v_{bat} - v_f$, where v_f denotes the voltage drop across the filter inductors L_{slc} or $L_{dlc,1}$ and $L_{dlc,2}$. The filters are designed to achieve the necessary attenuation at the frequency of the second harmonic component in i_d so that the desired value for r_I in the battery current is met. As an additional design constraint, a maximum permissible value for v_f must not be exceeded. The expression for the resonant frequencies of each filter are given in (2) and (3) [11]. Since the DLC has four energy-storage components, but the number of constraints (r_I and v_f) does not increase, it is assumed that both filter capacitors and inductors have the same component values $L_{dlc,1} = L_{dlc,2} = L_{dlc}$ and $C_{dlc,1} = C_{dlc,2} = C_{dlc}$. Thus, (3) can be simplified to give (4). The inductance and capacitance combination which meets the design constraints and results in the lowest mass is chosen.

Fig. 2 shows the qualitative bode plots of the transfer functions of an SLC (blue) and a DLC (red) with their corresponding asymptotic approximations (colored dashed lines). It might be assumed that the DLC always results in lower capacitance and inductance values, considering only the asymptotic approximations, since higher resonant frequencies are possible with the same attenuation. However, this does not apply for frequencies to be attenuated close to the resonant frequencies. The horizontal and vertical black dashed lines mark the design point for the filters, where the attenuation of A_z has to be achieved at the specified frequency $\omega_z = 2f_1$ to meet a given r_I factor. From the asymptotes, the resonant frequencies and the achieved attenuation of $-40\,\mathrm{dB/dec}$ for the SLC and $-80\,\mathrm{dB/dec}$ for the DLC for frequencies greater than the respective resonant frequencies can easily be deduced. Comparing the transfer functions to their asymptotic approximations makes it obvious that the DLC attenuation is still further away from its potential indicated by the asymptote than the SLC attenuation. This indicates that, for the given combination of ω_z and A_z in Fig. 2, the use of a DLC may not be the optimal choice for a mass-optimized filter. Depending on the required attenuation for a given ω_z, the positions of the filter resonant frequencies vary. With decreasing A_z, the filter resonances shift to lower frequencies at different rates for SLC and DLC. If A_z were to decrease, the resonant frequencies of the DLC $\omega_{0,\mathrm{dlc},1/2}$ would shift less to lower ω due to its higher attenuation than the SLC resonant frequency $\omega_{0,\mathrm{slc}}$. This means that for a certain value of A_z, $\omega_{0,\mathrm{slc}}$ would be smaller than both values $\omega_{0,\mathrm{dlc},1/2}$ and therefore the DLC might prove to be the lighter filter variant.

$$\omega_{0,\mathrm{slc}} = \frac{1}{\sqrt{L_{\mathrm{slc}} C_{\mathrm{slc}}}} \tag{2}$$

$$\omega_{0,\mathrm{dlc},1/2}^2 = \frac{A + B + C \pm \sqrt{(A + B + C)^2 - 4AC}}{2AC} \tag{3}$$
$$\text{and } A = L_{\mathrm{dlc},1} C_{\mathrm{dlc},1}, B = L_{\mathrm{dlc},1} C_{\mathrm{dlc},2}, C = L_{\mathrm{dlc},2} C_{\mathrm{dlc},2}$$

$$\omega_{0,\mathrm{dlc},1/2} = +\sqrt{\frac{3 \pm \sqrt{5}}{2 L_{\mathrm{dlc}} C_{\mathrm{dlc}}}} \tag{4}$$

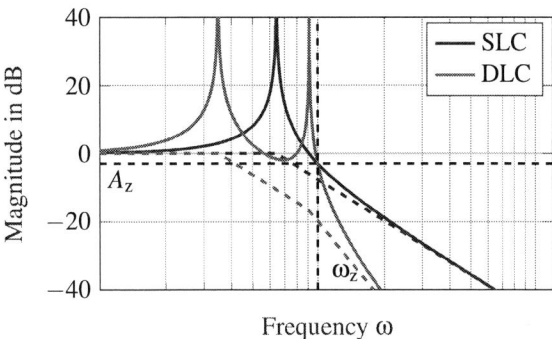

Fig. 2: Qualitative bode plots and their asymptotic approximations (colored dashed lines) for the LC filter variants for a given attenuation A_z and frequency ω_z

DC-DC Converter

For the DC-DC converter, an ideal control is assumed, which completely compensates for the second harmonic component of i_d. Only the current ripple Δi_{bat} originating from the switching of the DC-DC converter's semiconductors is still present in the battery current. Following the DC-DC converter design presented in [9] and [12], the capacitance of the module capacitor C_{dcdc} is designed for a voltage variation of $\pm 10\,\%$ and has to buffer the energy fluctuation caused by the second harmonic component of the inverter input current i_d. Since Δi_{bat} directly influences the RMS value of the battery current, the given value of r_I determines Δi_{bat} and thus the inductance

$$L_{dcdc} = \frac{v_{bat,min} D_{max}}{\Delta i_{bat} f_{dcdc}}.$$
(5)

In (5), D_{max} denotes the maximum duty cycle that may occur at the minimum battery voltage $v_{bat,min}$ and f_{dcdc} is the switching frequency of the DC-DC converter. Since the second harmonic component of the battery current is controlled by the DC-DC converter, f_{dcdc} needs to be sufficiently high. The influence of f_{dcdc} is subject to investigation in the following parameter study.

Interleaved DC-DC Converter

The CHBIL consists of multiple parallel DC-DC converters with a corresponding number of inductors, where the battery current is divided into the separate interleaved phase legs, leading to a reduced current in each single inductor. The inductance value of each interleaved phase is determined using (5) and is therefore the same as for the simple DC-DC converter. The module capacitor C_{il} is dimensioned the same as in the CHBDCDC. The MOSFETs in each interleaved phase leg are controlled with phase-shifted carrier signals at a frequency of f_{dcdc}. As depicted in Fig. 3, the additional interleaved phases further reduce the battery current ripple Δi_{bat}, and therefore the r_I value, due to phase-shifted control of the separate interleaved phase legs [13]. The resulting factor $r_{I,n_{IL}}$ for a CHBIL is determined using (6), where n_{IL} is the number of interleaved phase legs and D is the duty cycle. For all DC-DC converter variants, only continuous conduction mode is considered.

$$r_{I,n_{IL}} = \sqrt{\frac{1}{12}\left(\frac{\Delta i_{bat,n_{IL}}}{\bar{i}_{bat}}\right)^2 + 1} \quad \text{and} \quad \Delta i_{bat,n_{IL}}(D) = \frac{1}{n_{IL}}\frac{|\sin(n_{IL}\pi D)|}{|\sin(\pi D)|}\Delta i_{bat,n_{IL}=1}$$
(6)

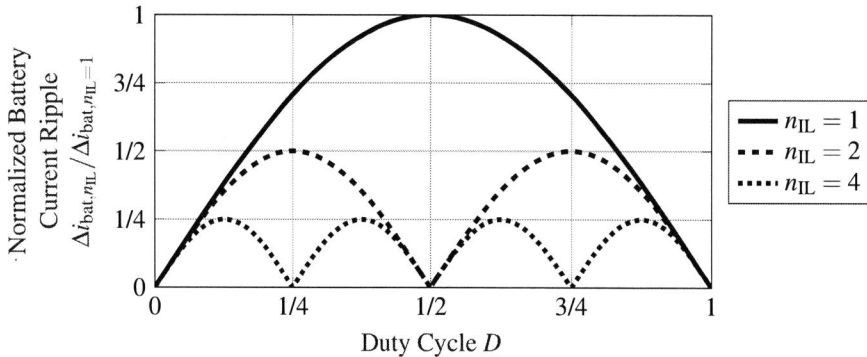

Fig. 3: Normalized battery current ripple vs. duty cycle as a function of the number of interleaved phase legs [13]

CHB Loss Calculation

The losses in the CHB are dissipated by the semiconductors in the H-bridge and the DC-DC converter(s) as well as by the passive components. The conduction loss calculation for the H-bridge semiconductors follows the approach detailed in [14] with minor adjustments, so that the losses can be assigned to a single MOSFET or body diode. The switching loss calculation is implemented according to [4] with slight modifications regarding the switching losses of the DC-DC converter MOSFETs. The switching losses of the H-bridge semiconductors are calculated according to [4], neglecting the current ripple of the machine current i_{ph}, because a sufficiently high machine inductance is assumed. For the DC-DC converter, however, the current ripple can be up to twice its DC value on the verge of discontinuous current mode. Hence, the calculation in [4] is modified to take the current ripple of the DC-DC converter into account. As for the passive components, the scalable component models from [4] are used to obtain the equivalent series resistance of the capacitor and the ohmic and magnetic losses of the inductor.

Parameter Studies

The parameter studies will give an impression of how certain design parameters influence the respective CHB inverter designs with respect to the resulting mass, efficiency, and power density. Power losses and efficiencies are always calculated for the worst-case operating point, which occurs at the lowest battery SoC. To determine n_{mod}, the inverter design is calculated for $2 \leq n_{mod} \leq 20$ and the design yielding the highest efficiency is chosen. The switching frequency of the H-bridge MOSFETs in each module f_s is adjusted according to n_{mod}, so that the resulting switching actions in v_{ph} always correspond to an effective switching frequency $f_{s,eff}$, leading to $f_s = \frac{f_{s,eff}}{2n_{mod}}$. Although the CHBDLC has been studied, it will not be discussed in the parameter studies, since it proved to be less efficient than the CHBSLC and did not reach such high power densities. As assumed before, it confirms that the higher DLC attenuation could not be used to benefit mass reduction because the total mass of all passive components of the DLC exceeds the SLC passive components' mass.

The parameter studies are based on the reference aircraft mission profile from [4], where two PMSMs are assumed, resulting in a maximum electrical apparent power of 2.1 MVA for one inverter. Additional parameters assumed for the case studies are listed in Table I. Note that the module capacitor voltage ripple of the CHBSLC and the DC-DC converter variants differ. This is due to the different effects a voltage ripple increase has on the resulting module mass. Since for variants containing a DC-DC converter, the capacitance decreases with rising voltage ripple, an increase in voltage ripple reduces the overall module mass. A voltage ripple of $\pm 10\%$ is considered acceptable here. As for the CHBSLC, an increase in $v_{f,max}$ both leads to an increased filter inductance and a decreased filter capacitance. The resulting mass decrease of the capacitor does not balance the mass increase of the inductor, therefore a smaller capacitor voltage ripple gives a mass advantage in the case of the CHBSLC.

Table I: Parameters for both case studies

Inverter power	S_{el}	2.1 MVA
Machine phase voltage (RMS)	$V_{ph,N}$	1.15 kV
Power factor	$\cos(\varphi_1)$	0.84
Derating factor	k_{der}	60 %
Effective switching frequency	$f_{s,eff}$	20 kHz
Max. filter voltage drop	$v_{f,max}$	2 % of v_{bat}

The contour plots in Fig. 4 show the variation of r_{uc} and r_l and the effects on efficiency, mass, and power density for the CHBSLC, CHBDCDC and CHBIL with $n_{IL} = 2$ or $n_{IL} = 4$ interleaved phase legs. For this case study, a fundamental machine frequency of $f_1 = 750$ Hz and a DC-DC converter switching frequency of $f_{dcdc} = 40$ kHz have been used.

A number of $n_{mod} = 2$ modules is calculated to be the most efficient choice for all CHBSLC design points. For the DC-DC converter variants, there are some design points that turn out to be more efficient with $n_{mod} = 3$. For the CHBDCDC, this is only the case for $r_{uc} = 1.17$ and $r_l = \{1.0006, 1.0062, 1.0073\}$. For the CHBIL with $n_{IL} = 2$, this is true for $r_{uc} < 1.37$ at most r_{IL} values, while for the CHBIL with $n_{IL} = 4$, this applies to all points.

The value of r_{uc} has a large impact on the CHBSLC mass, because the battery voltage is directly linked to the H-bridge input voltage via the LC filter. Since a large value of r_{uc} leads to higher maximum battery and capacitor voltages and the capacitor volume and ESR are proportional to the capacitor voltage, the capacitor mass increases accordingly. Therefore, a smaller r_{uc} leads to higher efficiency and lower mass. The r_l value marginally affects the efficiency, but a minimum for the CHBSLC mass is identified at $r_l \approx 1.015$. Two different effects increase the filter inductor mass for higher and lower values of r_l.

Lower r_I values lead to significantly larger inductance values and higher r_I values result in increased peak and RMS current values, both leading to an increased inductor mass. Therefore, at the design point $r_\mathrm{uc} = 1.17$ and $r_\mathrm{I} = 1.015$ a peak efficiency of 99.4 % and a maximum power density of $11.7\,\mathrm{kVA\,kg^{-1}}$ are realized for the CHBSLC.

Looking at the CHBDCDC, its efficiency increases at larger r_I values as well as at smaller r_uc values, since both indirectly influence the inductance value L_dcdc (see (5)). For the CHBDCDC mass, a similar relation can be observed: it is lightest at the highest r_I and lowest r_uc values. As introduced before, the effective r_I values of the CHBIL variants are reduced (see (6)) and therefore reach a maximum of $r_\mathrm{I} = 1.01$ and $r_\mathrm{I} = 1.0006$ for $n_\mathrm{IL} = 2$ and $n_\mathrm{IL} = 4$, respectively. Comparing the efficiency, both the CHBDCDC and CHBIL inverter variants are less efficient than the CHBSLC. In terms of mass, the CHBDCDC and CHBIL are lighter than the CHBSLC for almost all parameter combinations, especially for larger r_I values. One reason for using the CHBIL variant was the presumed mass advantage compared to the CHBDCDC. Comparing the CHBDCDC and CHBIL with $n_\mathrm{IL} = 2$, the assumption is confirmed: the CHBIL has a mass of 154 kg. However, a higher number of interleaved phase legs does not lead to a further mass reduction. This is due to the increased n_mod for the CHBIL with $n_\mathrm{IL} = 4$. The CHBDCDC reaches a maximum power density of $12.8\,\mathrm{kVA\,kg^{-1}}$ at the smallest value for r_uc and $r_\mathrm{I} = 1.019$, while the CHBIL with $n_\mathrm{IL} = 2$ has a power density of $13.6\,\mathrm{kVA\,kg^{-1}}$ for the same r_uc and $r_\mathrm{I} = 1.008$. Again, the CHBIL with $n_\mathrm{IL} = 4$ does not reach such high power densities because of its higher n_mod. To verify

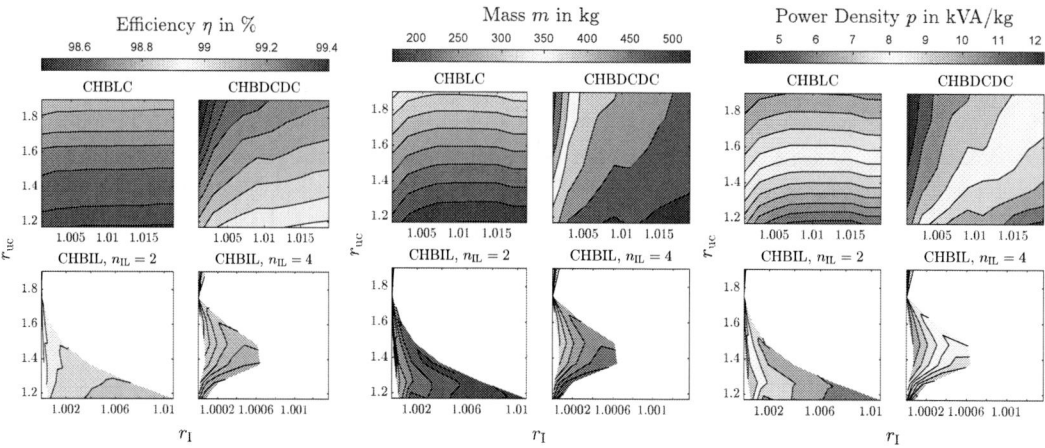

Fig. 4: Efficiency, mass, and gravimetric power density as a function of r_uc and r_I for a complete CHB and an apparent power of 2.1 MVA. The contour plot grid resolution in the r_I and r_uc direction is 10 by 7.

the analytically calculated results, a simulation of the CHBSLC was performed in Matlab Simulink, as it achieved the highest efficiency of 99.4 %. The semiconductors are modeled in PLECS and the PMSM is modeled as an independent sinusoidal current source. The simulated MOSFET voltages and currents together with the conduction losses and switching energies over one fundamental period of one H-bridge are shown in Fig. 5. Switching energies only occur in the positive half-wave of the MOSFET current, because it is assumed that during the negative half-wave, the integrated body diode fully conducts the current during interlock time, leading to zero-voltage switching. A comparison of the calculated and simulated losses of the H-bridge is shown in Table II. The main reason for the deviation between the calculated and simulated switching losses is the relatively low switching frequency f_s of the single semiconductors. Nevertheless, the analytical error for the total MOSFET losses compared to the simulated losses is below 4 %. At higher values of $f_\mathrm{s,eff}$, each MOSFET performs more switching operations, resulting in greater agreement between the simulated and calculated losses.

Analogous to Fig. 4, the contour plots in Fig. 6 show the variation of the fundamental frequency f_I and the DC-DC converter switching frequency f_dcdc and their effects on the efficiency, mass, and power density of each inverter type. A lithium-sulfur battery cell is assumed to be used, and a value of $r_\mathrm{uc} = 1.37$ is taken from the manufacturer's datasheet [15]. Also, a maximum value of $r_\mathrm{I} = 1.01$ is set, which

Table II: Comparison of the calculated and simulated H-bridge losses averaged over one fundamental period of one CHBSLC module with $r_\mathrm{I} = 1.015$, $r_\mathrm{uc} = 1.17$, and $f_\mathrm{s,eff} = 20\,\mathrm{kHz}$

| | Conduction Losses in W | | Switching Losses in W | |
	calculated	simulated	calculated	simulated
T1	484	497	39	34
T2	484	477	39	49
T3	484	500	39	42
T4	484	473	39	61

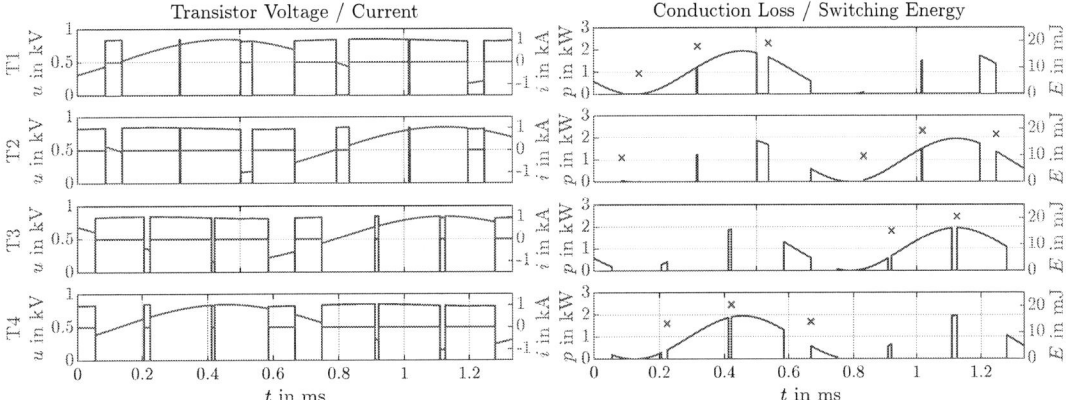

Fig. 5: Simulated MOSFET voltages and currents (left) and conduction losses and switching energies (right) for one H-bridge in a CHBSLC inverter with $n_\mathrm{mod} = 2$, $r_\mathrm{I} = 1.015$, $r_\mathrm{uc} = 1.17$, $f_\mathrm{s,eff} = 20\,\mathrm{kHz}$ and a switching frequency $f_\mathrm{s} = 5\,\mathrm{kHz}$ for each MOSFET

leads to reduced r_I values for the CHBIL variants of $r_{\mathrm{I},n_\mathrm{IL}=2} = 1.0012$ and $r_{\mathrm{I},n_\mathrm{IL}=4} = 1.0007$. While the CHBSLC is most efficient for $n_\mathrm{mod} = 2$ at all considered points, the CHBDCDC and CHBIL with $n_\mathrm{IL} = 2$ have $n_\mathrm{mod} = 2$ for $f_\mathrm{dcdc} < 45\,\mathrm{kHz}$ and $n_\mathrm{mod} = 3$ for $f_\mathrm{dcdc} > 45\,\mathrm{kHz}$. The CHBIL with $n_\mathrm{IL} = 4$ only uses $n_\mathrm{mod} = 2$ for $f_\mathrm{dcdc} < 35\,\mathrm{kHz}$ and is more efficient with $n_\mathrm{mod} = 3$ at all other frequencies. This is indicated by the dashed blue lines on the contour plots.

Efficiency, mass, and power density of the CHBSLC are not affected by f_dcdc since no DC-DC converter is used, but at higher values of f_1, the CHBSLC mass is significantly reduced from $508\,\mathrm{kg}$ at $f_1 = 200\,\mathrm{Hz}$ to $183\,\mathrm{kg}$ at $f_1 = 1\,\mathrm{kHz}$. Consequently, the power density increases accordingly. This is because the resonant frequency of the LC filter in each module rises with f_1 and therefore the filter component mass decreases.

The efficiencies of the CHBDCDC and CHBIL variants are only marginally affected by a variation of f_1. In contrast, the CHBDCDC efficiency strongly depends on f_dcdc. Higher values for f_dcdc lead to a decreased efficiency for all types containing a DC-DC converter. The abrupt changes in the course of the contour plots for the CHBIL variants, which are most pronounced on the mass and power density plots, are due to the increase of n_mod at values for f_dcdc of $35\,\mathrm{kHz}$ and $45\,\mathrm{kHz}$, which leads to an increased inverter mass and a decreased power density. The overall inverter mass decreases with both increasing f_1 and increasing f_dcdc for all CHBDCDC and CHBIL variants because of two effects. First, the capacitor mass is inversely proportional to f_1, which explains the mass reduction with rising f_1. Second, the inductor mass is directly related to f_dcdc, which leads to decreasing inverter mass for higher values of f_dcdc. It can be observed that, for values of $f_1 \lesssim 400\,\mathrm{Hz}$, a further increase in f_dcdc does not result in significant mass savings, since the capacitor mass is already dominant over the inductor mass. The same effect can be anticipated for values of $f_1 > 1\,\mathrm{kHz}$ and decreasing f_dcdc. In this case, a further increase in f_1 would not save significant mass in the capacitor, since the inductor mass is dominant over the capacitor mass at low f_{dcdc}.

Irrespective of the frequencies f_1 and f_{dcdc}, the CHBSLC is the most efficient inverter type, since there are no additional losses from a DC-DC converter. Especially for low f_1, CHBDCDC and CHBIL inverter variants are lighter alternatives. When using the CHBIL variants at high f_{dcdc}, they are superior to the CHBSLC in terms of mass and power density.

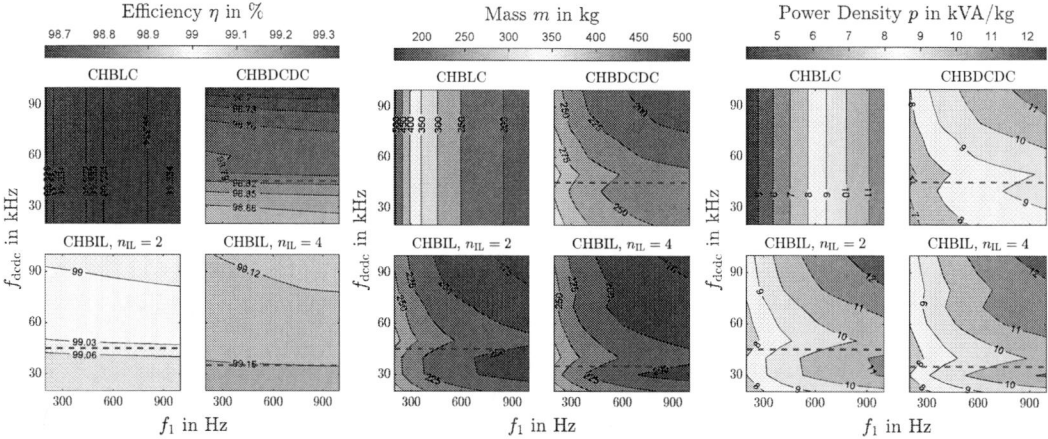

Fig. 6: Efficiency, mass, and gravimetric power density as a function of f_1 and f_{dcdc} for a complete CHB and an apparent power of 2.1 MVA. The grid has a resolution of 9 steps for each frequency. The dashed blue lines indicate the value of f_{dcdc} above which $n_{mod} = 3$, otherwise $n_{mod} = 2$.

The mass distribution of the different components in the inverter variants is shown in the pie charts in Fig. 7 for $f_1 = 600\,\text{Hz}$ and $f_{dcdc} = 60\,\text{kHz}$. The slice *Semiconductors* includes chip, cooling, and driver board masses. The dominant mass in the CHBSLC is by far the capacitor, whereas in the inverter variants including a DC-DC converter, the inductors account for the majority of the mass. With rising numbers of phase legs in the CHBIL, the relative inductor mass decreases while the semiconductor mass increases due to the additional half-bridges. Given that the absolute capacitor mass is constant and the inductor mass decreases with rising n_{IL} for all CHBIL, the relative share of the capacitor mass increases.

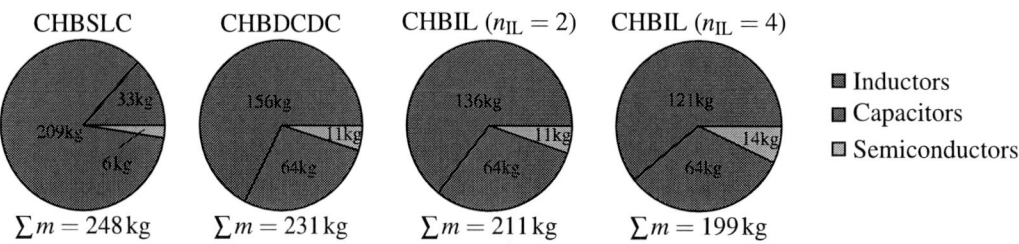

Fig. 7: Mass distribution of the components in a CHBSLC, CHBDCDC, and CHBIL with $n_{IL} = 2$ and $n_{IL} = 4$ at $f_1 = 600\,\text{Hz}$ and $f_{dcdc} = 60\,\text{kHz}$. Slice *Semiconductors* includes chip, cooling, and driver board masses.

Conclusion

In this paper, different module variants for a CHB have been studied in order to find suitable solutions for an SAEA. The focus of the module variant designs is the compensation of the second harmonic component in the battery current. In two parameter studies, the influences of the battery cell voltage window r_{uc}, the ratio of the RMS to DC battery current r_I, the fundamental machine frequency f_1 and the DC-DC converter switching frequency f_{dcdc} on the inverter's efficiency, mass, and power density have been investigated. To calculate the module components' losses and masses, scalable component models have been employed. The parameter studies show that a narrow cell voltage window is beneficial to the efficiency and power density in all module variants considered. Furthermore, increases in the fundamental frequency and the DC-DC converter switching frequency reduce mass and improve power density. The CHBIL reaches the highest power density but is inferior to the CHBSLC in terms of efficiency.

For a power train design process, the given parameter studies give an idea of how the named parameters of adjacent elements, such as a battery or machine, influence the resulting inverter. The results can also be used to choose the most suitable CHB variant for given battery and machine parameters. In combination with battery and machine models, the inverter model presented here makes an overall system optimization for mass and efficiency of SAEA propulsion systems possible, and this is planned for future studies.

References

[1] M. Filipenko, J. Kaiser, K. Plötner, and A. Strohmayer, "Nachhaltig durch die Luft," *Physik Journal*, no. 12, pp. 34–40, Dec. 2020.

[2] European Commission, *Flightpath 2050: Europe's Vision for Aviation.* LU: Publications Office, 2012.

[3] S. Sahoo, X. Zhao, and K. Kyprianidis, "A Review of Concepts, Benefits, and Challenges for Future Electrical Propulsion-Based Aircraft," *Aerospace*, vol. 7, no. 4, p. 44, Apr. 2020.

[4] J. Ebersberger, M. Hagedorn, M. Lorenz, and A. Mertens, "Potentials and Comparison of Inverter Topologies for Future All-Electric Aircraft Propulsion," *IEEE Journal of Emerging and Selected Topics in Power Electronics*, Apr. 2022.

[5] M. J. Brand, M. H. Hofmann, S. S. Schuster, P. Keil, and A. Jossen, "The influence of current ripples on the lifetime of lithium-ion batteries," *IEEE Transactions on Vehicular Technology*, vol. 67, no. 11, pp. 10 438–10 445, Nov. 2018.

[6] S. Karpuk and A. Elham, "Influence of Novel Airframe Technologies on the Feasibility of Fully-Electric Regional Aviation," *Aerospace*, vol. 8, no. 6, p. 163, Jun. 2021.

[7] Great Power, "Li-Ion 21700 4.7Ah Battery Datasheet," 2018.

[8] F. Jenni and D. Wüest, *Steuerverfahren für selbstgeführte Stromrichter.* vdf Hochschulverlag AG an der ETH Zürich, 1995.

[9] L. Baruschka and A. Mertens, "Comparison of Cascaded H-Bridge and Modular Multilevel Converters for BESS application," in *2011 IEEE Energy Conversion Congress and Exposition.* Phoenix, AZ, USA: IEEE, Sep. 2011, pp. 909–916.

[10] H. Schefer, L. Fauth, T. H. Kopp, R. Mallwitz, J. Friebe, and M. Kurrat, "Discussion on Electric Power Supply Systems for All Electric Aircraft," *IEEE Access*, vol. 8, pp. 84 188–84 216, 2020.

[11] A. Gruber, "Vergleich von Halbleitertechnologien und Schaltungstopologien zur Realisierung von Wechselrichtern für den Einsatz in Photovoltaiksystemen mit 1500 V Systemspannung," Ph.D. dissertation, Kassel University Press, Kassel, 2018.

[12] J. Specovius, "Gleichspannungswandler," in *Grundkurs Leistungselektronik.* Springer Fachmedien Wiesbaden, 2018, pp. 395–434.

[13] S. Zhang, "Analysis and minimization of the input current ripple of Interleaved Boost Converter," in *2012 Twenty-Seventh Annual IEEE Applied Power Electronics Conference and Exposition (APEC)*, Feb. 2012, pp. 852–856.

[14] X. Shen, H. Lin, B. Li, J. Liu, J. I. Leon, L. Wu, and L. G. Franquelo, "Loss Evaluation of Cascaded H-bridge and Modular Multilevel Converter for Motor Drive Applications," in *IECON 2018 - 44th Annual Conference of the IEEE Industrial Electronics Society*, Oct. 2018, pp. 1299–1304.

[15] Oxis Energy, "Ultra Light Lithium Sulfur Pouch Cell Datasheet," 2019.

Overview and Evaluation of Energy Balancing Techniques for MMCs with Various Input and Output Frequencies

Gyanendra Kumar Sah, Michael Schütt, Hans-Günter Eckel

UNIVERSITY OF ROSTOCK
Albert-Einstein-Str.2
D-18059 Rostock, Germany
Tel.: +49 / (0) 381 – 498 7137.
Fax: +49 / (0) 381 – 498 7102.
E-Mail: gyanendra.sah@uni-rostock.de
URL: http://www.iee.uni-rostock.de

Acknowledgments

This paper was made within the framework of the research project *Netz-Stabil* and financed by the European Social Fund (ESF/14-BM-A55-0015/16). This paper is part of the qualification program *Promotion of Young Scientists in Excellent Research Associations - Excellence Research Programme of the State of Mecklenburg-Western Pomerania.*

Keywords

«Modular Multilevel Converters (MMC)», «Capacitor Voltage Balancing», «AC-AC Converter», «Grid-Connected Converter», «Energy Balancing»

Abstract

This work provides an overview of all possible arm energy regulation techniques for the *Modular Multilevel Converter* (MMC). This research aims to identify all degrees of freedom (manipulated inputs) and possible combinations to regulate the arm energies of the MMC. Further, it offers a new generalized analytical toolchain to compare and evaluate the stability and performance of these balancing techniques. This toolchain includes normal operation, failure operation, and different three-phase and single-phase frequency conditions such as equal and unequal frequency applications (AC/AC, DC/AC). The common-mode voltage plays an essential role in regulating arm energies of the MMC for equal frequency operation. The proposed toolchain can also help to validate dimensioning of the converter analytically. Any existing energy balancing technique for MMC could be linked directly or indirectly to all derived energy balancing methods.

Introduction

The MMC is a promising topology for high-power applications, especially in the medium-voltage range [1, 2]. The MMC provides various degrees of freedom (manipulated inputs) to regulate its arm energies. Hence, multiple methods can be derived using a combination of these manipulated inputs to control the arm energies of the MMC. Some arm energy regulation techniques may not work for equal and unequal single-phase and three-phase frequency operations, which will be discussed later. Therefore, stability and performance analysis of the arm energy regulation techniques, especially for all expected (normal and abnormal) operating conditions, should be investigated before actual hardware implementation. It would further help if this investigation could be performed analytically, saving time, staffing, and money.

The common-mode voltage introduces an additional degree of freedom to regulate the arm energies of MMC [3]. The main objective of this work is to present a detailed mathematical toolchain for the investigation of various possible degrees of freedom to regulate arm energies of the MMC for different single-phase and three-phase frequency conditions. Further, the proposed toolchain facilitates analytical validation of the dimensioning of the converter during normal and abnormal operating conditions.

MMC Topology and Transformed Arm Powers in the αβ0-Frame

An accurate average electrical circuit diagram plays a vital role in deriving governing plant equations, which are later used for designing a suitable controller. Fig. 1.a illustrates the average equivalent electrical circuit diagram of direct MMC topology [1]. Here, v_{a123}, v_b, and v_{cm} represent three-phase, single-phase, and common-mode voltages. v_{cm} can be either DC or AC. The frequency of AC v_{cm} (f_{cm}) is generally chosen differently than the single-phase (f_b) and three-phase (f_a) frequencies, the proposed relation is $f_{cm} = 3\text{Max}(f_a, f_b)$. The neutral point (N) of the three-phase voltage source can be either floating or grounded. Each controlled arm voltage source ($v_{p/ny}$) represents a mathematical model of various submodules (SMs) connected in series. Depending upon the single-phase and three-phase frequencies, an SM is realized using either a full-bridge or half-bridge converter. Full-bridge-based SMs can be used universally, whereas half-bridge-based SMs cannot be used for equal frequency operation. Each branch current ($i_{p/ny}$) can be decomposed into two components: (a) circulating current (i_{by}) and (b) three-phase current ($0.5i_{ay}$). The expression of the summation and difference arm currents ($i_{\Sigma/\Delta y}$) and arm voltages ($v_{\Sigma/\Delta y}$) for phase y of MMC are shown in Fig. 1. The circulating current is the same for both the upper and lower arms in a phase. Fig. 1b represents the MMC topology and visualizes the energy components of the structure after ignoring all arm impedances (Z_{arm}). (1) illustrates the summation and difference arm powers of phase y of MMC. (2) to (7) represent the expression for the instantaneous summation and difference arm powers in the αβ0-frame [3, 4].

$$P_{\Sigma y} = p_{py} + p_{ny} \; ; \; P_{\Delta y} = p_{py} - p_{ny} \tag{1}$$

$$p_{\Sigma \alpha} = v_b \, i_{b\alpha} + v_{cm} \, i_{a\alpha} + \frac{v_{a\alpha} \, i_{a\alpha} - v_{a\beta} \, i_{a\beta}}{2} \tag{2}$$

$$p_{\Sigma \beta} = v_b \, i_{b\beta} + v_{cm} \, i_{a\beta} - \frac{v_{a\alpha} \, i_{a\beta} + v_{a\beta} \, i_{a\alpha}}{2} \tag{3}$$

$$p_{\Sigma 0} = v_b \, i_{b0} + \frac{v_{a\alpha} \, i_{a\alpha} + v_{a\beta} \, i_{a\beta}}{2} \tag{4}$$

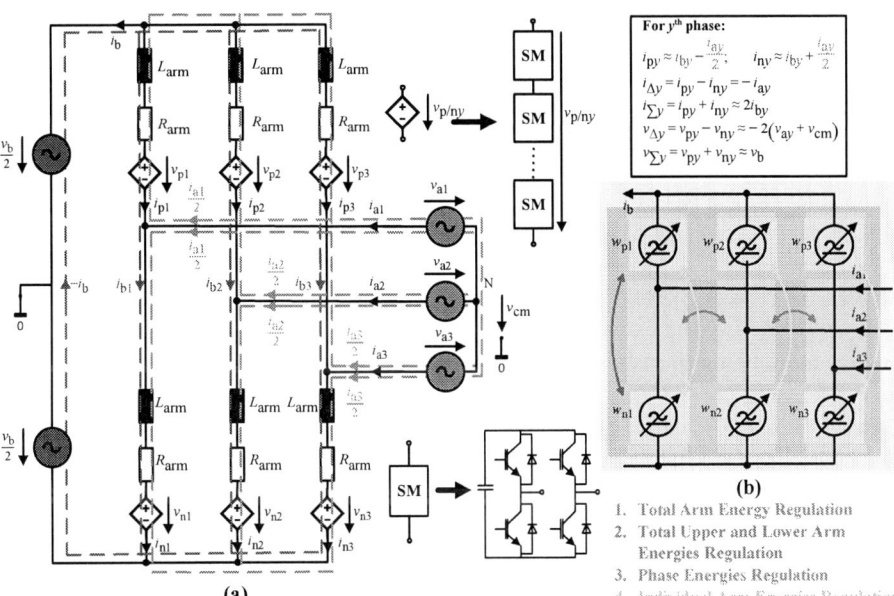

Fig. 1: Average equivalent electrical circuit diagram and arm energy regulations of MMC.

Table I: Different *Manipulated Inputs* (MIs) for regulating corresponding average arm energies of MMC.

1. Total Arm Energy Regulation (MI_1)	2. Total Upper and Lower Arms Energies Regulation (MI_2)
Positive sequence three-phase currents in the d-axis $\left(i_{ad+}\right)$.	Positive sequence circulating currents in the d-axis $\left(i_{bd+}\right)$.
Zero sequence circulating current in phase with $v_b \left(i_{b0}^b\right)$.	Zero sequence circulating current in phase with $v_{cm} \left(i_{b0}^{cm}\right)$.
3. Phase Energies Regulation (MI_{34})	**4. Individual Arm Energies Regulation (MI_{56})**
Circulating currents in phase with $v_b \left(i_{ba\beta}^b\right)$.	Three-phase currents in phase with $v_b \left(i_{aa\beta}^b\right)$.
Three-phase currents in phase with $v_{cm} \left(i_{aa\beta}^{cm}\right)$.	Single-phase currents in phase with $v_{a\alpha\beta}$ $\left(i_{b0}^{a\text{-}\alpha\beta}\right)$.
Negative sequence three-phase currents $\left(i_{adq-}\right)$.	Circulating currents in phase with $v_{cm} \left(i_{ba\beta}^{cm}\right)$.
Note: $i_{b0} = -\dfrac{i_b}{3}$	Negative sequence circulating currents $\left(i_{bdq-}\right)$.

$$p_{\Delta\alpha} = -\frac{v_b\, i_{a\alpha}}{2} - 2v_{a\alpha}\, i_{b0} - 2v_{cm}\, i_{b\alpha} - v_{a\alpha}\, i_{b\alpha} + v_{a\beta}\, i_{b\beta} \tag{5}$$

$$p_{\Delta\beta} = -\frac{v_b\, i_{a\beta}}{2} - 2v_{a\beta}\, i_{b0} - 2v_{cm}\, i_{b\beta} + v_{a\alpha}\, i_{b\beta} + v_{a\beta}\, i_{b\alpha} \tag{6}$$

$$p_{\Delta 0} = -2v_{cm}\, i_{b0} - v_{a\alpha}\, i_{b\alpha} - v_{a\beta}\, i_{b\beta} \tag{7}$$

All Possible Manipulated Inputs for Regulation of Arm Energies of MMC

Knowing all degrees of freedom can help derive all possible arm energy regulation techniques for the MMC. The arm energies ($W_{\Sigma/\Delta\alpha\beta 0}$) of the MMC can be regulated by varying the corresponding arm powers ($p_{\Sigma/\Delta\alpha\beta 0}$), which can be achieved by changing the *Manipulated Inputs* (MIs): circulating currents ($i_{ba\beta}$) and three-phase currents ($i_{aa\beta}$) (or single-phase current (i_{b0})).

The equations (2) and (3) represent the influence on the phase energies (MI_{34}) – see Fig.1(b). Similarly, equation (4) shows the influence on the total arm energy (MI_1), (5) and (6) the individual arm energies (MI_{56}), and (7) the total upper and lower arm energies (MI_2).

(8) represents the currents in the αβ-frame only due to the positive and negative sequence currents in the dq-frame. The Park and inverse Park transformation angle (θ_a) can be obtained from the output of the three-phase voltage *Phase-Locked Loop* (PLL). Since the same frequency components of current and voltage components make power, negative sequence currents also produce average power and thus represent one of the manipulated inputs according to (2) – (6), and (10). Table I lists all possible manipulated inputs for regulating corresponding arm energies (see also (2) to (10), and Fig. 1b). i_x^y represents current component (manipulated input) i_x in phase with the voltage v_y.

Table. II: Stability analysis of various average arm energies regulation techniques for direct AC-AC and AC-DC MMC topologies. Here, n = odd integers and $f_{cm} = 3\text{Max}(f_a, f_b)$.

Method	Manipulated Inputs $(MI_1, MI_2, MI_{34}, MI_{56})$	Determinant ($\|A\|$)	
		$f_a = f_b$	$f_a = nf_b, f_b = nf_a$ & $f_b = \text{DC}$
1	$i_{ad+}, i_{bd+}, i_{ba\beta}^b, i_{bdq-}$	0	$2V_a^4 V_b^2$
2	$i_{ad+}, i_{bd+}, i_{ba\beta}^b, i_{aa\beta}^b$	$\dfrac{V_a^2}{16}\left(\cos^2(3\phi_b)V_a^2 V_b^2 - \left(2V_a^2 - V_b^2\right)^2\right)$	$-\dfrac{1}{4}V_a^2 V_b^4$
3	$i_{ad+}, i_{bd+}, i_{adq-}, i_{bdq-}$	$-V_a^2\left(V_a^2 - \dfrac{V_b^2}{4}\right)^2$	$-V_a^6$
4	$i_{ad+}, i_{bd+}, i_{adq-}, i_{aa\beta}^b$	0	$\dfrac{1}{8}V_a^4 V_b^2$
5	$i_{b0}^b, i_{bd+}, i_{ba\beta}^b, i_{bdq-}$	0	$2\sqrt{2}V_a^3 V_b^3$
6	$i_{b0}^b, i_{bd+}, i_{ba\beta}^b, i_{b0}^{a-\alpha\beta}$	0	$-4\sqrt{2}V_a^3 V_b^3$
7	$i_{ad+}, i_{bd+}, i_{ba\beta}^b, i_{ba\beta}^{cm}$	$-2V_a^2 V_b^2 V_{cm}^2$	$-4V_a^2 V_b^2 V_{cm}^2$
8	$i_{ad+}, i_{bd+}, i_{aa\beta}^{cm}, i_{aa\beta}^b$	$-\dfrac{1}{8}V_a^2 V_b^2 V_{cm}^2$	$-\dfrac{1}{4}V_a^2 V_b^2 V_{cm}^2$
9	$i_{ad+}, i_{bd+}, i_{aa\beta}^{cm}, i_{ba\beta}^{cm}$	$-4V_a^2 V_{cm}^4$	$-4V_a^2 V_{cm}^4$
10	$i_{ad+}, i_{bd+}, i_{aa\beta}^{cm}, i_{bdq-}$	$2V_a^4 V_{cm}^2$	$2V_a^4 V_{cm}^2$
11	$i_{ad+}, i_{bd+}, i_{adq-}, i_{ba\beta}^{cm}$	$2V_a^4 V_{cm}^2$	$2V_a^4 V_{cm}^2$
12	$i_{b0}^b, i_{bd+}, i_{ba\beta}^b, i_{ba\beta}^{cm}$	$-2\sqrt{2}V_a V_b^3 V_{cm}^2$	$-4\sqrt{2}V_a V_b^3 V_{cm}^2$
13	$i_{b0}^b, i_{b0}^{cm}, i_{ba\beta}^b, i_{b0}^{a-\alpha\beta}$	0	$-8V_a^2 V_b^3 V_{cm}$
14	$i_{b0}^b, i_{b0}^{cm}, i_{ba\beta}^b, i_{ba\beta}^{cm}$	$-8V_b^3 V_{cm}^3$	$-8V_b^3 V_{cm}^3$
15	$i_{b0}^b, i_{b0}^{cm}, i_{ba\beta}^b, i_{bdq-}$	$V_a^2 V_b^3 V_{cm}$	$4V_a^2 V_b^3 V_{cm}$

In summary, the components of the circulating current ($i_{ba\beta}$), the three-phase current ($i_{aa\beta}$), and the single-phase current ($i_b = -i_{b0}/3$)) which are in phase with the single-phase voltage (v_b), three-phase voltage ($v_{aa\beta}$), and common-mode voltage (v_{cm}) represent the manipulated inputs of the power equations. Any combination of the six manipulated inputs (MI_1, MI_2, MI_{34}, MI_{56}) from Table. I represent a possible energy balancing technique for the MMC. Any existing energy balancing technique for MMC could be linked directly or indirectly to all derived energy balancing methods from Table. I.

$$i_{\alpha\beta} = D(\theta_a)\, i_{dq+} + \overline{D}^{-1}(\theta_a)\, i_{dq-} \rightarrow \begin{pmatrix} i_\alpha \\ i_\beta \end{pmatrix} = \begin{pmatrix} \cos(\theta_a)(i_{d+} + i_{d-}) + \sin(\theta_a)(i_{q-} - i_{q+}) \\ \cos(\theta_a)(i_{q+} + i_{q-}) + \sin(\theta_a)(i_{d+} - i_{d-}) \end{pmatrix} \tag{8}$$

$$D(\theta_a) = \begin{pmatrix} \cos(\theta_a) & -\sin(\theta_a) \\ \sin(\theta_a) & \cos(\theta_a) \end{pmatrix}; \quad \overline{D}^{-1}(\theta_a) = \begin{pmatrix} \cos(\theta_a) & \sin(\theta_a) \\ -\sin(\theta_a) & \cos(\theta_a) \end{pmatrix} \tag{9}$$

$$\overline{\left\{ v_{a\alpha}\, i_{x\alpha} - v_{a\beta}\, i_{x\beta} \right\}} = \hat{V}_{a+}\, i_{xd-}; \quad \overline{\left\{ v_{a\alpha}\, i_{x\beta} + v_{a\beta}\, i_{x\alpha} \right\}} = \hat{V}_{a+}\, i_{xq-}; \quad \overline{\left\{ v_{a\alpha}\, i_{x\alpha} + v_{a\beta}\, i_{x\beta} \right\}} = \hat{V}_{a+}\, i_{xd+} \tag{10}$$

Average Arm Energies Regulation Techniques for Direct MMC Topology

According to Table I, various energy balancing methods for the MMC can be derived by the combination of the manipulated inputs (MI_1, MI_2, MI_{34}, MI_{56}). (11) illustrates the definition of the three-phase voltage angle (θ_a), single-phase voltage angle (θ_b), and common-mode voltage angle (θ_a). Further, (12) represents the three-phase voltage ($v_{a\alpha\beta}$), single-phase voltage (v_b) and common-mode voltage (v_{cm}) expressions. (13) – (17) show the full expressions of the currents, including the previously mentioned definitions.

These expressions decompose the three-phase currents ($i_{a\alpha\beta}$), the circulating currents ($i_{b\alpha\beta}$), and the single-phase currents ($i_b \approx i_{b0}$) into components that are in phase with the voltages $v_{a\alpha\beta}$, v_b, and v_{cm}. $i_{b0}^{\perp b}$ and i_{aq+} are used to realize reactive powers in single-phase and three-phase networks, respectively.

These equations change slightly for AC-DC MMC topology. The DC value of the voltages and currents for AC-DC MMC application is equal to the corresponding RMS voltage and RMS currents during

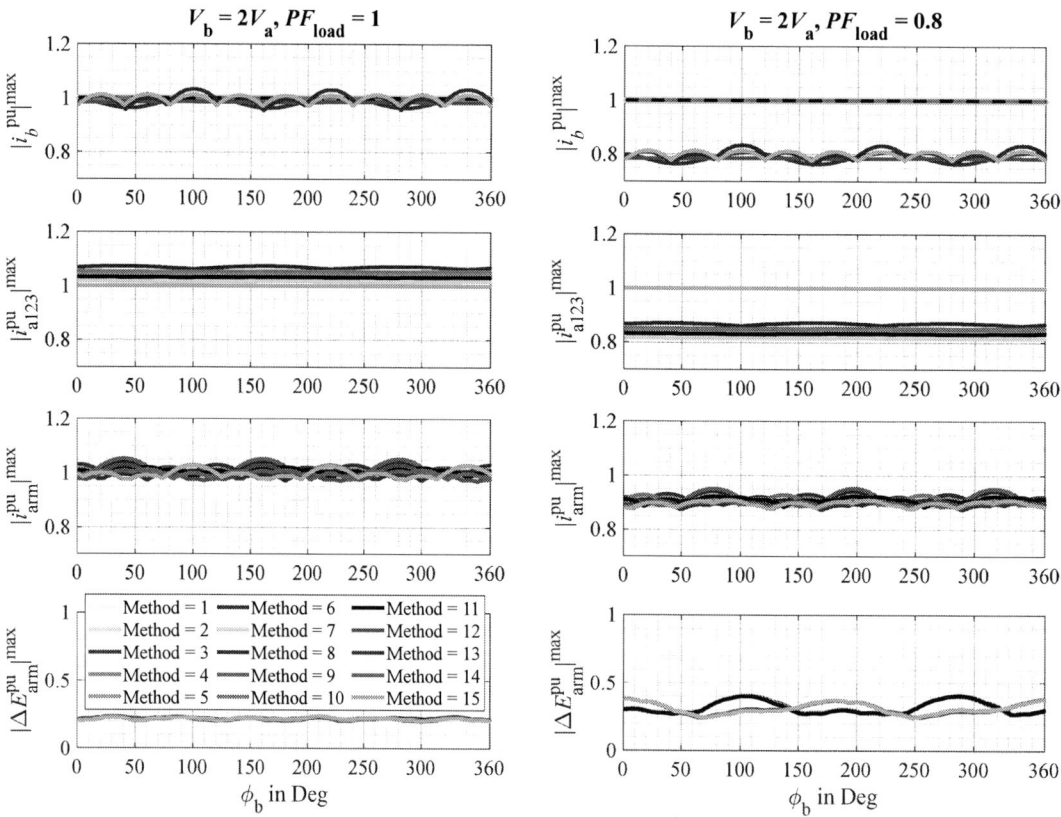

Fig. 2: Performance analysis of direct AC-AC MMC via current and energy values for all stable arm energy regulation techniques for $f_b = f_a/3$ application during normal operations at various load power factor. Here, $V_{cm} = 0.55\, V_a$.

AC-AC MMC application. If the single-phase side provides power, the three-phase currents are used to regulate the overall arm energies, and when the three-phase side acts as a power source, the single-phase currents are used to regulate the arm energies. All possible methods to regulate arm energies using six manipulated inputs (MI_1, MI_2, MI_{34}, MI_{56}) are listed in Table II. Method 1 to 6 represent the energy balancing methods that do not require common-mode voltage. The common-mode frequency (f_{cm}) equals $3\text{Max}(f_a, f_b)$ to avoid couplings with the single-phase and the three-phase networks. It should be noted that harmonics are injected into the grid of the MMC for eleven of the shown energy balancing techniques (Method 2, 3, 4, 6, 8, 9, 10, 11, 13, 14, and 15). No harmonics flow into the grid for four energy balancing methods (Method 1, 5, 7, and 12). One may choose any suitable energy balancing method based on the desired system requirements and controller's performance.

$$\theta_a = \omega_a t \; ; \; \theta_b = \omega_b t + \phi_b \; ; \; \theta_{cm} = \omega_{cm} t + \phi_{cm} \tag{11}$$

$$v_{a\alpha} = \sqrt{2}V_a \cos(\theta_a), \; v_{a\beta} = \sqrt{2}V_a \sin(\theta_a), \; v_{cm} = \sqrt{2}V_{cm} \cos(\theta_{cm}), \; \text{and} \; v_b = \sqrt{2}V_b \cos(\theta_b) \tag{12}$$

$$i_{a\alpha} = \cos(\theta_a)(i_{ad+} + i_{ad-}) + \sin(\theta_a)(i_{aq-} - i_{aq+}) + \sqrt{2}I_{a\alpha}^{cm} \cos(\theta_{cm}) + \sqrt{2}I_{a\alpha}^{b} \cos(\theta_b) \tag{13}$$

$$i_{a\beta} = \cos(\theta_a)(i_{aq+} + i_{aq-}) + \sin(\theta_a)(i_{ad+} - i_{ad-}) + \sqrt{2}I_{a\beta}^{cm} \cos(\theta_{cm}) + \sqrt{2}\,I_{a\beta}^{b} \cos(\theta_b) \tag{14}$$

$$i_{b\alpha} = \cos(\theta_a)(i_{bd+} + i_{bd-}) + \sin(\theta_a)\,i_{bq-} + \sqrt{2}I_{b\alpha}^{cm} \cos(\theta_{cm}) + \sqrt{2}I_{b\alpha}^{b} \cos(\theta_b) \tag{15}$$

$$i_{b\beta} = \cos(\theta_a)i_{bq-} + \sin(\theta_a)(i_{bd+} - i_{bd-}) + \sqrt{2}I_{b\beta}^{cm} \cos(\theta_{cm}) + \sqrt{2}\,I_{b\beta}^{b} \cos(\theta_b) \tag{16}$$

$$i_{b0} = \sqrt{2}I_{b0}^{\perp b} \sin(\theta_b) + \sqrt{2}I_{b0}^{b} \cos(\theta_b) + \sqrt{2}I_{b0}^{cm} \cos(\theta_{cm}) + \sqrt{2}I_{b0}^{a-\alpha} \cos(\theta_a) + \sqrt{2}I_{b0}^{a-\beta} \sin(\theta_a) \tag{17}$$

Stability and Performance Analysis

Validation of controller stability and performance analysis is vital while designing any controller. Arm currents, arm energies, and source current magnitude could be used to examine the performance of the MMC for the selected energy regulation method. For a chosen energy regulation method (see Table II), (18) represents the average of the equations (2) to (7) using matrix notation. Here, $P_{6\times1}$ represents the

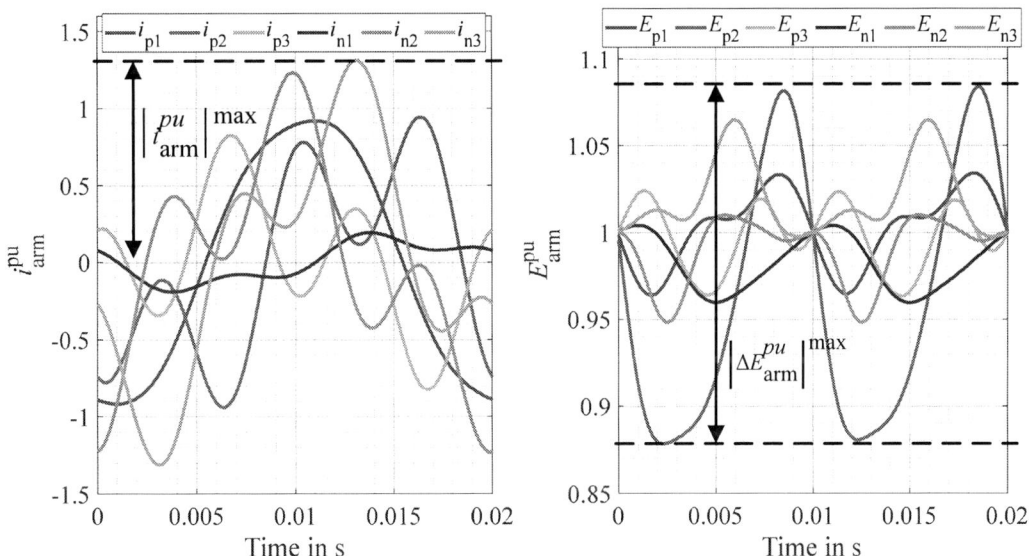

Fig. 3: Definition of *Maximum Arm Current* (left image) and *Arm Energy Ripple* (right image) using an example test case ($f_b = f_a$).

residual power matrix which consists of reactive power and coupling power components. In the steady-state, the average arm energies are constant; hence $\overline{p_{\Sigma/\Delta\alpha\beta0}} = 0$. The frequency for computing average power ($\overline{p_{\Sigma/\Delta\alpha\beta0}}$.) is selected using $f_{avg} = \text{Max}(0, \text{Min}(f_a, f_b))$. (19) represents the solution of (18) with $\overline{p_{\Sigma/\Delta\alpha\beta0}} = 0$. The determinant of the matrix A ($|A|$) for various methods and operating conditions are also listed in Table II. V_b represents the DC magnitude and RMS magnitude of the single-phase voltage for the AC-DC MMC topology and direct AC-AC MMC topology, respectively. If $|A| = 0$, the method is unstable. For unequal frequency operations ($f_a = n f_b$, $f_b = n f_a$ & $f_b = \text{DC}$), where n = odd integers, all methods are stable, and the manipulated inputs are naturally decoupled (A = diagonal matrix).

In contrast, methods 1, 4, 5, 6, and 13 are unstable for equal frequency operation ($f_a = f_b$). Further, Method 2 is unstable at $V_b = 2V_a$, and Method 3 is unstable at $V_b = 2V_a$ and $\phi_b = .m\pi/6$, where $m = 0$ & even integers for $f_a = f_b$. Hence, all methods without manipulated inputs in phase with v_{cm} (Method 1 to 6) are unstable at $V_b = 2V_a$ and $f_a = f_b$ because of the coupling between the manipulated inputs. These findings suggest that the common-mode voltage plays a vital role in balancing arm energies of direct AC-AC MMC for equal frequency operation.

$$\left(\overline{P_{\Sigma/\Delta\alpha\beta0}}\right)_{6\times1} = A_{6\times6} \, MI_{6\times1} + P_{6\times1} \tag{18}$$

$$MI_{6\times1} = -A_{6\times6}^{-1} P_{6\times1} \tag{19}$$

Solving (19) gives either RMS or amplitude of the selected manipulated input currents, which are substituted in (13) to (17) to obtain the instantaneous three-phase ($i_{a\alpha\beta}$), circulating ($i_{b\alpha\beta}$) and single-

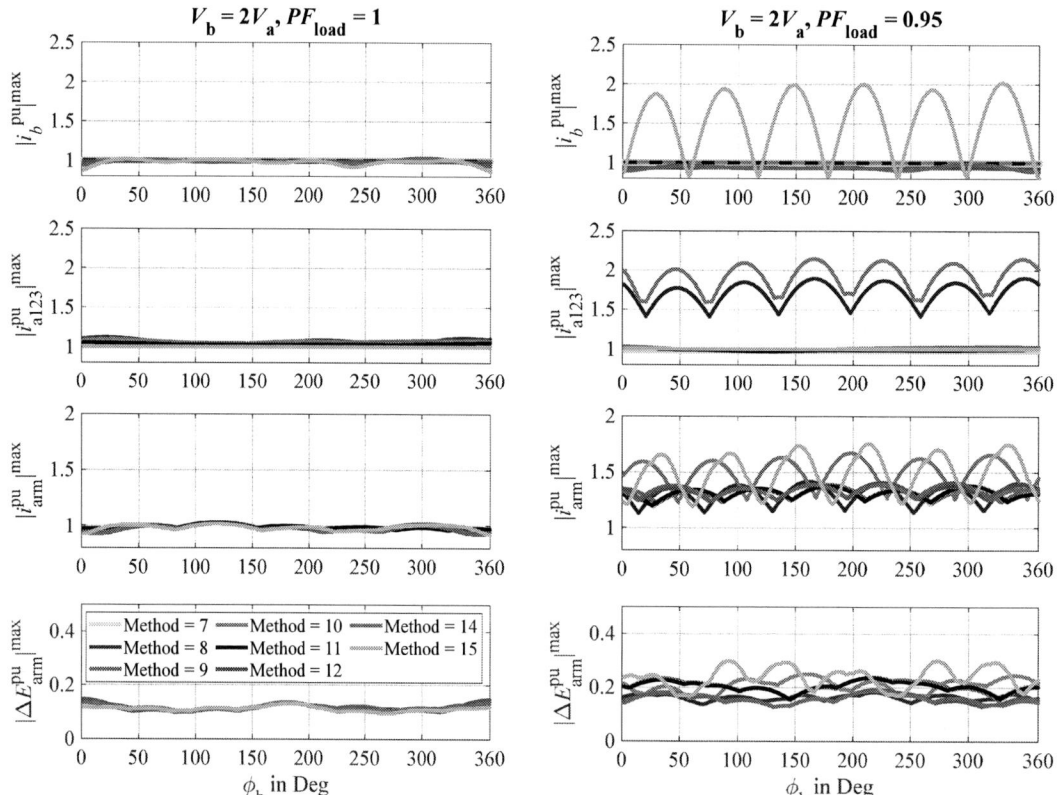

Fig. 4: Performance analysis of direct AC-AC MMC via current and energy values for all stable arm energy regulation techniques for $f_b = f_a$ application during normal operations at various load power factor. Here, $V_{cm} = 0.55 \ V_a$.

phase ($i_b \approx i_{b0}$) currents. It should be noted that each energy balancing method uses only six manipulated inputs. Hence, all remaining manipulated input's magnitude is considered zero while computing instantaneous currents using (13) to (17). The three-phase (i_{a123}) and circulating (i_{b123}) currents in the abc-frame are obtained by performing inverse Clarke transformation on $i_{a\alpha\beta}$, and $i_{b\alpha\beta0}$, respectively. The instantaneous arm currents ($i_{p/n123}$) are obtained from i_{a123} and i_{b123} (see Fig. 1). Further, $i_{a\alpha\beta}$, i_{b0}, and $i_{b\alpha\beta}$ are substituted in (2) to (7) to get instantaneous summation and difference arm powers ($p_{\Sigma/\Delta\alpha\beta0}$) in the $\alpha\beta0$-frame, which are transformed to the abc frame using inverse Clarke transformation.

The instantaneous arm powers ($p_{p/n123}$) can be obtained by solving (1). Assuming each arm's energies were initially at the reference value, the instantaneous arm energies ($E_{p/n123}$) are obtained by integrating computed instantaneous arm powers ($p_{p/n123}$).

The ripple in the arm energies ($\Delta E_{p/n123}$) can be computed from the computed instantaneous arm energies. If the maximum arm currents and energies are within the safe limits, then it can be said that the MMC is operating safely. Therefore, the maximum arm currents and arm energies could be investigated to validate the dimensioning of the converter for all operating conditions.

Analytical Result

Fig. 2 shows the solution of the equations for all stable methods at $V_b = 2V_a$ and various load power factors (PF_{load}) for $f_a = 3 f_b$ operation. The definition of maximum arm currents ($|i_{arm}^{pu}|^{max}$) and maximum ripple arm energy ($|\Delta E_{arm}^{pu}|^{max}$) is illustrated in Fig. 3. The nominal load current with the power factor (PF_{load}) is supplied for all the investigated cases, and no reactive power is drawn from the source of the MMC. The results show that the maximum arm current ($|i_{arm}^{pu}|^{max}$) is reduced when PF_{load} is reduced for $f_a = 3 f_b$ operation, and the maximum ripple arm energy $|\Delta E_{arm}^{pu}|^{max}$ increases when PF_{load} is reduced. The single-phase reactive power cannot be compensated by the three-phase reactive power and vice versa.

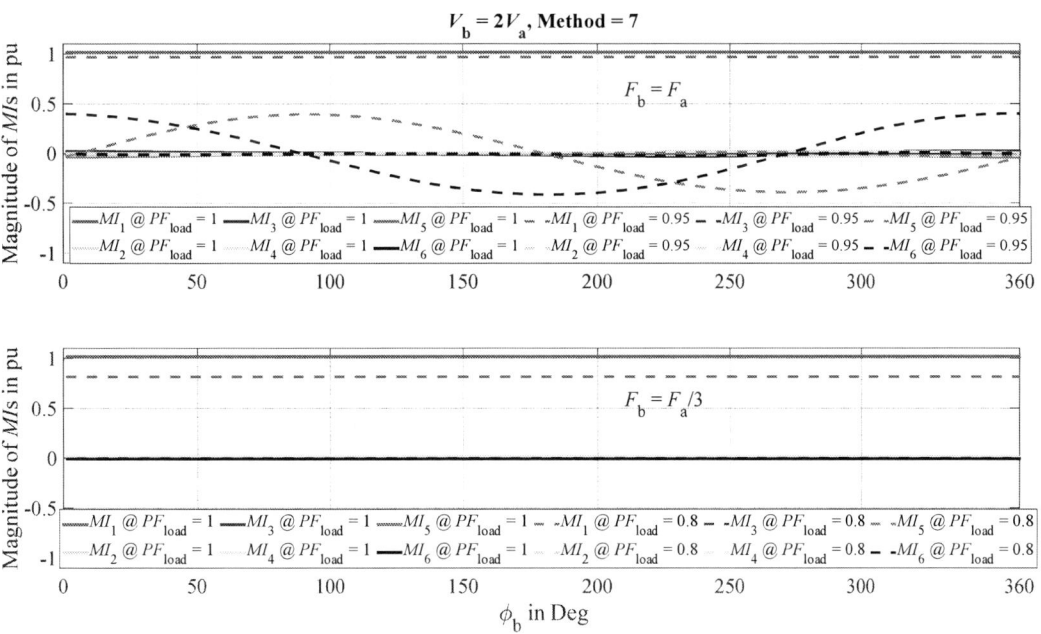

Fig. 5: *Manipulated Inputs* (MI) variation due to *Power Factor* (PF$_{load}$) change for energy regulation technique using Method-7 ($f_b = f_a/3$ and $f_b = f_a$) during normal operations. Solid line: PF$_{load} = 1$ and dotted line: less PF$_{load}$

Hence increase in $|\Delta E_{arm}^{pu}|^{max}$ with reduction in PF_{Load} is because the load reactive power oscillates inside MMC, which increases oscillation of the arm energies.

Similarly, Fig. 4 represents the solution of the equations for all stable methods at $V_b = 2V_a$ and various load power factors (PF_{load}) for $f_a = f_b$ operation. Method 8, 10, and 15 are stable, but their performance for $f_b = f_a$ operation is poor compared to other stable methods. In contrast with $f_a = 3 f_b$ operation, $|i_{arm}^{pu}|^{max}$ increases with a lower power factor at $f_a = f_b$ operation.

The first two terms ($- 0.5\ v_b\ i_{a\alpha\beta} - 2\ v_{a\alpha\beta}\ i_{b0}$) on the right-hand side of (5) and (6) represent coupling between the single-phase and three-phase systems. The *Manipulated Inputs* (MIs) only need to compensate for the losses at $f_a = 3 f_b$ because the three-phase and single-phase systems are naturally decoupled: the third harmonic component does not make power with the fundamental component. For $f_a = f_b$ operation, however, the MIs also compensate for couplings ($- 0.5\ v_b\ i_{a\alpha\beta} - 2\ v_{a\alpha\beta}\ i_{b0}$) present in the system. Fig. 5 shows the magnitude of manipulated inputs for both $f_b = f_a$ and $f_b = f_a/3$ operations at unity power factor and low power factor. For equal frequency operation, the coupling is predominant, and its effect also increases with a reduction in power factor. On the other hand, the magnitude of manipulated inputs does not increase with a low power factor for $f_b = f_a/3$ operation.

Fig. 6 represents the analytical results during the failure conditions on the single-phase side for well-behaved stable methods at $f_a = f_b$, and $V_b = 2V_a$. More arm current and more ripple arm energy are observed when the fault location is far away from the MMC because the load power factor is poor, increasing the coupling between manipulated inputs.

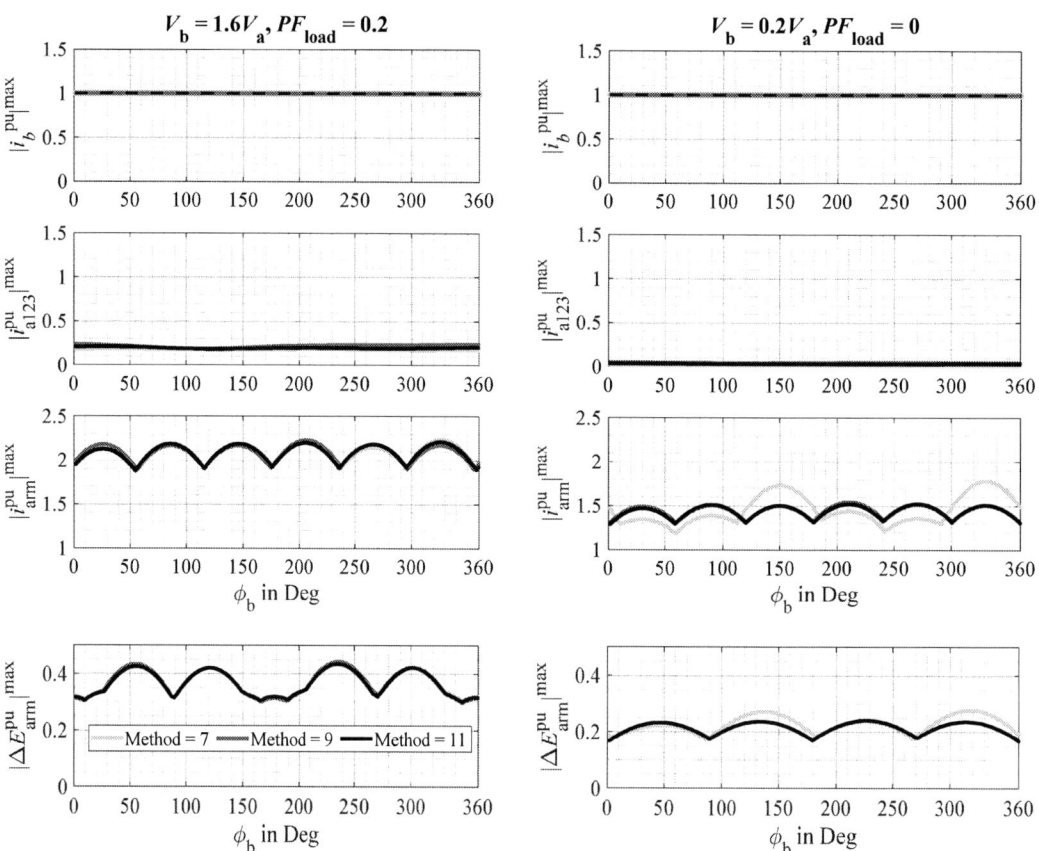

Fig. 6: Performance analysis during faulty conditions on the single-phase side of MMC for few well-behaved stable energy regulation techniques at $f_b = f_a$. Left: fault far away from MMC and Right: fault near MMC.

Conclusion

A new generalized mathematical toolchain is presented to investigate the stability and performance of all possible average arm energy regulation techniques for direct AC-AC and AC-DC MMC topologies. All proposed methods are stable for unequal frequency operations ($f_a = n f_b, f_b = n f_a$ & $f_b = $ DC), where $n = $ odd integers because the three-phase and single-phase systems are naturally decoupled. On the other hand, some methods are unstable for equal frequency operation ($f_a = f_b$) due to the coupling between the manipulated inputs. The common-mode voltage is found to be vital for the arm energy regulation techniques for equal frequency operation. For unequal frequency operation, fewer arm currents are required when the power factor of the load is small. On the other hand, more arm currents are required for equal frequency operation when the load power factor is low. Further, more ripple in the arm energies is observed with a low power factor because the reactive load power flows inside the MMC, which could not be compensated by the reactive power on the other side. The proposed toolchain can also validate the dimensioning of the direct AC-AC and AC-DC MMC topologies by analyzing the MMC energies and currents for all operating conditions of the converter (power factor, fault cases). Any existing MMC energy regulation technique could be linked directly or indirectly to all derived energy balancing methods.

In the future, it is intended to develop a similar analytical toolchain to further investigate the performance of the discussed energy balancing techniques for non-integer three-phase and single-phase frequency operation. This paper avoids this operation due to its increased complexity in computing the average mean value.

References

[1] J. Rodriguez, Jih-Sheng Lai and Fang Zheng Peng, "Multilevel inverters: a survey of topologies, controls, and applications," in IEEE Transactions on Industrial Electronics, vol. 49, no. 4, pp. 724-738, Aug. 2002, doi: 10.1109/TIE.2002.801052.

[2] L. Harnefors, S. Norrga, A. Antonopoulos and H. Nee, "Dynamic modeling of modular multilevel converters," Proceedings of the 2011 14th European Conference on Power Electronics and Applications, 2011, pp. 1-10.

[3] F. Kammerer, M. Gommeringer, J. Kolb and M. Braun, "Energy balancing of the Modular Multilevel Matrix Converter based on a new transformed arm power analysis," 2014 16th European Conference on Power Electronics and Applications, 2014, pp. 1-10, doi: 10.1109/EPE.2014.6910939.

[4] J. Kolb, F. Kammerer, M. Gommeringer and M. Braun, "Cascaded Control System of the Modular Multilevel Converter for Feeding Variable-Speed Drives," in IEEE Transactions on Power Electronics, vol. 30, no. 1, pp. 349-357, Jan. 2015, doi: 10.1109/TPEL.2014.2299894.

[5] M. Schütt and H. Eckel, "Design and analysis of complex vector current regulators for modular multilevel converters," 2017 19th European Conference on Power Electronics and Applications (EPE'17 ECCE Europe), 2017, pp. P.1-P.10, doi: 10.23919/EPE17ECCEEurope.2017.8099183.

[6] L. Angquist, A. Antonopoulos, D. Siemaszko, K. Ilves, M. Vasiladiotis and H. Nee, "Open-Loop Control of Modular Multilevel Converters Using Estimation of Stored Energy," in IEEE Transactions on Industry Applications, vol. 47, no. 6, pp. 2516-2524, Nov.-Dec. 2011, doi: 10.1109/TIA.2011.2168593.

[7] A. Antonopoulos, L. Angquist and H. Nee, "On dynamics and voltage control of the Modular Multilevel Converter," 2009 13th European Conference on Power Electronics and Applications, 2009, pp. 1-10.

[8] D. Siemaszko, A. Antonopoulos, K. Ilves, M. Vasiladiotis, L. Ängquist and H. Nee, "Evaluation of control and modulation methods for modular multilevel converters," The 2010 International Power Electronics Conference - ECCE ASIA -, 2010, pp. 746-753, doi: 10.1109/IPEC.2010.5544609.

[9] A. Lesnicar and R. Marquardt, "A new modular voltage source inverter topology," in *Proc. 10th EPE*, Toulouse, France, 2003, pp. 1–10.

[10] A. Lesnicar and R. Marquardt, "An innovative modular multilevel converter topology suitable for a wide power range," 2003 IEEE Bologna Power Tech Conference Proceedings,, 2003, pp. 6 pp. Vol.3-, doi: 10.1109/PTC.2003.1304403.

Comparative Lifetime Estimations for IGBT Modules in Wind Turbine Converters

Christian Neumann, Hans-Günter Eckel
UNIVERSITY OF ROSTOCK
Albert-Einstein-Str. 2
D-18059 Rostock, Germany
Tel.: +49 381 / 498-7138
E-Mail: christian.neumann@uni-rostock.de
URL: http://www.iee.uni-rostock.de/

Acknowledgements

The project WindVolt (0324256B) is supported by the Federal Ministry for Economic Affairs and Energy on the basis of a decision by the German Bundestag.

Keywords

«IGBT», «Lifetime», «Reliability», «Thermal cycling», «Wind energy».

Abstract

Comparative IGBT lifetime estimations for a generic wind turbine are presented. A thermal impedance matrix is utilized for thermal modeling, addressing thermal cross-coupling effects within the IGBT modules. For the estimation procedure, all temperature swings are calculated with a high temporal resolution, allowing the correct superposition of each swing.

Introduction

The overall thermal stress for IGBT modules in wind power applications consists of a superposition of fundamental frequency cycles correlated to the converter's output frequency and low-frequency cycles, caused by changing operation points due to fluctuations in wind speed [1]. Thus, the resulting mission profiles are complex, and it is mandatory to consider them as accurately as possible for realistic lifetime estimation.
The lifetime estimation procedures typically perform calculations on different time frames and merge the identified life consumptions to an overall result [2]. A lower time resolution is chosen to deal with wind-induced load changes, which results in averaged temperature profiles. Furthermore, it neglects the fundamental frequency ripple on top of them, as this is not displayed in the considered time frame. Consequently, the estimation procedures will operate on trivialized loading conditions, which do not represent the realistic stress an IGBT module has to endure. The lower the fundamental frequency of the converters, the higher this issue becomes. Moreover, the thermal model of the converter is often described as a Y-model [3]. This only addresses thermal cross-coupling effects between the IGBTs and *Free-Wheeling Diodes* (FWDs) on the case or heatsink level, but they already appear on the substrate level due to lateral heat spreading [4]. Both points mentioned above will lead to an overestimated lifetime of the IGBT modules.

This paper presents a lifetime estimation procedure that tackles the mentioned issues. The necessary calculations of the semiconductor losses and temperatures are performed on a high-resolution time basis. Hence, the mission profiles as inputs are transferred into temperature profiles, which contain thermal cycles from operation point changes and fundamental frequency components. Moreover, utilizing a thermal impedance matrix instead of a Y-model considers thermal cross-coupling effects way earlier. Hence, the loading conditions for the IGBT modules are mapped more realistically.
The investigations are carried out for the IGBT modules in the *Machine-Side Converter* (MSC) of a generic full-scale converter wind turbine (type 4) in the 3 MW class. Fig.1 illustrates the lifetime estimation procedure, which takes place in three steps.

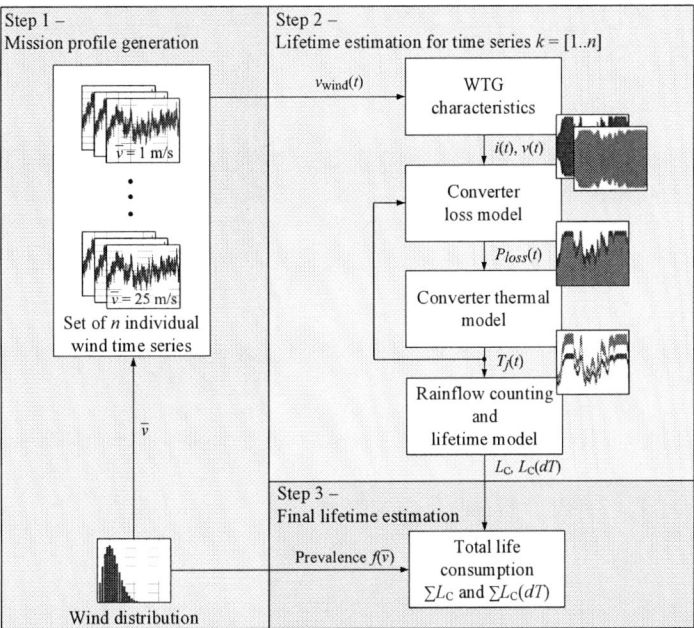

Fig. 1: Lifetime estimation procedure

A set of n synthetic wind time series at various average wind speeds will represent the mission profiles. They are used to calculate the thermal cyclings $T_j(t)$ of IGBTs and FWDs based on the converter's power losses and thermal models. A rainflow counting algorithm [5] is used to evaluate the thermal cyclings of the individual wind time series k regarding their height and quantity. By applying those thermal cycles to the lifetime model of the manufacturer and considering Miner's rule for damage accumulation, the relative life consumption (L_c) of the kth wind time series can be estimated. However, thermal cycles below 10 K are assumed to cause no harm to the module. Based on the prevalence of each average wind speed, which is defined by a Rayleigh wind distribution, those individual life consumptions are accumulated to an overall result ($\sum L_c$). The targeted lifetime in this paper is 20 years.

The converters for the comparative study are equipped with 1700 V half-bridge IGBT modules from Infineon. The upper IGBT and FWD are referred to as I_1 and D_1, whereas I_2 and D_2 is used for the lower ones. As the losses and thermal models for the upper and lower IGBT / FWD are identical, only the results for I_1 and D_1 are presented in this paper. Usually, a parallel connection of N IGBT modules is used for the converters. An equal current distribution among the paralleled modules is assumed, so the collector current I_C of a single IGBT or FWD shares $1/N$ of the total MSC current. As the main focus of the generic approach is to identify and investigate influencing factors on the estimated lifetime, the number of paralleled modules N does not necessarily have to be an integer value – quite contrary to real applications. Instead, N is set to meet a defined $T_{j,max}$ of the switches during operation at nominal power.

The generic approach allows the investigation of several influencing factors on the estimated lifetime. This paper covers different design and modeling aspects, like the thermal utilization of the IGBT modules or the base frequency of the generator, and the influence of the input's temporal resolution.

General model description

The following subsections explain the essential parts of the lifetime estimation procedure in more detail.

Wind profiles and wind distribution

In this study, 10-minutes stochastic wind time series are generated according to IEC 61400-3 using the Kaimal *Power Spectral Density* (PSD) to model wind turbulence [6]. The PSD is calculated for several mean wind speeds $\bar{v}_{hub}$ on the wind turbine's hub height (1). A Rayleigh distribution, with a wind site-specific mean wind speed $\bar{v}$, states the prevalence for $\bar{v}_{hub}$ over one year in bars of 1 m/s (2).

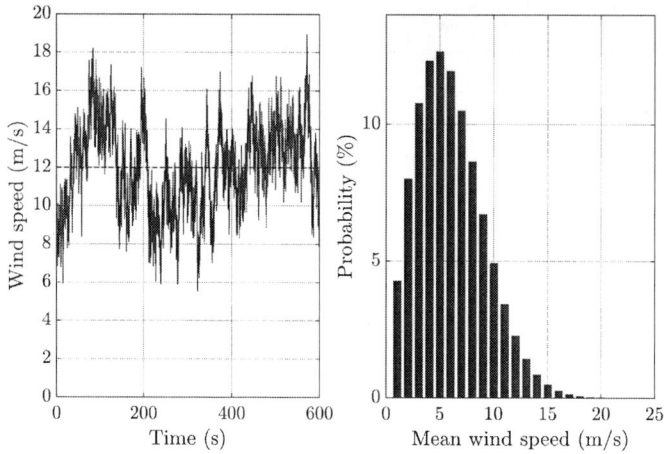

Fig. 2: Wind time series with a mean wind speed of 12 m/s (left); Rayleigh distribution for a wind site with an average wind speed of 6 m/s (right)

$$S_k(f) = 4\sigma_k^2 \frac{\dfrac{L_k}{\overline{v}_{hub}}}{\left(1 + 6f\dfrac{L_k}{\overline{v}_{hub}}\right)^{\frac{5}{3}}} \tag{1}$$

$$f(v) = \frac{\pi v}{2\overline{v}^2} e^{-0.25\pi(v/\overline{v})^2} \tag{2}$$

In the equations, $L_k = 340.2$ m denotes a turbulence length scale parameter for hub heights above 60 m [6] and σ_k is the standard deviation of the wind turbulence determined by the *Turbulence Intensity* (TI). The lifetime estimation is consecutively performed for all n wind time series. This study has two individual wind time series for each mean wind speed between 1 m/s and 25 m/s ($n = 50$).

Fig. 2 shows an exemplary wind time series for IEC class A, represented by a TI of 18%, and a Rayleigh distribution for an IEC class III wind site, having less than 7.5 m/s average wind speed on hub height.

Generator and MSC control structures

A *Permanent Magnet Synchronous Generator* (PMSG) is utilized for the type 4 wind turbine concept. The generator and converters are modeled with characteristic curves and equations to simulate their quasi-stationary behavior. At this calculation stage, an interpolation of the kth input wind time series to a higher time resolution is performed, which is chosen to represent the expected fundamental frequency cycles properly. Relations between the input wind speed, generator speed, and generator torque can be derived from the generic wind turbine's characteristics, shown in Fig. 3. However, the wind speed can not be directly translated into the generator quantities, as the dynamics would be way too fast. Instead, the wind speed is filtered by a first-order lowpass filter with the turbine's inertia constant $H = 4$ s to respect the mechanical dynamics of the drivetrain.

The mechanical generator quantities are transferred into the instantaneous MSC electrical quantities by using the necessary equations for the PMSG (3)-(5) and assuming an MSC *Field Oriented Control* (FOC) with *Zero d-axis Current* (ZDC) control strategy [7].

$$v_{sd} = R_{sd}i_{sd} + L_{sd}\frac{di_{sd}}{dt} - \omega_e L_{sq}i_{sq} \tag{3}$$

$$v_{sq} = R_{sq}i_{sq} + L_{sq}\frac{di_{sq}}{dt} + \omega_e L_{sd}i_{sd} + \omega_e \psi_m \tag{4}$$

$$T_e = \frac{3p}{2}\left(\psi_m i_{sq} - (L_{sd} - L_{sq})i_{sd}i_{sq}\right) \tag{5}$$

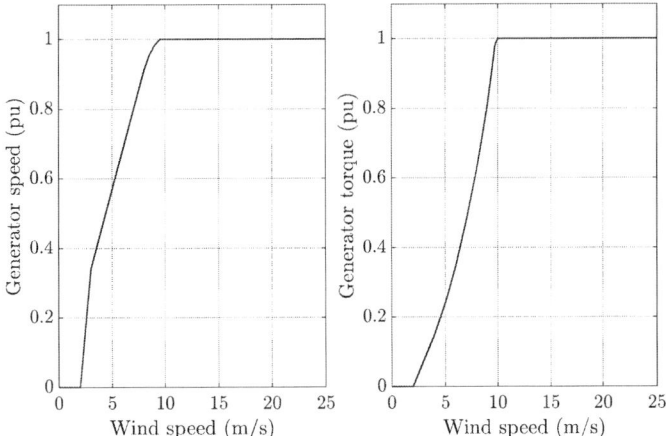

Fig. 3: Characteristic curves for generator speed (left) and torque (right) of the generic wind turbine

Based on the generator torque gathered from the characteristics (see Fig. 3) in combination with the input wind speed, dq-phasors for current and voltage of the MSC can be calculated. Hence, the instantaneous MSC currents and voltages can be derived, as shown for phase a, when the d-axis is chosen as phasor reference.

$$i_a(t) = i_{sq}\sin(\omega_e t - \frac{\pi}{2}) \tag{6}$$

$$v_a(t) = \sqrt{v_{sd}^2 + v_{sq}^2}\sin(\omega_e t + \varphi_v) \tag{7}$$

Where φ_v denotes the angle of the dq-voltage phasor regarding the d-axis.

Power loss and thermal model

The calculation of conduction and switching losses is based on datasheet values of the investigated IGBT modules. A curve-fitting approach is utilized to find the corresponding data regarding currents and junction temperatures T_j for different operation points, and instantaneous power losses are calculated for the upper and lower IGBT / FWD. The switching losses are determined by the junction temperature T_j, collector current I_C, DC voltage V_{dc}, and switching frequency f_{sw}. The conduction losses are defined as a function of T_j, I_C, and the relative duty cycle τ. A Pulse-Width Modulation (PWM) with a third harmonic injection is assumed for the latter one.

Because half-bridge IGBT modules are used, thermal cross-couplings between the upper and lower switches have to be considered in the lifetime estimation [4]. A matrix approach for thermal modeling is used to address this issue. All transient thermal impedances in (8) are given from junction to ambient and describe the thermal transfer between two nodes, e.g., the portion of passive heating seen at I_1 while D_1 is producing power losses (see $Z_{\text{th,ja }D_1 I_1}$).

$$\begin{bmatrix} T_{j\,I1} \\ T_{j\,D1} \\ T_{j\,I2} \\ T_{j\,D2} \end{bmatrix} = \begin{bmatrix} Z_{\text{th,ja }I1I1} & Z_{\text{th,ja }D1I1} & Z_{\text{th,ja }I2I1} & Z_{\text{th,ja }D2I1} \\ Z_{\text{th,ja }I1D1} & Z_{\text{th,ja }D1D1} & Z_{\text{th,ja }I2D1} & Z_{\text{th,ja }D2D1} \\ Z_{\text{th,ja }I1I2} & Z_{\text{th,ja }D1I2} & Z_{\text{th,ja }I2I2} & Z_{\text{th,ja }D2I2} \\ Z_{\text{th,ja }I1D2} & Z_{\text{th,ja }D1D2} & Z_{\text{th,ja }I2D2} & Z_{\text{th,ja }D2D2} \end{bmatrix} \begin{bmatrix} P_{I1} \\ P_{D1} \\ P_{I2} \\ P_{D2} \end{bmatrix} + T_a \tag{8}$$

The thermal impedances of the IGBT modules ($Z_{\text{th,jc}}$) and the generic heatsink ($Z_{\text{th,ca}}$) are given for Foster-type equivalent networks. Before connecting them in series, they have to be transformed into Cauer networks first. Cauer-type equivalent networks have a physical meaning and can be used for constructing a complete thermal system [8]. The resulting thermal impedances ($Z_{\text{th,ja}}$) are validated against real-world data. The junction temperatures can be estimated by calculating (8) with the instantaneous power losses. They contain ripples from changes in the wind conditions and the fundamental frequency component. A constant ambient temperature $T_a = 40\,^\circ\text{C}$ is used in this case study.

Rainflow counting, lifetime model, and lifetime estimation

The derived junction temperatures of the kth wind time series are further processed to a rainflow counting algorithm, which analyzes the thermal cycles regarding their cycle height, duration, number, and maximum junction temperature at which those cycles occur [5]. The results are applied to the lifetime model, which is not given in the form of an analytical function but the supplier's power cycling capability diagrams. The diagrams used here account for bond wire failures or wear out of the chip solder joints and represent the B_5 lifetime of the IGBT modules [9]. With this, the relative life consumption of the kth wind time series can be estimated by opposing the actual cycles against the cycles to failure from the curves. Finally, the individual results for all n calculations are accumulated using the prevalence of each average wind speed given by the wind distribution.

The following example might help for a better understanding of the procedure. Assuming $k = 9$ and $k = 10$ represent both the 10-minutes wind time series with a mean wind speed of 5 m/s, resulting in a relative life consumption of 3e-7 (pu) and 5e-7 (pu), respectively. Furthermore, the average wind speed of 5 m/s shall have a yearly prevalence of 12%. Hence, 126144 of these 10-minutes cycles will be seen over 20 years. The share of the overall life consumption is calculated to:

$$\Sigma L_c(5 \text{ m/s}) = \frac{126144}{2}(3\text{e-}7 + 5\text{e-}7) = 0.1 \text{ (pu)} \tag{9}$$

Meaning these 5 m/s wind bins will consume 10% of the 20-years lifetime.

Besides showing absolute numbers, cumulative life consumption plots are used for visualizing the results. They indicate which percentage of the 20-years lifetime is consumed for cycles up to a certain dT. Moreover, they show which dT has the highest contribution to overall life consumption (see Fig. 4).

Case studies with influencing factors on the estimated lifetime

This section presents various lifetime estimation results for the MSC IGBT modules according to the influencing factors mentioned in the beginning. A base case scenario is defined to establish a reference for the following comparison:

- Weak wind location with an average wind speed of 6 m/s on hub height (IEC IIIA, TI = 18%)
- Wind time series with a temporal resolution of 0.018 s
- Infineon 4th generation IGBT modules with 1400 A nominal current (FF1400R17IP4)
- Thermal design $T_{j,max}$ = 125 °C resulting in N = 3.9 parallel IGBT modules
- High-speed PMSG with 50 Hz fundamental frequency at nominal speed

First of all, a lifetime estimation for the base scenario is performed. Fig. 4 shows the cumulative life consumption as a function of the cycle height dT for the IGBT and FWD. On the one hand, this form of the presentation shows the overall life consumption on a 20 years basis, where any value higher than one results in a lifetime of fewer than 20 years. On the other hand, it also reveals which share of the overall life consumption different cycle heights will cause.

As expected, the FWD is the lifetime limiting factor on the MSC due to the rectifier operation of the converter. The life consumption of the FWD sums up to 275%, representing an estimated lifetime of only seven years. However, the IGBT reaches 28% or 71 years, respectively. Additionally, two regions can be identified on the FWD curve that show a significant share of life consumption. The first step can be seen in lower thermal cycles around 12 K. These cycles have their origin in the fundamental frequency of the converter output currents. In this example, they consume about 13% of the 20-year lifetime. Higher cycles arise from changes in the wind turbine operation points induced by fluctuating wind speeds. They result in a much higher life consumption of roughly 260 %, and hence it is crucial to consider them in lifetime estimation procedures in any case. By only investigating the fundamental frequency cycles of the IGBT module, based on the prevalences of the operation points, the potential lifetime issue would not have been identified in this case.

The following sub-sections present detailed lifetime estimation results for selected influencing factors. Table I gives a complete overview of all simulated aspects within this work.

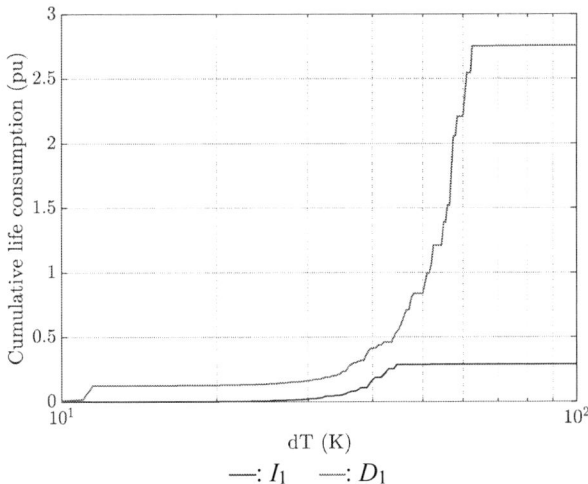

Fig. 4: Cumulative life consumption of the MSC IGBT modules for the base scenario

The thermal utilization level of the IGBT modules

The manufacturer's power cycling curves reveal higher $T_{j,max}$ results in fewer allowed cycles for the same thermal cycle height dT until the IGBT module might fail [9]. Hence, the thermal design temperature severely influences the estimated lifetime as cycles happening at higher maximum junction temperatures will lead to a higher life consumption and vice versa. Therefore, a comparative MSC design with a maximum junction temperature of 100 °C is utilized. In contrast to the base case, 5.83 instead of 3.9 modules are used in parallel to achieve lower junction temperatures on the devices.
Fig. 5 shows the cumulative life consumption curves for both cases. The increased number of IGBT modules results in lower losses on the individual IGBT module and smaller thermal cycles dT overall. Moreover, these cycles will happen at lower maximum junction temperatures, so the estimated lifetime is immensely increased from approximately seven to 76 years. Fundamental frequency cycles now occur below 10 K and do not cause any life consumption on the IGBT module as per the initial assumption.

The fundamental frequency of the converter

A notably challenging condition for the MSC IGBT modules arises with low-speed generator systems. In this case, the fundamental frequency of the converter's output currents is relatively low, resulting in

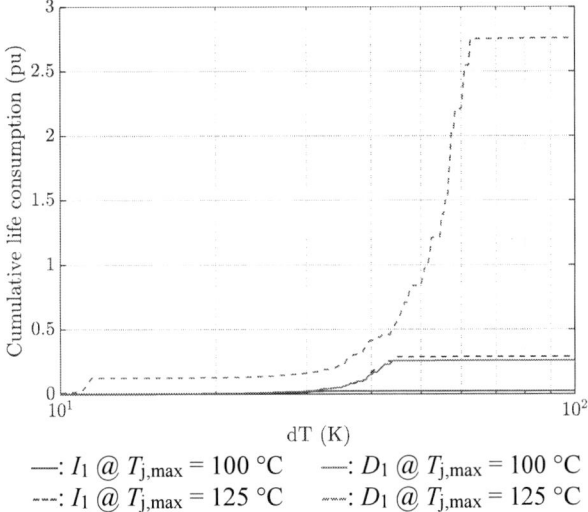

Fig. 5: Cumulative life consumption of the MSC IGBT modules for different thermal design $T_{j,max}$

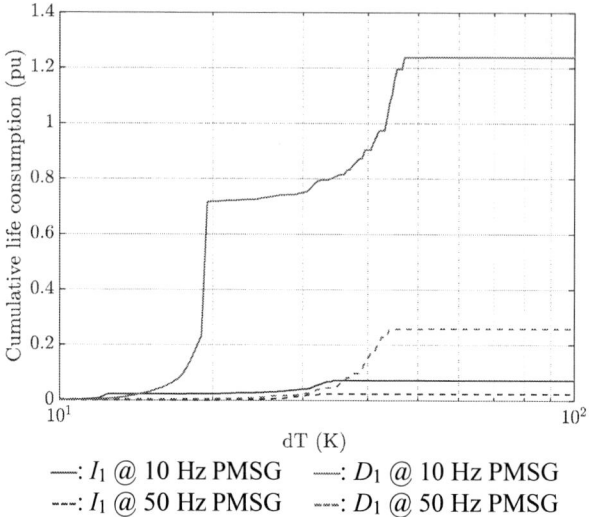

—: I_1 @ 10 Hz PMSG   —: D_1 @ 10 Hz PMSG

---: I_1 @ 50 Hz PMSG   ----: D_1 @ 50 Hz PMSG

Fig. 6: Cumulative life consumption of the MSC IGBT modules for different fundamental frequencies of the PMSG and reduced $T_{j,max}$ = 100 °C

extended heating periods for the power semiconductors and higher thermal cycles dT in general [1]. This sub-section compares a 50 Hz indirect-driven and a 10 Hz direct-driven PMSG. As the base scenario with a high-speed generator was already not fulfilling the lifetime requirements of 20 years, the estimated lifetime on a low-speed generator will be even lower. Therefore, the comparison in Fig. 6 will be shown for the variant with an already reduced maximum junction temperature $T_{j,max}$ = 100 °C from the previous sub-section.

Fig. 6 shows a significant step in the cumulative life consumption for cycles up to 20 K, representing the fundamental frequency cycles for the low-speed generator case. Due to the increased cycle height, 72 % of the lifetime is consumed already. With the mission profile-based cycles included, a life consumption of 124% is calculated, resulting in an estimated lifetime of only 16 years. The target lifetime could not be met even though 100 °C is a cautious design temperature for the IGBT modules.

The temporal resolution of the input wind time series

As highlighted in the beginning, a lower temporal resolution or averaged wind speeds are often used in lifetime estimation procedures to deal with wind-induced load changes [2]. The original wind time series inputs, having a temporal resolution of 0.018 s, are averaged in 10 s and one-minute intervals to replicate this effect.

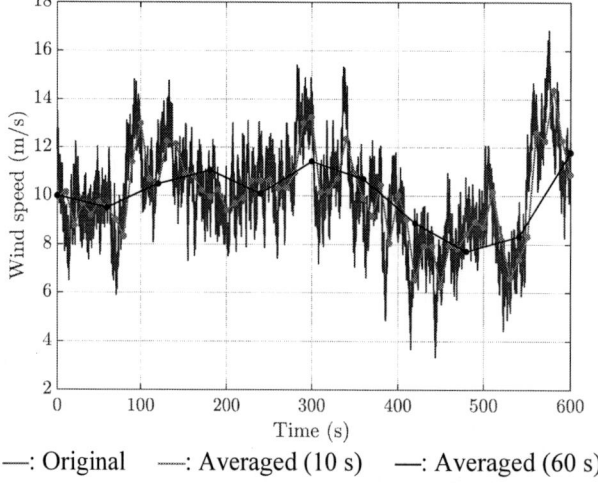

—: Original   —: Averaged (10 s)   —: Averaged (60 s)

Fig. 7: Wind time series with a mean wind speed of 10 m/s (TI = 18%) with and without averaging

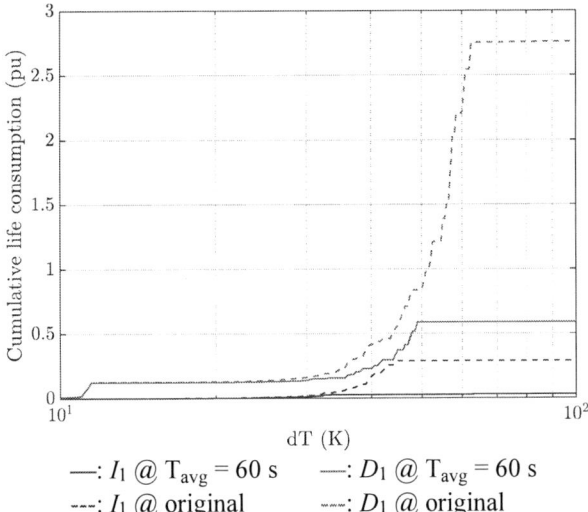

—: I_1 @ T_{avg} = 60 s   —: D_1 @ T_{avg} = 60 s

---: I_1 @ original   ---: D_1 @ original

Fig. 8: Cumulative life consumption of the MSC IGBT modules for 60 s averaged and non averaged wind time series inputs

Fig. 7 shows an exemplary wind time series with a mean wind speed of 10 m/s. A short-term averaging removes the turbulence from the signal, which has quite the same effect as the filtering with the inertia constant H. However, the local peaks and dips are minimized the more prolonged the averaging period becomes. Hence, averaging the wind data results in understated load scenarios for the semiconductors, which will affect the lifetime estimation results - especially for the partial load cases where more operation point changes are about to happen due to changes in wind speed.

Fig. 8 compares the one-minute averaging to the base scenario. The fundamental frequency cycles of the diode show the same life consumption in both cases, which was to be expected. Still, a significant difference can be seen in the wind-induced cycles in higher dT regions. On the one hand, the maximum cycle height of the lifetime relevant cycles is reduced. On the other hand, lesser wind-induced cycles are occurring due to the averaging, which results in a lower life consumption and can be seen in the separation of both FWD curves starting at 20 K cycles. The estimated lifetime has increased from seven to over 34 years. However, the averaging with 10 s has not changed the estimated lifetime as the averaging and filtering effect is quite similar (see Table I).

Other influencing factors

Table I shows the results of all lifetime estimations performed on the MSC. The colors indicate how well the targeted lifetime of 20 years is achieved – not reached (red), attained but with a low margin (orange), or entirely accomplished (green).

Besides the points highlighted in the previous sub-sections, variations in the wind conditions were conducted. Three different locations, represented by their annual mean wind speeds, are assumed for the same wind turbine with a nominal wind speed of 10 m/s. Sites with a lower mean wind speed will increase the amount of partial load operation points, where thermal cycles are generally happening at lower maximum junction temperatures. Therefore, the estimated lifetime increases with lower mean wind speeds. Interestingly, the portion of the relative life consumption caused by wind-induced cycles strongly increases when the mean wind speed of the site approaches the steep part of the wind turbine's power curve because changes from partial to full load are about to happen more frequently. In this case, many thermal cycles from lower temperatures to the designed maximum junction temperature will be seen, leading to a higher life consumption. Fig. 9 illustrates the described effects showing the three Rayleigh distributions against the wind turbine's power curve and the corresponding life consumption curves for the FWD D_1.

Furthermore, a wind class B with a lower turbulence intensity than class A is considered. The reduced turbulence in the wind data also reduces the wind-induced cycles and positively affects the estimated lifetime in all cases.

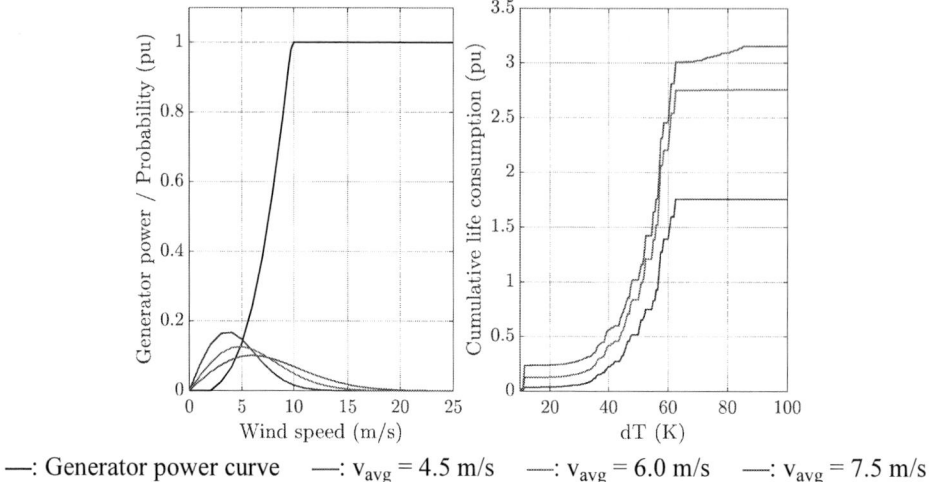

—: Generator power curve   —: v_{avg} = 4.5 m/s   —: v_{avg} = 6.0 m/s   —: v_{avg} = 7.5 m/s

Fig. 9: Power curve versus Rayleigh distributions for three different locations (left); Cumulative life consumption of the MSC IGBT modules (D_1) for FF1400R17IP4 with designed $T_{j,max}$ = 125 °C (right)

Finally, different generations of the power modules are considered. In [10], the superior power cycling capability of Infineon's 5th generation power modules (FF1800R17IP5) over the 4th generation is shown. For the same maximum junction temperature, the estimated lifetime is much higher in all cases. Even for the challenging low-speed PMSG with $T_{j,max}$ = 125 °C, the estimated lifetime is close to or above the targeted value. The main reason, in this case, is the much higher power cycling capability for the fundamental frequency cycles.

Table I: Results of the MSC lifetime estimation

PMSG	High-speed PMSG (50 Hz)				Low-speed PMSG (10 Hz)			
IGBT modules	IGBT4 FF1400R17IP4		IGBT5 FF1800R17IP5		IGBT4 FF1400R17IP4		IGBT5 FF1800R17IP5	
$T_{j,max}$ / Wind class	100 °C N = 5.83	125 °C N = 3.9	100 °C N = 5.62	125 °C N = 3.7	100 °C N = 6.89	125 °C N = 4.56	100 °C N = 6.5	125 °C N = 4.22
IEC IIIA 4.5 m/s	113 (0 %)	11 (4 %)	544 (0 %)	69 (0 %)	33 (24 %)	2 (442 %)	279 (0 %)	34 (3 %)
IEC IIIA 6.0 m/s	77 (0 %)	7 (13 %)	400 (0 %)	48 (0 %)	16 (72 %)	1 (1350 %)	195 (1 %)	23 (9 %)
IEC IIIA 6.0 m/s (10 s avg.)	-	7 (13 %)	-	-	16 (72 %)	-	-	-
IEC IIIA 6.0 m/s (60 s avg.)	934 (0 %)	34 (12 %)	>1000 (0 %)	524 (0 %)	26 (67 %)	1 (1270 %)	936 (1 %)	103 (8 %)
IEC IIIB 6.0 m/s (TI = 16%)	220 (0 %)	14 (12 %)	>1000 (0 %)	129 (0 %)	20 (70 %)	1 (1320 %)	389 (1 %)	44 (8 %)
IEC IIIA 7.5 m/s	68 (0 %)	6 (24 %)	367 (0 %)	43 (0 %)	10 (125 %)	<1 (2400 %)	176 (2 %)	20 (14 %)

The table shows the estimated lifetimes in years, limited by the diodes of the MSC. Cases marked with "-" were not considered. The explanation for the base-case scenario with 7 (13 %): The relative life consumption on a 20-year basis is 275 %, leading to a round-off lifetime of 7 years. Only 13 % of the 20-years lifetime was consumed by the fundamental frequency and 264 % by wind-induced cycles.

Conclusion

This paper presents an IGBT module lifetime estimation procedure based on synthetic wind time series inputs and a generic type 4 wind turbine. The parameters used in this paper are matched against real-world data for wind turbines, generators, and wind turbine converters.

The approach considers the superposition of fundamental frequency and wind-induced thermal cycles, as it performs necessary calculations with a high temporal resolution. Hence, it differs from known lifetime estimation procedures, which usually indicate no lifetime issues in their case studies [2]-[3]. However, there is still ongoing work to answer how these superimposed thermal cycles affect the IGBT module's lifetime, which is not part of typical power cycling tests. Additional research with the testbench introduced in [11] might help answer this question and validate the proposed method.

The generic approach allows for case studies with comparisons for controlled changes of the inputs to investigate various influencing factors on the estimated lifetime. This comparative MSC analysis might help to identify potential IGBT lifetime issues in real applications if certain conditions are fulfilled.

Parameters

Table II: Parameters of the generic wind turbine and PMSG

Parameter	Symbol (Unit)	High-speed version	Low-speed version
Wind turbine			
Rated power	S (MVA) / P (MW)	3.3 / 3.0	
Hub height	h_{hub} (m)	140	
Rotor diameter	d_R (m)	126	
Inertia constant H	H (s)	4	
Nominal rotor speed	n_R (rpm)	10	
Nominal wind speed	v_R (m/s)	10	
Wind speed operation range	v_{min} (m/s) / v_{max} (m/s)	3 / 25	
Gearbox ratio	i	1:100	-
PMSG			
Nominal voltage	V_n (V)	710	
Back EMF at nominal speed	V_p (V)	373	
Nominal speed	n_G (rpm)	1000	10
Number of pole pairs	p	3	60
Permanent flux	Ψ_m (Vs)	1.19	5.94
Inductances in d- and q- axis	L_d / L_q (mH)	0.145 / 0.200	0.745 / 1.0

Table III: Foster thermal impedance parameters of the generic heatsink

i Z_{th}	1		2		3		4	
	R_{th} (K/kW)	τ (s)	R_{th} (K/kW)	τ (s)	R_{th} (K/kW)	τ (s)	R_{th} (K/kW)	τ (s)
$Z_{th,ja}$ $I1I1$	0.2	0.004	4.62	0.695	3.54	1.046	21.84	6.107
$Z_{th,ja}$ $D1I1$	0.15	0.004	3.465	0.695	2.655	1.046	16.38	6.107
$Z_{th,ja}$ $I2I1$	0.078	0.008	1.797	1.39	1.377	2.092	8.493	12.214
$Z_{th,ja}$ $D2I1$	0.1	0.008	2.31	1.39	1.77	2.092	10.92	12.214
$Z_{th,ja}$ $I1D1$	0.117	0.004	2.695	0.695	2.065	1.046	12.74	6.107
$Z_{th,ja}$ $D1D1$	0.317	0.004	7.315	0.695	5.605	1.046	34.58	6.107
$Z_{th,ja}$ $I2D1$	0.078	0.008	1.797	1.39	1.377	2.092	8.493	12.214
$Z_{th,ja}$ $D2D1$	0.1	0.008	2.31	1.39	1.77	2.092	10.92	12.214
$Z_{th,ja}$ $I1I2$	0.078	0.008	1.797	1.39	1.377	2.092	8.493	12.214
$Z_{th,ja}$ $D1I2$	0.1	0.008	2.31	1.39	1.77	2.092	10.92	12.214
$Z_{th,ja}$ $I2I2$	0.2	0.004	4.62	0.695	3.54	1.046	21.84	6.107
$Z_{th,ja}$ $D2I2$	0.15	0.004	3.465	0.695	2.655	1.046	16.38	6.107
$Z_{th,ja}$ $I1D2$	0.078	0.008	1.797	1.39	1.377	2.092	8.493	12.214
$Z_{th,ja}$ $D1D2$	0.1	0.008	2.31	1.39	1.77	2.092	10.92	12.214
$Z_{th,ja}$ $I2D2$	0.117	0.004	2.695	0.695	2.065	1.046	12.74	6.107
$Z_{th,ja}$ $D2D2$	0.317	0.004	7.315	0.695	5.605	1.046	34.58	6.107

References

[1] D. Weiss, H.-G. Eckel, "Fundamental Frequency and Mission Profile Wearout of IGBT in DFIG Converters for Windpower", 15th European Conference on Power Electronics and Applications (EPE), pp. 1-6, 2013

[2] K. Ma, M. Liserre, F. Blaabjerg, "Lifetime estimation for the power semiconductors considering mission profiles in wind power converter", IEEE Energy Conversion Congress and Exposition, pp. 2962-2971, 2013

[3] G. Zhang, D. Zhou, F. Blaabjerg and J. Yang, "Mission profile resolution effects on lifetime estimation of doubly-fed induction generator power converter," *2017 IEEE Southern Power Electronics Conference (SPEC)*, 2017

[4] T. Hunger, O. Schilling, "Numerical investigation on thermal crosstalk of silicon dies in high voltage IGBT modules", PCIM Europe 2008

[5] A. Nieslony, Rainflow counting algorithm (v 1.2.0.0) available on MathWorks File Exchange, https://www.mathworks.com/matlabcentral/fileexchange/3026

[6] E. Cheynet, J. B. Jakobsen, C. Obhrai, "Spectral characteristics of surface-layer turbulence in the North Sea", Energy Procedia, Volume 137, pp. 414-427, 2017

[7] B. Wu, Y. Lang, N. Zargari, et al., Power Conversion and Control of Wind Energy Systems, Wiley-IEEE Press, 2011.

[8] Y. C. Gerstenmaier, W. Kiffe and G. Wachutka, "Combination of thermal subsystems modeled by rapid circuit transformation," *2007 13th International Workshop on Thermal Investigation of ICs and Systems (THERMINIC)*, 2007, pp. 115-120

[9] Infineon AG, "PC and TC Diagrams", Infineon AG Application Note AN2019-05, Edition 2021-02-1919

[10] T. Methfessel, F. Sauerland, K. Mainka and O. Schilling, "Enhanced lifetime and power-cycling modelling for PrimePACK™ .XT power modules", PCIM Europe digital days 2020, International Exhibition and Conference for Power Electronics, Intelligent Motion, Renewable Energy and Energy Management, 2020, pp. 1-8.

[11] T.-M. Plötz, J. Fuhrmann, H.-G. Eckel, „Powercycling Test Bench with Realistic Loss Distribution and Temperature Ripples", 2022 24th European Conference on Power Electronics and Applications (EPE'22 ECCE Europe), 2022

Single-phase, Five-level Inverter with SPWM-Based Neutral Point Voltage Balancing Scheme

Dmytro Kondratenko[1], Arkadiusz Lewicki[1] and Charles Odeh[1]
[1]GDAŃSK UNIVERSITY OF TECHNOLOGY
Gabriela Narutowicza 11/12
80-233 Gdansk, Poland
Tel.: +48 58 347 14 02
Fax: +48 58 347 18 02
E-Mail: dmytro.kondratenko@pg.edu.pl
URL: https://eia.pg.edu.pl/kaneike/main

Acknowledgements

This research project was funded by National Science Centre, Poland 2021/41/N/ST7/01968. For the purpose of Open Access, the author has applied a CC-BY public copyright license to any Author Accepted Manuscript (AAM) version arising from this submission.

Keywords

«Pulse Width Modulation (PWM)»», «T-type inverter», «Capacitor voltage balancing», «Voltage Source Inverter (VSI)».

Abstract

Multilevel inverter topologies provide several advantages over two-level inverter configuration. These benefits are the reason for the growing interest in multilevel topologies among research society. One of the most popular topological concepts (diode and active switch clamping) requires neutral-point potential balancing due to series-connected capacitor banks across the input dc link in such derived inverter configurations. This paper presents a Sinusoidal PWM (SPWM) scheme that ensures balanced and reduced input capacitor voltages' variations in single-phase T-type inverter. Comparison analyses with existing carrier-based pulse-width modulation methods are provided. Simulation results are provided which showcase the effectiveness of the control approach. Experimental validation of these results was provided with a five-level single-phase T-type VSI; that supplies single-phase RL load.

Introduction

In last decade, there is growing interest in multilevel inverter (MLI) topologies. Recently, MLI topologies have been widely used in various applications such as: photovoltaic (PV) generations, wind turbine systems, electric vehicle (EV) and telecommunication power systems, etc [1]-[3]. The multilevel topologies provide some advantages such as: reduced voltage on the switches and lower total harmonic distortion (THD) of the output voltages [3], [4].

In general, MLI can be classified into five main group: the flying capacitor, diode-clamped, cascade H-bridge (CHB), matrix converters and hybrid topologies, [5]. The imbalance of the neutral point voltage is an inherent problem of multilevel diode-clamped-inverter-derived topologies (neutral-point clamped – NPC [6], T-type [7] and F-type [8] converters). NPC and T-type inverters are the most prominent, with three-level configuration. NPC is the most widely used; nevertheless, T-type topology has low conduction losses when compared to NPC inverter, [9]. Operationally, all variants of the diode-clamped topologies are equivalent. Numerous modulation techniques have been proposed to improve the output waveforms quality and achieve the dc-link voltage balance for various operational conditions. The most popular modulation techniques can be divided into two groups: sinusoidal or carrier-based pulse-width modulation (SPWM or CBPWM), [10]-[14] and space vector pulse-width modulation (SVPWM), [15]-[19]. Several studies propose different modification of SVPWM and SPWM to solve the problem of input split dc-link voltage balancing. For space vector technique, the most popular technologies are different variations of a virtual space vector PWM (VSVPWM). This modulation concept leads to effective control of the dc-link voltage and reduced common-mode voltage (CMV), [20]. The main disadvantages of the balancing scheme based on virtual vector approach are various characteristics under different modulation index and power factor values. The higher the modulation impact is, the slower the DC-link voltage balancing is. Another approach to control the DC-link voltage is to use different types of controllers. This approach could be utilized by both modulation techniques, SVPWM and SPWM. The basic solution is a proportional, P, controller utilized as an additional stage in the PWM generation process, proposed in [21]. In the article presented in [22], the proportional resonant (PR) controller was developed. This control approach was aimed at reduction of voltage drifts and as well as voltage ripples. Due to the proposed structure, PR-controller utilizes two coefficients; similar to the conventional PI controller, but with an additional integrator block. These parameters depend on the duty cycle of small vectors and neutral point current, [22]. Just as different modifications of space vector modulation algorithms were presented, many variations of carrier-based modulation algorithms were proposed for the multilevel diode-clamped inverters, [23]-[25].

Natural balancing effect can be used for dc-link voltage control without any additional control scheme. However, it can be applied only in proper conditions. Two approaches for natural balancing exist: use of additional RLC balancing circuit, [26], and utilization of hybrid modulation method, [27].

This paper presents the concept of natural balancing in a single-phase T-type inverter shown in Fig. 1. The natural balancing of the split dc-link capacitor voltages for this inverter is achieved by the injection of balancing signal to the modulating signal. This work is organized as follows. Modulation method and natural balancing of the neutral-point voltage is presented in section 2. The comparative analysis with existing modulation method, simulation and experimental verifications are presented in section 3. Conclusions are drawn in section 4.

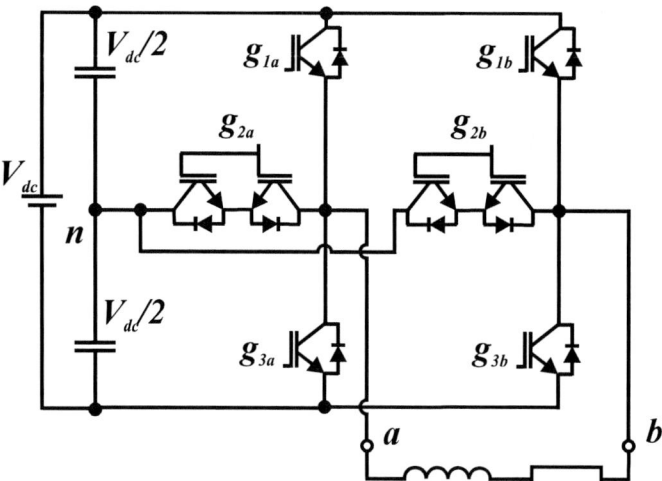

Fig. 1: Single-phase, five-level T-type inverter power circuit

Sinusoidal PWM with natural DC-link balancing

The simplified control block scheme of the modulation method is shown in Fig. 2. The index i is the phase-leg number/notation, $i=a,b$. Sinusoidal reference waveforms, SIN, and balancing injection signal, NPB, are obtained using (1) and (2).

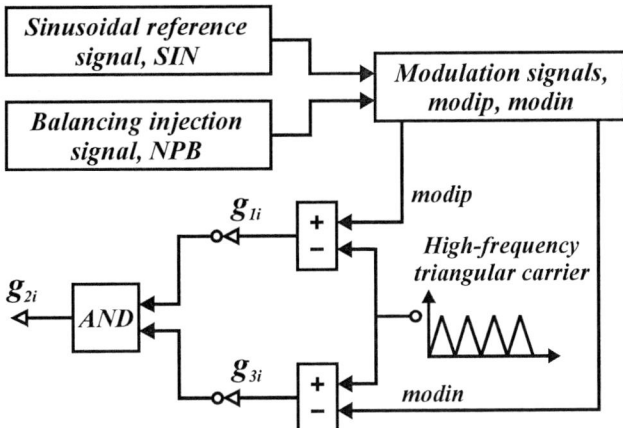

Fig. 2: Proposed SPWM control block scheme

$$SIN = m \cdot \sin(\omega \cdot t) \tag{1}$$

where m is a modulation index and ω is the fundamental output frequency.

$$NPB = B_p \cdot i_o \tag{2}$$

where i_o is an output current and B_p is a balancing component of the injected signal. The structure of the B_p signal is similar to proportional regulator and value of B_p could be obtain using (3).

$$B_p = k_p \cdot (V_{C1} - V_{C2}) \tag{3}$$

where k_p is a gain coefficient and V_{C1} and V_{C2} are lower and upper capacitor voltages.

The modulation signals block receives reference sinusoidal signal, SIN, and balancing component, NPB. This block computes two signals, $modip$ and $modin$, per each inverter leg. Modulation signal generating at this step is expressed in (4).

$$\begin{aligned}
&if\,(SIN > 0)\\
&modap = SIN + NPB;\ modan = 1;\\
&modbp = 0;\ modbn = 1 - SIN - NPB;\\
&else : modap = 0;\ modan = 1 + SIN + NPB;\\
&modbp = -SIN - NPB;\ modbn = 1;
\end{aligned} \tag{4}$$

At the last step, modulation signals are compared with triangular carrier signal, T; whose peak values are 0 and 1. The comparison equations that generate the gating signals are expressed in (5).

$$g_{1i} = modip > T;\ g_{3i} = modin < T;\ g_{2i} = !g_{1i}\ AND\ !g_{3i}; \tag{5}$$

Finally, three gating signals per one leg are generated. The switching sequence generated in such way provides a high-quality output voltage and current waveforms; and efficient neutral-point balancing.

Simulation and experimental verification

To verify the effectiveness and performance of the proposed modulation technique, simulation studies were conducted and experimental prototype was built. The power and logic circuit models of the inverter topology in Fig. 1 and the proposed control scheme were derived in PLECS software simulation environment. Capacitances of the DC-link capacitors are 750 µF, the switching frequency is 5 kHz and the dc-link voltage is 400 V. An RL load, 20 Ω and 20 mH, is connected at the inverter output terminals. Using the proposed modulation scheme, Fig. 3 shows the simulated inverter output voltage and current waveforms; along with the input split capacitor voltages' variations. With the same circuit parameters, classical control method was deployed in the inverter control; corresponding input and output waveforms are also shown in Fig. 3. Therein, the dynamic performances of these control approaches are displayed.

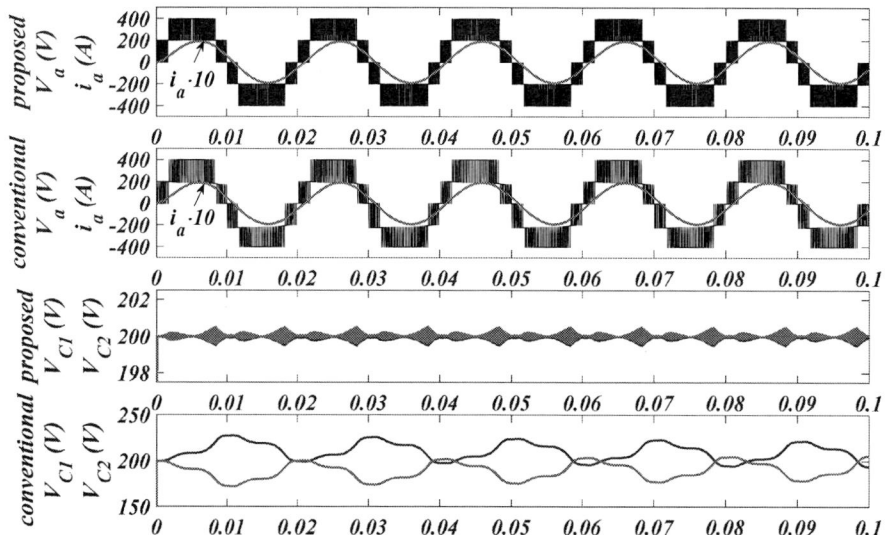

Fig. 3: Simulated output voltage, current and dc-link capacitor voltages waveforms for proposed and conventional modulation techniques.

The FFT analysis of the output current and voltage waveform is shown in Fig. 4. The harmonic performance of the proposed modulation method provides better THD value of the current, 0.34% compare to 1,1% for conventional method. The THD values of the output voltage are relatively similar, 26.55% for proposed and 27,0% for conventional method.

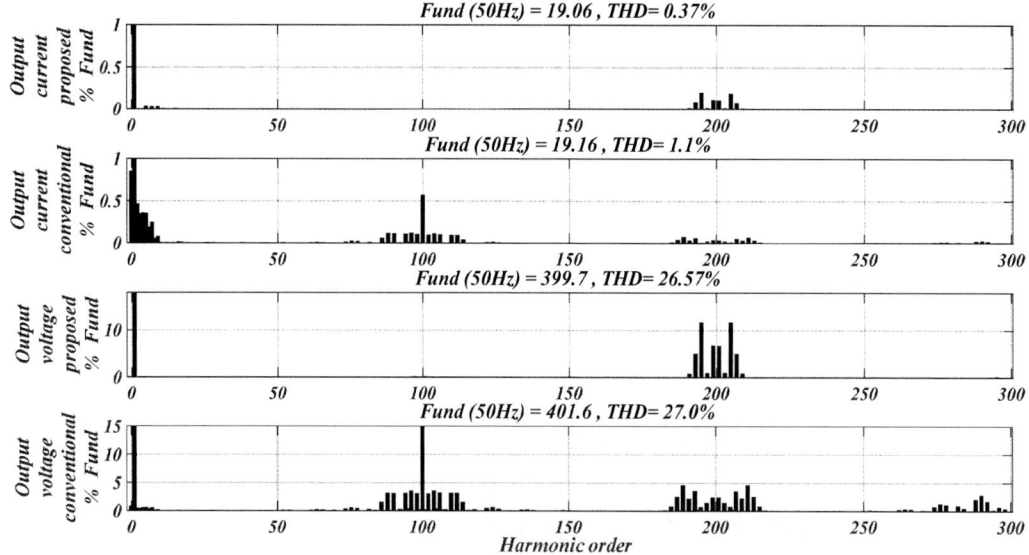

Fig. 4: Simulated output voltage, current and dc-link capacitor voltages waveforms for proposed and conventional modulation techniques.

To validate the effectiveness of the proposed modulation strategy, experimental study was conducted using single phase T-type experimental prototype. Table I gives the main prototype parameters. The Altera Cyclone II FPGA and ADSP21363L DSP processor were utilized to implement the proposed modulation scheme.

Table I: Prototype specification

	Specification
Capacitor bank	1100μF, 600V
RL load	40Ω, 20mH
Switching frequency	5kHz
Fundamental frequency	50Hz
DC-link voltage	400V
Power switches	AIKW50N60C

Using the proposed modulation scheme, Fig. 5 shows the experimental inverter output voltage and current waveforms; along with the input split capacitor voltages' variations. Modulation index was set to 0.8 and dc-link voltage value is 400 V.

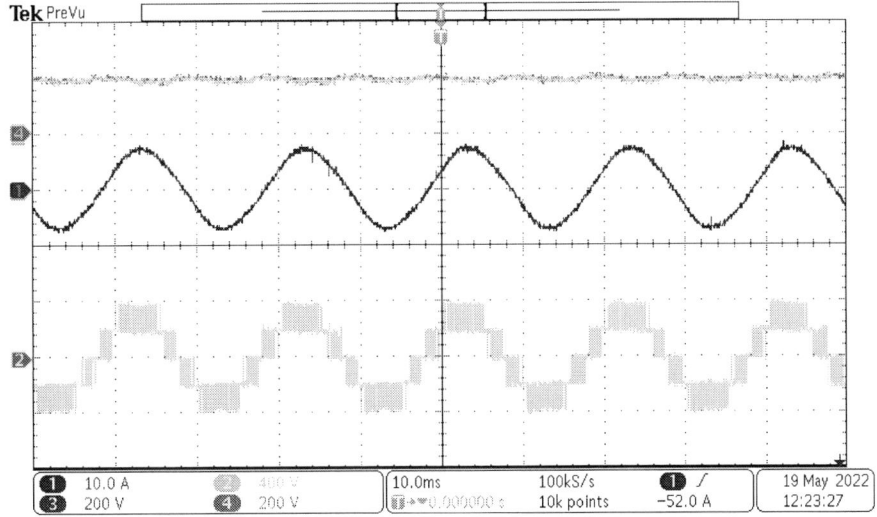

Fig. 5: Experimental output voltage, current and dc-link capacitor voltages waveforms for proposed modulation technique.

The dynamic performance of the proposed control approach, under step changes in the modulation index, is shown in Fig. 6. The modulation index value was changed from 0.8 to 0.4 and back to 0.8.

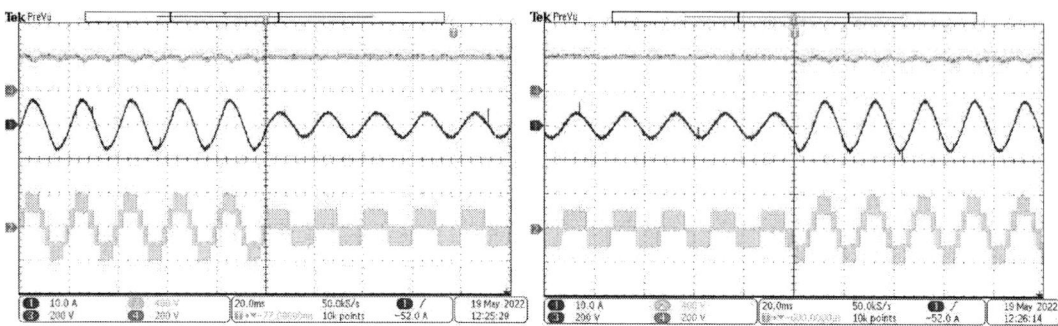

Fig. 6: Simulated output voltage, current and dc-link capacitor voltages waveforms for proposed and conventional modulation techniques.

Fig. 7 shows the balancing effectiveness of the proposed modulation strategy during the experimental verification studies.

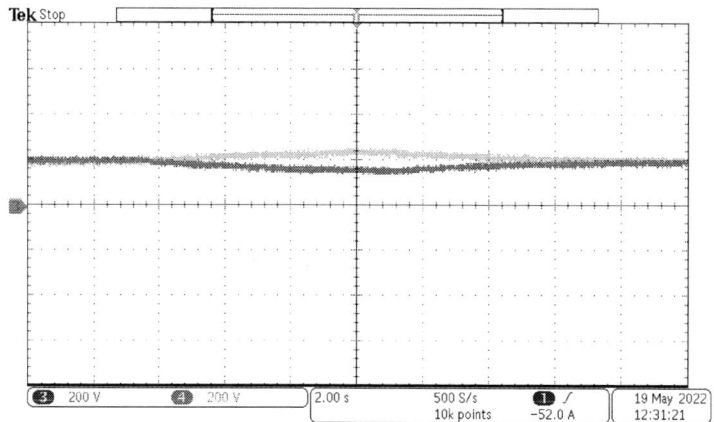

Fig. 7: DC-link capacitor voltages' variation without and with the neutral-point voltage balancing component.

Conclusion

Presented in this paper is a novel SPWM technique for single-phase, T-type inverter characterized by effective input split capacitor natural voltage balancing and reduced capacitor voltage variations. The proposed modulation strategy allows the minimization of number of regulators in the neutral-point voltage control scheme. The good performances shown in comparison with conventional modulation strategies validate the effectiveness and prospects of this control approach. Simulation results have been provided that show the effectiveness of the natural balancing in the proposed modulation technique for three-level, single-phase, T-type inverter. Experimental validations have been performed on a prototype inverter. Obtained experimental output waveforms matched in all respect with those of simulation studies.

References

[1] S. Amamra, K. Meghriche, A. Cherifi and B. Francois, "Multilevel Inverter Topology for Renewable Energy Grid Integration," in IEEE Transactions on Industrial Electronics, vol. 64, no. 11, pp. 8855-8866, Nov. 2017, doi: 10.1109/TIE.2016.2645887.

[2] T. D. Nguyen, Phan Quoc Dzung, D. N. Dat and N. H. Nhan, "The carrier - based PWM method to reduce common-mode voltage for three - level T - type neutral point clamp inverter," 2014 9th IEEE Conference on Industrial Electronics and Applications, 2014, pp. 1549-1554, doi: 10.1109/ICIEA.2014.6931415.

[3] R. Stala, "Application of Balancing Circuit for DC-Link Voltages Balance in a Single-Phase Diode-Clamped Inverter With Two Three-Level Legs," in IEEE Transactions on Industrial Electronics, vol. 58, no. 9, pp. 4185-4195, Sept. 2011, doi: 10.1109/TIE.2010.2093477.

[4] J. Chen, C. Zhang, A. Chen, X. Xing and F. Gao, "A Carrier-Based Fault-Tolerant Control Strategy for T-Type Rectifier With Neutral-Point Voltage Oscillations Suppression," in IEEE Transactions on Power Electronics, vol. 34, no. 11, pp. 10988-11001, Nov. 2019, doi: 10.1109/TPEL.2019.2900496.

[5] S. Kouro et al., "Recent Advances and Industrial Applications of Multilevel Converters," in IEEE Transactions on Industrial Electronics, vol. 57, no. 8, pp. 2553-2580, Aug. 2010, doi: 10.1109/TIE.2010.2049719.

[6] A. Nabae, I. Takahashi and H. Akagi, "A New Neutral-Point-Clamped PWM Inverter," in IEEE Transactions on Industry Applications, vol. IA-17, no. 5, pp. 518-523, Sept. 1981, doi: 10.1109/TIA.1981.4503992.

[7] B. Fuld, "Aufwandsarmer Thyristor-Dreistufen-Wechselrichter mit geringen Verlusten," in etzArchiv, vol. 11, pp. 261-264, VDE-Verlag, Berlin, Germany, 1989.

[8] C. I. Odeh, A. Lewicki, M. Morawiec and D. Kondratenko, "Three-Level F-Type Inverter," in IEEE Transactions on Power Electronics, vol. 36, no. 10, pp. 11265-11275, Oct. 2021, doi: 10.1109/TPEL.2021.3071359.

[9] M. Schweizer and J. W. Kolar, "Design and Implementation of a Highly Efficient Three-Level T-Type Converter for Low-Voltage Applications," in IEEE Transactions on Power Electronics, vol. 28, no. 2, pp. 899-907, Feb. 2013, doi: 10.1109/TPEL.2012.2203151.

[10] J. Pou, J. Zaragoza, S. Ceballos, M. Saeedifard and D. Boroyevich, "A Carrier-Based PWM Strategy With Zero-Sequence Voltage Injection for a Three-Level Neutral-Point-Clamped Converter," in IEEE Transactions on Power Electronics, vol. 27, no. 2, pp. 642-651, Feb. 2012, doi: 10.1109/TPEL.2010.2050783.

[11] W. Srirattanawichaikul, Y. Kumsuwan, S. Premrudeepreechacharn and B. Wu, "A carrier-based PWM strategy for three-level neutral-point-clamped voltage source inverters," 2011 IEEE Ninth International Conference on Power Electronics and Drive Systems, 2011, pp. 948-951, doi: 10.1109/PEDS.2011.6147369.

[12] A. Videt, P. Le Moigne, N. Idir, P. Baudesson and X. Cimetiere, "A New Carrier-Based PWM Providing Common-Mode-Current Reduction and DC-Bus Balancing for Three-Level Inverters," in IEEE Transactions on Industrial Electronics, vol. 54, no. 6, pp. 3001-3011, Dec. 2007, doi: 10.1109/TIE.2007.907001.

[13] N. Li, Y. Wang, R. Niu, W. Guo, W. Lei and Z. Wang, "A novel neutral point voltage automatic balancing carrier-based modulation strategy of three-level NPC converter," 2014 International Power Electronics Conference (IPEC-Hiroshima 2014 - ECCE ASIA), 2014, pp. 475-479, doi: 10.1109/IPEC.2014.6869626.

[14] S. Bo, L. Xinyi, L. Wanting, Z. Hui and L. Ning, "A New Carrier Based Neutral Point Potential Control Strategy for Three-Level NPC Converter," 2018 IEEE 4th Information Technology and Mechatronics Engineering Conference (ITOEC), 2018, pp. 427-431, doi: 10.1109/ITOEC.2018.8740588.

[15] S. Busquets-Monge, J. Bordonau, D. Boroyevich and S. Somavilla, "The nearest three virtual space vector PWM - a modulation for the comprehensive neutral-point balancing in the three-level NPC inverter," in IEEE Power Electronics Letters, vol. 2, no. 1, pp. 11-15, March 2004, doi: 10.1109/LPEL.2004.828445.

[16] H. Lin et al., "A Simplified 3-D NLM-Based SVPWM Technique With Voltage-Balancing Capability for 3LNPC Cascaded Multilevel Converter," in IEEE Transactions on Power Electronics, vol. 35, no. 4, pp. 3506-3518, April 2020, doi: 10.1109/TPEL.2019.2938606.

[17] F. Guo, T. Yang, S. Bozhko and P. Wheeler, "A Novel Virtual Space Vector Modulation Scheme for Three-Level NPC Power Converter with Neutral-Point Voltage Balancing and Common-Mode Voltage Reduction for Electric Starter/Generator System in More-Electric-Aircraft," 2019 IEEE Energy Conversion Congress and Exposition (ECCE), 2019, pp. 1852-1858, doi: 10.1109/ECCE.2019.8913186.

[18] C. Xia, G. Zhang, H. Shao and Y. Zhang, "Hybrid space vector PWM strategy for three-level NPC inverters with optimal extension mode," IECON 2014 - 40th Annual Conference of the IEEE Industrial Electronics Society, 2014, pp. 4649-4655, doi: 10.1109/IECON.2014.7049203.

[19] B. Fan, W. -g. Zhao, W. Yang and R. -q. Li, "A simplified SVPWM algorithm research based on the neutral-point voltage balance for NPC three-level inverter," 2012 IEEE International Conference on Automation and Logistics, 2012, pp. 150-154, doi: 10.1109/ICAL.2012.6308188.

[20] C. Hu et al., "An Improved Virtual Space Vector Modulation Scheme for Three-Level Active Neutral-Point-Clamped Inverter," in IEEE Transactions on Power Electronics, vol. 32, no. 10, pp. 7419-7434, Oct. 2017, doi: 10.1109/TPEL.2016.2621776.

[21] H. Kitidet and Y. Kumsuwan, "A CB-SVPWM control strategy for neutral-poin voltage balancing in three-level NPC inverters," TENCON 2017 - 2017 IEEE Region 10 Conference, 2017, pp. 1766-1771, doi: 10.1109/TENCON.2017.8228144.

[22] X. Lin, S. Gao, J. Li, H. Lei and Y. Kang, "A new control strategy to balance neutral-point voltage in three-level NPC inverter," 8th International Conference on Power Electronics - ECCE Asia, 2011, pp. 2593-2597, doi: 10.1109/ICPE.2011.5944742.

[23] U. A. Sipai and P. N. Tekwani, "Capacitor Voltage Balancing for Five-Level Neutral-Point Clamped Inverter using Modified Carrier Waves for Carrier-Based PWM Technique," 2019 IEEE 16th India Council International Conference (INDICON), 2019, pp. 1-4, doi: 10.1109/INDICON47234.2019.9028994.

[24] Logapriya J. and S. Mohamedyousuf, "Improved multilevel inverter with neutral point potential balancing for high power application," 2015 International Conference on Innovations in Information, Embedded and Communication Systems (ICIIECS), 2015, pp. 1-6, doi: 10.1109/ICIIECS.2015.7192987.

[25] Y. Li, Y. W. Li, H. Tian, N. R. Zargari and Z. Cheng, "A Modular Design Approach to Provide Exhaustive Carrier-Based PWM Patterns for Multilevel ANPC Converters," in IEEE Transactions on Industry Applications, vol. 55, no. 5, pp. 5032-5044, Sept.-Oct. 2019, doi: 10.1109/TIA.2019.2928240.

[26] R. Stala, "A Natural DC-Link Voltage Balancing of Diode-Clamped Inverters in Parallel Systems," in IEEE Transactions on Industrial Electronics, vol. 60, no. 11, pp. 5008-5018, Nov. 2013, doi: 10.1109/TIE.2012.2219839.

[27] M. Aly, E. M. Ahmed and M. Shoyama, "Modulation Method for Improving Reliability of Multilevel T-Type Inverter in PV Systems," in IEEE Journal of Emerging and Selected Topics in Power Electronics, vol. 8, no. 2, pp. 1298-1309, June 2020, doi: 10.1109/JESTPE.2019.2898105.

Magnetic Core Evaluation Kit for the Comparison of Core Losses

Wilmar Martinez[1], Xiaobing Shen[1], Siqi Lin[2], and Jens Friebe[2]
[1]Electrical Engineering Department (ESAT)
[1]KU Leuven - EnergyVille [2]Leibniz Universitat Hannover
[1]Diepenbeek - Genk, Belgium [2]Hannover, Germany
[1]wilmar.martinez@kuleuven.be [2]friebe@ial.uni-hannover.de

Keywords

≪Core Losses≫, ≪Magnetic≫, ≪Evaluation Kits≫, ≪Power Electronics≫, ≪Measurements≫

Abstract

Within an initiative of different scientific and industrial partners, a comparison of measurement, analytical, and numerical methods for core losses is conducted. Consequently, a core evaluation kit was developed and used in different parts of the world to evaluate the effects of size, shapes, and measurement practices on the core loss models that we use for power conversions systems. In this paper, the fluxometric four-wire characterisation method was used to characterise ferrite and nanocrystalline materials using different sizes, flux densities, supply voltages, and switching frequencies. It was found that using the same magnetic material, magnetic components have different iron loss properties when the size of the core is changed. These results are the very first of many procedures and tests with this initiative of the magnetic core evaluation kit. And it is expected to contribute to have a common and better understanding of core/iron losses of magnetic components in power electronics.

Introduction

The current energy sector is in transition from fossil-based to zero-carbon emission systems. Modern smart grids involve an increasing amount of dc technologies that have recently become possible due to rapid advancements in semiconductor devices required in power electronic energy conversion systems. Silicon-based semiconductors have reached their maturity as well as their performance limits [1], and only incremental improvements are expected in the future. However, rapid advancements have still been made in semiconductor devices with the introduction of wide bandgap (WBG) semiconductors like Gallium Nitride (GaN) and Silicon Carbide (SiC), which are faster, more efficient, and smaller than most of the conventional Silicon counterparts [2]. In this context, despite great efforts, the progress in magnetic design and its modelling is still lagging behind the advancements in semiconductor technology, especially after the developments of WBG semiconductor devices [3][4].

It is then widely said that magnetics is still the bottleneck of power electronics due to scaling laws that could affect the operation of high-frequency (HF) power converters [5] [6]. There is not a clear definition of the frequency range of HF in power electronics. Nevertheless, considering the current trend of converters and material, HF usually comprehends frequencies higher than 300kHz and higher than 30MHz in case of very HF. The bottleneck point of view has evolved during the last decade but there is still no consensus about the effects of the interaction between magnetic components and semiconductor devices.

One of the reasons of the lack of such consensus in the field is the difficulties to access to reliable characteristics of the magnetic materials and the cores that power electronics designers use. This is mainly due to the large amount of degrees of freedom that can play a role in magnetic characteristics and their measurements [7][8].

[7] indicated that the impact of size on the core can be divided into dimensional resonance- and skin effect in the core. The dimensional resonance effect was first proposed to explain the dielectric properties of ferrite materials [9]. Early literature pointed out that some ferrite materials have a high dielectric constant [10][11]. This leads to the effect of electric field cannot be neglected. A relevant preview is given in [12], especially focusing on the field of power electronics. [13] then discusses the possible influences of dimensional resonance on large hydroelectric generators. [5] analysed dimensional resonance effect on permeability and proposes a solution by segmentation and verifies it experimentally by a segmented common choke. Since the electric field cannot be neglected anymore, a practical method for measuring the electrical properties of manganese-zinc ferrites is given in [14]. The skin effect in the core is the concentration of flux on the core surface due to the influence of the magnetic field induced by the eddy current field in the core on the original magnetic field. As shown in Fig. 1(a), flux density is subject to dimensional resonance- and skin effect, so it concentrated at the flux surface and center.

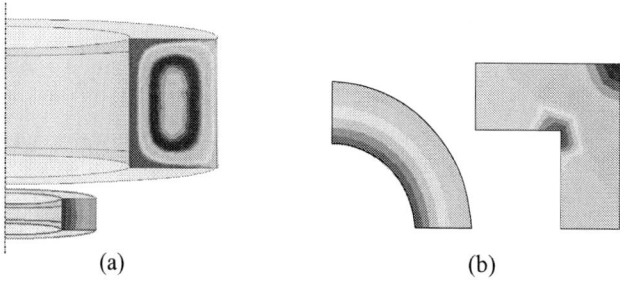

(a) (b)

Fig. 1: Flux density distributions for different core sizes and for different shapes. (a) Cross-sectional flux distributions for two ring cores of different sizes. The size of the upper core is 50 mm× 30 mm× 20 mm; the size of the lower core is 16 mm× 9.6 mm× 6.3 mm. The excitation source of the FEM simulation is a sine wave with 600 kHz. (b) The flux density distribution of two different shaped cores. The left side is a toroidal core with the size of 50 mm× 30 mm× 20 mm and the right side is a U-type core with the size of 50 mm× 25 mm× 20 mm.

In this context, a worldwide initiative towards the unification of core loss measurement procedures and experiences has been set up in the last year. As a result of this initiative, the development of a core evaluation kit available for several research groups in the world has been required.

The objective of such magnetic core evaluation kit is to evaluate the core loss effects of characteristics that are most of the time given for granted such as size, shape, dimensions, measurement procedures, sensors accuracy, etc.

This paper is divided as follows: First the core loss characterisation methods are summarised; and procedures and post-processing techniques are discussed. Then, the factors that influence core losses are discussed such as: materials, core shapes, core sizes, measurement techniques, among others. Finally, some preliminary results and conclusions are given.

Core Loss characterisation method

characterisation Procedure

In order to obtain the core loss characteristics of certain magnetic material, samples are constructed in ring shapes for better magnetic flux distribution. Then, two windings are wound in each ring and are excited by a sinusoidal power source. In the case of this evaluation procedure, a high frequency GaN Inverter is being used to emulate all the harmonic content generated by switching devices in practical implementations. Fig. 2 shows a schematic of the experimental setup used to obtain the measurements with which core losses are obtained. The Gallium Nitride inverter is constructed with 650V e-mode GaN HEMTs as it is depicted in Fig. 3. As devices under tests for this characterisation and evaluation procedure, several ring cores were constructed and wounded. Fig. 4 shows some of them. Different ring sizes were used to evaluate the size effect on core losses and also to achieve higher frequencies with DC voltages within the limits of the GaN devices (650V). The same happens with extremely low voltages, with certain sizes, high magnetic fluxes are easy to achieve, and the measurement procedure becomes compromised as the voltage levels land in the tolerance of the measurement equipment.

For the specific case of this paper, Ferrites and Nanocrystalline materials, intended to be used in high frequency applications, were selected, characterised, and evaluated. Table I shows the test specifications, as well as the ring and the GaN inverter parameters.

The GaN inverter is controlled with a unipolar modulation programmed in a 150MHz DSP. Such inverter is operated with a wide range of switching frequencies between 5kHz and 2MHz. Due to the nature of the unipolar modulation, carrier frequencies in the ring are in a range between 10kHz and 4MHz. With 4MHz being sufficient for most of the power electronic applications where WBG semiconductors are expected to be operated at high frequencies. Moreover, the fundamental frequencies are not applied for the test. The main purpose of the paper aims at analysing the magnetic loss of the cores with high frequencies.

Voltages in the secondary windings and currents through the first windings in the ring cores are determined with an oscilloscope capable of getting 10k points per measurement with a maximum sampling frequency of 50MS/s real time. With this scope, the primary-coil current is measured with a wideband current probe (Hall effect-clamp type). Voltage in the secondary coil is measured with a wideband differential voltage probe. Deskewing (synchronisation) of the current and voltage probes is conducted before the measurement procedure [15].

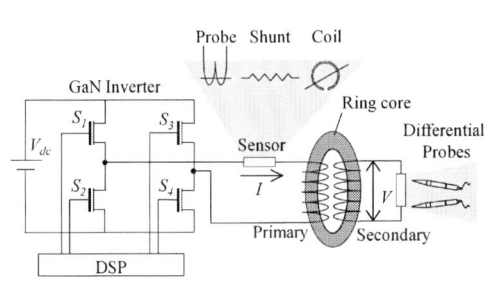

Fig. 2: Measurement circuit

Fig. 3: GaN Inverter Setup

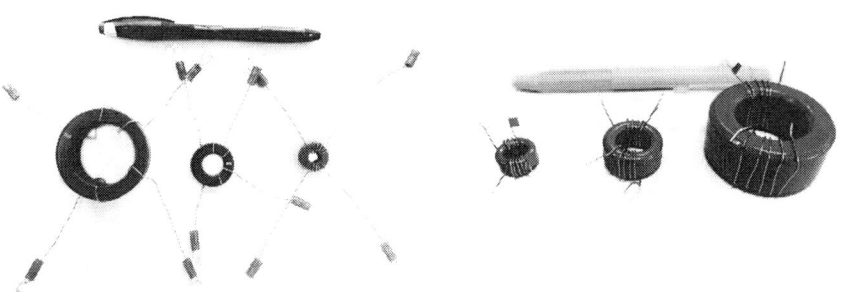

Fig. 4: Samples used in the evaluation procedure

Table I: Test setup specifications

Deadtime Td [ns]	10
DC voltage Vdc [V]	0-120
Switching frequency fs [kHz]	10-300

Post-processing

Once the tests are conducted and data from the current and voltage are obtained, the mathematical post-processing is conducted to find the BH operation and thereby the core loss calculation. Several core loss measurements are conducted in this study: Fluxometric (four-wire), calorimetric, and resonant processing are used to find the BH curves and to calculate the core loss density of the material under test. In the case of the fluxometric analysis, the following expressions are used to post-process the voltage and current data sets [16].

$$H_s(t) = \frac{N_1}{l_f} i(t) \tag{1}$$

$$B_0(t) = \frac{1}{N_1 A_{Fe}} \int v_2(t)dt \tag{2}$$

$$P_{Fe} = \frac{N_1 f}{N_2 A_{Fe} l_{Fe} \rho} \int_0^{1/f_0} i(t)v_2(t)dt \tag{3}$$

Fig. 5 and Fig. 6 show the reconstruction of one set of BH loops when the inverter is operating with a deadtime of 10ns and a sweep between 10 and 300kHz of switching frequency. In this test set, the comparison between three different maximum flux densities is done (100, 200, and 300mT). In the figures, the effect of the GaN switching frequency on the hysteresis pattern is evident.

In addition, it is possible to see the effect of increasing the magnetic flux on the output voltage. With the steps of hundreds between the three magnetic fluxes, the voltage to generate the targeted flux densities has a similar order of change, as shown in Fig. 7.

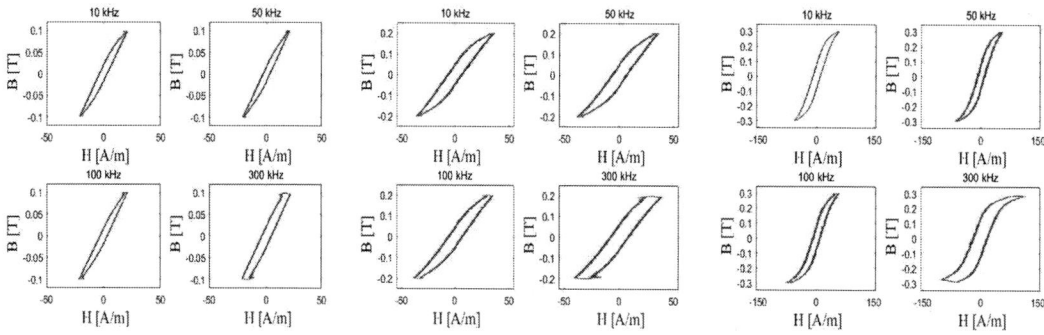

Fig. 5: BH loops for N87R25 with 100mT, 200mT, and 300mT of maximum flux density

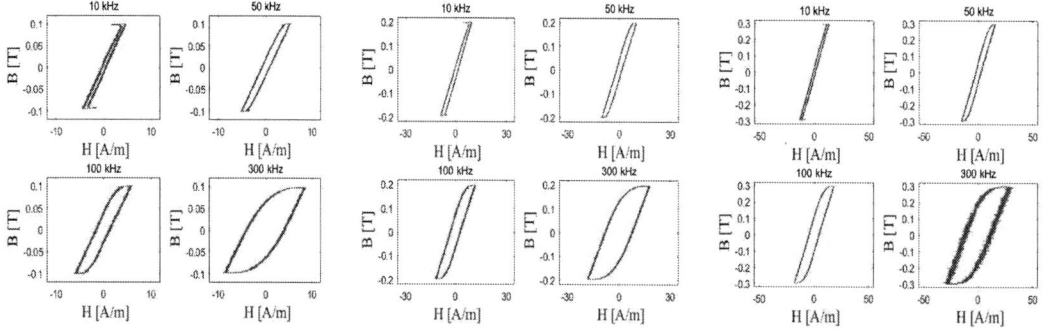

Fig. 6: BH loops for VP500R25 with 100mT, 200mT, and 300mT of maximum flux density

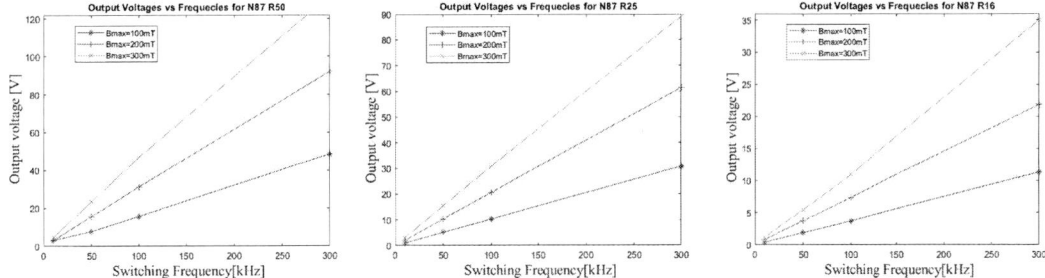

Fig. 7: Output Voltages VS Frequencies for N87 Cores with 100, 200, and 300mT of maximum flux density

Core losses - Influences

Material

For the characterisation procedure, Ferrite Powders, High Flux, and Sendust Nanocrystalline (high permeability and low permeability) were selected as materials to be compared in size, shape, and measurement procedure. These materials are widely used in power electronics systems. Being ferrite cores the more used for inductors and filters.

Nanocrystalline alloys have not only a high saturation flux density (over 1.2T), but also a high relative permeability ($1.2 \cdot 10^4$ - $1.2 \cdot 10^5$) and a very narrow magnetisation curve (with coercivities smaller than 0.5 A/m). However, due to their low resistivity (up to $1.15 \cdot 10^{-6}$ Ωm), these materials often require lamination to reduce eddy current losses. Because of the limited ability to store energy in the ungapped core, they are often used as materials for transformers, magnetic amplifiers, or sometimes also electrical machines [17][18][19].

The saturation flux density (up to 520 mT) in the ferrite cores is the lowest among the three, and the relative permeability (around 2200) is in the middle. Meanwhile, the static magnetic loss of these ferrites is the highest, but due to highest resistivity ($8 - 10$ Ωm), it has lowest eddy current losses at medium to high frequencies. Therefore, it is often used for power transformers, resonance inductors, etc. [17][18][19].

Table II: Comparison of different magnetic materials [20] [21] [22] [23] [24] [25]

	Alloys			Ferrites	
	Nanocrystalline alloys			Mn-Zn	
	VP500	VP712	VP550	N87	Fi335
Flux density (mT)	1240	1230	1210	490	520
Rel. Permeability	$9 \cdot 10^4$ (20 kHz)	12000 (12 kHz)	$1.2 \cdot 10^5$ (20 kHz)	2200 (25° C)	2100 (25° C)
Coercivity (A/m)	< 0.5	< 3	< 2	21	< 20
Resistivity (Ωm)	$1.15 \cdot 10^{-8}$	$1.15 \cdot 10^{-8}$	$1.15 \cdot 10^{-6}$	10	8
Core loss (kW/m³) at 100 kHz; 0.2 T	< 80 W/kg (0.3 T)		< 50 W/kg (0.3 T)	375 (100 °C)	310 (100 °C)
f (MHz)	...0.1	...0.1	...0.1	...0.3	...1

Size

The size effect on magnetic components and specially on the effective permeability has been discussed in the past [26] [27]. For this study, different sizes and shapes were selected. Specifically, three core sizes of the N87 and VP500F materials were selected: R16, R25, and R50.

Shape

One determining factor in core loss behaviour, that many times is neglected, is shape. Although, magnetic components can be constructed and realised with the same ideal characteristics (inductance, max flux, etc.) using different shapes and materials, shape can have a determining factor in the operation of the magnetic component. As shown in Fig. 1(b) even though the cross-sectional areas of the cores are the same, the different core shapes can cause differences in flux density distribution.

For this study, ring shapes are considered. Also, their flux distribution differences and the calculated excitation vs. flux distribution inside the core are considered.

Measurement Results

As mentioned before, the initial step of the use of the measurement kit was the four-wire characterisation procedure using ring shaped cores using N87 and VP500F materials in three different sizes: R50, R25, and R16.

After executing the characterisation method of section 2 and the post-processing, the following core loss characteristics were obtained for the three core sizes and the aforementioned materials. Fig. 8 and Fig. 9 show the measured core loss densities when the core is excited with a wide range of switching frequencies. As it can be seen from the figures, using the same size and material, core losses increase when the switching frequency is increased. Compared with the Nanocrystalline cores, Ferrites have a disadvantage of low thermal immunity. With the increasing frequency from 100kHz to 300kHz on N87 R50, its temperature goes above 90 °C and the voltage to get 300mT flux density exceeds the rated voltage 120V of DC voltage source. Therefore, in Fig. 8, the results of the N87 R50 do not show the data of the scenario of 300mT flux density under 300kHz.

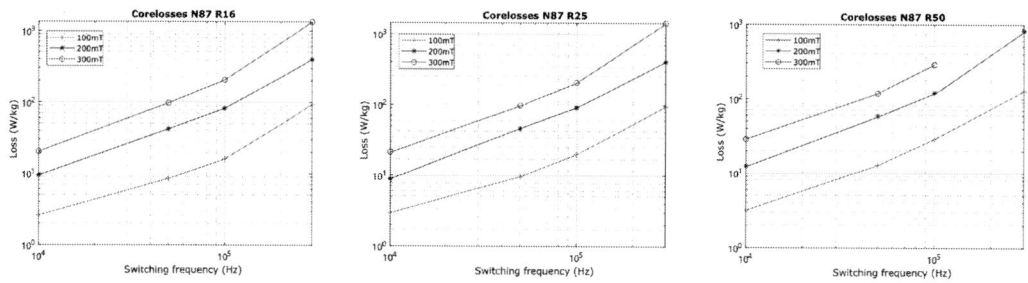

Fig. 8: Core losses Comparison of N87 R50, R25,R16 under different maximum flux densities

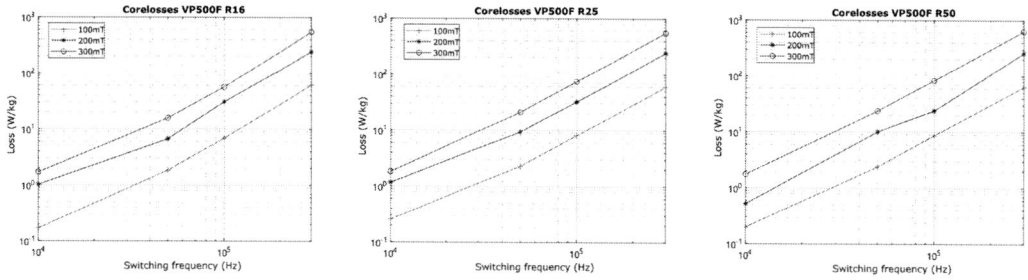

Fig. 9: Core losses Comparison of VP500F R50, R25,R16 under different maximum flux densities

Using the same material, the results of comparing different sizes at a certain flux density (100mT, 200mT, 300mT) are visualized in Fig. 10 and Fig. 11. The cases for N87 for a certain maximum density scenario are evidently different, although consistent with the trend of the fact that larger sizes generate more iron losses. But some fluctuations can be analysed in the cases with the VP500F material, due to the laminated structure which causes larger eddy current effects in the cross-sectional direction of the core to be less pronounced.

Fig. 12 gives an overview of the comparisons between these the materials at a certain size and same shape within different maximum flux densities. With the same size and shape, the VP500F cores show

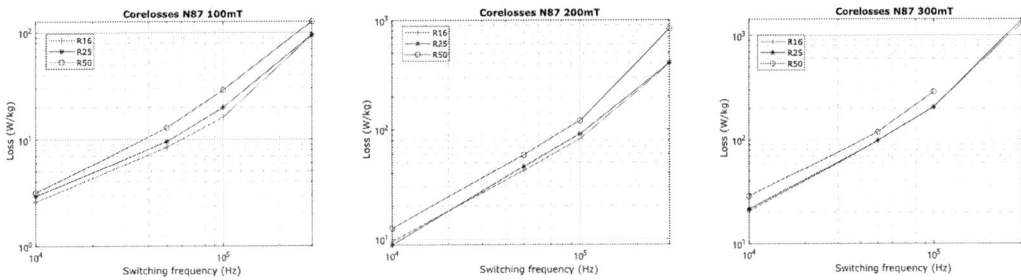

Fig. 10: Core losses Comparison of different N87 core sizes under a certain maximum flux density

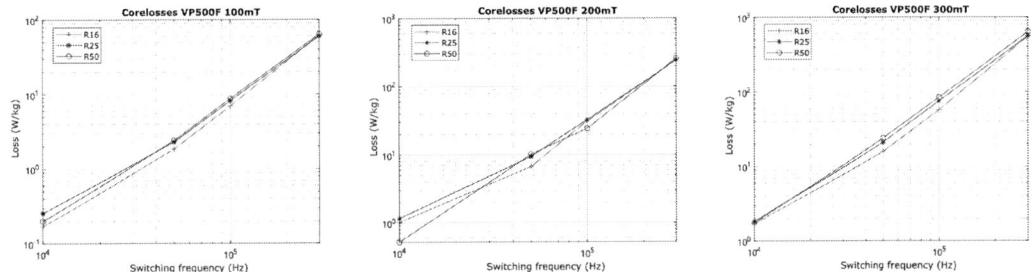

Fig. 11: Core losses Comparison of different VP500F core sizes under a certain maximum flux density

smaller core losses than the case of the N87. Among all these sizes, the N87 R50 shows the highest core losses, and the VP500F cores have the lowest core losses. During the measurements, nanocrystalline cores showed lower sensitivity to high temperature and they showed a relatively stable performance with higher frequencies.

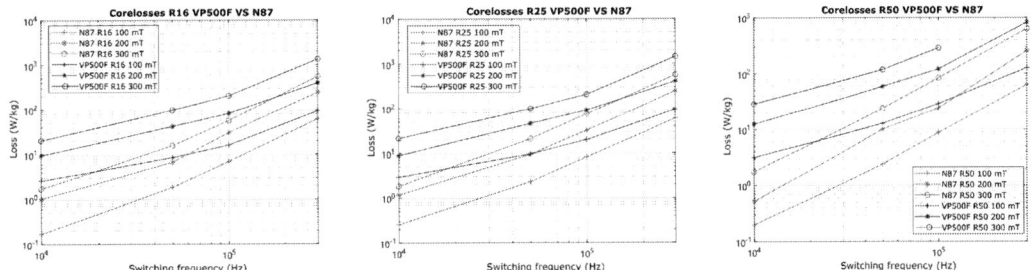

Fig. 12: Core losses comparison between N87 and VP500F with same sizes

Conclusion

In modern energy systems, power electronics is playing a huge role due to its advantages of high efficiency and adaptability. Nevertheless, still there are many components that need to be improved to make power electronics more reliable and to offer higher performance (e.g., high efficiency and high power density). One of these components are Magnetic components, being one of the most complex due to the large amount of degrees of freedom in their design and the lack of reliable characteristic, mainly because of the lack of unified characterisation methods.

Therefore, this paper presented an initiative to set up a core evaluation kit for the unification and experience sharing of magnetic characterisation procedures. For that, a magnetic core evaluation kit for comparison of core losses was defined and shared among several research labs around the world. In this paper, the four-wire method using N87 ferrite and VP500F nanocrystalline materials was presented.

It was verified that with the same material, the iron loss increases with a larger size of the core. It was found that a higher maximum flux density leads to a higher core loss, as well. VP500F nanocrystalline cores, with high permeability and high saturation, show lower iron losses and a stable temperature performance. They can offer a good operation at a wider range of temperature. Meanwhile, N87 ferrites cores responded well to frequency changes. Although, they are vulnerable to loss of performance at high temperature.

In this paper, the higher frequency effects on iron losses of magnetic cores were explored and analysed. These results give an insight into further research on iron loss modelling, and into common references and solid ground for high frequency magnetic component design.

References

[1] S. Madhusoodhanan, A. Tripathi, D. Patel, K. Mainali, A. Kadavelugu, S. Hazra, S. Bhattacharya, and K. Hatua, "Solid-state transformer and MV grid tie applications enabled by 15 kV SiC IGBTs and 10 kV SiC MOSFETs based multilevel converters," *IEEE Transactions on Industry Applications*, vol. 51, no. 4, pp. 3343–3360, jul 2015.

[2] G. Liu, K. Li, Y. Wang, H. Luo, and H. Luo, "Recent advances and trend of HEV/EV-oriented power semiconductors – an overview," *IET Power Electronics*, vol. 13, no. 3, pp. 394–404, feb 2020.

[3] A. J. Hanson and D. J. Perreault, "Modeling the magnetic behavior of n-winding components: Approaches for unshackling switching superheroes," *IEEE Power Electronics Magazine*, vol. 7, no. 1, pp. 35–45, mar 2020.

[4] W. Martinez and C. Cortes, "Design a dc-dc converter for a high performance electric vehicle," in *2012 International Conference on Connected Vehicles and Expo (ICCVE)*. IEEE, 2012, pp. 335–340.

[5] M. Kacki, M. S. Ryłko, J. G. Hayes, and C. R. Sullivan, "A study of flux distribution and impedance in solid and laminar ferrite cores," *2019 IEEE Applied Power Electronics Conference and Exposition (APEC)*, pp. 2681–2687, 2019.

[6] W. Martinez, J. Imaoka, M. Yamamoto, and K. Umetani, "High step-up interleaved converter for renewable energy and automotive applications," in *2015 International Conference on Renewable Energy Research and Applications (ICRERA)*, vol. 4. IEEE, 2015, pp. 809–814.

[7] C. R. Sullivan, "Core loss initiative: Technical," *APEC 2017 Industry Session - High Frequency Magnetics: Transforming the Black Magic to Engineering*, 2017.

[8] IMA, "Introduction to the ima working group," *APEC 2017 Industry Session - High Frequency Magnetics: Transforming the Black Magic to Engineering*, 2017.

[9] E. BLECHSCHMIDT, "Dielektrische eigenschaften von manganferriten," *Phys. Zeitschr*, pp. 212–216, 1938.

[10] F. G. Brockman, P. H. Dowling, and W. G. Steneck, "Dimensional effects resulting from a high dielectric constant found in a ferromagnetic ferrite," *Physical Review*, vol. 77, p. 8593, 1938.

[11] R. S. Glenn, "High-frequency dimensional effects in ferrite-core magnetic devices," *Virginia, USA, Dissertation of Virginia Polytechnic Institute, Blacksburg*, 1996.

[12] S. Lin, A. Ebrahimi, and J. Friebe, "Impact of electric field on magnetic flux distribution in electrical machines with very large size," *2021 IEEE Energy Conversion Congress and Exposition (ECCE)*, pp. 3969–3975, 2021.

[13] S. Lin, T. Brinker, L. Fauth, and J. Friebe, "Review of dimensional resonance effect for high frequency magnetic components," *2019 21st European Conference on Power Electronics and Applications (EPE '19 ECCE Europe)*, pp. P.1–P.10, 2019.

[14] M. Kacki, M. S. Rylko, J. G. Hayes, and C. R. Sullivan, "A practical method to define high frequency electrical properties of mnzn ferrites," *2020 IEEE Applied Power Electronics Conference and Exposition (APEC)*, pp. 216–222, 2020.

[15] W. Martinez, S. Odawara, and K. Fujisaki, "Iron loss characteristics evaluation using a high-frequency gan inverter excitation," *IEEE Transactions on Magnetics*, vol. 53, no. 11, pp. 1–7, 2017.

[16] P. Rasilo, W. Martinez, K. Fujisaki, J. Kyyrä, and A. Ruderman, "Simulink model for pwm-supplied laminated magnetic cores including hysteresis, eddy-current, and excess losses," *IEEE Transactions on Power Electronics*, vol. 34, no. 2, pp. 1683–1695, 2018.

[17] A. Goldman, "Handbook of modern ferromagnetic materials," *The Springer International Series in Engineering and Computer Science*, 2012. [Online]. Available: https://books.google.de/books?id=StbgBwAAQBAJ

[18] S. Tumanski, "Handbook of magnetic measurements (1st ed.)." *CRC Press.*, 2011.

[19] V. Valchev and A. Van den Bossche, "Inductors and transformers for power electronics (1st ed.)." *CRC Press.*, 2005.

[20] Magnetics, "Powder core catalog," 2020.

[21] Vacuumschelze, "Data sheet Vitroperm 500 F," 2021.

[22] Vacuumschelze, "Vitroperm 712 F new nanocrystalline cores to cover full range of common mode currents," 2017.

[23] Vacuumschelze, "Vitroperm 550 HF new nanocrystalline cores offering volume , weight cost optimized HF - Designs," 2021.

[24] EPCOS, "Ferrites and accessories-siferrit material N87," 2017.

[25] SUMIDA, "Magnetic material-Fi335," 2018.

[26] Snelling, "Soft ferrites – properties and applications," *ILIFFE Books, London*, 1969.

[27] M. Albach, "Induktivitäten in der leistungselektronik," *Springer Vieweg*, 2017.

Multi-Objective Optimization of Modular Multilevel Converter Systems

Nikolaus Patzelt, Christian Schlegel, Michail Vasiladiotis
Hitachi Energy
Spinnereistrasse 3
5300 Turgi, Switzerland
Email: nikolaus.patzelt@hitachienergy.com, christian.schlegel@hitachienergy.com,
michail.vasiladiotis@hitachienergy.com

Keywords

≪Modular Multilevel Converters (MMC)≫, ≪Multi-Objective Optimization≫, ≪Optimization Method≫,
≪Railway Power Supply≫.

Abstract

This paper investigates a multi-objective optimization method for modular multilevel converter systems.
The method is applied to the AC-AC MMC topology, which connects the three-phase utility grid to the
single-phase railway grid. The presented paper aims at a joint optimization of all system components
providing a set of pareto-optimal system designs. The optimization space encompasses core system
design parameters, such as the number of cells in a branch, the sizing of cell capacitor and branch inductance, but also the switching frequency and other parameters. The objectives studied in this paper are
the overall system cost and losses. The considered constraints comprise limitations on system components as well as the power capability of the converter, compliance with grid codes and system efficiency.
The obtained optimized design improved the total system cost by 7 % compared to a manually derived
benchmark design. It has been verified by offline and real-time simulations.

Introduction

The considered AC-AC static frequency converter is based on the modular multilevel converter topology [1][2]. An integrated gate-commutated thyristor (IGCT)-based MMC for railway power supply is
introduced in [3] and [4]. It connects the three-phase 50 Hz utility grid to a single-phase 16.7 Hz railway
supply and consists of an arbitrary number of cells N connected in series along one branch and a branch
inductor as shown in Fig. 1. The branch is the sub-circuit connecting a three-phase line with the single-
phase line. Every bipolar cell consists of a full-bridge circuit, a capacitor, a clamp circuit and a bypass
thyristor. The entire system includes the converter as well as the equipment interfacing to the utility and
railway grid such as transformers and possibly filters.

In state-of-the-art converter designs, typically, a set of individually optimized components are derived.
The combination of these components then serves as a design candidate. However, as the converter
system is highly non-linear, an analytical system optimization is challenging and the trade-off between
the different design parameters is not obvious. This paper proposes a method to jointly optimize various
design parameters for different objectives simultaneously to achieve an improved overall system design.

In the past years, there have been other approaches trying to optimize MMC system design. In [5],
an optimization with parameter variation is suggested, based on which the passive components are dimensioned. In [6], a minimization of the capacitor sizing is achieved by optimizing the arm circulating
current as well as the common mode voltage, leading to decreased energy storage requirements of the
MMC. A multi-objective optimization for the converter design has been performed in [7] and [8]. These
studies focus on footprint rather than cost of the converter and consider the design of the converter itself.

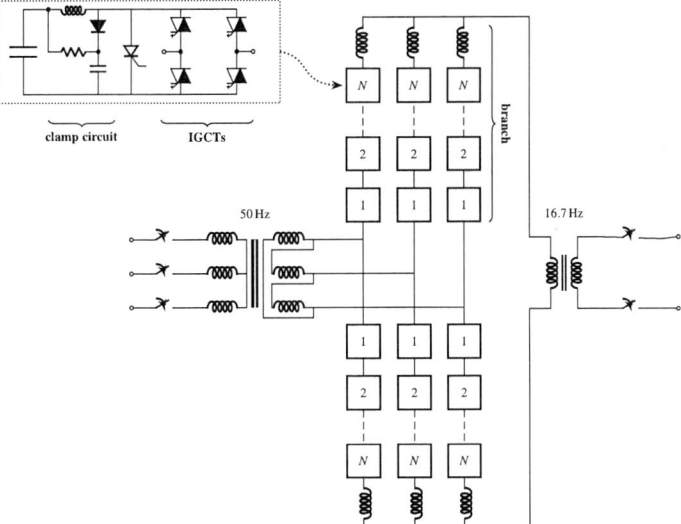

Fig. 1: AC-AC Modular Multilevel Converter for Railway Interties.

In contrast, this paper focuses on the steady-state and dynamic behaviour of the converter system including grid interface components such as transformers and filters. Moreover, the interaction with the system environment is studied considering different grid impedances and operating points.

To perform an optimization it is necessary to define an objective function vector with objectives as well as a set of constraints. The multi-objective optimization problem formulation is shown in (1) [9].

$$
\begin{aligned}
\min \quad & \mathbf{f}(\mathbf{x}) \\
\text{subj. to} \quad & g_j(\mathbf{x}) \leq 0 \quad \forall j \in [1, m] \\
& h_k = 0 \quad \forall k \in [1, p]
\end{aligned}
\tag{1}
$$

The objective function is given by $\mathbf{f} : V \rightarrow \mathbb{R}^n$ with $\mathbf{f}(\mathbf{x}) = (f_1(\mathbf{x}), f_2(\mathbf{x}), \ldots, f_n(\mathbf{x}))$.

Here n is the number of different objectives for a given optimization problem. The domain V with $\mathbf{x} \in V$ represents the set of design parameters, which, in the presented application, is a finite and closed set.

Structure of the System Design Optimization

The optimization method consists of four parts: (a) The parameter input, (b) the design derivation, (c) the design evaluation and (d) the optimization. The schematic of the optimization approach is shown in Fig. 2. The algorithm is based on a defined set of variable design parameters, which shall be optimized.

During each iteration step i, a set S_i of design parameters is chosen (a). During the design derivation (b), a design d_i is determined in two steps. Step 1 defines internal design parameters and derives the converter parameters. Step 2 calculates the system behavior in the frequency domain considering the system environment, such as grid impedances. The resulting design is checked for constraint violations.

As a next step, the suggested designs must be evaluated (c). Here, cost functions are used for evaluation, assigning a cost vector to the design d_i. In Fig. 2 the material cost and the losses are shown as an example. However, any design objective can be used here. Finally, the design is assessed according to the objectives in comparison to the previously analyzed designs (d). Depending on the optimization algorithm, a new set of parameters S_{i+1} is generated in order to evaluate the newly calculated design d_{i+1}.

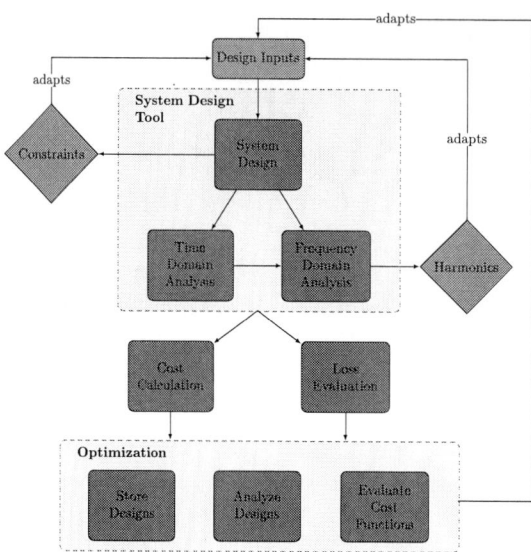

Fig. 2: Schematic Description of Proposed Multi-Objective Optimization Method.

Convergence

In order to ensure feasibility of the optimization, the convergence of the multi-objective optimization has been studied for the AC-AC MMC topology. Convergence is defined according to the stopping criterion given in [10] and a generation shall consist of 120 designs. Particularly relevant are the number of design parameters and the number of objectives. Fig. 3a shows several performed optimizations, relating the number of design parameters to the iterations until convergence. The number of parameters influences the convergence. In red the linear regression is shown to underline the trend.

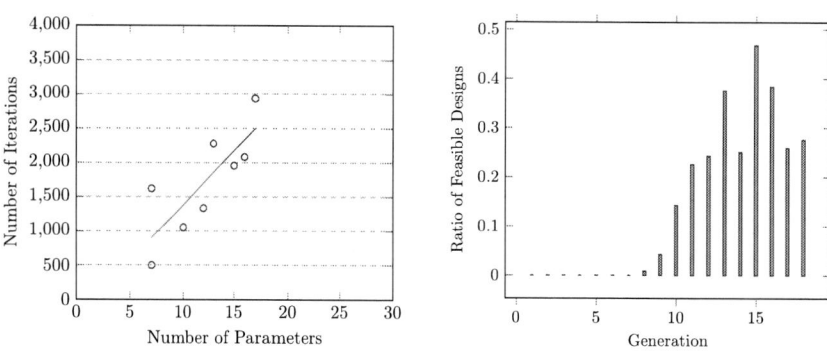

(a) Number of iterations until full convergence of multi-objective optimization.

(b) Discovery of feasible designs, given by the ratio of feasible to all evaluated designs.

Fig. 3: Convergence of the Multi-Objective Optimization.

The ratio of discovered feasible designs versus unfeasible designs shifts towards feasible designs in the course of the optimization as shown in Fig. 3b. Particularly in the first generations few feasible designs are being found, while after nine to ten generations, feasible designs are being discovered constantly.

Optimization Problem

Objective Function

This paper suggests a multi-objective optimization focusing on material cost and system losses. To quantify the material cost, individual cost functions for all parts of the converter system have been derived.

The sum of these costs defines the first objective. The components considered are

- MMC cells including base frame
- Branch inductances
- Single- and three-phase transformer
- Single- and three-phase filter
- Cooling unit.

The second objective are the system losses. They are the main contributor to the life cycle cost of a converter system and a key parameter defining the performance of a design. The system losses are calculated according to (2).

$$f_2(\mathbf{x}) = \mathbf{w}^\top \cdot \mathbf{L} \tag{2}$$

Here, $\mathbf{L}$ represents losses for different operating points weighted according to a pre-defined weighting function $\mathbf{w}$. The considered losses represent the system losses instead of the converter losses alone. Thus, the losses in the filters, transformers and cooling units are considered as well.

Constraints

The optimization objectives, constraints and design parameters need to be chosen carefully. To obtain a feasible solution, the mathematical formulation must contain all constraints that are attached to the optimization problem. Some constraints are inherent to the design. Other constraints are imposed by the system environment. The constraints that have been considered are shown in Table I. They encompass constraints on the components such as branch inductance, cell capacitance and the number of cells, power capabilities, the compliance with grid codes considering different frequency-dependent grid impedances as well as the system efficiency.

Table I: List of Constraints.

Description	Variable	Constraint
Branch inductance	L_{br}	$L_{br,min} < L_{br}$
Cell capacitance	C_{cell}	$C_{cell,min} < C_{cell} < C_{cell,max}$
Number of cells	N	$N_{min} < N$
Set of power ref. points (3-ph)	$\vec{S}_{3ph}$	$\vec{S}_{3ph,req} < \vec{S}_{3ph}$, diff. V levels
Set of power ref. points (1-ph)	$\vec{S}_{1ph}$	$\vec{S}_{1ph,req} < \vec{S}_{1ph}$, diff. V levels
V THD (3-ph, 1-ph)	THD_V	$THD_V < THD_{V,max}$
I THD (3-ph, 1-ph)	THD_I	$THD_I < THD_{I,max}$
V harm. content (3-ph, 1-ph)	$\vec{V}_{harm}$	$\vec{V}_{harm} < \vec{V}_{harm}^{max}$
I harm. content (3-ph, 1-ph)	$\vec{I}_{harm}$	$\vec{I}_{harm} < \vec{I}_{harm}^{max}$
Total system efficiency	η	$\eta_{req} < \eta$

Design Parameters

There is a high number of design parameters which can be taken into account when designing the converter system. However, while some may have a large impact, others only affect the objectives marginally. The most important design parameters are introduced in the following:

- **Switching frequency**: The switching frequency is selected based on the modulation technique and depending on the project specification. While higher switching frequencies reduce the harmonic content produced in the system, they increase the losses through switching.
- **Cell utilization**: The chosen number of cells could differ from the calculated value for the minimum required number of cells. Thus, the actual power capability of the converter may exceed the minimum required power capability. This power margin can be used to reduce either the current in the branch or the average cell voltage. The cell utilization defines whether the current or the voltage is minimized and has been used as a design parameter in the optimization.

- **Cell capacitor ripple**: The cell capacitor ripple (Δv_{cell}) directly influences the sizing of the capacitor itself. If a higher capacitor voltage ripple is allowed, the capacitor can be sized smaller. The cell capacitance is calculated by

$$C_{\text{cell}} = 2 \cdot \frac{\Delta E_{\text{br}}^{\max}}{N((V_{\text{cell}}^{\min} - V_{\text{cell}}^{\max})^2)} \tag{3}$$

where $\Delta E_{\text{br}}^{\max}$ is the maximum energy variation in the branch. Due to the dependence on the maximum ($V_{\text{cell}}^{\max}$) and minimum cell voltage ($V_{\text{cell}}^{\min}$), the cell capacitance can be steered by defining the cell capacitor ripple.
- **Transformer impedance**: A transformer impedance (X_{trafo}) is present in the design for the grid transformers on the single- and three-phase side as shown in Fig. 1.
- **Number of cells**: The number of cells per MMC branch (N).
- **Filter type**: On single- and three-phase side different filter types are evaluated.
- **Filter power**: The filter power determines the dimensioning of the individual filter components depending on the filter type.

Optimization Results

With the multi-objective optimization procedure, it is possible to analyze the relevance of different design parameters with respect to the selected objectives. This enables the designer to reduce the set of parameters to the most influential ones. To identify these, an analysis of the respective variable relevance has been performed. The correlation of different design parameters with the defined objectives and constraints has been assessed with the Pearson correlation coefficient $\rho_{X,Y}$, relating the random variables X and Y with

$$\rho_{X,Y} = \frac{1}{\sigma_x \sigma_y} Cov(X,Y), \tag{4}$$

where $Cov(X,Y) = \mathbb{E}\left[(X - \mathbb{E}[X]) \cdot (Y - \mathbb{E}[Y])\right]$ and σ_x and σ_y are the standard deviation of X and Y, respectively. Random variable X represents a design parameter, where Y represents an objective or constraint. The correlation can take values in the interval $[-1, 1]$, where a Pearson correlation of -1 indicates fully negative linear correlation and a value of 1 indicating total positive correlation. If the value is in the area around 0, there is no correlation detected. While this correlation gives good results for linear correlations, it may fail to detect other types of correlations. Fig. 4 illustrates the impact of five different design parameters on the cost, the losses, and the harmonics of the converter system.

Fig. 4a shows that the main driver of the material cost is the number of cells. This correlates with the intuition that the most cost-efficient design is a design with a minimum number of cells. There are, however, also other parameters that influence the cost. Most notably, this is the cell capacitor voltage ripple. The lower Δv_{cell} becomes, the lower is the required number of cells in the branch. The losses are mainly influenced by the switching frequency (Fig. 4b). This follows the argument of dependence of the switching losses to the switching frequency with $P_{loss,sw} \propto f_{sw}$. There are also many other factors influencing the loss calculation. This makes it hard to evaluate, which parameters are the most relevant. However, it can be said that due to the complexity and non-linearity of the topology, many parameters should be regarded when it comes to the loss evaluation of the system. It should be noted that the influence of the number of cells on the losses is only of an indirect nature. Increasing the number of cells allows to decrease the switching frequency and either cell voltage or branch current, leading to lower losses. Increasing the admissible cell capacitor ripple generally decreases the losses. While it is hard to take all the non-linear effects into account, this can be rooted in the positive correlation of the cell ripple with the number of cells. As the number of cells increases, the switching frequency can be decreased leading to lower losses. The harmonics are influenced by the respective transformer impedance. Moreover, the number of cells and again the switching frequency influence the harmonics.

These results are based on the analysis of various designs and their performance in terms of fulfillment of objectives or constraint violations. Another example of variable correlation is shown in Fig. 5. It

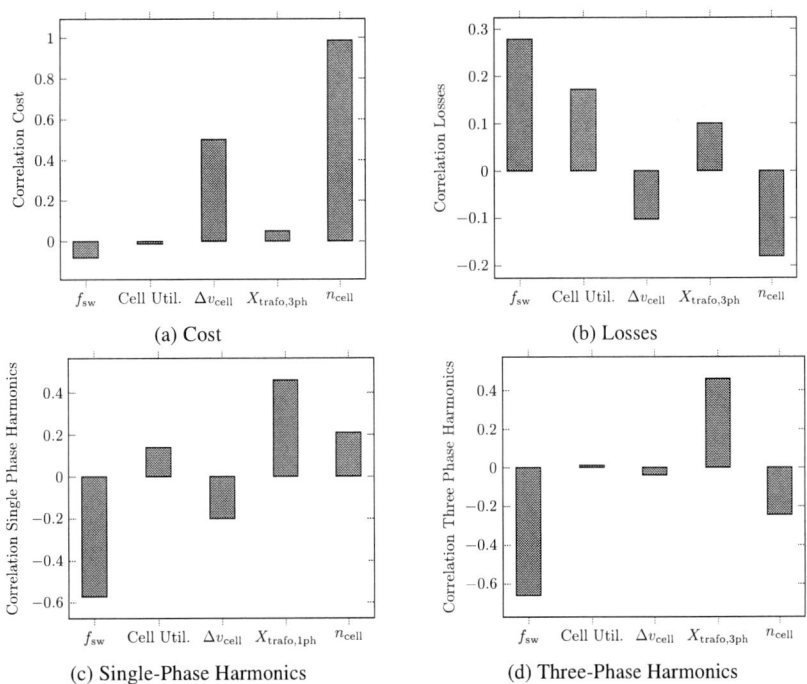

(a) Cost

(b) Losses

(c) Single-Phase Harmonics

(d) Three-Phase Harmonics

Fig. 4: Influence of various converter variables.

gives an example of a correlation between cell capacitor ripple and the required number of cells. There is a clear linear trend between capacitor ripple and the required number of cells. This also impacts the actual number of required cells and, thus, influences costs and losses of the design. Using such input-to-objective correlations, the design parameters can be analyzed and reduced to the most significant ones.

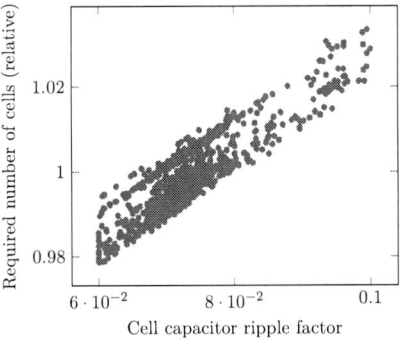

Fig. 5: Correlation between cell capacitor voltage ripple and the number of cells with the minimum number of cells N_{calc} as a function of the cell capacitor voltage ripple factor.

The optimization evaluates the design space, considering the introduced constraints optimizing for the objectives of total system cost and losses. The results of the optimization are compared with a benchmark design, which has been manually derived and which is verified and tested. The pareto-front of optimal designs is shown in Fig. 6. The proposed design, which has been derived from the optimization, is indicated as blue circle, and the existing manually derived benchmark design is shown as a blue rectangle. Two main clusters can be found that show a considerable cost difference. This is mainly due to the different number of cells, which is also considered in the design space of the optimization. As such, the optimization method can find improved designs reducing the required number of cells per branch.

During the verification of the dynamic behavior of designs, i.e. offline computer simulations, further constraints are introduced to the objective space. In the steady-state analysis of the converter system these are not considered yet as this would lead to an unacceptable run-time of the optimization. Therefore, in Fig. 6 pareto-optimal designs exist which become unfeasible in the verification step. This is not yet considered in the automated optimization shown in Fig. 2. As such, the currently proposed design appears to be sub-optimal in Fig. 6.

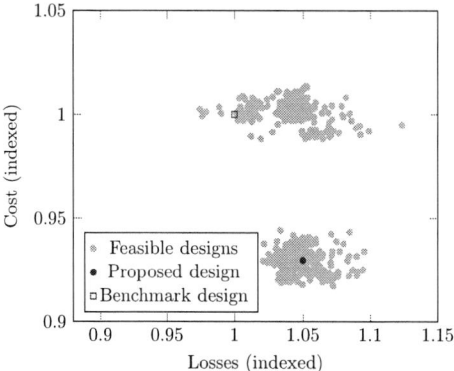

Fig. 6: Trade-off between cost and losses. Two clusters of evaluated designs are visible (green) with the benchmark design (blue square) in the cluster with higher system cost and the proposed design (blue circle) in the lower-cost cluster. Cost and losses are normalized based on the benchmark design.

Design Verification

In order to verify the obtained designs, different simulations are executed considering the full closed-loop control performance of the converter system. In a first step, offline computer simulations are conducted to verify that the dynamic behavior of the design does not violate any constraints. In a second step real-time simulations (RTS) with hardware-in-the-loop (HIL) are performed on different transient scenarios to confirm that the chosen designs fulfill the defined system requirements.

Offline Computer Simulations

Current Reference Step

The response of the converter to a current reference step on the three-phase side can be seen in Fig. 7. A reference step from 0.1 p.u. to -0.4 p.u. is performed. The reference is followed by a quick transient and without oscillations after the step. The closed-loop control performance of the suggested system design is comparable to the one of the benchmark design.

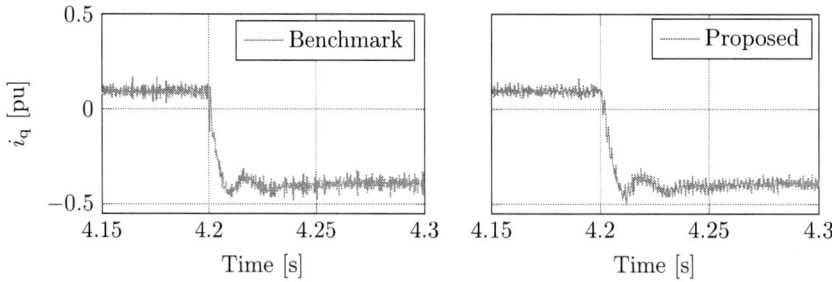

Fig. 7: Behavior of i_q after a reference current step. Comparison of offline computer simulations for the benchmark and the proposed design.

Three-Phase Grid Fault

A critical test for the performance of a system design is a grid fault scenario. A short-circuit is simulated and the resulting behavior is analyzed. In Fig. 8 the dynamic response in case of a fault on the three-phase utility grid side is shown. Visible is the behavior of the three-phase grid side voltages and currents. The benchmark design is shown in green while the proposed optimized design is shown in red. For the same trajectory of the primary grid voltage $u_G(t)$ it is visible that the transient behavior of the grid current $i_G(t)$ is comparable, both during the fault as well as after voltage recovery. Thus, the results confirm the appropriate behavior of the suggested design during a three-phase grid fault.

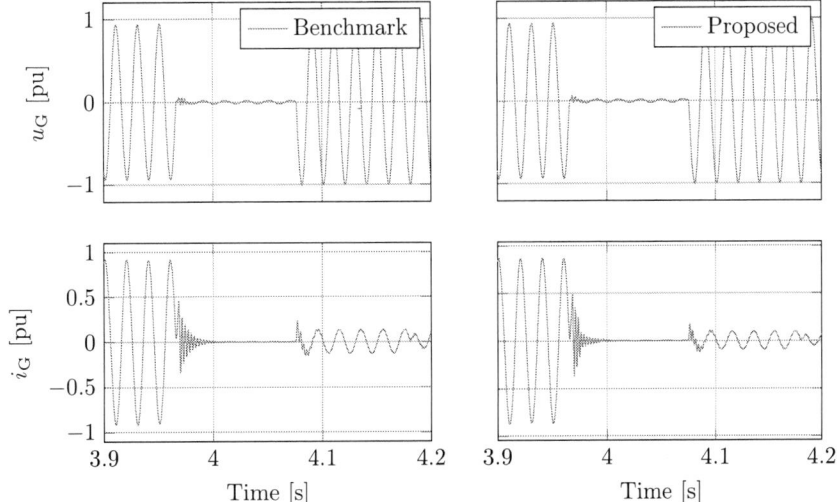

Fig. 8: Three-phase grid voltages and currents during a grid fault in offline computer simulations.

Real-Time Simulations

In a second step, real-time simulations are conducted. In this case, the real-time software and firmware are running on the actual control hardware, while the power system including MMC, transformers, filters and the AC grids is emulated in a commercial HIL system. To verify the design, various scenarios have been simulated. They include a current reference step, a grid fault, a power reference step, and a load step. The results are compared with the benchmark design.

Current Reference Step

In Fig. 9 the real-time simulation results of a current reference step are shown. The behavior is comparable both in the transient behavior as well as the recovery after the step. It is noted that the step change is different between the two designs (-0.4 p.u. for the benchmark and -0.5 p.u. for the proposed design, respectively) without loss of generality. Both designs show an acceptable and comparable current trajectory.

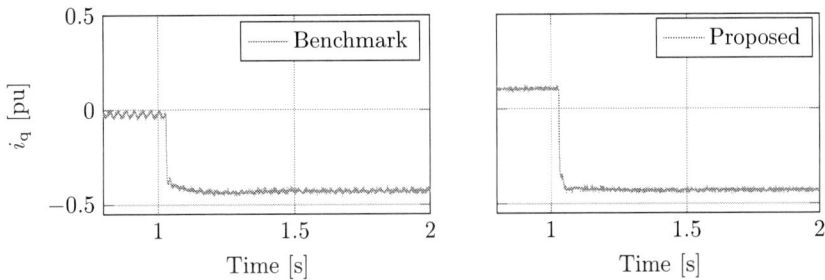

Fig. 9: HIL simulations of i_q after a reference current step for the benchmark and the proposed design.

Three-Phase Grid Fault

The real-time simulation of a fault in the utility grid is shown in Fig. 10. The upper plot shows the grid voltages for the benchmark and proposed design, respectively. The left plots show the occurrence of the fault while the right plots show the voltage return. The voltage behavior matches between the two designs.

The lower plots show the grid currents. Here, a different behavior is observed. In the proposed design, after return of the voltage an inrush-current is visible. The latter occurs due to saturation in the transformer. The differently designed transformer in the benchmark design does not saturate and, thus, leads to lower inrush-currents. On the other side, the benchmark design has to bear higher filter currents as the filter gets excited. The proposed design is robust enough to meet the design requirements taking into account the non-linear behaviors of the system. The different designs lead to different dynamic behavior during a grid fault, with both behaviors fulfilling the specification.

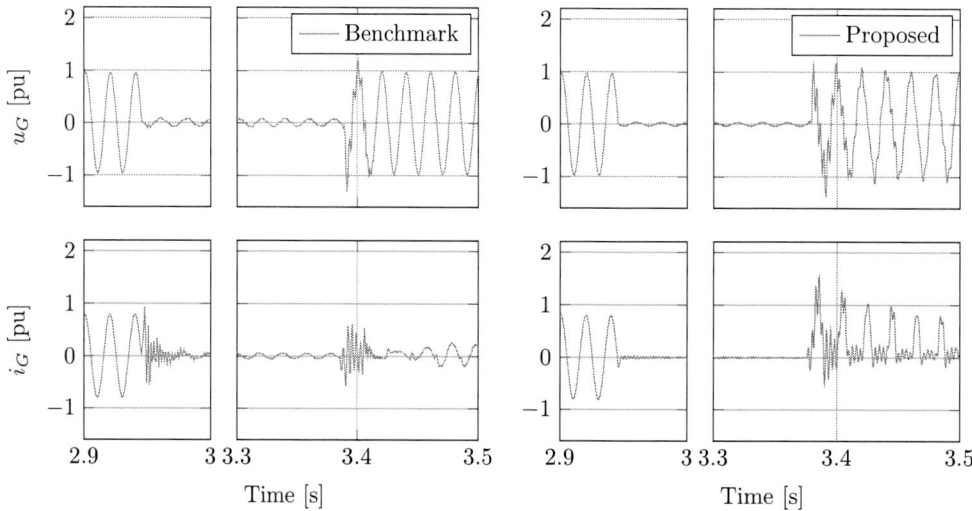

Fig. 10: HIL simulation of grid voltages and currents after a fault in the utility grid.

Power Reference Step and Load Step

As a further test, a power reference step is applied to the converter and the active power trajectory $P_{G,\text{filt}}$ is monitored and shown in Fig. 11. The trajectories of benchmark and proposed design are comparable and are complying with the design requirements, both in terms of time response and reference tracking.

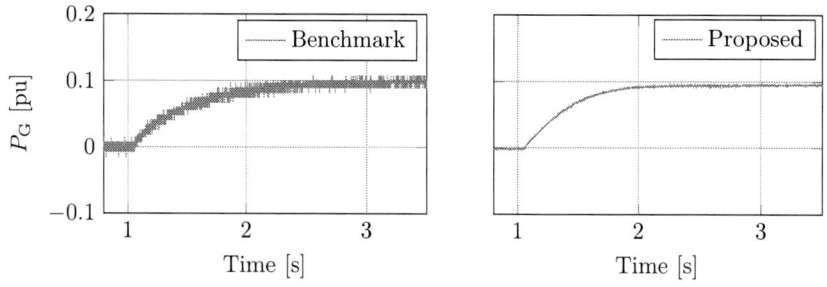

Fig. 11: HIL simulation of active power trajectory after a power reference step.

The reaction of the converter to a load step from 0 p.u. to 0.5 p.u. as shown in Fig. 12 is similarly compliant and comparable.

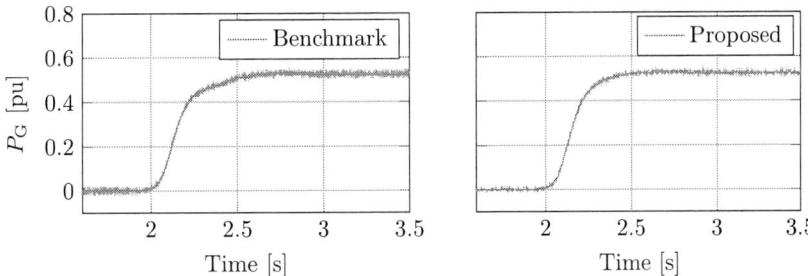

Fig. 12: HIL simulations of active power trajectory after a load step.

Conclusion

This paper suggests a multi-objective optimization procedure for power converters. The procedure has been derived and tested on an MMC topology. The suggested multi-objective optimization method enables an automated system design and aims to optimize an entire converter system for a set of objectives and constraints instead of an individual optimization of components. An analysis of the design space has been demonstrated, evaluating various design parameters and assessing their impact. This reduces the design space and allows a targeted optimization. The pareto-optimal system designs can be compared and assessed by the designer. Thus, the performance can be maintained with a reduced number and size of components, which leads to an improvement of the system in terms of cost or losses. The concept has been verified by offline computer and real-time simulations of steady-state and transient behavior of derived designs, which confirm the benefits of a derived candidate. As future work, the integration of dynamic behavior in the automated optimization would complement the optimization.

References

[1] A. Lesnicar, R. Marquardt: A new modular voltage source inverter topology, In Conf. Rec. EPE, 2003.

[2] S. Debnath, J. Qin, B. Bahrani, M. Saeedifard and P. Barbosa: Operation, Control, and Applications of the Modular Multilevel Converter: A Review, in IEEE Transactions on Power Electronics, vol. 30, no. 1, pp. 37-53, 2015.

[3] T. Thurnherr, P. Maibach, B. Buchmann and E. Baerlocher: Static Frequency Converters for Railway Power Supply based on IGCT High Power Semiconductors, PCIM Europe digital days 2020; International Exhibition and Conference for Power Electronics, Intelligent Motion, Renewable Energy and Energy Management, 2020, pp. 1-8.

[4] D. Weiss, M. Vasiladiotis, C. Banceanu, N. Drack, B. Odegard and A. Grondona: IGCT based Modular Multilevel Converter for an AC-AC Rail Power Supply, PCIM Europe 2017; International Exhibition and Conference for Power Electronics, Intelligent Motion, Renewable Energy and Energy Management, 2017, pp. 1-8.

[5] A. Hillers and J. Biela: Optimal design of the modular multilevel converter for an energy storage system based on split batteries, 15th European Conference on Power Electronics and Applications, 2013, pp. 1-11.

[6] S. Fuchs, M. Jeong and J. Biela: Reducing the Energy Storage Requirements of Modular Multilevel Converters with Optimal Capacitor Voltage Trajectory Shaping, 22nd European Conference on Power Electronics and Applications, 2020, pp. P.1-P.11.

[7] R. Sahu and S. Sudhoff: Design Paradigm for Modular Multilevel Converter-Based Generator Rectifier Systems, in IEEE Open Access Journal of Power and Energy, vol. 7, pp. 130-140, 2020.

[8] S. Am, P. Lefranc, D. Frey and M. Ibrahim: A generic virtual prototyping tool for multilevel modular converters (MMCs), IECON 2015 - 41st Annual Conference of the IEEE Industrial Electronics Society, 2015, pp. 1489-1494.

[9] G. Chiandussi, M. Codegone, S. Ferrero, F.E. Varesio: Comparison of multi-objective optimization methodologies for engineering applications, Computers and Mathematics with Applications, vol. 63, Issue 5, 2012, pp. 912-942.

[10] Luis Martí, Jesús García, Antonio Berlanga, José M. Molina: A stopping criterion for multi-objective optimization evolutionary algorithms, Information Sciences, vol. 367–368, 2016, pp. 700-718.

Sizing of Hybrid Energy Storage System for Residential PV Applications

Xiangqiang Wu, Zhongting Tang, Tamas Kerekes
Aalborg University
Pontoppidanstræde 101, 9220 Aalborg East
Aalborg, Denmark
Tel.: +45 91868836
Fax: +45 99408540
E-Mail: xiwu@energy.aau.dk
URL: https://www.en.aau.dk/

Keywords

« Photovoltaic», « Energy storage», « Battery », « Supercapacitor».

Abstract

In this paper, a sizing methodology is proposed for a grid-connected PV and hybrid energy storage system, which is used to determine the capacity share ratio of the Li-ion battery and the supercapacitor. Based on one case of the PV and the load mission profiles, the proposed sizing scheme can reduce the cycle numbers of the Li-ion battery and utilize the supercapacitor most, hence improving the self-consumption of the PV system with hybrid energy storage.

Introduction

Photovoltaic (PV) energy has grown rapidly in the last few decades due to global environmental awareness. However, managing PV production is challenging due to its intermittency and un-prediction, which means a huge amount of fluctuating power will be injected into the distribution systems. Therefore, solutions to flexibly control become mandatory tasks for PV systems [1].

In order to provide a reliable operation of PV systems, batteries are seen as a promising solution due to their decreasing prices, easy scalability, and flexibility. In addition, with the decreasing feed-in tariffs and increasing prices of grid electricity, maximum self-consumption operation of the PV generated electricity is becoming more attractive than feeding energy into the grid. In this case, optimal sizing of the energy storage system in grid-connected PV systems is a crucial issue, which relates to the reliable and economical system operation, and contributes to the more widespread use of batteries in PV applications. In this regard, much literature has proposed various sizing methods based on different evaluation objectives, which can be briefly categorized into three types: levelized cost of energy (LCOE) [2]-[4], power autonomy [5], and battery lifetime [6]-[8].

In addition, due to the changing solar irradiance and ambient temperature, there exist fluctuating PV power, which may decrease the lifespan of batteries and challenge the stable operation of the grid. Thus, that should be paid more attention to the dynamic response and battery lifetime. In recent years, hybrid energy storage systems (HESS) have drawn more attention since they can provide more control flexibility than that with battery only. Besides, HESS can extend the lifetime of batteries to some extent, which may make it a more cost-effective solution in energy storage applications [9]-[11]. Nevertheless, the optimal sizing criteria are not clear in prior-art literature, leading to the oversize of some storage elements and poor cost-effectiveness of the system.

This paper focuses on the optimal sizing of the hybrid energy storage system (i.e., Li-ion battery and supercapacitor) for residential PV applications. Based on given PV and load profiles, a sizing scheme is proposed to maximize the utilization of the supercapacitor and improve the self-consumption of the PV system. Firstly, the PV system configuration is presented, and a set of full-year PV and load profiles are adopted in this case. Secondly, the cycles of the battery and supercapacitor are presented and discussed considering the influences of the time constant of the low-pass filter. Then, the optimal sizing method is proposed based on the utilization of the supercapacitor and self-consumption of the system. The conclusion is presented in the last section.

System Configuration

Schematic

Fig. 1 shows the integrated PV and energy storage system, which consists of PV arrays, a hybrid energy storage system (i.e., Li-ion battery and supercapacitor), load, and the grid. In Fig. 1, the black arrows represent electrical links, while blue dashed arrows represent power flow directions. The battery and supercapacitor are parallel to each other in an active connection [12], which provides more control degrees of freedom. In this system, PV, battery, and supercapacitor connect the same DC bus through separate DC-DC converters, and P_{PV}, P_{bat}, P_{SC} are respectively output power of PV, battery, and supercapacitor. ΔP_{grid} is the power difference between PV production and energy storage, and P_{grid} is the power injected or absorbed from the grid.

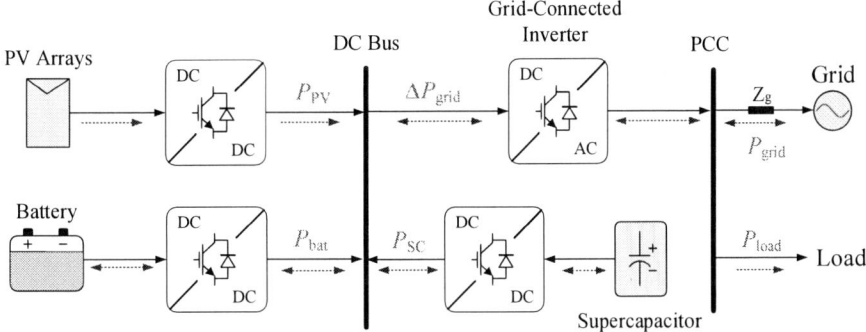

Fig. 1: Structure of the PV-HESS system.

Input Data

To analyze the power flows in the PV-HESS system of a typical household, a simulation model is developed. Both PV profile and load data sets are used as input for the simulation, which are shown in Fig. 2. In Fig. 2(a), the ambient temperature and solar irradiance are represented by a blue curve and orange curve, respectively, and these data are recorded for a full year with one-minute resolution in Aalborg, Denmark, located at 57° N, 9° E. The load profile of a typical Danish household is shown in Fig. 2(b), whose annual electricity consumption is about 3 MWh.

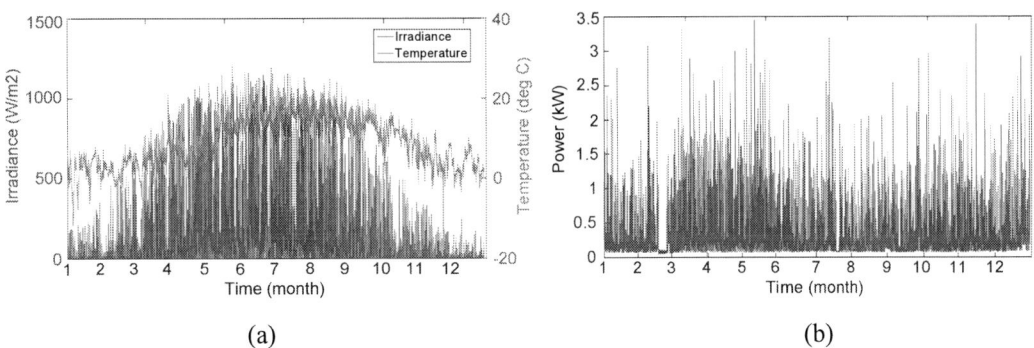

Fig. 2: PV and load profiles: (a) PV data; (b) load data.

In this paper, the PV model is based on the power temperature coefficient model [13], which is used solar irradiance and ambient temperature to estimate the generated power from the PV panels. The PV power P_{mp} at the maximum power point can be expressed as:

$$P_{mp}\left(G_e, T_c\right) = \frac{G_e}{G_{STC}} P_{mp,STC}\left[1 + \gamma_{mp}\left(T_c - T_{STC}\right)\right] \tag{1}$$

where G_e is the effective solar irradiance and G_{STC} is the solar irradiance under standard test condition (STC). $P_{mp,STC}$ and γ_{mp} are the measured peak power under STC and the normalized temperature coefficient of peak power, respectively. T_c and T_{STC} are the solar cell temperature and the temperature at STC conditions.

In this paper, the temperature coefficient is set to -0.35%/°C, and PV generator losses are considered as 8%. The efficiency of power electronic devices in this system is set to 95% at their nominal power. In addition, considering the power distribution between different power sources, the throughput model [14] is adopted to represent the Li-ion battery and supercapacitor, where the number of storage cycles n_c can be calculated as follows:

$$n_c = \frac{E_c + E_d}{2E_s} \tag{2}$$

where E_c and E_d are the energy charged and discharged to the battery/SC, respectively. E_s is the capacity of the battery/SC.

Based on the input data and simulation model described above, the PV and battery capacity can be sized. Here, the economic evaluation is performed based on the degree of self-sufficiency (d), which is used to qualify the energy independence (i.e., energy autonomy) of the microgrid from the utility grid. It represents the ratio of the load consumption (E_l) supplied by the PV-HESS system (including the directly used energy E_{du} and energy discharged for the load E_d), which can be expressed as:

$$d = \frac{E_{du} + E_d}{E_l} \times 100\% \tag{3}$$

Limited by the total investment, the PV and battery capacities are restricted to 10 kWp and 10 kWh respectively. In Fig. 3, the relationship among the self-sufficiency (i.e., power autonomy) of the system, PV, and battery capacity is presented. It can be observed that the ramp rate of self-sufficiency becomes much smaller when it is over 75%, which means a lot of extra investment can gain only a little power autonomy improvement. Therefore, it is reasonable to set the self-sufficiency to 75% by trading off the cost and self-sufficiency rate. In this case, the PV capacity is set to 10 kWp and the battery capacity is set to around 6 kWh.

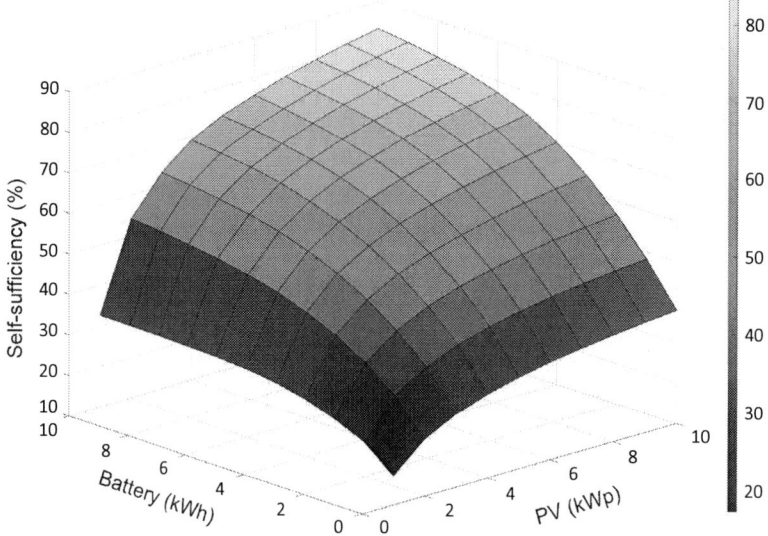

Fig. 3: Trend of self-sufficiency according to the PV and battery capacity

Control Strategy

In this paper, the DC link voltage is controlled by the inverter, and HESS is responsible to maintain the power level injected into the grid. The control framework of power distribution scheme is shown in Fig. 4, where the difference between PV production P_{pv_r} and load power P_{load_r} serves as the reference power of HESS. Through a first-order low-pass filter (LPF) and ramp rate limiter, the reference power of battery P_{bat_r} can be obtained, and the reference power of supercapacitor P_{SC_r} can be obtained by subtracting P_{bat_r} from the reference power of HESS P_{HESS_r}.

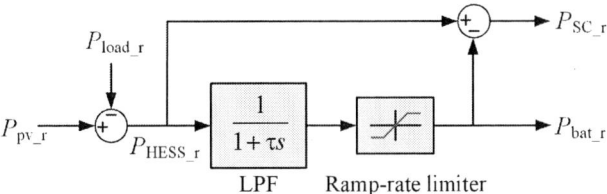

Fig. 4: Power distribution scheme

LPF-based power distribution strategy has been widely used due to its simplicity and superior performance. By changing the LPF time constant τ, the frequency of the current flowing through the battery can be changed. As a consequence, the low-frequency components in the input power will serve as the reference power for the battery, and the high-frequency components become the reference power for the supercapacitor, which accords with the characteristics of different energy storage elements. Considering that high-frequency components are harmful to batteries, the battery lifetime and efficiency of batteries can be improved by adopting supercapacitors with good dynamics. Therefore, an appropriate time constant of LPF and proper supercapacitor capacity are essential for the optimal HESS operation.

Simulation Results and Discussion

In order to size the supercapacitor in the HESS, the relationship among battery and supercapacitor cycles and control parameters (time constant of LPF) should be investigated. Due to the fact that supercapacitors have a much longer lifetime (e.g., over 500000 cycles) than batteries, the optimal supercapacitor size should be where the supercapacitor can gain the most cycles so that the supercapacitor can be utilized to the greatest extent.

Based on the above data and analysis, the simulation model was developed in MATLAB 2021b. The battery capacity is 6144 Wh in the simulation, consisting of twelve 12.8 V/40 Ah Li-ion battery cells. Considering that the battery response time (i.e., the time for a battery to provide energy at its full rated power) is 30 seconds or more, we mainly focus on the time constant over 30 seconds. In the simulation, a series of time constants are studied, which are 1, 30, 60, 120, 360, 600, 1200, 3600, and 6000 seconds. And the state of charge (SOC) of the battery and supercapacitor are both restricted to a range between 10% and 90% of their nominal capacities.

The relationship among supercapacitor cycles, capacities, and time constants is shown in Fig. 5(a). It can be seen that when the capacity of the supercapacitor is fixed, the cycles will increase at first and decrease later with an increasing time constant of LPF. For example, in the case of a 64 Wh supercapacitor, the supercapacitor cycles will increase from 2431 (τ =30 s) to 4236 (τ =360 s), then decrease to 2111 (τ =6000 s). This is because when the time constant increases, the power reference for the supercapacitor will increase as well. However, once the time constant reaches a certain value, the supercapacitor will be fully charged or discharged quickly, so that more energy will flow into the grid instead of the supercapacitor. In Fig.5(a), we can also observe that the time constant for peak supercapacitor cycles will change according to different supercapacitor capacities, and the bigger the supercapacitor size, the larger time constant is needed for peak cycles. Specifically, when the supercapacitor size is between 64 Wh to 128 Wh, the time constant is 360s for peak cycles, while 600s is for the supercapacitor ranging from 128 Wh to 448 Wh. And when the supercapacitor is larger than 448 Wh, a 1200s or higher time constant is needed.

Fig.5(b) shows the total energy injected into and absorbed from the grid. With an increasing time constant of the low-pass filter, the energy flowing through the inverter is increasing as well, which illustrates that there is more energy exchanged between the PV system and the grid. Considering the different prices of feed-in energy and procurement of electricity from the grid, the increasing energy throughput will lower the cost-effectiveness of the system. On the other hand, under a certain time constant, higher supercapacitor capacity leads to lower energy throughput, which means that more energy is consumed on-site, hence the self-consumption will be improved.

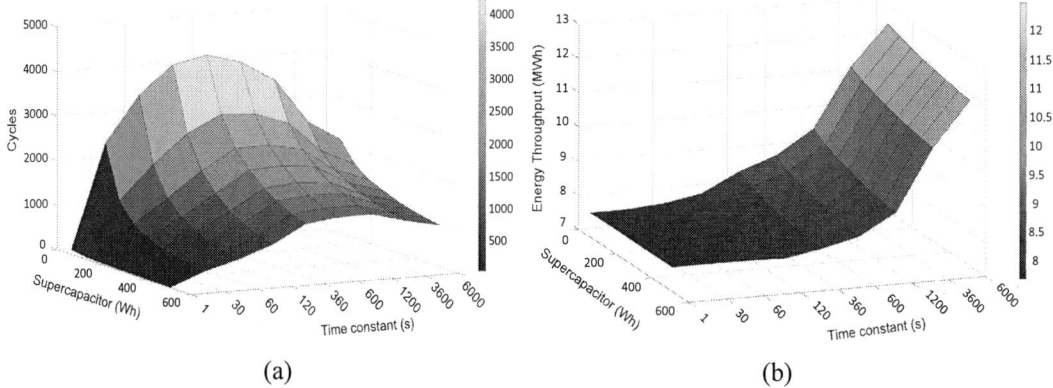

(a)   (b)

Fig. 5: Simulation results under different capacities and time constants: (a) supercapacitor cycles; (b) energy throughput to/from the grid.

Based on the above analysis, in order to achieve the best balance between the utilization of the supercapacitor and self-consumption of the system, the time constant is designed to 360 s, and the supercapacitor can be set to 128 Wh (i.e., 1/48 of Li-ion battery). Fig. 6(a) shows the cycles of the Li-ion battery and the energy to/from the grid under different time constants. With a time constant of 360 s, the battery cycles can be reduced by 1.89% compared to that of 1 s time constant. Considering that the high-frequency power fluctuations are filtered, the Li-ion battery lifetime can be prolonged more significantly.

In this case, Fig. 6(b) shows a one-day simulation, where the red curve represents the reference power for the HESS, while the blue and green curves represent the reference power for the battery and the supercapacitor, respectively. The result demonstrates that the high-frequency power fluctuations are charged and discharged through the supercapacitor while the battery mainly deals with the low-frequency power fluctuations. In that case, the battery can obtain some relief effects from the high-frequency harmonics, which can extend the battery lifetime.

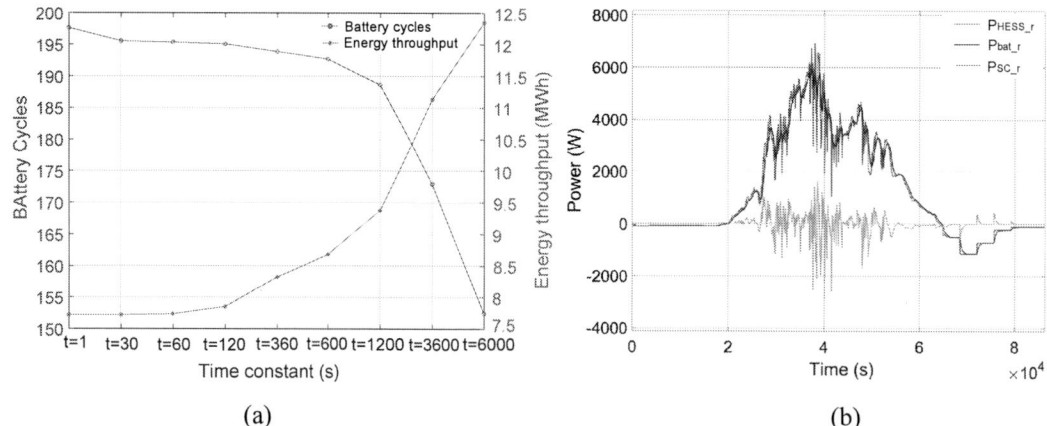

(a)   (b)

Fig. 6: Power distribution of the HESS and battery cycles under different time constants: (a) battery cycles and energy throughput; (b) power distribution.

Conclusion

In this paper, the sizing optimization of a grid-connected PV-HESS system is presented. An analysis of LCOE of the PV and the battery is made to size the PV and Li-ion battery capacities firstly. Then, according to the relationship between supercapacitor cycles and the time constant of the LPF, the optimum scaling of the supercapacitor can be determined to achieve maximum utilization of the supercapacitor. Meanwhile, the self-consumption of the PV system can be improved. The results obtained from the simulation show that the sizing parameters are appropriate and have good dynamic performance. Besides, the battery cycles can be reduced to some extent.

This sizing method is easy to realize and can be extended to other applications, including different types of energy storage technologies. For example, in dry and sunny areas, the appropriate time constant of LPFs may be different due to the decreasing fluctuating power. Then the capacity of the supercapacitor can be determined to improve the self-consumption by considering the greatest utilization of the supercapacitor. Therefore, this method has a good prospect of engineering application.

References

[1] Tran Q., Cong Pham M., Parent L., Sousa K.: Integration of PV systems into grid: from impact analysis to solutions, 2018 IEEE International Conference on Environment and Electrical Engineering and 2018 IEEE Industrial and Commercial Power Systems Europe (EEEIC / I&CPS Europe), 2018, pp. 1-6.

[2] Weniger J., Tjaden T., Quaschning V.: Sizing of residential PV battery systems, Energy Procedia 2014, Vol. 46, pp. 78-87.

[3] Torkashvand M., Khodadadi A., Sanjareh M. B., Nazary M. H.: A life cycle-cost analysis of Li-ion and lead-acid BESSs and their actively hybridized ESSs with supercapacitors for islanded microgrid applications, IEEE Access 2020, Vol. 8, pp. 153215-153225.

[4] Wee K. W., Choi S. S., Vilathgamuwa D. M.: Design of a least-cost battery-supercapacitor energy storage system for realizing dispatchable wind power, IEEE Transactions on Sustainable Energy 2013, Vol. 4, no. 3, pp. 786-796.

[5] Javidsharifi M., Pourroshanfekr H., Kerekes T.: Optimum sizing of photovoltaic and energy storage systems for powering green base stations in cellular networks, Energies 2021, Vol. 14, no.7, pp. 1895.

[6] Gee A., Robinson F., Dunn R.: Analysis of battery lifetime extension in a small-scale wind-energy system using supercapacitors, IEEE Transactions on Energy Conversion 2013, Vol. 28, no. 1, pp. 24-33.

[7] Dulout J., Jammes B., Alonso C., A. Luna A., Guerrero J. M.: Optimal sizing of a lithium battery energy storage system for grid-connected photovoltaic systems, 2017 IEEE Second International Conference on DC Microgrids (ICDCM), 2017, pp. 582-587.

[8] Alramlawi, M. Li P.: Design optimization of a residential PV-battery microgrid with a detailed battery lifetime estimation model, IEEE Transactions on Industry Applications 2020, Vol. 56, no. 2, pp. 2020-2030.

[9] Roy P., Karayaka H., Yan Y., Alqudah Y.: Size optimization of battery-supercapacitor hybrid energy storage system for 1MW grid connected PV array, 2017 North American Power Symposium (NAPS), 2017, pp. 1-6.

[10] Ferreira K. Santos W. M., César Rueda Medina A.: Sizing of supercapacitor and BESS for peak shaving applications, 2019 IEEE 15th Brazilian Power Electronics Conference and 5th IEEE Southern Power Electronics Conference (COBEP/SPEC), 2019, pp. 1-6.

[11] Mandal S., Mandal K. K., De M., Das, G.: A new improved algorithm for optimal sizing of battery-supercapacitor based hybrid energy storage systems, 2018 Emerging Trends in Electronic Devices and Computational Techniques (EDCT), 2018, pp. 1-6.

[12] Jing W., Lai C. H., Wong S.: Battery-supercapacitor hybrid energy storage system in standalone DC microgrids: a review, IET Renewable Power Generation 2017, Vol. 11, no.4, pp. 461-469.

[13] Nichinte A. S., Vyawahare, V. A., Magare D. B.: Estimation and comparison of module temperature model coefficient for different PV technology module, 2020 3rd International Conference on Communication System, Computing and IT Applications (CSCITA), 2020, pp. 13-17.

[14] Rosewater D. M., Copp D. A., Nguyen T. A., Byrne R. H., Santoso S.: Battery energy storage models for optimal control, IEEE Access 2019, Vol. 7, pp. 178357-178391.

DC Bias Currents in Full-Bridge DC-DC Converters in Context of WBG Semiconductors and High Switching Frequencies

Niklas Badenhop, Lukas Fräger, Dennis Kampen, Sascha Langfermann, Michael Owzareck
BLOCK TRANSFORMATOREN-ELEKTRONIK GMBH
Max-Planck-Straße 36-46
Verden, Germany
Tel.: +49 / (0) 4231 678 374
Fax: +49 / (0) 4231 678 277
E-Mail: niklas.badenhop@block.eu
URL: http://www.block.eu

Acknowledgements

Parts of this work were funded by the German Federal Ministry for Economic Affairs and Energy (BMWi) under grant number 03EN2010A-G in the project STIM. The authors are responsible for the content of this publication.

Keywords

DC-DC converter, Dual Active Bridge (DAB), Wide bandgap devices, Silicon Carbide (SiC), Non-identical devices

Abstract

This paper aims to address the challenges of ever higher switching frequencies and better semiconductors with respect to DC currents in DC-DC full-bridge converters. The causes of DC offset currents as non-identical components and higher switching frequencies in combination with wide bandgap (WBG) semiconductors are explained and discussed.

1 Introduction

Many papers have been published to address DC bias currents in DC-DC converters recent years (e.g. [1], [2]). The focus in most papers is eliminating the DC bias current (e.g. [1], [3], [4], [5], [6]). Little was published about the exact causes and influences due to better semiconductors and higher switching frequencies in combination with high DC-link voltages and possible solutions in detail until now. [7] presents a prediction method to calculate the maximum dc bias for a given transformer. This paper aims to present an overview about the causes of DC bias currents, especially in context of high switching frequencies, WBG semiconductors and high DC-link voltages. It provides analytic quantization of the effects proven by simulations and measurement results. Furthermore, it shows solutions to prevent and eliminate these currents to an acceptable level.

2 System Overview

The investigated system of this paper is a dual active bridge (DAB) DC-DC converter, see Fig. 1. However, other topologies without series capacitor show the same behavior. Also, the results can be adapted and used for half-bridge converter.

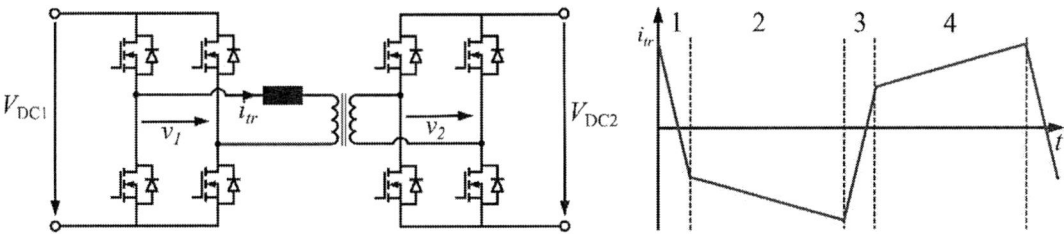

Fig. 1: Typical circuit of a dual active bridge (DAB) left and typical transformer current waveform right

There are two full-bridges, consisting of Silicon Carbide (SiC) MOSFETs and an MF-transformer with integrated leakage inductance. The two full-bridges generate square wave voltages v_1 and v_2. By controlling the phase shift between these voltages, the power transfer and output voltage V_{DC2} can be controlled.

3 Theoretical Considerations

As shown in Fig. 1, the transformer current i_{tr} has different slopes depending on the states of the full-bridges. In interval 1, secondary voltage v_2 is positive and primary voltage v_1 is negative. In interval 2, both voltages are negative. In intervals 3 and 4 the voltages are correspondingly vice versa. Equations 1-4 describes the behavior, $\frac{di_{tr}}{dt}$ is the transformer current slope, V_{DC1} the primary voltage, V_{DC2} the secondary voltage and L_σ the stray inductance of the transformer. The magnetizing inductance L_M is assumed to be large and the influence on the current slope is neglected [8]. Here a transformer transformation ratio of unity is assumed.

$$\left(\frac{di_{tr}}{dt}\right)_1 = -\frac{V_{DC1} + V_{DC2}}{L_\sigma} \tag{1}$$

$$\left(\frac{di_{tr}}{dt}\right)_2 = -\frac{V_{DC1} - V_{DC2}}{L_\sigma} \tag{2}$$

$$\left(\frac{di_{tr}}{dt}\right)_3 = \frac{V_{DC1} + V_{DC2}}{L_\sigma} \tag{3}$$

$$\left(\frac{di_{tr}}{dt}\right)_4 = \frac{V_{DC1} - V_{DC2}}{L_\sigma} \tag{4}$$

Intervals 1 and 3 or 2 and 4 respectively must be the same in terms of current slope $\frac{di_{tr}}{dt}$ and time t, otherwise there occur a DC bias current. This can cause saturation of transformers core material, can be cause higher core losses [9] and asymmetrical losses and stress of the semiconductors.

Due to many effects, the intervals are always slightly different. DC bias currents can be generated by one or both of the full-bridges. This happens when the semiconductors are turned on for different times. This causes a DC offset of the bridge voltages v_1 or v_2. The DC-impedance of the circuit R_{loop} consists of two MOSFETs with On-Resistance $R_{ds,on}$ and the transformer winding resistance R_{tr}:

$$R_{loop} = 2 \cdot R_{ds,on} + R_{tr} \tag{5}$$

In combination with the offset voltage between v_1 or v_2 a DC bias current results. The offset voltage ΔV_{Offset} can be quantified by the switching interval error ΔT_{error}, the switching frequency f_s and the DC-link voltage V_{DC}:

$$\Delta V_{Offset} = 2 \cdot \Delta T_{error} \cdot f_s \cdot V_{DC} \tag{6}$$

In context of high-power density, high switching frequencies and high efficiency there are a lot of challenges to keep the DC offset as small as possible [3].

With the offset voltage ΔV_{Offset} and the circuit DC-impedance R_{loop}, the current offset can be calculated:

$$\Delta I_{Offset} = \frac{\Delta V_{Offset}}{R_{loop}} \tag{7}$$

With higher switching frequencies, differences of the intervals become more critical. Even small timing errors cause large DC bias currents. Furthermore, high DC-link voltages amplify the effect, just like a low resistance of the power loop.

DC bias current is caused by dynamical and steady state timing errors. While earlier papers often describe and handle dynamical time errors caused by sudden phase shifts of control (e.g. [2], [10]), higher switching frequencies and WBG semiconductor amplify the impact of steady state timing error so that these become more important.

Errors of the switching times are mainly caused by steady state tolerances of components: the resolution of the PWM unit, variation of the propagation delay of digital isolators or gate driver ICs, signal runtime on PCB, tolerances of components e.g. capacitors in signal filters, tolerances of switching levels of digital inputs and variation of the MOSFET's threshold voltage. There are further small effects, which also influence the switching behavior. Most of the effects are in picosecond or small nanosecond range, but for high switching frequencies, these effects become important and must be paid attention to. The deviations are component related and therefore not reproducible. Thus, the worst case must be assumed for the analysis and design.

Furthermore, there are also DC currents which are caused by the control, e.g. by changing the phase-shift. This dynamical, transient caused offset occurs only for a short time because of the damping due to the impedance of the power loop. The converter, especially the transformer must be robust enough to handle the sum of static and transient dc bias. Multiple methods are published to overcome the transient, dynamical phase shift with resulting dc bias current (e.g. [2]).

4 System Evaluation

As an example, some practical static component tolerances of a real DAB converter were analyzed:

- High resolution PWM of Microcontroller   195 ps
- Skew time of digital isolator channel to channel   4 ns [11]
- Propagation delay of gate driver part to part   7 ns [12] or 14 ns [13]

Note, that all values can be positive or negative. So, the tolerances are doubled. By using a switching frequency of 100 kHz and a DC-link voltage of 800 V using the equation 2 this results in a DC offset voltage of

$$\Delta V_{Offset} = 2 \cdot \frac{11{,}2\text{ ns}}{10{,}0\text{ µs}} \cdot 800\text{ V} = 1{,}79\text{ V} \tag{8}$$

when two of the MOSFETs are turned off with the delay. When also the other two transistors are turned on with a delay, the effect will be doubled. If other effects like threshold voltages or filter are considered, the offset increases further.

With two SiC MOSFETs (Infineon IMZ120R090M1H [14]), each with a $R_{ds,on}$ of 90 mΩ and a resistance of transformers primary winding of approx. 20 mΩ, the current offset and the current offset per time difference can be calculated:

$$\frac{1{,}79\text{ V}}{200\text{ mΩ}} = 8{,}95\text{ A} \rightarrow 0{,}8\text{ A}/_\text{ns} \tag{9}$$

The results show that even small time errors combined with high switching frequency, high voltages and low power loop resistance result in unacceptable DC bias current. Thus, saturation of the transformer is possible and asymmetrical stress of the semiconductors are implied. When using high performance WBG MOSFETs with low $R_{ds,on}$ and increasing frequencies, the effect will continue to intensify.

In contrast, by using a state of the art IGBT converter with switching frequency of 8 kHz and a DC-link voltage of 800 V the DC Offset voltage can be calculated to:

$$\Delta V_{Offset} = 2 \cdot \frac{11{,}2 \text{ ns}}{125 \text{ μs}} \cdot 800 \text{ V} = 0{,}14 \text{ V} \tag{10}$$

when two of the IGBTs are turned off with the same delay.

With two IGBT (Infineon IKW50N120CS7 [15]), each with a $R_{ce,on} = \frac{2 \text{ V}}{15 \text{ A}} = 133 \text{ m}\Omega$ and a resistance of transformers primary winding of approx. 20 mΩ the bias current is

$$\frac{0{,}14 \text{ V}}{286 \text{ m}\Omega} = 0{,}49 \text{ A} \rightarrow 43{,}8 \text{ mA}/_{\text{ns}} \tag{11}$$

This DC bias current is much smaller and can be easily handled by using a transformer with small air gap to prevent core saturation, which will be shown in the next section. No DC bias control is needed. The example shows the influence of switching frequency and power loop resistance to the DC bias current. So, the problem becomes more relevant when using WBG semiconductors and high switching frequencies.

In recent literature, there are multiple solutions to overcome this problem. [1] gives an overview about flux measurement and flux balancing methods. In [16] the AC current is measured, and the DC part of the current is separated to then control the duty cycles of the MOSFETs. [17] shows a solution with measurement of transformers flux via a hall sensor and control of the duty cycle. Both methods need an accurate and expensive measurement with a high bandwidth and an additional control loop to adjust the duty cycle. In [3] the average current is measured, and then full-bridges duty cycle is controlled. This method does not need a high bandwidth but also an additional control loop. [18] also control the DC bias by changing the duty cycle. By using current injection [19] remove the DC bias. Another often used method is to place DC blocking caps on primary and secondary side. This is simple to implement, but the capacitors must handle high currents, become large and expensive and decrease the power density. By using MOSFETs with higher $R_{ds,on}$ values and transformer windings with higher resistance, the effect of DC currents also can be reduced but losses increase. Another option is to design the transformer to be resistant to DC currents up to the maximum appearing DC-bias. With this the transformer size may increase since the dc and ac flux density must be considered, but no extra control loop is needed. Depending on the maximum DC bias current, this can be the simplest solution to overcome core saturation. However, the disadvantage is the asymmetric stress of the semiconductors, so that this method is only suitable for small dc bias currents.
Since the currents can never be completely eliminated, DC blocking caps or DC resistant transformer are always required.

5 Design of Transformer

For a first estimation of the flux density in the core a linear magnetization characteristic of the BH-curve can be made if the operation point is below the saturation of the core material. In this case the resulting flux density is the addition of the flux density of the rectangular voltage $\hat{B}_{AC}$ and the flux density of the DC current $\hat{B}_{DC}$.

For a rectangular voltage the flux density of the transformer can be easily calculated with the peak Voltage $\hat{U}$, the turns N and the cross-sectional area of the core A_{Fe}, whereas the DC flux density is calculated by the DC-current i_{DC}, the magnetic resistance of the transformer R_m and the cross-sectional area of the core.

$$\hat{B} = \hat{B}_{AC} + \hat{B}_{DC} = \frac{\hat{U}}{4f_s N A_{Fe}} + \frac{N i_{DC}}{R_m A_{Fe}} \tag{12}$$

For small air gaps, neglecting the fringing flux, the magnetic resistance is calculated through the mean magnetic path length l_m of the core, the magnetic permeability μ_r, the magnetic constant μ_0 and the air gap δ.

$$\hat{B} = \hat{B}_{AC} + \hat{B}_{DC} = \frac{\hat{U}}{4f_s N A_{Fe}} + \frac{N i_{DC} \mu_0}{\frac{l_m}{\mu_r} + \delta} \tag{13}$$

As seen in equation (13) the DC flux density can be a problem even with small DC current offsets when a core has a large permeability and no air gaps. An uncut nanocrystalline core tends to have these properties. The DC flux density can be limited through air gaps, however the main inductance also decreases with larger air gaps which has to be considered in the design.

6 Simulation Results

A simulation in PLECS is taken to verify the theoretical considerations, see Fig. 2. Here, the simulation results confirm the theoretical assumptions.

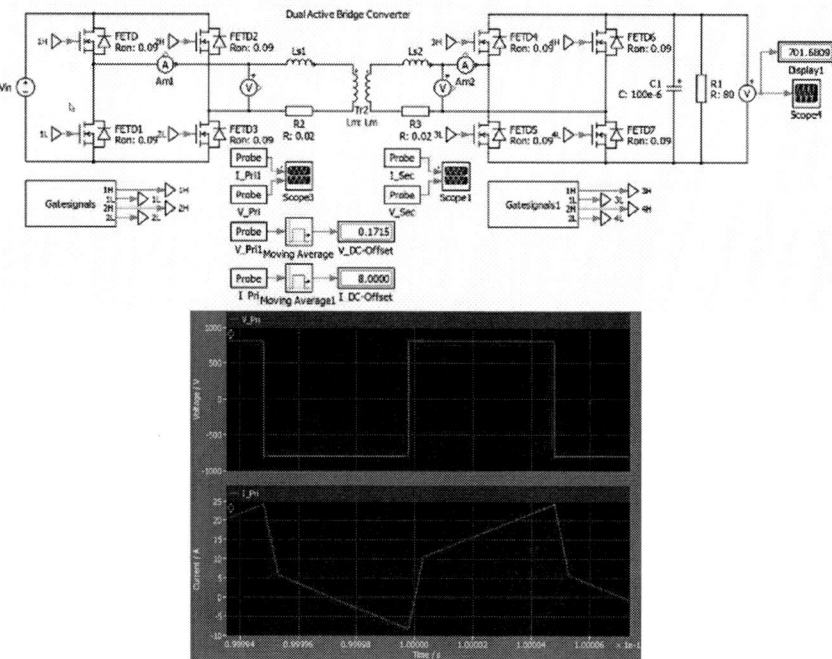

Fig. 2: Simulation setup in PLECS top and primary waveforms bottom

An open loop DAB converter is simulated with a fixed phase-shift at 800 V input voltage. The positive halfwave of primary voltage is about 10 ns longer than the negative (MOSFET 0 and 3 turned off 10 ns later). Therefore, there is a DC-bias voltage of 1,6 V which generates a DC-bias current of 8,0 A with an ohmic impedance of the power loop of 200 mΩ. Multiple simulations with different delays of individual MOSFETs on primary and secondary side verify the theoretical calculations.

7 Test Hardware and Measurements

To verify the theoretical considerations and simulations, a test hardware was built up and measurements were taken. The specifications of the test hardware are shown in Table I, the used measurement equipment in Table II. Here, a transformer with air gap is used to prevent core saturation and allow the measurement of the dc bias current.

Table I: List of relevant components

MOSFETs	IMZ120R090M1H [14]
Gate driver	1EDC60I12AH [13]
Gate-resistor	10 Ω
Isolator	ISO7740 [11]
Core	6x N87 E55/25
Air gap	0.2 mm
Main inductance	410 μH
Leakage inductance	15 μH @ each side
Winding	10 turns of 1400x0,05mm HF-litz-wire

Table II: Measurement equipment

Oscilloscope	Tek MSO Series 4, 500MHz, 12bit, 6Ch
Voltage probes	THDP0200
Current probe	TCP0150A

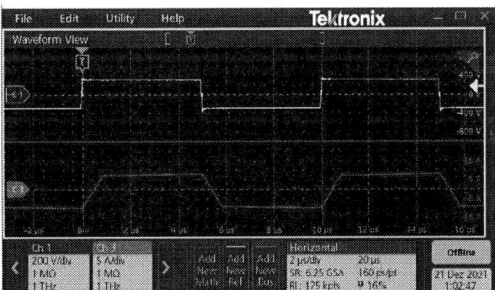

Fig. 3: Test hardware and measurement of DC bias current @ 300 V and 2,5 kW

In Fig. 3 the test hardware and measurement results are shown. To prevent saturation of the transformer only smaller voltages and loads are tested. First the imbalance of full-bridge voltage and then the DC-bias current is measured. Here, only the primary full-bridge is considered.

Table III: Measurement results

	Setup 1 @ 300 V and 2,5 kW	Setup 2 @ 300 V and 2,5 kW
T_{bias}	-9 ns	-12 ns
$I_{bias,measured}$	-1,10 A	-1,39 A
$I_{bias,calculated}$	-1,69 A	-2,25 A

Table III shows the measurement results of two setups. The PWM unit generates 100 kHz signal without an offset. Due to the component variations, the full-bridge voltage shows a DC bias. Time measurements show a time offset of -9 ns and -12 ns between positive and negative half-wave. These timing errors generate DC bias currents of -1,10 A and -1,39 A. With a gate-source voltage of 15 V and an assumed junction temperature of $T_j = 50\,°C$ the $R_{ds,on}$ is about 150 mΩ [14]. The winding resistance of the transformer was measured and is 20 mΩ. With these values the comparison to the theoretical considerations and simulation confirms the assumption with about 50-60 % higher values. These can be explained by a higher loop resistance due to the PCB and component tolerances. The calculations give the worst-case bias current.

8 Conclusion

This paper presents an overview of the main causes of static DC bias currents in DC-DC converters with respect to high switching frequencies and WBG semiconductors. An analytic approach to quantify the DC-offset is given and evaluated with measurements. Solutions to prevent or eliminating the issue are discussed and a preferred method is suggested.

References

[1] G. Ortiz, L. Fässler, J. W. Kolar and O. Apeldoorn, "Flux Balancing of Isolation Transformers and Application of "The Magnetic Ear" for Closed-Loop Volt–Second Compensation," *IEEE Transactions on Power Electronics,* vol. 29, no. 8, pp. 4078-4090, 2014.

[2] M. Wattenberg, U. Schwalbe and M. Pfost, "Impact of DC-Bias on Dual Active Bridge Control and How to Avoid it," *2019 21st European Conference on Power Electronics and Applications (EPE '19 ECCE Europe),* pp. P.1-P.8, 2019.

[3] Y. Yao, S. Xu, W. Sun and S. Lu, "A Novel Compensator for Eliminating DC Magnetizing Current Bias in Hybrid Modulated Dual Active Bridge Converters," *Journal of Power Electronics,* vol. 16, no. 5, pp. 1650-1660, 2016.

[4] D. Costinett, D. Seltzer, D. Maksimovic and R. Zane, "Inherent volt-second balancing of magnetic devices in zero-voltage switched power converters," *2013 Twenty-Eighth Annual IEEE Applied Power Electronics Conference and Exposition (APEC),* pp. 9-15, 2013.

[5] Q. Bu, H. Wen, J. Wen, Y. Hu and Y. Du, "Transient DC Bias Elimination of Dual-Active-Bridge DC–DC Converter With Improved Triple-Phase-Shift Control," *IEEE Transactions on Industrial Electronics,* vol. 67, no. 10, pp. 8587-8598, 2020.

[6] Z. Ji, Q. Wang, D. Li and Y. Sun, "Fast DC-Bias Current Control of Dual Active Bridge Converters With Feedforward Compensation," *IEEE Transactions on Circuits and Systems II: Express Briefs,* vol. 67, no. 11, pp. 2587-2591, 2020.

[7] L. a. C. W. Shu, Z. Lin, D. Ma, X. He and W. A. Syed, "DC Bias Study for DC-DC Dual-Active-Bridge Converter," *2018 IEEE 4th Southern Power Electronics Conference (SPEC),* pp. 1-5, 2018.

[8] L. Fraeger, N. Badenhop, S. Langfermann, M. Owzareck, D. Kampen and J. Friebe, "Dual Active Bridge Converter: Simple Peak Current Limitation by Dual Phase Shift Control," *PCIM Europe 2022,* 2022.

[9] J. Mühlethaler, J. Biela, J. W. Kolar and A. Ecklebe, "Core Losses Under the DC Bias Condition Based on Steinmetz Parameters," *IEEE Transactions on Power Electronics,* vol. 27, no. 2, pp. 953-963, 2012.

[10] B. Zhao, Q. Song, W. Liu and Y. Zhao, "Transient DC Bias and Current Impact Effects of High-Frequency-Isolated Bidirectional DC–DC Converter in Practice," *IEEE Transactions on Power Electronics,* vol. 31, no. 4, pp. 3203-3216, 2016.

[11] Texas Instruments, "Datasheet ISO7740," 2020. [Online]. Available: https://www.ti.com/product/ISO7740.

[12] Infineon, "Datasheet EICEDRIVER 1ED31... Compact," 2021. [Online]. Available: https://www.infineon.com/cms/en/product/power/gate-driver-ics/1ed3124mu12h/.

[13] Infineon, "Datasheet EICEDRIVER 1EDC Compact," 2017. [Online]. Available: https://www.infineon.com/cms/de/product/power/gate-driver-ics/1edc60i12ah/.

[14] Infineon, "Datasheet IMZ120R090M1H SiC MOSFET," 2020. [Online]. Available: https://www.infineon.com/cms/en/product/power/mosfet/silicon-carbide/discretes/imz120r090m1h/.

[15] Infineon, "Datasheet IKW50N120CS7," 2021. [Online]. Available: https://www.infineon.com/cms/de/product/power/igbt/igbt-discretes/ikw50n120cs7/.

[16] B. Zhang, S. Shao, L. Chen, X. Wu and J. Zhang, "Steady State and Transient DC Magnetic Flux Bias Suppression Methods for a Dual Active Bridge Converter," *IEEE Journal of Emerging and Selected Topics in Power Electronics,* vol. 9, no. 1, pp. 744-753, 2021.

[17] F. Dawson, "DC-DC converter interphase transformer design considerations: volt-seconds balancing," *IEEE Transactions on Magnetics,* vol. 26, no. 5, pp. 2250-2252, 1990.

[18] S. Han, I. Munuswamy and D. Divan, "Preventing transformer saturation in bi-directional dual active bridge buck-boost DC/DC converters," *2010 IEEE Energy Conversion Congress and Exposition,* pp. 1450-1457, 2010.

[19] S. Dutta and S. Bhattacharya, "A method to measure the DC bias in high frequency isolation transformer of the dual active bridge DC to DC converter and its removal using current injection and PWM switching," *2014 IEEE Energy Conversion Congress and Exposition (ECCE),* pp. 1134-1139, 2014.

Parameter tuning method for class Φ_2 converters for high-frequency wireless power transfer applications

Yining Liu, Prasad Jayathurathnage, and Jorma Kyyrä,
Aalto University
Maarintie 8, 02150
Espoo, Finland
Phone: +358504366005, +358504477981, +358505639146
Email: yining.1.liu@aalto.fi, prasad.jayathurathnage@aalto.fi, jorma.kyyra@aalto.fi

Acknowledgments

We would like to thank Professor Sergei Tretyakov for the guidance and useful discussions.

Keywords

≪High frequency power converter≫, ≪resonant converter≫, ≪parasitics≫, ≪soft switching≫, ≪zero-voltage switching≫

Abstract

This paper presents a method to compensate detuning of class Φ_2 push-pull converters due to practical disparities such as component values tolerances, parasitic effects, and manufacturing errors. A simple and effective tuning method is proposed based on only four steps. An example 100 W, 6.78 MHz wireless power transfer system is presented, that reaches 82.3% efficiency after tuning using the proposed method.

Introduction

High-frequency power conversion at MHz frequencies has become a very important research topic since the advent of wide-band-gap switching devices including Gallium-Nitrite (GaN) and silicon-carbide (SiC) devices. With the increasing of switching frequency, the size of the magnetic components can be reduced significantly, and the power density can be increased. However, the key challenges in MHz-power converters are dominance of switching losses, effects of parasitics, and high voltage/current stresses on the switching components. There have been several research attempts to address theses challenges. For example, realization of zero-voltage-switching (ZVS) and zero-voltage-derivative-switching (ZVDS) can reduce the switching losses significantly [6]. Different converter topologies have been extensively studied including class E [3], class EF [2], [5], class Φ_2 [4], and push-pull [1] topologies to realize different characteristics and advantages. Among these topologies, class Φ_2 push-pull topology has the advantages of increased power density with reduced DC current ripple as well as the inductance value. In [1], a theoretical design method is proposed to calculate optimal component parameters. With a proper design, a T-network is able to remove the second harmonic voltage from the power switches and greatly reduce their voltage stress. Meanwhile, its differential branch can provide required inductive current for ZVS, which simplifies coil tuning because no more residual inductance is needed in the load branch. Such structure also helps with an intrinsic constant voltage output, because the total voltage drop on the coil and its series tuning branch is always zero at the operating frequency.

Despite all these advantages, one of the main challenges in all MHz power converters is that the circuit components need to be precisely tuned to their designed values to achieve full advantages. In practical implementations, effects of parasitics and manufacturing tolerances may significantly deteriorate performance parameters, such as efficiency and load-independent constant output features. Converters can easily lose their soft-switching operations, which results in a much higher voltage and current stress on the components. On the other hand, the large number of design parameters and their coupling effects to circuit performance make it hard to find the detuned parameter(s). In practice, it is almost impossible to tune the circuit to its theoretically designed working point only through simulations or trail-and-error based approaches. Unfortunately, these detuning effects and corresponding parameter design methods are not discussed in the literature, and converters often work at sub-optimal operation points due to afore-mentioned practical disparities. Therefore, it is important to investigate parameter detuning effects and find a systematic way to tune the system to the desired working point. This paper investigates parameter detuning effects of class Φ_2 converters, and introduces a systematic approach for tuning the system to the proper working condition.

This paper is organized as follows. The working mode analysis of the selected class Φ_2 inverter is presented in Section II, the parameter tuning is presented in Section III, followed by experimental results.

Working mode analysis

The class Φ_2 inverter circuit structure is shown in Fig. 1. The definition of the components, the branch currents, and node voltages are depicted in Fig. 1. Considering the $180°$ phase shift between the left and right legs in push-pull operations, the current and voltage components can be divided into differential and common modes:

$$i_{\mathrm{odd}}\left(\omega_s t\right) = i_{\mathrm{odd1}} + i_{\mathrm{odd2}} = \left(i_{\mathrm{L1a}}\left(\omega_s t\right) - i_{\mathrm{L1b}}\left(\omega_s t\right)\right) + \left(i_{\mathrm{L2a}}\left(\omega_s t\right) - i_{\mathrm{L2b}}\left(\omega_s t\right)\right) \tag{1}$$

$$i_{\mathrm{diff}}\left(\omega_s t\right) = i_{\mathrm{o}}\left(\omega_s t\right) + i_{\mathrm{odd}}\left(\omega_s t\right) \tag{2}$$

$$v_{\mathrm{o}}\left(\omega_s t\right) = v_{\mathrm{ds1}}\left(\omega_s t\right) - v_{\mathrm{ds2}}\left(\omega_s t\right) \tag{3}$$

$$i_{\mathrm{even}}\left(\omega_s t\right) = \frac{i_{\mathrm{L2a}}\left(\omega_s t\right) + i_{\mathrm{L2b}}\left(\omega_s t\right)}{2} \tag{4}$$

$$I_{\mathrm{DC}} = \frac{i_{\mathrm{L1a}}\left(\omega_s t\right) + i_{\mathrm{L1b}}\left(\omega_s t\right)}{2} \tag{5}$$

Based on these definitions, differential currents i_{diff}, i_{odd} and i_{o} circulate around push and pull legs, which results in a $180°$ phase difference between voltage components on two legs. i_{odd} represents the inductive part of the total differential current i_{diff}, while i_{o} represents the resistive part with the reference to the output voltage v_{o}. In comparison, the common-mode current components I_{DC} and i_{even} flow through both legs in parallel, and the voltage components at the same position on left or right legs have also the same phase. Note that the class Φ_2 rectifier can be realized by replacing the AC output and DC source by an AC source and DC load resistance, respectively. The switching components for a rectifier can be either diodes (passive rectifiers) or FETs (active rectifiers). Therefore, the analysis and the experimental work presented in this work are also valid for the class Φ_2 rectifier.

The main working waveforms for inverters are shown in Fig. 1. Four working modes (as depicted in Fig. 1(b) and (c)) are contained in one full switching cycle as detailed in the following.

- **Mode 1** between $[\mathbf{0}, \theta_s]$ starts when switch Q_1 is turned off. Both switches Q_1 and Q_2 remain OFF during this period, the circulating current i_{diff} therefore charges C_{1a} and discharges C_{1b}, which contributes to the variation in differential voltage, i.e. makes the output AC voltage v_{o} to go from negative to positive. At the end of mode 1, $i_{\mathrm{C1b}} = 0$, v_{ds2} is discharged to zero by i_{C1b} with a zero derivative which prepares Q_2 for turning on with both ZVS and ZVDS in the next mode.
- **Mode 2** lasts between $(\theta_s, \pi]$. Q_2 is ON in this mode, which clamps v_{ds2} to zero. The output voltage v_{o} equals to v_{ds1}. Switch Q_2 is turned OFF at the end of this mode, which is at half of the switching period.
- **Mode 3** between $(\pi, \pi + \theta_s]$ and **Mode 4** between $(\pi + \theta_s, \mathbf{2\pi}]$ work in a similar way as Modes 1 and 2. During Mode 3, both switches are OFF, i_{diff} charges C_{1b} and discharges C_{1a} to prepare Q_1-ON in Mode 4, v_{o} falls from positive to negative. At $(\pi + \theta_s)$, Q_1 turns ON with ZVS and ZVDS, v_{ds1} is clamped to zero, and v_{o} equals to $(-v_{\mathrm{ds2}})$ during Mode 4.

The components L_{2a}, L_{2b}, and $2C_2$ constitute a T-network. From the definition in (1), in the differential-mode equivalent circuit, the series-connected horizontal branches, L_{2a} and L_{2b}, form an inductive branch between the push and pull legs to provide a path for the differential current i_{odd2}. In the common-mode, $L_{2a} - C_2$ and $L_{2b} - C_2$ form two separate branches in parallel with Q_1 and Q_2, respectively. By making L_2 and C_2 resonate at $2f_s$, these two branches provide low-impedance paths for i_{even}, which help to filter out the second harmonic component from v_{ds} and greatly reduce voltage stress.

$2Z_{eq}$ represents the equivalent load impedance reflected from the receiver side of the wireless power transfer (WPT) network, for a series-series compensation network [7], it can be calculated as

$$2Z_{eq} = \frac{\omega_s^2 M^2}{Z_{rec}} \tag{6}$$

where M is the mutual inductance between WPT coils, and Z_{rec} is the input impedance of the rectifier. To ensure the highest possible power transfer efficiency and load-independent characteristics, the impedance Z_{rec} is usually tuned with only a resistive part, therefore, $2Z_{eq}$ can be simplified as $2R_{eq}$ in Fig. 1. The output branch is assumed to have a high quality factor, which ensures approximately sinusoidal AC current, i_o, as the load branch acts like a high-Q resonant filter.

As defined in (3), v_o only exists in the differential mode equivalent circuit ($n\omega_s$, $n = 1,3,5,...$), where $L_{2a,b}$ and $L_{1a,b}$ composed two inductive branches in parallel with the load branch. The currents flowing through $L_{2a,b}$ and $L_{1a,b}$ contribute together to the total inductive current i_{odd} which is $90°$ lagging from i_o at fundamental frequency. Considering only the dominant harmonics $n = 1,3$, the total differential current can be written as

$$i_o(\omega_s t) = I_o \sin(\omega_s t + \phi_1), \tag{7}$$

$$i_{odd}(\omega_s t) = -I_{L,1} \cos(\omega_s t + \phi_1) - I_{L,3} \cos(3\omega_s t + \phi_3), \tag{8}$$

$$i_{diff}(\omega_s t) = i_o(\omega_s t) + i_{odd}(\omega_s t) = I_{diff} \sin(\omega_s t + \theta) - I_{L,3} \cos(3\omega_s t + \phi_3), \tag{9}$$

where I_o and ϕ_1 are the amplitude and phase of the load current. $I_{L,1}$ and $I_{L,3}$ represent the amplitudes of inductive current i_{odd} at the fundamental and third harmonic frequencies, respectively. $\frac{\phi_3}{3}$ provides the phase angle at $3\omega_s$. I_{diff} and β are the amplitude and phase of i_{diff}. All these parameters are shown in

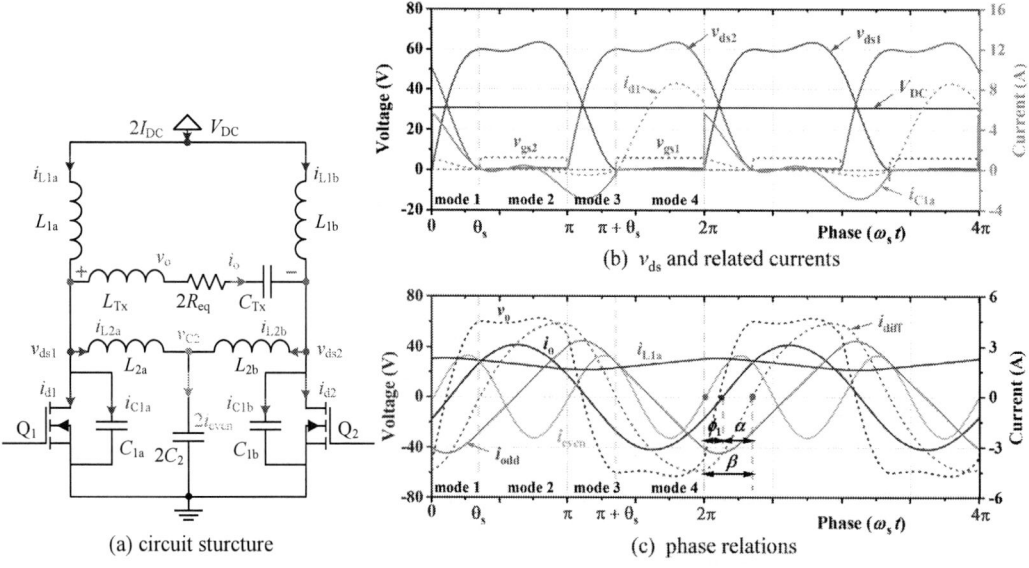

Fig. 1: Class Φ_2 push-pull inverter (a) circuit structure, (b) v_{ds} and its related current waveform, (c) main current waveform and phase relations.

Fig. 1(c), and they are related as

$$\beta = \phi_1 + \alpha, \tag{10}$$

$$\alpha = \arctan\left(\frac{I_{L,1}}{I_o}\right), \tag{11}$$

$$I_{\text{diff}} = \sqrt{I_o^2 + I_{L,1}^2}. \tag{12}$$

According to the working principles and design specifications, all the component values of the push-pull class Φ_2 circuit can be obtained through the design equations given in [6].

Parameter tuning due to parasitic effects

Even though a set of design parameters can be theoretically calculated, the manufacturing tolerances and parasitic effects cause inaccuracy of the component values and shift the circuit away from the optimal working point (with ZVS and ZVDS achieved at the nominal power). Therefore, actual implementations based on commercially available components should be properly tuned to the optimal working point at its nominal working status with full load and the specified mutual inductance.

Parameter detuning

Parameter detuning effects are analyzed based on a practically achievable inductor value L_2, when considering inaccuracies of C_1, C_2, Z_{eq} (or M) and R_{Load}. System power can be further tuned by adjusting V_{DC}. These different practical imperfections will have different effects on either the waveform or the system power level. In Fig. 2 and Fig. 3, v_{ds1}, i_{C1a} (or v_{in}, i_{in}) waveforms as well as the DC output power are given according to $\pm 20\%$ inaccuracies in the above parameters.

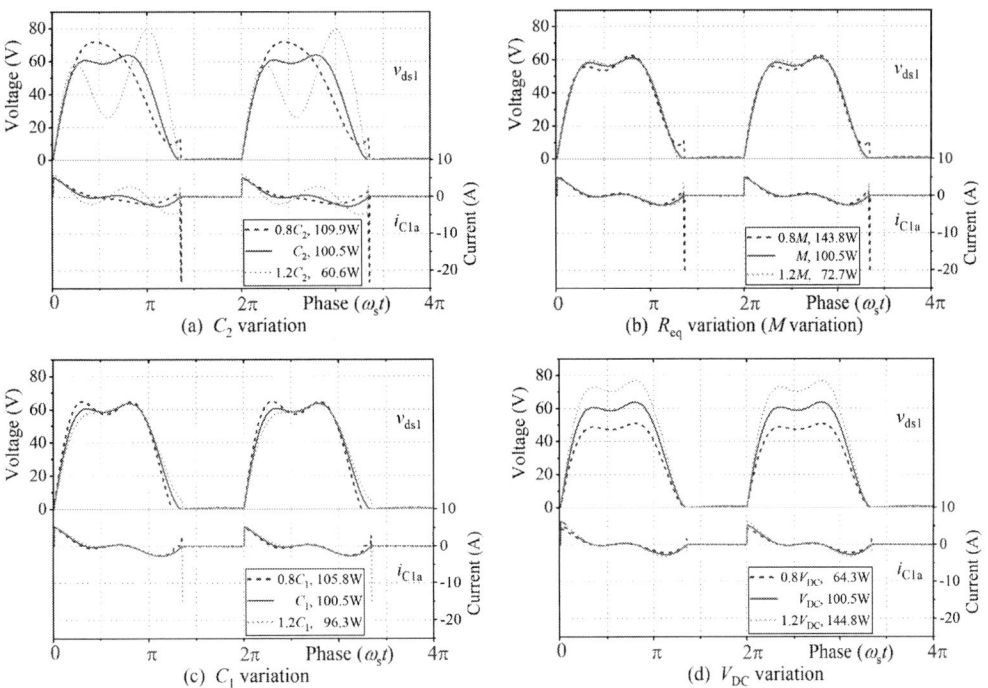

Fig. 2: Main waveform of a inverter with (a) C_2 variation, (b) R_{eq} variation (brought by M inaccuracies), (c) C_1 variation, (d) V_{DC} variation.

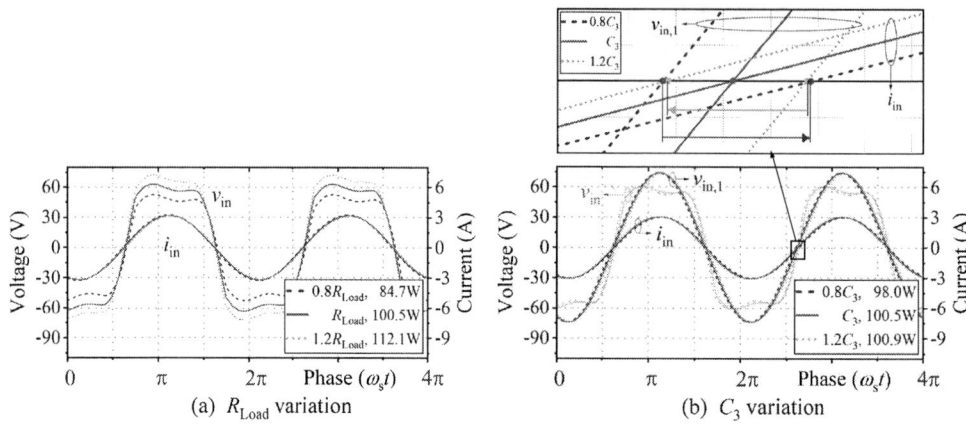

Fig. 3: Main waveform of a rectifier with (a) R_{Load} variation, (b) C_3 variation.

Parameters related to T-network resonance L_2, C_2, L_4, C_4

If each component value has a tolerance of $\pm 20\%$, the inaccuracy of the T-network resonance will have the most significant effect on both v_{ds} waveform and the power level, as seen in Fig. 2. Since the T-networks work in a same way at both inverter and rectifier sides, here the T-network in the inverter (with C_2 and L_2) is taken as an example to analyse parameter detuning effects. When the resonance frequency of the T-network is different from $2f_{\text{s}}$, the amplitude I_2 and phase ϕ_2 of i_{even} shift away from their designed values, which greatly changes the shape of i_{C1a} as well as v_{ds}, according to

$$i_{\text{C1a}}(\omega_{\text{s}}t) = I_{\text{DC}} - i_{\text{even}}(\omega_{\text{s}}t) - i_{\text{diff}}(\omega_{\text{s}}t), \quad \omega_{\text{s}}t \in [0, \pi + \theta_{\text{s}}] \tag{13}$$

$$v_{\text{ds1}}(\omega_{\text{s}}t) = \frac{1}{\omega_{\text{s}}C_1} \int i_{\text{C1a}}(\omega_{\text{s}}t)\,\mathrm{d}(\omega_{\text{s}}t). \tag{14}$$

In addition, ZVS and ZVDS conditions are not met because the instantaneous value of current i_{C1a} becomes non-zero at the switching moment (refer to i_{C1a} waveform in Fig. 2(a)). Detuning of the resonance frequency of T-network also creates a negative effect on voltage stress.

Parameters related to soft switching operations Z_{eq}, R_{Load}, M, C_1, C_3

Detuning of Z_{eq} leads to similar effects as detuning of R_{eq}, because it brings some additional reactance to the load branch and causes detuning of the load resistance along its whole variation range. Considering (6) and comparing to the inaccuracy on Z_{rec}, detuning of M brings higher inaccuracy on the equivalent load R_{eq} as its effect is squared in the equation. Both R_{eq} and C_1 detuning result in v_{ds} with a similar shape but lost soft-switching characteristics in the inverter. Although the variation in v_{ds} is small at the switching moment, it can lead to high spike current in the circuit which degrade converter efficiency and component lifetime. However, by comparing Fig. 2 (b) and (c), C_1 inaccuracies have little effect on the output power P_{o}, its negative effect is limited to soft-switching. On the other hand, R_{eq} variation changes the system power significantly, but it can be easily identified through the DC output voltage V_{Load} or the inverter output current i_{o}.

Similarly for diode rectifiers, the load resistance R_{Load} affects the system power as well as the output DC voltage, while the capacitance in parallel with the switching components (i.e. C_3 on the rectifier side) mainly affects the rectifier input impedance Z_{rec}, as shown in Fig. 3. Since the diodes have naturally ZVDS turning-off, an unsuitable C_3 will result in a reactance in Z_{rec} which can be observed easily through the phase difference between the rectifier input voltage and current. By observing the fundamental voltage component $v_{\text{in},1}$, the zoomed curves in Fig. 3 (b) indicate the phases of input impedance Z_{rec} based on $\pm 20\%$ C_3 inaccuracies. As seen in the figure, only the properly tuned C_3 can provide a purely resistive Z_{rec}, which is important for high WPT efficiency and load independent operations.

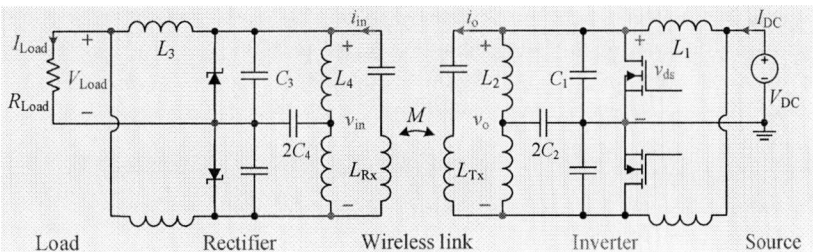

Fig. 4: Schematic of the WPT system.

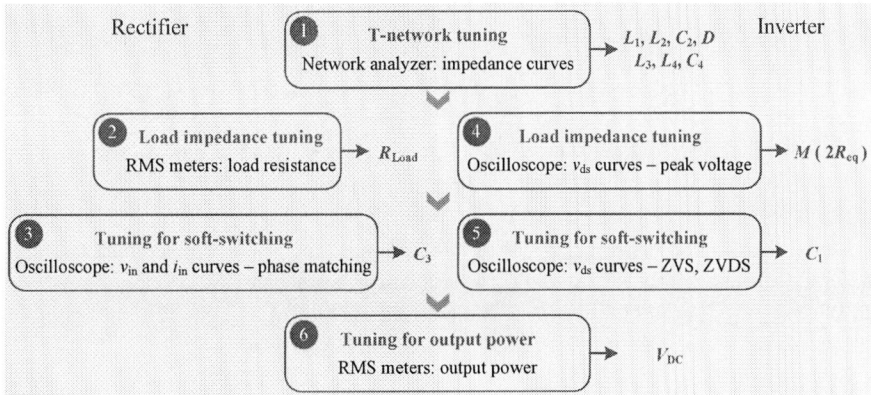

Fig. 5: System tuning steps.

All the inaccuracies in $L_2, L_4, C_{1\sim4}$ come from manufacturing tolerances, while R_{eq} inaccuracy can be a combination of multiple factors, as given in (6). Rectifier-side inaccuracies can bring in reactance for Z_{rec} as well as Z_{eq}, both Z_{rec} and Z_{eq} are not so straightforward to measure during the test. The mutual inductance is also hard to measure after implementation. Considering practical aspects, the system needs to be built for its optimal working status with the full load and the specified mutual inductance as in the design case. But in practical tests, it can be hard to find the specified M and R_{Load}, which even makes the tuning process targeting at a sub-optimal point.

Finally, variations of input DC voltage V_{DC} do not affect soft-switching characteristics. v_{ds} shape, ZVS and ZVDS are maintained while changing the input voltage value. Thus, V_{DC} can be used to adjust the output power based on the load requirement.

Parameter tuning steps

As analyzed above, parameters which have most complex effects on converter operations should be tuned firstly. In addition, similarly to the design procedure, the tuning direction should be from the load to the supply to ensure the required power at the specified load of the whole system. Therefore, based on the system schematic in Fig. 4, it is reasonable to make parameter tuning in steps shown in Fig. 5. T-networks are tuned first, as they have effects on the waveform shapes, power level, and soft-switching. The load impedances are tuned next with effects on the last two components. The next tuning step is about the capacitance in parallel with the switching components, which have effects mostly on soft switching, with negligible effects on the power level. Finally, the system power can be tuned separately by adjusting the input DC supply voltage. The tuning steps for the entire system can be explained as below:

- In **Step 1**, the T-network parameters for both converters are first tuned to resonance at $2f_s$ by adjusting the value of C_2, C_4 using a network analyzer. The designed duty cycle can also be properly adjusted according to the measured inductance value following the design equations in [6].
- In **Step 2**, considering the rectifier circuit, the DC resistance R_{Load} can be easily calculated and then adjusted to their specified full load by measuring the output DC voltage and current as $\frac{V_{Load}}{I_{Load}}$.
- As shown in Fig. 3, the passive class Φ_2 rectifiers have naturally ZVDS-off and ZVS-on because

of the diode operations. Therefore, in **Step 3**, a suitable capacitance C_3 that contributes to the required purely resistive rectifier impedance Z_{rec} can be found based on its related voltage and current waveforms. By tuning C_3 to its optimal value, the fundamental component of input voltage and current of the rectifier $v_{in,1}$ and i_{in} will become in phase. On the other hand, the capacitance in an active rectifier with controllable FETs can be tuned by observing v_{ds} waveforms until they show ZVDS operations.

- After finishing the tuning process of the rectifier, the reactance in its input impedance Z_{rec} is tuned to zero. The remained resistance is the value that contributes to the pre-defined load resistance in (6) for the inverter design. Thus, the inverter load impedance tuning can be simplified to the tuning of mutual inductance M as in **Step 4**. From the current relation of a series-series compensation network [8], the inverter and rectifier ac currents and voltages are related as

$$\frac{i_o}{i_{in}} = \frac{v_{in}}{v_o} = \frac{Z_{rec}}{\omega_s M}. \tag{15}$$

Therefore, with the pre-defined M and Z_{rec} from rectifier parameter design, the ratio between the current (or voltage) at the inverter and rectifier sides is a known value. The required M is reached by tuning the v_{ds} or ac current waveforms on both sides to the given ratio.

- Next, soft-switching on the inverter side is affected only by parallel capacitance C_1. In **Step 5**, the value of C_1 is tuned by observing the v_{ds} waveforms to ensure the ZVDS operations.
- Finally at **Step 6**, desired output power can be easily achieved by adjusting the input DC voltage, as the variation in V_{DC} does not affect soft-switching characteristics.

Experimental verification

A prototype WPT system has been set up with a class Φ_2 inverter, a wireless link, and a class Φ_2 rectifier, as shown in Fig. 6. The corresponding structure is shown in Fig. 4 and specifications are given as Table I.

Table I: Design specifications and component values

Inverter			Rectifier	
Input Voltage, V_{DC}	30	V	Power, P_{Load}	100 W
Duty cycle, D	31.4 %		Load, R_{Load}	7.5 Ω
Input inductor, L_1	2.7	μH	Output inductor, L_3	2.7 μH
T-network L_2	278	nH	T-network L_4	278 nH
T-network $2C_2$	992	pF	T-network $2C_4$	992 pF
External capacitor, C_1	1	nF	External capacitor, C_3	1 nF

The initial waveforms in Fig. 7(a) lose ZVDS switching and shows lower power and v_{ds} peak voltage

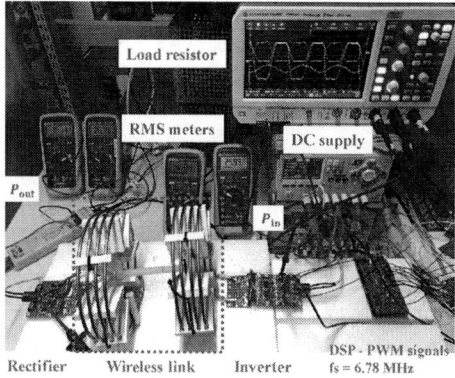

Fig. 6: Experimental system setup.

than designed. Through the tuning steps outlined in Fig. 5, the system is tuned to its designed operation point as in Fig. 7(b). The final v_{ds} and V_{Load} waveforms show that after tuning both ZVS and ZVDS are fully achieved at the designed output power. The achieved ZVDS operations help to reduce fluctuation in v_{ds} waveforms during its ON-period. The system reaches 82.3% efficiency at 100 W output power at 90 mm transfer distance.

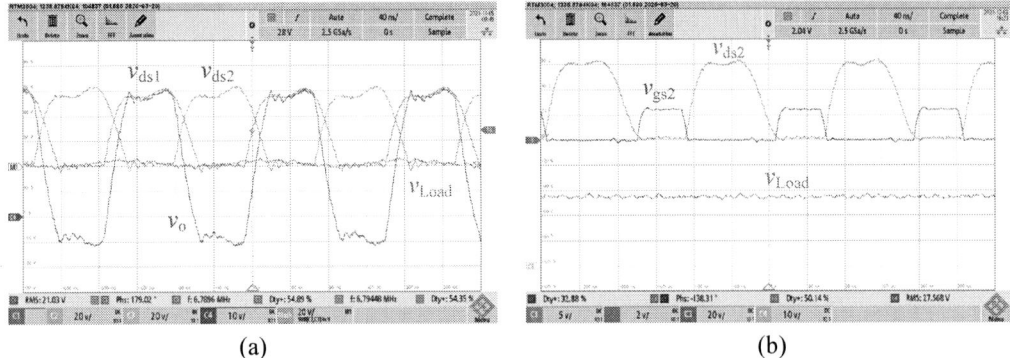

(a)   (b)

Fig. 7: v_{ds} and V_{Load} waveform. (a) Before and (b) after parameter tuning.

Conclusion

This paper analyzed parameter detuning effects on class Φ_2 push-pull converters. Detailed steps of the proposed tuning method are given based on decoupled features from experimental observation. By implementing the proposed tuning method, the system can be tuned to the designed operation point with both ZVS and ZVDS switching. An experimental realization demonstrated around 100 W output power regardless of existing parasitic effects, with the end-to-end efficiency of 82.3%.

References

[1] L. Gu, G. Zulauf, Z. Zhang, S. Chakraborty and J. Rivas-Davila: Push–Pull Class Φ_2 RF Power Amplifier, IEEE Transactions on Power Electronics vol 35 no 10, pp. 10515-10531

[2] S. Aldhaher, D. C. Yates and P. D. Mitcheson: Load-Independent Class E/EF Inverters and Rectifiers for MHz-Switching Applications, IEEE Transactions on Power Electronics Vol 33 no 10, pp. 8270-8287

[3] G. Kkelis, D. C. Yates and P. D. Mitcheson: Class-E Half-Wave Zero dv/dt Rectifiers for Inductive Power Transfer, IEEE Transactions on Power Electronics vol 32 no 11, pp. 8322-8337

[4] X. Zou, Z. Zhang, Z. Dong, Y. Zhou, X. Ren and Q. Chen: A 10-MHz eGaN FETs based isolated class-$\Phi2$ DCX, 2016 IEEE Applied Power Electronics Conference and Exposition (APEC), pp. 2518-2524

[5] Z. Kaczmarczyk: High-Efficiency Class E, EF_2, and E/F_3 Inverters, IEEE Transactions on Industrial Electronics vol 53 no 5, pp. 1584-1593

[6] H. Tebianian, Y. Salami, B. Jeyasurya and J. E. Quaicoe: A 13.56-MHz Full-Bridge Class-D ZVS Inverter With Dynamic Dead-Time Control for Wireless Power Transfer Systems, IEEE Transactions on Industrial Electronics vol 67 no 2, pp. 1487-1497

[7] S. Y. R. Hui, W. Zhong and C. K. Lee: A Critical Review of Recent Progress in Mid-Range Wireless Power Transfer, IEEE Transactions on Power Electronics vol 29, no 9, pp. 4500-4511

[8] Chwei-Sen Wang, O. H. Stielau and G. A. Covic: Design considerations for a contactless electric vehicle battery charger, IEEE Transactions on Industrial Electronics vol 52, no 5, pp. 1308-1314

Inductor Design Optimization Using FEA Supervised Machine Learning

D. Cajander[1,3], I. Viarouge[2], P. Viarouge[3], D. Aguglia[3,4]

[1]ENERGY Inst., Haute Ecole d'Ingénieurie et d'Architecture Fribourg, HES-SO University of
Applied Sciences and Arts Western Switzerland, Boulevard de Pérolles 80
CH-1700 Fribourg, Switzerland
Tel. : +41 26 429 65 57
Email: david.cajander@hefr.ch
https://www.heia-fr.ch

[2] ELECTROTECHNOLOGIES Selem Inc, 2610, Rue Gérard-Lajoie, GIP 3G1, Quebec (QC),
Canada
Email: selem@oricom.ca

[3]LEEPCI Lab., Electrical and Computer Eng. Dept., Laval University, GIK 7P4, Quebec
(QC), Canada
Email: philippe.viarouge@gel.ulaval.ca

[4]CERN - European Organization for Nuclear Research, Accelerator Systems Dept., Electric
Power Converter Group
CH- 1211 Geneva 23, Switzerland
Email: davide.aguglia@cern.ch

Acknowledgements

DC thanks Beat Wolf[5] for his collaboration on data analysis and machine learning ([5]Institute of Complex Systems, Haute Ecole d'Ingénieurie et d'Architecture Fribourg, HES-SO University of Applied Sciences and Arts Western Switzerland, Boulevard de Pérolles 80, CH-1700 Fribourg, Switzerland).

Keywords

«Magnetic device», «Machine learning», «Deep Learning», «Design optimization», «Finite-element analysis», «Data analysis», «Neural network», «Passive component»

Abstract

An optimal inductor design methodology using dimensioning models derived from Finite Element Analysis (FEA) supervised Artificial Neural Networks (ANN) is presented. The efficiency of such trained ANN dimensioning models in terms of compromise between precision and computing time is demonstrated for the cylindrical inductor topology with air and magnetic material core including saturation.

Introduction

Inductors are widely used in all kinds of power converters like SVC, AFE, and Multilevel Converters for energy storage, AC & DC filtering and magnetic coupling functionalities. Such magnetic components are often specified by the power electronics engineer considering them as simple elements of an electrical circuit. However, their practical feasibility, size, thermal or electro-mechanical (e.g. short circuits) behaviors can easily be overlooked. In the more and more demanding quest for compact and economical solutions, these aspects shall be carefully considered since the early steps of the converter design. For each application, inductors are subject to several requirements, such as electrical insulation, thermal and mechanical aspects, inductance linearity over a specified current range, internal resistance, fault currents to withstand, and lifetime [2]. The inductor design is a multi-variable non-linear optimization problem where the performance objectives to minimize are the volume and/or the mass and the cost. Several constraints can also be defined depending on the requirements. Several different inductor topologies presented in Fig. 1 can be evaluated by the designer to make an optimal choice for each application. A versatile and generic design methodology is mandatory.

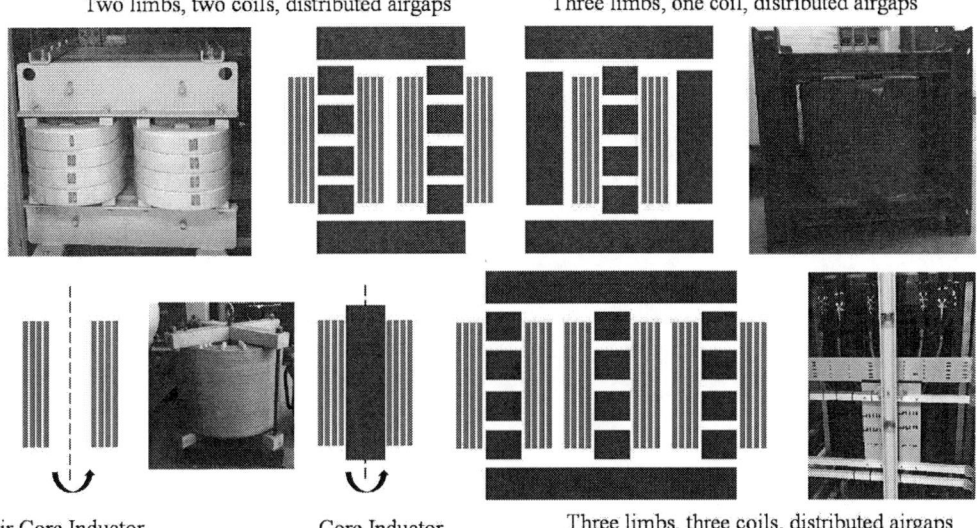

Fig. 1: Main inductor topologies used in Power Converters

Cylindrical Inductor topologies and application

The FEA supervised ANNs optimal design methodology presented in this paper has been applied to the following cylindrical inductor topologies: the cylindrical air core inductors and the cylindrical inductors with a magnetic core as shown in Fig.2. These kinds of inductors are used in several AC & DC filtering applications like the air core inductors used in SVC with thyristor-controlled reactor (TCR).

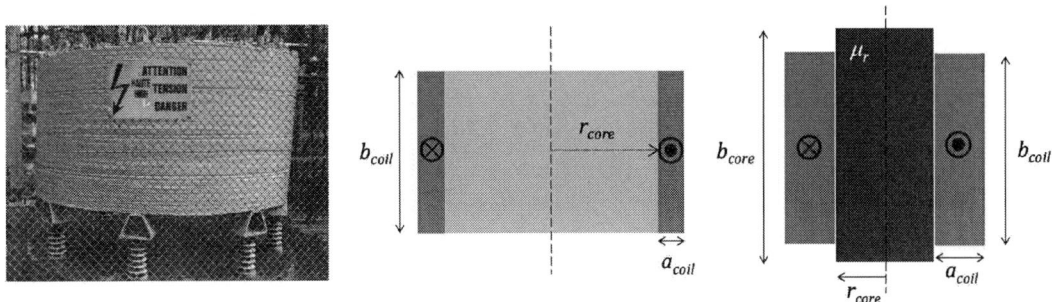

Fig. 2: Cylindrical air core inductor used in SVC with thyristor-controlled reactor (TCR) and Cylindrical inductor with magnetic material core

Cylindrical Inductor Design Optimization Using FEA

The inductor CAD environment is presented in Fig. 3. The optimal design methodology is using an inverse problem approach where a Non-Linear Optimization procedure (NLO) is associated to a device dimensioning model to minimize an objective function while respecting the specification constraints.

There are three different components in the dimensioning model: an electromagnetic model, a thermal model and a mechanical model.

In this modular CAD environment, the user can choose models with different levels of complexity according to a suitable compromise between accuracy and computing time. For example, in the electromagnetic dimensioning model, the inductance can be computed by a simplified analytical model [1] or a complex model based on 2D or 3D Finite Element Analysis (FEA) including non-linear B(H) characteristics of the core magnetic material.

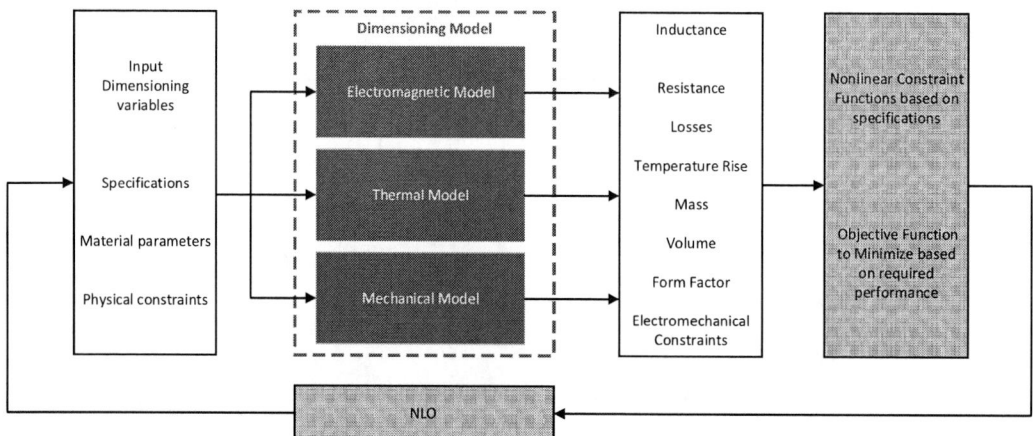

Fig. 3: Cylindrical inductor optimal design environment

Input dimensioning variables

There are 3 independent geometric dimensions for the air core inductor and 4 independent geometric dimensions for the magnetic core inductor as shown in Fig. 2.
By introduction of dimensionless form factors, the geometric variables can be naturally scaled to the coil inner radius r_{core} (m):

$$FF_{a_{coil}/r_{core}} = \frac{a_{coil}}{r_{core}} \qquad FF_{b_{coil}/r_{core}} = \frac{b_{coil}}{r_{core}} \qquad FF_{b_{core}/b_{coil}} = \frac{b_{core}}{b_{coil}} \qquad (1)$$

It has been shown that the use of dimensionless form factors is providing an efficient preliminary scaling to improve the NLO method efficiency.
There are 5 independent dimensioning variables for the optimal air core inductor design problem: 3 geometric dimensions, the number of turns N and the current usually imposed by the specifications.
There are 6 independent dimensioning variables for the optimal magnetic core inductor design problem: 4 geometric dimensions, the number of turns N and the current usually imposed by the specifications.

Objective function of the optimization method

Several objective functions can be used in the environment but the usual function to minimize during the cylindrical inductor design is the mass of the copper coil or the total mass of the core and the coil. The value of the objective function to minimize is computed from the input variables by use of the mechanical dimensioning model at each iteration of the optimization process.

Constraint functions of the optimization method

The constraint functions of the design optimization problem are directly derived from the main inductor specifications like the required inductance value, the peak and RMS current, the inductance linearity over a specified current range in the case of a saturable magnetic core inductor, the DC resistance, the quality factor in AC applications, the ambient temperature, the maximum temperature rise, the electrical isolation specs and the overall dimensions or space constraints. As shown in Figure 3, the values of the different constraint functions are computed from the input variables by use of the electromagnetic, thermal and mechanical dimensioning models at each iteration of the optimization process.

Electromagnetic dimensioning model

The main challenge in the dimensioning models used in the optimal design environment is the accuracy of the inductance computation for each set of the input variables. In order to meet the requirements with these inductor topologies, the inductance must be computed by a model based on Finite Element Analysis (FEA). According to their cylindrical symmetry, a 2D Finite Element Analysis (FEA) is preferred. With such a method, the non-linear characteristic B(H) of the core magnetic material can also be taken into consideration in the case of a saturable magnetic core inductor. Depending on the specified inductance linearity over the imposed current range and on the characteristics of the magnetic material in use, it is also possible in some applications to consider that the material relative permeability μ_r is constant over the specified current range.

In the case of the cylindrical air core inductor, the inductance to be computed by 2D FEA at each iteration of the NLO process is independent of the current and a function of 4 input variables only:

$$L = N^2 . L_1(r_{core}, FF_{a_{coil}/r_{core}}, FF_{b_{coil}/r_{core}}) \tag{2}$$

In this case, the specific or one-turn inductance ($N=1$) L_1 is a function of 3 variables only.

In the case of a cylindrical inductor with a magnetic material core of fixed linear relative permeability μ_r, the inductance to be computed by 2D FEA at each iteration of the NLO process is independent of the current and a function of 5 input variables:

$$L = N^2 . L_1(r_{core}, FF_{a_{coil}/r_{core}}, FF_{b_{coil}/r_{core}}, FF_{b_{core}/b_{coil}}) \tag{3}$$

In this case, the specific or one-turn inductance (N=1) L_1 is a function of 4 variables only.

In the case of the cylindrical inductor with a core made of saturable magnetic material with a non-linear characteristic B(H), the inductance to be computed by 2D FEA at each iteration of the NLO process is dependent of the current density and a function 6 input variables:

$$L = N^2 . L_1(r_{core}, FF_{a_{coil}/r_{core}}, FF_{b_{coil}/r_{core}}, FF_{b_{core}/b_{coil}}, J) \tag{4}$$

One can notice that for a fixed number of turns N, the product of the current density J (A/m^2) by the copper coil section imposed by the geometric dimensions is equal to the total coil magnetomotive force MMF. This MMF and the geometric dimensions are then fixing the magnetic field H distribution and consequently the core induction B distribution through the core according to the magnetic material characteristic $B(H)$.

In this case, the specific or one-turn inductance (N=1) L_1 is a function 5 variables only.

Non-Linear Optimization procedure (NLO)

The NLO procedure used in the cylindrical inductor optimal design environment presented in Fig. 3 is the Generalized Reduced Gradient (GRG) method [13].

Air core optimal design example using FEA

An application example of the optimal design methodology is illustrated in Fig. 4 in the case of a specified 10mH 100A air core inductor with 95°C temp rise. The objective function is the coil mass minimization. The convergence and precision characteristics of the NLO process are detailed in Table I. At each iteration of the NLO process, the mass, the losses, the resistance and the temperature rise are derived from the thermal and mechanical dimensioning models and the inductance is computed by 2D FEA. One can notice the number of iterations of the GRG method that is needed to converge towards the optimal solution. The inductance computation by the 2D FEA at each iteration is slowing the execution time.

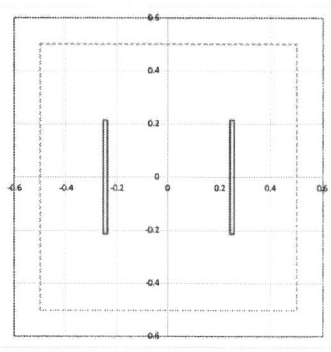 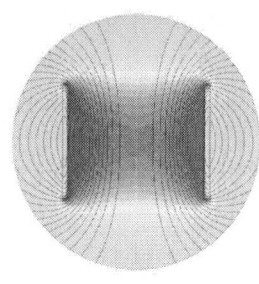

Fig. 4: Application example of FEA optimal design: 10mH 100A Cylindrical air core inductor with 95°C temp rise (r_{core}= .225m, $FF_{a_{coil}/r_{core}}$ = .0904 , $FF_{b_{coil}/r_{core}}$= 1.6463 , N =165)

Table I: Air core inductor optimization example using FEA

Air core inductor	Specifications			Optimal solution			NLO process	
Inductance (H)	Current (A)	Temp Rise (°C)	Copper Mass (kg)	Inductance (H)	Relative error to specs (%)	GRG Iterations number	Execution time (s)
1.00E-02	100	95	99.9	9.76E-03	2.4	1115	239

This example demonstrates that despite their better performance in terms of precision, the FEA techniques can become heavy and time consuming in an iterative optimization process. Hybrid methods like space mapping techniques combining coarse and fine models in the same optimization process have been proposed in the literature [3][4] to reach a feasible optimal solution with a limited number of finite element computations. This solution is validated by FEA with a suitable compromise between precision and computing time but its unicity and optimality are not guaranteed [5]. In this paper, another alternative is investigated to solve the preceding compromise by using the accurate FEA method to train Artificial Neural Networks (ANNs) dimensioning models and improve the computing and convergence efficiency of the NLO method [6] [7].

Supervised Artificial Neural Networks (ANNs) Dimensioning Models

Artificial Neural Networks (ANNs) are fast becoming a new viable approach to replace FEA based dimensioning models for optimal design, thanks to their gain in computing speed and simplicity. For instance, pre-trained ANN models are essentially smooth functions of tuned parameters, thus, the gradient is readily computed exactly compared to FEA methods. The ANN approach can been extended to saturated core inductors [8] [9] because it is possible to train and build a neural network to model nonlinear relationship by a suitable choice of the neural network architecture, the learning rate, the learning algorithm and the activation function.

In this paper, the ANNs or models are trained in a supervised setting and their creation is following the associated typical workflow described in Fig. 6. In the supervised setting, a model is trained on a dedicated database of labeled input/target pairs to minimize the error between its predictions computed from the inputs and the associated targets, in this case inductances [11].

Generation of inductance databases with FEA

Three databases presented in Table II and Table III have been generated from 2D FEA magnetostatic computation of specific inductances L_1 (N=1) using the FEA tool integrated in the inductor CAD environment of Fig. 3: Database 1 for cylindrical air core inductors, Database 2 for cylindrical inductors with a magnetic material core of fixed linear relative permeability μ_r =1000, Database 3 for cylindrical inductors with a magnetic core made of saturable material presenting a non-linear characteristic *B(H)*.

In Database 3, the range and distribution of the current density J is adjusted for each set of dimensions and form factors in order to operate the core magnetic material on its whole $B(H)$ nonlinear range (the product of J by the copper coil section imposed by the dimensions is equal to the total coil magnetomotive force that is fixing the core induction level B irrespective of the turn number). This approach is illustrated in Fig. 5 by the $L_1(J)$ and $B(J)$ characteristics of one inductor of Database 3 with a given set of dimensions and form factors.

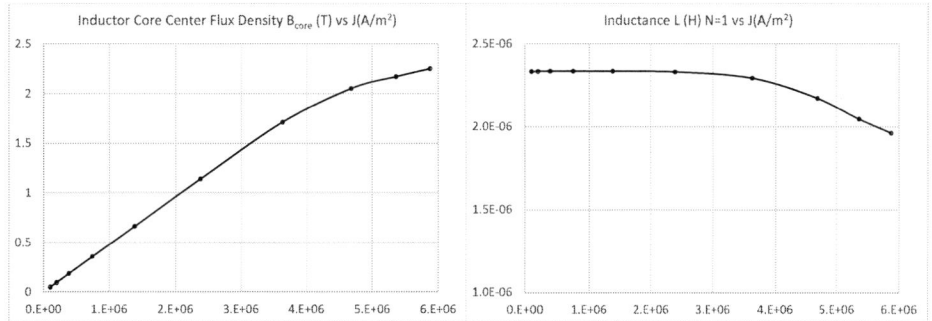

Fig. 5: Non-linear $L_1(J)$ and $B(J)$ characteristics of one specific inductance L_1 of Database 3 with a given set of dimensions & form factors (M19 core magnetic material used)

Table II: Inductance Databases for Supervised learning models

Inductor Database	Output: specific or one-turn inductance (N=1) computed by FEA	Number of inputs	Number of samples & Generation time
Database 1 Air core inductor	$L_1(r_{core}, FF_{a_{coil}/r_{core}}, FF_{b_{coil}/r_{core}})$	3	125000 8h 18 mn
Database 2 Magnetic core inductor with fixed permeability material μ_r=1000	$L_1(r_{core}, FF_{a_{coil}/r_{core}}, FF_{b_{coil}/r_{core}}, FF_{b_{core}/b_{coil}})$	4	104976 7h 23mn
Database 3 Saturable core inductor with M19 Laminated Silicon Steel	$L_1(r_{core}, FF_{a_{coil}/r_{core}}, FF_{b_{coil}/r_{core}}, FF_{b_{core}/b_{coil}}, J)$	5	190080 19h 33mn

Table III: Database variables mean and standard deviation

Database	r_{core} (m)		$FF_{a_{coil}/r_{core}}$		$FF_{b_{coil}/r_{core}}$	
	Mean	Std	Mean	Std	Mean	Std
Database 1	0.395	0.119	1.235	0.432	1.235	0.432
Database 2	0.400	0.111	1.208	0.432	1.208	0.432
Database 3	1.411	0.460	1.187	0.431	1.187	0.431

Database	$FF_{b_{core}/b_{coil}}$		J (A/m^2)		L_1 (H)	
	Mean	Std	Mean	Std	Mean	Std
Database 1	--	--	--	--	1.00E-06	3.50E-07
Database 2	0.761	0.403	not applicable	not applicable	1.50E-05	5.30E-06
Database 3	1.458	0.287	8.70E+05	9.40E+05	7.00E-06	2.00E-06

Workflow of ANN supervised training

The workflow for creating a model starts with the preprocessing of the database (see Tables II & III for the details of the database used). It is more efficient to train ANNs when the inputs and targets are all scaled to be in the interval [0, 1]. Thus, the inputs were standardized by computing their z-scores and a logarithm was applied to the target air core inductors and the magnetic core inductors with fixed permeability material of Database 1&2. The preprocessing for the database of saturable core inductors (Database 3) is slightly different: instead of applying a logarithm to the target inductances, a Z-standardization was applied similarly to the inputs.

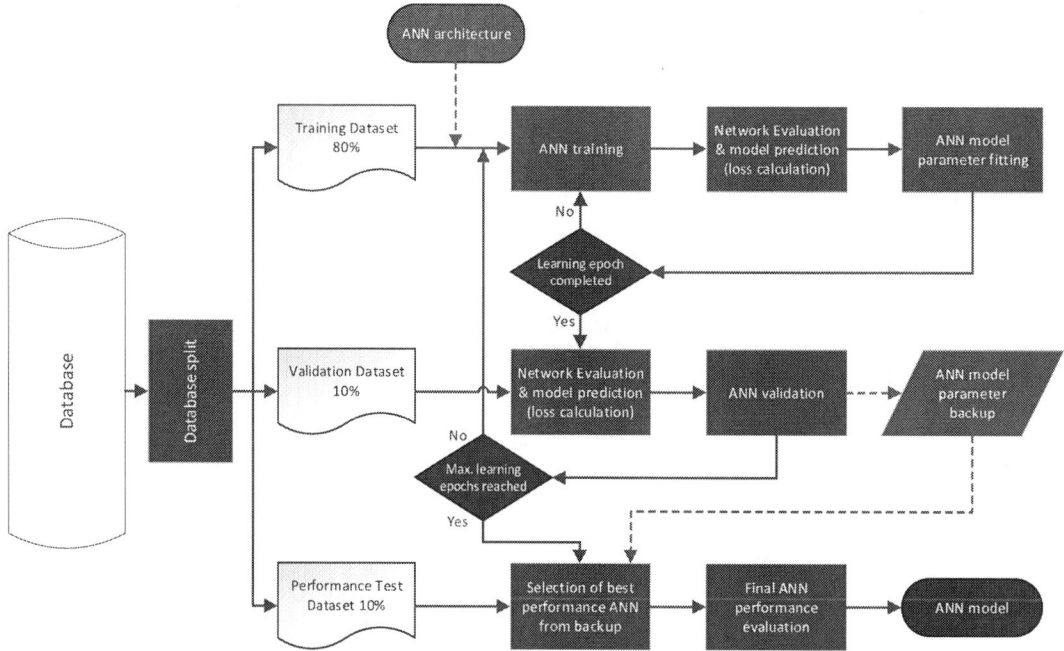

Fig. 6: Workflow of ANN supervised training

Before training, the database is shuffled and divided into 3 subsets: a training set (80%), a validation set (10%) and an evaluation set (10%).

During the training stage, the model predicts the inductance of all the training set inputs, then the error between the predictions and the targets is evaluated using a loss function, finally the parameters of the model are updated through a gradient descent algorithm relying on backpropagation to compute the gradient of the parameters [11]. One training iteration or epoch is completed after all the labeled pairs in the training set have been explored. In this paper, the mean squared error between prediction and target was used as the loss function. The first models were trained for 100 epochs and the third model for 500 epochs.

While training is underway, the performance of the model is evaluated using the validation set to identify overfitting on new data at a set interval of epochs. The loss is computed using the same loss function used on the training set. A backup of the model is created.

After the set number of epochs is reached, the models that were saved during validation are evaluated on the evaluation set composed of data unseen during training. The best performing model is then selected as the final model.

Parameters pertaining to model architecture such as number of layers, number of neurons per layer and the learning rate, must be specified before training and will be kept fixed thereafter. In this article, these hyper-parameters were optimized using the Optuna [12] framework and the machine learning was handled using the Pytorch [10] library.

ANN Architecture and Learning Optimization

In order to enhance the general performance of an ANN, it is important to determine the best network architecture and learning rate. Therefore, the ANN architecture optimization is done ahead of the training in a separate cycle of trials. The process starts with a random set of hyperparameters that are each comprised in predefined ranges set by the user (i.e. number of layers, number of hidden neurons and the learning rate). After a training cycle of 20 epochs the performance and the hyperparameters are stored and a new trial is started with a different set of hyperparameters. The Optuna framework determines the next hyperparameters to be tested based on the performance of the past recorded trials. This iterative process allows to refine the ANN architecture and learning rate according to the latest results. After a series of 100 trials, each made of 20 training epochs, the best performing ANN hyperparameters are stored to allow the training process to start directly with an optimized set of hyperparameters. This enhances the learning and forecasting abilities of the ANN model, while reducing the training effort for a given performance as well.

Description of the three Trained ANNs

The architecture of the three trained models and their performance on their test set are reported in table IV. Every model is a fully connected ANN whose activation function after every hidden layer is the Rectified Linear Unit (ReLu). The architecture of Model 1 is illustrated in Fig. 7.

Table IV: Test Set Performance of the Trained Models

	Database	Number of Inputs	Number of Hidden layers	Number of neurons per layer	Average Absolute error	Min Absolute error	Max Absolute error
Model 1 Air core inductor	**Database 1**	3	2	396	1.147e-08	1.085e-13	4.826e-07
Model 2 Magnetic core inductor μ_r=1000	**Database 2**	4	3	162	5.441e-09	1.054e-15	4.550e-08
Model 3 Saturable core inductor	**Database 3**	5	5	400	1.226e-07	3.183e-12	8.771e-07

All models are trained with a mini-batch gradient descent and the optimizer was Pytorch's Adam with an optimal learning rate according to the described learning optimization method. Model 1 and 2 were trained in 100 epochs while Model 3 was for 500 epochs. The reported models were the best ones found during training with the smallest validation loss.

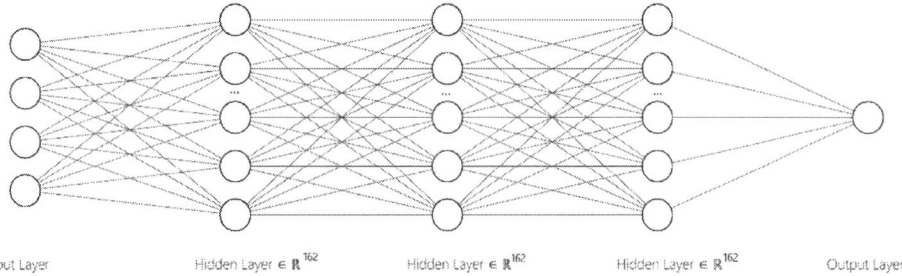

Fig. 7: Architecture of the fully-connected ANN Model 1

Cylindrical Inductor Design Optimization Using ANNs

The three ANN Models have been integrated in the inductor CAD environment of Fig. 3 to be used for the optimal design of the three kinds of cylindrical inductors: air core type (Model 1), magnetic core type with constant $\mu_r=1000$ (Model 2), saturable magnetic core type (Model 3). At each iteration of the NLO process, the mass, the losses, the resistance and the temperature rise are still derived from the thermal and mechanical dimensioning models but the inductance is no longer computed from 2D FEA but from one of the trained ANN model. Three specific examples are presented to validate the proposed optimal inductor design methodology using supervised ANNs and to compare its performance to the optimal inductor design methodology using the direct FEA approach that was presented in the first part of this paper. In order to precisely quantify the relative performances of both methods, the initial guess of the NLO process are identical.

Air core optimal design example

Table V presents the comparative analysis of both optimal design methodologies in the case of the same 10mH 100A air core inductor with 95°C temp rise of Table I and Figure 4. The same objective and constraint functions are used. The dimensions of both optimal solutions are very close $r_{core}= 0.239$m vs .225m , $FF_{a_{coil}/r_{core}}= 0.067$ vs .0904, $FF_{b_{coil}/r_{core}}= 1.7902$ vs 1.6463, $N=170$ vs 165. The comparative analysis of the optimal solutions found in terms of optimal mass, relative error, convergence and precision characteristics of the NLO process are detailed in Table V. The number of iterations of the GRG algorithm is 9% lower with the ANN Model 1. The computation time of the NLO process with the use of the ANN Model 1 is 66 times faster than with the direct use of FEA.

Table V: FEA vs ANN Optimization for an air core inductor

	Specifications					Optimal solution	
	Inductance (H)	Current (A)	Copper Mass (kg)	Inductance (H)	Relative error to specs (%)	Iterations number	Execution time (s)
Optimization with FEA	1.00E-02	100	99.9	9.76E-03	2.4	1115	239
Optimization with ANN Model 1	1.00E-02	100	96.3	9.99E-03	0.1	1020	3.6

Constant permeability magnetic core inductor optimal design example

Table VI and Fig. 8 present the comparative analysis of both optimal design methodologies in the case of a 10mH 100A magnetic core inductor with $\mu_r=1000$ and 95°C temp rise. The objective function is the total mass minimization. The comparative analysis of the optimal solutions found in terms of optimal mass, relative error, convergence and precision characteristics of the NLO process are detailed in Table VI. The number of iterations of the GRG algorithm is 23% lower with the ANN Model 2. The computation time of the NLO process with the use of the ANN Model 2 is 54 times faster than with the direct use of FEA.

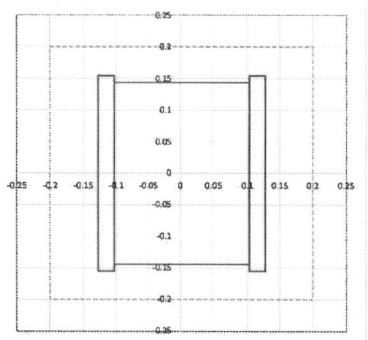 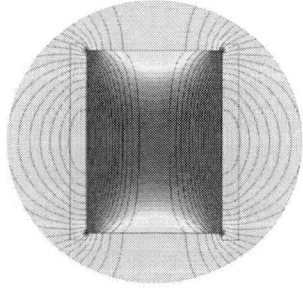

Fig. 8: Cylindrical inductor Magnetic core inductor with constant permeability (dimensions in m)

Table VI: FEA vs ANN Optimization for a magnetic core inductor with μr=1000

	Specifications					Optimal solution	
	Inductance (H)	Current (A)	Total Mass (kg)	Inductance (H)	Relative error to specs (%)	Iterations number	Execution time (s)
Optimization with FEA	1.00E-02	100	106.8	9.68E-03	3.2	1206	155
Optimization with ANN Model 2	1.00E-02	100	119.2	9.99E-03	0.1	930	2.85

Saturable magnetic core inductor optimal design example

Table VII and Fig. 9 present the comparative analysis of both optimal design methodologies in the case of a 44.8mH 100A saturable magnetic core made of M19 Laminated Silicon Steel inductor with 95°C temp rise. The objective function is the copper mass minimization in this example. The comparative analysis of the optimal solutions found in terms of optimal copper mass, relative error, convergence and precision characteristics of the NLO process are detailed in Table VII. The number of iterations of the GRG algorithm is 7% higher with the ANN Model 3. The computation time of the NLO process with the use of the ANN Model 3 is 214 times faster than with the direct use of FEA.

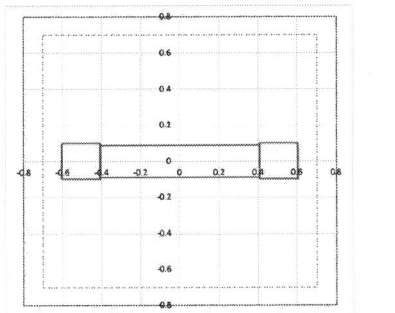 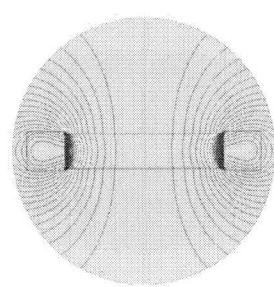

Fig. 9: Cylindrical inductor with a saturable magnetic core inductor including non-linear material B(H) characteristic (dimensions in m)

Table VII: FEA vs ANN Optimization for a saturable magnetic core inductor

	Specifications		Optimal solution			NLO process	
	Inductance (H)	Current (A)	Copper Mass (kg)	Inductance (H)	Relative error to specs (%)	Iterations number	Execution time (s)
Optimization with FEA	4.48E-02	100	925	4.62E-02	3.0	2496	1872
Optimization with ANN Model 3	4.48E-02	100	1092	4.51E-02	0.5	2676	8.76

One can notice in the three examples that the precision of the ANN models is highly acceptable for inductor design purpose. The number of GRG iterations is usually lower with the ANN model for a better relative precision. The ANN model inductance computation time is 54 to 66 times faster than in the FEA case without saturation. In the case of a saturable inductor nonlinear FEA become heavy and time consuming and the ANN model inductance computation time is 214 times faster. FEA supervised Artificial Neural Networks dimensioning models can replace the time consuming FEA method with the same accuracy in the optimal design environment and highly improve the computing and convergence efficiency of the NLO method.

Conclusion

It has been clearly demonstrated on this example of inductance topology that the use of FEA supervised trained ANNs dimensioning models is a very efficient alternative in terms of compromise between precision and computing time in an optimal inductor design environment. The systematic use of dimensionless form factors as input variables of the supervised ANN model also provides inherent preliminary scaling that improve the convergence of the training process. This methodology can be efficiently extended to the optimal design of the other inductor topologies presented in Fig. 1 and to other parts of the dimensioning model used in the environment like the computation of electro-magnetic forces or the heavy FEA computations of HF copper losses in AC applications.

References

[1] H. A. Wheeler: Formulas for the Skin Effect, Proceedings of the IRE, vol. 30, no. 9, pp. 412-424, Sept. 1942, doi: 10.1109/JRPROC.1942.232015.

[2] Sudhoff S. D.: Introduction to Inductor Design, Power Magnetic Devices: A Multi-Objective Design Approach, Wiley- IEEE Book Chapter, 2022,

[3] Choi H.-S. et al.: A new design technique of magnetic systems using space mapping algorithm, IEEE Transactions on Magnetics, Vol 37, no 5, p 3627 3630

[4] Leal-Romo F. et al.: Design optimization of a planar spiral inductor using space mapping, IEEE 26th Conference on Electrical Performance of Electronic Packaging and Systems (EPEPS), 2017, 3 p.

[5] S. Candolfi, P. Viarouge, D. Aguglia and J. Cros: Hybrid design optimization of high voltage pulse transformers for klystron modulators 2014 IEEE International Power Modulator and High Voltage Conference (IPMHVC), 2014, pp. 197-200, doi: 10.1109/IPMHVC.2014.7287242

[6] Guillod T. et al.: Artificial Neural Network (ANN) Based Fast and Accurate Inductor Modeling and Design, IEEE open journal of Power Electronics, Vol. 1, pp. 284-299

[7] Arnoux P.-H. et al.: Modeling Finite-Element Constraint to Run an Electrical Machine Design Optimization Using Machine Learning, IEEE Transactions on Magnetics, Vol. 51, no 3, 4 p.

[8] Burrascano P. et al.: Neural Models of Ferrite Inductors Non-Linear Behavior, IEEE International Symposium on Circuits and Systems (ISCAS), 2019, 5 p.

[9] Cincotti S. et al.: A neural network model of parametric nonlinear hysteretic inductors, IEEE Transactions on Magnetics, Vol 34, no 5, 3040-3043

[10] "Pytorch," PyTorch. [Online]. Available: https://pytorch.org/. [Accessed: 08-Dec-2021].

[11] Bengio, Yoshua; LeCun, Yann; Hinton, Geoffrey (2015). "Deep Learning". Nature. 521 (7553): 436–444. Bibcode:2015Natur.521..436L. doi:10.1038/nature14539. PMID 26017442. S2CID 3074096

[12] "Optuna" [Online]. Available: https://optuna.org/. [Accessed: 08-Dec-2021].

[13] Lasdon L.S. et al.: Design and Testing of a Generalized Reduced Gradient Code for Nonlinear Programming, ACM Transactions on Mathematical SoftwareVolume 4 Issue 1March 1978 pp 34-50. doi:10.1145/355769.355773

Enabling large-scaled MMC EMT-RMS co-simulation by data exchange in the loop (DXiL)

Xiong Xiao, Soham Choudhury, Martin Coumont, Jutta Hanson
Department of Electrical Power Supply with Integration of Renewable Energy (E5)
Technical University of Darmstadt
Landgraf-Georg-Straße 4, Darmstadt, Germany
Phone: +49 (0) 6151/16-24657
Fax: +49 (0) 6151/16-24665
Email: xiong.xiao@e5.tu-darmstadt.de
URL: https://www.e5.tu-darmstadt.de

Keywords

≪HVDC≫, ≪co-simulation≫, ≪MATLAB/Simulink≫, ≪electromagnetic-transients≫, ≪phasor simulation≫.

Abstract

This paper presents a comprehensive co-simulation method of electro-magnetic (EMT) and electrome-chanical (RMS) dynamic models in MATLAB/Simulink to investigate transient stability in large scale hybrid AC/DC grids. The modeling of modular multilevel converter (MMC) as well as co-simulation of EMT-RMS programs in MATLAB/Simulink is first described. For further reducing the unnecessary data exchange time between EMT and RMS programs, data exchange in the loop interaction method is introduced. With this method, the frequent data exchange through MATLAB workspace is avoided, and the co-simulation is thus accelerated significantly.

Introduction

Electro-magnetic transient (EMT) simulation and electro-mechanical transient (RMS) simulation are two main modeling and simulation tools for dynamic studies of power systems. EMT models use differential equations for describing the electro-magnetic transient processes and are suitable for simulating a relative small power grid. However, they are not suitable for studying large-scale systems because of huge calculation expense. In the contrast, RMS models, also called positive sequence models or transient simulation models, are implemented in the phasor-frame without considering any harmonics. They are suitable for investigating large-scale power system dynamics, such as rotor-angle stability of large AC grids. e As more and more HVDC grids with modular multilevel converters (MMC) are coming into application in the future transmission grids, it is necessary to investigate dynamic stability for large-scale hybrid AC/DC transmission grids, which calls for suitable dynamic models not only for AC grids, but also for DC grids and MMC. In standard power system simulation softwares such as MATLAB/Simulink and DIgSILENT/PowerFactory, there are already well developed RMS models for AC grids, and they have been proven accurately and rapidly since years. Even though MMC RMS models have been developed in recent years, such as [1]-[6], they are still not able to replace EMT models in some cases, such as long-term dc-voltage drops [2].

Co-simulation of EMT-type programs and RMS-type programs are another opportunity for simulating hybrid AC/DC power systems, in cases when MMC RMS models do not have identical results as EMT models. In such simulations, the AC system is simulated in RMS-type programs, while the DC grid is simulated in EMT-type programs. As in [7], an EMT-type averaged converter model is combined with a RMS-type AC grid model. Researches about interfacing techniques between the EMT-and RMS-type programs have been also carried out in [8]-[10].

MATLAB/Simulink provides both EMT-and RMS-type simulation environments, nevertheless, there are few literature about implementation the co-simulation into MATLAB/Simulink. A hybrid EMT-RMS simulation method for MATLAB/Simulink for studying distribution grids is developed in [11], where the standard voltage source converter (VSC) model in Simulink-library is used for the EMT part. However, there are still no standard models for MMC in MATLAB/Simulink. Based on the method of [11], this paper provides a MATLAB/Simulink-based EMT-RMS hybrid modeling method for simulating large-scale hybrid AC/DC transmission grids. Moreover, date exchange in the loop (DXiL) method is introduced for accelerating the co-simulation.

Principle for Co-simulating EMT and RMS models

As is mentioned in the introduction, AC and DC systems are respectively simulated in RMS- and EMT-type programs, it is necessary to define the interface of both types for each HVDC converter. This section explains the generic co-simulation principle adopted in this paper, including equivalent circuits of RMS and EMT model in each other, interaction protocol and the time interpolation method. It is worth noting that the co-simulation principle is applicable to all simulation softwares, not only MATLAB/Simulink.

Selecting interface bus

The point of common coupling (PCC) is selected as the interface bus, where the currents, voltages as well as power can be measured in both EMT and RMS models.

Subsystem equivalent circuits

As EMT and RMS models are simulated with quite different types of solver, it is necessary to define the equivalent models of them in each other, enabling them independently to be solved.

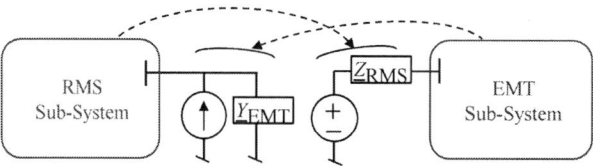

Fig. 1: EMT and RMS subsystem equivalent models [9].

Figure 1 shows the boundary condition of a converter consisting of EMT and RMS parts, including the Norton equivalent circuit of the EMT subsystem in the RMS model, as well as the Thevenin equivalent circuit of the RMS subsystem in the EMT model.

-Thevenin equivalent circuit of the RMS subsystem in the EMT model

The phasor voltage $\underline{V}_{\mathrm{PCC}}$ and current $\underline{I}_{\mathrm{PCC}}$, as well as the AC grid equivalent impedance $\underline{Z}_{\mathrm{RMS}}^{\mathrm{eq}}$ are measured at the PCC. In this way, the Thevenin equivalent voltage source of the RMS subsystem in the EMT model can be calculated as

$$\underline{V}_{\mathrm{RMS}}^{\mathrm{eq}} = \underline{V}_{\mathrm{PCC}} + \underline{I}_{\mathrm{PCC}} \cdot \underline{Z}_{\mathrm{RMS}}^{\mathrm{eq}}. \tag{1}$$

-Norton equivalent circuit of the EMT subsystem in the RMS model

The three-phase complex power is computed as

$$P + \mathrm{j}Q = \frac{1}{2}(\underline{V}_{\mathrm{a}} \cdot \underline{I}_{\mathrm{a}}^* + \underline{V}_{\mathrm{b}} \cdot \underline{I}_{\mathrm{b}}^* + \underline{V}_{\mathrm{c}} \cdot \underline{I}_{\mathrm{c}}^*), \tag{2}$$

where $\underline{V}_{\mathrm{a,b,c}}$ and $\underline{I}_{\mathrm{a,b,c}}$ are three-phase voltage and current phasor signals, respectively. The complex power $P + \mathrm{j}Q$ is provided by the EMT system, and the voltages $\underline{V}_{\mathrm{a,b,c}}$ are measured locally in the RMS system. As this work concentrates only on symmetric situations, it is adopted that all three-phase voltages and currents are symmetric, so that the power is equal in each phase, namely

$$\underline{V}_a \cdot \underline{I}_a^* = \underline{V}_b \cdot \underline{I}_b^* = \underline{V}_c \cdot \underline{I}_c^*. \tag{3}$$

Thus, the three-phase currents in the Norton equivalent circuit are derived as

$$\underline{I}_i = \frac{2}{3}\left(\frac{P+jQ}{\underline{V}_i}\right)^*, \quad i = a, b, c. \tag{4}$$

Interaction protocol

The interaction protocol between EMT and RMS subsystems can be parallel or serial. In this work, series protocol is adopted for its accuracy, which is shown in Figure 2(a).

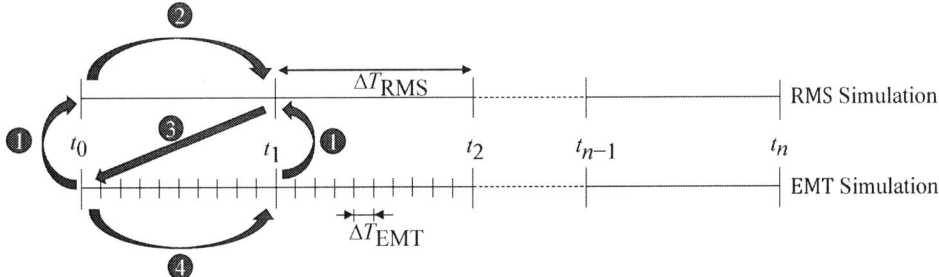

(a) Series interaction protocol [11].

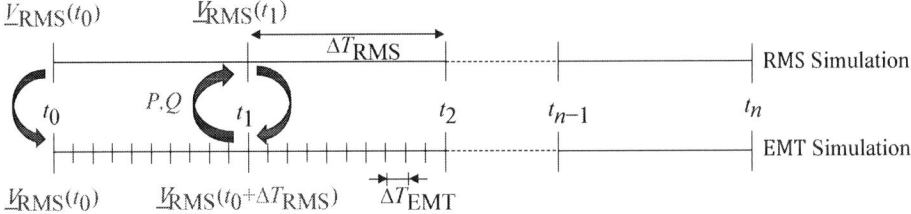

(b) Data exchange between EMT and RMS models.

Fig. 2: Interaction protocol of EMT-RMS co-simulation.

The series interaction protocol is explained as follows:

- At the very beginning of the co-simulation t_0, both EMT and RMS models are initialized according to power flow calculation results.

- Step 1: At the beginning of the i-th co-simulation cycle, the active power P and reactive power Q in the EMT subsystem, derived from currents and voltages measured at PCC, are transferred into the RMS subsystem.

- Step 2: Divided by the measured voltages, P and Q are converted into currents, and the Norton equivalent circuit in the RMS model is updated according to (4). The RMS program runs for one step ΔT_{RMS} till t_i.

- Step 3: The starting and end value of RMS equivalent voltage throughout the time step ΔT_{RMS}, $\underline{V}_{RMS}^{eq}(t_{i-1})$ and $\underline{V}_{RMS}^{eq}(t_i)$, are transferred to the EMT subsystem.

- Step 4: With $\underline{V}_{RMS}^{eq}(t_{i-1})$ and $\underline{V}_{RMS}^{eq}(t_i)$ as start and end value, the momentary value of RMS subsystem equivalent voltage V_{RMS} can be derived through the interpolation method in [9]. The EMT model is simulated for a RMS time step ΔT_{RMS} till t_i with the time step ΔT_{EMT}.

- Step 1-4 are repeated until the end of co-simulation.

Co-simulation in MATLAB/Simulink

The EMT and RMS parts are modeled first separately in two Simulink files, whose data exchange is carried out through the MATLAB workspace, just like in [11]. Standard models in the library of „Simscape/Electrical/Specialized Power Systems" are adopted for modeling the AC system, whereas the MMC is user defined as it is not available in the Simulink model library.

As there are hundreds of sub-modules (SM) in each phase of the MMC, it is difficult to model all of them exactly. Moreover, for investigating the dynamic behavior of a power grid in system-level, it is not necessary to consider the inner states of each SM. Thus, the averaged value modeling method for MMC in [12] is adopted, which is shown in Figure 3. Assuming that in each arm, the arm voltage is always

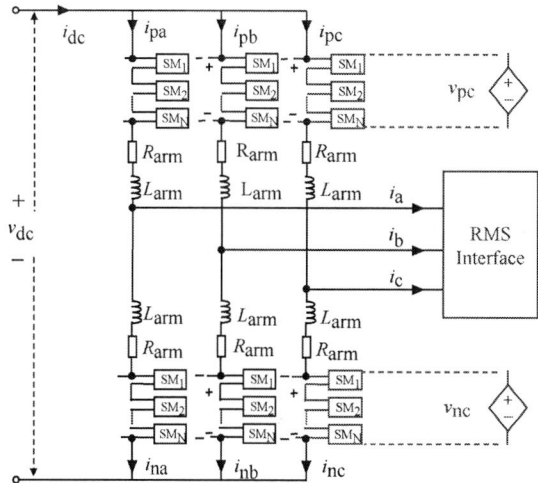

Fig. 3: Averaged value model of modular multilevel converter.

equally distributed to inserted SMs, all SM voltages in each arm are considered as an equivalent voltage source, so that

$$v_{pi} = \sum_{j=1}^{N} v_{SM}^{pj}, \quad i = a, b, c \tag{5}$$

and

$$v_{ni} = \sum_{j=1}^{N} v_{SM}^{nj}, \quad i = a, b, c, \tag{6}$$

where v_{pi} and v_{ni} are respectively equivalent voltages of upper and lower arm in each phase; v_{SM}^{pj} and v_{SM}^{nj} are sub-module voltages. As the transformer is modeled in RMS subsystem, the MMC in Figure 3 is directly connected to the RMS interface, where the three-phase voltages and currents are measured and an equivalent RMS AC grid is established.

In the i-th co-simulation cycle, the RMS simulation is first executed for a time step before it is paused, voltage result $\underline{V}_{RMS}(t_i)$ is logged to the workspace after ΔT_{RMS}. The interface voltage in EMT simulation $\underline{V}_{EMT}(t_{i-1})$ is interpolated in the workspace and updated in the EMT Simulink file. Executing EMT simulation for ΔT_{RMS}, power data $P(t_i)$ and $Q(t_i)$ are transferred to the RMS file through MATLAB workspace. In the above-mentioned steps, **set_param** handle is used for updating parameters from workspace, pausing and stepwise running EMT and RMS programs.

Different from normal Simulink blocks, the impedance block representing $\underline{Z}_{RMS}$ in Figure 1 can not be updated when the simulation is running. That means, the impedance value can not be changed with **set_param** handle, as long as the Simulation is running.

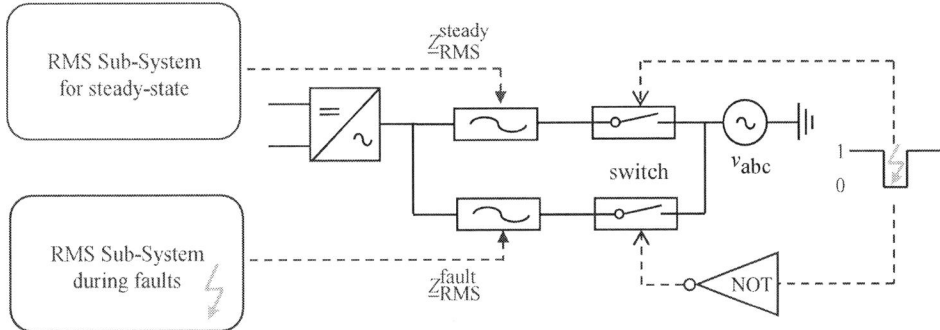

Fig. 4: Update impedance value in EMT model during co-simulation.

To overcome this problem, a back-up fault impedance block is introduced for representing the equivalent AC grid impedance during faults. As is shown in Figure 4, the value of the fault impedance is measured off-line in the RMS system before running the co-simulation. The fault impedance block is bypassed in steady-state and comes into service only if a fault takes place.

Simulation acceleration through data exchange in the loop (DXiL)

It can be seen from Figure 5 that in the i-th co-simulation cycle, there is extra time cost for pausing on program and starting the other, saving data to the workspace and importing data from it. The longer the co-simulation is carried out, the more such extra time expanse must be paid.

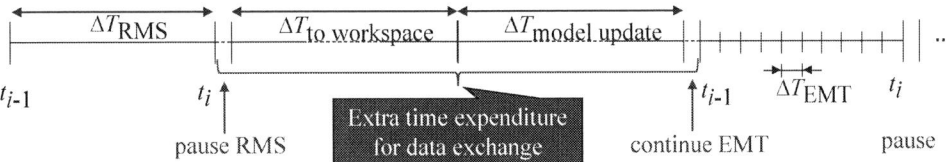

Fig. 5: Pausing time and time for data exchange for co-simulation.

One method for accelerating the co-simulation is to increase the RMS time step ΔT_{RMS} and reduce the cycle number. In some cases, ΔT_{RMS} can be reduced to 10-20 ms, just like in [11], where AC transmission grids are represented as voltage sources, instead of generators and loads. However, for simulating other AC grids with generators whose elements have relative tiny time constants, ΔT_{RMS} can not be as long as 10 ms, but maximum 1-2 ms.

Another thought is to avoid data exchange through the MATLAB workspace, which is the main contribution of this work. The main idea is to put both RMS and EMT models in one Simulink file and run them in rotation. Instead of extra programming, the „enabled subsystem" block can already realize rotational running different systems.

Enabled subsystem in MATLAB/Simulink

An example is shown in Figure 6, where a three-phase signal is generated in a enabled subsystem block. The subsystem is deactivated from 0.1 s to 0.2 s. It can be seen that the state variables in the subsystem are kept unchanged, when the subsystem is deactivated. This character can be applied to EMT-RMS co-simulation in one Simulink file.

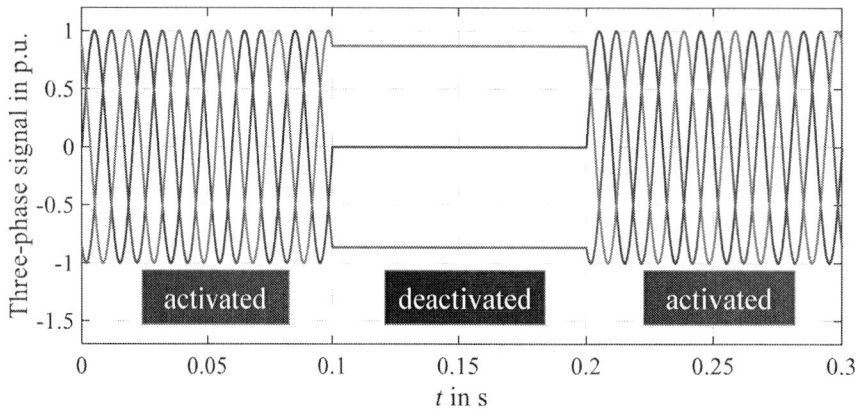

Fig. 6: Rotational simulation with enabled subsystem block.

Hybrid EMT-RMS model in MATLAB/Simulink

As is shown in Figure 7, the EMT and RMS models are now put in two enabled subsystem blocks in one Simulink file. The subsystem is activated under a pulse signal **1** and deactivated under a signal **0**. At

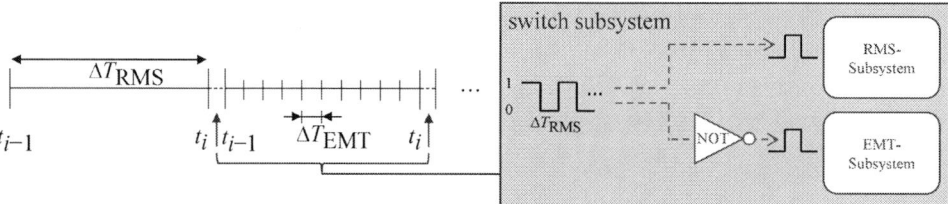

Fig. 7: Hybrid EMT-RMS model in different enabled subsystems.

the beginning of the i-th co-simulation cycle, a pulse signal **1** is generated to activate RMS subsystem, whereas EMT subsystem is deactivated under the inverse signal **0**. The pulse signal is overturned as long as the RMS simulation is finished, and the simulation mode is switched to EMT. In this way, the two subsystems are alternatively activated and deactivated in every co-simulation cycle, enabling EMT and RMS simulations to be executed rationally.

Dealing with simulation time

Theoretically, EMT and RMS systems should be simulated parallel at the same time. However, they have to be executed alternatively in order, when they are put in one Simulink file. Thus, the recorded time in Simulink is twice as the simulation time. Following remedies are taken for dealing with this problem:

- The total simulation time is doubled.

- The duration of all events in the grid, such as shout circuits, is doubled.

- For interpolating voltages in the EMT subsystem according to [9], „Clock" block is used to get the actual simulation time. In the DXiL co-simulation mode, however, the recorded simulation time should be halved, before it is used for voltage interpolation.

- All data are saved in form of „structure with time" after the co-simulation. The time in such structures should be halved before plotting the simulation results.

Case study and simulation results

Test bench and case

The test bench is shown in Figure 8. The AC system is represented as a voltage source in series with

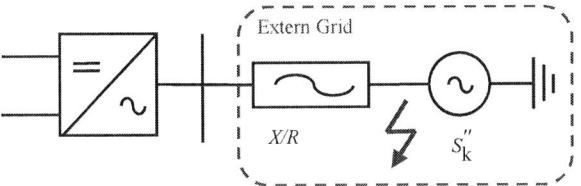

Fig. 8: Test bench.

an impedance. The short circuit ratio (SCR) as well as the X/R ratio are both 10. The co-simulation is executed for 0.3 s. A three-phase AC voltage drop to 50% is simulated at the PCC from 0.1 s to 0.2 s, representing a symmetrical AC fault somewhere in the AC network.

Co-simulation of EMT-RMS models in separate Simulink-files

The single phase voltage V_a as well as three-phase voltages in the direct-quadrature-zero coordinate $V_{d,q}$ are shown in Figure 9.

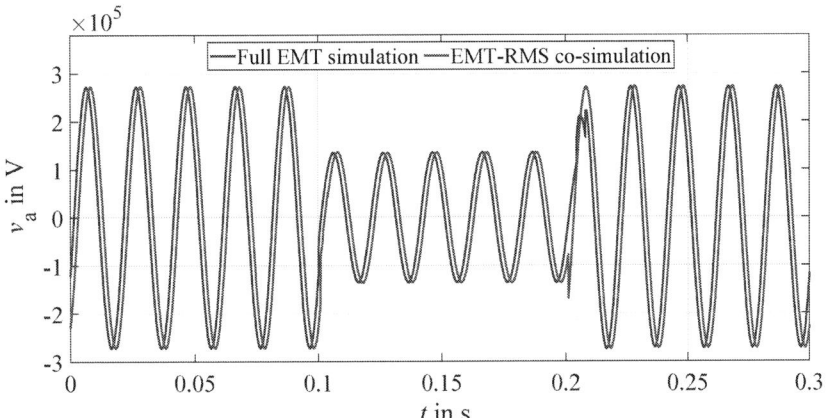

(a) Single phase voltage for a three-phase short circuit in the test bench.

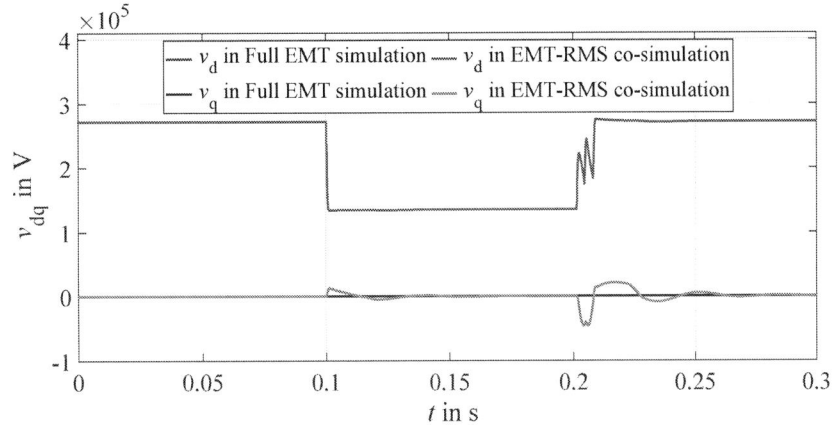

(b) Three phase voltage in the direct-quadrature-zero coordinate for a three-phase short circuit in the test bench.

Fig. 9: Simulation results for EMT-RMS co-simulation.

With acceptable errors, the co-simulation can bring out fast the same results as full EMT simulation not only by steady state, but also under transient processes.

However, the co-simulation takes a very long time for 134 seconds, which is regarded as very long for such a small test bench. Even though this co-simulation method can achieve fast the same results as full EMT simulation, it can not be applied for simulating large-scale power systems because of the huge time cost. As is already explained in Figure 5, the long duration is attributed to frequent data exchanges between both sub-modules through MATLAB work space.

Effect of acceleration

The same test bench as well as the same event as above is now studied with the DXiL acceleration method. The simulation results are shown in Figure 10. As is shown in Figure 10(a), the simulation time

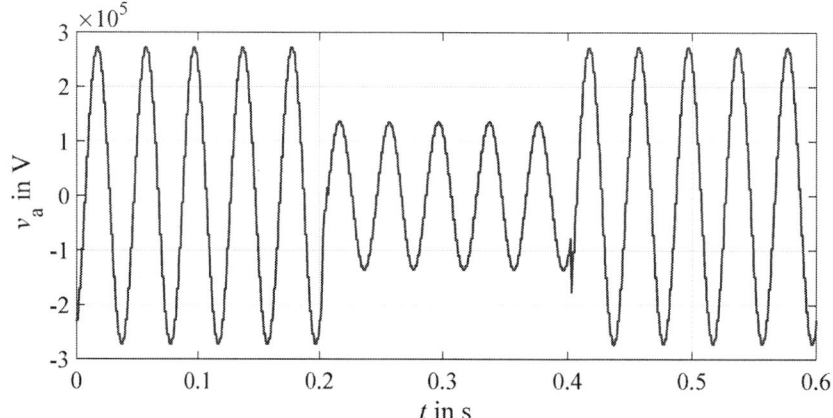

(a) Single phase voltage in EMT subsystem with doubled simulation time.

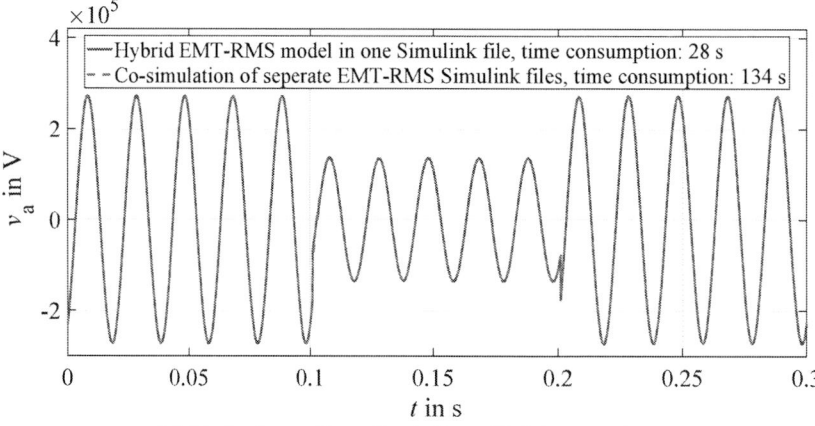

(b) Single phase voltage after dealing with data processing.

Fig. 10: Simulation results with DXiL acceleration for conditions.

is doubled from 0.3 s to 0.6 s, and the fault duration is 0.2 s instead of 0.1 s. Interspaces can be observed in the sinusoidal wave, once the EMT subsystem is deactivated. The processed simulation result is shown in Figure 10(b), where the simulation time and fault duration are reset as 0.3 s and 0.1 s, respectively. The co-simulation with DXiL acceleration method can achieve complete the same results as co-simulation of separate EMT and RMS Simulink files, whereas the simulation duration is reduced from 134 seconds to 28 seconds.

Conclusions

EMT-RMS co-simulation aims at keeping the exact electro-magnetic transient process in DC network and avoiding unnecessary simulation electro-magnetic transient process in AC network, when the transient stability of large-scale hybrid AC/DC power systems is investigated. If the data exchange process brought about by co-simulation takes too much time, however, the original intention of co-simulation is violated. This work has figured out that the EMT and RMS programs are not worth to be paused by data exchange, especially when the RMS time step is set as low to 1-2 ms. With the DXiL interaction method, the data exchange is always going on during simulation running, liberating both EMT and RMS simulations from frequent pausing and restarting throughout the co-simulation.

Acknowledgments

This work was supported by the German Federal Ministry for Economic Affairs and Climate Action (BMWK) under theproject OVANET 2.0 (0350037C).

References

[1] Liu, S., Xu, Z., Hua, W., Tang, G. & Xue, Y. Electromechanical transient modeling of modular multilevel converter based multi-terminal HVDC systems. *IEEE Transactions On Power Systems*. **29**, 72-83 (2013)

[2] Xiao, L., Xu, Z., Xiao, H., Zhang, Z., Wang, G. & Xu, Y. Electro-mechanical transient modeling of MMC based multi-terminal HVDC system with DC faults considered. *International Journal Of Electrical Power & Energy Systems*. **113** pp. 1002-1013 (2019)

[3] Beerten, J., Cole, S. & Belmans, R. Modeling of multi-terminal VSC HVDC systems with distributed DC voltage control. *IEEE Transactions On Power Systems*. **29**, 34-42 (2013)

[4] Longze, K., Lin, Z. & Fangyuan, L. Electromechanical modelling of modular multilevel converter based HVDC system and its application. *2017 IEEE Conference On Energy Internet And Energy System Integration (EI2)*. pp. 1-6 (2017)

[5] Wang, S., Zhao, X., Xue, F., Li, W., Peng, H., Shi, D., Wang, S. & Wang, Z. Electromechanical transient modeling of modular multilevel converter based HVDC network. *2019 IEEE Sustainable Power And Energy Conference (iSPEC)*. pp. 1845-1850 (2019)

[6] Du, W., Tuffner, F., Schneider, K., Lasseter, R., Xie, J., Chen, Z. & Bhattarai, B. Modeling of grid-forming and grid-following inverters for dynamic simulation of large-scale distribution systems. *IEEE Transactions On Power Delivery*. **36**, 2035-2045 (2020)

[7] Meer, A., Gibescu, M., Meijden, M., Kling, W. & Ferreira, J. Advanced hybrid transient stability and EMT simulation for VSC-HVDC systems. *IEEE Transactions On Power Delivery*. **30**, 1057-1066 (2014)

[8] Jalili-Marandi, V., Dinavahi, V., Strunz, K., Martinez, J. & Ramirez, A. Interfacing techniques for transient stability and electromagnetic transient programs IEEE task force on interfacing techniques for simulation tools. *IEEE Transactions On Power Delivery*. **24**, 2385-2395 (2009)

[9] Plumier, F., Aristidou, P., Geuzaine, C. & Van Cutsem, T. Co-simulation of electromagnetic transients and phasor models: A relaxation approach. *IEEE Transactions On Power Delivery*. **31**, 2360-2369 (2016)

[10] Meng, X. & Wang, L. Interfacing an EMT-type modular multilevel converter HVDC model in transient stability simulation. *IET Generation, Transmission & Distribution*. **11**, 3002-3008 (2017)

[11] Athaide, D., Qin, J. & Zou, Y. MATLAB/Simulink-based electromagnetic transient-transient stability hybrid simulation for electric power systems with converter interfaced generation. *2019 IEEE Texas Power And Energy Conference (TPEC)*. pp. 1-6 (2019)

[12] Peralta, J., Saad, H., Dennetière, S., Mahseredjian, J. & Nguefeu, S. Detailed and averaged models for a 401-level MMC–HVDC system. *IEEE Transactions On Power Delivery*. **27**, 1501-1508 (2012)

Advanced Low-Voltage System-in-Package Half-Bridge MOSFET

with Added Protection Features

S. Musumeci, V. Barba
POLITECNICO DI TORINO Corso Duca
degli Abruzzi 24. Torino, Italy
Tel.: +39 / (0) –11 090 7127 E-Mail:
salvatore.musumeci@polito.it,
vincenzo.barba@polito.it

F. Scrimizzi, C. Mistretta
STMICROELECTRONICS Stradale
Primosole 50, Catania, Italy
Tel.: +39 / (0) – 95.7404401 E-Mail:
filippo.scrimizzi@st.com
carmelo.mistretta@st.com

Keywords

« System in Package », «MOSFET», «Interleaved Buck Converter», «Advanced Power Supply», «Integrated Half-Bridge»

Abstract:

This paper deals with a smart System in Package (SiP) synchronous MOSFET half-bridge with internal current and thermal sensing circuits devoted to low-voltage supplies for autonomous driving applications. This kind of advanced application requests that Buck, Boost, or Buck-Boost converters can supply and stabilize the voltage. Furthermore, the temperature and the load current must be continuously monitored and controlled to increase the system's reliability. The SiP half-bridge described is experimentally evaluated in a Buck converter operation. Finally, a Buck interleaved configuration to improve the output current ripple and increase the current fed is experimentally investigated. The experimental results investigation demonstrates the effectiveness of the proposed SiP integrated half-bridge solution for an enhanced and reliable low voltage power supply.

Introduction

In the field of point of load power supplies for low voltage critical load such as microprocessors for new frontiers of high performance and reliability CPU cores, and in the section DDR (Double Data Rate) of chipset in the Fully Autonomous Driving (FAD) vehicles, the smart integrated synchronous half-bridge converter is a very attractive solution [1]. In this kind of application, the auxiliary power supplies play a crucial role in the reliability of the electronic systems used [2], [3]. In particular, considering the advanced automotive applications such as FAD, power supplies are a critical circuit system to ensure the safety of autonomous driving. For these power supply applications is necessary to monitor the main quantities such as maximum current and temperature thus that the limitation and protection systems can act promptly to avoid damage and dangers to the user and the vehicle. To take full advantage of the use of different technologies for systems that integrate power and complex signal circuits, SiPs represent a very viable solution. The advantages of the System in Package (SiP) are well known [4]

- ➤ Increasing of the switching transients
- ➤ Parasitic components (inductances and capacitances) reduction
- ➤ System compactness
- ➤ High power density

From point of view of the technology solution, the new wide-bandgap devices such as GaN FETs allow very high performance, and the SiP or full monolithic integration of power devices and gate drivers are the promising approaches to obtain advanced both switching and thermal conductivity features for power converters at high temperatures over 200 °C such as requested in electrical vehicle application [5]. However, at low-voltage application under 30Vthe pure silicon, MOSFETs (beyond the strongly reduced costs) a SiP technology approach is still very competitive especially as regards the direct resistance (R_{DSon}) obtainable compared with low-voltage GaNs available in the market [6]. In an integrated solution of a half-bridge and driver, GaN technology features a drain-source voltage of 80V [7]. For these motivations, a SiP device with low voltage MOSFETs is a very effective way to achieve low-cost and high-performance integrated power supply systems. Furthermore, the MOSFET technology allows obtaining a switch for converter system with a simply driver circuit and quite satisfactory thermal behaviour [8]. The half-bridge circuit is a flexible topology to obtain the basic converter circuit such as Buck, Boost useful in voltage regulator circuit for power supplies system. In the paper, an asymmetric low-voltage power MOSFET-based half-bridge with gate protection, and sensing circuits integrated device is investigated. The operative conditions of the proposed devices for power supplies in advanced applications are extended in the range of 600 to 2000 kHz. The main switching results are carried out to demonstrate the effectiveness of the SiP solution in protected half-bridge converter applications. The added temperature and current sensing performance are described and tested. Furthermore, the presented SiP solution in the interleaved configuration are evaluated.

Integrated MOSFET Based Synchronous Half-Bridge Converter

The SiP in half-bridge configuration is composed of a MOSFET power stage in the last generation of advanced trench-gate Strip-FET technology [9]. While the gate control and sensing circuits are developed in the latest Bipolar, CMOS, and DMOS (BCD9) technology arrangement [10]. The SiP simplified circuit schematic is depicted in Fig. 1a, beyond the classic gate drivers, protection, and control logic circuits, the device has temperature and current sensing circuits. These device features are essential for the reliability of FAD systems. In the paper, these added sensing circuits are investigated in terms of fidelity and accuracy of the electrical and thermal quantities measurement.

The devices are encapsulated in a standard three-island power Quad Flat No-lead (QFN) package. The power QFN package is designed for medium power applications. It integrates low on-resistance and high-speed switching MOSFETs. Furthermore, the power QFN is a flexible solution showing a highly efficient space-saving package and very low parasitic inductances. Nowadays QFN package is used in a wide range of power applications Integrated Circuits (IC), with power stage and signal circuits in the same package. The package Power QFN in this SiP configuration is depicted in Fig. 1b. The Figure of Merit (FOM) is the parameter to evaluate the benefit of the MOSFET power stage. The FOM are related to the mathematical multiplication of on-resistance (R_{DSon}) and total Gate Charge (Q_G). In Table 1 the FOM parameters for the high-side (HS) and the low-side (LS) devices are described. The R_{DSon} reported in Table I are relative to the maximum value.

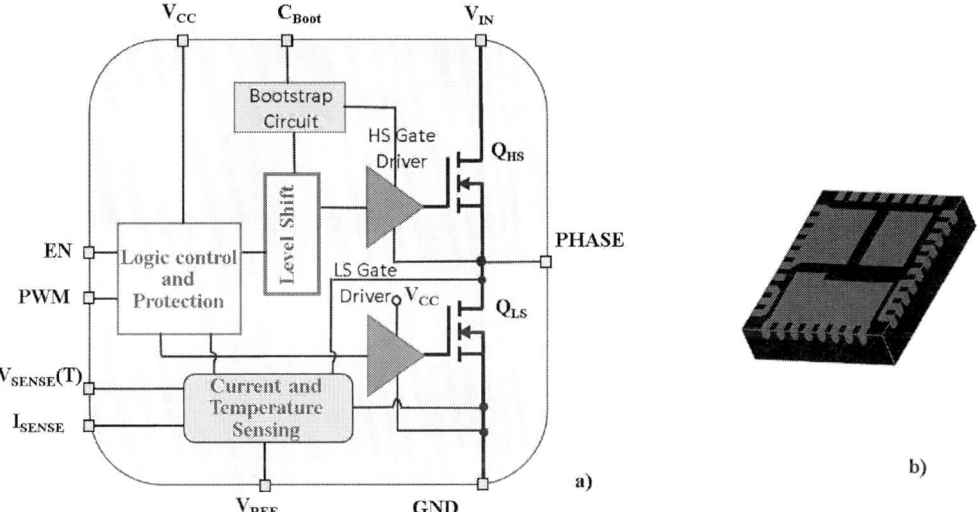

Fig. 1 a) Simplified schematic of half-bridge SiP power stage, gate control and sensing circuits. b) Power QFN package used.

The half-bridge integrated circuit can be used as Buck or Boost converter to supply a wide range of voltage load requests. To investigate the proposed SiP solution, a suitable evaluation board is developed. By the use of the experimental board, the sensing circuit behaviour is investigated. Furthermore, the converter switching performance and efficiency measurement at different switching frequencies are analyzed.

Current and Thermal Sensing Circuits

The thermal and current sensing circuits allow measuring the phase current by a dedicated circuit. The board design leads to an experimental relation to computing the actual current value by the following design equation.

$$I_{SENSE} = K_I \cdot (V_{ISENSE} - V_{REF}), \; K_I = \left(\frac{1000}{5} \cdot \frac{A}{V} \right) \tag{1}$$

An external current probe (based on the Hall effect) to measure the low side MOSFET average current (through the R_{DSon} and the phase node voltage processing) as the load varies is considered to compare the effectiveness of the proposed sensing approach.

The layout of the experimental board is reported in Fig. 2a. While the comparison between sensing current I_{SENSE} evaluated by (1) and the actual measured current I_{Lmean}. Is depicted in the graph in Fig.2b.

Table I Main MOSFETs Parameters

Parameters	HS MOSFET	LS MOSFET
R_{DSon} @25°C [mΩ]	4	1.2
R_{DSon} @100°C [mΩ]	6	1.8
Total gate charge: $Q_{G(max)}$ @10V [nC]	11	29.4

Thermal Sensor Arrangement and Operation

The temperature is measurement is obtained by a voltage signal proportional to the thermal variation. The voltage V_{TSENSE} linked with the device temperature is obtained by the equation

$$V_{TSENSE}(T) = V_{REF} + K_T \cdot T_{SENSE}, \; K_T = \left(8 \frac{mV}{°C} \right) \tag{2}$$

The thermal sensor was placed inside the smart system in the package of the device. It was positioned in the bottom right corner of the driver, (see a green rectangle in Fig. 3a), to be as close as possible to both power switches. Therefore, the sensing placement choice allows a better thermal derating behavior. It is a Proportional to Absolute Temperature (PTAT) Bandgap Current sensor, a conventional architecture implemented in the IC to link the sensor current to the actual temperature of the chip. To check the fidelity and mismatch between the real MOSFET devices' junction temperatures and the one detected by the thermal sensor, a dedicated thermal measurement was run. The diode placed in the chip structure of the driver, used as a thermal sensor (reachable through two pins of the IC device), and the body-drain diodes of the high side/low side MOSFETs were pre-characterized to define their forward voltage drop dependence versus the temperature.

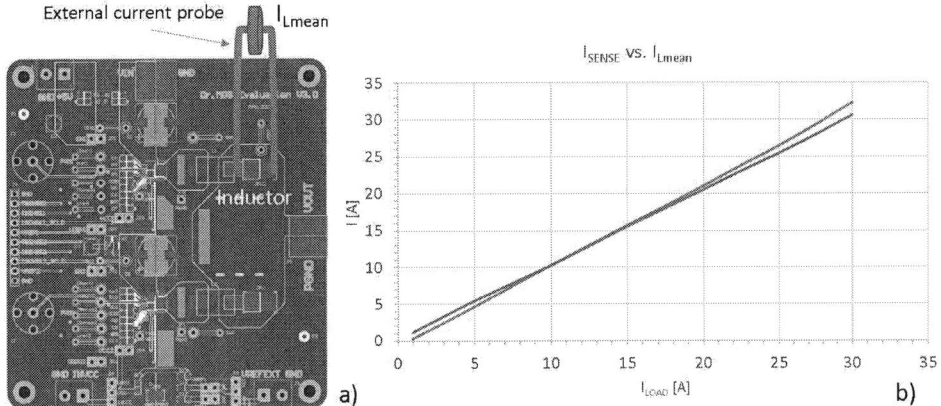

Fig. 2 a) Layout of half-bridge SiP evaluation board with external current probe. b) I_{SENSE} versus I_{Lmean} current comparison.

The two diodes on the power switches and the one on the driver were initially calibrated by heating them to a known temperature, their voltage drop V_F for MOSFET was measured at different temperatures and then their calibration was fixed. They were used as thermal sensors, according to this calibration methodology. It provides a fixed current, once the thermal steady state is achieved. The V_{FD} voltage (voltage variation on sensing driver diode) was measured, allowing to compute the temperature, through (2) of the chip tracking back their V_F versus, T_j characteristics considering the dependence previously set during the characterization itself. This method applied to the two power MOSFETs' diodes and the driver's one defined a measured mutual thermal impedance. This parameter allows us to sense the heat transferred by the power devices to the driver chip. Thus the temperature increase of every single die when others are managing power can be estimated. The temperatures were estimated following the above-reported methodology either on the driver or on the MOSFETs and so was defined the maximum temperature gap among the power devices and the driver itself that, by a simple matching through the copper frame and package construction features, is heated up indirectly. In Fig. 3b is reported a typical waveform to physically demonstrate the process previously described with the relevant current, voltage and energy loss values. Where V_{FD} is the voltage drop waveforms for driver diode, and V_{FS} is the voltage transient waveforms for Body diodes of the MOSFETs at the imposed current source–drain I_{FS}. In Fig. 3b, the diode placed in the driver circuit (Fig. 3a) is pre-turned on in forward bias with a small current of few mA, flowing through it, which is not able to raise its temperature. The waveform shows the V_F variation of both driver and body-drain diodes when a fixed current of 4.2 A flows through the body-drain diode of the power MOSFETs.

Temperature estimation and validation

To demonstrate the effectiveness of the thermal sensing approach a temperature estimation at different load current conditions I_{load} is carried out. The operative conditions are V_{in}=12V, V_{out}=1V, inductive load, L=150nH, at switching frequency f_{sw}=600kHz, and output current variation I_{LOAD} in the range $5 - 30$ A. The temperatures evaluated by the thermal sensor housed in the SiP are compared with the measurements achieved with an infrared thermal camera on the top of the device. In Table II the main results of the sensing temperature and the measured one are reported. Considering the estimation and measurement results at 20 A, the difference between the temperature from the thermal sensing T_{SENSE} and the temperature detected by the infrared thermal camera T_{ITC} is around 10°C.

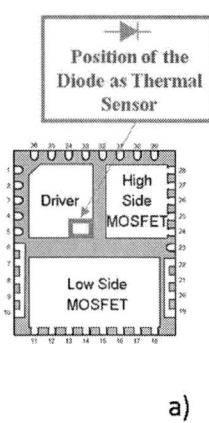 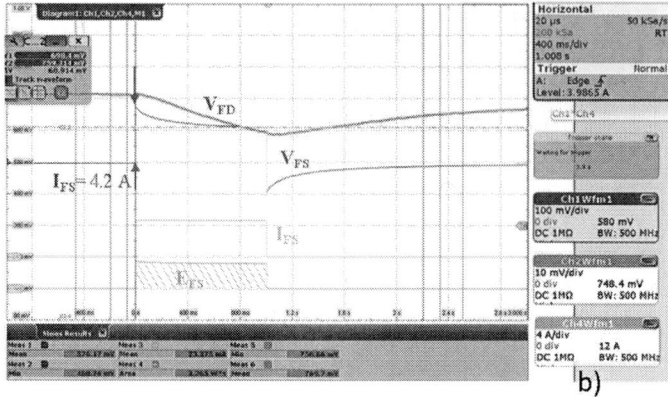

a) b)

Fig. 3 a) Thermal sensor position in the SiP device. b) V_{FD} variation on driver diode due to thermal mutual coupling body diode heating, body-drain diode V_{FS}, I_{FS} body-drain diode current imposed and the energy losses calculated in the area defined by the current pulse shape, E_{FS}=3.265W·s.

The graphics of Fig. 4 show the behavior of the temperature rising with the increase of the I_{LOAD} and the comparison with the measurements obtained with the infrared thermal camera on the top of the device. The differences in the two curves displayed are related to the equivalent thermal resistance between the package and the die R_{THeq}.

Table II. Voltage measured and Temperature Estimation with thermal sensor and infrared thermo camera versus load current

Parameters	I_{out} (A)	V_{TSENSE} (V)	T_{SENSE} (°C)	T_{ITC} Thermo camera (°C)
Operative Condition V_{in}=12V, V_{out}=1V, f_{sw}=600kHz	5	0.895	36.88	43.3
	10	0.965	45.63	55.0
	15	1.031	53.88	62.8
	20	1.132	66.50	76.3
	25	1.267	83.38	96.5
	30	1.451	106.38	123.6

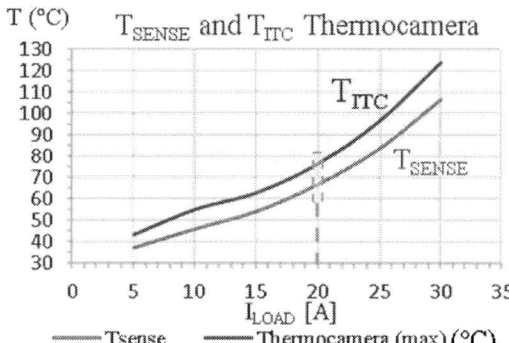

Fig. 4. Sensing temperature T_{SENSE} comparison with the temperature measured with the infrared thermal camera T_{ITC} at different load current I_{LOAD}.

In the same working condition at 20A the the power losses in the high and low side MOSFETs were estimated. In a single switch, both switching and conduction losses were considered. At I_{LOAD}= 20 A, in the low side (Q_{LS}) MOSFET the power losses are 1.06 W while in the high side (Q_{HS})are 0.67 W. We used a 4layer board and the results of a pre-characterization show the high side, the low side and the mutual Z_{TH} curves reported in the picture of Fig. 5.

At steady state the R_{TH} are:

- $R_{TH\,LS-4layer}$= 24.6 °C/W
- $R_{TH\,HS-4layer}$= 29.6 °C/W
- $R_{TH\,HSLS-4layer}$= 16.8 °C/W (mutual thermal resistance)

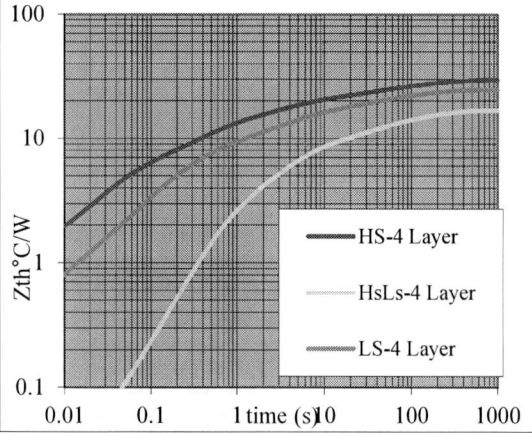

Fig. 5. Thermal impedance of the junction to 4 layer copper (Cu) board curves of the high side switch, the low side switch, and the mutual Z_{TH}.

Thus it is possible to calculate the single contribution to the junction temperature variation at $I_{LOAD}= 20$ A in a single switch. The temperature difference ΔT_{jLS} due to the power losses (P_{LSTOT}) of the Q_{LS} switch is

$$\Delta T_{JLS} = R_{TH\ LS-4layer} \cdot P_{LSTOT} = 24.6 \cdot 1.06 = 26.1°C \tag{3}$$

The ΔT_{jHS} due to the power losses (P_{HSTOT}) of the Q_{HS} switch is

$$\Delta T_{JHS} = R_{TH\ HS-4layer} \cdot P_{HSTOT} = 29.6 \cdot 0.67 = 19.8°C \tag{4}$$

The $\Delta T_{j\ mutual}$ due to the power losses due to the mutual impedance is

$$\Delta T_{J\ mutual} = R_{TH\ HSLS-4layer} \cdot (P_{LSTOT} + P_{HSTOT}) = 16.8 \cdot (1.06 + 0.67) = 29\ °C \tag{5}$$

The total temperature difference $\Delta T_{j\ TOT}$ is the sum of the every contribute.

$$\Delta T_{j\ TOT} = \Delta T_{JLS} + \Delta T_{JHS} + \Delta T_{J\ mutual} = 26.1 + 19.8 + 29 = 74.8\ °C \tag{6}$$

The temperature evaluated by (6) is in agreement with the corresponding results obtained in Table II.

Furthermore, the validation of the thermal sensing features is carried out in comparison with both thermocouple and infrared thermal camera measurements with a different heating approach. An external thermal gun to heat up the component up to a fixed temperature is used, without imposing a current I_{LOAD}. This passive validation methodology allows for a more accurate temperature evaluation demonstrating the effectiveness of the thermal sensor solution proposed. At thermic steady-state, the heat propagation reaches the MOSFETs and driver circuit in the SiP. Thus, the V_{TSENSE} furnishes a temperature closer to the actual one set by the heat gun. The thermal sensor temperature T_{SENSE} is obtained by reading V_{TSENSE} and applying (2). The package temperatures set by the thermal gun have been measured both through the infrared thermal camera and the thermocouple (considering the thermocouple results as a reference for the thermal gun action). The T_{SENSE} is achieved by (2) detecting the corresponding V_{TSENSE}. The V_{TSENSE} and the temperatures achieved by thermal gun heating on the SiP device are reported in Table III. In Fig. 6 a graphic comparison of the three temperature measurements is reported with the voltage V_{TSENSE} detected at 70°C highlighted.

As shown in Fig. 6, at 70°C the V_{TSENSE} is 1.16V. This V_{TSENSE} value compared to the value achieved in the previous test at 20A (1.16 V) highlighted in Table II shows a discrepancy of 0.03V (around 2.6%) the $\Delta V = 1.16 - 1.13 = 0.03$ V matches with a $\Delta T = 3.75$ °C mismatch.

TABLE III Results of V_{TSENSE} and the Temperatures Achieved by Thermal Gun Heating

V_{TSENSE} (V)	T_{SENSE} (°C)	T_{ITC} Thermo camera (°C)	T_{TC} Thermo couple (°C)
1.030	53.75	50.6	50
1.290	86.25	84.7	80
1.485	110.63	103.4	100
1.640	130	132.8	130
1.778	147.25	150.2	150

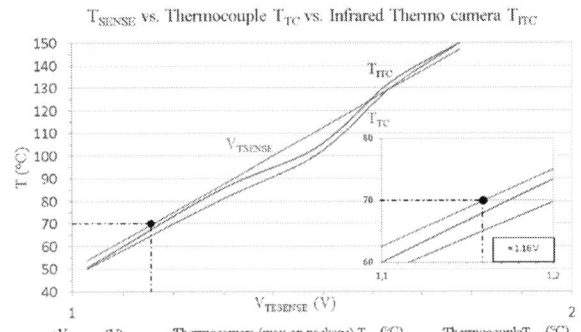

Fig. 6. Sensing temperature T_{SENSE} comparison with the temperature measured with the infrared thermal camera T_{ITC} and a thermocouple T_{TC}.

Experimental Buck Converter Evaluation

The switching behaviour analysis of a single-phase buck converter is based on the following test conditions reported in Table IV.

The switching waveforms of the synchronous switching leg in Buck configuration (see Fig. 7a) at 600kHz are reported in Fig. 7b. The current paths during the different stages of the transient behaviour are depicted in Fig. 7a. Furthermore, in Fig. 7b the devices' current path transients are also indicated. Moreover, the voltage trace related to the current sensing (V_{ISENSE}) are reported with the I_{out} and V_{in}.

Table IV. Buck Converter Test Conditions

V_{in} [V]	12-16
V_{out} [V]	1-1.1
I_{load} [A]	5-30
f_{sw} [kHz]	600-2000
L [H]	150n

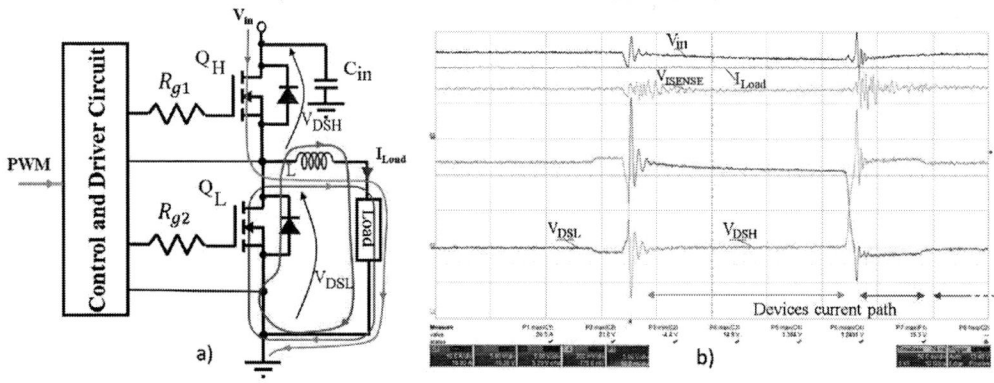

Fig. 7 a) Synchronous switching leg for Buck converter topology and current path during the switching transients. b) Switching waveforms of the Drain-Source of both low-side and high-side. The input voltage V_{in} with the output current (I_{out}) and the voltage trace related to the current sensing are reported too. I_{LOAD}=10A/div, V_{DHS}=V_{DLS}=5V/div, V_{in}=5V/div, t=50ns/div.

In the switching waveforms reported, the ringing voltage and the spike are due to high di/dt on the parasitic inductances in the switching power mesh of the evaluation board layout. The steady-state switching waveforms in the operative condition of Table IV considering a I_{LOAD}=20A with the temperature sensing (T_{SENSE}) displayed are reported in Fig. 8a. In Fig. 8b a zoomed view during the switching transients is shown. The peak value appears during the switching transients of the low side switch and high side switch due to the parasitic inductances of the power loop lead a maximum voltage peak up to 20V on the low side MOSFET. T_{SENSE} waveform maintain a clean voltage information also during the transients switching conditions as shown in Fig. 8b.

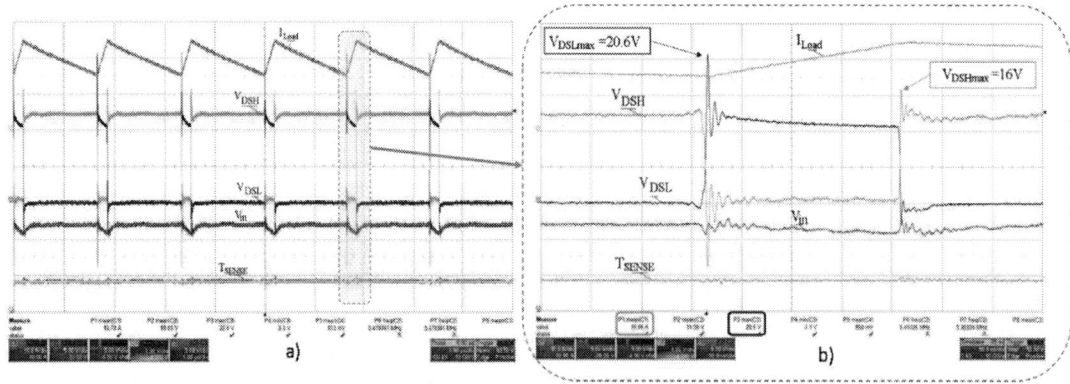

Fig. 8 a) Synchronous switching leg for Buck converter switching waveforms in steady-state. b) Zoomed view of the switching transient. I_{LOAD}=10A/div, V_{DHS}=V_{DLS}=5V/div, V_{in}=5V/div, T_{SENSE}=1V/div. a)t=100μs/div, b)t=50ns/div.

The over current protection action is described in Fig. 9. During normal working (at I_{load}=20 A, V_{in}= 12V and f_{sw}= 600kHz) the device was externally heated to bring it to an over temperature condition (OTP) as described in Fig. 9a. During the OTP the PWM stopped and it restarted once the OTP state was removed. The zoomed view of Fig. 9b details the return of standard PWM operating conditions when the over-temperature is removed.

The efficiency of the Buck converter developed on the SiP system is satisfactory and in line with the requirements of the applications considered. The device has the best performance at a switching frequency of 600kHz. However, the efficiency is also remarkable at the frequency of 1.2 MHz and features lower efficiency at 2MHz. In Fig. 10 the efficiency results versus the load current (I_{Load}) are reported at the three switching frequencies in the range reported in Table 4.

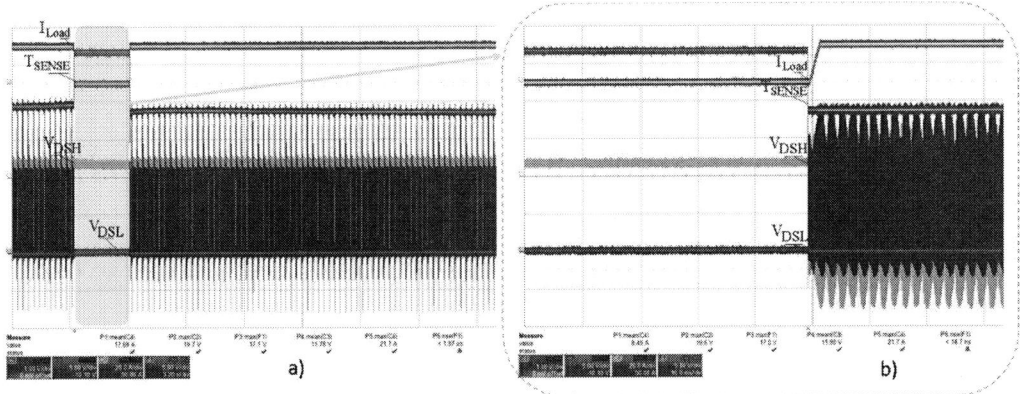

Fig. 9 a) Steady-state operating condition with an over-temperature condition caused through a time-limited externally heating. b) Zoomed view of the return of standard PWM operating conditions when the over-temperature is removed. I_{LOAD}=10A/div, V_{DHS}=V_{DLS}=5V/div, T_{SENSE}=1V/div, I_{LOAD}=20A/div a) t=10ms/div, b)t=1ms/div.

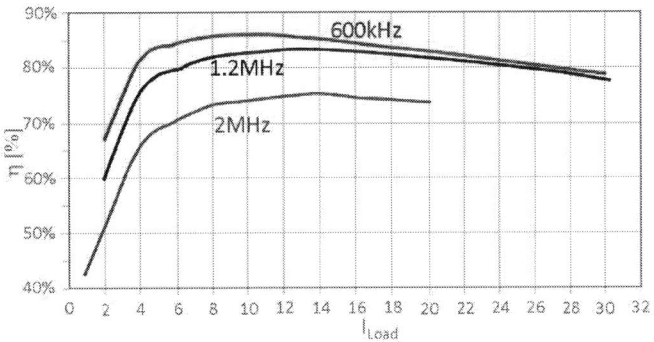

Fig. 10 Power converter efficiency versus the load current at different switching frequency

Interleaved Buck Converter Application

In FAD application the interleaved buck and boost configuration is a viable solution to improve the current output waveforms quality reducing the ripple. Furthermore, the output-filter sizes decrease in an interleaved implementation due to the lowered current in the power switching leg for each phase [11]. The number of the operating switching legs can be activated according to the load request optimizing the device's current share. Selecting the maximum number of the switching legs (generally 3 in this kind of application) the current in the single device is reduced acting in the MOSFETs heating reduction. A simplified block scheme of SiP MOSFET in interleaved configuration implementing 3 switching legs is reported in Fig. 11.

Therefore, the interleaved solution features several advantages which can be summarized in
- Reduction of ripple on the output current;
- Reduced input capacitance;
- Reduced output capacitance;
- Better thermal performance and efficiency at high load currents.

The current ripple of the output current depends on the duty-cycle and the number of switching legs adopted in the interleaving technique [12]. Considering two phases of interleaved Buck converter with non-coupled inductors, the experimental operative waveforms at Continuous Current Mode (CCM) are reported in Fig. 12a.

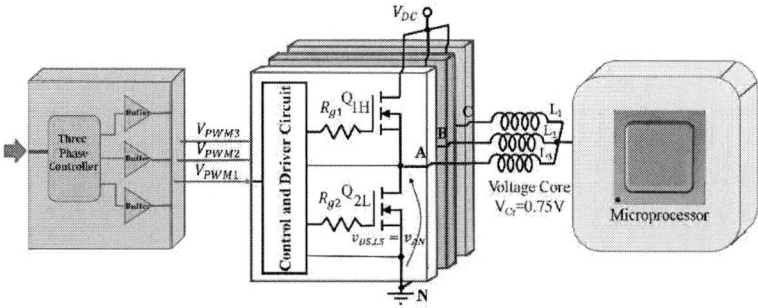

Fig.11 Power supply based on the SiP MOSFET half-bridge converter in interleaved configuration

The total output current $I_{Loadtot}$ is the sum of the single leg current I_{Load1} and I_{Load2}. In Fig 12b a single switching leg voltage are zoomed with the maximum peak voltage ($V_{DSL1, peak}$) of 17.8V. The reduction of the output load current ripple peak ($\Delta I_{Loadtot}$ =0.5A) is depicted in Fig.12b. Finally, the dead time imposed to avoid cross conduction is also highlighted. In the actual two phase interleaved converter design a dead time of 45ns was chosen.

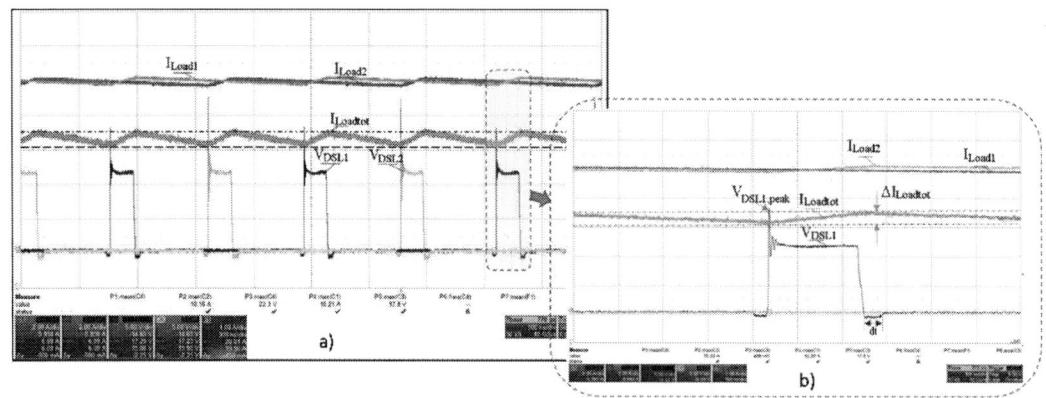

Fig.12. a) Low side voltages and currents switching waveforms of the two switching legs in interleaved configuration with the total load current. b) Single switching voltage V_{DSL1} with the total output current waveform $I_{Loadtot}$ and the underlined dead time chosen. $I_{Load1}=I_{load2}$=2A/div, $V_{DLS1}=V_{DLS2}$=5V/div, $I_{Loadtot}$=1A/div, a) t=500ns/div, b) t=100ns/div

Conclusions

In the proposed paper an advanced cost-effective SiP half-bridge MOSFETs is presented. The integrated device is oriented to develop a flexible and reliable power supply to power the FAD IC system. The half-bridge SiP solution can be used to achieve Buck, Boost, or Buck-Boost converters. In the paper, a Buck converter switching evaluation is carried out to demonstrate the usefulness of the proposed hybrid integrated power switch and control gate circuits. The inner sensing circuits related to the leg current and the power device temperature are focused. In particular, the temperature sensing approach is tested and validated by several measurement experimental tests and numerical analysis to demonstrate the effectiveness of the proposed sensing solution. Furthermore, a two-phase interleaved buck converter application is experimentally evaluated to demonstrate the effectiveness of the SiP solution. In the interleaved configuration, the control circuit allows activating the switching legs according to the load demand to modulate the IC supply current request. In this way, the achievable Buck, Boost or Buck-Boost converters are able to supply and stabilize the voltage at the value required by the active safety system in automotive application.

References

[1] J. A. Baxter, D. A. Merced, D. J. Costinett, L. M. Tolbert and B. Ozpineci, "Review of Electrical Architectures and Power Requirements for Automated Vehicles," *2018 IEEE Transportation Electrification Conference and Expo (ITEC)*, 2018, Long Beach, CA, USA, pp. 944-949, doi: 10.1109/ITEC.2018.8449961.

[2] T. Mio, et al. "Auxiliary Power Supply System for Electric Power Steering (EPS) and High-Heat-Resistant Lithium-Ion Capacitor," in *World Electr. Veh. J.* 2019, *10*, 27. https://doi.org/10.3390/wevj10020027.

[3] E. Armando, F. Fusillo, S. Musumeci and F. Scrimizzi, "Low Voltage Trench-Gate MOSFETs for High Efficiency Auxiliary Power Supply Applications," in *2019 International Conference on Clean Electrical Power (ICCEP)*, 2-4 July 2019, Otranto, Italy, pp. 165-170, doi: 10.1109/ICCEP.2019.8890217.

[4] C. O'Mathuna, "PwrSiP power supply in package power system in package," *2016 International Symposium on 3D Power Electronics Integration and Manufacturing (3D-PEIM)*, 13-15 June 2016, Raleigh, NC, USA, pp. 1-21, doi: 10.1109/3DPEIM.2016.7570569.

[5] R. Hou, Y. Shen, H. Zhao, H. Hu, J. Lu and T. Long, "Power Loss Characterization and Modeling for GaN-Based Hard-Switching Half-Bridges Considering Dynamic on-State Resistance," in *IEEE Transactions on Transportation Electrification*, vol. 6, no. 2, pp. 540-553, June 2020, doi: 10.1109/TTE.2020.2989036.

[6] S. Musumeci, E. Armando, F. Mandrile, F. Scrimizzi, G. Longo and C. Mistretta, "Experimental Evaluation of an Enhanced GaN-Based Non-Symmetric Switching Leg Integrated Module for Synchronous Buck Converter Applications," *2021 23rd European Conference on Power Electronics and Applications (EPE'21 ECCE Europe)*, 6-10 Sept. 2021, Ghent, Belgium, pp. 1-10.

[7] S. Musumeci, F. Mandrile, V. Barba, M. Palma, "Low-Voltage GaN FETs in Motor Control Application; Issues and Advantages: A Review," in *Energies* 2021, *14*, 6378. https://doi.org/10.3390/en14196378

[8] A. Raciti, et al., "A bi-dimensional model for power MOSFET devices accounting for the behavior in unclamped inductive switching conditions," *IECON 2013 - 39th Annual Conference of the IEEE Industrial Electronics Society*, 10-13 Nov. 2013, Vienna, Austria, pp. 134-139, doi: 10.1109/IECON.2013.6699124.

[9] F. Scrimizzi and F. Fusillo, "MOSFET Technologies for Auxiliary DC-DC Converters," *PCIM Europe 2018; International Exhibition and Conference for Power Electronics, Intelligent Motion, Renewable Energy and Energy Management*, 5-7 June 2018, Nuremberg, Germany, pp. 1-4.

[10] B. Murari, F. Bertotti, G. Vignola, Smart Power ICs, 2nd Edition, Springer Verlag, Berlin, 2002.

[11] S. Musumeci, R. Bojoi, E. Armando, S. Borlo, F. Mandrile, "Three-Legs Interleaved Boost Power Factor Corrector for High-Power LED Lighting Application," in *Energies* 2020, *13*, 1728. https://doi.org/10.3390/en13071728

[12] X. Yang, S. Zong and G. Fan, "Analysis and validation of the output current ripple in interleaved buck converter," in *IECON 2017 - 43rd Annual Conference of the IEEE Industrial Electronics Society*, 29 Oct.-1 Nov. 2017, Beijing, China, pp. 846-851, doi: 10.1109/IECON.2017.8216146.

Evaluation of common-mode leakage current of Aalborg-type transformerless PV inverters

Georgios I. Orfanoudakis
HELLENIC MEDITERRANEAN
UNIVERSITY
Department of Electrical & Computer Eng.
Estauromenos, 71004
Heraklion, Crete, Greece
Tel: +30 2810 379817
E–mail: gorfas@hmu.gr

Eftychios Koutroulis
TECHNICAL UNIVERSITY OF CRETE
School of Electrical & Computer
Engineering, University Campus
Chania, Greece
Tel.: +30 28210 37233
E–mail: efkout@electronics.tuc.gr

Georgios Foteinopoulos
TECHNICAL UNIVERSITY OF CRETE
School of Electrical & Computer
Engineering, University Campus
Chania, Greece
E–mail: georgefwt@gmail.com

Weimin Wu
SHANGHAI MARITIME UNIVERSITY
Department of Electrical Engineering,
Shanghai 201306, China
E–mail: wmwu@shmtu.edu.cn

Acknowledgements

This work was performed within the framework of the project "eSOLAR: Principle and control of high-efficiency Buck-Boost type Photovoltaic inverter" of the program "Bilateral and Multilateral Research & Technology Co-operation between Greece and China", funded in Greece by the Operational Program "Competitiveness, Entrepreneurship and Innovation 2014-2020" (co-funded by the European Regional Development Fund) and managed by the General Secretariat of Research and Technology, Ministry of Education, Research, and Religious Affairs under the project eSOLAR/T7ΔKI-00066. In China, this work was supported by the National Key Research and Development Project of China under Grant 2017YFGH001164. This support is gratefully acknowledged.

Keywords

«Grid-connected inverter», «Transformerless PV inverter», «Dual-mode time-sharing inverter», «Buck in Buck, Boost in Boost inverter», «Step-up inverter», «Rectified sine wave DC-link voltage», «Aalborg inverter», «Common-mode current», «Ground leakage current».

Abstract

Single-phase transformerless photovoltaic (PV) inverters with voltage step-up capability are widely employed for integrating PV generation to the electric grid. Transformerless PV inverter topologies include special features to suppress common-mode (CM) leakage currents that can flow through parasitic capacitances that appear between the PV array and the ground. The Aalborg inverter belongs to the family of dual-mode time-sharing PV inverters, in which a voltage step-up (Boost) stage operates alternatively to a step-down (Buck) stage to create a rectified sine wave DC-link voltage, which is then unfolded to the grid. This paper analyses, quantifies and improves CM leakage current generation for Aalborg-type transformerless PV inverters. It highlights the factors that can lead to high peak and RMS values of leakage current for these topologies and proposes an output filter modification to reduce them by up to 70%. The analytical results are supported by simulation in MATLAB-Simulink and are applicable to other topologies with rectified sine wave DC-link voltages.

1. Introduction

Single-phase transformerless inverter topologies have found great acceptance in the field of integration of photovoltaic (PV) systems to the electric grid, due to their high efficiency and the benefits they provide in terms of size, weight and cost [1, 2]. The transformerless inverter (that is, the DC/AC conversion) stage in relevant commercial products is normally preceded by one (or more) voltage step-up (i.e. Boost) stage, which is responsible for performing the PV array Maximum Power Point Tracking (MPPT) process and providing the inverter with an adequate DC-link voltage, even at low solar irradiance and/or high ambient temperature operating conditions.

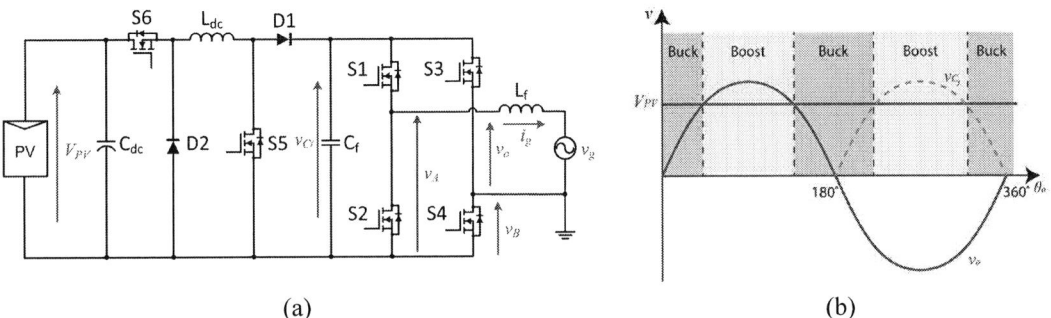

Fig. 1: (a) Aalborg full-bridge inverter topology, (b) Regions of Buck/Boost mode [5].

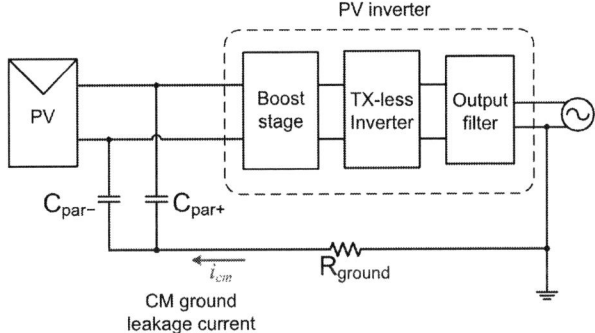

Fig. 2: Common-mode ground leakage current circulation path.

The Boost and the inverter stages are traditionally controlled with the aim of maintaining a stable DC-link voltage. Nevertheless, according to a different concept for two-stage transformerless PV inverters, the DC-link voltage waveform has the form of a rectified sine wave [2]. This waveform is generated by a DC-DC conversion stage, while an H-bridge simply "unfolds" it and supplies it to the grid. A family of PV inverters operating according to this concept are the dual-mode time-sharing inverters [3, 4]. These inverters operate in Buck or Boost mode depending on the instantaneous value of their output voltage, v_o, in relation to the PV array voltage, V_{PV}, as shown in Figure 1(b). Among them, the full-bridge Aalborg inverter, shown in Figure 1(a) [5] and its variants presented in [6, 7] exhibit high efficiency due to the elimination of one inductor as compared to other dual-mode time-sharing topologies, and the operation of only one power switch at high switching frequency at each moment.

In addition to high efficiency, a significant requirement for PV inverters relates to the suppression of common-mode (CM) ground leakage currents. Ground leakage currents can flow in transformerless inverter systems due to the existence of parasitic capacitances between the solar cells of the PV array and the ground. They circulate through the normally grounded electric grid neutral, as illustrated in Figure 2, and cause deterioration of the PV cells and safety hazards. Transformerless PV inverters apply a number of different techniques to suppress CM currents. These techniques and their effectiveness have been studied in the literature [1, 8 − 11] for other transformerless PV inverter topologies, but not for dual-mode time-sharing topologies.

This paper makes the following contributions: (a) analyses the CM leakage current generation mechanism for the full-bridge Aalborg inverter, which also applies to other dual-mode time-sharing topologies, (b) proposes an output filter modification which can lead to significant reduction of CM leakage current, (c) derives analytical expressions for the RMS value of the CM currents, for the original and the modified filter configurations, (d) quantifies the generated CM current under different inverter operating conditions in comparison with a well-known transformerless topology used as a benchmark, and (e) assesses the effect of each filter parameter on the generated CM current. The presented analysis is supported by simulations in MATLAB-Simulink.

2. CM current generation mechanism

2.1 Characteristics of generated CM voltage

The CM voltage, v_{cm}, of a single-phase inverter topology is calculated as:

$$v_{cm} = \frac{v_A + v_B}{2} \tag{1}$$

where v_A and v_B are the voltages of the two output terminals with respect to the negative DC-bus terminal, shown in Figure 1(a). In a conventional H-bridge inverter, these voltages have a Pulse Width Modulated (PWM) waveform, which results in fast variation of the CM voltage and flow of leakage currents through the parasitic capacitances of the PV array. For the topology of Figure 1(a), however, the H-bridge consisting of S1 to S4, simply switches at the grid frequency to unfold the voltage of capacitor C_f. The H-bridge can only be at two different switching states (not four, since the zero states are not used), namely S1-S4 or S2-S3. It can be shown that for both states, the CM voltage generated by the topology is equal to $v_{Cf} / 2$. Given that C_f is the output filter capacitor for the preceding Buck/Boost converter, the CM voltage of this inverter is not pulsating. Instead, it (ideally) has the form of a rectified sine wave, with half the amplitude of v_{Cf}, which is shown in Figure 1(b). Since a rectified sine wave voltage does not exhibit sudden variations (i.e., high values of dv/dt), the topology is fundamentally expected to generate minimal values of CM leakage current. Nevertheless, v_{Cf} also contains different types of distortion that do result in the generation of CM leakage current. The appearing types of distortion depend on whether the inverter operates in Buck or Boost mode, as it will be discussed below.

Buck mode

When the inverter operates in Buck mode, the following two types of distortion appear:

I. Capacitor voltage ripple: Since capacitor C_f acts as an LC filter capacitor for the Buck converter, an amount of voltage ripple is expected to appear at the converter switching frequency. The amplitude of this voltage ripple depends on the values of L_{dc} and C_f, as well as the switching frequency and duty cycle of the Buck converter. In the present topology, a typical peak-peak value for this ripple is in the range of 1% of the peak grid voltage.

II. Voltage distortion at grid voltage zero-crossings: The topology of Figure 1(a) is designed to operate only with a unity power factor. In practice, though, due to the phase lag introduced to the output current by the filter inductors, the voltage of capacitor C_f does not exactly follow a rectified sine wave in the area of zero-crossings of the grid voltage. Namely, fast voltage transients appear as a result of the capacitor current changing direction before the capacitor voltage reaches the value of zero.

Boost mode

When the inverter operates in Boost mode, or transitions between Buck and Boost mode, the following types of distortion appear:

III. High capacitor voltage ripple: Capacitor C_f acts as an output capacitor for the Boost converter, too. The same capacitor will have several times higher voltage ripple when used with a Boost converter as compared to a Buck converter, due to the pulsating form and higher peak values of the current that the Boost converter supplies it with. Hence, when the inverter operates in Boost mode, the variation of the capacitor voltage is significantly higher, reaching up to approximately 15% of the peak grid voltage.

IV. Voltage distortion at the transitions between the two modes: Capacitor voltage distortion is observed at the points of the transitions between the Buck and Boost modes. The characteristics (e.g. magnitude and duration) of this type of distortion depend on the inverter control strategy and the method (if any) applied to ensure a smooth transition between the two modes.

2.2 Equivalent circuits for CM current generation

With reference to Figure 1(a), during state S1-S4, the grounded neutral of the grid is connected to the negative terminal of the PV array through S4, thus fixing the voltages of the parasitic capacitors C_{par-} and C_{par+} to 0 and V_{PV}, respectively. This direct connection eliminates the flow of leakage currents during state S1-S4, that is, approximately for the duration of each positive half cycle of the grid voltage. On the other hand, during state S2-S3, the neutral of the grid is connected to the positive terminal of C_f, rendering the voltages of the parasitic capacitors C_{par-} and C_{par+} dependent on v_{Cf}. The CM equivalent circuits for the two states are presented in Figure 3(a) and 3(b), respectively.

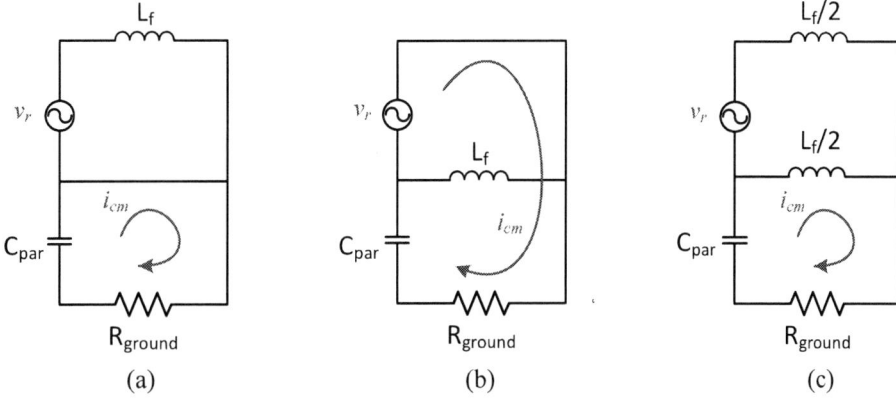

Fig. 3: Common-mode equivalent circuits for (a) state S1-S4, (b) state S2-S3, (c) proposed filter configuration.

Regarding distortion types I and III, if the RMS value of the switching-frequency ripple on v_{Cf} is equal to $V_{r,RMS}$ (abbreviation of $V_{Cf,ripple,RMS}$), then:

$$I_{cm,r,RMS,1-4} = 0, \text{ during state S1-S4} \tag{2}$$

$$I_{cm,r,RMS,2-3} = V_{r,RMS} / \sqrt{R_{ground}^2 + 1/(\omega_s C_{par})^2} \text{ , during state S2-S3} \tag{3}$$

where $\omega_s = 2\pi f_s$, with f_s being the switching frequency of the Buck/Boost stage, and C_{par} is the total parasitic capacitance to ground.

Given that the duration of each state is equal to half of the fundamental period, the overall RMS value of the switching-frequency CM current is as follows:

$$I_{cm,r,RMS} = V_{r,RMS} / \sqrt{2 \left[R_{ground}^2 + 1/(\omega_s C_{par})^2 \right]} \tag{4}$$

Moreover, it is noted that the CM current contains a component at the fundamental frequency of the grid, due to the relevant (i.e. rectified sine wave) variation of v_{Cf}, which can be shown to also appear only during state S2-S3. The RMS value of this component is given by Eq. (5), and is normally low or even negligible as compared to $I_{cm,r,RMS}$:

$$I_{cm,f,RMS} = V_{g,RMS} / \sqrt{2 \left[R_{ground}^2 + 1/(\omega_g C_{par})^2 \right]} \approx V_{g,RMS} \, \omega_g C_{par} / \sqrt{2} \tag{5}$$

where $\omega_g = 2\pi f_g$, with f_g being the fundamental frequency of the grid (50/60Hz), and $V_{g,RMS}$ is the RMS value of the grid phase voltage. The effect of R_{ground} can always be neglected in this expression.

3. Proposed output filter modification

With the aim of reducing the CM current generated by the full-bridge Aalborg inverter, a simple modification of its output filter is proposed in this paper. Namely, the inverter output filter inductor, L_f, is split to two inductors connected as shown in Figure 4, each having half of the original inductance value.

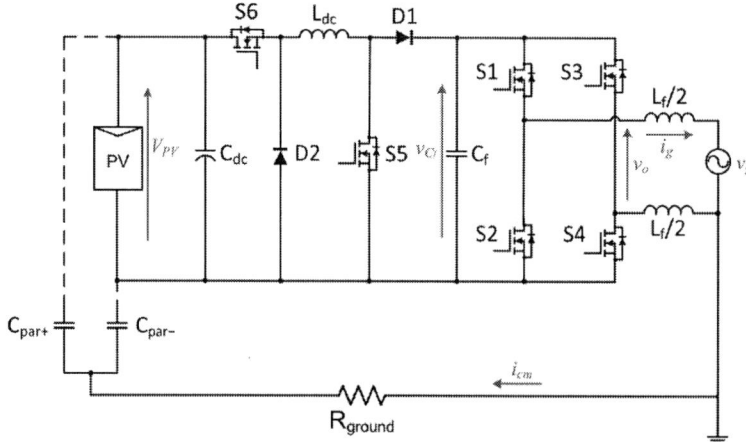

Fig. 4: Topology with the proposed modification, shown with the leakage ground current circulation path.

This configuration for the output filter was found to drastically reduce the peaks of the leakage currents at the zero-crossings of the grid voltage (distortion type II), since it inserts an inductor in a former zero-inductance CM current circulation path. With reference to distortion types I and III, due to the insertion of this inductor, the neutral of the grid is no longer connected to the negative DC-link terminal during H-bridge state S1-S4. Thus, CM current flows in this topology during both grid voltage half-cycles.

The modification of the output filter does not affect the value of $V_{r,RMS}$. The RMS value of the CM current can be derived based on the equivalent circuit of Figure 3(c), which holds for both inverter states. Neglecting the effect of R_{ground} with the aim of simplifying the analytical derivation, the RMS value of the switching-frequency CM current for the proposed filter configuration is as follows:

$$I'_{cm,r,RMS} = V_{r,RMS} \cdot 2\omega_s C_{par} / \left| 4 - \omega_s^2 L_f C_{par} \right| \tag{6}$$

According to Eq. (6), resonance will occur if L_f takes the following value:

$$L_{f,res} = \frac{4}{\omega_s^2 C_{par}} \tag{7}$$

Moreover, comparison of $I'_{cm,r,RMS}$ from Eq. (6) with $I_{cm,r,RMS}$ from Eq. (4), neglecting R_{ground}, reveals that the ratio between the two currents is:

$$\frac{I'_{cm,r,RMS}}{I_{cm,r,RMS}} = \frac{2\sqrt{2}}{\left| 4 - \omega_s^2 L_f C_{par} \right|} \tag{8}$$

The RMS value of the fundamental-frequency component of the CM current is again given by Eq. (5). Consequently, inductor L_f must have at least a certain value, $L_{f,min}$, below, for the modified filter configuration to result in a lower RMS value of CM current than the original one:

$$L_{f,min} = \frac{4 + 2\sqrt{2}}{\omega_s^2 C_{par}} \tag{9}$$

4. Simulation results

Simulations results are presented in this section to illustrate the discussed CM current characteristics and verify the analytical expressions derived for the original and the proposed filter configurations.

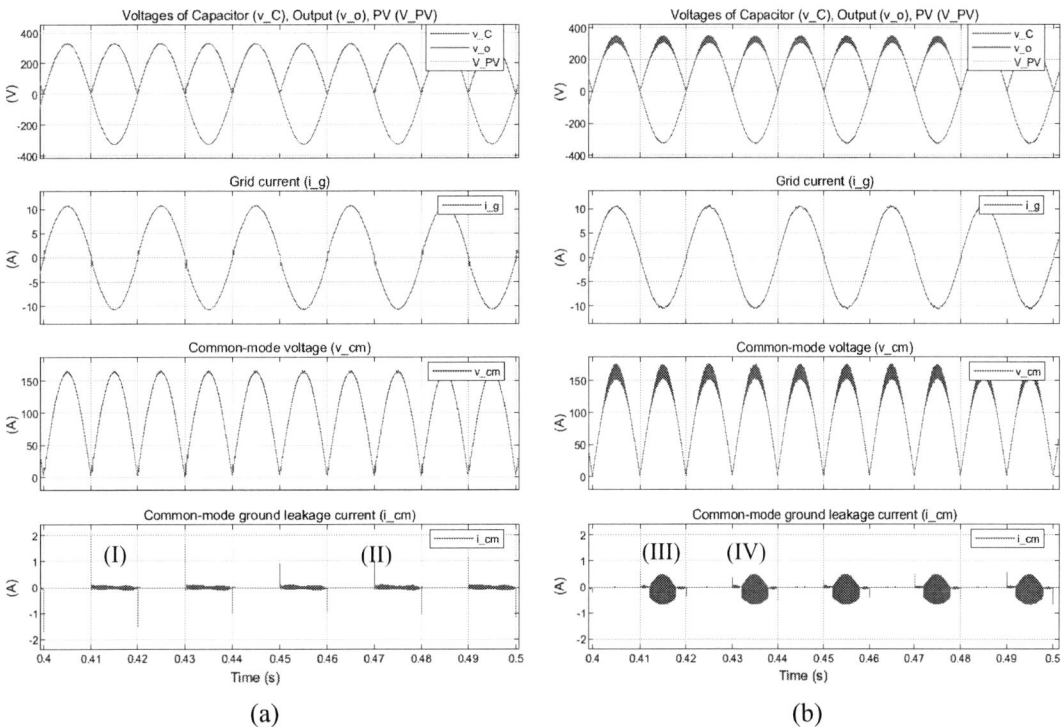

Fig. 5: Simulation results illustrating the operation of the full-bridge Aalborg inverter with: (a) Buck only operation, and (b) Buck-Boost operation. The RMS values of the CM ground leakage current are 38mA/218mA.

The topology of Figure 1(a) was modeled in MATLAB/Simulink as a 2kVA PV inverter switched at 30kHz, with L_{dc} = 1mH, L_f = 1.2mH and C_f = 3μF. The PV array was simulated as a DC source with V_{PV} = 400V for inverter operation in Buck only mode and V_{PV} = 200V for operation in Buck-Boost mode. The parasitic capacitances to ground were given values of C_{par+} = C_{par-} = 100nF (giving a total of C_{par} = 200nF per PV array kW), while the ground resistance R_{ground} was assumed to have a value of 0.5Ω. The inverter was connected to a 230V/50Hz grid.

Figure 5(a) illustrates a set of waveforms during Buck operation ($v_{o,pk}$ < V_{PV}), while 5(b) illustrates the respective waveforms during Buck-Boost operation. The CM current resulting from the four types of capacitor C_f voltage distortion is marked in the two bottom graphs, using symbols I – IV. The following main observations can be noted:

- Although CM voltage distortion of types I and III appears uniformly during both grid voltage half-cycles, CM current flows only during the negative half-cycles. As explained earlier, this is because of the direct connection of the grid neutral to the negative DC-link terminal during the positive half-cycles.

- High peak values of CM current appear at the zero crossings of the grid voltage. These CM currents originate from the CM voltage distortion of type II and exhibit high peak values due to the lack of inductance in their circulation path (which includes the capacitor C_f, the connection to the grid neutral, the ground resistance and the parasitic capacitances).

- The CM voltage distortion of type IV and the resulting CM current are unnoticeable, which is owing to the fact that the applied control strategy [12] is designed to smoothen the transients appearing when switching between the Buck and Boost modes.

The RMS values of the CM current in these results are 38mA for the Buck mode and 218mA for the Buck-Boost mode, while its peak values are in the order of 0.5 to 2A.

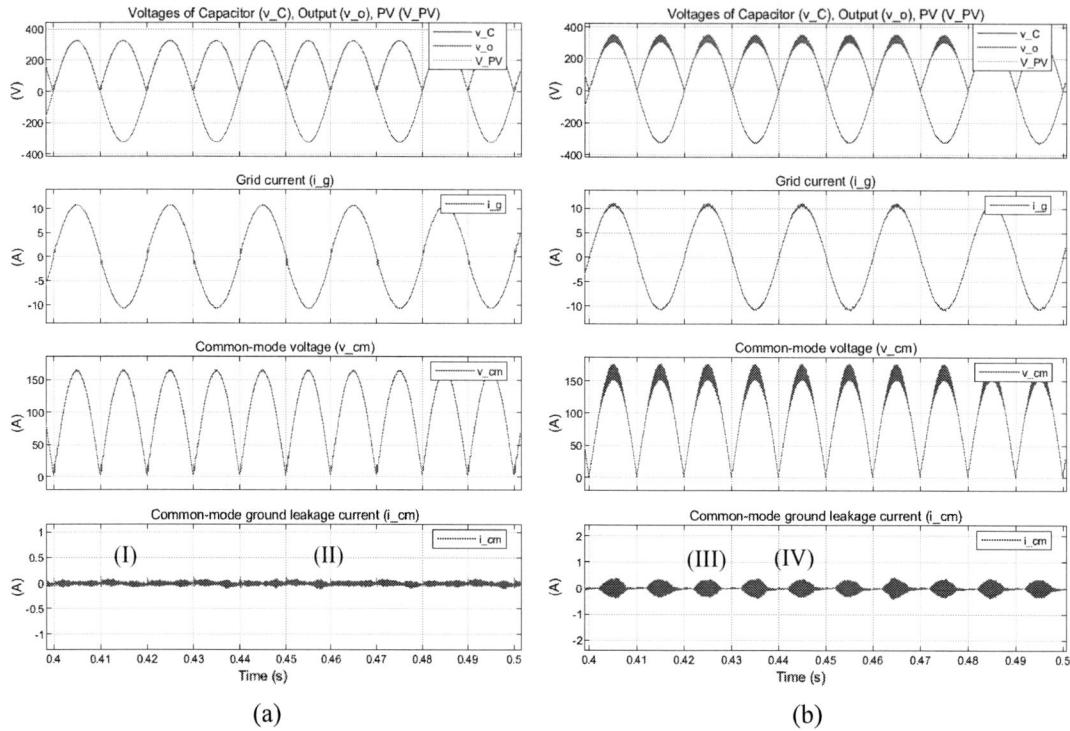

(a)   (b)

Fig. 6: Comparative simulation results illustrating the operation of the topology with the proposed output filter modification for: (a) Buck only operation, and (b) Buck-Boost operation. The RMS values of the CM ground leakage current are 27mA/129mA.

Figure 6 presents the respective waveforms for the proposed filter configuration. The following observations can be made in this case:

- CM current flows during both grid voltage half-cycles, as shown in the bottom graphs of Figure 6. As mentioned in Section 3, this is because the neutral of the grid is not directly connected to the negative DC-link terminal during H-bridge state S1-S4.

- No high peak values of CM current appear at the zero crossings of the grid voltage, due to the existence of inductance in both the line and neutral connections to the grid.

The resulting RMS values of CM current under the same conditions are reduced to 27mA for the Buck mode and 129mA for the Buck-Boost mode. This reduction agrees with the ratio of Eq. (8), which is equal to 0.62 for the given values of ω_s, L_f and C_{par}.

Table I: CM current for the two filter configurations

PV voltage [V]	PV power [W]	Original [mA]	Modified [mA]	Ratio
200	2000	247.8	142.6	0.58
	1000	124.6	71.3	0.57
	500	64.2	37.7	0.59
	250	36.7	21.8	0.60
400	2000	37.7	22.9	0.61
	1000	35.6	22.9	0.64
	500	34.6	22.9	0.66
	250	34.6	22.9	0.66

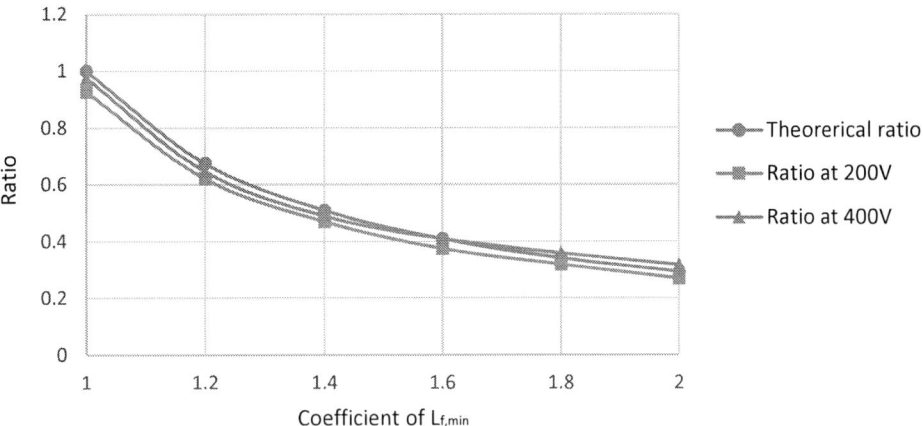

Fig. 7: Comparison of the theoretical ratio from Eq. (8), with ratios from simulations at 200 and 400V, for different values of L_f, ranging from $L_{f,min}$ to $2 \cdot L_{f,min}$.

Additional simulations were performed to compare the RMS value of the CM current for the original and the modified filter configuration under different inverter operating conditions, namely PV array voltage and power (current). The results are presented in Table I. The calculated ratio of $I'_{cm,r,RMS}$ over $I_{cm,r,RMS}$ is shown in the last column, and can be observed to remain within −5/+8% from the theoretically calculated value of 0.62. The deviations are due to the neglection of R_{ground} in the derivation of Eq. (6).

To provide a comparison, according to further simulations, a HERIC inverter [13] supplied by a Boost converter generates 18mA of RMS leakage current under all the conditions of Table I. Nevertheless, this value refers to a HERIC inverter with an L output filter, in the form of two separate inductors (at the line and neutral connections to the grid), each with inductance $L_f / 2 = 600\mu H$. If, with the aim of reducing the grid current distortion to the level of the topology in Figure 4 ($THD_I < 2\%$), an LCL filter is used instead (with four separate inductors of $L = L_f / 2$, and $C = C_f = 3\mu F$), the generated CM current RMS value increases significantly, to around 100mA.

Again with reference to the topologies of Figures 1(a) and 4, other values of L_f, from $L_{f,min}$, which has a value of 961μH according to Eq. (9), up to $2 \cdot L_{f,min}$, were tried in simulation for inverter operation at 2000W and 200V/400V, to further validate the derived expressions. The resulting ratios of generated CM currents are plotted as a function of L_f in Figure 7. It can be observed that both curves closely follow the expected ratio, derived from Eq. (8).

Finally, the effect of the values of L_{dc} and C_f on the generated CM current was examined by simulation. As it was expected, for both filter configurations, the RMS value of the CM current is inversely proportional to the values of L_{dc} and C_f. This is due to the fact that the switching-frequency ripple on v_{Cf}, which generates CM current according to Eq. (4) and (6), is primarily determined by these two parameters and reduces proportionally to them.

5. Discussion

It becomes apparent from the results presented in Section 4 that the proposed filter modification can significantly reduce the generated CM current, if a grid-side inductor with adequately high inductance is selected. Alternatively, CM current reduction could be achieved by increasing the value of the Buck/Boost inductor, L_{dc}, and/or the Buck/Boost output filter capacitor, C_f.

Nevertheless, the increment of L_{dc} is not preferable for the following reasons: a) L_{dc} is rated at a higher current (typically around 2 − 3 times) than L_f, as it supports the operation in Boost mode, b) L_{dc} is subject to higher copper and core losses, because it carries the switching-frequency ripple current generated by the Buck and Boost stages (whereas the current through L_f contains minimal switching-frequency ripple), c) Due to the switching-frequency ripple, L_{dc} must be wound on a low-loss (e.g. ferrite) core, while L_f can be manufactured as a common, line-frequency inductor, and d) Increment of

L_{dc} by a factor of 2 will reduce the CM current to 0.5 of its original value, while increment of L_f by a factor of 2 reduces the CM current to less than 0.3 of its original value, as shown in Figure 7. Thus, increment of L_{dc} instead of L_f would incur higher cost, volume and weight, and result in more losses and lower CM current suppression.

On the other hand, the alternative of increasing the capacitance of C_f could be valid, since it can lower $V_{r,RMS}$ and thus the RMS value of the CM current with minor additional cost. However, in the full-bridge Aalborg inverter as well as in other topologies with rectified sine wave DC-link voltage, the increment of C_f increases the grid current distortion around the zero-crossings of the grid voltage. The distortion relates to the voltage "unfolding" and the restriction for unity power factor, discussed in [12]. It becomes more evident when the inverter operates at low power and practically imposes a limit on the size of C_f.

Given the value of C_f, the proposed filter configuration therefore offers a preferred method for reducing the CM current generated by the full-bridge Aalborg inverter or similar topologies. It should be noted, however, that application of the method requires knowledge (approximate, obtained based on PV panel characteristics or measurement) of the parasitic capacitance C_{par}, since this is required to select a suitable value for L_f, according to Eq. (9). Finally, the grid impedance, which has not been considered in this study, will add to L_f and act favourably with respect to CM current suppression.

6. Conclusion

In this paper, the CM leakage current generation mechanism for an Aalborg-type PV inverter was analysed and an output filter modification was proposed which can reduce the generated CM leakage ground current RMS value by approximately 70%. Analytical expressions were derived and verified by extensive simulations in MATLAB-Simulink, over a wide range of inverter operating conditions and filter parameter values. The presented results are significant because they can also be applied to other topologies belonging in the family of rectified sine-wave DC-link voltage transformerless PV inverters.

References

[1] M. N. H. Khan, M. Forouzesh, Y. P. Siwakoti, L. Li, T. Kerekes and F. Blaabjerg, "Transformerless Inverter Topologies for Single-Phase Photovoltaic Systems: A Comparative Review," in *IEEE Journal of Emerging and Selected Topics in Power Electronics*, vol. 8, no. 1, pp. 805-835, March 2020.

[2] W. Liu, K. Niazi, T. Kerekes and Y. Yang, "A Review on Transformerless Step-Up Single-Phase Inverters with Different DC-Link Voltage for Photovoltaic Applications," in *Energies*, MDPI, vol. 12 (19), pp. 1-17, September 2019.

[3] W. Wu and T. Tang, "Dual-Mode Time-Sharing Cascaded Sinusoidal Inverter," in *IEEE Transactions on Energy Conversion*, vol. 22, no. 3, pp. 795-797, Sept. 2007.

[4] Z. Zhao, M. Xu, Q. Chen, J. Lai and Y. Cho, "Derivation, Analysis, and Implementation of a Boost–Buck Converter-Based High-Efficiency PV Inverter," in *IEEE Transactions on Power Electronics*, vol. 27, no. 3, pp. 1304-1313, March 2012.

[5] W. Wu, J. Ji and F. Blaabjerg, "Aalborg Inverter - A New Type of "Buck in Buck, Boost in Boost" Grid-Tied Inverter," in *IEEE Transactions on Power Electronics*, vol. 30, no. 9, pp. 4784-4793, Sept. 2015.

[6] H. Wang, W. Wu, H. S. Chung and F. Blaabjerg, "Coupled-Inductor-Based Aalborg Inverter with Input DC Energy Regulation," in *IEEE Transactions on Industrial Electronics*, vol. 65, no. 5, pp. 3826-3836, May 2018.

[7] S. Zhang, W. Wu, H. Wang, Y. He, H. S. Chung and F. Blaabjerg, "Voltage Balance Control Based Aalborg Inverter with Single Source in Photovoltaic System," *2018 IEEE International Power Electronics and Application Conference and Exposition (PEAC)*, Shenzhen, 2018, pp. 1-4.

[8] Z. Özkan and A. M. Hava, "A survey and extension of high efficiency grid connected transformerless solar inverters with focus on leakage current characteristics," *IEEE Energy Conversion Congress and Exposition (ECCE)*, Raleigh, NC, 2012, pp. 3453-3460.

[9] H. F. Xiao, and S. J. Xie, "Leakage Current Analytical Model and Application in Single-Phase Transformerless Photovoltaic Grid-Connected Inverter," *IEEE Transactions on Electromagnetic Compatibility*, vol. 52, no. 4, pp. 902-913, 2010.

[10] D. Zografos, E. Koutroulis, Y. Yang and F. Blaabjerg, "Minimization of leakage ground current in transformerless single-phase full-bridge photovoltaic inverters," *17th European Conference on Power Electronics and Applications (EPE'15 ECCE)*, Geneva, 2015, pp. 1-10.

[11] G. I. Orfanoudakis, E. Koutroulis and G. Foteinopoulos, "The role of diodes in the leakage current suppression mechanism of decoupling transformerless PV inverter topologies", *10th International Conference on Modern Circuits and Systems Technologies (MOCAST)*, Thessaloniki, 2021, pp. 1-4.

[12] G. I. Orfanoudakis, E. Koutroulis, G. Foteinopoulos and W. Wu, "Synchronous Reference Frame current control of Aalborg-type PV inverters," *2021 23rd European Conference on Power Electronics and Applications (EPE'21 ECCE Europe)*, 2021, pp. 1-10.

[13] S. Heribert, S. Christoph and K. Jurgen, "Inverter for transforming a DC voltage into an AC current or an AC voltage". Europe Patent 1 369 985 A2, 2003.

Multi-Frequency Traction-to-Auxiliary Integrated EV Drivetrain: Eliminating the Need for an Auxiliary Power Module

1st Caniggia Viana
Department of Electrical Engineering
University of Toronto
Toronto, Canada
caniggiadiniz@gmail.com

2nd Mehanathan Pathmanathan
Department of Electrical Engineering
University of Toronto
Toronto, Canada
meha.pathmanathan@mail.utoronto.ca

3rd Peter W. Lehn
Department of Electrical Engineering
University of Toronto
Toronto, Canada
lehn@ecf.utoronto.ca

Abstract

Leveraging multi-frequency power transfer in the drivetrain, a solution is presented to eliminate the auxiliary power module in electric vehicles. The concept exploits energy harvesting from the drivetrain switching to achieve traction-to-auxiliary power transfer. Only a compensation capacitor, high-frequency transformer, diode rectifier, and a CL filter are added to the drivetrain. Simulations and experimental verification are conducted to validate the proposed system.

ACKNOWLEDGEMENT

This work was supported by the Natural Sciences and Engineering Research Council of Canada (NSERC) under Grant CRDPJ 513206-17.

Index Terms

Auxiliary power module, traction-to-auxiliary, electric vehicle, drivetrain, switching harmonic energy harvesting modeling.

I. INTRODUCTION

Electric vehicle (EV) adoption is significantly impacted by vehicle weight and production cost. While the former indirectly impacts range, a critical buyer concern, the latter directly affects the consumer choice [1]. The auxiliary power module (APM), included in EVs to supply auxiliary loads, contributes significantly with weight and cost, given its substantial and increasing power rating in modern EVs [2].

To mitigate these issues, two main strategies have previously been explored. Increasing the switching frequency of the APM allows for the use of smaller passive components, largely reducing the APM's total size [3], [4]. This approach has been proposed in the literature, although it may require special, and often expensive, wide-bandgap switches. Integration, wherein power electronic or passive components already present in the vehicle are leveraged to implement the APM, provides an alternative route to decrease the APM contribution to vehicle cost and weight [5]. In the literature, solutions have been investigated where a third winding and power electronic interface (PEI) are added to the OBC transformer [6], [7], [8]. Other solutions leverage the entire OBC as the primary side PEI of an isolated dc/dc converter, implementing the APM [9], [10]. A solution is proposed in [11], capable of implementing an APM without the need for a dedicated OBC. However, that solution is only suitable for dual inverter drivetrains, and cannot be utilized along with more conventional single inverter drives.

This work proposes a fully integrated traction-to-auxiliary (T2A) power supply, implemented by leveraging and controlling the zero-sequence component of the voltage applied to the machine, which eliminates the need for an APM. The proposed circuit possesses the following unique set of characteristics:

1) It can be implemented without the addition of any active switches, providing a cost-effective T2A solution. The additional circuitry is limited to a compensation capacitor, an isolation transformer, a diode rectifier, and a CL filter.
2) It does not require any specific OBC topology, making this a widely applicable solution.
3) When driving, no additional active switching action is required to operate the T2A system, resulting in no additional switching loss.
4) It harvests energy from the switching frequency component of the zero-sequence voltage and current in the drivetrain, a largely underutilized degree of freedom in the drivetrain. This increases the asset utilization ratio, allowing vehicle manufacturers to get more functionality out of the already deployed components.

II. PROPOSED SOLUTION

The proposed solution consists of connecting the neutral point of the traction motor to the primary side of the T2A transformer, followed by the compensation capacitor, C_r. On the secondary side of the transformer, a center tapped diode rectifier is connected to a CL filter, followed by the low voltage (LV) battery.

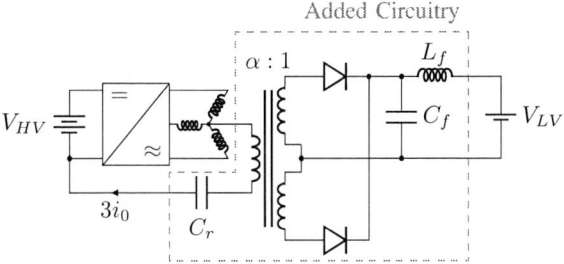

Fig. 1: Proposed T2A solution.

1) Operating Principle: The zero-sequence voltage, v_0, produced by the drivetrain, as described by the Clarke transformation. This voltage is defined as

$$v_0 = \frac{V_{HV}}{3} \sum_{i=a}^{c} g_i, \tag{1}$$

where g_i is the gating pulse associated with the top switch of leg "i" of the inverter, $i \in \{a, b, c\}$. The equivalent circuit representing the associated power transfer in Fig. 1 is shown in Fig. 2.

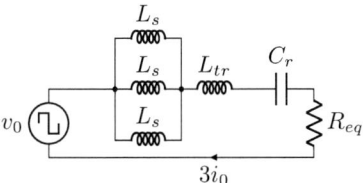

Fig. 2: Equivalent zero-sequence model of the proposed circuit.

In Fig. 2, L_s represents the zero-sequence (leakage) inductance of the machine, L_{tr} represents the leakage inductance of the transformer, referred to the primary side, and C_r is chosen to resonate, at the switching frequency, f_{sw}, with the loop inductance.

$$C_r = \left(\frac{1}{2\pi f_{sw}}\right)^2 \left(\frac{1}{L_{tr} + \frac{L_s}{3}}\right) \tag{2}$$

With placement of C_r defined by Fig. 2, the resonant nature of the circuit ensures the switching frequency component of the current dominates. Therefore, the power transfer in the circuit is related to the magnitude of the phasor describing the harmonic cluster around switching frequency of v_0, $\underline{V}^{(f_{sw})}$.

$$\underline{V}^{(f_{sw})} = \frac{2}{T_{sw}} \int_0^{T_{sw}} v_0(t) e^{-j2\pi f_{sw}} dt \tag{3}$$

In (3), T_{sw} is the switching period, $\frac{1}{f_{sw}}$. The objective of the control system is to control the magnitude of the voltage described in (3) without interfering with the driving operation.

A. Single Frequency Approximation

Note that (3) abstracts the effects of the cluster of frequency disturbances as a single-frequencies. While not exactly mathematically descriptive of the Fourier content of the zero-sequence voltage produced by this approach, this equation allows for power transfer control, provided that the impedance of the transformer is chosen such that

$$\frac{R_{eq}}{2\pi \left(L_{tr} + \frac{L_s}{3}\right)} < f_{sw} + 5f_r, \tag{4}$$

where f_r is the maximum electric frequency of the rotor. This design condition ensures that the LC filter hs wide enough bandwidth to allow the all frequency components in the cluster to produce significant currents.

III. Control System

In this paper, a SPWM modulator synthesizes the machine voltage requested by the drive control to track, for instance, a torque or speed reference. Sufficiently covered in the literature, the drive control is outside the scope of this work. Note, that the same approach applies even if injection of third harmonic is used to enable higher higher voltage synthesis.

As in regular drive systems, the space-vector representation of the voltage requested by the drive control is given, as a function of the modulation index, M, and modulation angle, θ, by

$$\vec{v}^* = \frac{V_{HV}}{2} M e^{j\theta}. \tag{5}$$

The modulator defines the gating pulses by comparing the modulating signals with an appropriate carrier. To implement the T2A controller, three carriers c_a, c_b, and c_c are defined, such that

$$\begin{bmatrix} g_a \\ g_b \\ g_c \end{bmatrix} = \begin{bmatrix} M\cos(\theta) > c_a \\ M\cos(\theta - 120°) > c_b \\ M\cos(\theta + 120°) > c_c \end{bmatrix}, \tag{6}$$

where a controllable phase-shift, $\delta \in [0°, 120°]$, is established between carriers, as shown in Fig. 3.

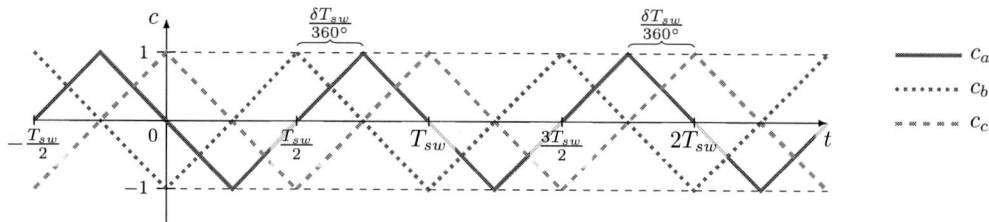

Fig. 3: Carrier definition and phase-to-phase carrier phase-shift representation.

The magnitude of the switching frequency component of the zero-sequence voltage is dependent on carrier phase-shift, δ, and can be shown to be approximately

$$\left| \underline{V}^{(f_{sw})} \right| \approx \left(\frac{2V_{HV}}{3\pi} \right) J_0 \left(\frac{M\pi}{2} \right) |1 + 2\cos(\delta)|, \tag{7}$$

where J_0 is a Bessel function of the first kind. Equation (7) implies that the T2A power transfer can be controlled by varying δ. A value of $\delta = 0°$ leads to maximum voltage and consequent maximum power transfer, whereas $\delta = 120°$ brings T2A power output to 0. Fig. 4 shows how the magnitude of the combines frequencies clustered around the switching frequency vary as a function of the modulation index, M, and the carrier phase-shift, δ. Moreover, changing the carrier phase-shift does not affect the line-to-line voltages applied to the machine, hence, the system does not have any effect on the drive operation.

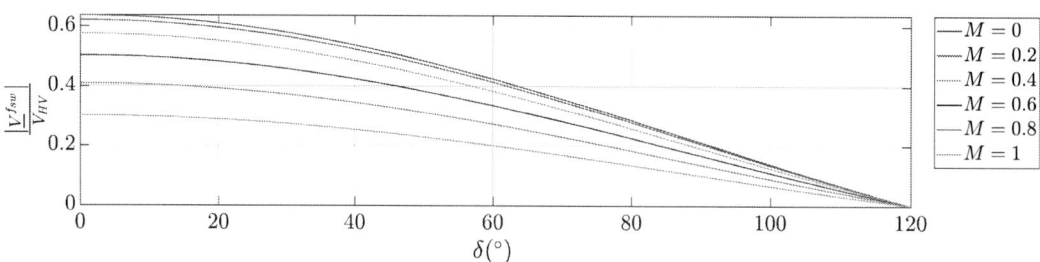

Fig. 4: Magnitude of the zero-sequence voltage produced around the switching frequency as a function of the modulation index and carrier phase-shift.

A control system is proposed using a simple PID controller to correct the value of delta to ensure i_{LV} tracks the associated reference, i_{LV}^*. This control loop runs in parallel with the traditional drive control. The T2A control diagram is shown in Fig. 5. The saturation block ensures $0° \leq \delta \leq 120°$.

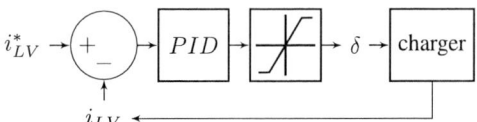

Fig. 5: Proposed T2A control loop, to be run in parallel with the drive control.

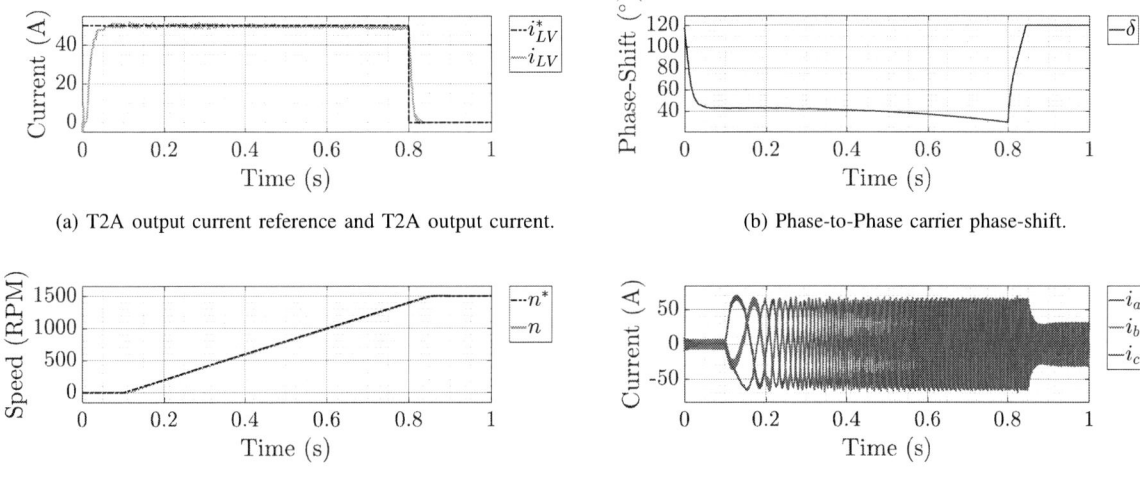

(a) T2A output current reference and T2A output current.

(b) Phase-to-Phase carrier phase-shift.

(c) Reference speed and speed.

(d) Phase currents through the machine windings.

Fig. 6: Simulation results, showing simultaneous T2A and drive operation.

IV. SIMULATIONS

Using the control architecture described in Fig. 5, along with a traditional speed-tracking drive control, the system in Fig. 1 is simulated. The simulation results are shown in Fig. 6.

At time $t = 0$, the T2A output current reference is set to $i_{LV*} = 50\ A$. The system satisfactorily tracks the reference, as shown in Fig. 6a. To achieve T2A power transfer, the T2A controller reduces δ, from $120°$, which results in the lowest voltage $\left|\underline{V}^{(1)}\right|$, toward $0°$, which results in the highest voltage, as shown in Fig. 6b.

At time $t = 0.1\ s$, the speed reference is ramped up to $1500\ RPM$. The traditional drive control increases the current reference and accelerates the system, tracking the speed reference, as shown in Fig. 6c. Importantly, the current output of the system, shown in Fig. 6d, is unremarkable and similar to what is expected in a regular drive system, demonstrating that the presence and operation of the T2A system do not affect the drive system in any significant way. As the speed increases, the modulation index, M, increases, leading to a requirement of a somewhat lower value of delta to maintain the same power level. This relationship is best seen in Fig. 6b, at $0.5\ s < t < 0.8\ s$.

At time time $t = 0.8\ s$ the T2A output current reference is stepped down to $0\ A$, which the T2A control tracks. Once again, both the T2A operation and transient are demonstrated to not impact the drive control.

V. EXPERIMENTAL RESULTS

An experimental circuit is setup to verify the analytical claims. The experimental setup has three main parts:

1) The power electronics and controls.
2) The PMSM and dynamometer, enabling driving operation at a specifiable speed and torque.
3) The T2A added circuitry, including transformer, rectifier, and compensation capacitor.

Pictures of the experimental setup are shown in Fig. 7. The parameters used in this experiment are shown in Table I.

The auxiliary voltage used in this experiment is $V_{LV} = 6V$, enforced by an electronic load, emulating the auxiliary battery. Note that this is a scaled representation of the 12 V typically used in EVs. This voltage level was selected based on laboratory component availability.

Four experiments are conducted to demonstrate the operation of the circuit.

(a) Inverter and DSP.

(b) PMSM & dynamometer.

(c) Transformer & rectifier.

(d) Capacitor C_r.

Fig. 7: Experimental setup to be used for experimental verification.

TABLE I: Parameters used in experimental verification.

Description	Symbol	Value
System		
Battery Voltage	V_{HV}	400 V
Auxiliary Voltage	v_{LV}	60 V
Switching Frequency	f_{sw}	10 kHz
Machine		
0-Seq. Inductance	L_S	0.53 mH
q-axis Inductance	L_q	0.94 mH
d-axis Inductance	L_d	0.73 mH
Series Resistance	R_S	45 $m\Omega$
Pole Pairs	$p/2$	5
Flux Linkage	R_S	0.127 Wb
Rated Current (RMS)	I_{max}	200 A
T2A		
Transformer Inductance	L_{tr}	700 μH
Compensation Capacitor	C_r	280 nF
Turns Ratio	α	20 V/V
Filter Capacitance	C_f	4 mF
Filter Inductance	L_f	5.37 μH

1) T2A transient from 0 to 30 A output current reference at standstill.
2) T2A transient from 0 to 30 A output current reference while the drivetrain is operated at 500 RPM and 10 Nm output.
3) Steady-state T2A operation at 0 A output current at standstill.
4) Steady-state T2A operation at 30 A output current at standstill.

During the first experiment, the drivetrain is set to idle, i.e., to operate at 0 RPM and 0 Nm. With the T2A system initially outputting 0 current, a transient in reference output current, i_{LV}^*, is applied from 0 to 30 A. Auxiliary voltage and currents resulting from this experiment are shown in Fig. 8. The during the transient, the current rises to track the reference. The auxiliary voltage increases slightly, in the presence of current.

During the T2A transient, the current flowing through phase a of the machine is measured. The result is shown in Fig. 8. During the transient, the current remains approximately 0.

Using a torque transducer, the torque on the axle connecting the PMSM under test and the dynamometer is measured. The result is shown in Fig. 8. No significant transient can be observed in the torque during the T2A transient. This result suggests that the T2A system, despite leveraging the machine and drive inverter, has no significant impact on the mechanical behavior of the system.

The second experiment explored herein, similarly to the first one, applies a T2A transient in output current from 0 to 30 A. The difference between tests lies in the fact that the drive, here, is set to 500 RPM and 10 Nm. In this setup, the dynamometer is responsible for setting the system speed, while the drivetrain operates in torque control mode. The auxiliary current and voltage waveforms resulting from the aforementioned procedure are measured and shown in Fig. 9. As was observed in the test at standstill, the current rises to meet the reference, while the auxiliary voltage displays a slight increase.

The current flowing through phase a of the machine is measured during this operation. The result is shown in Fig. 9. It can be seen the the current is primarily sinusoidal, as a result of the drive operation. The ripple can be observed to be higher in the portion of the curve where the T2A system is processing zero power. This is the case because to produce $i_{LV} = 0$ the system applies approximately $\delta = 120°$ phase-shift, producing a significant line-to-line voltage component at $10kHz$. In contrast, to output $i_{LV} > 0$ the phase-shift is reduced, thereby reducing the line-to-line voltage component, but aligning the line-to-neutral

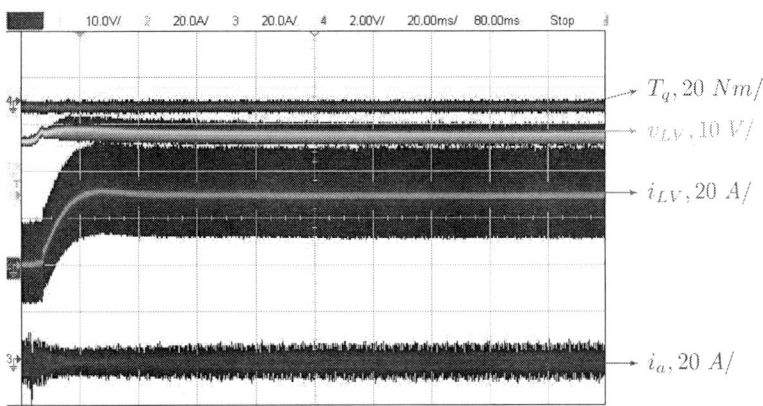

Fig. 8: Measured auxiliary voltage and currents, stator phase current, and drivetrain torque during T2A transient at standstill.

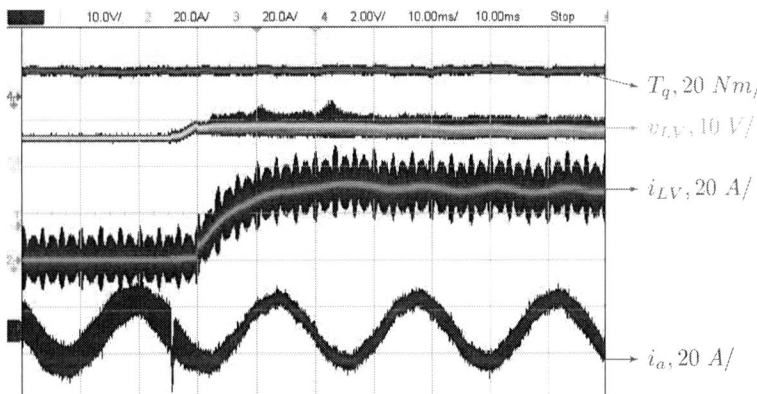

Fig. 9: Measured auxiliary voltage and currents, stator phase current, and drivetrain torque during T2A transient at $500\ RPM$ and $10\ Nm$.

currents and voltages of all phases, thus increasing the current through neutral.

While the transient is applied, the measured torque remains at approximately 10 Nm. Once again, no appreciable transient is seen in the torque produced by the system. This result corroborates the analytical conclusion that the T2A system does not interfere with the drive mechanical behavior.

Two more experiments are conducted with the drive system at standstill. These tests aim to show that the system operates with less stator phase current ripple when outputting significant T2A power.

Firstly, the T2A system is set to operate at zero output current, $i_{LV}^* = 0$. The machine phase a stator current is measured. As shown in Fig. 10, the peak current value is around 7.5 A.

The previous experiment is repeated, but this time with the T2A system outputting $i_{LV} = 30\ A$. Fig. 11 shows the measured stator phase a current. In comparison with the no power case, this scenario produces less phase current ripple, with peak current below 5 A. This phenomenon is explained by the alignment of the phase-shift, decreasing the line-to-line voltage seen by the stator. In other words, to output power, the T2A system increases the 10 kHz component of the 0-axis current through the machine, but it reduces the 10 kHz component of $\alpha-$ and $\beta-$axis currents more than commensurately, thus reducing the phase current ripple in comparison to the zero output case.

VI. CONCLUSION

A traction-to-auxiliary power supply integrated to the drivetrain is proposed for electric vehicles. The solution leverages the switching frequency component of the zero-sequence current through the drive machine, a usually underutilized degree of freedom in the drive system, to facilitate power transfer. As a result, the electric vehicles requirement of an auxiliary power module, a significant component both in weight and in cost, is eliminated. In lieu of a standalone auxiliary power

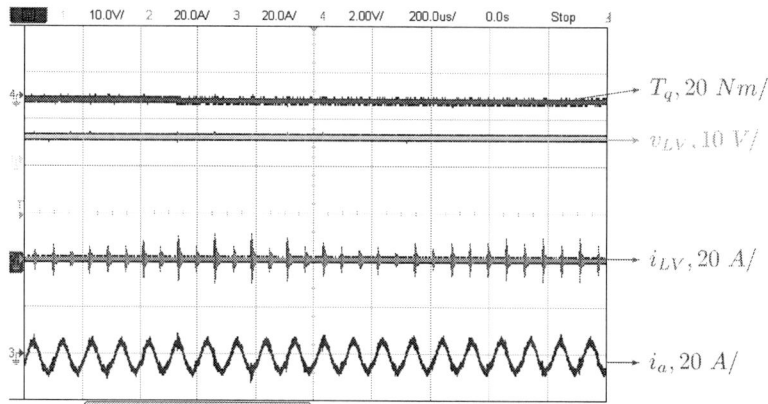

Fig. 10: Measured auxiliary voltage and currents, stator phase current, and drivetrain torque during steady-state operation at standstill, with zero T2A output power.

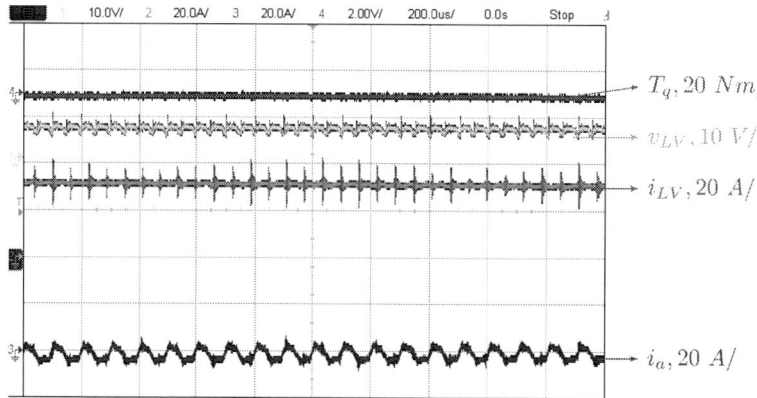

Fig. 11: Measured auxiliary voltage and currents, stator phase current, and drivetrain torque during steady-state operation at standstill, with zero T2A output power.

module, a small transformer and other small passive components are included. Despite using the zero-sequence component, this multi-frequency technique does not impede techniques that facilitate higher drive voltage synthesis, such as third harmonic injection.

Simulation results are presented to corroborate the analytical results, wherein the proposed auxiliary charger is demonstrated to operate simultaneously with the drive system. The operation is achieve by virtue of controlling the phase-shift between carriers in the drive system modulator, while not interfering with drive controls. An experimental setup is constructed and used to demonstrate not only satisfactory traction-to-auxiliary operation, but also that the additional system does not cause any noticeable interference with the mechanical behavior of the drivetrain, despite leveraging both the traction machine and the traction inverter during driving operation.

REFERENCES

[1] M. Coffman, P. Bernstein, and S. Wee, "Factors affecting EV adoption: A literature review and EV forecast for Hawaii," *Electric Vehicle Transportation Center*, no. April 2015, pp. 1–36, 2015.

[2] S. M. Hasan, M. N. Anwar, M. Teimorzadeh, and D. P. Tasky, "Features and challenges for Auxiliary Power Module (APM) design for hybrid/electric vehicle applications," *2011 IEEE Vehicle Power and Propulsion Conference, VPPC 2011*, 2011.

[3] A. M. Naradhipa, S. Kim, D. Yang, S. Choi, I. Yeo, and Y. Lee, "Power Density Optimization of 700 kHz GaN-Based Auxiliary Power Module for Electric Vehicles," *IEEE Transactions on Power Electronics*, vol. 36, no. 5, pp. 5610–5621, 2021.

[4] H. Moradisizkoohi, N. Elsayad, and O. A. Mohammed, "Experimental Demonstration of a Modular, Quasi-Resonant Bidirectional DC-DC Converter Using GaN Switches for Electric Vehicles," *IEEE Transactions on Industry Applications*, vol. 55, no. 6, pp. 7787–7803, 2019.

[5] A. Khaligh and M. Dantonio, "Global Trends in High-Power On-Board Chargers for Electric Vehicles," *IEEE Transactions on Vehicular Technology*, vol. 68, no. 4, pp. 3306–3324, 2019.

[6] L. Zhu, H. Bai, A. Brown, and L. Keuck, "A Current-fed Three-port DC/DC Converter for Integration of On-board Charger and Auxiliary Power Module in Electric Vehicles," *2021 IEEE Applied Power Electronics Conference and Exposition (APEC)*, pp. 577–582, 2021.

[7] Y. Tang, J. Lu, B. Wu, S. Zou, W. Ding, and A. Khaligh, "An Integrated Dual-Output Isolated Converter for Plug-in Electric Vehicles," *IEEE Transactions on Vehicular Technology*, vol. 67, no. 2, pp. 966–976, 2018.

[8] B. Farhangi and H. A. Toliyat, "Modeling and Analyzing Multiport Isolation Transformer Capacitive Components for Onboard Vehicular Power Conditioners," *IEEE Transactions on Industrial Electronics*, vol. 62, no. 5, pp. 3134–3142, 2015.

[9] J. G. Pinto, V. Monteiro, H. Gonçalves, and J. L. Afonso, "Onboard reconfigurable battery charger for electric vehicles with traction-to-auxiliary mode," *IEEE Transactions on Vehicular Technology*, vol. 63, no. 3, pp. 1104–1116, 2014.

[10] S. Kim and F. S. Kang, "Multifunctional onboard battery charger for plug-in electric vehicles," *IEEE Transactions on Industrial Electronics*, vol. 62, no. 6, pp. 3460–3472, 2015.

[11] C. Viana, M. Pathmanathan, and P. W. Lehn, "Auxiliary Power Module Elimination in EVs Using Dual Inverter Drivetrain," *IEEE Transactions on Power Electronics*, pp. 1-1, 2022.

Potentials to Improve the Post-Fault Performance of a Fault-Tolerant Inverter System in Electrified Aircraft Propulsion System

Yongtao Cao[1,2], Leon Fauth[1,2], Jens Friebe[1,2] and Axel Mertens[1,2]
[1]Cluster of Excellence SE²A - Sustainable and Energy-Efficient Aviation
Technische Universität Braunschweig
Universitätsplatz 2
38106 Braunschweig, Germany
[2]Institute for Drive Systems and Power Electronics
Leibniz Universität Hannover
Welfengarten 1
30167 Hannover, Germany
Phone: +49 511 762 14121
Email: Yongtao.Cao@ial.uni-hannover.de
URL: https://www.ial.uni-hannover.de/

Acknowledgments

We would like to acknowledge the funding by the Deutsche Forschungsgemeinschaft (DFG, German Research Foundation) under Germany's Excellence Strategy – EXC 2163/1 - Sustainable and Energy-Efficient Aviation – Project-ID 390881007. The authors are responsible for the content of this publication.

Keywords

≪Fault tolerance≫, ≪Reliability≫, ≪All Electric Aircraft≫

Abstract

Due to the increased exposure to cosmic radiation at higher flight altitude, a significant voltage derating of power semiconductors is required to reduce the failure rate of the inverters designed for electrified aircraft propulsion systems. By utilizing the necessary oversized block-voltage capability of the power semiconductors while employing the concept of variable DC-link voltage, post-fault performance of the inverter system can be improved.

Introduction

To achieve the ambitious goal of net-zero carbon emissions in Europe by 2050, several projects have been proposed to investigate sustainable and energy-efficient aviation through electrified propulsion systems. However, electrification of aircraft propulsion systems is still confronted with several technical challenges, especially in terms of reliability. For the application of power electronic converters in aircraft propulsion systems, the increased exposure to cosmic radiation at high altitude, changing ambient temperature during a flight mission and low air pressure at high altitude all pose new challenges. These working conditions should be carefully taken into account during the design phase. For example, compared to an application on the ground, a greater voltage derating of the power semiconductors due to increased exposure to cosmic radiation at high altitude needs to be considered to reduce the device failure rate in aircraft application [1].

The system reliability can be further improved by using a twin-motor propulsion architecture. However, the loss of one motor is still serious for large passenger aircraft, especially when it happens at low altitude and low airspeed, where maximum power is needed during take-off, as it could potentially result in a catastrophic accident [2]. In order to avoid the consequences of motor shutdown due to power semiconductor failure and improve the overall system reliability, fault-tolerant inverter systems can be implemented in the electric aircraft propulsion system. In [3], multilevel inverters with potential fault-tolerance such as active neutral point clamped (ANPC) inverter, neutral point clamped (NPC) inverter, modular multilevel converter (MMC) and T-Type inverter are investigated for future electric aircraft propulsion systems. Based on the classic fault-tolerant inverter concepts, this paper introduces the potential to improve their post-fault performance in aircraft applications by utilizing the necessary oversizing of the power semiconductors due to cosmic radiation. These strategies are especially practical when the fault happens at take-off with low flight altitude (less cosmic radiation) and high power demand. Combined with the concept of variable DC-link voltage, the post-fault performance can be improved yet further.

Mission profile and voltage utilization

In Fig. 1, the mission profile of an electric short-range aircraft derived from [4] is shown. Its cruising altitude is at 6 km above sea level. For mid- or long-range aircraft, the cruising altitude could be up to 12 km. As can be seen from the figure, maximum power and maximum output voltage are required at take-off, where the flight altitude is still low. Due to the single-event burnout effect (SEB), necessary voltage derating of power semiconductors needs to be considered to reduce the failure rate for aircraft applications.

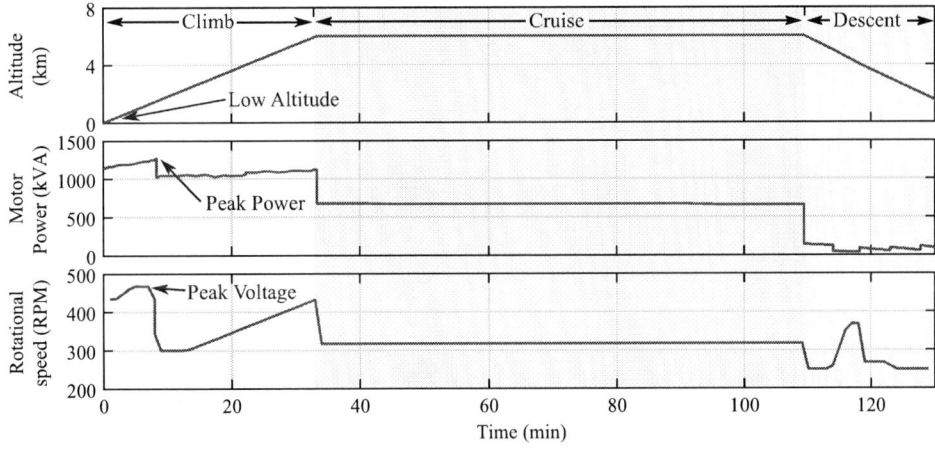

Fig. 1: Typical mission profile for a short range flight, with data based on [4]. The flight altitude (top), motor power (middle) and rotational speed (bottom), which is also proportional to the motor voltage. Most critical for cosmic radiation induced failures is the cruise phase because of the high altitude.

The SEB like all single-event effects (SEE) is a random failure which can not be predicted and can cause a sudden destruction of a semiconductor device. It is generated by a radiation particle interacting with the active area of the semiconductor. These are mainly neutrons in typical aircraft applications, which are generated by spallation from cosmic radiation. The number of available particles increases exponentially with higher altitudes [5]. Due to the stochastic nature of the failure mechanism, a linear dependency of the failure rate on the number of particles is assumed, resulting in an exponential dependency of the failure rate on the flight altitude. Additionally, the failure rate is increased if the device is operated closer to its breakdown voltage level [6]. Based on this, the failure rate of a 1700 V SiC-MOSFET per square centimeter of chip area is depicted as a function of the voltage utilization at different flight altitudes in Fig. 2.

Since the failure rate of the current airliner engine is around 1000 FIT (FIT: failure in time, 1 FIT = 1 Failure in 10^9 device hours) to meet the aircraft safety requirement [7], 1000 FIT can be set as the reference failure rate of one electrified aircraft propulsion system composed of battery, power electronic systems, electric motor and propeller (or turbofan). The required failure rate of the inverter system in an electrified aircraft propulsion system can therefore be expected to be around hundred or several hundred FIT. Considering the fact that capacitors, driver board, PCB and power semiconductors all contribute to the inverter system failure rate, for a megawatt-level (3 MW to 20 MW) multilevel inverter with large chip areas (200 cm² to 1200 cm², chip areas are estimated according to the method introduced in [3]), the maximum acceptable failure rate of the power semiconductor is considered to be at least less than 0.1 FIT per square centimeter in this paper. In practical applications, the failure rate requirement of the power semiconductors may vary slightly, depending on the failure rate allocation in the electric propulsion drive system. However, the aim of this paper is only to show the potential to improve the post-fault performance in electrified aircraft applications. As can be seen from Fig. 2, 45% voltage utilization at cruise altitude of 6 km above sea level is required for a 1700 V SiC-MOSFET to maintain this required failure rate. Compared to aircraft applications, 67% voltage utilization of the power semiconductor is typical for the drive systems in electric vehicle on the ground, a 1200 V SiC-MOSFET is usually used to block the 800 V DC-link voltage. In this paper, 67% will be assumed to be the maximum allowed voltage utilization of the power semiconductors for an emergency situation at low altitude.

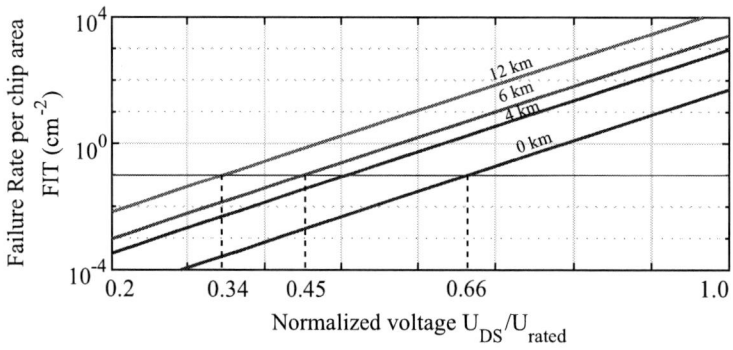

Fig. 2: Example failure rate for a 1.7 kV SiC-MOSFET at different altitude. Failure rate data based on results from [6].

Fixed DC-link voltage while utilizing the necessary oversizing

A multilevel inverter such as an MMC has multiple power semiconductors connected in series per arm (see Fig. 3). By applying an appropriate fault-tolerant control strategy and fault-tolerant isolation methodology in the design, the inverter system can still provide full performance despite the failure of a submodule. However, if the DC-link voltage of the inverter system is fixed, the other healthy submodules in the faulty arm must be able to withstand a higher voltage stress for fault-tolerant operation, meaning that oversized or redundant devices are necessary for classic fault-tolerant design. Nevertheless, by utilizing the necessary oversizing resulting from power semiconductor voltage derating implemented to withstand cosmic radiation and choosing a suitable number of submodules, unnecessary oversizing or redundancy can be avoided for fault-tolerant operations.

In Fig. 3, a fault-tolerant MMC with a fixed 3 kV DC-link voltage U_{DC} is depicted as an example. If SiC-MOSFET with a rated blocking voltage $U_{DS} = 1700$ V is chosen for this application, at least four submodules per arm are required to block the full DC-link voltage U_{DC} due to their required maximum voltage utilization $u_1 = 0.45$ at the cruise altitude of 6 km:

$$\frac{U_{DC}}{U_{DS} \cdot n} = \frac{3000 \text{ V}}{1700 \text{ V} \cdot 4} = 0.44 < u_1 \tag{1}$$

In case of a submodule failure, the failed submodule can be isolated with a bypass switch. If voltage utilization $u_{max} = 0.67$ is allowed for power semiconductors at take-off and emergencies, the rest of the three submodules can still block the full DC-link voltage U_{DC} with the voltage utilization u_{fault}.

$$\frac{U_{DC}}{U_{DS} \cdot (n-1)} = \frac{3000 \text{ V}}{1700 \text{ V} \cdot (4-1)} = u_{fault} < u_{max} \tag{2}$$

In this case, no redundant submodule or extra oversizing is required for this fault-tolerant operation. However, additional by-pass switches for fault isolation in the submodules will still bring additional costs.

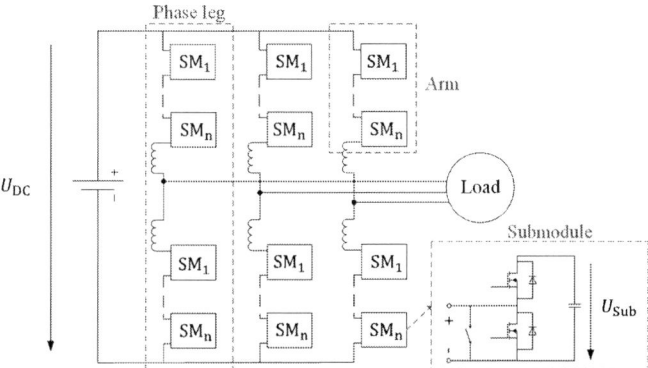

Fig. 3: Fault-tolerant MMC

Optimized variable DC-link voltage for fault-tolerant operation

In the following section, variable DC-link voltage strategies will be introduced to improve the post-fault performance of the power converter by utilizing the block-voltage capability of its power semiconductors. A NPC inverter will be analysed in a case study.

Optimized reduced DC-link voltage for fault-tolerant operation of NPC inverter

As can be seen from Equation (2), for fault-tolerant operation of a multilevel inverter with fixed DC-link voltage and multiple power semiconductors connected in series, a higher maximum voltage utilization u_{max} is required as n decreases. However, u_{max} is limited due to the overvoltage protection requirement. Such a limitation is especially critical for the fault-tolerant operation of a three-phase three-level NPC.

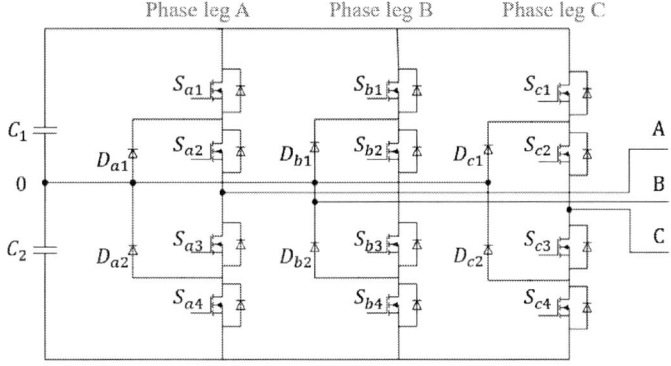

Fig. 4: Three-phase three-level NPC inverter

For example, if the outer power semiconductor S_1 has a short circuit failure (see Fig. 5 left), the inner power semiconductor S_2 has to withstand the full DC-link voltage during '-' state (S_3 and S_4 on). However, power semiconductors of the NPC inverter are usually only designed to be able to block half of the DC-link voltage U_{DC}, the increased voltage of S_2 could lead to a secondary failure and eventually result in system shutdown. Furthermore, due to the short circuit of S_1, the '0' state can also not be realized (the current flow through D_5 and S_2 is not possible since S_2 cannot be turned on). As can be seen from Fig. 5, only the white area of the space vector diagram can be reached in this case, a symmetric circular vector trajectory therefore cannot be generated [8].

However, if the DC-link voltage of the NPC inverter can be adjusted (reduced) for this fault scenario in such a way, that the voltage stress on S_2 is reduced in an acceptable range, the '-' state is available again, the NPC inverter can at least work in a 2-level mode allowing the use of a full modulation index in this case (see Fig. 5 right). In an emergency, especially when the fault happens during take-off of the electric aircraft, the necessary oversizing of the power semiconductors due to cosmic radiation can be utilized to maximize the post-fault performance. For example, in Fig. 5, 1700 V SiC-MOSFETs are employed in an NPC inverter with 1500 V U_{DC}. The voltage utilization of the power semiconductors is 44%, which corresponds to the required voltage derating in aircraft applications with a cruise altitude of 6 km (see Fig. 2). When employing the strategy of variable DC-link voltage, 1139 V U_{DC} (1700V $\cdot$ 67% = 1139V) can be applied to the DC-link in this fault scenario, which corresponds to 76% (1139V/1500V = 76%) of the original maximum output voltage of the NPC inverter with fixed DC-link voltage. System shutdown can thus be avoided with this strategy. In Fig. 6, the fault-tolerant strategy is validated in a simulation with constant load parameter, where the fault occurs at 0.05 seconds followed by the proposed fault-tolerant strategy with decreased DC-link voltage.

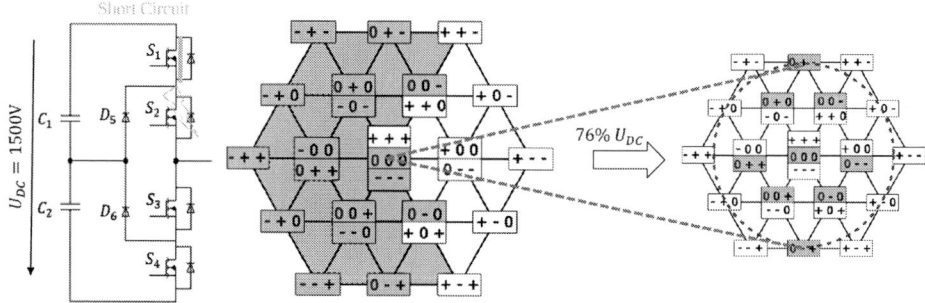

Fig. 5: Improved post-fault performance of an NPC using an optimized decreased DC-link voltage when S_1 suffers a short circuit. Red switching states are not available under this kind of fault, grey areas of the space vector diagram are therefore not reachable anymore. With reduced DC-link voltage (76% U_{DC}), the '-' state is possible again due to the reduced voltage stress on S_2 (67% voltage utilization) to realize fault-tolerant operation (right space vector diagram).

The variable DC-link voltage can either be adjusted with a DC-DC converter at the battery side of the DC distribution system or using the converter system in a serial-hybrid electric aircraft propulsion system.

It should also be mentioned that, such kind of strategy is under the assumption of power semiconductors with safe short circuit capability (in this case: S_1), allowing carrying the current further under its short circuit failure. One of the solutions is to use press-pack packaging. Press-pack packaging also brings the benefit of better cooling capability compared with wire-bonded power semiconductors. However, the higher cost of press-pack technology is one of the main reasons that limited its widespread use [9].

Optimized increased DC-link voltage for fault-tolerant operation of NPC inverter

For a single-event fault scenario such as an open circuit of S_1, or short circuit of S_2, an NPC with fixed DC-link voltage can use its clamping path to connect one faulty phase to the DC mid-point, so that the inner hexagon of the space vector diagram can still be used for fault-tolerant operation with reduced output (50 % degradation) (see Fig. 7). An ANPC inverter has more flexibility in the use of this strategy due to its clamping active controllable power semiconductors in the clamping path [10].

With the help of a variable DC-link voltage in the NPC or ANPC inverter system, the inner hexagon of the space vector can be further increased in the emergency case of take-off by utilizing the necessary

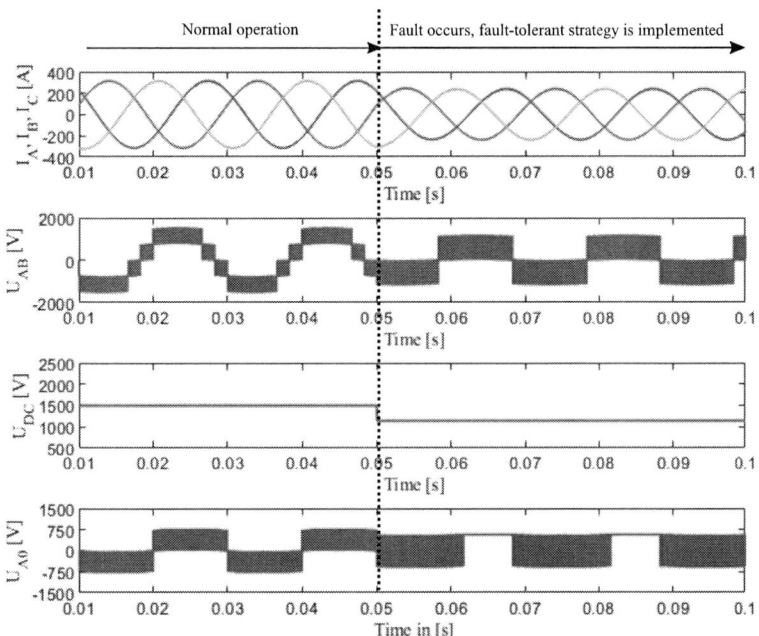

Fig. 6: Improved post-fault performance with proposed strategy under short circuit failure of the outer power semiconductor S_1 in a three-phase three-level NPC inverter (fault occurs at 0.05 seconds)

oversizing of the power semiconductors due to cosmic radiation. To give an example in Fig. 7, 1700 V SiC-MOSFETS are employed on a NPC inverter with 1500 V DC-link voltage, which corresponds to the required 44 percent voltage utilization of the power semiconductor at cruise. Under the condition of open circuit of S_1, 50 percent of the maximum output voltage can still be generated for fault-tolerant operation (see Fig. 7 middle). If the DC-link voltage is adjustable during take-off in this scenario, it can be increased to 2278 V. The voltage utilization on the power semiconductors is increased from 44 % to 67 % (2278 V / 2 / 1700 V = 67 %), allowing the necessary oversizing to be utilized. Compared to the maximum output of the NPC with fixed DC-link voltage during fault-tolerant operation, 76 % (2278 V / 2 / 1500 V = 76 %) of its maximum output voltage can still be provided, instead of 50 %. In Fig. 8, the fault-tolerant strategy is validated in a simulation with constant load parameter, where the fault occurs at 0.05 seconds followed by the proposed fault-tolerant strategy with increased DC-link voltage.

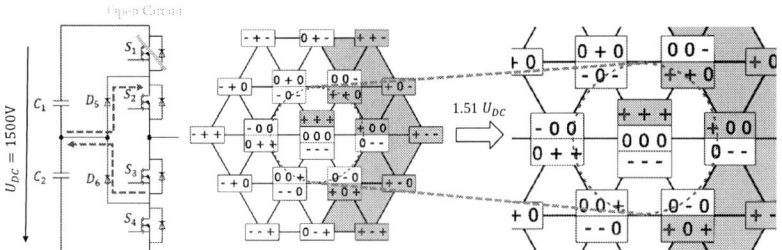

Fig. 7: Improved post-fault performance of an NPC using an optimized increased DC-link voltage strategy under the condition of the short circuit of S_1. Red switching states are not available, grey areas of the space vector diagram are therefore not reachable anymore, a smaller symmetric circular vector trajectory can still be generated. With increased DC-link voltage (151% U_{DC}), the voltage utilization of the power semiconductors can be maximized to increase the inner hexagon of the space vector diagram.

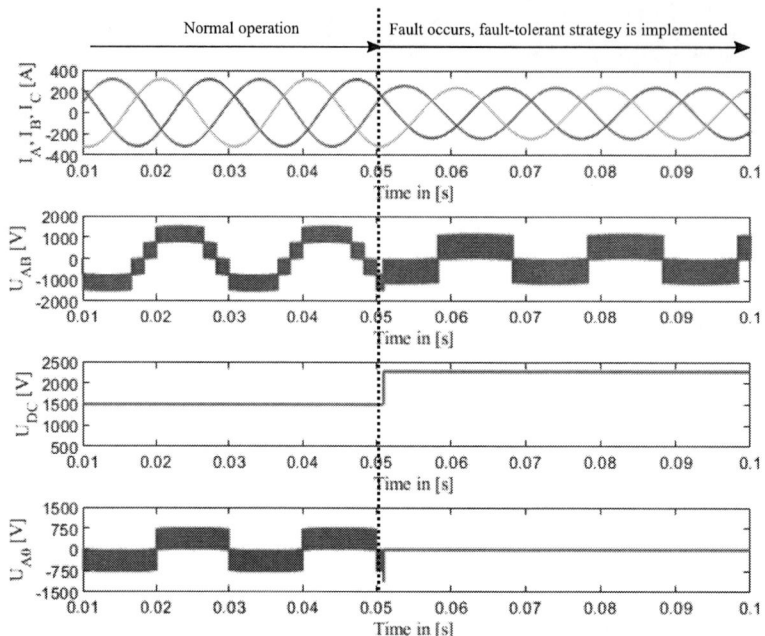

Fig. 8: Improved post-fault performance with proposed strategy under open circuit failure of the outer power semiconductor S_1 in a three-phase three-level NPC inverter (fault occurs at 0.05 seconds)

Conclusion

In this paper, the effect of cosmic radiation on the voltage derating of the power semiconductors in aircraft applications was discussed. By utilizing the necessary oversizing and combining this with proposed fault-tolerant design methodologies, a modular multilevel inverter with a fixed DC-link voltage can still provide full performance if one submodule fails. Considering a conventional 3-level inverter like NPC or ANPC, the post-fault performance of the inverter system can be further improved with the proposed concept of a variable DC-link voltage. The improved post-fault performance can contribute to aircraft safety, especially during take-off, when the maximum power is required and the flight altitude is low.

References

[1] H. Schefer, L. Fauth, T. H. Kopp, R. Mallwitz, J. Friebe and M. Kurrat, "Discussion on electric power supply systems for all electric aircraft," in IEEE Access, vol. 8, pp. 84188-84216, 2020.

[2] D. Kritzinger: "Aircraft system safety - military and civil aeronautical applications," Woodhead Publishing, pp.76-77, 2006.

[3] J. Ebersberger, M. Hagedorn, M. Lorenz and A. Mertens, "Potentials and Comparison of Inverter Topologies for Future All-Electric Aircraft Propulsion," in IEEE Journal of Emerging and Selected Topics in Power Electronics.

[4] S. Karpuk and A. Elham, "Influence of novel airframe technologies on the feasibility of fully-electric regional aviation," Aerospace, vol.8, no.6, pp.163, 2021.

[5] Measurement and Reporting of Alpha Particle and Terrestrial Cosmic Ray-Induced Soft Errors in Semiconductor Devices, JESD89A, JEDEC Solid State Technology Association, Oct. 2006. [Online]. Available: https://www.jedec.org/sites/default/files/docs/jesd89a.pdf

[6] C. Felgemacher, S. V. Araujo, P. Zacharias, K. Nesemann, and A. Gruber, "Cosmic radiation ruggedness of Si and SiC power semiconductors," in 2016 28th International Symposium on Power Semiconductor Devices and ICs (ISPSD), Prague, Czech Republic, Jun. 2016 - Jun. 2016, pp. 51–54.

[7] A. H. Epstein and S. M. O'Flarity, "Considerations for reducing aviation's CO2 with aircraft electric propulsion, " Journal of Propulsion and Power, vol.35, no.3, pp.572–582, 2019.

[8] Y. Cao, L. Fauth, A. Mertens and J. Friebe, "Comparison and Analysis of Multi-State Reliability of Fault-Tolerant Inverter Topologies for the Electric Aircraft Propulsion System," 2021 24th International Conference on Electrical Machines and Systems (ICEMS), 2021, pp. 766-771.

[9] K.-B. Lee and J.-S. Lee, "Reliability Improvement Technology for Power Converters," Power Systems, 2017.

[10] J. Li, A. Q. Huang, Z. Liang and S. Bhattacharya, "Analysis and Design of Active NPC (ANPC) Inverters for Fault-Tolerant Operation of High-Power Electrical Drives," in IEEE Transactions on Power Electronics, vol. 27, no. 2, pp. 519-533, Feb. 2012.

Model Predictive Control-enabled Fault Ride Through Operation Strategy for High Power Wind Turbine

Pedro Catalán[1], Yanbo Wang[2], Zhe Chen[2], Joseba Arza[3]

[1]*Ingeteam Power Technology S.A.,* Zamudio, Spain
[2]*AAU Energy, Aalborg University,* Aalborg, Denmark
[3]*Ingeteam R&D Europe,* Zamudio, Spain

E-mail: pedro.catalan@ingeteam.com, ywa@et.aau.dk, zch@et.aau.dk,
joseba.arza@ingeteam.com

Keywords

«Model Predictive Control», «Wind energy», «Grid-connected converter», «Fault ride-through», «Converter control».

Abstract

The increasing penetration of wind power poses new requirements in terms of ancillary services to support the stable operation of power system. Based on the flexible positive and negative sequence control (FPNSC) method for current reference generation, this paper presents a model predictive control (MPC)-enabled fault ride through operation strategy for high power wind turbine complying with next-generation grid code requirements. The proposed strategy calculates the current references considering the converter current constraint, and further exploits the dynamic response of MPC to optimize the voltage support capability during balanced and unbalanced grid faults. Simulation results are given to validate the operation performance of the proposed fault ride through operation strategy.

Introduction

The offshore wind energy is promoting the penetration of wind power generation into power system. To exploit the offshore wind energy in a cost-effective way, the high-power wind turbine is being paid the increasing concerns. Consequently, the application of high-power Wind Energy Conversion System (WECS) has become an important trend in modern wind power system. The back-to-back neutral-point-clamped (NPC) converter [1]–[4] has been presented as an attractive option for high-power wind turbine in terms of power quality and system costs, which is able to perform medium voltage operation by three-level output.

Fault-ride through (FRT) operation capability of wind turbine is one of important concerns. The grid codes require that the wind turbine remains the uninterrupted grid-connected operation under a temporary voltage drop, where reactive power is injected into power grid to support voltage restoration. The requirements about reactive current injection are specified using voltage-time profiles. Also, the FRT operation must be accomplished within 20 ms once the grid voltage falls below the 90% value. In [5], the effect of current injection into power grid on transient stability is analyzed. The reactive power requirements have traditionally involved the positive sequence. However, in 2015 the German standard VDE-AR-N 4120 indicates the requirements for the negative sequence current injection for the first time [6]. In [7], the performance of a type-4 WT during unbalanced grid faults is analyzed, showing the positive effects of the negative current injection. According to the latest revisions from VDE, the next-generation grid codes will demand the injection of positive and negative-sequence reactive current in the presence of asymmetrical faults. Consequently, dual-sequence current reference control will be required. The Flexible Positive and Negative Sequence Control (FPNSC) is presented as a flexible structure in [8], which can accommodate different control objectives dependent on two independent gains for the reactive current injection in positive-sequence (k^+) and negative-sequence (k^-).

The advanced control strategies have been frequently developed for wind power converter. For instance, the application of finite control set model predictive control (FCS-MPC) in 3L-NPC converter has been paid increasing concerns [9]. In [10], the MPC of power converters applied in variable-speed WECS is reviewed. The application of FCS-MPC can improve the operation performance of wind power converters. In [11], a direct model predictive control scheme for NPC converter in permanent magnetic synchronous generator (PMSG)-based wind turbine for unbalanced grid is presented. In [12], a robust FCS-MPC control method for a direct-drive PMSG wind turbine system with NPC power converter is proposed. This work improves the robustness against parameter uncertainties by means of revised predictions of the state variables. In [13], a FCS model predictive direct torque and power control for 3L-NPC converter in PMSG-based wind turbine is developed. In [14], a model predictive direct power control for high-power NPC converters is presented. The instantaneous active and reactive power are regulated. In [15], a self-tuning FCS-MPC is proposed to control switching frequency of the 3L-NPC converter, which improves the power conversion efficiency. The method demonstrates its effectiveness against variations of grid impedance and the operating point.

However, the fault-ride through capability of NPC converter equipped with MPC in the presence of grid faults is merely addressed. Therefore, this paper presents an MPC-based control structure with FPNSC to operate with different reactive current injection profiles during grid faults. The effectiveness of the proposed control strategy is validated by simulation verification. Further, the capability of the 3L-NPC power converter to deal with both balanced and unbalanced voltage faults is investigated.

System Description

The 3L-NPC converter is widely used in high-power applications due to the good performance with low switching frequency. The aim of MPC is to regulate the currents into power grid according to the reference values. Also, the MPC is employed to regulate the neutral point potential of NPC converter, which are related with the switching state and converter currents. The outer control loops such as the grid synchronization and the DC-link voltage controller are adopted to generate the current references. Fig. 1 shows the diagram of 3L-NPC converter with FCS-MPC, where the discrete-time model is adopted to predict the future behaviors of the plant and implement the optimal control actions based on predefined control objectives.

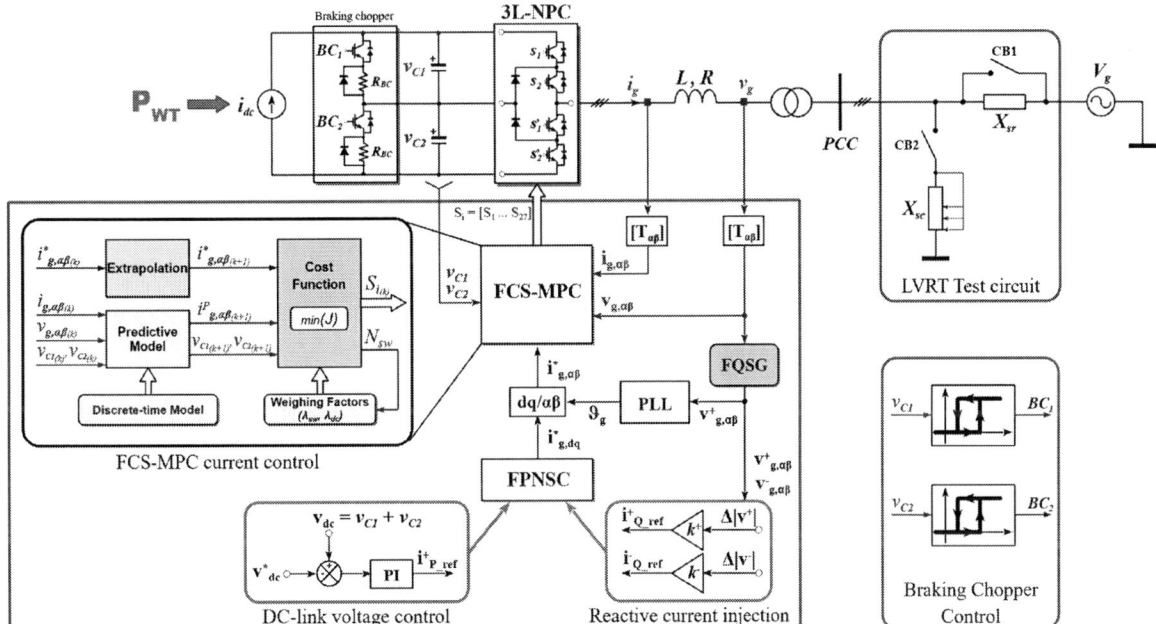

Fig. 1   The diagram of 3L-NPC converter with FCS-MPC.

The cost function considering the finite set of possible switching states of the power converter is established, where the given cost function is minimized by selecting the optimal state combination. Then, the optimal control law is obtained. The one-step horizon is proposed for the FCS-MPC. Although long prediction horizon usually leads to improved performance, it can be negligible for converters with an L-filter, whose transfer function is a first-order system. Moreover, long prediction horizons significantly increase the computational burdens, which affects the real-time implementation.

A. Operation Principle of Grid-Connected NPC Converter

The discrete-time model of injected grid current is established as (1).

$$i_{g(k+1)} = i_{g(k)} + \frac{T_s}{L}\left[v_{(k)} - Ri_{g(k)} - v_{g(k)}\right] \tag{1}$$

where R and L are the output resistance and inductance of the 3L-NPC converter, v is voltage generated by the converter, v_g is the grid voltage, T_s is the sampling time and i_g is the grid-injected current. The discrete model (1) is used to predict dynamic behaviors for the future value of grid current, considering all possible voltage vectors v generated by the NPC converter.

Currents in stationary reference frame can be controlled by defining a cost function based on the error between the predicted currents and the extrapolated references. To estimate the grid current error at (k+1) sampling instant, the reference currents are extrapolated to (k+1) state. The reference extrapolation technique can be used to compensate the sampling delay (T_s), and decrease the average reference tracking error [16].

$$J_i = \left(\left[i^*_{g,\alpha(k+1)} - i_{g,\alpha(k+1)}\right]^2 + \left[i^*_{g,\beta(k+1)} - i_{g,\beta(k+1)}\right]^2\right) \tag{2}$$

B. DC-link Voltage Balance of NPC Converter

The current through the clamping diodes and neutral point causes voltage imbalance between capacitors. When the output voltage is set to 0, the voltage of DC-link midpoint may fluctuate. Hence, a neutral point balancing control strategy is required to mitigate voltage imbalance. Considering NPC converter switching states, the discrete-time model of the DC-link capacitors is established as (3).

$$v_{C1(k+1)} = v_{C1(k)} + \frac{T_s}{C_1}i_{C1(k)}$$
$$v_{C2(k+1)} = v_{C2(k)} + \frac{T_s}{C_2}i_{C2(k)} \tag{3}$$

where v_{C1}, v_{C2} and i_{C1}, i_{C2} are the voltages and currents in DC-link capacitors $C1$ and $C2$, respectively.

Currents through the capacitors are obtained based on the load currents and the present switching state. Thus, no additional measurements are required. The DC-link capacitors voltage balance is added to the cost function as a secondary goal J_{dc} as (4).

$$J_{dc} = \lambda_{dc} \cdot \left[v_{C1(k+1)} - v_{C2(k+1)}\right]^2 \tag{4}$$

where v_{C1} and v_{C2} are the voltages of DC-link capacitors and λ_{dc} is the weighting factor.

C. Switching Frequency Control

The absence of modulator in FCS-MPC results in random switching frequency. A new element J_{sw} is incorporated into cost function to reduce switching frequency by the simple penalty strategy [17].

$$\Delta s_i(k) = |s_i(k) - s_i(k-1)| \tag{5}$$

$$J_{sw} = \lambda_{sw} \cdot \frac{1}{N} \sum_{i=1}^{N} \Delta s_i(k)$$

where $s_i(k)$ is the state of each i switching element, N is the number of switching elements and λ_{sw} is the weighting factor.

The overall reduction of the switching frequency can be performed since the change of state is penalized, so that the controller is encouraged to remain in the same state. The reduction of switching frequency will decrease the switching losses, which improves the power conversion efficiency.

The cost function of the MPC controller is defined by combining the following control objectives. (1) the AC current tracking error in $\alpha\beta$-domain, (2) the DC-link neutral point balance, (3) the switching frequency by the simple penalty approach.

$$J = J_i + J_{dc} + J_{sw} \tag{6}$$

The proposed fault ride through operation strategy based on MPC

As described previously, both positive and negative-sequence currents need to be regulated in a dual-decoupled control framework. This includes injection of active power in the positive-sequence, injection of capacitive reactive power in the positive sequence to support positive-sequence voltage, and injection of inductive reactive power in negative sequence to attenuate negative-sequence voltage. To address this, a synchronization structure that may extract the positive sequence and negative sequence of grid voltage is needed. In this paper, the FQSG-PLL [18] is employed to estimate the voltage sequences and the MPC is used to regulate the grid currents in the $\alpha\beta$-frame.

This section presents the proposed strategy for generating the current references according to the FPNSC structure. First, the grid code requirements for reactive power injection are described. Second, the limiting values for the current references and the remaining capacity for active power are calculated considering the converter current constraint. Third, the operation of braking chopper to protect DC-link during grid faults is described. Finally, the dual-decoupled current references are referred to the $\alpha\beta$-frame to be regulated by the MPC.

A. Reactive Current Reference

In the presence of voltage fault events, the low-voltage ride-through requirements consider reactive current injection both in the positive-sequence and in the negative-sequence. The reactive current references needed for compliance are given as (7).

$$i_{Q\_ref}^+ = k^+ \cdot \Delta|v^+|$$
$$i_{Q\_ref}^- = k^- \cdot \Delta|v^-| \tag{7}$$

where $i_{Q\_ref}^+$ is the positive-sequence reactive current reference, $i_{Q\_ref}^-$ is the negative-sequence reactive current reference, k^+ and k^- are the gains applied to the grid voltage deviations $\Delta|v^+|$ and $\Delta|v^-|$. The proportional gains are defined by grid operators in the range of $0 \le k \le 6$, in order to optimize the grid voltage support capability.

B. Current Limitation

The selection of the current references should consider the converter current constraint I_{max}, which is determined by operating conditions such as cooling temperature, switching frequency or supported overload. The positive-sequence and the negative-sequence current references are defined within the current constraint of the power converter as (8) and (9).

$$|I^+| + |I^-| \le I_{max} \tag{8}$$

$$|I^+| = \sqrt{\left(I^+_{P\_ref}\right)^2 + \left(I^+_{Q\_ref}\right)^2}$$

$$|I^-| = \sqrt{\left(I^-_{P\_ref}\right)^2 + \left(I^-_{Q\_ref}\right)^2} \tag{9}$$

where $|I^+|$ is the amplitude of the current reference in the positive sequence, defined by its components for active current ($I^+_{Q\_ref}$) and reactive current ($I^+_{Q\_ref}$), and $|I^-|$ is the amplitude of the current reference in the negative sequence, defined by its components for active current ($I^-_{P\_ref}$) and reactive current ($I^-_{Q\_ref}$).

If active current in negative sequence is set as zero ($I^-_{P\_ref}$ = 0), the equations (8) and (9) can be rewritten as (10).

$$\sqrt{\left(I^+_{P\_ref}\right)^2 + \left(I^+_{Q\_ref}\right)^2} + I^-_{Q\_ref} \le I_{max} \tag{10}$$

The current references should be prioritized to satisfy converter current constraint. Considering that $I^+_{Q\_ref}$ and $I^-_{Q\_ref}$ are determined for grid-code compliance, the cascade-limits for current references are given as (11)-(13).

$$limI^-_Q = I_{max} \tag{11}$$

$$limI^+_Q = I_{max} - I^-_{Q\_ref} \tag{12}$$

$$limI^+_P = \sqrt{\left(I_{max} - I^-_{Q\_ref}\right)^2 - \left(I^+_{Q\_ref}\right)^2} \tag{13}$$

where $limI^-_Q$ is the limitation of the negative-sequence reactive current reference $I^-_{Q\_ref}$, $limI^+_Q$ is the limitation of the positive-sequence reactive current reference $I^+_{Q\_ref}$, and $limI^+_P$ is the limitation of the positive-sequence active current reference $I^+_{P\_ref}$.

C. DC-link voltage controller

The DC-link voltage controller generates the positive-sequence active current reference $I^+_{P\_ref}$ to regulate the active power output of the power converter.

In the presence of grid fault, the active power absorption capability of power grid is decreased. And reactive current is injected to support grid voltage restoration, which further reduces the limit for the active current reference as (13). Additionally, the active current injection is null during severe grid faults to avoid instability system. However, the mechanical system of wind turbine still maintains the active power injection in transient period. This power imbalance between the generated power and power injected to the grid should be dissipated by employing a braking chopper, which avoids the voltage boost in DC-link. In the NPC converter, each half bus is protected by a braking chopper as shown in Fig. 1.

The braking chopper is operated by the hysteresis control of the voltage at the capacitor terminals. Considering the nominal voltage V_{dc}, the chopper switch-on voltage is fixed at V_{BC_on} = 1.15 V_{dc} and the switch-off voltage is fixed at V_{BC_off} = 1.05 V_{dc}. The switch-on voltage must ensure a safety margin with regards to the maximum working voltage of the components. Further, the voltage difference between both hysteresis thresholds must be sufficient to moderate the switching frequency. The braking resistor R_{BC} must be sized to dissipate rated active power, and its minimum value is established to keep the current value and the power losses of the braking semiconductor below the maximum ratings.

D. Current controller references

The MPC is responsible to regulate the grid currents in the αβ-frame. Then, the current references should be transferred from the dual-decoupled control framework to the αβ-frame by means of the positive-sequence voltage angle ϑ_g ($v^+_q \approx 0$) defined by the FQSG-PLL as (14)-(16).

$$i_\alpha^* = i_{\alpha\_ref}^+ + i_{\alpha\_ref}^-$$
$$i_\beta^* = i_{\beta\_ref}^+ + i_{\beta\_ref}^- \tag{14}$$

$$i_{\alpha\_ref}^+ = \frac{v_d^+}{|v^+|} i_{P\_ref}^+ \cdot \cos\vartheta_g + \frac{v_d^+}{|v^+|} i_{Q\_ref}^+ \cdot \sin\vartheta_g$$
$$i_{\beta\_ref}^+ = \frac{v_d^+}{|v^+|} i_{P\_ref}^+ \cdot \sin\vartheta_g - \frac{v_d^+}{|v^+|} i_{Q\_ref}^+ \cdot \cos\vartheta_g \tag{15}$$

$$i_{\alpha\_ref}^- = \frac{-v_q^-}{|v^-|} i_{Q\_ref}^- \cdot \cos\vartheta_g + \frac{v_d^-}{|v^-|} i_{Q\_ref}^- \cdot \sin\vartheta_g$$
$$i_{\beta\_ref}^- = \frac{v_q^-}{|v^-|} i_{Q\_ref}^- \cdot \sin\vartheta_g + \frac{v_d^-}{|v^-|} i_{Q\_ref}^- \cdot \cos\vartheta_g \tag{16}$$

where $i_{\alpha\_ref}^+$ and $i_{\beta\_ref}^+$ are the αβ-frame representation for the positive-sequence current references, $i_{\alpha\_ref}^-$ and $i_{\beta\_ref}^-$ are the αβ-frame representation for the negative-sequence current references, and i_α^* and i_β^* are the αβ-frame current references for the predictive controller. v_d^+, v_d^- and v_q^- are the grid voltage components obtained from the FQSG-PLL.

Simulation Verification

In this section, time-domain simulation in MATLAB/SIMULINK is performed to validate the proposed fault-ride through operation strategy for 3L-NPC converter in the presence of grid faults. The detailed parameters of the system are given in Table I.

TABLE I: SIMULATION PARAMETERS

Parameter	Value
Rated Power P_N	4 MW
LV Rated Voltage V_g	3100 V
HV Rated Voltage V_{HV}	66000 V
DC-link voltage V_{dc}	5600 V
Switching frequency f_{sw}	1 kHz
DC-link capacitors C_1, C_2	20 mF
Braking chopper resistor R_{BC}	1.0 Ω
Filter Inductance L	400 µH
Filter Resistor R	1.3 mΩ
Sampling Time T_s	50 µs

A. Test system description

Fig. 1 shows the FRT Test circuit in simulation verification. The FRT test system is designed based on the following aspects. (1) the rapid generation of two- or three-phase voltage dips at the instant of the fault can be performed, (2) the constant voltage during the fault duration at a reduced LVRT value can be produced, (3) the rapid voltage recovery to the original voltage value at the instant of fault clearing can be performed.

The variable series reactance (Xsr) is used to limit the short circuit current, so that only a small reduction in voltage at the grid side is observed during a fault simulation. The short circuit reactance (Xsc) is used to obtain the required voltage dip. The short circuit reactance is variable, so that dips with various magnitudes of the applied voltage can be obtained. The series reactance circuit breaker ($CB1$) is used to bypass the series reactance (Xsr) during normal operation of the power converter. The circuit breaker is opened at 1s before the introduction of the short circuit and closed sometime after the short circuit is removed. The short circuit breaker (*CB2*) is used to generate and end the time-controlled voltage dips.

B. Simulation results analysis

The power converter is operated at rated active power in steady state. The simulation tests include 100% voltage faults at PCC at 2s. The active power input to the DC-link remains unchanged during the whole test. All of the results are obtained from a post-processing of the voltages and currents measured at PCC to extract the positive and negative sequence.

Fig. 2 shows the simulation results of 3L-NPC converter system during a 3-phase 100% voltage fault at PCC. In Fig. 2 (a), the reactive current injection profile has been configured as $k^+ = 0$ and $k^- = 0$, which implies that positive-sequence reactive current is set as zero. In Fig. 2(b), the reactive current injection profile has been specified as $k^+ = 2$ and $k^- = 0$, which implies that positive-sequence reactive current will reach rated values for those voltage faults over 50%. In both cases, the negative-sequence reactive current is null since the voltage fault is balanced.

Fig. 3 shows the simulation results of 3L-NPC converter system during a 100% 2-phase voltage fault at PCC. In Fig. 3 (a), the reactive current injection profile has been specified as $k^+ = 2$ and $k^- = 0$, which implies that negative-sequence reactive current is set as zero. In this case, due to the lack of negative-sequence current injection, the power converter can be operated with active current because of the current limit remainder. The minor effect is caused by the fact that the voltage fault has been directly generated at PCC. In Fig. 3(b), the reactive current injection profile has been specified as $k^+ = 2$ and $k^- = 2$. In this case, both positive- and negative-sequence reactive current are injected into the grid. The current imbalance slightly reduces the voltage imbalance.

The simulation results show the quick dynamic response of the MPC-based current controller can be implemented since the reactive current is injected within less than 20ms. Additionally, the stable operation of the power converter during the fault events has been demonstrated. The power imbalance between the generated power and power injected to the grid is dissipated by the braking chopper, which maintains the voltage boost under safety margins. After the fault clearance, the active current returns to its original value in a ramped manner to continue the steady state operation.

Conclusion

This paper presents an MPC-based fault ride through operation strategy for 3L-NPC converter applied in high-power wind turbine. The FPNSC method is proposed to generate current references, and the limiting values are derived according to converter current constraint. The FCS-MPC is used to regulate the currents into the power grid in the αβ-frame, to balance the neutral point potential, and to control the switching frequency of NPC converter. Furthermore, the dynamic response of MPC is exploited to optimize the voltage support capability during balanced and unbalanced grid faults. The operation performance of the proposed strategy is analyzed and validated against symmetrical and asymmetrical faults. The effectiveness of the MPC-based control strategy is validated for high-power wind converter operation during grid faults, which is applicable for modern grid code requirements of offshore wind power system.

Model Predictive Control-enabled Fault Ride Through Operation Strategy for High Power Wind Turbine

CATALÁN Pedro

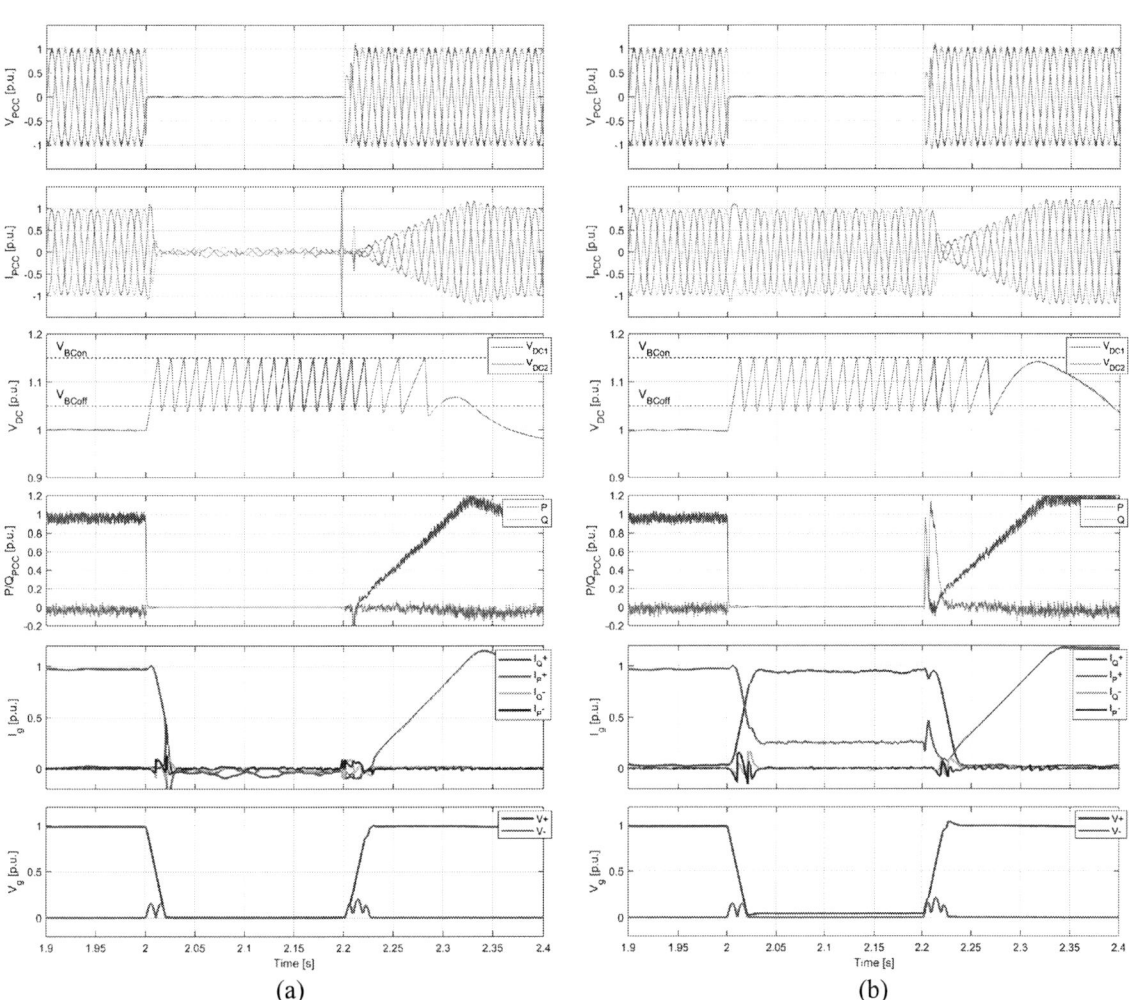

Fig. 2 The simulation results of the 3L-NPC converter under balanced grid fault.
(a) $k^+ = 0$, $k^- = 0$, (b) $k^+ = 2$, $k^- = 0$.

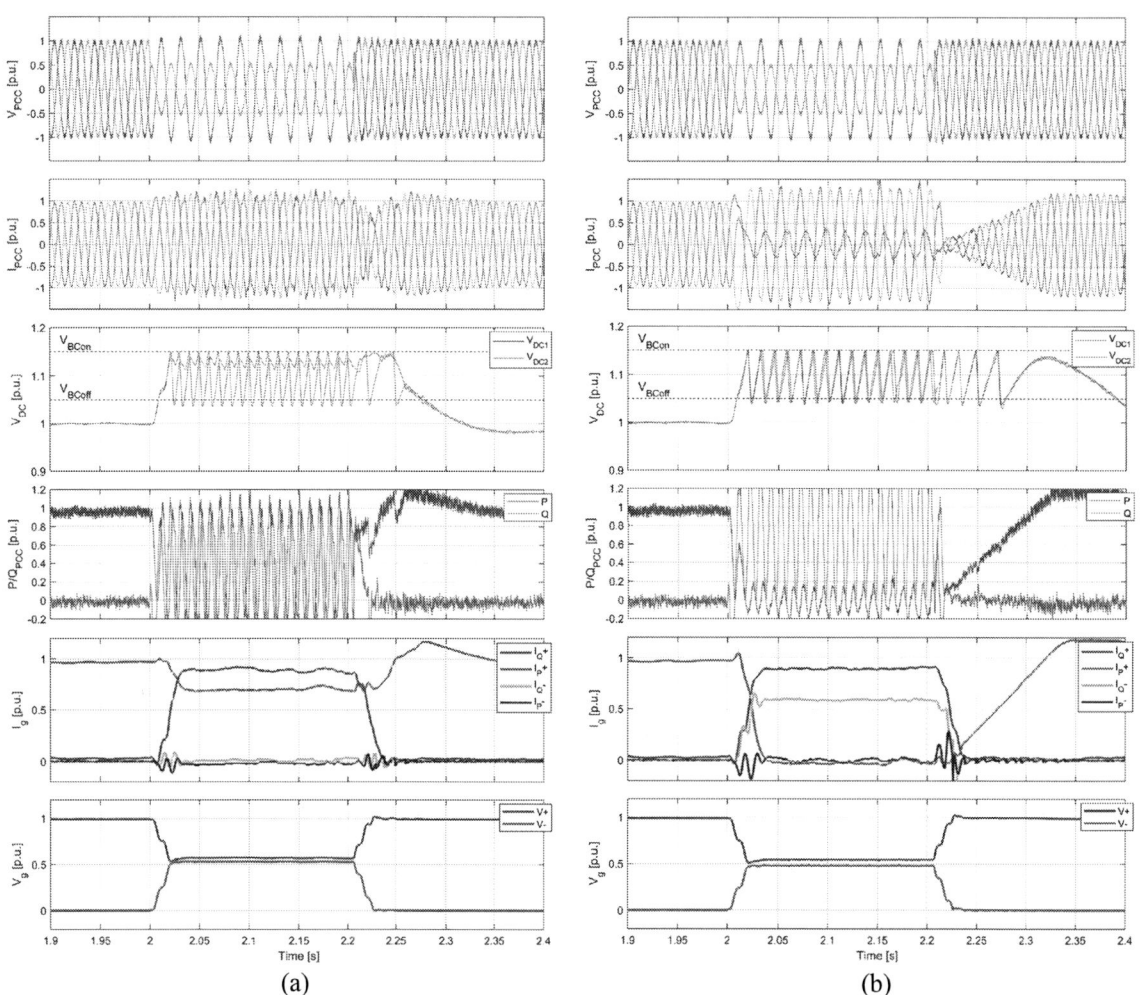

Fig. 3 The simulation results of the 3L-NPC converter under unbalanced grid fault.
(a) $k^+ = 2$, $k^- = 0$, (b) $k^+ = 2$, $k^- = 2$.

References

[1] A. Nabae, I. Takahashi, and H. Akagi, "A New Neutral-Point-Clamped PWM Inverter," *IEEE Trans. on Ind. Applicat.*, vol. IA-17, no. 5, pp. 518–523, Sep. 1981.

[2] J. Rodriguez, S. Bernet, P. K. Steimer, and I. E. Lizama, "A Survey on Neutral-Point-Clamped Inverters," *IEEE Trans. Ind. Electron.*, vol. 57, no. 7, pp. 2219–2230, Jul. 2010.

[3] J. Chivite-Zabalza, I. Larrazabal, I. Zubimendi, S. Aurtenetxea, and M. Zabaleta, "Multi-megawatt wind turbine converter configurations suitable for off-shore applications, combining 3-L NPC PEBBs," in *2013 IEEE Energy Conversion Congress and Exposition*, Denver, CO, USA, 2013, pp. 2635–2640.

[4] J. Li, Z. Hu, Y. Wang, and Z. Chen, "H14 Three-Level Inverter for Common-Mode Voltage Suppression," *IEEJ Trans. Elec Electron. Eng.*, vol. 16, no. 2, pp. 315–323, Feb. 2021.

[5] I. Erlich, F. Shewarega, S. Engelhardt, J. Kretschmann, J. Fortmann, and F. Koch, "Effect of wind turbine output current during faults on grid voltage and the transient stability of wind parks," in *2009 IEEE Power & Energy Society General Meeting*, Calgary, Canada, 2009, pp. 1–8.

[6] VDE-AR-N 4120, "Technical requirements for the connection and operation of customer installations to the high voltage network (TAB high voltage)," Jan. 2015.

[7] T. Neumann, T. Wijnhoven, G. Deconinck, and I. Erlich, "Enhanced Dynamic Voltage Control of Type 4 Wind Turbines During Unbalanced Grid Faults," *IEEE Trans. Energy Convers.*, vol. 30, no. 4, pp. 1650–1659, Dec. 2015.

[8] M. Graungaard Taul, X. Wang, P. Davari, and F. Blaabjerg, "Current Reference Generation Based on Next-Generation Grid Code Requirements of Grid-Tied Converters During Asymmetrical Faults," *IEEE J. Emerg. Sel. Topics Power Electron.*, vol. 8, no. 4, pp. 3784–3797, Dec. 2020.

[9] J. Rodriguez *et al.*, "State of the Art of Finite Control Set Model Predictive Control in Power Electronics," *IEEE Trans. Ind. Inf.*, vol. 9, no. 2, pp. 1003–1016, May 2013.

[10] V. Yaramasu and B. Wu, *Model predictive control of wind energy conversion systems*. New York: Wiley, 2017.

[11] Z. Zhang and R. Kennel, "Direct Model Predictive Control of three-level NPC back-to-back power converter PMSG wind turbine systems under unbalanced grid," in *2015 IEEE International Symposium on Predictive Control of Electrical Drives and Power Electronics (PRECEDE)*, Valparaiso, 2015, pp. 97–102.

[12] Z. Zhang, Z. Li, M. P. Kazmierkowski, J. Rodriguez, and R. Kennel, "Robust Predictive Control of Three-Level NPC Back-to-Back Power Converter PMSG Wind Turbine Systems with Revised Predictions," *IEEE Trans. Power Electron.*, vol. 33, no. 11, pp. 9588–9598, Nov. 2018.

[13] Z. Zhang, F. Wang, J. Wang, J. Rodriguez, and R. Kennel, "Nonlinear Direct Control for Three-Level NPC Back-to-Back Converter PMSG Wind Turbine Systems: Experimental Assessment With FPGA," *IEEE Trans. Ind. Inf.*, vol. 13, no. 3, pp. 1172–1183, Jun. 2017.

[14] J. Scoltock, T. Geyer, and U. K. Madawala, "Model Predictive Direct Power Control for Grid-Connected NPC Converters," *IEEE Trans. Ind. Electron.*, vol. 62, no. 9, pp. 5319–5328, Sep. 2015.

[15] P. Catalan, Y. Wang, Z. Chen, and J. Arza, "Model Predictive Control Strategy for NPC Converter-based Wind Turbine with Switching Frequency Control," in *2021 IEEE Southern Power Electronics Conference (SPEC)*, Kigali, Rwanda, 2021, pp. 1–7.

[16] P. Cortes, F. Quiroz, and J. Rodriguez, "Predictive control of a grid-connected cascaded H-bridge multilevel converter," in *Proceedings of the 2011 14th European Conference on Power Electronics and Applications*, 2011, pp. 1–7.

[17] R. Vargas, P. Cortes, U. Ammann, J. Rodriguez, and J. Pontt, "Predictive Control of a Three-Phase Neutral-Point-Clamped Inverter," *IEEE Trans. Ind. Electron.*, vol. 54, no. 5, pp. 2697–2705, Oct. 2007.

[18] J. I. Garcia, J. I. Candela, A. Luna, and P. Catalan, "Grid synchronization structure for wind converters under grid fault conditions," in *IECON 2016 - 42nd Annual Conference of the IEEE Industrial Electronics Society*, Florence, Italy, 2016, pp. 2313–2318.

AUTHOR INDEX

Abdalrahman, Adil 2241, 3282, 3757
Abdullah, Ahmed.. 554
Abedini, Hossein... 865
Aceña, Javier Cañas.. 484
Adabi, Jafar... 2537
Addin, Ali Sharaf... 1824
Afonso, Luciana C. 4018
Aganza-Torres, Alejandro................................. 1328
Agarwal, Ritika.. 3615
Agirrezabala, Eneko 3327
Aguglia, D. 1955
Ahmed, Emad M.. 1015
Aiello, Giuseppe 2628
Aillerie, Michel.. 315
Aizpuru, I... 2903
Aizpuru, Iosu 325, 3327, 3574, 3750
Akuru, Udochukwu B... 2958
Al-Haddad, Kamal 1025
Alaluss, Mohamed 1424
Alatise, Olayiwola 1497, 2477
Albrecht, Fabian 2726
Aldarmon, Mohamed... 2574
Ali, Mohammad.. 2392, 3022
Ali, Ramy 390
Ali, Rana Asad.. 698
Allard, Bruno 169, 3862
Allioua, Abdelmoumin 2835
Alvarez, Asier... 279
Alvarez-Herault, Marie-Cecile 1147
Alves, Wendell Da Cunha.................................. 1046
Alvi, Muhammad H 1692
Aly, Mokhtar 1015
Andersen, Michael A. E..................................... 1561
Ando, Y. 1785
Andresen, Jan... 1684
Ansari, Sajad A... 3440
Antonopoulos, Antonios 297, 432
Anzola, J.. 2903, 2967
Anzola, Jon 3574
Apostolidou, Nena 1796
Appel, Tobias... 1121
Apte, Pramod 2773
Arabsalmanabadi, Bita...................................... 1025
Arias, Manuel 152
Arrizabalaga, Antxon.. 325
Arrozy, Juris 681
Arruti, Asier.. 3574, 3750
Artal-Sevil, J. S...................................... 2903, 2967

Arza, Joseba 484, 2011
Asllani, Besar... 2515
Asoodar, Mohsen 2843
Atzler, Frank 3391
Aunsborg, Thore Stig 825
Aviñó, Oriol 2715
Ayarzaguena, Ibán...................................1765, 3336
Aztiria, Jon 325
Baars, Nico.. 2788
Babin, Anthony 3696
Baburske, Roman 1424
Bacha, S. 2422, 3179
Bacha, Seddik... 3140, 3928
Bacheti, Gabriel Gaburro 421
Bachmann, Matthias... 3501
Badenhop, Niklas 145, 1939
Baek, Seung-Hyuk 2877
Bagaber, Bakr.. 3037, 3711
Baimel, D. 3254
Baimel, N. 3254
Bak, Claus Leth 2504
Bakhos, Gianni 3928
Bakran, Mark-M......................................805, 1036, 2744
Bakri, Reda 1046
Balachandran, Arvind 1456
Balasubramanian, Sridhar 2030
Ballestín-Bernad, V....................................2903, 2967
Banana, Shady.. 1064
Banavath, Satish Naik 730
Banda, Joseph..187, 289
Barba, V. 1975
Barbi, Eli 3254
Barg, Sobhi 361
Barman, Subhranil... 2462
Barón, Kevin Muñoz... 2698
Bashar, Erfan 2477
Basic, Duro 125
Basler, Michael 242
Basler, Thomas............................... 1424, 1713, 1733, 3373
Bauer, Luca 971
Bauer, Pavol 1319, 3607, 3729
Baumann, Michael 1167
Baumann, Timm Felix 2355
Bäumler, Christian 1733
Bayer, Markus 115
Bayhan, Sertac 3518
Bayram, Islam Safak 3518
Beck, Simon 1434, 2038

Beckemeier, Christian..2327
Beczkowski, Szymon Michal...2661
Beineke, Stephan 3501
Beiranvand, Hamzeh....................833, 3092, 3846, 3966
Belhaouane, Mohamed Moez 582
Benchaib, Abdelkrim 3928
Bendfeld, Christian 1620
Benech, Philippe...169
Bensetti, Mohamed 3883
Bergmann, Lukas..1036
Bergveld, Henk Jan...3796
Bermejo, Jose Manuel.......................................1765, 3336
Bernal, Carlos 3327
Bernal-Agustín, J. L..2967
Bernal-Ruiz, Carlos 3750
Bernichon, Thomas 3920
Bertilsson, Kent 361
Bertin, Matthieu..534
Beukes, Johan 3112
Beye, Mamadou Lamine...2736
Beza, Mebtu 1187
Bezerra, Vinicius Freire...2689
Bhatnagar, Pallavee 3804
Bhattacharya, Arghyadip 178
Bhoi, Sachin Kumar..3031
Biadene, Davide...865
Biela, Juergen 1402
Biela, Jürgen 651, 662, 933, 1391, 1434, 2038, 2544
Bieler, Arne 1121
Bier, Anthony 922, 2736
Billa, Laxma R..2301
Bimmel, Luc 2736
Binder, Andreas..2316
Bitsi, Konstantina 3246
Blaabjerg, Frede..................2110, 2182, 2496, 2504, 2939
Blanes, J. M. 3382, 3401
Blank, Thomas...232
Blanquez, Francisco R. 2189, 2451
Blasco-Gimenez, Ramon 2189, 2451
Blasuttigh, Nicola 3846
Blatsi, Zoe..2824, 3813
Blömeke, Alexander 4025
Böcker, Joachim 2276, 2432, 2754, 3625, 3686
Bockholt, Jan 1286
Boettcher, Norman..1128
Bohllaender, Marco 4016
Bohne, David..514
Boige, Francois 944
Boisson, Guillaume Piquet...960
Bolzoni, A...1371
Bongiorno, Massimo...1187
Bonten, Remco 634

Böorngen, Hannes 1754
Borcherding, Holger..2852
Börngen, Hannes 3362
Boroyevich, Dushan 2806
Bosch, Swen 2219
Bosga, Sjoerd G. 3246
Bouscayrol, Alain..2175
Boutleux, Emmanuel 251
Boutry, Arthur..2515
Brabetz, Ludwig...2383
Branco, Cesar Augusto Santana Castelo 2948
Braun, Gerrit 2205
Braz, Cesar 1445
Briff, Pablo 451
Bringezu, Thilo 662
Brinker, Tobias...2977
Brogioli, Doriano Constantino 833
Brommer, Volker 2726
Bronstein, S. 3254
Brooks, Michael 279
Brückner, Thomas 1824
Brulin, Pierre-Yves 3831
Brunner, Andreas 593
Brunner, Frank 3775
Brüns, Michael 474
Bruyere, Antoine 1046
Bruyere, Paul 960
Bucarey, Victor 1074
Budo, Kohei..213, 351
Bueno, Emilio José. 421
Bueno-Mariani, Guilherme 3272
Bugarski, Stevan 2334
Bünte, Andreas..380
Burgos, Rolando..1692, 2806
Burgos-Mellado, Claudio 1074, 3429
Burkart, Ralph M. 203
Burke, Richard 3696
Bushra, Rehnuma...2392, 3022
Busquets-Monge, Sergio 2715
Buticchi, Giampaolo 3014
Buttay, Cyril..2049, 2515
Byen, Byengjoo...1207
Caarls, Esin Ilhan 681
Cabrera, Michel...169
Cacciato, Mario 2628
Caillierez, Antoine 3883
Cajander, D. 1955
Cakal, Gokhan...3947
Caldognetto, Tommaso 865
Camargo, Renner Sartório...421
Camurca, Luis 3101
Can, Görkem..3092

Cano, Tania C. 335
Cao, Jingming 3215, 3225
Cao, Yongtao 2003
Cappelle, Jan 1300
Cárcamo, Alberto 1083
Carcouet, S. 843
Carpita, Mauro 1543
Carrasco, Miguel 370
Casado, P. 3382, 3401
Castellazzi, Alberto 689, 2156, 2285, 2402, 2893, 3084
Castelli-Dezza, Francesco 1476
Castro, Ignacio 335
Catalán, Pedro 2011
Catellani, Stéphane 922, 990
Ceccarelli, Lorenzo 681
Chakraborty, Sajib 2101, 3031
Chang, Che-Wei 1692
Charkaoui, Abdelmouneim 442
Chatterjee, Kishore 178, 2462
Chen, Zhe 2011
Chen, Zhu 3235
Chevalier, Florian 3582
Chida, Makoto 1580
Chinthavali, Madhu Sudhan 344
Chiumeo, Riccardo 3206
Choksi, Kushan 344
Choudhury, Soham 1966
Chub, Andrii 730
Cimetiere, Xavier 1046
Clerc, Guy 251
Clerici, Alessio 3206
Cobaleda, Diego Bernal 2581
Cogitore, Bruno 1216
Colmenero, Manuel 2189, 2451
Cosso, Simone 2919
Coumont, Martin 1966
Crovetti, Paolo 554
Cui, Yi 3986
Czerwenka, Philipp 593
Dahmen, Christopher 1824, 1855
Damian, Ioan Catalin 2266
Damm, Gilney 3590
Danielsson, Christer 2843
Dargahi, Vahid 2073
Davari, Pooya 2496
Davidson, Jonathan N. 3440
De Bernardinis, Alexandre 315
De Carne, Giovanni 3014
De Cesaris, Ivan 223
De Donato, Giulio 1569
De Doncker, Rik W. 709, 1266, 2119, 3599, 3676,
....3740, 3766, 3893

De Lillo, Liliana 3450
De Matos, Jose Gomes 2948
De Oliveira, Eduardo Facanha 2441
De, Dipankar 689
Deb, Arkadeep 1497
Deblecker, Olivier 504
Deboy, Gerald 3984
Deck, Patrick 514
Deckers, Martijn 2795
Degaa, Laid 3696
Delette, Gérard 922
Deng, Kai 3235
Dennetiere, Sébastien 582
Derammelaere, Stijn 3344
Despouys, Olivier 2486
Dick, Christian P. 514
Dickmann, Stefan 758
Dieckerhoff, Sibylle 1466, 2596, 2607, 2644, 3775
Dieng, A. 2092, 2930
Dierks, Rebecca 1533
Dietrich, Tim-Hendrik 1094
Disselkamp, Simon 2912
Domae, Shinichi 3084
Domes, Daniel 2744
Domes, Konrad 1137
Dong, Chaoyu 3215, 3225
Dong, Dong 1692, 2515
Dong, Jianning 1319
Dong, Tenghui 3084
Dorner, Oscar 1177
Dos Santos, Pedro Leal 604
Dragicevic, Tomislav 2496, 2939, 3429
Drexler, Christoph 411, 1167
Driesen, J. 3655
Driesen, Johan 2795
Drimizi, Youssef 2869
Drissi, Khalil El Khamlichi 3786
Duarte, Jorge L. 681, 798
Duarte, Jorge 2788
Duchamp, Jean-Marc 169
Dujic, Drazen 2049
Dumtzlaff, Jacob 1865
Duquesne, Thierry 3582
Dürbaum, Thomas 88, 307
Duun, Sune Bro 825
Dworakowski, P. 2422
Dworakowski, Piotr 2049
Ebel, Thomas 3130
Ebner, Kathrin 4015
Eckart, Martin 3646
Eckel, Hans-Guenter 3460

Eckel, Hans-Günter 11, 59, 70, 980, 1294, 1703,
.... 1744, 1885, 1895, 2308, 4003
Eckstein, Mattea 1277
Effenberger, Thomas 1754
Eggers, Malte 1466
Ehlich, Martin 2852
El Baghdadi, Mohamed 2101, 2293, 3031
El Sherif, Alaa 3796
El-Refaie, Ayman 719, 1692
Ellinger, Thomas 2885
Emmers, G. 3655
Emmers, Glenn 2795
Empringham, Lee 3450
Encarnação, Lucas Frizera 421
Endo, Yusuke 2285
Epping, Daniel 749
Erckrath, Tobias 1350, 1620
Eremia, Mircea 2266
Eriksson, Lars 1456
Erlbacher, Tobias 1128
Ernst, Alexander 3149, 3159
Es-Seghaie, Hajar 922
Escoffier, René 990
Etoz, Burhan 1497
Faber, Samuel 307
Falchi, Daniele 2486
Faramehr, Soroush 3822
Farhangi, Shahrokh 787
Fauth, Leon 2003, 2638, 3838
Fayolle-Lecocq, Murielle 990
Fazli, Nastaran 11
Fehr, Hendrik 49, 3391
Felgemacher, Christian 442, 4004
Fernández, Arturo 152
Ferreyra, Fabio 554
Festerling, Tobias 1237
Finney, Stephen 80, 3470, 3813
Fischer, Katharina 1674, 1804
Fischer, Manuel 749
Fischer-Baeumer, Rico 1137
Fölkel, Lorandt 279
Formentini, Andrea 2919, 3975
Forouzesh, Mojtaba 1590, 1601
Forsstrom, Ville 3301
Förster, Nikolas 2432
Foster, Martin P. 3353, 3440
Foteinopoulos, Georgios 1985
Fräger, Lukas 145, 641, 1939, 2588, 2773
Frank, S. R. 3411
Franzki, Jonas 261
Frey, David 1147
Fricke, Tobias 1247

Fricke, Torben 1381
Friebe, Jens 1914, 2003, 2327, 2392, 2588, 2638,
.... 2655, 2689, 2773, 2977, 3022, 3059, 3545, 3838
Fritze, Eric 758
Fröhling, Sören 1674
Fuchs, Simon 1434, 2038
Fuhrmann, Jan 980
Fukunaga, Shuhei 108
Ganeshpure, Dhanashree Ashok 3729
Gao, Xiang 3014
Gaona, Daniel 2441
Garces, Santiago Ramos 3344
Garcia, Raul Murillo 2355
Garrigós, A. 3382, 3401
Gaubert, Jean-Paul 1525
Gauthier, Jean-Yves 3862
Gavelle, Mathieu 2618
Gehl, Adrian 2912
Geiss, Michael 2554
Gemma, Filippo 3975
Geng, Weiwei 3722
Geng, Xiaomeng 2596, 2644, 3775
Gennaro, Francesco 2628
Gensior, Albrecht 49, 370, 3391
Gerges, Tony 169
German, Ronan 2175
Germishuizen, J. J. 3318
Geury, Thomas 2101
Gholami, M. 3179
Gholami, Mehrdad 3140
Ghumman, Sukhjit S 2763
Gieraths, Antje 767
Gierschner, Magdalena 1294
Gierschner, Sidney 11
Gillon, Frédéric 1046
Girona-Badia, Jaume 3704
Glaser, Martin 4020
Gleissner, Michael 805, 1036
Gnärig, Lasse 370
Goetz, Stefan 1025, 1064, 1197, 3636, 3665
Gohler, Katherina 1804
Gohrmann, Kai 1137
Golev, Victor 1286
Goller, Maximilian 1733
Gomes, Lucas Vinícius De Araújo 3059
Gomes, Zariff Meira 3590
Gómez, Alexis A. 1765, 3336
Gomis-Bellmunt, Oriol 2486, 3704
Gonzalez, Jose Ortiz 1497
Gonzalez-Hernando, Fernando 3938
Gonzalez-Torres, Juan-Carlos 3928
Götz, Georg Tobias 709

Gräber, Hendrik	2977
Grabs, Volker	97
Gradinger, Thomas B.	203
Grant, Thomas	2301
Grass, Norbert	2366
Grau, Vivien	854
Gremme, Florian	4021
Griepentrog, Gerd	160, 2780, 2835
Grodnichev, Anton	624
Groke, Holger	3169
Groon, Fabian	3092
Groten, Jonas	279
Gruson, François	582
Guerrero, Bruno	944
Gui, Qiuye	49
Guillaud, Xavier	582
Günes, Ece Olcay	1361
Gupta, Kirti	2110
Gupta, Krishna Kumar	3615, 3804
Gutierrez, Alonso	2618
Haag, Felix	2726
Haake, Daniel	624
Haarer, Jörg	971, 1237, 1277
Habersetzer, Antoine	4015
Hably, A.	3179
Hably, Ahmad	3140
Hackl, Philipp	39
Haederli, Christoph	3282
Häfner, Ying-Jiang	2241, 3282, 3757
Hagedorn, Maximilian	1875
Hajar, K.	3179
Hajar, Khaled	3140
Hajian, Masood	468
Hakkila, Akseli	297
Hald, Alex	380
Hameyer, Kay	3005, 3235
Hammes, David	11
Handt, Karsten	2607
Hanf, Michael	3169
Hanisch, Lucas Vincent	261
Hanisch, Lucas	1094
Hänsel, Stefan	572
Hansen, Sandra	3966
Hanson, Alex J.	1722
Hanson, Jutta	1966
Hao, Chuantong	80, 3470
Hardan, Faysal	468
Harmand, Souad	2996
Hasan, Md. Mahamudul	3031
Hasler, J. P.	1371
Hassan, Tayssir	1466
Hatori, K.	1785

Hatori, Kenji	777
Hattori, Takato	739
Hauenschild, Philipp	1506
Haug, Martin	279, 698
Hayes, John G.	2470
Hegazy, Omar	2101, 2293, 3031
Heide, Daniel	3711
Heien, Christian	1294
Heimler, Patrick	1713
Hein, Yves	1294
Helmholdt-Zhu, Ting	97, 854
Hembel, Ahmed	3947
Henke, Markus	261, 1094, 2030
Henkenjohann, Jonas	1684
Henn, Jochen	3599
Henneberg, Dustin	2885, 3491
Herbold, Johannes	749
Hernando, Marta M.	1083, 1765, 3336
Herzog, Hans-Georg	952
Heydari, Rasool	2682
Hikihara, Takashi	108
Hiller, M.	3411
Hiller, Marc	115, 999
Hillmer, Hartmut	2383
Hilt, Oliver	2596, 2644, 3775
Himker, Niklas	1631
Himmelmann, Patrick	999
Hiraki, Eiji	2164
Hirning, David	971, 1237, 1277, 3536
Hissel, Daniel	315
Hjerrild, Jesper	2504
Hoerner, Michael	1754
Hofer, Heimo	1445
Hofer, Matthias	2251
Hoff, Bjarte	3198
Hoffmann, Klaus F.	758, 2726, 3188
Hoffmann, Madlen	3262
Hoffstadt, Thorben	1157
Hofmann, Viktor	195, 400
Hofmann, Wilfried	3957
Hofstetter, Patrick	195, 400
Hölscher, Jonas	2432
Holtje, Pauline	1665
Holzke, Wilfried	3149, 3159, 3169
Horn, Markus	2383
Hortans, Magnus	3309
Hoshi, Nobukazu	1776, 1844
Hosseinabadi, Farzad	3031
Hosseini, Elham	1025
Hou, Jingning	3722
Houwen, Simon	3344
Hridya, I	187

Hu, Anliang	651
Hu, Bin	2182
Hu, Xiaowei	3722
Huang, Jiasheng	1561
Huerta, Gabriel Ramos	1226
Huesgen, Till	2230
Huisman, Henk	634, 673, 681
Hutzler, Michael	1445
Idir, Nadir	2996, 3582, 3822
Igic, Petar	3822
Iida, Masaki	2164
Iman-Eini, Hossein	787
Imgart, Paul	1187
Incurvati, Maurizio	223, 268
Inoue, Michiko	3420
Iraola, Unai	3327
Ishihara, Mastaka	2164
Itoh, Jun-Ichi	902, 1104, 2127
Ittamveettil, Hridya	289
Izurza, Pedro	484
Jaber, Hamzeh J.	2156, 2285, 3084
Jacques, Dries	3344
Jagannath, Sriram	3362
Jahdi, Saeed	1497, 2477
Jain, Anekant	3615
Jain, Sanjay K.	3615, 3804
Jamal, Adeel	2780
Jaman, Shahid	3031
Jankovic, Marija	442
Jayathurathnage, Prasad	1947
Jena, Kasinath	3804
Jenhani, Firas	1343
Jeong, Byunghwang	1207
Jeschke, Sebina	3235
Jha, Kapil	187, 289
Jia, Hongjie	3215, 3225
Jia, Ming	1266
Joebges, Philipp	1266
Johansson, N.	1371
Johnson, C. Mark	3450
Jonsson, Tomas	1456
Jordà, Xavier	2715
Jørgensen, Asger Bjørn	825, 1641, 2661
Jöst, Dominik	4025
Jovanovic, Raka	3518
Juchem, Ralf	4023
Judge, Paul	80
Junemann, Lennart	1665
Jung, Marco	624, 1515, 1611, 1620
Junghans, Christoph	3460
Junyent-Ferre, Adria	2574
Kabbara, Wassim	3883

Kacetl, Jan	1197, 3636, 3665
Kacetl, Tomáš	1197, 3636, 3665
Kacki, Marcin	2470
Kadem, Karim	3590
Kaerst, Jens Peter	544
Kaiser, Jeremias	307
Kallfass, Ingmar	2698, 3565
Kamel, Tamer	468
Kaminski, Nando	2230, 3149, 3169
Kamm, Simon	2698
Kampen, Dennis	145, 1939, 2588
Kamper, Maarten J.	2958
Karakasli, Vefa	2835
Karamanakos, Petros	297, 1476, 1754
Karau, Fabian	3292
Karnehm, Dominic	767
Karwatzki, Dennis	195
Kasten, Henning	3501
Kayser, Felix	59, 4003
Keilmann, Robert	891
Kempchen, Malte	2912
Kemper, Philipp	749
Kennel, Ralph	1754, 2366, 3362
Kerekes, Tamas	1933
Keshavarzi, Davood	1064
Khader, Meriem	2655
Khan, Basit Ali	2537
Khan, Mohammed Ali	135
Khan, Nameer	3796
Khan, Siam Hasan	484
Khanzadeh, Babak	2344
Khenfri, Fouad	3831
Kiehnle, Philip	999
Kiffe, Axel	1157
Kikuchi, Naoto	1104
Kim, Dong-Uk	1207
Kim, Sungmin	1207, 2877
Kinzer, Dan	3987
Kirsch, Andreas	380
Kitagawa, Wataru	739
Kjærsgaard, Benjamin Futtrup	825
Klee, Matthias	1515
Klever, Severin	3676
Klötzer, Sebastian	4011
Knebusch, Benjamin	1665, 3048
Ko, Youngjong	3014
Kobayashi, Hiroyasu	1580
Kocewiak, Lukasz	2504
Koch, Jan-Niklas	2852
Koczy, Dawid	3149
Kohlhepp, Benedikt	88, 307
Kojima, Tetsuya	3740

Kondo, Keiichiro 1580
Kondratenko, Dmytro 1906
Kopp, Tobias 912
Kormska, Tomáš 1114
Körner, Patrick 2021
Korthauer, Bastian 3625
Kosesoy, Yusuf 634
Kostka, Benedikt 1649
Kostynski, Daniel 3855
Koteich, Mohamad 534
Kouro, Samir 1015
Koutroulis, Eftychios 1985
Kowal, Julia 4014
Kragl, Robert 2554
Krick, Alexander 3989
Krigar, Tim 2375
Krishnamoorthy, Harish Sarma 730
Krüger, Helge 3966
Krümpelmann, Marcel 1631
Kubulus, Pawel Piotr 2661
Kuder, Manuel 767
Kumar, Amit 451
Kumar, Kaushik Naresh 1486
Kumar, Manish 3511
Kuperman, A. 3254
Kuprat, Johannes 3067
Kuring, Carsten 2596, 2644, 3775
Kurrat, Michael 912
Kurukuru, V S Bharath 135
Kusaka, Keisuke 1104, 2127
Kusche, Stephan 3704
Kusebauch, Manuel 3491
Küster, Pierre 411
Kwak, Jaedon 2893
Kyyrä, Jorma 1947
La Mantia, Fabio 833
Labonne, A. 3179
Labonne, Antoine 3140
Labrousse, D. 843
Lacerda, Vinícius Albernaz 3704
Laclaverie, Julien 944
Laforet, David 1445
Lamar, Diego G. 335, 1083, 1765, 3336
Lange, Jarren 2276
Lange, Yannic 2644
Langfermann, Sascha 1939, 2588
Lanzarotto, D. 2564
Larrañaga, Uxue 3938
Larrazabal, Igor 1765, 3336
Larsson, Anders 1456
Lataire, Philippe 2293
Laumen, Michael 3766

Lauri, Andrea 865
Laza, Saioa Burutxaga 370
Lazkano, Markel Zubiaga 484
Le Leslé, Johan 2526
Le Métayer, Pierre 2049
Lee, Jaehong 2877
Lee, Seung-Hwan 2877
Lee, Yonghwa 2402
Lefebvre, Bruno 2515
Lefevre, Guillaume 2526
Legay, Florian 3529
Lehn, Peter W. 1995, 2084, 2145, 2763
Leifert, Torsten 4013
Lemaire-Semail, Betty 2175, 2996
Lembeye, Yves 1216
Lenz, Kevin 442
Lenzen, Patrick 2413
Leuer, Michael 3292
Leuzzi, Riccardo 3975
Lévy, PE 843
Lewicki, Arkadiusz 1906
Lexow, Daniel 1744
Li, Feifei 3235
Li, Ke 3822
Li, Marui 3215, 3225
Li, Qiang 3722
Li, Weihan 4025
Li, Xiang 2301
Li, Xupeng 3373
Li, Zheming 2744
Liang, Mincui 3786
Lichtenstein, Timo 1674
Liebfried, Oliver 2726
Liegmann, Eyke 1754, 3362
Lievre, Aurelien 2175
Lin, Siqi 1914, 2638
Lin-Shi, Xuefang 3862
Lindemann, Georg 3555
Linder, Stefan 3992
Lippold, Florian 1506
Liserre, Marco 421, 833, 3014, 3067, 3092, 3101,
.... 3846, 3966
Liu, Chao 1561
Liu, Steven 604
Liu, Xing 1733, 3373
Liu, Yan-Fei 1590, 1601
Liu, Yining 1947
Llanos, Jacqueline 3429
Löfgren, Jonas 3920
Lombard, Philippe 169
López, Abraham 152
Lorenz, Andreas 814

Lorenz, Erwin 1167
Lorenz, Malte 1875
Lorenz, Oscar 873
Loudot, Serge 3883
Lu, Xuyang 3822
Lu, Yizhou 883
Luan, Shaokang 3309
Luckert, Franz 2706
Luecke, Stefan 3075
Luh, Matthias 232
Luo, Fang 344, 2860
Lusardi, Federico 3975
Lutsch, Michael 88
Lutz, Josef 1713
Lutzen, Hauke 2230
Ma, Wenhao 80
Maamri, Nezha 1525
Maibach, Philippe 3282
Maier, Robert W. 2744
Maitra, Abhishek 1424
Mallwitz, Regine 891, 912, 1094, 1247, 1506
Mambetow, Arthur 145
Manthey, Tobias 2655, 2689, 3059
Marca, Ygor Pereira 798
Marcaide, Inko 3920
Marcault, Emmanuel 2618
Marchesoni, Mario 2919
Margreiter, Thomas 223
Margueron, Xavier 1046
Marks, Hendrik 2030
Marquardt, Rainer 1855
Marroquí, D. 3382, 3401
Martin, Jérémy 990, 2736
Martinez, Wilmar 1914, 2197, 2581
Martinez-Garcia, Herminio 1056
Martinez-Padron, Daniel S. 1256
Martnez, Wilmar 2638
Marx, Philipp 1237, 1277, 3536
März, Martin 493, 3262
Mashaly, Aly 442
Mashayekh, Ali 767
Mathúna, Cian Ó 4006
Mattavelli, Paolo 865
Matthies, David 3159
Maussion, Pascal 2869
Maynard, X. 843
Mazuela, Mikel 325, 3327, 3574
Meddour, Aissam Riad 3696
Mehran, Kamyar 614, 3353
Mehrasa, M. 3179
Mehrasa, Majid 3140
Meier, Hans 2021

Meinert, Janus Dybdahl 825
Meissner, Michael 758, 3188
Mellor, Phil 2477
Mendoza-Araya, Patricio 1177, 1226
Meng, Qingchao 933
Menzel, Steffen 3169
Merlin, Michael M. C. 2824, 3813
Merlin, Michael 80, 3470
Mersche, Stefan 115
Mertens, Axel 641, 1350, 1533, 1631, 1649, 1665,
..1684, 1865, 1875, 2003, 2066, 2392, 2706, 3022, 3037, 3048,
3075, 3555, 3711
Miaja, Pablo F. 152
Mijatovic, Nenad 2496, 2939
Miller, T. J. E. 3318
Minami, Masataka 2285
Mir, Tabish Nazir 468
Mirza, Abdul Basit 344
Mirzadeh, Mina 1350
Mirzaeva, Galina 3903
Miskiewicz, Rafal 1486
Mistretta, C. 1975
Mita, Salvatore 2628
Mo, Wai Keung 3130
Möckel, Andreas 3391
Moench, Stefan 242
Mogorovic, Marko 203
Mohanta, MK Kharabela 689
Möhlenkamp, Georg 3993
Mohsenzade, Sadegh 614, 3353
Moldenhauer, Deniz-Heinz 2205
Mondal, Gopal 572
Mondzik, Andrzej 3804
Monmasson, Eric 1256
Mönninghoff, Sebastian 3005
Montero, E. Rodriguez 1834
Morales-Paredes, Helmo K. 1074
Morand, Julien 2526
Morel, F. 2422, 2564
Morey, Philippe 1543
Morshed, Muhammad 2301
Motte-Michellon, Denis 1216
Mouselinos, Theodoros P. 1551
Moussa, Hassan 3590
Movagharnejad, Hedieh 3048
Mu, Yunfei 3215
Müller, Jonas 2230
Müller, Tankred 474
Munk-Nielsen, Stig 825, 1641, 2661, 3309
Muñoz-Carpintero, Diego 1074, 3429
Muruaga, Endika Bilbao 3529
Musolino, Francesco 554

Mustafeez-Ul-Hassan 2860
Musumeci, S. 1975
Muyllaert, Koenraad 2383
Mysore, Madhu Lakshman 1424
Naeve, Tomasz 1445
Nagayasu, Kiwa 2164
Naghibi, Javad 614, 3353
Nahalparvari, Mehrdad 2843
Najjar, Mohammad 2682
Nakamura, Keiichi 777
Nakamura, Taketsune 3084
Nami, Ashkan 2241, 3757
Nannen, Hauke 160
Nassurdine, B. Mohamed...843
Nayak, Khirod Kumar..2241, 3757
Nayampalli, Vishwas Acharya..1703
Nazeri, Ahmad Ali 1309, 1336, 1343, 2670, 3871
Neal, Harley..2301
Nee, Hans-Peter 2843
Nehmer, Dominik 1036
Neira, Sebastian 2824, 3813
Neuland, Tanja..3991
Neumann, Christian 1895
Neumann, Ingmar 1445
Neumeister, Matthias..572
Nguyen, Allen..1722
Nguyen, Khanh-Hung 562, 1309
Nguyen, Van-Sang 922, 990
Nguyen, Xuan Viet Linh 169
Nian, Heng..2182
Niasar, Mohamad Ghaffarian..3729
Nie, Shuang 2145
Niedernostheide, Franz-J. 2744
Niedernostheide, Franz-Josef.......................................1424
Nielebock, Sebastian...493, 2607
Niemetz, Michael..2021
Niggemann, Oliver 3545
Nikowitz, Mario 2251
Nishio, Atsushi 351
Nishitani, Yota..3420
Nishizawa, Shin-Ichi...1128
Noboru, Wakana 777
Noisette, Philippe..3910
Nooshabadi, Morteza Tadbiri 787
Nordström, Lars 883, 1006
Nymand, Morten 2682
O'Donnell, Terence 390
O'Driscoll, Seamus 4006
Obernolte, Urs 854
Odeh, Charles 1906
Okada, Ryohei 1776, 1844
Olbrich, Markus..2912

Oliveira, Hercules Araujo 2948
Orbay, Raik 3920
Orchard, Marcos..3429
Orfanoudakis, Georgios I. 1985
Örgüt, Osman 1361
Orlik, Bernd 3149, 3159, 3169
Ortega, David 1765, 3336
Ortiz-Gonzalez, Jose 2477
Orts, C...3382, 3401
Oshnoei, Arman 2939
Ota, Ryosuke...1776, 1844
Ouyang, Ziwei...1413, 1561
Owzareck, Michael 1939, 2588
Oyarbide, Estanis 3327
Paasch, Kasper M..3130
Pace, Loris 3582
Páez, J. D. 2422
Pagnani, Daniela 2504
Panigrahi, Bijaya Ketan...2110, 3511
Papadopoulos, Georgios..1391
Papadopoulos, Theofilos 432
Papafotiou, George..2788
Papanikolaou, Nick...1796, 2257
Papastergiou, Konstantinos 2355
Pascal, Yoann..3067
Pasquier, Christophe 3786
Passalacqua, Massimiliano 2919
Passmore, Brandon 4005
Pathmanathan, Mehanathan 1995, 2084, 2145, 2763
Patin, Nicolas 1256
Patti, Dario 2628
Patzelt, Nikolaus 1923
Paul, Arup Ratan 178
Pauls, Denis..2441
Pavone, Mario 554
Pedroso, Douglas 335
Peftitsis, Dimosthenis...1486, 2355
Pelletier, Sebastien 223
Penczek, Adam..3804
Peng, Hujun..3235
Péra, Marie-Cécile...315
Pereda, Javier 2824
Pereira, Thiago 3014, 3092, 3101, 3846
Perez, Gaëtan 960
Perez-Cebolla, Francisco Jose...............................3574, 3750
Peroutka, Zdenek 1114
Perpiñá, Xavier...2715
Perrin, Rémi 2526
Perrin, Remi 3272
Petritz, Andreas 279
Petzoldt, Jürgen...2885, 3491
Peyghami, Saeed 2939

Pfeiffer, Jonas	411, 1167
Pfost, Martin	2375, 2413
Phanse, Ajinkya	1722
Phulpin, Tanguy	3883
Pichon, Pierre-Yves	2526
Pickert, Phil Leon	1381
Piepenbrock, Till	2432
Pietrzak-David, Maria	2869
Pigott, John	3796
Pinheiro, José Renes	3590
Piqué, Gerard Villar	3796
Piróg, Stanislaw	3804
Placzek, Julius M.	833
Plat, Arnaud	3862
Plötz, Till-Mathis	980
Pogulaguntla, Aditya	730
Pohlmann, Sebastian	767
Polezhaev, Vladimir	2230
Ponick, Bernd	1381, 1665, 3048, 3711
Poormohammadi, Fereshteh	2795
Pöschke, Florian	3704
Pouresmaeil, Edris	2537
Pouresmaeil, Kaveh	2788
Pouresmaeil, Mobina	2537
Pramanick, Sumit	1658, 3511
Pree, Elias	1445
Prenleloup, Pierre	3529
Prieto-Araujo, Eduardo	2486, 3704
Puls, Simon	2852
Puschmann, Frank	749
Qin, Zian	3607
Quabeck, Stefan	3893
Quade, Katharina Lilith	4025
Quay, Rüdiger	242
Rabkowski, Jacek	1486, 3938
Rädel, Uwe	2885, 3491
Radha, Krishna Moorthy	344
Rafiq, Aamir	1658
Raggini, Diego	3206
Raghavendra, I Venkata	730
Rahmani, Mehdi	2496
Raison, Bertrand	1147
Rajabian, Amir Azam	614
Ramdane, Brahim	1216
Ramirez, Fernando	289
Rasekh, Navid	3120
Rasool, Haaris	2101, 2293
Raßmann, Rando	1286
Rathjen, Kai-Uwe	758
Rault, Pierre	582
Ravyts, Simon	1300
Raya, Mariana	2715

Razi, R.	3179
Razi, Reza	3140
Regnat, Guillaume	2526
Rehlaender, Philipp	2432, 2754, 3625
Reimann, René	3159
Reincke-Collon, Carsten	370, 3391
Reindl, Andrea	2021
Reiner, Richard	242
Reißenweber, Lukas	525
Reitmeier, Dominik	2211
Remón, Daniel	1083
Rettner, Cornelius	4019
Reyes-Chamorro, Lorenzo	3429
Reynaud, Jean-François	3529
Ribeiro, Luiz Antonio De Souza	2948
Richard, Lucas	1147
Rickert, Kai	115
Rigbers, Klaus	4023
Rigogiannis, Nick	2257
Ringbeck, Florian	4025
Risch, Raffael	651
Rizoug, Nassim	3696, 3831
Robinson, Jonathan	572
Rocha, Gabriel Silva	2948
Roche, Jan-Philipp	3545
Rodríguez, Alberto	335, 1083, 1765, 3336
Rodriguez, Daniel C.	3893
Rodriguez, Joan Marc	2574
Rodriguez, José	1015
Roes, Maurice G. L.	798
Roes, Maurice	2788
Roß, Tilo	3391
Rossi, Mattia	1476
Rothenburger, Max	2383
Roth-Stielow, Jörg	971, 1237, 1277, 3536
Rouphael, Rosalie	1525
Rudolph, Christian	474
Rueß, Manuel	3565
Rufer, Alfred	30
Ruppert, Lukas A.	3766
Ruthardt, Johannes	971
Rylko, Marek S.	2470
Sadarnac, Daniel	3883
Saeidi, Mahmoud	1336, 1343, 3871
Safdarzadeh, Omid	2316
Sah, Gyanendra Kumar	1885
Sahan, Benjamin	1137
Sahin, Ilker	1361
Sahoo, Subham	2110, 2182
Sahu, Malaya Kumar	2241, 3757
Sahu, Silpashree	689
Said, Nasri	2618

Saito, Wataru 1128
Sakai, J. 1785
Salehi, Navid 1056
Samples, Ben 4005
Sanchez, Juan........................... 873
Sanchez-Ruiz, Alain 484
Santos, Francisco 3101
Sanusi, Bima Nugraha 1413
Sanz-Alcaine, José Miguel......... 3750
Sarlioglu, Bulent 3947
Sato, Kota 1580
Sato, Takashi 3420
Sauer, Dirk Uwe 4012, 4025
Sauerland, Henning................... 3159
Sawicki, Jean–paul 315
Scarcella, Giuseppe 1569
Scelba, Giacomo 1569, 2628
Schäffner, Philipp 279
Schafmeister, Frank 2432, 2754, 3625, 3686
Schanen, Jean-Luc 787
Schanen, JL.............................. 843
Schefer, Hendrik 891, 912, 1094
Schellekens, Jan 634
Schierle, Guido 3188
Schiestl, Martin 223, 268
Schillinger, Tobias 3646
Schillingmann, Henning 2030
Schlegel, Christian 1923
Schlegel, Ludwig 3957
Schmid, Markus 268
Schmidhuber, Michael 411, 1167
Schmies, Dominik..................... 2276
Schmitz, Laurids 3599
Schnabel, Fabian................. 624, 1515
Scholjegerdes, Moritz 3005
Schön, André 814
Schrödl, Manfred 2251
Schueltzke, Jens 1167
Schuerhuber, Robert 39
Schuhmann, Thomas 3646
Schullerus, Gernot 593, 2334
Schulte, Horst 3704
Schulz, D. 3411
Schulze, Gerold........................ 2383
Schulze, Hans-Joachim 1424
Schumann, Christian 2058
Schumann, Sven 4022
Schümann, Ulf.......................... 1286
Schupp, Jan.............................. 3309
Schütt, Michael 1885, 2308
Schwarz, Babette 1381
Schwendemann, R. 3411

Scohier, Martin......................... 504
Scrimizzi, F. 1975
Sebastián, Javier................. 1765, 3336
Seibel, Axel....................... 1515, 1620
Seitz, Arne............................... 4015
Seliger, Norbert 22
Semail, Eric 2996
Sen, Paresh C. 1590, 1601
Sepehr, Amir 2537
Serdyuk, Yuriy 2344
Sergentanis, Grigorios 3450
Serra, Amiron Wolff Dos Santos ... 2948
Seybold, Felix 3536
Shahparasti, Mahdi................... 2682
Sharma, Kanuj.......................... 2698
Shawky, Ahmed 1015
Shen, Chengjun 2477
Shen, Xiaobing...................1914, 2197
Shinoda, Kosei 3928
Shintani, Michihiro 3420
Shousha, Mahmoud...............279, 698
Shuqin, Wang........................... 1815
Siala, Sami 125
Siemaszko, Daniel 3910
Siemieniec, Ralf 1445
Sievers, Markus 3855
Singh, Rupam 135
Singh, Shashank Shekhawat 279
Singh, Sukhjit 2084
Skala, Aleksander..................... 3804
Skibin, Stanislav....................... 3301
Soeiro, Thiago Batista..........1319, 3729
Solomentsev, Michael 1722
Solovyov, Vyacheslav............... 2860
Soltau, N. 1785
Soltau, Nils.............................. 777
Sönmez, Ertugrul................593, 2334
Soundararajan, Ajeeth Phrassanna ... 3729
Soupremanien, Ulrich 922
Spieler, Matthias 1692
Sprunck, Sebastian 1611
Sreekanth, T 730
Stadler, Alexander.................... 525
Stadlober, Barbara.................... 279
Staiger, Jochen 2219
Stala, Robert............................ 3804
Stalleicken, Frederik 2607
Stallmann, Frederik 641
Stärz, Ronald.....................223, 268
Stathis, Spyridon 1402
Staubach, Christian 1137
Steckler, P. B........................... 2564

Stefanski, L.	3411
Steffen, Jonas	1515
Steinhart, Heinrich	2219
Štengl, Josef	1114
Stevic, Marija	2985
Stewart, Joshua	2806
Steyn, Kyle	3112
Stille, Karl Stephan	2276
Stock, Alexander	1
Stöckl, Thomas	952
Stone, David A.	3440
Strunk, Robin	1350
Stul, Koen	1300
Stutz, Christian	493
Suberski, Martin	2885, 3491
Sujeeth, Arjun	2628
Sullivan, Charles R.	2470
Svensson, Jan R.	1187
Tabrizi, Gholamreza	1611
Takamori, Taro	1128
Takayama, Hajime	108
Takeshita, Takaharu	213, 351, 739
Talla, Jakub	1114
Tang, Chengjun	2813
Tang, Zhongting	1933
Tashakor, Nima	1025, 1064, 1197, 3636, 3665
Tatakis, Emmanuel C.	1551
Tegtmeier, Bernd	1674
Teske, Peter	1466
Thiringer, Torbjörn	2344, 2813, 3920
Thoma, Jürgen	2554
Thönelt, Nick	1713
Thönnessen, André	3676
Tian, Fanghao	2581
Tillmann, Philipp	3740
Tiwari, Arvind Kumar	289
Tiwari, Arvind	187
To, Pham Ha Trieu	59, 70, 4003
Tornello, Luigi Danilo	1569
Torres, C.	3382, 3401
Torrico, Grover	361, 1815
Tournez, Florian	2175
Tran, Dai Duong	2293
Tran, Manh Tuan	2101, 2293
Tresca, Giulia	3975
Trescases, Olivier	3796
Tricoli, Pietro	468
Trochimiuk, Przemyslaw	1486
Tschepp, Andreas	279
Turrisi, Gaetano	1569
Tzanakis, Athanasios	3920
Uicich, Simon	3862

Ulbing, Alexander	3855
Ulmer, Sabrina	593, 2334
Ulrich, Burkhard	459
Umetani, Kazuhiro	2164
Unruh, Peter	1620
Unruh, Roland	3686
Urkizu, June	325
Vaccaro, Luis	2919
Vaessen, Peter	3729
Vagg, Christopher	3696
Vagnon, Eric	2515
Vahid, Sina	719
Vala, Sama Salehi	344
Valderrama, Carlos	504
Valenzuela, Rodrigo Alonso Alvarez	814
Van Cappellen, Leander	2795
Van Mierlo, Joeri	2101
Van Oosterwyck, Nick	3344
Van Tuan, Mai	351
Vandenbussche, Thomas	1300
Vanfretti, Luigi	3928
Vanwalleghem, Bart	3344
Vasiladiotis, Michail	1923
Vatamanu, Lucian	1046
Vázquez, Aitor	1083
Vázquez, Francisco	1765, 3336
Velasco-Quesada, Guillermo	1056
Velazco, Diego	251
Vellvehi, Miquel	2715
Venkataramanan, Giri	3480
Venugopal, Ravinder	2985
Verdier, Jacques	169
Vermeerch, Pierre	582
Veroni, Alessandro	3206
Vershinin, K.	2564
Viana, Caniggia	1995, 2084
Viarouge, I.	1955
Viarouge, P.	1955
Vidal-Albalate, Ricardo	2189
Videau, Nicolas	944
Videt, Arnaud	3822
Villar, Irma	3529, 3938
Vitorino, Montiê Alves	2689, 3059
Vogelsberger, M.	1834
Volzer, Benjamin	2554
Von Hoegen, Anne	3740
Wada, Keiji	1128
Wagner, Valentin	514
Wakelin, Bruce	3309
Wallart, Francois	251
Wallscheid, Oliver	2276, 2432
Waltereit, Patrick	242

Wang, Chu	3722
Wang, Jun	2136, 3120
Wang, Kangan	3014
Wang, Rui	673, 1641
Wang, Xiaoya	3722
Wang, Xin	315
Wang, Yanbo	2011
Wang, Yangang	2301
Waradzyn, Zbigniew	3804
Watanabe, Hiroki	1104
Wattenberg, Martin	873
Weicker, Martin	2316
Weires, Jonas	604
Weiser, Mathias C. J.	3565
Weiss, Xavier	1006
Wenzel, Johannes C.	2066
Werlig, Christian	3966
Weyh, Thomas	767
Wicht, Bernhard	2912
Wieczorek, Nick	3775
Wiemer, Adrian	2544
Wiesemann, Julius	1865
Wiesner, E.	1785
Wiesner, Eugen	777
Wijnands, Korneel	673, 798, 2788
Wilkowski, Matt	4008
Willer, Felix	3838
Willich, Viktor	3555
Wohlrath, Fritz	525
Wolbank, T.	1834
Wolf, Mihaela	2596, 2644, 3775
Wolfstädter, Simon	4017
Wölk, Alexander	279
Wouters, Hans	2197
Woywode, Oliver	758
Wu, Weimin	1985
Wu, Xiangqiang	1933
Wu, Yuxuan	2860
Wunsch, Bernhard	3301
Würfl, Joachim	2596, 2644, 3775
Würsig, Andreas	3966
Xia, Peizhou	3470
Xiao, Qian	3215, 3225
Xiao, Xiong	1966
Xie, Jun	2885, 3491
Xie, Lihong	2136
Xu, Huihui	709
Xu, James	3796
Xu, Qianwen	883, 1006
Xu, Wei	2136
Xu, Zhongqing	912
Xu, Zixiao	2182

Yadav, Sachin	3607
Yamaguchi, Masamichi	2127
Yamashita, Shota	213
Yamauchi, Kohei	2119
Yang, Huoming	1466
Yang, Jiajun	3014
Yang, Juefei	2477
Yang, Yinghui	3993
Yang, Yongheng	2257
Yaqoob, M.	1815
Yasuda, Takumi	902
Yeganeh, Mohammad Sadegh Orfi	2496, 2939
Yu, Guangyao	1319
Yu, Xiao	562, 1309, 2383
Yu, Xiaodan	3225
Yuan, Xibo	2136, 3120
Zacharias, Peter	411, 562, 1309, 1328, 1336, 1343, 2383, 2670, 3871
Zacher, Benjamin H.	2058
Zampardi, Giorgia	833
Zanchetta, Pericle	3975
Zatocil, Heiko	160
Zdanowski, Mariusz	3938
Zhang, Bo	1733
Zhang, Shimin	709
Zhang, Yaqian	2182
Zhang, Zhe	1561
Zhang, Zhuoqi	1776
Zhang, Ziqian	39
Zhao, Hongbo	1641, 3309
Zheng, Zhixue	315
Zhetessov, Aidar	3480
Zhu, Zi-Qiang	2958
Ziani, Adel	944
Ziegler, Philipp	971, 1237, 1277, 3536
Zilic, Rufad	1336
Zocher, Markus	2366
Zolfi, Pouya	719
Zou, Zhixiang	3014
Zsurzsan, Tiberiu Gabriel	1561

IEEE
445 Hoes Lane
Piscataway, NJ 08854-4141

ISBN 978-1-6654-8700-9